AF684770

THIN FILMS IN TRIBOLOGY

TRIBOLOGY SERIES, 25

THIN FILMS IN TRIBOLOGY

edited by

D. DOWSON, C. M. TAYLOR, T. H. C. CHILDS, M. GODET and G. DALMAZ

Proceedings of the 19th Leeds-Lyon Symposium on Tribology
held at the Institute of Tribology, University of Leeds, U.K.
8th - 11th September 1992

ELSEVIER

Amsterdam – London – New York – Tokyo 1993

For the Institute of Tribology, Leeds University
and
The Institut National des Sciences Appliquées de Lyon

LIBRARY COPY
RETURN TO
LEWIS LIBRARY
CLEVELAND, OHIO

TJ
1075
.A2
L43
1992
c.1

ELSEVIER SCIENCE PUBLISHERS B.V.
Sara Burgerhartstraat 25
P.O. Box 211, 1000 AE Amsterdam, The Netherlands

Library of Congress Cataloging-in-Publication Data

Leeds-Lyon Symposium on Tribology (19th : 1992 : University of Leeds)
Thin films in tribology : proceedings of the 19th Leeds-Lyon Symposium on Tribology held at the Institute of Tribology, University of Leeds, U.K. 8th-11th September 1992 / edited by D. Dowson ... [et al.].
p. cm. -- (Tribology series ; 25)
Includes bibliographical references.
ISBN 0-444-89789-5 (alk. paper)
1. Tribology--Congresses. 2. Thin films--Congresses. I. Dowson, D. II. University of Leeds. Institute of Tribology. III. Institut national des sciences appliquées de Lyon. IV. Title. V. Series.
TJ1075.A2L43 1992
621.8'9--dc20 93-5673
CIP

ISBN: 0 444 89789 5 (Vol. 25)

© 1993 Elsevier Science Publishers B.V. All rights reserved.

No part of this publication may be reproduced, stored in a retrieval system or transmitted in any form or by any means, electronic, mechanical, photocopying, recording or otherwise, without the prior written permission of the publisher, Elsevier Science Publishers B.V., Copyright & Permissions Department, P.O. Box 521, 1000 AM Amsterdam, The Netherlands.

Special regulations for readers in the U.S.A. – This publication has been registered with the Copyright Clearance Center Inc. (CCC), Salem, Massachusetts. Information can be obtained from the CCC about conditions under which photocopies of parts of this publication may be made in the U.S.A. All other copyright questions, including photocopying outside of the U.S.A., should be referred to the copyright owner, Elsevier Science Publishers B.V., unless otherwise specified.

No responsibility is assumed by the publisher for any injury and/or damage to persons or property as a matter of products liability, negligence or otherwise, or from any use or operation of any methods, products, instructions or ideas contained in the material herein.

This book is printed on acid-free paper.

Printed in The Netherlands

INTRODUCTION

The nineteenth Leeds-Lyon Symposium on Tribology was held at Bodington Hall, University of Leeds, from 8th to 11th September 1992. The focus of the meeting and its title was 'Thin Films in Tribology'. Some one hundred and fifty delegates from twenty countries gathered to listen to and discuss seventy papers presented in nineteen sessions. This number of papers is an increase on numbers at previous meetings and indicates the continuing popularity of the Symposium.

It was again pleasant to welcome Professor Maurice Godet and a group from our sister institution, INSA de Lyon. On this occasion we also particularly welcomed Professor K L Johnson from the University of Cambridge, both to mark our respect for his wide-ranging achievements in tribology and on his retirement, and as Guest of Honour and after dinner speaker at the Queens Hotel, Leeds: these Proceedings are dedicated to him.

The nineteenth Symposium was special for another reason too, as the last occasion on which Professor Duncan Dowson, co-founder with Maurice Godet of the Symposia, could attend the Symposium at Leeds as a serving member of staff. In addition to the traditional Thursday afternoon tour which this year was to Chatsworth House and The Abbeydale Industrial Museum and culminated with dinner and a display of the silver at the Cutlers' Hall, Sheffield, the UK champion Grimethorpe Colliery Brass Band was invited to play in the University's Parkinson Court on the Wednesday evening. One could well believe that the walls of Jericho once fell to trumpet blasts, although it must be reported that Leeds survived.

Professor Dowson's impending retirement to some extent determined the theme of the meeting. From Victorian days when working lubricant films were calculated in tens of micrometres, to today when molecular dynamics simulations and even experiments are beginning to look at nanometre, single molecule thick, films, the study of surfaces which is the tribologists' challenge has moved to finer and finer scales. It was felt appropriate to invite reviews from across the tribological field with thin films as their theme. Thus, in addition to the Keynote Lecture, delivered by Professor Dowson on Thin Films in Tribology, most half-day working periods were started by an invited review session. The meeting finished with a panel discussion. We would like to thank the following for their reviews and panel contributions: K L Johnson, T Bell, C M Myllerup and B J Hamrock, H S Cheng, T H C Childs, H A Spikes, M O Robbins, R C Coy, M Godet and B Jacobson.

The further working sessions of the Symposium were as lively as ever, concentrating on contact, friction and traction (session 3), coatings and surface modification (Sessions 4 and 12), analysis and design of highly stressed components (Session 5), elastohydrodynamics (Sessions 7 and 11), lubricated wear (Session 8), Hardness and metal cutting (Session 9), molecular scale thin films (Session 13), bio-tribology (Sessions 15 and 18), compliant surfaces (Session 16) and coating wear (Session 17). We would like to thank the Chairmen of Sessions and the papers' authors, as recorded in these proceedings, for the success of the Symposium.

To mount such a Symposium in an effective style needs the help of many and we would like to record sincere appreciation of the great contribution of ours colleagues - Mrs Cath Goulbourn, Mr Ron Harding, Mr Brian Jobbins, Mr David Jones, Mrs Sheila Moore, our technicians and research fellow/research students. It is once again a pleasure to record our thanks to Elsevier Science Publishers B.V., Amsterdam for their impressive production of the volume of proceedings. We were fortunate, as in the past, to attract the financial support of a number of concerns, support which helps to encourage younger researchers to attend the meeting and reduce the registration fee. We gratefully acknowledge the following organisations in this regard.

BP International Ltd
Castrol Ltd
Mitchell Bearings
Shell Thornton Research Centre
SKF Engineering and Research Centre
US Army Research, Development and Standardisation Group (UK)

The wide range of topics covered by the Leeds-Lyon series of tribology symposia is evidenced by the following sequential list:-

[1]	1974	(Leeds)	Cavitation and Related Phenomena in Lubrication
[2]	1975	(Lyon)	Super Laminar Flow in Bearings
[3]	1976	(Leeds)	The Wear of Non-Metallic Materials
[4]	1977	(Lyon)	Surface Roughness Effects in Lubrication
[5]	1978	(Leeds)	Elastohydrodynamics and Related Topics
[6]	1979	(Lyon)	Thermal Effects in Tribology
[7]	1980	(Leeds)	Friction and Traction
[8]	1981	(Lyon)	The Running-In Process in Tribology
[9]	1982	(Leeds)	Tribology of Reciprocating Engines
[10]	1983	(Lyon)	Numerical and Experimental Methods in Tribology
[11]	1984	(Leeds)	Mixed Lubrication and Lubricated Wear
[12]	1985	(Lyon)	Mechanisms and Surface Distress
[13]	1986	(Leeds)	Fluid Film Lubrication - Osborne Reynolds Centenary
[14]	1987	(Lyon)	Interface Dynamics
[15]	1988	(Leeds)	The Tribological Design of Machine Elements
[16]	1989	(Lyon)	Mechanics of Coatings
[17]	1990	(Leeds)	Vehicle Tribology
[18]	1991	(Lyon)	Wear Particles: From the Cradle to the Grave
[19]	1992	(Leeds)	Thin Films in Tribology

We look forward with anticipation to the twentieth meeting in Lyon from 7th to 10th September 1993. The Symposium will focus on energy dissipation processes and the origins of friction in tribology at all scales, from all points of view and the title will be **"Disspative Processes in Tribology".**

Chris Taylor
Duncan Dowson
Tom Childs

CONTENTS

SESSION I
KEYNOTE ADDRESS

Chairman : Professor L Flamand

Paper I (i) **Thin Films in Tribology**

Thin Films in Tribology / D. Dowson et al. (Editors)
1993 Elsevier Science Publishers B.V.

Thin Films in Tribology

D Dowson

Institute of Tribology, Department of Mechanical Engineering, The University of Leeds, Leeds LS2 9JT, United Kingdom

In this Keynote Address the author reviews briefly the full spectrum of thin film tribology, ranging from surface modification and coatings to the behaviour of thin, protective lubricating films. Attention is drawn to the remarkable way in which the effective minimum film thickness in highly loaded, reliable and efficient lubricated machine elements has fallen from tens of microns to tens of nano-metres throughout the 20th century. The near coincidence of studies of 'solidified' lubricants and 'boundary' lubricants in tribology is noted.

1. INTRODUCTION

The short definition of Tribology, introduced more than a quarter of a century ago, is that it is ...."**The science and technology of interacting surfaces in relative motion and the practices related thereto....**". Since the subject was defined there has been much research and not a little success in understanding the response of the bulk materials, both solids and fluids, which form the tribological conjunctions, to the severe loading, kinematic, thermal and environmental conditions which they experience in performing their tribological functions. Much of this progress has been developed from firm foundations in continuum mechanics and concepts of fracture mechanics and fatigue in the case of solid mechanics. Analysis has nevertheless generally rested upon idealisations of the materials considered and in many ways it is remarkable that the predictions of performance have been so consistent with practical experience.

Notwithstanding the impressive progress recorded in the understanding of the tribological performance of bulk materials, the most spectacular developments in recent times have resulted from the recognition of the distinctive roles of thin films, both solid and fluid, in tribology. The operating surfaces of solids might exhibit physical properties quite different from those of the underlying bulk materials, while the fluid nature of the lubricant itself becomes questionable in the exceptionally thin films which appear to determine success or failure in many lubricated machine components. Some illustrations of thin film dominance in tribology include;

- chemical reactions with the lubricant or the environment which might create surface films on the solids with properties which dominate the tribological properties of the conjunctions.

- lubricants might solidify and cease to respond as Newtonian fluids.

- the effective thickness of the lubricating film might fall to the level of the roughness scale of the solid surfaces, and might even approach molecular proportions.

- the surfaces of the solids might have been modified, either deliberately or by pre-treatment, or as a result of exposure to the operating environment of the tribological conjunction.

- wear debris (third bodies) might accumulate within the conjunction in thin layers and dominate the tribological performance of the system.

- Molecular rather than continuum mechanics might be necessary to account for the detailed solid-to-solid and solid-to-fluid interactions in thin film tribology.

Many of these topics form the subject of review and research papers in the 19th Leeds-Lyon

Symposium on Tribology, while the recognition of their importance accounts for the title of the Symposium itself. In the present paper brief mention will be made of some of the major fields of progress reported fully elsewhere in the Symposium, but particular attention will be given to studies of thin fluid film lubrication.

2. THE SPECTRUM of THIN FILM TRIBOLOGY

Initially the selection of sliding pairs of materials for machinery was guided by consideration of the properties of the bulk materials. However, in due course it was recognised that in many components, like plain bearings, it was adequate to provide a low shear strength, low melting point bearing material on the bearing surface to support the relatively hard and expensive journal or shaft. As the unit loads on machinery increased, the relatively thick layers of soft material, like white metal, gave way to the thin multiple layers of materials capable of resisting fatigue and operating in the ever more hostile environments that are familiar today in many plain bearings. General reviews of these developments were published in [1,2,3], while valuable guidance to wear resistance now appears in handbooks [4,5].

It can be argued that the spectacular developments in plain bearing technology did much to promote interest in many aspects of thin film tribology. It is, however, the emphasis on wear reduction, particularly in unlubricated conjunctions, in recent years that has been responsible for much of the progress evident today.

Perhaps the greatest advance followed the recognition that the bulk materials could be selected for their structural integrity, general mechanical properties and relatively low cost while the surfaces of rubbing solids could be modified by a variety of techniques to enhance their resistance to corrosion or wear. In general wear resistance is enhanced by increasing the hardness of the surface, through such processes as;

Heat Treatment. Perhaps the best known method of preparing relatively thick layers of harder material in low and medium carbon steels is **Case hardening**, or carburising, based upon the diffusion of carbon into the surface when the metals are heated in the presence of a carbonaceous material. Other thermal processes for the enhancement of hardness based upon induction flames, lasers and electron beams are widely employed, depending upon the materials and sizes of the specimens.

Diffusion. Thermo-chemical action can be used to prepare harder surfaces in relatively thin layers through the use of elements such as carbon, sulphur and nitrogen.

An alternative to the modification of the surface layers of the bulk material is to apply a coating of either a very hard material or a solid lubricant.

Electroplating is a well known and established procedure for coating materials such as iron, chromium and nickel.

Welding is used to provide protective alloy coatings based upon materials such as cobalt, iron, nickel and tungsten carbide.

Flame-spraying is widely used where metals, alloys or ceramics are to be applied to enhance the resistance to wear. The coatings are relatively thick and difficulties can be encountered with awkwardly shaped components.

Ion Plating and Plasma deposition are the coating processes that have attracted the most attention in tribological circles in recent years. They are used for hard and soft coatings alike and generally yield dense structures with good tribological properties. The two major categories of coating systems under these headings are **Chemical Vapour Deposition (CVD) and Physical Vapour Deposition (PVD)**. In the former, gaseous reagents provide the source of the coating materials, while in the latter the coating material is generally evaporated from solid placed within the vacuum chamber. The CVD process generally involves high temperatures, which may be disadvantageous, whereas PVD minimises this problem. The coatings are very thin, such that the original surface topography is not impaired and adequate bond strength is often claimed. The assurance of adequate bonding is nevertheless a matter of the utmost importance in many tribological situations, such as in medical implants, and much effort is currently being applied to the evaluation and improvement of coatings produced by ion plating techniques.

2.2 The Integrity of Coatings

If it is true that protective coatings, both hard and soft, have been responsible for much of the progress in wear protection under rubbing conditions [6], it is equally true that much of the concern about their use has been related to the issue of their integrity on the substrates.

In some cases it is not a disaster if the coating eventually wears away, since many substrates are themselves reasonably good wear resisting materials. However, in other cases it might be disastrous if the coating detaches itself from the substrate, since the debris might be hard, abrasive and destructive to the tribological system. It is the responsibility of the tribologist to ensure that the protective coatings which he or she recommends have adequate bond strength at the interface with the underlying material throughout the lifetime of operation of the sliding pairs of materials.

The point was brought to my attention most vividly almost ten years ago, when we were concerned with the development of improved counterfaces for the ultra-high molecular weight polyethylene acetabular cups used in total replacement hip joints [7]. The wear studies had already indicated that high density alumina ceramic provided an excellent counterface to the polyethylene, with wear of the latter taking place at a lower rate than when well prepared metals were used [8,9]. The mechanical strength of metallic femoral stems and heads nevertheless presented desirable features for the severe loads and bending moments encountered by implants. Interest thus developed in the possibility of retaining the mechanical strength of metallic stems and heads, and the tribological advantages of ceramic counterfaces by applying alumina coatings to the metallic heads.

Plasma sprayed alumina ceramic coatings having thicknesses in the range 200-300 μm were applied to the surfaces at velocities up to 500 m/s using argon with a small amount of hydrogen as the carrier and plasma gas. The bond strength specimens were prepared on the circular ends of cylindrical metallic specimens of diameter 25 mm. The ends of the metal cylinders were cleaned by pressure grit blasting and the ceramic coatings were applied either directly or by interposing a Ni, Cr, Aℓ bond coat some 75-100 μm thick.

Normal loads were applied to the coating in a tensile testing machine by sticking the flat face of a carefully prepared stainless steel cylinder to the alumina with Araldite. A constant strain rate of 0.5 mm/min was adopted and some specimens were tested in air and others after soaking for various periods in deionised water.

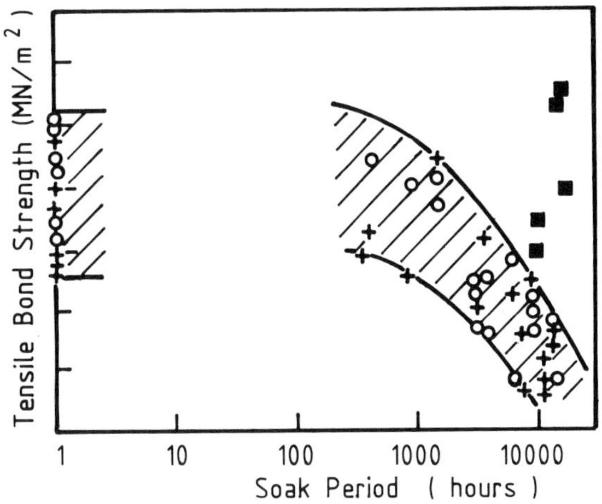

Fig. 1. Variation of Tensile Bond Strength of Coatings of Alumina on Metals. (■ control; + with bond coat; o without bond coat)

The findings are shown in Figure 1. The initial bond strengths are recorded in the ordinate and it is evident that little deterioration took place during the first 1,000 hours or so. However, the bond strength of the specimens soaked in water decreased rapidly thereafter, in some cases by almost 90% after 10,000 hours (say 14 months). There were no significant differences between the results for the two different metallic substrates and the bond coat had little influence on the bond strength.

While survival times of about 1,000 hours may be satisfactory in many engineering situations, they are clearly inadequate for implant applications. It is only now, with the considerable improvements in coating technology and the development of alternative surface treatments, that surface modifications are once again attracting the attention of bio-engineers engaged in the advancement of total replacement joints.

3. FLUID FILM LUBRICATION

In the previous section surface modification of the solids, often in thick films of micron on sub-micron proportions, was discussed in relation to developments in thin film tribology. We now move from the solids to the fluids in tribological conjunctions to find evidence of equally spectacular advances in thin film tribology.

3.1 The Thinning Film

I have outlined elsewhere [10] the manner in which advances in analysis, design and manufacture have enabled fluid-film lubricated systems to operate with films of ever decreasing thickness throughout the 20th century. At the beginning of the century minimum film thicknesses in reliable fluid film bearings were probably of order 10^{-5} m and maybe even as high as 10^{-4} m. Development of the internal combustion engine provided the spur for improvements of big end and main crankshaft bearings which regularly operate today with minimum film thicknesses of order 10^{-6} m.

In the second half of the 20th century the remarkable tenacity of fluid-films in highly stressed, lubricated machine elements was revealed through the development of understanding of the phenomenon now known as elasto-hydrodynamic lubrication. The initial measure of such films was the micron, but as experience grew we began to talk confidently of minimum film thicknesses of sub-micron proportions; say 10^{-7} m. Analytical procedures for smooth solids were developed for the prediction of such films in conjunctions lubricated by either Newtonian or Non-Newtonian fluids. The analyses became increasingly sophisticated and in recent times a number of the initial restrictions on the analysis have been relaxed to permit solutions to be obtained which take account of the surface roughness of the solids; point (elliptical) and line contacts; isothermal and thermal conditions and the realistic rheological characteristics of the lubricants. As the available solutions multiplied, and the learned society literature on tribology still devotes a good deal of space to the subject, empirical film thickness equations were developed, verified by experiment and incorporated into design procedures.

Experimental techniques based primarily upon interferometry and capacitance systems not only verified the theoretical predictions but also revealed the effectiveness of very thin elasto-hydrodynamic films. Roberts and Taylor [11] had revealed water films having thicknesses of about 10^{-8}m between rubber and glass as early as 1971, while Spikes [12] more recently developed an ultra thin film interferometry system capable of measuring effective elastohydrodynamic films between glass and metals of similar film thicknesses. Flying heads float above computer discs with great effectiveness and remarkable reliability on films of air generated by hydrodynamic actions which have thicknesses in the range 10^{-9}m to 10^{-7}m.

We now recognise that the thin, tenacious and very effective films of fluid developed beneath the rugosities on surfaces as they pass through elasto-hydrodynamic conjunctions can maintain fluid film lubrication between many real (rough) surfaces which might otherwise be expected to touch each other. This appears to explain why many elasto-hydrodynamic conjunctions are effective even though the predicted film thicknesses, based upon smooth surface analysis, are but modest compared to the composite surface roughness of the solids. This is known as micro-elasto-hydrodynamic action or simply asperity lubrication and it appears to be particularly effective in low elastic modulus conjunctions such as elastomeric bearings, seals and synovial joints. Much remains to be done on the analysis and experimental study of micro-elasto-hydrodynamic lubrication, but it is already clear that such studies are enhancing our understanding of the lubrication of highly deformed rough surfaces and also exposing the inadequacy of the simple (Λ) ratio concept as a guide to the performance and long term survival of certain elastohydrodynamic conjunctions. The essentials of micro-elasto-hydrodynamic actions are reviewed in the next section.

3.2 Micro-Elasto-Hydrodynamics

The well known features of smooth surface elasto-hydrodynamically lubricated conjunctions are shown in Figure 2. In this illustration the moving plane surface represents a rigid solid while a cylinder of appropriate radius and elasticity is shown as the deformable solid.

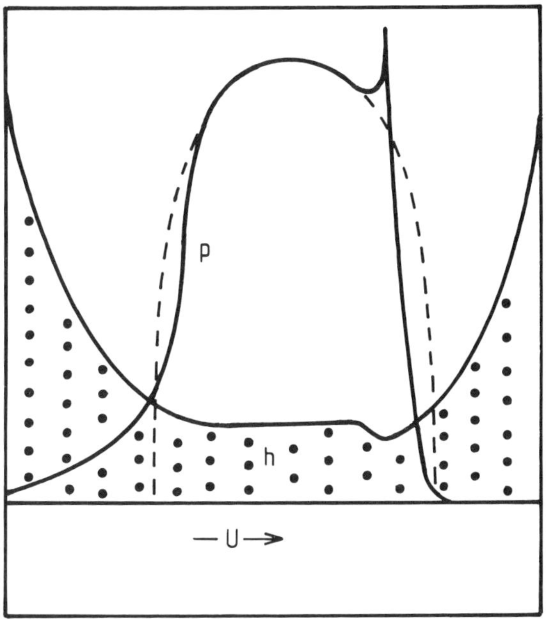

Fig. 2. Elasto-Hydrodynamic Conjunction

If transverse roughnesses modelled by a sinusoidal profile are considered on the deformable surface, the usual smooth surfaced elasto-hydrodynamic conjunction pressure curve will be perturbed by the local asperity hydrodynamic action associated with the peaks and troughs as shown in Figure [3]. Several numerical solutions to the micro-elastohydrodynamic lubrication problem have now been reported [13,14], with attention being focused upon the influence of the amplitude, wave length and orientation or lay of the roughness profiles upon pressure distribution, film thickness and stresses within the solids.

The important outcome of micro-elasto-hydrodynamic action is that the surface rugosities are partially flattened by the enhanced local pressures which they generate. This may allow the surfaces to be separated by an effective film of lubricant when they would otherwise be expected to make contact, thus extending the regime of fluid film lubrication to very thin film situations.

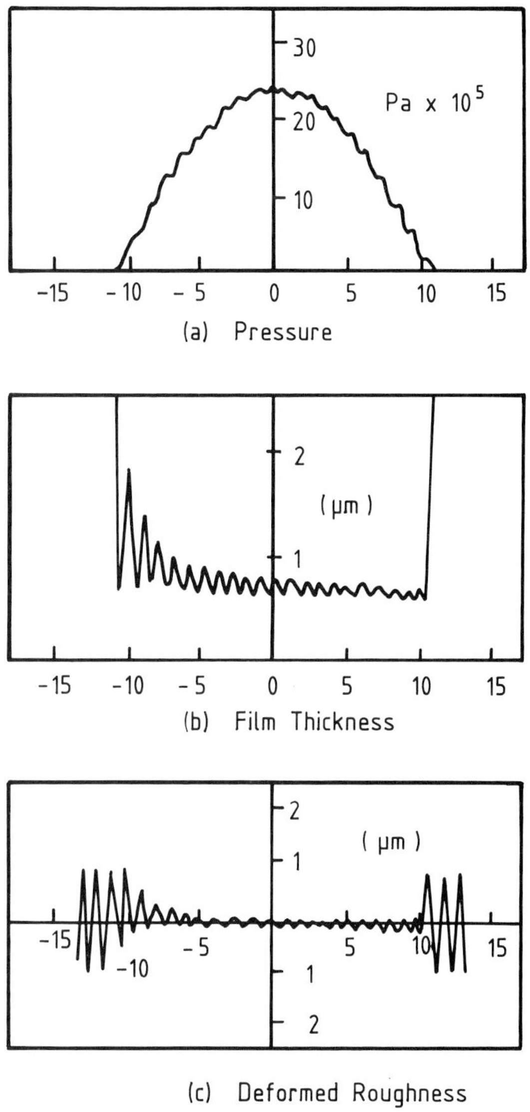

Fig. 3. Features of Micro-Elastohydrodynamic (Asperity) Lubrication.

3.3. The Lambda (Λ) Ratio.

It has been useful in the past to ascertain the likely effectiveness of fluid film lubrication by calculating the ratio of the calculated elastohydrodynamic film thickness (h) for smooth surfaces to the composite surface roughness $[(\sigma_1)^2 + (\sigma_2)^2]^{1/2}$ of the solids. Experience has shown that if the resulting lambda (Λ) ratio exceeds about three, the conjunction enjoys effective fluid film action as indicated by Figure 4. While a value of unity will inevitably lead to substantial surface-to-surface contacts, wear and a reduced fatigue life, there is increasing evidence that conjunctions with intermediate lambda values can survive much better than expected due to micro-elasto-hydrodynamic effects.

This raises doubts about the real value of the lambda ratio. It is clearly a useful initial guide, to conjunction survival, but if circumstances permit micro-elastohydrodynamic action, a more detailed investigation of the actual minimum film thickness should be undertaken. A modified lambda ratio in which the composite roughness term is adjusted to reflect the remarkable elastic smoothing action which sometimes takes place might still be useful, but as the resulting minimum film thicknesses become smaller and smaller and approach molecular proportions, account must also be taken of the absolute separation of the solids.

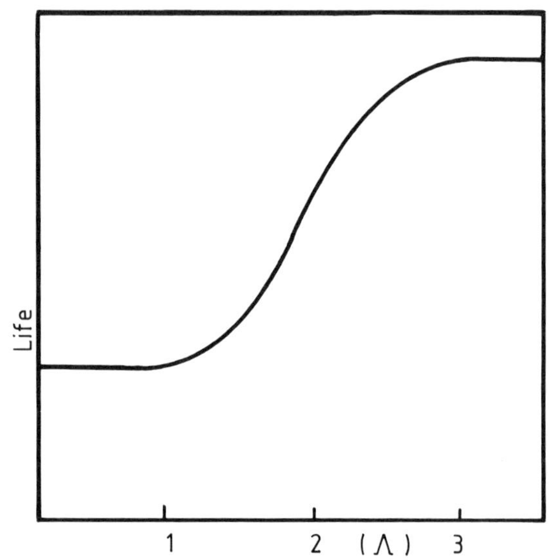

Fig. 4. Influence of Lambda Ratio (Λ) Upon Life.

4. THE STRIBECK CURVE

The transitions from 'fluid film' to 'boundary' lubrication via an intermediate zone known as 'mixed' lubrication are often represented by a Stribeck [15] curve. The first President of the American Society of Mechanical Engineers, Robert H Thurston [16] reported in 1885 that the friction in a journal bearing passed through a minimum as the load increased; a transition now associated with a change from fluid-film to mixed lubrication. Gümbel [17] unified the observations of Thurston and Stribeck when he represented journal bearing friction in a given bearing as a function of a single dimensionless group ($\eta\omega/p$), where (η) is the lubricant viscosity, (ω) the angular velocity of the shaft and (p) the projected load per unit area.

The general form of the Stribeck curve is shown in Figure 5. It was one of the earliest forms of lubrication regime charts, distinguishing as it did between fluid-film; mixed and boundary lubrication. The minimum coefficient of friction is of order 10^{-3}, while the level representative of boundary lubrication in engineering systems is about 10^{-1}. The mixed lubrication curve connecting these extremes is steep, thus offering attractive possibilities for a significant reduction in the coefficient of friction by the promotion of a little hydrodynamic action, or a severe penalty on fluid-film bearing performance when asperity interaction occurs.

The possibility of extending fluid-film lubrication into regimes that would appear to be mixed or boundary lubricated, through micro-elastohydrodynamic action has already been noted in the previous section. These suggestions are by no means recent, since Michell [18] reviewed some of the issues in his excellent book published in 1950. He recalled that Professor Heidebroek [19] had discussed a paper by Tenot, published in 1937, and declared that the rising friction in the mixed lubrication regimes was not necessarily due to dry, or solid, friction. He pointed out that if regard was paid to the rugosities on the bearing surfaces, convergent films could be envisaged between them which would contribute to the local load carrying capacity and greatly to an increase in the friction.

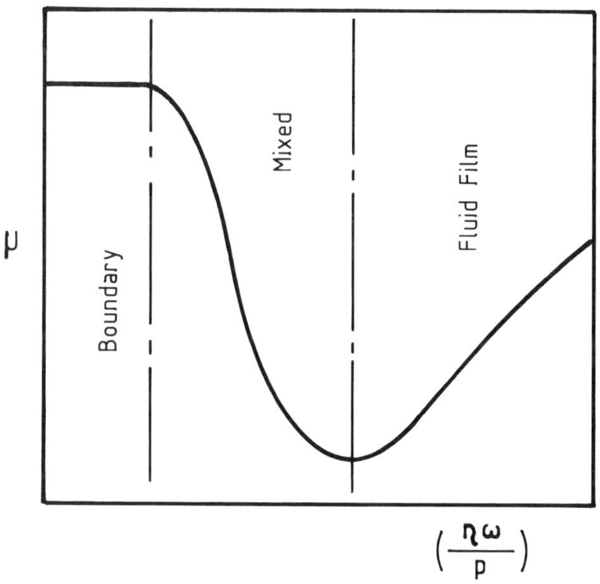

Fig. 5. Stribeck Curve.

In 1940 Muskat and Evinger [20] demonstrated that viscosity-pressure effects in the lubricant would have the same effect upon the performance of a plane inclined surface bearing.

Michell [18] himself contributed to the debate by presenting an early and quite remarkable analysis of the influence upon load carrying capacity of longitudinal corrugations of sine-wave form on a pad adjacent to a smooth slider. He showed that this form of roughness alone could yield a minimum in the friction trace of similar form to that of the Stribeck curve. It provided firm theoretical support for the concept outlined by Professor Heiderbroek. Michell [18] preserved a balanced view of the factors responsible for the minimum friction characteristic in the Stribeck curve. In 1950 he wrote;

> ... **"In actual bearings of course the rugosities will not be regular corrugations, nor of equal height, as assumed in this calculation, and they will occur on both surfaces, not on the pad only. The more prominent of the irregularities on the opposed surfaces will come into contact with one another, while the less prominent are still separated by comparatively thick films of lubricant. The fluent film will then become broken by discontinuities progressively increasing in number and extent as (h) is reduced, and the resistance to sliding, instead of being solely the resistance of a viscous film of diminishing thickness, will be partly of that nature and will consist partly of frictional and abrasive resistance at the solid contacts, the latter finally predominating".**

Current work confirms the significant role of asperity lubrication in protecting machine components like gears and low elastic modulus bearings like synovial joints, from the levels of friction and wear which might initially be expected from lubrication regime charts or Stribeck diagrams. It is interesting to reflect on the way in which engineers have enhanced the performance of lubricated machine elements throughout the 20th century, by operating equipment safely and reliably at ever decreasing values of the minimum film thickness.

5. MINIMUM FILM THICKNESS TRENDS

In Section 3.1 attention was drawn to the evidence of decreasing effective minimum film thickness in lubricated components throughout the 20th century. This trend is summarised in Table 1 and a number of practical illustrations of the 'design' minimum film thickness are noted.

The move from one or two orders of magnitude of micron sized film thicknesses to nano meters throughout the century is clearly evident and this is represented on a time base in Figure 6. Present day students of lubrication in many realistic machine elements are truly tackling common problems, whether they be from a consideration of even thinner fluid films or thicker boundary lubricating films.

Table 1

Minimum Film Thickness Trends in the 20th Century.

Date and Practical Examples	Film Thickness (m)
Late-19th century (1900) - Plain Bearings	10^{-4} - 10^{-5}
Mid-20th century (1950) Steadily Loaded Bearings	10^{-5}
Late-20th century (1980) - Dynamically Loaded Bearings.	10^{-5} - 10^{-6} 10^{-6} - 10^{-7}
End-20th century (1990 -2000) - Asperity Lubrication; low elastic modulus solids; magnetic recording, plasto-elasto hydrodynamic lubrication	10^{-7} - 10^{-8} (Or even 10^{-9})

It is no accident that studies of lubricant rheology reveal evidence of effective solidification and a limiting shear stress corresponding to a traction coefficient of about 0.1, while the coefficients of friction in boundary lubrication are also of this magnitude. The gap is really closing between the two sides of mixed lubrication and in due course students of the subject of tribology might well wonder why it took so long to recognise the continuity of physical phenomena at the lubricated interface between machine elements.

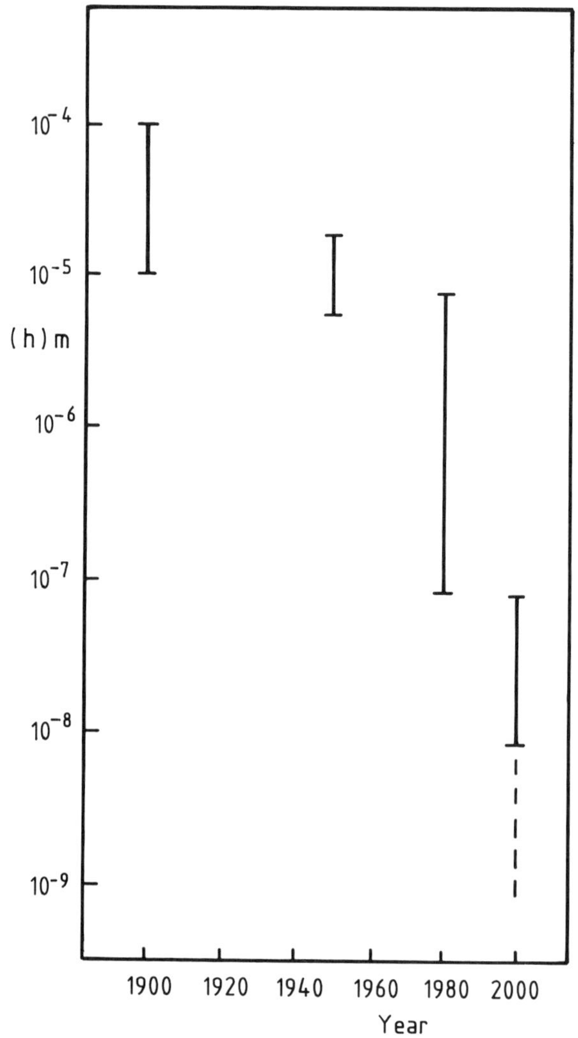

Fig. 6. The Thinning Film

The trends outlined above have opened the door for mathematically based simulations of solid-to-solid and solid-to-lubricant interactions to be tacked on the basis of molecular dynamics. Landman et al [21] have actively tackled the problems of dry contact on this basis, while Robbins [22] has contributed to the fluid interface problem in the present volume.

The predictions of atomic force modelling and the findings from atomic force microscopes can be expected to reveal a new understanding of many of the basic physical phenomena which control the surface interactions we now call 'tribology'. One of the most interesting and comforting findings to date is that the predictions from molecular dynamic simulations are remarkably consistent with the predictions of continuum mechanics down to asperity scale phenomena. The continuum mechanics approach to tribology still has much to offer!

REFERENCES

[1] "Proceedings of the Conference on Lubrication and Wear", The Institution of Mechanical Engineers, London, 1957.

[2] "Lubrication and Wear; Fundamentals and Application to Design", Proceedings of The Institution of Mechanical Engineers, Volume 182, Part 3A, 1967.

[3] Dowson, D. "A History of Tribology", Longman, London, pp 677, 198.

[4] Peterson, M B and Winer, W O (Eds), "Wear Control Handbook", ASME, 1980, pp 1358.

[5] Neale, M J. (Ed), "Tribology handbook", Butterworths, 1973.

[6] Dowson, D, Taylor, C M and Godet, M (Eds) "Mechanics of Coatings", Proceedings of the 16th Leeds-Lyon Symposium on Tribology, Tribology Series 17, Elsevier Amsterdam, pp 498.

[7] Randall, J E and Dowson, D, "The Bond Strength and Ageing of Plasma-Sprayed Coatings of Alumina on Metal Substrates", Engineering in Medicine, Volume 13, No. 2, 1984, pp 67-71.

[8] Dowson, D and Linnett, L W, "A Study of the Wear of Ultra-High Molecular Weight Polyethylene Against a High Alumina Ceramic", in Mechanical Properties of Biomaterials, Eds. Hastings, G W and Williams, D F, Advances in Biomaterials, Volume 2, Part One: Orthopaedics. Tribological Properties in Joint Replacecment, John Wiley, Chichester, Chapter 1, 1980, pp 3-26.

[9] Dowson, D and Harding, R T, "The Wear Characteristics of Ultra High Molecular Weight Polyethylene Against a High Density Alumina Ceramic Under Wet (Distilled Water) and Dry Conditions", WEAR, Vol. 75, (1982), pp 313-331.

[10] Dowson, D. "Developments in Lubrication - the Thinning Films", J. Phys. D: Appl. Phys. 25 (1992), 334-339.

[11] Roberts, A D and Tabor, D, "The Extrusion of Liquids Between Highly Elastic Solids", Proc. Roy. Soc. London, A325, (1971), 323-345.

[12] Spikes, H A, "The Behaviour of Lubricants in Contacts: Current Understanding and Future Possibilities", 7th BP Tribology Lecture (1992), IMechE, London, (to be published).

[13] Dowson, D and Jin, Z M, "Micro-Elasto-hydrodynamic Lubrication of Synovial Joints", Engineering in Medicine, (1986), Volume 15, No. 2, pp 63-65.

[14] Dowson, D and Jin, Z M, "Micro-elasto-hydrodynamic Lubrication of Low-Elastic-Modulus Solids on Rigid Substrates", Journal of Physics D: Applied Physics 'Frontiers of Tribology', (1992), Vol. 25, 1A, 116-123.

[15] Stribeck, R, 'Die Wesentlichen Eigenschaften der Gleit und Rollenlager', Z. Ver. dt. Ing. (1902), 46, No. 38, 1341-8; No. 39, 1463-70.

[16] Thurston, R H, "A Treatise on Friction and Lost Work in Machinery and Millwork", (1995), Wiley, New York.

[17] Gümbel, L, "Das Problem der Lagerreibung", Mbl, berl, Bez Ver, dt. Ing. (VDI) (1914), 87-104; 109-120.

[18] Michell, A G M, "Lubrication - Its Principles and Practice", (1950), Blackie, London and Glasgow.

[19] Heiderbroek, E, "General Discussion on Lubrication and Lubricants", (1937), IMechE, Vol. 1, 133-137.

[20] Muskat, M and Evinger, H H, J. App. Phys. (1940), 739-748.

[21] Laudman, U, Luedtke, W D and Ringer, E M, "Molecular Dynamic Simulations of Adhesive Contact Formation and Friction", (1992), in Singer, I L and Pollock, H M eds., Fundamentals of Friction: Macroscopic and Microscopic Processes, Kluwer Academic Publishers, The Netherlands, 463-510.

[22] Robbins, M, Proceedings of the 19th Leeds-Lyon Symposium on Tribology, (1993), 'Thin Films in Tribology', Elsevier Science and Technology.

SESSION II
INVITED REVIEWS

Chairman :	Professor M Godet
Paper II (i)	**Non-Newtonian Effects in Elastohydrodynamic Lubrication**
Paper II (ii)	**Engineering the Surface to Combat Wear**
Paper II (iii)	**Local Effects in Thin Film Lubrication**

Thin Films in Tribology / D. Dowson et al. (Editors)
© 1993 Elsevier Science Publishers B.V. All rights reserved.

NON-NEWTONIAN EFFECTS IN ELASTOHYDRODYNAMIC LUBRICATION

K.L. Johnson

Cambridge University Engineering Department
Trumpington Street, Cambridge CB2 1PZ, UK

Researches during the last two decades into the rheology of typical EHD lubricants have establised that they exhibit several features of non-Newtonian behaviour at the pressures and shear rates encountered in concentrated contacts. Viscoelasticity and non-linearity (shear thinning) in the shear stress-strain rate relation, including rate independent shear in the manner of a plastic solid, have been identified. It is well known that the film thickness in a conventional EHD contact, having smooth surfaces, is established in the entry zone where the pressure is relatively low. Since non-Newtonian effects generally only manifest themselves at high pressure, the effect on smooth-surface film thickness is small. However those aspects of EHL such as traction, which are governed by the viscosity of the oil in the centre of the nip, are strongly influenced by non-Newtonian properties. Much current interest centres on the micro-EHL of rough surfaces, which it is hoped will throw more light on the mechanism of film breakdown and 'scuffing' failure. Since the asperities extend through the nip where the pressure is high, micro-EHL mechanics involves the non-Newtonian properties of the lubricant.

1. INTRODUCTION

Researchers in elastohydrodynamic lubrication have long appreciated that lubricants which are Newtonian at ambient pressure exhibit non-Newtonian properties at the extreme pressures encountered in concentrated contacts. This fact seems less clearly understood by the wider tribological community largely, I suspect, on account of the success of Newtonian theory in predicting EHL film thicknesses. The reason for this apparent paradox is also clear: the film thickness in an EHD contact is largely established in the entry zone where the pressure is comparatively low and the fluid is more or less Newtonian. Serious non-Newtonian behaviour would only be expected in the central zone where the pressure is high, and where it might be expected to influence the resistance to sliding - the traction - and micro-EHL of surface asperities as they pass through the nip. Both of these important aspects will be considered below.

Boundary lubrication which roughly obeys Amontons' law, with a coefficient of friction around 0.1, was thought to be something different from either hydrodynamic or elastohydrodynamic lubrication. It can now be largely explained as a manifestation of the non-Newtonian behaviour of lubricants at high asperity pressures.

2. NON-NEWTONIAN CHARACTERISTICS

2.1. Observations of non-Newtonian behaviour

The earliest and most direct way of observing non-Newtonian response in EHL is by carrying out a 'traction test' in a rolling contact disc machine at a high (constant) load. Measuring the variation of traction force with sliding speed, making the appropriate correction for viscous heating and knowing the film thickness, gives a plot of mean shear stress $\bar{\tau}$ against shear strain rate $\dot{\gamma}$: an isothermal 'traction curve' (Fig. 1). Typically this comprises a linear region (which would be expected with a Newtonian oil), followed at high stress by a non-linear (shear thinning) zone, followed at even higher stress (if the pressure is high enough) by a zone in which the fluid shears at a constant stress independently of shear rate. The

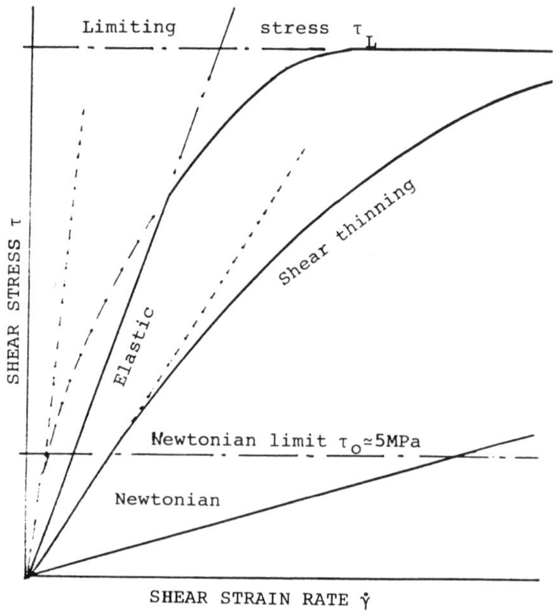

Figure 1. Non-Newtonian effects in EHL at constant temperature and pressure.

behaviour at high shear stress is clearly non-Newtonian.

The disc machine in which the pressure varies through the nip from zero to a maximum and to zero again in a time of 1.0 ms is not an ideal way of studying lubricant properties which are a strong function of pressure, particularly since only the *mean* shear stress can be measured. It is clearly desirable to supplement the disc machine results by more conventional rheometric measurements under conditions of constant pressure and temperature, but the high pressures exerienced in EHD contacts makes this a daunting challenge. Important contributions to the study of the non-Newtonian properties of lubricants using high-pressure rheometers have been made by Lamb and co-workers [1], Bair & Winer [2,3] and Jacobson [4]. High pressure measurements have been made using impact or stress wave techniques by Jacobson [4], Clyens [5] and Ramesh [6].

From these researches there is now general agreement that non-Newtonian properties of common EHD lubricants are manifested in three fairly distinct ways:

(i) **Viscoelasticity**, in which an increase in shear rate at small shear strains causes the response of the fluid to change from viscous (liquid-like) to elastic (solid-like).
(ii) **Plasticity**, in which the fluid shears at a limiting stress τ_L which is independent of shear strain and shear rate in the manner of a perfectly plastic solid.
(iii) **Shear thinning**, in which the relationship between shear stress and shear rate is non-linear with decreasing effective viscosity.

2.2. Interpretation

Although there is general agreement that the above aspects of non-Newtonian behaviour arise in EHD contacts, depending on the conditions, there are features of their interpretation which remain to be resolved. *Viscoelastic response* of lubricating fluids has been demonstrated (1) by high-frequency oscillating shear experiments on a cooled and/or pressurised sample of the fluid. The transition from viscous to elastic response is governed by the Deborah number D, which is defined by

$$D = \frac{\text{Relaxation time}}{\text{Deformation time}}$$

where the relaxation time λ may be expressed

$$\lambda = \frac{\text{Viscosity } \eta}{\text{Elastic shear modulus G}}$$

Viscoelastic effects may be expected when D > 1.0. They arise in EHD contacts through the transient nature of the deformation. The fluid passes through the contact in a time $t_1 = \ell / \bar{U}$, typically 0.1 - 1.0 ms, where ℓ is the length of the contact and $\bar{U}$ is the entrainment velocity. The Deborah number may therefore be written:

$$D = \frac{\eta}{G} / t_1 = \frac{\eta \bar{U}}{G\ell} \tag{1}$$

The oscillating shear tests reveal that the shear modulus of typical oils varies linearly with pressure with a value of ~ 10^9 Pa at EHD pressures. Since viscosity increases exponentially with pressure, D will increase with pressure and will exceed unity when η exceeds ~ 10^6 Pas, which is commonly reached at EHD pressures. The viscoelastic effect influences the gradient of the linear part of the traction curve at very small sliding speeds and high contact pressures [7,8].

All methods of measurement: disc machine, high-pressure viscometer or impact reveal a limiting shear stress at large strains which is related to pressure by a linear law of the form

$$\tau_L = A + Bp \tag{2}$$

where A is generally small compared with Bp at EHD pressures and B is a constant for each fluid whose value (dependent on temperature) generally lies in the range 0.05 - 0.12. Measurements of elastic shear modulus G show that it varies with pressure according to a linear relationship similar to equation (2) such that, over a wide range of pressure, the ratio of G to τ_L is a constant ≈ 30 (Fig. 2). This result

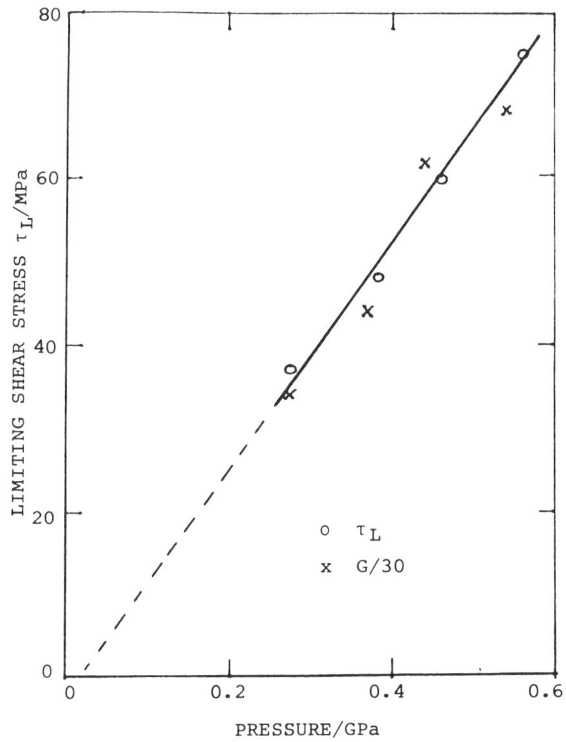

Figure 2. Comparison of limiting shear stress τ_L and elastic shear modulus G for 5P4E from [12].

strongly suggests that the limiting shear stress τ_L at large strains is governed by the same intermolecular (van de Waals) forces which govern the elastic shear modulus G at small strains [9], and represents the ultimate shear strength of the material.

There is less agreement about the interpretation of the shear thinning zone. Two different approaches have been taken. The first taken by Hirst & Moore [10] and Johnson & Tevaarwerk [11] is based on the Eyring thermal activation theory of fluid flow; the second, taken by Bair & Winer [12] and Jacobson [4] is based on the 'free volume' concept of polymeric fluids.

The Eyring model starts from the idea that in a fluid nominally 'at rest', molecules (or segments of larger molecules), are randomly moving relative to each other when their individual thermal energy $k\Theta$ exceeds the activation energy E, i.e. the irreversible work which has to be done against intermolecular (van de Waals) forces in moving between adjacent positions. If the pressure is increased by p the activation energy is increased by $v_p p$, where v_p is the pressure 'activation volume', i.e. the local expansion which is necessary to accommodate the movement. If now a shear stress τ is applied, the random thermal motion in the direction of the stress is assisted and that in the opposite direction resisted by the work $v_\tau \tau$, where v_τ is the shear activation volume, i.e. the displaced volume associated with the molecular movement. Its magnitude is that of the molecular segment undergoing the movement, say 10^{-27} m^{-3}. By standard thermal activation theory the flow equation relating the shear rate $\dot{\gamma}$ to the shear stress τ may be written

$$\dot{\gamma} = Ak\Theta \exp\left(-\frac{E + v_p p}{k\Theta}\right) \sinh\left(\frac{v_\tau \tau}{k\Theta}\right) \tag{3}$$

where A is a constant, k is Boltzmann's constant and Θ is the absolute temperature. Equation (3) may be rewritten:

$$\eta\dot{\gamma} = \tau_o \sinh(\tau / \tau_o) \tag{4}$$

where the viscosity η is given by

$$\eta = \frac{1}{Av_\tau}\exp\left(\frac{E + v_p p}{k\Theta}\right) = \eta_o(\Theta)\exp(\alpha p) \qquad (5)$$

and the pressure coefficient of viscosity $\alpha = v_p/k\Theta$. Taking v_p to be about $v_\tau/10$ gives $\alpha \approx 2.6\times10^{-8}\ Pa^{-1}$ which is typical for EHD lubricants. When $\tau << \tau$ equation (4) reduces to the Newtonian form:

$$\tau = \eta\dot{\gamma} = \eta_o\dot{\gamma}\exp(\alpha p) \qquad (6)$$

and when $\tau >> \tau_o$, $\sinh(\tau/\tau_o) \approx \frac{1}{2}\exp(\tau/\tau_o)$ so that equation (4) becomes

$$\tau = \tau_o Ln(2\eta\dot{\gamma}/\tau_o) \approx \tau_o Ln\left(\frac{2\eta_o\dot{\gamma}}{\tau_o}\right) + \alpha\tau_o p \qquad (7)$$

The transition from Newtonian (linear) to non-Newtonian (logarithmic) behaviour occurs when $\tau \approx \tau_o = k\Theta/v_\tau \approx 4\,MPa$. This model received good support from the disc machine traction tests. For the three fluids tested the non-Newtonian transition was observed at a shear stress in the range 2-8 MPa as predicted and plots of mean shear stress $\bar{\tau}$ against log (shear rate $\dot{\gamma}$) gave straight lines (see Fig. 3), as expected from equation (7).

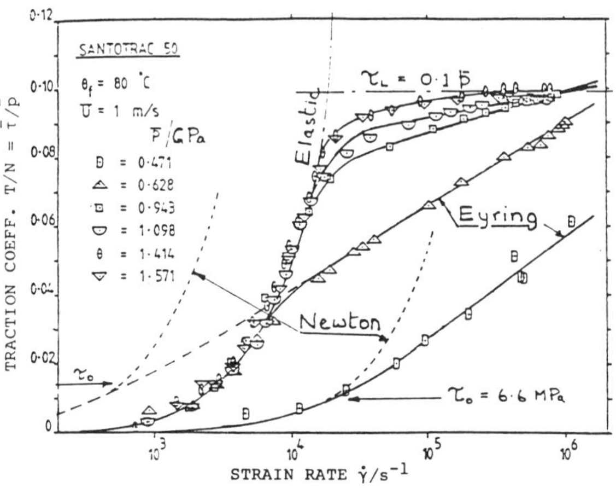

Figure 3. Traction curves for Santotract 50 at 80°C showing shear thinning (eq. 4,9) the viscoelastic effect and limiting stress, from [7].

It has not proved easy to reproduce the conditions of Fig. 4 in a high-pressure viscometer, but the transition to shear thinning at a shear stress > 1.0 MPa has been observed by Jacobson & Höglund [13]. In their tests the fluid is sheared at a constant rate and the shear stress measured as a function of pressure (see Fig. 4). At low pressure, when $\tau << \tau_o$, τ varies exponentially with p according to equation (6). At high pressure, where $\tau >> \tau_o$, and at constant strain rate $\dot{\gamma}$, τ varies linearly with p according to equation (7), with gradient

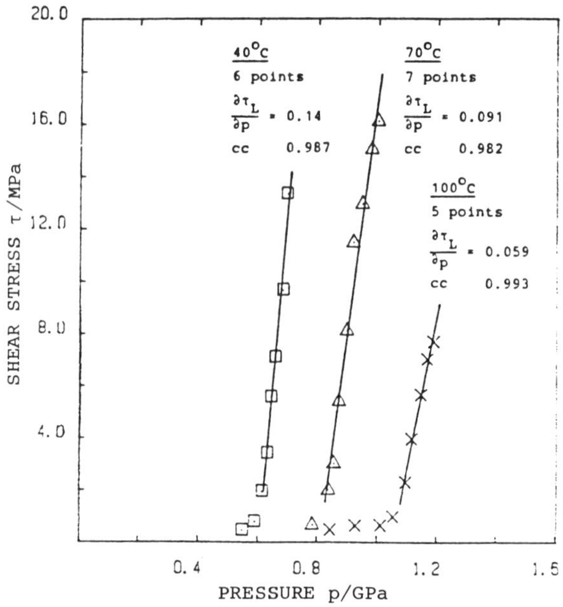

Figure 4. Measurements of shear stress τ as a function of pressure p at constant shear rate, from [13].

$\partial\tau/\partial p = \alpha\tau_o$. Values of $\alpha\tau_o$ obtained in the disc machine are compared wth $\partial\tau/\partial p$ from Jacobson & Hogland (Fig. 5) are given below.

	40°C	70°C	100°C
$\partial\tau/\partial p$ from ref. [13]	0.14	0.091	0.059
$\alpha\tau_o$ from ref. [7]	0.17	0.15	0.14

Although the Eyring theory appears to model well the transition from Newtonian to shear thinning flow, it does not explain the transition from non-linear flow to limiting plastic shear.

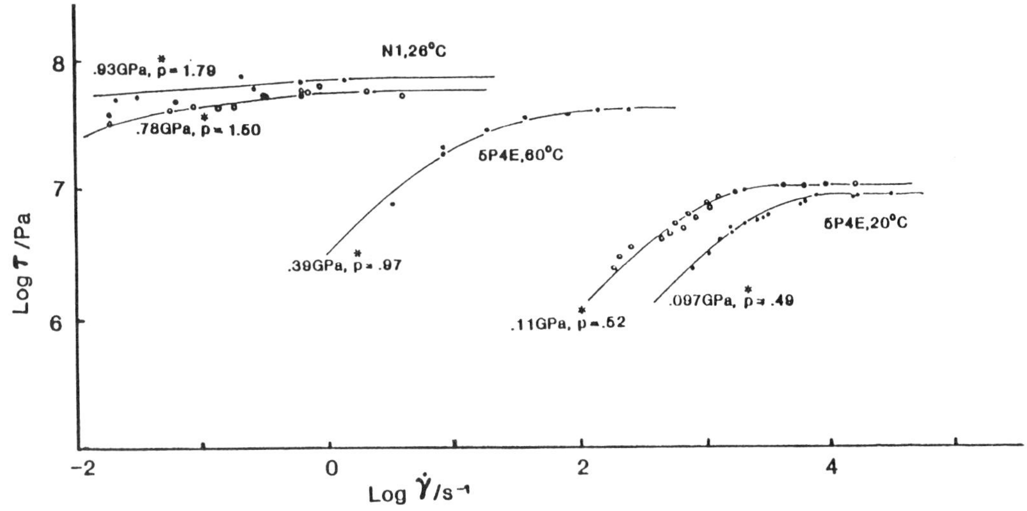

Figure 5. Transition from Newtonian to plastic flow (eq. 8), from [12].

The alternative approach invokes the property of amorphous polymers known as the 'glass transition'. This is defined as the state (temperature and pressure) at which a discontinuity in thermal expansion coefficient or bulk modulus is observed. It is modelled by imagining the specific volume to be made up of two parts: the 'free volume' and the 'occupied volume'. The glass transition occurs when, either by thermal contraction or by compression, the free volume is eliminated leaving only the occupied volume. In this state the material deforms like an amorphous solid: elastic at small strains and plastic at large strains. If liquid-like rate dependent flow takes place, it is with a greatly enhanced relaxation time. The viscosity at the glass transition is of order 10^7 Pas or greater [14].

The free volume model of Williams et al. [15] has been modified by Yasutomi et al. [16] to included effect of pressure on the glass transition temperature Tg and has been shown to fit the variations with temperature and pressure of the Newtonian (low shear stress) viscosity of several lubricants. However this model does not address the onset of shear thinning at high shear stress. This has been done by Bair & Winer [12] though an equation derived from their high pressure viscometer experiments:

$$\eta\dot{\gamma} = \tau_L \exp(p^*\tau^*)^3 \ \mathrm{Ln}\{1/(1-\tau^*)\} \tag{8}$$

where $\tau^* = \tau/\tau_L$ and $p^* = p/p_g$ (p_g = glass transition pressure). This equation is fitted to the observations in Fig. 5.

2.3 Constitutive equations

The manifestation of non-Newtonian behaviour discussed above: viscoelasticity, shear thinning and limiting stress, can be conveniently combined into a non-linear Maxwell law which, for one dimensional flow, may be written as the sum of a linear elastic term and a non-linear viscous term:

$$\eta\dot{\gamma} = \frac{\eta(p,\Theta)}{G(p,\Theta)} \cdot \frac{d\tau}{dt} + F(\tau) \tag{9}$$

If the Eyring model is adopted, the non-linear viscous term $F(\tau)$ takes the form of the right-hand side of equation (4) provided $\tau < \tau_L$. Otherwise $\tau = \tau_L$ = constant independent of $\dot{\gamma}$. If the glass transition model is adopted, $F(\tau)$ takes the form of the right-hand side of equation (8). In either case equation (9) can be generalised by the von Mises hypothesis to accommodate simultaneous shear in two directions such as arises in a rolling contact with spin [11,12].

At this juncture it is important from an engineering point of view to emphasise the similarity between these two models rather than their differences. In circumstances where the elastic term is negligible (Deborah no. << 1), the relationship between shear stress τ and shear rate $\dot{\gamma}$ is given by equations (4) and (8). They both show the onset of shear thinning (i.e. limit of Newtonian behaviour) at a shear stress in the range 1.0 - 10 MPa and a limiting stress in the range 0.07 p - 0.10 p for most fluids. Temperature has only a minor effect.

Indeed useful engineering calculations have been carried out by further simplified constitutive equations. If the viscous term in equation (9) is taken to be Newtonian up to the limiting shear stress, i.e. $F(\tau) = \tau$ for $\tau < \tau_L$, equation (9) represents a linear viscoelastic (Maxwell) fluid, and when D << 1, it reduces to a Newtonian liquid.

Another simple constitutive law which includes shear thinning and limiting stress τ_L has been used by Lee & Hamrock [17]. Described as a 'circular' non-Newtonian model, it puts

$$F(\tau) = \tau\left\{1-(\tau/\tau_L)^2\right\}^{-1/2} \qquad (10)$$

2.4 Non-Newtonian Reynolds' equation

In order to perform EHD calculations for a non-Newtonian fluid it is convenient to derive an appropriate form of Reynolds' equation. This has been done for the one-dimensional flow of an Eyring fluid, specified by equation (4), by Houpert & Hamrock [18] and independently by Conry et al. [19]. In the form suggested by Greenwood & Johnson [10] it may be written:

$$1-\frac{h^*}{h} = \left[S^2 + \left(\frac{\sinh t_p}{t_p t_v}\right)^2\right]^{\frac{1}{2}}\left(\coth t_p - \frac{1}{t_p}\right) \qquad (11)$$

where $t_p = \dfrac{(h/2)(dp/dx)}{\tau_o}$, $t_v = \dfrac{\eta(u_1+u_2)}{h\tau_o}$,

$S = \dfrac{u_2-u_1}{u_2+u_1}$; u_1 and u_2 are the surface speeds, h is the film thickness and h^* its value where the pressure gradient (dp/dx) = 0.

Note that the shear stresses at the wall for a Newtonian fluid are $\tau_{1,2} = St_r \pm t_p$ so that t_p and st_r may be interpreted as the ratios of the Poiseuille shear t_p and Couette shear t_r to t_o. Thus non-Newtonian action would be expected when either t_p or St_r exceeds unity. Since t_v is very high (>100) in the high pressure region of an EHL contact, the Newtonian limit is reached at very small values of slide-roll ratio (S < 0.01). At greater values of S the second term in the square bracket of equation (11) becomes negligible whence

$$1-\frac{h^*}{h} = |S|\left(\coth t_p - \frac{1}{t_p}\right) \approx \frac{1}{3}|S|t_p \qquad (12)$$

From this we see that, in the non-Newtonian regime, the film shape is sensitive to the slide roll ratio, whereas in Newtonian conditions the film shape is independent of S.

A comparable non-Newtonian Reynolds' equation based on the 'circular' constitutive equation (10) (neglecting viscoelastic effects) has been derived by Lee & Hamrock [17].

Difficulties arise in developing a Reynolds' equation for two dimensional flow due to the von Mises quadratic interaction between the two components of shear stress τ_{xz} and τ_{yz}.

3. INFLUENCE OF NON-NEWTONIAN FLOW ON EHL CONTACTS

3.1 Film thickness

As mentioned at the outset, the success of Newtonian analysis in predicting EHD film thicknesses suggests that non-Newtonian properties of the lubricant exert a second order effect. In pure rolling, where the flow is Poiseuille the shear stress is given by (h/2)(dp/dx), which does not reach the critical level in the entry region where the film thickness is determined. Film thickness is therefore only influenced by *sliding* as implied by equation (12). Even so the effect is generally small.

A Grubin-type analysis of the entry region by Gecim & Winer [21] indicated a reduction in film thickness by about 30% at a slide-roll ratio S = 2. More complete numerical analyses of line contacts by Lee & Hamrock [17] using the 'circular' law of equation (10) and by Conry et al. [22] using the Eyring equation (11) came to broadly the same conclusion.

Interferometric measurements by Jacobson [4] of the non-Newtonian influence of sliding on the film thickness in circular point contacts are shown in Fig. 6. The central film thickness is reduced somewhat by sliding, but the minimum thickness is hardly affected.

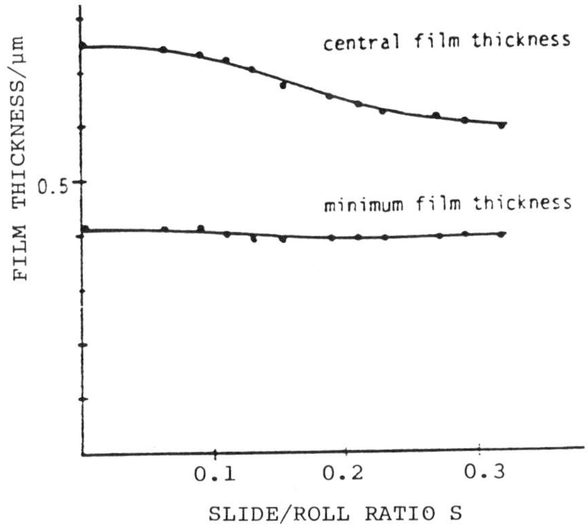

Figure 6. Non-Newtonian effect on film thickness of a smooth point contact, from [4].

3.2 Traction

The profound effect of non-Newtonian properties on traction forces in rolling/sliding contact has already been mentioned.

The viscoelastic effect, which arises when the relaxation time of the fluid (η/G) exceeds the time of passage through the contact, influences the linear part of the traction curve at small strain rates (concave upwards in the $\bar{\tau}/\bar{p}-\log\dot{\gamma}$ plot in Fig. 3). Shear thinning according to the Eyring hypothesis appears as straight lines in Fig. 3 (equation (4)). The limiting stress τ_L is only reached at the highest pressure ($\bar{p}=1.57$ GPa) in traction tests on Santotrac 50 at 100°. Extrapolating from the data of Alsad et al. [14] suggests that the glass transition at 100°C occurs at a pressure of about 1.4 GPa.

A traction map for EHL contacts lubricated with Santotrac 50 is presented in Fig. 7. The coordinates ($\alpha_o\bar{p}$) and ($\alpha_o\eta_o\bar{U}R'$) are known if the operating conditions of the contact are specified. In regime I the behaviour is Newtonian in which equation (6) applies. In regime II shear thinning occurs and equation (4) applies. The isothermal traction curves will be of the form shown in Fig. 4 for $\bar{p}$ = 0.47 and 0.62 GPa. In regime III the viscoelastic effect occurs at small strain rates, as seen in the curves at pressures 0.78 - 1.09 GPa. In regime IV the behaviour is elastic at small strain rates and plastic at the limiting shear stress at higher strain rates, as found when $\bar{p}$ = 1.57 GPa.

The above discussion and the data of Figs. 4, 6 and 8 refer to full-film EHL where the ratio Λ of the nominal film thickness to combined surface roughness exceeds 3. At lower values of Λ local variations in pressure caused by the surface roughness begin to influence the traction [23,24,25]. With well run-in surfaces micro-EHL maintains thin films even though $\Lambda < 1.5$. In these circumstances it is most likely that at the asperity pressures the lubricant will be operating in the elastic-plastic regime, in which the traction coefficient is in the range 0.05 - 0.10 independently of the load and speed. These are the circumstances which were described in the past as 'boundary lubrication'. It may now be seen to be a manifestation of non-Newtonian behaviour at high pressure.

3.3 Micro EHL: (a) Transverse roughness

The breakdown of EHD films in concentrated contacts normally occurs when operating in the 'mixed' regime in which Λ is less than 1.5 say. This has led to the study of micro-EHL, i.e. EHL theory applied to the asperities within a bulk contact. Since this action takes place where the bulk pressure is high and the local asperity pressures are even higher, it is evident that non-Newtonian properties of the oil should be taken into account.

Most rough finishing operations of metal surfaces, e.g. turning or grinding, leave the surface with a pronounced 'lay'. In an EHL contact the lay may be transverse to the direction of rolling and sliding, as in hobbed and ground gear teeth, or longitudinal as in turned and ground discs. It is not unrealistic therefore to model the surface roughness one-dimensionally and to treat the two cases separately. We shall consider transverse roughness first.

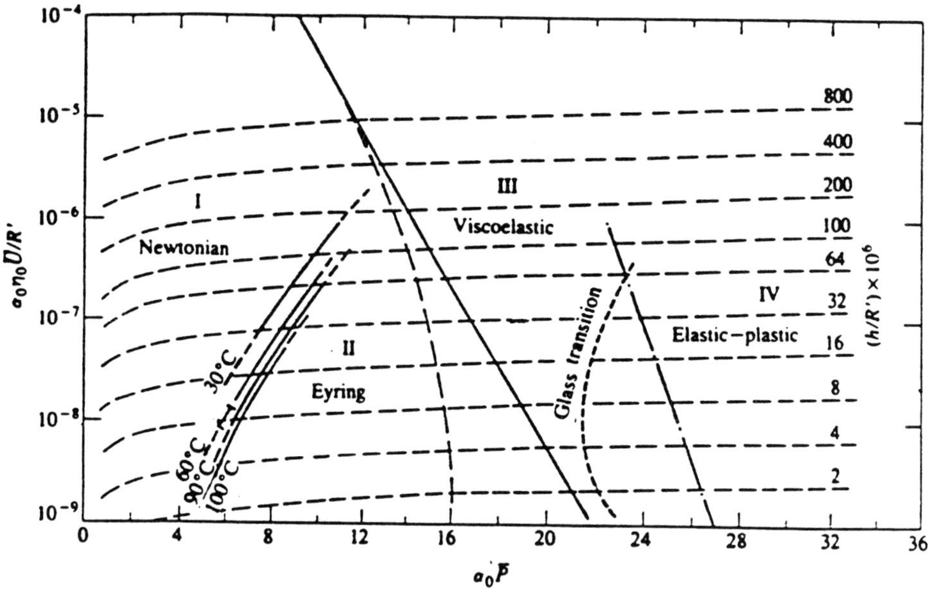

Figure 7. Traction map for Santotrac 50, from [8].

Distinction must be made between cases of steady and unsteady flow. Only for a sliding contact in which the rough surface is stationary and the moving surface is smooth, is the flow pattern steady and the steady-flow Reynolds' equation applicable. This is the case which has received most attention e.g. [26]. If one or both rough surfaces are moving an unsteady EHD analysis is required even if the bulk geometry is unchanging. This situation has been studied by Venner [27] for a Newtonian lubricant and Chang et al. [28] for an Eyring lubricant.

The situations which have been investigated include a regular stationary sinusoidal surface, a surface defect such as a bump or a groove both moving and stationary. In all these cases with a Newtonian lubricant the conclusion is the same: the bump and ripples are flattened by micro-elastohydrodynamic action to be replaced by a large fluctuation in pressure. An example of such a calculation, taken from Kweh et al. [26], is shown in Fig. 8. The undeformed sinusoidal stationary surface is shown dotted. The deformed surface is seen to be almost flat with a film thickness not very different from that for a smooth surface. The corresponding pressure ripple has an amplitude which is close to that required to flatten a surface having the amplitude Δ ($\pm$ 0.25 μm) and wavelength λ (0.2 mm) shown, i.e.

$$p^* = \pi E'\Delta / 2\lambda = 0.45 \text{ GPa} \tag{13}$$

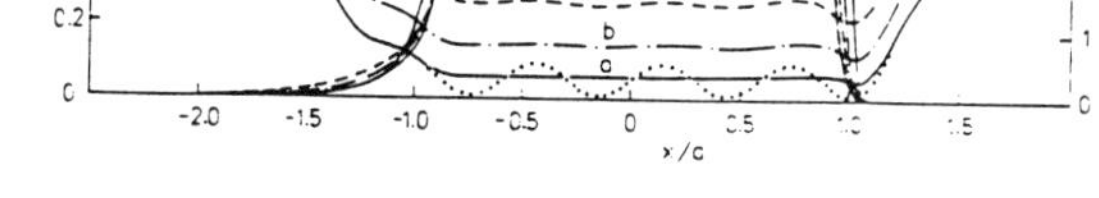

Figure 8. Newtonian analysis of transversely rough EHD contact, from [26].........Initial profile.

This result suggests that the EHD effect of the asperities can be examined separately from the bulk contact, with a considerable saving in computational complexity. Thus Greenwood & Johnson [20] have made a simple linearised Newtonian analysis of a flat surface in contact with an infinite stationary surface of sinusoidal profile. At a mean pressure representative of EHL contacts the surface ripples were squashed flat; the small residual ripple was almost entirely due to lubricant compressibility.

However, we must now test the results of Newtonian theory against the shear stress criteria of non-Newtonian flow. By way of example, consider the results of Kweh et al. shown in Fig. 8, taking the pressure fluctuation to be represented closely enough by equation (13). The Poiseuille shear stress $\tau_p = (h/2)(dp/dx)$ and the Couette shear $\tau_r = \eta(u_1 - u_2)/h$ are given in Table 1. If we take the onset of shear thinning to be at τ_o = 5 MPa and the limiting stress to be at $\tau_L = 0.08\ \bar{p}$ = 66 MPa, we note that the Poiseuille stress due to the pressure gradient only just reaches the onset of non-Newtonian behaviour. On the other hand, the Couette stress, due to the high sliding speed greatly exceeds the limiting stress of any mineral oil. It is clear from this simple 'shear stress test' that the results of the Newtonian calculations must be suspect at the very least.

Table 1

Speeds: u_1 = 24.9 m/s ; u_2 = 0

Mean Hertz pressure $\bar{p}$ = 0.83 GPa

Temperature/°C	50	80	140
Viscosity at $\bar{p}$/Pas	3879	204	6.93
Film thickness $h_o/\mu m$	1.43	0.78	0.33
Poiseuille stress τ_p/MPa	10.2	5.5	2.3
Couette stress τ_v/MPa	6.7×10^4	6.7×10^4	6.7×10^4

In order to assess the influence of shear thinning Greenwood (unpublished note) carried out a linearised analysis of a flat surface in sliding contact with a stationary infinite sinusoidal surface, using the approximate Reynolds-Eyring equation (12). The result is very different from that found with a Newtonian lubricant. The pressure fluctuation is much smaller and the amplitude of the deformed profile is close to its undeformed value; the main change being a small phase-shift.

Full computer studies of EHL contacts typical of steel gear teeth, using either the Reynolds-Eyring equation [26,28] or the Hamrock equation [17], come to the same conclusion. Surface irregularities, whether individual 'bumps', sinusoidal ripples or random roughness, are generally flattened by a Newtonian lubricant. Shear thinning, by restricting the shear stress which can be developed, leads to smaller fluctuations in pressure and much reduced deformation of the asperities. This is clearly an important area for further research, particularly into the circumstances of unsteady flow when two rough surfaces are moving.

3.4 Micro-EHL: (b) Longitudinal asperities

Several computer analyses have been made [27,28] of EHL contacts having regular longitudinal roughness. This situation is simpler than transverse roughness since it is reasonable to assume steady conditions. Also two rough surfaces can be reduced to one surface in rolling/sliding contact with a plane. At low values of Λ, the pressure over the ridges greatly exceeds that in the furrows, thereby inducing lateral pressure gradients $(\partial p/\partial y)$. With a Newtonian lubricant, however, a film of significant thickness is generated beneath the ridges which is unaffected by sliding.

With a non-Newtonian fluid the condition of high pressure and thin films at the asperity crests ensure that in longitudinal sliding the shear stress reaches its limiting value τ_L. In these conditions the longitudinal and lateral shear stresses τ_{xz} and τ_{yz} respectively are coupled by

$$\tau_{xz}^2 + \tau_{yz}^2 = \tau_L{}^2 \qquad (14)$$

If, due to longitudinal sliding τ_{xz} approaches τ_L a much reduced value of τ_{yz} is available to resist side leakage. Oil therefore leaks laterally from the crests of the ridges into the troughs, leading to a thinning of the film as it passes through the contact. This non-Newtonian effect was recognised by Jacobson et al. [29] and independently by Johnson & Higginson [30]. A simple analysis for a contact of rectangular planform (2a & 2b), assuming the fluid to be Newtonian up to the limiting shear stress, gave the following expression for the ratio of film thickness h at exit (x = 2a) to that at inlet (x = 0):

$$\frac{h_1}{h_o} = \frac{1}{1+AS} \tag{15}$$

where S = slide/roll ratio, $(U_1-U_1)/(U_1+U_1)$
$A = h_o a p_o / 2\tau_L b^2$
and p_o = centre line (y=0) pressure assumed constant.

It is clear from equation (15) that the effect is activated by the slide/roll ratio S. Electrical conductance tests were carried out in a disc machine in which one disc had a specially prepared circumferential sinusoidal roughness of wavelength 270 μm and peak-to-valley height 15 μm. The results for varying slide/roll ratios are shown in Fig. 9. The conditions change from almost zero conductance in pure rolling to complete conductance at S = 15%.

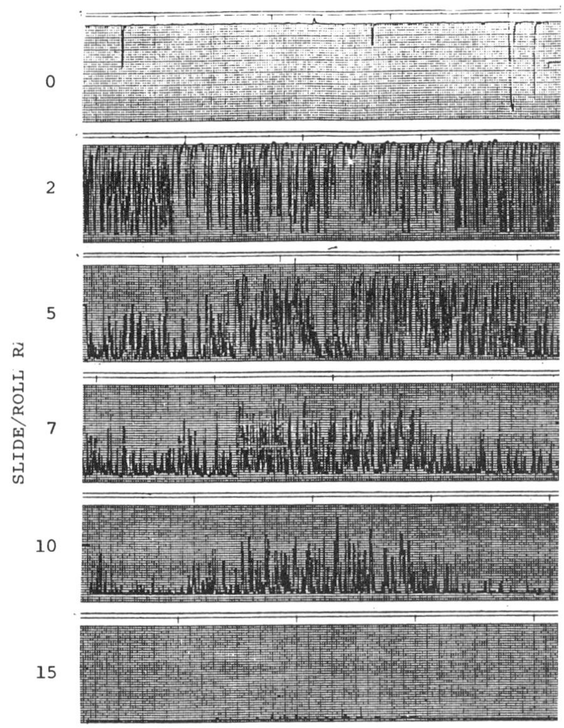

Figure 9. Conductance of a circumferentially rough disc with slide/roll ratio, from [32].

4. CONCLUSIONS

(1) Three aspects of non-Newtonian rheology have been observed in EHL contacts: (i) viscoelasticity, (ii) shear thinning and (iii) a limiting shear stress independent of shear rate.

(2) Viscoelastic effects occur when the relaxation time of the fluid (η/G) exceeds the time of passage of the fluid through the contact, which normally requires viscosities in excess of 10^6 Pas. Viscoelasticity is only important in the linear zone of small slide/roll ratios.

(3) Shear thinning sets in when the shear stress in the fluid reaches a value ~ 5 MPa. Different models of the process have been suggested leading to somewhat different $\tau - \dot{\gamma}$ relationships. From the engineering point of view the choice of model is not critical.

(4) The shear stress reaches a limiting value $\tau_L \approx Cp$, where the fluid flows at constant stress independently of shear rate. C is a constant for a particular fluid, which decreases slightly with temperature and lies generally in the range 0.06 - 0.12

(5) Non-Newtonian properties have a negligible effect on the film thickness between smooth surfaces ($\Lambda >> 3$), except at very high slide/roll ratios.

(6) The traction in highly loaded EHL contacts is greatly reduced below its Newtonian value by shear thinning and limiting shear stress.

(7) Micro-EHD action between damaged or rough surfaces operating at low Λ ratio is strongly influenced by non-Newtonian behaviour. This is an important area of on-going research.

(8) Transverse asperities or surface defects are deformed by EHD action to a much smaller extent that Newtonian analysis would suggest.

(9) The limiting shear stress permits side leakage of oil from the crests of longitudinal asperities into the furrows during rolling with longitudinal sliding.

REFERENCES

1. Barlow, A.J. et al. The effect of pressure on the viscoelastic properties of liquids, Proc. Roy. Soc. Lond. A327 (1972) 403.
2. Bair, S. & Winer, W.O. Shear strength measurements of lubricants at high pressure. Trans. ASME, J. Lubric. Technol., Ser. F, 101 (1979) 251.
3. Bair, S. & Winer, W.O. Some observations in high pressure rheology of lubricants. Trans. ASME, J. Lubric. Technol., Ser. F, 104 (1982) 357.
4. Jacobson, B.O. *Rheology and EHL*, Elsevier, 1991.
5. Clyens, S., Evans, C.R. and Johnson, K.L. Measurement of viscosity of supercooled liquids at high shear rates with a Hopkinson torsion bar. Proc. Roy. Soc. Lond., Ser. A, 381 (1982) 195.
6. Ramesh, K.T. and Clifton R.J. A pressure shear plate experiment for studying the rheology of lubricants at high pressures and high shearing rates. ASME Journal of Tribology, 109, No. 2 (1987).
7. Evans, C.R. and Johnson, K.L. Rheological properties of EHD lubricants. Proc. Inst. Mech. Engrs, Pt C, 200 (1986) 303.
8. Evans, C.R. and Johnson, K.L. Regimes of traction in elastohydrodynamic lubrication. Proc. Instn. Mech. Engrs, Part C, 200(C5) (1986).
9. Tabor, D. in *Microscopic aspects of adhesion and lubrication*, ed. J.M. Georges, Elsevier (1981) 651.
10. Hirst, W. and Moore, A.J. Non-Newtonian behaviour in EHL. Proc. R. Soc. Lond., Ser. A, 337 (1974) 101.
11. Johnson, K.L. and Tevaarwerk, J.L. Shear behaviour of elastohydrodynamic oil films. Proc. R. Soc. Lond., Ser. A., 356 (1977) 215.
12. Bair, S. and Winer, A.O. The high pressure high shear stress rheology of liquid lubricants. Trans. ASME Ser. F, J. of Tribology, 114 (1992) 1.
13. Jacobson, B.O. and Höglund, E. Experimental investigations of the shear strength of lubricants subjected to high pressure. Trans. ASME, Ser. F, J. Lub. Tech. 108 (1986) 571.
14. Alsaad, M., Bair, S., Sandborn, D.M. and Winer, W.O. Glass transition in lubricants: its relation to EHL. Trans. ASME, J. Lub. Tech., Ser. F, 100 (1978) 404.
15. Williams, M.L., Landel, R.F. and Ferry, J.D. Relaxation mechanism in amorphous polymers and glass forming liquids. J. Am. Chem. Soc. 77 (1955) 3701.
16. Yasutomi, S., Bair S. and Winer, W.O. An application of a free volume model to lubricant rheology. Trans. ASME, Ser. F, J. of Tribology, 106 (1984) 291.
17. Lee, R-T. and Hamrock, B.J. A circular non-Newtonian fluid model. Trans. ASME, Ser. F, J. of Tribology, 112 (1990), 486.
18. Houpert, L. and Hamrock, B.J. EHL calculations used as a tool to study scuffing. In *Proc. 12th Leeds-Lyon Symp.*, Butterworths, London (1986) 146.
19. Conry, T.F., Wang, S. and Cusano, C. A Reynolds'-Eyring equation for EHL in line contacts. Trans. ASME, Ser. F, J. of Tribology, 109 (1987) 648.
20. Greenwood, J.A. and Johnson, K.L. The behaviour of transverse roughness in sliding EHL contacts, Wear, 153 (1992) 107.
21. Gecim, B. and Winer, W.O. Lubricant limiting shear stress effect on EHD film thickness. Trans. ASME Ser. F, J. of Tribology, 102 (1980) 213.
22. Wang, S., Cusano, C. and Conry, T.F. Thermal analysis of EHL of line contacts using the Ree-Eyring fluid model. Trans. ASME, Ser. F, J. of Tribology, 113 (1991) 232.
23. Bair, S. and Winer, W.O. Regimes of traction in concentrated contact lubrication. Trans. ASME, Ser. F, J. of Tribology, 104 (1982) 382.
24. Evans, C.R. and Johnson, K.L. The influence of surface roughness on EHD traction. Proc. Inst. Mech. Engrs. Pt C. 201 (1987) 45.
25. Schipper, D.J., Vroegop, P.H., de Gee A.W.J. and Bosma, R. Micro-EHL in lubricated concentrated contacts. Trans. ASME, Ser. F, J. of Tribology, 112 (1990) 392.

26. Kweh, C.C., Evans, H.P. and Snidle, R.W. Micro-EHL of an elliptical contact with transverse and three-dimensional sinusoidal roughness. Trans. ASME, Ser. F, J. of Tribology, 111 (1989) 577.
27. Venner, C.H. Multilevel solution of the EHL line and point contact problems. Ph.D. dissertation, University of Twente, Enschede, NL, 1991.
28. Chang, L. Cusano, C. and Conry, T.F. Effects of lubricant rheology and kinematic conditions on micro-EHL. Trans. ASME, Ser. F, J. of Tribology, 111 (1989) 344.
29. Karami, G., Evans, H.P. and Snidle, R.W. EHL of circumferentially finished rollers having sinusoidal roughness. Proc. Inst. Mech. Engrs. Pt C, 201 (1987), 29.
30. Barragan de Ling, F. dM., Evans, H.P. and Snidle, R.W. Micro-EHL of circumferentially finished rollers: the influence of temperature and roughness. Trans. ASME, Ser. F, J. of Tribology, 111 (1989) 730.
31. Jacobson, B.O., Ioannides, E. and Tripp, J.H. Redistribution of solidified films in rough Hertzian contacts. Proc. 14th Leeds-Lyon Symp., Lyon (1987) 51.
32. Johnson, K.L. and Higginson, J.G. A non-Newtonian effect of sliding in micro-EHL. Wear, 128 (1988) 249.

Thin Films in Tribology / D. Dowson et al. (Editors)
1993 Elsevier Science Publishers B.V.

Engineering the Surface to Combat Wear

T. Bell

Wolfson Institute for Surface Engineering, School of Metallurgy and Materials, University of Birmingham, PO Box 363, Birmingham, B15 2TT, UK

The disciplines of tribology and surface engineering are outlined as a basis for the design of engineering surfaces to combat wear. It is demonstrated that a deep metallurgical insight is essential when engineering the surface of stainless steel for protection against both corrosion and wear and that a new hybrid technology involving plasma enhanced ion implantation appears to show great promise in this field. Liquid phase alloying of titanium using high energy beam surface engineering technologies is shown to be of value in heavy load contact situations, whereas modest load wear conditions are readily combated using thermochemical conversion technology. In complex design situations two surface technologies can be combined - duplex surface engineering - to give very substantial improvements in wear and load bearing capacity. This duplex approach is illustrated through the industrially accepted TiN/plasma nitrided combination technology. Recent contact mechanics modelling of this duplex system is reviewed showing that thin ceramic coatings can act as "stress barrier" coatings on realistic industrial engineered surfaces.

1. INTRODUCTION

The concept of tribology was introduced in 1966(1) to cover the overall subject of interacting surfaces in relative motion. This multi-disciplinary subject embracing engineering, physics, chemistry and metallurgy has been developed since on an international basis in an attempt to offset the great economic losses, mainly of an avoidable nature, associated with the wear of engineering components. The early 1980's saw the development of the related interdisciplinary subject of surface engineering.(2,3) The impetus for this development was stimulated, at least in part by the growing commercial maturity of a wide range of surface technologies, such as laser and electron beam processing, plasma thermochemical techniques, innovative engineering coating and ion implantation surface modification methods. Surface engineering is not simply about using one of a number of techniques to coat or otherwise enhance surface properties, including tribological properties. Indeed, the new surface technologies, together with traditional surface treatments provide a complex portfolio of treatments which make possible the design and manufacture of lamellae composite engineering components. By changing the structural characteristics or the composition of the surface of a component, new desirable engineering properties can be achieved. Implicit in the decision to engineer the properties of the surface means that some existing properties of the substrate material are adequate for the component to perform satisfactorily from a technical standpoint and to be economically viable. This is reflected in the concept that Surface Engineering is(4):

"The design of surface and substrate together as a system to give a cost effective performance enhancement of which neither is capable on its own."

The portfolio of surface technologies which are available to the design engineer is summarised in Fig. 1, where it can be seen that the thickness of the engineered surface can vary from several millimetres for weld overlays, to a few micrometres for PVD and CVD coatings,

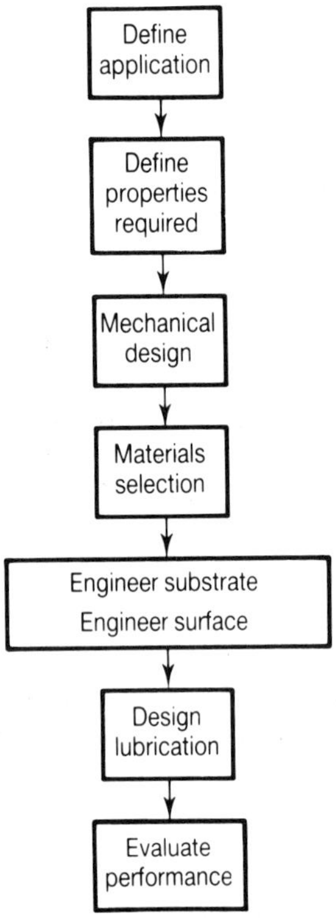

Figure 1. Sequence of steps in designing a tribological component.

whilst the surface modification depth brought about by ion implantation is 0.1 μm or less. Likewise, coating hardnesses span a wide range from about 250-350 H_v for some spray coatings, to 1000 H_v for nitrided steels, and to 1300-1600 H_v for detonation gun (D-gun) carbide-metal cermet coatings, while up to 3500 H_v can be achieved for PVD (TiN) coatings; diamond-like films can achieve hardnesses approaching that of natural diamond.

2. DESIGNING ENGINEERED SURFACES

The sequence of events involved in the manufacture of components for use in the machine building sector of industry, and which involve surface engineering procedures are summarised in Fig. 2.(5,6) The first step in selecting a surface modification treatment is to determine the surface and substrate engineering requirements. Frequently, they involve but one of the sets of conditions listed in Table 1.(7) However, there can be many challenging design situations where there is a requirement for a combination of several properties. Although procedures are available(8,9) to assist the designer in the complex task of selecting optimum and economically viable surface technologies to meet the required wear/friction/corrosion/fatigue properties, they have usually been developed for individual technologies, and frequently are lacking in detail due to an absence of specific experimental data. In the absence of a complete comprehensive design system, which is unlikely to be available for several years, a logical analysis of several surface engineering case studies relating to wear-resistant surfaces would seen to be a worthwhile approach for examining the impact of surface engineering on wear. Several such case histories are detailed in the literature(10-18) and can be illustrated through recent developments in the surface

Table 1
Properties required by engineering components

(1)	Abrasive wear resistance under conditions of low compressive loading
(2)	Abrasive wear resistance under conditions of high compressive loading
(3)	Resistance to scuffing or seizure
(4)	Bending or torsional strength
(5)	Bending or torsional fatigue strength
(6)	Resistance to mechanical pitting (surface contact fatigue)
(7)	Resistance to case crushing (surface collapse)
(8)	Resistance to corrosion and wear

THICKNESS OF COATINGS AND DEPTH OF DIFFUSION TREATMENTS

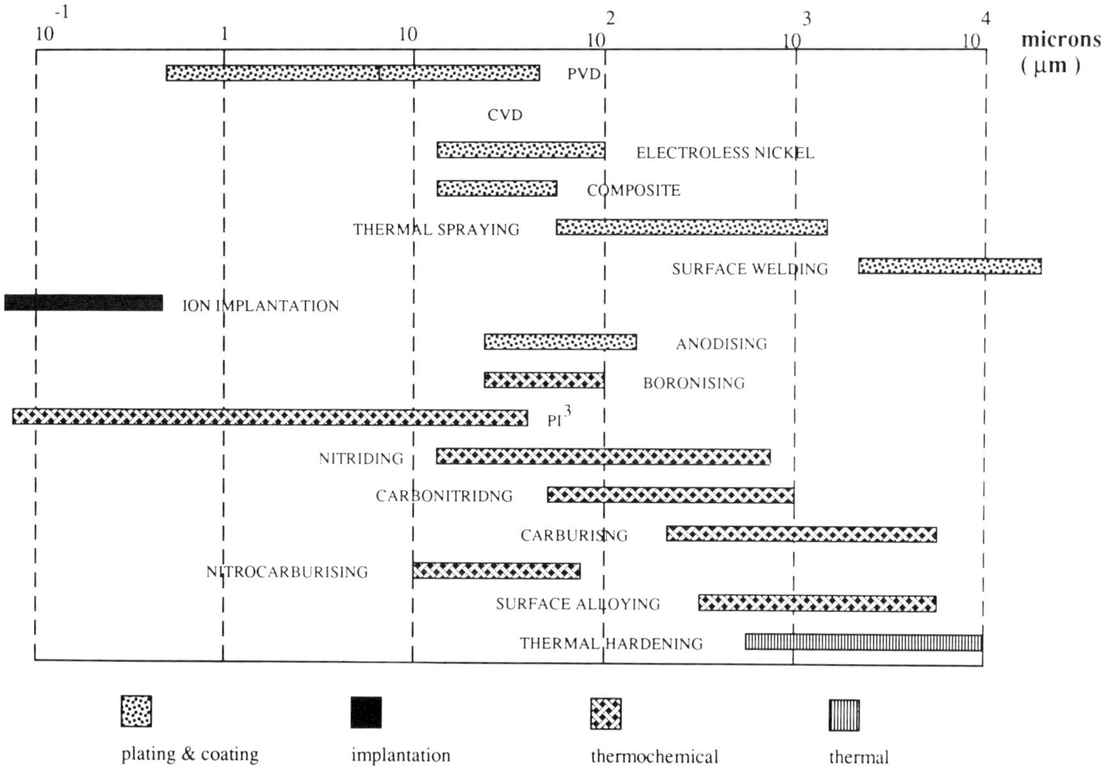

Figure 2. Classification of the Surface Engineering Technologies as a function of surface modification thickness.

engineering of stainless steel and titanium alloys.

3. SURFACE ENGINEERING OF STAINLESS STEEL

There is frequently an engineering requirement to improve the wear resistance of austenitic stainless steels. Plasma nitriding is particularly suitable for the surface modification of such high alloy steels as stainless steels, where the presence of oxide scales (Cr_2O_3) on the surface may prevent nitrogen mass transfer reactions during conventional gas nitriding. This advantage of plasma nitriding over conventional nitriding lies in the sputtering effects of energetic positive ions, which effectively remove such contaminants from the component surface, thus accelerating nitrogen mass transfer from the plasma to the component. A variety of austenitic stainless steels have been successfully nitrided for significantly improved wear resistance in a glow discharge plasma over a wide range of temperatures.

The most prevalent wear damage in austenitic stainless steel is gross plastic deformation.(19) Another factor influencing its wear performance is stacking fault energy(20) which varies from a low 20 m Jm^{-2} to an intermediate value of 80 m Jm^{-2} for type 316.(19) This is important since a low stacking fault energy indicates a faster work hardening rate. Conventional plasma nitriding of autenitic stainless steel at temperatures in excess of 500°C leads to the formation of a surface layer of the Fe_4N phase and precipitation of CrN in the nitrided case(21,22) and as a consequence the corrosion resistance is considerably reduced, accordingly in some circumstances the surface engineer is reluctant to specify plasma nitriding due to a requirement for combined wear and corrosion resistance (i.e., condition 8 in Table 1).

Over the past few years low temperature plasma nitriding in the temperature range 400-500°C has resulted in the formation of an expanded austenite phase at the surface, i.e., nitrogen is in solid solution, a condition which is known to increase pitting corrosion resistance.(23) Accordingly, there has been much interest in low temperature plasma nitriding to produce this expanded austenite which is known as the "S-phase". The "S-phase" can provide corrosion resistance equivalent to that of untreated material(21,22,24,25) and hardnesses in the range 980-1350 $H_{v0.025}$(21,22). Hannula et al(22) have suggested that the high hardness values are caused by nitrogen supersaturation of austenite, a high dislocation density and fine e nitride platelets in the "S-phase".

Recent work in Australia has led to a novel method of low temperature plasma nitriding. This is referred to in Fig.1 as PI^3, (Plasma Immersion Ion Implantation) from where it can be seen that material can be treated to depths in the range 0.1 μ to 10 μ. The process which is still in the development stage is, in fact, a hybrid process utilising the ballistic aspects of conventional ion implantation and the diffusive thermochemical character of plasma nitriding.

The hybrid low temperature plasma nitriding technology involves immersing the components, which are biased to a high negative potential, in a low-pressure rf nitrogen plasma, as distinct to the conventional glow discharge plasma. The high negative bias accelerates nitrogen ions which are implanted into all the exposed surfaces of the components, thus eliminating the line-of-sight restrictions of conventional ion beam implantation.(26) The ion bombardment causes significant heating and also results in the diffusion of nitrogen well beyond the implantation range(27) giving rise to large increases in surface hardness with AISI-316 stainless steel (Fig.3(28)) and improvements in the load bearing capacity of the implanted layer.

Samples treated at temperatures in the range 150°C to 520°C have been examined in detail by Samandi et al(28) using a CSEM Pin on Disc tribometer employing single crystal ruby balls with alcohol as the lubricant. Their results are summarised in Fig. 4. It can be seen, from an overall standpoint that, as the treatment temperature increased, so did the wear resistant properties of the stainless steel. Wear resistance improved by a factor of between 200 and 350 at loads of 10 and 20N. This could be attributed to the creation of conditions favourable to the formation of a protective oxide layer which was maintained throughout the wear test and thus inhibits the severe metallic wear regime. At all temperatures below 520°C substantial quantities of the "S-phase" were found to be present, by X-ray diffraction methods, thus showing promise of good corrosion resistance.

The potentiodynamic curves for anodic polarisation of the plasma nitrided samples processed by the hybrid technology are shown in Fig. 5. At the lowest treatment temperatures, there was only slight improvement over the untreated steel but as the temperature increased to 350°C both the critical current density for passivation and the passive current density decreased. For material treated at 350°C a 20 times decrease in the passive current was achieved reflecting a significant increase in corrosion resistance. The curves indicate that the corrosion resistance of material nitrided at 450°C was slightly worse than the untreated steel. Further deterioration in corrosion performance occurred for the sample treated at 520°C thus reflecting the earlier observations of decreased corrosion resistance as a result of conventional glow discharge plasma nitriding.

It can be seen therefore, that the hybrid, implantation assisted rf plasma nitriding technology has considerable potential for the treatment of austenitic stainless steels at temperatures in the range 350-450°C. It provides for the possibility of developing a hard, highly wear resistant surface without compromising the corrosion resistance.

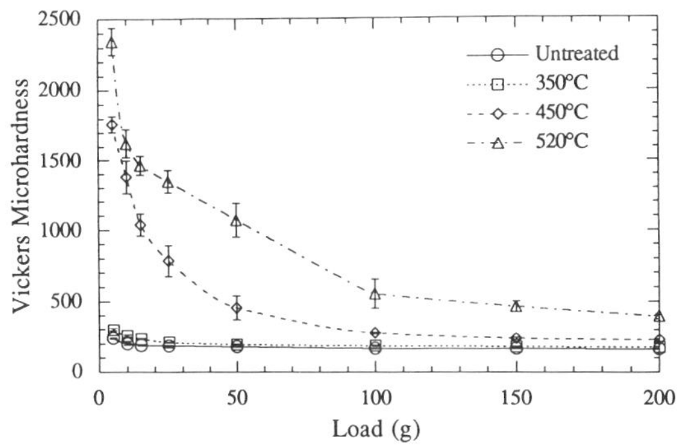

Figure 3. Effect of PI³ treatment temperature on microhardness.

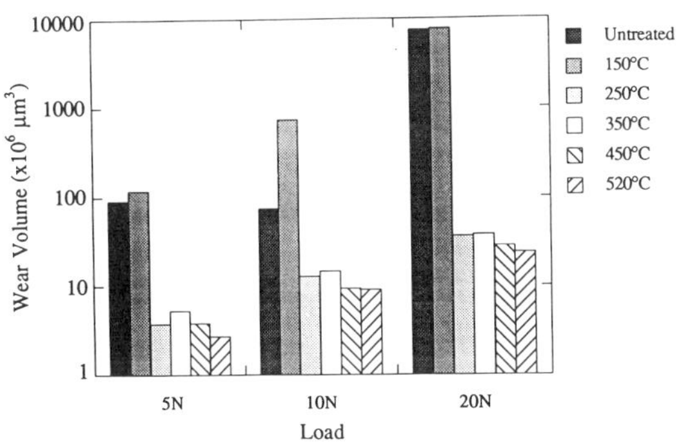

Figure 4. Wear volumes of stainless steel discs at different loads after 10^3 revolutions.

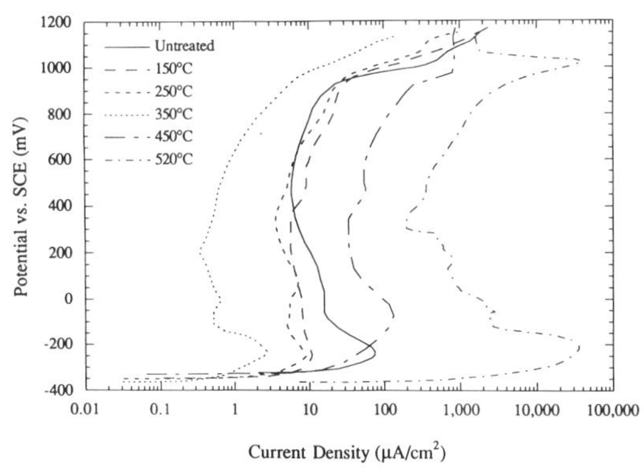

Figure 5. Potentiodyamic curves obtained for anodic polarisation of PI³ treated samples.

4. SURFACE ENGINEERING OF TITANIUM ALLOYS

The high strength to weight ratio of titanium and its excellent corrosion resistance make it the designer's first choice material for many applications in the fields of aerospace and chemical engineering. Its poor tribological properties have, however, frequently inhibited its use not only in these areas but, more particularly, in the wider field of general engineering. When an untreated titanium alloy surface is in rubbing contact with another material, even when lubricated, there is a strong tendency for galling to take place leading to severe wear and/or seizure. Traditional methods of surface treatment, such as anodising and electroplating, have had only limited success in overcoming the problem and surfaces treated in such ways are not able to withstand more than light surface stresses. However, in recent years, technological advances in plasma processing and physical vapour deposition (PVD) techniques and in surface treatment by high energy lasers or electron beams have opened up opportunities for greatly enhancing the tribological performance of titanium alloys. In particular, it has become possible to apply a much thicker hardened "case" to titanium alloys thereby enabling them to withstand much higher contact stresses. In this respect, current technological advances will be considered in two categories:

- modest loads - conditions of lubricated sliding wear without high normal contact stresses
- heavy contact loads - lubricated contact where there is a substantial element of rolling contact and, consequently, high subsurface stresses are induced.

Modest loads — Advances in methods of physical vapour deposition have led to increasing use of TiN coatings on titanium components. Typically, a 3 µm TiN layer is formed by processing at about 500°C. The layer is very hard and smooth with a low coefficient of friction.

Its industrial use has recently increased and, indeed, it is being considered for protection against dust erosion as, for instance, on aircraft or helicopter gas turbine compressor blades which may be impacted by sand or volcanic ash.(29) Another development has been in the thermochemical treatment of titanium alloys using glow discharge plasma technology. In this technique nitrogen ions are accelerated towards the titanium surface where they react with it to form an in-situ titanium nitride layer. The process is typically performed at 700-850°C and the accompanying solid-state diffusion leads to a build up of a multi-layer coating. Beneath a thin gold-coloured layer of TiN is a somewhat thicker layer of predominately Ti_2N phase which is less hard but still has good wear resistance. Typical microhardness profiles of plasma nitrided IMI Ti 318 are shown in Fig. 6.

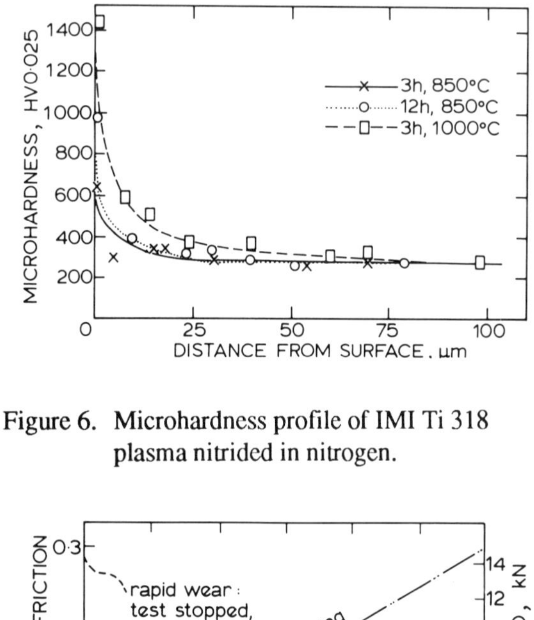

Figure 6. Microhardness profile of IMI Ti 318 plasma nitrided in nitrogen.

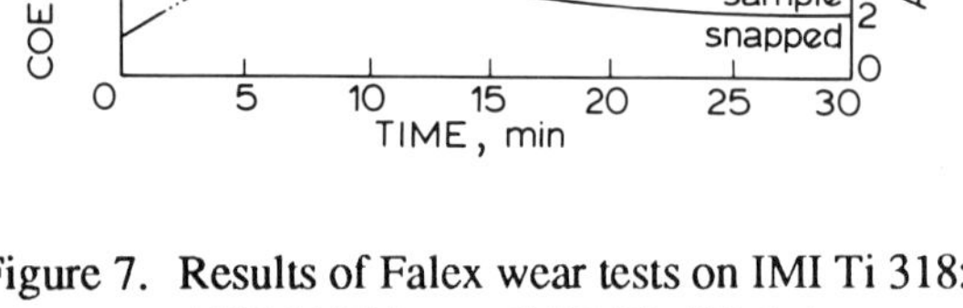

Figure 7. Results of Falex wear tests on IMI Ti 318: AISI 1137 jaws; SAE 30 oil lubricant; plasma nitriding treatment was 12h at 850°C in nitrogen.

It can be seen that the "case" extends into the substrate to give an overall depth of hardening of circa 30 µm. The resultant improved wear performance is illustrated in Fig. 7, which shows the results of Falex wear tests in which a cylindrical titanium pin was rotated, with oil lubrication, between two steel V-blocks under a progressively increased clamping force.

Heavy Contact Loads — In order to withstand heavy loads with rolling contact where substantial subsurface Hertzian stresses are generated, a greater depth of hardening is required and this is impossible to achieve by solid state metallurgical techniques. In order to produce a deep hardened layer in a titanium alloy surface it is necessary to carry out liquid phase thermochemical alloying. The simplest method of liquid phase alloying of selected areas of titanium surfaces involves focussing a laser beam so as to melt the surface of the metal in a stream of nitrogen or a nitrogen-argon mixture. The nitrogen becomes alloyed with the molten titanium by convection and diffusion, and the self-quenching effect of the substrate cools the molten metal on the surface at a rate of around 10^5°C s^{-1}. The component surface is fed across below the laser beam so as to produce a melt track of nitrogen rich hard metal. In order to cover a substantial area of the component surface, the process is repeated to give a series of overlapping runs. By varying the processing parameters of a 1.2 kWCO_2 laser, the depth and hardness of the melted alloy can be varied over a wide range as is illustrated in Fig. 8.

The lubricated wear behaviour of laser-nitrided Ti-6Al-4V under high loads is compared with that of the untreated alloy in Fig. 9. An Amsler wear tester was used in which the titanium disc was tested against an EN19 steel counterface under conditions of rolling with 30% slip. The outerface of the laser-nitrided titanium disc was alloyed to a hardness of about 800 H_v and surface ground before wear testing at various loads. It can be seen that there is an initial period of running in wear which is followed by a steady state condition when the wear rate is constant. At all stresses, the untreated titanium wore extremely rapidly, whereas the wear rate of the laser nitrided surface was greatly reduced. Under an applied load of 13 kg the wear rate of the laser nitrided alloy was reduced by a factor of 100 to a rate about the same as that of a steel (EN19) surface tested under the same conditions. On testing at 54 and 120 kg applied load (500 and 750 MPa contact pressure) rather rapid initial wear took place, but a steady state condition was nevertheless established at which the wear rate was a factor of ten less than for the untreated alloy.

Although the laser-nitriding process has thus been shown to result in a marked improvement in wear properties at much higher stress levels than that obtainable from the plasma nitriding technology, the wear rate is still considerably

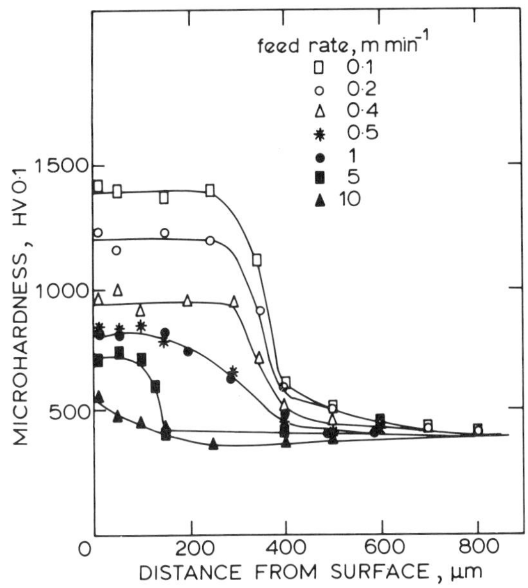

Figure 8 Effect of specimen feed rate on microhardness profile of IMI Ti 318 laser melted in nitrogen.

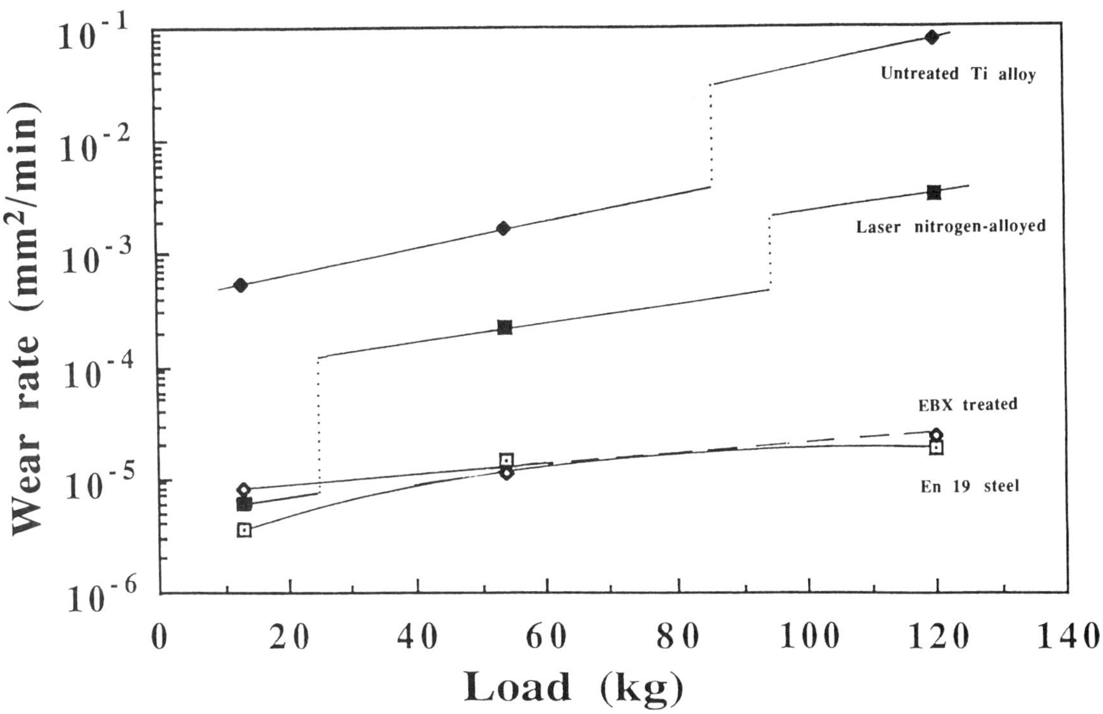

Figure 9. Steady state wear rate for lubricated rolling/30% sliding at various loads. Loads of 13, 53 and 120 kg corrspond to Hertzian stresses of 250, 500 and 750 MPa.

Table 2
Surface Engineered Components Which Are Used for Are Undergoing Trials in High Performance Cars

In Regular Use	On Trial
Plasma nitrided steering racks	PVD coated steering racks
Plasma nitrided screw fasteners	PVD coated exhaust and inlet valves
Plasma nitrided clutch withdrawal bobbins	Laser nitrided gear selector fingers
Plasma nitrided reverse gears	Laser nitrided con rod side faces
Plasma nitrided lay-shaft and drive shaft spacers	EB alloyed drive shaft nuts
PVD coated brake pistons	EB alloyed power differential drive pins
PVD coated wheel hubs	Plasma nitrided valve spring retainer caps

greater, as can be seen in Fig. 9, than that achieved by steel against steel. Recent work has shown(29) that by alloying with an element or compounds, like silicon carbide, a surface hardened alloy of considerable toughness can be produced and such alloying has been investigated using the alternative high energy beam thermochemical process of electron beam alloying (EBX). As a result wear characteristics have been achieved, as can be seen from the data in Fig. 9, which are comparable to steel on steel. Table 2, summarises surface engineered titanium components which are used or are undergoing trials in high performance cars.

5. DUPLEX SURFACE ENGINEERING

It is through the ever-increasing market acceptance of existing first-generation surface engineering technologies, and their incorporation into the design and manufacture of economically viable, technically enhanced engineering components that the future of the discipline of surface engineering is assured, especially in its contribution to improvements in wear, corrosion and related tribological environments. Accordingly, in examining the future of surface engineering, it is important to examine today's research in this and associated fields.

There can be no doubt that the scope for duplex or multiple-surface treatments to provide an array of novel surface engineered products in a wear-related environment is virtually unlimited. Several of the multiple-surface treatments which are currently being investigated for the design of components for operation under complex conditions are listed in Table 3 and are examples of "designer surfaces", the development and application of which will prevail in the future.

Since these duplex technologies are still in the relatively early stages of development the present paper will now discuss the basic principles and metallurgical aspects of duplex surface engineering. In addition, the TiN coated-nitrided steel system will be used to consider the contact problems of real multi-layer systems using the numerical model recently developed by Cole and Sayles.(40,41)

Duplex surface engineering involves the sequential application of two surface engineering technologies, their metallurgical interactions play a very important role in determining the properties of the resultant composites. By way of example, consideration is given in this Section to the PVD deposition of the ceramic TiN on top of plasma nitrided steel. As a result the composite comprises a thin coating of TiN on the surface, with a relatively thick subsurface layer produced by the thermochemical process of plasma nitriding, and a ductile core: The composite possesses properties characteristic of both processes, i.e., a combination of low friction, low wear, improved coating-substrate adhesion strength, superior load bearing capacity and high fatigue strength. Accordingly, it is essential to carefully control

Table 3
Typical Duplex Surface Engineering Technologies

(1) Laser treatment of thermal and plasma spray coatings (13)
(2) Ion beam mixing and ion-assisted coatings (30,31)
(3) Hot isostatic pressing of overlay coatings (32,33)
(4) Thermochemical treatment of precarburised steels (34,35)
(5) Thermochemical treatment of pre-laser-hardened steels (36)
(6) CVD treatment of precarburised steels (37)
(7) PVD treatment of pre-nitrided steels (38,39)
(8) PVD treatment of high energy beam alloyed aluminium (5)

both processes so that the structures and properties resulting from the first process, i.e., plasma nitriding, are not deteriorated by the second process, i.e., TiN deposition. Improper process control and/or an inadequate appreciation of the metallurgical reactions involved, will lead to softening, decomposition and structural changes of the subsurface layers resulting in undesired coating/substrate combinations.(38,42,43) The combined mechanical, metallurgical and chemical properties resulting from properly executed duplex surface engineering have proved, on the basis of information available, to be better than those produced by any individual

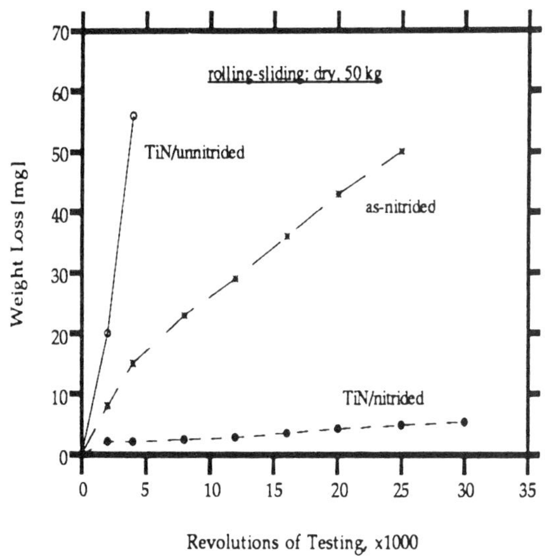

Figure 10. Wear curves for nitrided, TiN coated and duplex treated EN40B.

process.(44,38,42,43,45) Fig. 10, for example, illustrates rolling sliding wear test results for TiN/non-nitrided steel, plasma nitrided steel and the TiN/nitrided steel systems. Clearly, the duplex "TiN/nitrided steel system" shows by far the highest wear resistance. Pure sliding tests using a ball-on-wheel configuration also demonstrated that the duplex system exhibited a much higher load bearing capacity than the single TiN/unnitrided system.(38) Such duplex systems are rapidly finding applications within the engineering component sector of manufacturing industry(46) as a result of the combined effect of the thin ceramic coating and the strengthened subsurface that provides remarkable tribological and load bearing properties. Due to the very thin nature of the surface coatings, the substrate must bear the majority of the applied load, and if it has insufficient strength to bear the contact loading, and thus support the coating, plastic deformation will occur, leading to premature failure of the hard surface coating.(38,44) The contact mechanics necessary to obtain an accurate knowledge of the contact stresses in such multi-layered structures has been analysed recently by Cole and Sayles(40,41), and is currently being examined in depth as part of BRITE/EURAM Project BE 4242. From preliminary results which have been published(47) concerning contact stresses in the duplex TiN/nitrided system and from a further paper in the present Conference volume by Olver et al, the concept of a "stress barrier coating" has been developed.(48) These recent publications conclude that for real engineered surfaces, where surface roughness is taken into account:

(1) Surface roughness gives rise, in many circumstances to a thin, highly-stressed layer. Reinforcing this layer (e.g., with a TiN coating) can reduce plastic deformation and improve rolling contact fatigue life.(48)

(2) The stress distribution in the bulk of the substrate under a TiN coating is little different from that in an untreated surface but in the zone near the TiN-plasma nitrided substrate interface the highest stresses are only half as large as the values which occur at the same depth in the untreated case. This massive reduction in stress helps explain the improved behaviour of the surface engineered components in tribological situations.(47)

(3) The increased strength of the substrate due to plasma nitriding, and which extends for approximately 500 μm from the surface, is also most beneficial in helping the substrate survive the stresses for which it is exposed.

REFERENCES

1. Lubrication (Tribology) Education and Science (Jost Report), HMSO, London, 1966.
2. T. Bell, Editorial, Surface Engineering, 1985, Vol. 1, p. 1.
3. T. Bell, Surface Engineering: A Rapidly Developing Discipline, European Journal of Engineering Education, 12, No. 1, 1987, pp. 27-32.
4. D.J. Vaughan, Future of Surface Engineering in the UK, CEST Report, 1991.
5. T. Bell, Towards Designer Surfaces, Metals and Materials, 7, 1991, 478-485.
6. I.M. Hutchings, Tribology: Friction and Wear of Engineering Materials, Edward Arnold, p. 204, 1992.
7. T. Bell, Surface Engineering: Its Current and Future Impact on Tribology, J. Applied Phys.,1992, 5, , A297-A306.
8. W.H. Roberts, Surface Engineering, Int. Tribol. Conf., Melbourne, 1987, Institution of Engineers, pp. 445-451
9. C.S. Syan, A. Matthews and K.G. Swift, Coating and Treatment Selection, Surface Engineering, 1986, 2, pp. 249-255,
10. T. Bell, Surface Heat Treatment of Steel to Combat Wear, Metallurgia, 1982, 49, pp 103-111.
11. D.N.H. Trafford, T. Bell, J.H.P.C. Megaw and A.S. Bransden, Proc. Heat Treatment '79, The Metals Society, 1979, pp 32.1-10.
12. D.N.H. Trafford, T. Bell, J.H.P.C. Megaw and A.S. Bransden, Proc. Heat Treatment '81, The Metals Society, London, 1981, pp. 198-206.
13. R. Sivakumar and B.L. Mordike, Laser Melting of Plasma Sprayed NiCoCrALY Coating, Surface Engineering, 1987, 3, pp. 299-309.
14. T. Bell. H.W. Bergman, J. Lanagan, P.H. Morton and A.M. Staines, Surface Engineering with Nitrogen, Surface Engineering 1986, 2, pp. 133-143.
15. P.H. Morton and T. Bell, Surface Engineering ofTitanium, Mem. Etudes Sci. Rev. Metall., 1989, No. 10, pp. 639-646.
16. A.S. Bransden, S.T. Gaggard, B.C. Inwood, and J.H.P.C. Megaw, Laser Hardending of Ring Grooves in Medium Speed Diesel Engine Pistons, Surface Engineering, 1986, 2, pp 107-113.
17. C.G. John, Modern Engineering Plating Processes and Techniques, Wear Resistant Treatments and Coatings - Course Notes, NCT, Risley, UK, 1986.
18. Industrial Case Study: Niflor-autocatalytic Nickel-PTFE Composite Coating, Surface Engineering, 1987, 2, pp 16-18.
19. K.L. Hsu, T.M. Ahn and D.A. Rigney, Friction, Wear and Microstructure of Unlubricated Austenitic Stainless Steels.
20. G.E. Dieter, Mechanical Metallurgy, McGraw Hill, 1988, pp 212-219.

21. P.A. Dearnley, A. Namvar, G.G.A. Hibberd and T. Bell, Proc. Conf. Plasma Surface Engineering, Garmisch-Partenkirchen, FRG, Sept. 1988, Vol. 1, pp. 219-226.
22. S.P. Hannada, P. Nenonen and J.P. Hirvonen, Surface Structure and Properties of Ion Nitrided Austenitic Stainless Steel, Thin Solid Films, 181, 1989, pp. 343-350.
23. J.E. Truman, M.J. Coleman and K.R. Pirt, British Corrosion Journal, 1977, 12, p. 236.
24. K. Ichii, K. Fugimura and T. Takese, Structure of the Ion-Nitrided Layer of 18-8 Stainless Steel, Technol. Rep. Kansai Univ., 27, 1986, pp. 135-144.
25. Z.L. Zhang and T. Bell, Structure and Corrosion Resistance of Plasma Nitrided Stainless Steel, Surface Engineering, Vol. 1, No. 2, 1985, pp. 131-136.
26. J. Conrad, Plasma Pulse Technique for Ion Implantation, 3rd International Conference on Plasma Surface Engineering, Garmisch-Partenkirchen, Oct. 1992.
27. G.A. Collins, R. Hutchings and J. Tendys, Plasma Immersion Ion Implantation - The Role of Diffusion, ibid.
28. M. Samandi, B.A. Shedden, D.I. Smith, G.A. Collins, R. Hutchings and J. Tendys, Microstructure, Corrosion and Tribological Behaviour of Plasma Immersion Ion Implanted Austenitic Stainless Steel, ibid.
29. T. Bell, P.H. Morton and B.L. Mordike, Titanium Alloys: Surface Treatment Encyclopedia of Materials Science and Engineering, Ed. Robert W. Cahn, Supplementary Volume 3, Pergamon Press, pp. 2083-2087.
30. G. Carter, Ion Implantation and Ion Beam Mixing, Ion Assisted Surface Treatment Techniques and Processes (University of Warwick), 1982, p. 92.
31. G. Dearnley, Ion Implantation and Ion Assisted Coatings for Wear Resistance in Metals, Surface Engineering, 1986, 2, pp. 213-22.
32. D.M. Anderson and B.A. Rickinson, Hipack Coatings,, Proc. lst Int. Conference on Surface Engineering, (Cambridge, The Welding Institute), 1985, pp. 16.1-10.
33. H.D. Steffens, R. Dansmer and U. Fisher, Post-treatment of Composites by Hot Isostatic Pressing, Surface Engineering, 1988, 4, pp. 39-43.
34. A. Olver, Private Communication, 1989.
35. H. Yasbanda, M.Sc.(Eng) Thesis, University of Birmingham, 1987.
36. T. Bell and A. Bloyce, Nitriding Laser Treated Titanium Bearing Low Alloy Steels, Proc. Heat Treatment '84, The Metals Society, London, 1984, paper 36, pp. 36.1-7.
37. S. McLean, CVD Treatment of Carburised Low Alloy High Speed Steel, M.Phil. Thesis, University of Birmingham, 1990.
38. Y. Sun and T. Bell, Combined Plasma Nitriding and PVD Treatments, Trans. Inst. Metal Finishing, 1992, 70 (1), p. 38.
39. M. Zlutanovic, Deposition of (TiAl)N Coatings on Plasma Nitrided Steel, Surface and Coatings Technology, 48, 1991, pp. 19-24.
40. S.J. Cole and R.S. Sayles, ASME paper No. 91, Trib. 11, 1991, STLE and ASME Trib. Conf.
41. S.J. Cole and R.S. Sayles, Microstructural Science, Vol. 20, Feb. 1992.
42. D.P. Davis, Heat Treatment of Metals, 18(4), 1991. p. 97.
43. D.K. Chen and Z. Chen, Heat Treatment of Metals (in Chinese), 154(6), 1990, p. 39.
44. Y. Sun and T. Bell, Mater. Sci. Eng., A140, 1991, p. 419.
45. T. Bell and M. H. Sohi, Aluminium Today, May 1990, p. 44.
46. A.M. Staines, Trends in Plasma Assisted Surface Engineering Processes, Heat Treatment of Metals, 1990, 4, p. 85.
47. T. Bell, Y. Sun and S.J. Cole, Duplex Surface Engineering, Heat and Surface '92, 8th International Congress on Heat Treatment of Materials, Kyoto, Japan, 1992, p. 1-6.
48. A.V. Olver, S.J. Cole and R.S. Sayles, Contact Stresses in Nitrided Steels, 19th Leeds-Lyon Symposium on Tribology - Thin Films in Tribology, Leeds, 8-11 September 1992.

Thin Films in Tribology / D. Dowson et al. (Editors)
1993 Elsevier Science Publishers B.V.

Local Effects in Thin Film Lubrication

C. M. Myllerup[a] and B. J. Hamrock[b]

[a]Department of Machine Elements, Technical University of Denmark, 2800 Lyngby, Denmark

[b]Department of Mechanical Engineering, Ohio State University, Ohio 43210, USA

The basic mechanisms involved in elastohydrodynamic lubrication are presented. Machine elements operate with effective fluid film lubrication even when the films are of micrometer and submicrometer dimensions. These thin films have caused a rethinking of some of the basic assumptions that have been used in the past. The first part of the paper focuses on recent developments in considering non-Newtonian fluid rheology, thermal effects, and nonsmooth surfaces. The latter part of the paper considers research just being initiated such as obtaining a basic understanding of the various terms in the Navier-Stokes equations as applied when singularities exist within the lubricated conjunction. Also, the modelling of how a solid particle travels through an elastohydrodynamically lubricated conjunction was briefly considered.

1.0. INTRODUCTION

Many lubricated machine elements operate with effective fluid film lubrication even when the films become very thin. Present applications have film thicknesses of micrometer and submicrometer dimensions. Film thicknesses have been steadily decreasing since the start of this century, when they were believed to be of the order of 10^{-5} m. It is anticipated that this trend will continue. The main reasons for the decreasing film thicknesses since the turn of the century are the ability of surfaces to be manufactured to much higher accuracy and the more severe operating demands.

These decreasing film thicknesses have caused researchers to rethink some earlier assumptions. The present review paper focuses on four issues that are important in thin fluid film lubrication. These are lubricant rheology, thermal effects, surface topography, and surface damage due to debris. The approaches in this paper are to consider these effects in thin fluid films, to focus on the assumptions being imposed in present considerations, and to comment on the validity of these approaches. Not only will the level of current understanding be discussed, but questions the authors recently have been groping with will be presented.

For the section on surface topography not only will an overview of the state of the art be given and problems still needing solution be discussed, but an analysis will also be performed focusing on the importance of higher order effects in nonsmooth surfaces. The objective of the surface topography section is to provide a measure of the validity of the assumptions imposed in deriving the Reynolds equation as well as an understanding of the magnitude of the terms neglected in the Navier-Stokes equations in arriving at the Reynolds equation. The surface topography section will therefore be much longer than the other three sections.

2.0. LUBRICANT RHEOLOGY

The goal of fluid film lubrication is to establish an effective fluid film that will support the load while reducing friction between the moving surfaces. In nonconformal surfaces, which are normally found in elastohydrodynamic lubrication, the full burden of the load must be carried by a very small lubricated area. The load per unit area in elastohydrodynamically lubricated (EHL) conjunctions generally exceeds 0.7 GPa and may be as large as 4.0 GPa. These fluid film pressures result in elastic deformation of the solid surfaces and also considerably increase the fluid viscosity. Inasmuch as viscosity is a measure of the fluid's resistance to flow, this increase of pressure greatly enhances the lubricant's ability to support a load without being squeezed out of the lubricated conjunction. The ability of the lubricant to generate a lubricating film that is adequate to maintain separation between the moving surfaces determines the effectiveness of elastohydrodynamic lubrication.

In elastohydrodynamic lubrication the lubricant experiences rapid and very large pressure variations, a rapid transit time, possibly large temperature changes, elongational shear rates, and particularly in sliding contacts, high shear strain rates. As a result of high fluid film pressures and shear heating, the viscosity changes drastically within the lubricating film. In addition, the existence of very high shear rates results in shear stresses that are well beyond the limiting shear stress of the lubricant. Furthermore, the transit time required for the fluid to go through the high-pressure zone is between 10^{-3} and 10^{-4} s. Thus the conditions prevailing in an EHL conjunction are quite severe.

2.1. Non-Newtonian Rheology

In most recent elastohydrodynamic lubrication analyses it is assumed that the fluid behaves in a non-Newtonian manner. That is, as the shear stress approaches the limiting shear stress, the relationship between shear stress and shear strain ceases to be linear and non-Newtonian behavior results.

The non-Newtonian property of limiting shear stress implies that the lubricant does not experience shear stress exceeding this limit. The limiting shear stress is usually considered as an asymptotic limit. Should the fluid model, however, allow the shear stress to reach the limiting shear stress, slippage will occur, either within the oil film or at the interface between the oil and one of the bounding surfaces. A fluid model that allows for such a slip phenomenon will be referred to as "a visco-plastic fluid model." The distinction between asymptotic and viscoplastic limiting shear strength behavior is important because the latter requires the replacement of the traditional velocity boundary conditions by stress boundary conditions. This means that there will be transitions in the type of boundary conditions imposed if the shear stress reaches the plastic limit inside the conjunction, a condition that may introduce singularities in the solution.

Bair and Winer (1979) and Gecim and Winer (1980) proposed alternative expressions for the relationship between shear stress and shear strain rate that exhibit asymptotic limiting shear stress behavior. Both Bair and Winer (1979) and Gecim and Winer (1980) incorporated their models into the analysis of EHL line contacts by using a Grubin-like inlet analysis. In attempts to obtain a full numerical solution of the problem of EHL line contacts, Bair and Winer's model has been approximated into simple forms such as the linear model proposed by Trachman (1971), which was used in elastohydrodynamic lubrication by Wang and Zhang (1988) and Iivonen and Hamrock (1989), or the circular model proposed by Lee and Hamrock (1990a). Both the linear and the circular fluid models allow a modified Reynolds equation to be established by using an approach similar to the one that applies for a Newtonian fluid. The circular model is, however, not as severe a non-Newtonian fluid model as is the linear model, and is therefore in better agreement with experimental observations. The modified Reynolds equation for the circular fluid model was solved by Lee and Hamrock (1990a) by using the fast system approach developed by Houpert and Hamrock (1986). This approach was later refined by Hsiao and Hamrock (1992) and used to investigate thermal effects in non-Newtonian elastohydrodynamic lubrication. The existence of the solution for a circular fluid model allows the most significant non-Newtonian feature to be incorporated into the analysis of elastohydrodynamic lubrication, namely the limiting shear stress. It is, however, restrictive not to be able to model more freely the transition between Newtonian behavior at low shear strain rate and limiting shear stress behavior at high shear strain rate.

Elsharkawy and Hamrock (1991) proposed a general nonlinear viscous model where the limiting shear stress has already been incorporated as

$$\frac{\partial u}{\partial z} = \frac{\tau}{\eta}\left[1 - \left[\frac{\tau}{\tau_L}\right]^n\right]^{-1/n} \tag{1}$$

where $$\tau_L = \tau_{atm} + \gamma\, p \tag{2}$$

Equation (1) can be written as

$$\frac{\tau}{\tau_L} = \frac{\dfrac{\eta}{\tau_L}\dfrac{\partial u}{\partial z}}{\left[1 + \left[\dfrac{\eta}{\tau_L}\dfrac{\partial u}{\partial z}\right]^n\right]^{1/n}} \tag{3}$$

Equation (3) is plotted in figure 1 for different values of n. In figure 1(a) the conventional way of describing the various models is used. An alternative representation of these non-Newtonian models is obtained when using the effective viscosity first introduced by Bair and Winer (1979).

$$\eta_{eff} = \tau\left[\frac{\partial u}{\partial z}\right]^{-1} = \eta\left[1 - \left[\frac{\tau}{\tau_L}\right]^n\right]^{1/n} \tag{4}$$

This relationship between shear stress and effective viscosity is shown in figure 1(b) for various values of the exponent n.

The main advantage of the general model is that all the other nonlinear viscous models with the asymptotic limiting shear stress property are readily derived or closely approximated by appropriate choice of the exponent n. As indicated in figure 1, n = 1 results in the linear model and n = 2 in the

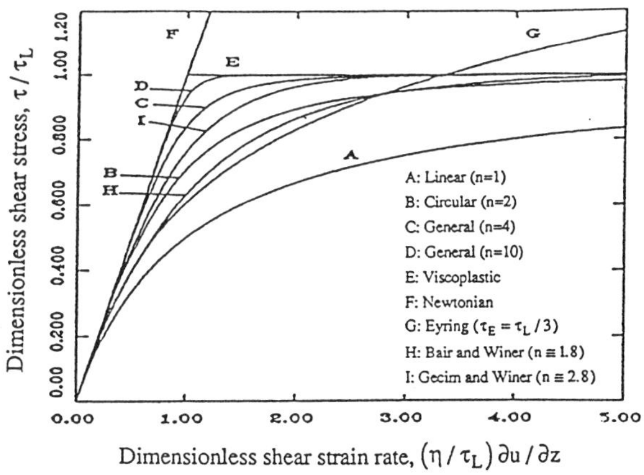

Figure 1a. Non-Newtonian rheological models represented by the effect of shear strain rate on shear stress. (Myllerup et al., 1992).

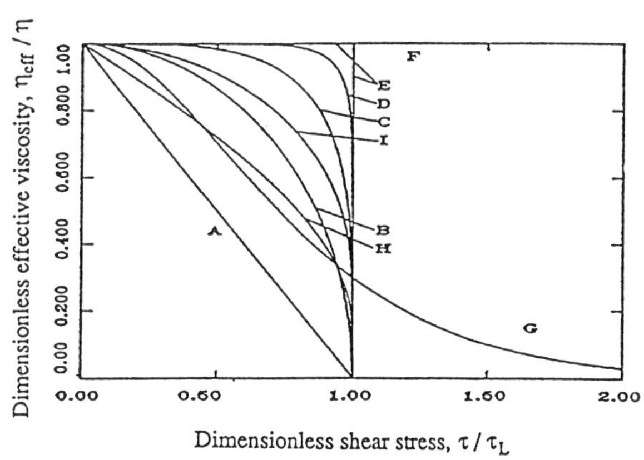

Figure 1b. Non-Newtonian rheological models represented by the effect of shear stress on effective viscosity. (Myllerup et al., 1992).

circular model while n = 1.8 approximates the Bair and Winer (1979) model and n = 2.8 the Gecim and Winer (1980) model. The general fluid model will in fact, if incorporated into the analysis of elastohydrodynamic lubrication, allow a given asymptotic relationship between shear strain rate and shear stress to also be simulated in the transition zone. In the general case of arbitrary n it is not yet possible to derive a modified Reynolds equation directly as was the case for n = 2. Myllerup et al. (1992) obtained an approximate solution to the general problem of non-Newtonian elastohydrodynamic lubrication by using the fact that for $\tau/\tau_L << 1$ equation (1) is Newtonian to a leading order of τ/τ_L and that Couette dominance exists in the high-pressure zone. These results indicate that the effects of n are relatively small for n > 1; the fundamental rheological parameters η and τ_L are adequately represented by a circular fluid model.

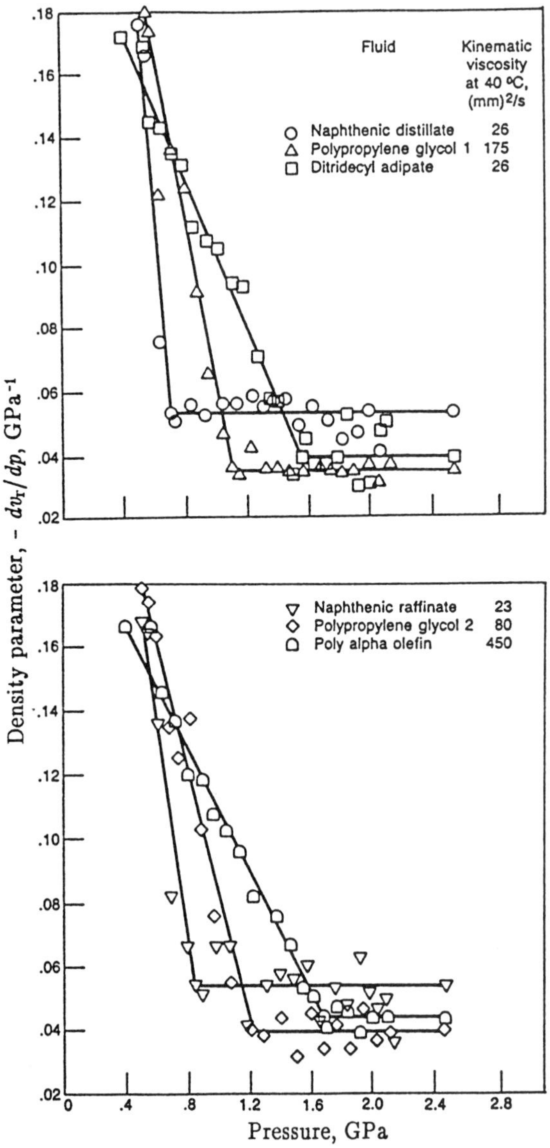

Figure 2. Effect of pressure on density parameter for six base fluids. (Hamrock et al., 1987).

2.2. Density

Besides the non-Newtonian fluid properties, another important property of the fluid is the variation of the fluid density with pressure. Hamrock, et al. (1987) obtained experimental data showing the effect of pressure on density for a range of pressure from 0.4 to 2.2 GPa for six base fluids at a constant temperature of 20°C. An important parameter used to describe the results is the change of relative volume with change of pressure dv_r/dp. The relative volume (v_r) is the ratio between actual and initial volume. For pressures less than the solidification pressure ($p < p_S$), a small change in pressure results in a large change in dv_r/dp. For pressures greater than the solidification pressure ($p > p_S$), there is little change in dv_r/dp.

In elastohydrodynamic lubrication the rate of pressure increase is extremely high, typically 10^{13} Pa/s. The lubricant under these conditions will not have time to crystallize but will be compressed to an amorphous solid, the distance between the molecules of the lubricant becoming smaller and smaller. There exists a point where the molecules are not free to move and any further compression will result in deformation of the molecules. The pressure where this first starts to occur is the solidification pressure. Thus, the solidification pressure represents a transition in compressibility stiffness. For $p > p_S$ the lubricant behavior is more like that of an incompressible substance.

Figure 2 shows the change in relative volume with changing pressure for six base fluids. The solidification pressure varied considerably for the different base fluids tested. Furthermore, for $p < p_S$ the experimental data assumed different slopes for the different base fluids. One can thus conclude that molecular packing has a dominant effect on fluid rheology. These same observations were made by Granick (1992) when discussing rheology in ultrathin fluid films (films approaching molecular dimensions).

2.3. Unanswered Questions

With films in EHL conjunctions becoming thinner and approaching the molecular scale, the first basic question is whether or not the lubricant should be considered to be a continuum. This is a topic of current interest and one that present researchers are attempting to understand.

Granick (1991) asks a number of questions relating to ultrathin films. Some of these provocative questions are:

a. What determines the effective viscosity of an ultrathin film?
b. What makes ultrathin liquids turn into solid when they are sufficiently thin?
c. Why is solidification suppressed by continuous motion?

Or more generally, what are the dynamics of liquids in intimate contact with a solid boundary when the film is ultrathin? In the material presented in this paper it was assumed that a continuum applies and that the film is at least two orders of magnitude larger than the molecules of the lubricant and thus much larger than those discussed by Granick (1991).

Do lubricating oils exhibit a limiting shear stress? Hoglund and Jacobson (1986), Bair and Winer (1979), and Bair (1990) indicate that indeed lubricating oils do have limiting shear stress. But a question as to what is the limiting shear stress of a fluid still exists. Figure 3 from Bair (1990) indicates that Hoglund and Jacobson (1986) label their results as limiting shear stress whereas Bair believes these results are still in the Newtonian region. Hoglund and Jacobson (1986) observed a transition in the relationship between shear stress and pressure. The transition occured at approximately 0.6 GPa at 40°C and 0.8 GPa at 70°C and was taken to indicate solidification. As would be expected for a Newtonian fluid of exponential viscosity-pressure dependence the relationship between shear stress and pressure was exponential below the solidification pressure, whereas for $p > p_s$ this relationship was in good agreement with the linear limiting shear stress model from equation (2). Typical values of the proportionality constant

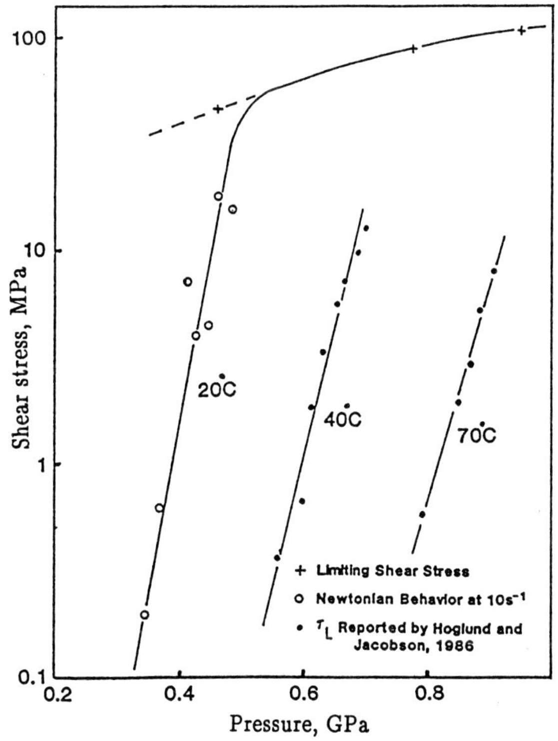

Figure 3. Comparision of the shear stress measurements of Hoglund and Jacobson (1986) with measured viscosity and limiting stress. (Bair, 1990).

$\gamma = \partial\tau_L/\partial p$ for a wide range of fluids are between 0.02 and 0.08. Bair (1990), however, found that when comparing his experimental results on a logarithmic scale with the ones of Hoglund and Jacobson (1986) the transition observed by the latter was replaced by another at higher pressures, see figure 3. Bair (1990) suggests that it is only at this stage that the limiting shear stress of the fluid is reached.

For a lubricating fluid that exhibits a limiting shear stress, the fundamental physical phenomenon which results in this property remains to be established. The concept of solidification pressure which was used in section 2.2 to explain the transition in compressibility has been mentioned; when a critical normal stress is reached within the lubricant it no longer behaves as a fluid in the traditional sense. There are a number of questions that are related to this basic problem:

d. When plotting shear stress against shear strain rate is the limiting shear stress an asymptotic or is it a plastic limit? That is, will the limiting shear stress ever actually be reached?

e. At atmospheric pressure and room temperature but at very large shear strain rates ($\partial u/\partial z > 10^6$/s), do lubricating fluids experience a limiting shear stress? Applications of tapes and floppy disks operate with these shear strain rates.

Answering any of these questions may help to clarify whether solidification occurs. Should the limiting shear stress be a plastic phenomenon, either as shear bands as presented by Bair (1992) or as shear localization along the film (inside the fluid or at the solid surfaces), we must incorporate a plastic transition into the general rheological model from equation (1) by imposing $\tau = \tau_L$ under given conditions. A final question then arises:

f. What happens at the transition from no slip to the necessary stress boundary condition in the case of localized shearing?

Figure 4 from Hsiao and Hamrock (1992) illustrates why this is an important issue. Figure 4 gives the distribution of stress boundaries at

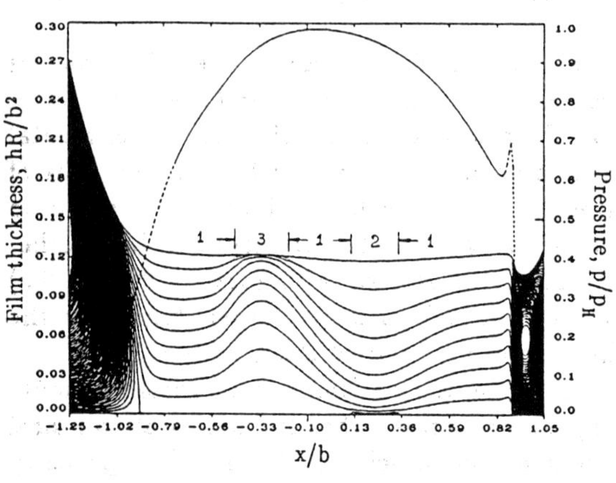

Figure 4. Velocity contours indicating flow-zone distribution. ("1" designates flow with velocity boundaries, "2" designates flow-zone with stress boundary at $z = 0$, and "3" designates flow zone with stress boundary at $z = h$). (Hsiao and Hamrock, 1992).

which the local limiting shear stress is imposed. The number "1" indicates that an ordinary no-slip boundary condition was used in this region. The number "2" denotes the flow zone where the limiting shear stress is reached at solid surface "a". The number "3" denotes the flow zone where the limiting shear stress is reached at solid surface "b." Does the lubricant actually move through the conjunction as indicated by these velocity contours, and is the solution regular or singular at the transitions in boundary conditions?

These are some of the major questions as related to lubricant rheology that are, in the authors' opinion, still unanswered.

One would be remiss if a discussion of lubricant rheology omitted a new book on the subject by Jacobson (1991). The book summarizes the nearly 30 years of research the author has conducted on lubricant rheology.

3.0. THERMAL EFFECTS

Recently Hsiao and Hamrock (1992) used the fast approach and solved the thermal Reynolds equation for EHL conjunctions with a circular non-Newtonian fluid model. The use of fictious Reynolds equations ensured the validity of the solutions obtained and helped identify the flow zones with boundary shear stresses that reached the local limiting shear stress of the lubricant. Imposing shear-stress boundary conditions enabled the simulation of an EHL conjunction lubricated by a circular non-Newtonian fluid when the limiting shear stress of the lubricant is reached. The transition from circular to viscoplastic fluid model occured when the actual shear stress differed from the limiting shear stress by less than the numerical representation of the computer.

Using the thermal influence coefficients and embedding the solid-liquid interface temperature into the discrete energy equations enables the application of a straightforward successive-overrelaxation-by-lines finite difference method. It solves for the field temperatures and the interface temperatures simultaneously.

Using a conservative energy equation avoids the need for a difficult numerical procedure to obtain a converged solution for a thermal EHL conjunction that includes strong reverse flows. Applying a two-dimensional heat flow model validates the evaluation of the solid-liquid interface temperatures for an EHL conjunction with a solid surface near the stationary state.

The slide-roll ratio is introduced when incorporating non-Newtonian fluid rheology effects into an elastohydrodynamic lubrication analysis and is defined as

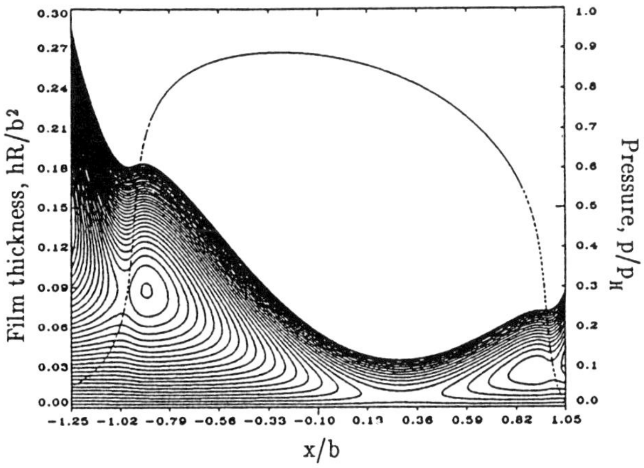

Figure 5a. Streamlines with film shape and pressure profile for $A = 10$, isothermal results. (Hsiao and Hamrock, 1992).

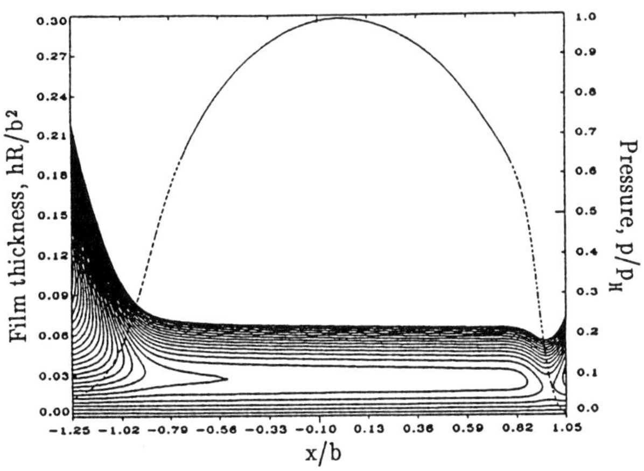

Figure 5b. Streamlines with film shape and pressure profile for $A = 10$, thermal results. (Hsiao and Hamrock, 1992).

$$A = \frac{2(u_b - u_a)}{(u_b + u_a)} \tag{5}$$

Note that $A = 0$ implies that $u_a = u_b$ while $A = 2$ implies that $u_a = 0$; when $0 \lesssim A < 2$, both solid surfaces move in the same direction through the conjunction. Also, $A = 10$ implies that $u_a = -2u_b/3$ and $A \longrightarrow \infty$ implies that $u_a \longrightarrow -u_b$; when $A > 2$ the bounding surfaces move in opposite directions inside the conjunction.

3.1. Recent Results

Some of the results from Hsiao and Hamrock (1992) are presented here, focusing on the difference between isothermal and thermal solutions as well as on local effects.

Figures 5(a) and 5(b) compare isothermal and thermal results in showing the formation of the film shape, pressure profile, and streamlines for identical operating conditions. At a slide-roll ratio of 10, with temperature effects not considered, figure 5(a) predicts the possibility of film collapse in an EHL conjunction. Figure 5(b) shows that the temperature rise in the flow field prevents a sharp decrease in film thickness near the center of the Hertzian zone. Furthermore, the pressure profile is more Hertzian. It implies that at high slide-roll ratios ($A = 10$) two seperate lubricant flows moving in opposite directions prevent metal-to-metal contact from occuring in the Hertzian zone.

Figure 6 shows the viscosity profiles at slide-roll ratios of 0, 1, 2, and 10 for isothermal and thermal results. The viscosity profiles from the isothermal analysis reflect the distributions of pressure, such as shown in figure 5(a) for $A = 10$. It also shows that the viscosity profiles remain almost unchanged when the surfaces move in the same direction or are in the stationary state ($A \lesssim 2$). When the bounding surfaces start to move in opposite directions ($A > 2$), clearly the magnitude of the viscosity (influenced by pressure) decreases and the maximum viscosity ratio shifts towards the inlet region for the isothermal results.

On the other hand, the viscosity profiles

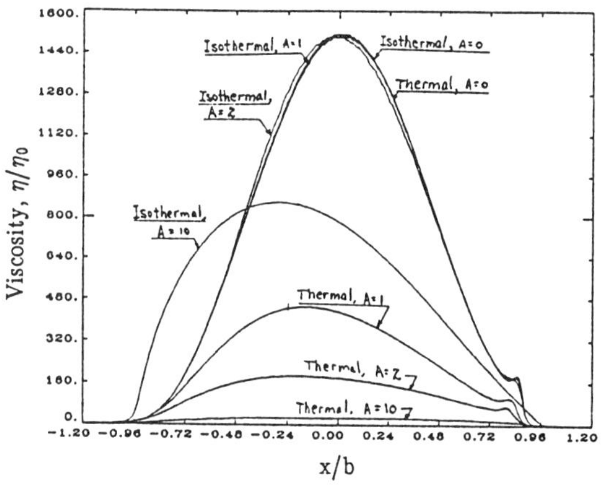

Figure 6. Viscosity profiles for various slide-roll ratios, isothermal and thermal results. (Hsiao and Hamrock, 1992).

from thermal analysis indicate the synergistic effects of pressure and temperature. The decrease in viscosity with increasing slide-roll ratio is continuous and dramatic. The location of maximum viscosity remains almost unchanged as A is changed. It reflects the fact that the pressure profiles are more Hertzian and that the temperature profiles follow more or less the same shape as the pressure profiles in the Hertzian zone along the contact.

Hsiao and Hamrock (1992) also reported that when the surfaces are in counter motion ($A > 2$) the friction coefficient is only one-half that predicted by an isothermal analysis if temperature effects are considered.

Thus the conclusion to be drawn from thermal considerations is that indeed they are important if an accurate assessment of performance is to be obtained.

3.2. Unanswered Questions

A question concerning thermal effects is, do thermal spikes exist within the conjunction? Thermal spikes do occur at a comparable location to the pressure spikes in an EHL conjunction. This can be seen in figure 7 which shows a temperature distribution obtained by Hsiao and Hamrock (1992) for near rolling conditions at $A = 0.1$. The cause of these temperature spikes is not completely understood, although the fluctuation of the boundary shear stress difference at this location may explain their existence.

Another consideration is that if a critical normal stress (the solidification pressure) exists beyond which the rheological behavior of the lubricant changes, will this also imply transitions in the thermal properties of the lubricant?

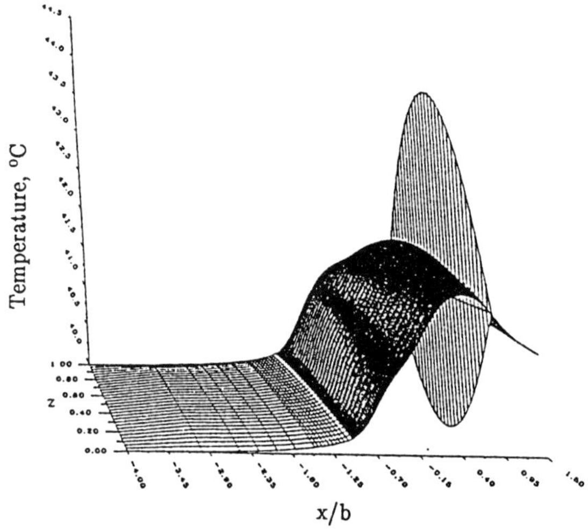

Figure 7. Temperature distribution for A = 0.1. (Hsiao and Hamrock, 1992).

4.0. SURFACE TOPOGRAPHY

As films become thinner, deviations of surface topography significantly affect the lubrication of the surfaces. The performance parameters in an EHL conjunction are modified by the amplitude and texture of the roughness patterns. Two extra parameters seem a priori to be necessary in describing the performance parameters:

a. The rms surface height relative to the film thickness
b. The ratio of the correlation lengths in the two principal roughness directions

This second parameter indicates the directional orientation of the roughness.

Considerable research has been devoted to the lubrication of surfaces where the surface texture is considered. For example, Houpert and Hamrock (1985) considered the presence of bumps and dents while assuming Eyring's rheological fluid model. Later Lee and Hamrock (1990b) investigated microelastohydrodynamic lubrication while assuming the circular fluid model.

These references treat the solutions of rough EHL surfaces such that the asperities are underformable objects being pressed into the surfaces by the pressure from the film generation. It has also been assumed that the thin film approximation applies. The thin film approximation implies that the pressure is constant across the film and allows the continuity equation to be expressed in terms of pressure and velocity (or stress) boundary conditions. Depending on the rheological model, a modified version of the Reynolds equation results.

The main question is, when dealing with rough surfaces in nonconformal conjunctions, should the full Navier-Stokes equations be used, or can some terms be neglected? The distinction between Reynolds and Stokes roughness was proposed by Elrod (1973). If b' is the characteristic wavelength of the asperity and h is the local film thickness, the ratio of these is used to describe two types of roughness:

a. Reynolds roughness $h/b' << 1$, where the thin film assumption applies and the Reynolds equation is locally valid
b. Stokes roughness $h/b' >> 1$, where the Stokes equations must be used for any local analysis

The region between was not dealt with in any definite way by Elrod (1973). Considering that the Reynolds equation is derived from a reduced form of the Stokes equations, it is probably more correct to say that if Reynolds roughness (the most restrictive condition) cannot be applied, then Stokes roughness should be used.

In any case, if terms are neglected in the full Navier-Stokes equations, how do we assess that these actually are small in a given application? Also, in attempting to get higher order effects, what should the perturbation parameter be in improving conventional lubrication theory?

This section is different from the other three major sections in that we have chosen to attempt to answer the questions we posed.

4.1. Historical Overview of Perturbation Approaches

A fundamental question that still deserves some attention is, what does the thin film assumption, which leads to the formulation of the Reynolds equation, really mean? Wannier (1950) suggested that it was a coordinate perturbation in terms of the cross-film coordinate z. His work was based on the Stokes equations, so that inertial terms had already been discarded. Wannier (1950) proposed that once the Reynolds solution was obtained, higher orders could be derived by regression. He failed to point out that when higher orders of z are taken into account for the pressure and velocity equations, the same should be done in the expression for the velocity boundary condition. When this is done properly, the coordinate perturbation results in a replacement of the second-order Reynolds equation by a sixth-order differential equation for pressure when attempting an increase in the accuracy of z.

Elrod (1959) proposed instead of a perturbation on the z-coordinate, a perturbation in terms of the film aspect ratio

$$\epsilon = h_0/L \tag{6}$$

where L = length, m
h_0 = general film thickness level, m

Elrod has applied this to conformal surfaces, such as those in a slider bearing. The dimensionalization by Elrod (1959) included using the Reynolds number in the dimensionless pressure and velocity. Only one parameter (ϵ) of small magnitude therefore appeared in the analysis.

Diprima and Stuart (1972) introduced the Reynolds number Re as an additional parameter of small magnitude in applying their perturbation approach to lubrication theory of conformal surfaces. A detailed review of the method applied to cartesian coordinates is given in Hamrock (1991). In this perturbation theory it is assumed that the pressure and the velocity can be expanded in terms of composite series of ϵ and Re, giving the Reynolds equation as the first-order approximation. This is the approach that we will use in applying our perturbation analysis.

4.2. Dimensionalization

An attempt will be made to better understand the importance of higher order effects in dealing with surface topography. Because of the added complexity we will confine our discussions during the remainder of section 4 to rigid surfaces and isoviscous Newtonian fluids with constant properties. As stated earlier, we will be dealing with nonconformal surfaces and will be neglecting side leakage.

The first major hurdle is that for rigid nonconformal surfaces there is no characteristic bearing length and no general film thickness as there is for conformal surfaces. It is incorrect to use the film aspect ratio defined in equation (6) for nonconformal considerations. Thus the proper expressions for both ϵ and Re need to be established.

By using conformal mapping it is possible to transform the nonconformal film region into a quadratic region of well-defined proportions. The appropriate curvilinear coordinate system for this purpose is the dipolar system having coordinates (ξ, ζ) defined by

$$\xi + \hat{\imath}\,\zeta = \ln\left[\frac{x + \hat{\imath}\,(z + c)}{x + \hat{\imath}\,(z - c)}\right] \tag{7}$$

where c = constant, m

This system of coordinates was used by Jeffery (1922a) to obtain an exact solution to the Stokes equations for rotating cylinders in a viscous fluid. The singularity in the dipolar coordinate system at the z-axis implies that the solution gives antimetric pressure distribution about the the line generated by connecting the center of cylinders, or, in other terms, that $p(-x) = -p(x)$. Thus, the solution compares with solving the Reynolds equation with Sommerfeld boundary conditions. In the following we consider the special case where one of the cylinders in contact is replaced with a plane. The range of the dipolar coordinates for our nonconformal conjunction is therefore $-\pi < \zeta < \pi$ and $0 < \xi < \xi_b$, where

$$\cosh\xi_b = 1 + h_c/R \tag{8}$$

$$\sinh\xi_b = \sqrt{(2+h_c/R)h_c/R} \tag{9}$$

where h_c = central film thickness, m

R = radius of cylinder, m

The film aspect ratio is measured by ξ_b/π in dipolar coordinates, which to a leading order of h_c/R is proportional to

$$\epsilon = \sqrt{h_c/R} \tag{10}$$

If we therefore choose a cartesian dimensionalization for the cross-film direction to be

$$z = h_c\,Z \quad \text{and} \quad h = h_c\,H \tag{11}$$

and at the same time assume that the accuracy of the perturbation approach (magnitude of ϵ and Re) is not altered by a change of the coordinate system, the characteristic length L for a nonconformal conjunction must be $\sqrt{h_c R}$. Thus the dimensionalization in the x-direction is

$$x = \sqrt{h_c R}\;X \tag{12}$$

Martin (1916), while solving the pressure in the Reynolds equation for a rigid, nonconformal conjunction, and Jeffery (1922a), while solving the Stokes equations, found that 90 % of the load (integrated pressure) is carried within a region of about 4 $\sqrt{h_c R}$. These results are consistent with the dimensionalization from equation (2) in that X is of order-one throughout the load carrying region.

The dimensionalization of the velocity in the direction of surface motion is scaled with respect to twice the mean surface velocity or

$$u = \tilde{u}\,\bar{u} \tag{13}$$

where $$\tilde{u} = u_a + u_b \tag{14}$$

Having chosen this dimensionalization, in order to ensure balance in the continuity equation, the velocity in the direction of the fluid film must be dimensionalized as

$$w = \epsilon \tilde{u}\, \bar{w} \tag{15}$$

The dimensionalization of the pressure is obtained by balancing the dominant velocity derivative against the pressure gradient in the Navier-Stokes equations, or

$$p = \frac{\eta_0 \tilde{u}}{\epsilon^3 R} P \tag{16}$$

The appropriate Reynolds number in the dimensionalization becomes

$$Re = \frac{\tilde{u} \rho_0 R}{\eta_0}\, \epsilon^3 \tag{17}$$

The lubricant is considered incompressible. The stream function is made dimensionless by the following equation:

$$\psi = \epsilon^2 R \tilde{u}\, \Psi \tag{18}$$

The stream function satisfies continuity, thus giving

$$\bar{u} = \frac{\partial \Psi}{\partial Z} \quad \text{and} \quad \bar{w} = -\frac{\partial \Psi}{\partial X} \tag{19}$$

For this dimensionalization the Navier-Stokes equations become

$$Re \left[\frac{\partial \Psi}{\partial Z} \frac{\partial^2 \Psi}{\partial X \partial Z} - \frac{\partial \Psi}{\partial X} \frac{\partial^2 \Psi}{\partial Z^2} \right] = -\frac{\partial P}{\partial X} + \epsilon^2 \frac{\partial^3 \Psi}{\partial X^2 \partial Z} + \frac{\partial^3 \Psi}{\partial Z^3} \tag{20}$$

$$-Re\, \epsilon^2 \left[\frac{\partial \Psi}{\partial Z} \frac{\partial^2 \Psi}{\partial X^2} - \frac{\partial \Psi}{\partial X} \frac{\partial^2 \Psi}{\partial X \partial Z} \right] = -\frac{\partial P}{\partial Z} - \epsilon^2 \frac{\partial^3 \Psi}{\partial X \partial Z^2} - \epsilon^4 \frac{\partial^3 \Psi}{\partial X^3} \tag{21}$$

These equations are identical to those obtained by Hamrock (1991) for conformal surfaces, the only difference is in the dimensionalization.

4.3. Perturbation Approach

Even though the film aspect ratio (ϵ) for nonconformal geometry is not quite as small as that for a conformal geometry, it still seems appropriate to use the parametric perturbation suggested by Diprima and Stuart (1972). Therefore, assuming *regular* expansions of the dimensionless pressure and stream functions in terms of both ϵ^2 and Re gives

$$P = P_{00} + \epsilon^2 P_{01} + Re\, P_{10} + \ldots \tag{22}$$

$$\Psi = \Psi_{00} + \epsilon^2 \Psi_{01} + Re\, \Psi_{10} + \ldots \tag{23}$$

Substituting these into equations (20) and (21) while collecting terms of the same order and following the procedures given in Hamrock (1991) for pure sliding ($A = -2$) gives

O(1):
$$\frac{dP_{00}}{dX} = 6\left[\frac{H - H_m}{H^3}\right] \tag{24}$$

$$\Psi_{00} = \frac{1}{6} \frac{dP_{00}}{dX} Z^3 - \frac{1}{4H}\left[2 + \frac{dP_{00}}{dX} H^2\right] Z^2 + Z \tag{25}$$

O(ϵ^2):
$$\frac{\partial P_{01}}{\partial X} = -\frac{\partial^3 \Psi_{00}}{\partial X^2 \partial Z} + \frac{d\tilde{A}_{01}}{dX} \tag{26}$$

$$H^3 \frac{d\tilde{A}_{01}}{dX} = \tilde{E}_{01} - \frac{4H}{5}\left[\frac{dH}{dX}\right]^2 - \frac{H^2}{5}\frac{d^2 H}{dX^2} + \frac{H^3}{10}\left[6\left[\frac{dH}{dX}\right]^2 + \frac{d^2 H}{dX^2} H\right] \frac{dP_{00}}{dX} \tag{27}$$

O(Re):
$$H^3 \frac{dP_{10}}{dX} = \tilde{D}_{10} + \frac{H^2}{280}\left[3\left[\frac{dP_{00}}{dX}\right]^2 H^4 - 34\frac{dP_{00}}{dX} H^2 + 56\right]\frac{dH}{dX} \tag{28}$$

It is observed that new arbitrary parameters (A_{01}, E_{01}, and D_{10}) appear at each increase in perturbational order. The reason is that these equations are an integrated form of the basic differential equations. These parameters are determined by imposing the appropriate boundary conditions.

For smooth nonconformal surfaces the solution of the Stokes equations given by Jeffery (1922a) provides us with an exact solution for P_{01} when imposing the Sommerfeldt boundary conditions. No exact solutions exist that will enable us to verify the equations for P_{10}.

In present applications the order-one Reynolds equation, equation (24), is solved by using the Reynolds boundary condition. As discussed by Berthe (1974), multiplicity in solutions for the exit boundary exist when the characteristic wavelength of the asperity (b') becomes smaller than a critical value. In such cases the solution giving the largest film reaction was chosen as suggested by Berthe (1974).

The integration of the inertial pressure correction P_{10} is straightforward when imposing zero correction at X_{start} and X_{end}. For the solution of the viscous pressure correction P_{01} the Z-dependence from the $\partial^3\Psi_{00}/\partial X^2\partial Z$ term causes a problem of how to interpret X_{start} and X_{end} at an $O(\epsilon^2)$ accuracy. Choosing $X_{start} \longrightarrow -1/\epsilon$ and setting $P_{01}(X_{start}, 0) = 0$ corresponds to requiring the solution to be bounded. The second boundary condition for P_{01} was $P_{01}(X_{end}, 0) = 0$, and as a result the film rupture boundary will also have a Z-dependence.

4.4. Results

First of all it should be pointed out that for smooth surfaces all three pressure contributions P_{00}, P_{01}, and P_{10} are of order one. This implies that the error introduced by using the Reynolds equation, that is, by retaining P_{00} only, is of order ϵ^2 for the neglected viscous terms and of order Re for the neglected inertial terms. Depending on the operating conditions it will be either the viscous or the inertial correction that dominates the error, but for smooth surfaces and $\epsilon << 1$ both are expected to be small and P_{00} will thus suffice for most practical purposes.

Two types of surface texture were studied, namely harmonic roughness patterns and single asperities, either as bumps or dents. In all cases the rough features were superimposed on a stationary cylinder of radius R so that the conditions were steady-state. Let h_c denote the central film thickness for the cylinder that serves as the line of reference, or in other terms, let the plane and the center of this cylinder be a distance $R + h_c$ appart.

When microlubrication conditions exist the dimensions h_c and R are no longer characteristics of the load carrying film because each asperity constitutes a separate lubricated region of given curvature and minimum film thickness. In all cases to be presented, however, film rupture occurs beyond the position of h_c and although film reformation may be possible under subsequent asperities, the bulk of the load will be carried by a coherent film. It is therefore expected that the film aspect ratio ϵ as defined by equation (10) still has relevance and that the above dimensionalization therfore applies. If z_m is the amplitude of the harmonic roughness pattern and b' its wavelength then the dimensionless film thickness becomes

$$H = \frac{1}{\epsilon^2}(1+ \epsilon^2 - \sqrt{1 - \epsilon^2 X^2}) - Z_m \cos\left[2\pi \frac{X - X_0}{B'}\right] \qquad (29)$$

where
$$z_m = h_c\, Z_m \qquad (30)$$

$$b' = \sqrt{h_c R}\, B' \qquad (31)$$

and X_0 denotes the dimensionless distance from the position of h_c to the closest minimum in film thickness. By variation of Z_m and B' it was found that the surface texture has an important effect on the viscous correction but hardly any effect on the inertial correction. In nonconformal rough surfaces it is therefore likely that the viscous correction dominates the error even when the conditions are such that $\epsilon > \eta_0/u\rho_0 R$ so that Re is larger than ϵ^2. Also it was found that the parameter z_m/b' is much more important than the parameter h/b' in determining when the thin film assumption applies. These considerations are illustrated in figures 8 and 9.

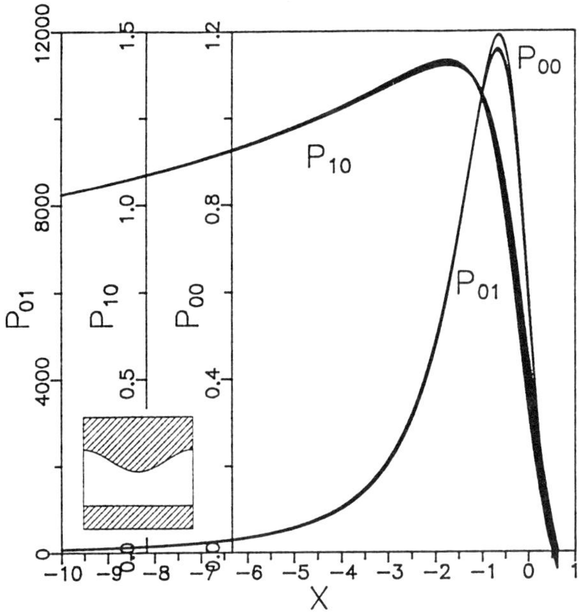

Figure 8a. Pressure contributions for a harmonic roughness pattern with minimum film thickness of 7.5×10^{-2} μm thus resulting in operating parameters of $\epsilon^2 = 1.0\times10^{-5}$, $Re = 3.2\times10^{-6}$, $Z_m = 0.25$ and $B' = 7.9\times10^{-3}$.

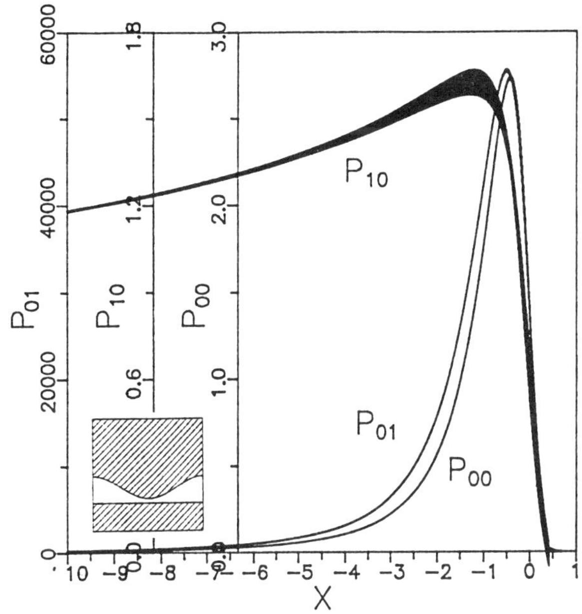

Figure 8b. Pressure contributions for a harmonic roughness pattern with minimum film thickness of 1.0×10^{-2} μm thus resulting in operating parameters of $\epsilon^2 = 3.5\times10^{-6}$, $Re = 6.5\times10^{-7}$, $Z_m = 0.714$ and $B' = 1.34\times10^{-2}$.

Figure 8 shows the effect of a reduction in film thickness on the three pressure contributions P_{00}, P_{01}, and P_{10}. The surface texture remains constant in figures 8 and 9. The operating parameters used for these results were $R = 1.0$ cm, $u\rho_0 R/\eta_0 = 1.0\times10^2$, $x_0 = 0$, $z_m = 2.5\times10^{-2}$ μm and $z_m/b' = 0.1$. In figure 8(a) the minimum film thickness in the conjunction was 7.5×10^{-2} μm ($= 3z_m$) at $x = 0$ so $h_c = 1.0\times10^{-1}$ μm and thus $\epsilon^2 = 1.0\times10^{-5}$, $Re = 3.2\times10^{-6}$, $Z_m = 0.25$ and $B' = 7.9\times10^{-3}$. In figure 8(b) the mini- mum film thickness was reduced to 1.0×10^{-2} μm ($= 0.4z_m$), which gives $h_c = 3.5\times10^{-2}$ μm, implying $\epsilon^2 = 3.5\times10^{-6}$, $Re = 6.5\times10^{-7}$, $Z_m = 0.71$ and $B' = 1.34\times10^{-2}$. Because the change in film thickness also has the effect of reducing the film aspect ratio ϵ, the pressure distributions in figure 8(a) are not directly comparable with those in figure 8(b). The respective magnitudes of the corrections $\epsilon^2 P_{01}$ and ReP_{10} relative to P_{00} are, however, comparable and it transpires that the reduction of h_{min}/b' from 0.3 to 0.04 only changes $\epsilon^2 P_{01}/P_{00}$ from about 10% to about 7%. In this case, as in all the other cases we have evaluated, the inertial correction remained very small relative to the first order pressure.

In figure 9 a reduction in minimum film thickness was obtained by increasing the amplitude of the roughness pattern while at the same time keeping h_c the same value used in figure 9(a). Again using $R = 1.0$ cm, $u\rho_0 R/\eta_0 = 1.0\times10^2$, $x_0 = 0$, and $h_c = 1.0\times10^{-1}$ μm, but now with the amplitude was set to $z_m = 5.0\times10^{-2}$ μm compared to 2.5×10^{-2} μm used in figure 9(a) so that $z_m/b' = 0.2$, $\epsilon^2 = 1.0\times10^{-5}$, $Re = 3.2\times10^{-6}$, $Z_m = 0.5$ and $B' = 7.9\times10^{-3}$. Comparing the results in figure 8(a) to those in figure 9 we observe an increase in viscous correction $\epsilon^2 P_{01}/P_{00}$ from about 10% to approximately 40%. Such results are in contradiction with the proposed condition $h/b' << 1$ for Reynolds roughness.

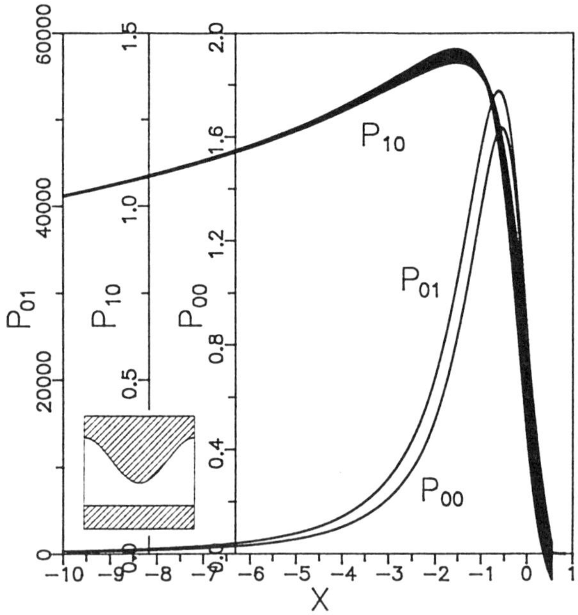

Figure 9. Pressure contributions for a harmonic roughness pattern with minimum film thickness of 7.5×10^{-2} μm and amplitude of roughness of 5.0×10^{-2} μm thus resulting in operating parameters of $\epsilon^2 = 1.0\times10^{-5}$, $Re = 3.2\times10^{-6}$, $Z_m = 0.50$ and $B' = 7.9\times10^{-3}$.

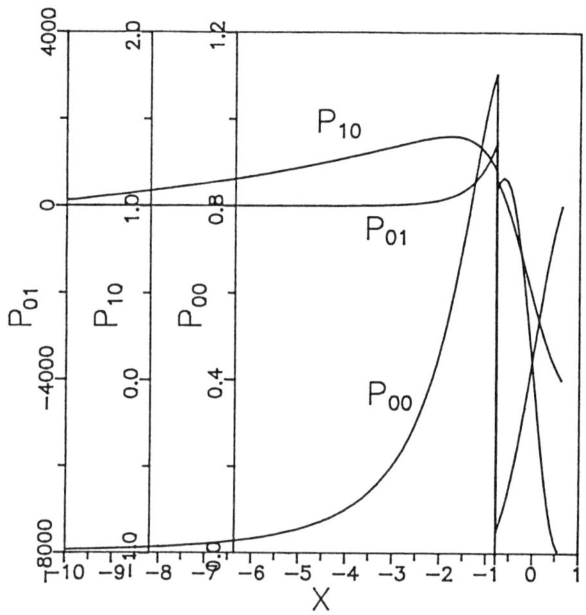

Figure 10. Pressure contributions for a single bump located at $X_0 = -0.78$ with operating parameters $\epsilon^2 = 1.0\times10^{-5}$, $Re = 3.2\times10^{-6}$, $Z_m = 0.50$ and $B' = 7.9\times10^{-3}$.

The magnitude of the viscous corrections the amplitude and b' as the wavelength the presented in figures 8 and 9 are substantial when compared with the order ϵ^2 obtained for smooth surfaces. The values of z_m/b' used were indeed such that the resulting viscous corrections no longer should be considered negligible. The maximum asperity slope was larger than 30^0 in the three cases which is somewhat in excess of what can be expected for real surfaces in nonconformal applications. That slopes of this size can occur locally on damaged surfaces is, however, undeniable. Consider for example the case of a solid particle adhering to the surface, or that of a surface crack. It is interesting to know how the presence of single bumps or dents will affect the higher orders of pressure. In order to avoid discontinuities in either film thickness, film gradient, or film curvature, the single asperity was modeled as one wavelength of a harmonic. Still using z_m as film equation becomes

$$H = \frac{1}{\epsilon^2}(1 + \epsilon^2 - \sqrt{1 - \epsilon^2 X^2}) \qquad (32)$$

for $X \notin [X_0 - B'/2,\ X_0 + B'/2]$, and

$$H = \frac{1}{\epsilon^2}(1 + \epsilon^2 - \sqrt{1 - \epsilon^2 X^2}) - Z_m\left[1 + \cos\left[2\pi\,\frac{X - X_0}{B'}\right]\right] \qquad (33)$$

under the asperity; $X \in [X_0 - B'/2,\ X_0 + B'/2]$. For the bump in figure 10 the operating conditions were $h_c = 1.0\times10^{-1}$ μm, $x_0 = -25$ μm, $u\rho_0 R/\eta_0 = 1.0\times10^2$, $z_m = 5.0\times10^{-2}$ μm, $R = 1.0$ cm, and $z_m/b' = 0.2$. This implies that the dimensionless variables had the values $\epsilon^2 = 1.0\times10^{-5}$, $Re = 3.2\times10^{-6}$, $Z_m = 0.5$, $B' = 7.9\times10^{-3}$, and $X_0 = -0.78$. To model the dent in figure 11 the sign of Z_m was changed to give a local increase in film thickness. In order to accentuate the influence of the dent we used $z_m = -5.0\times10^{-1}$ μm. Other than that, the

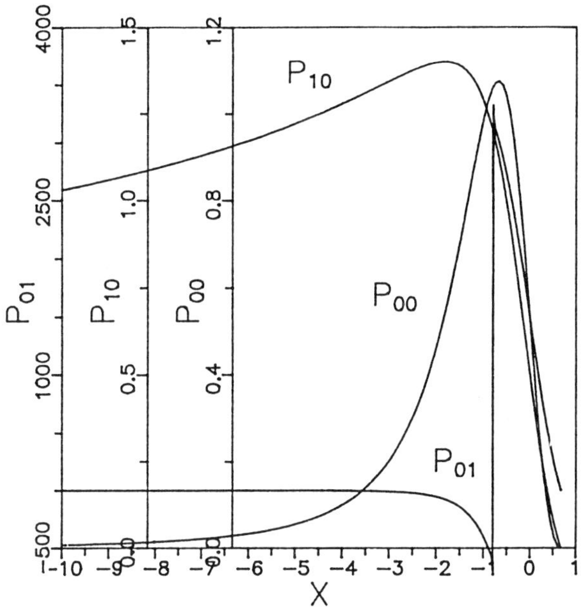

Figure 11. Pressure contributions for a single dent located at $X_0 = -0.78$ with operating parameters $\epsilon^2 = 1.0\times10^{-5}$, $Re = 3.2\times10^{-6}$, $Z_m = -5.0$ and $B' = 7.9\times10^{-3}$.

conditions were identical for figure 11 as they were for figure 10, and thus in the latter $Z_m = -5.0$.

For single asperities we observe very little variation in the inertial correction, as was also the case for harmonic roughness patterns. We can thus conclude that, for surfaces where the roughness cannot be considered as Reynolds roughness, the inertial contribution to the momentum equations remains negligible and the Stokes equations are indeed sufficient to describe the lubricant flow. Also, figures 10 and 11 indicate that not only does the single asperity result in a large pressure contribution of the viscous correction $\epsilon^2 P_{01}$ in the vicinity of the asperity, but throughout the conjunction as well.

The presence of a single surface defect can cause the higher order effects to become significant and thus cause the Reynolds equation to be inadequate. But as to now no alternative exists when the surface texture is Stokes roughness.

5.0. SURFACE DAMAGE DUE TO DEBRIS

It is well recognized that inertial and on-line lubricant filtration is important for long life in nonconformal machine elements. Furthermore it is recognized that elastohydrodynamic lubrication films are much thinner than typical oil film ratings and that inevitably particles are left in the lubricant that are several times larger than the minimum surface separation.

The presence of solid particles in the lubricant for nonconformal machine elements in the hard elastohydodynamic lubrication regime reduces life considerably. The transit of solid particles between surfaces in sliding motion (e.g., gear teeth) results in abrasive wear. Where the surfaces are in rolling motion (e.g., rolling element bearings), it causes fatigue of the surface materials and eventual failure. Today because of improved steel quality the primary source of fatigue crack propagation is surface, or near-surface, defects resulting from solid particles that roll in. In both rolling and sliding surface motion the initial failure produces more solid particles, thereby accelerating the rate of surface deterioration.

Although there has been experimental work on the effects of particles in EHL conjunctions (see, e.g., Sayles and Ioannides, 1988; or Wan and Spikes, 1988), little attempt has been made to model this event. Before discussing the modeling of a particle in a lubricated conjunction, singular features of conventional lubrication theory need to be considered.

5.1. Singularity Features of Lubrication Theory

Consider a fluid film slider bearing consisting of two parallel regions connected by means of a fixed-incline region as shown in figure 12. The usual approach to obtaining a solution consists in assuming thin film conditions, that is, to solve equation (24) in each region, and in equating both pressure and mass flow rate at the common boundaries. Because the second derivation of the film thickness with the coordinate in the direction of sliding (d^2h/dx^2, in essence the film

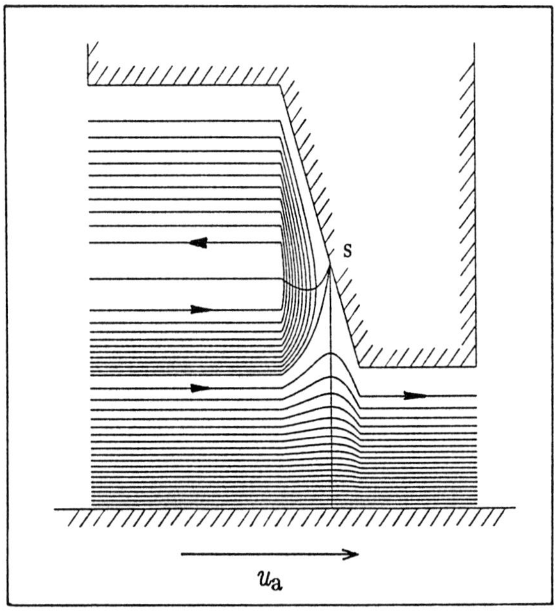

Figure 12. Streamline pattern predicted by using matched solutions to the Reynolds equation in a film region containing discontinuities in curvature.

curvature) is singular at the common boundaries, no correction to the flow can be obtained by using equations (26) and (28). The approximation is *irregular.* If we now allow the length of the fixed-incline region to decrease without changing the shoulder height, a very steep local film gradient results and another type of singularity appears. In fact, in any film geometry where locally $dh/dx > 1$ or $d^2h/dx^2 > 1$ the perturbation expansion should be singular, at least in terms of the film aspect ratio (ϵ). The singularity arising from the curvature singularity is less severe than the one arising from the film gradient, but in both cases using the Reynolds equation will not allow a proper matching between the fluid regions (the coupling between up- and downstream flow fails). Seen in this light, the increase in viscous correction obtained above for rough surfaces must be taken as an indication that the regular perturbation approach is about to fail because the film gradient and the film curvature approach their singular values. In such cases the actual value of the higher order solution P_{01} should not be used to improve the leading order solution P_{00}, but rather taken as a measure of the validity of the assumptions leading to the Reynolds equation.

5.2. Particle Modeling

Depending on the ratio of the particle diameter to the film thickness (d_p/h), different considerations apply for the particle motion. If the particle diameter is to be of comparable dimension to the minimum film thickness, the condition $d_p/h << 1$ corresponds to a particle far from the center of the conjunction and located well into the inlet region. There the particle will move as predicted by the theory of low-Reynolds-number hydrodynamics (see, e.g., Jeffery, 1922b; Happel and Brenner, 1965; or Leal, 1980). Only one wall needs to be accounted for at a time, and it transpires that the particle will tend to move as the fluid it replaces. The influence of the particle on the flow will only extend to one or two times the particle diameter.

The case where $d_p/h > 1$ (particle entrapment) has been considered by Sayles and Ioannides (1988). The problem is basically one of solid mechanics because no lubricant separates the particle from the solid surfaces.

Haber et al. (1987) proposed an approximate solution in the intermediate flow region where d_p/h is only slightly smaller than one. This solution assumes quasi-parallel surfaces and is thus restricted to conformal geometries. In nonconformal applications the lubricant would experience backflow, and none of the simplifying assumptions (which made the preceding solutions possible) would be valid. It is under these conditions that the considerations concerning singular features in fluid film lubrication covered in the last section become important. Because the flow partly occurs in the small region between the particle and either surface, it is tempting to resort to some type of local Reynolds solution. This is in essence what Languirand and Tichy (1983) did for a particle of high aspect ratio (thus increasing the regions above and underneath

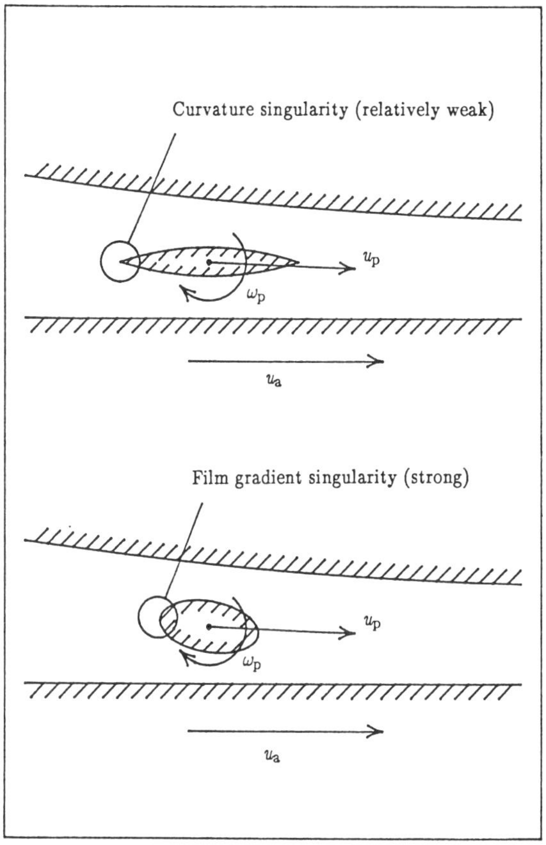

Figure 13. Nature of the singularities arising when attempting to model particle motion by means of the thin film approximation.

the particle where the Reynolds solution would be valid). Because the aspect ratio was chosen to be large in their work (the particle length was between 5 and 10 times the particle thickness), it is possible that the leading and trailing film gradient and curvature do not have a considerable effect. It does remain that these are singularities in the solution, as illustrated in figure 13. Thus it is believed that a proper singular perturbation expansion to replace the regular one of Diprima and Stuart (1972) is needed for the particle consideration.

6.0. CONCLUSIONS

The basic mechanisms involved in elastohydrodynamic lubrication were presented. Elastohydrodynamic lubrication, as characterized by solutions of the coupled elasticity, rheology, and Reynolds equations, has revealed the fundamental understanding of film thickness, pressure, film temperature, surface temperatures, and stress fields required for design and failure predictions of lubricated nonconformal conjunctions. Machine elements operate with effective fluid film lubrication even when the films are of micrometer and submicrometer dimensions. The results until recently focused on Newtonian fluid, smooth surfaces, and isothermal conditions. With film thicknesses in EHL conjunctions decreasing (still being able to adequately lubricate the surfaces), earlier assumptions have had to be reconsidered.

The first part of this paper focused on the developments that have taken into account non-Newtonian fluid rheology, thermal effects, and nonsmooth surfaces. The review presented was an attempt to describe the important aspects of these additional considerations. Since it is a review paper, we also attempted to pose questions that the authors are groping with, hoping that open discussion may result in better understanding. It must be emphasized that these conjectures are those of the authors and that others might see things differently.

The latter part of the paper focused on research just being initiated. A basic understanding of various terms in the Navier-Stokes equations as applied when singularities exist within the lubricating conjunction was attempted. A perturbation analysis was performed in hopes of better understanding the importance of higher order effects in dealing with surface topography. The modelling of how a solid particle travels through an EHL conjunction was also briefly discussed.

7.0. AKNOWLEDGEMENTS

The authors wish to thank the Danish Research Academy (Denmark) for funding the visit for one year of C.M. Myllerup at the Ohio State University, as well as SKF Engineering and Research Center (The Netherlands) and the Mobil Foundation for their support of this effort.

8.0. REFERENCES

Bair, S. (1990) High Shear Stress Rheology of Liquid Lubricants. Ph.D. Thesis, Georgia Institute of Technology, USA.

Bair, S. (1992) "Shear Bands in EHL Conjunctions," Presented at the Open Forum of the ASME-STLE Conference, St. Louis, MO., USA, Oct. 1992.

Bair, S. and Winer, W.O. (1979) "Shear Strength Measurements of Lubricants at High Pressure," J. Lubr. Tech., vol. 101, no. 3, pp. 251-257.

Berthe, D. (1974) Les Effets Hydrodynamiques sur la Fatigue des Surfaces dans les Contacts Hertziens. Ph.D. Thesis INSA Lyon, France.

Diprima, R.C. and Stuart, J.T. (1972) "Flow Between Eccentric Rotating Cylinders," J. Lubr. Tech., vol 94, no. 3, pp. 266-274.

Elrod, H.G. (1959) "A Derivation of the Basic Equations for Hydrodynamic Lubrication with a Fluid Having Constant Properties," Report I-A2049-5, Franklin Institute, USA.

Elrod, H.G. (1973) "Thin-Film Lubrication Theory for Newtonian Fluids With Surfaces Possessing Striated Roughness or Grooving," J. Lubr. Tech., vol. 95, no. 4, pp. 484-489.

Elsharkawy, A.A. and Hamrock, B.J. (1991) "Subsurface Stresses in Micro-EHL Line Contacts," J. Tribology, vol. 113, no. 3, pp. 645-656.

Gecim, B. and Winer, W.O. (1980) "Lubricant Limiting Shear Stress Effect on EHD Film Thickness," J. Lubr. Tech., vol. 102, no. 2, pp. 213-219.

Granick, S. (1991) "Motion and Relaxations of Confined Liquids," Science, Sept. 1991, vol. 243, pp. 1374-1379.

Granick, S., (1992), Private communications

Haber, S., Shapira, M., and Etsion, I. (1987) "The Effect of Two-Phase Lubricant on Bearing Performance, ASLE Trans., vol. 30, no. 1, pp. 34-40.

Hamrock, B.J. (1991) Fundamentals of Fluid Film Lubrication, NASA RP-1255.

Hamrock, B.J., Jacobson, B.O., and Bergstrom, S.I. (1987) "Measurement of the Density of Base Fluids at Pressure up to 2.2 GPa," ASLE Trans., vol. 30, no. 2, pp. 196-202.

Happel, J. and Brenner, H. (1965) Low Reynolds Number Hydrodynamics. Englewood Cliffs, N.J., Prentice Hall.

Hoglund, E. and Jacobson, B.O. (1986) "Experimental Investigations of Shear Strength of Lubricants Subjected to High Pressure and Temperature," J. Tribology, vol. 108, no. 4, pp. 571-578.

Houpert, L. and Hamrock, B.J. (1985) "Elastohydrodynamic Lubrication Calculations Used as a Tool to Study Scuffing" Proc. 12th Leeds-Lyon Symp. on Tribology, Mechanical Engineering Publications, Bury St. Edmunds, Suffolk, England.

Houpert, L. and Hamrock, B.J. (1986) "Fast Approach for Calculating Film Thicknesses and Pressures in Elastohydrodynamically Lubricated Contacts at High Loads," J. Tribology, vol. 108, no. 3, pp. 411-420.

Hsiao, H.S. and Hamrock, B.J. (1992) "A Complete Solution for Thermal Elastohydrodynamic Lubrication of Line Contacts Using the Circular Non-Newtonian Fluid Model," J. Tribology, vol. 114, no. 3, pp. 540-552.

Iivonen, H.I. and Hamrock, B.J. (1989) "A Non-Newtonian Fluid Model for Elastohydrodynamic Lubrication of rectangular Contacts," Proc. 5th Int. Congress on Tribology, Helsinki, Finland, June 1989, pp. 178-183.

Jacobson, B.O. (1991) Rheology and Elastohydrodynamic Lubrication, Elsevier, Amsterdam, Tribology Series, 19.

Jeffery, G.B. (1922a) "The Rotation of Two Circular Cylinders in a Viscous Fluid," Proc. Roy. Soc. London A101, pp. 169-174.

Jeffery, G.B. (1922b) "The Motion of Ellipsoidal Particles Immersed in a Viscous Fluid," Proc. Roy. Soc. London, vol. A102, pp. 161-179.

Languirand, M.T. and Tichy, J.A. (1983) "The Effect of a Translating High Aspect Ratio Particle in a Plane Slider Bearing," J. Lubr. Tech., vol. 105, no. 3, pp. 396-405.

Leal, L.G. (1980) "Particle Motions in a Viscous Fluid," Ann. Rev. Fluid Mech., vol. 12, pp. 435-476.

Lee, R.T. and Hamrock, B.J. (1990a) "A Circular Non-Newtonian Fluid Model — Part I — Used in Elastohydrodynamic Lubrication," J. Tribology, vol. 112, no. 3, pp. 486-496.

Lee, R.T. and Hamrock, B.J. (1990b) "A Circular Non-Newtonian Fluid Model — Part II — Used in Microelastohydrodynamic Lubrication," J. Tribology, vol. 112, no. 3, pp. 497-505.

Martin, H.M. (1916) "Lubrication of Gear Teeth," Engineering (London), vol. 102, pp. 119-121.

Myllerup, C.M., Elsharkawy, A.A., and Hamrock, B.J. (1992) "Couette Dominance Used for Non-Newtonian Elastohydrodynamic Lubrication," Submitted to the J. Tribology for publication.

Sayles, R.S. and Ioannides, E. (1988) "Debris Damage in Rolling Bearing and Its Effect on Fatigue Life, J. Tribology, vol. 110, no. 1, pp. 26-31.

Trachman, E.G. (1971) "The Rheological Effects on Friction in EHL," Ph.D. Thesis, Northwestern University, Evanston, IL., USA.

Wan, G.T.Y. and Spikes, H.A. (1988) "The Behavior of Suspended Solid Particles in Rolling and Sliding Elastohydrodynamic Contacts," ASLE Trans., vol. 31, 1, pp. 12-21.

Wang, S.H. and Zhang, H.H. (1988) "The Surface Strength of Gears and Rheological Characteristics of Lubricants, Wear, vol. 127, pp. 1-14.

Wannier, G.H. (1950) "A Contribution to the Hydrodynamics of Lubrication," J. Appl. Math., vol. 8, no. 1, pp. 1-32.

SESSION III
CONTACT, FRICTION, TRACTION

Chairman : Dr J M Greenwood

Thin Films in Tribology / D. Dowson et al. (Editors)
© 1993 Elsevier Science Publishers B.V. All rights reserved.

The Effect of oil film thickness on the rolling friction under line contact in the range from 0.01 to 10μm

K. Matsubara, T. Deguchi* and S. Sasanuma

Department of Production Mechanical Engineering, Tokai University,
1112, Kitakaname Hiratsuka-shi Kanagawa 259-12, Japan

In this paper, rolling experiments were done on condition that a couple of rollers rolls reciprocally on the flat specimen of steel having different surface roughnesses. The rolling members were coated with thin oil film thicknesses that were made to desiccate a solvent in oil diluted solutions on the surfaces. The rollers were made of steel, ceramics and nylon.

The experimental results show there is a transition phenomenon in which the coefficients of rolling friction for steel and ceramic rollers suddenly increase from an oil film thickness. For nylon rollers it does not show any transition phenomena.

The estimated rolling velocities corresponding to the oil film thicknesses were examined by a formula in EHL. For example, the maximum film factor (ratio of film thickness to surface roughness) of 10 is equivalent to a velocity of 350 cm/s.

1. INTRODUCTION

On the mechanisms of rolling friction a considerable literature exists on this subject [1-6]. The roughness theory has been put forward by Bikerman [7], Drutowski [8], Soda [9], Sasaki [10] and so on, and the adhesion theory by Tomlinson [11], Hisada [12] and Schwartzkoff [13].

Hysteresis losses on rolling friction have been studied by Tabor [14] as an important factor. After that, many a research has been developed to a high degree of maturity based on the hysteresis losses theory for over three decades.

An interesting point among them is that the rolling members coated with thin soft metals have little effect on rolling. As an example, four kinds of the thin metals coated on some plastics from 0.1 to 1μm have been examined by Matsubara et al [15]. According to the report, there were no effects on the rolling friction.

From experimental evidences it is known that thin oil films coated have little effect on the rolling resistance as well as the thin metal films. If thicker oil films were applied to the rolling members, the rolling resistance must be increased by the resistance to flow of the lubricant. It is supposed there is a critical film thickness of lubricant between the thick and thin film thickness.

In this paper, the effects of oil film thickness on the rolling friction have been examined by a couple of rollers rolling pendulum system [16] instead of a ball on a concave track reported [17,18]. Test rollers were chosen S45C, Al_2O_3, ZrO_2, Si_3N_4 and nylon.

Flat surfaces of rolling track were

*Graduate Student of Tokai University (Present Electronic Systems and Components Division, Komatsu Ltd),
2597 Shinomiya, Hiratsuka-shi Kanagawa 254, Japan

made of S45C having the surface roughnesses from 3 to 0.1 μm R_{max}. The film thicknesses required were made by an artificial means to desiccate a solvent in the oil diluted solutions on the surfaces in the same way independently described elsewhere [19, 20].

The rolling friction were measured in a range of oil film thickness from 0.01 to over 10 μm. The experimental results show that the coefficients of rolling friction for steel and ceramic rollers increase suddenly from an oil film thickness on a half way of its increasing. For nylon rollers it does not show any transition phenomena. The reasons for differences will be discussed later.

In this paper, a relation between both film thicknesses made artificially and calculated by EHL are also examined.

2. APPARATUS AND EXPERIMENTAL PROCEDURE

2.1 Principle of apparatus

The principle of the apparatus for measuring rolling friction in this paper is shown in Fig. 1 on the basis of Offner's report [16]. A testing couple consists of two rollers of the same radius which is keeping a gap with a rod making in alignment of their centres. The rod is wound by a thread hanging a suspended weight w.

The construction of the apparatus is drawn schematically in Fig. 2. The dimensions are 42 cm in length, 33 cm

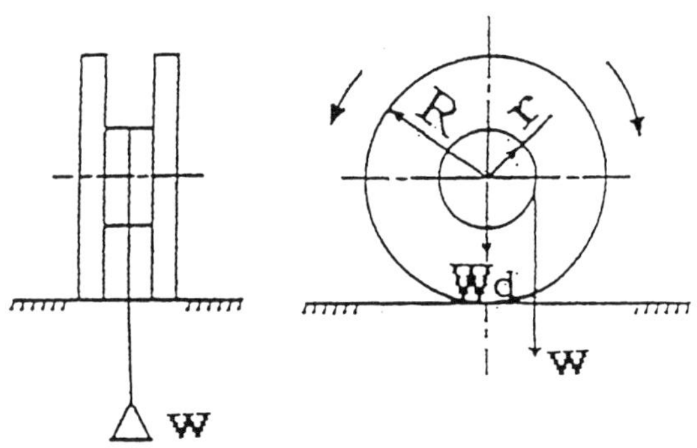

Fig. 1 The principle of apparatus

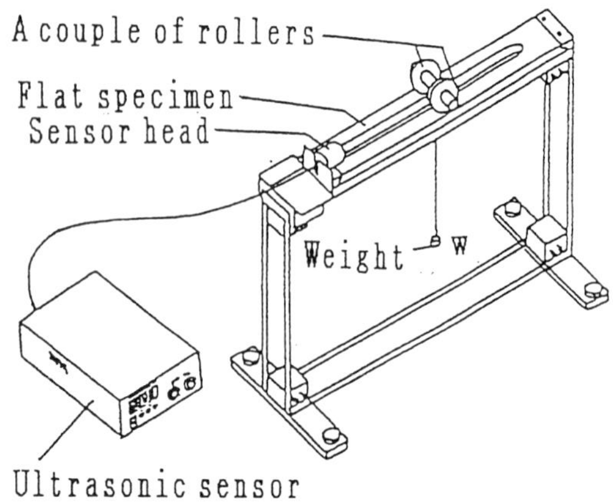

Fig. 2 Layout of the apparatus

in height and 5 cm in width respectively. Side plates that keep rigidity of the apparatus were made of PMMA with also object of seeing inside of the apparatus by eyes. The top of the apparatus was made of a base plate on which a flat specimen is set.

The roller rolling pendulum system has a long rolling distance. When the rollers released from a set position of about 10 cm apart from the centre of the track it rolls reciprocally on the flat specimen. The resistance to rolling may depend on an amount of the total rolling distance L of the rollers. The distances L accumulated from the first run to the end are obtained through a personal computer in which measured data are inputted from an ultrasonic sensor as shown in Fig. 2.

A sum of weight of the couple of rollers W_a and the suspended weight w would be $W_t = W_a + w$. For calculation of the coefficient of rolling friction, Offner and Tomita [18] have proposed an equation of potential energy for the rolling system as follows;

$$w \cdot h = \rho W_t \cdot L \qquad (1)$$

where, $w \cdot h$ = potential energy of the weight w in the height h between the beginning and the end of rolling over

Table 1 Roughness (σ_1) of rollers

Sample	S45C	Al_2O_3	ZrO_2	Si_3N_4	6N
σ_1 μm	0.2	2.0	0.14	0.25	0.5

total distance L of rolling.

The coefficient of rolling friction ρ is given by

$$\rho = w \cdot h / W_t \cdot L \qquad (2)$$

here, $w \cdot h$ and $W_t \cdot L$ are the loss of potential energy and the work of rolling.

2.2 Samples for experiments

2.2.1 Rollers

A carbon steel, ceramics (Al_2O_3, ZrO_2, Si_3N_4) and 6 nylon were chosen as the materials for the rollers. The dimensions for each roller are 50 mm in diametre and 10 mm in thickness. The rollers of carbon steel were machined from the machine structural use of 0.45 % C, S45C, and were treated by carborization and then by hardening. After the heat treatments were done, the rolling surfaces were ground keeping the surface hardness. The roughness σ_1 of the rollers are shown in Table 1.

2.2.2 Flat specimens

Flat specimens of steel were made in the same way as the roller made. The dimensions of flat specimen are 50 mm in width and 300 mm in length. The specimens have a narrow space that was made in the middle along the rolling track as shown in Fig. 2. Four kinds of the roughness of 3.0, 0.5, 0.3 and 0.1 R_{max} were made in perpendicular to the rolling direction.

2.3 Method for polishing and cleaning surfaces

The rolling members were carefully cleaned with a 125 W ultrasonic washing machine in the trichloroethane for 10 minutes each and the cleaning process was repeatedly three times.

2.4 Methods for making thin oil film thickness

A thin oil film thickness is made as follows: First, a concentration of the diluted solution from mineral oil SAE #10 and trichloroethane as a dilutent is made. Next, the oil diluted solution is coated on the surfaces. Then, the dilutent is desiccated from the solution. For different coating areas keeping a constant film thickness the volume of the solution must be changed. For different film thicknesses the concentrations must be changed with the proportion of the coating area.

2.5 Experimental methods

The couple of testing rollers was put on the flat specimen, then the weight w suspended by a thread was adjusted with winding up on the shaft of the couple, and a wedge was put in front of the rollers to keep the suitable set position.

Previously, the height of the suspended weight w in the vertical was measured. When the wedge was taken off quietly, the couple of rollers start to roll reciprocally with damped oscillations. The curves are shown in Fig. 3: (A) under dry and (B)the track (surface roughness of 3.0 μm) coated with the oil film thickness of 5.2 μm.

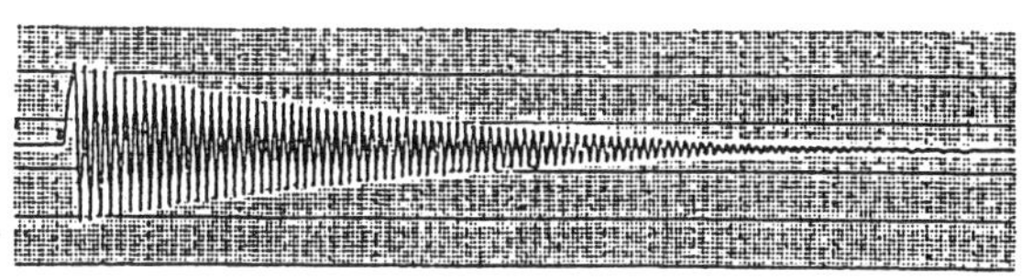

(A) Dry

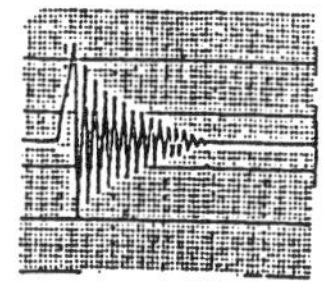

(B) 5μm in film thickness

Fig. 3 Damped oscillation curves

Table 2 mean values and standard deviations of dry rolling friction

Specimen	S45C		Al_2O_3		ZrO_2		Si_3O_4		6N	
σ_2 μm	3.0	0.1	3.0	0.1	3.0	0.1	3.0	0.1	3.0	0.1
m ,$\times 10^{-3}$	0.45	0.45	0.35	0.28	0.18	0.13	0.22	0.19	2.3	3.1
σ_{n-1}, $\times 10^{-4}$	0.17	0.38	0.18	0.19	0.19	0.07	0.23	0.07	2.48	4.08

The total rolling distance L is calculated by the personal computer, and the vertical displacements h of the weight w from the start to the last run is obtained from the readings.

The experiments were made at a room temperature of 25℃ and about 50 % R.H.

3. EXPERIMENTAL RESULTS AND DISCUSSIONS

3.1 The coefficient of rolling friction under dry condition

The coefficients of rolling friction for S45C, ceramics and 6 nylon were obtained under the dry condition as shown in Table 2. The experiments were done on tracks made with two roughnesses(σ_2) of 3.0 and 0.1 μm.

The mean values and standard deviations show that the effect of surface roughness on the rolling friction seems to be almost independent.

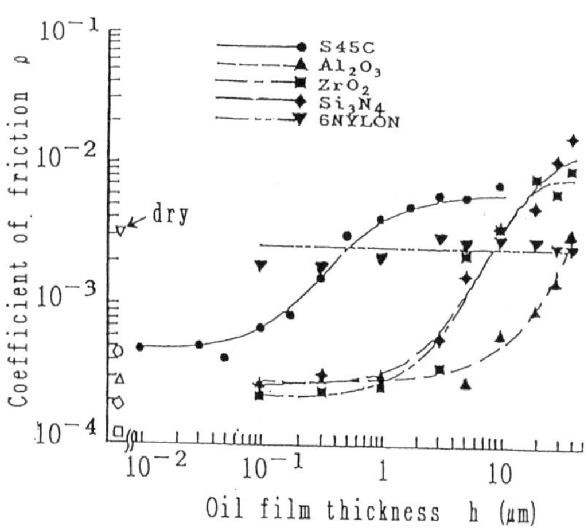

Fig. 4 The effect of film thickness

3.2 The effect of oil film thickness

When a surface roughness (σ_2) of 0.1 μm on the track kept constant, the relation between the coefficient of rolling friction and oil film thickness is shown in Fig. 4. The graphs show that in the case of steel S45C the coefficient of rolling friction keeps constant under the boundary lubrication of the film thickness of 0.1 μm.

When exceed the value of 0.1 μm higher the coefficient increases with increasing film thickness. Thus, the transition phenomena of the coefficient of rolling friction have been shown clearly in these experiments. In the case of Si_3N_4 and ZrO_2 the same phenomena have been observed, but a starting point of the phenomena on the film thickness was 1 μm and it was increased by ten times than that of steel. The other ceramic of Al_2O_3 became the starting point as larger as about 3 μm.

In the case of 6 nylon there are no transition phenomena. These experimental results would suggest that the mechanism of rolling friction of nylon differs with other materials of testing rollers.

These phenomena under the line contact show substantially the same tendency to the results of the point contact of rolling reported [17,18].

If the film thicknesses were over about 10 μm every couple of rollers begins to slide without rolling and the direction of motion is lost and run out into a different direction.

3.3 The surface roughness of flat surface of rolling tracks

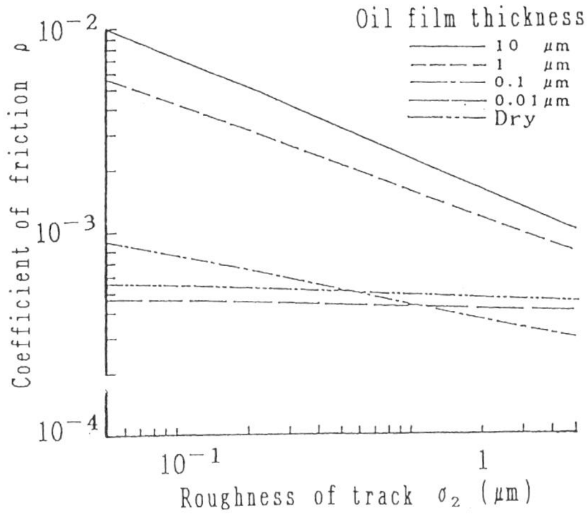

Fig. 5 The effect of surface roughness

The graphs have been complicated with many experimental results plotted, so the lines are drawn with omitted them, using a least squared method. Figures from Fig. 5 will be made in the same way.

The effect of surface roughness of the track on the rolling friction for couples of steel rollers is shown in Fig. 5 as a function of film thickness. In a case of film thickness of 10 μm the coefficient of rolling friction decreases with a tendency to reach the value of dry friction. Under the condition of dry friction and boundary lubrication of the film thickness of 0.1 μm, the coefficient of rolling friction keeps constant with increasing the surface roughness. The behaviours of ceramics are the same as that of steel. The nylon did not show any change with surface roughness.

3.4 Film factor ξ and the coefficient of rolling friction ρ

The effect of thin oil film thickness on the rolling friction has been considered to be little [1, 18]. Under such a condition the rolling may be situated in the boundary lubrication, thus some parts of the contact are supported by oil film the other by solid contacts.

The roughness of ceramic rollers was larger than that of steel as shown in Table 1, thus the ceramic rollers may require more thick oil film than that of steel to show the same behaviour. Accordingly, the transition phenomena may also depend on the surface roughness.

From the point of view the relation between roughness and film thickness must be discussed. Here, the nondimensional film factor $\xi(h/\sigma)$ defines as ratio of the film thickness h to the relative roughness σ:

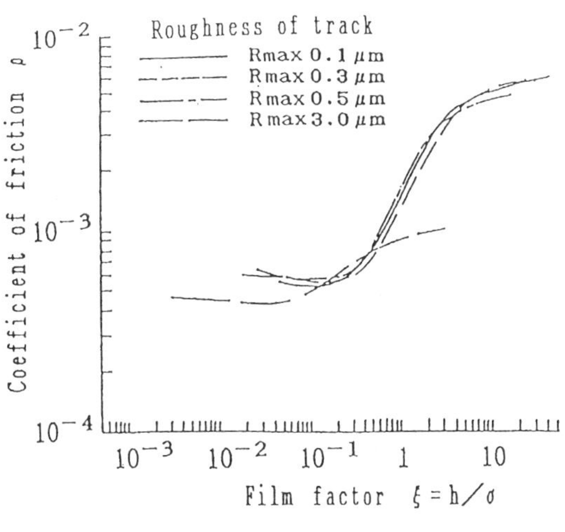

Fig. 6 Nondimensional film factor for steel S45C

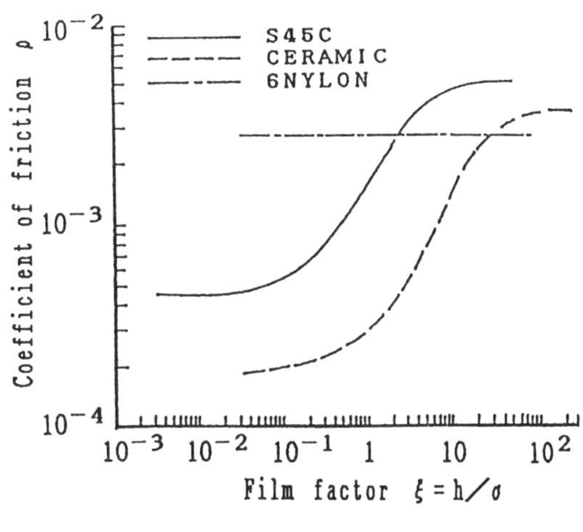

Fig. 7 Nondimensional film factor for each material of rollers

Table 3 Minimum film thickness by EHL

Specimen	S45C	Al_2O_3	ZrO_2	Si_3O_4	6N
h_{min} μm	0.13	0.11	0.10	0.12	0.15

Table 4 Film factor ξ' by h_{min}

Roughness σ_2 μm	Relative roughness σ μm	Film factor ξ' (h_{min}/σ) ,$\times 10^{-1}$	ρ ,$\times 10^{-3}$
3.0	3.01	0.43	0.47
0.5	0.54	2.41	0.59
0.3	0.36	3.61	0.66
0.1	0.22	5.91	0.89

{$\sigma=(\sigma_1{}^2+\sigma_2{}^2)^{1/2}$: σ_1 and σ_2 are roughness of the roller and rolling track}. The value of 1 of the film factor means that the thickness of the oil film corresponds to the surface roughness or 10 is ten times higher the one.

For steel rollers the relation between the coefficient of rolling friction and the film factor are shown in Fig. 6 as a parameter of the surface roughness. The figure presents that each graph seems to include in one graph adapting the film factor. The other materials of ceramics and 6 nylon are also shown in Fig. 7 including the results of steel S45C in Fig. 6.

As introduced the idea that the film factor is used for arrangement of the experimental results instead of independent factors of roughness and film thickness, the relations became clear as shown in Fig. 7. From this arrangement a rolling regime under lubrication is classified into three parts of boundary lubrication, fluid lubrication and transition zone between them.

3.5 The results examined by EHL

The velocities of rolling must be generally keep constant under the rolling condition of EHL. Then, an excessive attempt using EHL computation will try to apply for the experimental results obtained in the damped reciprocal motions of rolling.

3.5.1 Computation of h_{min}

On the contact of rolling surfaces the minimum film thicknesses h_{min} are calculated by an equation of Dowson-Higginson [21], afterward, the equation have revised as follows [22]:

$$H = 2.65\, G^{0.54} U'^{0.7} W^{-0.13} \qquad (3)$$

here,

$H = h_{min} / R$

G : Material parameter

W : Load parameter

R : Equivalent radius of roller

U' : Velocity parameter

Here, it is assumed that a mean value of rolling velocity in this experiment can be applied for the equation (3).

The values of minimum film thickness using 3.5 cm/s of the mean rolling velocity calculated are shown in Table 3. The minimum film thicknesses for each roller are almost the same.

3.5.2 Film factor ξ' by h_{min} and ρ

Film factors $\xi'(=h_{min}/\sigma)$ calculated from the minimum film thickness h_{min} by the equation (3) and surface roughnesses of rolling tracks are shown in Table 4. The coefficients of rolling friction for the film factor have been obtained from the graph in Fig. 7.

3.5.3 ξ and ratios of minimum and maximum coefficient of rolling friction τ'

Ratios of the coefficients of rolling friction (the observed values to the maximum value) for steel S45C and ceramics against the film factors are shown in Fig. 8. Almost the same behaviours have been reported by Tallian et al [23] examined with the electrical resistance, by Dowson et al [24] with rolling friction including slip, by Aihara and Sawamoto [25] with the torque and by Johnson et al [26] with the theoretical treatments.

They have examined a relation between the film factor and the non-contact time and or torque (or rolling friction).

According to their data, in the case of the film factor $\xi < 1$, the contact is metallic and $\xi > 3$ completely liquid lubrication. The values of film factor ξ' and ρ in Table 4 have been included in Fig. 8. The film factor ξ' is less than 1 mainly due to the low velocity of 3.5 cm/s.

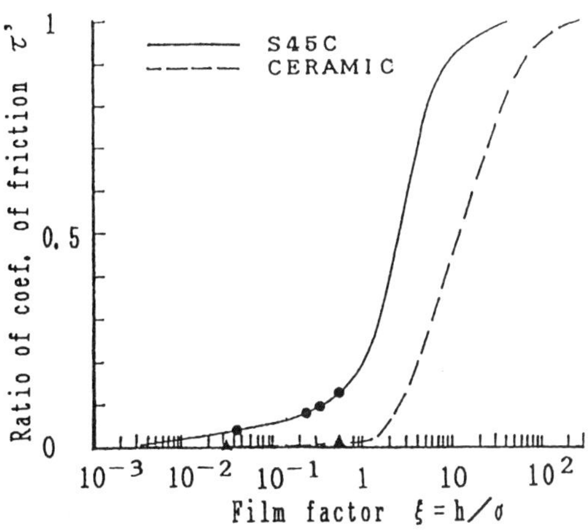

Fig. 8 Film factor and ratio of the rolling friction

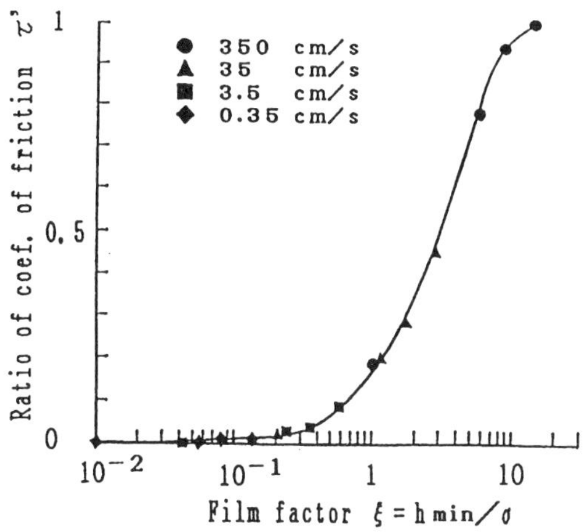

Fig. 9 Film factor and velocities calculated from EHL

3.5.4 Rolling velocities calculated by EHL

In the case of steel S45C, the film factors ξ' were obtained from the equation (3) using the film thickness h_{min} against the assumed rolling velocities of 0.1, 1, 10 and 100 times the mean velocity of 3.5 cm/s in this experiment. The values are shown in Fig. 9.

From the relation between the ratio of the coefficient of rolling friction and the film factor, a maximum film factor of 10 made by the artificial method in this experiment correspond to a velocity of 350 cm/s by EHL. For ceramics the treatment is the same as steel.

3.6 Hertzian contact pressure

In these experiments the contact configuration is the line contact that tries to compare with the point contact reported already [17,18]. The maximum and mean contact pressures for line and point contacts are shown in Table 5 for each sample.

From these data both the maximum and mean contact pressures of the line contact are clearly much less than that of point contacts and ranging from steel

Table 5 Hertzian contact pressures

Sample \ P	P_{max}, ×10^{-2} GPa		P_{mean}, ×10^{-2} GPa	
Contact	Line	Point	Line	Point
S45C	1.85	8.26	1.45	5.51
Al_2O_3	1.68	12.5	1.32	8.36
ZrO_2	1.62	12.5	1.27	8.32
Si_3N_4	1.56	—	1.23	—
SIALON	—	11.5	—	7.67
6N	0.206	0.461	0.162	0.304

※ — No data

S45C to nylon. However, in the case of point contact the values of the rollers range from ceramics to steel and nylon. On the other hand, under the same film thickness, the coefficients of rolling friction of the line contact show the lower values than that of point contacts. In addition to it, the starting point of the transition of the line contacts is late compared with the point contact.

4. CONCLUSIONS

With increasing oil film thickness in this experiment and with a small amount of the oil lubricated by Ishikawa et al [27], both behaviours of rolling friction undergo a remarkable difference. In the former the coefficient of rolling friction keeps almost constant initially then increasing drastically and in the latter decreases surely.

An interesting difference has occurred both rolling experiments based on the experimental conditions. The Hertzian pressure and velocity are as follows: from 1.08 to 1.65 GPa and a constant velocity of 2.5×10^{-3}cm/s in Ishikawa's and 5.62×10^{-2} GPa and a mean value of 3.5 cm/s in this experiment.

The other things affecting the rolling friction should be surface energy on the rolling surfaces. The ceramic rollers have lower coefficient of rolling than that of steel under the dry condition as shown in Table 2. The ceramics have also the higher starting point of transition phenomenon than that of steel. The reason is not clear but surface energy of ceramics especially small [29] compared with that of steel.

It is very characterized that comparing behaviours of rolling friction between the steel and ceramic rollers and the nylon are different. The nylon rollers do not show any transition phenomena keeping a nearly constant coefficient of rolling friction of about 8.0×10^{-3} within the range chosen within these experiments. The behavior of nylon is mainly governed by the large hysteresis losses of itself that is superior to the effect of oil film covered on the surfaces of steel and ceramic rollers.

If the experiments were done under the condition over a thicker oil film thickness the couple of every testing roller begun to slide without rolling. Afterward, the couple of rollers were lost the direction of motion and run out into a different way. Under the conditions in the perfect fluid lubrication the rollers had chosen an ease way to move themselves.

The rolling friction under the fully flooded condition will be calculated by a formula (24) by Goksem et al [28]. Under such conditions the resistance to flow would be compared with the rolling resistance. However, in this paper such discussion will be left in the near future.

In conclusions, the mechanism of rolling friction will be divided into two types: one is surface problems such as adhesion, roughness or cleanness and so on; the other is bulk problem such as hysteresis losses. In the case of steel and ceramic rollers will be concluded in the surface problems, whereas the nylon is clearly included in the category of the bulk properties.

In this experiment, the rolling is

reciprocal damped motions. If it is assumed that the fixed direction of rolling with a constant velocity was decided the EHL equation of Dowson-Higginson is considered to be applied. A mean rolling velocity of 3.5 cm/s was calculated based on the concept. To consider the assumed rolling velocity corresponding to the oil films is important. For example, the maximum film factor of 10 was obtained as to correspond to the velocity of 350 cm/s calculated by the EHL equation.

REFERENCES

[1] F.P. Bowden and D.Tabor : The friction and Lubrication of Solids(II), Oxford (1964), Chap.XX.
[2] K. Matsubara, Tribology (Japanese), Sangyo-Tosho, Chap.5,and,Chap.7-4.
[3] M.D. Hersey, Trans. ASME, JOLT, 91 (1969) 260.
[4] M.D. Hersey and M.S. Downes, Trans. ASME, JOLT, 91(1969) 264.
[5] M.D. Hersey, Trans. ASME, JOLT, 91 (1969)269.
[6] M.D. Hersey and P.L. Golden, Trans. ASME, JOLT, 92(1970)83.
[7] J.J.Bikerman, J.Appl. Phys., 20 (1949)971.
[8] R.C.Drutowski, Trans. ASME, 81(1959) 233.
[9] K. Yakushiji, S. Kumagaya and N. Soda, Tokyo Univ. Riken Hokoku, 3, No.3-4(1949)104.
[10] T. Sasaki, N.Okino, Trans. JSME, 27 (1961)1456.
[11] G.A. Tomlinson, Phil. Mag., 7(1929) 905.
[12] T. Hisada, K. Tsugawa, J. Mech. Lab., 4(1950)9.
[13] K. Schwartzkoff, Wear, 9(1966)349.
[14] D.Tabor, Proc. Roy. Soc. A, 229 (1955) 229.
[15] K. Matsubara and M. Watanabe, Proc. JSLE-ASLE Int. Lub. Tokyo, Elservier,(1976)119.
[16] D.H. Offner and N.Tomita, Trans. ASME, JOLT, 100(1978)25.
[17] K. Matsubara, et al:Tribology Conf. London, I Mech E.(1987)261.
[18] K. Matsubara and S. Sasanuma, Proc. Jap.Int.Trib. Conf. Nagoya, (1990) 941.
[19] H.M. Briscoe, Trib. Int. 23, No.2 (1990)67.
[20] F.P. Bowden and D. Tabor, Friction, Heinenn Educational Books Ltd. (1974)120.
[21] D. Dowson and G.R. Higginson, Engg. (Lond.), 192(1961)158.
[22] D. Dowson, Proc.IME, 182(1967-1968) Conf. on Lubri. and Wear(London) (1967)151.
[23] T.E. Tallian, J.I. McCool and L.B. Sibley, Proc. IME,180,Pt3B(1965-66) 169
[24] D. Dowson and T.L. Whomes, Proc. IME, 182, Pt 1, 14(1967-68)292.
[25] D. Dowson, et al(edited), Mixed Lubrication and Lubricated Wear (Proc. 11th Leeds-Lyon Symp. on Tribology), S. Aihara and T. Sawamoto, Butterworths (1985)302.
[26] K.L.Johnson, J.A. Greenwood and S.Y. Poon, Wear,19(1972)91.
[27] Y. Ishikawa and M. Suda, Seimitsukikai, 49(1983)1065.
[28] P.G. Goksem and R.A. Hargreaves, Trans. ASME, JOLT, 100(1978)436.
[29] R. Warren, Act. Meta., 26 (1978) 1789.

© 1993 Elsevier Science Publishers B.V. All rights reserved.

Contact Stresses in Nitrided Steels

by A.V. Olver, *S.J. Cole and R.S. Sayles

Department of Mechanical Engineering
Imperial College, London, SW7 2BX.

A recently developed elastostatic model which includes the effects of roughness and layered microstructures, is applied to the selection and processing of materials, with the objective of increasing resistance to rolling fatigue and related forms of damage. The effect of nitriding and of thin ceramic coatings is investigated.

1. INTRODUCTION

The resistance of materials to rolling fatigue and related types of damage, has been widely studied and is an important consideration in selecting materials and processes for engineering components such as gears, cams, and rolling bearings. The mechanism is poorly understood but is usually thought of in terms of cyclic contact stresses leading to crack initiation (1, 2). Because the stress field is predominantly compressive, rolling fatigue is often preceded by substantial plastic deformation. This paper forms part of a quest to improve the rolling fatigue performance of engineering materials and is based on the proposition that if plastic deformation can be avoided by reducing stresses or increasing strength, performance is likely to be substantially enhanced. Use is made of a recently developed contact model which enables the realistic features of roughness and of layered microstructures to be handled.

2. THEORETICAL MODELS

The theoretical models used in this study have been described in detail in earlier papers (3,4,5,6). The models are elastostatic in character and are therefore relevant to rolling fatigue only when any fluid lubricant film is thin; this is usually the most severe condition and hence is most relevant to materials engineering. Important features of the theoretical treatment are that both arbitrary roughness and elastically inhonomogeneous layers can be included and stresses in the bodies calculated. Tangential loading is neglected (figure1).

3. ROUGHNESS

Early work (4,7) showed that the roughness most associated with practical engineering production processes gave rise to a very thin, highly stressed layer, within which plastic deformation was likely. Whether this plastic deformation ceased after a short initial period of running was a sensitive function of the hardness difference and the distribution of roughness between the contacting bodies (1). The most severe situation, in which plastic deformation was persistent and cyclic in character, was associated with a hard, rough counterface and smooth, softer specimen and led rapidly to crack initiation.

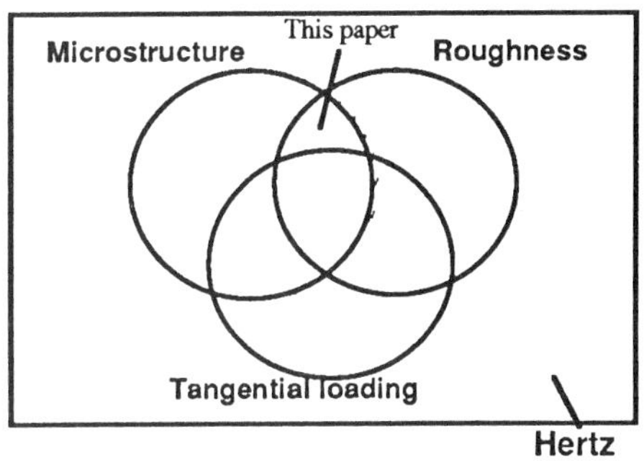

Figure 1. Elastostatic contact models.

Some typical engineering surfaces are shown in figure 2. Figure 3 shows the distribution of shear stress produced under typical normal contact conditions, using the method described by Bailey et al (4). A thin, very highly

* Present Address: School of Metallurgy and Materials, University of Birmingham

TABLE 1 Layer Properties Used for the Analysis

PHASE	Young's Modulus, E/GPa	Poisson's ratio	Layer thickness /μm
Steel (diffusion layer)	212*	0.30	-
Epsilon iron nitride	219	0.45	10
Gamma prime nitride	204	0.29	10
TiN (compliant)	300	0.30	5
TiN (stiff)	640	0.30	5

*for the results presented in figures 10-14, E=200 GPa.

TABLE II Contact Conditions

Figure No.	Mean radius of curvature, mm	Load per unit length, N/μm	Hertz Pressure*, GPa	Contact Half-Width, mm
8-9	5	0.35	1.61	0.1318
10-14	5	0.35	1.565	0.1424

* Calculated for smooth counterface pressing on substrate.

stressed layer is present for all except the polished surfaces. It was consequently decided to investigate methods of reinforcing this near-surface region. One method used was a short-duration nitriding treatment applied to a temper resistant carburised steel. This process has been described as 'duplex hardening' by Davies (8) but the present authors prefer the term 'binary,' reserving 'duplex' for processes which carry out both surface hardening treatments simultaneously. The shear yield stress distributions arising from various surface hardening treatments are shown in figure 4.

Comparison with the applied stresses in figure 3 shows that the extent of near-surface plastic deformation is considerably reduced in the case of the binary hardened surface. Note also that the conventional nitrided steel is close to yield at the case-core interface as well as at the surface. Rolling fatigue tests on specimens treated in this way were carried out using the apparatus (figure 5) described by Graham et al (9). The results have been made public by Davies (8) and are described in detail by Olver (7). A substantial increase in performance was obtained when the binary treatment was applied to a ground specimen in Vasco X2 carburised steel. The tests were run in contact with a (softer) conventional carburised steel.

Plastic deformation on an asperity scale was thus confined to the counterface, a feature confirmed by subsequent examination of the specimen surfaces; this probably accounts for

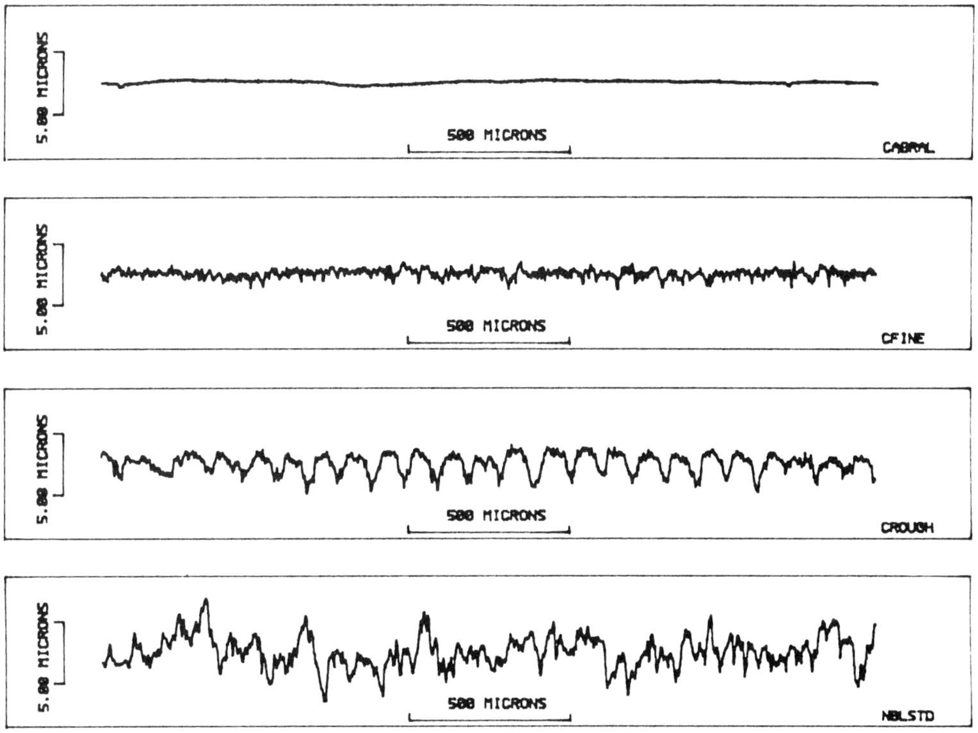

Figure 2. Typical surface roughness profiles. From top to bottom: polished, fine ground, rough ground, shot blasted.

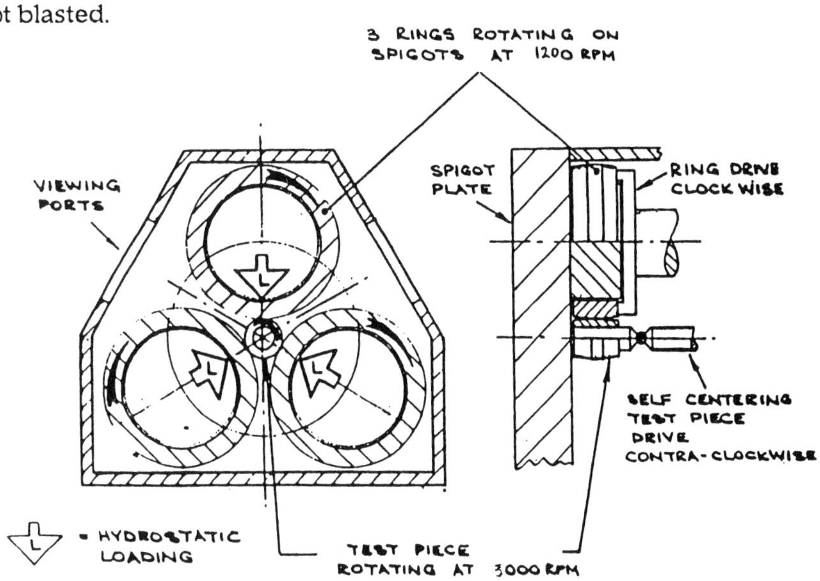

Figure 5. Test rig for rolling fatigue experiments. The test specimen is a roller of 18.7 mm diameter.

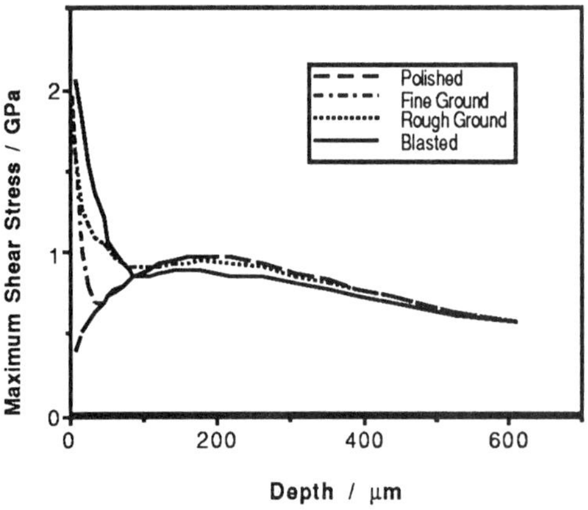

Figure 3. The effect of some typical surface roughnesses on the distribution of principal shear stress in a heavily loaded contact.

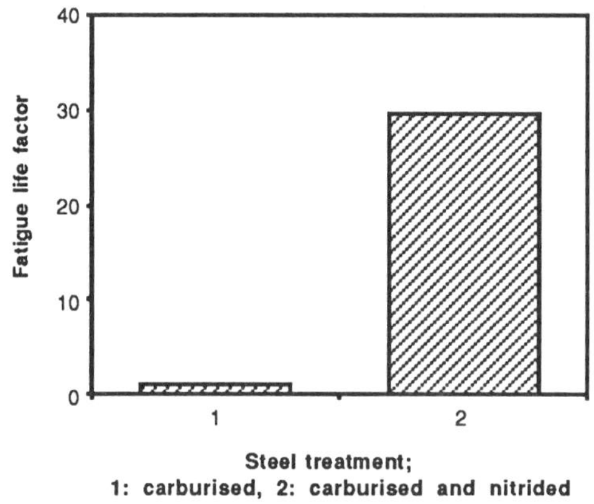

Figure 6. Comparison between rolling fatigue lives of carburised and binary hardened Vasco X2 steel.

the large life increase of about 30-fold, which is illustrated in figure 6. Despite this improvement, it is apparent from the theoretical results shown in figures 3 and 4 that the near-surface plastic deformation would not have been entirely suppressed had the specimens not been tested against a softer counterface. The nitriding treatment did not really provide adequate surface hardness; its depth, on the other hand, was more than sufficient - a reversal of the usual view of nitriding. This has prompted a closer look at nitriding itself and at the possibility of further reinforcement using thin ceramic layers. Both involved use of a recently-developed theoretical treatment in which two thin surface layers of different elastic modulus are present. The results are given in the remainder of the paper.

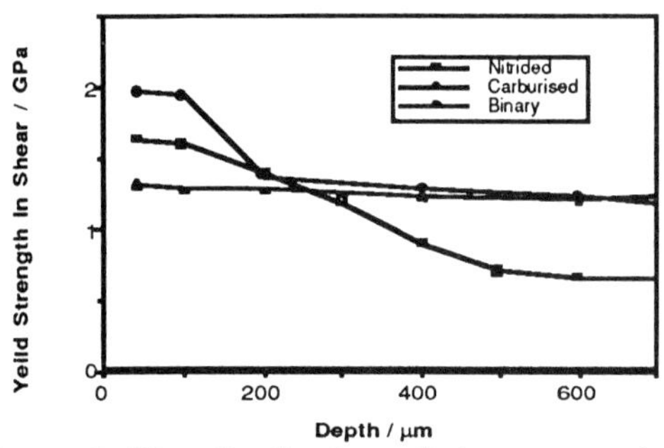

Figure 4. The distribution of shear strength produced by some surface hardening processes.

4. IRON NITRIDE

Nitriding and related processes such as nitrocarburising, produce surface iron nitride compound layers which are elastically different from the substrate. These are sometimes removed leaving only a hardened diffusion layer. In the case of gas nitriding a double layer is produced (figure 7) which is nearly always - at least partly - removed by grinding,blasting or chemical treatment. An outer layer of epsilon-iron nitride of typically 10μm thickness overlies a bottom 10μm layer of gamma-prime nitride. The double layer model was used to investigate the contact of this structure under conditions typical of commercial nitrided gears and contact conditions derived appropriately. A major difficulty was the lack of accurate elastic properties for these compounds; values estimated by Rozendaal et al (10), Hanabusa et al (11) and Somers et al (12) were used. How accurate these estimates, which are based on data for structurally similar carbides, are, is not known. Results of the contact analysis shown as contour maps of various stress components are given in figures 8-10.

The stresses are very high in the compound layers particularly in the outer epsilon layer. In addition substrate stresses still approach yield magnitude most notably below the large

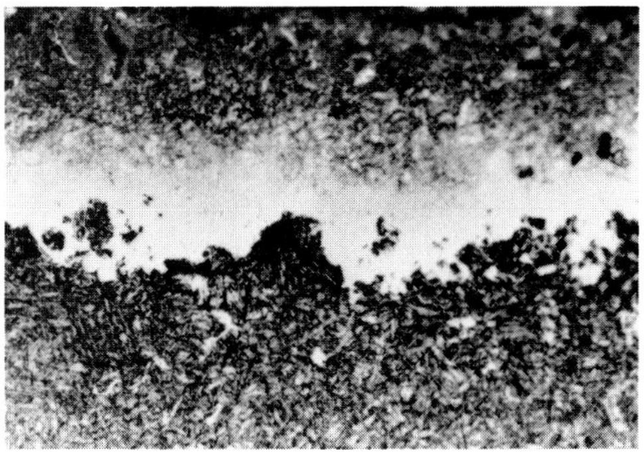

Figure 7. Taper microsection of nitrided 3%Cr-Mo-V steel showing the light etching compound nitride layer (centre) consisting of epsilon and gamma-prime iron nitrides. The compound layer is about 20μm thick in total.

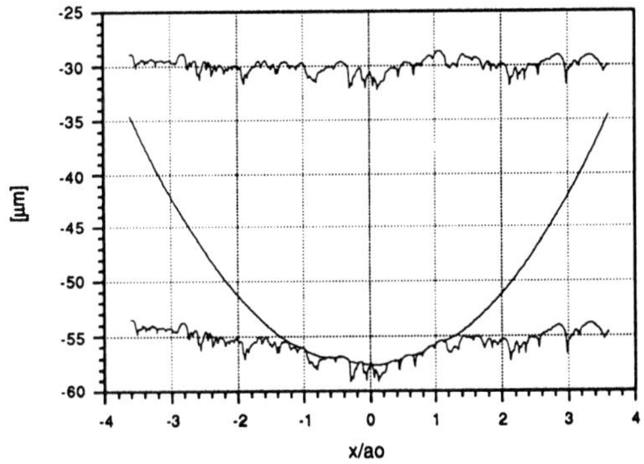

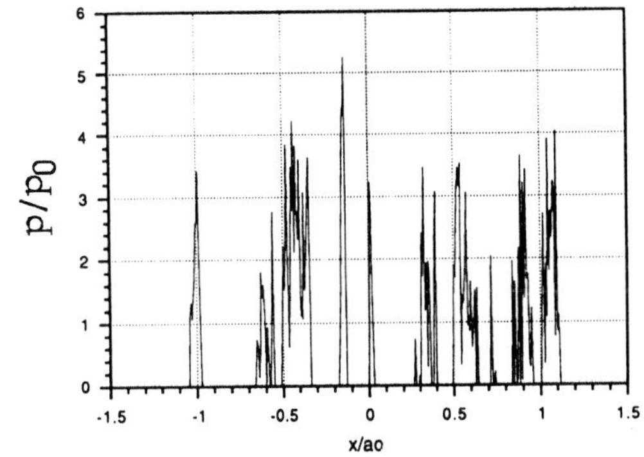

Figure 8. The pressure distribution and deflected shape resulting from the contact of the surface of a nitrided gear tooth and a smooth cylindrical counterface. a_o=half width, and p_o=maximum pressure of a reference Hertzian contact between counterface and (smooth) substrate.

Figures 9-14.

Figures 9-14 show contour plots of stress components in layered structures, the surface of the body being at the top of the illustration. The scale used is non-uniform in order to display the details of the stresses in the layers. The layer thicknesses are shown on the plots and in each case the region of substrate shown is $2a_o$ deep and $4a_o$ wide.

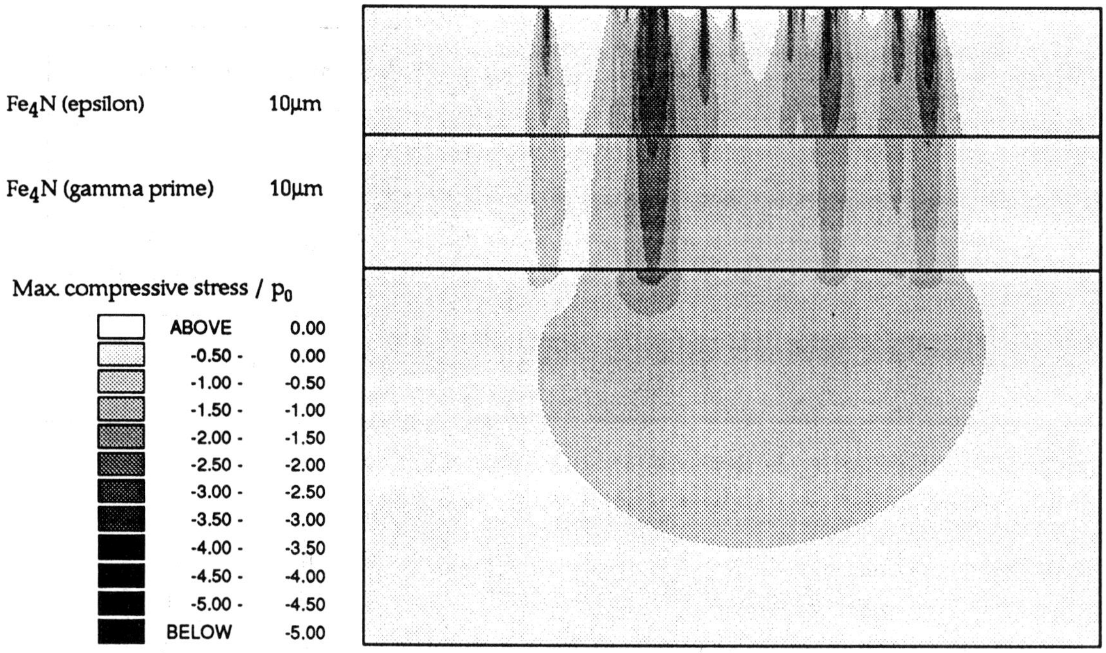

Figure 9 Distribution of principal compressive stress in nitrided steel, double compound layer. Surface roughness produces significant stresses in the compound layers.

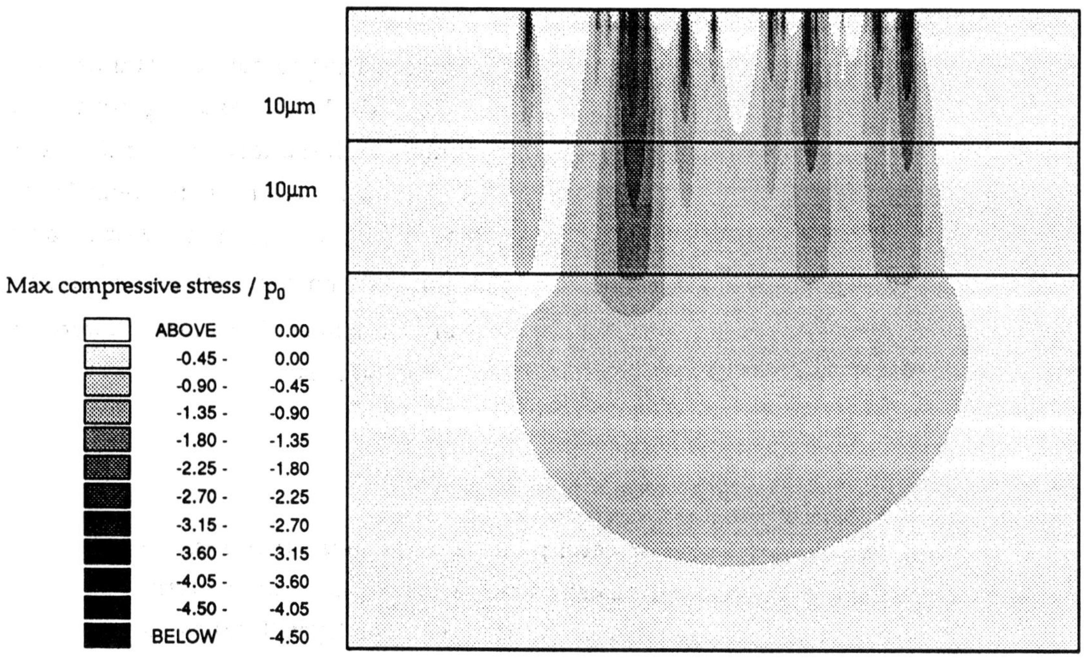

Figure 10 As fig. 9, but after removal of compound layer, same surface roughness.

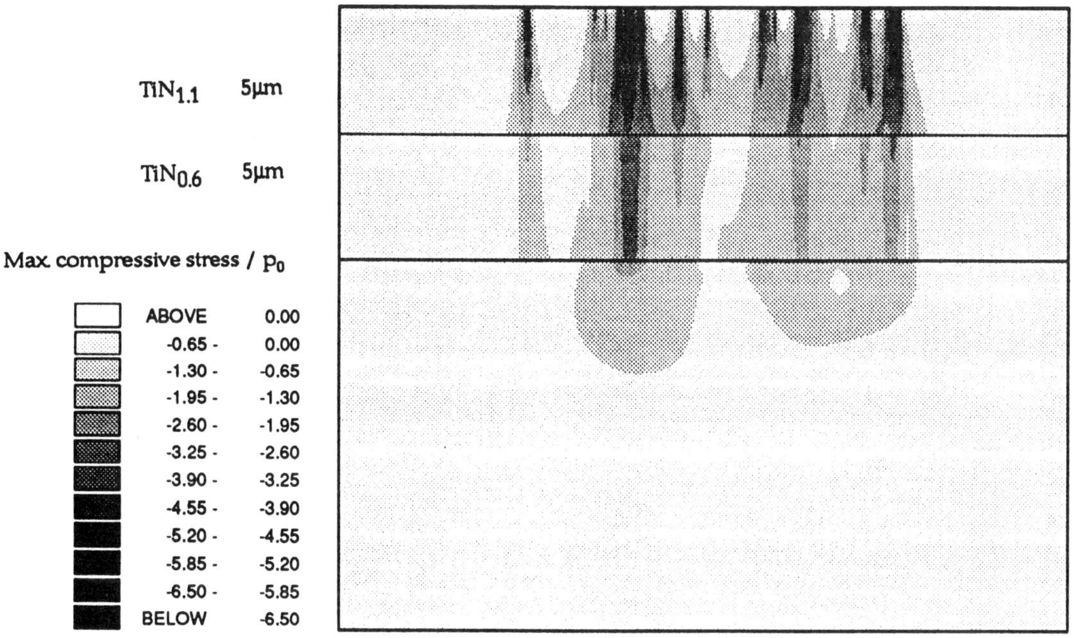

Figure 11 As fig. 10, but including effect of double layer of TiN. Upper layer E= 640 GPa, lower layer E= 300 GPa.

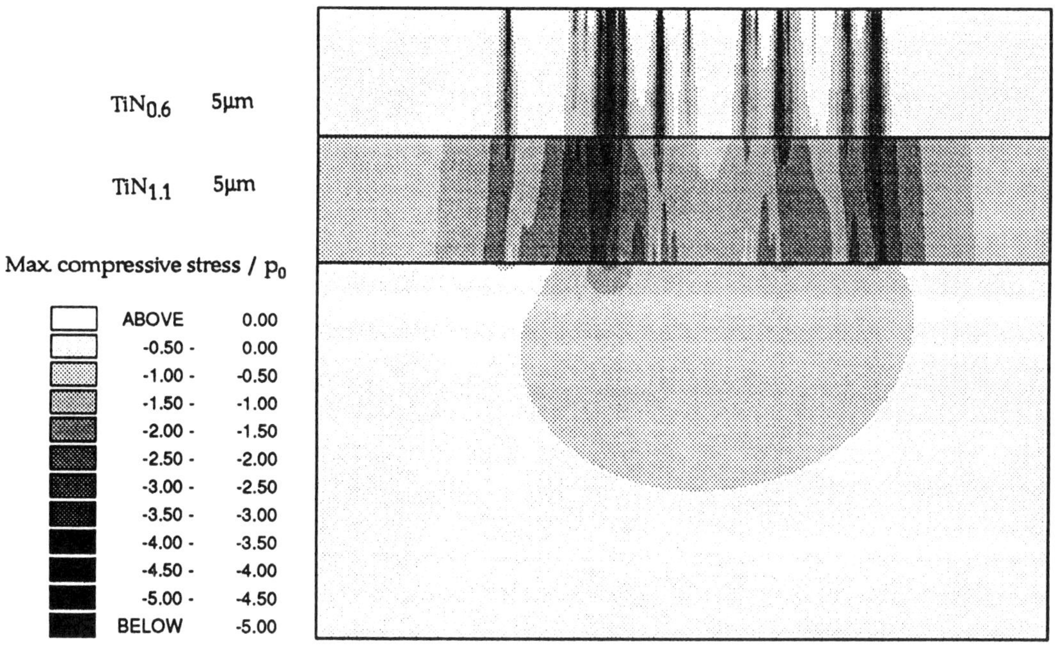

Figure 12 As figure 11, but with layers reversed.

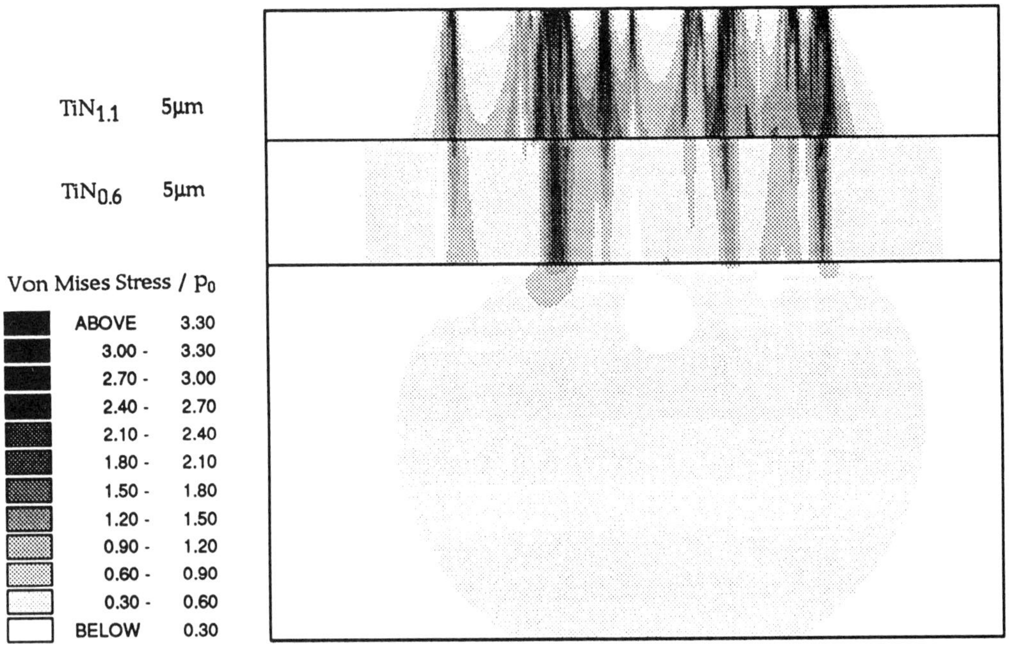

Figure 13 As fig. 11, but showing distribution of Von Mises stress.

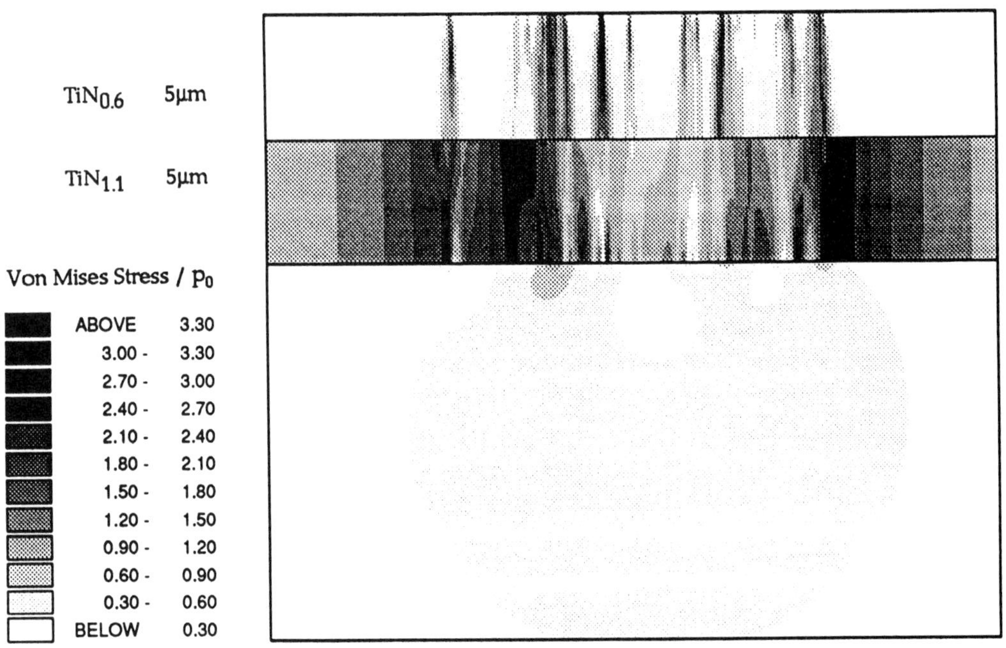

Figure 14 As fig. 13, but with layers reversed, i.e. less stiff (low N) layer outermost.

asperity at x/a_0 = -0.5. Taken together these show the necessity to remove the compound layer; either rupture of the epsilon layer or yield of the substrate would cause considerable damage, the latter because it would be most likely to cause layer debonding. Removal of the layers gives rise to the half-space distribution shown in figure 10. Although, as discussed earlier, surface yield is still expected, this is less important in the absence of a coating.

In conclusion, it appears that the compound layers produced during nitriding offer little substrate protection and can be expected to fail rapidly when heavily loaded against a rough surface. This is indeed a common experience and is often considerably worsened by the presence of porosity in the extreme surface (12).

5. TITANIUM NITRIDE

An alternative strategy is to remove any compound layer - virtually none is produced in the binary process described earlier - and use a thin ceramic coating to carry the 'roughness stresses'. Titanium nitride is particularly attractive because it can be deposited with good adhesion and without producing excessive roughness. In addition, the elastic properties are a function of the nitrogen content and hence can, in principle, be manipulated by suitable process control.Values of Young's modulus given in the literature vary from 300 GPa to around 640 GPa (13-16). In order to explore some of the possibilities presented by this, we have modeled a double layer of stiff (N:Ti = 1.1) and compliant (N:Ti = 0.5) titanium nitride using the same roughness as for the compound layer studies. Results are shown in figures 11-14.

If the stiff layer is placed outermost, very high direct stresses result. Bearing in mind that any sliding would produce corresponding tension (of about 1200 MPa) this configuration might be thought detrimental. On the other hand, the substrate stresses are very much reduced and yield appears to be avoided (figure 13). If the compliant layer is placed outermost, the surface stresses are reduced at the expense of higher shear stresses in the lower layer. the shear stresses are especially high in the regions each side - rather than directly below - the contact (figure 14). The reason for this lies in the stress conditions at the stiff-compliant boundary (figure 15). Continuity of strain requires that the horizontal compressive stress is greater on the stiffer side of the boundary (Xx in figure 15). This increases the maximum shear stress in the stiffer layer only if the normal compression is less than that acting horizontally, a condition encountered below the periphery rather than the centre of the contact.

Which arrangement of coated layers would be most damage-resistant in practice cannot be predicted with confidence most particularly because the morphology, adhesion and other properties might also be expected to depend on nitrogen content, although it is noted that the hardness is not significantly affected (16). The residual stresses might also be of importance.

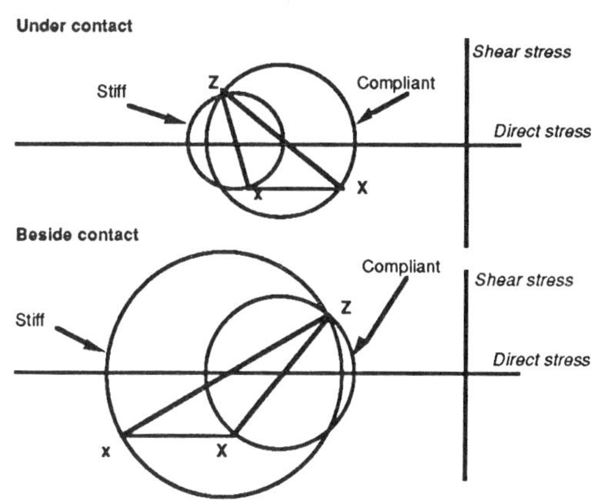

Figure 15. Stress conditions at the stiff/compliant interface shown as a Mohr's circle plot.

6. CONCLUSIONS

Surface roughness gives rise, in many circumstances to a thin, highly-stressed layer. Reinforcing this layer can reduce plastic deformation and improve rolling fatigue life.

Theoretical results suggest that considerable potential exists for further improvements by using thin, hard coatings; these can protect the substrate from damage produced by roughness: a 'stress barrier coating'.

The variable modulus of TiN can be exploited to change the stresses caused by roughness.

REFERENCES

1. A. V. Olver, H. A. Spikes and P. B. Macpherson, Wear in Rolling Contacts, Wear 112 (1986) p121

2. J. C. Hamer, J. M. Hutchinson, A. V. Olver, R. S. Sayles, E. Ioannides, Fatigue Life Modelling of Gear Contacts, (Proc. Conf.) British Gear Association Annual Congress, Sheffield 1991.

3. M. N. Webster, R. S. Sayles, A Numerical Model for the Elastic, Frictionless Contact of Real Rough Surfaces, Trans ASME, J Trib 108 (1986) p314.

4. D. M. Bailey and R. S. Sayles, Effect of Roughness and Sliding Friction on Contact Stresses, Trans ASME J Trib 113 (1991) p729.

5. S. J. Cole and R. S. Sayles, A Numerical Method for th Contact of Layered Elastic Bodies with Real Rough Surfaces, Trans ASME J Trib 114 (1991) p334.

6. S. J. Cole and R. S. Sayles, Stresses in and Beneath a Surface Coating Due to a Rough Surface Contact, Microstructural Science, 20 (1991).

7. A.V. Olver, Fatigue in Helicopter Transmission: Rolling Fatigue Studies, Westland Helicopters Ltd Research Paper RP758 (1989).

8. D.P. Davies, Duplex Hardening, an Advanced Surface Treatment Technique Combining Carburising and Nitriding Heat Treatment of Metals, 4 (1991) p97.

9. R.C. Graham, A.V. Olver, P.B. Macpherson, An Investigation into the Mechanism of Pitting in High Hardness Carburised Steels, Am. Soc. Mech. Engrs. Pamphlet 80-C2/DET.

10. H.C.P. Rozendaal, P.F. Colijn and E.J. Mittemeijer, Morphology Composition and Residual Stresses of Compound Layers of Nitrocarburized Iron and Steels, Surface Engineering, 1 (1985) p30.

11. I. Hanabusa, S. Fubura and H. Fujurera, Bull, Jap. Soc. Mech. Eng. 12 (1969) p931.

12. M.A. Somers and E.J. Mittemeijer, Formative and Growth of Compound Layers on Nitrocarburising Iron: Kinetics and Microstructural Evolution, Surface Engineering, 3 (1987) p123.

13. R. Whiting, M.A. Angadi, Young's Modulus of Thin Films Using a Simplified Vibrating Reed Method, Measurement Science and Technology 1 (1990) p662.

14. M.Y. Al-Jaroudi, Reactive Magnetron Sputtering of Tin on Steel, Thin Solid Films 182 (1989) p153.

15. W.C. Oliver, C.J. Mettargue, Characterising the Hardness and Mechanical Properties of Thin Films Using a Mechanical Properties Microprobe, Thin Solid Films 101 (1988) p117.

16. M. Ohern, R.H. Parrish, W.C. Oliver, Evaluation of the Mechanical Properties of TiN by Ultra-Low Load Indentation. Thin Solid Films, 181 (1989) p357.

17. Gupta, P.K., Walowit, J.A., "Contact Stresses between an Elastic Cylinder and a Layered Elastic Solid", Jnl. of Lub. Tech.,Trans. ASME,p250-257, April 1974.

Thin Films in Tribology / D. Dowson et al. (Editors)
© 1993 Elsevier Science Publishers B.V. All rights reserved.

STEADY STATE TOPOGRAPHY OF SURFACES IN REPEATED BOUNDARY LUBRICATED SLIDING

A. Kapoor and K.L. Johnson

Cambridge University Engineering Department
Trumpington Street, Cambridge CB2 1PZ, UK

The present paper considers the repeated sliding contact of a rough surface which has 2-dimensional cylindrical asperities of equal radius but of random (Gaussian) height over a hard regular surface. Boundary lubrication ensures that there is only a little or no wear. During the contact high asperities undergo plastic deformation - residual stresses develop, asperity height and profile change and the material may strain harden. Making the hypothesis that in the steady state the load is carried purely elastically the steady state topography of the deformed rough surface has been found. The height distribution is found to change from the Gaussian to one which is bounded. The dimensionless nominal pressure is also obtained as a function of the normalised separation (d/σ). Theoretical predictions of the change in surface topography are in excellent agreement with the experimental results of Williamson et al. [1].

1. INTRODUCTION

Peak heights of a virgin surface generally follow a Gaussian distribution [1]. But if the surface slides against another surface of comparable or greater hardness its topography undergoes a change [2-5]. This change results from either plastic flow when it is termed 'burnishing', or wear when it is called 'running-in', or more probably both.

There have been many attempts to model the process of the change in topography due to sliding. Thomas [6] simulated on a computer the removal of the asperity tops by abrasive wear. As successive layers of the rough surface are sliced away parameters such as roughness, mean peak curvature, mean slope, etc. decrease in a systematic way. His model amounted to truncating the Gaussian height distribution at a height h and replacing the lost segment by an impulse of equal area at h. Though simple to analyse, wear rarely occurs in this way. Instead King et al. [7] consider truncating functions of triangular and Gaussian shape, to obtain the worn height distribution, but there is no theoretical basis for the choice of truncating function or its standard deviation. Golden [8] overcame this problem by considering a wear mechanism such that the rate of decrease in height of an asperity with time is a function of the degree of its penetration, so that if the height of the peak is z then the wear function will be $W(z-d)$ and will be zero for $z \leq d$, d being the separation of the plane. He suggests determining W experimentally and analysed a simple case assuming W to be linear. He found that the resulting topography is that of the original surface up to a height d. Above this height it is again Gaussian but smoother. As time progresses it remains Gaussian but becomes smoother still. Using a similar approach Sugimura et al. [9] derived a 'wear-in' equation which describes changes in the height probability density function of the composite profile due to wear using size distribution of wear particles.

Models by Thomas and Sugimura rely on wear to change the surface topography and therefore do not apply where the change is primarily due to plastic flow as in boundary lubricated contacts. The models of Golden and King allow change in asperity heights by plastic flow, but require an experimental determination of the truncating function or the wear function.

To develop the present model we make the hypothesis that due to plastic flow in the early passages of a slider, the shape and height of asperities in the surface will be modified such that, in steady state, the load will be carried purely elastically. This approach has been used previously by Kapoor and Johnson [10] to analyse the geometry changes and the groove formation during rolling and sliding of a torroid on an elastic-plastic half space, and by Johnson and Shercliff [11] (later referred to as JS) to analyse shakedown of 2-d asperities.

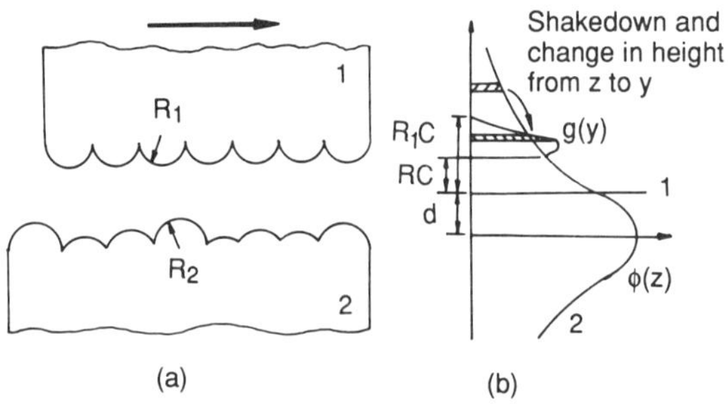

Figure 1. Shakedown of an elastic regular and elastic-plastic rough surface

1.1. Physical model

We consider a rough surface of 2-d cylindrical asperities of equal radii and of heights following a Gaussian distribution. This is being traversed by a hard regular surface having 2-d cylindrical asperities of equal height and radii (Fig. 1). During an 'encounter', in which a hard asperity slides past a softer one, the size of the contact patch and the maximum contact pressure change. For some asperities the maximum contact pressure exceeds the elastic limit and plastic flow occurs causing a change of height and shape This process continues until the height and shape have changed such that the contact pressure does not cause any further plastic flow. This process is an example of 'shakedown'. In the following we shall refer to this shakedown of a single pair of asperities as the 'unit event'.

During sliding all the asperities above a certain height will develop a contact pressure high enough to cause plastic flow. Such asperities after interaction will have a modified height and shape which will carry the load elastically. The other asperities either do not make contact or do not reach the elastic limit. By correlating the original to its modified height we shall obtain the modified height distribution of asperities in the rough surface. Also by integrating the load carried by each asperity in the shakedown state, in a manner similar to Greenwood & Williamson [12] we shall obtain the load carried by the surface in steady state.

Two somewhat idealised cases will be examined: (i) lateral asperities in which the sliding motion is perpendicular to the axes of the asperities, as in hobbed gear teeth, and (ii) longitudinal asperities aligned in the direction of sliding as in a lathe-turned journal bearing.

2. LATERAL ASPERITIES

2.1. The unit event

The shakedown of a pair of lateral asperities was analysed by JS [11]. They considered plain-strain deformation of 2-d asperities with their axes perpendicular to the direction of sliding, Fig. 2a. The unit event comprises repeated sliding at a pre-specified separation Δ_o until a steady (shakedown) state is reached. During an elastic encounter, the contact starts at the leading edge, with semi-contact width a and contact pressure p rising from zero, at point A to a maximum at B, and falling to zero again at C (Fig. 2b). Above a critical value of Δ_o, the pressure p_o exceeds the elastic limit p_o^e whereupon the following changes occur ;

(i) protective residual stresses develop

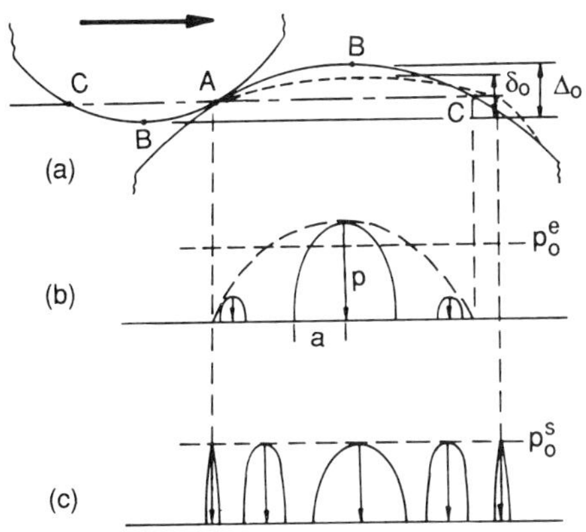

Figure 2 (a) The unit event
(b) Variation of pressure on undeformed profile.
(c) Variation of pressure on shakedown profile.

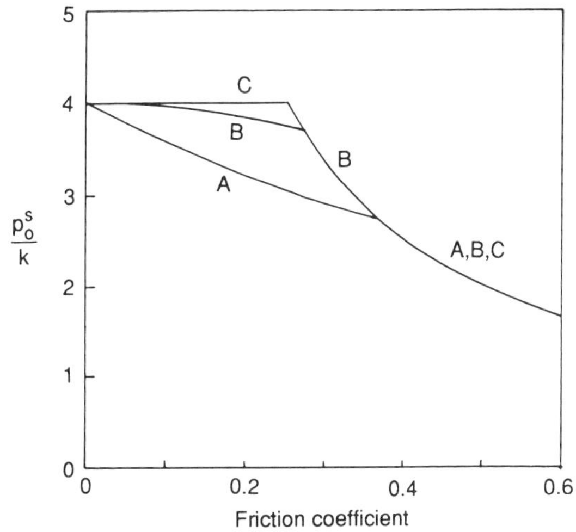

Figure 3. Shakedown limits to be used in eq. (1)
A - Lateral asperity (perpendicular to sliding).
B - Longitudinal asperity (parallel to sliding).
C - Kinematic hardening for both the lateral and longitudinal asperity.

(ii) plastic flow causes the asperity shape and height to change and
(iii) the material strain hardens.

JS made the hypothesis that the deformed asperity adopts a shape which maintains the contact pressure at the shakedown limit throughout the interaction (Fig. 2c). The shakedown limit is defined as the maximum pressure which can be supported purely elastically in the steady state; it takes into account the above-mentioned changes. Two cases were considered: (i) one hard and one soft asperity in which the plastic deformation is confined to the softer asperity and (ii) asperities of equal hardness in which the plastic deformation was assumed to be shared equally between the two. The results of case (i) will be used in the analysis. JS define the constant C by

$$C = \left(\frac{p_o^s}{E^*}\right)^2 \left[2\,\ell n\left(4\frac{E^*}{p_o^s}\right) - 1\right] \tag{1}$$

where p_o^s is the shakedown pressure for the given friction conditions (Fig. 3) and the equivalent elastic modulus of the two surfaces $E^* = [(1-\nu_1^2)/E_1 + (1-\nu_2^2)/E_2]^{-1}$. ν being the poisson's ratio.

The shakedown response of individual asperities to a non-dimensional interference (Δ_o/RC) is presented in Figs. 8 and 9 of ref. (11). With a little rearrangement these plots express the relationships

$$\frac{\delta_o}{R_1 C} = F_1\left(\frac{\Delta_o}{RC}, t\right) \tag{2}$$

and

$$\frac{\overline{P}/R_1 E^*}{\left(p_o^s/E^*\right)^2} = F_2\left(\frac{\Delta_o}{RC}, t\right) \tag{3}$$

where δ_o is the interference in the steady (shakedown) state and $\overline{P}$ is the normal contact force averaged over the period of an encounter. The equivalent radius R is defined by

$$\frac{1}{R} = \frac{1}{R_1} + \frac{1}{R_2} = \frac{1}{R_1}(1+t) \tag{4}$$

where t is the ratio of the radius R_1 of the non-deforming asperity to the radius R_2 of the deforming asperity. For (Δ_o/RC) < 0.7 the encounter is entirely elastic, so that $\delta_o = \Delta_o$ and $\overline{P}$ is given by elastic analysis. For

$(\Delta_o/RC) > 1.0$ the shakedown equations (2) and (3) above apply; in the range $0.7 < (\Delta_o/RC) < 1.0$ shakedown is achieved but the shakedown limit is not reached. In this small range $\overline{P}$ is estimated by interpolation. When (Δ_o/RC) is large (>>1), (δ_o/R_1C) approaches the value 1.0 and L approaches π. Above this mean load the contacting asperity is flattened and elastic shakedown is not possible. Plastic deformation then takes place repeatedly in the steady state.

2.2. Random rough surface

Let there be N asperities per unit length and the radii R_1 and R_2 be constants for the respective surfaces 1 and 2 in Fig. 1.

The initial peak height distribution of the deforming surface 2 is Gaussian and the probability of finding an asperity between heights z and z + dz is $\phi(z)dz$, where the probability density function $\phi(z)$ is given as

$$\phi(z) = \frac{1}{\sqrt{2\pi}\cdot\sigma} e^{-\frac{z^2}{2\sigma^2}} \equiv \frac{1}{\sqrt{2\pi}\cdot\sigma} \text{Exp.}\left[\frac{-z^2}{2\sigma^2}\right] \qquad (5)$$

where σ is the standard deviation of the peak heights.

Since the asperities of surface 1 are of the same height, its height distribution function is an impulse of area unity. We fix the origin of our axis system at the mean peak height of surface 2. If the separation between the surfaces is d, the interference Δ_o as defined in Fig. 2 is

$$\Delta_o = z - d \qquad (6)$$

After shakedown (in the steady state) it is modified to g(y), where y is the new height of the deformed asperity, Fig. 1b. The interference δ_o of the deformed asperity is

$$\delta_o = y - d \qquad (7)$$

Since the number of asperities remains constant during shakedown we can write,

$$\phi(z)\,dz = g(y)\,dy$$

$$\text{or,} \quad g(y) = \phi(z) / \frac{dy}{dz} \qquad (8)$$

This equation relates the modified peak height distribution to the original one in a fundamental way and does not assume the original distribution to be Gaussian. In the present analysis, however, we assume $\phi(z)$ to be Gaussian and that for $t = 1$, F_1 can be represented by

$$\frac{\delta_o}{R_1C} = 1.05 - \frac{0.277}{(\Delta_o / R_1C)} \qquad (9)$$

The error involved in eq. (9) and the plot represented by eq. (2) is negligible for $(\Delta_o / RC) < 5$, about 2% at $(\Delta_o / RC) = 12$ and 5.5% at $(\Delta_o / RC) = \infty$. Since there are only a few asperities in the tail of Gaussian distribution the actual error in the form of g(y) should be even smaller. Using equations (5) to (7) and (9), substituting in eq. (8) and standardising the distribution [13], we obtain

$$g(y) = 1.46\,\eta^2\,\text{Exp}\left[-\frac{1}{2}\left(\frac{d}{\sigma} + \frac{R_1C}{\sigma}\cdot\eta\right)^2\right] \qquad (10)$$

$$\text{where} \quad \eta = \frac{0.277(R_1C/\sigma)}{1.05(R_1C/\sigma) + (d/\sigma) - (y/\sigma)}$$

remembering that the modification of heights is possible only when $(\Delta_o / RC) > 1$ and for $(\Delta_o/RC) >> 1$, $(\delta_o/R_1C) = 1$ which implies that (δ_o / RC) can vary only between 1 to 2. The range of y for which eq. (10) is valid is

$$\left(\frac{d}{\sigma}\right) + \frac{1}{1+t}\left(\frac{R_1}{\sigma}C\right) \le \left(\frac{y}{\sigma}\right) \le \left(\frac{d}{\sigma}\right) + \left(\frac{R_1}{\sigma}C\right) \qquad (11)$$

The asperities remain undeformed and the distribution unchanged in the range y

$$\left(\frac{y}{\sigma}\right) < \left(\frac{d}{\sigma}\right) + \frac{1}{1+t}\cdot\frac{R_1C}{\sigma} \qquad (12)$$

whereupon,

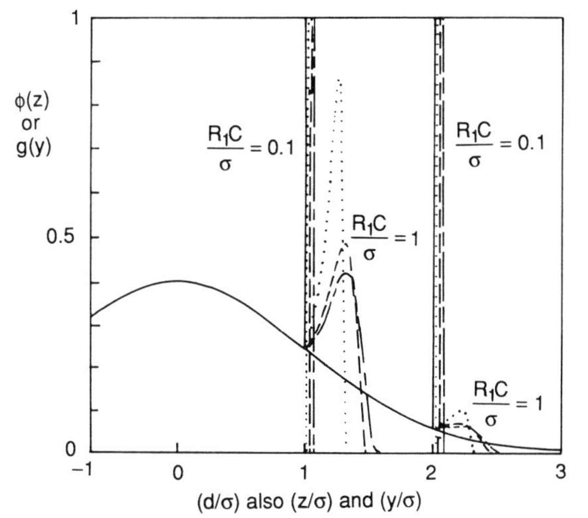

Figure. 4. Modification of asperity heights
——— Original distribution (Gaussian)
— — — Modified distribution for $R_1/R_2 = 2$
- - - - - Modified distribution for $R_1/R_2 = 1$
· · · · · · Modified distribution for $R_1/R_2 = 0.5$

$$g(y) = \frac{1}{\sqrt{2\pi}} \cdot \mathrm{Exp}\left[-\frac{1}{2}\left(\frac{y}{\sigma}\right)^2\right] \tag{13}$$

Eq.(10 to 13) completely define the new distribution of asperity heights which depends only on the non-dimensional separation between the surfaces (d/σ) and the non-dimensional group (R_1C/σ). The above analysis was repeated for $t = 0.5$ and 2.0. The modified distributions of height so found are plotted in Fig. 4. The cumulative height distributions are shown in Fig. 5. In this diagram the undeformed Gaussian distribution appears as a straight line. As expected, the variance in height of the deformed asperities is much reduced. Soft asperities $(R_1C/\sigma < 0.1)$ are almost flattened to the same height, so that the distribution reduces to a vertical line.

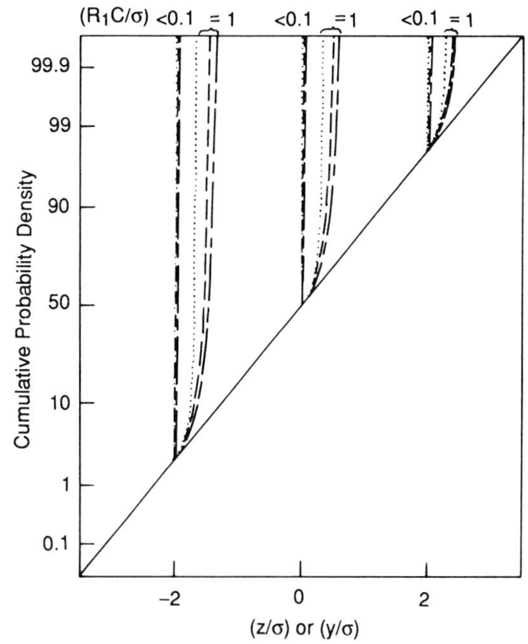

Figure 5. Modification of asperity heights
——— Original distribution (Gaussian)
— — — Modified distribution for $R_1/R_2 = 2$
- - - - - Modified distribution for $R_1/R_2 = 1$
· · · · · · Modified distribution for $R_1/R_2 = 0.5$

2.3. Load carried in steady state

Eq. (3) gives the mean load per unit length of the asperity during an encounter in the steady state. As before we assume N asperities per unit length of the surface and a separation d between the surfaces (Fig. 1). Only those asperities which are higher than d will make the contact and carry load. Let us consider asperities between heights z and $z + dz$. Their number will be $N \cdot \phi(z)dz$, the interference $\Delta_o = z - d$ and the mean load carried will be

$$F_2\left(\frac{z-d}{R_1C}, t\right) . N . \phi(z)dz \tag{14}$$

And the load carried by a unit square can be found by summing over all the asperities in unit length, which make contact, i.e. for whom $\Delta_o > o$ or $z > d$. The load carried by the unit square is the nominal pressure $\bar{p}$, given by

$$\frac{(\bar{p}/R_1E^*)}{(p_o^s/E^*)^2} = N \int_{z=d}^{\infty} F_2\left(\frac{z-d}{R_1C}, t\right) \cdot \phi(z)dz \tag{15}$$

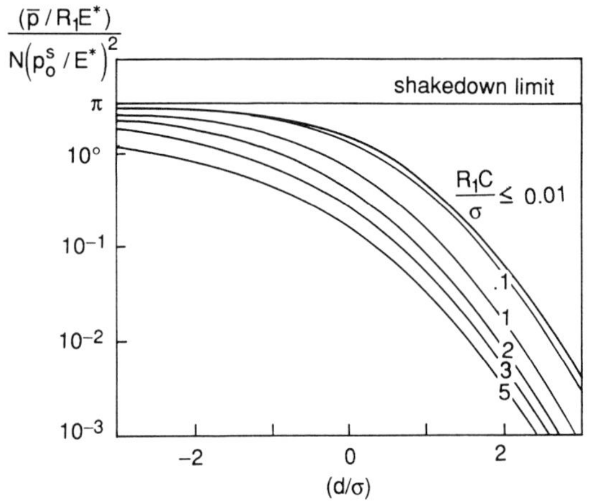

Figure 6. Normalised mean nominal pressure at steady state

The function F_2 is found from the data of ref. ([11]. Since, after the rearrangement as in eq. (3) the curves for $t = 0.5$, 1 and 2 are very close, the following function may be used to represent all of them to good approximation:

$$F_2 = 2\left(\frac{\Delta_o}{R_1C}\right) \quad , \quad 0 \le \frac{\Delta_o}{R_1C} \le 1 \tag{16a}$$

$$F_2 = 3 - \frac{1}{(\Delta_o / R_1C)} \quad , \quad \frac{\Delta_o}{R_1C} > 1 \tag{16b}$$

Substituting in eq. (15), writing $s = z/\sigma$, $\alpha = (d/\sigma)$, $\beta = (R_1C/\sigma)$ and taking $\phi(z)$ to be Gaussian gives

$$\frac{(\bar{p}/R_1E^*)}{N(p_o^s/E^*)^2} = \sqrt{\frac{2}{\pi}} \int_{\alpha}^{\alpha+\beta} \frac{s-\alpha}{\beta} e^{-\frac{s^2}{2}} ds + \int_{\alpha+\beta}^{\infty} \left(3 - \frac{\beta}{s-\alpha}\right) e^{-\frac{s^2}{2}} ds \tag{17}$$

This equation gives the relationship between the mean pressure $\bar{p}$ and the separation (d/σ) for different values of (R_1C/σ). It has been evaluated numerically and is plotted in Fig. 6.

3. RATIO OF ASPERITIES IN SHAKEDOWN

We assume N asperities per unit length and the height distribution to be Gaussian. At a normalised separation (d/σ), the number of asperities in contact is,

$$n = N \int_{d/\sigma}^{\infty} \phi(z)\, dz \tag{18}$$

From section 2.1 we know that shakedown occurs when $(\Delta_o/RC) > 0.7$. So the number of asperities which undergo shakedown is,

$$n_s = N \int_{d/\sigma + 0.7RC/\sigma}^{\infty} \phi(z)\, dz \tag{19}$$

The fraction of asperities undergoing shakedown to those in contact, i.e. (n_s/n) is a function of (RC/σ) and (d/σ) and is plotted in Fig. 7.

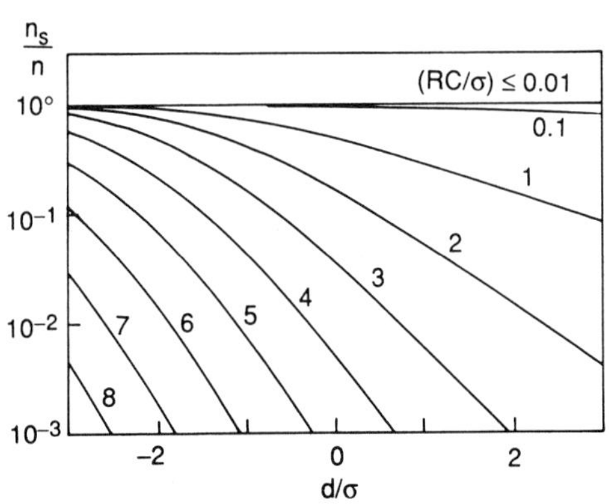

Figure 7. Fraction of asperities in shakedown.

4. LONGITUDINAL ASPERITIES

In this section we consider sliding parallel to the axes of the asperities. If the transverse profile of the asperities is unchanging and the sliding direction is exactly parallel to their axes, the hard surface will impart a pattern of conforming grooves onto the softer mating surface. The contact area will grow until the shakedown pressure is reached, as analysed in ref. [10]. A more realistic situation is that autocorrelation between transverse profiles decays with axial distance. This situation would arise, even with regular turned cylindrical surface, since the feed of the cutting tool would incline the axis of the ridges slightly to the circumferential direction. With uncorrelated transverse profiles repeated contact will be made in positions which vary randomly in the lateral direction. In the limit of many contacts the effect will be the same for the unit event as repeated passes in the *lateral* direction, leading to the same expressions for the change in height and curvature as for lateral asperities. In frictional sliding, however, the value of the shakedown pressure p_o^s is slightly different. It has been evaluated and is shown in Fig. 3.

A further difference between lateral and longitudinal asperities lies in the deformed shape of the asperity. As shown in Fig. 2, material displaced from the crests of lateral asperities appears in a bulge on the downstream side. The same might be expected with regular longitudinal asperities slightly inclined to the direction of sliding (e.g. turned surface). But with surfaces having randomly uncorrelated transverse profiles the deformation is likely to be more symmetrically disposed. This feature does not, however, influence significantly the deformed height and load in the steady state given by equations (2) and (3).

5. EXPERIMENTS

To check the validity of the proposed theory the predictions were compared with the experimental results of Williamson et al. [1]. In these experiments a bead blasted mild steel pad was pressed against a rotating hard steel table which had been finely ground. The system was flooded with 20 SAE lubricating oil and the surface topography was monitored during the test. Initially the height distribution was truly Gaussian (curve A in Fig. 8) but, as running-in proceeded, the height distribution

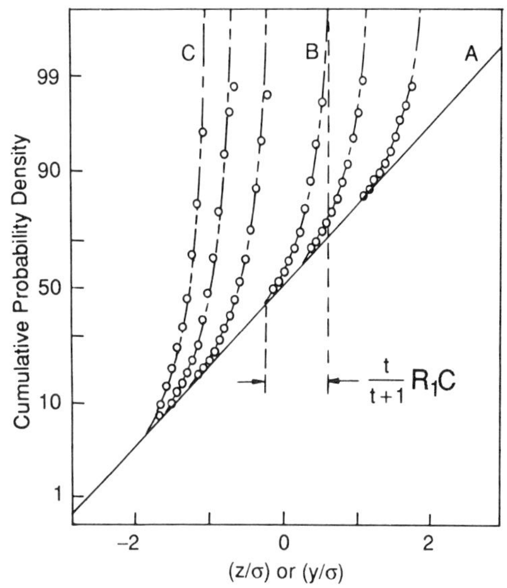

Figure 8. Modification of peak heights
——— Original distribution (Gaussian)
— — — Theory for t = 0.5, 1, 2
o Experiments by Williamson et al. [1]

was progressively modified.

To predict the modified height distribution values of (R_1C/σ) and (d/σ) are required in equation (10). The specimen had a roughness of 50 μin. c.l.a. so σ = (c.l.a.)/0.8 = 62.5 μin. = 1.588 μm . However, the hardness of the specimen and the mean curvature R_1 of the hard surface are not specified, so that the parameter (R_1C/σ) was found by fitting the theoretical curve B to the experimental points. We note that, by equation (11), the spread in height of the asperities which have been deformed is restricted to $CR_1t/(1+t)$. This was equated to the experimental variation in height (60 μin.) in the set of results marked B in Fig. 8. It remains to fix t , the ratio of hard and soft asperity curvature. Values of t = 0.5, 1.0 and 2.0 were used to obtain theoretical curves which differed little, e.g. t = 1 gives (CR_1/σ) = 2.0 to be used in equation (10) to plot all the theoretical curves for different separations (d/σ). The overall fit is very satisfactory, in spite of the fact that the experimental surface was bead-blasted giving it an *isotropic* roughness, while the theory was developed for linear roughness (longitudinal or lateral).

6. DISCUSSION

The analysis presented in this paper applies to the running-in (shakedown) during sliding of a hard (elastic) and a soft (elastic-plastic) rough surface. Both have been assumed to have 2-d cylindrical asperities with a Gaussian height distribution, but the hard surface is highly smooth compared with the soft surface. Machining operations like turning, grinding, shaping, etc. produce this type of 2-d surface roughness. Two cases have been analysed: (i) when asperities slide perpendicular to their axes - a practical example being hobbed gear teeth which have surface roughness lay perpendicular to the sliding and, (ii) when asperities slide parallel to their axes, which happens when a turned shaft rotates in a turned bushing. In case (ii), provided sucessive transverse profiles are not exactly correlated along the direction of sliding, the same analysis applies to both longitudinal and lateral asperities, with a small change in the value of the shakedown pressure p_o^s . When exact correlation occurs a series of permanent grooves are formed as described in ref. [10].

During the initial contact some asperities experience a contact pressure greater than the shakedown limit. They undergo plastic deformation and their height and shape changes. Protective residual stresses also develop and further sliding is purely elastic with the contact pressure at the shakedown limit. It is recognised that wear in early sliding also assists the process of shakedown by reducing asperity heights. In lubricated sliding wear reduces significantly after running in and so the predictions of the above theory might be expected to apply, even if there was some wear in the early cycles of sliding.

The ratio of asperities undergoing shakedown to those in contact, Fig. 7, is a function of (RC/σ) and the load or separation (d/σ). If $(RC/\sigma) < 0.01$ almost all of the contacting asperities are plastically deformed to shakedown state, however low the load. On the other hand if $(RC/\sigma) = 5$ most of the asperities are elastic at low loads but as load increases and (d/σ) decreases more and more asperities become plastic and achieve shakedown.

$$\frac{CR}{\sigma} = \left(\frac{p_o^s}{E^*}\right)^2 \left(\frac{R}{\sigma}\right)\left[2\ell n\left(\frac{4E^*}{p_o^s}\right) - 1\right] \tag{20}$$

(CR/σ) is a function of the surface topography (R and σ) and the material properties (E^* and p_o^s , which also depends on the friction coefficient) and governs the extent of plastic deformation. In this respect it is analagous to the plasticity index ψ of Greenwood & Williamson [12], which governs the extent of plastic deformation in the normal contact of rough surfaces. If we write $(1/\psi^2) = (H/E^*)^2 (R/\sigma)$ the comparison with eq. (20) is obvious, with the shakedown pressure p_o^s replacing the hardness H . The results of a sliding experiment are reported in Fig. 9 of ref. [12], where it is noted that the value of ψ decreased from about 10 to 1 during sliding, indicating that initially there was plastic flow but later the contact became elastic. In the shakedown of surfaces, this is exactly what the analysis predicts - initially some asperities deform plastically but continued sliding causes them to shakedown and ultimately become elastic in the steady state.

If the initial asperity height distribution is known, the modified distribution is given by eq. (8). It has been solved assuming the initial distribution to be Gaussian. In contrast with the view that the modified distribution remains Gaussian with a smaller deviation, we find that shakedown by a hard surface results in a distribution which has a finite bound with a spread of tRC and may be non-dimensionalised by dividing by σ , to give $S = (tRC/\sigma)$. If the modified distributions are plotted for the same value of S , but different values of t (= 0.5, 1, 2) they are indistinguishable (see Fig. 8) showing that the modified distribution is primarily governed by S. If S is small (< 0.01, say) the asperities are effectively crushed flat with only a small spread in height.

The variation in mean nominal pressure $\bar{p}$ with normalised separation (d/σ) is shown in Fig. 6. For a given load (i.e. nominal pressure $\bar{p}$) and (R_1C/σ) the steady state separation may be read from the plot and used in conjunction with Fig. 5 to determine the steady-state profile. As the load is increased shakedown of the surfaces takes place at a reduced separation, i.e. the surfaces move closer. At (d/σ) less than about -1 the hollows of the surface will be completely filled with the plastically displaced material and the asperities would be expected to interact. The present model, based on asperities behaving independently will then be unreliable. For a hypothetical case of the softer surface having been squashed completely flat the normalized mean pressure may be written as

$$\frac{\bar{p}_{max}/R_1E^*}{N\left(p_o^s/E^*\right)^2} = \pi \tag{21}$$

This is the maximum pressure which the surfaces can suppot in the eastic state and the curves in Fig. 6 can be seen to approach it asymptotically. If this value is exceeded, repeated plastic deformation will take place in the steady-state, with an expected marked increase in wear rate. This result suggests a possible mechanism for a transition from mild to severe wear.

The surface layers are known to harden in repeated sliding which influences the value of the shear yield stress k in Fig. 3. With the kinematically hardening model, curve C should be used and ideally k found from a cyclic stress-strain curve. In the absence of this data, $k \approx$ (hardness)/6.

In this paper the asperities on the hard surface have been assumed to all have the same height (i.e. zero variance). In reality some asperities will be higher than others, and it will be the highest hard asperity which will give rise to the maximum plastic deformation as it traverses the softer surface. For lateral asperities the modified profile will still be given by the present analysis if the separation d between the mean peak heights of the two surfaces is replaced by the separation d' between the mean height of the soft surface and the *maximum height* h of the hard surface

$$\text{i.e.} \quad \frac{d'}{\sigma} = \frac{d}{\sigma} - \frac{h}{\sigma} \tag{22}$$

The height h of the tallest asperity may be estimated by a method given in appendix 3 of ref. [14]. However in steady-state contact with this modified profile, lower asperities on the hard surface will not reach the shakedown pressure, so that the nominal pressure given in Fig. 6 will be an upper bound to the true mean pressure.

With longitudinal asperities in which the hard asperities only move through a small lateral distance relative to the soft surface, the present theory applies provided that σ is taken to the combined rms roughness of *both* surfaces. If σ_1 and σ_2 are the rms roughnesses of the hard and soft surfaces,

$$\sigma^2 = \sigma_1^2 + \sigma_2^2 \tag{23}$$

CONCLUSIONS

The steady state topography and nominal pressure of an elastic-plastic rough surface sliding against an elastic regular surface have been analysed, with the following results:

(i) The modified asperity height distribution is bounded and depends on the parameter $S = (R_1/R_2)RC$, which is inversely proportional to (plasticity index)^2. The lower the values of S, the steeper is the modified distribution.

(ii) The predictions of the theory agree remarkably well with experiments done by Williamson et al. [1].

(iii) The normalised mean nominal pressure increases as the separation (d/σ) is reduced and approach a limiting value of π. Above this critical value, the steady state is not completely elastic and some high asperities continue to undergo repeated plastic flow. Since continuous plastic flow is likely to result in a larger wear rate, this critical value could be assocated with a transition from mild to severe wear.

(iv) If the hard surface is rough the actual nominal pressure is less than the one predicted by theory. However, for a surface sliding along the axis of the asperities with negligible or no sideways motion the difference is very small.

(v) The modified height distribution is not affected by roughness of the hard surface for sliding perpendicular to the axis of asperities or along the axis with large sideways motion. For sliding along the axis with very small sideways motion the combined rms roughness of both surfaces should be used, eq. (23).

ACKNOWLEDGEMENT

The work was supported by a co-funded Research Grant from SERC and Nuclear Electric Berkeley Laboratories.

REFERENCES

1. Williamson, J.B.P., Pullen, J. and Hunt, R.T., 'The shape of solid surfaces', in Ling F. (ed.) 'Surface Mechanics', ASME, New York (1969).
2. Thomas, T.R. 'The characterisation of changes in surface topography in running-in', Proc. 4th Leeds-Lyon Symp., MEP, London (1978) p.99-108.
3. Rowe, G.W., Kaliszer, H., Trmal, G. and Cotter, A., 'Running-in of plain bearings', Wear 34 (1975) p.1-14.
4. Morimoto, T. and Tamamura, K., 'Burnishing process using a rotating ball-tool-effect of tool material on the burnishing process', Wear, 147 (1992) p.185-193.
5. Ostvik, R. and Christenson, H., 'Changes in surface topography with running-in', Proc. Inst. Mech. Engrs., 183, Part 3P (1968/69) 59-67.
6. Thomas, T.R., 'Computer simulation of wear', Wear, 22 (1972) p.83-90.
7. King, T.G., Watson, W. and Stout, K.J., 'Modelling the micro-geometry of lubricated wear', Proc. 4th Leeds-Lyon Symp., MEP, London (1978) p.333-43.
8. Golden, J.M., 'The evolution of asperity height distributions of a surface subjected to wear', Wear, 39 (1976) p.25-44.
9. Sugimura, J., Kimura, Y. and Amino, K., 'Analysis of the topographic changes due to wear - geometry of the running-in process', JSLE, n8 (1987) p.69-74.
10. Kapoor, A. and Johnson, K.L., 'Effect of changes in contact geometry on shakedown of surfaces in rolling/sliding contact', Int. J. Mech. Sci., vol. 34, No. 3 (1992) p.223-239.
11. Johnson, K.L. and Shercliff, H.R., 'Shakedown of 2-dimensional asperities in sliding contact', Int. J. Mech. Sci., vol. 34, No. 5 (1992) p.375-394.
12. Greenwood, J.A. and Williamson, J.B.P., 'Contact of nominally flat surfaces', Proc. Roy. Soc. Lon., A295 (1966) p.300-19.
13. Meyer, P.L., 'Introductory Probability and Statistical Applications', Addison Wesley Publishing Company (1978) p.185-186.
14. Greenwood, J.A. and Tripp, J.H., 'The contact of nominally flat surfaces', Proc. I. Mech. E., vol. 185, 48/71 (1970-71) p.625-633.

Thin Films in Tribology / D. Dowson et al. (Editors)
1993 Elsevier Science Publishers B.V.

Towards a better surface finish for bearing materials.

by A.A.Torrance and A.Parkinson

Department of Mechanical and Manufacturing Engineering,
Trinity College,
DUBLIN 2, Republic of Ireland.

Recent work suggests that the slopes of asperities may provide a much better guide to the quality of bearing surfaces than more traditional parameters such as R_a. This paper describes how these slopes can be measured and analyzed with a PC to allow the prediction of friction and wear in certain types of contact. The technique is applied to lubricated ground steel surfaces rubbing against copper and aluminium sliders., and to the same surfaces polished to varying degrees. It is found that within experimental error, the predictions agree with the measurments, and that even light polishing can markedly reduce friction and wear, provided that it reduces asperity slopes.

1. INTRODUCTION.

Although it is well known to practising engineers that the friction of boundary lubricated contacts depends on the texture of the rubbing surfaces, most scientific research into the subject has neglected this important topic. However, in recent years, several workers have published models of asperity contacts which could allow the effects of surface roughness on friction to be predicted (1 - 4). These models assume that an asperity contact can be represented as a hard wedge which slides over a soft ductile surface, pushing a plastic wave ahead of it. Slip line fields can then be developed to predict the stress and strain in the softer material due to the passage of a hard asperity. An example is shown in figure 1. In all models, the stress and strain depend on the interfacial shear strength ratio (*f*) and the angle which the hard wedge makes with the soft surface. If a real hard surface can be represented as an array of such asperities, then its friction can be calculated successfully from its *slope* distribution and the value of *f* imposed by the lubricant (5,6). It is also possible to calculate wear coefficients for the softer member of the rubbing pair (7,8), but less surely, since some of the mechanical data needed by the model is not available.

Many combinations of materials used in bearing contacts fulfil this requirement. Examples are soft polymer and non-ferrous bushes on hard steel shafts, crankshaft bearings under mixed lubrication, human joint prostheses and many types of seal. Indeed, there is much metallographic evidence (9) that even harder, more brittle materials also deform plastically in rubbing contacts. An example is the spheroidal iron used in camshafts sliding against a hard steel follower.

In all these applications, it is customary to assess and specify surface finish in terms of standard parameters such as R_a and R_q. While this practice helps to ensure that bearing surfaces meet certain minimum quality standards, it falls short

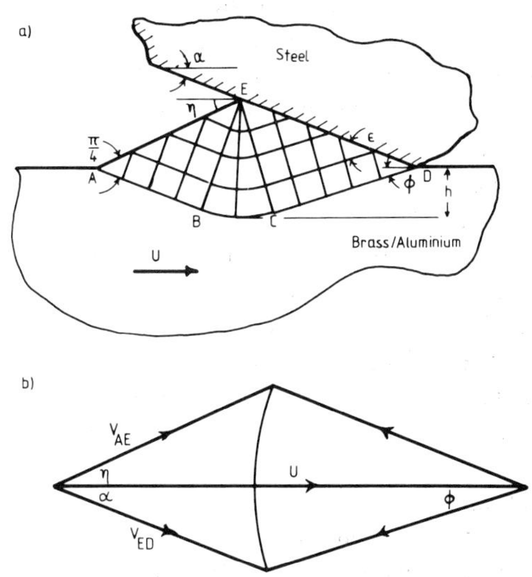

Figure 1a) Slip line field for wave formation.
b) Hodograph.

of providing an accurate assessment of fitness for purpose.

According to the work just cited, asperity slope would be a much better measure of the quality of such bearing surfaces. However if useful measurements of slope are to be made, it is important to define the scale of surface features which will be significant for determining friction. This will allow a bandwidth to be defined over which slopes should be measured. Slope is sensitive to both the long and short wavelength cut-offs chosen for the filter (10,11), so if these are not correctly specified, measurements will have no meaning.

Previous work (6,8,11,12) has shown how bandwidths should be chosen. It is necessary to distinguish four scales of roughness on the hard surface:

1. Longer than the contact length (l_c).
2. Between l_c and the asperity length (l_a).
3. Shorter than l_a but thicker than the boundary film.
4. Thinner than the boundary film.

Scales 1 and 4 do not influence the friction directly.

Previous work describes the influence of the intermediate scales of roughness (6,8,11,12). Scale 2 influences the attack angle (α) which the hard surface makes with the soft one (figure 1). Scale 3 affects the roughness of the asperity interface ED (figure 1) and through this the effective shear strength of that interface. If the contact pressure in the bearing is changed l_a will also change, as the real area of contact rises. This alters the average slopes of scales 2 and 3, and with some surfaces, causes the coefficient of friction to be load-dependent (6,8). Thus it is no simple task to measure surface slopes in a way which will allow the prediction of friction. However, if surface slopes can be measured thus, it is possible to obtain a direct quantitative assessment of the fitness of a bearing surface for its purpose. It is also possible to investigate how modifications to surface finish will affect friction and wear coefficients.

In earlier work (6,8,11,12), the main emphasis was on demonstrating the feasibility of such an approach. However progress was slowed by the fact that very slow-running software was used to analyze the profiles of bearing surfaces. In some cases, 12 hours were needed to finish an analysis with an IBM PC. For the work to become useful, it was necessary to develop software which would perform the analysis in a reasonable time (<60 s) and be easy to use. In the first part of this paper, the development of such software is briefly described. It is then applied to the measurement of the effects of various smoothing processes on steel surfaces, and the prediction of their friction and wear coefficients, which are verified experimentally.

2. DATA ACQUISTION AND SOFTWARE .

Surface profiles were acquired with a Talysurf 4 using a stylus with a 2.5 micron tip radius at a speed of 0.0508 mm/s. The signal from the instrument was fed through a low-pass filter-amplifier unit set to pass frequencies of < 15 Hz so as to avoid aliasing of frequencies around or above the sampling frequency of 40 Hz. The vertical magnification was set to give a signal of

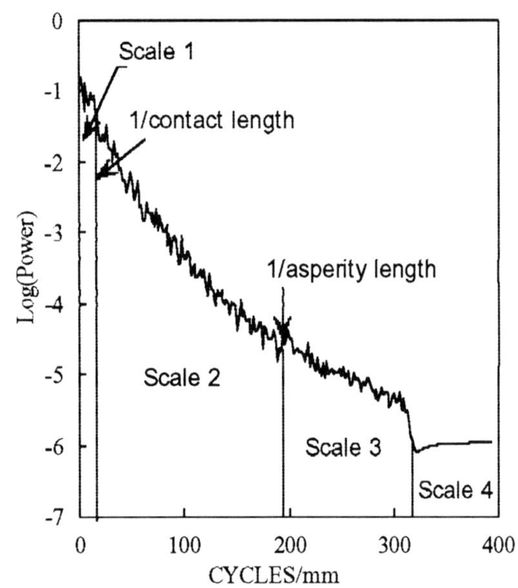

Figure 2. Power spectrum of profile of ground steel surface.

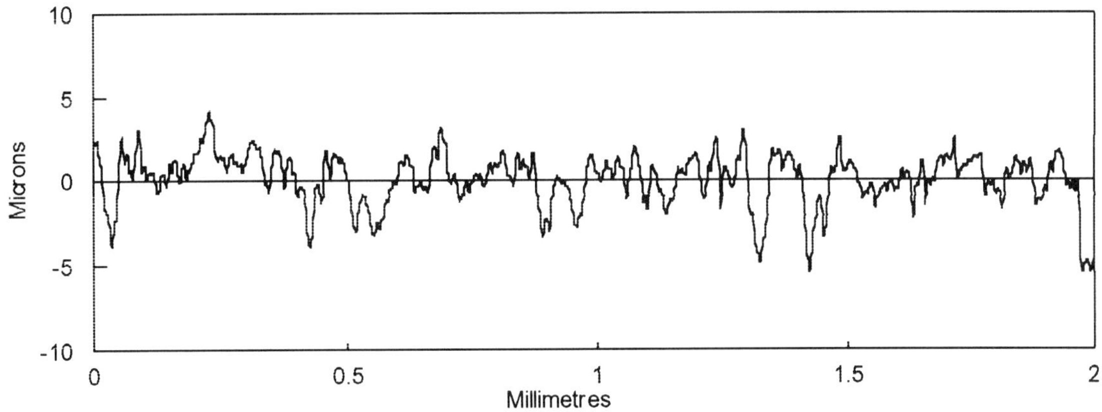

Figure 3. Profile of ground steel surface used in friction experiments. R_a = 1.04 microns.

± 3 volts to match the range of the 12 bit Analogue to Digital convertor (± 5 volts) used in the system. The digital data was recorded with an IBM 88 PC in blocks of 8192 points. For the most part it was analyzed using an IBM 386 fitted with a co-processor. The analysis is based on the spectral methods described in previous work (12,13) which allow profiles to be filtered and differentiated in an efficient manner and permit their power spectra to be calculated. An example of such a spectrum is shown in figure 2 with the four different scales of roughness indicated. A section of the filtered profile from which it was calculated is shown in figure 3.

In order to calculate the average slopes of the two significant roughness scales (2 and 3) it is convenient to use a property of the power spectrum: if it is integrated between two frequency limits, the result is equal to the square of the rms roughness of the wavelengths lying between those limits. If the thickness of the lubricant film is known, this can be used to find the boundary between scales 3 and 4. The power spectrum is integrated stepwise from the high frequency end. When the frequency is reached which makes scale 4 roughness equal to the lubricant film thickness, we have reached the upper frequency limit of scale 3, which can be used with the contact length l_c to set a bandwidth for the analysis.

The slopes of the profile lying within scales 2 and 3 can now be found by passing the filtered profile through a differentiator filter. Also, the asperity contact length (l_a) can be found as a function of % bearing area by scanning the filtered profile (fig 3) from its highest point to determine the height distribution and the number of level crossings. The average slope can also be found as a function of bearing area by scanning simultaeneously the slope record.

To get a complete picture of the frictional performance of the surface, we must now find how the slopes in scales 2 and 3 vary with bearing area. One way of doing this would be to pass the profile through a differentiator filter, once for scale 2 and once for scale 3, for every different value of l_a corresponding to each level of the bearing area curve. Unfortunately this would be very wasteful of computer time and would limit the usefulness of the software. A much quicker way of achieving the same thing is to make use of the power spectrum of surface slopes. This is shown in figure 4. Each value of l_a is used as a limit for the integration of the slope spectrum. Two integrations are performed: from zero to $1/l_a$ gives the square of the rms slope of scale 2; from $1/l_a$ to the upper cut-off of the spectrum gives the rms slope of scale 3 squared. Calculating the slopes in this way takes only a few milliseconds and keeps the overall run time within reasonable bounds. The distribution of slopes is then assumed to be gaussian for the purposes of friction analysis. Also, since friction depends on the slopes of *asperities in contact*, it is necessary to apply a correction factor to the rms slope calculated from the spectrum, which is for

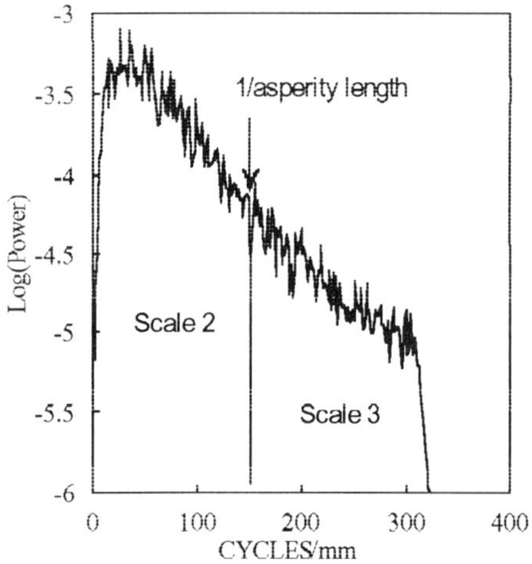

Figure 4. Power spectrum of slopes.

the whole profile within the bandwidth. For a given % bearing area, this is taken to be the rms slope of asperities within the bearing area divided by the rms slope of the whole profile. Admittedly this method involves some assumptions and approximations which would not be needed were the profile to be passed repeatedly through the filter. However, when this latter approach was used in previous work (6,8) it required run times of up to 12 hours. The approach adopted in this paper gave run times of about 30 seconds using a 386 PC fitted with a co-processor and software written in C.

Once the scale 2 and scale 3 slopes had been found as a function of % bearing area, friction coefficients could also be calculated as a function of % bearing area provided *f* was known. Given the hardness of the softer material, contact pressure could be related to % bearing area. as described previously (6,8,10,11). Also, wear coefficients can be found from the hodograph (fig. 1) provided the reduced modulus of the contact is known and the strain needed to produce a wear particle in one cycle of the asperity stress field.

3. FRICTION AND WEAR EXPERIMENTS.

There are many different effects which could be studied using the software just described. For instance, increasing the lubricant film thickness will move the upper frequency limit of scale 3 to lower levels, thereby reducing the slope and hence the friction. In this work, however, we have tried to find what effect surface modifications might have on friction. Since slopes are predicted to determine friction coefficients, it was decided to find how polishing a ground surface would affect its friction coefficient. Two degrees of polishing were used: light and thorough. Light polishing was achieved by spinning the ground surface shown in figure 3 in a lathe and rubbing it with a cloth soaked in Brasso for 1 minute. This smoothed the asperities, but hardly changed the surface roughness. The resulting profile is shown in figure 5. Thorough polishing was achieved by first using silicon carbide paper, then 6 μm diamond paste. This gave the profile shown in figure 6. As can be seen, light

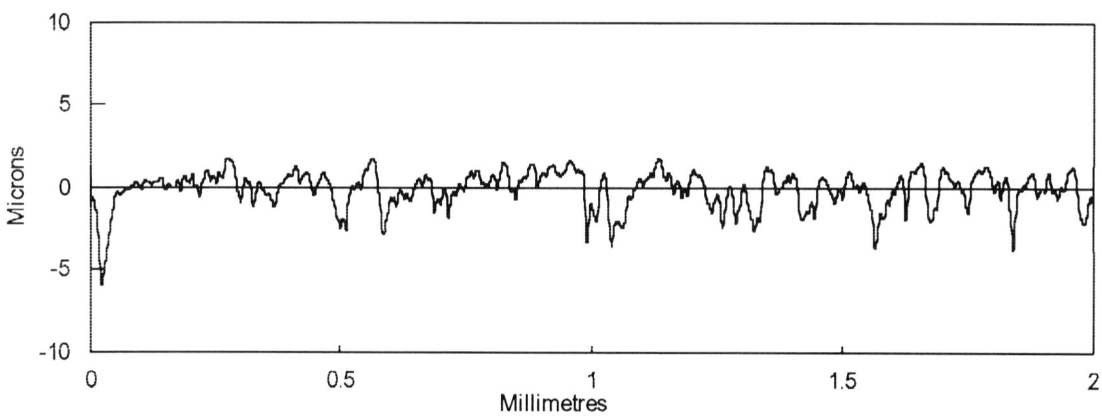

Figure 5. Profile of a ground surface after polishing with Brasso. $R_a = 0.9$ microns.

polishing reduces the surface roughness by only a small amount, whereas thorough polishing reduces it greatly. However, a close comparison of figures 3 and 5 reveals that the light polish has made the asperities more rounded. Hence we may expect them to have lower slopes. For the thoroughly polished surface, the slopes will be lower still.

The steel surfaces shown in figures 3,5 and 6 were rubbed against copper and aluminium sliders using the Bowden-Leben machine shown in figure 7. The harder of the two surfaces in contact *(1)* is one of the mild steel surfaces just mentioned, the specimen itself being a 17 mm diameter bar. The softer specimen *(2)* was a square-sided block of either copper or aluminium, with its bearing surface turned to a 140 mm radius as shown in the inset. This gave a crossed-cylinder contact, allowing high contact pressures to be reached at modest loads. The steel bar was moved beneath the soft slider at a speed of 0.33 mm/s with a leadscrew *(7)* driven by a stepper motor *(8)*. Its axis was inclined at 13.5° to its direction of motion so that the contact point moved steadily across the slider during the test. A known load is placed in *(3)* and

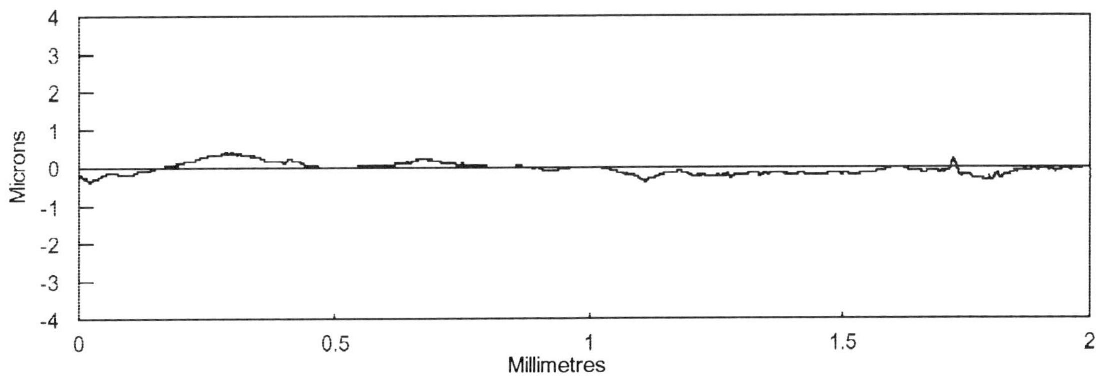

Figure 6. Profile of surface polished with 6 micron diamond paste, R_a = 0.1 microns.

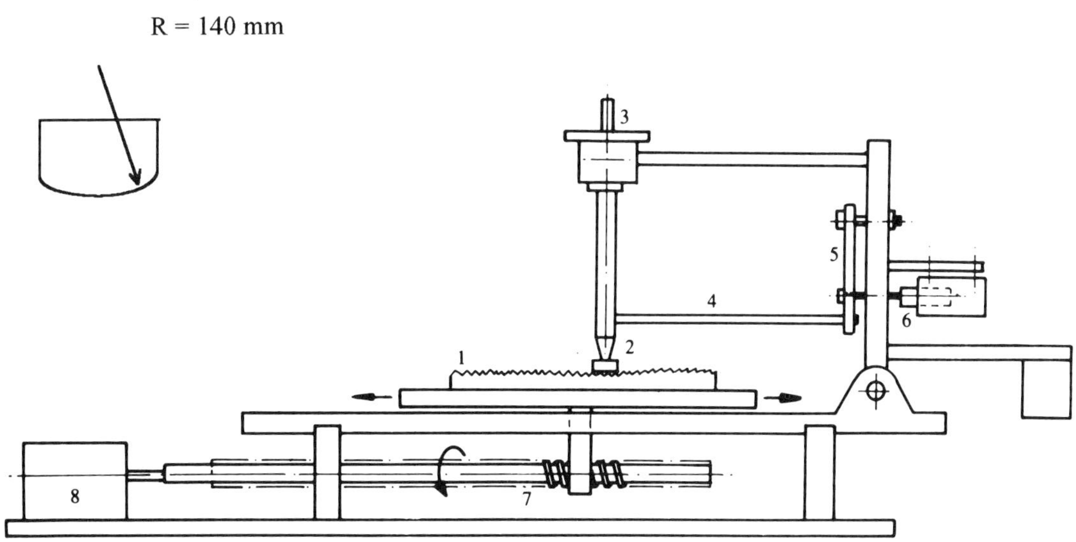

Figure 7. Diagram of Bowden-Leben Machine: inset shows slider in section.

the frictional force recorded by monitoring with the transducer *(6)* the displacement of the bar *(4)* which is attached to the slider and restrained by the cantilever *(5)*. The system was calibrated with deadweights. Loads of 5N - 40 N were used giving average hertzian contact pressures from 80 MPa - 290 MPa.

At the start of each test, the steel surface was covered with hydraulic oil. The load was applied, and the motor started. The ground steel surface was driven under the slider a distance of 20 mm, then the motor was stopped and the slider removed. An attempt was made to measure the wear of the slider by passing a profilometer over the wear scar and calculating the volume of material removed from the shape of the scar. Unfortunately the results were not very accurate as the depth of the scar produced in this way was not much deeper than the roughness of the slider. With the lightly polished steel surface, the wear was much less, so the tests were prolonged by reversing the machine and performing five reciprocations: a distance of 200 mm in all. However, even after this, the wear scars were no bigger than those produced by the ground steel, showing that the wear coefficients were an order of magnitude less. With the thoroughly polished surface, there was no detectable wear even after thirty strokes of the machine. The results of these measurements are shown in table 1. Also shown are the wear coefficients predicted using the methods described in a previous publication (11). Although their seems to be good quantitative agreement between predicted and measured values, both are subject to considerable uncertainty. The measurements themselves were not very accurate, and the predictions depend on the strain needed to produce a wear particle in one strain cycle (C), a quantity which is estimated rather than measured directly (11,12). What is significant about these results is that both the theory and the measurements show wear decreasing by an order of magnitude after light polishing and by a further 3 - 5 orders of magnitude after thorough polishing.

The measurements of friction were more precise, and are shown as the experimental points in in figures 8 and 9. Three theoretical lines are also shown in each figure. They were calculated

Table 1.
Wear Coefficients.

	Predicted	Measured
Ground / Al	$\sim 10^{-3}$	$\sim 10^{-3}$
Lt. Polished / Al	$\sim 10^{-4}$	$\sim 10^{-4}$
Th. Polished / Al	$\sim 5 \times 10^{-8}$	N/A
Ground / Cu	$\sim 2 \times 10^{-3}$	$\sim 10^{-3}$
Lt. Polished / Cu	$\sim 10^{-4}$	$\sim 10^{-4}$
Th. Polished / Cu	$\sim 8 \times 10^{-8}$	N/A

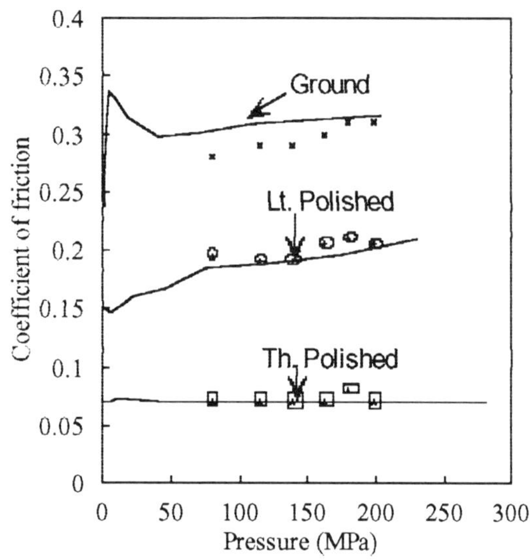

Figure 8. Friction *vs* contact pressure for aluminium sliders. Theoretical lines and experimental points.

Table 2.
Mechanical and physical properties.

Material	Hv (MPa)	E (GPa)	*f*	C
Mild steel	2500	207	N/A	N/A
Copper	970	110	0.37	50
Aluminium	670	70	0.32	57

using the software described in the previous section from profiles of the three steel surfaces used. and from the material properties shown in table 2. The boundary film thickness was assumed to be 10 nm.

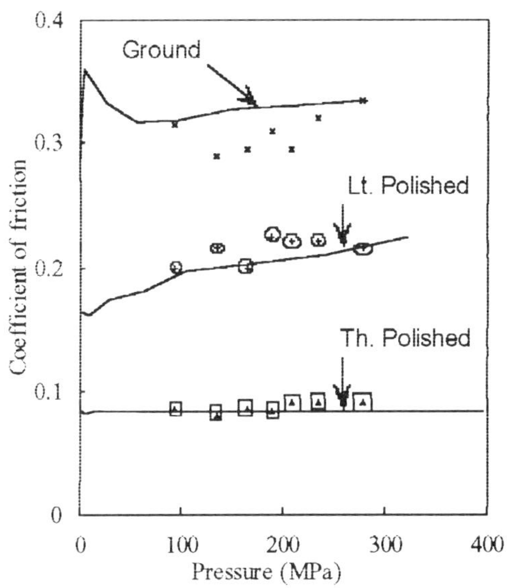

Figure 9. Friction *vs* contact pressure for copper sliders. Theoretical lines and experimental

Hardness was measured with a Vickers tester at 20 Kgf load. Young's modulus (E) was taken from tables. The value of f was found by choosing the value which correlated best the friction measurements made on the thoroughly polished surface. It depends on both the lubricant and the slider material. The agreement between predicted and measured values is good, indicating that for the type of contact studied here, the software should provide a good assessment of the fitness of the steel surface for a bearing.

4. DISCUSSION.

The results reported above show that asperity slopes can exert a powerful effect on boundary friction. Comparing the friction of the lightly polished steel bar with that of the ground steel bar makes this especially evident. The light polishing has had very little effect on the roughness itself, but it has reduced significantly the asperity slopes. The magnitude of the reduction is well predicted by the software for both copper and aluminium sliders. A further reduction in friction can be achieved by thoroughly polishing the steel bar to a near mirror finish. This reduces the roughness greatly, and also the asperity slopes. Once again, the magnitude of the reduction is well predicted by the software. It is remarkable that in all cases, the predicted values are within 10 % of the measurements, that is to say, within experimental error.

For the ground surface, and for the lightly polished surface, the software predicts that friction should be load dependent. This is borne out by the measurements, with the reservation that the predicted trend lies within their error band. It is interesting to observe that the trends become more marked at pressures below the measurement range, the estimated friction of the ground surface rising to 0.35 at low loads, whilst the friction of the polished surface falls to 0.16. This suggests that at the pressures found in many practical bearing assemblies, light polishing may reduce friction much more than in the measurements made here. This has important implications for several tribological processes. The most obvious is perhaps running-in. This is known to modify surface texture in a way which reduces friction and wear (14). Our work suggests that successful running in must involve a reduction in asperity slope, and offers a convenient way of monitoring it. The same can be said of some of the finishing processes used for bearing surfaces: plateau honing, superfinishing and polishing for example. We believe that each of these processes is best assessed by its ability to reduce asperity slopes and suggest that the most appropriate way of controlling them is also through their slopes.

Reducing friction reduces mechanical wear (15,16), so the smoothing of asperities should also have an effect on wear rates. Moreover, using methods similar to those described here, a link between friction and wear can be predicted theoretically (12,17). Qualitatively, our measurements and predictions confirm this and agree with each other, but the measurements must be improved if the wear predictions of the software are to be tested quantitatively.

There is much more work which could be done on finishing processes and running-in using the techniques described here. However, there are also other areas of tribology where this approach could be useful. An example is the influence of film thickness on boundary friction. Our work predicts that an increase in boundary film thickness should remove the effects on friction of some of the smallest asperities. This is accounted for by the software through a reduction of the high frequency cut-off of the filter which in turn reduces the effective slopes of the asperities. It could therefore be a useful tool for the assessment and development of low friction boundary lubricants.

For the time being, the experimental verification of the software has been confined to a fairly narrow range of contacts. However, work is now in hand to extend it to other situations, and to develop better techniques of measuring all the mechanical and physical properties it requires.

5. CONCLUSIONS.

1. Software, convenient to use on a PC has been developed to predict friction and wear coefficients from asperity slopes

2. When asperity slopes are deliberately reduced by polishing, large reductions in friction and wear coefficients result, even when the polishing is too light to influence surface roughness very much

3. A better way of controlling the surface finish of bearing surfaces is to measure their asperity slope, rather than the usual roughness parameters.

REFERENCES.

1. J.M.Challen and P.L.B.Oxley, An explanation of the different regimes of friction and wear using asperity deformation models, Wear, 53 (1979) 229-243
2. J.M.Challen, L.J.McLean and P.L.B.Oxley, Plastic deformation of a metal surface in sliding contact with a hard wedge, Proc. Roy. Soc. Ser. A, 394 (1984) 161-181.
3. J.M.Challen and P.L.B.Oxley, A slip line field analysis of the transition from local asperity contact to full contact in metallic sliding riction, Wear, 100 (1984) 171-193.
4. K.Komvopoulos, N.Saka and N.P.Suh, Plowing friction in dry and lubricated metal sliding, ASME J. Trib., 108 (1986) 301-313.
5. H.Moalic, J.A.Fitzpatrick and A.A.Torrance, The correlation of the characteristics of rough surfaces with their friction coefficients, Proc. I. Mech. E., 201 (1987) 321-329.
6. P.Lacey, A.A.Torrance and J.A.Fitzpatrick, The relation between the friction of lubricated surfaces and apparent normal pressure, ASME, J.Trib., 111 (1989) 260-264.
7. J.M.Challen, P.L.B.Oxley and B.S.Hockenhull, Prediction of Archard's wear coefficient for metallic sliding friction assuming a low cycle fatigue wear mechanism, Wear, 111 (1986) 275-288.
8. P.Lacey and A.A.Torrance, The calculation of wear coefficients for plastic contacts, Wear 145 (1991) 367-383.
9. D.A.Rigney, The role of characterization in understanding debris generation, in 18th Leeds-Lyon Symposium Elsevier (1991) ,Ed. D.Dowson and M.Godet.
10. H.Moalic, J.A.Fitzpatrick and A.A.Torrance, A spectral approach to the analysis of rough surfaces, ASME, J. Trib., 111 (1989) 359-363.
11. H.Moalic, Spectral analysis of rough surfaces, Ph.D. Thesis, University of Dublin (Trinity College) 1986.
12. A.A.Torrance, Debris detachment in plastic contacts, in 18thLeeds-Lyon Symposium Elsevier (1991) , Ed. D.Dowson and M.Godet.
13. A.A.Torrance, The statistics of the extrema of surfaces as a means of assessing their mechanical performance, Wear 1992, in the press.
14. T.R.Thomas, Rough surfaces, Longman (1982).
15. W.H.Roberts, Some current trends in tribology in the U.K. and Europe, JSLE International Tribology Conference, Tokyo (1985).
16. E.Rabinowicz, The dependence of the adhesive wear coefficient on the surface energy of adhesion, in Wear of Materials, Ed. W.A.Glaeser, ASME, New York (1977) 36-40.
17. J.M.Challen, E.M.Kopalinsky and P.L.B.Oxley, An asperity deformation model for relating the coefficients of friction and wear in sliding metal friction, in Tribology - Friction, Lubrication and Wear fifty years on, Vol II, I.Mech E., London (1987) Paper C156/87.

SESSION IV
COATINGS AND SURFACE MODIFICATION (1)

Chairman : Dr B J Briscoe

Thin Films in Tribology / D. Dowson et al. (Editors)
1993 Elsevier Science Publishers B.V.

Wear and Friction Behavior of Ni-SiC Composite Coatings

E.A. Rosset, S. Mischler, D. Landolt

Materials Department, Ecole Polytechnique Fédérale de Lausanne, 1015 Lausanne, Switzerland

The sliding wear and frictional behavior of Ni metal matrix composite coatings have been studied by using a reciprocating motion wear test rig. The effect of the heat treatment on structure and wear behavior is investigated. An attempt is made to elucidate the role of the interface between the particles and the matrix. Atomic Force Microscopy (AFM) and Scanning Electron Microscopy (SEM) have been used to characterize the worn surfaces.

1. INTRODUCTION

Electroless and galvanic deposition techniques are cheap and flexible methods to deposit functional metal films on a wide range of substrates. They offer an easy way to produce metal matrix composite coatings. For example, one can obtain self-lubricating and anti-wear coatings by embedding either solid lubricant particles (i.e. PTFE) or hard particles (i.e. carbides or nitrides) in a metallic matrix.

This paper presents a study of NiP - SiC composite coatings. It is well known that their tribological properties can be significantly improved with an appropriate thermal treatment. The influence of the thermal treatment on wear properties has been widely studied in terms of performance improvement but the wear mechanisms are not been yet fully understood.

The aim of the present paper is to compare the sliding wear and frictional behaviour of a NiP-SiC composite before and after a thermal treatment. Wear rates and frictional coefficients are determined under dry sliding wear conditions against alumina. The morphology of the worn surfaces and of the wear debris are observed by Scanning Electron Microscopy and Atomic Force Microscopy in order to understand the wear mechanisms. Simple fracture test were carried out in order to observe the dominant mechanical failure mode of the composites.

2. EXPERIMENTAL

2.1 Test Material

Sliding wear conditions were established by rubbing an alumina pin against a metal matrix

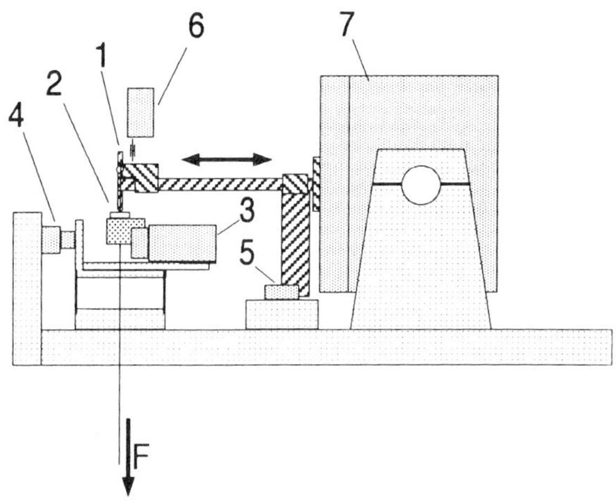

Figure 1. Wear test rig.

composite coating plated on steel. Technical pure alumina rods of 4 mm diameter were supplied by Metoxit AG, Thayngen, Switzerland.

SiC particles, 1.8 μm mean diameter, supplied by Lonza - Werke Gmbh (UF-10), Waldshut, Germany, are embedded by co-deposition in an electroless NiP metal matrix 40 μm thick. The phosphorus content is 7-8 wt%. The volumic ratio of SiC particles is 27%. Thermal treatment consists in heating the coating for 5 hours at 290°C in a nitrogen/hydrogen atmosphere and increase the hardness to 1200 HV0.05.

For fracture surface analysis the coating was separated from the substrate by dissolution of the steel substrate in hydrochloric acid. Fracture surfaces were produced by bending the isolated coating in air.

2.2 Wear test

The wear tests were carried out in the reciprocating pin-on-plate rig depicted schematically in Fig. 1. Friction tests are done in air, 50-60 %RH, 20-23 °C.

Sliding wear conditions are established by rubbing the end of a vertically mounted pin, figure 1 (1), against a flat sample (2). The sample holder is located above a load cell (3) allowing for the measurement of the normal force. The frictional force is measured with a piezoelectric force transducer (4). An IR optical sensor (5) measures the horizontal pin displacement and a lasermeter (Keyence LC-2100)(6) measures the vertical pin displacement with a 0.2 μm resolution. Reciprocal pin motion is provided by an electrodynamic vibration exciter (Brüel Kjaer 4809)(7).

During the wear test the frictional and the normal forces as well as the horizontal and vertical pin displacements transients were monitored using a Macintosh IIfx computer equipped with a National Instrument general purpose 16 bits I/O board controlled by the LabView 2 data acquisition and control software. Real time data analysis provides the wear rate and coefficient of friction evolution.

The oscillating pins were prepared by machining the ends of the alumina rods in the shape of truncated cones (120° included angle). The diameter of the flat end was 0.5 mm giving an apparent contact area of 0.2 mm^2. The pin was oscillating at a frequency of 5 Hz. The stroke length of 5 mm corresponded to an average sliding velocity of 50 mm/s. The applied normal load was of 5 N resulting in a nominal contact pressure of 25 MPa. The samples were tested just after a 5 minutes ultrasonic cleaning in ethanol and were dried with a warm air jet. The initial roughness, Ra, was 1.8 μm for the non heat treated (HT) samples and 1.6 μm for the HT samples, respectively, over a cut-off length of 0.8 mm.

2.3 Wear analysis

The surface topography was measured by use of a Taylor-Hobson talysurf 10 profilometer and a Park Scientific Instruments SFM-BD2 Atomic Force Microscope (AFM). The morphology was observed with a Cambridge Stereoscan 650 SEM. EDAX analysis was performed using a Cambridge Stereoscan 250 SEM. The structure was analysed with a Siemens D500 Kristalloflex RX diffractometer.

3. RESULTS

3.1 Structure

The X-Ray spectrum recorded before heat treatment (Fig. 2a) exhibits well defined peaks corresponding to the α-SiC as well as a broad peak

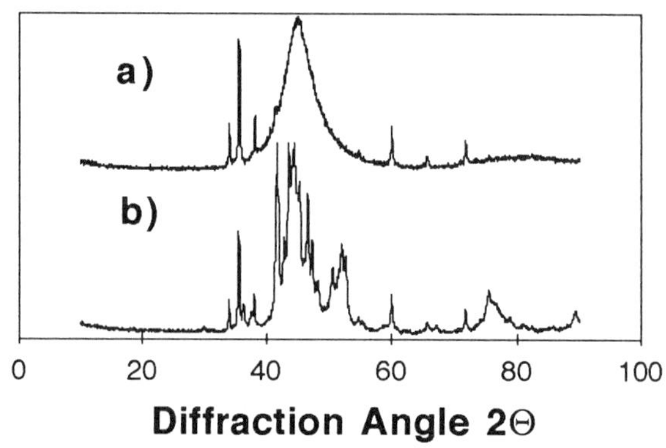

Figure 2. X-Ray Spectrum of a) Non HT sample, b) HT sample.

at 45°. The presence of this peak indicates an amorphous structure of the Ni-P matrix.

After heat treatment several X-Ray lines (Fig. 2b) appear superimposed to the broad peak corresponding to the amorphous state. This indicates that a partial recristallisation of the matrix has occurred. The lines observed in Fig. 2b correspond to the compounds α-SiC; Ni_3P and Ni_3Si. It is known that, by thermal activation, amorphous NiP undergoes an eutectic reaction leading to the formation of two phases: pure Nickel and Ni_3P. For a phosphorous content of 7.5 % the ratio of the volume fractions of the two phases is 50:50 [1].

The Ni_3Si phase forms by diffusion of Si from the SiC particles in the Ni matrix. Kloos and co-workers [2] observed a complete SiC dissociation in a NiP composite with 5% SiC after a heat treatment of 2 hours at 600°C.

3.2 Wear tests

The evolution of the coefficient of friction (μ) and of the vertical pin displacement with the number of cycles is plotted in Fig. 3. The value of the coefficient of friction scatters considerably between different tests on the same sample and assume the value of 0.3 ± 0.1 independently on the heat treatment.

The wear coefficients were determined by profilometry after the wear test and correspond to $3.8\ 10^{-6}$ and $8.8\ 10^{-6}$ mm^3/Nm for samples with and without heat treatment respectively. The deviation of the wear coefficients was 20%.

The wear track was generally flat independent of heat treatment. Numerous pits were observes on the non HT samples, whilst only few were present on the HT sample.

The morphology observed in figure 4 top exhibits geometrical features. Cracks, probably due to fatigue, were present at the ends of the pits in the sliding direction.

The surface morphology of the pits shown in figure 4 bottom corresponds to the original surface. These pits resulted from the valleys which have not interacted mechanically with the pin yet and were due to the original sample roughness. Very few cracks, similar to the non HT samples, were observed.

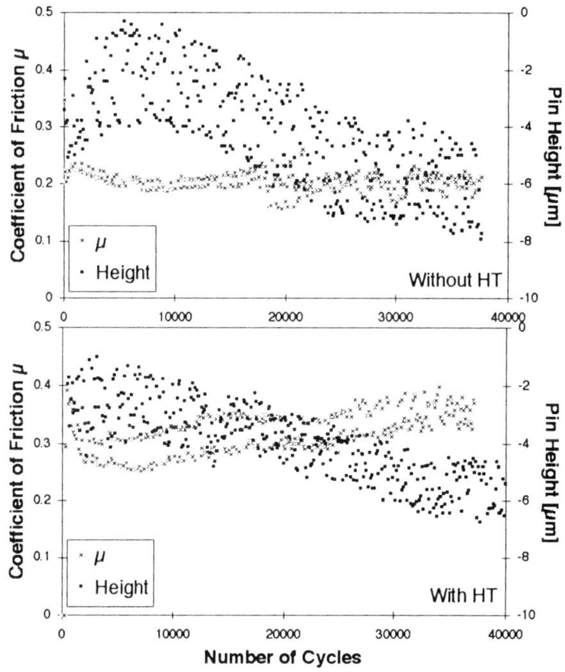

Figure 3. Coefficient of friction and vertical pin displacement evolution, without HT (top), with HT (bottom).

The AFM images recorded on the flat parts of the wear track surface (Fig. 5) shows the presence of grooves on both the matrix and the SiC particles. The observed grooves indicate abrasion by sub-micron ploughing. The SiC particles protrude slightly (about 20 nm) from the matrix.

The AFM image shown in figure 6 has been recorded in a pit of the non HT sample and indicates a faceted surface morphology.

Wear debris were found aggregating at the end of the wear tracks on both samples. Significant differences in debris size and size distribu-

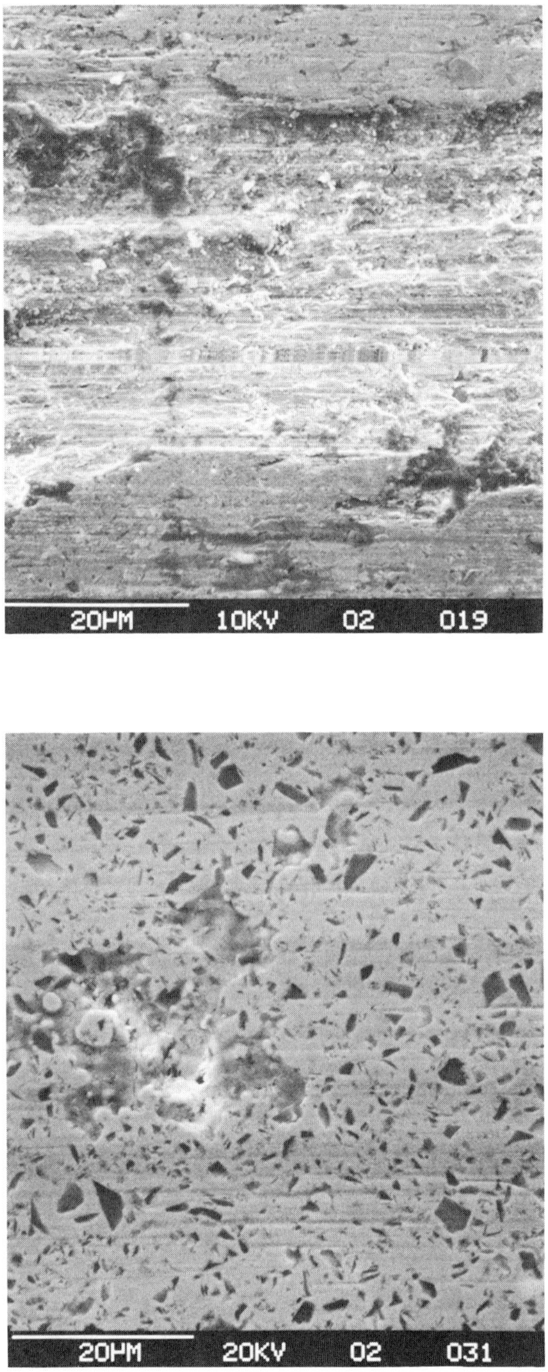

Figure 4. Wear track surfaces, Non HT sample (Top), HT sample (Bottom).

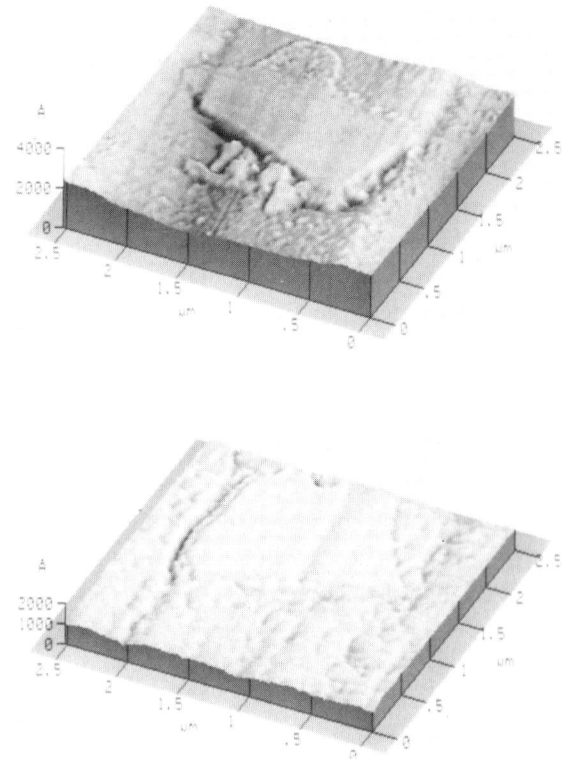

Figure 5. Imbedded particles in the tracks. AFM topography of non HT sample (top) and HT sample (bottom).

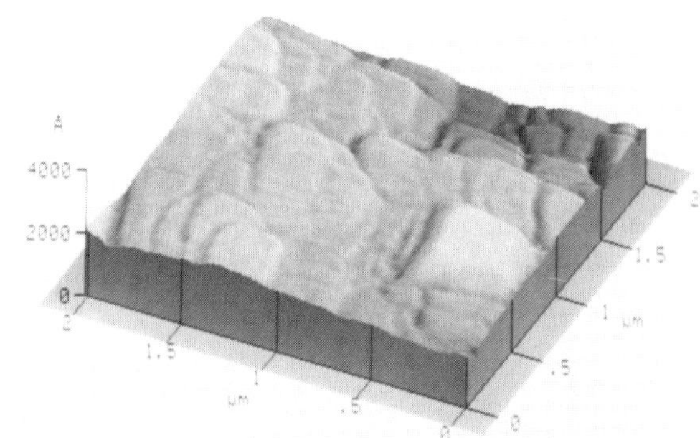

Figure 6. AFM typical topography of a pit in a non HT sample.

tion are found by scanning electron microscopy (Fig. 7). On HT samples wear debris exhibit a very fine (submicron) and uniform size distribution. The size of wear debris generated on non HT samples ranges from 5 μm to 0.2 μm for most of the debris, the larger debris exhibiting a faceted geometry. Using EDAX analysis several large debris were identified as being SiC particles.

Considerable metal transfer on the alumina pin was found during sliding on the non HT samples. In this case the pin flat end was almost completely covered with a dark film. EDAX analysis showed that this film was composed of Ni, P and Si. The irregular morphology of this adhesion film is clearly illustrated by the SEM image (Fig. 8 top).

The flat end of the pin rubbed against the HT sample was only locally covered with a dark films. EDAX analysis showed that the dark stains observed on the SEM images (Fig. 8 bottom) contain Ni, P and Si whilst the rest of the surface exhibits only an EDAX signal corresponding to the Al from the Alumina pin.

3.3 Fracture surfaces

SEM images of fracture surfaces are shown in Fig. 9. The morphology of these surfaces depends on thermal treatment of the sample: the non HT sample exhibits a faceted fracture surface with a facet size in the range 1 to 5 μm. Several SiC particles were identified on the non HT fracture surface (dark features in the centre of Fig. 9 top). The shape and the average size of the facets observed in Fig. 9 top correspond very well with the geometry and dimensions of the SiC particles. This indicates that in the non HT sample the fracture propagates preferentially at the interface between particle and matrix.

The fracture surface features observed on the HT sample are much finer than the average size of the SiC particles. This indicates that the fracture path goes, at least partially, through the SiC particle.

Figure 7. Wear debris. Non HT sample (Top), HT sample (Bottom).

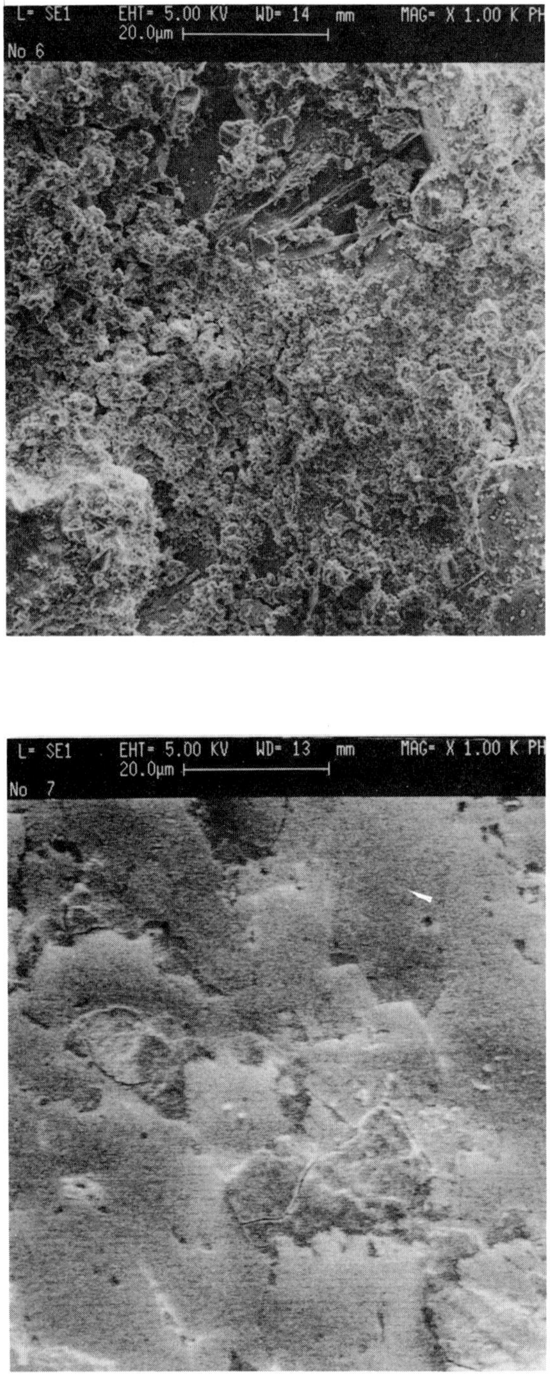

Figure 8. Pin surfaces after sliding, on non HT sample (Top), on HT sample (Bottom).

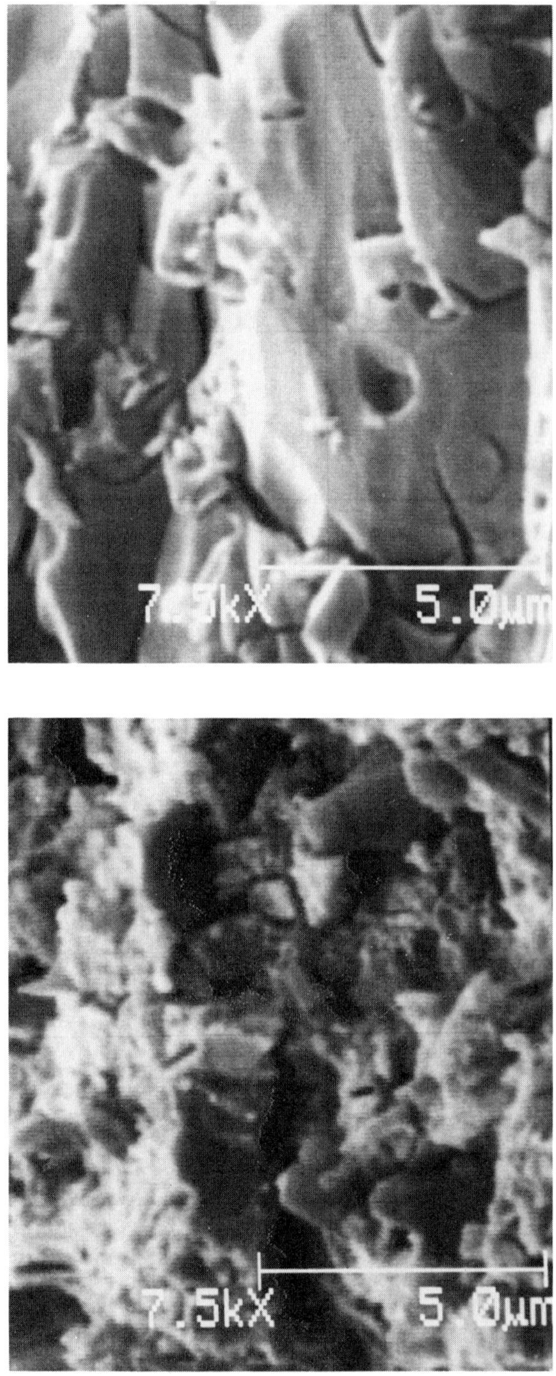

Figure 9. Fracture surface on non HT sample (Top) and on HT sample (Bottom).

4. DISCUSSION

The present results show that the heat treatment leads to a reduction of the wear coefficient of a factor of 2. The dominant wear mechanisms are also affected by the heat treatment : surface fatigue pits are observed only in non HT samples. On the same samples the extent of metal adhesion to the pin is more important. Wear debris found on HT samples are finer and exhibit a more homogeneous size distribution. Abrasion (Fig. 5) occurs on both samples.

The vertical pin displacement scattering (Fig. 3) relates directly to the size of the particles observed (Fig. 7) and thus to the heat treatment. The 2 μm elevation of the pin on the non HT sample (Fig. 3, top) corresponds to the formation of the transfer coating observed on the pin (Fig. 8, top). Such an elevation is scarcely observable for the HT sample (Fig. 3, bottom) and correlate with the rather clean surface of the pin (Fig. 8, bottom).

The observed differences in wear behaviour are a consequence of the structure modification induced by the heat treatment.

Before heat treatment the composite is composed of SiC particles dispersed in an amorphous Ni-P matrix. The X-Ray analysis shows that during heating phase transitions and chemical reactions occur which lead to the formation of Ni_3P and Ni_3Si. According to the literature formation of nickel phosphide takes place in the Ni-P matrix [1].

The formation of Ni_3Si is controlled by the diffusion of Si out from the SiC particles into the matrix and is therefore expected to take place at the interface between matrix and particle.

Additional compounds formed at the interface in composite materials influence the strength of matrix-particle cohesion and thus the mechanical properties of the material [3]. If the cohesion is too weak no load transfer between matrix and reinforcing particle occurs. In this case no improvement of the material properties results from the combination matrix-particle and cracks will propagate at the interface. A stronger interaction leads to a significant load transfer from the matrix to the reinforcing particle and thus cracks can be deviated or stopped.

The fracture surfaces (Fig. 9) indicate that the formation of nickel silicide improves the cohesion between SiC particles and the matrix. Prior to the heat treatment the composite cracks along the particles/matrix interfaces whilst after heat treatment the cohesion is much stronger and leads to a crack propagation through the SiC particles.

The capability of the interface to transfer load may play an important role in wear. The SiC particles protrude from the matrix in the wear track and therefore carry most of the applied load during sliding.

On non HT samples the load is not distributed to the surrounding matrix because of the weak particle-matrix cohesion. Highly concentrated stresses arise at the sharp edges of the SiC particles. This leads to microcracking with subsequent crack propagation at the interface caused by repeated alternating loading of the surface (surface fatigue). As consequence flaking of material occurs and pits are formed.

On HT samples the sliding forces applied on a SiC particle are distributed by the interface to the surrounding matrix so that no stress concentration appears. No flaking of material is then expected.

The predictions of this simple model agree well with the observed presence of pits only on the non HT sample. The faceted surface morphologies in the pits and the non HT fracture surface correspond to the model. Considering only the increase in the hardness of the matrix could lead to the opposite prediction. Because of the weak cohesion SiC particles are likely to be teared from the non HT composite during sliding. This can explain the presence of large SiC particles among the wear debris on the non HT sample.

Adhesion depends on physical and chemical properties of the materials in contact, the mode and value of loading and on the state of the surfaces (roughness, contamination). No major difference in roughness, surface chemistry or in chemical properties of the material is to be expected between HT and non HT samples. Instead an increase of a factor of 2 in hardness of the Ni-P phase was observed by Kloos [2] after a similar heat treatment. According to the adhesive law of wear an increase in hardness of the Ni-P matrix leads to a reduced material loss.

5. CONCLUSIONS

The microstructure and the wear behaviour of the studied Ni-P / SiC composite coating are modified by the applied heat treatment.

A mechanistic relationship between microstructure and wear process has been proposed. The role played by the matrix-SiC interface in the wear mechanism has been shown.

REFERENCES

[1] J. Osowiecki, W. Form, J. Forchelet Mémoires et Etudes Scientifiques Revue de Métallurgie 11, 581 (1985)

[2] K.H. Kloos, E. Wagner, E. Broszeit Z. Werkstofftech 9, 305 (1978)

[3] R.W. Davidge Interfaces in Materials, Proceedings of the Colloqium, Brussels, 9 December 1988

Thin Films in Tribology / D. Dowson et al. (Editors)
© 1993 Elsevier Science Publishers B.V. All rights reserved.

STUDY ON SOLID LUBRICATION PROPERTY OF MoS_2–CONTAINING Ni BASE AND Ni–P BASE COMPOSITE BRUSH PLATING LAYERS

Xian–Hua Zhang, Jia–Jun Liu, Bao–Liang Zhu
Tribology Research Institute, Tsinghua University
Solid Lubrication Lab, Lanzhou Institute of Chemical Physics

ABSTRACT

Using brush plating technique①to realize solid lubrication is one of new methods of solid lubrication with great vitality. The newly developed Ni / MoS_2 and Ni–P / MoS_2 composite brush plating layers have been widely applied on numerous machine parts with success. In order to evaluate their solid lubrication property and approach the mechanism of thier effect this paper conducted systematic investigations and comparisons of their tribological behaviors on a ″ball–on–disc″ testing machine under lubrication, and discussed their mechanism through microscopic anlayses of TEM, SEM, and X–ray diffraction. The results of research showed that the friction coefficient of both composite brush plating layers can reach generally 0.05~0.06, and the wear–resistance of Ni–P / $MoS_2$is 4~6 times that of Ni / MoS_2.

1.INTRODUCTION

MoS_2 is a widely used excellent solid lubricant. There are numerous methods of its application suiting different working conditions[1]. Along with the popularization of solid lubrication the investigations for searching more simplistic and effective methods of its application become more and more attractive.

Brush plating is quite similar to conventional vat plating in principle, but it shows very distinctive advantages beyond the latter, such as: high current density, rapid deposition rate, relative motion between anode (brush) and cathod (workpiece) during operation etc. All these characteristics are favorable for obtaining a sound composite plating layer with higher concentration of MoS_2 (can reach 80 vol%) and larger thickness (can reach 100 μm) compared with that by conventional vat plating method.

In order to evaluate the solid lubrication property of newly developed Ni base and Ni–P base MoS_2–containing composite brush plating layers, this paper conducted a comprehensive

①The brush plating technique, differing from the conventional vat plating, is an electrochemical process conducted with an electrolyte supplied to substrate by so–called brush (or tampon pad) to form the adherent deposit.

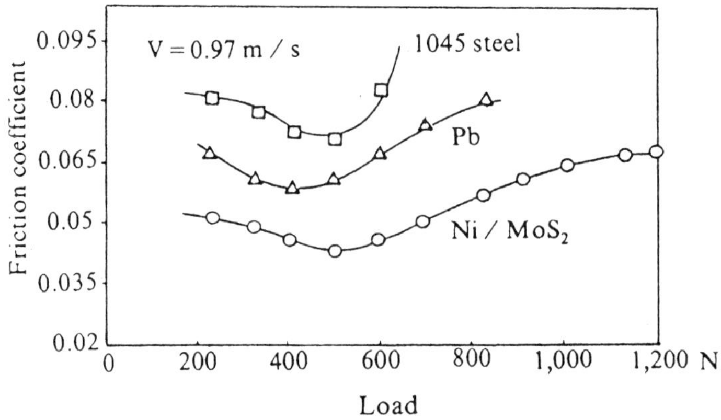

Fig.1 The variation of friction coefficient of Ni / MoS_2 composite plating layer with load

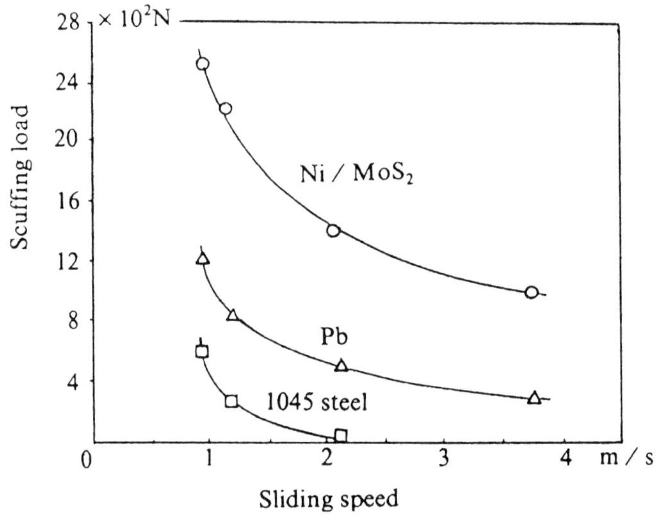

Fig.2 The variation of scuffing load of Ni / MoS_2 composite plating layer with sliding speed.

measurement and comparison of friction and wear behavior of two optimized Ni/MoS_2 and $Ni-P/MoS_2$ composite plating layers, and made a deep mechanism analysis of their effect.

2. EXPERIMENTAL METHODS

2.1 Basic compositions of brush plating solutions and parameters

Composition of Ni/MoS_2 plating solution: Nickel sulphate 200~280 g/l, Nickel chloride 40~70 g/l, complexing agent 10~20 g/l, additive 5~10 g/l, $MoS_2$10 g/l.

Parameters: Size of $MoS_2$1μm, voltage: 12v Current density: 20–40 A/dm^2.

Composition of $Ni-P/MoS_2$ plating solution: Nickel sulphate 200~250 g/l, Nickel chloride 40~60 g/l, Sodium hypophosphite 10~50 g/l, Additive I 60~100 g/l, Additvie II 20~30 g/l, Additive III appropriate amount, $MoS_2$10g/l.

Parameters: Size of MoS_2 1 μm, Voltage: 15V Current density: 8~45 A/dm^2

2.2 Friction and wear tests

A "ball–on–disc" testing machine was adopted. The upper ball specimen made from 52100 bearing steel was stationary. Radius of ball R=6.35mm. Surface roughness Ra=0.01μm. Hardness was HV770. The lower disc specimen made from 1045 steel was rotating. Its dimensions were Φ56×5mm. Hardness was HV450. The Ni/MoS_2 amd $Ni-P/MoS_2$ composite plating layers were brush plated on its surface. A non–cyclic drop–feed lubrication mode with $20^{\#}$ engine oil (kinematic viscosity was 17~23 cSt without additives) was employed. A 2–minute running–in period under 110N load was allowed for every new pair of specimens, and then the stepwise loading process was started with a rate of 100N per 10 sec. until scuffing. In this test the friction coefficient, scuffing load and wear life could be obtained.

2.3 Analyses of microstructure of plating layers.

The Scanning Electron Microscope of Model OPTON CSM 950 with Energy Dispersive X–ray analyser (EDX) and X–ray Diffractometer of Model RU–3014 (Cu target, Voltage 40 KV Beam Current 20mA) were employed.

3. EXPERIMENTAL RESULTS AND DISCUSSIONS

3.1 Friction and wear behavior of Ni/MoS_2 composite plating layer.

The concentration of MoS_2 in composite plating layer obtained by the method mentioned above was 11 vol.%. The thickness of plating layer was 10~12μm.

Fig.1 shows the variation of friction coefficient with load. In this figure the variations of friction coefficient of Pb plating layer and 1045 steel are also put up for comparison. It can be seen that Ni/MoS_2 composite plating layer ex-

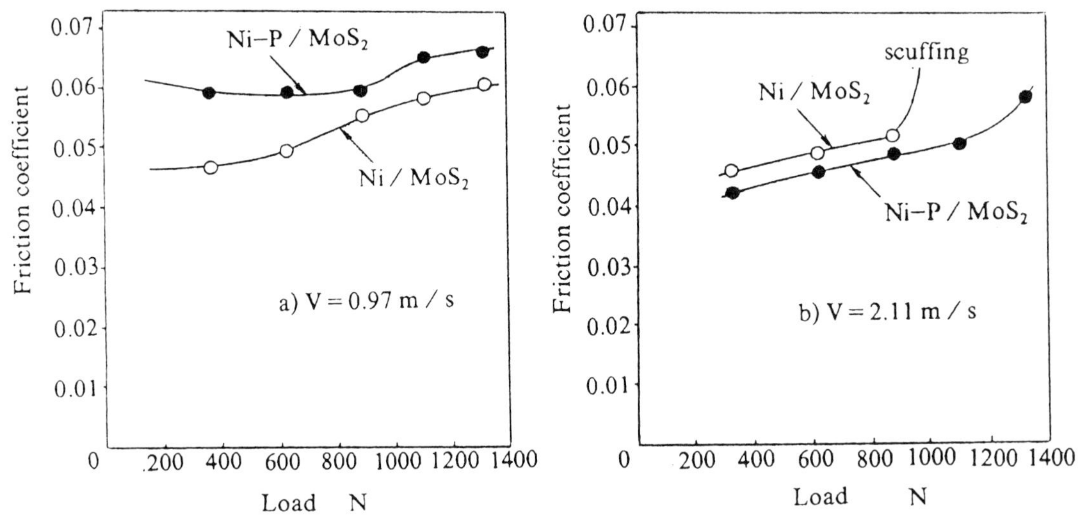

Fig.3 The varaition of friction coefficient of two composite plating layers with load

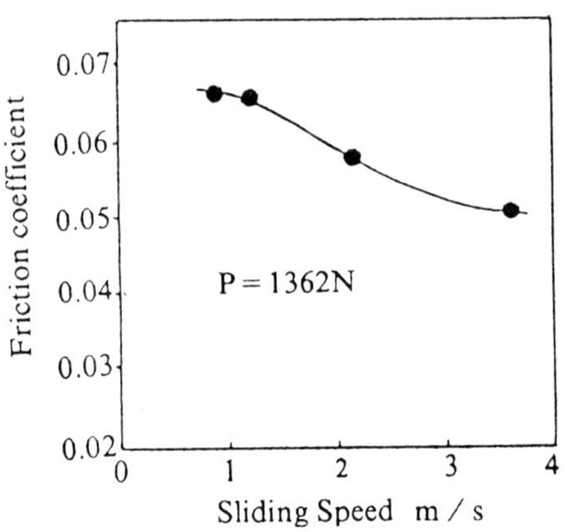

Fig.4 The variation of friction coefficient of Ni–P / MoS_2compostite plating layer

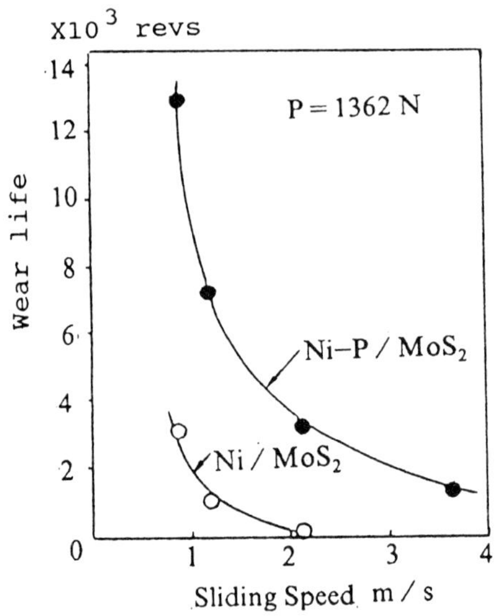

Fig.5 Th variation of wear life of two composite plating layers with sliding speed.

hibited obvious friction–reducing effect, its friction coefficient was is the range of 0.04~ 0.06, much lower than that of 1045 steel and Pb plating layer.

When the load reached 400~ 500N a valley of friction coefficent appeared. The similar phenomenon was also present for 1045 steel and Pb plating layer. This is probably due to that the load of 400~ 500N was just a transition point from mixed lubrication (partial EHD lubrication) to boundary lubrication. Before this point, the lubrication state was going from lower to higher percentage of boundary lubrication in mixed lubrication state, the slight decrease of friction coefficient with load might be related with more boundary films starting to act. After this point the oil film had lost fully its function, the friction coefficient was controlled mainly by the plastic deformation of plating layer (or substrate). i.e. the greater the deformation work was dissipated, the higher the friction coefficient would be exhibited[2].

Fig.2 shows the variation of scuffing load (load bearing capacity) of Ni / MoS_2 plating layer with sliding speed. The currves of 1045 steel and Pb plating layer are also put up for comparison. It can be found that the scuffing load of Ni / MoS_2 composite plating layer was much higher than that of two others, about 4 times that of 1045 steel and 2 times that of Pb plating layer. The reasons may be explained as follows: Before scuffing occured, the contact surface was in the state of boundary lubrication, the direct contact between metals was present at some local micro–regions, if the contact surface was fully covered by MoS_2, the adhesion between metals and plastic deformation of substrate beneth surface could be prevented effectively, thus the development of strain fatigue, which is the basic mechanism of delamination, could be inhibited[3]. The Pb plating layer showed the same function, but its effect was far from that of MoS_2.

3.2 The comparison of tribological behavior of composite plating layers

The concentration of MoS_2 in Ni–P / MoS_2 composite plating layer was also 11 vol.%

Fig.3 shows separately the variation of friction coefficient with load of Ni / MoS_2 and Ni–P / MoS_2 composite plating layers at 0.97 m / s and 2.11 m / s sliding speeds. It can be seen that the friction coefficient of Ni–P / MoS_2 was a little higher than that of Ni / MoS_2 due to its higher hardness (~ HV500, while hardness of Ni / MoS_2 was about HV250). However, it could still be kept at a stable and lower level of 0.06~0.07.

The scuffing of Ni / MoS_2 plating layer occured at 900N load when sliding speed was increased to 2.11 m / s. It indicates that the friction–reducing function of Ni / MoS_2 plating layer was of a certain limitation. i.e. it could be used availably only at the condition of lower sliding speed of <1 m / s. While the Ni–P / MoS_2 plating layer exhibited the

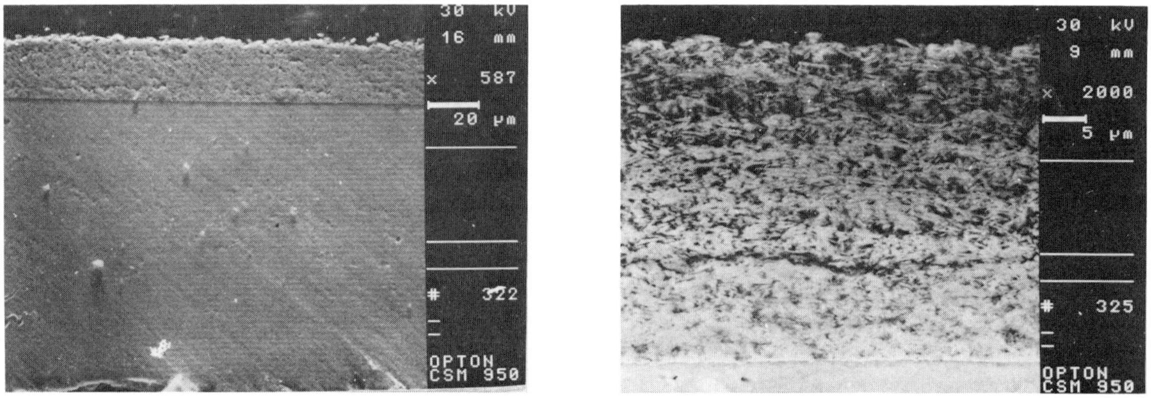

a) × 587 b) × 2000

Fig.6 The SEM morphology of section view of Ni / MoS_2 composite plating layer

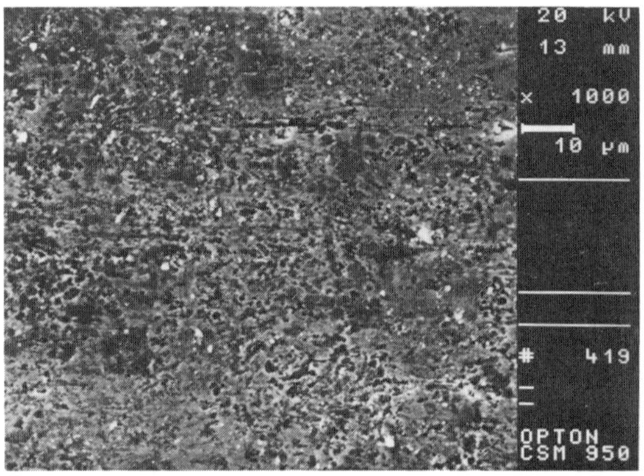

Fig.7 The SEM morphology of wear surface of Ni / MoS_2 composite plating layer. × 1000

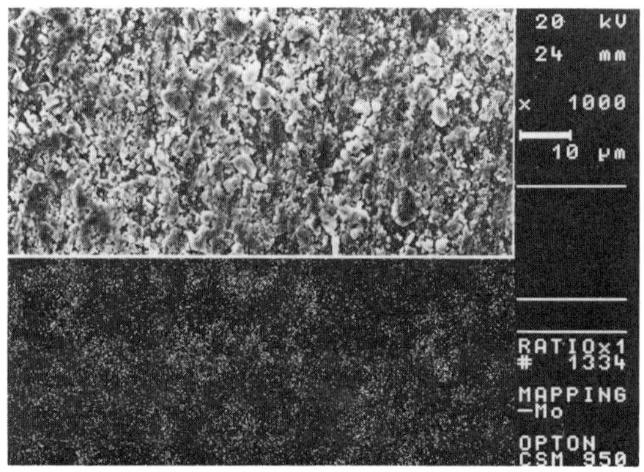

Fig.8 The distribution of Mo element on wear surface of Ni / MoS_2 composite plating layer

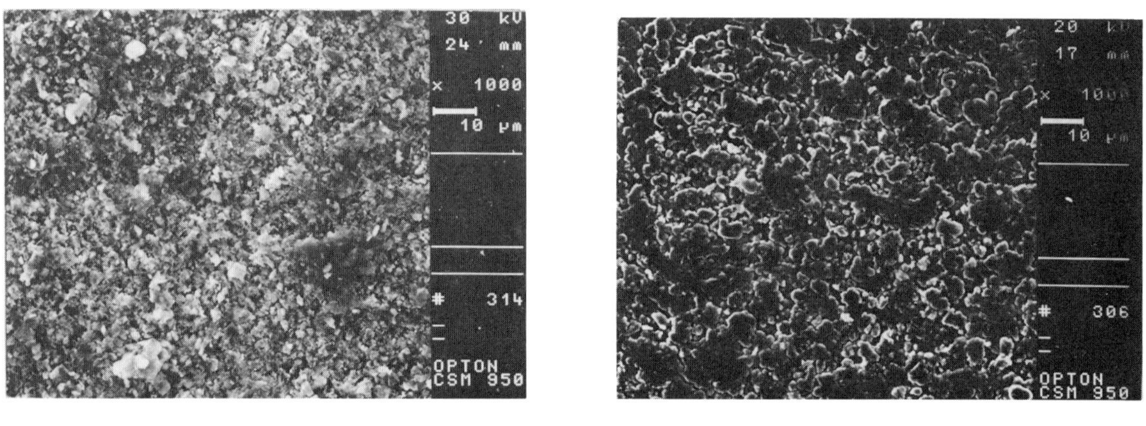

a) Ni / MoS_2 b) Ni–P / MoS_2

Fig.9 The SEM surface morphology of two composite plating layers × 1000 advantage of broader availability, its friction coefficient was still kept below 0.06 at 2.11 m / s sliding speed and 1362N. load condition.

Fig.4 shows the variation of friction coefficient of Ni–P / MoS_2 plating layer with sliding speed at 1362N load. Its reducing tendency at such high load condition was related obviously with the frictional temperature and changes of structure and property of Ni–P matrix. i.e. the high friction temperature could result in the decrease of surface hardness of plating layer, thus its solid lubrication property could be further improved.

Fig.5 shows the comparison of wear life of Ni / MoS_2 and Ni–P / MoS_2 composite plating layers varied with sliding speed at 1362N load. It can be seen that the wear resistance of Ni–P / MoS_2 plating layer was increased significantly about 4~ 6 times that of Ni / MoS_2 plating layer in the range of sliding speed <2m / s. When the sliding speed was over 2m / s, the Ni / MoS_2 plating layer was already failed completely, while the Ni–P / MoS_2 plating layer could still work as a solid lubrication film.

3.3 The charactersitic of MoS_2 structure in composite plating layer.

The TEM analysis results of thin film sample of Ni / MoS_2 plating layer revealed that there were two types of MoS_2 grains, strip–like and plate–like. The former represented the (0001) basal plane of MoS_2 grains perpendicular to the surface of plating layer, and the latter represented its parallel orientation. An obvious preferred orientation of MoS_2 grains parallel to the surface was further proved by the pole diagram of X–ray diffraction analysis[4].

Fig. 6 a) and b) show the SEM morphology of section view of Ni / MoS_2 composite plating layer at low and high magnification. From Fig.6 b) it can be found clearly the special laminary structure, which was formed in the process of co–deposition of MoS_2 grains covered by Ni^{2+} ions.

Fig.7 shows the SEM morphology of wear track of Ni / MoS_2 plating layer after dry friction. The black MoS_2 particles became visible when the Ni covers were worn off. The distribution of MoS_2 was quite homogeneous as shown in the face distribution of Mo element analysed by EDX in Fig.8.

The preferred orientation of MoS_2 grains in Ni / MoS_2 plating layer was determined by the laminary structure of original MoS_2 and the technological characteristic of brush plating, and was the main reason why it showed the better solid lubrication property than other MoS_2–containing coatings.

As for the Ni–P / MoS_2 composite plating layer, the similar morpholgy of MoS_2 was formed as in Ni / MoS_2 plating layer due to the same technology of brush plating. Fig.9 shows their SEM morphology of original plating layers.

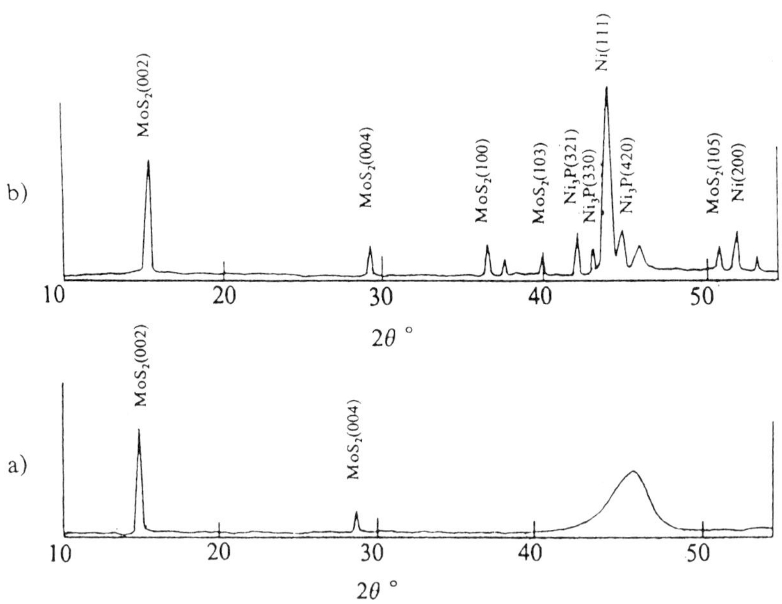

Fig.10 The X-ray diffraction spectra of Ni-P / MoS_2 composite plating layer at a) as plating state and b) after heating at 400℃

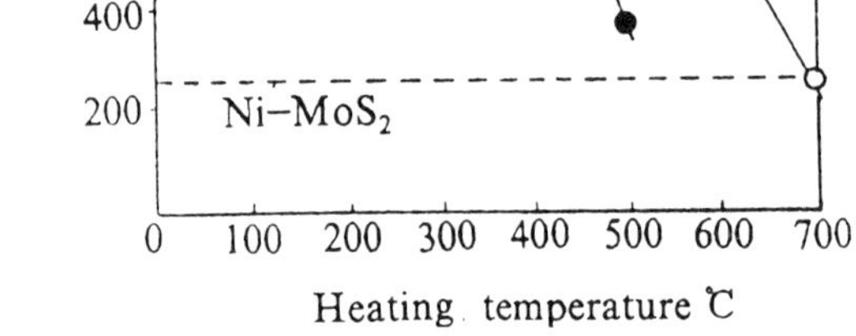

Fig.11 The variation of hardness of plating layers with heating temperature

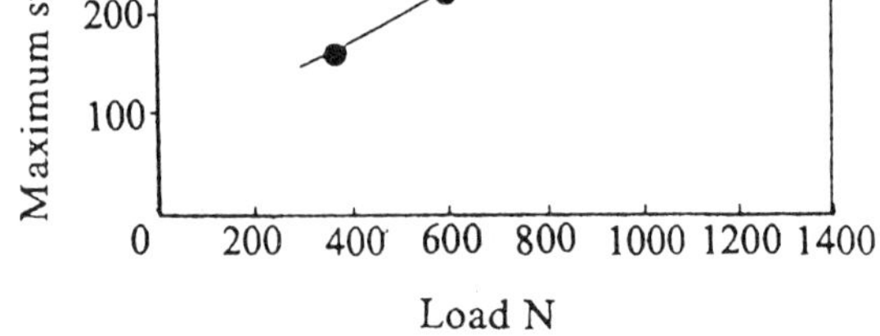

Fig.12 The variation of maximum surface temperature (calculated) of Ni-P / MoS_2 composite plating layer with load

3.4 The structural characteristic of Ni–P matrix in compositie plating layer.

Differing from pure Ni the Ni–P alloy could form an amorphous structure at plating state. The addition of MoS_2 didn't alter its morphology. Fig.10 shows the X–ray diffraction spectra of Ni–P / MoS_2 composite plating layer at different temperature. It indicates that the Ni–P matrix showed a complete amorphous structure at room temperature. Heating at a certain temperature, the amorphous structure could be transformed to a crystalline state with obviously increased hardness. This characteristic was just the starting point of using Ni–P to replace Ni matrix for improving the wear–resistance of composite plating layer.

Fig.11 shows the variation of hardness of Ni–P and Ni–P / MoS_2plating layers with heating temperature. Both of them exhibited the similar crystallization process from amorphous structure to growth of crystalline grains at higher temperature. It can be seen that the Ni–P plating layer showed the highest hardness about 1,000 HV, while the hardness of Ni–P / MoS_2 composite plating layer became decreased due to the addition of MoS_2, reaching about 700HV. maximally. However, it was obviously higher than that of Ni / MoS_2 whose maximum hardness was only 250 HV.

Considering the important influence of friction temperature on tribological behaviour of composite plating layer, and the difficulty of its measurement, this paper calculated the mean friction temperature of contact surface using a mathematic expression derived on a ball–on–disc contanct model[5]:

$$\theta = \frac{4fPVg}{J\pi R^2 K}\sum_{n=1}^{\infty} \frac{\int_0^{\pi} \sin^2\alpha F[\lambda_n(R_o - C\cdot Cos\alpha),0]d\alpha}{F^2(\lambda_n R,1)\lambda_n} J_o(\lambda_n r)$$

where θ is temperature rise $\theta = t - t_c$

t is practical temperature

t_c is ambient temperature

f is friction coefficient

P is load

V is sliding speed

g is gravity acceleration

J is mechanical equivalent of heat

R is radius of disc

R_o is radius of contact ring

K is thermo–conductivity

λ is constant

$J_o(\lambda_n r)$ is Bessel function

C is radius of contact spot

α is angular coordinate of contact spot

The calculated variation of maximum friction temperature of Ni–P / MoS_2 composite plating layer with load at a constant sliding speed of 0.97m / s is shown in Fig.12. The friction temperature was increased almost linearly, it could reach about 500℃ when the load increased to 1362N. It can be known from Fig.11 that its hardness was still higher than that of Ni / MoS_2 in the condition of such high surface temperature. This is the basic reason why Ni–P / MoS_2 composite plating layer possesses much better wear–resistance than Ni / MoS_2 plating layer.

4.CONCLUSIONS

1.The Ni Base and Ni–P Base MoS_2–containing composite brush plating layers all show an excellent friction–reducing property. Their friction coefficient is generally in the range of 0.04~ 0.06 in the lubricated condition.

2. The wear–resistance of Ni–P / MoS_2 plating layer is much higher than that of Ni / MoS_2 plating layer. Its wear life is about 4~ 6 times that of the latter. Therefore, it can be used in the more severe working conditions.

3. The MoS_2 grains in composite brush plating layer show a distinct preferred orientation of its (0001) basal plane parallel to the surface of plating layer. This is the basic reason why the composite brush plating layers show better solid lubrication property than other MoS_2–containing films do.

4. The maximum friction temperature of Ni–P / MoS_2 composite plating layer can reach about 500℃. At this temperature level its hardness is higher than that of Ni / MoS_2 plating layer. So its wear–resistance is better obviously than that of the latter.

REFERENCES

1M.Matsunaga, Y. Tsuya, Handbook of Solid Lubrication (Japanese) 1978

2P. Heilman, D.A.Rigney Wear 72(1981)195

3Liu Jia–Jun, Lu Zhi–Qiang, Cheng Yin–Qian Wear 140(199)135

4Xian Hua Zhang, Jia–Jun Liu, Bao–Liang zhu Lubrication Engineering vol.48, No.1 (1992)37.

5Liu Jia–Jun, Zhang Xian–Hua, Zhu Bao–Liang Tribology (Chinese) Vol.12, No.1 (1992)27.

Thin Films in Tribology / D. Dowson et al. (Editors)
© 1993 Elsevier Science Publishers B.V. All rights reserved.

Effect of Thin Solid Film Thickness on Friction of Metal

K.Hiratsuka

Tokyo Institute of Technology
2-12-1, Ookayama, Meguro-ku
Tokyo 152 JAPAN

5 kinds of Sn films from 0.07 to 110μm in thickness was rubbed against steel ball (SUJ2) in air and in vacuum in scanning electron microscope. The mean value and the variation of coefficient of friction were less in vacuum than in air. This is because the growth of transfer particle was enhanced by the atmosphere. The minimum coefficient of friction was obtained at the film thickness of 1.1μm in vacuum and of 12.7μm in air. The increase in the coefficient of friction was due to the transfer and wear of Sn film.

1. Introduction

Theory of adhesive friction explains the low friction of a soft metal coated on a hard substrate[1]. Bowden and Tabor insisted that, as for the friction of thin solid film, the worn track directly correlates with the contact area, so that, if the worn track is small, the coefficient of friction is low[1]. Moreover, as far as the friction of thin solid film was concerned, they made focus on the adhesive term only; they did not take into consideration of the ploughing or transfer effect on friction.

The author found that the coefficient of friction was increased with increase in the rate of transfer, and pointed out that the transfer of material has an great influence on the friction of metal[2]. This experiment made clear of the importance of the transfer of surface material as a controlling factor of the coefficient of friction. Fig.1 shows the result; the effect of growth rate of transfer particle on the friction of various metals[2]. As can be seen in this figure, the more the transfer particle grows up, the higher the coefficient of friction exhibits. In that experiment, the transfer was inhibited by rubbing the thin film. However,the film thickness was only one kind(0.1mm). In the present study, thin films which have some kinds of film thicknesses were tested. It enabled the study on the effect of film thickness on friction.

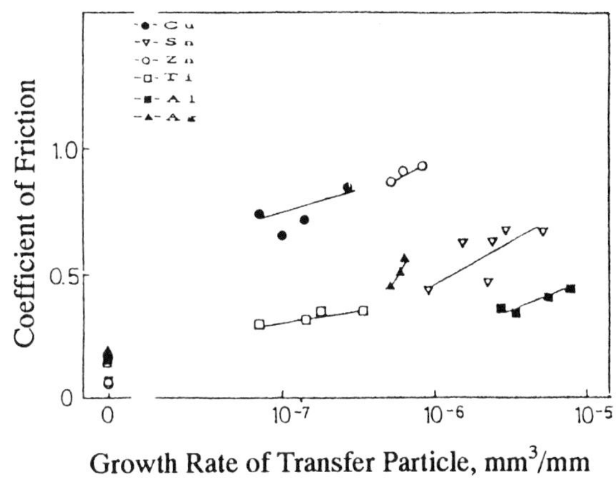

Fig.1 Effect of growth of transfer particle on coefficient of friction[2]

Moreover,the author demonstrated that the atmospheric oxygen changed the coefficient of friction by enhancing the transfer[3]. It was expected that the atmospheric oxygen changed the friction of thin film by changing the transfer phenomenon. From this point of view, wear test was conducted in air and in vacuum.

2. Experimental Procedure

Pin–on–Disk type wear test rig, which was set in the chamber of scanning electron microscope (SEM) was used (Fig.1). Wear test was conducted in vacuum of $4x10^{-3}$Pa and in laboratory air. Experimental conditions throughout the test were; Load=2.9N, Sliding Velocity=0.865mm/s, Sliding Distance=3.14m (Number of disk rotation=50). Sn was chosen as a material of thin film, because it exhibits a growth of transfer particle in a large scale[4]. The disk specimen (SS41 Steel, Hv=180) was polished by #4000 emery paper. After it was degreased and dipped in acid, Sn was electroplated on the surface from $SnSO_4$ solution. The purity of Sn was 99.5%. By changing the plating time, 5 kinds of solid films (0.07, 0.12, 1.1, 12.7, 110μm) were obtained. Film thickness of 110μm were measured by the weight difference between before and after plating. Those of others were identified by X–ray fluorescence coating thickness meter. These disk specimens were rubbed against steel ball (SUJ2, diameter=1mm). The mean friction diameter was 20mm, so that the sliding distance was 62.8mm per revolution.

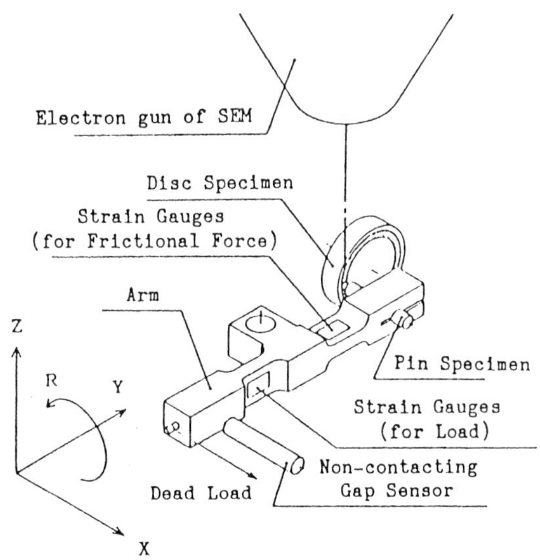

Fig.2 Schematic of test rig in SEM

3.Results

The mean, maximum and minimum coefficient of friction during the first revolution (sliding distance=62.8mm/s) is shown in Fig.3. In general, both mean value and its variation was less in vacuum than in air. The minimum coefficient of friction was obtained at the film thickness of 1.1μm in vacuum and of 12.7μm in air.

The variation of coefficient of friction as a function of sliding distance in the rubbing between SUJ2 and Sn of 1.1μm in thickness is shown in Fig.4. As shown in vacuum (a), the low and a little variation in the coefficient of friction continued until the end of rubbing. (However, the mean value changed to a little higher (0.15) at the end.)

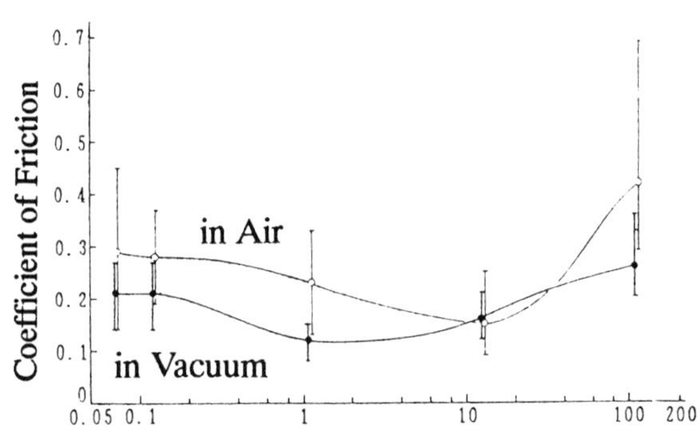

Fig.3 Effect of Sn film thickness on the friction between bearing steel (SUJ2) and Sn film in air and in vacuum

Fig.5 shows Sn and Fe map in the wear track of disk surface produced in vacuum, which was obtained by X-ray micro analyzer (XMA). Detected Fe in the wear track seems to mean that Sn was worn out. However Sn is present partly in the grooves which were abraded during the preparation of substrate steel surface. It is difficult for XMA to detect the elements less than 1μm in depth. Probably very thin Sn layer still remained on the wear track, because the coefficient of friction was still low. XMA study showed that Sn was completely present on the wear track after the rubbing of Sn film of 12.7 and 110μm in thickness. On the contrary, in the wear track of Sn films of 0.07 and 0.12μm, no Sn was detected.

As can be seen in Fig.4(b), in air, the coefficient of friction increased gradually and

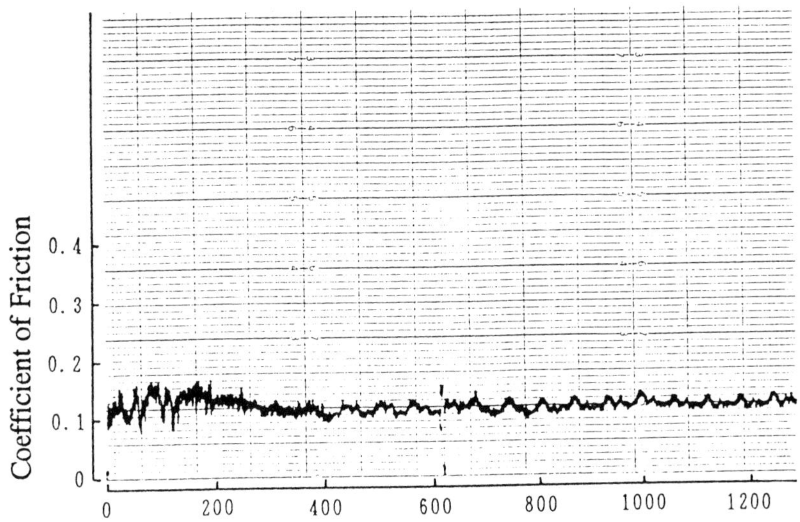

(a) in Vacuum

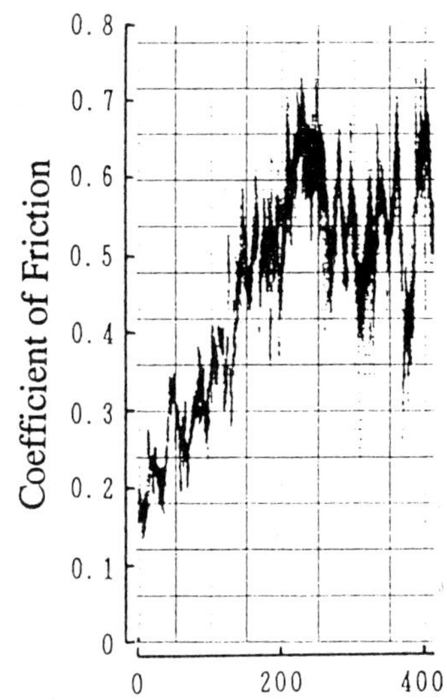

(b) in Air

Fig.4 Coefficient of friction as a function of sliding distance in vacuum and in air (Film thickness=1.1μm)

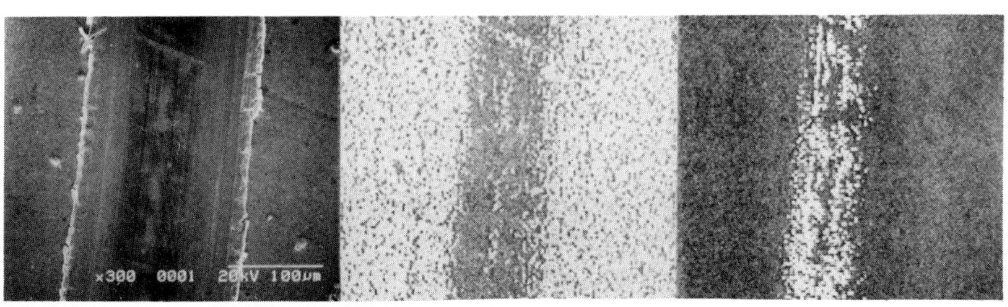

SEM Image **Sn** **Fe**

Fig.5 Sn and Fe map in worn track of disk in vacuum

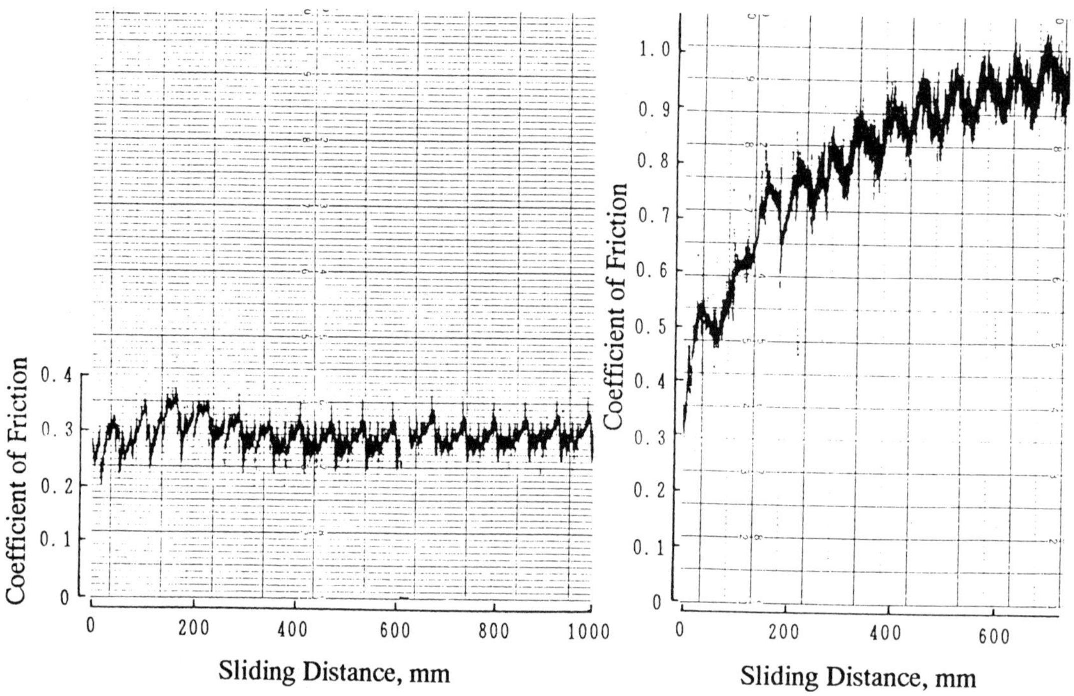

(a) in Vacuum (b) in Air

Fig.6 Coefficient of friction as a function of sliding distance in vacuum and in air (Film thickness=110μm)

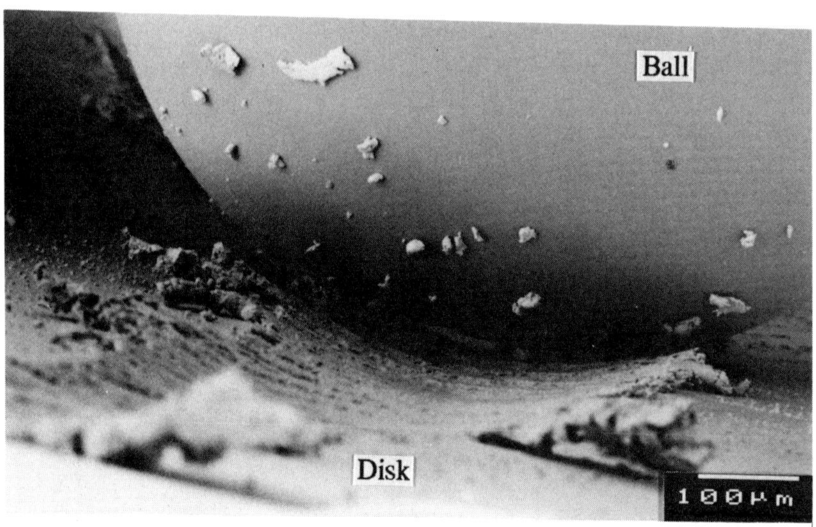

Fig.7 Contact point between ball and disk (in air, sliding distance=3.14m, Sn film thickness =110μm)

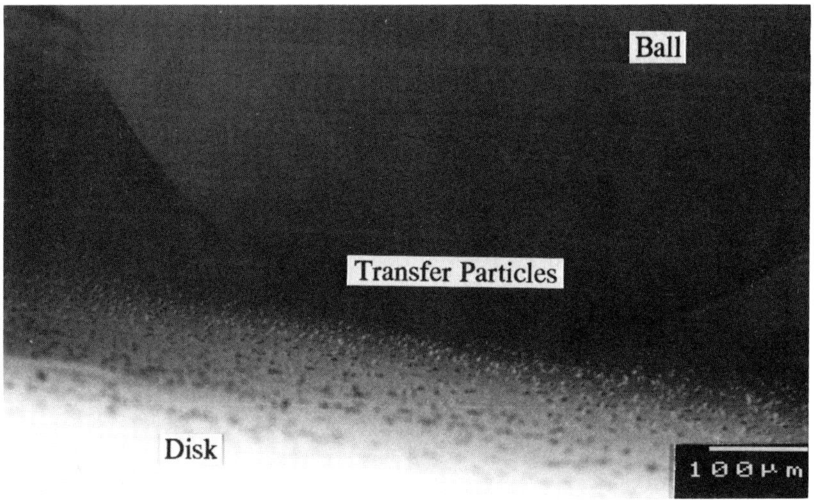

(a) Low magnification

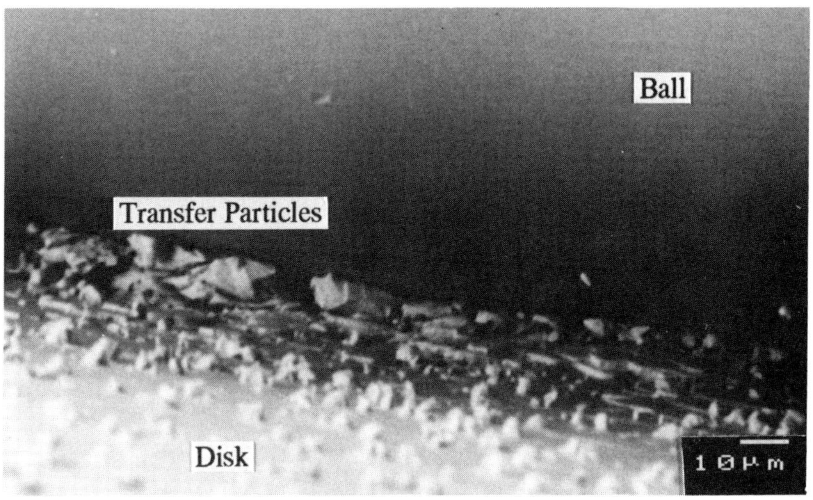

(b) High magnification

Fig.8 Contact point between ball and disk (in air, sliding distance=37mm, Sn film thickness =110μm)

reached a steady value after about 3 revolutions of the disk. Even in the first revolution (rubbing), the coefficient of friction increased. This increase is associated with the wear of Sn. As shown in Fig.6(a), in case that Sn film of 110μm thick was rubbed in vacuum, the coefficient of friction varied a little. Whereas in air, it increased(b). The reason for this increase was transfer and wear. These results were also observed in the case of 12.7μm in thickness. When SUJ2 steel was rubbed against Sn of 110μm thick, the increased coefficient of friction reached the steady value of 1.1 after the sliding distance of 1.4m. In this case, Sn film

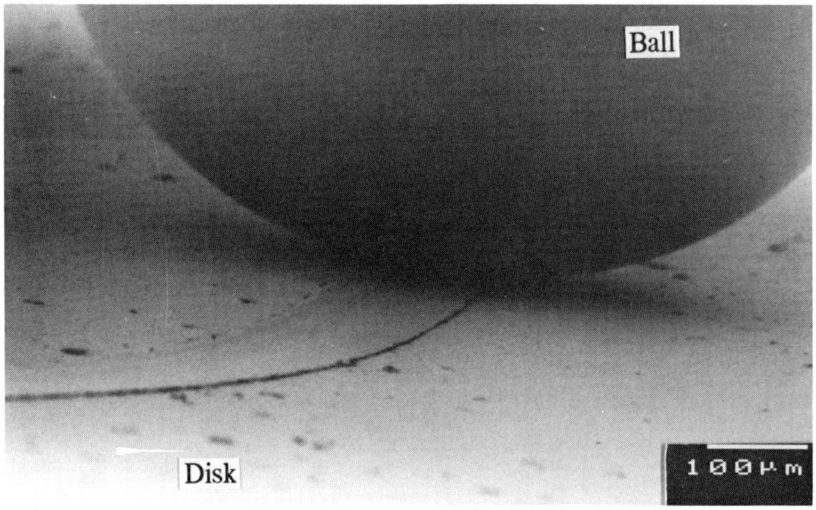

Fig.9 Contact point between ball and disk (in air, sliding distance=3.14m, Sn film thickness =1.1μm)

Fig.10 Contact point between ball and disk (in air, sliding distance=3.14m, Sn film thickness =12.7μm)

was too thick to be worn out, which was confirmed by XMA study.

Fig.7 shows the contact point between ball and disk of 110μm thick at the end of the rubbing in air. Sn disk surface was rough, and a lot of wear particles were generated. On the contrary in vacuum, wear particles could hardly observed, but surface Sn was protruded outside of the wear track.

Fig.8(a) and (b) show the contact point at

the 0.6th revolution (sliding distance=37mm). From the high magnification (b), it is understood that the transfer particle existed even in the first run.

Figs.9 and 10 show the contact point at the end of experiment in which the ball was rubbed in vacuum against Sn of 1.1 and 12.7μm thick, respectively. In vacuum, Sn was protruded to the both sides of wear track. Sn of 110μm thick showed the same feature. The widths of the wear tracks of Sn of 1.1, 12.7, 110μm in film thicknesses were 160, 312, 328μm, respectively., The mean coefficients of friction of those widths were 0.15, 0.21, 0.31, respectively.

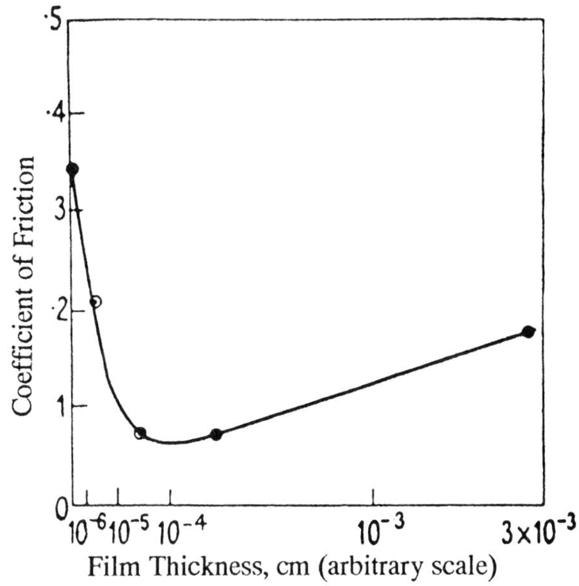

Fig.11 Effect of film thickness of indium coated on steel on coefficient of friction rubbed against steel ball[1]

4. Discussion

Bowden and Tabor found out that the minimum coefficient of friction exists in the friction between steel ball and indium film of about 0.1μm thick deposited on steel, as shown in Fig.10. The reason for the increase in coefficient of friction in the cases of film thicknesses less than 0.1μm was the nonuniformity of indium on steel; indium did not deposit all the surface of steel. In the present study, the time of electroplating of 0.07 and 0.12μm in thicknesses was 5 and 10 seconds, respectively. Thus it is considered that the Sn atoms were not uniformly covered on substrate steel. That is why the higher coefficient of friction was obtained in these films.

The effect of oxygen is also interesting. In vacuum, the coefficient of friction has almost constant value from the start if the film is not broken. On the contrary, conducted in air, the coefficient of friction increased. As shown in Fig.8, the transfer particle is generated in air. Those transfer particle is also observed in the rubbing of the films of 1.1 and 12.7μm thick. These results indicate that the transfer particle grows at the interface in air, which results in the increase in friction.

As mentioned at the end of the section of results, the width of wear track and the coefficient of friction in vacuum correlated; the coefficient of friction increased with increasing the width of wear track. However, there was a fact that the coefficient of friction was almost constant until the end of rubbing in vacuum, although the width of wear track increased during the rubbing. This result means that the coefficient of friction does not depend on the width of wear track. On the other hand, however, in air, as the width of wear track increased with sliding distance, the coefficient of friction also increased. The difference between vacuum and air was transfer and wear. It can be deduced that transfer and wear is one factor controlling the friction in the rubbing of thin film.

5. Conclusions

From the wear test in air and in vacuum between bearing steel (SUJ2) and 5 kinds of Sn films electroplated on steel, the following conclusions are drawn.

There exists a film thickness which exhibits the minimum coefficient of friction in both air and vacuum.

The coefficient of friction increased with the growth of transfer particle in air.

Acknowledgment

The author is grateful to Mr.T.Kawashima in Meltex Co.Ltd. for the preparation of Sn specimens and XMA study of worn tracks.

References

1. F.P.Bowden and D.Tabor, The Friction and Lubrication of Solids, Oxford Univ. Press (1954).
2. K.Hiratsuka, M.Kogure, T.Nakano and T.Sasada, Japanese J. Tribologist, 37, 10 (1992) 870.
3. K.Hiratsuka, Proc. 2nd Symp. Ferrography (1988) 1.
4. K.Hiratsuka, Y.Ando and A.Sugahara, J.JSLE Intern. Ed. 10 (1989) 33.

Thin Films in Tribology / D. Dowson et al. (Editors)
© 1993 Elsevier Science Publishers B.V. All rights reserved.

Synthesis of Tungsten Disulfide on Tungsten-tip Directly inside FIM

N.Ohmae, N.Mori, M.Tagawa and M.Umeno

Department of Precision Engineering,Faculty of Engineering, Osaka University, 2-1 Yamadaoka, Suita, Osaka,565, Japan

In-situ observation of the formation of tungsten disulfide (WS_2) was carried out by depositing sulfur on a W-tip at room temperature and by heating at varied temperatures. Atomic arrangements of synthesized WS_2 were imaged by FIM. The formation of WS_2 predominated at temperatures around 600~700K. Epitaxial growth of the (0001)WS_2//(001)W was found for WS_2 film heat-treated at 670K, but the difference in thermal expansion coefficients between tungsten and its disulfide caused considerable misfit defects at the interface. Atomic structures of bulk WS_2 and WS_2 powder embedded onto the W-tip also were imaged. Aligned short bright lines observed for layered solid lubricants can be explained by an extremely small atomic step existing on the apex even after field evaporation procedures.

1. INTRODUCTION

Tungsten disulfide (WS_2) is known as a lamellar solid lubricant which gives good stability at high temperatures and satisfactory service life in vacuum. Unlike MoS_2, WS_2 is rarely found as natural minerals, so that synthetic production of WS_2 has been made industrially. One of the synthesizing methods of WS_2 is a thermal decomposition of WO_3 or $(NH_4)_2WO_4$, i.e., a completely chemical synthesis. The other method is the direct synthesis by heating a mixture of tungsten powder and sulfur. However, an optimum heating temperature for manufacturing layered WS_2 has not yet been fully understood. Chemically produced WS_2 gives smooth appearance, while directly synthesized one shows slightly rough appearance.

The high voltage field ion microscope (HV-FIM)[1] successfully imaged lamellar solid lubricants as MoS_2, natural graphite and WS_2, and the aligned short bright lines indicated agglomerated basal planes which protruded above the apex of the tip on a sub-Å order[2]. When the field evaporation procedure produced an atomically smooth surface, net plane rings as in the case of metal-tip resulted. Atomic structure of lamellar materials thus is very complicated. WS_2 seems to be one of most proper materials for HV-FIM studies of the synthesis of solid lubricants, since atomic structure of W-tip surface has extensively been studied by many FIM works

[3-5]. The aim of this study was to analyze atomic structures of WS_2 synthesized by vacuum-depositing sulfur onto W-tips directly inside FIM.

2. EXPERIMENTAL PROCEDURE

Figure 1 schematically shows the tip-screen configuration of the HV-FIM used in this experiments. An auxiliary electrode is placed between the tip and the screen, and a negative high voltage up to 60kV is applicable to this electrode. The tip naturally is kept at ground potential, and at temperatures below 100K. The ultimate vacuum pressure of 10^{-7}Pa was obtained with use of a turbomolecular pump, a Ti-getter pump and a liquid-N_2 cooled cold finger.

Figure 2 diagrammatically presents the arrangement of W-tip, S-source and other major components of the HV-FIM. The tip was installed in a tip holder and attached to the bottom of the liquid N_2-cooled cold finger. The tip temperature was varied by resistively heating W-loop onto which the FIM tip was spot welded, and the temperature was measured by C-A thermocouple. The sample introducing rod was used as an electric current feedthrough for heating the tip. Sulfur was evaporated from a small hole ($2mm^{\phi}$) of the crucible which was resistively heated by a W-wire. Evaporation of water vapor is one of the serious problems, so that the S-source was pre-heated at the location away from the tip with use of a rotary vacuum feedthrough. The evaporation of S was monitored by a Q-mass spectrometer. The tip can be replaced through an

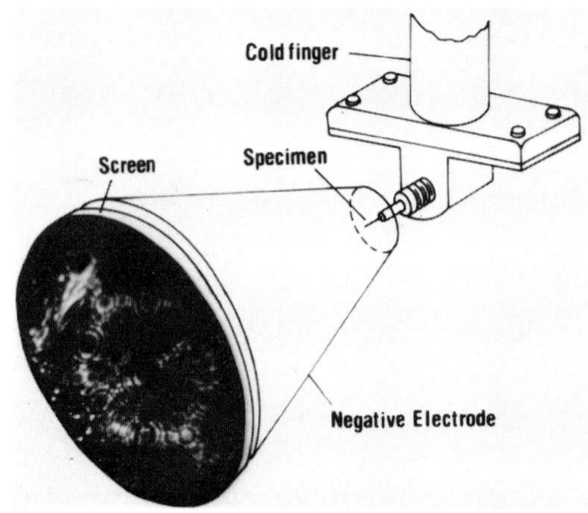

Figure 1. Schematic drawing of a tip-screen configuration of HV-FIM.

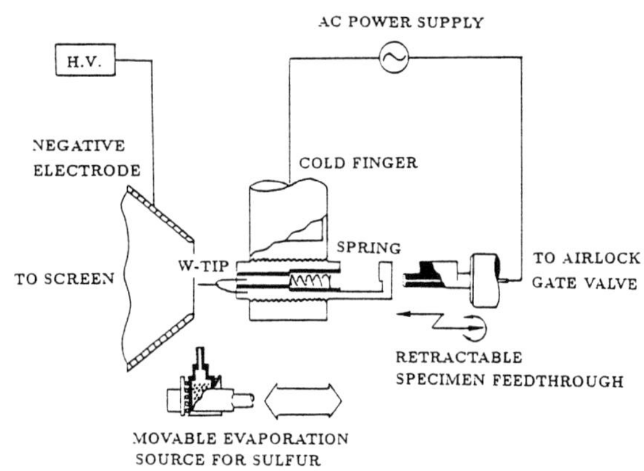

Figure 2. Experimental arrangements of W-tip and S-source.

airlock system without breaking the whole vacuum.

One end of the W wire of 60μm in diameter was electropolished using a solution of 3% KOH into a shape of micro-pointed tip whose radius of curvature was a few tens of a nanometer[6]. Before depositing S, an atomic structure of the W-

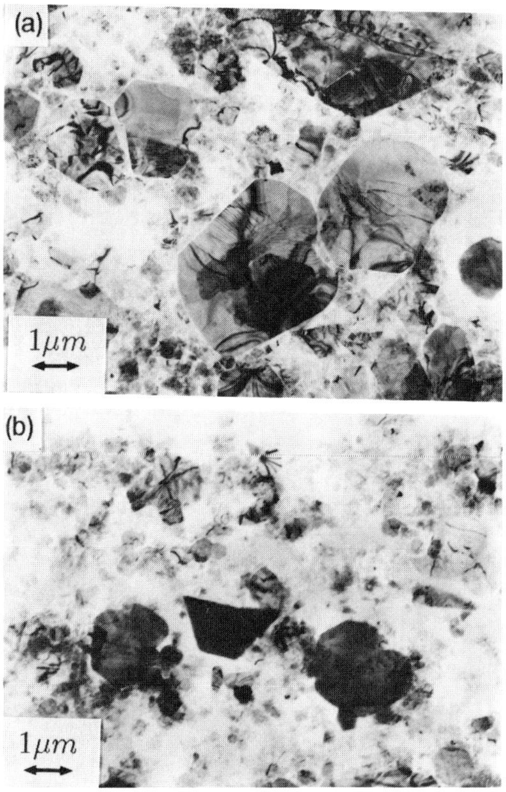

Figure 3. HV-TEM photographs of (a) chemically processed WS_2 and (b) directly synthesized WS_2.

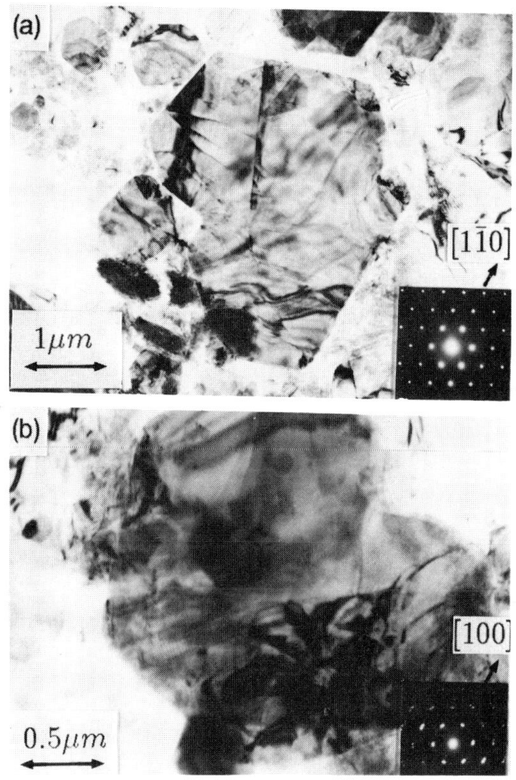

Figure 4. WS_2 powders observed at high magnifications by HV-TEM; (a) chemically processed WS_2, and (b) directly synthesized WS_2.

tip was photographed. Vacuum-deposition of S was carried out at room temperature, the exposure being a few hundreds Langmuires. The formation of WS_2 was studied by raising temperatures up to 1050K. The atomic structures of WS_2 as well as W were imaged with He gas.

3. RESULTS AND DISCUSSION

3.1.HV-TEM of WS_2

Figure 3 presents HV-TEM photographs of chemically processed WS_2 (a) and directly synthesized WS_2 (b). Polygonal thin flakes with the size of 10nm$\sim$2μm can be seen in these photographs. Notable difference in the size of WS_2 flakes between (a) and (b) was not found, though the directly synthesized WS_2 looks rough in appearance, as described earlier. Figure 4 shows the WS_2 flakes observed at higher magnifications. Chemically processed WS_2 in (a) shows sharp sides of the hexagonal shaped flake, and the selected area electron diffraction pattern signifies that the observed flake orients in the basal (001) plane. Figure 4(b) shows the HV-TEM photograph of directly synthesized WS_2. Comparing with Fig.4(a), hexagonal shape of a flake is not remarkable, and the stack of several layers with Moiré fringes is observed.

As is well known, WS_2 has a hexagonal crystal structure, and the stacking sequence is S-W-S-S-W-S. There exist relatively strong covalent bonds in each group of S-W-S, while weak van der Waals forces between S atoms and adjacent planes. Lattice constants, a_0 and c_0 are 0.332nm and 1.236nm, respectively. It is possible to calculate a_0 for WS_2 shown in Fig.4 by measuring the lattice spacings, and the values of a_0 were calculated at 0.32~0.33nm for both (a) and (b). Therefore it is difficult from HV-TEM to find out notable differences in the structure of WS_2 depending on the manufacturing processes, except the sharpness of hexagonal flakes.

3.2.Reaction of W/S at 78K and room temperature

The deposition of sulfur at 78K resulted only in the enhanced contrast of FIM image near the (112) planes. The field evaporation re-produced bulk W structure quickly. Therefore W and S do not react at 78K, but field adsorption of S onto W occurs.

Figure 5(a) shows the FIM photograph of W-tip before depositing S at room temperature. As for FIM observations of W-tip surface, Fig.5(a) is a rare example that the grain boundary is incorporated at the location indicated by an arrow, and it appears as if two (010) planes exist at the righthand side of the figure. Fig.5(b) indicates that relatively large spots begin to form imperfect net plane rings on the W (001), (112) and (010) planes. This may indicate that deposited S adsorbs on these crystallographic planes. However, numerous dark bands co-exist, suggesting an incorporation of many lattice imperfections. In Fig.5(c), bulk W atoms begin to appear, and atomic arrangements of W is clear in (d). The FIM photographs after depositing S contains considerable imperfection. This may be due either to the difference in thermal expansion coefficients of W and S or to defect generation during the field desorption processes of S.

Vacuum deposition of S at room temperature and at 78K thus was unsuccessful in producing WS_2 and the formation of WS_2 was studied by raising temperatures of W-tip on which S was deposited at room temperature.

3.3.W/S reaction at 550K

The S-coated tip heated at 550K showed a localized formation of S clusters as shown in Figure 6(b). By removing several layers using field evaporation, isolated net plane rings, though they appeared imperfect concentric circles, co-existed with bulk W, see Fig.6(c). The continuation of field evaporation up to 8.5kV gave the bulk W structure. Therefore the WS_2 formation at 550K is not perfectly stable.

3.4.W/S reaction at 670K

When the S-coated tip was heated at 670K, the formation of WS_2 was observed. Figure 7(a) presents the FIM photograph

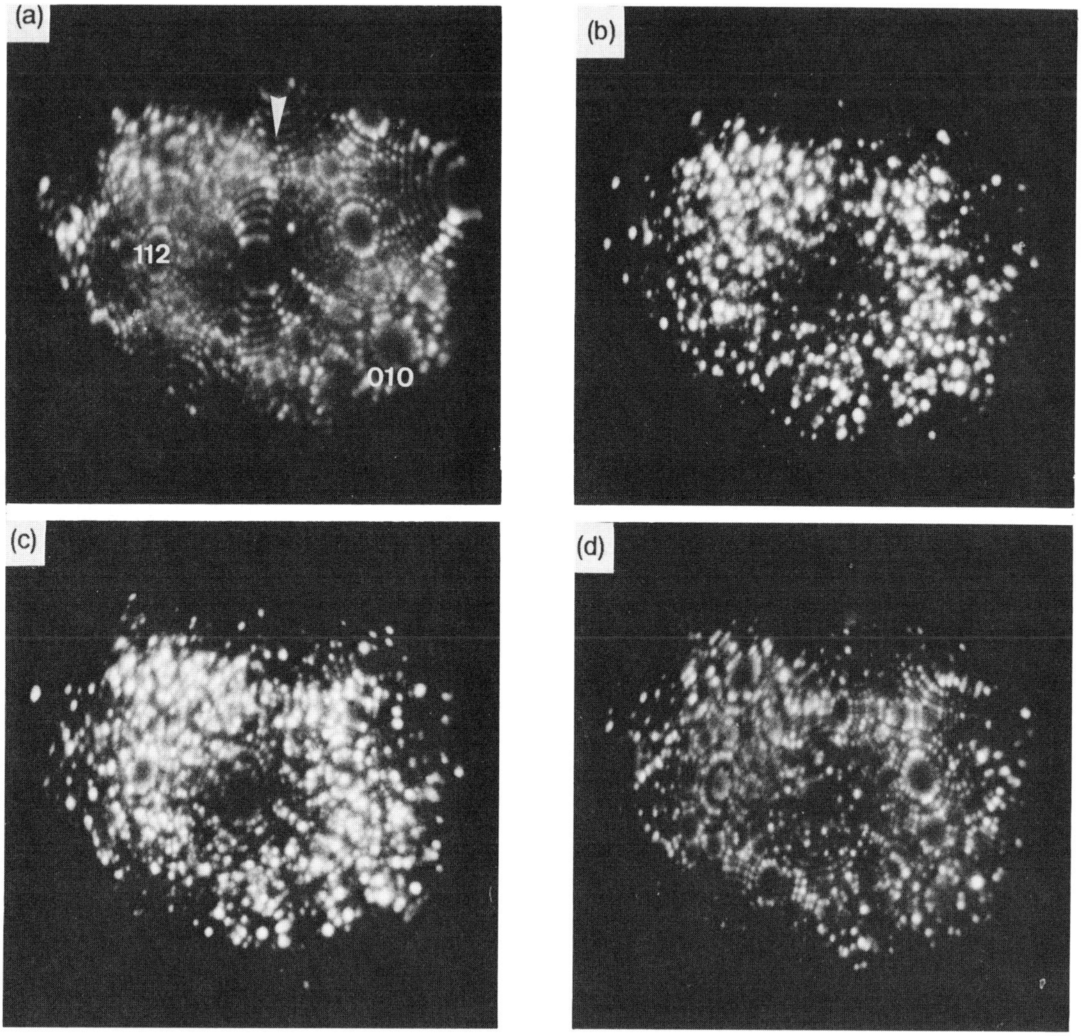

Figure 5. HV-FIM photographs of (a) W-tip prior to S-deposition, (b) after S-deposition at room temperature, and (c) and (d) after field evaporation; (a),(b),(c) and (d) were imaged at 7.9kV, 7.5kV, 7.6kV and 7.9kV with 0.36-Pa He.

of bulk W. The FIM photograph imediately after depositing S at room temperature is shown in Fig.7(b) which resembles in appearance to Fig.5(b). When the tip was heated at 670K, a completely different FIM image from Fig.7(c) resulted. The large net plane ring at the upper right-hand side in Fig.7(c) was identified as the WS_2 (0001) plane by comparing the computer-simulated pattern. The WS_2 (0001) showed the crystal growth on the W (001).

By calculating the misfit value of the (0001)WS_2//(001)W, the possibility of this epitaxial growth was well identified. The structure of bulk W begins to appear in (d), though localized WS_2 and S are still observed. The FIM photograph in (e) shows the structure of bluk W. However, the grain boundary which runs from the (001) plane to the (010) plane exists.

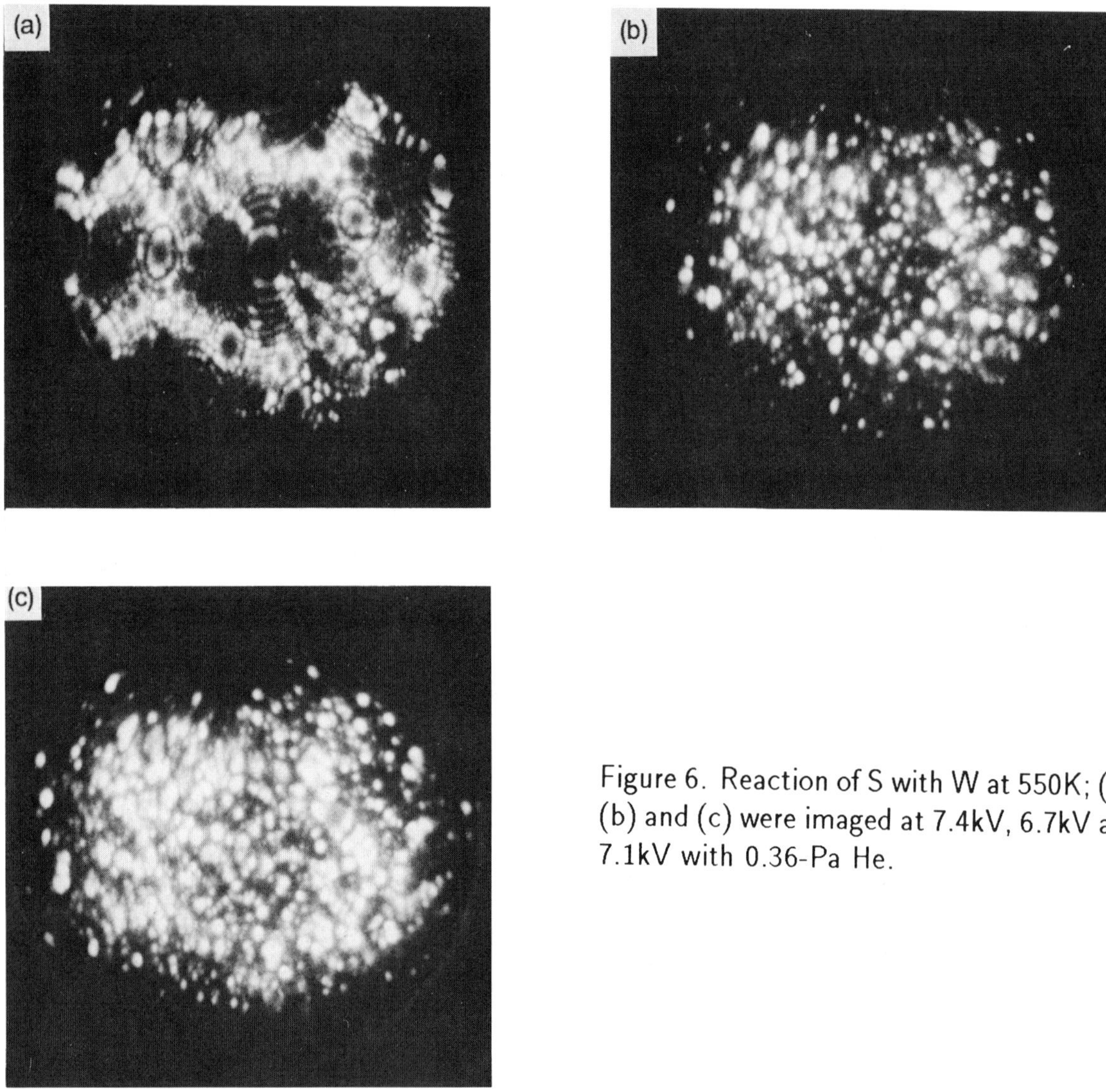

Figure 6. Reaction of S with W at 550K; (a), (b) and (c) were imaged at 7.4kV, 6.7kV and 7.1kV with 0.36-Pa He.

The growth of WS_2 on W thus predominated at a heating temperature of 670K, but the misfit between the two and the difference in thermal expansion coefficients caused grain boundary and many other defects imaged as dark bands.

3.5. W/S reaction at 900K and 1050K

The S-coated tip heated at 900K is shown in Figure 8 Removal of surface layer by field evaporation quickly revealed the co-existence of S clusters and W atoms. Dark bands corresponding to lattice defect also predominated. The vacuum pressure of 10^{-4}Pa at which the tip was heated was equivalent to the vapor pressure of WS_2 at 740K, i.e., $\sim 10^{-4}$Pa. Therefore, at 900K the WS_2 formation is not stable.

Owing to the same reason, the forma-

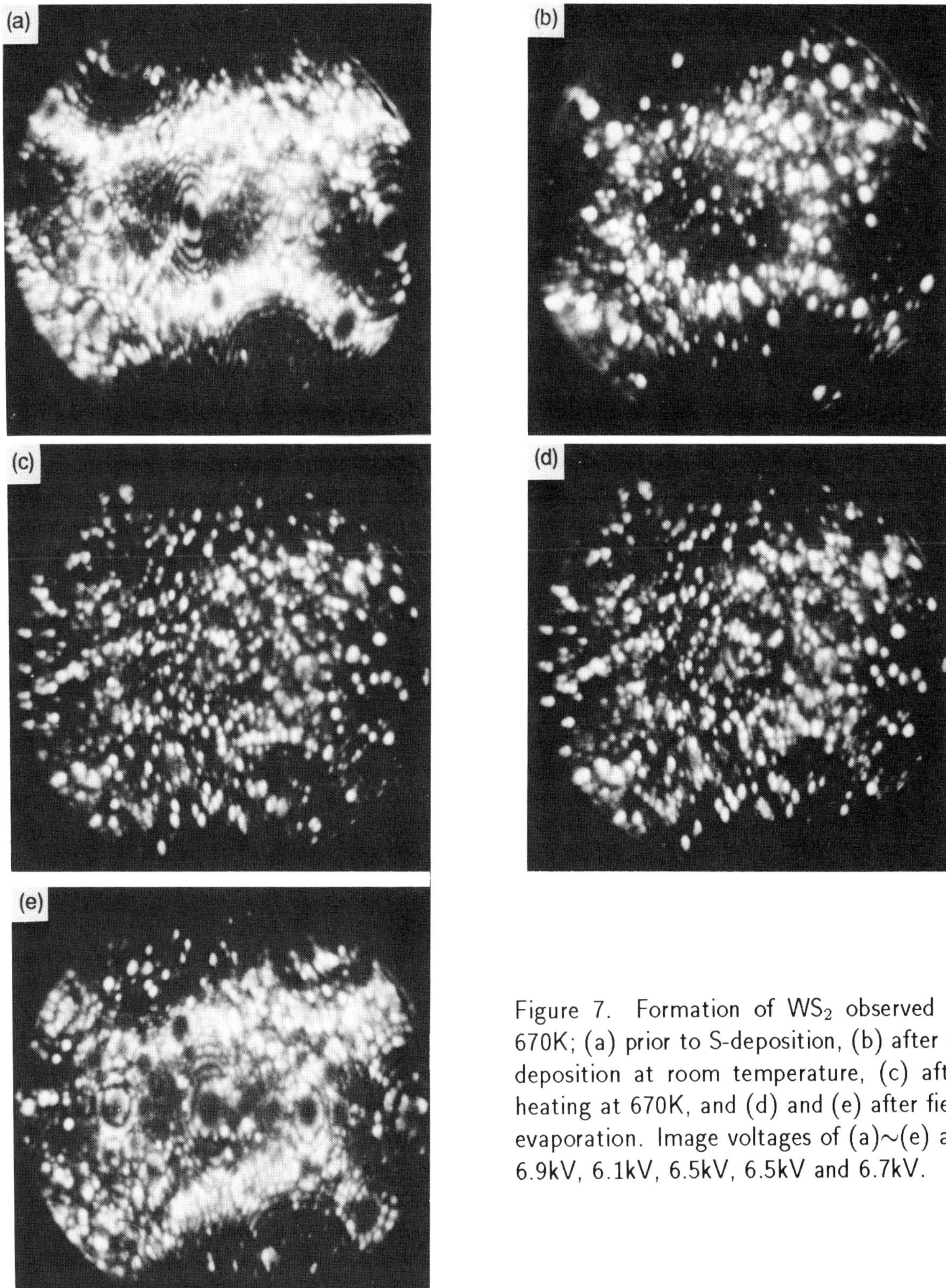

Figure 7. Formation of WS_2 observed at 670K; (a) prior to S-deposition, (b) after S-deposition at room temperature, (c) after heating at 670K, and (d) and (e) after field evaporation. Image voltages of (a)~(e) are 6.9kV, 6.1kV, 6.5kV, 6.5kV and 6.7kV.

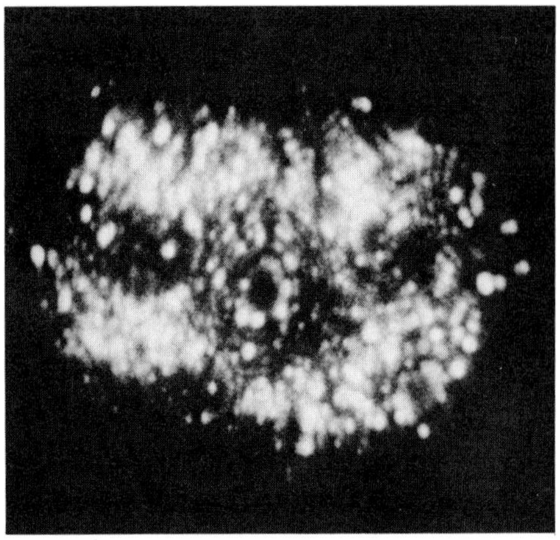

Figure 8. HV-FIM photograph of S-coated W-tip heated at 900K, imaged at 7.1kV with 0.36-Pa He after field evaporation at 7.4kV.

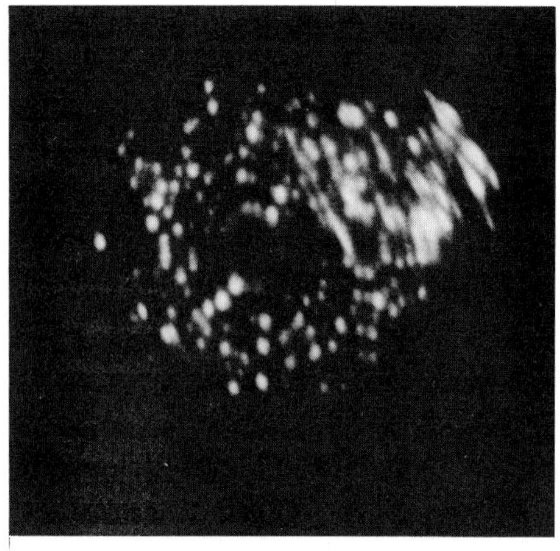

Figure 9. WS_2 powder embedded onto W-tip heated at 670K; imaged at 5.1kV with 0.34-Pa He.

tion of WS_2 was not recognized for the S-coated tip heated at 1050K. Clustered S atoms and bulk W atoms co-existed, the appearance of which resembled to Fig.8.

3.6.HV-FIM of WS_2 powder

Chemically processed WS_2 powder was embedded onto a W-tip and heated at 670K for 1 minute. Figure 9 indicates that aligned bright lines are formed at the righthand side of the W (001) plane. The alignment of bright lines is characteristic to lamellar material as WS_2, when viewed from the a-axis.

The tip shown in Figure 9 was heated again at 670K for 10 minutes. Extraordinary arrangements of large bright spots are observed as indicated by the arrows in Figure 10(a). Enhanced diffusion of S may cause partially the formation of polygonal ledge sites, so that this FIM photograph images WS_2 nearly from the c-axis. Further field evaporation resulted in the linear alignment of bright spots on the W (121) plane. The formation of short bright lines predominates at the upper righthand side of Fig.10(b). Fig.10(c) shows the structure of bulk W with numerous streaks where WS_2 powder adhered. Disturbances in the arrangement of of W-atoms and sub-boundaries are observable.

3.7.HV-FIM of bulk WS_2

Figures 11(a) and (b) show the HV-FIM photographs of bulk WS_2. Fig.11(a) is characterized by the aligned short bright lines in one direction. Unimaged dark area corresponds to intruded region at the apex of this tip. Fig.11(b) shows bright spots and bright lines when the tip surface was field evaporated up to approximately 50 kV. Because of the increased radius of curvature of this tip, large area of bulk WS_2 is observable. Localized net plane rings,

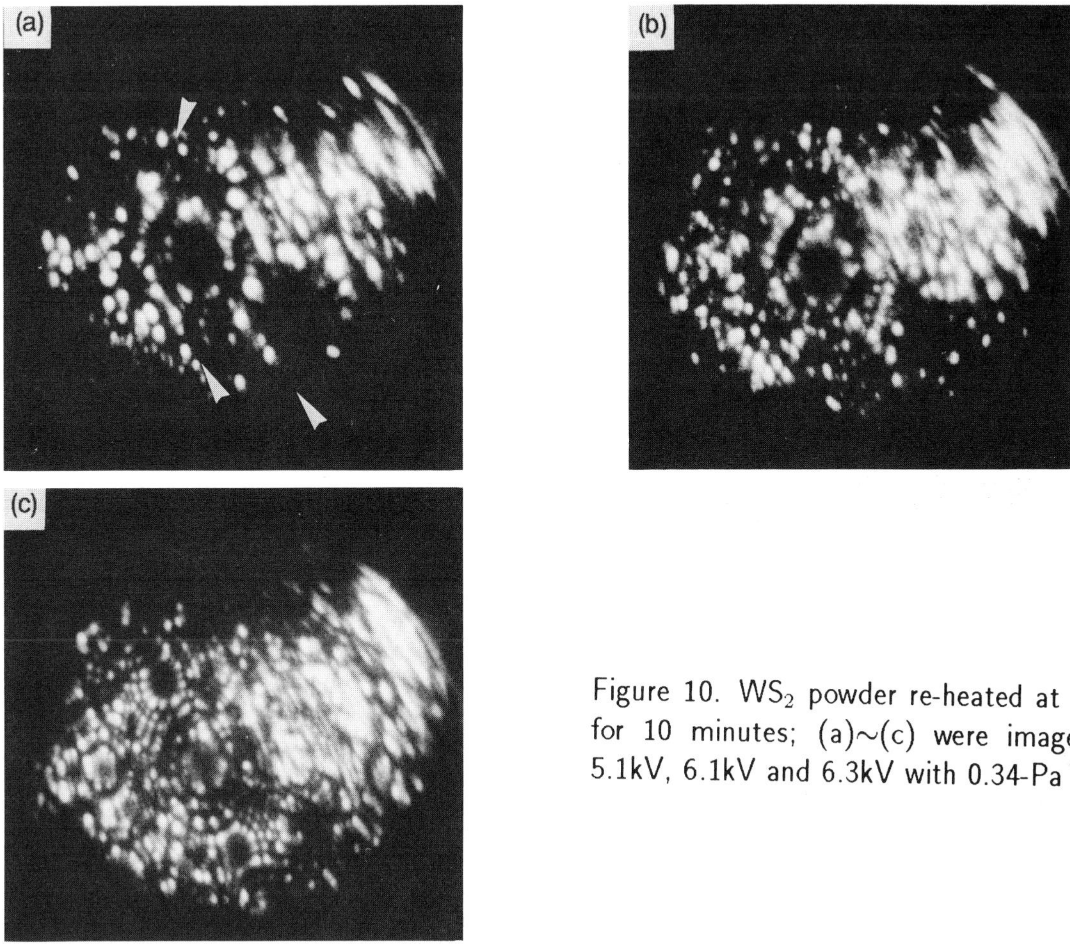

Figure 10. WS_2 powder re-heated at 670K for 10 minutes; (a)~(c) were imaged at 5.1kV, 6.1kV and 6.3kV with 0.34-Pa He.

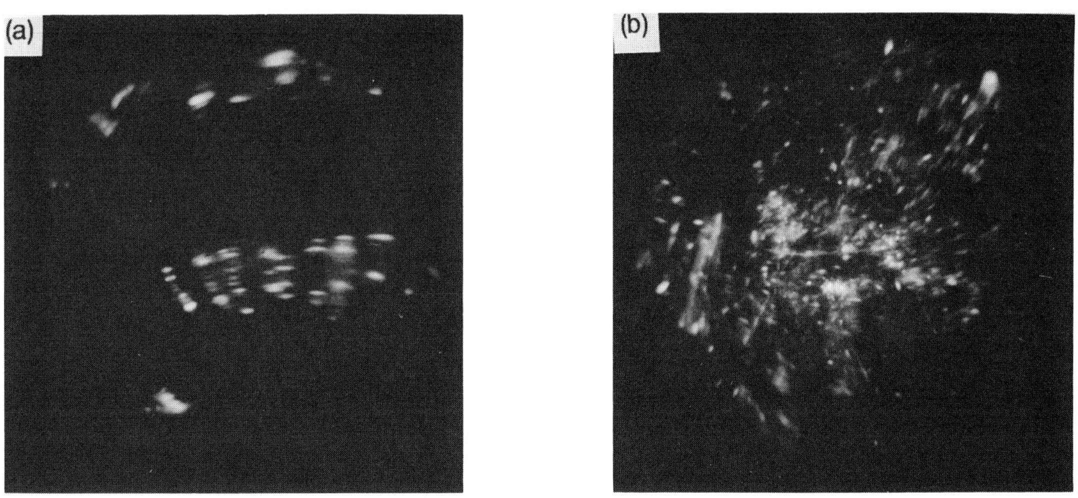

Figure 11. Aligned bright lines (a) and spots and lines (b) observed for bulk WS_2; (a) and (b) were imaged with 0.01-Pa H_2 at 36kV and 46kV.

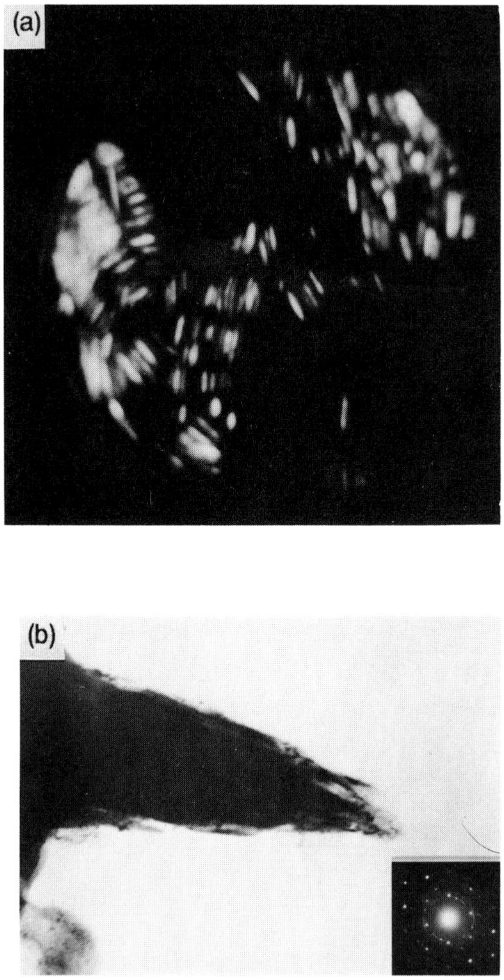

Figure 12. (a) HV-FIM of bulk WS_2, and (b) HV-TEM of the tip shown in (a); (a) was imaged with 0.05-Pa H_2 at 23.5kV.

aligned bright lines and streaks co-exist.

Figure 12 indicates the HV-FIM photograph of WS_2 (a) and the HV-TEM photograph of the tip (b). Aligned bright lines are imaged along several directions. Since selected area electron diffraction pattern shows spots and halo-rings, WS_2 in this figure is polycrystalline. Thus the bright lines align in the several directions depending on each agglomerated basal planes.

and by heating inside FIM at varied temperatures. A stable formation of WS_2 was found at 600~700K. The epitaxial growth of the $(0001)WS_2//(001)W$ was found. The HV-FIM observation of WS_2 powder embedded onto W-tip showed the mixture of aligned bright lines and polygonal ledge, suggesting that the powders orientated both in the a- and c-axes.

Bulk WS_2 showed only the aligned bright lines, corresponding to the polycrystalline nature of WS_2.

ACKNOWLEDGEMENTS

We are grateful to Mr. Shoichi Ohkage of Nihon Junkatsuzai Co.Ltd., for providing us with WS_2 powders.

REFERENCES

[1] M. Tagawa, S. Koike, N. Inoue, N. Ohmae and M. Umeno, J. Phys., C6-48 (1957) 583.

[2] N. Ohmae, M. Tagawa, M. Umeno and S. Koike, Proc. Jpn. Int'l. Tribol. Conf., Nagoya (1990) 1827.

[3] E. W. Müller and T. T. Tsong, Field Ion Microscopy-Principles and Aplocation, Elsevier, N.Y. (1969).

[4] R. Gomer, Field Emission and Field Ionization, Harvard Univ. Press (1961).

[5] R. Wagner, Field Ion Microscopy in Material Science, Springer-Verlag (1962).

[6] N. Ohmae, A. Nakamura, S. Koike and M. Umeno, J. Vac. Sci. Technol., A4 (1987) 1367.

SESSION V
ANALYSIS AND DESIGN OF HIGHLY STRESSED COMPONENTS

Chairman : Dr R C Coy

Thin Films in Tribology / D. Dowson et al. (Editors)
© 1993 Elsevier Science Publishers B.V. All rights reserved.

Micro-Wear of Tapered Roller Bearings Lubricated by Oil With and Without Foreign Particles

S.Noguchi, T.Sawamoto and S.Aihara

Tribology Research and Development, NSK Ltd.,
1-5-50 Kugenuma-Shinmei, Fujisawa, Kanagawa,
251 JAPAN

Tapered roller bearings were rotated with clean oil and with contaminated oil which contained a tiny amount of small hard particles. To evaluate minute wear at the contact, the change of surface profile at the same location was measured with a re-location technique. With clean oil, a running-in phenomenon was observed but with contaminated oil, the volume of wear continued to increase during entire test. Wear volume showed good correlation with P V value considering initial Λ.

1. INTRODUCTION

"Running-in" is an essential process to enhance performance properties, particularly with sliding bearings. This topic was selected as a symposium theme at one of the previous Leeds-Lyon symposia(1).

When running torque is measured with tapered roller bearings, we usually find that running torque decreases with time at the initial stage. This torque reduction is a result of the running-in effect at the roller end and inner-ring rib contact where contact is roll/slide which is shown in Fig.1. The main cause of running-in is the initial wear of surfaces in contact.

However, the amount of wear is very small in the case of rolling bearings and the studies dedicated to running-in process of rolling bearings are limited (2-5).

To evaluate the minute wear at the roller-end and rib contact, change of surface profile at the same location was measured by re-location technique after certain period of rotation. The running-in effect depends on the EHL film thickness/surface roughness ratios.

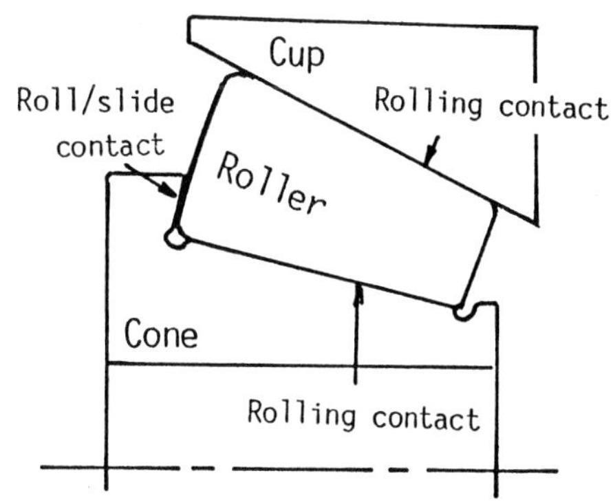

Fig.1 Tapered roller bearing

Running-in did not occur when oil contained contaminants, even if only a very tiny amount. And the sizes of wear particles were quite different.

2. SURFACE PROFILE MEASUREMENT

To examine the changes of surface profile after repeated rotation, it was necessary to measure the profile at the same location. A stylus type measuring apparatus was built to realize such requirements.

The detail of apparatus was shown elsewhere (6). Surface profile was measured while driving the X-stage after

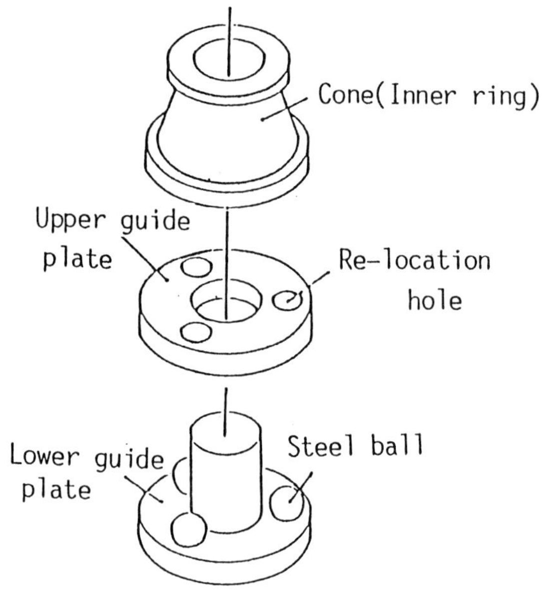

Fig.2 Re-locating plates

Table 1 Test condition for clean oil

Test bearing	HR30307C (Φ35mmxΦ80mmx22.75mm)
Axial load	4.9, 9.8, 13.72 kN
Rotatonal speed	Ajusted to obtain fixed initial Λ (below)
Initial Λ (roller end/rib)	0.5, 1.0, 1.5
Lubricating oil	Turbine oil (ISO VG32) 200 ml/min circulation
Oil temperature	30 + 3°C
Total revolutions	100,000 rev

adjusting the test specimen at the fixed position on the goniometer stage. The location detection was achieved by using the clock pulses of an A/D transducer. The signal was sent to a microcomputer through the A/D transducer and an amplifier.

Repeated measurements at the same location require the re-location of the test specimen with a sufficiently high accuracy. The method of re-location based on Williamson and Hunt(7) is shown in Fig.2. The upper guide plate with the inner ring of tapered roller bearing attached and the lower guide plate with the θ stage and provided with three holes distributed in a circle of the

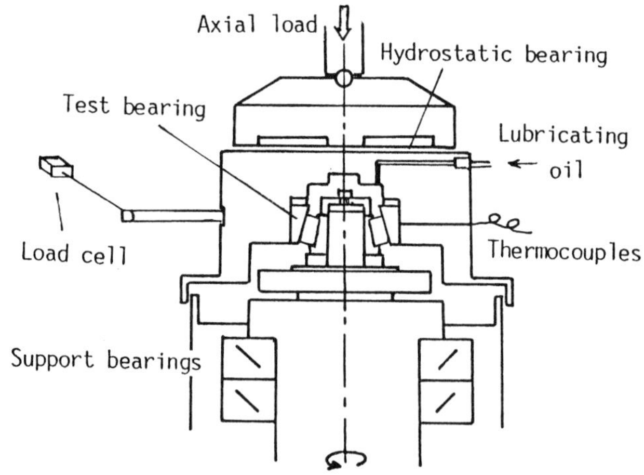

Fig.3 Test rig for clean oil

same pitch. The holes were aligned by using steel balls and fixed from above with an adhesive. This method ensured an accuracy of sub-micron for repeated re-locations at the same spot.

In the study the bearing surface was measured at many locations by driving the θ stage. Therefore, the re-location accuracy was affected by the accuracy of the θ stage. The resulting accuracy of repeated re-location was thought to be less than 5 μm.

3. EXPERIMENTS WITH CLEAN OIL

3.1 Test procedure

Inner ring of tapered ring was fixed with the upper guide plate as shown in Fig.2 during tests. Rotation and profile measurement were repeated.

Rotation test rig lubricated with clean oil is shown in Fig.3. Rotating shaft is vertical, a test bearing is fixed at the upper end and the load is applied axially via hydrostatic bearing. Running torque can be measured during rotation continuously.

Lubricating oil is supplied by a pump through 5 μm filter and temperature is controled by heater. Test bearing and lubricating oil are shown in Table 1 with the experimental conditions.

Total revolutions of the bearing for each test were 100,000. Measurements of surface profile were carried out for 2,000, 5,000, 10,000, 20,000, 40,000, 60,000 and 100,000 revolutions.

The rotational speed was determined from time to time by observing the oil film thickness which resulted in the fixed value of Λ(ratio of film thickness to composite surface roughness) at roller end / rib contact at the initial stage. The EHL film thickness was calculated with Hamrock-Dowson formula (8) for the roller end/rib contact and with Dowson-Higginson formula(9) for the roller /raceway contact.

3.2 Change of surface profiles at the roller end/ cone rib contact

Running torque of a tapered roller bearing usually decreases with rotation as shown in Fig.4 as an example. The torque decreases quickly at the initial stage but deems to reach certain value which depends on an operating condition.

The change of surface profile of the rib with rotation is shown in Fig.5. The change in the surface is clear by a gradual wearing away the surface profile irregularities. This is attributed to wear caused by sliding against the roller-end. The extent of surface profile change was determined quantitatively as a wear land surface area resulted from the difference between the initial surface profile and the surface profile measured after rotation at the same location.

As the unworn area exists and some roughness valleys remain after rotation, it was easy to correspond these profiles. The resulting difference in the surface was calculated by a microcomputer. 12 sites were measured and an average value of twelve datum was taken as the wear land surface area for these conditions. The volume of wear was calculated as the multiplication of the surface profile change (area) and the

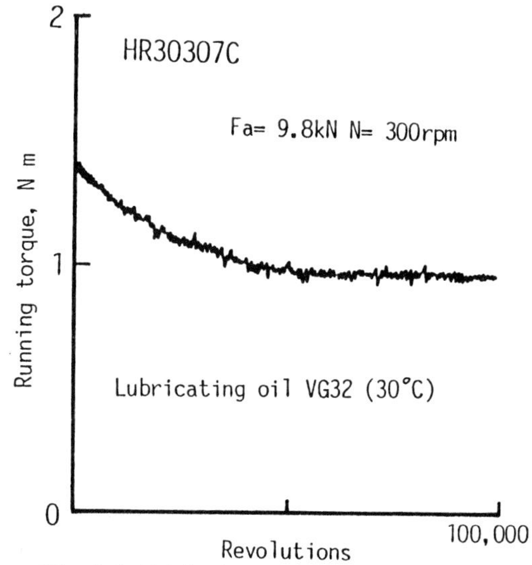

Fig.4 Initial change of running torque of tapered roller bearing

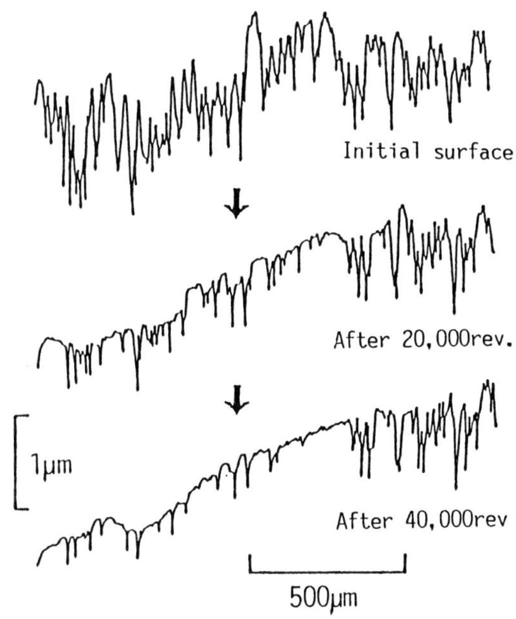

Fig.5 Change of surface profile on cone rib

length of contact (π x pitch circle diameter of contact).

The relationship between the sliding distance and the cumulative wear volume is shown in Fig.6. The sliding distance was calculated from the rotational speed, the total number of revolutions and the number of rollers. The length $8x10^4$ m, correspondded to the

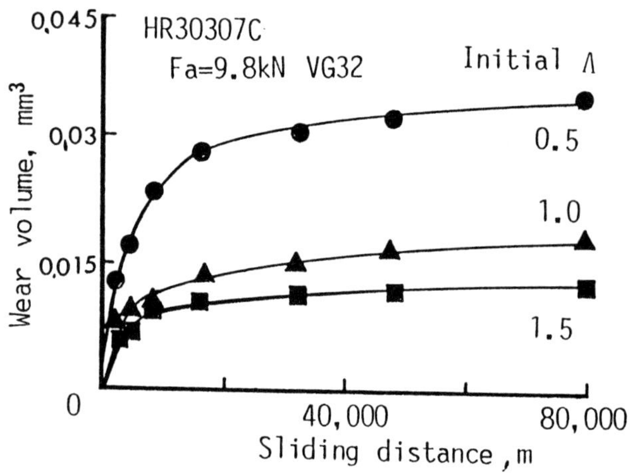

Fig.6 Relationship between sliding distance and cumulative wear volume

total number of revolutions, 100,000.

Wear at the initial stage increases extensively, yet hardly changed at the later stage and tended towards saturation. This phenomenon qualitatively coincides with the results obtained in the sliding tests(10). The results of Fig.6 are typical running-in curves.

Wear stops as the value Λ becomes relatively larger than the initial value because surface roughness reduced by the wear. Initial values Λ are 0.5 to 1.5 and the final value Λ ia thought to be around 3.

Fig.6 also shows that larger value Λ easily reaches the run-in stage with smaller wear amount. However, the specific wear rate observed on the cone rib just after the termination of the running-in process amounted to about 10^{-13} mm^2 /N, which is extremely small value compared with slide induced wear. And the massof maximum wear is less than 0.1mg.

The existing mass difference definition method could not be applied in this case. Therefore, a more efficient method was to be developed to quantitatively determine the wear from the change in the surface profile to examine wear on a microscopic scale(11).

The wear rate was also affected by the applied load. The wear volume after 100,000 revolutions was correlated to the P V value as shown in Fig.7. P is the initial maximum contact pressure at the roller-end /rib contact and V is the sliding speed.

As is evident from Fig.7 the wear volume on the cone rib after rotation is proportional to P V value and depend on the initial value Λ.

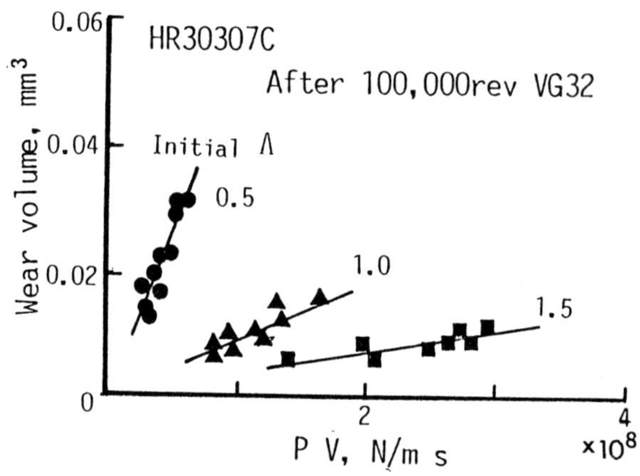

Fig.7 Relationship between P V and wear volume

3.3 Change of surface profiles at the roller/ raceway contact

Surface profiles of an inner-ring raceway were measured at the same time as the measurement of rib. The typical results are shown in Fig.8. The outlook of initial surface remained after rotattion, namely roughness peaks and valleys exist as before. However, the surface profile irregularities changed and became rounded. This is explained by an even plastic deformation of the surface roughness by rolling contact. The surface profile change is not as clear as on the rib.

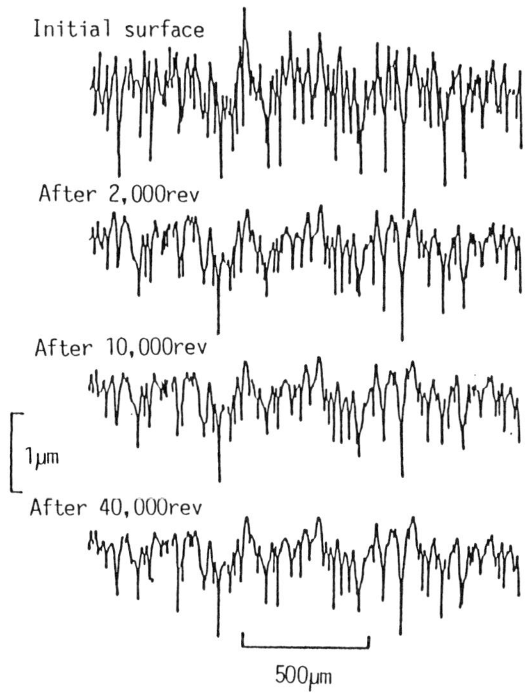

Fig.8 Change of surface profile on inner ring raceway

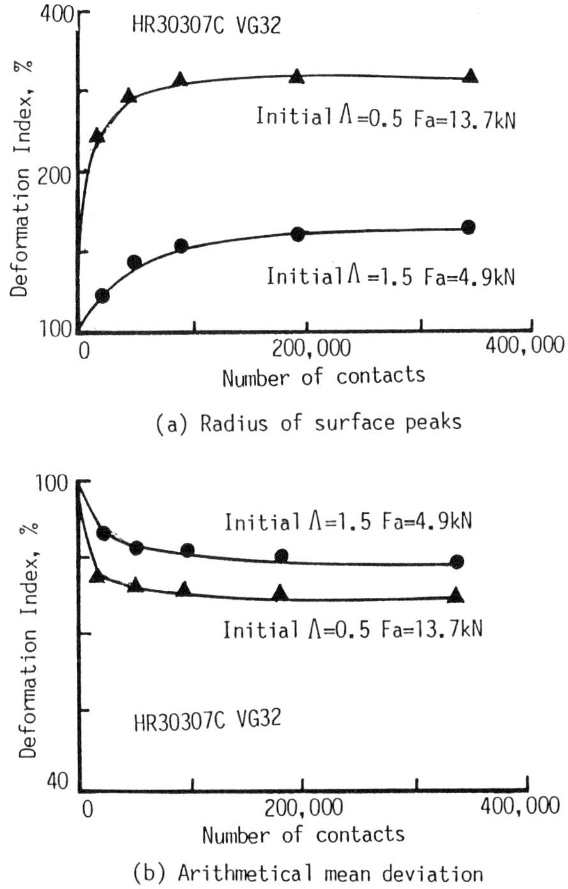

Fig.9 Change of Deformation Index with rotation

3.4 Deformation Index

To evaluate the degree of deformation two parameters were chosen. The radius of roughness peak was calculated for all the peaks to get average radius and the arithmetical mean deviation from the profile was determined as well.

Deformation index was defined as,

$$D.I. = 100 \times A/B \ (\%)$$

Where A is the measured value after rotation and B is the initial value.

Figs.9 are the change of deformation index for radius and arithmetic mean deviation with rotation respectively. Curves are similar to running-in curves obtained at the roller-end /rib contact, though mean deviation moves to opposite direction.

Figs.10 show the effect of load (max. contact pressure). Plastic deformation is clearly affected by the contact pressure and initial value Λ.

4. EXPERIMENTS WITH CONTAMINATED OILS

4.1 Test procedure

Hard particles in oil should affect wear of tapered roller bearings. Oil with particles(contaminated oil) can not be circlated at even concentration. The different test rig was used as shown in Fig.11. The bearing was set in oil and rotated from upper shaft. The bearing was fixed to cause particle circulation in oil. Temperature was not controled this time.

Test conditions are shown in Table 2. As hard particles, SiC was chosen whose

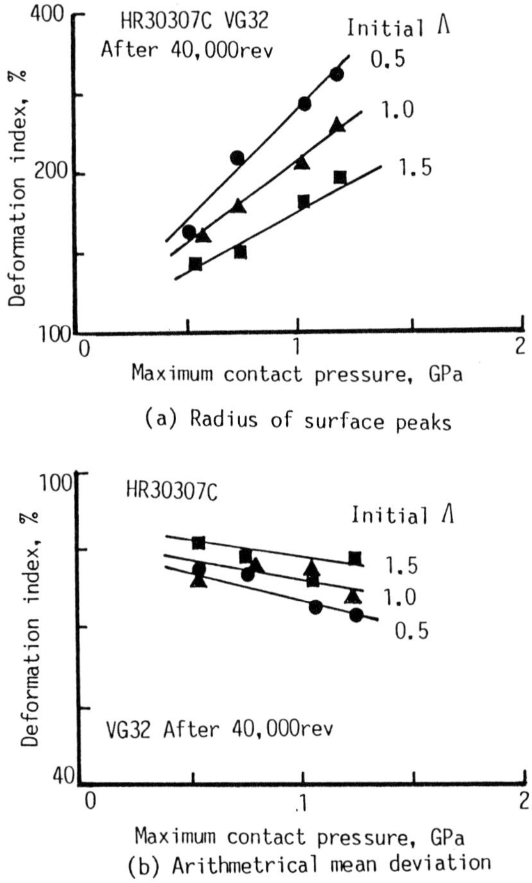

(a) Radius of surface peaks

(b) Arithmetrical mean deviation

Fig.10 Relationship between maximum contact pressure and Deformation Index

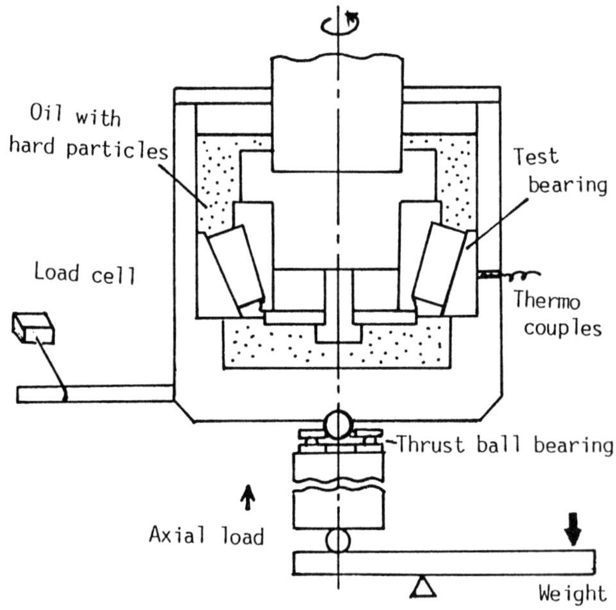

Fig.11 Test rig for oil with foreign particles

Table 2 Test condition for oil with hard particles

Test bearing	HR32307J (φ35mm×φ80mm×32.75mm)
Rotational speed	300, 500, 800 rpm
Axial load	2.0, 3.9, 5.9 kN
Lubricating oil	Turbine oil (ISO VG32) 200 ml resorvoir
Added particles	SiC #1000, #2000, #3000
Concentration	3, 5, 10, 30, 50 ppm
Total revolutions	100,000 rev

Knoop hardness is 2480 much harder than the bearing steel and average sizes are 11.5 μm (#1000), 6.7 μm (#2000) and 4.0 μm (#3000). Weight % of contaminant are 3, 5, 10, 30 and 50 ppm.

4.2 Experimental results at the roller-end and rib contact

Fig.12 shows the increase of volume of wear with sliding distance with the least contaminated oil at the roller-end cone rib contact. Wear does not stop even at 100,000 revolutions. The amount of wear increases with the contaminant percentage as shown in Fig.13. It is assumed that the wear is determined by the number of particles which enter the

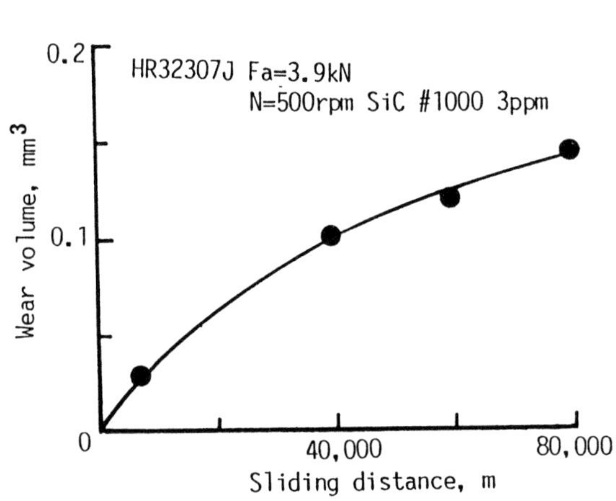

Fig.12 Relationship between sliding distance and cumulative wear volume

contact. As easily guessed, running-in effect does not occur with contaminated oils.

The wear volume was plotted with the parameters, with the particle sizes in Fig.14 and with the axial load in Fig.15.

Fig.14 shows that smaller particles cause more wear. The smaller particles can enter into the contact area easier than larger particles which are blocked from entering by the wedge shape of the inlet area of the roller/ rib contact. The oil film thickness at the rib contact is 0.1 - 0.3 μm and the size of particles is 4 -12 μm.

As the wear is abrassive wear, the cutting speed is effective. The wear increases with speed. To the contrary, the load does not affect the wear volume. This may be attributed to the fact that EHL oil film thickness is hardly dependent on applied load(Fig.15).

The volume of wear with contaminated oil increased drastically when compared with clean oil. The worn area after the rotation is shown in Fig.16. The difference is so evident. However, the wear rate is still smaller than the wear rate of sliding wear.

4.3 Wear at the roller/ raceway contact

With clean oil, the change of surface profile of raceway was thought mainly as the result of plastic deformation. However, in the case of contaminated oil, peaks of the surface roughness seem to be cut as shown in Fig.17. The surface profiles of initial surface and tested surface after 100,000 revolutions are compared. Uncontacted surface is almost identical but rotated surface shows the difference. Plastic deformation may exist but not so clear as before. Some peaks of profile are clearly cut by wear but only limited to the peaks. At the raceway contact is rolling contact with very small sliding usually less than 1%. Therefore, the wear volume is too small to make quantitative comparison.

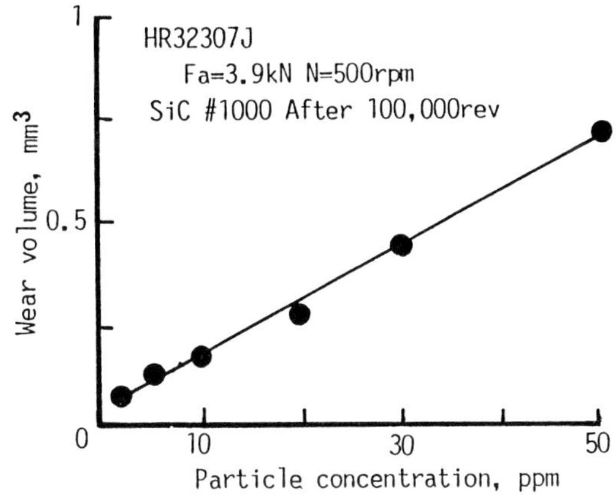

Fig.13 Effect of particle concentration on wear volume

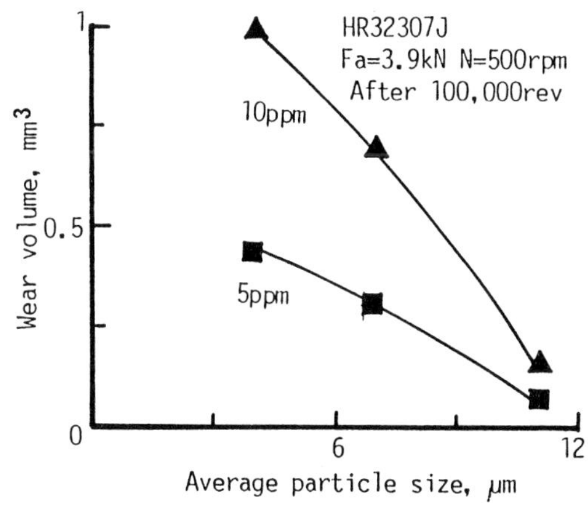

Fig.14 Effect of particle size on wear volume

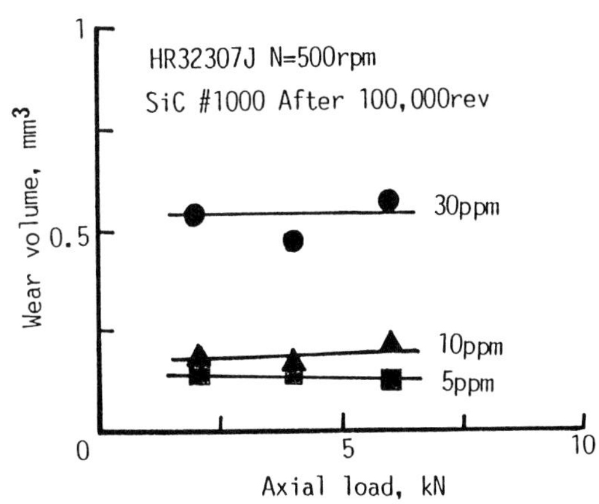

Fig.15 Relationship between axial load and wear volume

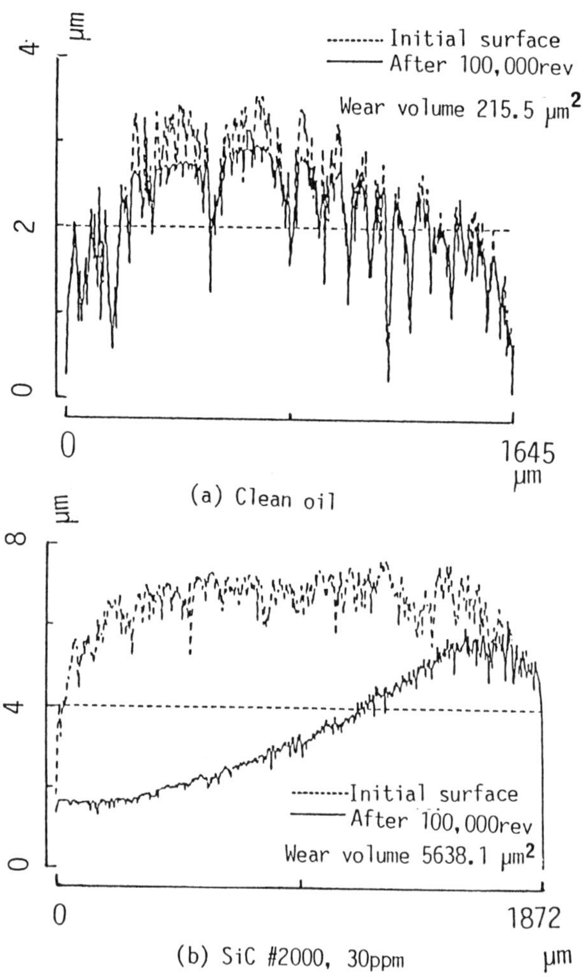

Fig.16 Comparison of wear with clean oil and with contaminated oil

4.4 Wear particles

To evaluate worn particles oils are examined by means of Ferrography. Test with clean oil was carried out by the test rig for contaminated oil to collect tested oil after rotation. Fig.18 is the photographs of worn particles. The wear volumes are quite different and the sizes of worn particles are different too. Contaminated oil produced small worn particles, because of possibly abrassive effect of hard contaminant.

Shapes of either particles are rather similar and observation by SEM could not clarify the difference.

Hard particles are said to make dents on the bearing surfaces(12) which may cause life reduction. Careful observation of the raceway surfaces was carried out but such holes were not found this time. The raceway surface was uniformly cut with small scratches and became rather shiny. This may be attributed to the size of hard particles which are very small in these experiments.

5. CONCLUSION

Measurements of the surface profile change with rotation in the process of running-in were carried out with tapered 9roller bearings. At the roller-end /

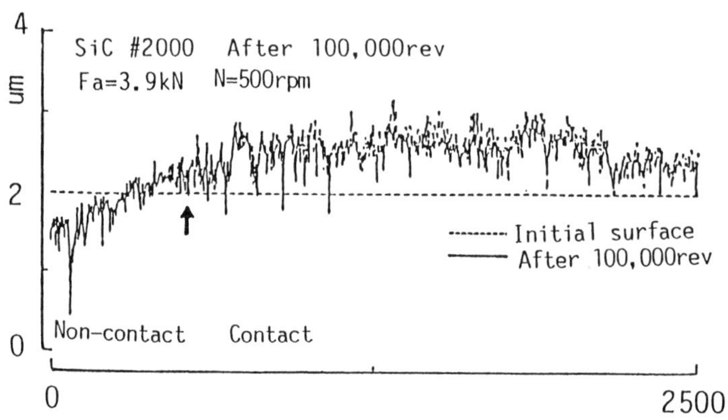

Fig.17 Change of surface profile on inner ring raceway for oil with foreign particles

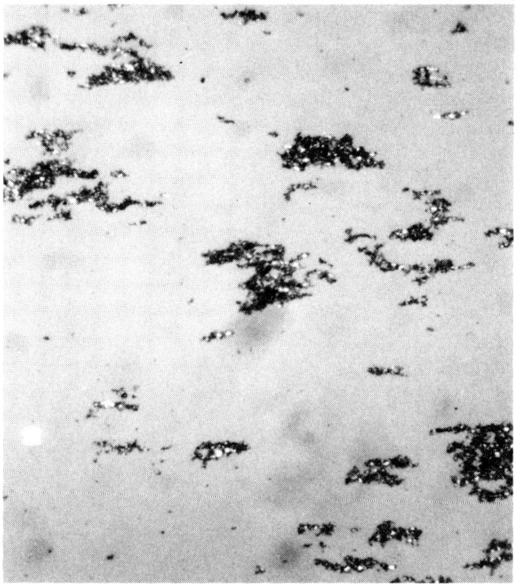

(a) Clean oil

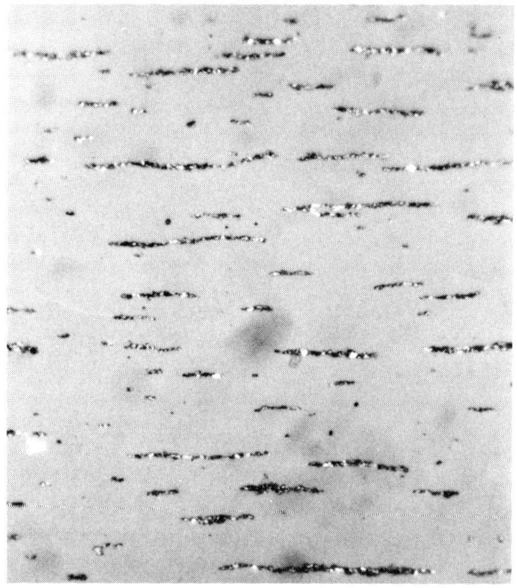

(b) Oil with hard particles

Fig.18 Collected wear particles by Ferrography

cone rib contact the surface profile change was due to sliding wear and at the roller and raceway contact the change was deemed to be due mainly to plastic deformation for the case of clean oil.

Running-in was seen only with clean oil but wear increased with contaminated oil. Wear volume during run-in stage isdepend on the operating conditions of bearings and showed good correlation with P V value considering initial A.

Hard particles in oil cause huge wear even though their number is limited, 3 ppm thistime. However, the wear rate is much smaller than the rate with sliding wear.

Sizes of worn particles are different between clean oil and contaminated oil. The latter produced a much greaternumber of smaller worn particles.

Though this might be a basic work, it gives some deeper understanding into tapered roller bearings in actual use.

REFERENCES

1. Proc. 8th Leeds-Lyon Symp.(1981) Butterworths.
2. NSK Bearing Journal, 607 (1957) 1.
3. Y.Yamamoto and M.Hashimoto, J.JSLE, 29-2 (1984) 137.
4. J.Sugimura et al, J.JSLE, 31-11 (1986) 813.
5. J.Kagami et al,J.JSLE, 33-12(1988)900.
6. S.Noguchi and T.Sawamoto, J.JSLE,37-2, (1992) 142.
7. J.B.P.Williamson and R.T.Hunt,J.Phys. E.,1 (1968) 749.
8. B.J.Hamrock and D.Dowson, Trans ASME, JoLT,99 (1977)264.
9. D.Dowson, Proc. IME, 182,Pt.3A(1968) 151.
10. For example: S.Norose and T.Sasada, J. JSLE, 24-4 (1978)231.
11. T.Tsukada et al, Preprints of JSLE, (1982)53.
12. R.S.Sayles and E.Ioannides, Trans. ASME, JoT, 110 (1988)26.

(Nos.2,3,4,5,6,9 & 10 in Japanese)

Thin Films in Tribology / D. Dowson et al. (Editors)
© 1993 Elsevier Science Publishers B.V. All rights reserved.

The Effect of Compliance On Peak Oil Film Pressure in Connecting Rod Bearings

D.N.Fenner Dept of Mechanical Engineering, King's College, University of London.
J.D.C.McIvor Structural Mechanics & Dynamics, Sydney, Australia.
J.M.Conway-Jones Glacier Vandervell Ltd, London.
H.Xu T&N Technology Ltd, Rugby.

An elastohydrodynamic study of a highly-loaded bearing fatigue test rig shows that elastic deformation significantly increases the extent of the oil film and leads to a marked reduction in the maximum oil film pressure. Predictions for the ratio of peak oil film pressure to maximum specific load are reduced from about 10 when the bearing is assumed rigid, to approximately 2 for an elastic housing. At maximum load there is a marked reduction on the oil film thickness towards the bearing edges and a flattening of the pressure variation in the axial direction. These effects should be taken into account in less rigid automotive big-end bearings when interpreting bearing fatigue test data.

1. INTRODUCTION

One of the criteria for the acceptance of a lubricated bearing material for a given duty is its fatigue strength under dynamic load. In the case of connecting-rod bearings maximum specific loads have increased over the years from 12-15 MPa for babitted white metal bearings to 35-50 MPa for current gasoline engines and 50-70 MPa for turbocharged truck diesel engines. In order to obtain fatigue data in less than the normal life expectancy of a modern engine, bearing fatigue test rigs may be run at even higher specific loads. When both bearing and shaft are assumed rigid the peak oil film pressures in these test rigs are typically predicted to be in the range 6-12 times the maximum specific load. However, it is well known that at such high pressures elastic deformation of the housing becomes significant and an elastohydrodynamic (EHD) analysis of the bearing is required.

In this work the effects of elasticity on the predicted performance of a highly-loaded bearing fatigue test rig are assessed using the finite element (FE) based method for EHD analysis developed by the first two named authors [1-3].

2. EHD ANALYSIS

A number of numerical methods, employing either a finite difference or finite element approach, have been developed for solving dynamically loaded EHD bearing problems. These methods are based either on the short bearing theory [4-8] or entail a two-dimensional solution of Reynolds equation [2,3,9-11]. The non-linearity of the EHD problem, resulting from the coupling of the elastic deflection of the bearing housing with the pressure derived from Reynolds equation, means that an iterative scheme is required for the solution of oil film pressures and thicknesses at each time step in the loading cycle of the bearing.

The Newton-Raphson (N-R) method has been successfully applied to the problem [10] and forms the basis of a robust and efficient solution algorithm. Even so, a prohibitive amount of computer CPU time may still be required for a converged solution. It has been shown recently [2,3], however, that use of higher-order rather than the more usually employed linear finite elements to represent the oil film, together with a fast N-R method can considerably reduce the CPU time required. The result is that EHD analysis, once felt to be an expensive luxury, can now be adopted as a practical tool for bearing design.

2.1 Governing equations

For an incompressible, isoviscous, Newtonian fluid the two-dimensional distribution of pressure in the oil film satisfies Reynolds equation in a (θ, z) co-ordinate system (see Fig.1) given by

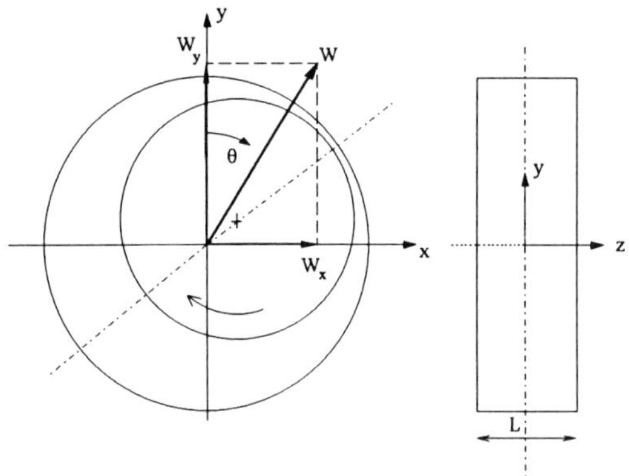

Fig.1 Bearing geometry

$$\nabla(h^3\nabla p) - 12\mu\left(\frac{U}{D}\frac{\partial h}{\partial\theta} + \frac{\partial h}{\partial t}\right) = 0 \qquad (1)$$

In this equation $p = p(\theta, z, t)$ and $h = h(\theta, z, t)$ are the oil film pressure and thickness at time t, respectively, μ is the viscosity, D is the bearing diameter and $U = U(t)$ is the relative velocity of the bearing surfaces. If the problem is symmetric about the bearing centre-line $z = 0$ then the solution domain can be restricted to $0 \le z \le L/2$, where L is the bearing width, and the pressure is required to satisfy the boundary conditions

$$p = 0 \text{ at the bearing edge, } z = \frac{L}{2} \qquad (2)$$

$$\frac{\partial p}{\partial z} = 0 \text{ at bearing centre-line, } z = 0 \qquad (3)$$

The oil film thickness, h, is given by

$$h = c - (e_x \sin\theta + e_y \cos\theta) + \delta \qquad (4)$$

where the first term, c, is the nominal radial clearance. The second term is due to the motion of the journal in the clearance space, where e_x and e_y are the x and y components of the eccentricity vector. The third term, δ, is the radial elastic displacement of the bearing housing resulting from the oil film pressure on the bearing surface.

In the FE formulation of the problem the film thickness at node i with a circumferential co-ordinate θ_i, is written as

$$h_i = c - (e_x \sin\theta_i + e_y \cos\theta_i) + \delta_i \qquad (5)$$

where δ_i is the radial displacement at node i. With the aid of a structural FE model of the bearing housing the radial components $\{F\}$ of the forces at nodes on the bearing surface are related to the displacements $\{\delta\}$ at these nodes through $\{F\} = [K]\{\delta\}$ where $[K]$ is the stiffness matrix. The nodal forces can be expressed in terms of the pressures $\{p\}$ at nodes on the bearing surface through $\{F\} = [M]\{p\}$ where the matrix $[M]$ depends upon the shape functions of the elements used [2]. Hence the nodal displacements can be written as

$$\{\delta\} = [C]\{p\} \qquad (6)$$

where $[C] = [K]^{-1}[M]$ is the pressure compliance matrix for nodes on the bearing surface, from which it follows that

$$\delta_i = \{C_i\}\{p\} \qquad (7)$$

where $\{C_i\}$ is the i 'th row of the matrix $[C]$.

The components W_x and W_y of the external load, W, on the journal at any instant in time must be exactly balanced by the forces found by integrating the pressure forces acting over the bearing surface area A. Thus

$$\begin{aligned} \int_A p\cos\theta \, rd\theta dz &= W_x \\ \int_A p\sin\theta \, rd\theta dz &= W_y \end{aligned} \qquad (8)$$

2.2 Newton-Raphson method

If $p^{(r)}$, $e_x^{(r)}$, $e_y^{(r)}$ denotes the r'th trial solution for the oil film pressure and journal eccentricity the N-R method can be used to obtain the $(r+1)$'th trial solution given by

$$\begin{aligned} p^{(r+1)} &= p^{(r)} + \Delta p^{(r)} \\ e_x^{(r+1)} &= e_x^{(r)} + \Delta e_x^{(r)} \\ e_y^{(r+1)} &= e_x^{(r)} + \Delta e_x^{(r)} \end{aligned} \tag{9}$$

where $\Delta p^{(r)}, \Delta e_x^{(r)}, \Delta e_y^{(r)}$ are the corrections. The Galerkin method of weighted residuals is used to obtain the finite element formulation of the problem. For a mesh containing n nodes the correction vector

$$\{X\} = \{\Delta p_1, \Delta p_2 \ldots\ldots \Delta p_n, \Delta e_x, \Delta e_y\}^T$$

is given by

$$[A]\{X\} = \{R\} \tag{10}$$

where the matrices $[A]$ and $\{R\}$ are given in [2]. If the elements of the correction matrix $\{X\}$ tend to zero as the sequence of iterations proceeds then the method is convergent. The solution is advanced to the next time step using implicit time-stepping.

One of the most time consuming operations in the N-R method is the reformulation of the matrix $[A]$ after each iteration. In an attempt to minimise computing time the first two named authors have developed a fast N-R method [2,3] in which a complete reformulation only takes place when the rate of convergence becomes unacceptably slow.

2.3 Cavitation

The boundary separating the lubricated and cavitated regions of the oil film is determined using the complementarity formulation [12] in which the net flow rate, f, at any point in the fluid, given by

$$f = \nabla(h^3 \nabla p) - 12\mu\left(\frac{U}{D}\frac{\partial h}{\partial \theta} + \frac{\partial h}{\partial t}\right) \tag{11}$$

is used to decompose the solution domain into :

i) Lubricated region where Reynolds equation (1) is satisfied and :

$$f = 0 \text{ and } p \geq 0 \tag{12}$$

ii) Cavitated region where :

$$f \leq 0 \text{ and } p = 0 \tag{13}$$

Using the Murty algorithm [13] a trial decomposition of the solution domain is set up by assigning an initial status to each node in the mesh, and the EHD problem is then solved for p and f. The status of any node where the complementarity conditions (12) and (13) are not satisfied is then altered and further iterations are performed until convergence is achieved.

2.4 Finite element meshes

A mesh of eight-node isoparametric elements (see Fig.2) is used to model the oil film, these having been shown [1] to be computationally more efficient than the linear elements normally used. The pressure compliance matrix $[C]$ is obtained using 20-node elastic brick elements for the bearing housing.

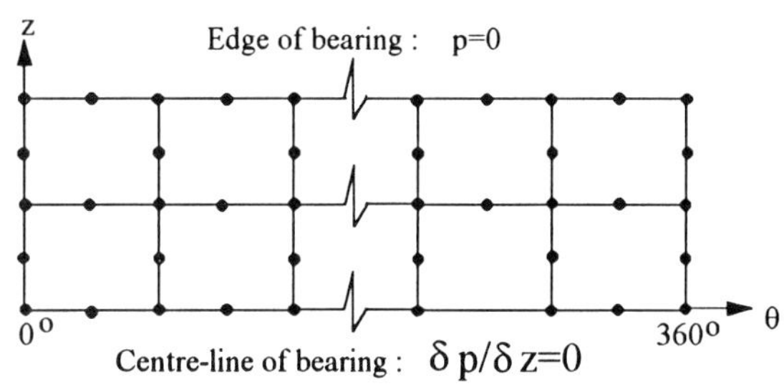

Fig 2. F.E. mesh for oil film

3. PREDICTIONS FOR BEARING FATIGUE TEST RIG

The Sapphire bearing fatigue test rig, details of which are given in the Appendix, is used for comparing the fatigue strength of bearing materials employed in reciprocating engines. The load case listed has a maximum specific load,

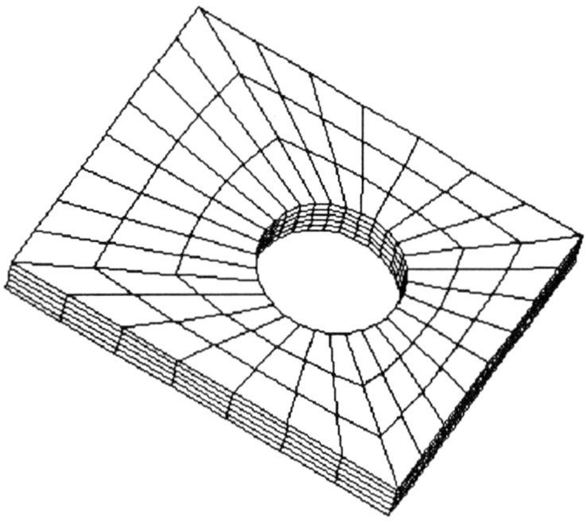

Fig 3 Half-width model of bearing housing

$P = W_{max} / (DL)$ of 109MPa and will cause fatigue cracking of aluminium bimetal bearings after about 25×10^6 load cycles.

The finite element mesh used to obtain the compliance matrix for the Sapphire bearing housing is shown in Fig 3 and the location of the constraints is shown in Fig 14. On the bearing surface there are 60 nodes in the circumferential direction and 11 across the half-width. The journal is assumed to be rigid.

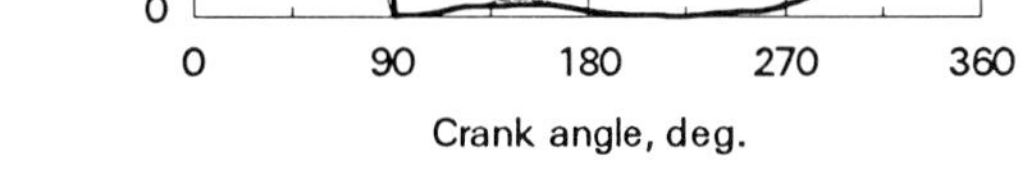

Fig. 4 Maximum oil film pressure variation

The predictions for the variations of the maximum oil film pressure, p_{max}, obtained using rigid and EHD analyses of the Sapphire bearing are shown in Fig.4. Elasticity reduces the predicted maximum oil film pressure from 1262MPa to 223 MPa. At maximum load the elastic bearing housing conforms with the journal over an arc of approximately 100 degree extent (Fig.5). This is comparable with the Hertzian contact condition modified by the classic calculation for EHD lubrication [14]. The extent of the load carrying

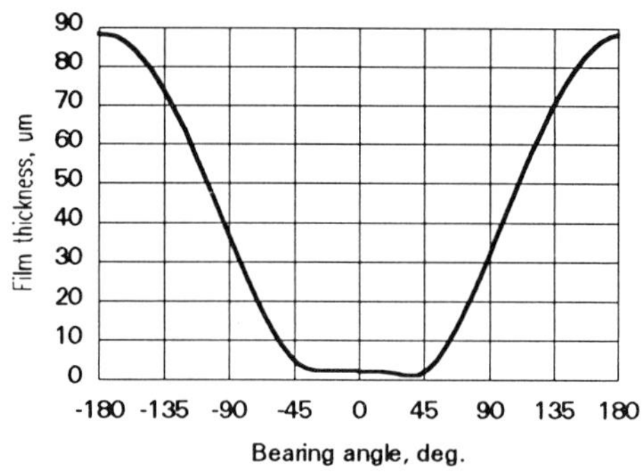

Fig.5 Oil film thickness variation along bearing centre-line at maximum load.

region of the oil film is approximately 5 times greater than that predicted for the rigid bearing (Fig.6). When the deformation of the journal is included it is expected that the reduction in p_{max} due to elasticity will be even more pronounced.

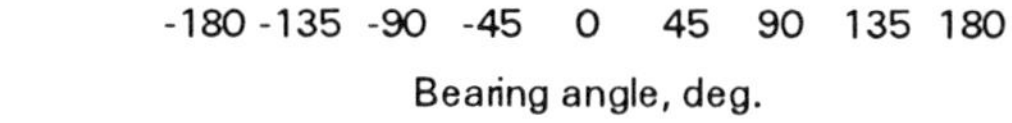

Fig 6 Oil film pressure variation along bearing centre-line at maximum load.

The variation of the ratio p_{max} / P with P is shown in Fig.7 where it is clear that, as well as the ratio having a much lower value, the trend is reversed for the elastic housing. Predictions for a rigid housing show an increase in the ratio with P

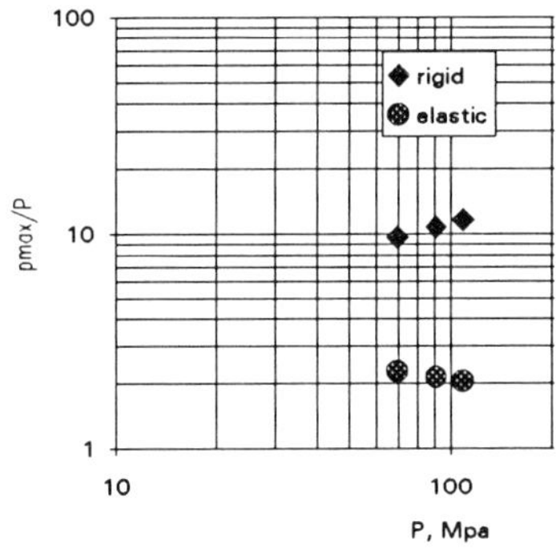

Fig 7 Variation of maximum oil film pressure with maximum specific load

whilst predictions for an elastic housing show a decrease with P as deformation spreads the load carrying film over a greater arc.

At maximum load the oil film thickness is relatively uniform (Fig.5) along the centre-line of the bearing within the deformed arc, and is more than 5 times that predicted for a rigid bearing. There is, however, a marked reduction in oil film thickness towards the edge of the bearing (see Fig 8) where the minimum is similar to that for the rigid case. The reduced film thickness at the bearing edge restricts oil flow and leads to a flattening of the pressure variation in the axial direction (Fig 9). This is in marked contrast to the parabolic pressure variation assumed in short bearing theory [4-8].

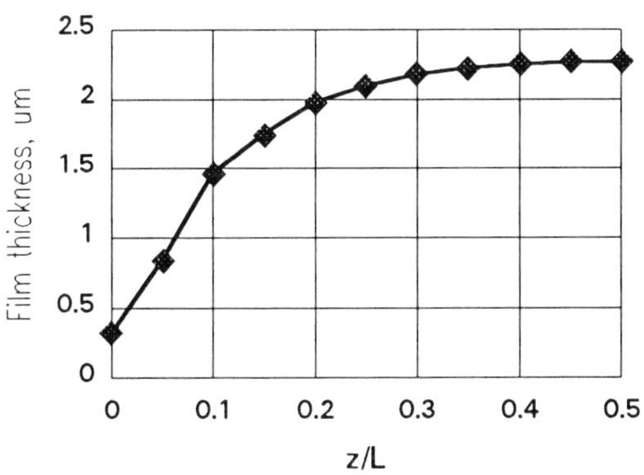

Fig. 8 Axial variation of oil film thickness at $\theta = 0$ deg. for maximum load.

The variation of the oil film thickness in the axial direction is found to be very sensitive to changes in the housing geometry. In some test procedures it is common practice to increase the maximum specific load by reducing the width of the bearing shell, whilst maintaining the original width of the bearing housing. This has the effect of concentrating the load around the centre-line of the housing and leads to an increase in the relative curvature of the axial profile since the outer part of

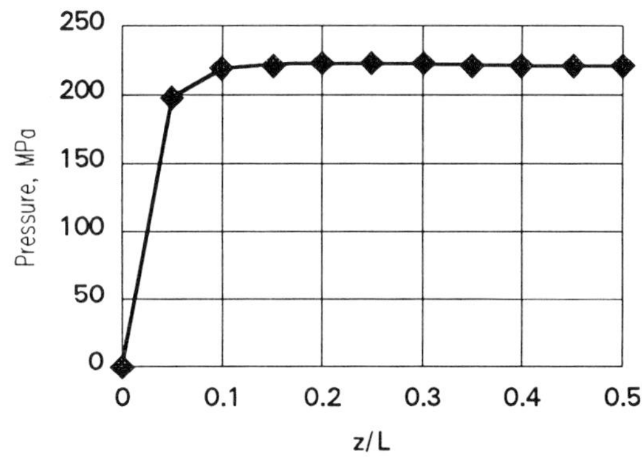

Fig. 9 Axial variation of pressure at $\theta = 0$ deg. for maximum load.

the housing is not loaded but can exert an exaggerated restraint on the effective edges of the bearing (Fig.10b). An EHD analysis of one such bearing predicts the occurrence of contact at the edges during the period of maximum load. This is in accordance with observations made during bearing tests which show clear evidence of metal-to-metal-contact at the edges.

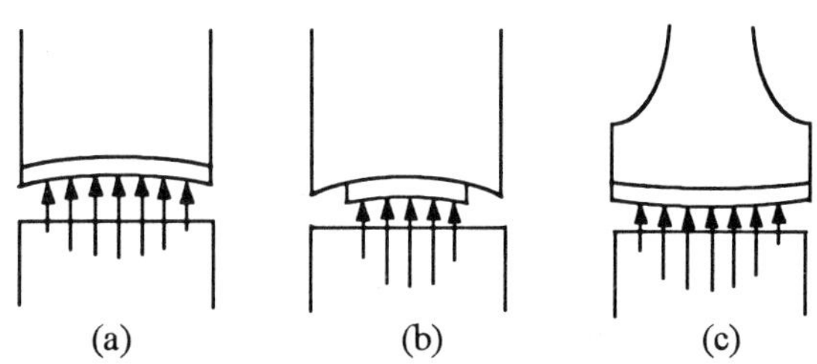

Fig. 10 Bearing housing geometry: (a) full-width bearing shell, (b) reduced-width bearing shell, (c) automotive big-end.

The empirical development of automotive connecting rods has resulted in relief of the back of the bearing housing on the rod side (see Fig.10c). This allows the edges of the bearing to deflect away from the journal thereby counteracting the tendency for the oil film thickness to diminish towards the edges. Excessive relief may, however, lead to unacceptably high stress concentrations in the central region.

4. PIEZO-VISCOUS EFFECTS

It is well known that at high pressures the viscosity of lubricating oil may increase by a factor of 10 or more. The relationship commonly used is the Barus equation

$$\mu = \mu_0 e^{\alpha p} \tag{14}$$

where μ is the viscosity at pressure p, μ_0 is the viscosity at atmospheric pressure $p = 0$ and α is the piezo-viscosity coefficient.

Data for typical engine oils [15] up to pressures of 200 MPa indicate that α reduces with oil film pressure, p (MPa), and with temperature, T (°C), in a manner which can be represented by

$$\alpha = 0.139 p^{-0.13} T^{-0.36} \tag{15}$$

Integration over the pressure contours predicted for the elastic Sapphire bearing at the instant of maximum load indicates an instantaneous mean viscosity of 41cP over the conforming area of the bearing and journal. Results obtained using this increased viscosity for the whole load cycle indicate that there is very little change in the extent of the deformed arc and the peak oil film pressure is only increased by about 3% as shown in Fig.11. However, Fig 12 shows that at maximum load the oil film thickness is more than doubled at the centre-line of the bearing and is more than 10 times thicker at the edge. Since the pressure is high for only about 30% of each cycle the effect is likely to be less than that indicated here. Incorporation of piezo-viscous effects into the solution procedure is currently in progress.

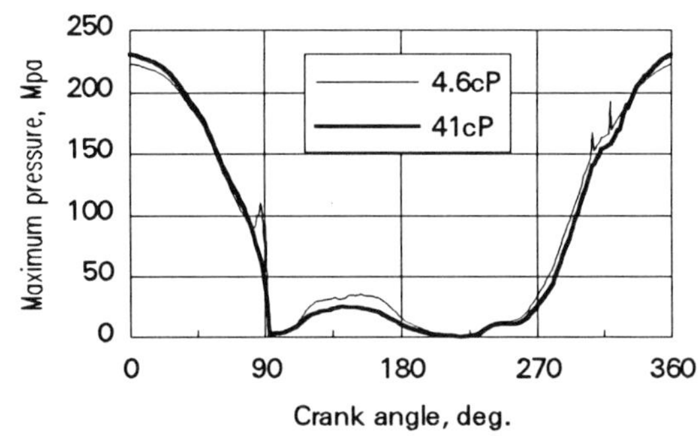

Fig. 11 Maximum oil film pressure variation

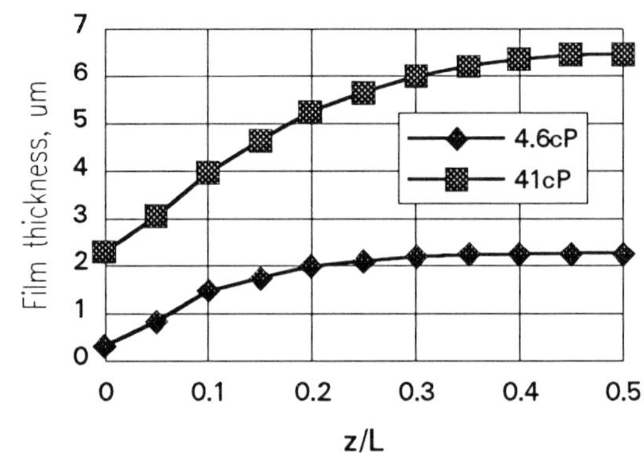

Fig. 12 Axial variation of oil film thickness at θ=0 deg. for maximum load.

5. CONCLUSIONS

The actual mechanism of cracking of the surface layer of lubricated bearings is not completely understood. It is becoming clear, however, that it arises from a combination of stresses which are primarily related to the condition of contact stress. The results presented in this work show the importance of using EHD analysis to predict the performance of bearings in a relatively rigid, highly-loaded test rig. It will be important to take into account the effect of flexibility in less rigid automotive type big-end bearings when interpreting results from fatigue test rigs and this matter is currently in hand.

Predictions for the peak oil film pressure, obtained assuming a rigid bearing, will not only greatly exaggerate the severity of conditions but will also give the wrong trends. Allowance for the flexibility of the shaft is likely to further reduce the severity of the peak oil film pressure and the stresses in the bearing.

Results obtained for the axial variations of pressure and film thickness clearly indicate the inadequacies of the assumptions in short bearing theory.

The omission of an allowance for piezo-viscous effects has a relatively small effect on the computed peak oil film pressure and consequently on the stresses likely to cause fatigue cracking. The effect on the predicted minimum oil film thickness is, however, significant.

REFERENCES

1. McIvor, J.D.C., & Fenner, D.N., An evaluation of eight-node quadrilateral finite elements for the analysis of a dynamically loaded journal bearing, Proc.I.Mech.E., Part C, 202(C2), 95-101, 1988.
2. McIvor, J.D.C., The analysis of dynamically loaded flexible journal bearings using higher-order finite elements. PhD thesis, University of London, 1988.
3. McIvor, J.D.C., & Fenner, D.N., Finite element analysis of dynamically loaded flexible journal bearings: a fast Newton-Raphson method, ASME J. Tribology, 111, 597-604, 1989.
4. Fantino, B.& Frene, J. Comparison of dynamic behaviour of elastic connecting rod bearings in both petrol and diesel engines, ASME J. Tribology, 107, 87-91, 1985
5. van der Tempel, L.,Moes, H. & Bosma, R., Numerical simulation of dynamically loaded flexible short journal bearings, ASME J. Tribology, 107, 396-401, 1985.
6. Goenka, P.K. & Oh, K.P., An optimal short bearing theory for the elastohydrodynamic solution of journal bearings, ASME J.Tribology, 108, 294-299, 1986.
7. Aitken, M.B. & McCallion, H., Elastohydrodynamic lubrication of big-end bearings Part1:theory, Proc.I.Mech E., Part C, 205(C2), 99-106, 1991.
8. Xu, H. & Smith, E.H, A new approach to the solution of elastohydrodynamic lubrication of crankshaft bearings, Proc.I.Mech.E.,204, 187-197, 1990.
9. LaBouff, G.A., & Booker, J.F., Dynamically loaded journal bearings: a finite element treatment for rigid and elastic surfaces, ASME J.Tribology, 107, 505-515, 1985.
10. Oh, K.P., & Goenka, P.K., The elastohydrodynamic solution of journal bearings under dynamic loading, ASME J.Tribology, 107, 389-395, 1985.
11. Kumar, A., Goenka, P.K. & Booker, J.F., Modal analysis of elastohydrodynamic lubrication: a connecting rod application, ASME J.Tribology, 112, 524-534, 1990.
12. Oh, K.P., The numerical solution of dynamically loaded elastohydrodynamic contact as a nonlinear complementarity problem, ASME J.Tribology, 106, 88-95, 1984.
13. Murty, K.G., Note on a Bard-type scheme for solving the complementarity problem, Opsearch, 11,123-130, 1974.
14. Dowson, D. & Higginson, G.R., Theory of roller-bearing lubrication and deformation, Proceedings of 1963 Lubrication and Wear Group Convention, 216-227, I.Mech.E., London, 1964.
15. Hutton, J.F., Jones, B. & Bates, T.W. Effects of isotropic pressure on the high temperature shear rate viscosity of motor oils, SAE Paper 830030, 1983.

APPENDIX

In the Sapphire hydrodynamic bearing fatigue test rig (see Fig 13) dynamic loading is applied to a connecting-rod bearing. The test bearing engages with a journal which rotates with a small (0.381mm) eccentricity at 3000 rev/min. The connecting rod is attached to a double-acting piston operating in an oil-filled cylinder with relief valves to control the pressure. Adjustment of the valves enables the forward and reverse loads on the bearing to be controlled, as indicated by a strain

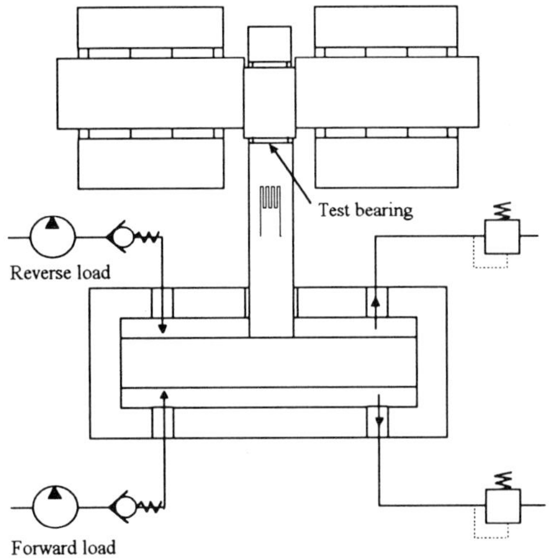

Fig. 13 Sapphire bearing fatigue test rig

gauge on the connecting-rod. The load is increased stepwise after inspections at 20 hour ($3.6x10^6$ load cycles) intervals until fatigue cracking is observed. Oil is supplied through a 6mm diameter hole in the bearing cap half at 45° from the split line. The oil used is Shell Rotella T 10W with no viscosity index improvers.

The dimensions of the bearing housing are shown in Fig. 14 and the bearing data and loads are given in Table 1.

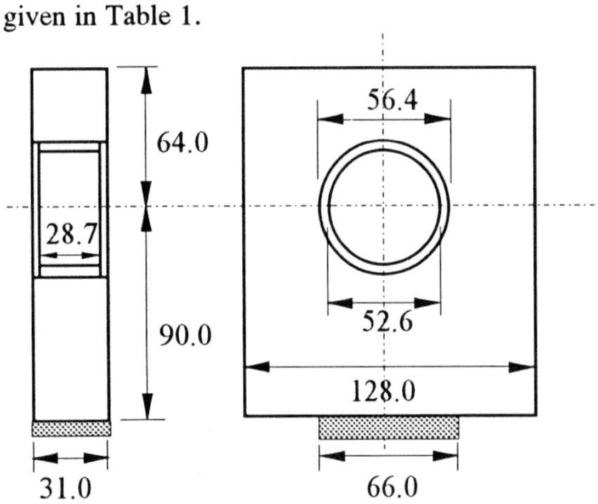

Fig 14 Sappire bearing housing
(all dimensions in mm)

Diameter	52.9 mm
Length	28.7 mm
Diametral clearance	0.07mm
Speed	3000 rev/min
Effective viscosity	4.6cP
Maximum specific load	109 MPa
Crank length	0.381 mm
Con-rod length	259 mm

Crank angle deg.	Wx N	Wy N	Crank angle deg.	Wx N	Wy N
0	0	165322	10	23	162676
20	45	158721	30	66	149474
40	85	133605	50	101	117751
60	114	8996	70	124	6616
80	129	41045	90	131	1461
100	129	-2816	110	123	-7229
120	114	-16294	130	101	-18201
140	84	-19455	150	66	-16741
160	45	-14043	170	23	-8707
180	0	-3920	190	-23	-2336
200	-45	-903	210	-66	-946
220	-84	455	230	-101	1842
240	-114	7730	250	-123	7637
260	-129	9130	270	-131	17256
280	-129	27772	290	-124	47581
300	-114	70052	310	-101	99168
320	-85	112368	330	-66	133546
340	-45	149430	350	-23	158694

Table 1 Bearing data and loads

Thin Films in Tribology / D. Dowson et al. (Editors)
1993 Elsevier Science Publishers B.V.

A Transient Elastohydrodynamic Lubrication Analysis for Spiral Bevel Gears

S.G. Lim,* D.E. Brewe, and R.F. Handschuh

National Aeronautics and Space Administration, Lewis Research Center, 21000 Brookpark Road, Cleveland, Ohio 44135, U.S.A.

A transient elastohydrodynamic lubrication (EHL) analysis for spiral bevel gears is presented. The time dependent Reynolds equation is solved numerically considering the effect of the rate of change of contact parameters. Some fundamental characteristics of the dynamic loading are investigated in detail and are compared with the Grubin's approximations. Next, this solution method is applied to the spiral bevel gear system. Because the ellipticity parameter is high, the line contact case is assumed.

1. INTRODUCTION

It is well known that the lack of a lubricant film at the tooth contact may lead to gear failure. Therefore it is important that reliable calculations of the film thickness can be made accounting for the conditions experienced by the contacting members. Elastohydrodynamic lubrication (EHL) has been used to predict the oil film thickness successfully utilizing the Dowson and Higginson [1] formula for line contact case. But this formula is based on a quasi-steady state assumption. That is, the contact parameters (load, speed, and curvature) during the operation can be represented by the instantaneous curvatures, rolling speed, and contact load, independent of past history and the rate of change of these contact parameters. In this assumption, the time-dependent term or the transient effect is neglected. Gu [2] has indicated that the transient effect might be important for heavily loaded gears.

In considering a transient analysis, the relative normal motion of the contacting members must be included as well as the rolling/sliding motion that leads to lubricant entrainment. The fundamental transient phenomena in EHL was investigated by Vichard [3]. In that paper, he used the Grubin's approximation to study the transient effect in an EHL line contact. A consequence of the transient/dynamic conditions revealed a damping phenomena reflected in the film thickness calculations. Dowson et al. [4] calculated the time-dependent film thickness of the piston rings and demonstrated the importance of the squeeze action near the top dead center. Lee and Hamrock [5] solved the transient Reynolds equation. But in that paper, the central film thickness rather than the dynamic load distribution was considered as an input data. Recently, using the Grubin's approximation, Tieu and Worden [6] investigated the transient film thickness of a pair of involute gears under dynamic loads. Even though the Grubin's approximation offers a considerable simplicity, it is limited in that the typical EHL film thickness at the end of the contact can not be described adequately, and accordingly the squeeze velocity distribution is affected.

Spiral bevel gears in helicopter transmissions are operated in heavily loaded situations where the lubricant film thickness becomes very thin. The transient EHL analysis in the highly loaded case has not yet been investigated. The problem is to find the film thickness distribution with a known loading function, in which the normal motion is accountable in the solution. In this paper, the time-dependent Reynolds equation is solved without the Grubin's approximation. Some fundamental characteristics of the dynamic loading are

*National Research Council—NASA Research Associate at Lewis Research Center.

investigated in detail. Next, this solution method is applied to the spiral bevel gear system. Because the ellipticity parameter is high (i.e., ~25 to 30) for the example under consideration, the line contact case was assumed. The results include the minimum and central film thickness distribution during the meshing period.

1.2 Notation

A	R/R_i
a	semi-minor axis of contact ellipse
b	semi-major axis of contact ellipse
e	outlet (cavitation) boundary
E	e/R_i
E′	reduced elastic modulus
f	line contact force
f_p	point contact force
f_s	quasi-steady-state load variation
G	dimensionless material parameter, $\alpha E'$
H	dimensionless film thickness, h/R_i
H_0	dimensionless minimum film thickness for Grubin's approximation
h	lubricant film thickness
h_e	film thickness at $x = e$
h_0	minimum film thickness for Grubin's approximation
h_{min}	minimum film thickness
h_{cen}	central film thickness
N	$12\mu_0 u_m/E'R$
N_p	number of pinion teeth
P	dimensionless pressure, p/E'
p	hydrodynamic pressure
p_h	maximum hertzian pressure
R	relative radius of curvature, $1/R_1 + 1/R_2 = 1/R$
R_1, R_2	radii of circular cylinders in contact
R_i	reference radius of curvature
S	squeeze number, $\omega_D R/u_m$
t	time
u_m	mean rolling velocity
W	dimensionless load, f/ER_i
X	dimensionless x coordinate, x/R_i
x_A	inlet boundary
z	exponent in Roelands viscosity equation
α	pressure coefficient of viscosity
β	ratio of dynamic load to mean or quasi-dynamic load
ω	angular velocity of cylinder or pinion
ω_D	angular velocity of dynamic loading
μ	viscosity
μ_0	reference viscosity

2. THEORETICAL BACKGROUND

2.1 Governing Equations

The pressure distribution is represented by the Reynolds equation;

$$\frac{\partial}{\partial x}\left(\frac{h^3}{\mu}\frac{\partial p}{\partial x}\right) = 12 u_m \frac{\partial h}{\partial x} + 12 \frac{\partial h}{\partial t} \quad (1)$$

The lubricant is assumed to be incompressible. The first term in the right hand side represents the entraining action and the second term the squeeze action.

The film thickness equation is;

$$h(x,t) = h_e(t) + \frac{x^2 - e^2(t)}{2R(t)} - \frac{4}{\pi E'} \int_{x_A}^{e(t)} p(x',t) \ln \left| \frac{x - x'}{e(t) - x'} \right| dx' \qquad (2)$$

The dynamic load is balanced by the hydrodynamic pressure;

$$f(t) = \int_{x_A}^{e(t)} p(x,t)\, dx \qquad (3)$$

The pressure-viscosity relation is modelled by the Roelands Eq. (5);

$$\mu = \mu_0 \exp \left\{ \frac{p_0 \alpha}{z} \left[\left(1 + \frac{p}{p_0} \right)^z - 1 \right] \right\} \qquad (4)$$

By introducing the dimensionless parameters indicated in the notation, the above equations become;

$$\frac{\partial}{\partial X}\left(\frac{H^3}{\bar{\mu}} \frac{\partial P}{\partial X} \right) = N \left(A \frac{\partial H}{\partial X} + S \frac{\partial H}{\partial \theta} \right) \qquad (5)$$

$$H(X,\theta) = H_e(\theta) + \frac{X^2 - E^2(\theta)}{2A(\theta)} - \frac{4}{\pi} \int_{X_A}^{E(\theta)} P(X',\theta) \ln \left| \frac{X - X'}{E(\theta) - X'} \right| dX' \qquad (6)$$

$$W(\theta) = \int_{X_A}^{E(\theta)} P(X,\theta)\, dX \qquad (7)$$

The Reynolds boundary condition is used for pressure at the cavitation boundary, i.e.,

$$P(E,\theta) = \frac{\partial P(E,\theta)}{\partial X} = 0 \qquad (8)$$

2.2 Grubin's Approximation

The simultaneous solution to Eqs. (5) to (8) is no easy task. Due to the nonlinearity, the iteration is required and the numerical stability needs to be considered. Realizing this fact, Vichard [3] used the Grubin's approximation and he was able to obtain the following nonlinear ordinary differential equation:

$$\frac{2W^{3/2}}{GNA^{3/2}} - G_p = \frac{S}{\sqrt{WA}} \left[\frac{2W\dot{A}}{3A} G_a + 2\dot{W} G_w - \dot{H}_0 \left(G_s + 4 \frac{W^3}{H_0^3} \right) \right] \qquad (9)$$

$$G_p = \int_0^1 \left[z^2 \sqrt{1 - z^2} - z^4 \log \left(\frac{1 + \sqrt{1 - z^2}}{z} \right) \right] \frac{dz}{\Delta} \qquad (10)$$

$$G_a = \int_0^1 z \left(1 - z^2 \right)^{3/2} \frac{dz}{\Delta} \qquad (11)$$

$$G_s = \int_0^1 z^3 \frac{dz}{\Delta} \tag{12}$$

$$G_w = \int_0^1 \times z^3 \left[\log \left(\frac{1 + \sqrt{1 - z^2}}{z} \right) - \sqrt{1 - z^2} \right] \frac{dz}{\Delta} \tag{13}$$

$$\Delta = \left[\frac{\beta z^2}{2} + \sqrt{1 - z^2} - z^2 \log \left(\frac{1 + \sqrt{1 - z^2}}{z} \right) \right]^3 \tag{14}$$

$$\beta = \frac{H_0}{W} \tag{15}$$

Note that z is a dummy variable for integration. Given the initial H_0, Eq. (9) can be solved by using a time marching technique, and with the periodic condition, the converged solution can be found. Vichard [3] used exponential approximations for G_i (where i = p,a,s and w), but the direct numerical integrations are used in this paper.

The squeeze velocity distribution is expressed by

$$\dot{H} = \dot{H}_0 + 2\dot{W}\left[-B \sqrt{B^2 - 1} - \log\left(-B + \sqrt{B^2 - 1} \right) \right] - 4W\dot{B}\left(B^2 - 1 \right)^{1/2} \tag{16}$$

where,

$$B = \frac{X}{2\sqrt{WA}} \tag{17}$$

The above formulation provides a considerable simplification in treating the transient EHL line contact problem so that the solution can be obtained quickly and without difficulty.

3. NUMERICAL METHOD

Even though the Grubin's approximation is good for quick calculations, it has limitations in accuracy. Thus, the full numerical solution is required to get a more realistic solution. To avoid iterating on the location of the outlet boundary, the following body-fitted coordinate transformation is used [7];

$$\xi = \frac{X - X_A}{E(\theta) - X_A} \tag{18}$$

which results in the following transformed equations,

$$\frac{1}{\left(E - X_A\right)^2} \frac{\partial}{\partial \xi} \left(\frac{H^3}{\bar{\mu}} \frac{\partial P}{\partial \xi} \right) = N \left(\frac{A}{E - X_A} \frac{\partial H}{\partial \xi} + S \frac{\partial H}{\partial \theta} - S \frac{\dot{E}\xi}{E - X_A} \frac{\partial H}{\partial \xi} \right) \tag{19}$$

$$H(\xi,\theta) = H_e(\theta) + H_g(\xi,\theta) + H_w(\xi,\theta) \tag{20}$$

$$W(\theta) = \left(E(\theta) - X_A\right) \int_0^1 P(\xi,\theta)\, d\xi \tag{21}$$

where,

$$H_g(\xi,\theta) = \frac{1}{2A(\theta)}\left\{\left(\xi\left(E(\theta) - X_A\right) + X_A\right)^2 - E^2(\theta)\right\} \tag{22}$$

$$H_w(\xi,\theta) = \frac{4\left(E(\theta) - X_A\right)}{\pi}\int_0^1 P(\xi,\theta)\ln\left|\frac{\xi - \xi'}{1 - \xi'}\right|d\xi' \tag{23}$$

The time derivative component is composed of three terms;

$$\frac{\partial H(\xi,\theta)}{\partial\theta} = \dot{H}_e(\theta) + \dot{H}_g(\xi,\theta) + \dot{H}_w(\xi,\theta) \tag{24}$$

The $\dot{H}_g$ can be obtained analytically, and $\dot{H}_e$ and $\dot{H}_w$ are treated numerically;

$$\dot{H}_e = \frac{H_e^n - H_e^{n-1}}{\Delta t} \tag{25}$$

$$\dot{H}_w = \frac{\left(H_w\right)_i^n - \left(H_w\right)_i^{n-1}}{\Delta t} \tag{26}$$

Equation (19) is central differenced, and the Newton-Raphson method is used to solve the system equations. The unknowns are pressures, cavitation boundary, and the film thickness at the exit point. This "system approach" was used by Houpert and Hamrock [8] for line contact and Lim and Brewe [9] for point contact. Since the details of the system approach are available in the above references, the derivations of the system matrix is not presented here. In general the converged solution is obtained within three iterations when the steady-state solution is used as the initial condition. The inlet boundary x_A is taken as three times the semi-Hertzian length. In the dimensionless parameter S, ω_D is dynamic frequency. It is usually not equal to the rotational speed, but it has to do with the system excitation frequencies. For example, the natural frequency of the gear system is one of the sources of excitation frequency.

4. RESULTS AND DISCUSSION

4.1 The transient solution of the cylinder-plate geometry.

In the governing equations described in section 2.1, the operating parameters (force, speed, and the radius of curvature) at the contact are time- dependent. To help understand some basic transient effects without the added complexity of a time dependent curvature and rolling speed, a dynamically loaded rolling cylinder on the plate is

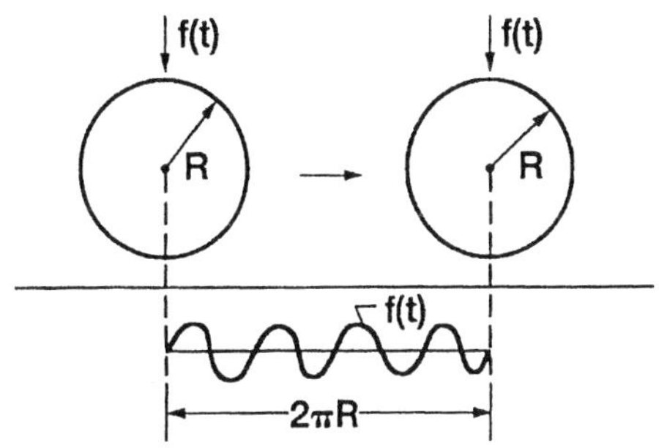

Figure 1. The dynamically loaded cylinder on the plate model.

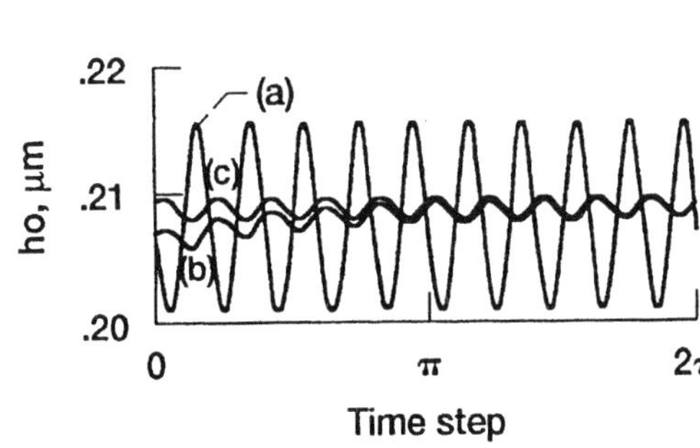

Figure 2. Mimimum film thickness distribution for $\gamma = 10$ and $\beta = 0.3$.

considered in Fig. 1. The dynamic load is mimicked by the following sinusoidal variation:

$$f(t) = f_0\left(1 + \beta \sin \omega_D t\right) \tag{27}$$

$$\omega_D = \gamma\omega \tag{28}$$

The period for one revolution of the cylinder is $2\pi/\omega\gamma$, and the period for the dynamic frequency is $2\pi/\omega\gamma$. The γ represents the intensity of the frequency in the dynamic loading. For example, when $\gamma = 10$, there are 10 fluctuations in one complete rotation of the cylinder. The β represents the ratio of the applied force relative to the mean load. According to the experimental results [10], the dynamic contact loading in gears is composed of many harmonic fluctuations during a meshing period. Thus, these sinusoidal variations with different γ can be used to mimic fundamental gear dynamics. The operating parameters for this section are $f_0 = 1.281\times10^6$ N/m, $R_i = 0.02025$ m, $u_m = 9.5365$ m/sec, $\mu_0 = 4.9\times10^{-3}$ Pa · sec, G = 2088.

Figure 2 shows the film thickness distribution for $\gamma = 10$ and $\beta = 0.3$ using the Grubin's approximation. Starting with the QSSS as an initial condition for the TS (curve b), the converged TS (curve c) is obtained utilizing the periodicity condition. Comparing with QSSS (curve a), the damping effect is apparent with some amount of phase-lag and attenuated amplitude. The fundamental assumption here is that the film thickness in the Hertzian contact region is flat so that the typical EHL film shape at the exit region and the possible film thickness variation in the Hertzian region due to the squeeze effect cannot be described properly.

The full numerical solution for $\gamma = 10$ and $\beta = 0.3$ with same operating parameters is presented in Fig. 3. There is no significant amplitude change in the minimum film thickness although some phase-lag is observed. But, there is small change in central film thickness distribution with

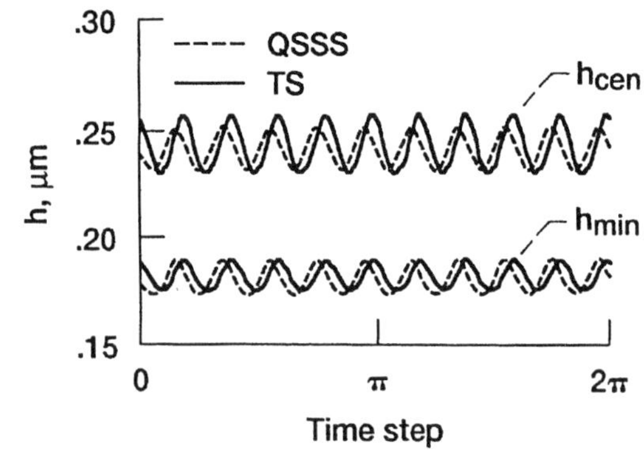

Figure 3. Minimum and central film thickness distribution for $\gamma = 10$ and $\beta = 0.3$.

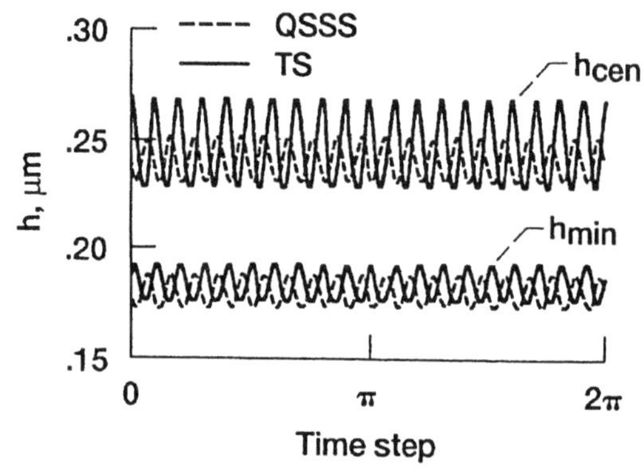

Figure 4. Minimum and central film thickness distribution for $\gamma = 20$ and $\beta = 0.3$.

phase-lag. The transient effects can be seen more vividly for $\gamma = 20$ in Fig. 4. The mean level for both the minimum and central film thickness distribution is higher for the TS and the QSSS. This is not what we have come to expect from the damping phenomena observed in Fig. 2. However, the amplitude of the TS central film thickness is nearly twice that for the QSSS. Thus the squeeze action inherent with the TS has a greater impact on the central film thickness as the dynamic frequency is increased.

To examine this phenomena, it is necessary to look into the detailed pressure distribution, film thickness profile, and the squeeze velocity distribution. The sequence of the film shape during one dynamic loading cycle (Fig. 5) reveals significant differences between the TS and QSSS. Compared

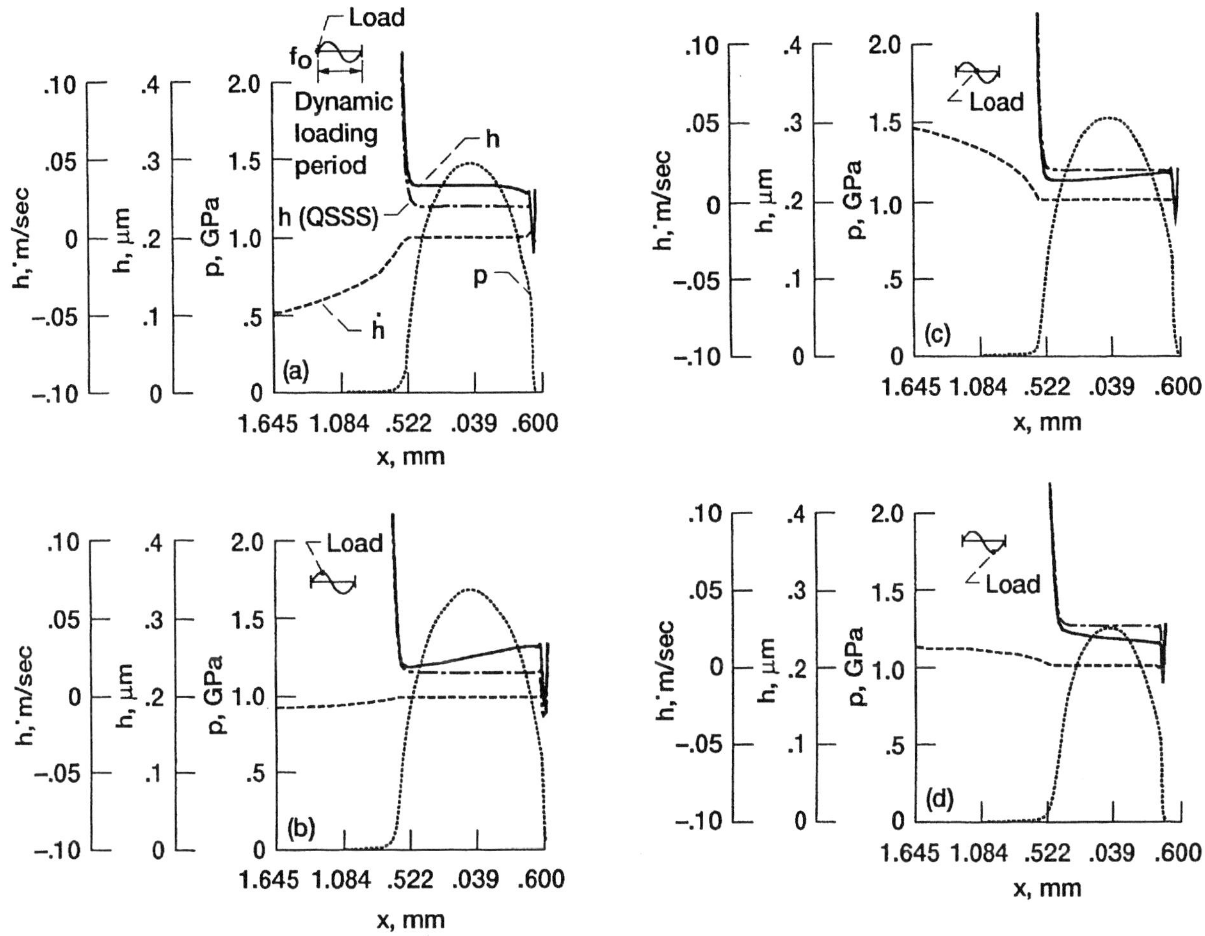

Figure 5. Film thickness, pressure and squeeze velocity distribution during one dynamic cycle for Fig. 4.

with the film thickness profiles for QSSS, the TS film thickness profiles are shifted and dimpled to varying degrees in time. Also, it is seen that the central film thickness is influenced more than the minimum film thickness. This fact results in a large change in the central film thickness distribution. In Ref. [11], the experimental results of transient EHL were presented, where this kind of dimple shape was detected in the point contact case but not in the line contact case. But, the loading in this paper is two orders of magnitude higher than the experimental condition for the line contact and the dimple shape is expected in the highly loaded case.

Next, since the excitation frequency is, in general, a combination of low frequency and high frequency, the following dynamic loading is considered for the cylinder-plate model:

$$f(t) = f_0\left(1 + \beta_1 \sin \omega_{D1} t + \beta_2 \sin \omega_{D2} t\right) \qquad (29)$$

Figure 6 depicts the film thickness distribution for $\beta_1 = 0.2$, $\beta_2 = 0.3$, and $\omega_{D1} = \omega_D$, $\omega_{D2} = \omega$. The film thickness distribution shows the response to high frequency variation as well as low frequency variation. Again, the most significant differences between the TS and QSSS are observed

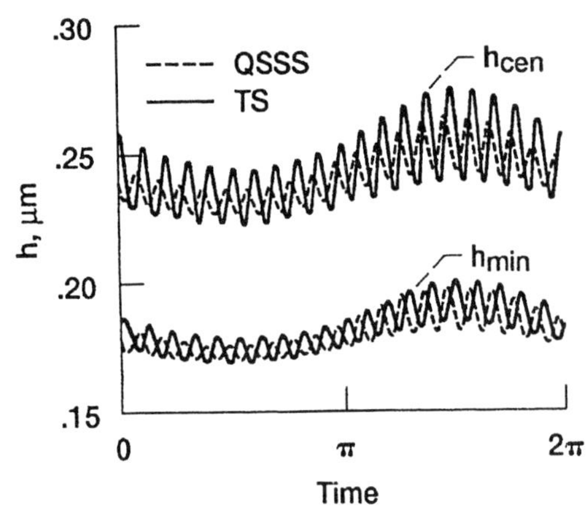

Figure 6. Minimum and central film thickness distribution for the dynamic loading of Eq. (29).

with the results of the central film thickness. The damping effect is manifest in the phase-lag whereas the increased amplitude of the central film thickness is due to the elastic behavior of the solid and resulting film-shape.

4.2 The Transient Solution for the Spiral Bevel Gear

Spiral bevel gears are widely used in power transmission systems. In the design of gears, the lubrication analysis is essential since inadequate lubrication causes the gear failure. The so-called λ ratio, the ratio of the oil film thickness to the magnitude of the surface roughness is one of the major factors which determine the gear life.

For the spiral bevel gears used in this paper, the contact parameters, load, speed, and the radius of curvatures are obtained by running the tooth contact analysis (TCA) computer program provided in ref. [12]. The numerical data are taken from the helicopter transmission which is being tested at NASA Lewis Research Center. According to the TCA output for the gear under consideration, the order of magnitude of the normal force at contact is 20 kN, where the maximum Hertzian pressure is about 1.5 GPa and the ellipticity parameter is greater than 25. Thus, the nature of the contact is almost line contact. That is, the ellipticity parameter is so high that the effect of the side leakage of the lubricant is negligible. Accordingly, the point contact TCA output can be converted to the line contact parameters using the effective length of the Hertzian contact. That is, using the relation

$$\frac{\pi p_h a \ell_e}{2} = \frac{2\pi a b p_h}{3} \tag{30}$$

the effective length

$$\ell_e = \frac{4b}{3} \tag{31}$$

Thus, the effective line contact force = f_p/ℓ_e N/m. The gear specifications, the operating conditions, lubricant properties, and the gear material properties are provided in Table 1. Fig. 7(a) shows the effective forces in three positions of the gear. The three positions refer to the contact patterns at the mean (centered), toe, and heel positions. The TCA is based on the assumption that the teeth are rigid and that only one pair of teeth are in contact at any time. The fluctuations of the loading due to the vibrations are not

Table 1
Gear operating parameters

Gear Data:	
Number of teeth of gear	62
Number of teeth of pinion	19
Young's modulus, GPa	207
Poisson's ratio	0.3
Operating condition:	
Torque, N-mm (lb-in.)	549219 (4862)
Pinion speed, rpm	6016
Lubricant:	
Temperature, °C (°F)	100 (212)
Lubricant	MIL-L-23699
Kinematic viscosity (cSt)	5.09
Specific gravity	0.9725
Pressure-viscosity coefficient, 1/GPa	9.50

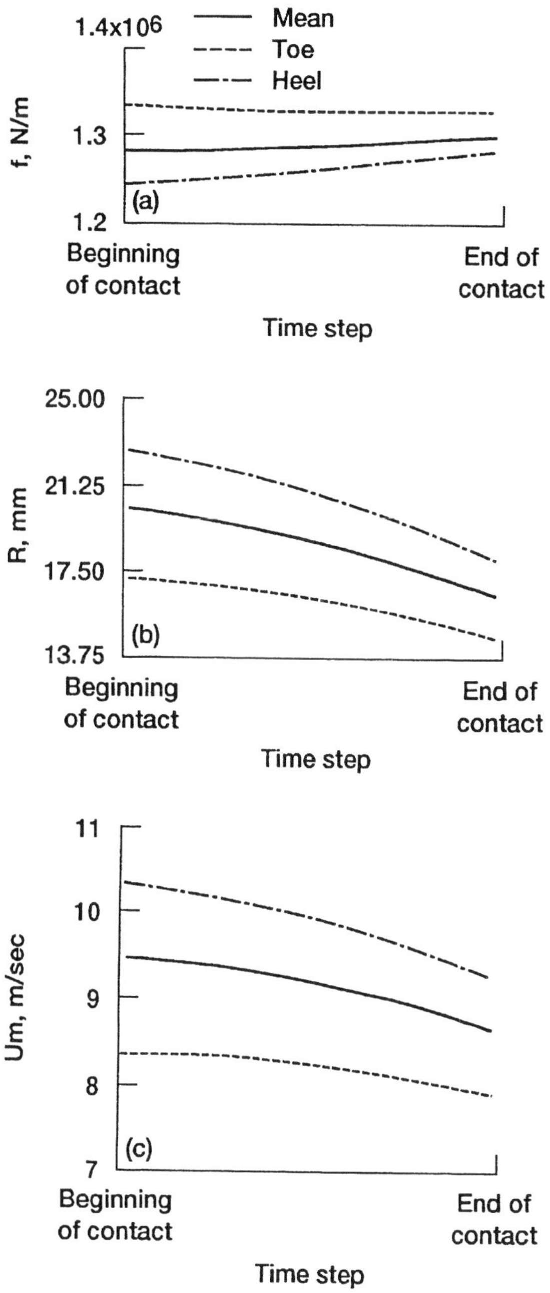

Figure 7. Contact parameter variation of the spiral bevel gear during one meshing period (a) load (b) radius of curvature (c) mean rolling speed.

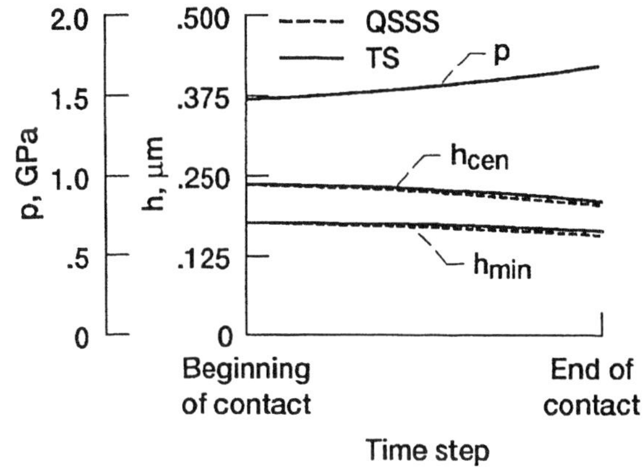

Figure 8. Film thickness and maximum pressure distribution for quasi-dynamic loading in spiral bevel gear.

considered, therefore the contact forces are approximations. In this sense, they are called quasi-dynamic loads in this paper. The effective radius of curvatures and the mean rolling speeds of contacts are provided in Figs. 7(b) and (c).

Figure 8 compares the steady-state solution and the transient solution of the mean position for the quasi-dynamic loading. The steady- state solution of the initial contact was used as an initial condition for the transient solution.It is seen that the difference is quite small.

Since the quasi-dynamic loading is different from the real situation, the following simulated dynamic loading is used for this transient analysis

$$f(t) = f_s\left(1 + \beta \sin \omega_D t\right) \qquad (32)$$

Again, γ represents the intensity of the dynamic loading. The meshing period is $2\pi/\omega_p N_p$ and the period for the dynamic frequency, t_D is $2\pi/\omega_p N_p \gamma$. Thus, the dynamic frequency, ω_D is $2\pi/t_D$.

In the gear problem, unlike the constant geometry and rolling speed in the cylinder on the plate model, all the contact parameters are changing with different dynamic frequencies. And, since the loading fluctuations are occurring in a meshing period which is a very short time compared with the rolling period of the cylinder on the plate, the intensity of the squeeze effect is higher than the simplified model.

Figure 9 shows the comparison between the QSSS and TS for $\beta = 0.1$ and $\gamma = 10$. The general trend is that, as discussed in section 4.1, the squeeze velocity causes a large amount of fluctuations in the film thickness distribution. But, the difference between the maximum pressure distribution is small. In this simulation, the meshing cycle with decreasing rolling speed and curvature and increasing quasi-dynamic load give rise to a low frequency variation. The load is composed of the low frequency and the mimicked high frequency combination. The initial condition at the beginning of the engagement affects the film thickness distribution. But, the regularity is established beyond the point A. In a gear problem, unlike the cylinder on the plate, the periodicity condition is unavailable since the condition at the beginning and the end of the contact are not the same. The nature of the initial contact is impact loading and its condition is different from the steady-state solution. The reasonable estimation of this initial condition will improve the TS and this is worth addressing in future research.

Increasing β to 0.2 and retaining the same γ, the variation in the central film thickness is increased (Fig. 10). The shape of the minimum film thickness distribution has changed reflecting the influence of the time varying curvature and speed. A significant portion of the minimum film TS is lower than the QSSS. But, the differences in the maximum pressures remain small.

In Fig. 11, the γ is increased to 20 with $\beta = 0.1$. It is seen that the central and minimum film thickness distribution is shifted upward and the mean levels are significantly higher than QSSS. Also, the differences in the maximum pressure are significant. The TS generally predicts higher maximum pressures than the QSSS.

In the above observations, the large β causes lower minimum film thickness distribution than the QSSS and high frequency (γ) leads to increased minimum film thickness distribution and increased maximum pressure distribution.

One of the important steps in the lubrication analysis is to have an accurate dynamic loading distribution during the meshing cycle. The TCA

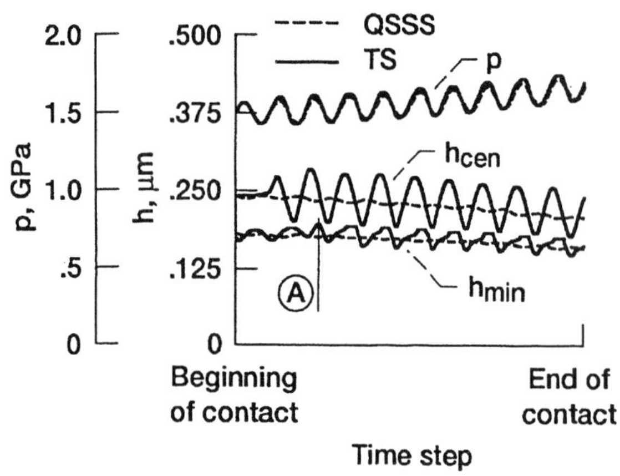

Figure 9. Film thickness and maximum pressure distribution for mimicked dynamic loading in spiral bevel gear ($\beta = 0.1$, $\gamma = 10$).

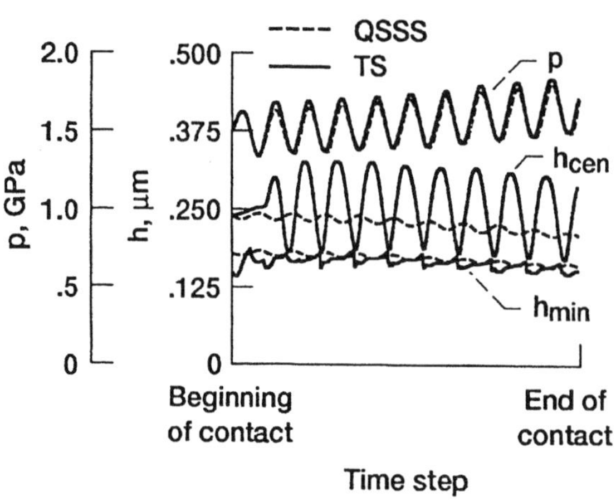

Figure 10. Film thickness and maximum pressure distribution for mimicked dynamic loading in spiral bevel gear ($\beta = 02$, $\gamma = 10$).

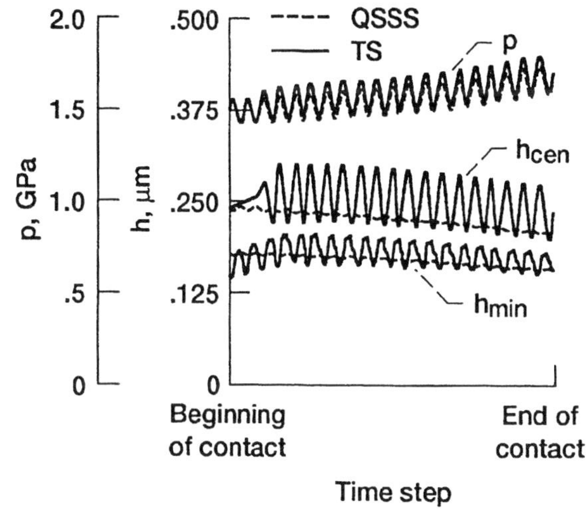

Figure 11. Film thickness and maximum pressure distribution for mimicked dynamic loading in spiral bevel gears ($\beta = 0.1$, $\gamma = 20$).

program used in this paper provides an approximation. Especially, since the contact ratio is generally greater than one, it is expected that there is an abrupt change in the contact force distribution. This may lead to a more severe squeeze effect than reported in this paper.

5. CONCLUSION

The transient effects of the spiral bevel gear lubrication problem were studied. To help understand some basic transient effects without the added complexity of a time-dependent geometry, a dynamically loaded rolling cylinder on a plate was analyzed first. Because the ellipticity parameter determined for the spiral bevel gear geometry is large (i.e., 25 to 30), the line contact model is used throughout. To begin, the Grubin's approximation, one could understand more readily the role of certain film shape details in the transient behavior. The transient behavior was illuminated by comparing solutions which included the squeeze motion (i.e., referred to as transient solutions (TS)) with quasi-steady-state solutions (i.e., QSSS) that did not include the squeeze motion. For the spiral bevel gear analyzed in this study, the contact parameters (load, speed, and the radius of curvatures) were obtained from the tooth contact analysis (TCA) computer program for a helicopter transmission being tested at this laboratory. Unlike the dynamic roller on a plate model, the curvatures and speed are time dependent. The effect of periodic loading as governed by the force-ratio, β, and the dynamic frequency factor, γ, was investigated to produce the following results:

1. The effect of doubling the dynamic frequency factor, γ, increased the mean film thickness (both minimum and central) levels significantly throughout the dynamic cycle. It also tended to increase the mean level of the maximum pressure.

2. The effect of doubling the dynamic force factor, γ, resulted in the amplitude of the central film thickness nearly doubled. However, the amplitude of the minimum film thickness decreased slightly. The amplitude of the maximum pressure doubled, although the mean level of the maximum pressure remained the same.

3. The decrease in the mean level of film thickness (both minimum and central) for one dynamic cycle of the spiral bevel gear was attributed to the decrease in speed and curvature for that period of time. While the film thickness was decreasing the mean level of the maximum pressure was increasing.

4. The time trace of the film thickness (both central and minimum) for the transient solution (TS) lagged-in-phase the quasi-steady-state solution (QSSS) throughout the dynamic cycle for both the cylinder- on-plate and spiral bevel gear.

5. The most significant differences between the TS and QSSS are observed with the results of the central film thickness. In addition to the TS lagging the QSSS in phase, the amplitude of the central film thickness was increased significantly due to the elastic behavior of solid and resulting film-shape.

6. ACKNOWLEDGMENT

This work was done while S.G. Lim held a National Research Council—NASA Lewis Research Center Research Associateship.

7. REFERENCES

(1) Dowson, D.; and Higginson, G.R.: Elastohydrodynamic Lubrication. Pergamon Press, 1966.

(2) Gu, A.: Elastohydrodynamic Lubrication of Involute Gears. J. Eng. Ind., vol. 95, no. 4, Nov. 1973, pp. 1164-1170.

(3) Vichard, J.P.: Transient Effects in the Lubrication of Hertzian Contact. J. Mech. Eng. Sci., vol. 13, no. 3, 1971, pp. 173-189.

(4) Dowson, D.; Ruddy, B.L.; and Economou, P.N.: The Elastohydrodynamic Lubrication of Piston Rings. Proc. Royal. Soc. London Ser. A, vol. 386, 1983, pp. 409-430.

(5) Lee, R.; and Hamrock, B.J.: Squeeze and Entraining Motion in Nonconformal Line Contacts Part II—Elastohydrodynamic Lubrication. ASME Paper 88-Trib-37, 1988.

(6) Tieu, A.K.; and Worden, J.: Transient Oil Film Thickness in Gear Contacts Under Dynamic Loads. Fluid Film Lubrication—Osborne Reynolds Centenary; Proceedings of the 13th Leeds-Lyon Symposium on Tribology, Dowson, D. et al., eds., Elsevier, New York, 1986, pp. 285-290.

(7) Lim, S.G.; Brewe, D.E.; and Prahl, J.M.: On the Numerical Solution of the Dynamically Loaded Hydrodynamic Lubrication of the Point Contact Problem. Tribol. Trans., vol. 34, no. 2, 1991, pp. 195-204.

(8) Houpert, L.G.; and Hamrock, B.J.: Fast Approach for Calculating Film Thickness and Pressures in the Elastohydrodynamically Lubricated Contacts at High Loads. J. Tribol., vol. 108, no. 3, July 1986, pp. 411-420.

(9) Lim, S.; and Brewe, D.: A System-Approach to the Elastohydrodynamic Lubrication Point-Contact Problem. Tribol. Trans., vol. 35, no. 2, 1992, pp. 367-373.

(10) Kasuba, R.; and Radzimovsky, E.I.: A Multi-Purpose Planetary Gear Testing Machine for Studies of Gear Drive Dynamics, Efficiency, and Lubrication. J. Eng. Ind., vol. 95, no. 4, Nov. 1973, pp. 1123-1130.

(11) Ren, N.; Zhu, D.; and Wen, S.Z.: Experimental Method for Quantitative Analysis of Transient EHL. Tribol. Int., vol. 24, no. 4, Aug. 1991, pp. 225-230.

(12) Litvin, F.L.; Zhang, Y.; and Chen, J.: Users Manual for Tooth Contact Analysis of Face-Milled Spiral Bevel Gears with Given Machine-Tool Settings. NASA CR-189093, 1991.

Thin Films in Tribology / D. Dowson et al. (Editors)
© 1993 Elsevier Science Publishers B.V. All rights reserved.

Surface contact in a symmetrical loading of a multileaf journal bearing.

I.Iordanoff[a], P Hermel[b] and P Stéfan[c]

[a]Student at the P.Sabatier university

[b]Ingineer at ABG.SEMCA·

[c]University Lecturer at P.Sabatier university.

This work is about the analytical and experimental study of the starting torque of a multileaf journal bearing with a very small radial clearance and a study of surface contacts.

INTRODUCTION

Replacement of the high-speed, high temperature turbine ball bearing may be attractive in the conception of little vehicular gaz turbine.Thus, the solution often proposed at low load is an expansible gaz bearing. The bearing studied in the report is a multileaf bearing. one of the most important points in the conception of the bearing is the knowledge of the starting torque: it will determine the maximum power necessary to make the turbine function. This work is about the analytical and experimental study of the leaves-to-shaft loading. When the shaft is installed in the bearing the relative radial clearance is about 10^{-3}. So one of the main problems of the research is to simulate the important surface contact between the different parts of the bearing. The solution we find is taken from a finite element method. Experimental study has been made by taking different bearings with different surface finishes in order to compare them with theoretical study. It is found that both give approximately the same starting torque, contact surface and final geometry. This paper is divided into three parts. First we give a description of the bearing after the leaves are installed. In the second part we research the bearing load when the leaf-to-leaf linking starts sliding . Then we impose a step by step loading till we reach the final shaft radius.

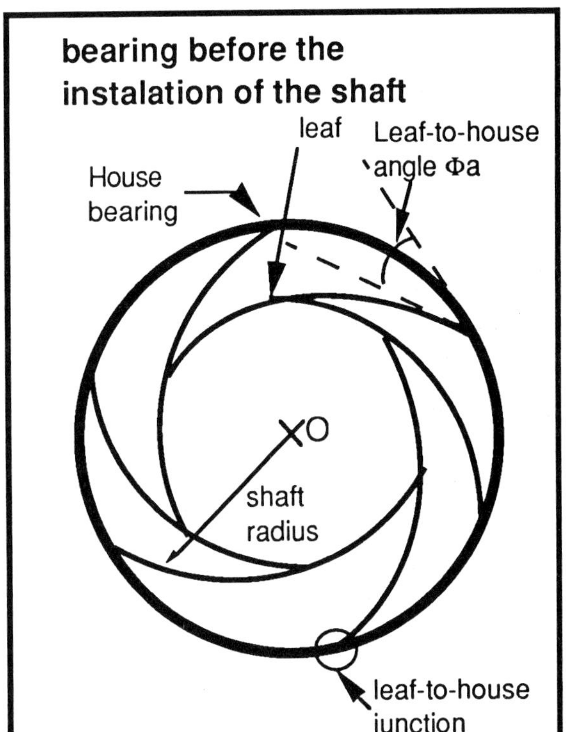

figure 1

1.DESCRIPTION OF THE MODEL

1.1The step by step loading

An initial study gives the geometry and preloading conditions of the instalation of the leaves in the bearing.Then, we find the points of the leaves which are closer to the bearing center. This gives us the initial shaft radius Ra. Then, at each incrementation of Ra, we find the new geometry of the bearing and all the boundary conditions. When The nominal radius is reached, the simulation ends.

1.2.Main problems to be solved

When the shaft is installed in the bearing, the relative radial clearance is about 10^{-3}. So contacts existing between the different parts of the bearing cannot be considered as punctual.

Leaf-to-leaf link gets bloked at the beginning of shaft instalation. The leaf-to-house link clearance makes the leaf-to-house angle Fa decrease during the loading.

1.3 Main hypothesis

We work with little deformation hypothesis.

The leaves are supposed to be circular beams at the beginning. At each step, we calculate U(a,b), V(a,b), V̲(a,b), w(a,b) denoting the linear radial-radial,tangential-radial, radial-tangential and tangential-tangential displacement(at point a)-force(at point b) relationships for a circular beam which approximates the real geometry of the leaf (reference 2 § 3).

The leaf-to-house link is supposed to be a unilateral pivot link (figure 2).When Mt>0, the leaf-to-house angle variation dFa is supposed to be very small at each step. Thus we will able to give linear relationships between the displacement vector DLn and DFa.

Leaves are considered tangentials at contact point.

The problem is supposed to be symmetrical and we will work on one leaf.

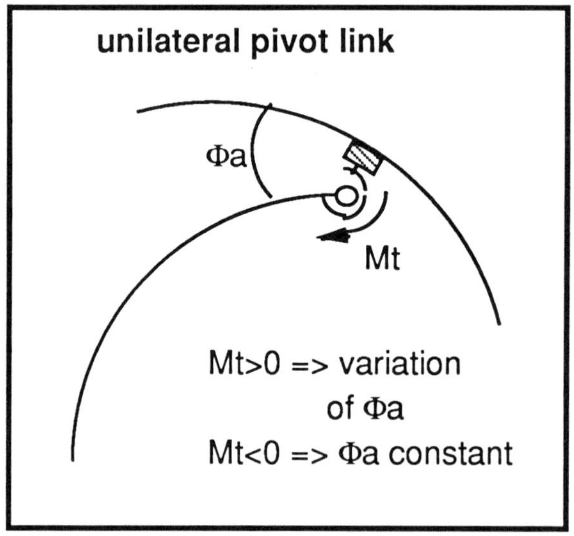

figure 2

1.4 Discretisation of the model.

Each leaf is discretised into n points. Each point is associated with a load Rj and a displacement DLj. Thus, linear relationships give us a matrix relation dRn=Kn,n*DLn, where Kn,n is a stiffness matrice, dRn is a loading vector and DLn is a displacement vector..

2.SIMULATION

2.1.Instalation of the leaves.(fig 3 § 4)

Point L1 has to be under point A2 (figure 3). If Fr0 is the load of the lower leaf, -Fr0 is the load of the upper leaf at the contact point. (figure 4) With

dL=A2L1+ Ep, we have the relation:

DL=Fr0(U(L,L)+U(A,A)-U(L,A)-U(A,L))

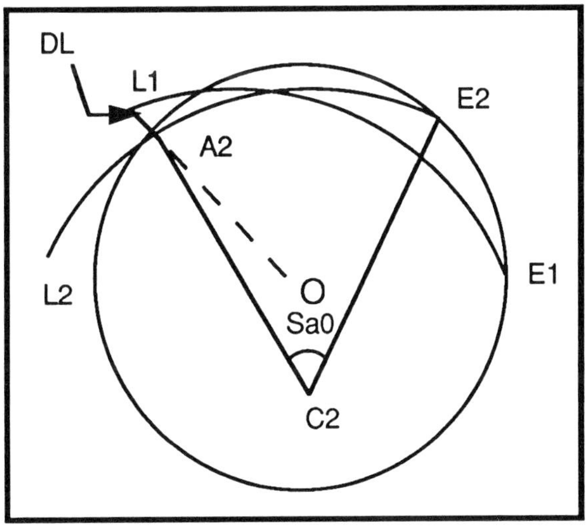

figure 3

Fr0 and the position of point A are obtained by an iterative calculation.

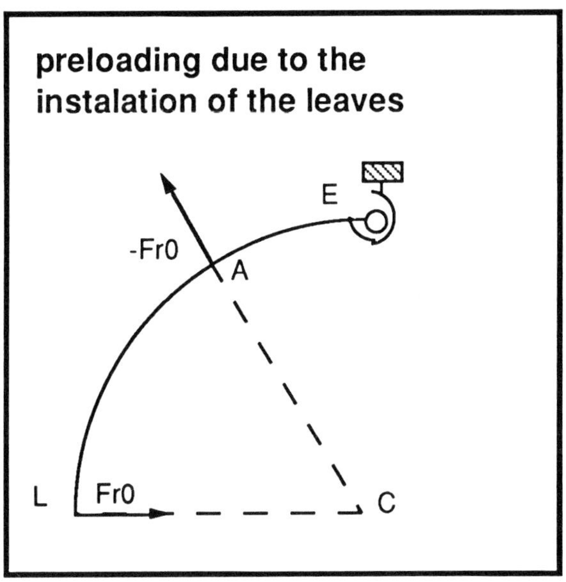

figure4

2.2.Beginning of the relative motion between leaves.

Contacts are supposed punctual and the leaf-to-house angle doesn't move in this part(Mt<0).We have four unknown elements

_ Fr: Normal reaction between leaves.
_ Ft: Tangential reaction between leaves
_ Fp: Radial shaft load.
_ Ra: Shaft radius at the beginning of relative motion.

We have 4 equations:

_ Equality of radial displacement at leaf-to-leaf contact.
_ Equality of tangential displacement.
_ Relationship between the leaf-to-beam contact point displacement and the load Fp.
_ Ft= mFr when relative motion begins.

Thus we are able to calculate the conditions when leaf-to-leaf linking starts sliding.

2.3.Simulation with contact areas and a variable angle Fa

We impose dRa, the incrementation of shaft radius. At each step, we know radial displacement at shaft-to-leaf contact point(1). Radial displacements are equal between leaf-to-leaf contact point(2). If a point is in contact with the house bearing, its displacement is 0(3).When the leaf-to-house angle Fa decreases, the pivot-link-balance makes it necessary for the moment Mt to be zero(4).The unknown elements are:

_ The reactions applied at each contact point.
_ The displacement of the points which are not contact points (in order to obtain the geometry of the leaf).
_ The variation of angle Fa

The conditions 1, 2, 3, 4 can be writtenDli=F(U,$\underline{V}$,V,W,dRi,dFa)

We can solve this first matrix equation and find dRm=Km,m*DLm, where dRm is the vector of the unknown contact loading. Then we can find the geometry of the leaf with the relation DLn= Sn,n * dRn where n is the number of the points of discretisation. (dRj=0 if j isn't a contact point and dRj=dRi if j is a contact point).

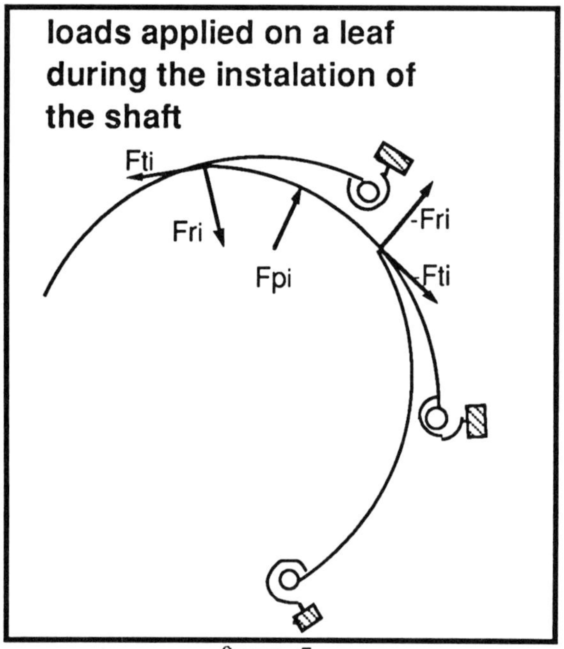

figure 5

Boundary conditions

We calculate for each leaf point the separation distance between the house bearing, the shaft and the other leaves. If this distance equals 0 we have a new contact point.

We also calculate, at each contact point the load Ri=S dRi. If Ri equals 0, the point is no longer a contact point.

2.4.Result and discussion

bearing geometry taken for the simulation visualised in figure 6 7 8 § 9

_ number of leaves:	_ 8
_ shaft radius:	variable
_ leaf thickness:	_ 0.17 mm

Figure 6 shows the saturation apparition as clearance approaches 0. We can see three different areas in this curve.

figure 6 : **Curve Torque=F(radial clearance)**

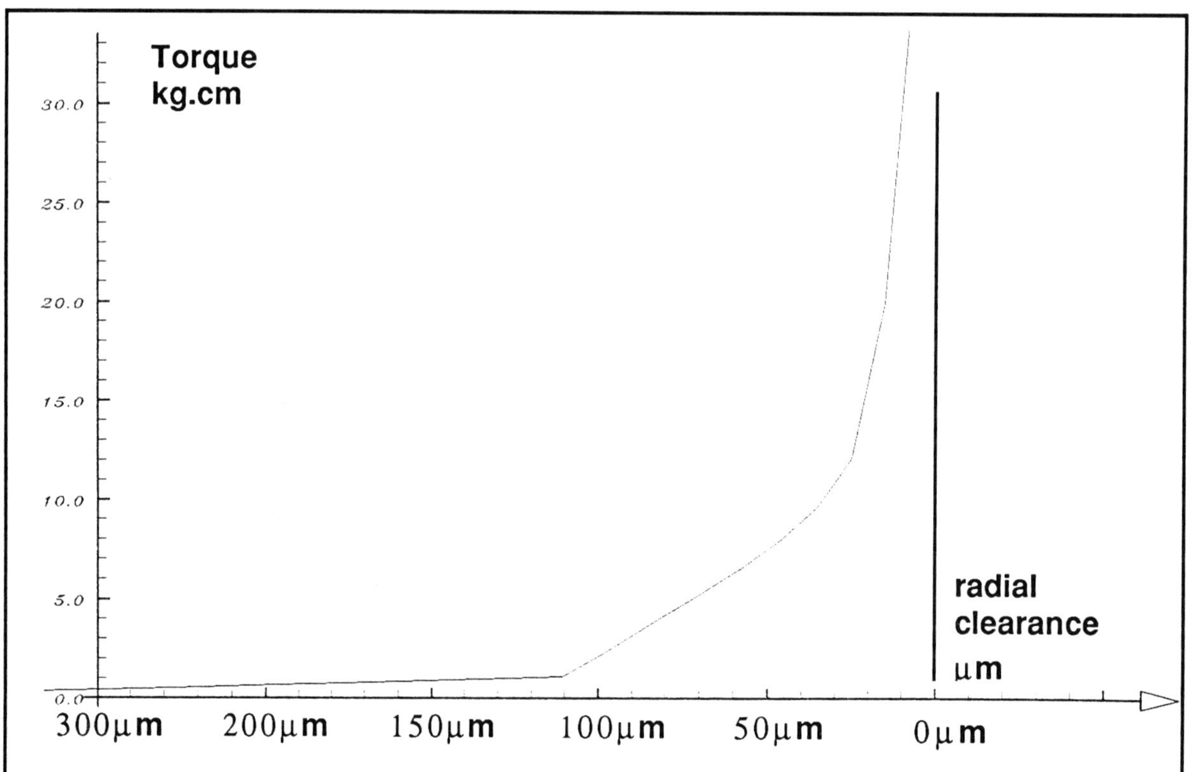

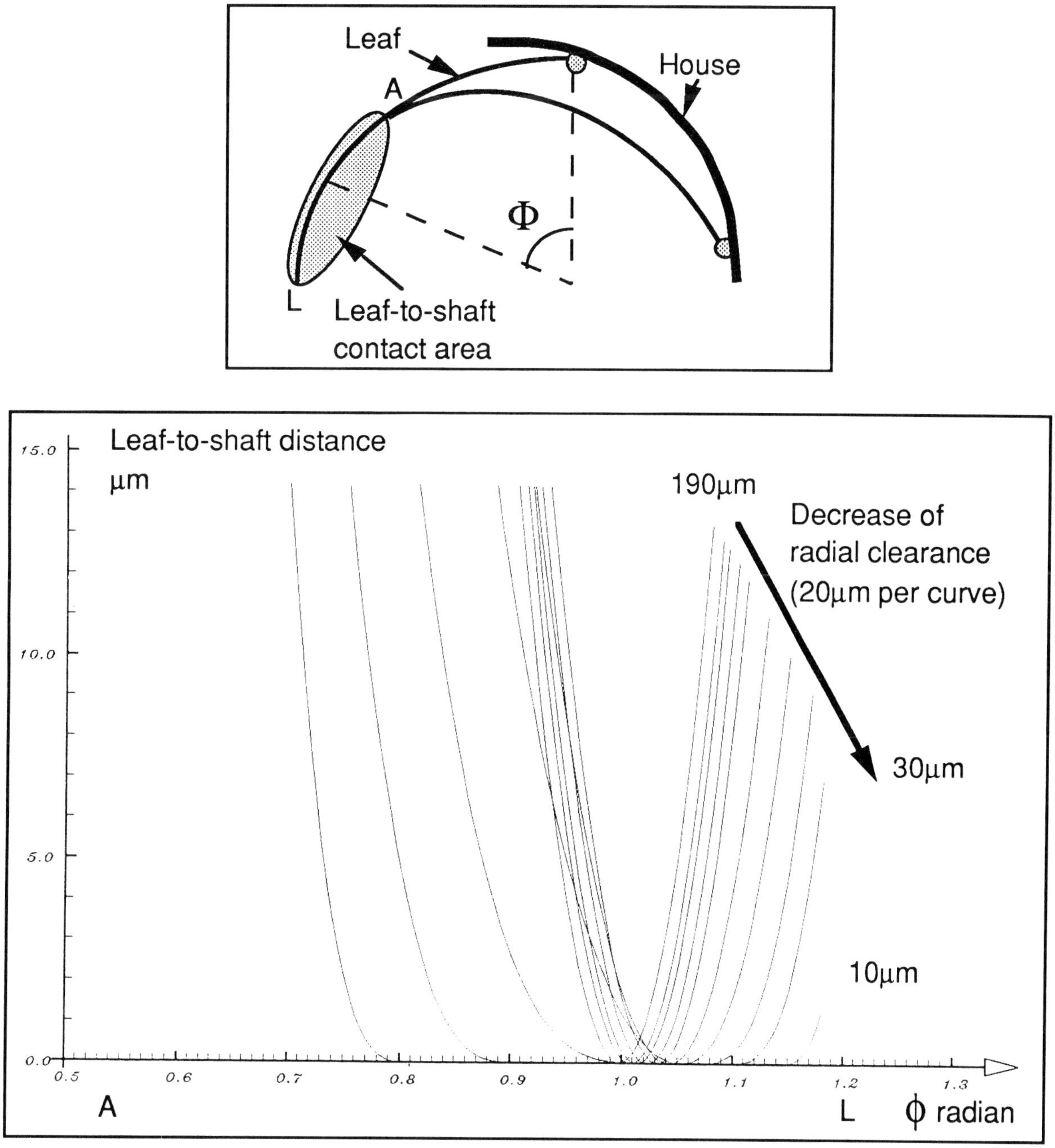

Figure 7

Figure 7 shows the development of the leaf-to-shaft contact zone near 80% of the leaf length. Experimentally, contact surface is developing in the same zone than in the simulation. The first area of the curve 6 is due to the development of this contact surface.

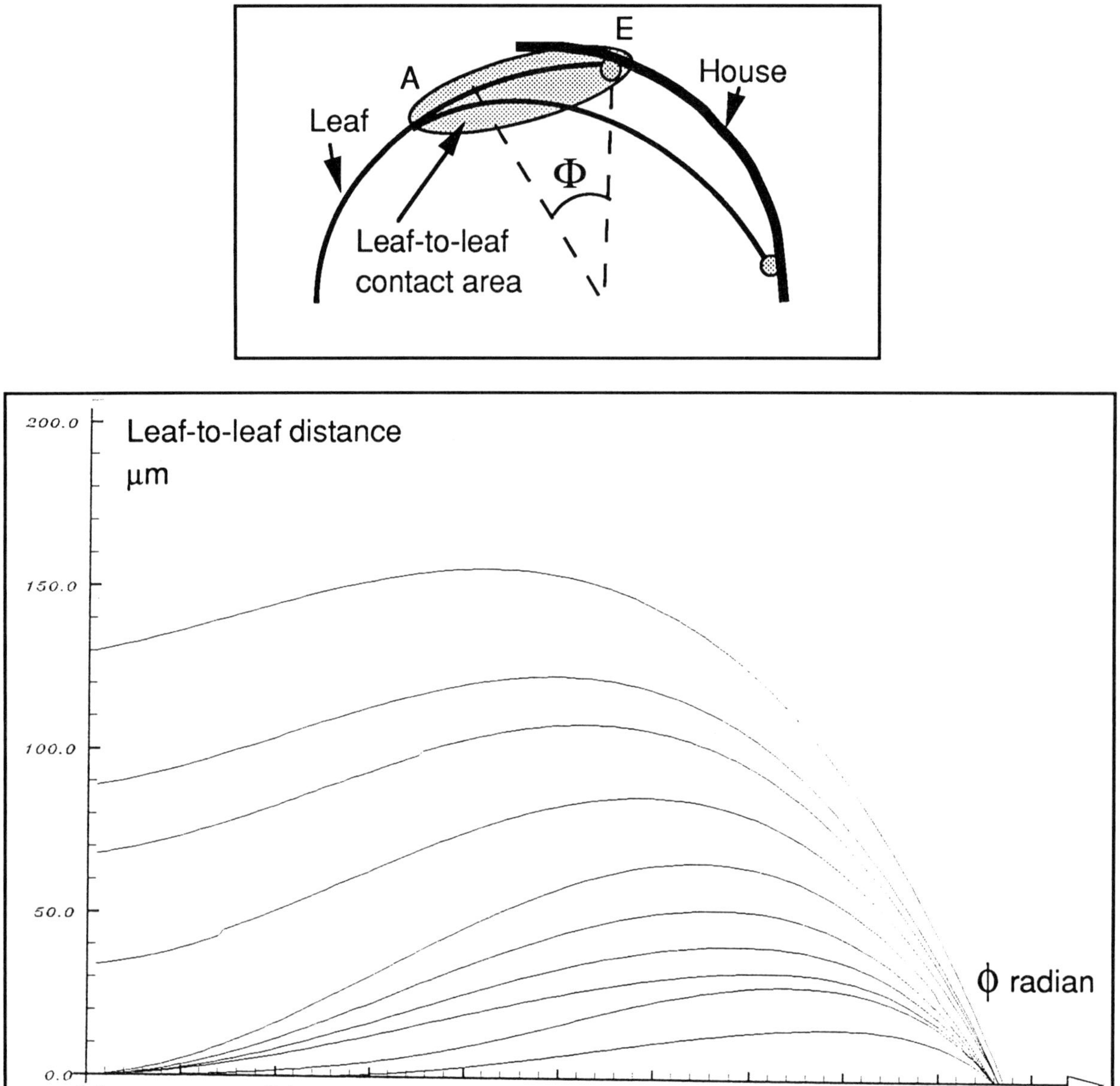

Figure 8

Figure 8 shows that there is the apparition of a second leaf-to-leaf contact surface near the leaf-to-house link. The second area of the curve 6 is due to the development of this contact surface.The apparition of this contact surface is confirmed by the experimental study made by pr.Glienicke (4). In his study, he found that the stiffness increases when a contact point appears near the leaf-to-house link.

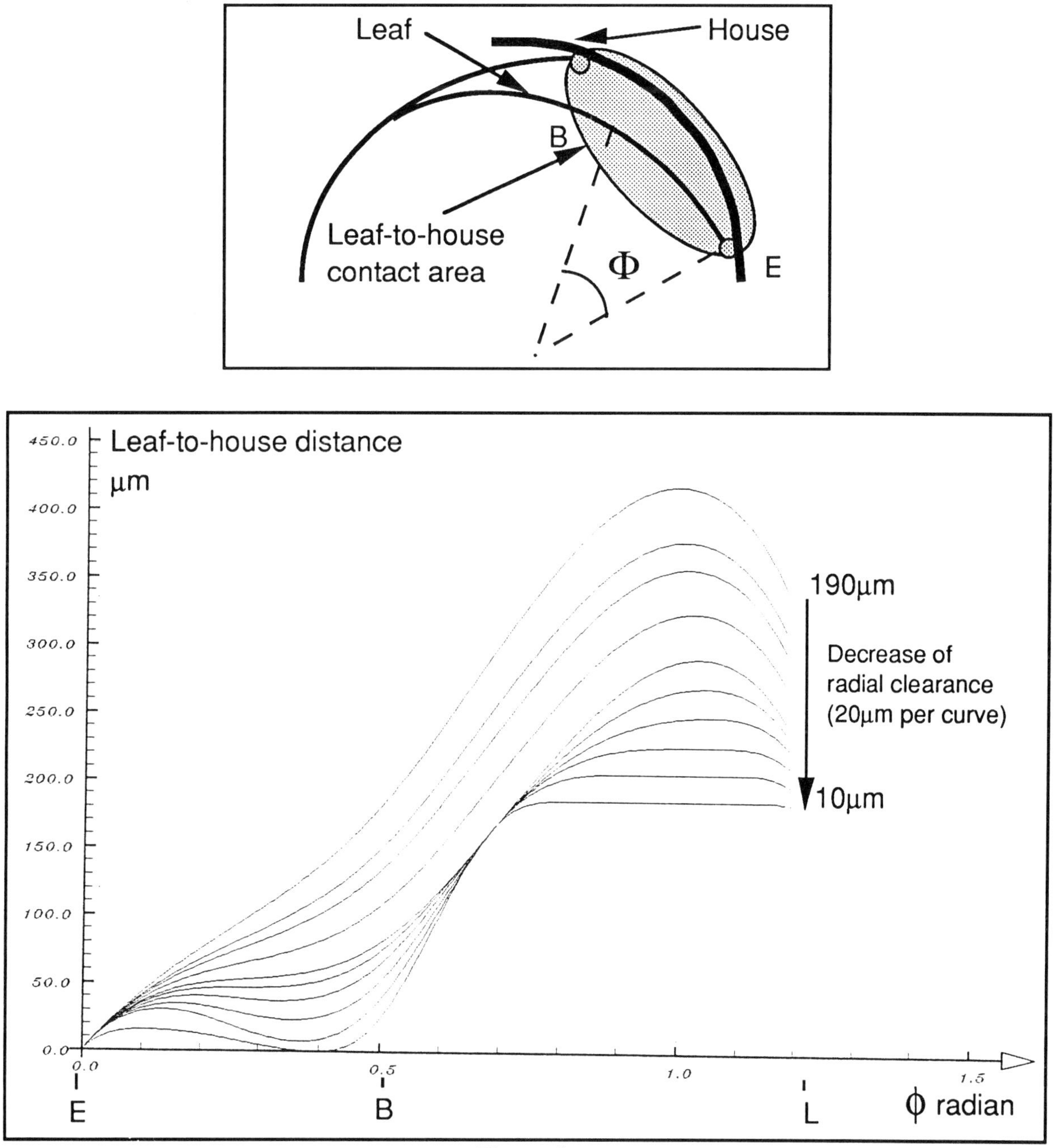

Figure 9

Figure 9 shows the apparition of a leaf-to house contact surface very close to the saturation. The third area of the curve 6 is due to this contact surface.

Pr Glienicke has made the same observation in his experimental study. The elastic part of the leaf is now reduced and the stiffness of each leaf becomes very high.

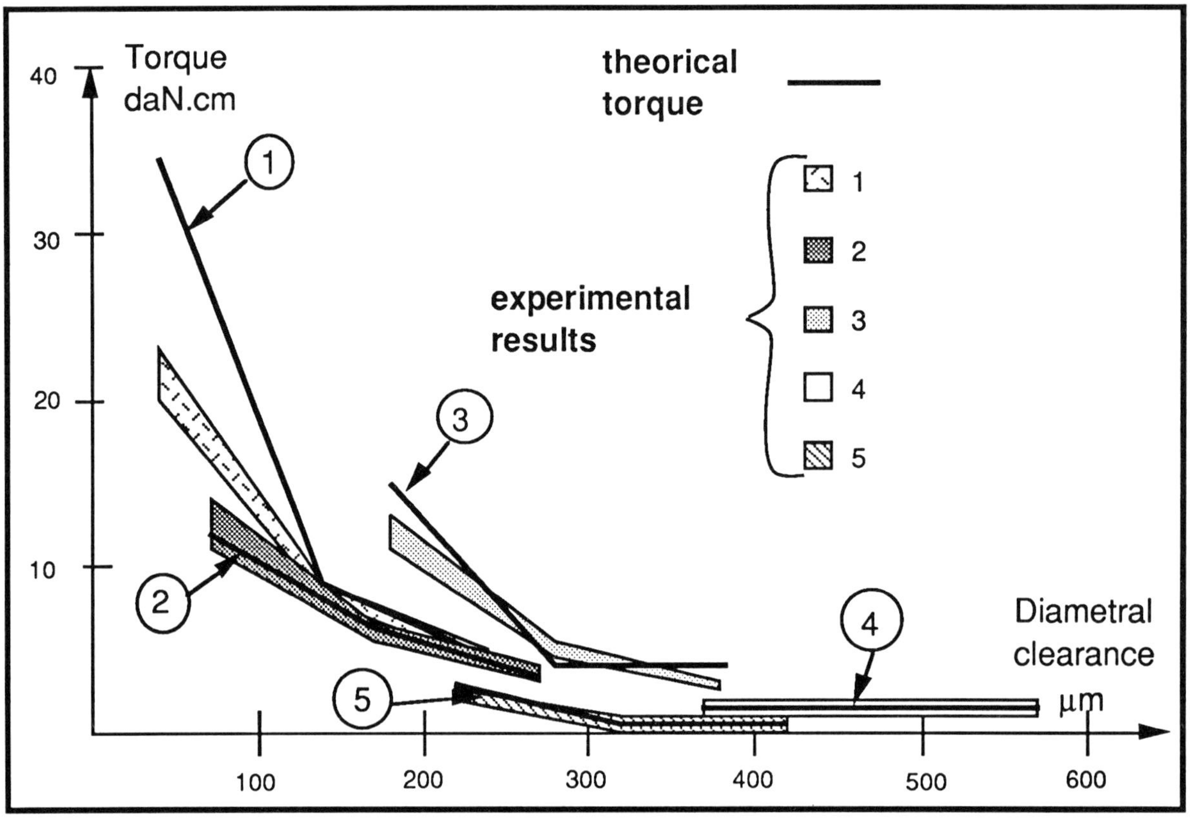

Figure 10

3 .EXPERIMENTAL STUDY

The study is made with 3 house bearings and five leaves of different thickness. Thus, we obtain different diametral clearances.

Leaf 1, 2, 5 are teflon coated.

Leaf 3 and 4 are uncoated.

We see in figure 10 that the starting torques we find are very good for journal radius reaching 99,85% of the theoretical saturation shaft radius (which corresponds to a clearance of 0). So, load and final geometry of the bearing after the installation of the shaft are well simulated by the model for this kind of bearing.

4.APPLICATION

We want to obtain a starting torque as little as possible. Thus, we will prefer to take a leaf covered with teflon and with a little thickness. But the dynamic properties of the bearing are better if the bearing has enough stiffness. In practice, for a bearing, we will choice a radial clearance corresponding to the second part of curve 6. In addition to that studies in a non symmetrical case and in a dynamic case have to be done in order to find the best bearing as possible.

5.CONTINUATION OF THIS WORK

This study could be continued in two

directions:
_ The study of bearings which have little geometrical differences with this bearing.

_ The study of this type of bearing in a non symmetrical case like the radial displacement of the shaft in the bearing. (this will give us the stiffness curve of the bearing)

References

1 Wahl,Mechanical spring

2 KP.OH and SM Rohde,a theoretical investigation of multileaf journal bearing, transaction of the ASME, Journal of applied Mechanics

3 RJ.Tripett, KP.OH, SM Rohde,theoretical and experimental load deflection studies of a multileaf journal bearing, trzansaction of the ASME.

4 Pr Glienicke, étude sur les paliers à feuilles séparées, special study for ABG SEMCA.

SESSION VI
INVITED REVIEWS

Chairman : Professor G Dalmaz

Paper VI (i) **Mixed Lubrication and Lubricated Wear**

Paper VI (ii) **Metal Machining Mechanics and Tribology from the Macro to the Nano Scale**

Thin Films in Tribology / D. Dowson et al. (Editors)
1993 Elsevier Science Publishers B.V.

Mixed Lubrication and Lubricated Wear

Herbert S. Cheng

Center for Engineering Tribology, Northwestern University
Evanston, IL 60208

This review attempts to update our current analytical understanding of the macro-scale lubrication variables such as the average film thickness, lubricant and contact pressures, shear stress, and surface temperature, as well as the micro-scale random variables induced by the randomly distributed asperity contacts.

Attention is first given to the all important average film thickness which controls the degree of asperity contacts, and the intensity and extent of the contacting pressure and temperature. Early developments of average Reynolds equation using average flow factors offer a convenient way to determine the effects of surface roughness height and the lay orientation on the average film thickness. However, recent numerical simulations of elastohydrodynamic line and point contacts with longitudinal or transverse rough ridges have shown these early models seem to yield a much larger roughness effect particularly for transverse roughness operating at a low lambda, the ratio of the average film thickness to the composite roughness. In the light of these numerical simulations and some recent experimental results of film thickness and friction in mixed lubrication, an assessment is given to the possible limitations of the average flow model and some possible reasons for the overestimate.

The development of accurate models for the asperity contact pressure and temperature as functions of lambda is important for determining the hydrodynamic to contact load and friction ratios as well as the asperity contact temperature distribution in mixed lubrication. A brief review is also given to recent advancements in this area.

Even though theories of mixed lubrication are not fully matured, there have been some limited successes in applying them to predict lubricated wear and wear transition to scuffing in elastohydrodynamic contacts. A recent wear model in mixed lubrication is also discussed

1. INTRODUCTION

When the lubricant pressure is insufficient to lift up all the asperities between two sliding lubricated surfaces, the normal bearing load is partly borne by the asperity tips. The lubrication is a mixed mode containing both boundary lubrication at the sliding asperities and thick film lubrication in the valleys.

Many lubricated contacts in rolling bearings, gears, cams as well as some heavily loaded sliding bearings can operate satisfactorily in this mode with a significant portion of the bearing load carried by the deformed asperities. The lubrication performance and breakdown of the entire contact are influenced by the average macroscopic as well as the locally fluctuated microscopic film thickness, pressure, surface temperature, and shear stress. The analysis, characterization and prediction of these macro and micro quantities in mixed lubrication appear to be a necessary road path towards developing better failure predictions of lubricated contacts.

Analyses in mixed lubrication involving rough surfaces are considerably more complex and difficult than hydrodynamic analyses of smooth surfaces. Methods for analyzing the effects of surface roughness and their lay orientation on the average film thickness and load capacity were reviewed earlier by Elrod (1978) and by Cheng (1984).

With recent advancements of more powerful numerical techniques such as the multigrid method, direct numerical simulations with real measured surface roughness profiles and topographies have shown to be feasible. These new methods are expected to accelerate the analytical understanding of mixed lubrication.

In this paper, a review is given to recent analytical and experimental results on contact modeling of rough surfaces, macro and micro lubricant film thickness, pressure, surface temperature, and friction. In addition, a recent wear model for sliding contacts in mixed lubrication is also reviewed.

2.CONTACT MODELING OF ROUGH SURFACES

One of the important quantities in mixed lubrication is the ratio of the total asperity contact area to the apparent bearing area, the contact area ratio. It is this ratio which determines the intensity of the asperity contact load in mixed lubrication.

The interelations among the contact area ratio, asperity contact load, and the ratio of the average gap to the composite roughness, h/σ, can be obtained by contact modeling. The most commonly used contact models include Greenwood and Tripp model (1971) for spherical asperity tips, Onions and Archard model (1973) for 2-D asperity ridges, and Bush, Gibson and Thomas model (1975) for 3-D roughness. However,these models do not include the interactive asperity deformation, and become less applicable at high contact loads.

A more direct method for contact modeling is numerical simulation which can account for interactive asperity deformation. Numerical simulations of rough contacts were accomplished by Lai and Cheng (1985) for 3-D roughness, and Webster et. al.(1986), and Lee and Cheng(1991) for 2-D roughness. The major disadvantage of this approach is the lengthy computation required for real rough surfaces.

More recently, a simple analytical contact model for 2-D roughness was developed by Aramaki et. al. (1993a). This model covers a

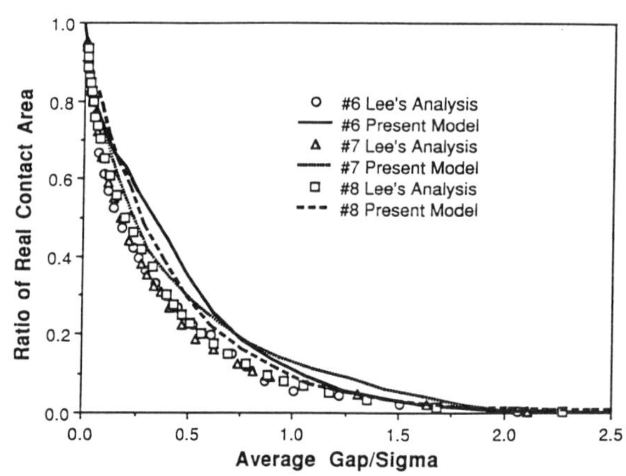

Fig. 1 Ratio of Real Contact Area (#6-#8 surfaces)

wide load range and requires very little computation time. Fig 1 shows that the relations of the contact area ratio and average gap ratio obtained from this simple contact model agree reasonably well with those obtained by Lee

Figs. 2(a) and 2(b) give comparison of the pressure flow factors, ϕ_x , and the shear flow factor, ϕ_u, recently compiled by Bhushan and Tondor (1989). With the exception of the wide load range and requires very little computation time. Fig 1 shows that the relations of the contact area ratio and average gap ratio obtained from this simple contact model agree reasonably well with those obtained by Lee and Cheng (1991) with full simulation.

3. AVERAGE FILM THICKNESS

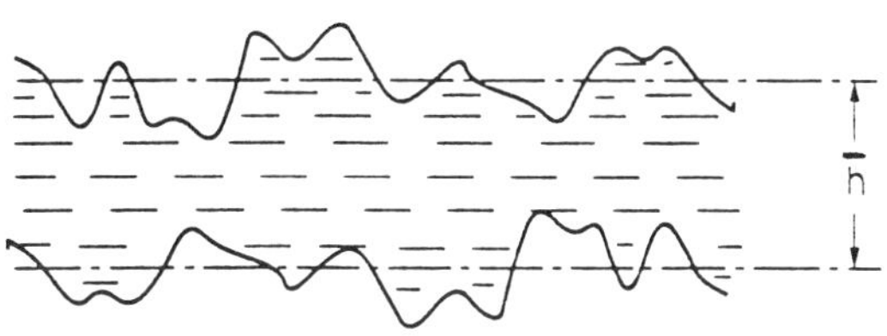

The average film thickness is important in mixed lubrication because its ratio to the composite roughness determines the extent of asperity contact in mixed lubrication. The influence of roughness and their lay orientation with respect to flow on the average film thickness can be determined by solving the Reynolds equation with average flow factors to account for the change of flow due to the roughness. These flow factors were introduced numerically by Patir and Cheng (1978) and analytically by Elrod (1979).

The method of flow factors offers a convenient way to determine the average hydrodynamic pressure in mixed lubrication. It has received considerable attention, re-examination, and refinement. Among these, Teale and Lebeck (1980) showed that the flow factors are sensitive to the grid schemes used to discretize the flow. Tripp (1983) generalized the concept of flow factors for extended applications. Tondor (1986) developed analytically the flow factors for unidirectional striated roughness oriented at different angles with respect to the direction of sliding. Hu and Zheng (1989) re-examined the boundary conditions and numerical schemes and partially explained the variance of the flow factors.

Figs. 2(a) and 2(b) give comparison of the pressure flow factors, ϕ_X , and the shear flow factor, ϕ_u, recently compiled by Bhushan and Tondor (1989). With the exception of the pressure flow factors by Tondor for isotropic roughness, $\gamma B=1$, which differ in trend with others, all the other results show reasonably satisfactory agreement.

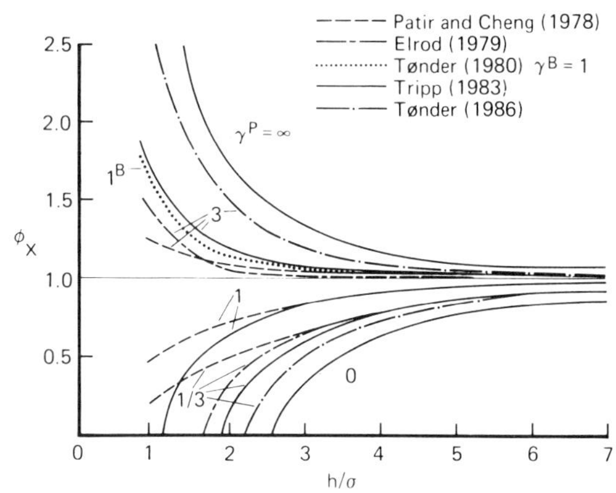

Fig. 2(*a*)

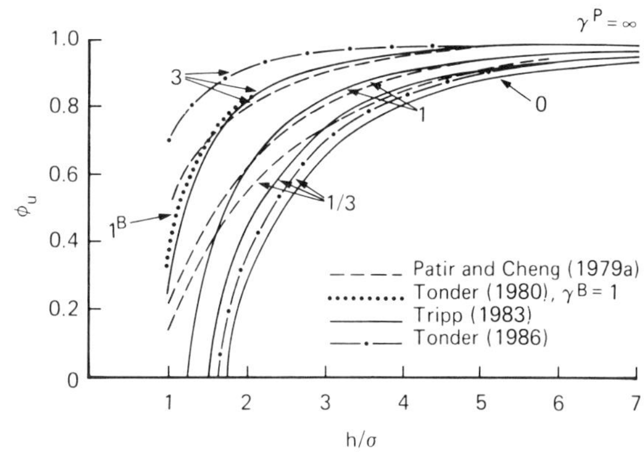

Fig. 2(*b*)

Fig. 2 Comparison of flow factors calculated from different theories (*a*) ϕ_X, (*b*) ϕ_u for stationary roughness. ϕ_u for moving roughness is $2-\phi_u$ for stationary roughness. Superscript *B* on 1 represents that definition for γ^B instead of γ^P was used for calculation of γ.

The flow factor method has been applied recently to determine the average film thickness, average hydrodynamic pressure, and contact load ratio by Jeng and Hamrock (1987), Zhu and Cheng (1988), Lin and Chu (1991) for elastohydrodynamic point contacts, by Sadeghi and Sui (1989) for line contacts with compressible lubricants, and by Bhushan and Tondor (1989) for gas lubricated sliders.

While the flow factors offer a convenient method to determine the average film and pressure in mixed lubrication, it must be kept in mind that they are calculated with asperity tips truncated geometrically. For heavily loaded sliding contacts where the asperity shapes are distorted severely due to micro-elastohydrodynamic lubrication, the results may not be applicable.

4. ASPERITY FILM THICKNESS AND PRESSURE

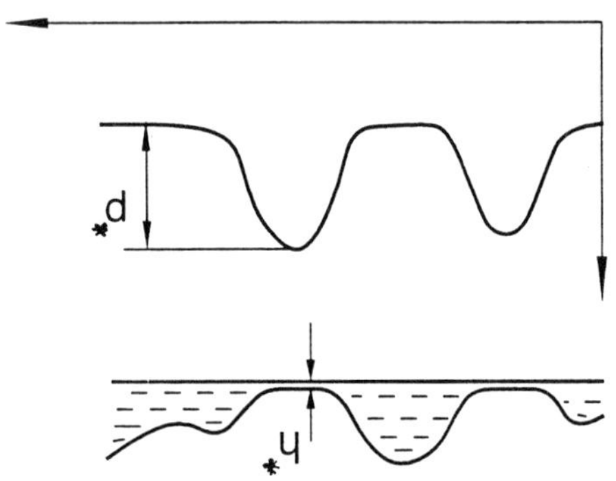

With the rapid increase in computer storage and speed, it is now feasible to simulate numerically the detailed film thickness and pressure variations around asperities, asperity ridges, furrows, dents, and other irregularities.The direct simulation offers a new way to study the effects of single or multiple surface irregularities on the lubrication behavior of rough contacts, and to characterize the mixed lubrication performance. Some selected results for line and point contacts are reviewed here.

4.1. Line Contacts

The lubrication behavior around a stationary or moving single asperity in line contacts is now well understood through work by Goglia, Cusano, and Conry (1984), and by Ai and Zheng (1989). These results for single asperity reveal major features of micro-EHL.

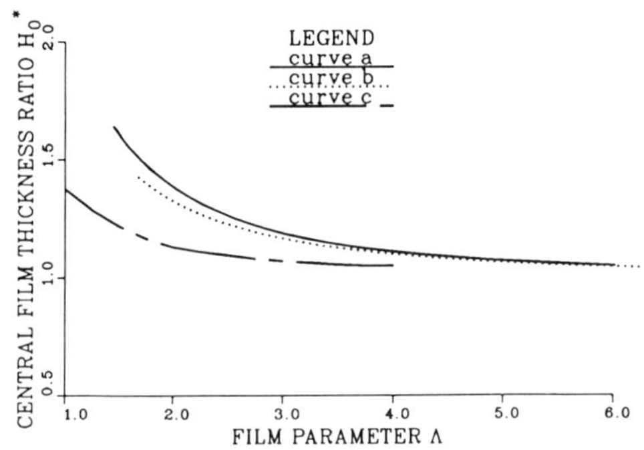

Fig. 3 Relation between the average central film thickness and the film parameter (*a*) $\lambda = 0.2$, $\phi = \pi$, (*b*) $\lambda_1 = 0.13$, $\phi_1 = 1.3\pi$, $\lambda_2 = 0.2$, $\phi_2 = \pi$, $A_1 = 2.33A_2$, and (*c*) taken from Fig. 4 in Zhu et al. (1990) for $\gamma = 1/6$)

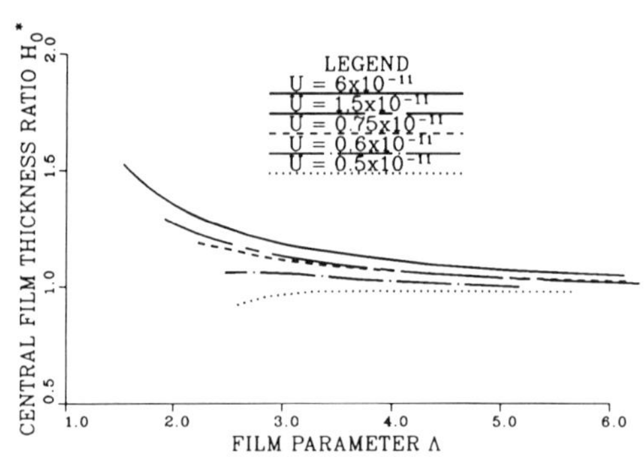

Fig. 4 Effects of the speed parameter on H_o^*-Λ relation of a sliding contact

For effects of multiple surface irregularities, Chang and Webster (1991) calculated the average central film thickness for wavy surfaces and found that the entraining velocity and the slide to roll ratio have a strong influence on the lubrication process. Fig. 3 shows the variation of H_O, the ratio of the central film thickness for the rough case to that for the smooth case, with the film parameter Λ, film to roughness ratio, for a moderately loaded case. The trend agrees with that predicted by using the average flow model. Fig. 4 shows the effect of entraining speed on the central film thickess ratio. The roughness influence diminishes as the entraining speed decreases. This was attributed to the change of phase of the waviness at the inlet and possibly to larger asperity deformation at lower speeds.

Results using measured real surface roughness for a simple sliding line contact were obtained by Kweh et. al. (1992). Fig. 5 shows typical pressure and film thickness profiles. The average film thickness for this rough contact was found to be approximately 20% higher than the smooth contact. This trend appears to be consistent with that found by the average flow method. The minimum film thickness for the rough contact is, however, slightly less than that for the smooth contact.

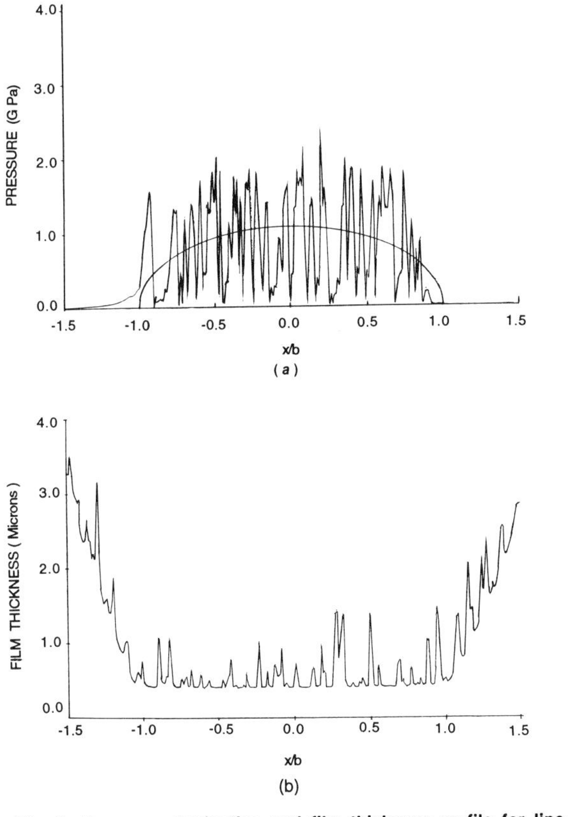

Fig. 5 Pressure distribution and film thickness profile for line contact with transverse, real roughness. (a) Pressure distribution. Also shown is corresponding Hertzian distribution for smooth surfaces. (b) Film thickness profile (temperature = 100 °C)

The multigrid method was used recently by Venner and ten Napel (1992) to study the effects of real roughness in a sliding line contact. Their results of pressure and film profiles are shown in Fig. 6 for a heavily loaded contact. The initial roughness is seen to be completely flattened out by the local elastic deformation in the conjunction. As a result, the film thickness profile is practically the same as that of the smooth case. There is no increase in the average film comparing to the smooth case. In contrast to the results in Fig. 6, the initial roughness in Fig. 5 by Kweh et.al. still contains many valleys which results in an average film thickness higher than the smooth case. In either case, the local asperity deformation for heavily loaded sliding contacts is significant. Results from average flow method should not be used for these conditions.

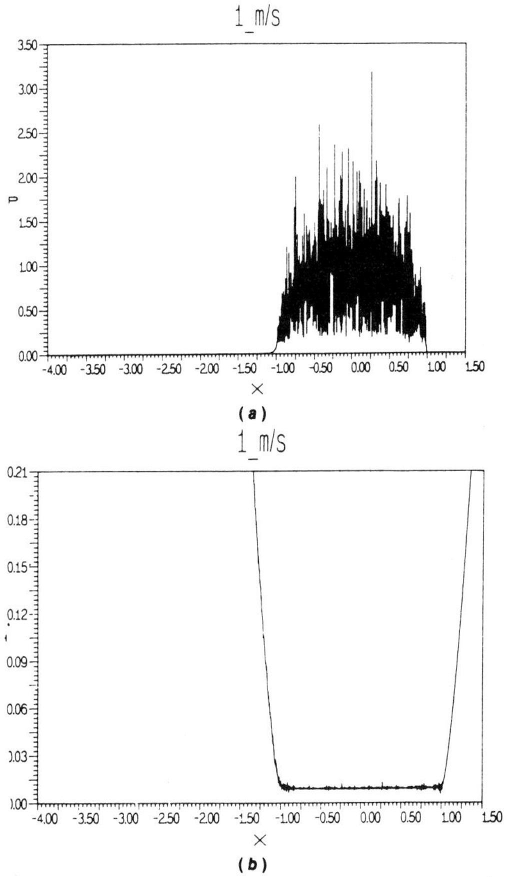

Fig. 6 **Pressure profile (*a*) and film shape (*b*) calculated with rough surface for a maximum Hertzian pressure of 2 GPa and surface velocity of 1 m/s**

4.2. Point Contacts

The multigrid method was also used by Lubrecht, ten Napel and Bosma (1988) to investigate the effects of sinusoidal transverse and longitudinal roughness on the film thickness and pressure in point contacts. They found that the average film ratio, the ratio of average rough film to smooth film, is sensitive to the asperity height parameter A, and the asperity radius parameter R_a. Fig. 7 for transverse roughness shows a very interesting variation of the average film ratio with the asperity radius or wave length. For large R_a, the film ratio is unity showing no roughness effect from a long wave length. It increases as the wave length decreases, but drops below unity for very small wave length roughness. This effect is enhanced with increasing asperity height.

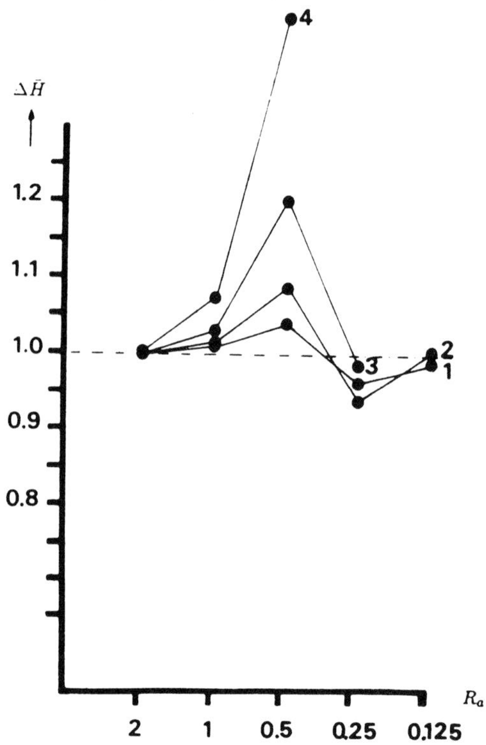

Fig. 7 Average film ratio, ΔH V.S. asperity radius parameter, transverse roughness.
1: A=0.05; 3: A=0.2
2: A = 0.1; 4: A = 0.4
A = asperity height parameter

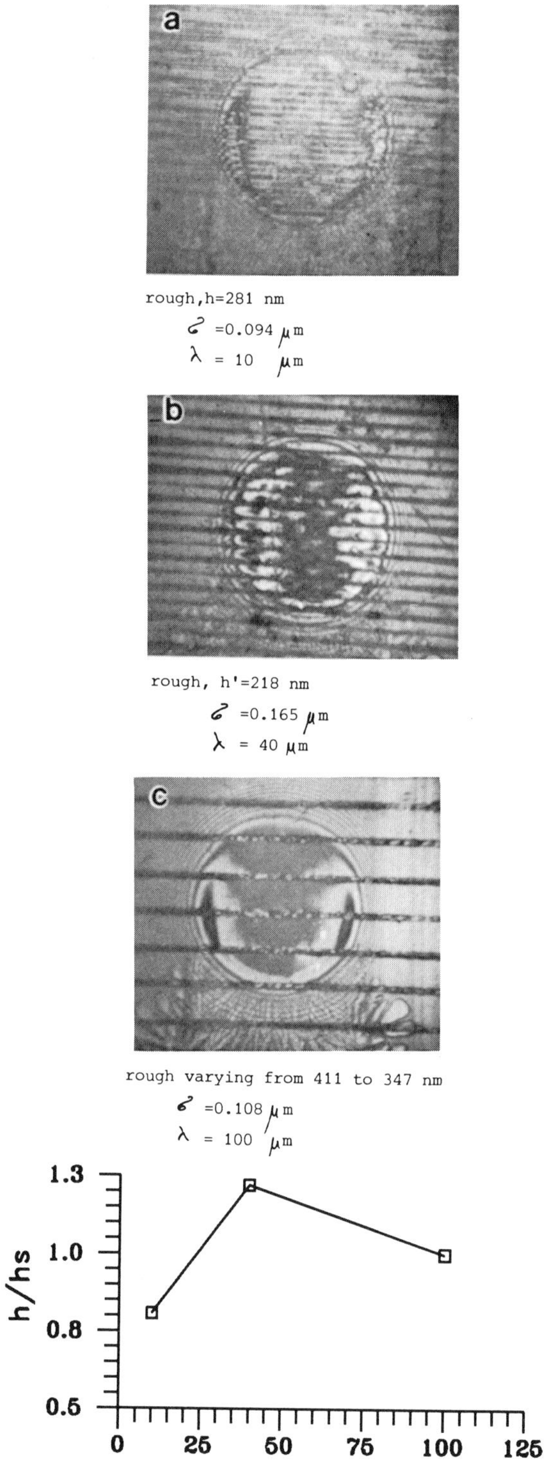

Fig. 8 Microinterferograms of ball contacts with transverse roughness pattern in rolling.
a: h/h_s =0.807; b: h/h_s = 1.22; c: h/hs= 1.0

Interestingly, the wave length effect predicted in sliding ball contacts by Lubrecht et.al. appears to have been observed optically by deSilva et.al. (1986) with artificial sinusoidal roughness in rolling ball contacts. As shown in Fig. 8, three wave lengths were used, λ = 10, 40, 100 μm. The film ratios corresponding to these cases are 0.807, 1.22, and 1.00 respectively. Plotting these values in the lower figure, one sees a similar pattern as that shown in Fig. 7.

The results for longitudinal roughness calculated by Lubrecht et.al. are shown in Fig. 9. For very large wave lengths, the film ratio is less than unity showing the expected reduction in average film due to longitudinal roughness. The unexpected feature is the increase in the film ratio for small wave lengths. This intriguing feature is worthy of further exploration.

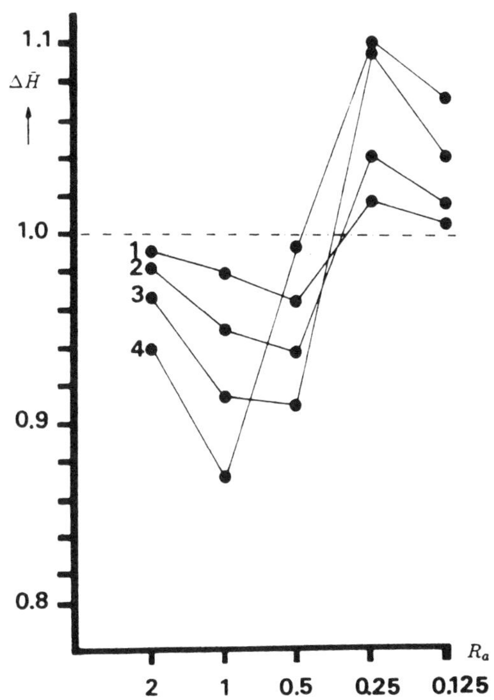

Fig. 9 Average film ratio, ΔH V.S. asperity radius parameter, longitudinal roughness
1: A=0.05; 3: A=0.2
2: A = 0.1; 4: A = 0.4
A = asperity height parameter

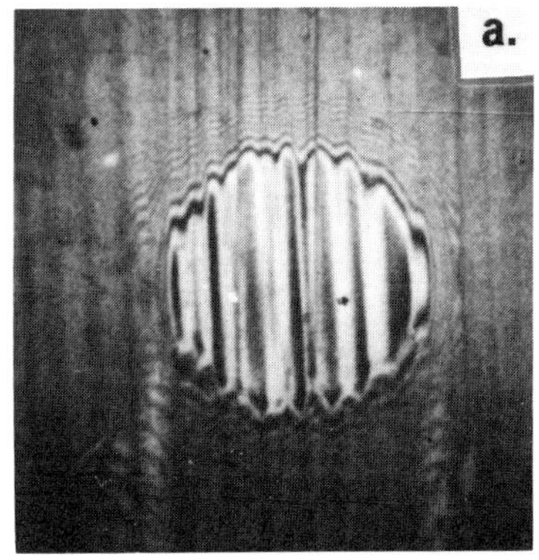

rough,h'= 94 nm

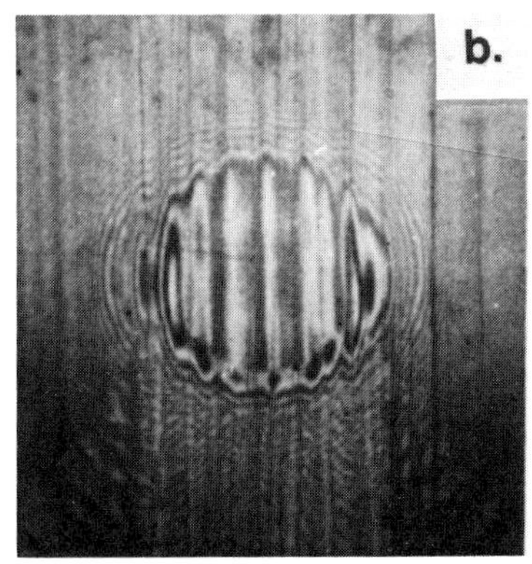

rough,h'= 368 nm

Fig. 10 Microinterferograms of ball contacts with longitudinal roughness pattern in rolling, λ = 30 nm for both.
a: h_s =149 nm; b: h_s =609 nm

Unfortunately, the data of optical film thickness for longitudinal roughness obtained by deSilva et.al. for rolling ball contacts is only for one wave length, as shown in Fig. 10. It does show a slight film reduction due to the longitudinal roughness.

5. AVERAGE FRICTION

The average friction in mixed lubrication consists of three components, the horizontal force from the lubricant pressure, shearing of the viscous lubricant, and shearing of the asperity contacts. Analyses of friction due to the first two components for rough contacts were given by Patir and Cheng (1978). They introduced two shear stress factors for calculating the frictional components contributed by the lubricant.

Coulumb model is usually used for calculating the asperity friction. A value of around 0.1 is usually chosen as the asperity friction coefficient for mineral oils in mixed lubrication. In actual asperity contacts, the shear stress depends on the asperity lubricant film which can be a micro-EHL film, a mono-layer boundary film, an oxide film, or no film. The use of a constant friction coefficient for the asperity contacts is an over simplification and should be revised.

Nakahara et.al. (1984) employed the shear stress factors by Patir and Cheng (1978) and studied the effects of surface roughness orientation and bearing slenderness ratio on the friction coefficient of a rough slider.The calculated friction coefficients are compared with the measured ones in Figs.11 and 12. For short sliders of bearing slender ratio of 0.5, Fig. 11 shows that the transverse roughness gives a higher friction because the resistance to side flow is smaller and the film is thinner. For bearings with a larger slender ratio, 2.0, the reverse is true. The longitudinal rough slider has a slightly higher friction because of a slightly lower film thickness. These measured friction coefficients match quite closely with those calculated with the average flow and shear factors.

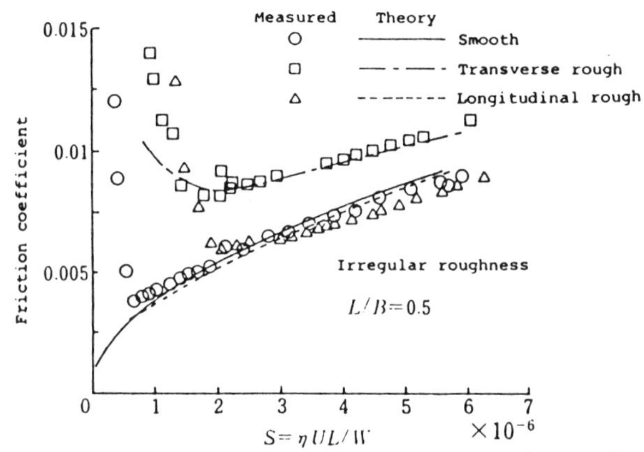

Fig 11 Friction characteristicse for irregular rough surface (L/B=0.5)

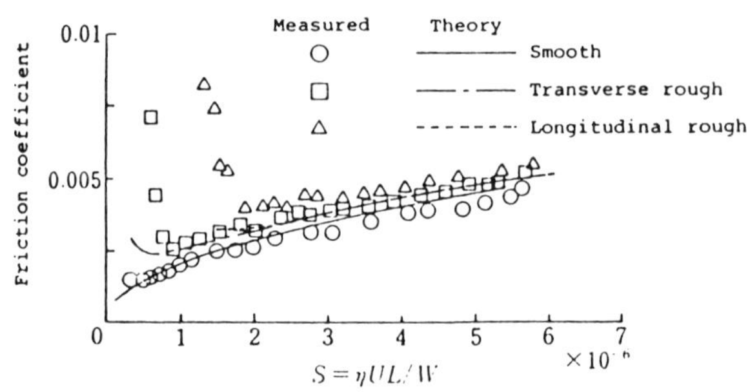

Fig 12 Friction characteristics for irregular rough surface (L/B=2)

6. AVERAGE AND ASPERITY SURFACE TEMPERATURE

The average surface temperature of contacts in mixed lubrication can be determined from classical theories using the average friction. Commonly used methods for the average surface temperature were reviewed by Winer and Cheng (1980).

For calculating the asperity contact temperatures and their distribution, numerical or statistical simulations are oftened used. For instance, Lai and Cheng (1985) performed simulation for a randomly generated 3-D roughness. McCool (1991) developed a software to generate the distribution of asperity contact temperatures for spherical asperity tips based on Greenwood and Williamson's contact model. For longitudinal roughness, Lee and Cheng (1991) has developed a method to simulate the asprity temperature distribution.

For estimating the asperity temperatures, one may use a simplified analytical model developed very recently by Aramaki, Chung and Cheng (1992). Fig. 13 shows typical asperity temperature distributions predicted from this model. A reasonably close agreeement was found between these analytical predictions and those obtained from numerical simulation.

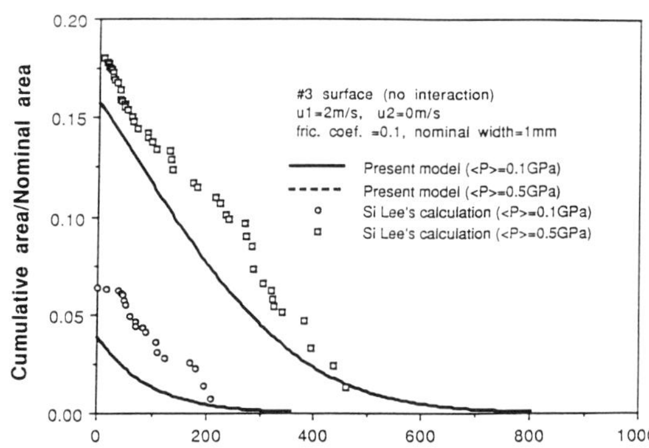

Fig. 13 Temperature distribution (#3 surface)

7. A SLIDING WEAR MODEL IN MIXED LUBRICATION

Techniques developed in mixed lubrication for determining the asperity temperature distribution can be used to refine the sliding wear model developed earlier by Rowe(1966) using Kingsbury's concept of thermal desorption (Kingsbury, 1958), and applied later by Stolarski (1979) to a simple sliding system.

In this model, both the thermal desorption wear mechanism at low asperity contact temperature and the oxidative wear mecha-nism at elevated asperity contact temperature.Digitized actual surface roughness profiles are used in simulating two contacting surfaces with longitudinal roughness, and in obtaining the asperity contact temperature distributions within a Hertzian contact region. Wear measurements in a two-disk machine have been made over a wide range of slide-to-roll ratios.

Fig. 14 shows a typical relation between the cumulative asperity contact area and the asperity contact temperature in a line contact with longitudinal roughness. These were calculated by Lee and Cheng's program (1991) under conditions used in the experiment.

In Fig. 15, both the analytically predicted wear rate and the experimentally observed wear rate show a continuous decrease in wear rate with the increase in slide-to-roll ratio for small slide-to-roll ratios. The wear rate increses drastically shortly after it passes through the minimum friction and followed by scuffing. The comparison between the predicted and measured wear rates is quite satisfactory.

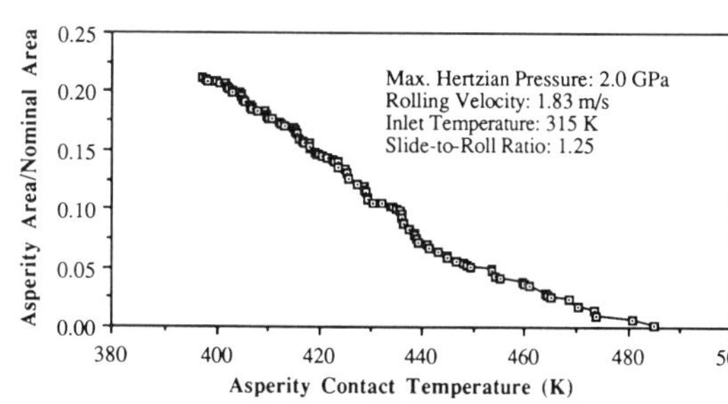

Fig. 14 Cumulative asperity contact area and temperature distributions

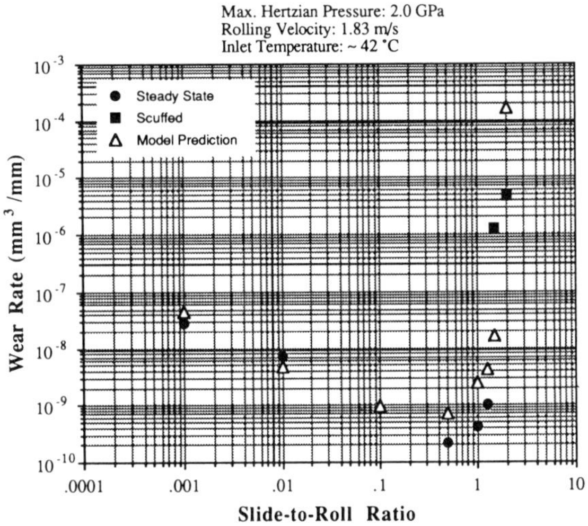

Fig. 15 Comparison of model prediction with experimental observation (soft specimen)

8. CONCLUDING REMARKS

1. Recent numerical simulations of line and point contact EHL using wavy surfaces and real rough surfaces have shown considerable promise as a powerful tool to determine the effects of major surface roughness parameters on the behavior of mixed-film EHL contacts.

2. Film thickness measured optically with balls of wavy surface show trends agree with those predicted by numerical simulations.

3. Friction predicted by average flow and shear factors show good agreement with measured value for rough hydrodynamic slider.

4. For heavily loaded EHL sliding contacts, numerical simulations show that the average flow method predicts a much too high average film due to severe asperity deformation.

5. Sliding wear and wear transition is strongly dependent on the asperity temperature distribution in mixed-film EHL, and can be predicted by the concept of thermal desorption.

REFERENCES

1. Elrod, H. G.,Proc. of the 4th L-L Symposium, pp.11, (1978)
2. Cheng, H. S., Proc. of the 11th L-L. Symposium, pp.11-20 (1984)
3. Greenwood, J.,A., and Tripp, J. H., Proc. Inst. Mech. Engrs. 185 (1970-1971) pp.625-633.
4. Onions, R. A., and Archard, J. F., J. Phys. D., Appl. Phys. Vol 6 (1973) pp. 289-304.
5. Bush, A. W., Gibson, R. F., and Thomas, T. R., Wear, Vol. 35 (1975), pp. 87-111
6. Lai, T. W. and Cheng, H. S., ASLE Trans., Vol. 28, No. 2. pp. 172-180.
7. Webster, M.N. Ioannides, E. and Sayles, R. S., Proc. of the 12 L-L. Symposium, pp.207-221 (1985)
8. Lee, S. C. and Cheng, H.S., Tribology Trans. Vol. 35 (1992) , No. 3, pp. 523-529.
9. Aramaki, H., Cheng, H. S., and Chung, Y. W., To appear in J. of Tribology (1993 a)
10. Patir, N, and Cheng H. S., J. Lubr. Tech., Trans. ASME, Vol. 100, No. 1 (1978), pp.12-17.
11. Elrod, H., J. Lubr. Tech., Trans. ASEM Vol. 101, jan (1979) pp.8-14.
12. Teale, J. L., and Lebeck, A. O., J. Lubr. Tech., Trans. ASME, Vol. 102, July (1980).
13. Tripp, J. H., J. Lubr. Tech., Trans. ASME, Vol. 105, No.3 (1983) pp. 458-465.
14. Tonder, K., Tribology and Mechanics of Magnetic Storage Systems, Ed. by Bhushan te al., SP-16, ASLE, Park Ridge, IL., pp. 183-192.
15. Hu, Y. Z., and Zheng, L. Q., J. of Tribology, Vol. 111 (1989) pp.525.
16.Bhushan, B., and Tonder, K., J. of Tribology, Vol.111 (1989), April, pp.220-226.
17. Jeng, Y. R., and Hamrock, B. J., ASLE Trans.Vol 30, No.4 , pp. 531-538.

18. Zhu, D., and Cheng, H.,S., J. of Tribology, Vol. 110, Jan. (1988), pp. 32-37.
19. Lin, J. F., and Chu, H. Y., J of Tribology, Vol. 113, Jan. (1991), pp.12-21.
20. Sadeghi, F., and Sui, P. C., J of Tribology, Vol. 111, Jan. (1989), pp. 56-62.
21. Bhushan, B., and Tonder, K., J. of Tribolgoy, Vol.111, April (1989), pp.228-237.
22. Goglia, P.,Conry, T. F, and Cusano, C., J. of Tribology, Vol. 106, No.1 (1984), pp.104- 112.
23. Ai, X., and Zheng, L., J. of Tribology, Vol. 111, No.4 (1989), pp. 569-576.
24. Chang, L., and Webster, M. N., J. of Tribology, Vol. 113, Jan. (1991), pp. 110-115.
25. Kweh, C. C., Patching, M. J., Evans, H. P., and Snidle, R. W., J. of Tribology, Vol. 114, July (1992), pp.412-419.
26.Venner, C. H., and ten Napel W. E., J. of Tribology, Vol. 114, July (1992), pp.616-622.
27. Lubrecht, A. A., ten Napel, W. E., and Bosma, R., J. of Tribology, Vol. 110, July (1988) pp.421- 426.
28. de Silva, G. W. S., Leather, J. A., and Sayles, R. S., Proc. of the 12th L-L. Symposium (1985), pp.258-272.
29.Patir, N., and Cheng, H. S., J. of Lubr. Tech., Trans. ASME,Vol. 101, April (1979), pp.220-230.
30. Nakahara, T., Takesue, M., and Aoki, H., J. of JSME International Edition, N0. 5 (1984), pp.65-70.
31. McCool, J. I., J. of Tribology, Vol. 110, Jan. (1988), pp. 106-111.
32. Lee, S. C., and Cheng, H. S., J. of Tribology, Vol. 113, April (1991), pp. 327-334.
33. Aramaki, H., Cheng, H., S., and Chung, Y. W., To appear in J. of Tribology (1993).
34.Wu, S., and Cheng, H. S., J. of Tribology, Vol. 113, Jan. (1991), pp.134-141.

Thin Films in Tribology / D. Dowson et al. (Editors)
© 1993 Elsevier Science Publishers B.V. All rights reserved.

Metal machining mechanics and tribology from the macro to the nano scale

T. H. C. Childs

Department of Mechanical Engineering, University of Leeds, Leeds LS2 9JT, United Kingdom

The paper briefly reviews observations and models of continuous chip formation at feeds in the sub-millimeter, sub-micrometer and nanometer ranges. In the sub-millimeter range, differences in chip flow among different materials are well known; differences in the friction interactions between chips and tools and in materials' flow at high strain rates and temperatures are of major interest. In the sub-micrometer range a change from dislocation to theoretical strength controlled plasticity may occur at feeds around 0.1 μm, leading to an increased importance of elasticity in chip formation at lower feeds. In the nanometer range, molecular dynamics simulations of the machining process do not yet agree with experiments: the question is why?

1. INTRODUCTION

Metal machining at feeds of mm. and sub-mm. has been a main stream engineering activity since the start of the industrial revolution. Interest is in a wide range of both tool and work materials: steels, carbides, cermets and ceramics as tools; ferrous and non-ferrous metals, including stainless steels, nickel and titanium alloys, as work materials. The materials have a strong influence on the characteristics of chip formation. Thousands if not tens of thousands, and tens if not hundreds of books have been written on the subject (1-5). Section 2 of this paper briefly sets out the objectives of chip flow modelling in the sub-mm. regime before concentrating on the importance of chip/tool friction and bulk mechanical properties in controlling chip flow in these conditions.

By contrast the study of chip formation at μm. and approaching nm. feeds is much more recent, associated with precision engineering - the manufacture of mirrors and computer disc substrates (6) - and has been almost exclusively concerned with the machining of copper or aluminium alloys with diamond tools. The rich variety of materials' variables of the sub-mm. feed range has not been and perhaps cannot be studied. Indeed only a very few papers have been published which give comprehensive information about chip formation in these conditions (7,8). It does seem that chip formation is significantly different at these low feeds from the sub-mm. range. Chip thickness relative to the feed is much greater and perhaps the shear stress on the primary shear plane rises to a theoretical strength level. This is considered in section 3 of this review.

Molecular dynamics and molecular statics simulations of metal machining are being carried out that conclude that sub-nm. layers, only a few atoms thick could be removed by machining (9-12). They certainly conclude that theoretical strength levels of shear stress are involved in chip formation at this level but they do not predict the very large chip thickness to feed ratios observed experimentally at slightly larger feeds. These findings are reviewed and discussed in section 4.

2. MACHINING AT SUB-MM. FEEDS

2.1. Theoretical preliminaries

Chips can form either as a continuous ribbon flowing directly over a tool rake face or separated from the rake face by a built-up-edge; or they can form discontinuously as equiaxed fragments broken from the workpiece (4). This paper is concerned only with the first type. Figure 1 shows schematically a section through such a chip. It is presented here as a way of introducing notation and ideas. A tool of rake angle α approaches work material at a cutting speed U to remove a feed depth f of material as a chip of thickness t and radius R. Chip formation is primarily by plastic shear over a zone OA so narrow that it may be regarded as a first approximation as a plane, although a secondary plastic shear also occurs above OA and along the rake face OB. The chip form is usually described by the shear plane angle ϕ rather than the chip thickness ratio t/f but the two are related by geometry:

$$f / \sin\phi = t / \cos(\phi - \alpha). \tag{1}$$

The resultant force FR acts between the chip and tool, inclined to the rake face at the friction angle λ.It has components FC and FT in the cutting and thrust directions and its size is constrained by the shear flow stress k on the primary shear plane, as its resolved component on that plane, divided by the area of the plane, must equal k:

$$FR \cos(\phi + \lambda - \alpha) = kfd / \sin\phi \tag{2}$$

where d is the depth of cut.

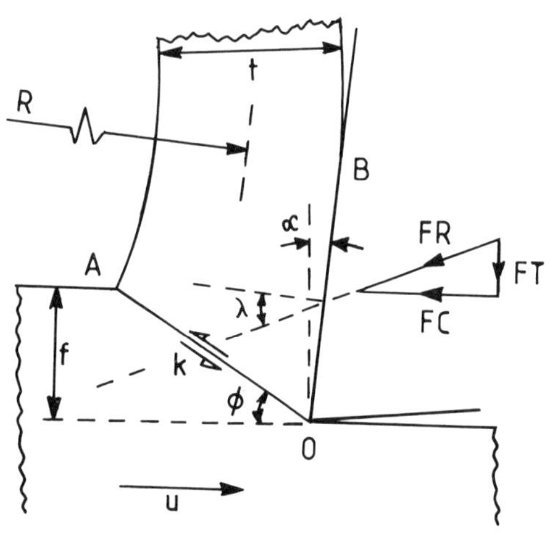

Figure 1. Continuous chip formation notation.

The main task of mechanical models of chip formation is to predict t/f or ϕ, although in some cases to predict the radius ratio R/f is also of value. Once ϕ is known, for example, and assuming λ is also known (see later), other quantities of interest such as FC and FT are readily calculated. Equations 3 and 4 record the non-dimensional specific forces derived from equation 2.

$$FC/(kfd) = \cos(\lambda - \alpha)/(\sin\phi \cos(\phi + \lambda - \alpha)) \tag{3}$$

$$FT/(kfd) = (FC/(kfd)) \tan(\lambda - \alpha) \tag{4}$$

It is well known that changing the lubrication between the chip and tool changes ϕ, for example (13). Early models of chip formation attempted to relate ϕ to λ alone. Thus Ernst and Merchant (14), assumed chip formation would occur with minimum energy and derived

$$\phi = \pi/4 - (\lambda - \alpha)/2, \tag{5}$$

while Lee and Shaffer (15) developed the first slip-line field model of machining, in which plasticity conditions in the secondary shear zone are assumed to control ϕ, to obtain

$$\phi = \pi/4 - (\lambda - \alpha). \tag{6}$$

Neither (5) nor (6) is obeyed experimentally. Although many materials show a linear relation between ϕ and λ, the constants of the relationship vary from material to material (16); and in some cases a linear relation is not observed. Subsequent slip-line field modelling, reviewed and extended in (17), demonstrates that a unique relationship between ϕ and λ is not to be expected. Experimental studies have demonstrated that friction force between the chip and tool is not proportional to the load, so that a friction coefficient does not best describe the chip/tool interaction. These points are developed in the next section.

2.2. Friction in metal machining

The small contact lengths, typically less than 1 mm., and high contact stresses, around 1 GPa, between the chip and tool make experimental studies of chip/tool friction interactions difficult. Apart from photoelastic studies, mainly with model materials (18-20), the main experimental method has been the split tool technique (21-22). Figure 2a shows the contact stresses between the chip and tool measured by the split tool technique when machining a low alloy steel with a cemented carbide tool of zero rake angle, at a feed of 0.2 mm. and a cutting speed of 100 m/min. (23). Figure 2b replots the results as friction stress versus normal stress. It also shows a second set of results, for a free cutting low alloy steel cut in the same conditions as the first. At low normal stress a Coulomb friction law applies,

$$\tau = \mu\sigma_n , \qquad (7a)$$

with μ = 1.3 and 0.75 for the standard and free cutting steels respectively. But at high normal stress the friction stress saturates to

$$\tau = mk . \qquad (7b)$$

m = 1 and 0.86 respectively if k is set at its value on the rake face close to the cutting edge.

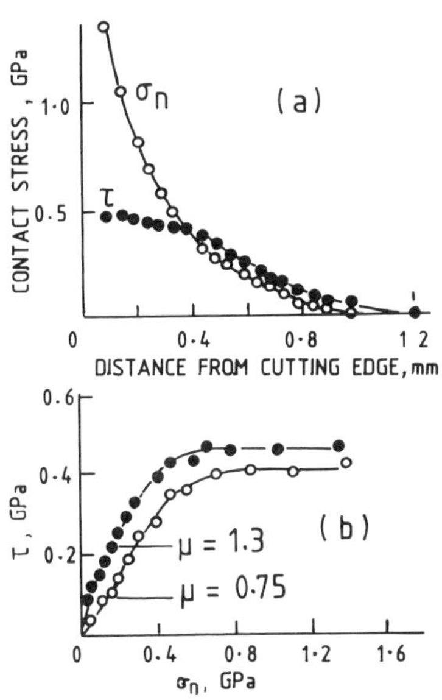

Figure 2. The variation of friction and normal stress (a) with distance from the cutting edge for a low alloy steel; and (b) with each other for the low alloy steel (•) and a free cutting version (o).

Table 1.
Collected m and μ values for metal machining

Tool	Work	m	μ	Refs.
Photo-elastic polymer	Pb.	1.0	0.8	18
	" "	1.0	0.5	19
High speed steel	Pb-Sn	0.8	0.4	21
	Al.	0.7-1.0	1.4-2.2	21
	Cu.	0.75	0.75	21
Cement'd carbide	α-brass	0.9	0.8	23
	steels[1]	0.9-1.0	1.3-2.5	24-26
	steels[2]	0.8-1.0	0.7-1.2	25-26
	steels[3]	0.75	0.28	26

steels[1] = carbon and low alloy
steels [2] = MnS and semi-free cutting
steels[3] = leaded free cutting

Table 1 summarises m and μ data from a wide range of sources. In no case is m < 0.7. For steels, with the exception of a leaded free cutting steel, 0.8 < m <1.0. With the exception again of the leaded steel, μ > 0.7 and can reach 2.5. All these reinforce the view of metal cutting as essentially a poorly lubricated process. It should be pointed out, however, that no data exists for ceramic or coated tools because split tools have not successfully been made from these.

Slip- line field theories of chip forming have been developed modelling friction according to equation 7b. Even though equation 7a is not a good model, an average friction angle λ can be calculated from the results. Figure 3 shows predicted variations of ϕ with λ that are found to agree with experiments (17). Instead of a one to one relation between ϕ and λ, as predicted by equations 5 and 6, bounded regions in which ϕ and λ lie are predicted for every combination of m and α. Figure 3 shows such regions for m = 0.82 and 0.99 and α = 0° and 30°. It also shows the lower bound of the region for m = 0.97 and the upper bound for m = 1.0. A conclusion of this modelling is that in the high friction conditions of metal cutting (m > 0.8), a small change of friction can have a large influence on chip flow. Much experimental and theoretical work remains to be done in studying friction in metal cutting.

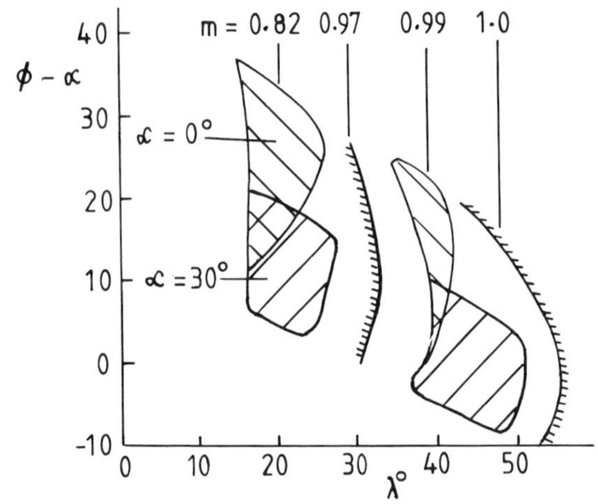

Figure 3. Slip-line field predictions of ϕ and λ.

Although experimental results fall within the expected regions of figure 3, and in that sense it can be claimed that the mechanics of metal machining is understood, this conclusion is not very useful. In figure 4, which includes experimental results, the bounds of figure 3 are converted to a dependence of specific forces on λ, using equations 3 and 4. In figure 4a the bounds are for $\alpha = 0°$ and are so wide that they only weakly constrain the sizes of the forces. A much stronger constraint is placed on the forces when $\alpha = 30°$, but the practical use of such a high rake angle tool is prevented by cutting edge failure.

Slip-line field theories assume a material's flow stress does not vary with strain, strain rate or temperature. Unique estimates of chip flow may be obtained if flow stress is modelled more realistically, but finite element analysis is then needed.

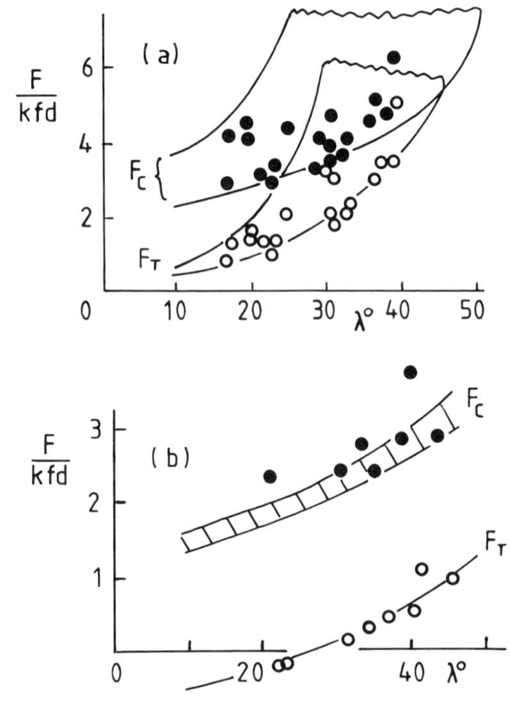

Figure 4. Slip line field and experimental variations of tool force with λ. $\alpha = 0°$ (a) and $30°$ (b).

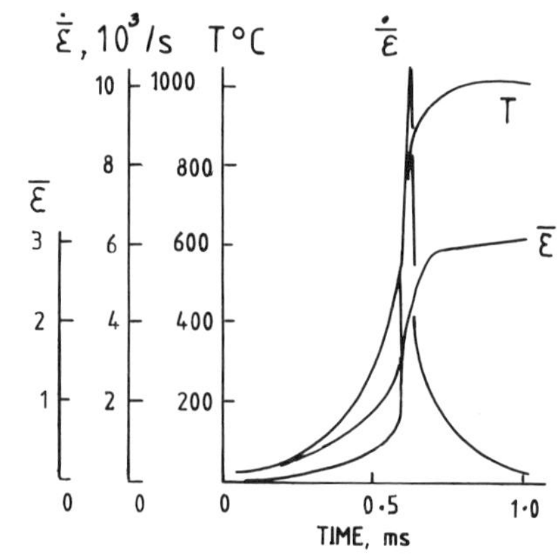

Figure 5. Strain, strain rate and temperature history example in metal machining.

2.3. Importance of mechanical properties

Figure 5 shows a finite element estimate of the strain, strain rate and temperature history of chip material flowing over the rake face in conditions close to those of figure 2. Strain rates up to 10^4 and temperatures of 1000°C are reached in milliseconds. Hopkinson bar testing (27) has been used to measure the variation in uniaxial flow stress σ_Y in such conditions but strain rates and temperatures only up to 10^3 and 700°C have been achieved. Data may be found in (23 - 27) and has been fitted to

$$\sigma_Y = A\,(\text{strain})^n(\text{strain rate})^m. \qquad (8)$$

When such mechanical property data and friction data as in figure 2 are input to an elastic-plastic finite element programme predictions of chip flow are being obtained (figure 6) which are good for free cutting steels, when friction is controlled by solid lubricant deposits on the rake face. But poor predictions, particularly of thrust force, are obtained for non-free cutting steels, when friction is controlled by the shear flow stress of the chip material itself (28). It may be that extrapolation of equation 8 to much higher temperatures and strain rates than those for which it was obtained is in error. The determination of flow stress properties in conditions relevant to machining is another field for materials research, in addition to the study of friction.

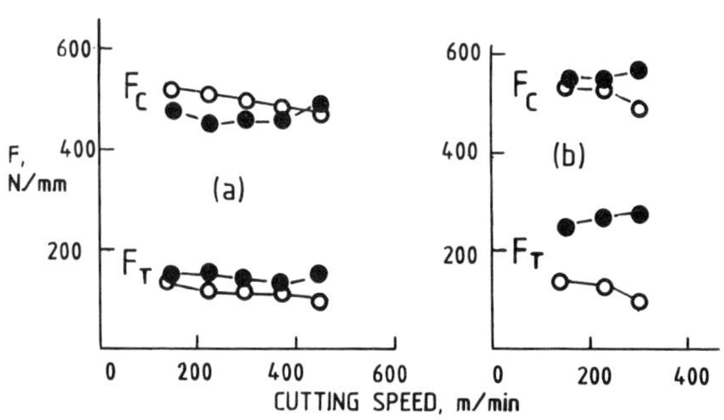

Figure 6. Measured (•) and predicted (o) forces per unit depth of cut for (a) a free cutting and (b) a standard low alloy steel, f = 0.25 mm., $\alpha = 6°$.

3. MACHINING AT SUB-MICRON FEEDS

In the previous sections it was assumed that tools are perfectly sharp. In experiments that have been carried out at sub-mm. feeds, either with artificially round-edged tools (29) or with tools with a negative chamfer (30), it has been observed that a stagnant zone of work material forms under the rounded edge or chamfer. Chip flow occurs over this zone without much change in the shear plane angle relative to that with a sharp edged tool: t/f ratios of about 2 are recorded in (29,30). However the stagnant zone gives rise to an increase in the thrust force relative to the cutting force. Figure 7 reports measured changes in non-dimensional specific forces as the ratio of feed f to cutting edge radius r was altered in experiments with artificially round-edged cutting tools (31). For f/r < 1, thrust force became greater than cutting force.

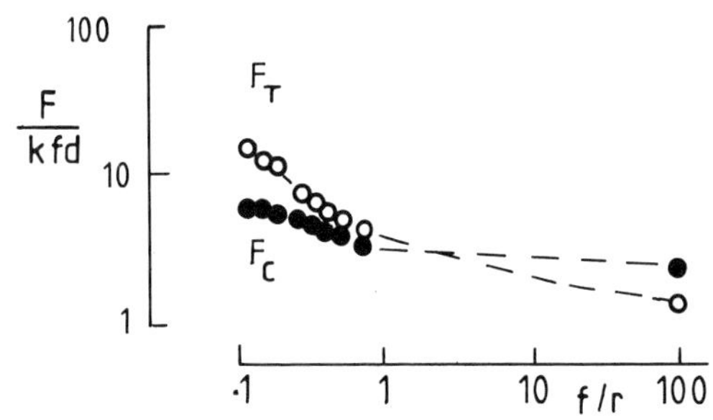

Figure 7. Specific forces versus f/r, f > 0.04 mm. derived from reference (31).

Qualitatively similar variations of force have been found with experiments at sub-μm. feeds, but the variations of force have been much larger than at higher feeds. Figure 8 is derived from experiments reported by Moriwaki and Okuda (7). r has been taken as 35 nm, their middle estimate, and k has been taken as 600 MPa, also their estimate. Non-dimensional specific forces rise to 100 to 1000 as f reduces to 0·0025 μm and f/r to 0.1, some 10 to 100 times larger than those shown in figure 7. The question is why?

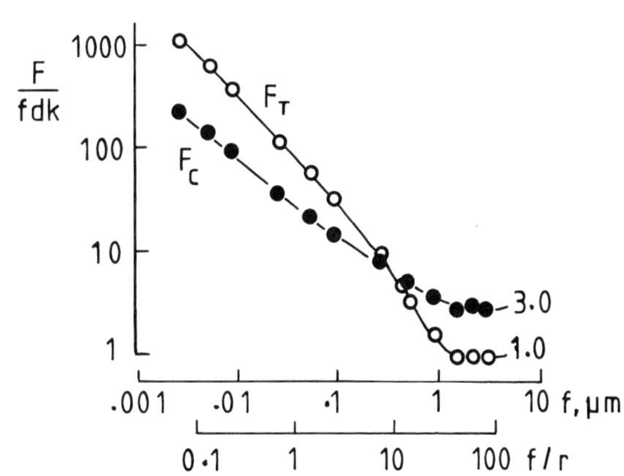

Figure 8. Specific forces versus f and f/r, derived from reference 7, copper machined by diamond.

Part of the answer lies in much larger chip thickness ratios. Figure 9a shows the measured variation of t/f with f and the corresponding variation of ϕ calculated from equation 1. t/f of 60 at f =0·0025 µm. leads to $\phi = 1°$. This is an extraordinary low value, judged by sub-mm. feed experience.

Further insight is obtained from figure 9b in which is plotted absolute values of t and forces per unit depth of cut. As the feed reduces below 0.01 µm., t, FC and FT all become almost independent of feed.

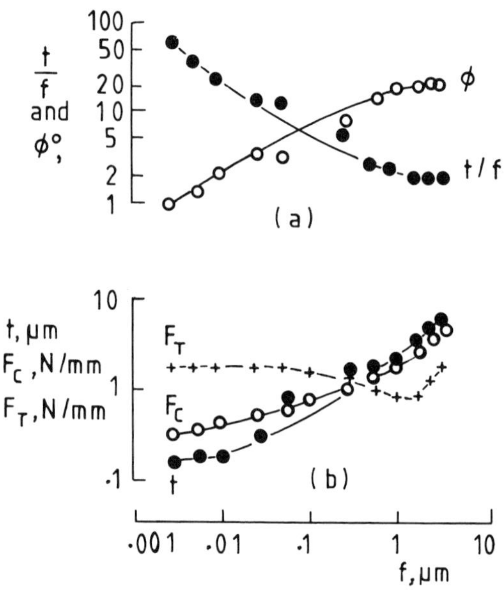

Figure 9. Further derived data from (7): the variation with f of (a) t/f and ϕ and (b) t, FC and FT.

A speculative explanation of the small estimated ϕ and the insensitivity of t, FC and FT to f at the smallest feeds concerns possible effects of elasticity at small feeds. Figure 10a draws to scale the chip form and tool edge during cutting at the feed of ·0025 µm., assuming the work material to have an infinitely large elastic modulus. Figure 10b suggests a possible effect of elasticity, namely that the tool sinks into the work by a depth w much larger than f. In that case the shear plane under load might rotate to a larger value of ϕ, 10° in figure 10b. If f << w, t, FC and FT might become insensitive to f.

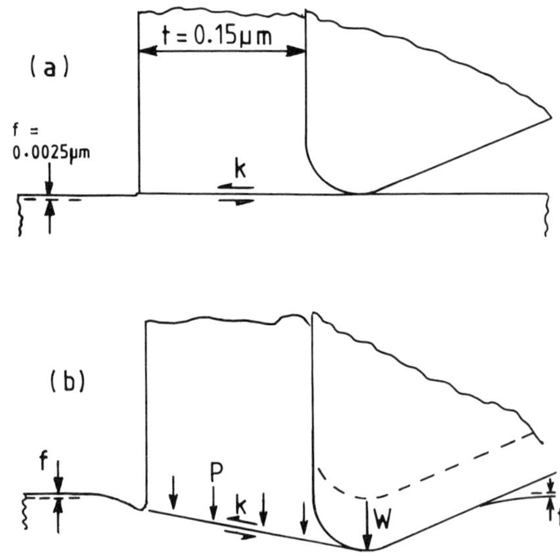

Figure 10. Speculations on elastic effects at almost nanometer feeds, discussed in text.

Whether this speculation is correct depends on whether it is plausible that w >> f. In the absence of a rigorous elastic analysis, an initial estimate of the possible size of w/f may be made by considering the average surface deflection w_{av} beneath a rectangular contact patch of width t and length d on a semi-infinite solid, acted on by a pressure p equal to k (figure 11). The result may be derived from (32)

$$w_{av} / f \cong (t/f)\,(k/E)\,(1 + \log(d/t)) \qquad (9)$$

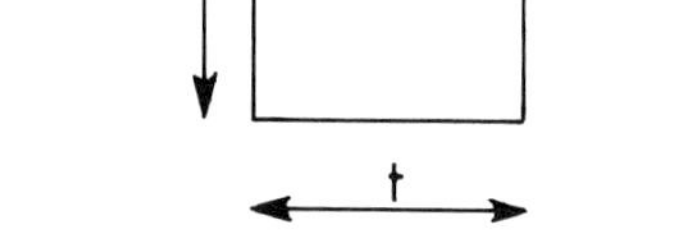

Figure 11. Elastic deflection due to p = k over a rectangular patch of dimensions chip thickness t x depth of cut d.

Taking E = 130 GPa and k = 600 MPa as before, and using values of d/t (about 1000) and t/f from Moriwaki's paper, the lower line of figure 12 is obtained. As f approaches 0·0025 µm., w_{av} / f approaches 1.0. Although this is not negligible, it is not >> 1.0 as needed for the figure 10b view. However if k is taken as 5 GPa, rounded up from the expected theoretical strength value of 4.6 GPa, the upper line of figure 12 results, in which w_{av} / f approaches 10.

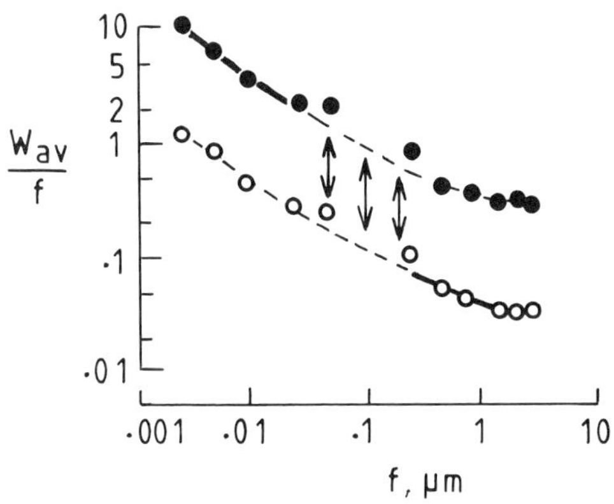

Figure 12. Calculated changes of w_{av} / f with f, for k = 600 MPa (o) and 5 GPa (•)

For elasticity to modify the flow as suggested in figure 10, it is at least necessary that the shear flow stress rises to its theoretical strength value. Perhaps, as suggested in figure 12, a transition occurs between dislocation controlled plasticity and theoretical strength values around f = 0.1 µm.. This suggestion was made by Moriwaki, but apparently rejected by him. He applied equation 2 to his data to calculate k. Figure 13 shows the result. At large feeds, k = 600 MPa , as assumed earlier here, but as f reduces below 0.1 µm., k rises, but only to 2 GPa. Moriwaki rejected this as not large enough for a theoretical strength value. However if elastic deflections were significant, application of equation 2 would be in error, both because of its neglecting the possible rotation of the shear plane under load and because some proportion of the tool force would have acted across the tool flank, rather than all on the rake face.

That f = 0.1 µm. may be a critical value also comes from a second look at figure 9b. At greater feeds, the variation of FC with f follows that of t with f, but below that value FC and t diverge. The suggestion that a change to theoretical strength values of plasticity occurs at feeds below 0.1 µm. is plausible, but more experimentation is needed before a clear conclusion is reached. It is critical that the observation of very large t/f values be confirmed. Experiments by Lucca et al. (8) show t/f only 3.7 at f = 0.05 µm., but do not extend to lower feeds. Recent experiments at the nanometer level do not report values of chip thickness or thrust force, but only estimate cutting force (9,11).

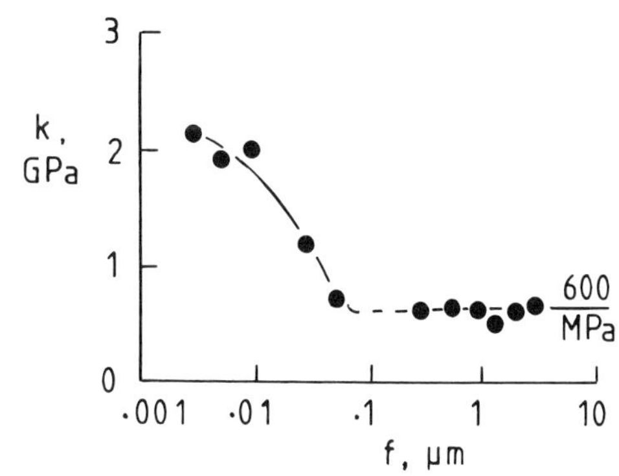

Figure 13. Calculated variation of k with f.

4. SIMULATED MACHINING AT NANO-METER FEEDS

A close-packed layer of copper atoms has a thickness of 0.27 nm.. The removal of such a thin layer has never been achieved by experiment (and may never be because of the difficulty of manufacturing sufficiently sharp tools), but molecular dynamics computer simulations have been performed to assess whether chip formation might occur at such a small feed, and the answer has been yes (9,11,12).

Simulation follows the displacements of atoms in a work material by time integration of Newton's second law applied to each atom in the model (copper and aluminium materials have been studied). The force on each atom is obtained from the summation of the interactions between it and every other atom in the model which are in turn obtained from the relative positions of the atoms and an assumed interatomic potential between them.

Motions have been calculated resulting from the displacement of a tool-shaped array of diamond-bonded carbon atoms into the work, and forces on the tool obtained from a further application of Newton's second law. The calculations are time consuming: 170 hours of computing time has been quoted for a 35 nm. tool displacement and a model consisting of 6000 atoms (11). An alternative approach, less intensive of computing time but only valid for modelling machining at 0°K, is molecular statics simulation, in which the atomic interactions are reduced to a non-linear finite element model (10).

The results of both molecular dynamics and molecular statics simulations predict chip formation with aluminium and copper, at feeds from 0.25 to 1 nm., to occur with specific cutting forces between 20 and 60 GPa. However in only one case (11), have thrust force and shear plane angle data been given, the latter deducible from visualisations of the chip formation process. Table 2 is derived from this.

Table 2
Molecular dynamics simulated results of copper machined by diamond, tool edge radius 0.5 nm.

f,nm.	FC/(fd), GPa	FT/(fd), GPa	ϕ, °	k, GPa
0.24	62.5	54.2	15	12.0
0.53	43.4	13.2	23	13.4

The shear plane angles ϕ are typical of sub-mm. feed values and are not as observed experimentally at sub-micron feeds (figure 9). The primary shear zone shear stress k derived by applying equation 2 is definitely of a theoretical shear strength size. In fact it is 2.5 to 3 times the theoretical shear strength, perhaps the result of forcing shear on a non-close-packed plane.

Other results in (11), for aluminium, may be used to deduce how non-dimensional specific forces are predicted to vary with f/r. Figure 14 is obtained if it is assumed that in these simulations k for aluminium is the same fraction of its Youngs modulus as is k for copper. A comparison with figure 7 shows FT/(kfd) to vary almost identically as it does at sub-mm. feeds, but FC/(kfd), although it changes from being greater than to less than FT/(kfd), peaks near to f/r = 0.1.

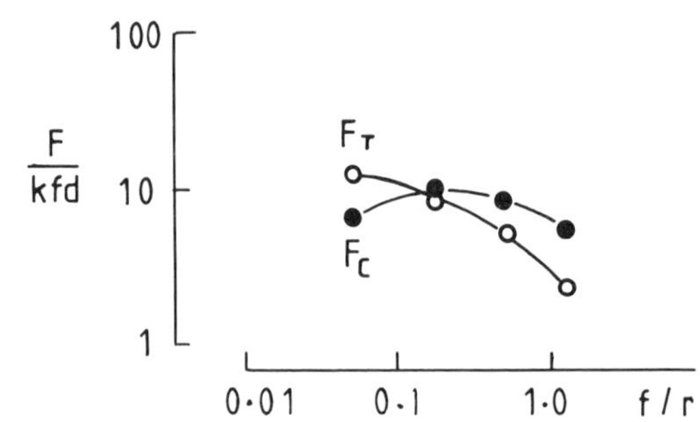

Figure 14. Molecular dynamics simulated non-dimensional specific forces: aluminium machined by diamond, f = 0.27 to 0.63 nm, r = 0.5 to 5 nm

The molecular dynamics simulations support the suggestion from the sub-micron experiments (section 3) that k takes a theoretical strength value at low feeds. However their high shear plane angles and reduction of FC/(kfd) with f/r < 0.1 do not agree with experiment. There are a number of ways in which they may not be realistic. (i) The friction between chip and tool, modelled by the interatomic potential between chip and tool atoms, may not be correct: molecular statics simulation has demonstrated that changing the chip/tool interaction does change chip thickness (10). (ii) The simulations do not show elastic spring back effects on the flanks of the tools: the boundaries of the models may be too stiff. Much work remains to be done in this exciting new area.

5.SUMMARY

Comparisons have been made between metal machining studies at sub-mm., sub-micrometer and nanometer feeds by considering non-dimensional specific forces and chip thickness ratios in these regimes.

Slip-line field theories provide bounds to the expected values of non-dimensional forces. At sub-mm. feeds experimental data fall within these bounds. The bounds are sensitive to the value of the friction between chip and tool, so that materials engineering factors that control this are important. For tools of near-zero or smaller rake angles, the bounds are so wide as not to be very constraining on the size of the forces. Then flow stress variations with strain, strain rate and temperature are found to provide additional conditions that give unique predictions of chip flow. So at sub-mm. feeds, mechanical modelling is well developed but its application is held back by lack of knowledge of materials' properties concerning chip/tool friction and bulk flow at extreme strain rates and temperatures.

Experiments at sub-micron, approaching nanometer, feeds show much higher non-dimensional specific forces than at higher feeds. This comment applies particularly when the shear flow stress k is given its bulk value, but less so if it is given a theoretical strength value: a number of reasons suggest that k might change from dislocation to theoretical strength controlled at a feed of around 0.1 μm.. A further cause of a high non-dimensional specific force is a large chip thickness ratio at low feeds. It is speculated in this paper that this might be caused by elastic effects that are negligible at sub-mm. feeds.

The results of molecular dynamics and molecular statics simulations of machining at sub-nanometer feeds give specific forces consistent with plastic shear at theoretical strength levels. They also show chip thickness ratios very like those at sub-mm. feeds. This is surprising as the experiments at nanometer feeds referred to in the last paragraph show much larger values. It is a matter of current research to resolve these differences between simulation and experiment.

REFERENCES

1. G. Boothroyd and W. A. Knight, Fundamentals of Machining and Machine Tools,Dekker, New York, 1989
2. P. L. B. Oxley, Mechanics of Machining, Ellis Horwood, Chichester, 1989.
3. M. C. Shaw, Metal Cutting Principles, Clarendon Press, Oxford, 1984.
4. E. M. Trent, Metal Cutting, Butterworths, London, 1984.
5. E. Usui, Modern Theory of Metal Cutting, Kyoritsu Shuppan, Tokyo,1990 (in Japanese).
6. N. Ikawa et al., 1991, Annals CIRP 40/2, 587-594.
7. T. Moriwaki and K. Okuda, 1989, Annals CIRP 38/1, 115-118.
8. D. A. Lucca et al., 1991, Annals CIRP 40/1, 69-72.
9. N. Ikawa et al., 1991, Annals CIRP 40/1, 551-554.
10. T. Inamura et al., 1991, Int. Jnl. JSPE 25, 259-266.
11. S. Shimada et al., 1992, Annals CIRP 41/1, 117-120.
12. I. F. Stowers, 1992, video presentation.
13. T. H. C. Childs, 1972, Proc. I. Mech. E. (Lond.) 186 Pt. 1, 717-727.
14. H. Ernst and M. E. Merchant, Chip Formation, Friction and High Quality Machined Surfaces pp 299-335, ASM, Ohio 1941.
15. E. H. Lee and B. W. Shaffer, 1951, J. Appl. Mech. 18, 405-413.
16. H. Ll. D. Pugh, Proc. Conf. Technology of Engng. Manufacture pp.237-254, I. Mech. E. Lond. 1958.
17. T. H. C. Childs, 1980, Int. J. Mech. Sci. 22, 457-466.
18. W. Kattwinkel, 1957, Industrieanzeiger 79, 525-532.
19. H. Chandrasekeran and D. V. Kapoor, 1965, Trans. ASME Jnl. Eng. Ind. 87, 495-502.
20 A. Bagchi and P. K. Wright,1987, Proc. Roy. Soc. Lond. A409, 99-113.
21. S. Kato et al., 1972, Trans. ASME Jnl. Eng. Ind. 94, 683-689.

22. G. Barrow et al. 1982, Int. Jnl. Mach. Tool Des. Res. 22,75-85.
23. T. H. C. Childs and K. Maekawa, 1990, Wear 139, 235-250.
24. T. Shirakashi and E. Usui, 1973, Jnl. JSPE 39, 966-972.
25. E. Usui et al., 1981, Bull. JSPE 15, 237-242.
26. K. Maekawa et al., Proc. 2nd. Int. Conf. Behaviour of Materials in Machining pp. 132-145, Inst. Metals, London, 1991
27. T. Shirakashi et al., 1983, Bull. JSPE 17, 161-166.
28. A. Otieno, current Ph.D. studies, University of Leeds.
29. P. J. Thompson and M. Sarwar, Proc. 15th Int. MTDR Conf. pp.217-226, Macmillan, London, 1975.
30. P. Wallen et al, 1988, Int.J.Mach.Tools Manufact. 28, 515-528
31. D. A. Tamianu and J. H. Dautzenberg, 1991, Annals CIRP 40/1, 65-67.
32. S. Timoshenko and J. N. Goodier, Theory of Elasticity, article 124, 2nd. edition, McGraw-Hill, New York 1951.

SESSION VII
ELASTOHYDRODYNAMICS (1)

Chairman : Professor W O Winer

Thin Films in Tribology / D. Dowson et al. (Editors)
© 1993 Elsevier Science Publishers B.V. All rights reserved.

ASPECTS OF TWO-SIDED SURFACE WAVINESS IN AN EHL LINE CONTACT

A.A. Lubrecht[a,b] and C.H. Venner[b]

[a] SKF Engineering & Research Centre B.V., Nieuwegein, the Netherlands
[b] University of Twente, Enschede, The Netherlands

Applying a multigrid solver to the transient Reynolds equation, this paper investigates the influence of two-sided waviness on pressure and film thickness in a medium loaded EHL line contact. The contact properties are studied for one set of operating conditions varying the ratio of the surface velocities, as well as the amplitude and wavelength of the waviness. The characteristic quantities used to describe the transient problem are straightforward extensions of the ones describing the stationary problem such as H_{min}, H_{ave} and P_{max}. Where possible physical explanations of the observed relations are presented.

1 Introduction

The research in the field of numerical Elasto Hydrodynamic Lubrication (EHL) is gradually moving away from the 'classical' idealized smooth surface problem and facing the complex reality in which engineering surfaces are non-smooth on the scale of thickness of the lubricating oil film. Presently we are puzzled by a problem that must have sounded familiar to a tribologist in the early decades of this century: How to understand, utilize and optimize the separating oil film in highly loaded contacts. Whereas the previous problem focussed on the so-called macro contact, the current one deals with the micro contact, the influence of surface roughness. The goal is to predict and optimize the operation of highly loaded contacts working with film thicknesses smaller than the undeformed surface roughness amplitude.

A number of investigations have addressed the steady state roughness problem [5, 6, 7, 9, 10, 12] and [13], however in general the roughness moves through the contact, thereby creating a transient problem. The time dependent problem was addressed in [1, 2, 3, 14, 17]. In [11] the transient analysis was used to explain an experimentally observed sensitivity to the overrolling direction of the location of the failure site with respect to the original indentation. In [19] a transient analysis of the influence of one-sided waviness on the film thickness was described. Several unexpected features were found, related to the velocity at which pressure and film thickness perturbations travel through the contact zone. It was shown that the pressure perturbations travel with the speed of the wavy surface (u_2 in that case), whereas the film thickness disturbances move with the average surface speed $(u_1 + u_2)/2$. These predicted features have recently been observed experimentally [8]. As was shown in the theoretical paper [19] the solution of the general non-smooth surface contact problem requires a transient solution of the EHL equations. In that paper as well as in the present one the authors have selected Multigrid solvers, since they meet the two requirements necessary to tackle this particular problem; they are fast and robust. Furthermore, and perhaps the most important argument is, that these techniques can straightforwardly be extended to more complete (and thus more complex) models of reality as for instance the point contact problem. This extension of the algorithm to point contact problems causes the computational complexity to increase only moderately. As a consequence the increase in computing time is not a large obstacle in view of the present increase in computer capacity. The main reason for addressing the line contact problem in this work is not so much the gain in computing time as the additional simplicity of the line contact problem when compared with the full non-smooth point contact problem.

In this paper the waviness extends to both surfaces, which introduces two additional degrees of freedom to the problem compared to the case studied in [19]. The complete problem is described by the contact condition parameters $(M, L$ or $W, U, G)$, the ratio of the surface velocities and for sinusoidal waviness the dimensionless amplitude and wavelength of both surfaces, and the initial phase difference. To reduce the number of parameters to some extent, the waviness wavelength on the upper and lower surface is assumed identical. By calculating either average values or extreme values over time, i.e. averages/extremes over all phase differences, the initial phase difference of the waviness becomes irrelevant.

1.1 Notation

A	amplitude
$\mathcal{A}$	dimensionless amplitude, $\mathcal{A} = AR/b^2$
b	halfwidth of Hertzian contact, $b = \sqrt{(8\mathrm{w}R)/(\pi E')}$
E	modulus of elasticity
E'	reduced modulus of elasticity, $2/E' = (1-\nu_1^2)/E_1 + (1-\nu_2^2)/E_2$
G	dimensionless materials parameter, $G = \alpha\ E'$
h	film thickness
h_{ave}	average film thickness, $h_{ave} = \frac{1}{2b}\int_{-b}^{+b} h(x)\ dx$
h_{inv}	inverse film thickness, $h_{inv} = (\frac{1}{x_b - x_a}\int_{x_a}^{x_b} h^{-1}(x)\ dx)^{-1}$
h_{min}	minimum film thickness
H	dimensionless film thickness, $H = hR/b^2$
H_{ave}	dimensionless average film thickness, $H_{ave} = h_{ave}R/b^2$
H^+_{ave}	maximum over time of H_{ave}
H^-_{ave}	minimum over time of H_{ave}
$\overline{H_{ave}}$	average over time of H_{ave}
H_{inv}	dimensionless inverse film thickness, $H_{inv} = h_{inv}R/b^2$
$\overline{H_{inv}}$	average over time of H_{inv}
H_{min}	dimensionless minimum film thickness, $H_{min} = h_{min}R/b^2$
H^+_{min}	maximum over time of H_{min}
H^-_{min}	minimum over time of H_{min}
H_0	integration constant
L	dimensionless material parameter (Moes), $L = G(2U)^{0.25}$
M	dimensionless load parameter (Moes), $M = W(2U)^{-0.5}$
n	number of discretisation points
p	pressure
p_h	maximum Hertzian pressure, $p_h = (2\mathrm{w})/(\pi b)$
P	dimensionless pressure, $P = p/p_h$
P_{max}	maximum dimensionless pressure
P^+_{max}	maximum over time of P_{max}
P^-_{max}	minimum over time of P_{max}
R	reduced radius of curvature, $1/R = 1/R_1 + 1/R_2$
$\mathcal{R}$	dimensionless deviations from the smooth profile
t	time
T	dimensionless time, $T = (tu_s)/(2b)$
u_1	velocity of lower surface
u_2	velocity of upper surface
u_s	sum velocity, $u_s = u_1 + u_2$
U	dimensionless speed parameter, $U = \eta_0 u_s/(2E'R)$
x	coordinate
x_d	location of surface feature
x_s	location of surface feature at $t = 0$
X, X'	dimensionless coordinate, $X = x/b$, $X' = x'/b$
X_d	dimensionless location of surface feature, $X_d = x_d/b$
X_s	dimensionless location of surface feature at $T = 0$, $X_s = x_s/b$
X_a, X_b	dimensionless inlet, outlet boundary domain, $X_a = x_a/b, X_b = x_b/b$
w	external load per unit width
W	dimensionless load parameter, $W = \mathrm{w}/(E'R)$
w	wavelength of surface feature
$\mathcal{W}$	dimensionless wavelength of surface feature, $\mathcal{W} = w/b$
z	viscosity index (Roelands equation)
α	pressure viscosity index
$\bar{\alpha}$	dimensionless parameter, $\bar{\alpha} = \alpha p_h$
Δ_T	dimensionless time increment
Δ_X	dimensionless space increment
ϵ	coefficient in Reynolds equation, $\epsilon = (\bar{\rho}H^3)/(\bar{\eta}\lambda)$
λ	dimensionless speed parameter, $\lambda = (6\eta_0 u_s R^2)/(b^3 p_h)$
η	viscosity

η_0	viscosity at ambient pressure
$\bar{\eta}$	dimensionless viscosity, $\bar{\eta} = \eta/\eta_0$
ρ	density
ρ_0	density at ambient pressure
$\bar{\rho}$	dimensionless density, $\bar{\rho} = \rho/\rho_0$

2 Theory

For completeness this section first presents the equations to be solved. Subsequently, the physical parameters describing the contact conditions are given in Table 1, together with different sets of dimensionless parameters. Table 2 lists the numerical parameters used in the calculations. The theory behind the Multigrid solvers is not repeated here and can be found in for instance [12, 16, 18], and in the references of these works.

2.1 Equations

The one-dimensional transient Reynolds equation is written in a dimensionless form:

$$\frac{\partial}{\partial X}\left(\epsilon \frac{\partial P}{\partial X}\right) - \frac{\partial(\bar{\rho}H)}{\partial X} - \frac{\partial(\bar{\rho}H)}{\partial T} = 0 \quad (1)$$

The boundary conditions are $P(X_a, T) = P(X_b, T) = 0$, $\forall T$ where X_a and X_b denote the boundaries of the domain. Furthermore, the cavitation condition $P(X,T) \geq 0$, $\forall X, T$ must be satisfied throughout the domain. ϵ and λ are defined according to:

$$\epsilon = \frac{\bar{\rho}H^3}{\bar{\eta}\lambda} \qquad \lambda = \frac{6\eta_0 u_s R^2}{b^3 p_h}$$

The density ρ is assumed to depend on the pressure according to the Dowson and Higginson relation [4] and the Roelands viscosity pressure relation [15] is used. The film thickness equation is made dimensionless using the same parameters and accounting for a moving surface feature reads:

$$H(X,T) = H_0(T) - \mathcal{R}_1(X,T) - \mathcal{R}_2(X,T) + \frac{X^2}{2} - \frac{1}{\pi}\int_{X_a}^{X_b} P(X',T)\ln|X - X'|\,dX' \quad (2)$$

where $\mathcal{R}_1(X,T)$ denotes the undeformed geometry of the waviness on the lower body at dimensionless time T and $\mathcal{R}_2(X,T)$ describes the waviness on the upper body. $H_0(T)$ is an integration constant.

$$\mathcal{R}_i(X,T) = \begin{cases} 0, & X \geq X_{d,i} \\ \mathcal{A}_i \sin\left(2\pi \frac{X - X_{d,i}}{\mathcal{W}_i}\right), & X < X_{d,i} \end{cases} \quad (3)$$

$$X_{d,i} = X_s + 2u_i/u_s\, T; \qquad i = 1, 2$$

At all times the force balance condition is imposed, i.e. the integral over the pressure must balance the externally applied contact load. This condition determines the value of the integration constant $H_0(T)$ in equation (2). Expressed in the dimensionless variables the force balance equation reads:

$$\int_{X_a}^{X_b} P(X,T)\,dX - \frac{\pi}{2} = 0 \quad \forall T \quad (4)$$

Parameter	Value	Dimension
E'	$2.26\ 10^{11}$	$[Pa]$
α	$2.2\ 10^{-8}$	$[Pa^{-1}]$
z	0.68	
η_0	$40\ 10^{-3}$	$[Pa\ s]$
u_s	1.94	$[ms^{-1}]$
R	$1.41\ 10^{-2}$	$[m]$
b	$5.0\ 10^{-4}$	$[m]$
p_h	$2.0\ 10^{9}$	$[Pa]$
h_{min}	0.232	$[\mu m]$
Dimensionless parameters		
M	100	
L	11.08	
$\bar{\alpha}$	44.0	
λ	$3.7\ 10^{-4}$	
W	$4.92\ 10^{-4}$	
U	$1.22\ 10^{-11}$	
G	$4.97\ 10^{3}$	
H_{min}	$1.316\ 10^{-2}$	
H_{ave}	$1.484\ 10^{-2}$	
H_{inv}	$2.674\ 10^{-2}$	

Table 1: *Different parameters and their values for the present loading condition.*

2.2 Conditions

The parameters used to describe the contact conditions in the calculations are given in Table 1, together with some dimensionless parameters. The numerical parameters used in the calculation are given in Table 2. These contact conditions are identical to the ones used in [19]. In that paper characteristics of the Reynolds equation are extensively described, especially the dominance of the shear flow over the pressure flow in the contact region. As a consequence the pressure perturbations become detached from the film thickness variations. For an extensive theoretical and numerical description of this effect the reader is referred to [19].

Parameter	Value
X_a	-2.5
X_b	1.5
Δ_T	0.005
Δ_X	$2.841\ 10^{-3}$
n	1409

Table 2: *Different parameters in the numerical solution.*

2.3 Time dependence

Because we are dealing with a time dependent problem, the familiar stationary contact quantities like H_{min}, H_{ave} and P_{max} now need another index to determine their relation with time. The superscript $..^{+}$ denotes the maximum value over time, the superscript $..^{-}$ denotes the minumum value over time, and the superscript $\overline{..}$ denotes the average value over time. These minima, maxima and averages are taken over the time interval when the values have become periodic, from $T = 4.0$ onwards in case of figure 1. This to exclude the initial period when the waviness is entering the contact area. The most interesting quantities are probably H^{-}_{min} denoting the minimum value of the minimum film thickness over time, $\overline{H_{ave}}$, the average value of the average film thickness over time and P^{+}_{max} the maximum value of the maximum pressure over time.

2.4 Numerical accuracy

In order to obtain a reasonable numerical accuracy for the extreme values like H^{-}_{min} a fine mesh had to be combined with a small time increment. The reader is reminded that H^{-}_{min} is the minimum value over time of the minimum film thickness H_{min}. The numerical accuracy in this quantity is better than 10% over the entire parameter range presented. The average film thicknesses $\overline{H_{ave}}$ and $\overline{H_{inv}}$ converge much more rapidly. Generally, the numerical error in these quantities is less than 2%. In order to eliminate the start-up effects and to obtain results from a fully periodic solution, the number of time steps was chosen inversely proportional to the velocity of the slowest surface. Thus 4000 timesteps were used for $u_2/u_s = 0.1$ and $u_2/u_s = 0.9$, 2000 timesteps were used for $u_2/u_s = 0.2$ and $u_2/u_s = 0.8$, 1500 timesteps were used for $u_2/u_s = 0.3$ and $u_2/u_s = 0.7$ and 1000 timesteps were used for $u_2/u_s = 0.4$ and $u_2/u_s = 0.6$.

3 General results

In figures 1, 2 and 3, three different ways of presenting the solutions are shown. Each has its own advantages and disadvantages, and each one sheds light on a particular detail of the results. The conditions were $w = 0.25b$, $A = 0.25$ μm, $u_2/u_s = 0.25$ (thus $u_1/u_s = 0.75$ and $u_1/u_2 = 3.0$), $X_{d,1} = X_{d,2} = -2.0$ at $T = 0.0$. Figure 1 shows the pressure (P) and film thickness (H) at a location ($X = 0$) as a function of the dimensionless time (T). At $T = 0$ the waviness on both surfaces is outside the contact and both pressure and film thickness have their smooth surface values. From $T = 2u_s/(2u_1) = 1.33$ onwards the waviness of the lower (faster) body reaches $X = 0.0$ and the pressure variations become sinusoidal. From $T = 2u_s/(2u_2) = 4.0$ onwards the upper body waviness reaches this location and the pressure becomes harmonic. Because the film thickness variations generated by the waviness of the upper and lower body travel at an average surface velocity ($u_s/2$) within the high pressure region ($|X| < 1.0$) the two variations in the film thickness reach $X = 0.0$ at $T = u_s/(2u_2) + 1.0 = 3.0$ and $T = u_s/(2u_1) + 1.0 = 1.66$ respectively. As a consequence the pressure and film thickness variations shown in figure 1 become harmonic at different times ($T = 4.0$ and $T = 3.0$ respectively). For locations $X > 0$ this time difference is even greater.

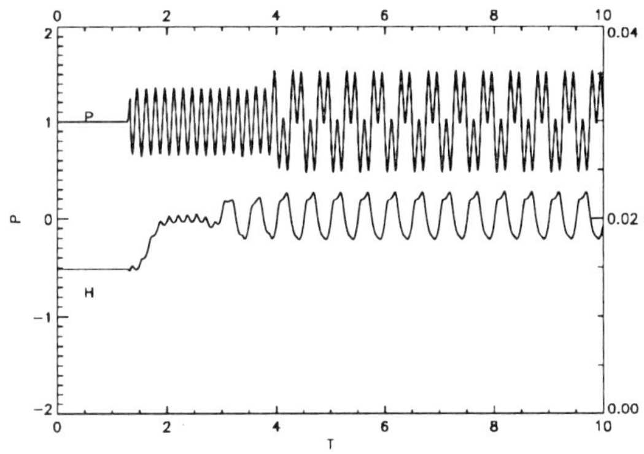

Figure 1 *Pressure and film thickness for $X = 0$ as a function of T, $u_2/u_s = 0.25$*

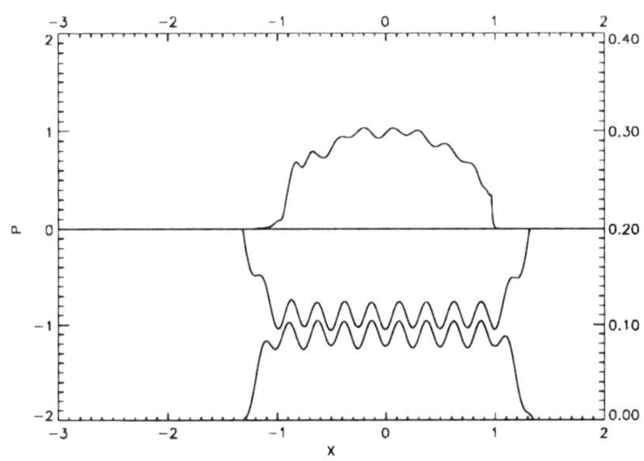

Figure 2a *Pressure and film thickness for $T =$ 6.88 as a function of X, $u_2/u_s = 0.25$*

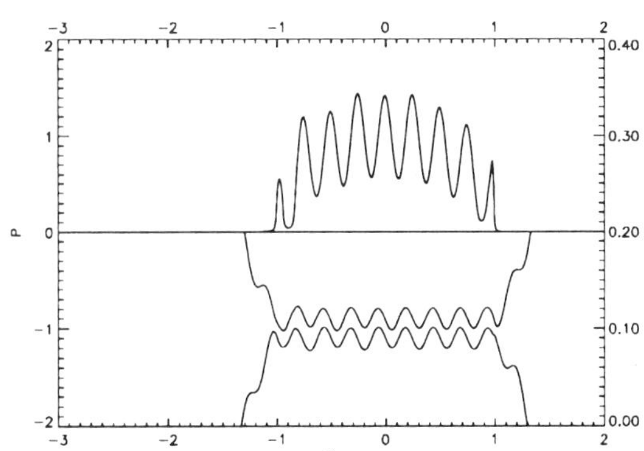

Figure 2b *Pressure and film thickness for $T =$ 6.94 as a function of X, $u_2/u_s = 0.25$*

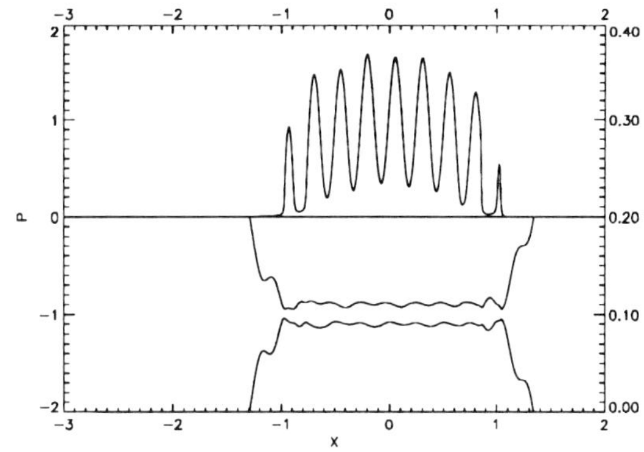

Figure 2c *Pressure and film thickness for $T =$ 7.00 as a function of X, $u_2/u_s = 0.25$*

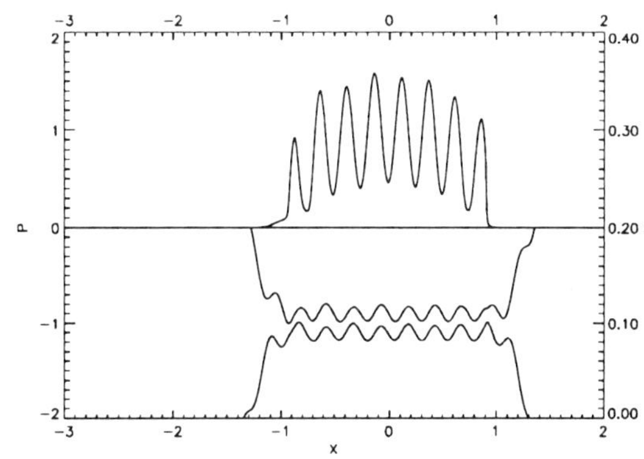

Figure 2d *Pressure and film thickness for $T =$ 7.06 as a function of X, $u_2/u_s = 0.25$*

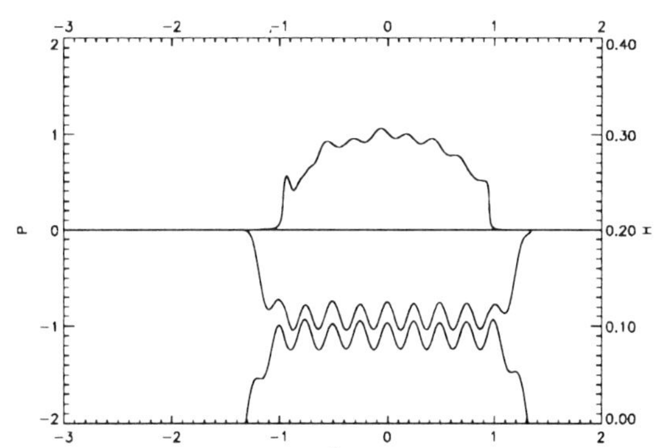

Figure 2e *Pressure and film thickness for $T =$ 7.13 as a function of X, $u_2/u_s = 0.25$*

A second way of visualizing the data is shown in figures 2a-e which displays the pressure and film thickness profiles for a certain timestep ($T = 6.88, 6.94, 7.00, 7.06, 7.13$) as a function of X. From these figures it is clear that the maximum pressure fluctuations coincide with the minimum film thickness variations, and vice versa. Because the lower body velocity is three times larger than the upper body velocity the lower body waviness has to "jump over" the upper body waviness, thereby causing a major flattening of the waviness and therefore requiring the large pressure fluctuations (figure 2c). The pressure fluctuations are minimal when the waviness of both surfaces is in phase (figure 2a and 2e). Clearly, this type of data is better presented as a movie.

The third way of presenting the pressure and film thickness variations is by considering them as a function of time for a certain location on the bodies. This description is important when one is interested in the stresses and stress cycles a certain volume element inside the material experiences. Figure 3a displays film thickness and pressure on a waviness top located on the upper (slow) surface, figure 3b does the same for a waviness top on the lower (fast) surface. Note that the time spent in the contact zone for the upper (slow) surface is three times as large as for the lower (fast) surface.

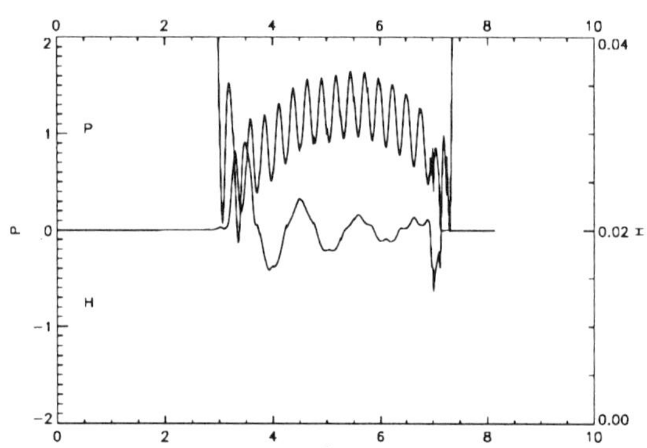

Figure 3a *Pressure and film thickness for a waviness top on the upper surface as a function of* T, $u_2/u_s = 0.25$

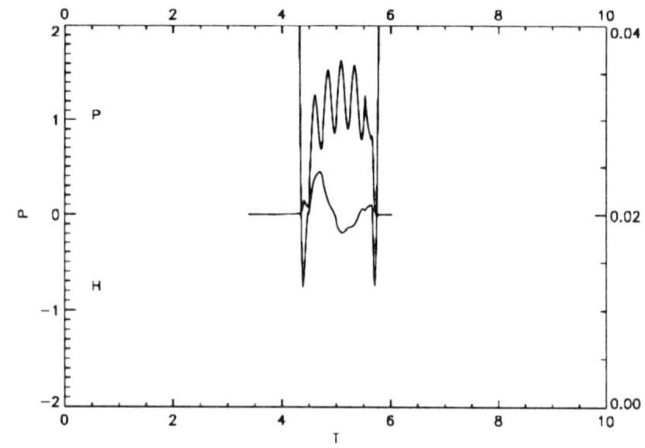

Figure 3b *Pressure and film thickness for a waviness top on the lower surface as a function of* T, $u_2/u_s = 0.25$

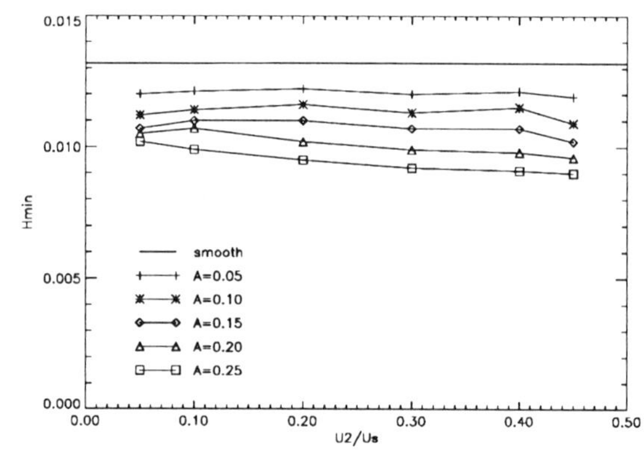

Figure 4a *Minimum film thickness* H^-_{min} *as a function of* u_2/u_s, $A_1 = A_2 = A$ *in* μm, $\mathcal{W}_1 = \mathcal{W}_2 = 0.25$

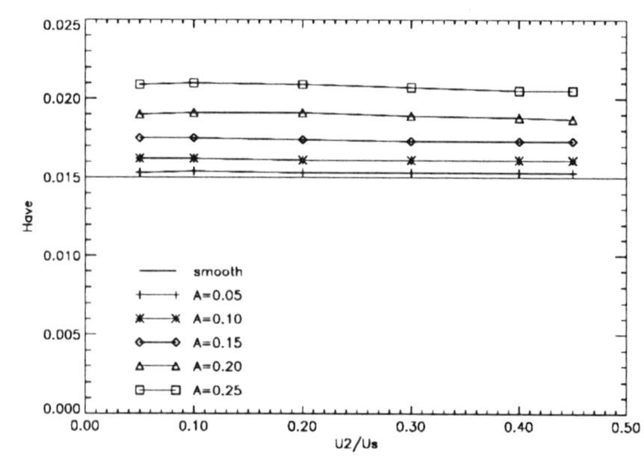

Figure 4b *Average film thickness* $\overline{H_{ave}}$ *as a function of* u_2/u_s, $A_1 = A_2 = A$ *in* μm, $\mathcal{W}_1 = \mathcal{W}_2 = 0.25$

4 Influence of u_2/u_s for equal waviness amplitude

From figure 1 it can be observed that the film thickness at $X = 0.0$ increases over its smooth value when the waviness enters the contact. This section explores how the different parameters like minimum film thickness, average film thickness and maximum pressure behave as a function of the surface speeds. Figures 4a and b show the influence of the parameter u_2/u_s on the minimum film thickness and on the average film thickness for the case that the waviness on both surfaces has the same amplitude. The definitions of H^-_{min} and $\overline{H_{ave}}$ are given in section 2.3. Only the first half of the graphs is shown since they are symmetric around $u_2/u_s = 0.5$. The points $u_2/u_s = 0.0$ and 0.5 are omitted since they do not represent an average over all phase differences of the waviness.

From these figures it can be observed that the minimum film thickness decreases with amplitude below the smooth surface value. The average film thickness increases with increasing amplitude over the smooth surface value. From these two figures it can also be concluded that the influence of the surface velocity difference on the minimum and average film thickness is small. Also the maximum pressure was found to be insensitive to the velocity difference.

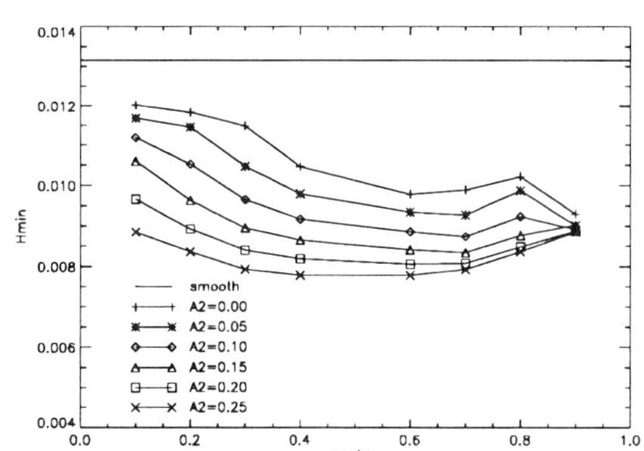

Figure 5a *Minimum film thickness H^-_{min} as a function of u_2/u_s and A_2, $A_1 = 0.25$ μm, $\mathcal{W}_1 = \mathcal{W}_2 = 0.25$*

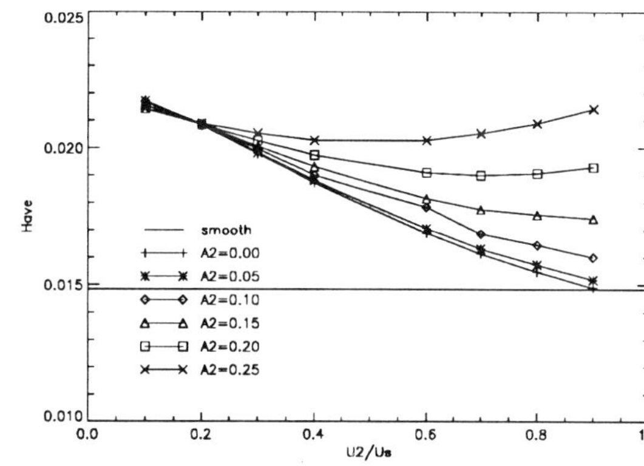

Figure 5b *Average film thickness $\overline{H_{ave}}$ as a function of u_2/u_s and A_2, $A_1 = 0.25$ μm, $\mathcal{W}_1 = \mathcal{W}_2 = 0.25$*

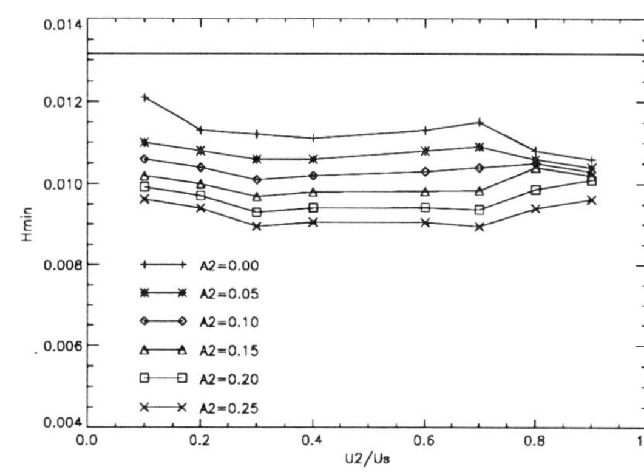

Figure 6a *Minimum film thickness H^-_{min} as a function of u_2/u_s and A_2, $A_1 = 0.25$ μm, $\mathcal{W}_1 = \mathcal{W}_2 = 0.50$*

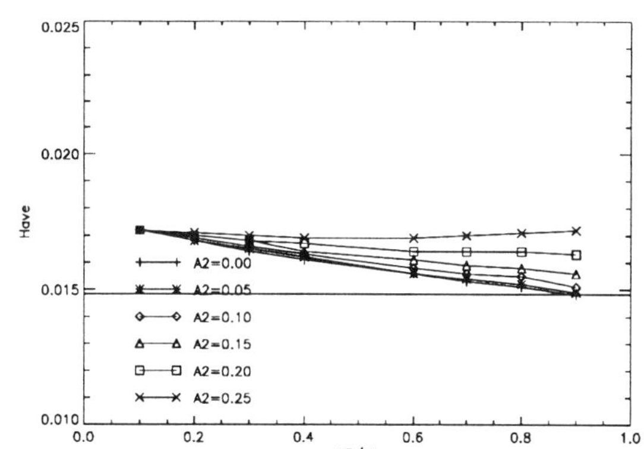

Figure 6b *Average film thickness $\overline{H_{ave}}$ as a function of u_2/u_s and A_2, $A_1 = 0.25$ μm, $\mathcal{W}_1 = \mathcal{W}_2 = 0.50$*

5 Influence of u_2/u_s, for different waviness amplitude

In this section the waviness amplitude of the lower surface is kept contant $A_1 = 0.25$, while the amplitude of the upper body waviness is varied from $A_2 = 0.0$ to $A_2 = 0.25$. Now the symmetry around $u_2/u_s = 0.5$ is broken of course, and the entire range from $u_2/u_s = 0.1$ to $u_2/u_s = 0.9$ is displayed. The points $u_2/u_s = 0.0, 0.5$ and 1.0 are again omitted since they do not represent an average over all phase differences.

From figure 5a it can be observed that the minimum film thickness decreases with increasing amplitude A_2 and with increasing values of u_2/u_s as was found in the previous section. For large values of u_2/u_s and A_2 it increases slightly again, in order to become symmetric for the case of $A_2 = 0.25$. From figure 5b it can be observed that the average film thickness increases with increasing amplitude A_2 and decreases with increasing values of u_2/u_s. For large values of u_2/u_s and A_2 it increases slightly again, becoming symmetric around $u_2/u_s = 0.5$ for the case that $A_1 = A_2 = 0.25$. The maximum pressure P^+_{max} was found to be virtually independent of the value of u_2/u_s, as was observed in the previous section.

In figure 6 the wavelength was doubled compared to figure 5. The influence of this doubling on the minimum film thickness is small, compare figures 5a and 6a. The maximum pressure is halved for a given amplitude, as is predicted by the dry contact theory. The increase of the average film thickness over its smooth value and its dependency on u_2/u_s and $\mathcal{A}_2$ are considerably reduced as can be observed by comparing the figures 5b and 6b. Roughly speaking, the increase in average film thickness is halved by doubling the wavelength of the waviness.

6 Influence of amplitude

This section analyses in detail the influence of the amplitude on the maximum pressure, the minimum and the average film thickness as observed in the previous section. From a dry contact analysis the maximum pressure can be expected to increase linearly with the waviness amplitude, a

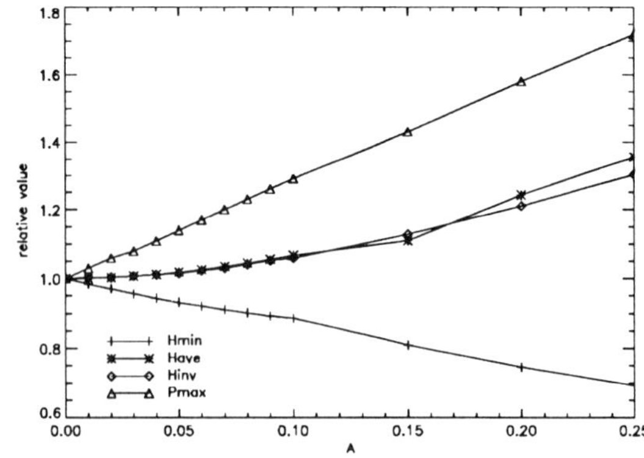

Figure 7a *Relative change in pressure P^+_{max} and film thickness H^-_{min}, $\overline{H_{inv}}$ and $\overline{H_{ave}}$as a function of the amplitude $A_1 = A_2 = A$ in μm, $\mathcal{W}_1 = \mathcal{W}_2 = 0.25$, $u_2/u_s = 0.3$*

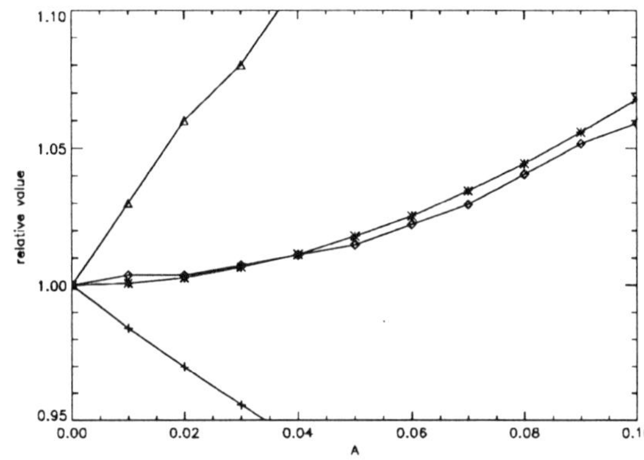

Figure 7b *Detail of figure 7a*

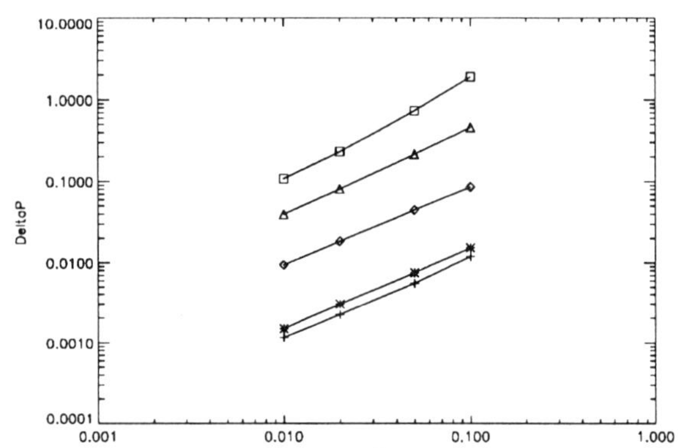

Figure 8 *Relative pressure increase in the inlet region as a function of the amplitude $A_1 = A_2 = A$ in μm, $\mathcal{W}_1 = \mathcal{W}_2 = 0.25$, $u_2/u_s = 0.3$, bottom to top for $X = -2.0, -1.5, -1.2, -1.1, -1.0$.*

trend that is observed from figure 7a and is analysed in detail in [7]. The minimum film thickness decreases roughly linear with the amplitude, a trend that is easily accepted. However, the average film thickness as well as the inverse film thickness seem to increase quadratically with the amplitude. This can be more clearly observed from figure 7b, which displays a detail of figure 7a. This trend, which was also observed in [19], needs further explanation.

The classical explanation of the smooth average film thickness and the nearly parallel film thickness in the contact area assumes that the pressure in the inlet reaches a certain value where the oil viscosity becomes so large that pressure-induced flow is nearly absent. From this (zero-pressure-flow) point onwards, all oil is transported through the contact.

This explanation can be extended to wavy contacts. Consider a point in the inlet where the pressure is such that the viscosity becomes sufficiently high to eliminate virtually all pressure induced flow. Now assume that in the neighbourhood of this point the pressure fluctuations depend linearly on the waviness amplitude, as happens in the contact region (see figure 7a). The location of this zero pressure-flow point then moves outwards in a first approximation proportionally with the height of the pressure fluctuations which are themselves proportional to the waviness amplitude. The gap height at this point increases roughly proportional with the square of this distance, since the cylinder can be accurately approximated by a parabola as the influence of deformation can be neglected. As a result the gap height at this point, and therefore the amount of oil transported through the contact, and thus the average film thickness in the contact area will increase with the square of the waviness amplitude. Figure 8 shows the relative increase of the maximum pressure in the inlet at various locations as a function of the amplitude. From this figure it can be concluded that the pressure increase in the inlet is indeed proportional to the waviness amplitude, as is the case in the Hertzian contact zone. Concluding this section it can be stated that the average film thickness increases quadratically with the waviness amplitude because the location of the zero-pressure-flow point moves linearly towards the inlet as a function of the waviness amplitude.

7 Discussion and conclusion

In this paper the influence of two-sided waviness on pressure and film thickness was investigated. The influence of the non-Newtonian fluid behaviour as well as thermal effects has been omitted. The authors are aware of the importance of these effects especially at large velocity differences. However, in order to be able to correctly interpret the results it was considered beneficial to study one effect at a time.

The observed dependence of P^{+}_{max}, H^{-}_{min} and $\overline{H_{ave}}$ on $\mathcal{A}_1$, $\mathcal{A}_2$, $\mathcal{W}_1$, $\mathcal{W}_2$ and u_2/u_s can be summarized as folows. The parameters with the additional subscript "s" refer to the parameter values for the smooth solution.

• The maximum pressure attained in the oil film is close to the dry contact pressure, is virtually independent of the ratio of the surface velocities and increases linearly with the ratio of amplitude and wavelength:
$P^{+}_{max} = P_{max,s} + c_1(\mathcal{A}_1/\mathcal{W}_1 + \mathcal{A}_2/\mathcal{W}_2)$.

• The minimum film thickness H^{-}_{min} attained in the oil film decreases linearly with increasing waviness amplitude and is independent of the wavelength $\mathcal{W}$. As a first approximation:
$H^{-}_{min} = H_{min,s} - c_2(\mathcal{A}_1 u_1 + \mathcal{A}_2 u_2)$.

• The average film thickness $\overline{H_{ave}}$ attained in the oil film increases quadratically with ratio of amplitude and wavelength for small amplitudes. As a first approximation:
$\overline{H_{ave}} = H_{ave,s} + c_3(\mathcal{A}_1 u_1/\mathcal{W}_1 + \mathcal{A}_2 u_2/\mathcal{W}_2)^2$.

8 Acknowledgement

The authors would like to thank Dr. H.H. Wittmeyer, Managing Director of the SKF Engineering & Research Centre for his kind permission to publish this work, part of which was supported by a fellowship of the "Royal Netherlands Academy of Arts and Sciences".

References

[1] **Chang, L., Cusano, C., and Conry, T.F.**, 1989, "Effects of Lubrication Rheology and Kinematic Conditions on Micro-Elastohydrodynamic Lubrication," *ASME JOT*, **111**, 344-351.

[2] **Chang, L., and Webster, M.N.**, 1991, "A Study of Elastohydrodynamic Lubrication of Rough Surfaces," *ASME JOT*, **113**, 110-115.

[3] **Chang, L.**, 1992, "Traction in Thermal Elastohydrodynamic Lubrication of Rough Surfaces," *ASME JOT*, **114**, 186-191.

[4] **Dowson, D., and Higginson, G.R.**, 1966, "Elastohydrodynamic Lubrication, The Fundamentals of Roller and Gear Lubrication," Pergamon Press, Oxford, Great Britain.

[5] **Goglia, P.R., Cusano, C., and Conry, T.F.**, 1984, "The Effects of Surface Irregularities on the Elastohydrodynamic Lubrication of Sliding Line Contacts. Part I - Single Irregularities," *ASME JOT*, **106**, 104-112.

[6] **Goglia, P.R., Cusano, C., and Conry, T.F.**, 1984, "The Effects of Surface Irregularities on the Elastohydrodynamic Lubrication of Sliding Line Contacts. Part II - Wavy Surfaces," *ASME JOT*, **106**, 113-119.

[7] **Greenwood, J.A., and Johnson, K.L.**, 1992, "The Behaviour of Transverse Roughness in Sliding Elastohydrodynamically Lubricated Contacts," *WEAR*, **153**, 107-117.

[8] **Kaneta, M.**, 1991, "private communication"

[9] **Kweh, C.C., Evans, H.P., and Snidle, R.W.**, 1989, "Micro-Elastohydrodynamic Lubrication of an Elliptical Contact with Transverse and Three-Dimensional Roughness," *ASME JOT*, **111**, 577-583.

[10] **Lee, R.T., and Hamrock, B.J.**, 1990, "A circular non-Newtonian Fluid Model: Part II - used in Micro-Elastohydrodynamic Lubrication," *ASME JOT*, **112**, 497-505.

[11] **Lubrecht, A.A., Venner, C.H., Lane, S., Jacobson, B., and Ioannides, E.**, 1990, "Surface Damage - Comparison of Theoretical and Experimental Endurance Lives of Rolling Bearings," *Proceedings of the 1990 Japan International Tribology Conference, Nagoya, Japan*, **1**, 185-190.

[12] **Lubrecht, A.A.**, 1987, "Numerical Solution of the EHL Line and Point Contact Problem Using Multigrid Techniques," PhD. Thesis, University of Twente, Enschede, The Netherlands. ISBN 90-9001583-3

[13] **Lubrecht, A.A., ten Napel, W.E., Bosma, R.**, 1988, "The Influence of Longitudinal and Transverse Roughness on the Elastohydrodynamic Lubrication of Circular Contacts", *ASME JOT*, **110**, 421-426.

[14] **Osborn, K.F., Sadeghi, F.**, 1992, "Time Dependent Line EHD Lubrication using the Multigrid/Multilevel Technique", *ASME JOT*, **114**, 68-74.

[15] **Roelands, C.J.A.**, 1966, "Correlational Aspects of the Viscosity-Temperature-Pressure Relationship of Lubricating Oils," PhD. Thesis, Technical University Delft, Delft, The Netherlands, (V.R.B., Groningen, The Netherlands).

[16] **Venner, C.H., ten Napel, W.E., and Bosma, R.**, 1990, "Advanced Multilevel Solution of the EHL Line Contact Problem," *ASME JOT*, **112**, 426-432.

[17] **Venner, C.H., Lubrecht, A.A., and ten Napel, W.E.**, 1991, "Numerical Simulation of the Overrolling of a Surface Feature in an EHL Line Contact," *ASME JOT*, **113**, 777-783.

[18] **Venner, C.H.**, 1991, "Multilevel Solution of the EHL Line and Point Contact Problems," PhD. Thesis, University of Twente, Enschede, The Netherlands. ISBN 90-9003974-0.

[19] **Venner, C.H., and Lubrecht, A.A.**, 1992, "Transient Analysis of Surface Features in an EHL Line Contact in the case of Sliding," *to appear in ASME JOT*

Thin Films in Tribology / D. Dowson et al. (Editors)
1993 Elsevier Science Publishers B.V.

Traction in a simulated EHD roller/ring contact

J. Seabra * , A. Sottomayor ** and A. Campos **

CETRIB - Unidade de Tribologia e Manutenção Industrial
DEMEGI* / INEGI**
Faculdade de Engenharia da Universidade do Porto
Rua dos Bragas - 4099 PORTO Codex - Portugal

ABSTRACT

Traction forces are an important parameter in the analysis of an elastohydrodynamic contact problem. This work is about the determination of traction forces and friction coefficients in an EHD roller/ring contact, in a roller bearing.

A simulation of the EHD contact is developed, which consists on a waved dry Hertzian contact, separated by a constant film thickness calculated using Dowson's expression, and corrected by a thermal factor. This procedure, although approximated, permits to develop a simple algorithm for the evaluation of the waved pressure distribution and of the waved contact geometry.

The results obtained concern the MOBIL Jet II oil, which is considered as a viscoelastic (Ree-Eyring) fluid, whose physical characteristics (viscosity, shear modulus and reference shear stress) are pressure and temperature dependent.

The variation of friction and traction with the operating conditions is studied, showing the influence of the rolling and sliding speeds, and the influence of the maximum Hertzian pressure. The influence of the surface waviness of the contacting solids is also analysed, showing how this parameter is important for the traction behaviour of the EHD contact.

*DEMEGI - Departamento de Engenharia Mecânica e Gestão Industrial
**INEGI - Instituto de Engenharia Mecânica e Gestão Industrial

INTRODUCTION

The traction force inside an EHD contact is a very important aspect of the contact behaviour, since it has a determinant influence on the energy dissipated and thus on the contact temperature. It also modifies the contact pressure distribution, by introducing a tangential component, and making the subsurface stresses more severe.

It is known that traction in an EHD contact depends on several parameters like the applied load or the rolling and sliding speeds. It also depends on the pressure distribution and in everything that might modify it, like the surface waviness, which becomes an important parameter in traction evaluation.

Thermoelastohydrodynamic rough models and their numerical solution, although already available at least for line contacts, might be rather complex and time consuming. In order to avoid these inconveniences, a simulation of the isothermal EHD waved contact was developed.

That simulation consists on a dry Hertzian contact, separated a constant film thickness calculated using the Dowson's expression [1], and corrected by a thermal factor [2]. The pressure distribution and deformed geometry are those of the dry Hertzian contact. This procedure, although approximated, permits to develop a simple algorithm for the evaluation of the waved pressure distribution and of the waved contact geometry.

These results can be used to perform EHD traction and friction coefficient calculations, and to test the influence of several parameters, like the rheological model, the operating conditions or the surface waviness [10, 11, 12].

1. SIMULATION OF THE EHD ROLLER/RING WAVED CONTACT

1.1. Geometry of the contact

In a roller/ring contact, the pressure distribution along the transverse direction depends very much on the roller geometry in that direction. Thus, this geometry must be defined with care, since it will effect not only the pressure distribution but also the film thickness and the tangential pressure distribution.

The non-deformed geometry of the contact is shown on figure 1.

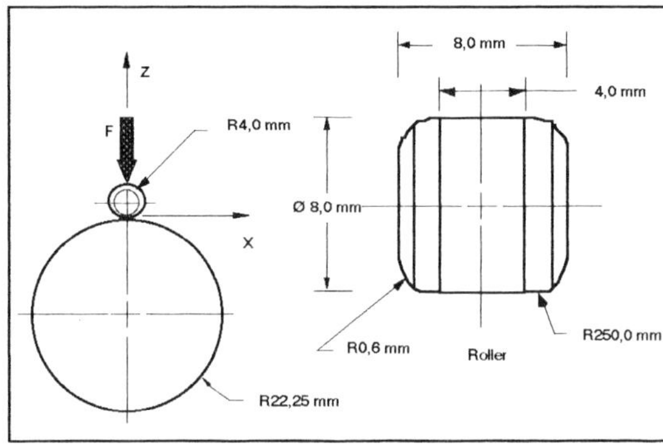

Figure 1 - Geometry of a roller/ring contact.

Considering the roller/ring contact as a line contact some parameters can be evaluated. The contact width was taken to be the length of the cylindrical part of the roller ($\ell = 2a = 4.00$ mm), and the equivalent radius of the contact, $R = [0.5\ (R_1^{-1} + R_2^{-1})]^{-1}$ [3], is equal to 6.79 mm.

Only one macrogeometry was considered.

1.2. Surface waviness

Four microgeometries were considered.

In the first one, both the roller and the ring have smooth surfaces. In the others, the roller is smooth and the ring has a longitudinally waved (sinusoidal) surface.

In table 1 the amplitude for each case is presented.

Surf. Waviness	case 1	case 2	case 3	case 4
Amplitude - μm	0.00	0.25	0.50	0.75

Table 1 - Amplitude of the surface waviness.

An amplitude of 0.25 μm, corresponds to an equivalent RMS roughness of the contacting surfaces, $\sigma = (\sigma_1^2 + \sigma_2^2)^{0.5}$, of 0.356 μm, which is a characteristic value for a roller bearing [4]. Thus, the amplitude of 0.75 μm, is three times larger than the usual value, being much more severe.

The wavelength is constant and was calculated supposing that within the cylindrical part of the roller there were 10 waves. Thus $\lambda = 2a / 10 = 0.4$ mm.

1.3. Pressure distribution and deformed geometry of the contact

The EHD pressure distribution and deformed geometry correspond to the Hertzian dry ones. The deformed geometry is shown on figure 2 for the case of a waved surface contact.

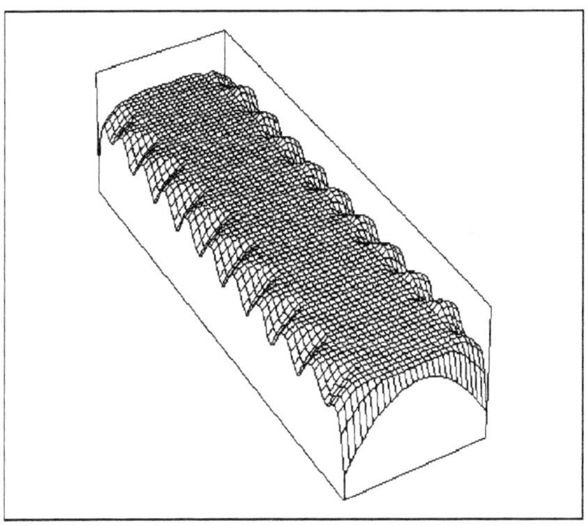

Figure 2 - The deformed surface.

There are two important differences between, the pressure distribution and the deformed geometry considered, and the correspondent EHD real ones. The pressure distribution doesn't show the characteristic pressure spike, and the contact geometry has no film thickness restriction at the divergent boundary.

Future work, will consider a real EHD contact, in order to obtain more realistic results, which will be compared with those presented in this report.

1.4. Film thickness

In the high pressure zone of the contact, the deformed surface is almost flat (see figure 2), meaning that film thickness is constant, and thus not much influenced by the surface waviness [5].

The two surfaces are separated by a constant film thickness, h_0, given by Dowson's formula for line contacts [1], which states that,

$$h_0 = 1.949\left(\eta_0 U \alpha\right)^{0.727} R^{0.364} \left(E \ell / F_n\right)^{0.091}$$

This isothermal, smooth and fully flooded film thickness is corrected by a thermal factor, ϕ_T, defined as [5]

$$\phi_T = \left\{ 1 + 0.1 \left[\left(1 + 14.8\, V_e^{0.83}\right) L^{0.64} \right] \right\}^{-1}$$

where

$$V_e = \frac{|U_1 - U_2|}{U_1 + U_2} \quad ; \quad L = \frac{\beta \left(U_1 + U_2\right)^2 \eta_0}{K}$$

The thermal factor, ϕ_T, due to the viscous shear heating in the convergent of the EHD contact, is very useful in this analysis, since the isothermal, smooth and fully flooded formulation of the EHD contact problem, doesn't take into account the sliding rate, V_e .

Beside this thermal correction, a surface waviness or roughness correction factor, could also be considered. Nevertheless, previous results [5] for a longitudinally waved (or rough) contact, showed that this correction is not very important.

2. OPERATING CONDITIONS

The friction force in an EHD contact, is a function of the operating conditions, mainly the load and the rolling and sliding speeds. It also depends on the surface roughness or waviness since it affects both the pressure and the film thickness in an EHD contact.

Four load cases were studied, corresponding to maximum Hertzian pressures of 0.5, 1.0, 1.5 and 2.0 GPa in a smooth dry line contact, $p_0 = [\ (2\ F_n\ E)\ /\ (\pi\ \ell\ R)\ ]^{0.5}$ [3].

This range of maximum Hertzian pressure is large enough to cover all practical cases of an EHD roller/ring contact, for a roller bearing in a jet engine.

The tangential load, F_t , is the integral of the tangential pressure distribution, τ, which depends on the lubricant rheology as it will be seen later on. The global friction coefficient, μ, represents the ratio between tangential and normal loads.

The rotation speed of the inner ring varies from 1000 rpm to 40000 rpm; thus the mean rolling surface speed $(U1+U2)/2$, goes from 2.33 m/s to 93.2 m/s. The sliding speed for each rolling speed, is such that the sliding rate, V_e, goes from 0.01 to 0.10. Thus, the sliding speed range goes from 0.02 to 0.93 m/s.

The material of the roller and of the inner ring, is steel, with a Young's modulus of 210 GPa and a Poisson's coefficient of 0.3 . The equivalent Young's modulus for the two contacting bodies is equal to 230.77 GPa, $E = \{\ 0.5\ [(1-\nu_1^2)/E_1 + (1-\ \nu_2^2)/E_2\ ]\ \}^{-1}$ [3]. Only one material was considered.

3. LUBRICANT RHEOLOGY

The friction force in an EHD contact, is a function of the rheological law considered, and of the lubricant pressure and temperature dependence (piezo and thermoviscosity).

Several rheological laws might be considered in order to describe the lubricant behaviour in an EHD contact. These equations always depend on some lubricant parameters, like the shear modulus G, the reference shear stress τ_r or the dynamic viscosity η. The lubricant properties are generally pressure and temperature dependent, and some times, they might even depend on other parameters, for instance the rolling speed.

In this work, the viscous and the viscoelastic rheological laws were considered, both for a Ree-Eyring fluid with a reference shear stress, τ_r. These models were applied to the MOBIL Jet II oil, and the corresponding equations for the shear stress were obtained.

The analysis of the results obtained [10], allowed to conclude that, for the lubricant and operating conditions considered, the viscoelastic model is more realistic producing smaller friction coefficients than the viscous model.

3.1. Viscoelastic model

In the Maxwell model the shear rate $\dot{\gamma}$ is the sum of an elastic component $\dot{\gamma}_e$ and a viscous component $\dot{\gamma}_v$. Assuming that the viscous component is the same of the Ree-Eyring fluid, the shear rate is defined as [6, 9]

$$\dot{\gamma} = \dot{\gamma}_e + \dot{\gamma}_v = \frac{\dot{\tau}}{G} + \frac{\tau_e}{\eta} \, sh\left(\frac{\tau}{\tau_r}\right)$$

When sliding occurs in two orthogonal directions, the previous expression must be replaced by the following system of equations:

$$\dot{\gamma}_x = \frac{\dot{\tau}_x}{G} + \frac{\tau_x}{\tau_{eq}} \frac{\tau_r}{\eta} \, sh\left(\frac{\tau_{eq}}{\tau_r}\right)$$

$$\dot{\gamma}_y = \frac{\dot{\tau}_y}{G} + \frac{\tau_y}{\tau_{eq}} \frac{\tau_r}{\eta} \, sh\left(\frac{\tau_{eq}}{\tau_r}\right)$$

,

where, $\tau_{eq} = \sqrt{\tau_x^2 + \tau_y^2}$, and

$$\tau_x = \tau_x\,(x,y,z) \Rightarrow \dot{\tau}_x = \frac{d}{dt}\tau_x = \frac{d}{dt}\tau_x\,(x,y,z)$$

$$\tau_y = \tau_y\,(x,y,z) \Rightarrow \dot{\tau}_y = \frac{d}{dt}\tau_y = \frac{d}{dt}\tau_y\,(x,y,z).$$

Using the derivative chain rule for functions of more than one variable, it can be written that,

$$\dot{\tau}_{x\,(x,y,z)} = \left[\frac{\partial}{\partial x}\tau_{x\,(x,y,z)}, \frac{\partial}{\partial y}\tau_{x\,(x,y,z)}, \frac{\partial}{\partial z}\tau_{x\,(x,y,z)}\right]\begin{bmatrix}\frac{d}{dt}x\\ \frac{d}{dt}y\\ \frac{d}{dt}z\end{bmatrix}$$

If rolling only exists in the xx direction, thus

$$\frac{d}{dt}x = U \neq 0\ ;\ \frac{d}{dt}y = V = 0\ ;\ \frac{d}{dt}z = W = 0$$

implying that

$$\dot{\tau}_x\,(x,y,z) = U\frac{\partial}{\partial x}\tau_x\,(x,y,z)$$

$$= \lim_{\Delta x \to 0}\frac{\tau_x\,(x,y,z) - \tau_x\,(x-\Delta x,y,z)}{\Delta x}.$$

For a small Δ_x the limit can be approximated by the finite ratio, and so

$$\dot{\tau}_{x\,(x,y,z)} = U\,\frac{\tau_x\,(x,y,z) - \tau_x\,(x-\Delta x,y,z)}{\Delta x}\ ;$$

by a similar procedure

$$\dot{\tau}_{y\,(x,y,z)} = U\,\frac{\tau_y\,(x,y,z) - \tau_y\,(x-\Delta x,y,z)}{\Delta x}$$

and the system of equations can be reduced to find the root X of the following equation:

$$A\,X + \frac{B}{C_1}\,sh\left(\frac{X}{B}\right) - C_2 = 0 \qquad (1)$$

with A, B, C, C_1, C_2, and X defined as:

$$A = \frac{U}{G\Delta x}\ ,\qquad B = \frac{\tau_r}{\sqrt{1+C^2}}\ ,$$

$$C = \frac{\tau_y}{\tau_x} = \frac{\dot{\gamma}_y + \frac{U}{G}\,\frac{\tau_y(x-\Delta x)}{\Delta x}}{\dot{\gamma}_x + \frac{U}{G}\,\frac{\tau_x(x-\Delta x)}{\Delta x}} \qquad (2),$$

$$C_1 = \eta,\ \ C_2 = \dot{\gamma}_x + A\ \tau_x\,(x-\Delta x)\ ,\ X = \tau_x$$

To find the root $\tau_x = X$ of the previous equation, the Newton method was used; the other shear stress component, τ_y, can be calculated by the definition of C once τ_x is known.

3.2. Lubricant parameters

The viscoelastic rheological model depends on the dynamic viscosity, η, and on the Ree-Eyring reference shear stress, τ_r. It also depends on the shear modulus, G. Each lubricant presents its characteristic values for η, G and τ_r.

In an elastohydrodynamic contact, the pressures and temperatures attained might considerably modify these parameters. The following relationships were considered for MOBIL Jet II oil.

3.2.1. Viscosity

Experience shows that at high pressures, the Barus law for piezoviscosity, might give excessively high values of viscosity, and thus produce non-reasonable values of the friction force. In film thickness calculations, the Barus law also gives higher values of film thickness than for instance the Roelands law, although the difference here is much smaller.

Previous results [10], where both Barus and Roelands piezoviscosity laws were considered, showed that the Roelands law is more adequate for the high level of pressures involved in this case.

Roelands piezo and thermoviscosity relationship states that [8]:

$$\eta=\eta_0 \exp\left[\alpha_\eta^* P\right]$$

where

$$\alpha_\eta^* P=(\ln\eta_0+9.67)\left[(1+5.1\times10^{-9}P)^Z\left(\frac{T-138}{To-138}\right)^{-So}-1\right]$$

$$Z=\frac{0.196\times10^9\,\alpha_\eta}{\ln\eta_0+9.67}\,,\qquad S_0=\frac{\delta_\eta(To-138)}{\ln\eta_0+9.67}$$

with $Z = 0.5115$ (at 373°K) and $S_0 = 1.015$.

The viscosity of the lubricant, η_0, at ambient pressure and 373°K, is equal to 0.05 Pas.

3.2.2. Shear modulus

The shear modulus varies with pressure and temperature as follows:

$$G=G_0 \exp\left[\alpha_G P-\beta_G\left(\frac{1}{T}-\frac{1}{T_0}\right)\right]$$

with $G_0 = 0.245 \times 10^6$ Pa

$\alpha_G = 2.1 \times 10^{-9}$ Pa^{-1} (at To=313 °K)

$\beta_G = 3800$ °K

3.2.3. Ree-Eyring reference shear stress

The Ree-Eyring reference shear stress varies with pressure as follows (τ_r is considered temperature independent):

$$\tau=\tau_{r0} \exp\left[\alpha_{\tau_r} P\right]$$

with $\tau_{r0} = 0.602 \times 10^6$ Pa

$\alpha_{\tau_r} = 2.57 \times 10^{-9}$ Pa^{-1}

These parameters and their variation with pressure and temperature were obtained from reference [9] by Houpert.

4. FRICTION COEFFICIENT RESULTS

The algorithm developed for the evaluation of the tangential tractions inside the lubricant film is based on the solution of expression (1).

The mid layer of the lubricant film is discretized into a grid of points (i,j) corresponding to the points used for the evaluation of the dry Hertzian pressure field.

The determination of the tangential traction in point (i,j) depends on the value obtained in point (i,j-1) as shows expression (2). An iterative procedure was developed, based on the initial value of $\tau_{i,j-1}$. The algorithm converges very well, and is completely independent of that initial value.

The results presented here, concern an EHD roller/ring contact in a roller bearing, lubricated with MOBIL Jet II oil.

The temperature was taken to be constant (373°K) along the contact.

4.1. Tangential pressure distribution

Figure 3 shows the tangential pressure distribution for a waved contact. This result was obtained considering:

Maximum Hertzian pressure - 1Gpa,
Rolling speed - 69.9 m/s (ω=30000 rpm),
Sliding rate - 0.50, and
Amplitude of the waviness - 0.25 μm.

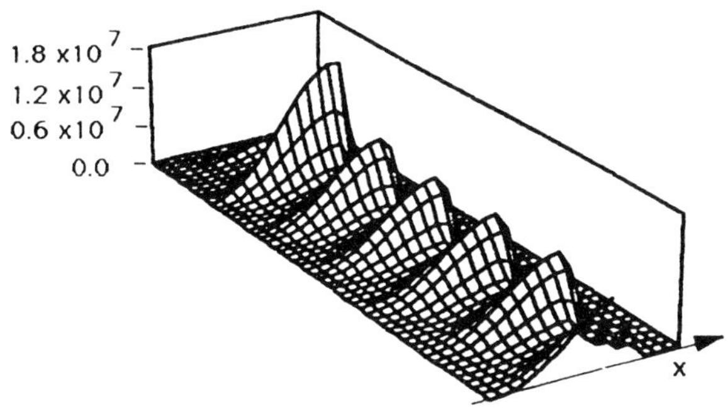

Figure 3 - Tangential pressure distribution for a waved contact.

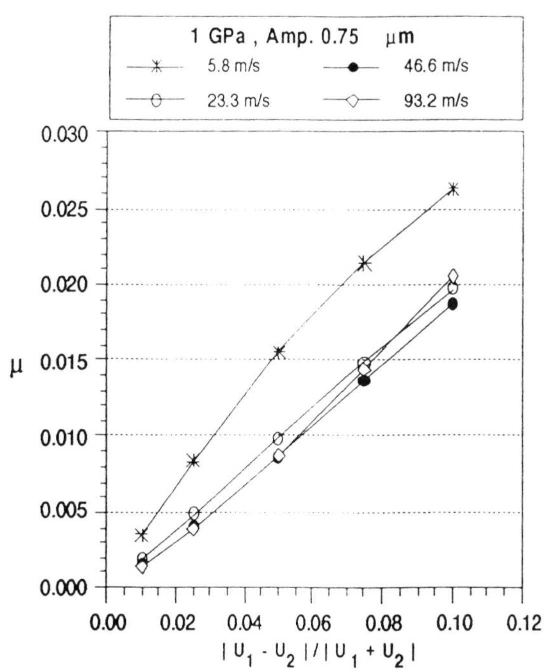

Figure 4 - Variation of the friction coefficient with the sliding rate for several rolling speeds.

Figure 3 shows that the tangential traction distribution is waved and in phase with the surface waviness.

In fact, the peaks of the surface waviness produce zones of high normal pressure which due to the lubricant pressure dependence give very high values of the dynamic viscosity, of the shear modulus and of the reference shear stress. As a consequence the tangential pressure presents high values in those zones.

The local friction coefficient, defined as $\mu_{loc} = \tau_{ij}/p_{ij}$, also presents a waved distribution. For the case under analysis, the maximum local friction coefficient is equal to 0.020, while the global friction coefficient is only 0.0045.

This local behaviour might be very important when analysing the thermal behaviour of the EHD contact, since it can be related to spots of very high temperature inside the lubricant film.

4.2. Influence of the operating conditions

Figures 4 shows the variation of the friction coefficient with the sliding rate, in a waved roller/ring contact, for several rolling speeds. These results were obtained considering a maximum Hertzian pressure of 1 GPa and an amplitude of the surface waviness of 0.75 μm.

It can be noticed that, as expected, increasing sliding rates produce increasing friction coefficients.

Figure 4 also shows that for constant sliding rates, the friction coefficient is not much affected by the rolling speed, and is almost constant. This is true within the narrow range considered for the sliding rate ($0.01 \leq V_e \leq 0.10$), and particularly for high rolling speeds.

The results obtained considering other maximum Hertzian pressures and other amplitudes of the surface waviness show the same behaviour.

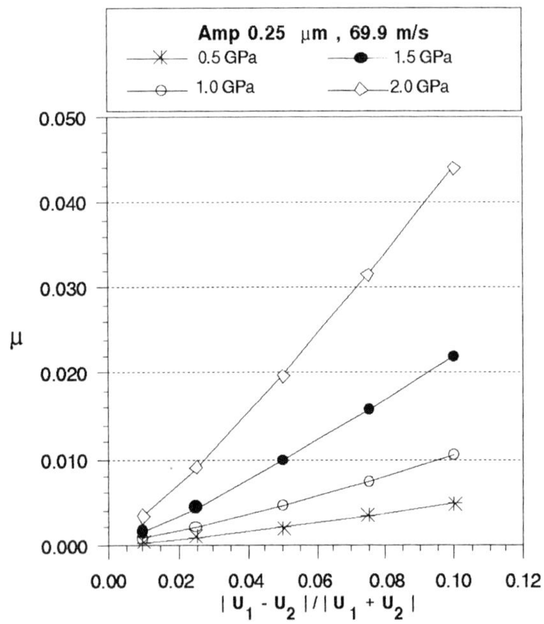

Figure 5 - Variation of the friction coefficient with the sliding rate for several maximum Hertzian pressures.

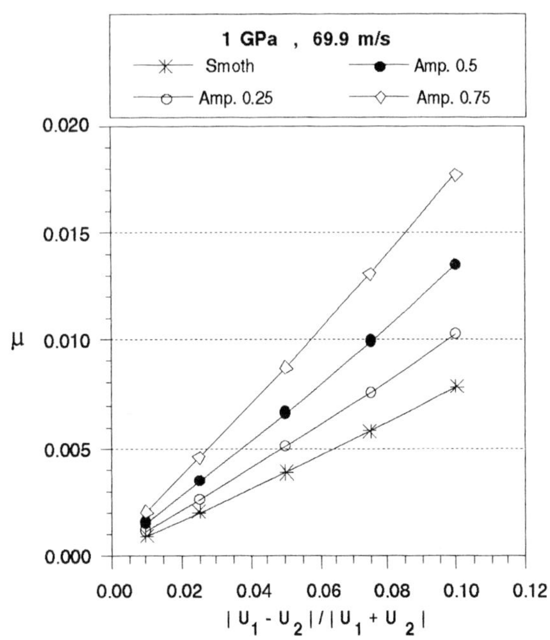

Figure 6 - Variation of the friction coefficient with the sliding rate for several amplitudes of the surface waviness.

Figures 5 shows the variation of the friction coefficient with the sliding rate, for different maximum Hertzian pressures. These results were obtained considering a rolling speed of 69.9 m/s (30000 rpm) and an amplitude of the surface waviness equal to 0.25 μm.

Increasing sliding rates produce increasing friction coefficients for each load (maximum Hertzian pressure) considered.

For a constant sliding rate, increasing the maximum Hertzian pressure produces a very important rise of the friction coefficient. For instance, considering a sliding rate of 5%, the friction coefficient goes from 0.004 to 0.020 (5 times more), when the maximum Hertzian pressure goes from 1 to 2 GPa.

Load and thus maximum Hertzian pressure is fundamental in the analysis of the tangential traction in an EHD contact. This effect becomes even more important as the surface waviness increases.

4.3. Influence of the surface waviness

Figures 6 shows the variation of the friction coefficient with the sliding rate, for the smooth and waved surfaces, for a maximum Hertzian pressure of 1 GPa and a rolling speed of 69.9 m/s (30000 rpm).

The results available show that the friction coefficient always increases when the amplitude of the surface waviness also increases. This augmentation is more important for the highest sliding rates (as shown in figure 6) and also for the highest maximum Hertzian pressures. On the opposite this augmentation is smaller when the rolling speed increases.

The considerable increase of the friction coefficient, when surface waviness is considered, is related to the normal pressure increase in some points within the contact zone, induced by the surface waviness itself.

This shows how important can be the local behaviour of the EHD contact, and all the

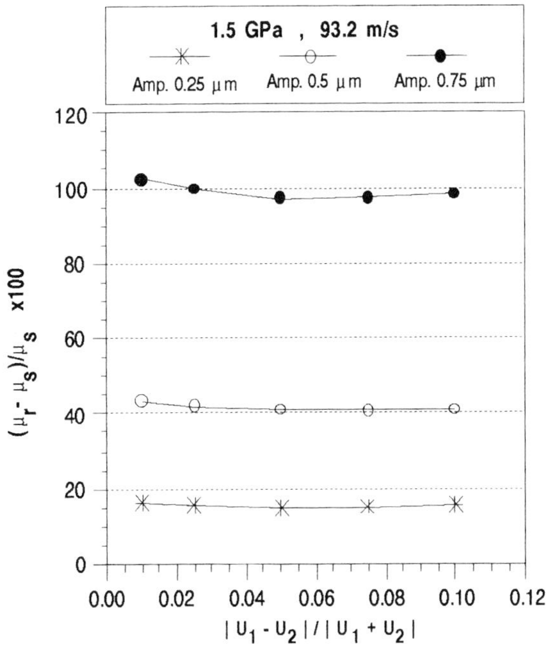

Figure 7 - Relative variation of the friction coefficient with surface waviness.

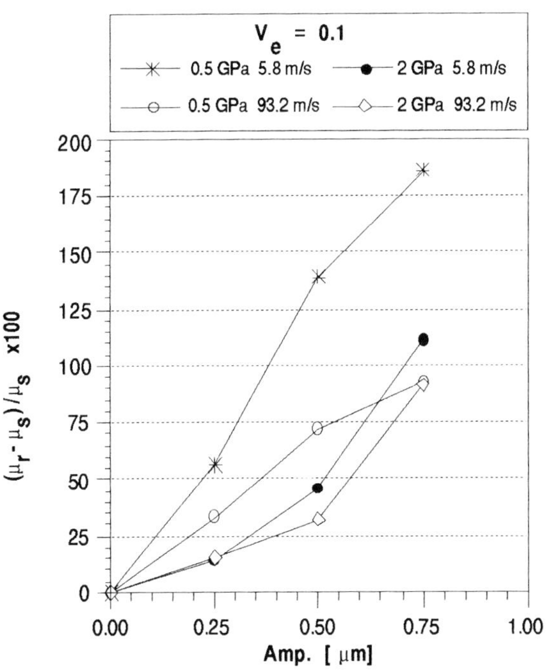

Figure 8 - Relative variation of the friction coefficient with the amplitude of the waviness.

aspects related to micro elastohydrodynamics and local friction coefficient.

It is also very interesting, to analyse the relative increase of the friction coefficient, comparing rough and smooth cases.

The parameter,

$$\mu_V = \frac{\left(\mu_{rough} - \mu_{smooth}\right)}{\mu_{smooth}} = \frac{\left(\mu_r - \mu_s\right)}{\mu_s}$$

represents the relative variation of the friction coefficient with the surface waviness.

Figure 7, shows the variation of the parameter μ_V with the sliding rate. It can be noticed that the relative increase of friction coefficient, is independent of the sliding rate, no matter the maximum Hertzian pressure, rolling speed or surface waviness. This is true at least for the range of the sliding rate considered ($0.01 \leq V_e \leq 0.10$).

Figure 8, shows the variation of the parameter μ_V, with the amplitude of the surface waviness, for constant sliding rate. μ_V always increases when the amplitude of the surface waviness also increases.

For each amplitude of the surface waviness, the relative variation of friction coefficient decreases with the maximum Hertzian pressure (for constant rolling speed), and decreases with the rolling speed (for constant maximum Hertzian pressure).

5. DISCUSSION

The results presented, clearly establish the variation of the friction coefficient with the operating conditions of the EHD contact, although they are not much different from others already published.

The results also show that the presence of surface waviness or roughness is no doubt an important parameter.

The surface waviness modifies the Hertzian normal pressure distribution, making it also wavy, and presenting several peaks of pressure higher than the maximum Hertzian one.

The lubricant under these normal pressure peaks is highly compressed, and due to its physical pressure dependence presents very high values of dynamic viscosity, shear modulus and reference shear stress, locally generating very high tangential pressures and friction coefficients.

These local friction coefficients can be several times greater than the global one for the EHD contact. This local behaviour is surely very important for a thermal micro EHD analysis of the contact.

These results are function of some lubricant parameters like η, G and τ_r, whose variation with pressure and temperature is perhaps not sufficiently well known. These parameters and their variation are very important, emphasizing the need for a serious and complete knowledge of the lubricant rheology.

CONCLUSION

The simulation of the EHD contact, using the waved dry Hertzian results, proved to be very useful, and allowed the development of a fast algorithm for the determination of the tangential pressure distribution and thus the friction coefficient.

The viscoelastic rheological model seems to be well adequate to represent the behaviour of the MOBIL Jet II oil.

The results presented here show the influence of the operating conditions on the friction coefficient in an EHD contact. It was also possible to study the influence of the surface waviness or roughness.

Generally, it can be said that the friction coefficient increases when:

a) the maximum Hertzian pressure (load) increases;
b) the sliding rate increases;
c) the rolling speed decreases.
d) the amplitude of the surface waviness increases.

The relative variation of the friction coefficient due to surface waviness, μ_v, also increases with the amplitude of the surface waviness. However, μ_v, seems to be almost independent of the sliding rate, at least for values of V_e till 0.10.

Acknowledgements

This work was developed within the BRITE-EURAM project "Bearing with minimum lubrication", Contract Aero - 0007 - A(c). The authors are grateful for the financial support conceded to this project by the European Community and for the collaboration of the other partners involved.

Nomenclature

a	- contact half width ($a = \ell / 2$)
Amp	- amplitude of the waviness
E	- equivalent Young's Modulus
F_n	- normal load applied to the contact
F_t	- tangential load applied to the contact
G	- shear modulus of the lubricant
h_0	- centre film thickness
K	- thermo conductivity of the lubricant
p_0	- maximum Hertzian pressure
R_i	- radius of each contacting surface
R	- equivalent radius of the contacting surfaces
S_0	- temperature exponent in Roelands law
U_i	- speed of surface i along OX direction
V_e	- sliding rate
X	- non-dimensional shear stress
Z	- pressure exponent in Roelands law
η	- viscosity of the lubricant
α_G	- shear modulus pressure coefficient
α^*_η	- Roelands law pressure viscosity coefficient
α_{τ_r}	- Ree-Eyring pressure coefficient
β_G	- Shear modulus temperature coefficient
ϕ_T	- correction factor for thermal effects
γ	- shear rate of the lubricant

γ_e - elastic shear rate
γ_v - viscous shear rate
λ - wavelength of the waviness
μ - friction coefficient
σ_i - RMS surface roughness
σ - equivalent RMS surface roughness
τ_e - equivalent shear stress
τ_{ro} - Ree-Eyring reference shear stress
τ_{xz} - shear stress in OX direction (τ_x)
τ_{yz} - shear stress in OY direction (τ_y)

References

[1] - Dowson, D. and Higginson, G.R., "Elastohydrodynamic lubrication". S.I. Edition, Pergamon Press Ltd., Oxford, 1977, 236p.

[2] - Gohar, R., "Elastohydrodynamics" - Ellis Horwood series in Mechanical Engineering, 1988.

[3] - Boresi, Sidebottom, Seely, Smith, "Advanced mechanics of materials" - Chp. 14: "Contact Stresses", John Wiley and Sons, New York, 1978.

[4] - Cheng, H. S., "Elastohydrodynamic Lubrication" in "Handbook of Lubrication, Theory and practice of tribology, Vol. II, Theory and design", edited by E. R. Booser, CRC Press, Inc., Boca Raton, Florida , USA, 1984.

[5] - Seabra, J. and Berthe, D., "Elastohydrodynamic point contacts. Part II - Influence of surface speeds, surface waviness and load on the contact behaviour". WEAR, Vol. 130, 1989, 319-335.

[6] - François Vergne, "Le contact elastohydrodynamic pontuel lubrifié par une fluid visquex lineaire ou non". Thése de Docteur nº 13-89, INSA de Lyon 1989.

[7] - Houpert, L. "Fast numerical calculation of EHD sliding traction forces; Application to rolling bearings". ASME Journal of Tribology, vol.107, April 1985.

[8] - Houpert, L. "New results of traction force calculations in elastohydro-dynamic contacts". ASME Journal of Tribology, vol.107, April 1985.

[9] - Houpert, L, "Contribution a l'etude du frottement dans un contact elastohydrodynamic". Thése de Docteur nº IDI 3-8019, I.N.S.A. Lyon 1980.

[10] - Campos, A., Sottomayor, J., Seabra, J. and Ferreira, L., "Analysis of a roller/ring contact in a roller bearing - Preliminary results for pressure, film thickness and friction force". BRITE - EURAM project: Bearing with minimum lubrication. Contract Aero - 0007 - A (c). Report nº 2, CETRIB, DEMEGI, Faculdade de Engenharia da Universidade do Porto, Jan. 1991, 17p.

[11] - Sottomayor, A., Campos, A., Seabra, J. and Ferreira, L., "Analysis of a roller/ring contact in a roller bearing - Influence of the lubricant rheology and of the surface waviness on the friction force". BRITE - EURAM project: Bearing with minimum lubrication. Contract Aero - 0007 - A (c). Report nº 3, CETRIB, DEMEGI, Faculdade de Engenharia da Universidade do Porto, July 1991, 16p.

[12] - Sottomayor, A., Campos, A., Seabra, J. and Ferreira, L., "Analysis of a roller/ring contact in a roller bearing - Results of friction coefficient in a simulated E.H.D. roller / ring contact". BRITE - EURAM project: Bearing with minimum lubrication. Contract Aero - 0007 - A (c). Report nº 4, CETRIB, DEMEGI, Faculdade de Engenharia da Universidade do Porto, Jan. 1992, 14p.

Thin Films in Tribology / D. Dowson et al. (Editors)
© 1993 Elsevier Science Publishers B.V. All rights reserved.

The behaviour of real transverse roughness in a sliding EHL contact.

J.A. Greenwood and G.E.Morales-Espejel.

University Engineering Department, Trumpington Street, Cambridge, CB2 1PZ, U.K.

A number of authors have shown theoretically that a sliding elastohydrodynamic contact, transverse imperfections, whether bumps, dents, sinusoidal ridges or even real (1-D) roughness may largely disappear, to be replaced by corresponding pressure variations. The complexity of full EHL calculations has limited the number of cases analysed, so that no criterion for smoothing emerged. By considering an 'infinite EHL contact' instead of the finite 'Hertzian' region of a real EHL contact, (but retaining the characteristic mean pressure and viscosity), Greenwood and Johnson (1992) developed a simple model to explain the behaviour: but their conclusion relied on the hypothesis that sinusoidal induced pressures, corresponding to a not-quite sinusoidal roughness, correctly represent a sinusoidal roughness.

Here we obtain the exact solution for an 'infinite EHL contact' with sinusoidal roughness, and confirm the criterion. However, the present method enables <u>any</u> (1-D) initial roughness to be studied: and in particular demonstrates that only the high wave number components of real roughness persist, so that again a rough surface becomes essentially plane.

The result throws doubt on models of 'partial' EHL contact in which the initial surface roughness is retained.

1. INTRODUCTION

It is widely believed that failure of lubricated contacts is associated with the roughness of the participating bodies; and it certainly appears from isothermal EHL theory that there will always be a lubricant film of acceptable thickness between smooth surfaces, because the rate of decrease of film thickness with increasing load is so slow. [In isothermal theory, increasing the speed is always beneficial: if, more realistically, we take into account the viscosity reduction close to inlet shear (Greenwood and Kauzlarich 1973 [1], Murch and W.R.D. Wilson 1975 [2], A.R. Wilson 1980 [3]), increasing the speed may reduce the film thickness, but this does not seem to be, of itself, the explanation of scuffing failures].

An elementary explanation of the role of roughness is to assume that it plays no part in the determination of the nominal film thickness: that is, the mean surfaces of rough bodies have the separation predicted for smooth bodies. More sophisticated analyses allow for the effect of roughness, either by a (relatively) simple statistical averaging (Christensen 1971 [4], Tønder 1984 [5]) or by the flow-factor method introduced by Cheng 1978 [6] and widely followed. Christensen clearly intended his theory for 'classical' bearings, in which the bodies are regarded as rigid, and so it was entirely appropriate to assume also that the surface roughness was rigid; but it appears that the stochastic method is invariably used with the initial surface roughness even when applied to EHL contacts. Cheng, however, made one of the earliest calculations of the passage of an asperity through an EHL contact (Cheng 1973) [7], and demonstrated that it certainly did not pass through unchanged: can one, then, reasonably use the initial roughness in either method for EHL calculations ?

A number of authors have made calculations for an EHL contact with a wavy surface, as a first attempt on the real problem of a rough surface. It appears that with a limited number of waves, the nominal film thickness is altered, but in a somewhat confusing manner related to the precise placing of the waves with respect to the end of the inlet (Chang, Cusano and Conry [8], Venner and Lubrecht [9]). Within the 'Hertzian' region, the waviness largely disappears, at least in the steady flow case. i.e. when there is pure sliding with a smooth, moving surface and a stationary rough surface. Greenwood and Johnson 1992 [10] have provided a simple analysis of the persistence/flattening, which appears to explain rather satisfactorily the results of full EHL calculations. However, Venner 1991 [11] has progressed beyond this, and succeeded in performing full EHL calculations for line contact of a surface which is rough, not merely wavy; and, again for the steady case, finds that the roughness largely disappears, while the central nominal film thickness is barely changed. The purpose of this note is to provide a simple analysis of the behaviour of transverse roughness in pure sliding.

2. ANALYSIS

We shall assume, like Greenwood and Johnson, that the behaviour of the roughness in the Hertzian region of an EHL contact can be modelled by considering an infinitely large, nominally plane contact: the nominal film thickness is assumed to have been determined by conditions in the inlet, and the nominal pressure will be assumed to be constant at a value related to the Hertz pressure. We consider the cases of a Newtonian lubricant and a non-Newtonian, Eyring lubricant for which the Reynolds' equation is:

$$1-\frac{h^*\rho^*}{h\rho}=(\coth t_p-\frac{1}{t_p})\sqrt{s^2+\left(\frac{\sinh t_p}{t_p t_v}\right)^2} \quad (1)$$

where :

$$s=\frac{u_2-u_1}{u_2+u_1},\ t_p=\frac{h}{2\tau_o}\frac{dp}{dx},\ t_v=\frac{\eta(u_2+u_1)}{h\tau_o}$$

$$\bar{u}=\frac{u_2+u_1}{2}$$

τ_o = lubricant Eyring stress. For pure sliding, $s=\pm1$.

For Newtonian fluid, $t_p\simeq 0,\ \ t_v\simeq 0$ giving the usual Reynolds' equation.

It will be assumed that the viscosity follows the Barus' law:

$$\eta=\eta_o e^{\alpha p} \quad (2)$$

and the density obeys the Dowson and Higginson equation.

$$\rho=\frac{1+\gamma\, p}{1+\beta\, p} \quad (3)$$

For known pressures p, the film thickness h can readily be determined from (1) by iteration. (Since the equation is only cubic in h, an explicit solution is available: but for a step-by-step solution iterating is more convenient).

We are ignoring the overall geometry of the contact: but we wish to allow for the elastic deformation of the surface roughness.

For our infinite contact, we can conveniently represent the pressures by a Fourier integral, or in practice, by a discrete Fourier transform:

$$p(x)=p_o+\sum_{n=1}^{N}p_n exp(2\pi nix/\lambda) \quad (4)$$

The deflection of a half-space by a sinusoidal pressure $p_1\cos 2\pi x/\lambda$ is well-known to be:

$$v=\frac{1-\nu^2}{\pi E}\lambda p_1\cos\frac{2\pi x}{\lambda}$$

Noting that we require the combined deflection of two bodies, and using superposition, the deflection due to $p(x)$ is:

$$v=v_o+\sum_{n=1}^{N}v_n exp(2\pi nix/\lambda) \quad (5)$$

with: $v_n=\frac{2\lambda}{n\pi E'}p_n$

where:

$$\frac{2}{E'}=\frac{1-\nu_1^2}{E_1}+\frac{1-\nu_2^2}{E_2}$$

Finally we require:

$$h(x)=h^*+z(x)+v(x) \quad (6)$$

where $z(x)$ is the initial, undeformed, roughness.

2.1 Wavy surface

Greenwood and Johnson argued that it was sufficient to consider a sinusoidal pressure distribution: if this proves to be consistent with an almost sinusoidal initial roughness, it may be concluded that an exactly sinusoidal roughness will be consistent with an almost sinusoidal pressure. The above analysis permits this hypothesis to be verified: we take the roughness $z(x)$ to be a single Fourier term, while $v(x)$, $p(x)$ and $h(x)$ all have as many terms as required. The pressure is indeed not sinusoidal, but the eye can certainly not detect this, see Fig. 1.

2.2 Rough surface

A convenient simulation of a typical engineering surface (Rice [12]) is:

$$z(x) = \sum_{n=1}^{n_{max}} z_n \cos(2n\pi x/\lambda + \varepsilon_n) \tag{7}$$

where ε_n are random phases $(0 \leq \varepsilon_n < 2\pi)$. For this surface the autocorrelation function is:

$$E\{z(x)z(x+\zeta)\} = \frac{1}{2}\sum_{n=1}^{n_{max}} z_n^2 \cos(2\pi n\zeta/\lambda) \tag{8}$$

Sayles and Thomas [13] suggest that typical engineering surfaces have a power spectral density:

$$G(k) = c/k^2$$

where k is the wave-number $1/\lambda$. By taking $z_n = A/n$ we obtain a discrete approximation to this. Note that this is equivalent to choosing a profile of fractal dimension 1.5 (Majumdar and Tien [14]).

The RMS roughness σ is then given by:

$$\sigma^2 = \frac{1}{2}\sum_{n=1}^{n_{max}} {z_n}^2 = \frac{A^2}{2}\sum_{n=1}^{n_{max}} \frac{1}{n^2} \tag{9}$$

3. SOLUTION

To solve equations (1),(4) and (6), we take the Fourier expansion of $z(x)$ and try as an initial guess $h(x) = h^*$, i.e. $v(x) = -z(x)$. Equations (4) and (5) give the corresponding pressures; equation (1) then determines $h(x)$ and finally equation (6) gives a new $v(x)$. It proves that this scheme is unstable for higher wave numbers: but application of Aitken's method of extrapolation to successive sets of iterates (Hamming [15]), improves the convergence.

The reason for the difficulties appears to be that when calculating pressures from displacements, the pressure coefficients are multiplied by the wave number: $p_n = \frac{\pi E' v_n}{2\lambda} n$ thus increasing the numerical error for high values of n.

An inverse scheme which obtains displacements from pressures necessitates the calculation of both pressures and pressure gradients from the Eyring-Reynolds equation. This involves the problem of finding the pressure at the beginning of the analysed interval, which turns out to be a very unstable process as well.

In cases of higher amplitudes of initial roughness and short wave lengths, despite the use of Aitken's extrapolation, instabilities may persist; for those cases the above sequence can be modified to improve convergence.

Defining:

$$f = h(\coth t_p - \frac{1}{t_p})\sqrt{s^2 + \left(\frac{\sinh t_p}{t_p t_v}\right)^2} \tag{10}$$

equation (1) becomes:

$$h - [\rho(p_o)/\rho]h_G = f \tag{11}$$

where:

$$h_G = \frac{h(p_{max})\rho(p_{max})}{\rho(p_o)}$$

now from (6):

$$v = \frac{\rho(p_o)}{\rho}h_G - h^* - z + f \tag{12}$$

finally from (3):

$$\frac{\rho(p_o)}{\rho} = \frac{1+\gamma\, p_o}{1+\beta\, p_o}\frac{1+\beta\, p}{1+\gamma\, p} \tag{13}$$

from which it is easy to find that:

$$\frac{\rho(p_o)}{\rho} = 1-(p-p_o)C+\frac{\gamma\, C(p-p_o)^2}{1+\gamma\, p} \tag{14}$$

where:

$$C = \frac{\gamma-\beta}{(1+\beta\, p_o)(1+\gamma\, p_o)} \tag{15}$$

substituting (14) into (12) one obtains:

$$v+h_G(p-p_o)C =$$

$$(h_G-h^*)+h_G\gamma C\frac{(p-p_o)^2}{1+\gamma\, p}+f-z \tag{16}$$

Since v_o is arbitrary one may ignore the constant terms. Thus from (4) and (5), and applying FFT to the right hand side of (16):

$$\frac{2\lambda}{\pi E'}\sum_{n=1}^{N}\frac{p_n}{n}exp\left(\frac{2\pi nxi}{\lambda}\right)+h_G C\sum_{n=1}^{N}p_n exp\left(\frac{2\pi nxi}{\lambda}\right)$$

$$=\sum_{n=1}^{N}q_n exp\left(\frac{2\pi nxi}{\lambda}\right) \tag{17}$$

therefore:

$$q_n = p_n\left(\frac{2\lambda}{\pi nE'}+Ch_G\right) \tag{18}$$

Equation (16) may be solved iteratively by using estimates of the pressure to find the right hand side and hence values of q_n: equation (18) then gives improved values of the pressure.

For an incompressible fluid we have $\gamma=\beta$ and therefore $C=0$.

4. RESULTS AND DISCUSSION

Three sets of data were used, they are given in table 1. Set 1 corresponds to the non- Newtonian compressible example with sinusoidal surface (of initial amplitude h_1) taken from Fig. 2 of Chang [16], to enable a comparison. Set 2 is used to make comparisons among Newtonian, non-Newtonian, compressible and incompressible wavy surfaces. Finally, set 3 corresponds to a non-Newtonian compressible real roughness example. From the EHL pressures of this example the subsurface stresses are also calculated.

a) Comparison with Chang's results.
Using data set 1 from table 1, the example shown in Fig.1 was solved. The variables plotted are the ones used by Chang [16] . The Chang's solution for the complete Hertzian geometry is shown in Fig. 2.

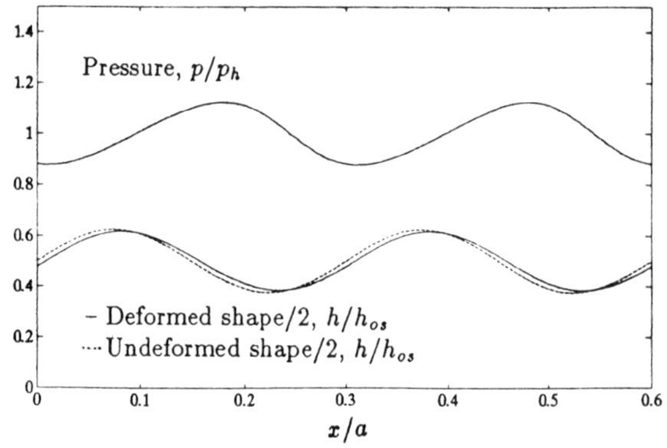

Figure 1: Results present analysis.

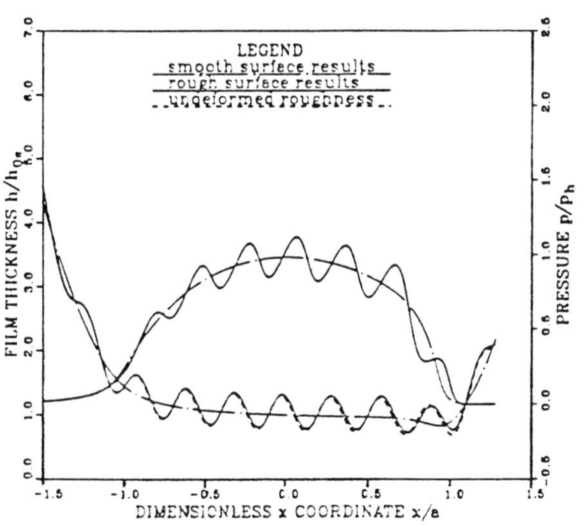

Figure 2: Chang's complete solution.

One can see that the amplitudes of pressure and shape ripples in both cases are in good agreement. For this example both the deformation of shape and the pressure ripples are small: note that the pressure ripples are 90° ahead of the shape ripples.

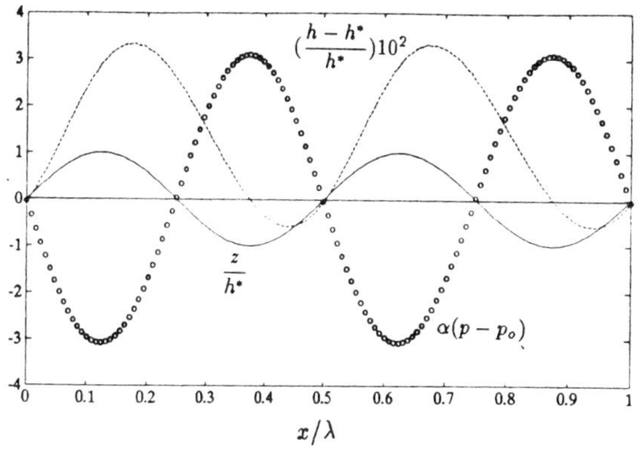

Figure 3: Non-Newtonian Compressible.

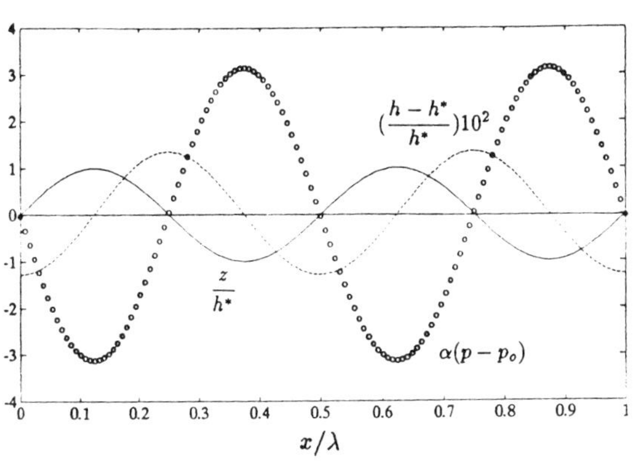

Figure 4: Non-Newtonian Incompressible.

b) Newtonian - non-Newtonian comparisons. With data set 2 from table 1, four solutions have been obtained, Figs. 3, 4, 5 and 6, to enable comparisons among Newtonian, non-Newtonian, compressible and incompressible cases starting with a undeformed sinusoidal roughness.

We should point out that the amplitude of the undeformed sinusoidal roughness (h_1) is greater than the central film thickness of the gap where it is supposed to pass through, but the elastic deformation occurring makes this possible.

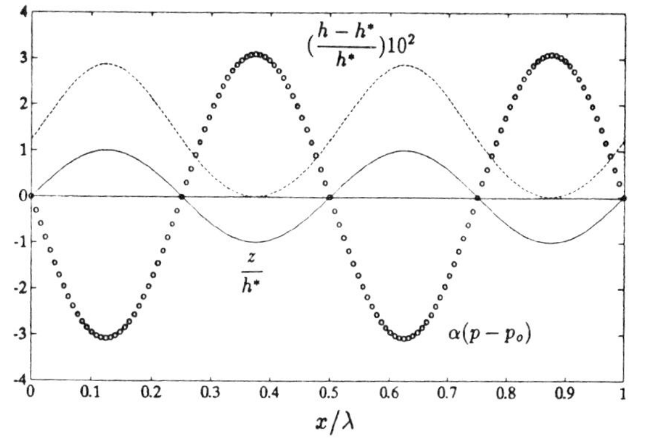

Figure 5: Newtonian Compressible.

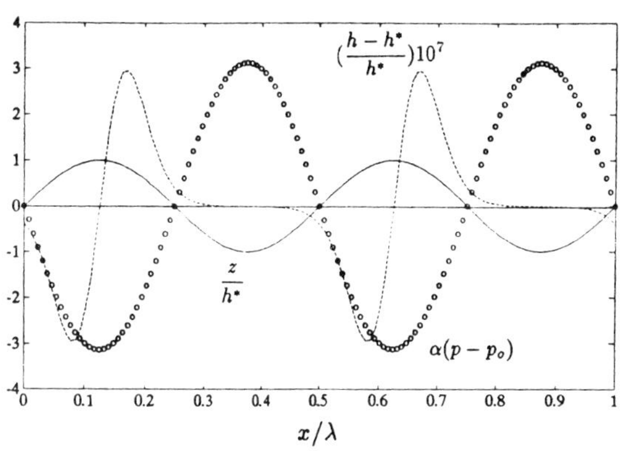

Figure 6: Newtonian Incompressible.

By comparing Fig. 3 with Fig. 5 and Fig.4 with Fig.6, one can see that non-Newtonian behaviour produces a bigger final amplitude of shape than the corresponding Newtonian case, while in contrast the amplitude of the final pressure ripples are very similar indeed.

With the Newtonian compressible fluid, Fig. 5, the shape is in phase with its initial shape and exactly 180° out of phase with the pressures.

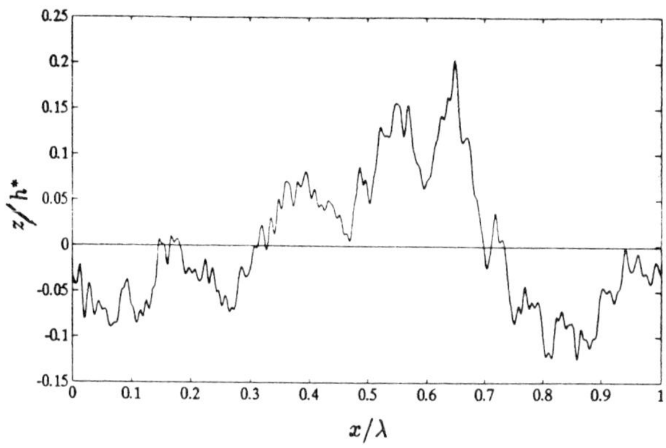

Figure 7: Initial shape, RMS=0.074.

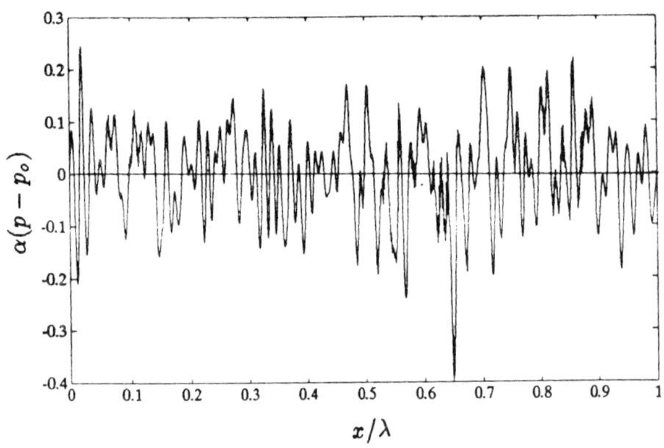

Figure 9: Pressure variations, RMS=0.089.

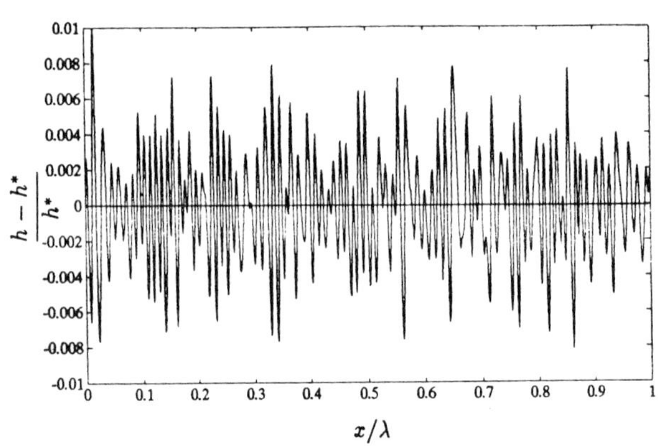

Figure 8: Final shape, RMS=0.003.

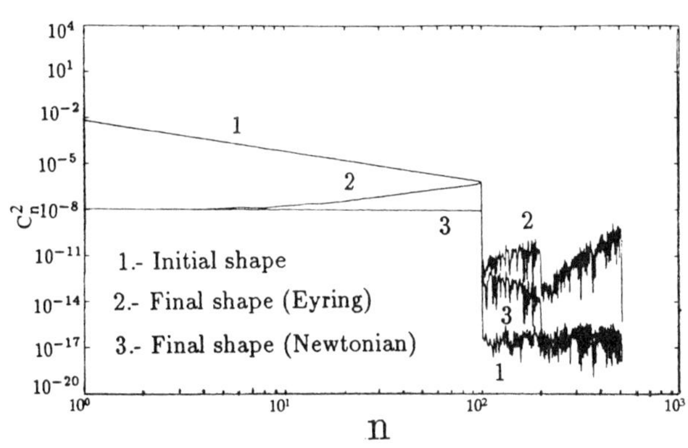

Figure 10: Square Fourier coefficients.

The compressible non-Newtonian case, Fig. 3, shows a final shape shifted in phase respect to its initial one.

c) Roughness deformation for an Eyring fluid. Real roughness can be analysed by the present scheme in exactly the same way as the wavy surface. Data set 3 from table 1 was used for this example, and the initial rough surface, Fig. 7, has been obtained by applying equation (7) with $n_{max} = 100$, and 1024 points along the abscissa x.

Fig. 8 shows the deformed shape and Fig.9 the corresponding pressure distribution, Fig.10 shows a comparison between the Fourier coefficients (amplitude of harmonics) of the undeformed (initial) shape and deformed (final) shape.

One can see that the low frequency amplitudes have been very considerably reduced, but the higher frequencies are relatively little changed. It must be admitted that the reduction of the fun-

Table 1
Sets of data

Set no.	τ_o	p_o	E'	$\bar{u}$	h^*	h_1	λ	α	η_o	h_1/h^*
	MPa	GPa	GPa	m/s	μ m	μ m	mm	GPa^{-1}	Pa s	-
1	5	1.1	219	4	0.60	0.15	0.071	15.9	0.04	0.25
2	10	1.0	227	25	0.7125	0.7125	1.2	14.8	0.0179	1.0
3	10	1.0	227	25	0.304	0.062	(5.7)	9.6	0.0024	0.203

damental by a factor of 1000 as shown occurs for a wave length of 5.7 mm, which is unrealistic: the reduction for $\lambda = 0.57mm$ is only a factor of 100. Note that the curve suggests that a roughness component with a wave number exceeding 100 will be magnified (the theory below does not suggest this): certainly when such components are included, the iteration fails.

Thus, a rough surface becomes essentially plane. This result suggests that the basic assumption of the 'partial' EHL models of retaining the initial surface roughness could be suspect.

Another feature shown in Fig. 10 is that the coefficients line for the deformed shape is slightly curved at high 'n' for all non-Newtonian cases. In contrast, Newtonian examples always show a completely horizontal pattern.

d) Analytical Theory.

Looking again the deformation of sinusoidal roughness; the question arises: why does Fig. 1 shows the roughness persisting, while Fig. 3, also for a non-Newtonian compressible lubricant, shows it largely disappearing ? and why is there a phase difference between pressure and shape of 90° in Fig. 1 but 180° in Fig. 3 ?

If it is assumed that both the pressure variations and the film thickness variations are small, and that the mean pressure and therefore the viscosity are high, equation (1) reduces to:

$$h - \frac{\rho^* h^*}{\rho} = \frac{h^2}{6\tau_o}\frac{dp}{dx} \simeq \frac{\bar{h}^2}{6\tau_o}\frac{dp}{dx} \tag{19}$$

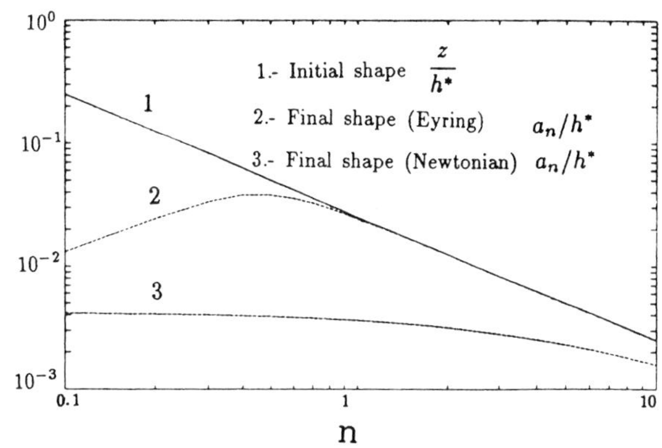

Figure 11: Variation of shape amplitudes with wave number.

If we also linearise the density variations, we can obtain an analytical solution to (19), from which the ratio of the amplitude of the film thickness variations and the initial roughness (now both sinusoidal) is:

$$\frac{a_n}{z_n} = \frac{\sqrt{B_n^2 + C^2}}{\sqrt{(C + A_n)^2 + B_n^2}} \tag{20}$$

where :

$$C = \frac{\gamma - \beta}{(1 + \beta\, p_o)(1 + \gamma\, p_o)}, \quad B_n = \frac{\pi \bar{h} n}{3\lambda\tau_o}$$

and:

$$A_n = \frac{2\lambda}{\pi E'\bar{h}n}$$

$\bar{h}$ = mean value of h (to the accuracy of the linear theory $h^* = \bar{h}$).

The effect of varying the wave number 'n' is shown in Fig. 11 (using data set 1). It is clear the critical dependence on the ratio $\lambda/\bar{h}$: the amplitude of the roughness in Fig. 1 is indeed unchanged; but this would no longer be the case if the wave length were increased by a factor 3. One can show by considering equation (20) that, for plausible values of the parameters, the critical wave length for which the amplitude is reduced by a factor of 2 is given by the simple relation:

$$\frac{\lambda}{\bar{h}} > 1.6\sqrt{\frac{E'}{\tau_o}} \qquad (21)$$

Thus, for Chang's nominal film thickness $\bar{h} = 0.60\ \mu m$, the critical wave length is $\lambda > 200\ \mu m$ in agreement with Fig. 11.

5. SUBSURFACE STRESSES

According to the shakedown criteria for elastic rolling contacts, the condition to avoid continously plastic deformation (see Johnson [17]), is:

$$\tau_{xz} \leq k \qquad (22)$$

where k is the yield stress in pure shear. So the attention is drawn in the orthogonal stress τ_{xz} .

It is well known that the orthogonal shear stress due to a cosinusoidal pressure distribution on the surface $p = p_n \cos(s_n x)$ is given by:

$$\tau_{xz} = -p_n(s_n z)e^{-s_n z}\sin(s_n x) \qquad (23)$$

where z represents distance below the surface. And for low values of z , the stress corresponding to the mean pressure is near zero. So it is easy to find the stress due to our pressure distribution by superimposition of the terms due to the separate Fourier terms.

Figs. 12 and 13 show the distribution of τ_{xz}/S_p at two depths below the surface (where S_p is the standard deviation of the pressures at the surface). τ_{xz} is clearly a random variable with zero mean, and as expected from the Central Limit Theorem is approximately a Gaussian variable for reasonably small depths. The standard $2S$ and $3S$ confidence limits (where S is the standard deviation of τ_{xz}) are also shown on the figures, and it is clear that the usual rules for assessing the expected largest values hold.

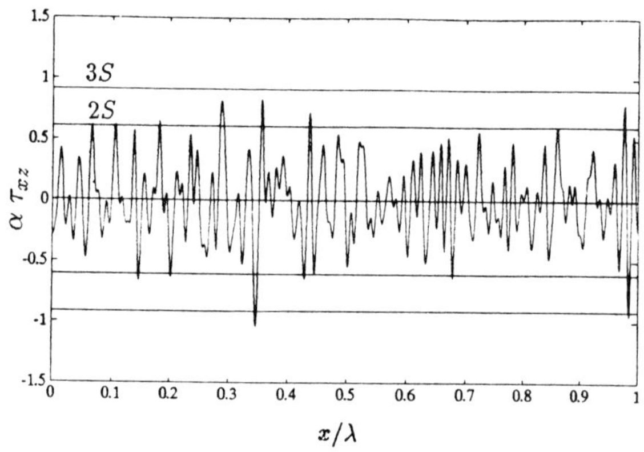

Figure 12: τ_{xz} at $z/\lambda = 0.0035$, $S = 0.030$.

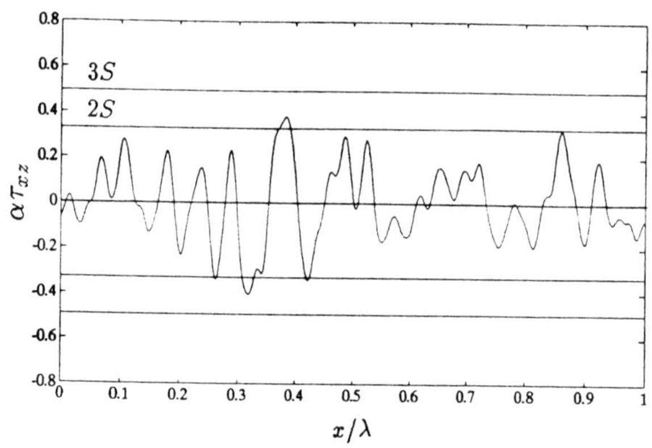

Figure 13: τ_{xz} at $z/\lambda = 0.020$, $S = 0.164$.

Fig. 14 shows the variation of $2S/S_p$ with depth z/λ and suggests that the maximum stress will probably occur at $z/\lambda = 0.0035$, with a 5 percent probability of the value 0.61 S_p being exceeded.

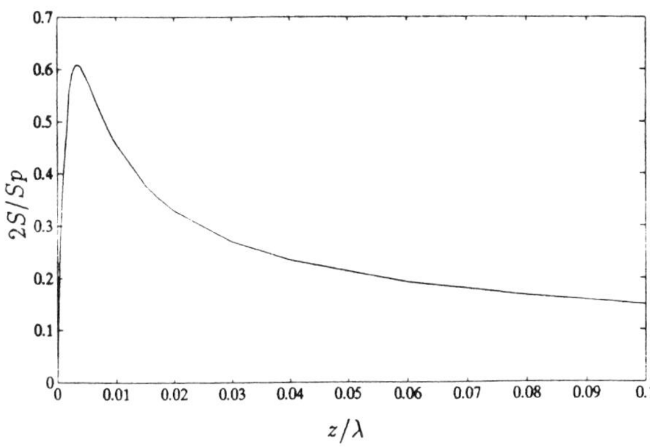

Figure 14: RMS ratios of τ_{xz} vs. z/λ.

6. SUMMARY AND CONCLUSIONS

A simple scheme to investigate the effects of 1-D real roughness and waveness in steady state EHL sliding contacts has been developed . The basic simplification of the scheme is to consider a small section of an infinitely long contact instead of the finite Hertzian region of a real EHL contact. Analysing the roughness and pressures into their harmonic components enables us to incorporate the simple sinusoidal roughness model developed by Greenwood and Johnson (1992).

Solutions have been obtained for wavy and real roughness surfaces. A comparison with another author's results is shown, giving good agreement with the complete micro EHL solution.

Comparisons between Newtonian and non-Newtonian cases were made as well as between compressible and incompressible, showing that non-Newtonian solutions will give higher final amplitudes of deformed shape than their Newtonian equivalents. It was also shown that the final shape in non-Newtonian cases is out of phase respect to the initial roughness (Figs. 3-6). In Fig. 10 was shown that for surfaces with real roughness the low frequency components are almost completely flattened after deformation which suggests that the basic assumption of 'partial' EHL models of retaining the original surface roughness could be suspect. An analytical solution of the linearised Eyring-Reynolds equation for sinusoidal roughness was presented showing that the deformation of the roughness is strongly dependent on the ratio $\lambda/\bar{h}$. For small values of this ratio, the roughness is expected to persist.

Finally results for two different depths of the subsurface orthogonal stress τ_{xz} under 'real' roughness EHL pressure were obtained, from which it is clear that increasing the depth removes the high frequencies in the stress. Since the stress τ_{xz} along x proved to be a Gaussian distribution for reasonably small depths, the standard confidence limits were used to assess the expected largest value.

REFERENCES

1. J.A.Greenwood and J.J. Kauzlarich, Inlet shear heating in elastohydrodynamic lubrication, ASME J. Lub. Tech.95 (1973)417.

2. L.E.Murch and W.R.D.Wilson, A thermal EHD inlet zone analysis, ASME J. Lub. Tech.97 (1975) 212.

3. A.R. Wilson, An experimental thermal conduction for predicted oil film thickness, 6th Leeds-Lyon Symposium on Tribology "Thermal Effects" I.Mech.E. (1979)179.

4. H.Christensen and K. Tønder, The hydrodynamic lubrication of rough surfaces, ASME J. Lub. Tech.93, No.3 (1971).

5. K. Tønder, The hydrodynamic lubrication of rough surfaces based on a new perturbation approach, ASME J. of Trib.106 (1984)440.

6. N.Patir and H.S. Cheng, Effect of surface roughness on the central film thickness in EHD contacts. 5th Leeds-Lyon Symposium on Tribology (1978).

7. Lee Kwan and H.S. Cheng, Effects of surface asperity on elastohydrodynamic lubrication, NASA report no. CR-2195 (1973).

8. L. Chang, C. Cusano, T.F. Conry, Effects of lubricant rheology and kinematic conditions on micro-elastohydrodynamic lubrication, ASME J. of Trib. 111 (1989).

9. C.H. Venner, A.A. Lubrecht, W.E. ten Napel, Numerical simulation of overrolling of a surface feature in a EHL line contact, ASME J. of Trib. 113(1991).

10. J.A. Greenwood and K.L. Johnson, The behaviour of transverse roughness in sliding elastohydrodynamic lubricated contacts, Wear 153 (1992)107.

11. C.H. Venner, W.E. ten Napel, Surface Roughness Effects in an EHL line contact, ASME J. of Trib. 114 (1992) 616.

12. S.O. Rice, The mathematical analysis of random noise, Bell Syst. Tech. J. 23 (1944)282.

13. R.S. Sayles and T.R. Thomas, Surface topography as a nonstationary random process, Nature, 271 (1978)431.

14. A. Majumdar and C.L. Tien, Fractal characterization and simulation of rough surfaces, Wear , 136 (1990)313.

15. R.W. Hamming, Numerical methods for scientists and engineers, Dover Publications, New York (1973).

16. L. Chang and M.N. Webster, On the pressure rippling and roughness deformation in elastohydrodynamic lubrication of rough surfaces, submitted to ASME J. of Trib. (1992).

17. K.L. Johnson, Contact Mechanics, Cambridge University Press (1987).

ACKNOWLEDGEMENT

The work was supported by a grant from ITESM and CONACYT, Mexico.

Thin Films in Tribology / D. Dowson et al. (Editors)
1993 Elsevier Science Publishers B.V.

Non-Dimensional Groups in Elastohydrodynamic Lubrication

D. DOWSON , E. S. SONG , C. M. TAYLOR

Non-dimensional groups which have been adopted in presenting EHL results are reviewed. It is shown that when the Roelands [1] viscosity-pressure relationship is adopted, the most popular set of the groups (G, U, W) do not determine the solution uniquely.

Therefore, it is suggested that the selection of lubricant properties and the definition of (G) be standardized. It is also concluded that two additional non-dimensional groups are needed for the unique representation of EHL operational conditions. For example,

$$H = f\ (\ Z,\ U,\ W,\ \eta_0/\eta_\infty,\ E'/\gamma\ )$$

for circular or line contact problems. (Z) can be replaced by (G). Buckingham's Π-theorem confirms this in terms of the maximum number of separate independent non-dimensional groups of the system.

NOTATION

E	elasticity modulus
E'	equivalent elasticity modulus, $1/E' = (1-\nu^2)/E$
f	a function
F	applied load
G	non-dimensional material group, $\alpha E'$
h	film thickness
h_{00}	fictitious separation at the origin
h_{cen}	central film thickness
h_{min}	minimum film thickness
H	non-dimensional film thickness, h/R_x
H_{00}	non-dimensional fictitious separation at the origin, h_{00}/R_x
H_{cen}	non-dimensional central film thickness, h_{cen}/R_x
H_{min}	non-dimensional minimum film thickness, h_{min}/R_x
k	ellipticity ratio, $k = (R_y/R_x)^{2/3}$
L, M	Moes non-dimensional groups, $L=G(2U)^{1/4}$ and $M=W(2U)^{-3/4}$
M, L, T	dimensions for Mass, Length and Time
p	pressure
p_h	maximum Hertzian pressure
p_{sp}	pressure spike
$p_{iv,as}$	asymptotic isoviscous pressure
P	non-dimensional pressure, p/E'
P_H	non-dimensional maximum Hertzian pressure, p_h/E'
P_{sp}	non-dimensional pressure spike, p_{sp}/E'
R_x, R_y	equivalent radii of curvature in the x- and y- directions
u_e	entraining velocity
U	non-dimensional velocity group, $\eta_0 u_e/E' R_x$
W	non-dimensional load group, $F/E' R_x^2$
X, Y	non-dimensional coordinates, x/R_x and y/R_x
Z	viscosity-pressure index of the Roelands relationship
α	viscosity-pressure coefficient
α'	defined by $1/p_{iv,as}$
β_1	constant of the density-pressure relationship, 1.68×10^{-9} Pa $^{-1}$
β_2	constant of the density-pressure relationship, 0.58×10^{-9} Pa $^{-1}$
γ	constant of the Roelands relationship, 1.9609×10^8 Pa
η	viscosity of the lubricant
η_0	atmospheric viscosity of the lubricant
η_∞	constant of the Roelands relationship, 6.31×10^{-5} Pa.s
$\bar{\eta}$	non-dimensional viscosity, η/η_0
$\bar{\eta}_h$	non-dimensional viscosity at the maximum Hertzian pressure
ν	Poisson's ratio
ρ	density of the lubricant
ρ_0	atmospheric density of the lubricant
$\bar{\rho}$	non-dimensional density, ρ/ρ_0

INTRODUCTION

In presenting EHL results, non-dimensional groupings incorporating the important physical variables have traditionally been adopted. For example, a set of (G, U, W) was adopted by Dowson and Higginson [2], while (L,M) have been used by Venner and Napel [3]. Film thickness formulae have also been developed and based upon these groups. However, it has occasionally been noticed that a comparison of EHL results for fixed values of the dimensionless groups shows differences between the results presented by various authors beyond allowable computational errors.

Since the development of the above non-dimensional groups were based upon the early assumptions of the EHL theory (e.g. incompressible fluid and the Barus viscosity-pressure relationship), these groups need to be reconsidered in general terms.

OBTAINING DIFFERENT RESULTS

The Roelands viscosity-pressure relationship has been widely adopted and shown to be more realistic than the Barus equation (eq.1).

$$\eta = \eta_0 \, e^{\alpha p} \tag{1}$$

The density-pressure relationship adopted has commonly been that proposed by Dowson and Higginson [2]. Therefore, the governing equations for an elliptical contact EHL problem are the Reynolds equation (2), the film shape equation (3), the viscosity-pressure relationship (4), the density-pressure relationship (5) and the force balance equation (6).

$$\frac{\partial}{\partial x}\left(\frac{\rho h^3}{12\eta}\frac{\partial p}{\partial x}\right) + \frac{\partial}{\partial y}\left(\frac{\rho h^3}{12\eta}\frac{\partial p}{\partial y}\right) = u_e \frac{\partial}{\partial x}(\rho h) \tag{2}$$

$$h = h_{00} + \frac{x^2}{2R_x} + \frac{y^2}{2R_y} + \frac{2}{\pi E'}\int_{-\infty}^{\infty}\int_{-\infty}^{\infty}\frac{p \; d\overline{x} \; d\overline{y}}{\sqrt{(x-\overline{x})^2+(y-\overline{y})^2}} \tag{3}$$

$$\eta = \eta_0 \left(\eta_\infty/\eta_0\right)^{\left[1-(1+p/\gamma)^Z\right]} \tag{4}$$

$$\rho = \rho_0 \left(1 + \frac{\beta_2 \, p}{1+\beta_1 \, p}\right) \tag{5}$$

$$F = \int_{-\infty}^{\infty}\int_{-\infty}^{\infty} p \; dx \; dy \tag{6}$$

When the mathematical modelling is the same, the results obtained should not be different beyond the computational error. However, the film thickness results for fixed values of (G, U, W) are proved to vary by changing the individual variables of the non-dimensional groups. This implies that the set of (G, U, W) does not fully define the operational condition. In fact, another source of the deviation in comparisons has been that lubricant properties are not input in a standard way. Prior to the main context in the next section, some examples are reported here detailing the determination of the lubricant properties.

In the Roelands viscosity-pressure relationship, eq.(4), two parameters (η_∞ and γ) are constants which neither vary for different lubricants nor for different pressures and temperatures. However, the atmospheric viscosity (η_0) and the viscosity-pressure index (Z) are inherent properties of a specific lubricant. While (η_0) can easily be obtained by experiment, this is not so for the viscosity-pressure index (Z), since it requires numerous tests followed by curve-fitting. Alternatively, as was shown by Roelands [1] and later by Houpert [4], (Z) can be determined somewhat more easily, if the viscosity-pressure coefficient (α) is known, using ;

$$Z = \alpha\gamma / \ln(\eta_0/\eta_\infty) \tag{7}$$

This equation provides with good accuracy the conversion of the lubricant properties adopted in the Barus and the Roelands relationships. Here, (α) needs to be measured at low pressures, which is not difficult.

The adoption of the three lubricant properties (η_0, α and Z) which satisfied equation (7) can be found in Lubrecht, Napel and Bosma [5], Chang, Cusano and Conry [6], Kim and Sadeghi [7], etc. On the other hand, the adoption of properties which did not satisfy equation (7) can be found in Hamrock and Dowson [8], Chittenden [9], Kweh, Evans and Snidle [10], etc. Each publication cited above has used the Roelands relationship for the viscosity.

Kweh et al [10] have obtained (η_0, α, Z) from the information published on lubricant properties, followed by curve fitting. However, Hamrock and Dowson [8] and Chittenden [9] utilized a different concept of (α') ;

$$\alpha' = 1/p_{iv,as} \quad (8)$$

in defining the material group ($G=E'/p_{iv,as}=\alpha' E'$). The parameter ($p_{iv,as}$) is a conceptual pressure described in Roelands [1], termed the asymptotic isoviscous pressure and defined by ;

$$p_{iv,as} = \int_0^\infty \eta_0/\eta(p)\ dp \quad (9)$$

When the Barus relationship is substituted into equation (9), ($1/p_{iv,as}$) reduces exactly to (α). Since the Roelands relationship was used by the above authors, the notation of (α') reflected in their film thickness formulae should not be confused with (α) appearing in the Barus relationship. When the Roelands relationship (eq.4) is substituted into equation (9), the value of (α') obtained becomes smaller than the (α) obtained from equation (7) with ($Z = 0.67$), up to about (10 %) according to the value of (η_0).

In addition, the integration in equation (9) yields 'an incomplete gamma function' for the Roelands relationship, hence ($p_{iv,as}$) needs to be read from a table in Roelands [1] with given values of (η_0) and (Z). For this, (Z) has been assumed by the above authors to be a constant (0.67), which is not always true for different lubricants. According to Roelands [1], (Z) is a lubricant property ranging from ($0.5 \sim 1.1$) for most mineral oils. Therefore, it would be beneficial to have a single definition for the non-dimensional material group to be ($G = \alpha E'$), where (α) is the viscosity-pressure coefficient rather than ($1/p_{iv,as}$). This study will adopt ($G = \alpha E'$). In due course, the determination of (Z) for a lubricant with measured (η_0) and (α) preferably needs to follow equation (7).

The first example given in Table 1 shows the variation of the results for fixed values of (G, U, W) by changing variables, where equation (7) is not satisfied. The results presented are for a circular contact EHL situation and the details on the numerical method and the accuracy are listed in Song [11].

Modulating (η_0) and (u_e) and keeping their product unchanged does not alter the maximum Hertzian pressure (p_h). From ambient pressure

Table 1 Variations in (H_{cen}) and (H_{min}) by different data for fixed values of (G, U, W) not following equation (7).
(G=5713 $U=4.69\times10^{-13}$ $W=1.82\times10^{-8}$)

R_x=0.0254 (m), R_y=0.0254 (m),
F=2.58 (N), $E'=2.1972\times10^{11}$ (Pa),
Z=0.7257, $\alpha=2.6\times10^{-8}$ (1/Pa)
(1*) : data for a real oil (LVI-260 at 60°C)

Data No.	1*	2	3	4	5
η_0 (Pa.s)	0.071	0.03	0.01	0.005	0.0005
u_e (m/s)	0.037	0.087	0.262	0.524	5.24
$H_{cen} \times 10^5$	0.343	0.321	0.290	0.270	0.205
$H_{min} \times 10^5$	0.227	0.214	0.196	0.185	0.149
$P_{sp} \times 10^2$	0.105	0.101	0.097	—	—
$\overline{\eta}_h$	133	73	34	21	4
Remarks	p_h=0.21 (GPa), pure sliding, isothermal				

to (p_h), the Roelands viscosity will show different dependencies upon the pressure if (η_0) and (Z) are selected differently. For smaller (η_0), the variation of the viscosity throughout the field becomes smaller as seen in Table 1 through ($\overline{\eta}_h$), the non-dimensional viscosity at p_h. Therefore, the results incline towards the isoviscous case for smaller (η_0). Such an inclination to isoviscous features can also be noticed in other parameters. The de-crease in the non-dimensional pressure spike(P_{sp}) combined with the decrease in the film thickness ratio (H_{cen}/H_{min}) is considered to reflect the weakened piezo-viscous influence, since changes in the opposite direction have been commonly acknowledged to be due to the piezo-viscous effect of the lubricant. Therefore, the maximum difference in the film thickness was (40 %) compared with data case 1, which could be still bigger for a data set with much smaller (η_0). By adopting a much smaller (η_0), it is also true that the numerical stability on convergence improves considerably. However, it should be emphasized again that such a set of (η_0, α, Z) would be quite unrealistic since most mineral oils would satisfy equation (7) with a high accuracy.

Up to this point, emphasis has been placed

on a standard way of determining the lubricant properties (η_0, α, Z) and the common definition for (G), to facilitate both the comparison and the utility of the EHL results. Even then, we cannot accomplish uniqueness in the results when assessed in terms of three fixed non-dimensional groups (G, U, W) as illustrated in Table 2. Compared with the results for the reference case 3, the maximum differences for the cases considered were (9 %) in (H_{cen}) and (2.6 %) in (H_{min}), though they could be still bigger with other sets of data utilizing much smaller or much bigger (η_0). In the meanwhile, it should be noted that adopting such data sets with much smaller or much bigger (η_0) brings about unrealistically large values for the elasticity modulus $(E^{'})$ in a fixed (G, U, W) structure.

Table 2 Variations in (H_{cen}) and (H_{min}) for different data for fixed values of (G, U, W)
— following equation (7).
$(G{=}5713 \quad U{=}4.69{\times}10^{-13} \quad W{=}1.82{\times}10^{-8})$

$R_x{=}0.0254$ (m), $R_y{=}0.0254$ (m),
$Z{=}0.7257$, Pure Sliding
(3*) : data for a real oil (LVI-260 at 60^oC)

Data No.	1	2	3*	4	5	6
η_0 (Pa.s)	319.0	0.335	0.071	1.08e-2	1.38e-3	2.95e-4
α (1/Pa)	5.71e-8	3.17e-8	2.60e-8	1.90e-8	1.14e-8	0.57e-8
$E^{'}$ (Pa)	1.00e11	1.80e11	2.20e11	3.00e11	5.00e11	10.0e11
F (N)	1.174	2.114	2.580	3.523	5.871	11.74
u_e (m/s)	3.73e-6	6.41e-3	0.0369	0.330	4.307	40.33
$H_{cen} \times 10^5$	0.355	0.346	0.343	0.337	0.326	0.312
$H_{min} \times 10^5$	0.228	0.227	0.227	0.227	0.223	0.221
$P_{sp} \times 10^2$	0.113	0.107	0.105	0.101	0.097	0.095
$\overline{\eta}_h$	177	136	133	121	89	57
p_h (GPa)	0.096	0.17	0.21	0.29	0.48	0.96

These variations stem from the fact that the non-dimensional material group (G) incorporates what is normally taken to be the Barus viscosity-pressure coefficient (α), which is never used in the numerical analysis if the Roelands relationship is adopted. The viscosity-pressure coefficient (α) is used only to present the results in terms of (G). Due to this fact, the numerical results for (H_{cen}) and (H_{min}) can be different for fixed values of (G, U, W) if different (α) values are assumed. Alternatively, when all the variables except (α) are the same between two cases, there cannot be a change in numerical results. However, the film thickness formulae presented in terms of (G, U, W) predict different values. Therefore, it is apparent that the presentation of the results in terms of (G, U, W) is not adequate for the adoption of the Roelands viscosity-pressure relationship. Nor is there a sufficient number of the non-dimensional groups to define fully the operating conditions. The latter point is detailed in the next section.

NON-DIMENSIONAL GROUPS FOR THE UNIQUENESS

When the equations (2) to (6) are non-dimensionalized, they reduce to ;

$$\frac{\partial}{\partial X}\left(\frac{\overline{\rho}H^3}{12\overline{\eta}}\frac{\partial P}{\partial X}\right)+\frac{\partial}{\partial Y}\left(\frac{\overline{\rho}H^3}{12\overline{\eta}}\frac{\partial P}{\partial Y}\right)=U\frac{\partial}{\partial X}(\overline{\rho}H) \tag{10}$$

$$H = H_{00} + \frac{X^2}{2} + \frac{Y^2}{2}\left(k^{-3/2}\right) + \frac{2}{\pi}\int_{-\infty}^{\infty}\int_{-\infty}^{\infty}\frac{P \ d\overline{X} \ d\overline{Y}}{\sqrt{(X-\overline{X})^2+(Y-\overline{Y})^2}} \tag{11}$$

$$\overline{\eta} = (\eta_0/\eta_\infty)^{[(1+PE^{'}/\gamma)^Z-1]} \tag{12}$$

$$\overline{\rho} = 1 + \frac{E^{'}\beta_2 \ P}{1+E^{'}\beta_1 \ P} \tag{13}$$

$$W = \int_{-\infty}^{\infty}\int_{-\infty}^{\infty} P \ dX \ dY \tag{14}$$

Although different normalisations may be adopted, they would not change the principle conclusions reached in this study.

These five equations have been solved for the five unknowns $(P, H, \overline{\eta}, \overline{\rho}, H_{00})$, given the values of the seven non-dimensional variables (k, Z, U, W, η_0/η_∞, $E^{'}/\gamma$, $E^{'}\beta_1$, $E^{'}\beta_2$). Here, of course, consistent application of the boundary conditions is assumed. Therefore, any specific result should be a function of the seven non-dimensional variables, for example ;

$$H_{cen} = f\ (\ k, Z, U, W, \eta_0/\eta_\infty, E'/\gamma, E'\beta_1, E'\beta_2\ ) \qquad (15)$$

Now, it is recalled that the density-pressure relationship (eq.5) has commonly been adopted since Dowson & Higginson [2], so as to assume (β_1) and (β_2) to be absolute constants. This may be accepted whilst considering mineral oils. With this assumption, both ($E'\beta_1$) and ($E'\beta_2$) become unnecessary in equation (15) since ($E'\beta_1 = (\gamma\beta_1)E'/\gamma$) and ($E'\beta_2 = (\gamma\beta_2)E'/\gamma$), where ($\gamma\beta_1$) and ($\gamma\beta_2$) are not variables any more. Hence, equation (15) reduces to ;

$$H_{cen}=f\ (k, Z, U, W, \eta_0/\eta_\infty, E'/\gamma\ ) \qquad (16)$$

Also, it will be shown for circular or line contact EHL problems that ;

$$H_{cen}=f\ (Z, U, W, \eta_0/\eta_\infty, E'/\gamma\ ) \qquad (17)$$

since the ellipticity ratio (k) is constant and equal to unity. In addition, the above equations may still incorporate the familiar materials group (G) replacing (Z), providing the conversion of equation (7) is satisfied at all times. Therefore, the variations of the results demonstrated in Table 2 for fixed (G, U, W) are interpreted as being due to the variations in (η_0/η_∞) and (E'/γ).

APPLYING THE Π-THEOREM

The Buckingham's Π-theorem [12] is applied to the EHL system described by equations (2) to (6). The procedure is to suppress the five simultaneous equations by a single functional relationship. If one unknown (e.g. η) is replaced with the corresponding equation (e.g. eq.4), there will remain four equations for four unknowns, and so on. In this manner, equations (2), (4), (5) and (6) can be substituted into equation (3) one by one to obtain the required relationship as follows ;

$$f\ (\ h, F, u_e, x, y, R_x, R_y, E', \eta_0, \eta_\infty, \gamma, Z, \beta_1, \beta_2\ ) = 0 \qquad (18)$$

Here, three points need to be mentioned. Since equation (6) is a global equation and the unknown (h_{oo}) is associated with the global system, it may be considered that equation (6) yields a solution for (h_{oo}). Therefore, (F) becomes a quantity of the relation instead of (h_{oo}) after the substitution. When equation (5) is substituted into equation (2), it is apparent that the ambient density (ρ_0) appears in every term, without influencing the results obtained. Hence, (ρ_0) disappeared in equation (18). Finally, (η_∞) and (γ) are not removed from the relation even if they are absolute constants. This is because they are not dimensionless constants.

After transformation, equation (18) can be presented in terms of five different kinds of quantities and nine ratios to them ;

$$f\ (\ R_x, E', \eta_0, F, u_e, H, X, Y, k, \eta_0/\eta_\infty, Z, E'/\gamma, E'\beta_1, E'\beta_2\ ) = 0 \qquad (19)$$

The nine ratios ($H, X, Y, k, \eta_0/\eta_\infty, Z, E'/\gamma, E'\beta_1, E'\beta_2$) should eventually remain in the non- dimensional expression of eq. (19).

For a relation which subsists among 5 quantities of different kind (R_x, E', η_0, F, u_e), the Π-theorem determines 'the maximum number of separate independent non-dimensional groups'. The dimensions of the 5 quantities are represented in brackets.

$$R_x[L], \quad E'[ML^{-1}T^{-2}], \quad \eta_0[ML^{-1}T^{-1}], \quad F[MLT^{-2}], \quad u_e[LT^{-1}]$$

Since the number of the fundamental dimensions is 3, (M, L, T), two (=5-3) non-dimensional groups will be obtained according to the theorem. That is,

$$\Pi_1 = \eta_0 u_e/E'R_x \quad (= U) \qquad \text{and}$$
$$\Pi_2 = F/E'R_x^2 \quad (= W) \qquad (20)$$

Thus, the relation (eq. 19) can be fully represented in the following non-dimensional form ;

$$f\ (\ U,\ W,\ H,\ X,\ Y,\ k,\ \eta_0/\eta_\infty,\ Z,\ E'/\gamma,\ E'\beta_1,\ E'\beta_2\ ) = 0 \qquad (21)$$

Therefore, the form of the relation for any specific result (H_{cen}) at a given point ($X=Y=0$) will adopt the form of equation (15). Subsequently, we can also reach equations (16) and (17) with the assumption that the fluid is incompressible or that a common density-pressure relationship is adopted.

Furthermore, it is worthy of note that the development of the non-dimensional groups starts from the set of the governing equations which correspond to a specific system. In due course, those groups necessary for other systems (e.g. non-Newtonian EHL, thermal EHL, or adopting other assumptions which alter the governing equations) may vary both in number and shape for the unique representation of the results.

CONCLUSIONS

1. In utilizing and comparing the EHL results, (α') defined by $(1/p_{iv,as})$ should not be confused with the viscosity-pressure coefficient (α).
2. The material group (G) preferably needs to be uniquely defined by $(G = \alpha E')$.
3. The determination of the lubricant properties (η_0, α, Z) needs to be standardized by satisfying ;

$$Z = \alpha\gamma / \ln(\eta_0/\eta_\infty)$$

4. Even the most familiar non-dimensional groups (G, U, W) do not fully define the operating conditions.
5. For the adoption of the Roelands viscosity-pressure relationship, two additional non-dimensional groups need to be introduced for the unique representation of the results, for example ;

$$H_{cen} = f\ (\ k, Z, U, W,\ \eta_0/\eta_\infty,\ E'/\gamma)$$

for elliptical contact EHL problems, and

$$H_{cen} = f\ (\ Z, U, W,\ \eta_0/\eta_\infty,\ E'/\gamma\ )$$

for circular or line contact EHL problems.
6. In the above equations, (G) can replace (Z) providing the relationship of equation (7) is satisfied.
7. The Buckingham's Π-theorem confirms that these equations comprise the maximum number of separate independent non-dimensional groups.

REFERENCES

[1] **Roelands C.J.A., 1966**, " Correlational aspects of the viscosity-temperature-pressure relationship of lubricating oils ", Ph.D. thesis, University of Delft.

[2] **Dowson D. and Higginson G.R., 1959**, " A numerical solution to the elastohydrodynamic problem ", J. Mech. Eng. Sci., 1(1), pp 7-15.

[3] **Venner C.H. and Napel W.E. ten, 1992**, " Multilevel solution of the elastohydrodynamically lubricated circular contact problem. Part II : Smooth surface results ", Wear, 152, pp 369-381.

[4] **Houpert L.G., 1985**, " New results of traction force calculations in elastohydrodynamic contacts ", J. of Tribology, Trans. ASME, 107, pp 241-248.

[5] **Lubrecht A.A., Napel W.E. ten and Bosma R., 1987**, " Multigrid, an alternative method of solution for two-dimensional elastohydrodynamically lubricated point contact calculations ", J. of Tribology, Trans. ASME, 109, pp 437-443.

[6] **Chang L., Cusano C. and Conry T.F., 1989**, " Effects of lubricant rheology and kinematic conditions on micro- elastohydrodynamic lubrication ", J. of Tribology, Trans. ASME, 111, pp 344-351.

[7] **Kim K.H. and Sadeghi F., 1991**, " Non-Newtonian elastohydrodynamic lubrication of point contact ", J. of Tribology, Trans. ASME, 113, pp 703-711.

[8] **Hamrock B.J. and Dowson D., 1976**, " Isothermal elastohydrodynamic lubrication of point contacts, Part (I) - Theoretical formulation ", J. of Lubrication Tech., Trans. ASME, 98 (2), pp 223-229.

[9] **Chittenden R.J., 1984**, " The elastohydrodynamic lubrication of elliptical contacts ", Ph.D. thesis, The University of Leeds, U.K.

[10] **Kweh C.C., Evans H.P. and Snidle R.W., 1989**, " Micro-elastohydrodynamic lubrication of an elliptical contact with transverse and three-dimensional roughness ", J. of Tribology, Trans. ASME, 111, pp 577-584.

[11] **Song E.S., 1992**, " The influence of asperity lubrication in hydrodynamic and elastohydrodynamic contacts ", Ph.D. thesis, The University of Leeds, U.K.

[12] **Buckingham E., 1914**, " On physically similar systems ; Illustrations of the use of dimensional equations ", Physical Review, 4, pp 345-376.

SESSION VIII
LUBRICATED WEAR

Chairman : Dr P M Cann

Effect of Corrosion on the Wear Behavior of Passivating Metals in Aqueous Solutions

S. Mischler, E.A. Rosset, D. Landolt

Materials Department, Ecole Polytechnique Fédérale de Lausanne, 1015 Lausanne, Switzerland

The corrosive wear behavior under sliding condition of a Fe25Cr alloy - a model material for stainless steel - in 1M H_2SO_4 has been studied by using electrochemical techniques. An attempt is made to correlate quantitatively corrosion parameters with material loss under corrosive wear conditions. Surface characterisation has been done by Auger Electron Spectroscopy (AES) and Atomic Force Microscopy (AFM).

1. INTRODUCTION

Corrosion products play a very important technological role. For example stainless steels and titanium alloys owe their good corrosion resistance to the presence of a thin surface oxide film (passive film) formed by reaction of the metal with the corrosive medium. In many practical applications stainless steels are subject to combined degradation by aqueous corrosion and wear and it is therefore of interest to understand the behaviour of passive films under corrosive wear (tribocorrosion) conditions.

The present study was initiated with the aim to elucidate the effect of passive films on the wear behaviour of a Fe-25% Cr alloy in 1 M sulphuric acid. Wear tests were carried out on samples polarised at a potential of 700 mV corresponding to the passive potential region (all potentials in this paper are given with respect to the normal hydrogen reference electrode). The wear rate, the friction coefficient and the current were recorded during the wear test. The morphology and the composition of the worn surfaces were characterised by Atomic Force Microscopy (AFM) and Auger Electron Spectroscopy (AES).

For reference, a wear test was carried out on a sample polarised cathodically (-800 mV) where no corrosion takes place.

The passivation kinetics of the Fe25Cr alloy polarised at 700 mV was determined electrochemically with the aim to find a correlation with the current measured during the wear test.

2. ELECTROCHEMICAL ASPECTS

In acidic solutions one can describe the corrosion of a metal M by the following electrochemical reactions involving oxidation of

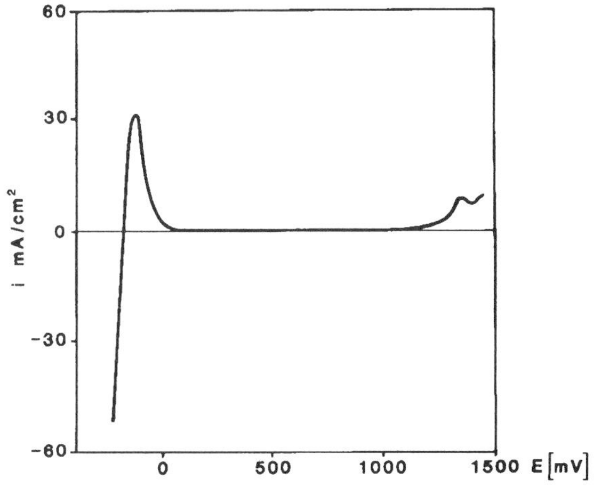

Figure 1. Polarisation curve of Fe25Cr in 1M H_2SO_4

the metal M and reduction of the protons via transfer of n electrons per metal atom oxidised.

$$\text{oxidation} \qquad M = M^{n+} + ne^- \qquad (1)$$

$$\text{reduction} \qquad nH^+ + ne- = n/2\ H_2 \qquad (2)$$

Electrochemical techniques are very helpful in studying corrosion mechanisms because they offer the possibility to control the chemical nature of the surface exposed to an electrolyte and to measure in-situ the material loss rate associated with corrosion. This possibilities are best illustrated by the polarisation curves.

To establish a polarisation curve in addition to the electrode being studied (working electrode) a reference electrode and an inert auxiliary electrode are employed. A potentiostat automatically maintains the desired potential between the working and reference electrodes by passing an appropriate current between the working and counter electrode. Polarisation curves are established in the potentiostatic mode by plotting the measured current versus the applied potential.

The polarisation curve of the Fe25Cr alloy in 1M H_2SO_4 is illustrated in Fig. 1. Two regions can be distinguished: the cathodic region were the current is negative and the anodic region where the current is positive.

At the corrosion potential (-200mV in Fig. 1) the current is zero and the reactions (1) and (2) occur at the same rate. In the cathodic region, at potentials well below the corrosion potential (about 100mV), negligible metal dissolution occurs and the current is determined by the kinetics of reaction (2). In the same way in the anodic region well above the corrosion potential the reaction (2) becomes negligible and the current is determined by the metal oxidation kinetics (reaction 1).

The relation between weight loss of the metal and current in the anodic region is given by Faraday's law:

$$\frac{dm}{dt} = \frac{IM_1}{nF} \qquad (3)$$

where m is the metal mass dissolved during the time t, I is the current, F is the Faraday's constant (96500 C/mole) and M_1 the atomic mass of the metal (55 g/mole for Fe25Cr)

The anodic region is divided in three parts corresponding to three different surface states of the metal. In the active region (-200 to about 0 mV) the metal dissolves directly in contact with the solution. In the range 0 to 1100 mV (passive region) the surface is covered with an approximately 2 nm thick oxide film (mainly trivalent chromium oxide), which inhibits the metal dissolution and leads to very low currents (10^{-3} mA/cm^2). Above 1100 mV (transpassive region) the current increases, the passive film being no more stable because of the oxidation of trivalent chromium oxide to soluble hexavalent chromium.

For a more detailed discussion of the electrochemical techniques the reader is referred to the literature [1].

3. EXPERIMENTAL

3.1 Test Materials

Sliding wear conditions were established by rubbing an alumina pin (chosen for its chemical inertness) against a high purity iron-chromium alloy (25% w Cr) plate. The hardness of the alloy was 160 HV30. The oscillating pins were prepared by machining the ends of 4 mm diameter alumina rods in the shape of truncated cones (120° included angle). The diameter of the flat end was 0.68 mm giving an apparent contact area of 0.36 mm^2. The pin was oscillating at a frequency of 5 Hz. The stroke length of 3.6 mm corresponded to an average sliding velocity of 36 mm/s. The applied normal load was 5 N resulting in a nominal contact pressure of 14 MPa. The metal samples were prepared by embedding 12x12 mm

Fe25Cr plates in Technovit mount. After polishing with 6 μm diamond paste and ultrasonic cleaning in ethanol the samples were mounted in the electrochemical cell of the wear test rig schematically presented in Fig. 2.

The 1M sulphuric acid solution was prepared using doubly distilled water and analysis grade 96% sulphuric acid (Merck). The solutions were not deaerated. All tests were carried out at room temperature (20-23°C).

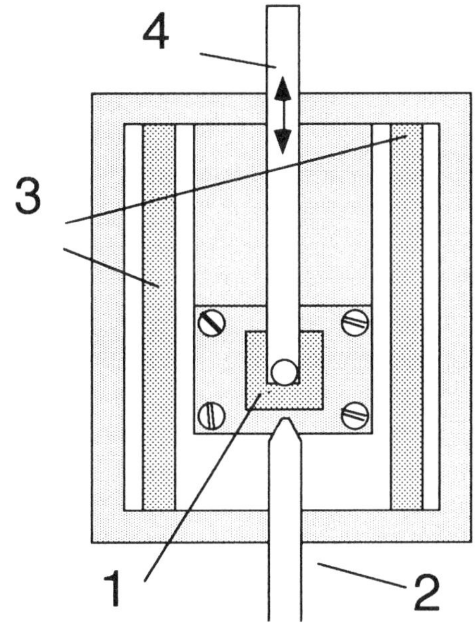

Figure 2. Top view of the electrochemical cell. 1- Working electrode. 2- Reference electrode. 3- Auxiliary electrodes. 4- Moving arm with pin.

3.2 Electrochemical tests

The polarisation curves were measured in the potentiodynamic mode by scanning the potential from -750 mV to 1400 mV at a scan rate of 5 mV/s.

The passivation kinetics were determined by the potential step method: after 5 minutes of cathodic polarisation at -800 mV the potential was stepped to 700 mV. The resulting current transient was digitally recorded at a sampling rate of 1 kHz.

3.3. Wear Test

The tribo-corrosion experiments were carried out in the reciprocating pin-on-plate rig described in more detail in [2]. Sliding wear conditions were established by rubbing the end of a vertically mounted pin against a fixed flat plate sample.

The electrode potentials are applied using a HEKA PG285 potentiostat connected to a graphite counter electrode and a mercury sulphate reference electrode (Fig. 2).

The electrochemical cell is located above a load cell allowing for the measurement of the normal force. The frictional force is measured with a piezoelectric force transducer. Reciprocal pin motion is provided by an electrodynamic vibration exciter (Brüel Kjaer 4809) driven by a triangular waveform signal. The pin motion was measured by a photoelectric sensor.

During the wear test the frictional and the normal forces as well as the current were monitored using a Macintosh IIfx computer (LabView2 software from National Instruments) equipped with a National Instruments general purpose 16 bit I/O board.

The test conditions involved cathodic polarisation at -800 mV for 5 min followed by passivation at 700 mV. After 25 minutes passivation the rubbing was started. At the end of the test the metal samples and the alumina pins were removed from the solution, rinsed with distilled water and dried with an argon jet.

The surface morphology resulting from tribocorrosion was characterised by optical microscopy and Atomic Force Microscopy (Park Scientific Instruments, SFM-BD2).

Surface analysis and secondary electron imaging were carried out in a Perkin Elmer 660 Scanning Auger Microscope using a 3 keV (25 nA) electron beam. Depth profile acquisition was performed by rastering a 2 keV Ar+ beam over an area of 1.5 x 1.5 mm. Under these conditions the sputter rate of Ta_2O_5 was 0.30 nm/min.

4. RESULTS

Figure 3 shows the evolution of the current with the time after switching the sample potential from the cathodic region to the passive region. At the beginning of the passivation the current is very high but decreases quickly as soon as the passive film is formed. The experimental curve plotted in Fig. 3 can be described by the empirical relation (4)

$$i = 10^3 . \exp(-0.139 . t) + 175 . t^{-0.614} \quad (4)$$

where t is expressed in ms and i in mA/cm^2. The curve calculated using (4) is plotted in Fig. 3 together with the experimental points.

The obtained friction and material loss parameters are listed in Table 1.

The tabulated coefficients of friction μ correspond to the average values recorded during the rubbing. Variations in μ were less than 10%.

A linear decrease of the vertical pin position with sliding time was observed after a running-in period of 60 s. Since no pin geometry change was observed after the wear test, the measured decrease can be attributed to wear of the metal. The

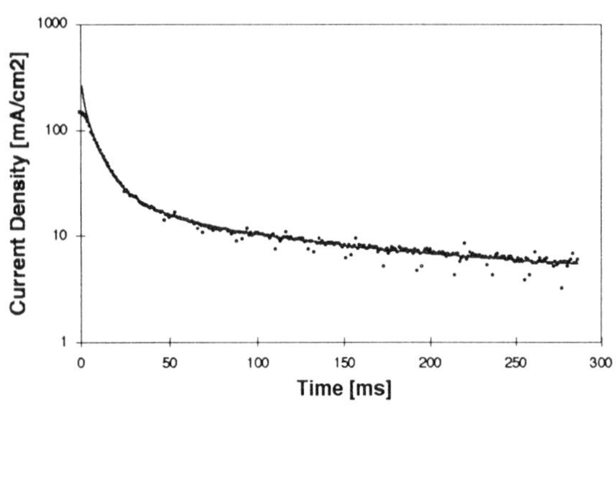

Figure 3. Experimental and fitted passivation current transients.

wear rates listed in table 1 correspond to the slope of the curve indicating the vertical pin position versus time. The average wear rate measured on passivated surfaces is 7.3 nm/s.

The start of rubbing on passivated samples corresponds to an increase of the current from 10^{-3} mA to 0.65 mA. After the running-in period the current remained within a scatter band of 5%. The rubbing did not affect the current of the sample polarised in the cathodic region.

Current densities were calculated by dividing the current by the wear scar surface. It is assumed that the current is due only to the dissolution occurring in the wear scar. The influence on the measured current of oxygen reduction is considered to be negligible.

In order to calculate the wear scar surface a rectangular geometry of the wear scar was assumed. Length and width of the scar were determined by optical microscopy after the test. The averaged wall area was estimated by taking half the depth measured after sliding.

The current densities were converted into material loss using (3). This yielded an average dissolution rate of (14.3/n) nm/s.

The variation of current and of pin horizontal position recorded during a stroke are shown in Fig. 4. At the end of each stroke, when the direction of motion changes, the pin remains motionless for about 25 ms. This is because the static frictional force exceeds the force transmitted by the exciter to the pin in order to generate the alternate motion. During the dead time the current drops to about 25% of its maximum value.

A low magnification image of the wear track obtained on a passivated sample is shown in Fig. 5. The geometry of the wear track exhibits a regular and well defined shape corresponding to the pin geometry.

More detailed three dimensional images of the bottom of the wear tracks are obtained by

Table 1
Wear tests results

Applied potential [V]	Sliding distance [m]	Sliding time [s]	μ	Total wear rate [nm/s]	Current density [mA/cm2]
0.7	126	3500	.24	7.5	21
0.7	146	3840	.24	8.1	19
0.7	355	10140	.25	6.4	17
-.8	225	6000	.22	1.5	--

using Atomic Force Microscopy (Fig. 6). Note that the relief in Figure 6 has been enhanced by using a vertical scale magnification factor of 50 with respect to the horizontal scale. Area roughness statistics can be extracted from AFM images by calculating the mean deviation from the average height. This value corresponds in principle to the R_a value determined in conventional two dimensional profilometry.

In this way we obtain roughness values of 61 and 39 nm for the wear track formed at cathodic potential and passive potential respectively. For comparison the roughness outside the wear track on a passivated sample corresponds to 28 nm.

Wear debris were found neither on the samples during rubbing nor after filtering the solution or the water used for washing the samples. Some wear debris were found accumulating at the edges of the flat end of the pin however. Wear debris apparently did not play a significant role in the wear process.

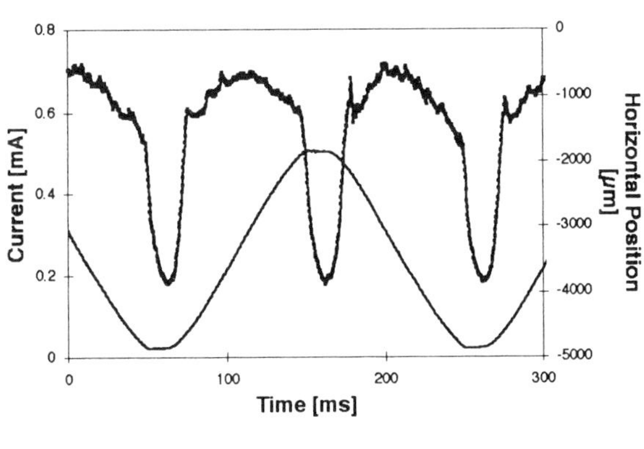

Figure 4. Variation of current and horizontal pin position during sliding.

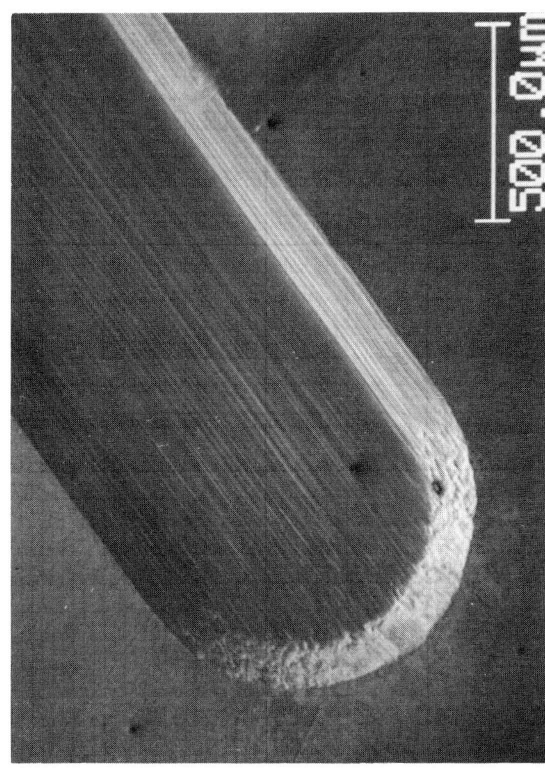

Figure 5. SEM image of the wear scar on passivated sample.

The grooves observed by AFM are oriented along the pin motion axis and are probably the result of the microploughing action of asperities present on the alumina surface. Abrasive wear is expected when one of the two moving surfaces is considerably harder than the other, as was the case in the present experiment.

The geometry of the pin as observed by optical microscopy did not change during the experiment. Under the conditions of this study alumina, therefore, was not subject to corrosion or wear.

The Auger depth profile measured on a area outside the wear track of a passivated sample is shown in Fig. 7. The oxygen profile exhibits a well defined oxide-metal interface at about 1.5 nm. Carbon is present as a surface contamination only, the carbon signal goes to zero at an apparent depth of 0.3 nm. Sulphur is found in the passive film probably as sulphate from the solution. A detailed surface analytical study of the composition of a passive film formed in similar conditions can be found in the literature [3].

The oxygen profile measured in the wear track (Fig. 8) does not show a sharp interface probably because the surface roughness induces shadowing effects of the ion beam used for sputtering. This implies that locally the passive oxide film is not removed by sputtering and still contributes to the overall Auger signal. Ion beam shadowing effects render very difficult the interpretation of depth profiles and therefore no quantification of the profiles of Fig. 7 and 8 were carried out in the present study. It is however possible to conclude that the passive film formed in the wear track has about the same thickness as the one formed outside the track but contains less sulphur.

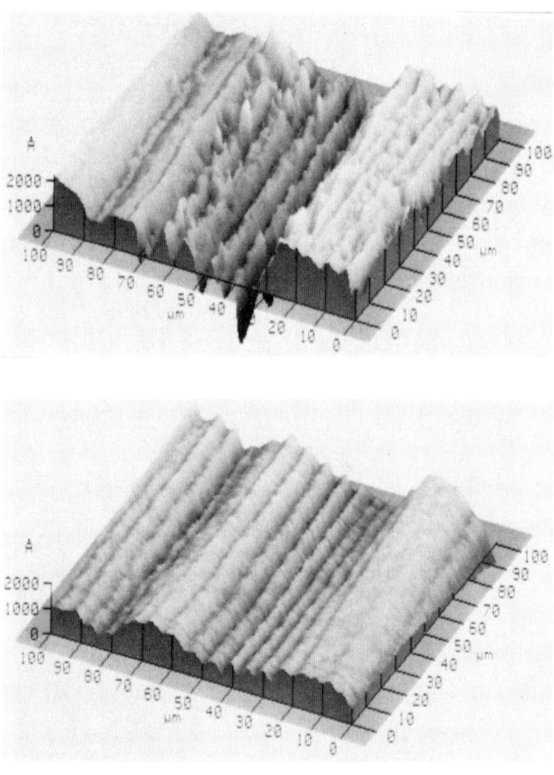

Figure 6. AFM images of the wear tracks formed at the cathodic potential (above) and at the passive potential (below).

5) DISCUSSION

Uhlig [4] has proposed a simple mechanistic model based on a chemical and a mechanical factor to describe the weight loss of a metal surface undergoing fretting corrosion in a gaseous environment. Uhlig considered that an asperity rubbing on a metal surface produces a track of clean metal which immediately oxidizes and therefore leads to a chemically controlled surface damage. The mechanical factor correspond to a certain amount of wear caused by the asperities digging below the surface.

In principle, the results of the present study can be interpreted by a similar model.

Under sliding conditions the passive film is periodically removed from the contact area exposing the nascent metal surface to the acid solution. A considerable amount of dissolution is required in order to repassivate the exposed surface as shown by the current transient of Fig. 3.

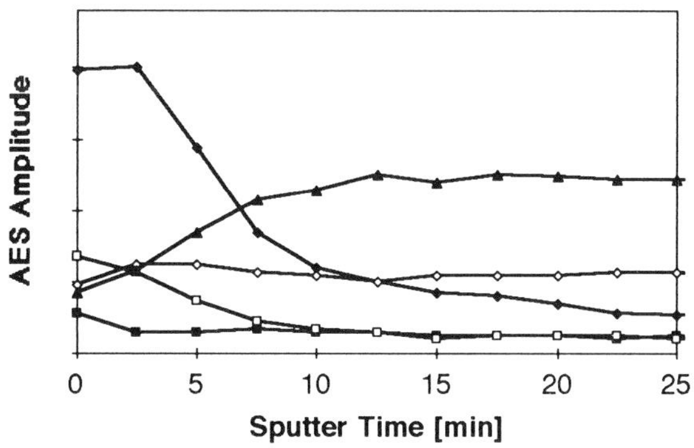

Figure 7. Auger depth profile measured outside the track on passivated sample.

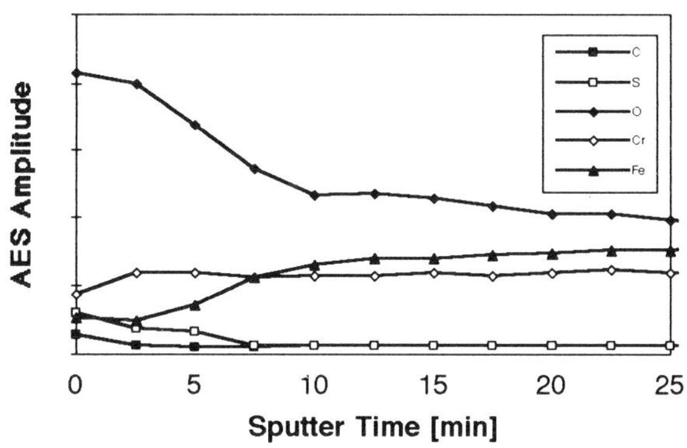

Figure 8. Auger depth profile measured inside the wear track on passivated sample.

Since at each stroke of the pin the passive film is removed and rebuilt the material loss due to corrosion is higher under sliding conditions than during static conditions. This can explain the increase of two orders of magnitude of the current observed on passivated samples when sliding starts.

An increased corrosion rate compared to static conditions would also be expected if rubbing leads only to film thinning. To periodically restore the original film thickness metal must be oxidised. In this case the total material loss should correspond to the amount of metal oxidized and hence the valence n in equation 3 can be obtained by (5).

$$n = \frac{w_t}{MI / F\rho} \qquad (5)$$

Where w_t is the total wear rate listed in table 1 and ρ the density (7.69 g/ml). Equation (5), for an average value of w_t=2.3 nm/s, yields a dissolution valence n=1.96. This value is to low for the considered passive potential, where chromium is expected to dissolve at a valence of 3 and iron at valences ranging from 2 to 3. This suggest that passive film thinning is not the predominent mechanism of material removal. However, a more exact determination of the dissolution valence is required in order to confirm this conclusion.

The presence of a passive film on the Fe25Cr surface can explain the smoother morphology of the wear track surfaces formed at the passive potential compared to the surfaces formed under cathodic conditions where no film is present. During the sliding, and at each stroke, the film covering the rough surface is removed together with some underlying metal from the contact areas, i.e. from the asperities. After the film removal the fresh metal surface exposed to the corrosive solution at the tips of the asperities undergoes dissolution until the passive film is rebuilt. On the non-contact areas, i.e. valleys, the passive film remains and thus no dissolution occurs. The sliding process thus causes a chemical selective material loss from the asperities which leads to smoothing. In absence of corrosion the relief is determined only by the ploughing action of the asperities on the alumina surface and no chemical induced smoothing occurs.

Assuming the validity of the wear mechanism involving passive film removal from the surface with subsequent repassivation it should be possible to calculate the current transient of Fig. 4 on the basis of the passivation kinetic.

For this purpose the length of the wear track can be divided in N segments of equal length. The total current I measured as a function of the time t is then given by (6)

$$I = S \cdot \sum_{k=1}^{N} i\,(t - t_A(k)) \qquad (6)$$

where S is the wear scar area and $t_A(k)$ corresponds to the time when the segment k was last activated by rubbing.

The activation time of each segment is determined by the displacement profile of the pin. Under the present conditions the motion of the pin can be divided into four steps involving a forward displacement lasting 75 ms, a dead time of 25 ms with no motion, a backwards displacement lasting again 75 ms and finally again a dead time of 25 ms. Thus for forward displacement the activation time of each segment k is defined by:

$$t_A(k) = (k-1) \cdot L/v \qquad [\mathrm{ms}] \qquad (7)$$

where L is the length of a segment and v is the displacement velocity (30 mm/s in our case). For the backwards displacement $t_A(k)$ is given by

$$t_A(k) = 75 + 25 + (N-k+1) \cdot L/v \qquad [\mathrm{ms}] \qquad (8)$$

Although the conditions in the wear scars may differ from the passivation conditions used to establish (4) this relation can describe the function $i(t-t_A)$ of (6). In particular slightly higher temperatures due to friction and changes in solution composition due to corrosion may play a role.

The values calculated numerically using (6) with 75 segments are compared to experimental results in Fig. 9. A reasonable good agreement is found in the shape of the curves. The absolute value of the calculated current is however about 150 times higher than the measured current. This can be interpreted by assuming than the real area of contact corresponds to 0.5 - 1 % of the apparent area of contact.

One may assume that the mechanical factor of corrosive wear of passivated samples corresponds to the total wear found on the sample polarised cathodically, where no corrosion occurs. In this case an average chemical wear rate of

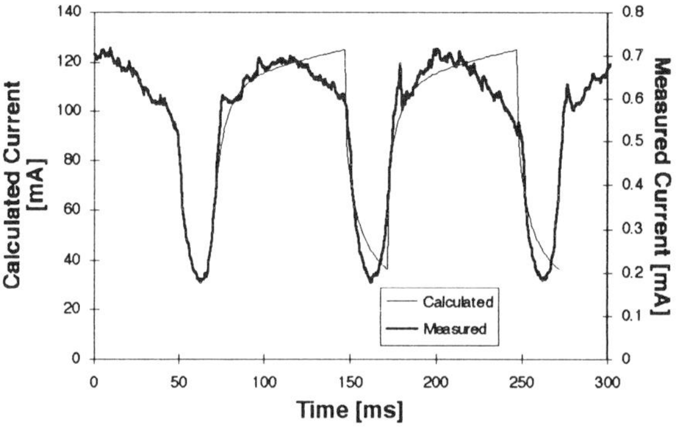

Figure 9. Calculated and measured current transients during sliding on passivated sample.

5.8 nm/s can be obtained by subtracting the wear rate of the cathodic sample (1.5 nm/s) to the average wear rate of passivated samples (7.3 nm/s). Comparing this value with the average current density leads to a valence of dissolution of 2.47. This value is reasonable and implies that, assuming chromium dissolves at valence 3, 30% of the iron atoms dissolve with a valence of 3 and 70% with a valence 2.

However, because of the presence of the passive film, it is not likely that mechanical wear at the passive potential corresponds to that at the cathodic potential. The observed differences in surface morphology and in the frictional coefficient should also influence the wear behaviour.

If the valence of dissolution were known one could calculate exactly the chemical material removal rate by using Faraday's law and therefore the mechanical wear rate by comparison with the total wear rate.

6) CONCLUSIONS

A simple mechanistic model considering a mechanical and a chemical factor may be used to describe weight loss and surface morphology of Fe25Cr alloy surfaces sliding against alumina in 1M sulphuric acid under applied passive potential of 700mV.

The current measured during sliding can be calculated by known displacement of the pin on the basis of the passivation kinetics measured under static conditions.

Knowledge of the exact dissolution valence of the metal during sliding is required in order to quantitatively interpretate corrosive wear measurements.

REFERENCES

[1] H.H. Uhlig and R.W. Revie
Corrosion and Corrosion Control
J. Wiley & Sons, New York (1985)

[2] E.A. Rosset, S. Mischler, D. Landolt
To be published in Thin Films in Tribology
Proceedings of the 19th Leeds-Lyon Symposium on Tribology

[3] S. Mischler, H.J. Mathieu and D. Landolt
Surface and Interface Analysis 11,182 (1988)

[4] H.H. Uhlig
Journal of Applied Mechanics 21, 401 (1954)

Thin Films in Tribology / D. Dowson et al. (Editors)
1993 Elsevier Science Publishers B.V.

LUBRICANT CAPABILITY IN SQUEEZE-SLIDING MOTION

Elisabet Kassfeldt, D. Sc. (Mech)
Jan Lundberg, D. Sc. (Mech)
Division of Machine Elements, Luleå University of Technology,
S-951 87 Luleå, Sweden

SUMMARY

Nine different types of commercial lubricants have been tested in an apparatus which can produce squeeze and sliding motion simultaneously. It is found that lubricants with the same viscosity show big differences in lubricant capability in terms of wear and oil film collapse, and that these differences are due to the additives.

1. INTRODUCTION

Gears in applications such as final drives in cars or elevator gearboxes are exposed to combined squeeze and sliding motion. It is shown by Lundberg et al [1] that these circumstances will cause increased possibility of oil collapse between the operating gear teeth.

In this investigation, nine different types of oil intended for use in final drives are tested in an apparatus which can produce both squeeze and sliding motion at the same time. Wear and oil film collapse are studied for each oil. The tests have been performed at two different temperatures (50° and 22°C).

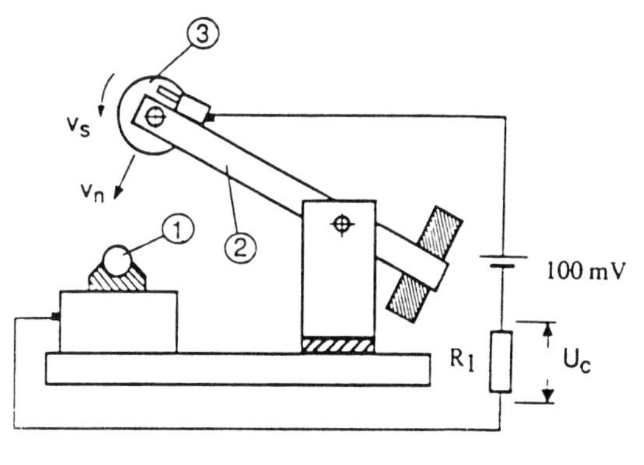

Figure 1. Sketch of the experimental set-up

2. MEASURING SYSTEM

The test rig is described in detail in ref [1] and is only briefly explained below. The principal design of the experimental apparatus is illustrated in Fig 1.The pendulum (2) with the steel roller (3) is free to fall on to the steel ball (1). The steel roller has a diameter of 69.5 mm and the steel ball a diameter of 25 mm.

The test oil is positioned both on the ball (1) and the roller (3).

If the oil film is broken at an impact, a current flows from the 100 mV power supply through the ball and the roller. The amplitude and time of the voltage U_c over the resistance R_1 = 100Ω are registered by a transient recorder. The normal velocity v_n, just before impact, is registered by an oscilloscope. The sliding velocity v_s, given by the angular speed of the air nozzle driven roller, is measured by an IR-detector connected to an adapter unit. This equipment gives ten pulses per revolution, which are registered by the same oscilloscope that reads the normal velocity. A tri-axial force transducer is located under the ball, giving the normal force F_n at impact. The registration of the normal force is linked together with the current registration through ball and roller.

The tests have been performed at two different temperatures (50°C and 22°C). In order to achieve the temperature 50°C, special heat equipment was developed, see Fig 2.

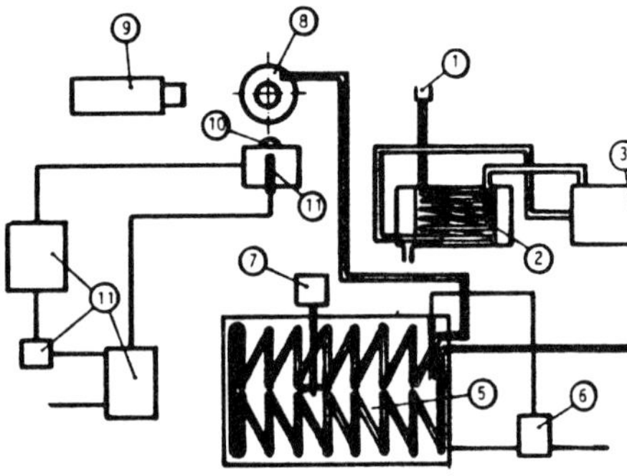

Figure 2 Heat equipment

The air which is used for driving the roller is initially taken from the air reservoir (1) of the laboratory. This air comes into an open heat exchanger (2) consisting of 2 m of copper tubes, Ø 8.5 mm, embedded in an open chamber with output water from a closed exchanger (3). This exchanger (3) is connected to the hot water supply (4) of the laboratory. After the air has passed through the open exchanger(2), the air then flows into the closed exchanger (3). The next step for the air is to flow through Ø 9 mm tubing measuring about 6 m in length, embedded in an open chamber (5) with a heating device with an effect of 2000 W. This chamber (5) is filled up with water which is heated up to boiling point. The temperature is held constant with the use of a temperature regulator (6). A circulator system (7) is responsible for the circulation of water in the chamber (5). The heated air now comes through the nozzle to the roller (8). Even more heat is needed to compensate for the cooling due to the rotation of the roller. A 330 W hot air fan (9) is used to ensure enough heat on the roller. The fan is placed 150 mm from the roller and blows hot air directly on to the roller. The ball (10) is heated by an electric heater and the temperature is regulated by a back linked temperature regulator (11). This system is capable of holding the temperature of the ball at 50 ± 1°C. The temperature of the roller at operating condition is measured with an infrared thermographic heat camera. This temperature is held at 50 ± 3°C.

3. TEST PROCEDURE

Four different types of tests have been performed

1) Wear tests at 50°C
2) Wear tests at 22°C
3) Oil collapse tests at 50°C
4) Oil collapse tests at 22°C

In the wear tests at 50°C, every oil is exposed to 105 impacts with normal velocity v_n = 0.13 ± 0.01 m/s and sliding velocity v_s = 6.0 ± 0.2 m/s. The maximum force during an impact reaches values up to F_n = 411± 30 N due to actual conditions. With the use of the theory from Hamrock and Dowson [2] this gives a contact pressure of p = 1970 ± 50 N/mm^2.

The amount of wear is estimated by measuring the wear spot on the ball and the total wear volume is calculated with the use of

$$V = \frac{\pi h^2}{3}(3R-h) \qquad (1)$$

and

$$h = R - \sqrt{R^2 - \frac{D^2}{4}} \qquad (2)$$

where V is the scraped - away volume of the segment of the ball, h is the height of this segment, R is the ball radius, and the average diameter D of the wear spot is measured through a microscope with four measurements with an angle of 45° between the measurements.

One example of typical wear spots after 105 impacts is shown in Fig 3.

Figure 3. Typical wear spots after 105 impacts

Finally a dimensionless wear number S is defined as

$$S = \frac{V \text{ with actual test oil}}{V \text{ at dry condition}} \qquad (3)$$

In the wear tests at 22°C the only difference apart from the temperature, compared to the tests at 50°C, is that the normal velocity has to be somewhat increased. The reason for a change of the velocity is that the increased viscosity, due to the decreased temperature, will give a more damped behaviour. Thus an increased normal velocity is needed in order to maintain the force and the pressure level. The normal velocity is now $v_n = 0.16 \pm 0.03$ m/s, giving normal force $F_n = 435 \pm 40$ N and pressure $p = 2000 \pm 30$ N/mm^2. One example of this condition can be seen in Fig 4. This figure shows the normal force F_n(A) and the electrical breakdown (B) as a function of time t. The fact that there is an electric current through the ball and the roller during almost the whole contact time indicates that the oil collapse is well established and thus a high amount of wear is developing.

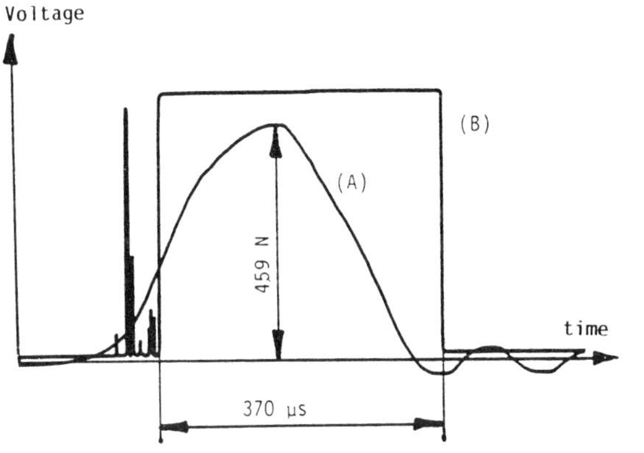

Figure 4. Typical impact

The procedure in the oil collapse tests was to step-by-step and at constant normal velocity approach the sliding velocity giving the first sign of asperity contact, like the peak on curve (A) shown in Fig 5.

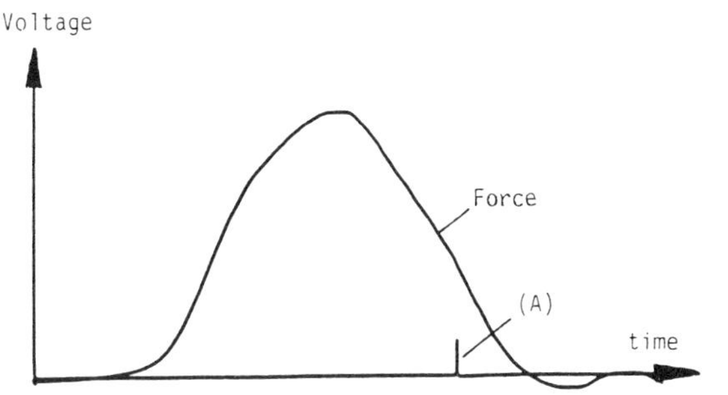

Figure 5. Typical case in oil collapse test

Table 1.
Test lubricant types

Lubricant	Type	Viscosity [c St] 22°C	50°C
1	Polyalphaolefine, ester	428	119
2	Mineral, HWI	182	57
3	Polyalphaolefine , ester	247	66
4	Polyalphaolefine, ester	276	87
5	Polyalphaolefine, ester	300	78
6	Mineral, conventional	425	84
7	Mineral	282	76
8	Mineral	458	89
9	Polyalphaolefine	325	80

This phenomenon is completely described in the works of Lundberg [3]. The little peak (A) in Fig 5 indicates asperity contact and thus the oil film is equal to zero somewhere in the contact. The normal velocity for both temperature cases is $v_n = 0.15 \pm 0.01$ m/s, normal force $F_n = 413 \pm 50$ N, giving pressure p = 2000 ± 90 N/mm^2.

4. OILS TESTED

Nine oils of different types have been tested, see Table 1.

5. TEST RESULTS AND DISCUSSIONS

The limit sliding velocity v_{sl} is presented as a function of the viscosity, with the temperature as a parameter in Fig 6 for the polyalphaolefins and in Fig 7 for the mineral oils.

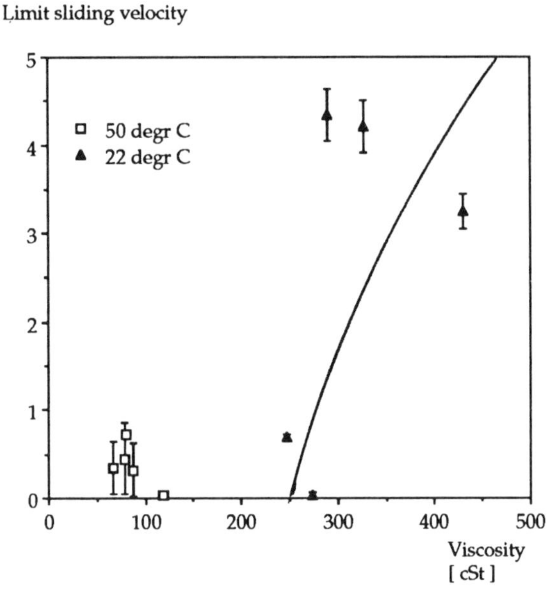

Figure 6 Limit slide velocity v_{sl} versus viscosity ν for polyalphaolefins

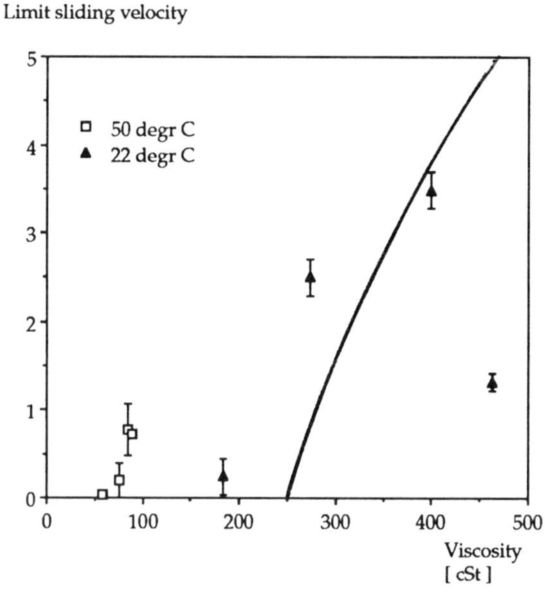

Figure 7 Limit slide velocity v_{sl} versus viscosity ν for mineral oils

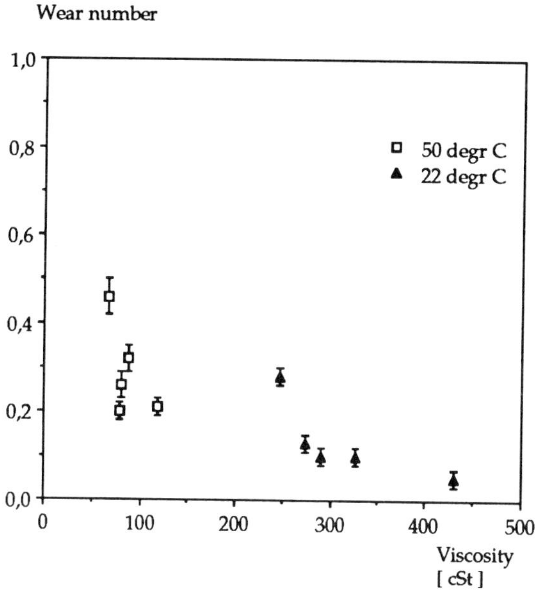

Figure 8 Wear number S versus viscosity ν for polyalphaolefins

In Figs 6 and 7 there is also a reference curve presented. This curve is the result of measurements of the limit slide velocity for different polyalphaolefins without any EP-additives. The reference measurements are, except for surface roughness, made with similiar conditions and are completely described in ref [3].

The intersection point between the reference curve and the ν-axis depends on the surface roughness for both the ball and the ring in each test. It is assumed that the shape of the curve does not change due to surface roughness.

The wear number S is presented as a function of the viscosity with the temperature as a parameter in Fig 8 for the polyalphaolefins, and in Fig 9 for the mineral oils.

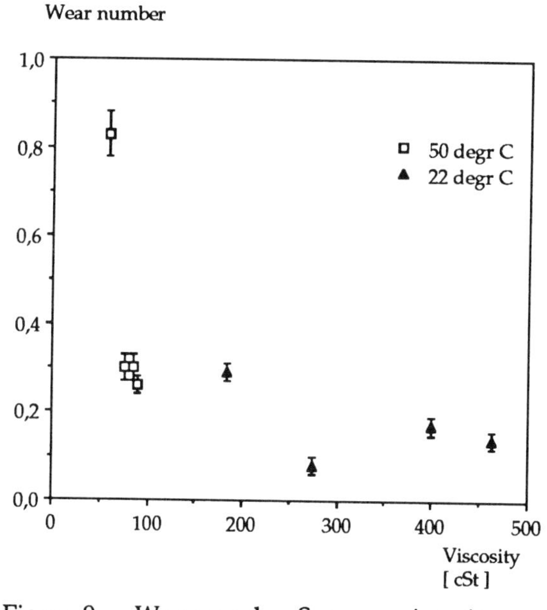

Figure 9 Wear number S versus viscosity ν for mineral oils

While the experiments in ref [3] show a clear relation between the limit slide velocity and the viscosity, if the oils have the same additives, the relation between the wear number and the viscosity, with the same additives, is also expected to be clear. The wear number decreases with increasing viscosity, and correspondingly the limit slide velocity increases with increasing viscosity.

For the lower temperature it can be seen that the wear number shows small dependence on the viscosity, $0.05 < S < 0.31$, over a rather wide range of viscosity, 180 cSt < ν < 470 cSt. In comparision with this it can be seen from the higher temperature that the wear number differs between $0.20 < S < 0.81$ in a small range of viscosity, 60 cSt < ν < 90 cSt.

6. CONCLUSIONS

From the test results the following conclusions can be drawn.

1. For both the polyalphaolefins and mineral oils at the higher temperature, the limit slide velocity v_{sl} as a function of viscosity is similiar and almost constant.

2. A comparison between the reference curve measured at 22 °C for a polyalphaolefin and the tested polyalphaolefins shows that 2 of 5 have a high limit slide velocity. This is due to the fact that the reference curve is valid for polyalphaolefins without any EP-additives, while the tested polyalphaolefins contain different additives.

3. A comparison between the reference curve and the tested mineral oils shows that 1-2 mineral oils have a lower limit slide velocity than the reference curve. This can, apart from the additives, also depend on the base oil.

4. For both the polyalphaolefins and the mineral oils the wear is decreased when the viscosity is increased.

5. Considering wear there is a similiar viscosity dependence for both the mineral oils and the polyalfaolefins.

7. ACCURACY OF THE METHOD

The accuracy of the method is illustrated in Figs 6-9 by means of the staples. The width in S-direction is due to the differences between each measurement of the diameter of the wear spots.

The width in v_{sl}-direction is not due to difficulties in measuring the velocity but difficulties in changing the velocity with proper steps.

The accuracy of the viscosity is dependent on the measurements of the oil bulk temperature. The errors in the ν-direction are represented by the symbols in the diagram. No consideration is maid to internal heating in the contact zone.

8. SURFACE ROUGHNESS

Surface roughness is very important both in order to reduce the wear and to get an oil film which does not collapse. In order to obtain knowledge about the actual surface, surface profiles have been measured using a stylus instrument. In Table 2 some standard statistical classifications of the surfaces are given. The values are averages from five measurements.

In the wear tests, the roller is also damaged because of the high amount of wear. The apparatus makes it possible to adjust the position of the roller side ways and to adjust a new angular position of the ball. This enables new undamaged surfaces for each oil. However, the width of the roller can only manage 13 oils, therefore the roller has to be machined between the two test temperatures 50°C and 22°C.

The oil collapse tests needed two levels of surface roughness on the ball and roller because of the fact that the higher temperature 50°C decreased the viscosity of the test oils so much that extreme polishing of the surfaces had to be done in order to get readable measurement values. On the other hand, in the 22°C test for some of the oils, the viscosity was so much increased that an oil collapse test was never achieved, even for very high sliding velocities. Therefore a more rough surface had to be manufactured for this temperature.

Table 2.
Surface parameters

Test/object		Wear test		Wear test		Oil collapse test		Oil collapse test	
Temperature		50°C		22°C		50°C		22°C	
Object		Roller	Ball	Roller	Ball	Roller	Ball	Roller	Ball
LT	[mm]	5.60	5.60	5.60	5.60	0.56	0.56	0.56	0.56
LM	[mm]	4.00	4.00	4.00	4.00	0.40	0.40	0.40	0.40
LC	[mm]	0.80	0.80	0.80	0.80	0.08	0.08	0.08	0.08
R_{max}	[μm]	6.71	7.74	7.93	7.74	0.054	0.082	0.284	0.508
R_z	[μm]	5.13	6.18	5.34	6.18	0.041	0.056	0.211	0.370
R_p	[μm]	2.84	5.15	3.28	5.15	0.025	0.031	0.095	0.266
R_{3z}	[μm]	3.18	-	3.32	-	0.026	0.035	0.133	0.249
Ra	[μm]	0.90	1.72	0.84	1.72	0.008	0.007	0.028	0.057
Rt	[μm]	6.94	8.47	8.43	8.47	0.058	0.086	0.302	0.553

9. ACKNOWLEDGEMENTS

The authors would like to thank Professor Erik Höglund for fruitful discussions and Mrs Elisabeth Nyström for her excellent work in typing this paper.

10. REFERENCES

1. J. Lundberg, E. Höglund and S-I Bergström: "Apparatus for squeeze/sliding lubrication studies". Proceedings of Nordtrib 90, Hirtshals, Denmark, June 1990, pp 83-94.

2. B. Hamrock and D. Dowson: "Ball bearing lubrication" John Wiley & Sons, 1981.

3. J.Lundberg: "Viscosity dependence on squeeze/sliding lubrication". Accepted for publication in Wear WEA 5997.

Thin Films in Tribology / D. Dowson et al. (Editors)
1993 Elsevier Science Publishers B.V.

Scuffing as a random phenomenon

T.A.Stolarski

Department of Mechanical Engineering,
Brunel University,
Uxbridge, Middlesex, UB8 3PH, United Kingdom

A model of scuffing phenomenon in lubricated contacts is presented in terms of a random process. The model identifies conditions that produce boundary or mixed lubrication with asperity contacts for which the plasticity index is greater than 0.6. It then imposes, as a necessary prerequisite for scuffing, the condition that the contact temperature is high enough to cause significant desorption of adsorbed lubricant film. For such conditions determination of the probability of asperity contact and of plastic asperity contact provide upper and lower bounds respectively to the probability of scuffing.

NOTATION

E effective elastic modulus
E_c heat of adsorption of the lubricant
μ coefficient of friction
h the mean thickness of the film between rough surfaces
h_0 film thickness with smooth surfaces
M molecular weight of the lubricant
Y flow hardness of the softer material
r average radius of curvature at peak of asperity
R gas constant
T_c critical temperature for a given lubricant material combination
T_s surface temperature at contact
V relative velocity of contacting surfaces
W load on the contact
λ specific film thickness
ψ plasticity index

1. INTRODUCTION

Scuffing originates primarily from an increase in the level of friction between contacting surfaces in relative motion which in turn is associated with a lack of, or a deterioration in the conditions of lubrication at those surfaces. The definition of scuffing reflects the fact that local welds - the evidence of scuffing - occur between rubbing surfaces and lead to gross damage of them. It is obvious that for metallic surfaces to weld together the films on the surface of at least one of them must become disrupted and direct metal-metal contact must occur.

The onset of scuffing depends upon the interplay between a number of variables. These include materials, load on contact, level of friction, lubrication, velocity, surface roughness and temperature within the contact. A number of models and scuffing hypotheses have been proposed and all of them treat the problem in deterministic way. Blok [1] advanced the hypothesis that the factor controlling the breakdown of lubrication is the local temperature within the contact which consists of two components The first component is the bulk temperature of the metal parts at positions remote from the contact. The second component is the instantaneous temperature rise as the surface passes through the heat source represented by the contact. It is postulated that lubrication will break down and scuffing process will be initiated if the actual contact temperature exceeds a certain critical value. As the critical temperature depends on a number of parameters characterizing the materials in contact and the lubricating medium it is understandable that it cannot be predicted in a deterministic way but rather in the form of a probability that it will assume a certain value.

The main assumption of the model proposed by Dyson *et al.* [2] is that scuffing can be avoided if a very thin film of lubricant of very high viscosity is separating pairs of interacting asperities. This high viscosity, however, cannot be provided effectively by the local microelastohydrodynamic interactions between asperities but only by the high pressures resulting from the operation of the macroelastohydrodynamic system. Again, the model is a deterministic one and does not take into account the inherent variability of parameters involved and neglects the chemical effects at the metal-lubricant interface.

Recognizing the complexity of the scuffing process and its dependence on a number of factors varying in a random way it is postulated in this paper that a probabilistic approach to scuffing is adopted.

2. THE MODEL

A number of steps are required to determine the probability of scuffing. They can be presented in the form of a logical sequence of steps and easily translated into computer code. The complete rationale for the system of analysis is developed in the subsequent sections of the paper.

2.1. Contact lubrication regimes

The first step in the analysis is to determine the lubrication regime within the contact. This can be done by examining the values of the viscosity parameter, g_v and the elasticity parameter, g_e [3],

$$g_v = \frac{\alpha w}{R}\left(\frac{w}{\eta_o V}\right)^{1/2} \tag{1}$$

$$g_e = \left(\frac{w}{ER}\right)^{1/2}\left(\frac{w}{\eta_o V}\right)^{1/2} \tag{2}$$

where α is the pressure-viscosity coefficient, R represents equivalent radius of curvature of contacting surfaces, w is the normal load per unit length of contact, and η_0 is absolute viscosity of lubricant. Other symbols used in equations (1) and (2) are defined in the notation.

For hydrodynamic lubrication to take place the following must be fulfilled:

$g_v < 1.5$ and $g_e < 0.6$

Operating conditions outside the above limits for g_v and g_e usually mean elastohydrodynamic lubrication (ehd). The practical ehd lubrication is limited by another pair of parameters, that is the speed parameter, g_s and the load parameter g_l;

$$g_s = \alpha\left(\frac{E^3 \eta_o V}{R}\right)^{1/4} \tag{3}$$

$$g_l = \alpha\left(\frac{wE}{2\pi R}\right)^{1/2} \tag{4}$$

For ehd lubrication to occur the following must be satisfied,

$1.8 < g_s < 100$ $\quad$ $1.0 < g_l < 100$

The next step in the procedure involves the estimation of the lubricating film thickness. This can be done by using appropriate expression available from reference [4, 5, 6]. Thickness of the film calculated for the contact between smooth surfaces must be related to the topography of engineering surfaces in order to determine the functional lubrication regime. Usually, the specific film thickness, λ, [7] is used for that purpose.

The value of specific film thickness is an important indicator. If $\lambda>3$, the probability of metal-metal asperity contact is very low and for practical reasons no further analysis is required. However, if $\lambda<1$, the operating conditions are characteristic for boundary lubrication and the mode of asperity contact ought to be examined. To find out the mode of asperity contact it is necessary to calculate the plasticity index [8],

$$\psi = \frac{E}{Y}(\frac{\sigma_p}{r})^{1/2} \tag{5}$$

When ψ <0.6, the contact between the asperities will be elastic and when ψ>1.0, the contact will be partially plastic even at the lightest loads. The contact mode is mixed when ψ is in the range between 0.6 and 1.0. The case of ψ<0.6 is unimportant from the probability of scuffing point of view, but when ψ>0.6 the probability of scuffing is significant and should be examined by comparing the surface contact temperature and the characteristic lubricant temperature.

2.2. Thermal effects

The surface contact temperature, T_s, can be estimated using one of a number of published models [9,10]. According to Bowden and Tabor [9],

$$T_s = T_o + \Delta T \tag{6}$$

where T_0 is the bulk temperature and ΔT is given by,

$$\Delta T = 0.443\frac{g\mu V(WY)^{1/2}}{J(k_1+k_2)} \tag{7}$$

where k_1 and k_2 are thermal conductivities of materials in contact, J is the mechanical equivalent of heat and g is the gravitational constant.

The characteristic lubricant temperature, T_{cr} can be estimated for a given combination of materials and lubricant. The expression for characteristic temperature is in the form [11],

$$T_{cr} = \frac{3(\frac{E_c}{RT_{cr}}-2)-\{5(\frac{E_c}{RT_{cr}})^2-12(\frac{E_c}{RT_{cr}})+12\}^{1/2}}{\frac{30.9x10^5}{V}(\frac{T_m}{M})^{1/2}\frac{E_c}{RT_{cr}^2}\exp(-\frac{E_c}{RT_{cr}})} \quad \text{............} \tag{8}$$

where T_m is melting point of the lubricant, and M is its molecular weight.

If the surface contact temperature is greater than the characteristic lubricant temperature, the probability of scuffing is high. If, however, $T_s < T_{cr}$ scuffing is practically not probable.

2.3. Probability of asperity contact

It is essential for further analysis to determine the probability of asperity contact for a given film thickness. If h_1 and h_2 represent random heights of asperities of contacting surfaces and h is the film thickness, then the clearance is given by,

$$\Delta h = h - h_1 - h_2 \tag{9}$$

Therefore, the probability that Δh is negative is synonymous with the probability of asperity contact and is given by,

$$P(\Delta h \le 0) = \frac{1}{(2\pi)^{1/2}\sigma_*}\int_0^\infty \exp\{\frac{-(\bar{h}-\Delta h)^2}{2\sigma_*}\}d(-\Delta h) \quad \text{...........} \tag{10}$$

In the foregoing, $\bar{h}$ is the mean value of the separation between contacting rough surfaces and

$$\sigma^* = (\sigma_{p1}^2 + \sigma_{p2}^2)^{1/2}$$

is the standard deviation of peak heights.

The contact between asperities must be further investigated in order to determine the probability of plastic contact. It is given by the expression,

$$P(\Delta h \le \delta_p) = \frac{1}{(2\pi)^{1/2}\sigma_*}\int_0^{\infty} \exp\{\frac{-(\bar{h}-\Delta h)^2}{2\sigma_*^2}\} d(-\Delta h)$$

............(11)

where δ_p is derived from the concept of plasticity index and is given by,

$$\delta_p = \frac{0.89\sigma_*}{\psi^2} \qquad (12)$$

2.4. Probability of scuffing

It is important to remember that probability of asperity contact for a given surface finish and other contact conditions does not mean imminent scuffing, although both these phenomena are related through the expression,

$$P(\text{scuffing}) < P(\text{asperity contact}) \qquad (13)$$

Probability of scuffing is mainly defined by two parameters, i.e. the plasticity index ψ and the temperature within the contact. The necessary condition for the scuffing is fulfilled when $\psi>0.6$, which means plastic asperity contact. The sufficient condition for scuffing is when $T_s > T_{cr}$. It is only when both conditions are simultaneously met that the scuffing ought to be seriously considered and then,

$$P(\text{pl.a.contact}) < P(\text{scuffing}) < P(\text{a.contact})$$

.... (14)

In the approach presented in this paper, the probability of scuffing is given not by a precise number but in the form of a range of values provided by upper and lower bounds defined by the probability of plastic asperity contact and the probability of asperity contact.

3. A SPECIMEN OF RESULTS

A computer code was developed in order to test the prediction of the model. The programme calculates the probability of asperity contact, probability of asperity plastic contact, critical temperature and surface temperature. These four parameters are required to estimate the probability of scuffing.

A standard set of input data used is given below:

load on contact = 100 N; length of contact = 0.01 m; relative velocity = 0.5 m/s; relative radius of contacting surfaces = 0.005 m; viscosity = $0.247\text{x}10^{-1}$ Ns/m^2; pressure-viscosity coefficient = $1.8\text{x}10^{-8}$ m^2/N; radius of asperity peak = $83\text{x}10^{-6}$ m; surface roughness (c.l.a.) = $16\text{x}10^{-8}$ m; elastic modulus = $2.1\text{x}10^{11}$ N/m^2; flow hardness of the softer material in contact = $22\text{x}10^8$ N/m^2; molecular weight of lubricant = 226 kg/kmol; heat of adsorption of lubricant = 48976 kJ/kmol.

Only two parameters influencing scuffing were considered, that is velocity of relative motion and heat of adsorption of a lubricant. Figure 1 shows the probability of scuffing of a fictitious line contact in a function of relative velocity of contacting surfaces.

The probability of asperity contact is decreasing with the increase in velocity of relative motion and thus reducing the danger of scuffing (Figure 1). This is due to the increase in film thickness separating the contacting surfaces with velocity.

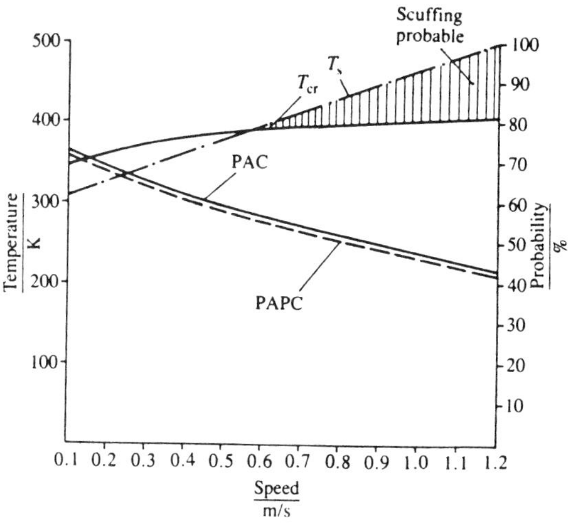

Figure 1. Probability of scuffing of line contact in a function of relative velocity of contacting surfaces.

Figure 2 shows the influence of the heat of adsorption of a lubricant on the probability of scuffing.

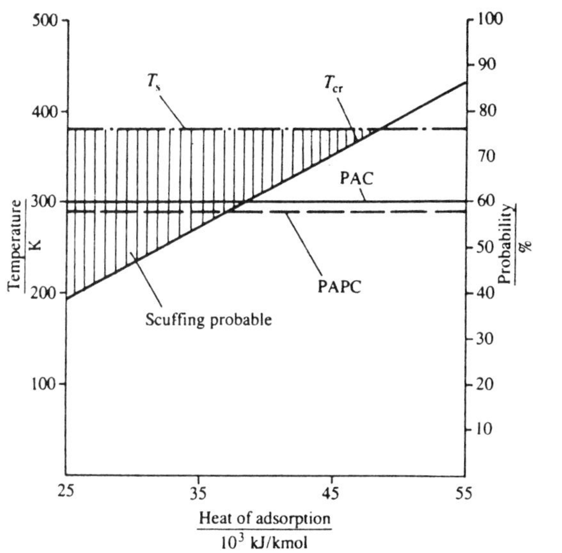

Figure 2. Probability of scuffing of a line contact in a function of heat of adsorption.

The influence of the heat of adsorption is mainly manifested through the critical temperature which, in turn, crucially depends on E_c. It is seen from Figure 2 that at low values of E_c the model predicts scuffing with a probability of 58-60 per cent. When E_c increases the danger of scuffing recedes because T_{cr} becomes greater than T_s. In other words, although the necessary condition for scuffing to occur is still fulfilled (asperity contact and $\psi>0.6$) the sufficient condition ($T_s> T_{cr}$) is not.

4. CONCLUDING COMMENTS

This paper presents an attempt to quantify a complex phenomenon called scuffing. The central feature of the model proposed is that the probability of scuffing is assessed only when two conditions are simultaneously fulfilled. The contact of surface asperities is regarded as a necessary condition and does not point to an imminent scuffing. Provided that the surface temperature does not exceed the critical temperature, direct metal-metal contact between asperities is prevented by the adsorbed boundary film of a lubricant.
A number of input data not readily available are required by the model proposed. It is thus apparent that further progress in mathematical modelling of a number of tribological problems depends crucially on the availability of information characterizing the system.

REFERENCES

1. H.Blok, The postulate about the constancy of scoring temperature, NASA SP-237,1970
2. R.W.Snidle, S.D.Rossides and A.Dyson, Proc.R.Soc., A395 (1984) 291.
3. K.L.Johnson, J.Mech.Engng Sci., 12 (1970) 9.
4. H.M.Martin, Engineering, 11 (1916) 119.

Thin Films in Tribology / D. Dowson et al. (Editors)
1993 Elsevier Science Publishers B.V.

Wear of Metals in a Magnetic Field in Boundary Lubrication

K. Hiratsuka

Tokyo Institute of Technology
2–12–1, Ookayama, Meguro–ku,
Tokyo, 152 JAPAN

Wear tests between S45C steel pin and S45C steel disk have been conducted in dry air and in 8 kinds of normal alkane (hexane, heptane, octane, nonane, decane, undecane, tridecane and tetradecane) and poly α–olefin with and without the magnetic field applied vertically to the rubbing surface. In all the lubricants except PAO, the magnetic field accelerated severe–mild wear transition, which resulted in a reduction in coefficient of friction in most lubricants and in a great reduction in total wear (1/10–1/1000) in all the lubricants. These phenomena are supposed to be due to the chemisorption of oxygen or oxidation activity enhanced by the magnetic field.

1. Introduction

The easiest way to magnetize a steel is to rub the magnet against it. At this point, it can be said that tribology and magnetism have been related for a long time. However, the effect of magnetic field on friction and wear has not been fully understood. This is because, depending on the sliding conditions such as rubbing material, sliding velocity and so on, wear was reported to be increased or decreased by the application of a magnetic field[1–5].

The author has also studied the effect of the magnetic field on the wear of metals in dry condition. Fig.1 shows the wear–sliding distance curve for the rubbing between Fe pin and Fe disk[6]. The magnetic field applied vertically to the rubbing surface (Hv) accelerated a severe–mild wear transition, which resulted in a great reduction in total wear. The same phenomenon was also found in S45C steel vs. S45C steel and Ni vs. Ni

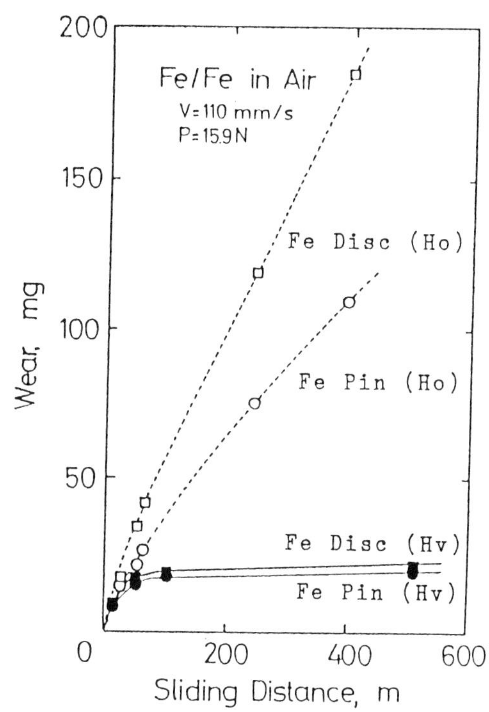

Fig.1 Effect of magnetic field on wear of Fe vs. Fe in air

rubbings[7].

These experiments were carried out in an air atmosphere. In order to apply this magnetic effect to the ordinary rubbing systems, metals in a magnetic field should be rubbed in boundary lubricated condition. As a first step, in the present study, wear tests between S45C steel and S45C steel were carried out without the lubricants and in 8 kinds of normal alkane and poly α-olefin in a magnetic field applied normally to the rubbing surface.

2. Experimental Procedure

The wear test was carried out using pin-on-disk type test rig. As shown in Fig.2, the lower portion of the disk specimen was dipped in a lubricant. By the rotation of disk, it was supplied to the friction interface due to the viscosity of liquid. The lubricants were 8 kinds of normal saturated hydrocarbon; hexane, heptane, octane, nonane, decane, undecane, tridecane, tetradecane and poly α-olefin (Kinetic viscosity is 5.7mm^2/s at 40°C and 29.9mm^2/s at 100°C. Viscosity index is 136.7). Both pin and disk specimen were S45C steel. Chemical composition is shown in Table1. The magnetic field normal to the rubbing surface was applied by attaching the permanent magnet(Nd-Fe-B alloy) on the other side of the disk. The outer diameter and mean friction diameter of disk was 25mm and 20mm, respectively. The pin diameter was 4mm. The outer and inner diameter of magnet was 25mm and 18mm, respectively. The thickness of it was 4mm. The magnetic field at the friction track of disk without the contact of pin specimen was from 400 to 500 gauss from the inner to the outer friction track. A magnetic attractive force of 6.9N operated between the contacted pin and disk. The load by the dead weight of 9.8N was also provided, so that the total load with a magnet was 16.7N. Without magnetic field, the load was also 16.7N by only dead weight. Sliding velocity was 100mm/s, and sliding distance was 360m. The surface of both pin and disk were polished by #1000 emery paper. The disk surface was flat. Whereas, the pin surface has a little curvature to prevent an edge contact. The friction force was detected by strain gauges. The displacement of pin position to the disk was also detected by non-contacting displacement sensor. This signal reflected how much pin and disk was worn and which wear mode was dominant, severe or mild. The shape, size and color of wear particles were analyzed by Ferrography.

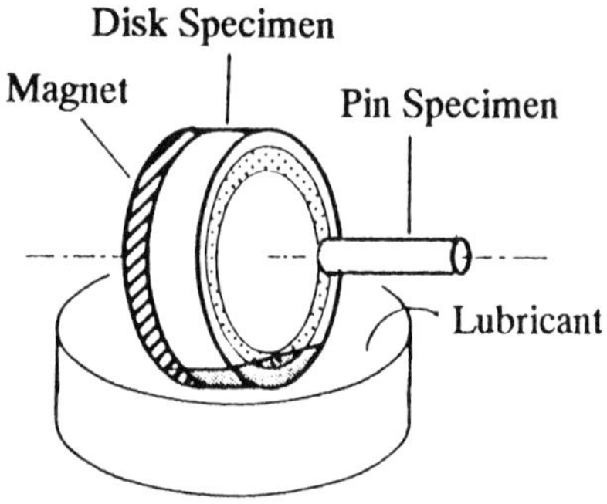

Fig.2 Schematic of wear test rig

Table1 Chemical composition of S45C steel

C	Si	Mn	P	S	Cr	Fe
0.45	0.23	0.73	0.026	0.014	0.10	Balance

3. Results

Fig. 3 shows the variation of the coefficient of friction and vertical displacement of the pin during the rubbing in hexane as a function of sliding distance, (a);without, and (b);with the magnetic field. Without the magnetic field, the

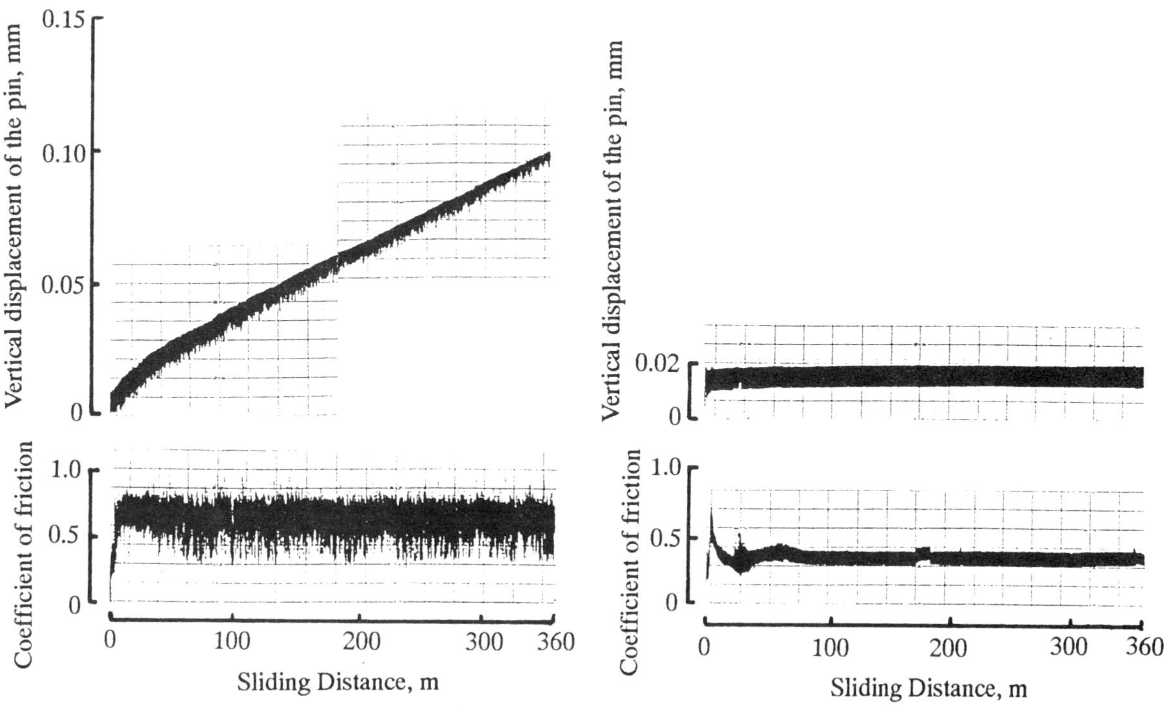

(a) without magnetic field (b) with magnetic field

Fig.3 Coefficient of friction and vertical displacement of the pin rubbed in hexane

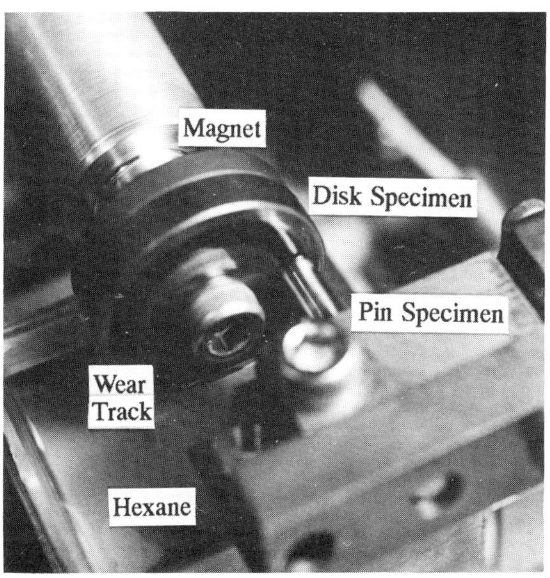

Fig.4 Contact point of pin and disk during rubbing in hexane in a magnetic field

coefficient of friction fluctuated very much and the mean value was almost constant throughout the test. Large displacement of the pin was observed which was due to the high rate of wear. By the application of a magnetic field, the coefficient of friction varied from high value (which was almost the same as without the magnetic field) to the low and stable value with less fluctuation than without the magnetic field.

Fig.4 is the photo of the pin and disk lubricated by hexane in a magnetic field at a sliding distance of about 255m. Hexane is observed around pin edge. The rubbing surface of disk turned to brown, which is a characteristic of mild wear. The wear particles are on and out of the friction track and showed the same color. From this observation, it is understood that the low value of coefficient of friction was due to the mild wear.

The friction surfaces of disk and the wear particles produced in heptane and tetradecane with and without the magnetic field ,which were analysed by Ferrography, are compared in Figs.5 and 6, respectively. When lubricated by hydrocarbons besides PAO, the surface and the wear particles showed metallic luster without magnetic field. Whereas with magnetic field, the surface and the wear particles were black or brown. This difference evidently shows that the wear mode changes from severe to mild by the magnetic field even if metals are lubricated.

Fig.7 shows the friction surfaces of disk produced by the rubbing in PAO with and without the magnetic field. Both surfaces showed black mild wear surfaces, with wider track in a magnetic field.

Fig.8 shows the coefficient of friction at the end of sliding with and without the magnetic field as a function of the number of carbon atoms of lubricant Alkane. (The values of dry friction are also plotted at the carbon number to

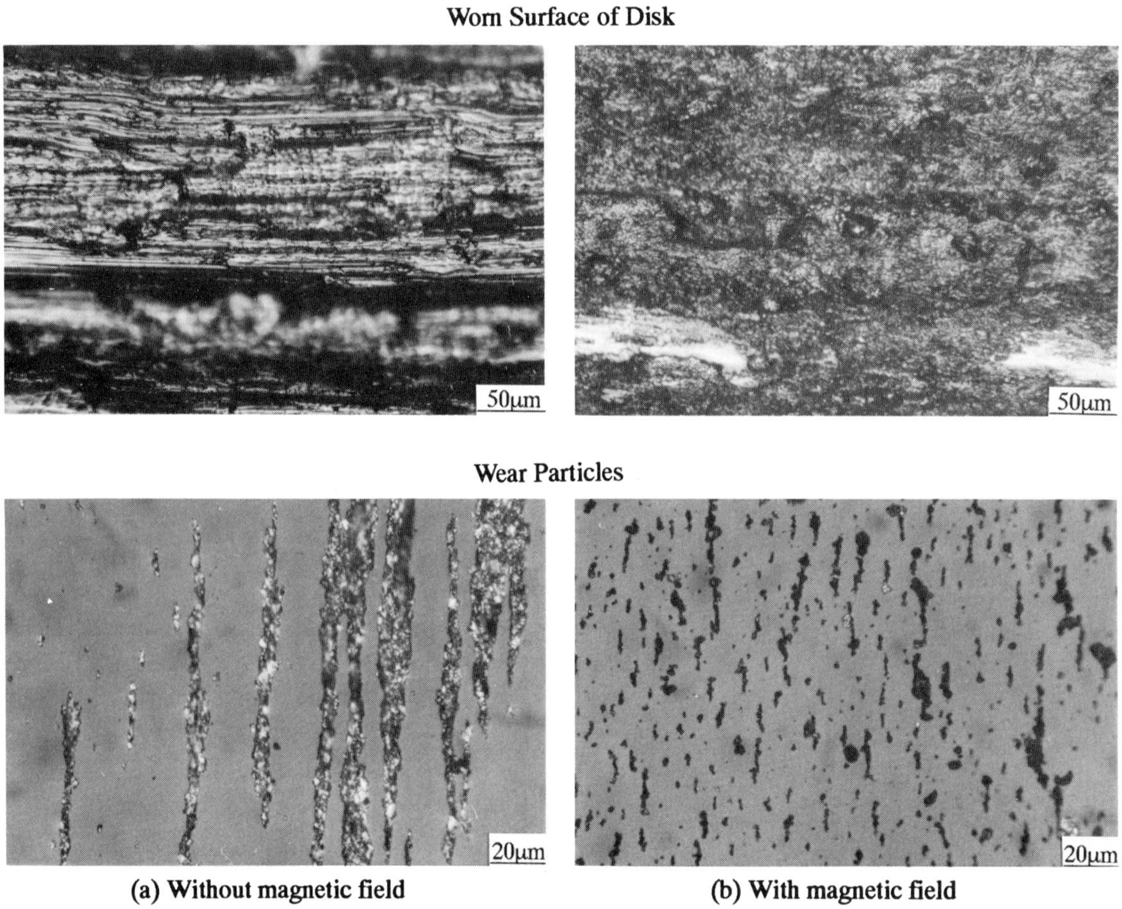

Fig.5 Worn surface of disk and wear particles produced in heptane with and without the magnetic field

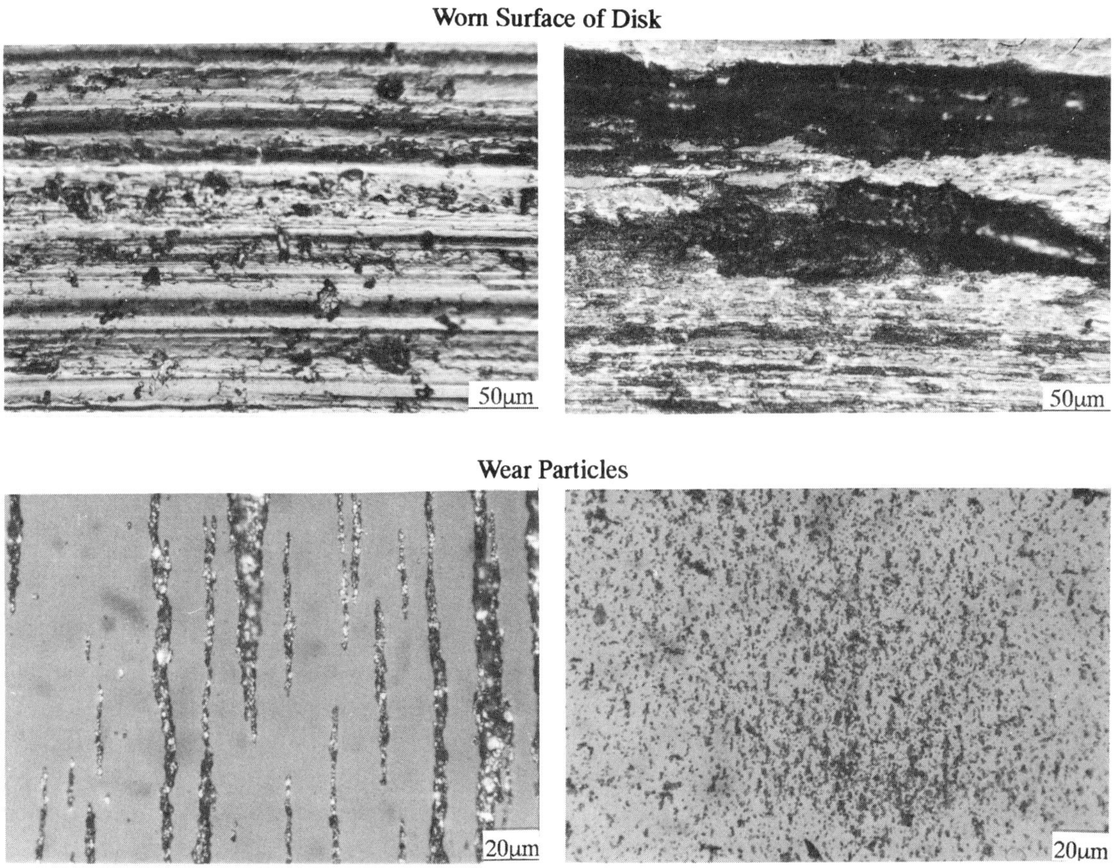

(a) Without magnetic field (b) With magnetic field

Fig.6 Worn surface of disk and wear particles produced in tetradecane with and without the magnetic field

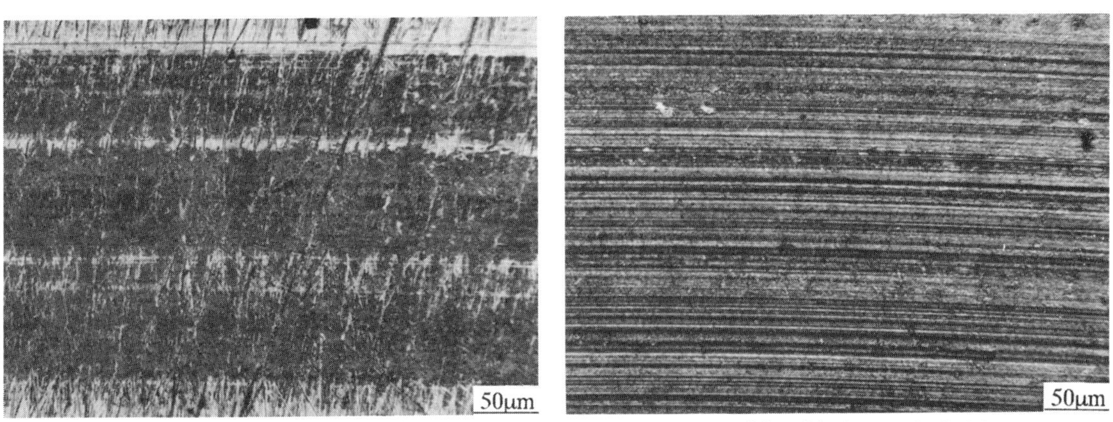

(a) Without magnetic field (b) With magnetic field

Fig.7 Worn surface of disk produced in PAO with and without the magnetic field

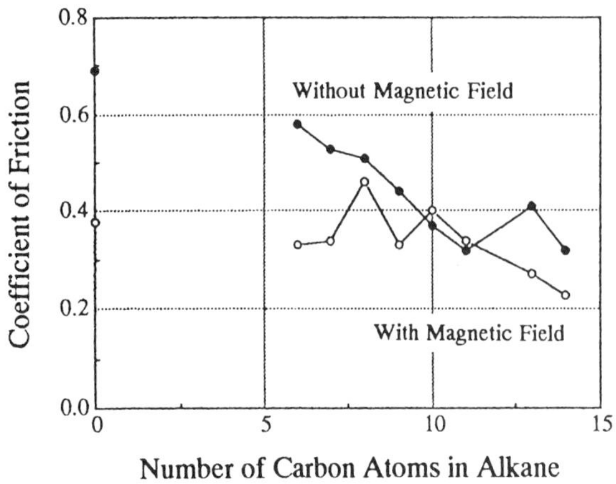

Fig.8 Coefficient of friction as a function of the number of carbon atoms of lubricant alkane with and without magnetic field

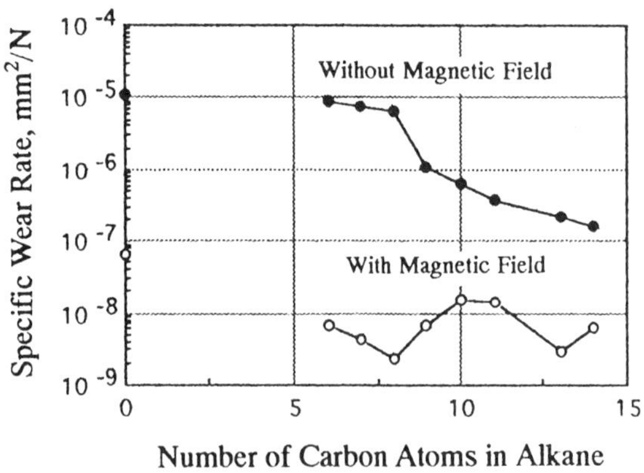

Fig.9 Specific wear rate as a function of the number of carbon atoms of lubricant alkane with and without magnetic field

be zero.) Generally, a coefficient of friction was decreased by the magnetic field. This is due to the transition to the mild wear. Without magnetic field, as the number of carbon atom decreased, the coefficient of friction increased, because hydrocarbons with lower carbon numbers had poor lubricity due to lower viscosity. Thus the coefficient of friction between presence and absence of a magnetic field differs more as the carbon number

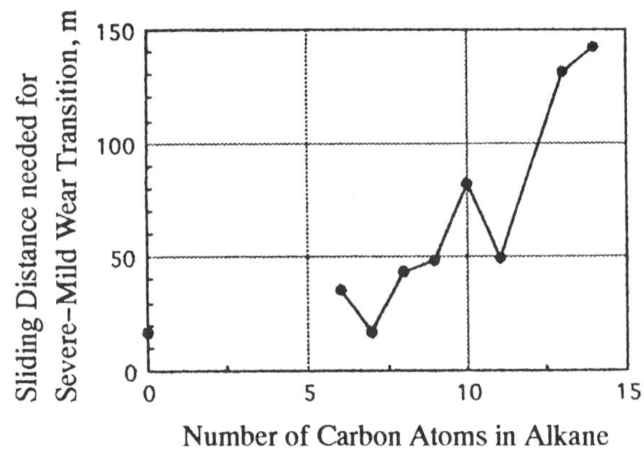

Fig.10 Sliding distance needed for severe–mild wear transition as a function of the number of carbon atoms of lubricant alkane with magnetic field

decreases. However, it is to be noticed that in some cases, the coefficient of friction was higher with the magnetic field. In PAO, the coefficient of friction with or without the magnetic field was 0.11 and 0.08, respectively.

Fig.9 shows the magnetic effect on specific wear rate in various lubricants. (The values of dry friction are also plotted at carbon number to be zero.) Without magnetic field, the specific wear rate decreases as the carbon number increases. This is due to the viscosity increase as the increase in carbon number. However, even when lubricated by tetradecane, the wear mode was still severe. On the contrary, under the magnetic field, the specific wear rate was decreased (1/10–1/100) than that without magnetic field and had almost constant value regardless of the number of carbon atoms of lubricants. This is due to the mild wear. In PAO, wear was too low to measure by electric balance both with and without the magnetic field.

The sliding distance needed for the severe–mild wear transition is summarized in Fig.10. As the number of carbon atom increases, it becomes longer.

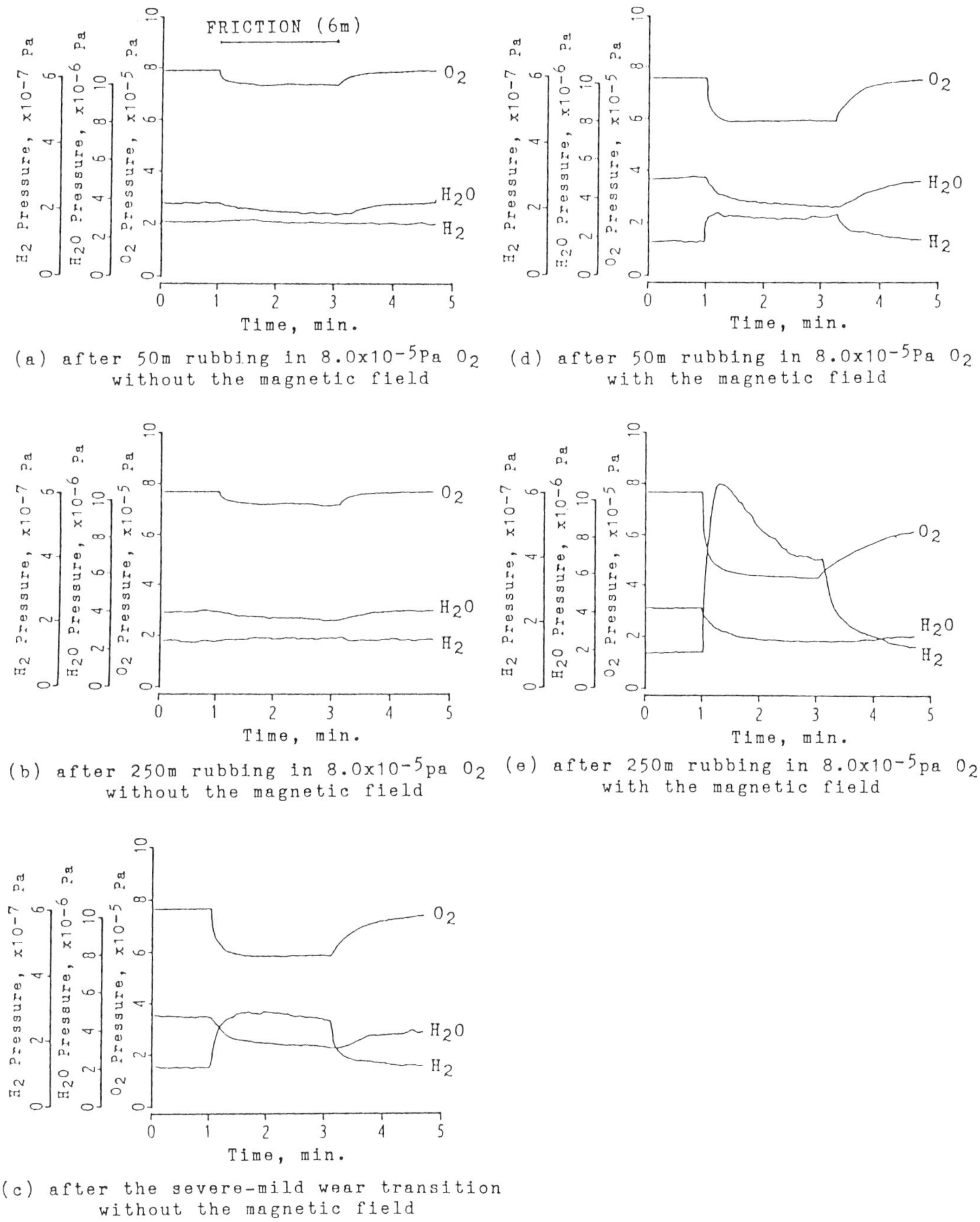

Fig.11 Magnetic effect on chemisorption of oxygen. (Fe/Fe, Sliding Velocity=52mm/s, Load=6.9N with the magnetic field (magnetic attractive force=5.9N), Load=6.9N without the magnetic field)

4. Discussion

In most of the lubricants, both the coefficient of friction and the wear were reduced by the magnetic field associated with mild wear. From the color of worn surface and wear particles, they are expected to be oxidized to Fe_3O_4 and $\alpha-Fe_2O_3$. The author demonstrated that the chemisorption of oxygen during friction was enhanced by the magnetic field in a very low oxygen pressure[8]. As shown in Fig.11, the pressure change of oxygen is greater with the magnetic field than without it. The mechanism of the acceleration of severe-mild wear transition in lubricants by the magnetic field seems to be the same; the chemisorption of oxygen dissolved in lubricants and the subsequent oxidation is enhanced by the magnetic field.

To accomplish the mild wear, the contact of two solids is needed. The reason for that the longer sliding distance is needed for the severe-mild wear transition in longer chain of alkane is the inhibition of solid contact due to the larger viscosity of longer alkane. Without the magnetic field, the lower coefficient of friction was obtained in longer alkane. This is also the result of the inhibition of solid contact associated with the lubrication by the longer alkane.

As shown in Fig.9, the specific wear rate in a magnetic field did not change very much. When lubricated by alkane with smaller carbon number, the reducing effect on wear by magnetic field is dominant. Whereas when lubricated by longer alkane, the reducing effect by viscosity is dominant. These two effects operated, which resulted in almost the same specific wear rate obtained in a magnetic field. It can also be said that the amount of severe wear needed for the severe-mild wear transition is almost the same regardless of the chain length of alkane.

5. Conclusions

S45C steel was rubbed against S45C steel without lubricants and in 8 kinds of normal alkane (hexane, heptane, octane, nonane, decane, undecane, tridecane, tetradecane) and poly α-olefin in a magnetic field applied normally to the rubbing surface. The following conclusions are dawn.

In most lubricants, the coefficient of friction was reduced by the magnetic field with the appearance of mild wear. In all the lubricants, the wear was greatly reduced associated with a mild wear (1/10–1/1000) by the magnetic field. Mild wear is supposed to be brought about by the chemisorption of oxygen and the oxidation being enhanced by the magnetic field in boundary lubricated condition.

References

1. D.K.Pal and N.C.Gupta, J. Inst. Eng. India, 53 (1973) 195.
2. V.Raghuram and M.K.Muju, Int. J. Mach. Tool Des. Res., 20 (1980) 87.
3. M.K.Muju and A.Ghosh, Wear, 41 (1977) 103.
4. M.K.Muju and A.Radhakrishna, Wear, 58 (1980) 49.
5. V.Ramakrishnan and A.Ghosh, Wear, 69 (1981) 71.
6. K.Hiratsuka, H.Saito and T.Sasada, Proc. 5th Intern. Cong. Tribology (EUROTRIB89) 2 (1989) 90.
7. K.Hiratsuka, T.Sasada and S.Norose, Wear 110 (1986) 251.
8. T.Sasada, K.Hiratsuka and H.Saito, Wear 135 (1990) 251.

Thin Films in Tribology / D. Dowson et al. (Editors)
© 1993 Elsevier Science Publishers B.V. All rights reserved.

THE BOUNDARIES OF ZDDP LUBRICATION

J.S. Sheasby and T.A. Caughlin

Department of Materials Engineering,
The University of Western Ontario
London, ON, Canada. N6A 5B9.

The antiwear performance of a commercial secondary zinc dialkyldithiophosphate (ZDDP) in base stock was examined over a wide range of conditions in a 4-ball machine. Particular attention was given to correlating the rate of wear with the appearance of wear scars. When the ZDDP was working effectively wear protection was due to the formation of thick glassy films on the rubbing surfaces. The films were transparent, tended to form into mounds, were composed of Zn, P, O, and S, and were essentially free of iron. Optimum wear protection required partial EHL conditions, a combined bulk and frictional temperature between 125-250 °C, and wearing surfaces of sufficient hardness and adhesiveness.

1. INTRODUCTION

Although the boundary lubrication properties of ZDDPs have been studied for many years a definitive mechanism for their action has not been developed. One barrier to progress has been the lack of conformity as to what constitutes effective boundary lubrication both in terms of wear rate and in the nature and appearance of worn surfaces. For instance in the IRG diagram [1] typical boundary lubrication wear rates are given as 5000 μm^3/N.m, whereas according to Rowe [2] boundary lubrication rates are in the range 0.13-13 μm^3/N.m, and for ZDDPs in particular, 0.13-1 μm^3/N.m. Similarly Cavdar and Ludema [3] report that the protective films formed from ZDDP have a dual structure with iron oxides next to the metal and organoiron compounds on top, and are only about 10 nm thick. Whereas Cann, Johnston and Spikes [4], and also Belin et al. [5] report the formation of 0.5 μm thick films of a material similar to an inorganic glass ($ZnO\ .P_2O_5.Fe_2O_3$) in which polyphosphates are linked by bridging Fe(3+) and Zn(2+) ions.

As Bell and Colgan [6] have established that the wear rate in boundary lubrication in an automobile valve train is typically 10 μm^3/N.m, the lower limits of acceptable behaviour were adopted in this work. Previous research in the authors' laboratory [7] has determined that whilst the ZDDP oil blend to be used in this work can readily attain this degree of protection, under other conditions the wear rate can be unacceptably high.

Near optimum wear protection was achieved in the 4-ball machine when operating at 100 °C, 15 kg load, 1200 rpm spindle speed, with standard EP grade 52100 steel balls. This work first explores these conditions further, and then a variety of other conditions, so that good and bad behaviour can be distinguished.

2. EXPERIMENTAL PROCEDURES

1/2" diameter balls were worn in a Shell 4-ball lubricant testing machine basically following the practise of ASTM D4172-82, 1987. AISI E-52100 EP balls, and a load of 147 N (15 kg) were used unless specified otherwise. Times of the tests varied from a few seconds to 3 days.

The base oil was Solvent 150 N (kinematic viscosity 28.31 Cst @ 40 °C) refined from Western Canadian crude. The antiwear agent was a commercial grade secondary zinc dialkyldithiophosphate applied at a treat level of 1.5 wt%.

The resulting wear scars were examined and measured by optical microscopy, by SEM, and by a Dektak profilometer. Scanning Auger analysis was performed on selected scars.

The profilometer traces were used to learn how the scars evolve, and to estimate the volume of wear. In the first seconds the balls actually gained height due to the accumulation of an antiwear film. Thereafter the profiles of the scars developed logically as the top ball wore into the 3 lower balls. The traces used for the calculation of wear volumes were performed parallel to the direction of sliding on balls on which the antiwear film had been removed by a 5% HF solution with the balls under cathodic protection. Volume calculations made the approximation of revolving the traces through 180°. The wear volumes are therefore not considered reliable to +/- 3000 μm^3.

For the more wear resistant conditions the rate of wear, and therefore the nominal wear constant, diminished with time. To facilitate comparison with other work the wear constants, K in $\mu m^3/N.m$, given in this paper are defined by the equation:

K = (wear vol at 60 mins)/0.408 x load x distance unless specified otherwise. Where the load was that applied to all three balls.

3. RESULTS

3.1 Standard Behaviour

The wear behaviour at 100 °C, 15 kg, 1200 rpm spindle speed (0.46 m/s) for 60 minutes was selected as the reference for good antiwear action. The wear constant was low, 0.1 $\mu m^3/N.m$, and the appearance of scars relatively uniform, Figure 1.

Such wear scars were surrounded by a fragile coloured material which following Stinton et al. [8], will be termed "friction film". This material was most clearly seen after a minimal rinse in hexane, and was easily removed by light rubbing. After the hexane rinse and rubbing, the scars were brightly coloured with yellows, browns and blues by the antiwear film proper. These antiwear films contained Zn, P, O, S and C, but no Fe. The films were probably inorganic glasses as they were hard (42 Rc), and their appearance was unchanged by heating to 440 °C.

In the SEM the films could be seen to be of uneven thickness, Figure 1a, forming mounds that tended to increase in size to about 30 μm diameter by 0.5 μm thick towards the trailing edge. (The leading edge of scars is on the left in all figures). The trailing side of scars also tended to have strips free of mounds as though some had detached sweeping the surface clean. The film between the thicker mounds, Figure 1b, had a characteristic porosity, and the sides of the larger mounds tended to be damaged, with the damage possibly extending into the metal.

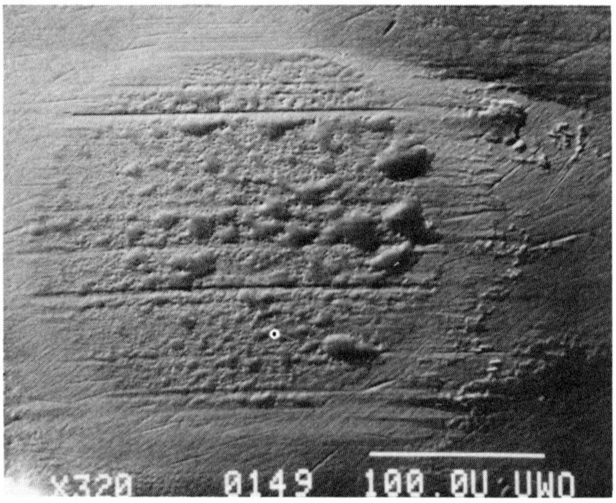

1a

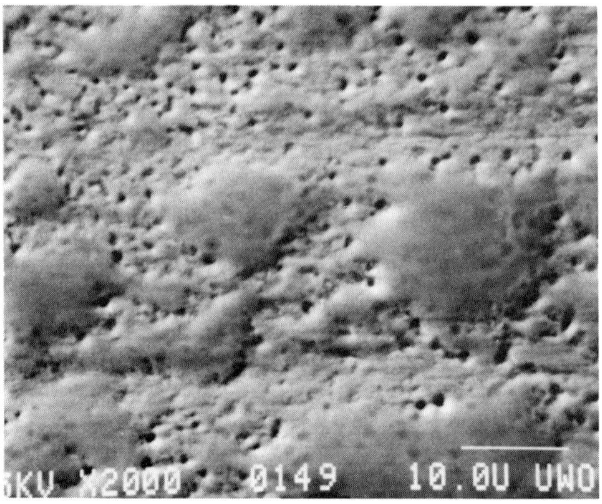

1b

Figure 1: SEM views of a scar after wear at 100 °C, 1200rpm, 147 N, for 1 hour, i.e. a standard scar. a) low magnification, some friction material is present in front of the leading edge, b) high magnification, showing porosity in the base film between the mounds.

The metal surface under the antiwear film and mounds had a random rough texture between light scratches, Figure 2. Such roughness implies that the antiwear film could only have been moving slowly ,if at all, over the wear scar. On the other hand, locations on wear scars under the strips that were free of mounds and porous antiwear film were relatively smooth, (arrowed in Figure 2b), implying that film had slid at these locations.

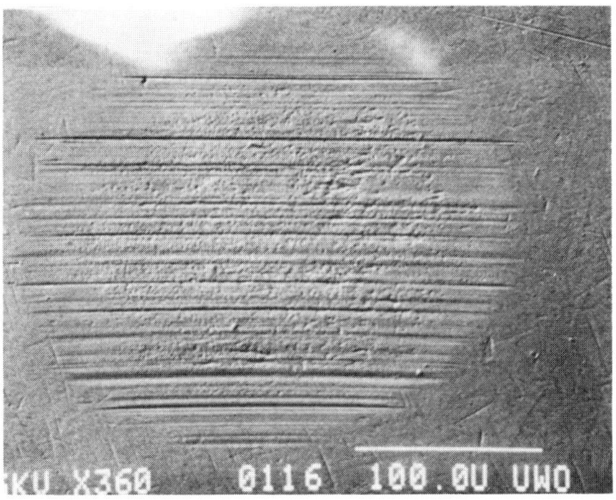

2a

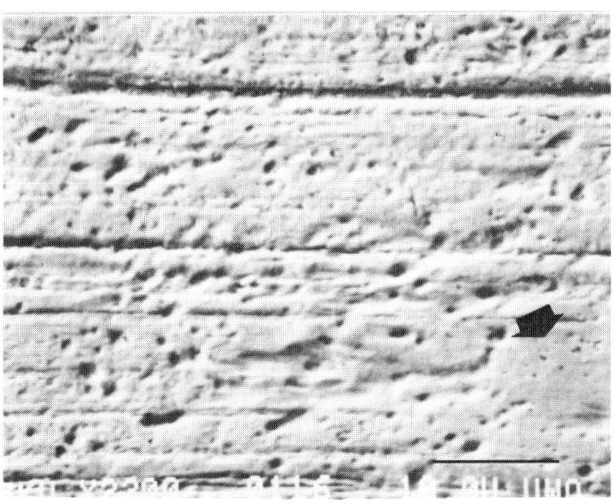

2b

Figure 2: SEM views of a wear scar similar to that in Figure 1 after removal of the antiwear film. a) low magnification, b) high magnification, a location smoothed by film sliding is arrowed.

3.2 Effect of Time at 100 °C, 15 kg, 1200 rpm.

By 3 seconds, i.e. stopping the 4-ball machine as soon as it reached speed, a thin uniform film had coated the scar. After 60 seconds the film had begun to thicken unevenly, and as wear continued the thicker parts developed into the mounds of the standard film, Figure 1. The thicker parts of the antiwear film on balls worn for several hours were frequently chipped around their edges, and were possibly separated from the base film by a boundary. Such thick regions may therefore more properly be described as pads, instead of mounds. After 6 hours the scars were very scratched apparently due to damage by detached film material.

Wear volumes as a function of distance are given in Figure 3. After about 100m of sliding the wear volume was approximately proportional to the square root of the distance.

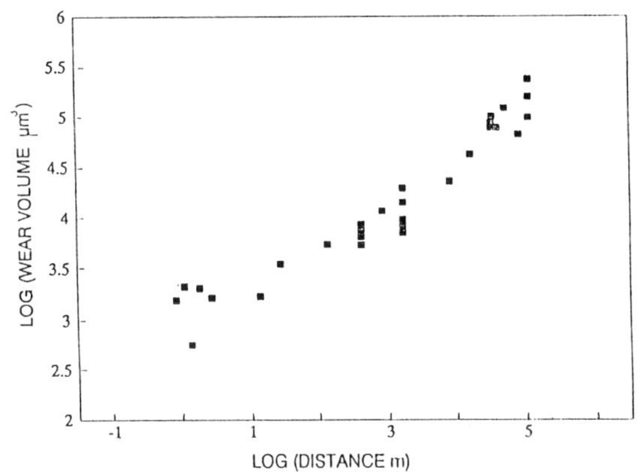

Figure 3: Dependence of wear volume on sliding distance at 100 °C, 1200 rpm, 147N.

3.3 Effect of Temperature at 15 kg, 1200 rpm.

The wear rate was insensitive to temperature from 0-150 °C, and had the same dependence upon distance as at 100 °C. Scars also had a similar appearance to the standard scar except that at the higher temperatures the antiwear pads were larger and appeared to be sliding either with, or over, the base film, Figure 4.

At 175 °C the wear rate was considerably higher than at the lower temperatures, Table I, and the antiwear film was different in character, Figure 5. In particular, antiwear pads that had apparently been detached from elsewhere accumulated on the leading edge of scars. Also the antiwear film on the leading part of the scar proper was quite smooth. Pads were present on essentially bare metal behind the smooth film, Figures 5b,c.

Wear was also faster below 0 °C, Table I. The antiwear film formed at low temperatures also appeared to be less adherent than at higher temperatures with patches accumulated on the leading edge of scars, Figure 6.

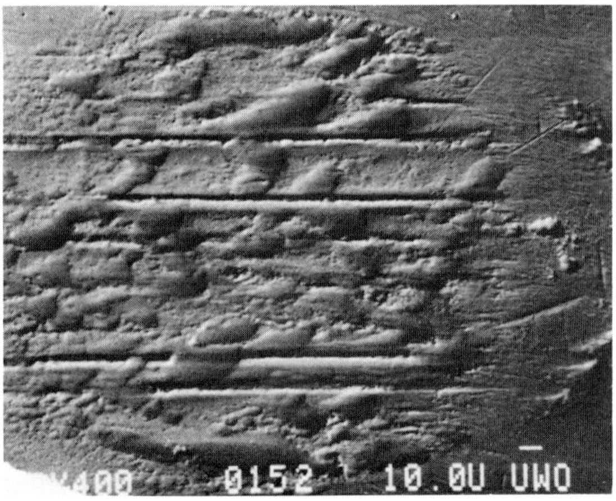

Figure 4: SEM view of a wear scar formed at 150 °C, 1200 rpm, 147 N, 60 min, showing larger pads and sliding on surface.

OIL	WEAR CONSTANT μm^3/N.m
1.5% Sec ZDDP in MCT 10	0.1
1.5% pure isopropyl ZDDP in MCT 10	0.1
1.5% Sec ZDDP in MCT 10 at 175°C	5.4
1.5% Sec ZDDP in MCT 10 at −10°C	0.4
1.5% Sec ZDDP in MCT 10 (N_2 atmosphere)	0.13
1.5% Sec ZDDP in PAO	0.2
1.5% Sec ZDDP in hexadecane	34
1.5% Sec ZDDP in oxidized hexadecane	12.6
1.5% Sec ZDDP in MCT 10 + 2% C black	0.21

Table I. Wear constants after 60 mins wear at 100 °C, 15 kg, 1200 rpm.

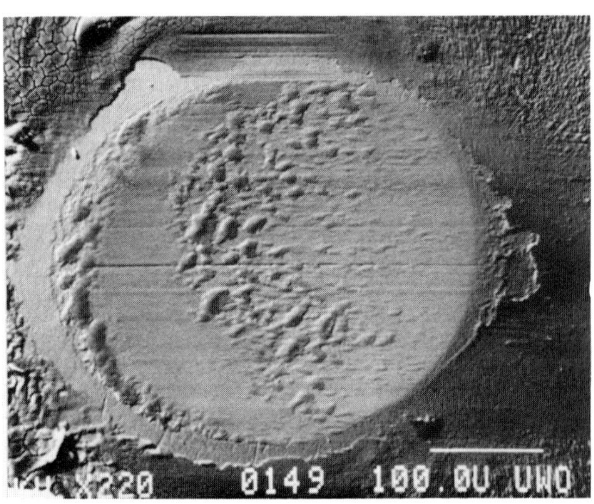

5a

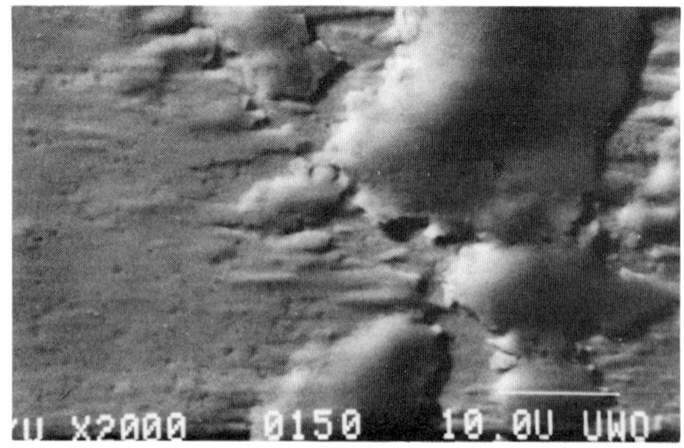

5b

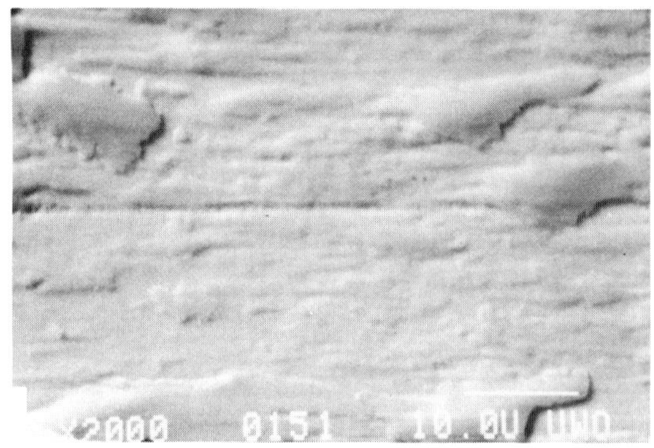

5c

Figure 5: SEM views of a wear scar formed at 175 °C, 1200 rpm, 147 N, 60 min, a) whole scar showing pads accumulated on the leading edge, b) junction of the filmed leading zone and the padded region, c) closer to trailing edge.

Figure 6: SEM view of a wear scar formed at 0 °C, 1200 rpm, 147N, 330 min.

3.4 Effect of Miscellaneous Factors at 100°C, 15 kg, 1200 rpm.

Wear constants after changing several factors are also given in Table I. Bubbling N_2 through the oil had no effect on the wear rate or appearance of scars. Similarly no effect was seen of a voltage applied between the ring ball and bottom balls of either + or - 1 V. The addition of 2% of C black increased the wear rate slightly but had no effect on the appearance of the wear scar. Changing the base oil to PAO made no difference; however the ZDDP in hexadecane caused more wear than in base stock alone. Some improvement was achieved when the hexadecane was preoxidized.

3.5 Effect of Second Additive at 75°C, 15 kg, 1200 rpm.

A range of commercial detergents, dispersants, and friction modifiers were added individually to the ZDDP containing oil. The resulting wear constants are shown in Table II. All degraded the performance by factors that ranged from 3 to 20 times. The degree of deterioration correlated well with changes in the appearance of scars, in the SEM in that the film became progressively thinner and smeared as the wear rate increased. This progression could not be discerned in the optical microscope even though there were considerable differences in both the scars and the friction material halos around the scars.

OIL 1.5 wt% SECONDARY ZDDP in MCT 10 WITH	WEAR CONSTANT μm^3/N.m
alone	0.1
6% dispersant #1	0.23
2% Ca Phenate (detergent)	0.23
2% Na Sulphonate 400 TBN (detergent)	0.31
2% Ca Sulphonate 400 TBN (detergent)	0.58
2% Na Phenate 400 TBN (detergent)	0.65
2% Ca Sulphonate 300 TBN (detergent)	0.73
6% dispersant #2	0.83
2% Mg Sulphonate 400 TBN (detergent)	0.92
0.1% friction modifier	1.84

Table II. Effect of second additives on wear constants after 60 mins wear at 100°C, 15 kg, 1200 rpm.

3.6 Effect of Ball Hardness at 100°C, 15 kg, 1200 rpm.

Standard balls were tempered to varying degrees to investigate the effect of metal hardness. The configurations: all balls soft, and hard top ball against softened bottom balls, gave similar wear rates, but only the latter will be described as these scars were less affected by severe wear.

The wear constants increased progressively as the balls were softened, Figure 7. Wear volumes measured on the softer balls had a large variability being affected by metal flow and the statistical presence or absence of large features on the wear scars.

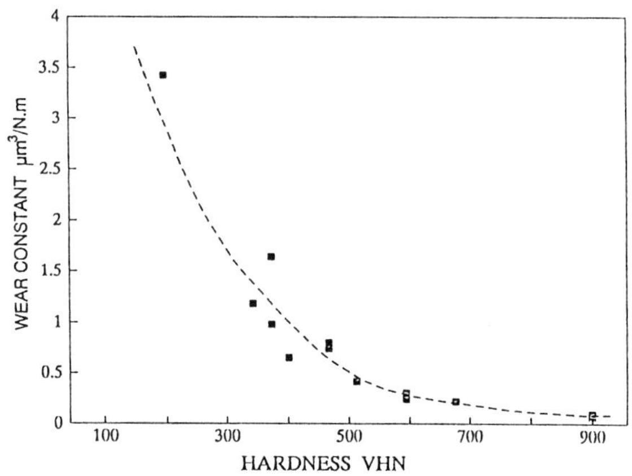

Figure 7: Dependence of the wear constant on wear ball hardness with ring ball fully hard. 100°C, 1200 rpm, 147 N, 60 min.

Exaggerated growth of several of the antiwear film mounds on each scar occurred on balls tempered to 680-595 VHN, Figure 8. These large pads of antiwear film could not be seen in the optical microscope due to their transparency. After stripping the antiwear film from such scars it was apparent from the flow pattern in the metal that the pads were being pushed into the ball, Figure 8b. The exaggerated pads were absent on balls softer than 400 VHN. The wear scars on these soft balls exhibited considerable metal flow and scratching, and were coated by a relatively even antiwear film. Wear scars on balls of hardness 470 VHN had both large pads and extensive metal flow.

3.7 Effect of Load at 100°C, 1200 rpm.

In the tests to observe the effect of load on the scar appearance and wear rate the balls were first run for 1 min at a load of 15 kg (147 N), and then for a further 5 min at the test load. Figure 9 shows that the wear rate increased rapidly with load above 25 kg.

Scars formed at 25 kg were similar to the standard 15 kg scar, while scars formed at 35 kg were quite different, being more scratched and irregular in outline. The antiwear film between the thicker pads on the latter scars was not porous as in Figure 1b, but was smeared and torn and did not cover the deeper grooves, Figure 10. Figure 10b shows the same location as Figure 10a, after removal of the antiwear film. Signs of thermal softening were absent.

With machine loads of 50 kg or greater, scars were grooved, irregular and deeply pitted, Figure 11. The flatter parts of the scar were flowed and torn, and lightly coated with antiwear film, Figure 11b. The appearance changed little with stripping.

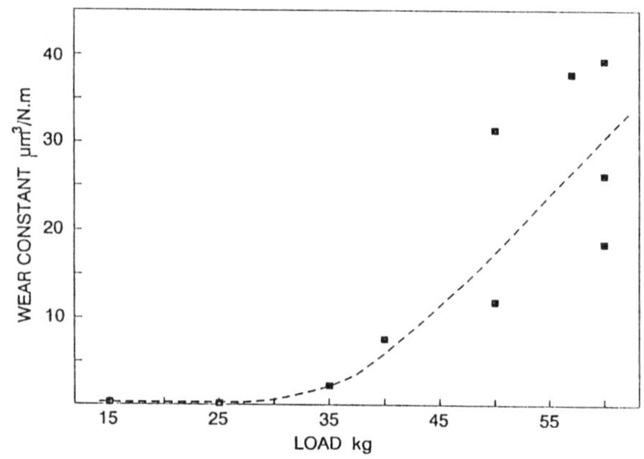

Figure 9: Dependence of the wear constant after 5 mins of wear on load, 100 °C, 1200 rpm.

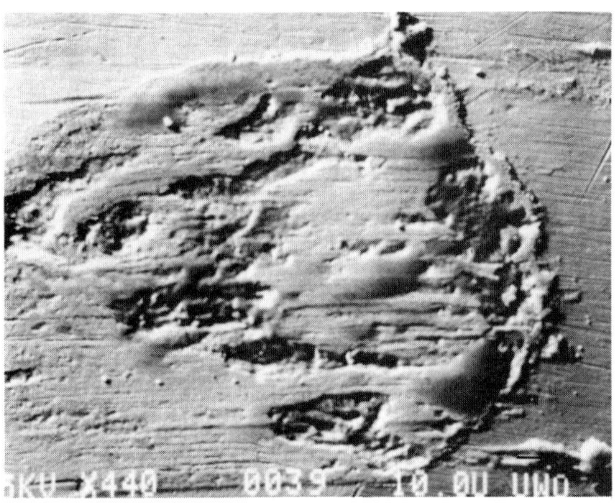

8a

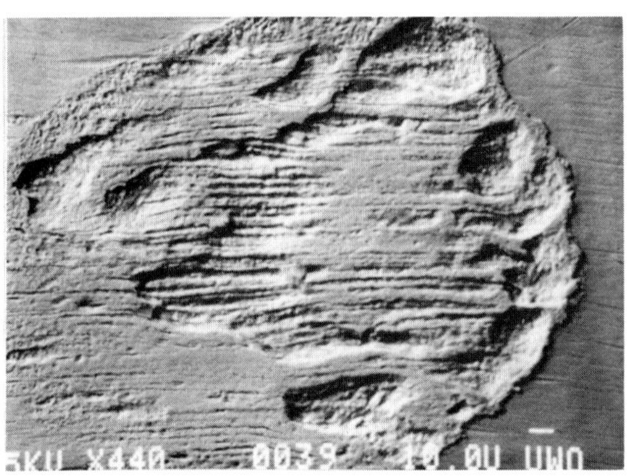

8b

Figure 8: SEM views of a wear scar formed on a ball of hardness 675 VHN after wear at 100 °C, 1200 rpm, 147 N, for 60 min. a) as worn, b) after removal of the antiwear film and pads.

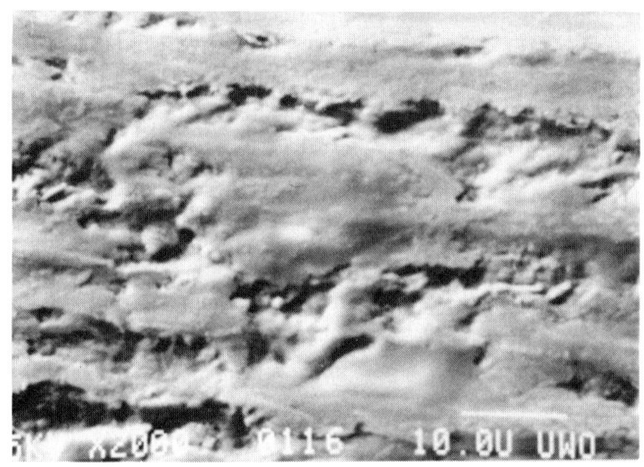

10a

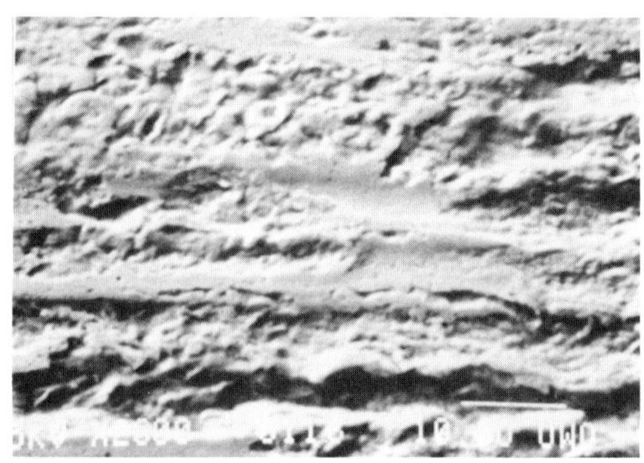

10b

Figure 10: High magnification SEM view of a scar formed on a ball worn for 5 min at 100 °C, 1200 rpm, 343 N. a) as worn, b) after removal of the antiwear film.

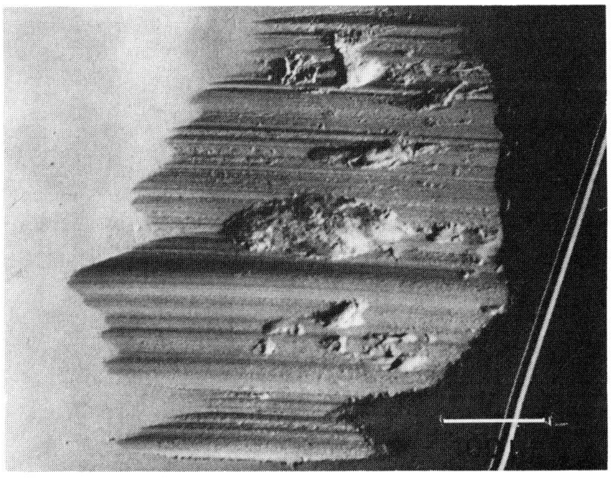

11a

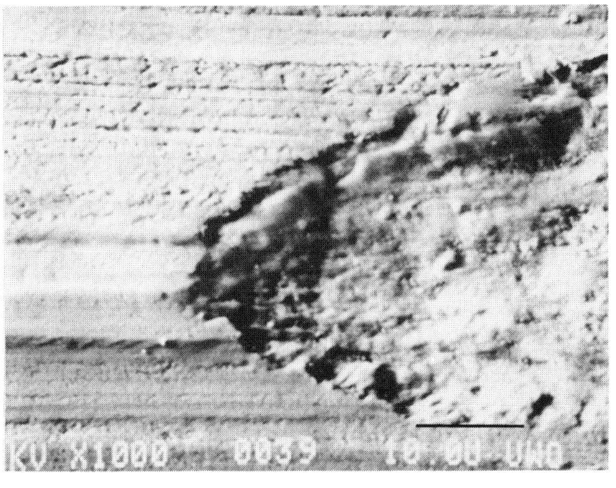

11b

Figure 11: SEM views of a scar formed on a ball worn for 5 min at 100°C, 1200 rpm, 588 N. a) full scar showing irregular shape and large gouges, b) high magnification of a gouge and surrounding filmed, but flowed, scar surface.

3.8 Effect of Ball Composition at 100°C, 15 kg, 1200 rpm.

Wear constants from several combinations of ball material are given in Table III. Self mated 440C stainless steel wore twice as fast as the standard 52100 steel, probably due to an increased movement of the antiwear film on the scar, as the overall film structure and appearance of the scar after stripping, was similar to scars formed on 52100 steel at 125-150°C.

The Co-Cr alloy was not lubricated effectively by the ZDDP, forming large bare scars, Table III. The scars were smaller, but still bare, when run against a steel ring ball. The scar on the steel ring ball was coated with antiwear film.

Table III. Wear constants with the indicated material combinations after 60 mins wear at 100°C, 15 kg, 1200 rpm. Hardness of 440C stainless steel Rc 66-67. Hardness of CO-Cr alloy Rc 63. Composition of Co-Cr alloy (Stoody 2N) C 2, Mn 1 max, Si 1, Cr 32, Ni 2.5 max, W 17, Fe 3 max, Co bal, Mo 1.

TOP BALL	BOTTOM BALLS	WEAR CONSTANT μm^3/N.m
440C	440C	0.18
Co-Cr	Co-Cr	12
52100	Co-Cr	0.53
Co-Cr	52100	0.13

3.9 Effect of Speed at 100°C, 15 kg Load.

Wear constants calculated from 1 hour tests at four temperatures as a function of sliding speed are shown by the open symbols in Figure 12. A transition from effective lubrication at high speeds to ineffective lubrication at low speeds was particularly evident at 100°C. At speeds above 15 mm/s wear rates and the appearance of scars was essentially independent of speed. However, at low speeds the ZDDP actually caused more wear than base stock alone.

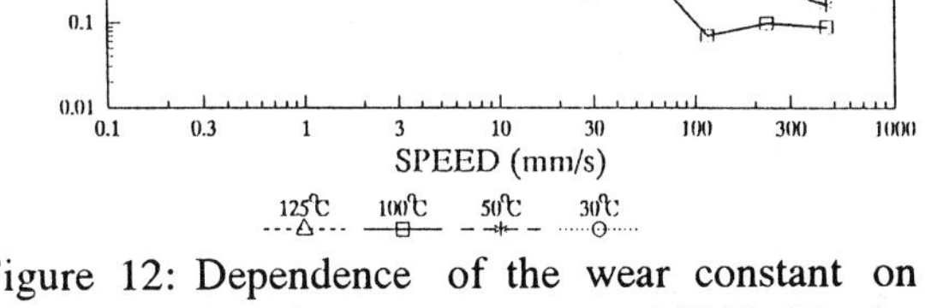

Figure 12: Dependence of the wear constant on sliding speed at four temperatures. 147 N, 60 min. Filled squares were started at 1200 rpm for a few seconds and then continued at the indicated speed. Filled circles were longer duration tests, demonstrating that wear was independent of time when protection was less than optimum.

Wear scars formed when the ZDDP was ineffective were characteristically covered and surrounded by a profuse black material, that was particularly packed on the leading third of scars, Figure 13a. After cleaning the scars were found to be nearly film free, and quite smooth under the packed mass, Figure 13b.

The transition from effective to ineffective lubrication was associated with initiation, as opposed to steady state, processes, as full wear resistance was achieved immediately when a test was continued at 1200 rpm after starting and running for 20 hours at 1 rpm. Similarly, if a test was started at the higher speed for just a few seconds so that an antiwear film formed, on continuation at low speeds the wear rate was low (but not as low as the 1200 rpm rate, filled squares in Figure 12).

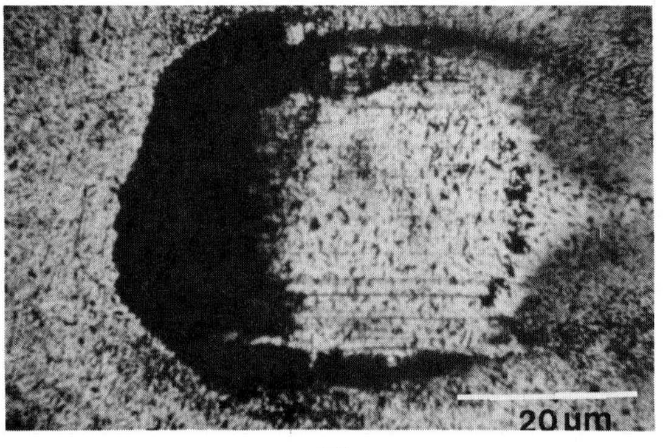

13a

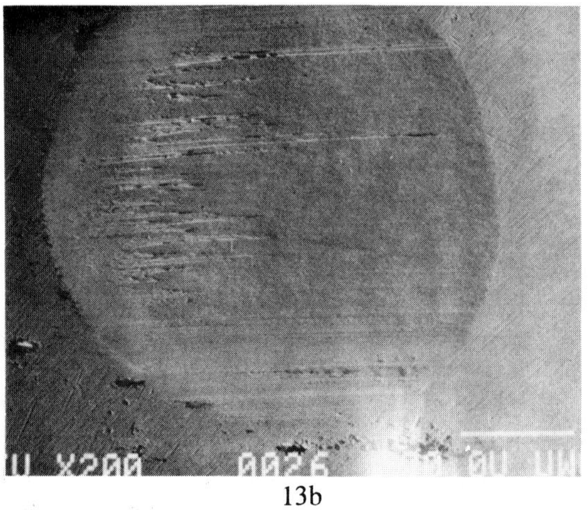

13b

Figure 13: Optical view of a wear scar formed at 1 rpm, 150 °C, 147 N, 240 min. a) with the fragile black debris in place, b) SEM view after light cleaning.

3.10 Effect of Speed at other Temperatures, 15 kg Load.

Figure 12 shows that the transition from effective to ineffective ZDDP action occurred at lower sliding speeds as the temperature was reduced. The transition also becomes less marked, so that whilst the runs at 30 °C, 1 rpm, had the profuse black debris associated with ineffective additive action, any change in wear rate was too small to be detected.

Wear rates at the lower temperatures and speeds were higher than at 100 °C, and increased as the speed was reduced. The reduced wear resistance correlated with reductions in the thickness and extent of the antiwear film, Figure 14. The wear rates associated with this film structure, unlike those of the standard film, did not decrease with wear time, (filled circles in Figure 12).

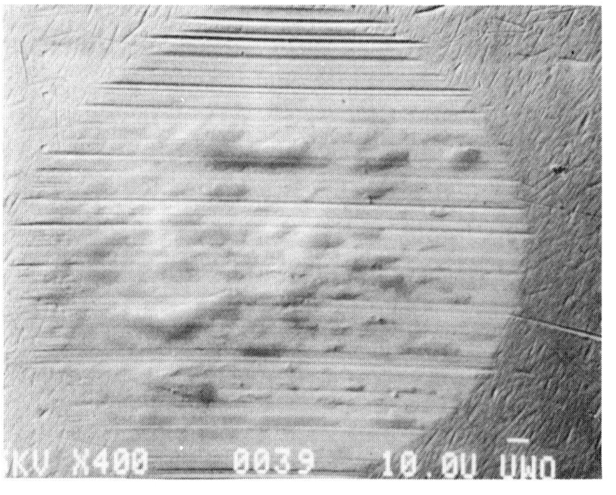

Figure 14: SEM view of a wear scar formed on a ball worn for 60 min at 50 °C, 41 rpm, 147 N. The scar was essentially bare between and around the pads of antiwear film.

3.11 Effect of Speed at other Loads, 100 °C.

Wear constants calculated from 1 hour tests at 100 °C, at four loads as a function of sliding speed are shown by the open symbols in Figure 15. The transition from effective to ineffective ZDDP action occurred at lower sliding speeds as the load was reduced. The transition at all loads was associated with the changes in appearance just described for 15 kg. The filled circle was for a test run for 1100 min, 10 kg load. The antiwear film on these balls was thicker and covered more of the scar than on the balls worn under the same conditions for 60 min.

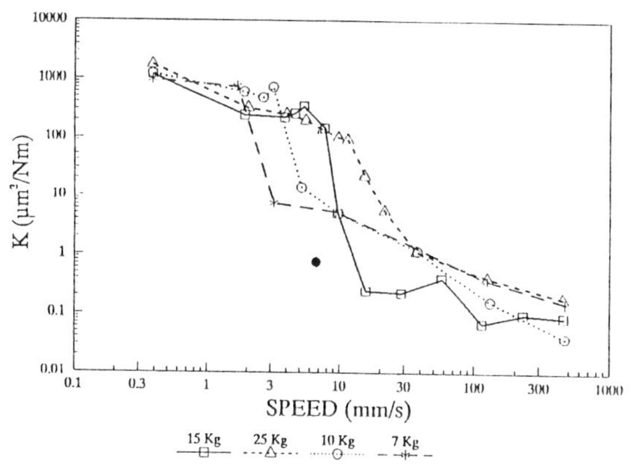

Figure 15: Dependence of the wear constant on sliding speed at four loads. 100°C, 60 min. The filled circle is for an 1100 min test, showing that the K(60 min) values for the slow speed tests are influenced by the short sliding distance.

3.12 Effect of oil viscosity

To examine the effect of oil viscosity on the extent of effective additive operation the ZDDP was blended into lighter and heavier base stocks. The tests were performed by incrementing the speed every 2 mins. The quality of the additive action was judged by the size of the fluctuations and value of the friction force. Table IV shows that the extent of optimum behaviour was increased when using the more viscous oil, and decreased when using the less viscous oil.

Table IV. Transitions speeds determined by friction behaviour, 15 kg load, 100°C, in oils containing 1.5 wt% secondary ZDDP.

OIL VISCOSITY AT 100°C cSt	SPEED TO GOOD AW ACTION mm/s	SPEED FROM GOOD AW ACTION mm/s
2	14.4	146
5	9.7	289
13	5.2	697

4. DISCUSSION

Under optimum conditions the 2nd ZDDP oil blend used in this work achieved wear protection as effective as any reported in the literature, i.e. K ≈0.1 μm/N.m. In other conditions wear rates could be unacceptably high, and even be higher than in base stock alone. In the latter situation, or when unacceptable damage to the wearing surfaces occurs, the limits to the usefulness of an additive are readily defined. When the deterioration is progressive, and the wear process does not roughen the surface unduly or form large debris particles, the Bell and Colgan value [6] for K of 10 μm/N.m is probably appropriate as a limit of acceptability.

Optimum wear resistance at 100°C was bounded at low and high speeds by failure of partial EHL, so that as noted by others [1, 9] EHL is a necessary condition for satisfactory boundary lubrication. However this work shows that EHL is not a sufficient condition, as wear can be unduly rapid within the EHL regime if conditions are not correct for a full antiwear film to form and to adhere. Using the Fein formula [10] antiwear films with the standard appearance, and performance, are estimated to require a minimum temperature to form, i.e. combined bulk and frictional heating, of about 125°C ; i.e. 100°C with 25 rpm, or 30°C with 1200 rpm. On the other hand the films appear to loose adherence on 52100 steel above about 250°C, (150°C with 1200 rpm). Adherence of the films was poor on Stoody 2N, and to a lesser extent on 440C, so that the oil gave unsatisfactory wear protection even though well within the partial EHL regime.

A very rapid form of wear occurred on EHL film collapse at low speeds in the presence of ZDDP, characterized by profuse amounts of carbonaceous material packed on and around essentially bare scars. Apparently the ZDDP had sufficient residual surface activity to prevent the formation of an oxide glaze that would normally give some wear protection. The ZDDP films require a certain stability of the wearing surfaces to become established: a property noted previously in the context of surface roughness [11].

Loss of effectiveness of the partial EHL film at high loads/speeds was evidently due to a thermal thinning of the oil film, as failure could be delayed or accelerated by appropriate changes in the viscosity of the oil. There was no evidence of overheating of the antiwear film. Consequently the increase in wear rate with load or speed was relatively gradual, making the cut-off for acceptable behaviour situation dependent.

The Dektak profiles showed that wear occurred in the presence of a thick nearly continuous antiwear film, albeit at a diminishing rate with time. As the antiwear films were essentially free of Fe it is not clear how this wear took place. When the protection was less perfect there was evidence of metal flow or abrasion, or at least of flow of the antiwear film over the surface that could remove metal as oxides. As the scar surface under the better films was microscopically rough, Figure 2, with only small areas smoothed by film sliding, the presumption has to be that when the film spalled locally some metal was taken as well.

Any change in the appearance of scars from those produced by the standard conditions of 100 °C, 1200 rpm, 147 N, was associated with an increase in the rate of wear. The type of condition that was changed, e.g. temperature, speed etc., caused characteristic changes in the film appearance. The terms used to describe these changes, such as softer, softened, worn, less adherent etc., were subjective, and should be validated by future work.

The role of the friction film in the antiwear process is not known. As suggested by others [4,8], it is possible that the film has EHL capability because of its high viscosity. Alternatively, the primary role of the friction film may be that of precursor material for the working glassy film. Future work will vary the quantity and composition of the friction film by the addition of second and third additives to the oil to see if there is a corresponding change in the glassy film.

5. CONCLUSIONS

This work found a clear correlation between the appearance of wear scars and the wear performance of balls worn in base stock containing 1.5 wt % of a commercial secondary ZDDP. When the ZDDP worked effectively wear protection was due to the formation of thick glassy films (aw film) on the rubbing surfaces. Optimum wear protection required partial EHL conditions and a combined bulk and frictional temperature between about 125-250 °C. Performance could be degraded by: temperature too low, temperature too high, metal too soft, use of hard Co-Cr alloy, and the presence of other additives.

ACKNOWLEDGMENTS

This work was funded by a Natural Sciences and Engineering Research Council of Canada Strategic Research Grant, an Imperial Oil University Research Grant, and by an Ontario Ministry of Colleges and Universities URIF award. The authors would like to express their appreciation to Z. Nisenholz, J. Rezek, W.A. Mackwood and M.P. Tangen who contributed to this study.

REFERENCES

1. A.W.J. de Gee, A. Begelinger and G. Salomon, Proceedings of the 11th L-L Symposium,1984, Paper V(i) 105.
2. R.S. Fein, in CRC Handbook of Lubrication Vol. II, CRC Press, Boca Raton1987, pp 49.
3. B. Cavdar and K.C. Ludema, Wear 148 (1991) 329.
4. P.M.E. Cann, G.J. Johnston and H.A. Spikes, Proc.Inst. Mech Eng., Int. Conf. Tribology London July (1987) pp. 543.
5. M. Belin, J.M. Martin, J.L. Mansot, Trib. Trans., 32 (1989) 410.
6. J.C. Bell and T.A. Colgan, Trib. Int. 24 (1991) 77.
7. J.S. Sheasby and Z. Nisenholz, To be published in Tribology Transactions.
8. Stinton, H.A. Spikes and A. Cameron, ASLE Trans., 25 (1981) 355.
9. E.S. Forbes, Tribology, 3 (1970) 145.
10. R.S. Fein, ASLE Trans., 8 (1960) 34.
11. J.S. Sheasby, T.A. Caughlin, A.G. Blahey and K.F. Laycock, Tribology Int. 23 (1990) 301.

SESSION IX
HARDNESS AND METAL CUTTING

Chairman : Professor T H C Childs

Thin Films in Tribology / D. Dowson et al. (Editors)
1993 Elsevier Science Publishers B.V.

Small scale effects in metal cutting

J A Williams

Cambridge University Engineering Department, Trumpington Street, Cambridge, CB2 1PZ.

Many metal deformation processes can be successfully modelled by analytical methods which treat the material as a uniform, homogeneous continuum. In these processes the plastically deforming zone has dimensions which are of the same order as those of the workpiece, i.e. large on the metallurgical or crystallographic scale. In metal cutting, on the other hand, the deformation zone at any instant is very small and may be comparable in size with the grain size of the workpiece: this is especially true in recently developed techniques of single-point micro-machining. Some experiments are described in which single crystals of two face centred cubic metals, copper and aluminium, and one hexagonal, magnesium, have been machined with carefully controlled crystal orientations in an effort to investigate the limitations of continuum ideas. The results indicate that there are circumstances in which both the direction of the primary shear plane and the flow stress of the material can indeed be influenced by such a fundamental parameter as the crystallographic stacking fault energy. It is now becoming feasible to model such processes on an atomic, rather than a continuum, basis; further experiments of this sort will be needed to validate these simulations.

1. INTRODUCTION

In machining, a sharp wedge shaped tool, whose geometry is defined by the value of its rake angle α, removes a thin surface layer from the workpiece by a process of intense plastic shear within the *primary shear zone*, Fig. 1.

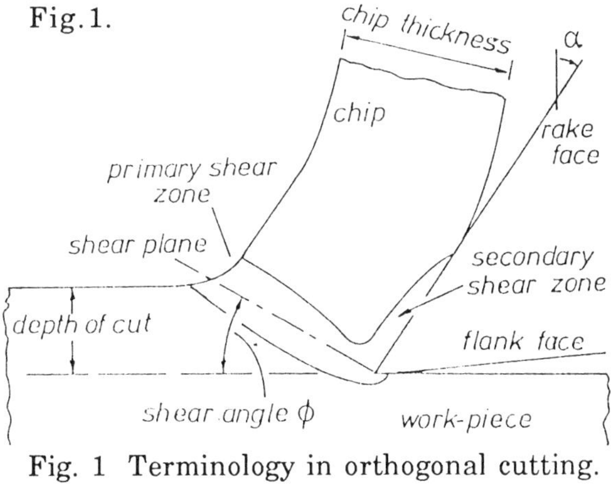

Fig. 1 Terminology in orthogonal cutting.

Many attempts have been made to apply the principles of continuum mechanics to this process, in the same way as has been accomplished for many other metal working processes, but thus far without the same degree of success. There are computational difficulties (even for the relatively simple geometry of plane strain orthogonal machining) because of the presence of the outer free surface and practical difficulties arising from the imperfectly understood friction or boundary conditions along the rake face of the tool. In addition the strains and strain rates involved in cutting, even at quite moderate workshop speeds, are truly remarkable. The material flowing through the primary shear zone suffers shear strains in the range of 1 to 5 at strain rates easily in excess of 10^6 s^{-1} and under these conditions the material constitutive relationship is likely to be problematic. One of the principal aims of any model of the cutting process is the prediction of the forces transmitted to the tool in terms of the tool geometry and material parameters and this essentially involves arriving at an estimate of the shear angle ϕ. Analyses all depend, in one way or another, on the adoption of a principle of least action: typically, it is postulated that the shear angle ϕ adopts that value which enables the movement of the tool to continue with the minimum overall expenditure of energy. This energy is dissipated principally in two zones, *viz* the primary zone referred to above, and a *secondary shear zone* which lies adjacent to the rake face of the tool and whose extent depends on the effects of friction and adhesion that act there. Lower forces of friction at the rake face give increased values of ϕ and reduce the work done, or energy consumed, per unit volume of material removed. Within the shear zones, the workpiece material is assumed to behave as a homogeneous isotropic continuum - although its flow stress may be influenced by local values of strain, strain rate and temperature. Most analyses make two further assumptions; firstly, that the tool remains 'sharp' (i.e. has a radius at its tip which is small compared to the depth of cut) and secondly, that the forces generated on the clearance or flank face of the tool (for example, as a result of the combination of the development of a wear land on the tool and the elastic spring-back between it and cut surface) are negligibly small. Under conditions when all the simplest idealisations hold geometric similarity will be preserved with changes of scale, so that the *specific* energy consumption would not be expected to vary greatly with changes in depth of cut.

In recent years the technology of the single point machining of non-ferrous materials on a scale much smaller than previously attempted has become established as a manufacturing route for engineering components with very precise dimensional tolerances and exceptionally fine surface finishes: in these circumstances some of these assumptions need to be re-examined. Micro-machining may involve significant sliding along the flank face of the tool and the contact forces generated here by the elastic recovery of the workpiece/tool combination, which are not significant at much greater depths of cut, can now be of the same order as the principal cutting forces. If the tool tip radius is comparable to, or even greater than, the depth of cut then the local, effective rake angle, experienced by the material just in front of the tool edge, may become very negative and bear little relation to the nominal value of the tool measured on the macroscale. A number of studies have brought out the importance of these effects [1-5]. Fig. 2 shows some data from Moriwaki using fine grain copper as the workpiece. When the layer removed is such that the ratio of the depth of cut to the tool tip radius more than about twenty then the specific energy consumption is almost constant; however, with diminishing depths of cut this quantity rises rapidly.

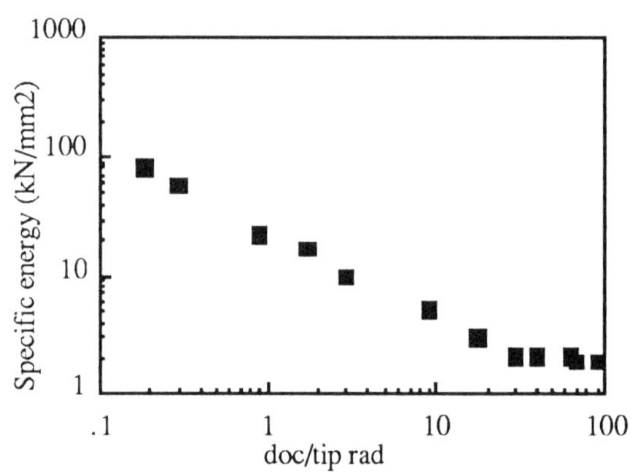

Fig.2 Variation in the specific energy of metal removal with depth of cut, from Moriwaki [5].

Once the depth of cut and the grain size of the workpiece material become of the same order of magnitude, it is not unreasonable to ask, even in the case of an infinitely sharp tool, whether the usual assumptions of homogeneity and isotropy still apply. All regular crystalline structures have preferred glide or slip planes and we could well imagine that the flow properties of individual grains would depend on their orientation relative to the cutting edge, as illustrated diagrammatically in Fig. 3.

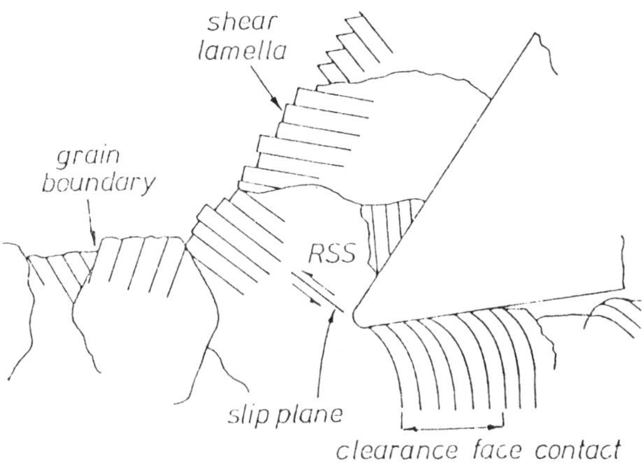

Fig. 3 Additional factors in orthogonal machining at small depths of cut.

The observation that the crystallographic macrostructure of metals may be developed by machining has been made noted for a number of face centred cubic metals [6-10] and is evidence of some form of non-isotropic behaviour. Fig. 4 shows the cut surface of a coarse grained specimen of commercially pure aluminium produced by orthogonal planing. It can be seen that certain grains are covered by strong surface markings which run perpendicularly to the direction of relative motion of the tool. These markings are regularly repeated within any given grain although their severity and spacing varies from one grain to another. Since the marks do not extend across the whole of the workpiece they cannot be attributed to tool movements, or chatter, but rather must have some direct crystallographic origin in the region close to the cutting edge. A similar pattern is seen on the outer free surface of the chip, see Fig. 5. and this morphology is also echoed by the dislocation structure within the bulk of the chip.

The origin of these features has been discussed elsewhere [7,11]. In addition the forces associated with the machining operation have been reported when cutting both single crystals of copper of known orientation [9] and coarse grained aluminium where the orientation of individual grains relative to the cutting tool was known [12]. In these latter experiments Ramalingam and Hazra obtained results on commercially pure aluminium, cut at a depth

Fig.4 The machined surface of a coarse grained aluminium workpiece.

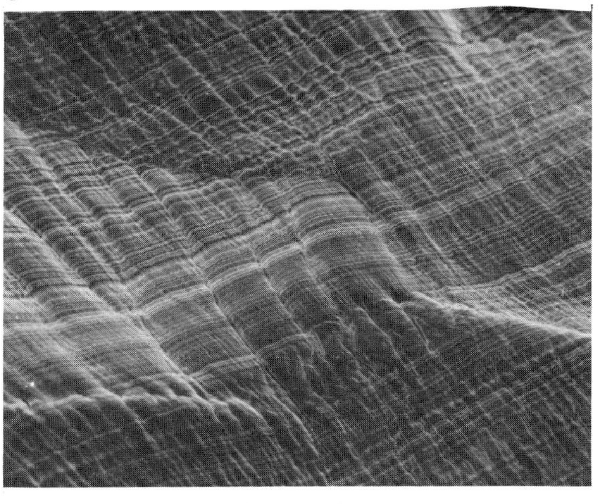

Fig.5 Evidence of the workpiece grain structure on the outer surface of a chip.

of cut of 125 μm at a speed of 16 mm/min, which indicated that the resolved shear stress (RSS) on the shear plane had a constant value, although the shear plane angle itself varied with crystal orientation over a wide range (from 14.6° to 41.7° in their data). They concluded that the shear in cutting occurred on a crystallographic plane of non-specific

Miller indices by the operation of multiple slip on a number of slip systems; the RSS was an invariant material property.

Williams and Gane [9] in their study of the microtoming of pure copper single crystals in two known orientations, with a typical depth of cut of 10 μm, conversely found that RSS was dependent on orientation and had a lower value when a (111) slip plane was aligned with the likely direction of the shear plane. The shear stress on the shear plane in the more favourable (110)[001] crystal orientation was significantly lower than in the (111)[112] position. Fig.6 illustrates the general cutting geometry; in the designation (lmn)[pqr] the cutting plane has Miller indices (lmn) and the cutting direction is [pqr]. The same crystal orientations have been used in the work of König and Spenrath [13] who found a similar dependence of cutting force on crystal orientation at similar depths of cut although at a much higher speed - unfortunately it is not possible from their published data to extract values of resolved shear stresses.

2. EXPERIMENTAL PROCEDURE AND RESULTS

In the experiments to be described, both aluminium and copper single crystals have been cut in the same experimental arrangement and in a number of orientations. Great care was taken to obtain an accurate value of the ratio of chip thickness to depth of cut and thus of the shear plane angle ϕ. In addition some results of the cutting behaviour of a single crystal of the hexagonal metal magnesium, which has fewer crystallographic slip systems and thereby much reduced ductility, are also presented.

The single crystal specimens were grown from a high purity melt using seed crystals and subsequently checked for orientation by X-ray diffractometry. Each workpiece was 3 mm thick, 20 mm wide by about 80 mm long.

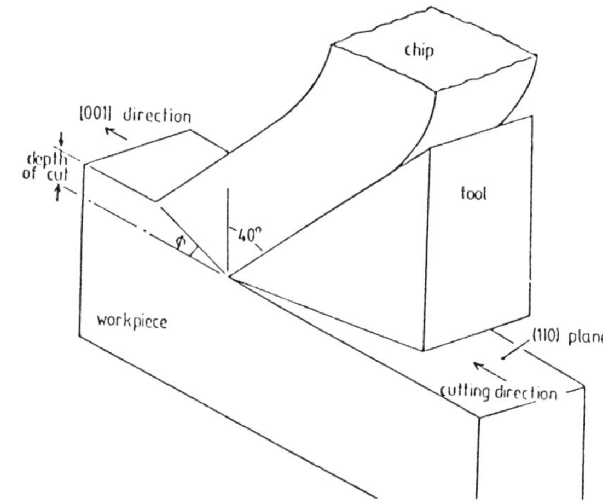

Fig. 6 General arrangement of the cutting geometry when machining copper and aluminium single crystals.

Copper crystals of ‹100›, ‹110› and ‹112› orientations and aluminium of ‹100› and ‹112› were prepared. The {111}‹112› crystals, being unsymmetrical, were cut in two directions. The general arrangement is illustrated in Fig. 6 and the various orientations and the positions of the most favourably inclined {111} slip planes in each situation are illustrated in Fig. 7. As a control a fine grained annealed polycrystalline specimen of similar purity of copper an aluminium were also included in the experimental schedule.

The experimental rig used for this investigation has been described elsewhere [14]. Cuts were made at a speed of 20 mm s^{-1} using high speed steel tools with a rake angle of 40° and a depth of cut of 100 μm (this was checked by weighing the resultant chip). The tools were ground to be as sharp as possible and it was estimated that the tip radius was no more than about 5 μm, i.e. very much smaller than the depth of cut. The values of the shear plane angle ϕ and the resolved shear stress (RSS) on the shear plane are shown in Tables 1 and 2. The errors quoted refer to scatter in the data - each test was repeated at least three times.

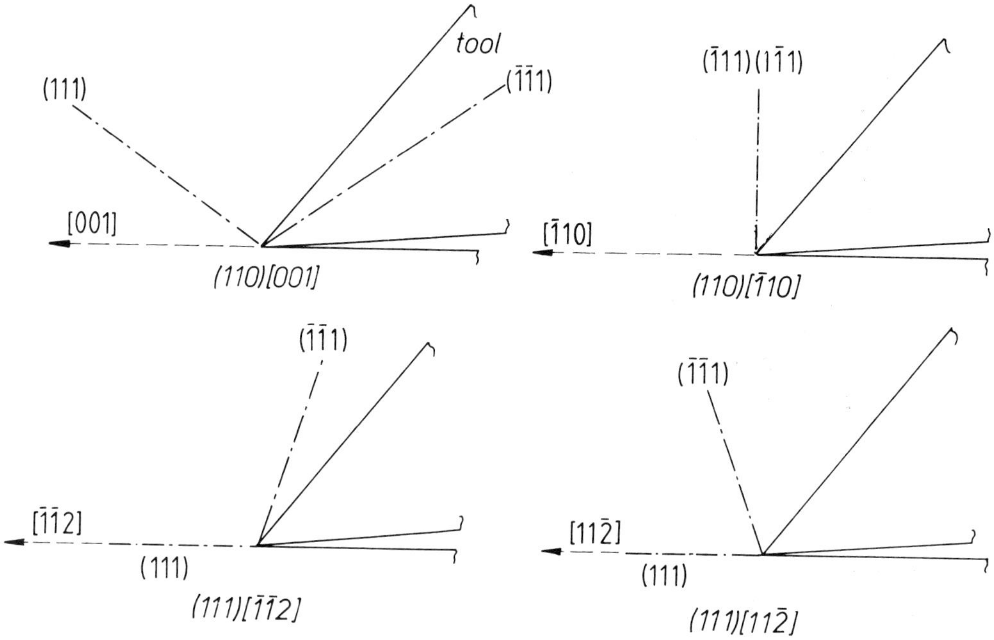

Fig. 7 Crystal orientations used when cutting single crystals of copper and aluminium showing the position of the {111} planes as seen from the side of the specimen.

Table 1

Copper single crystals. Shear angle ϕ and resolved shear stress RSS on the shear plane as a function of crystal orientation.

orientation	ϕ / °	RSS/ MPa
polycrystal	16.6 ±.5	298 ±7
(110)[001]	13.0 ±.7	288 ±9
$(110)[\bar{1}10]$(a)	5.9 ±.4	230 ±8
$(110)[\bar{1}10]$(b)	30.1 ±2.5	286 ±15
$(111)[\bar{1}\bar{1}2]$	22.2 ±1.3	320 ±5
$(111)[11\bar{2}]$	36.3 ±.9	261 ±11

Table 2

Aluminium single crystals. Shear angle ϕ and resolved shear stress RSS on the shear plane as a function of crystal orientation

orientation	ϕ /°	RSS/ MPa
polycrystal	9.2 ±.3	101 ±4
(110)[001]	9.2 ±.3	95 ±3
$(111)[\bar{1}\bar{1}2]$	7.3 ±.2	98 ±4
$(111)[11\bar{2}]$	8.6 ±.2	98 ±1

In the case of the copper $(110)[\bar{1}10]$ specimens the shear plane was able to take up one of two values with corresponding changes in chip thickness, indicated by entries (a) and (b) in Table 1: this switching from one geometry to another could occur two or three times in the course of a single cut. The higher shear angle was associated with a greater RSS on the shear plane. However, when allowance is made for the different *lengths* of the two shear planes the higher shear plane was found to correspond to a reduced shear force and thus a lower overall expenditure of energy.

The single crystals of magnesium were of similar dimensions to those of the copper and aluminium; each had a $\{10\bar{1}0\}$ face and were cut with the basal slip plane (0001) making various angles with the cutting direction as shown in Fig. 8. Again a fine grained polycrystalline specimen was included. In the case of this hexagonal close packed metal not only was the cutting force but also the mode of chip formation was found to vary dramatically with specimen orientation as indicated in Table 3. The behaviour varied from smooth continuous cutting, producing a tightly curled chip, case (d), to the situation in which no chip was formed and the tool dug into the surface as a fracture propagated along the (0001) plane, case (c).

Table 3

Cutting behaviour of magnesium single crystals.

orientation	(0001) plane angle	mode of chip formation	ϕ /°	RSS/MPa
(a)	0	discontinuous	54	89
(b)	45	continuous, segmented, no curl	43	53
(c)	-45	no chip, tool digs in	-	-
(d)	90	continuous, tight curl	79	43
polycrystal	varies	discontinuous, broken	54	91

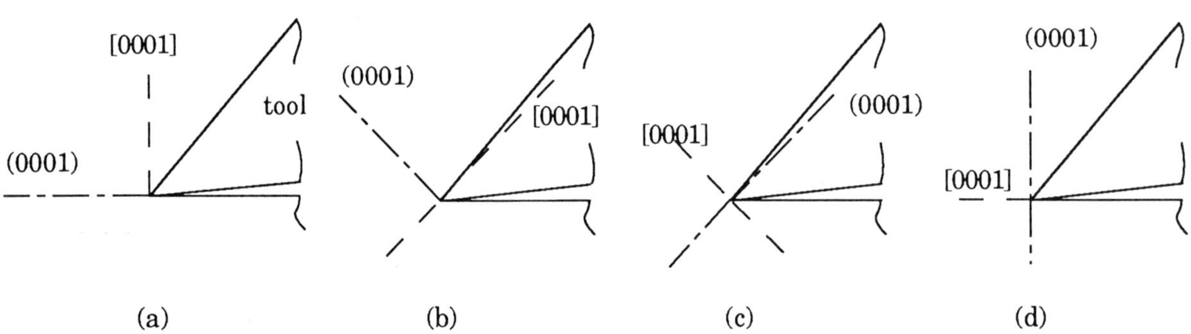

Fig. 8 Orientations used in studying the cutting behaviour of magnesium single crystals showing the position of the (0001) plane.

3. DISCUSSION

3.1 Copper and aluminium

Single crystals of copper were found to exhibit large variations in the resolved shear stress acting on the primary shear plane with an increase from 230 MPa in the most 'favourable' orientation to 320 MPa in the most difficult. The polycrystal had an RSS between the two single crystal values. In the case of the (110)[$\bar{1}$10] orientation there appear to be two possible cutting geometries. The shear plane with the higher value of shear angle operated at an RSS similar to the polycrystalline value. The effective resolved shear stress on the lower angled shear plane was significantly lower, but because the shear plane is longer, the resultant cutting forces in the two cases were similar. It seems probable that the switching behaviour between the two geometries was due to rake face frictional force. Increased resistance to chip flow up the face of the tool would favour the operation of the lower shear angle mode, but should the resistance to chip flow reduce - as a result of changes in the lubrication conditions - there is a return to the higher shear angle mode. The fact that shear angles of intermediate value were not observed, presumably because they involved the application of greater forces and the expenditure of more energy, offers further evidence of the concept of 'favourable' flow orientations. The {111}‹112› orientations give the most striking demonstration of a reduced RSS in a favourable crystal geometry. Rotating the specimen from the (111)[11$\bar{2}$] to the (111)[$\bar{1}\bar{1}$2] position reduced the RSS by more than 25% to a value some 20% less than that for the polycrystal.

The experimental results on similar specimens of aluminium, Table 2, show some differences when compared to the copper results. Both the variation in shear angle and those in the numerical values of RSS were very much less than with copper, in fact within the range of experimental error. In other words it appears that copper displays a dependence of RSS on {111} plane orientation while aluminium does not, thus validating the results of both references [9] and [12].

The difference between these two metals must involve the processes of dislocation movement that contribute to the shearing of the structure of the material as cutting proceeds. In particular it would seem that the ability of an individual dislocation to glide on more than {111} plane, i.e. the process of *cross-slip*, is of significance here. Consider the dislocation $\frac{1}{2}$[$\bar{1}$10](111) it is free to move on the (11$\bar{1}$)plane some 70° away. However, it may also dissociate into partials

$$\tfrac{1}{2}[1\bar{1}0](111) \Rightarrow \tfrac{1}{6}[1\bar{2}1] + \tfrac{1}{6}[2\bar{1}\bar{1}] \ .$$

These will repel each other as a result of their elastic interaction and as they separate a sheet of *stacking fault* is formed between them. The *stacking fault energy* associated with this region leads to an effective constant force pulling them back together; these two effects come to balance one another and so result in an equilibrium spacing. Glide, or movement, of the extended dislocation will be more difficult for a metal with a greater spacing. However, if a local constriction in this ribbon of stacking fault energy occurs, then the dislocation at the point of the restriction will find it easier to move out of the plane of the faulted region and to continue plastic straining by moving on other planes: this is the process of cross-slip. A certain amount of energy is needed to form such a constriction and this will depend on the width of the stacking fault. Materials with a high stacking fault energy, and so a low equilibrium spacing, should thus undergo cross-slip more readily than those of low stacking fault energy.

The stacking fault energies and equilibrium spacings for copper and aluminium are shown in Table 4 from which it can be seen that cross slip should occur much more readily in aluminium. Small local fluctuations in the stress field around a pair of partials (due to the presence of nearby dislocations, solute particles, precipitates or grain boundaries) frequently cause constriction and recombination. This is much less likely in copper where most of the dislocations would be confined to one glide plane. Thus the same processes of dislocation mobility that contribute to the low shear strength of pure aluminium also cause its flow stress in machining to be substantially independent of workpiece orientation. Since dislocations are less mobile in copper, favourable orientation of a glide plane reduces the resolved shear stress.

Table 4

Stacking fault energies and equilibrium spacings

metal	stacking fault energy $\times 10^{-3}$ J m^{-2}	equilibrium spacing atomic distances
Cu	55	10 - 15
Al	200	2

3.2 Magnesium

The experiments on the single crystals of magnesium show that in this case fracture as well as shear can play a part in cutting. Forces and chip morphology are highly orientation dependent. When the major slip plane is coincident with the cutting direction dislocation glide appears not to contribute significantly to material removal: chips are formed by fracture along the direction of maximum shear stress. Plastic flow becomes possible when the specimen is rotated so that the slip plane makes an acute angle with the cutting direction as in Fig. 8(b). The RSS required to produce the chip is reduced by this change of mechanism although fracture would still seem to be involved as the chip has a characteristic segmented or discontinuous form. However, if the slip plane is further rotated to the position of Fig. 8(c) then cleavage along this plane results in the tool digging into the surface and no steady state cutting condition is achieved but rather lumps of material are gouged from the surface. The basal plane can act both as a preferred slip or cleavage plane. If the slip plane is aligned perpendicularly to the cutting direction then, somewhat surprisingly, a tightly curled chip is produced which, when unrolled, is longer than the surface from which it was produced as illustrated in Fig 9. This involves an effective shear angle of more than (180-α)° and is quite incompatible with any of the continuum models of the machining process. The polycrystalline specimen of magnesium formed chips by a combination of shear and fracture at an RSS comparable to that of the 0° specimen.

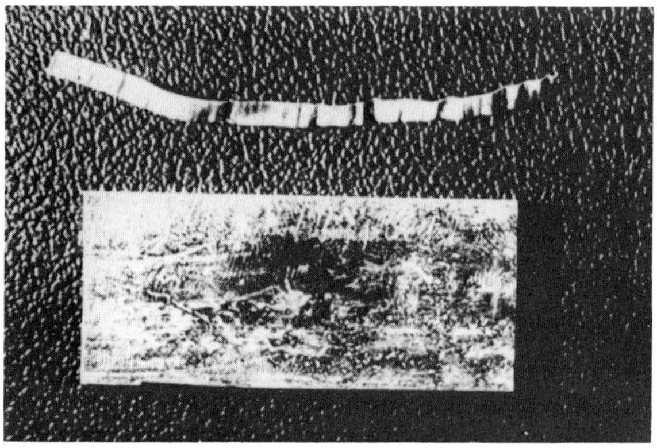

Fig. 9 The uncurled chip from the single crystal of magnesium cut under conditions (d) of Table 3 and Fig. 8. The chip is longer than the workpiece from which it has been removed.

3.3 Effects of scale

The values of the resolved shear stresses found in these experiments for single crystals of copper and aluminium were comparable with those of the polycrystalline specimens of the same metals. However, it is well known that when a single crystal of a pure metal is tested in simple tension the RSS is very much less than that for a similar polycrystalline specimen [16,17]. This is because existing dislocations within its structure can move across the crystal with little resistance and so accommodate the imposed strain by an accumulation of small, but finite, slip steps. In a polycrystal the dislocations are not able to emerge (and so disappear) so easily at a free surface; they become blocked, tangled into networks, which have a characteristic 'cell' size, and multiply. A single crystal produced by conventional means, as used in these experiments, is not initially free of defects but has an initial dislocation density: in metal cutting the edge of the cutting tool is a most effective dislocation multiplier and so this initial population of dislocations is rapidly amplified leading to the flow stress becoming similar to that of a polycrystal. If we had started with a perfect crystal, containing no initial stacking faults or imperfections, then the force required to cause the cutting tool to move would be very much increased because now it would involve actually *creating* dislocations in an otherwise perfect lattice.

There are clear similarities between the mechanism of cutting and that of indentation hardness, and for this process there is considerable experimental evidence that when carried out on the micro-scale, so that the depth of the hardness indentation is of the order of tens of nano-metres or less, there is a significant increase in hardness - by perhaps as much as an order of magnitude [18-20]. A number of possible explanations have been offered which involve complex interactions between the dislocations generated by the tip of the indenter and the existing dislocation networks in the upper surface of the specimen. Such explanations may also be relevant to the increase in cutting force and specific energy observed in sub-micron machining - although it is difficult to separate out these effects from those involving the finite tip radius and the region of elastic contact referred to in §1.

An alternative approach, which has only really become feasible in the last few years, is that of calculating the forces and energy requirements in these small scale processes by mathematically modelling the interatomic force relations within an array of atoms which is then forced to accommodate the movement of either the hardness indenter or the cutting tool; this is the technique of *molecular dynamics*. When it is remembered that a depth of cut of one micron is equivalent to several thousand atomic layers, so that the model would need to contain several million individual atoms, it is apparent that the computation effort required in such simulations is truly prodigious if anything approaching realistic scales is to be attempted. The calculation must be repeated at time increments which are a small fraction of the period of atomic vibration (say 10^{-12} s) so that many millions of iterations are needed to simulate even a fraction of a second of real time. Nevertheless, a start has been made [21-24] and Fig. 10 shows a two-dimensional computer generated graphic illustrating of the removal of a very thin surface layer from a perfect crystal lattice.

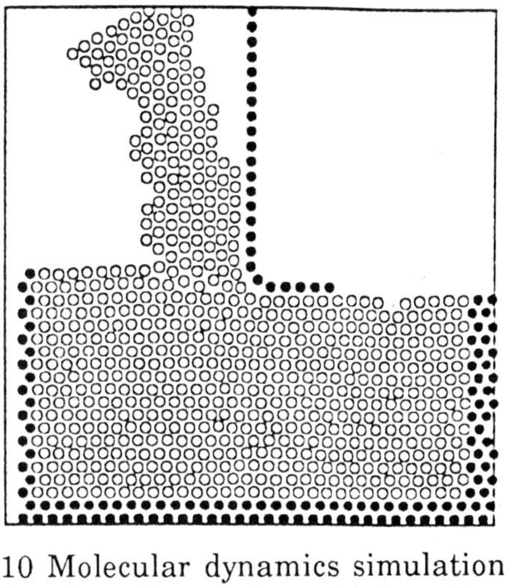

Fig. 10 Molecular dynamics simulation of small-scale orthogonal machining of copper. Frame contains about 1000 atoms and the depth of cut is 2 atomic layers: cutting speed 500 ms^{-1}; from [21].

When restricted in this way to two-dimensions there is a considerable saving in computational effort so that larger arrays of atoms can be dealt with; Belak [21] has shown details of the simulation of 2-D orthogonal cutting using as amany as 100000 atoms in 200 layers. The depth of cut was 32 layers (ca 8 nm), the effective cutting speed was 100 ms^{-1} and the tool tip radius was varied from 1.25 to 20 nm so that there was an overlap between these simulaions and the data of Moriwaki of Fig. 2. The same trend of rapidly increasing specific cutting energy at small ratios of depth of cut to tool tip radius was apparent.

4. CONCLUSIONS

When the grain size of the workpiece in orthogonal machining is comparable to the depth of cut, the crystallographic origin of the deformation process is apparent in both the magnitude of the cutting forces and the geometry of the shear deformation.

In the case of crystals, or grains, of face centred cubic metals of low stacking fault energy, both the direction of the primary shear plane, and the value of the resolved shear stress that acts on that plane, depend on the local crystallographic orientation relative to the cutting direction. In the case of fcc metals with higher stacking fault energies, the numerical value of the resolved shear stress is much less orientation dependent. This distinction arises from the differences in the ease of dislocation cross-slip between the two materials.

The response of a hexagonal metal is very dependent on the orientation of the basal plane relative to the direction of cutting. By varying this, the cutting behaviour may change from continuous chip formation, producing a chip which is thinner than the depth of cut, to gross fracture and cracking of the substrate.

The development of molecular dynamics offers the prospect of far more scientifically fundamental models of small scale processes than can be achieved with a classical continuum approach. Such models will be able to cope intrinsically with material anisotropy which has its basis in the crystal structure of the specimen. However, further experiments of the sort described will be needed to validate the predictions of such models.

REFERENCES

[1] W. R. Backer, E. R. Marshall and M. C. Shaw, Trans ASME, 74 (1952) 61.

[2] K. Nakayama and K Tamura, Trans ASME(B), 90 (1968) 119.

[3] D. Lucca, R. L. Rhorer and R. Komanduri, Annals of CIRP (1991) 69.

[4] N. Ikawa *et al*, Annals of CIRP, 40 (1991) 587.

[5] T. Moriwaki and K. Okuda, Annals of CIRP, 38 (1989) 115.

[6] L. M. Clareborough and G. J. Ogilvie, in *Machining Theory and Practice*, ASM, Cleveland, Ohio, 1950.

[7] J. T. Black, Trans ASME(B), 93 (1971) 507.

[8] J. T. Black, Trans ASME(B), 94 (1972) 307.

[9] J. A. Williams and N. Gane, Wear 42 (1977) 341.

[10] J. A. Williams and J. G. Horne, J. Mat. Sci. 17 (1982) 2618.

[11] W. M. Stobbs, J. S. Kallend and J. A. Williams, Acta Met., 24 (1976) 1083.

[12] S. Ramalingam and J. Hazra, Trans ASME(B), 95 (1973) 939.

[13] W. König and N. Spenrath, in *Progress in Precision Engineering*, Seyfried, Kunzmann, McKeown and Weck (eds), Springer-Verlag, Berlin, 1991.

[14] T. Wakabayashi, J. A. Williams and I. M. Hutchings, in *Tribology in Metal Cutting and Grinding*, IMechE, London, 1992.

[15] P. B. Hirsch (ed) *The Physics of Metals*, CUP, Cambridge, 1975.

[16] C. S. Barrett, Structure of metals, McGraw-Hill, New York, 1952.

[17] W. J. McGregor Tegart, *Elements of Mechanical Metallurgy*, Macmillan, London, 1966.
[18] N. Gane and F. P. Bowden, J. Appl. Phys. 39 (1968) 1432.
[19] G. M. Pharr and W. C. Oliver, J. Mater. Res. 4 (1989) 94.
[20] W. C. Oliver, R. Hutchings and J. B. Pethica in *Microindentation Techniques in Materials Science and Engineering,* ASTM STP 889, Philadelphia, 1986.
[21] J. Belak and I. F. Stowers, Proc. 1990 Annual Meeting of the Americal Soc. of Precision Engineers, 76.
[22] J. Belak, University of California Research Report (1991) UCRL-JC-106788.
[23] T. Inamura and N. Takezawa, in *Progress in Precision Engineering,* Seyfried, Kunzmann, McKeown and Weck (eds), Springer-Verlag, Berlin, 1991.
[24] J. Belak, private communication.

Thin Films in Tribology / D. Dowson et al. (Editors)
© 1993 Elsevier Science Publishers B.V. All rights reserved.

TOOL WEAR IN MACHINING OF ADVANCED CERAMICS

S.C. Laroiya*, M. Adithan* and Dell K. Allen**

*Technical Teachers' Training Institute, Sector 26, Chandigarh 160 019, India.

**Brigham Young University, Provo, Utah 84602, U.S.A.

Advanced Ceramics are extensively in use in modern technological world. A major breakthrough is required in lowering the production costs and improving producibility and reliability in use for advanced-ceramic components and products to achieve their predicted potential. Machining is needed to produce the required size, shape and surface finish not possible by forming methods. To accelerate the commercialization of Advanced Ceramics low cost machining is of interest to the researchers all over the world.

This paper is an attempt to study the tool wear problems associated with machining methods like Diamond machining, Electrodischarge Machining, Ultrasonic Machining and Rotary Ultrasonic Machining in the machining of advanced ceramics. Experiments have been conducted to machine High Alumina, Silicon Carbide and Steatite using ultrasonic drilling process and the data thus generated has been analysed and inferences drawn. The paper also discusses production accuracy, surface roughness, machining rates and tool wear values obtained while machining advanced ceramics.

1. INTRODUCTION

Ceramics have been in use since time immemorial. Advanced ceramics applications in the present time include mechanical components, hip joints, human body parts, computer chips and parts of spaceships. Their application is of special interest to the researchers all over the world because of their excellent properties like high strength-to-weight ratio and resistance to heat, wear and corrosion. During the sintering of these materials shrinkage cannot be entirely ruled out and therefore machining is needed to attain proper accuracy, shape, etc. Since the development of ceramic engines most of the ceramic components have been machined with diamond machining methods. The main aim of the scientists and researchers these days has been the development of low cost machining of advanced ceramics. Important properties of various ceramics are that they are weak in tension but strong in compression e.g Alumina Ceramics has tensile strength of 138 M.Pa and compressive strength of 240 M. Pa; homogeneous ceramics of single phase are usually stronger than those with several phases; ceramics possess good creep resistance, high corrosion resistance, high thermal shock resistance as well as the ability to withstand high service temperatures of the order of 1650°C. Amongst the new generation ceramics the most widely used ceramics in the industry are SiC, Si_3N_4, CBN, Al_2O_3, UCon. (a proprietary Nitrided refractory metal alloy developed by Union Carbide, USA).

Differences in the type of atomic bonding are responsible for hardness and Young's modulus of ceramic materials. Materials with metallic bonding exhibit high ductility, covalently bonded ceramics typically are hard, strong and have high melting temperature.

2. NEED FOR MACHINING OF ADVANCED CERAMICS

There are many reasons why we should go in for machining of advanced ceramics after their compacting and sintering:
a) during the sintering process shrinkage of the material cannot be entirely avoided, hence to achieve closer tolerances and dimensional accuracies of parts;
b) slots, grooves and side holes are required to be produced as a secondary operation after compacting and sintering since these configurations are difficult to obtain during compacting and moulding;
c) the production process has to accomodate limitless diversity of through holes, blind holes, slots shapes and odd contours for production of complex and intricate parts;
d) for slicing into small pieces and their wafers.

3. SURVEY OF EXISTING MACHINING METHODS AVAILABLE AND TOOL WEAR PROBLEMS ASSOCIATED

3.1 Diamond Turning:

Hidehiko Takeyama(1) has reported that while it is possible to machine the advanced ceramics with diamond tools, the tool wear is greatly affected by the cutting pressure and consequently limits the feed rate. He has further reported that tool shape should be so designed such that the cutting presure or the uncut chip thickness may be uniform along the cutting edge i.e. a pointed tool with a larger side cutting edge is desirable.

3.2 Diamond Grinding

For grinding of ceramics grinding machines with a high stiffness of 10 N/μm. between the tool holder and the work piece spindle are required. Machines designed to grind various types of steel with CBN wheels may also work in certain ceramic machining applications. For ceramic components machining higher wheel speeds upto 100 m/sec. are required to improve material removal rates and surface finish(2).

The mechanism of material removal in diamond grinding of ceramics is primarily due to the microbrittle fracture of the material(3). The grinding process is completed in two phases viz roughing and finishing. A high workpiece speed and a small depth of cut are desirable for roughing process (to reduce the grinding force), while the workpiece speed should be kept relatively low in the finishing process to overcome the chances of getting a damaged surface layer.

The problem of utmost importance in diamond grinding of ceramics is the excessive increase in the grinding force with time elapsed and the occurrence of chipping and cracking in the workpiece.

3.3 Creep feed grinding

Creep feed grinding has an important role in minimizing the chipping of the workpiece which quite often occurs at the entry and exit of the grinding wheel. It is therefore essential to introduce continuous dressing of grinding wheel to prevent the increase in the grinding force with time to take advantage of creep feed grinding.

Resin bonded wheels wear away faster but these produce much lower grinding forces while machining high Alumina Ceramics. Selection of an appropriate grinding wheel and cutting fluid is essential for efficient grinding of advanced ceramics.

From the point of view of surface roughness and wheel wear it is confirmed that emulsifiable oil shows better performance than other chemical fluids.

Toyota Machine Works at Japan (4) conducted several experiments to improve the quality of surface produced while grinding of ceramics. It was found that Nickle coated friable abrasives used in resinoid bonded grinding wheels produced better results while grinding high Alumina ceramics. However, while grinding Si_3N_4 components it was found that the material strength was reduced after grinding.

In creep feed grinding a combination of a stiff machine tool, which eliminates vibrations, with high positioning accuracy, and high rotational speed results in improved efficiency and dimensional accuracies.

W.Konig(5) has reported that in Si/SiC components surface grinding should be carried out based on an analysis of process kinematics and a study of the basic mechanisms of material removal because such methods will lead to considerable increase in material removal rate, component strength and tool life.

Sentoku and others(6,7) have reported that for milling of machinable ceramics, WC tools are required and the chip size is determined by the relationship between the angle of work corner and disengagement angle in cutting. They have further reported that cutting tool with large nose radius provides high efficiency cutting and lengthens tool life while decreasing the cross-sectional area of uncut chip increases specific cutting force.

Takete Nakano(8) has developed a new dressing technology for cutting off SIALON(Si-Al-O-N) ceramics which are difficult to cut off by conventional diamond machining. This is because the diamond wheel looses its cutting ability with time, due to pile up of ceramic chips around the diamond grains. Newly developed dressing technology consists of continuous removal of the chips during the cutting off process, in which a glass fiber reinforced plastic plate is used as a dresser. Sialon ceramics can be cut off under stable conditions at higher cutting speeds with a smaller cutting force. This also results in less wear of the diamond wheel.

T.Nakagawa(9) has found that Ultrasonic vibrations to a grinding wheel reduces grinding force and also a combination of vibrations and EDM is possible for machining certain types of ceramics.

4. ELECTRO-DISCHARGE MACHINING OF CERAMICS

Maccalous and Coppfer(10) have found that EDM process when used on complex Zirconium Diboride Ceramic components had higher material removal rates as compared to diamond grinding which can result in cost savings. They also found that fewer set ups and operations were required with the EDM process. W.Konig(11) has reported that it is possible to machine ceramic materials by EDM process if electrical conductivity is of the order of 100 ohm-cm. They verified that EDM achieves high removal rate as compared to traditional techniques like diamond machining methods. However, the metal removal mechanism for machining conductive ceramics differs considerably from those involved in metal machining.

5. LASER BEAM MACHINING OF CERAMICS

Laser beam's capability as a tool for machining of advanced ceramics has been reported by a number of scienti sts(12). The technique has high productivity and eliminates the problems of tool wear, but research in process optimization is required for its extensive application. H.K.Tonshoff (13) has reported a reduction or total absence of cracks and other defects in CO_2, Nd, YAG and excimer laser drilling of ceramics with short pulses and high beam intensities.

6. HOT MACHINING OF CERAMICS

Kunio Vehara and others(14) has reported that the cutting mechanism of the ceramics changes from brittle fracture to plastic deformation with the increase in temperatures, and at high temperatures the machining of ceramics is similar to conventional metal cutting. In hot machining the surface finish and tool life are improved remarkably. They have further reported that in hot machining ceramics can be machined with high removal rates and with good surface integrity.

7. WATER JET MACHINING OF CERAMICS

For ceramics machining a high velocity fluid jet either with or without abrasive particles in the cavitating or in continuous mode can be used. This type of machining is best for cutting slots and grooves or for trepanning large holes in ceramics. There are no heat effects nor large mechanical forces in the cutting process. Consequently very fragile ceramics can be cut without any damage. University of Rhode Island(4) has used a high pressure abrasive water jet for machining certain varieties of high alumina ceramics. The system consists of a robotically controlled water jet that handles pressures upto 200 M.Pa with water feed rates ranging from 0.1 to 1.30 kg/minute.

8. CHEMICAL ETCHING OF CERAMICS

Ceramics such as silicon wafers and quartz crystals can be machined with chemical etching and this results in the production of good quality holes. The following problems were, however, present: Strict and extensive controls were needed in the use of toxic chemicals and the process is very time consuming.

9. ULTRASONIC MACHINING OF ADVANCED CERAMICS

Since ultrasonic machining is a non-chemical, non-electrical and non-thermal process, it provides the user with a capability of machining complex shapes and contours in many hard and brittle materials.

This process is preferred over laser drilling as reported by Moreland(15). In laser drilling, there is always the drawback of microcracking and presence of laser slag. Hence ultrasonic machining is always preferable for high quality, high yield and high reliability applications.

For materials like structural ceramics and glasses,the structure of the material is such that large forces are required to machine these materials and these cause localized surface cracking which leads to reduce strength and load carrying capacity of the components. The properties of the components can be improved by careful machining. Ultrasonic machining provides means by which ceramics can be machined successfully preserving their surface integrity.

A recent application of ultrasonic machining is the machining of three dimensional shapes in new generation ceramics. This process is especially suitable for machining of complex three dimensional shapes up to 100 mm in diameter in brittle materials. USM has a number of important advantages for ceramic applications: conducting and non-conducting materials can be machined irrespective of their hardness; there are no chemical or electrical alterations on the workpiece surface and complex three dimensional contours can be machined as quickly as simple ones However, the main problems associated with successful application of ultrasonic machining are:

* Low material removal rates and hence low productivity;

* Likely production inaccuracies of the holes machined;

* Presence of longitudinal and lateral tool wear; and

* Gradual wear of the abrasive particles themselves.

Material removal rate in USM is influenced by a number of parameters such as amplitude and frequency of vibrations, static load,type and size of the abrasives, concentration of the abrasives in the slurry, method of feeding of abrasive slurry, the medium of the slurry,the work material properties tool geometry and tool material.

10. ROTARY ULTRASONIC MACHINING OF CERAMICS

Komaraiah(16) has conducted experiments on various ceramic material such as alumina, ferrite and percelain and studied surface integrity aspects with both conventional and rotary mode of ultrasonic machining process. The study reveals that rotary USM is superior to conventional USM in all respects. The rotary mode yields higher material removal rates, lower surface roughness and better accuracy. Rotary USM is always recommended for machining axisymmetric holes.

11. PRESENT INVESTIGATIONS

Experiments have been conducted on a vertical 600 watts ultrasonic drilling machine manufactured by Kerry's Ultrasonics, U.K. The advanced ceramics used in the study are Hexoloy Alpha Silicon Carbide supplied by Standard Oil and Engineering Company, U.S.A. and High Alumina (96%) and steatite ceramics supplied by Jyoti Ceramics, Nasik, India.

The abrasive: water ratio in the slurry used is 1:4 (by weight); SiC 280 Mesh abrasives are used in the investigations. For computing the tool wear and the production accuracy a profile projector is used.

12. RESULTS AND DISCUSSIONS

12.1 Production accuracy in the machining of advanced ceramics

Due to external feeding of the slurry and as the point of feeding is constant, gradual rounding off of hole at entry occurs and the hole size at entry tends to be enlarged. As drilling continues and proceeds further, there is a tapering effect on the hole due to the onset of lateral wear of the tool.

Conicity or taper in the hole is mainly due to lateral wear of the tool; use of tougher and harder tool materials such as stainless steel and tungsten carbide reduces lateral wear of the tool and the conicity of the holes produced. Conicity also depends on the nature of work material machined.

Out-of-roundness is caused by the non-uniform flow of abrasive particles in the gap between the tool and the workpiece; these particles scour the tool surface and produce lateral or side wear. The extent of lateral wear depends on the rate of abrasive flow and as this is not uniform around the circumference of the tool, tool wear is not uniform and out-of-roundness of the tool occurs first, followed by out-of-roundness of the hole produced. Out-of-roundness of the holes produced is assessed by diameter measurement.It is observed that out-of-roundness of the hole is more at entry. Production accuracyof holes machined is shown in Table 1 and 2. It has been observed that as the depth of the hole increases the machining rate decreases for a while and then remains almost a constant rate (Fig. 1 and 2). This is due to the non-availability of the fresh abrasive slurry at higher depths of hole.

12.2 Tool wear in the machining of advanced ceramics

Tool wear is an important factor which affects the performance of the ultrasonic drilling process. Tool wear results in frequent tool replacement. As the tool length and weight decrease the tool is not in resonance with the driving oscillator, amplitude of vibrations reduces and the machining rate decreases. A contoured tool profile results in dimensional inaccuracy of the holes machined. Tool wear also leads to taper formation in the holes resulting in poor circulation of the abrasive slurry and reduced machining rate. Tool wear in Ultrasonic drilling is influenced by a number of factors, such as cutting time, static load, type of abrasives used and tool material which in turn all affect the performance of the process considerably.

Nomoti(17) has observed that longitudinal tool wear is significant only at depths greater than 15 mm while machining ceramics. Rozenberg(18), has found that the relative tool wear was about 45 times more while machining tungsten carbide when compared to machining of glass. Tungsten carbide is difficult to machine because of its higher toughness when compared with that of glass. Adithan(19) has found that all the parameters which affect the material removal rate also affect the tool wear rate. In addition to longitudinal wear and lateral wear a third type of wear known as cavitation wear also exists. This is confirmed by SEM analysis of worn out tool surface (Fig. 3). Tool wear in the case of machining of Hexology Alpha Silicon Carbide was found to be several times more as compared to machining of high Alumina.

12.3 Tool Wear propagation

To study the tool wear propagation the shank was unscrewed from the transducer after the drilling operation and tool end viewed under a profile projector with a high magnification (x 100). The change in the profile of the tool after drilling every hole was traced. Both longitudinal and lateral wear values were measured from a preset reference surface to an accuracy of 1 micron. The changes of the tool contour from hole to hole become evident through the recording of the longitudinal sections into one another. From this the development of the wear profile could be seen (Fig.4). Fig. 5 shows the SEM of wear profile of tool. In the machining of steatite both longitudinal and lateral wear were present. When high Alumina was machined the longitudinal tool wear increased by several times as compared to steatite machining. Fig. 6 shows a plot between cumulative longitudinal tool wear and cumulative cutting time when Hexoloy Alpha Silicon Carbide is machined with different diameter holes. It is found that initially tool wear is proportional to cutting time and after certain time there is a rapid increase in tool wear. The curve shows following three distinct phases:

1. Primary Wear Region (Rapid initial Wear)

2. Linear Wear Region Up to start of tool break down (lower rate of wear during stable condition)

3. Unstable Wear Region (after start of tool breakdown until tool failure).

A sudden increase in tool wear values at the end of linear wear region confirms that the start of the tool breakdown has occurred after a certain definite time and further machining results in a steep increase in tool wear leading to Catastrophic tool failure.

The value of surface finish of the holes drilled in Hexoloy SiC varies from 0.5 to 0.7 μm(Ra).

13. ACKNOWLEDGEMENT

The authors wish to thank Dr. S.R. Gowariker, Director, Thapar Corporate Research and Development, Patiala for his valuable guidance in the preparation of this paper.

REFERENCES

1. Hidehiko Takeyama, "Machining of Hard Ceramics with Sintered Diamond Tools". Industrial Diamond Association of Japan's 30th Anniversary, May 1978, pp. 110-113.

2. 89th Ceramic Society meeting, Abstract, "Advanced Materials and Processes", 10/87, pp. 28.

3. I. Inasaki; "High Efficiency Grinding of Advanced Ceramics", Annals of the CIRP. Vol. 35/1/1986 pp. 211-214.

4. L.M. Shepard, "Machining of Advanced Ceramics", Advanced Materials and Processes. 12/87.

5. W. Konig, et al "Grinding of Si/SiC Ceramics: Achieving High Performance and Minimum Damage", Third International Symposium on Ceramic Materials, Las Vegas, NV, USA. Nov., 88

6. Sentoku, et al, "Machining Performance for Machinable Ceramics II" Milling. Journal Funtai Qyabi Funnatsuyak in. 1988 pp. 142-146.

7. Sentoku, et al, "Machining Performance for Machinable Ceramics III"(in Japanese),Journal Funtai Qyabi Funnatsuyk in 36(2), 1989, pp. 121-124.

8. Takate Nakano, et al "Newly developed dressing technology for cutting-off SIALON (Si-Al-O-N) Ceramics" Annals of the CIRP, Vol. 38/1/1989.

9. T. Nakagawa, et al "Novel Efficient and Precise grinding technology for Structural Ceramics" MRS International meeting on Advanced Materials, Vol. 5, Structural Ceramics/Fracture mechanics, Tokyo, Japan 30th May-June, 1988. pp. 157-169.

10. Maccalous, J.N; Coppfer, W.P. Electrical Discharge Machining of Zirconum Diboride" AIAA/ASME/SAL, Thirteenth Structural Dynamics and Materials Conference, San Antonio, Texas, April 10-11, 1972.

11. W. Konig, et al, "EDM Future steps towards the machining of Ceramics" Annals of the CIRP Vol. 37/1/1987 pp. 623-631.

12. Masaki Mori, et al, "Study on Ultrasonic Laser Machining" Annals of the CRIP. Vol. 25/1/76, pp. 185-189.

13. HK Tonshoff, et al, "Laser Cutting of Advanced Ceramics" Annals of the CIRP, Vol.38/1/1989.

14. Kunio Vehara, et al, "Cutting Ceramics with a technique of Hot machining" Annals of CIRP, Vol. 35/1/1986, pp. 55-58.

15. Moreland,Mary, A; "Ultrasonic Advantages revealed in the Hole Story," 1986, pp. 7.

16. Komaraiah, M; "Some studies on Ultrasonic Machining" Ph.D Thesis (1986) Osmania University, Hyderabad (India).

17. Nomoti, A; "Ultrasonic Machining by Low Power Vibration" Journal of Accoustic Society of America, 26(1964) pp. 10-81.

18. Rozenberg, et al, "Ultrasonic Cutting" Consultants Bureau, New York, 1964, Translation from the Russian.

19. Adithan, M. et al, "Parameter Influence on Tool Wear in Ultrasonic Drilling" Tribology International, Dec., 1974, pp. 260-68.

PRODUCTION ACCURACYOF HOLES MACHINED

TABLE 1 (Work Material: High Alumina)

Sr No	Diameter of hole (mm)	Oversize of hole (mm)	Out-of-roundness at entry (mm)	Out-of-roundness at exit (mm)	Conicity*
1.	2.90	0.15	0.18	0.32	0.05
2.	3.00	0.18	0.45	0.53	0.06
3.	3.50	0.22	0.19	0.20	0.09
4.	4.50	0.16	0.23	0.28	0.07
5.	5.00	0.10	0.36	0.41	0.08

*Conicity or Taper for 4 mm length

TABLE 2 (Work Material: Hexoloy Alpha Silicon Carbide)

Sr No	Diameter of hole (mm)	Oversize of hole (mm)	Out-of-roundness at entry (mm)	Out-of-roundness at exit (mm)	Conicity*
1.	2.500	0.255	0.150	0.230	0.385
2.	2.900	0.110	0.120	0.130	0.005
3.	3.500	0.140	0.210	0.210	0.020
4.	4.500	0.280	0.270	0.280	0.015
5.	6.500	0.185	0.125	0.175	0.075

*Conicity or Taper for 5 mm length

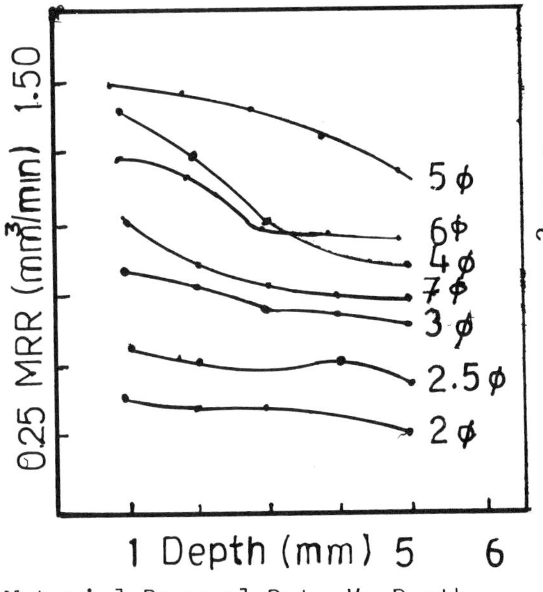

Fig 1: Material Removal Rate Vs Depth
Work Material: Hexoloy Alpha Silicon Carbide
Tool Material: Stainless Steel

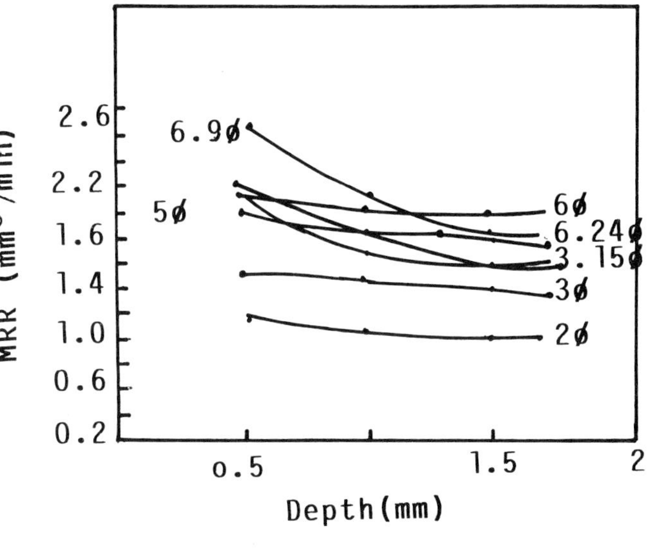

Fig.2: Material Removal Rate Vs Depth
Work Material : Steatite
Tool Material : Stainless Steel

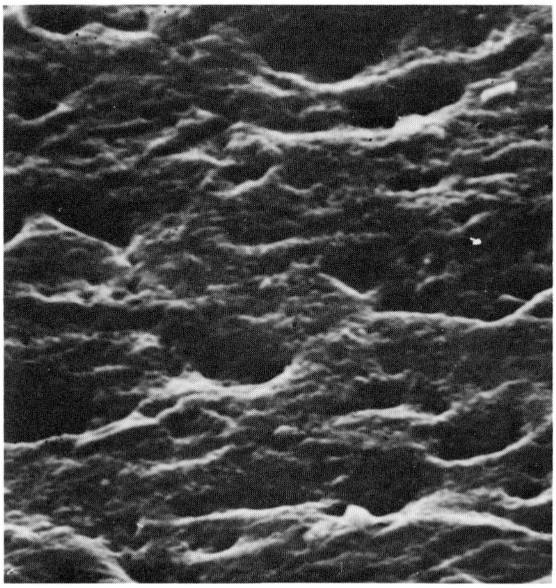

Fig. 3: Scanning Electron Micrograph of Tool Surface Produced after drilling holes in High Alumina. The Surface shows cavities formed due to cavitational wear (4000 x) Tool Material: Stainless Steel

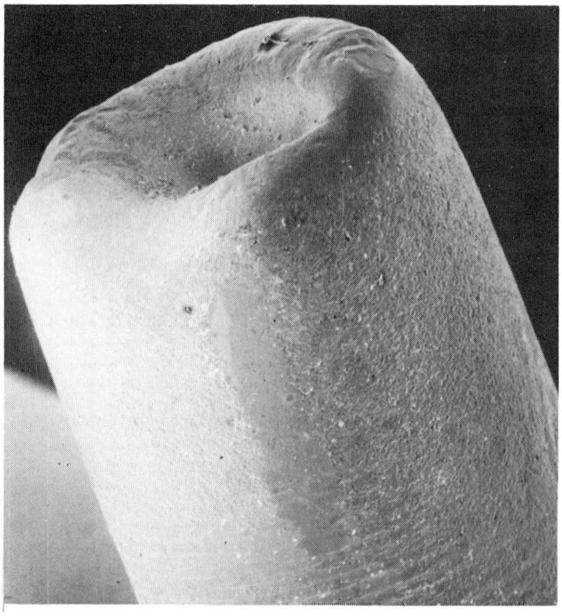

Fig. 5 : Scanning Electron Micrograph of Wear Profile of Tool (40 x) (Work Material: High Alumina) (Tool Material: Stainless Steel)

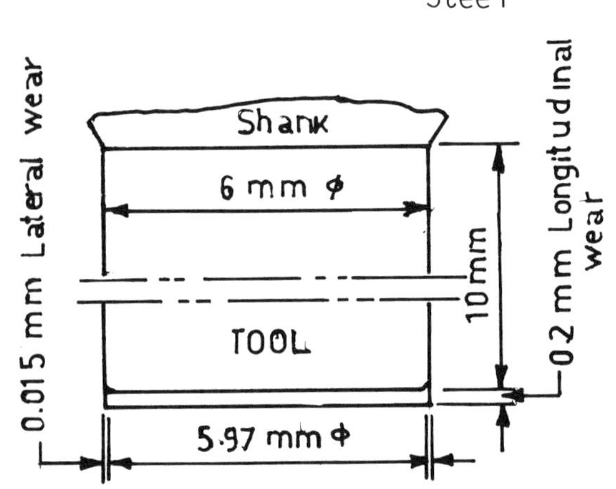

Fig. 4: Schematic representation of tool wear (Work material: Steatite (Tool material: Stainless Steel)

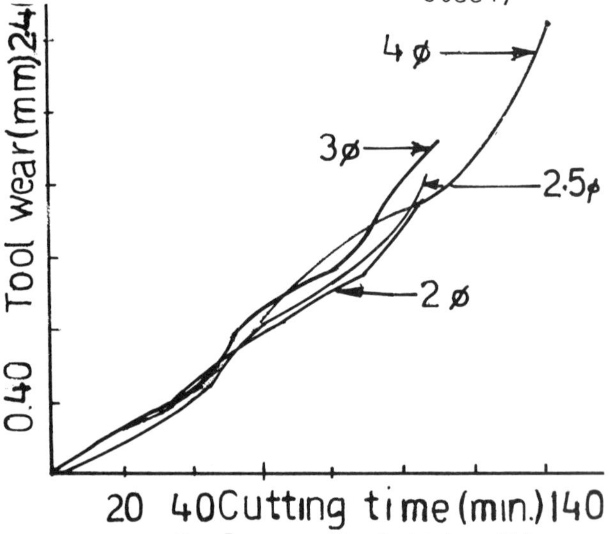

Fig. 6: Tool wear Vs Cutting Time (Work Material: Hexoloy Alpha Silicon Carbide) (Tool Material: Stainless Steel)

Thin Films in Tribology / D. Dowson et al. (Editors)
© 1993 Elsevier Science Publishers B.V. All rights reserved.

MECHANICAL PROPERTIES OF TiN, (Ti,Al)N, (Ti,Nb)N AND Ti(C,N) COATINGS MEASURED BY NANOINDENTATION

E. Vancoille, J.P. Celis and J.R. Roos [a]
[a] Department of Metallurgy and Materials Engineering
Katholieke Universiteit Leuven, de Croylaan 2 B 3001 Leuven
BELGIUM

Ternary alloyed TiN based PVD coatings are second generation coatings for high speed cutting of abrasive alloys. In this respect, hardness is certainly one of the most interesting mechanical properties, but evaluating it is a challenge because of the limited coating thickness (2 to 3 μm). Therefore, nanoindentation was used to determine hardness and Young's modulus of steered arc ion plated (Ti,Al)N and (Ti,Nb)N coatings with different compositions. It was found that the ternary coatings were harder and showed a higher Young's modulus compared to the simple TiN. Since high temperatures develop during machining at increased speeds, the effect of a heat treatment on the mechanical properties of these coatings is worth investigating. So the hardness of TiN, $(Ti_{50},Al_{50})N$, $(Ti_{85},Nb_{15})N$ and $Ti(C_{60},N_{40})$ coatings was determined after a heat treatment under argon at temperatures up to 900 °C. It was observed that the hardness decreased as the coatings are annealed at progressively higher temperatures. X-ray diffraction showed that the internal stress decreases and the crystallographic texture of the coatings is modified.

1. INTRODUCTION

Hardness in the broadest sense is defined as the resistance of a material against plastic deformation [1]. In a typical micro-Vickers hardness test the indentation diagonal is measured after lifting away the indenter. The hardness is then the ratio of the load applied to the area of the indent remaining. This hardness value is thus derived from a plastic deformation of the material and elastic effects are not taken into account. With this definition, hardness of bulk materials can be measured easily and is generally independent of the load applied. When performed on thin coatings on a substrate, hardness measurements imply the use of very low loads, since otherwise the plastic deformation field will extend into the substrate, and the hardness of the system rather than that of the coating would be measured. Since empirical rules [2] state that the indentation depth may not exceed one seventh or even one tenth of the coating thickness, the usual micro-Vickers load range (0.15 to 5 N) is not suited for characterizing 2 to 3 μm thick ceramic coatings with an expected hardness of 2000 to 3000 HVN. Low load indentation instruments not causing coating cracking or plastic deformation of the substrate are nowadays of interest. Pollock et al. [3] described an ultra low load penetration hardness tester having an in-depth displacement resolution of 5 nm. The ultra microhardness measurements were correlated with conventional optical readings of the indentation area and the results were found to be satisfactory. At small indentation depths a trigonal pyramidal Berkovich type indenter is most appropriate because such a indenter shape can be manufactured within narrow tolerances. Pethica et al. [4] described an instrument, nowadays called the Nanoindenter, for mechanical testing on the nanometre scale. The in-depth analyzing capabilities of the nanoindenter justify its use for investigating thin coatings where conventional Vickers indentations are too deep because of the high loads applied. The elastic behaviour of the material can be assessed by comparing the elastic recovery of the indentation pit with the total indentation depth [4]. Pethica discussed in great detail the rounding effect of the tip as the indentation depth is below 20 nm. In this case, the shape of the tip must be known exactly, e.g. by measuring the size of replicas of the indents in a transmission electron microscope. From the unloading part of the indentation curves Young's modulus can be computed [4]. This approach is a good alternative for the otherwise cumbersome beam deflection measurements [5]. A number of nanoindenter-like instruments were recently developed in different research laboratories. Yanagisawa and Motomura [6] described in a study about hard coatings an ultra microindentation apparatus with 1 μN load resolution and 4 nm displacement resolution. They found a satisfactory correspondence between the Young's modulus derived from a continuous indentation technique and the one measured with a beam bending method. Loubet et al. [7] discussed the theoretical background and formulas needed to determine hardness and Young's modulus from a continuous recording of load and displacement. Klages [8] reviewed the different mechanical properties of thin coatings such as internal stress, crack resistance and bending fatigue resistance in relation to hardness, and discussed the usefulness of nanoindentation for hardness and Young's modulus determination. Jehn [2] then compared the performance of different ultra low-load displacement instruments by comparing the measured hardness of TiN and (Ti,Al)N coatings. The Al or Nb alloyed TiN coatings are meant to be used in high performance cutting operations, where high temperatures develop at the cutting edges [9,10]. Although the contact length of the chip on (Ti,X)N coated tools and thus the friction are reduced [11], the heat generated may still degrade the mechanical properties of the coatings. Therefore the influence of heat on the mechanical properties of TiN and (Ti,X)N coatings has to be known.

The purpose of this work is to determine the nanohardness and Young's modulus of as-plated ternary coatings, such as (Ti,Al)N, (Ti,Nb)N and Ti(C,N) with different compositions. Additionally, the effect of a heat treatment on the nanohardness of (Ti,X)N coatings has been investigated.

2. EXPERIMENTAL

The substrate material used was High Speed Steel, Werkstoffnr. 1.3343 (S6-5-2 or AISI M2) with composition 0.90% C, 0.41% Cr, 5.0 % Mo, 1.9% V, 6.4% W, produced by Thyssen Edelstahlwerke AG under brand name THYRAPID 3341. The samples were mechanically polished before coating to a smooth surface finish (R_a = 0.05 µm). (Ti,Al)N, (Ti,Nb)N and Ti(C,N) coatings were produced by steered arc ion plating from targets with composition Ti, $Ti_{83}Al_{17}$, $Ti_{67}Al_{33}$, $Ti_{50}Al_{50}$, $Ti_{85}Nb_{15}$, and $Ti_{70}Nb_{30}$ (all in atom%). The coatings were all stoichiometric with regard to nitrogen. The ratio of the metals was found the same in the nitride coatings as in the metal cathodes [12]. Ti(C,N) coatings with a 2/3 ratio of C to N were also produced. A load or/and displacement controlled indentation instrument from Nano Instruments Inc., Knoxville TN was used. This "NANO I" is capable of applying small loads in the order of 0.2 µN and measures the resulting depth displacements down to 0.2 nm accurate. A computer controlled XY table with a 1 µm resolution allows indentations to be made on specified spots on the surface. The diamond indenter used is a triangular base Berkovich pyramid with an apex angle of 65.3° so that the same relationship holds for the indentation area vs penetration depth as for a standard Vickers indenter. A standard indentation cycle was programmed and the maximum load was chosen so that an indentation depth of 70 to 100 nm was reached. The ratio coating thickness to indentation depth is thus 1/20 to 1/30, so that the critical depth criterion is fulfilled [2]. On each sample 10 to 15 valid indentations were done to evaluate the scatter in the measurements.

For the hardness after annealing experiments, TiN, $(Ti_{50},Al_{50})N$, $(Ti_{85},Nb_{15})N$ and $Ti(C_{40},N_{60})$ coatings were encapsulated in a quartz tube pumped to 10^{-6} torr vacuum, flushed with argon, and then hermetically sealed. The remaining O_2 partial pressure did not exceed 10^{-8} torr. The samples were then heat treated in their quartz sealings for 24 hours at temperatures of 300, 500, 700, and 900 °C. The coating hardness was not determined at high temperature but at room-temperature by nanoindentation. The loading/unloading cycle applied was identical to the one used for the as-deposited coatings. X-Ray diffractograms of the annealed coatings were taken with a Philips diffractometer operating with CuKα radiation. SEM of the surface and of fractured cross sections was done. Raman spectroscopy was used to detect a possible oxygen contamination on the surface of the heat treated samples.

3. RESULTS and DISCUSSION

3.1. As-deposited coatings

A standard loading/unloading sequence to a maximal force of 4.5 mN was used. The cycle (Figure 1) consists of an approach segment in which the indenter approaches the surface from a height of 1 µm at a rate of 5 nm/sec and determines the point of contact by measuring the stiffness response of the system. The load is applied at a rate of 50 µN/sec until 4500 µN is reached. Then the indenter is held for ten sampling points (about three seconds) at this load before unloading is done at the same rate (50 μN/sec) to 80% of the maximum load. Then the indenter is again held stationary during 60 seconds. The displacement drift then recorded is assumed to be of thermal origin. Finally the load cell is reset and the indenter is lifted away from the surface. The total cycle duration is about seven minutes including the approach segment. The resulting maximum indentation depth for (Ti,X)N coatings is in the range 70 to 100 nm. These indentations are too small to be seen in the SEM because of the relative poor surface conductivity of ceramics. Only the indents where the approach segment effectively brought the indenter within 2 nm to the real surface were accepted as valid hardness readings. About one quarter of all indents does not meet this criterion making smooth surfaces a necessity for using nanoindentation.

Derivation of hardness and Young's modulus from load/ displacement data was done using the standard software routines based on the work by Doerner and Nix [13]. Figure 2 summarizes the hardness for steered arc (Ti,Al)N and (Ti,Nb)N coatings of different compositions. The hardness of the (Ti,Al)N coatings increases slightly as the aluminium content in the coating increases, while the hardness of the (Ti,Nb)N coatings decreases as the niobium content increases although the scatter is quite large (± 500 HVN). In any case, the ternary coatings are harder than the simple TiN. This agrees with the findings of Ives et al. [14] who reported Knoop hardness values ranging from 1900 to 3900 HK for their TiN steered arc coatings while no clear relationship between bias voltage, nitrogen pressure composition and hardness was observed. The present measurements done on the Ti(C,N) coatings displayed a hardness of 2500 to 3200 HVN depending on their composition. This indicates that Ti(C,N) is one of the hardest ternary coatings. As the carbon content of the steered arc ion plated Ti(C,N) coatings was decreased from 60 to 40 at%, the hardness was found to increase. This agrees with literature data on the Vickers hardness of TiC and Ti(C,N) coatings [15]. For easy reference to literature values, Vickers measurements interpreted with the Jönnson-Hogmark model [16] are also reported in Figure 2. The correspondence to the nanoindenter measurements is satisfactory, although in general Vickers measurements tend to give higher values for the hardness. Jehn [2] reported somewhat lower Vickers hardness values (1900 - 2200 HVN) for TiN and different (Ti,Al)N coatings on M2 steel than found here. Kennedy and Tang [17] reported a nanohardness value of 2930 HVN for TiN which is higher again. The hardness increase as the Al increases is in contradiction with the findings of Knotek and Prengel [18] for arc ion plated (Ti,Al)N coatings. They reported 1800 HVN values at -50V bias voltage for coatings with a high Al content, and 2200 HVN at -400V bias voltage for coatings with a lower Al content [18]. It seems from these data that hardness values as found in literature, tend to show a considerable scatter and no clear relationship between coating composition and hardness is seen.

Especially the spread in the nanoindentation measurements is worth some discussion. Because the calculation of the hardness relies on an accurate measurement of the displacement, it is easy to imagine that sample surface roughness plays an important role. Arc coatings tend to be rough due to the droplets or so-called macros's, but these inhomogeneities are also found across the whole thickness of the coating. One has in fact a composite material of a hard ceramic coating with softer metal droplets. Depending on the measuring location a high, low or intermediate hardness can be found. Because Young's modulus is also derived from the load/displacement cycle of the nanoindentation, the same remark concerning the spread in the experiments applies. Figure 3 shows Young's modulus for TiN, (Ti,Al)N and (Ti,Nb)N steered arc ion plated coatings for different compositions. The modulus is only slightly influenced by the insertion of Al or Nb in the TiN lattice. A very weak maximum is noticed for the (Ti,Nb)N coatings at 5 at% niobium. For the (Ti,Al)N coatings, Young's modulus can reach high values (350 GPa at 20 at% Al), but the scatter band is also considerable (10 %). The range of elastic moduli found here is in good agreement with most literature values reported. Oliver et al. [19] reported Young's moduli ranging from 300 to 450 GPa for sputtered TiN coatings on AISI304 whereas Kennedy and Tang [17] reported values of 280 GPa for the same TiN. Chollet et al. [20] initially found 640 GPa for the Young's modulus of TiN as measured by flexural resonance frequency measurements, although more precise measurements based on the same principle gave values of 384 to 446 GPa for CVD TiN depending on the type of substrate material [21]. Recent work on nanoindentation of TiN coatings on stainless steel was done by Sproul et al. [22] who reported a Young's Modulus of 400 to 500 GPa. The measurements presently done on Ti(C,N) coatings gave a Young's modulus of 300 GPa. So far no comparative measurement for Young's moduli of PVD Ti(C,N) could be found, but Chollet et al. [21] reported Young's moduli of 580 GPa for CVD TiC and of 264 GPa for CVD Ti(C,N), so that, considering the latter, our measurements seem to be in fair agreement.

3.2. Heat treated coatings

A comparison between the indentation cycles done on steered arc ion plated TiN and $(Ti_{85},Nb_{15})N$ coatings before and after heat treatment for 24 hours at 900 °C under argon is shown in Figure 4. Because the same load was applied in all experiments it is clear that the heat treated coatings have softened. Figure 5 gives the hardness as a function of the heat treatment temperature for the different types of coatings. The hardness of TiN decreases from 2200 HVN at room-temperature to 771 HVN after annealing at 900 °C, whereas the $(Ti_{50},Al_{50})N$ hardness decreases from 2400 HVN to 733 HVN after annealing at 900 °C. After annealing at 300 °C $(Ti_{50},Al_{50})N$ shows a hardness increase to 2900 HVN. For the $(Ti_{85},Nb_{15})N$ coatings the hardness decreases from 2800 HVN to 1210 HVN after annealing at 900 °C. For these samples, the nanoindentation values were cross-checked with Vickers hardness measurements calculated according to the Jönsson-Hogmark approach [16]. These experiments are in agreement with the results of Quinto et al. [23] who measured the hardness of PVD and CVD hard coatings deposited on cemented carbide with a Vickers indenter. A decrease in hardness from 2000 HVN at room-temperature down to 500 HVN at 1000 °C was found for TiN. They stated that the stored elastic energy within the metastable PVD microstructure provides a considerable driving force for "thermal recovery" at higher temperatures. (Ti,Al)N coatings with 75/25 and 50/50 at% Ti/Al (2500 HVN as deposited) were found to maintain a higher hot hardness than TiN up to 800 °C (1000 HVN), which seems to agree with the present findings. The relationship between hardness, internal stress and the peak width at half height as measured by X-ray diffractograms was also demonstrated [23]. Knotek et al. [24] measured the hot hardness of CVD TiN, PVD TiN, and PVD (Ti,Al,V)N coatings and found similar results to those of Quinto [23]. The Ti(C,N) coatings shows a similar behaviour (Figure 5d). The hardness decreases from 3000 HVN to 1100 HVN after heat treating at 900 °C. Tamura and Kibo [15] investigated the properties of TiC, TiN and $Ti(C_x,N_{1-x})$ sputtered coatings for annealing effects in air as well as in argon. The microhardness they measured for TiC by the Vickers method decreased from 5000 HVN to 3700 HVN, which corresponds well to the observed hardness decrease here of 2000 Vickers units for the Ti(C,N). These authors remarked however that more carbon in the coatings induced severe oxidation. Even under argon atmosphere, the residual oxygen originating from e.g. the uncoated sample substrate side forms a thin oxygen layer on the coating. Tamura and Kibo attributed the hardness decrease to this oxide [15]. This is an important point and so the quality of the inert blanket and the presence of a soft oxide layer on the sample must be checked. With Raman spectroscopy no titanium oxide signal could be found on any of the at 900 °C treated (Ti,X)N, though the Raman technique is known to be very sensitive to oxides [25]. The explanation for this can be sought in the specific nature of the arc plated coatings. The surface of the samples is covered with metal macro-droplets consisting of Ti, TiAl or TiNb metal. These macros can oxidize much more easily than their corresponding nitrides. In fact, such metal splats would function as oxygen getters and even diminish the oxygen partial pressure above the sample so that an oxidation of the nitride would be even more unlikely. The nanoindentation measurements were done on locations in between the surface defects were chosen. Consequently, one can be confident that the hardness decrease is not some artefact due to an oxygen contamination layer. The explanation for the observed decrease in hardness as the heat treating temperature increases has to be sought in the microstructural changes induced. Mader et al. [26] investigated the defect structure of ion-plated TiN during heating in a transmission electron spectroscopy hot-stage. Grain growth was not observed as the sample was heated for several hours at 900 °C in vacuum. So a softening or "annealing" mechanism such as is usually seen with metals could therefore not account for the observed hardness decrease [23,24,27]. Only at 1300 °C some dislocations were seen, while grain boundaries started to move. The hardness decrease must therefore be explained as a massive emission of point defects [27]. A more accessible way to confirm the hardness decrease is the peak shape and peak width at half

height on an XRD. Broad peaks and peak shift indicate high stresses and small crystallite size, and these features were reported to be responsible for the high hardness of TiN [23,27]. X-Ray diffractograms of TiN and Ti(C,N) coatings before and after annealing at 900 °C are shown in Figures 6. It can be seen from these qualitative measurements that after annealing at higher temperatures the diffraction peaks become narrower, indicative of a stress relief and/or grain growth. Perry [27] suggested that the usual internal stress of 3 to 5 GPa in the coatings could be reduced to 0.5 GPa.
For the Ti(C,N) coatings (Figure 6b), the (111) peak, characteristic for a steered arc coating decreases in height. The crystallographic texture of the coating upon tempering was found to change from the strong (111) to a mixed (111), (200) and (220) texture in accordance with [27]. But even a heat treatment at 900 °C for 24 hours is not long enough for attaining the full equilibrium texture. This is not so surprising because the heat treatment temperature is still low in comparison to the expected melting temperature (>3000 °C) of the ternary nitrides [26,27].

4. CONCLUSION

Ternary steered arc ion plated (Ti,Al)N coatings with compositions up to 50/50 at % Ti/Al and (Ti,Nb)N coatings with compositions up to 70/30 at% Ti/Nb were found to be harder than simple TiN. Steered arc Ti(C,N) and $(Ti_{50},Al_{50})N$ were shown to be the hardest ternary steered arc ion plated coatings with higher Young's moduli than steered arc TiN. Nanoindentation was also used to measure hardness and Young's modulus (Ti,X)N coatings heat treated in an inert atmosphere at temperatures up to 900°C. It was found that the hardness of the heat treated coatings decreased, which was confirmed by peak narrowing observed in X-Ray measurements.

5. ACKNOWLEDGMENTS

The authors wish to thank Dr. J. Ebberink of Gühring K.G. (Albstadt, Germany) and ir. S. Boelens[1] of Hauzer HTCE (Venlo, The Netherlands) for their help in preparing the coated samples and for the drilling tests. This research was carried out in the framework of the Interuniversity Networks for Fundamental Research founded by the Belgian Government (Contract IUAP nr. 4) and supported by the Commission of the European Community under Contract BRITE-RI1B-0201-C(AM).

----
[1] presently at Philips N.V., Eindhoven, The Netherlands

REFERENCES

1. D. Tabor, The Hardness of Metals, Clarendon Press, Oxford 1951
2. H. Jehn, Galvanotechnik, 80 (1989) N°4, 1193
3. D. Newey, M.A. Wilkins and H.M. Pollock, Phys. & Sci. Instrum., 15 (1982), 119
4. J.B. Pethica, R. Hutchings and W.C. Oliver, Philosophical Magazine A, 48 (4) (1983), 593
5. M.F. Doerner, D.S. Gardner and W.D. Nix, J. Mater. Res. 1 (6) (1986), 845
6. M. Yanagisawa and Y. Motomura, Lubrication Engineering, 43, 1 (1987), 52
7. J.L. Loubet, J.M. Georges and P. Kapsa, Proc. Leeds-Lyon Symp., Lyon 1989, Session F2, 1
8. C.P. Klages, Vortrage VDI Symposium-Haus Düsseldorf, 5/12/1989
9. J.M. Molarius, A.S. Korhonen, E. Harju and R. Lappalainen, Surf. Coat. Technol., 33 (1987) 117
10. H. Kunz, Vortragstext, Symposiums der Deutschen Gesellschaft für Metallkunde, 1986, 221
11. W. König and D. Kammermeier, VDI Berichte N°702, 1988, 291
12. J.R. Roos, J.P. Celis, E. Vancoille, H. Veltrop, S. Boelens, F. Jungblut, J. Ebberink and H. Homberg, Thin Solid Films, 193/194, (1990), 547
13. M.F. Doerner and W.D. Nix, J. Mater. Res. 1 (4), (1986), 601
14. M. Ives, J.S. Brooks, J. Cawley and W. Burgmer, Surf. Coat. Technol., 49 (1991), pp 244 - 252
15. M. Tamura, H. Kubo, Surf. Coatings Technol., 49 (1991) 194
16. O. Vingsbo, S. Hogmark, B. Jönsson and A. Ingemarsson, STP 889, ASTM (1986) 257
17. F. E. Kennedy and L. Tang, Proc. 16th

Leeds-Lyon Symposium, 1989, Ed. Elsevier, 40
18. O. Knotek and H.G. Prengel, Surface Modification Technologies IV, Paris 1991, pp 507 - 521
19. W.C. Oliver, C.J. Mchargue and S.J. Zinkle, Thin Solid Films, 153 (1987), 185
20. E. Török, A.J. Perry, L. Chollet and W.D. Sproul, Thin Solid Films, 153 (1987), 37
21. L. Chollet and C. Biselli, C., Proc. Leeds-Lyon Symp. 1989, 217
22. D.S. Stone, K.B. Yoder and W.D. Sproul, J. Vac. Sci. Technol. A 9 (4), (1991), 2543
23. D.T. Quinto, G.J. Wolfe and P.C. Jindal, Thin Solid Films, 153 (1987) 19
24. O. Knotek, R. Elsing, M. Atzor and H.G. Prengel, Proc. Wear of Materials Conference, Denver, Colorado, 1989, 557 - 561
25. Nakamoto and Kazuo, "Infrared and Raman spectra of inorganic and coordination compounds", Wiley, New York (N.Y.), 1986.
26. W. Mader, H.F. Fischmeister and E. Bergmann, Thin Solid Films, 182 (1989) 141
27. A.J. Perry, Thin Solid Films, 146 (1987), 165

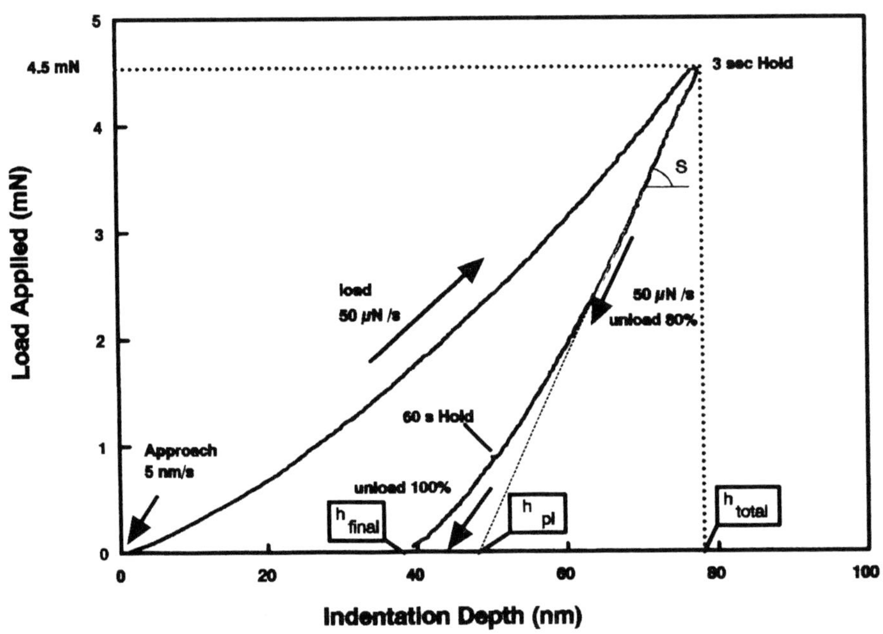

Figure 1 :

Typical nanoindentation experiment done for determining hardness and Young's modulus of (Ti,X)N coatings. The most important cycle parameters and the different segments are indicated in the figure.

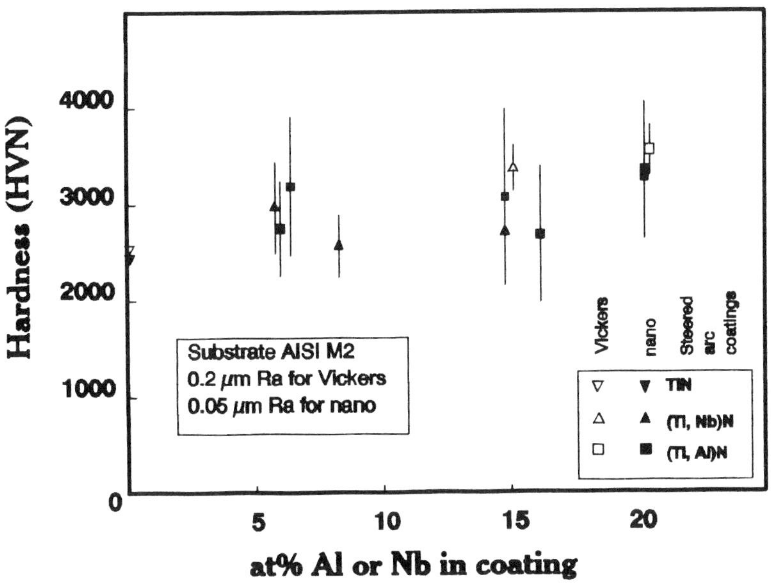

Figure 2 :

Hardness as a function of composition for TiN, (Ti,Al)N and (Ti,Nb)N coatings, measured by nanoindenter as well as Vickers measurements using Jonsson-Hogmark approach.

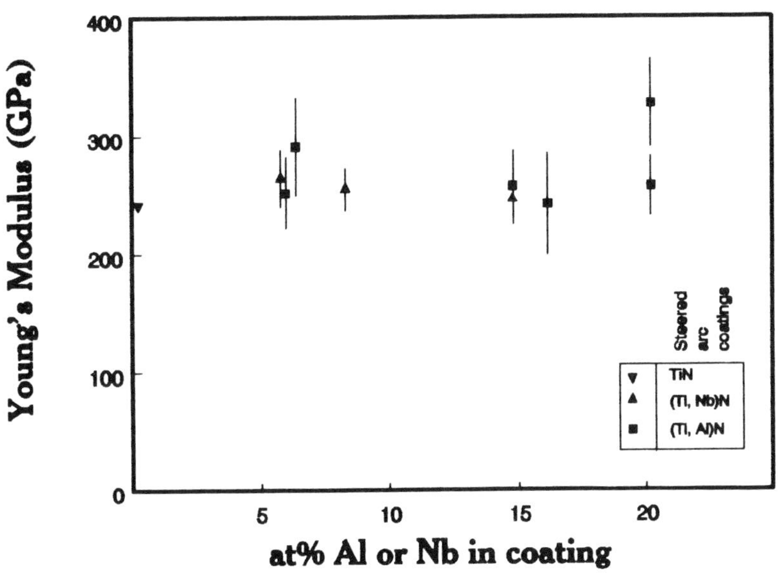

Figure 3 :

Young's modulus as a function of composition for TiN, (Ti,Al)N and (Ti,Nb)N coatings, measured by nanoindentation.

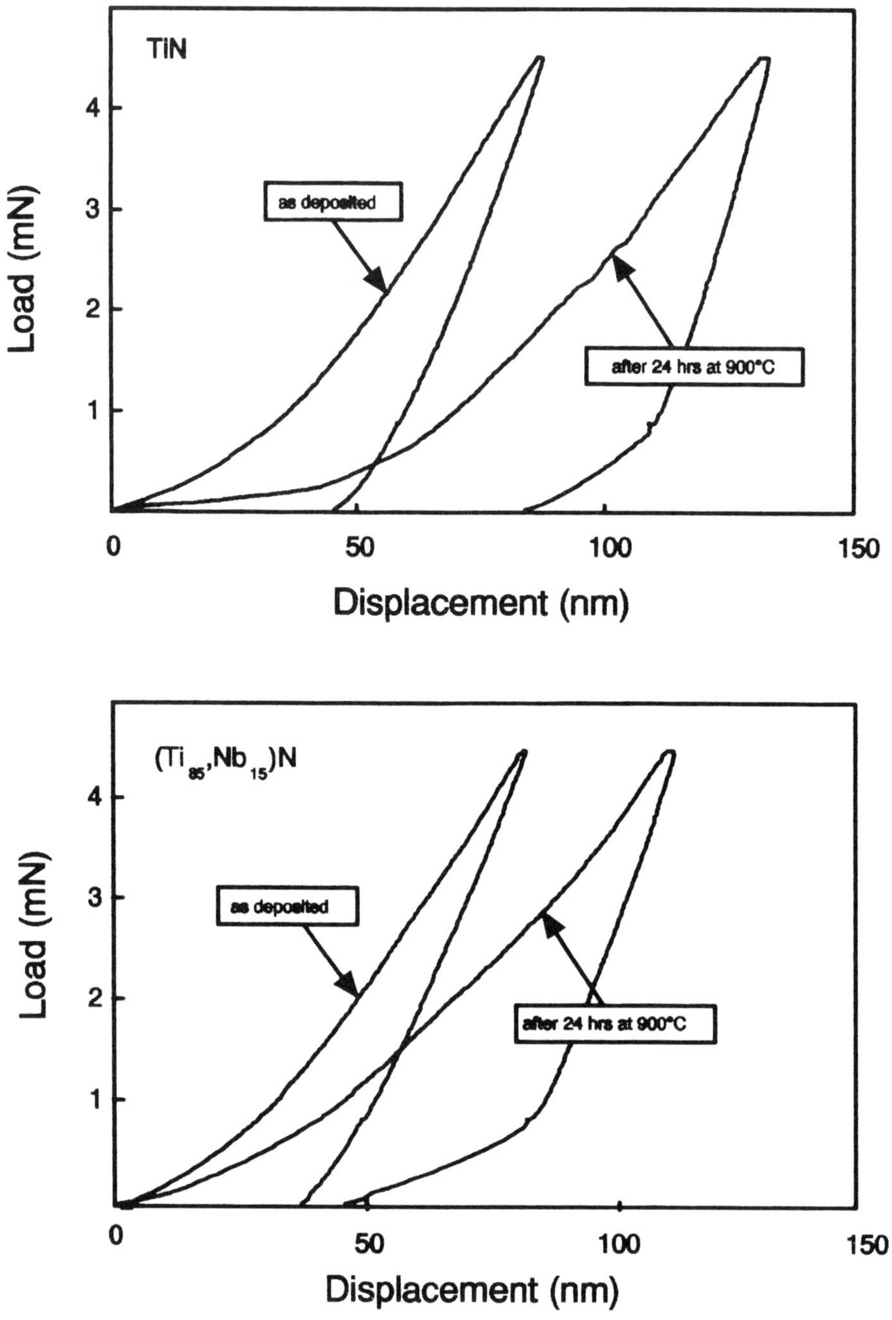

Figure 4 :

Typical indentation cycles on TiN and (Ti,Nb)N before and after heat treating. The softening is clearly observed as a shift of the curves to the right.

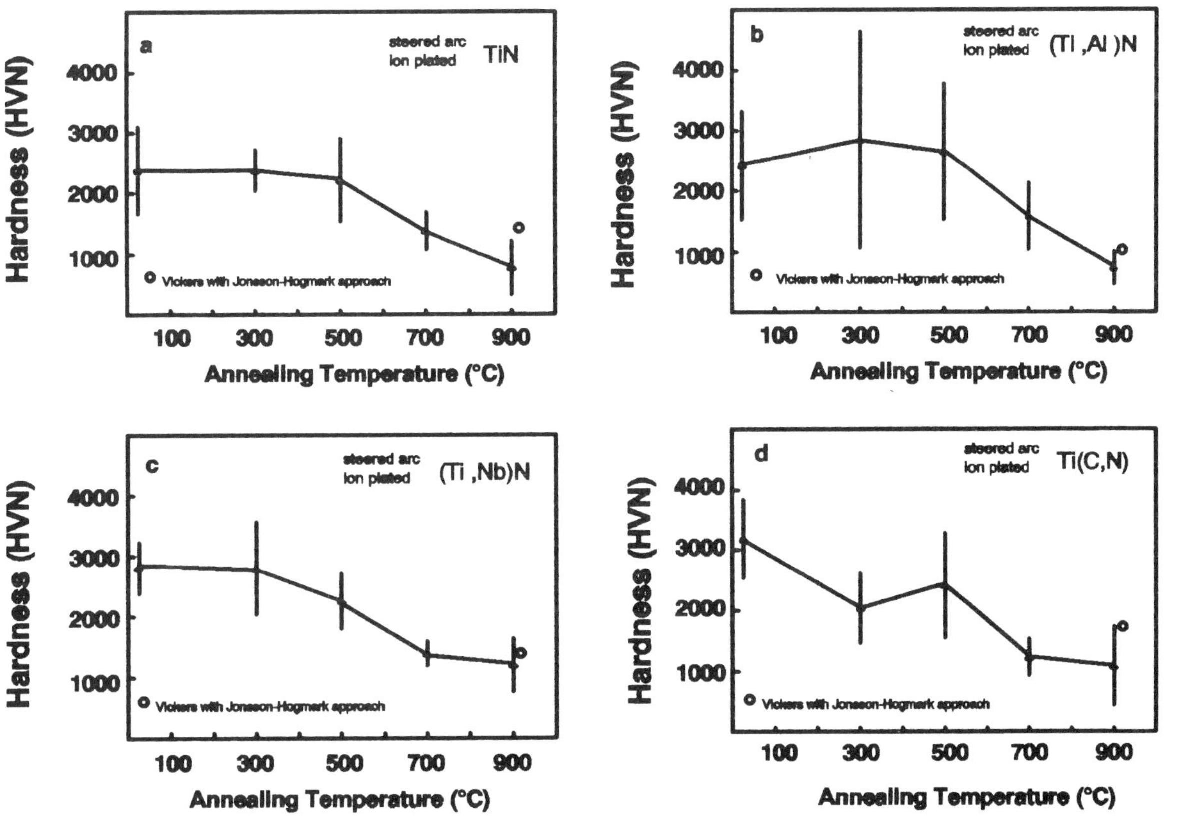

Figure 5 :

Hardness after heat treatment in inert atmosphere at different temperatures, for respectively a TiN (a), a (Ti,Al)N (b) , a (Ti,Nb)N (c) and a Ti(C,N) (d) coating.

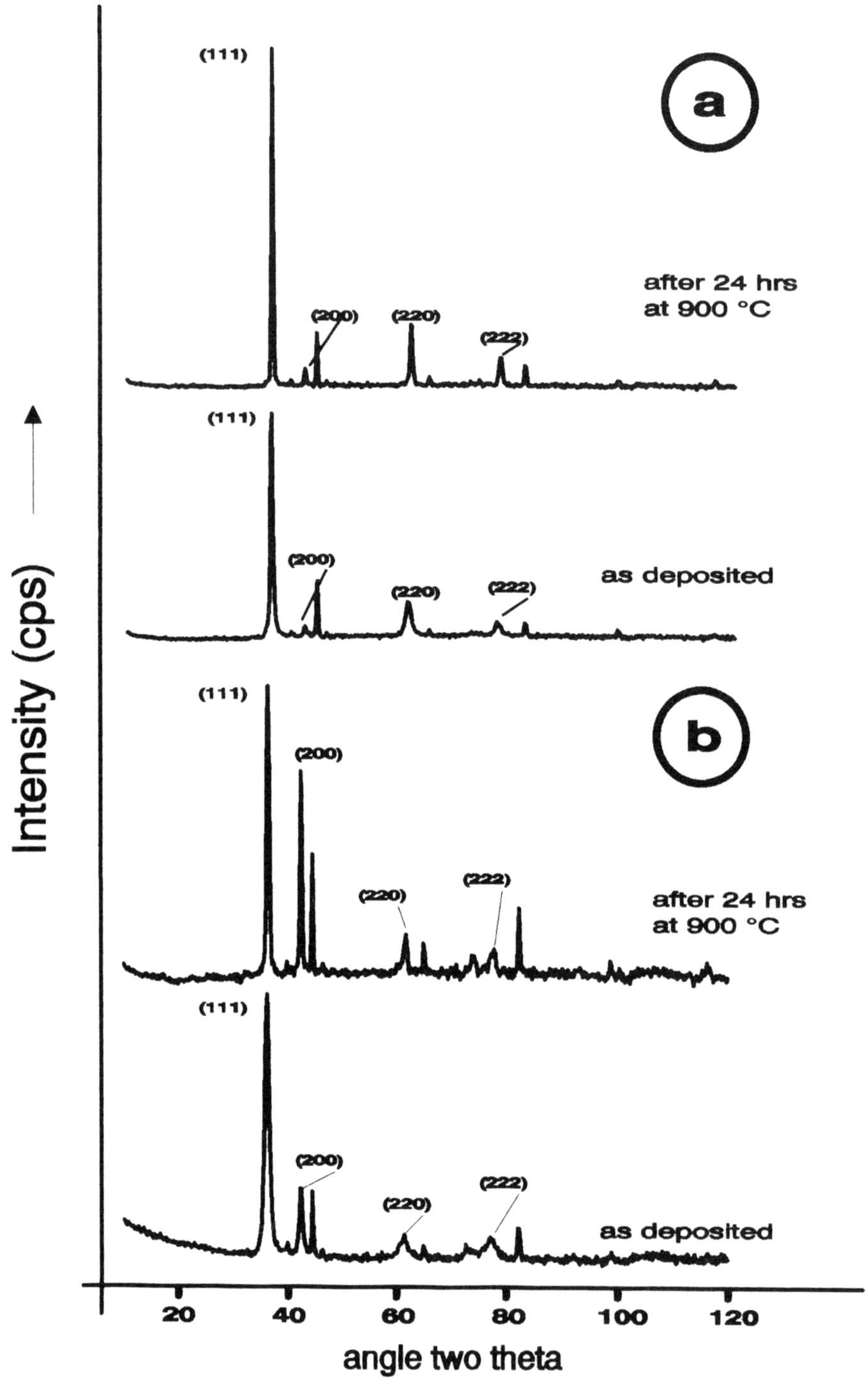

Figure 6 :

X-Ray diffractograms of a TiN (a) and a Ti(C,N) coating (b) in the as-deposited and after heat treating at 900 °C for 24 hours

Thin Films in Tribology / D. Dowson et al. (Editors)
1993 Elsevier Science Publishers B.V.

Controlling the performance of cutting operations - The critical micro-meters

L. Rozeanu and D. Pnueli
Technion, Haifa 32000, Israel

1. INTRODUCTION

The following text describes some factors which operate in the upper layer of the parts engaged in a cutting process. In the "tool" case the active upper layer is less than one μm thick, while in the "work-piece" case the thickness can reach hundreds of μm. Proper handling of these factors is expected to:

- minimize power consumption;
- improve quality of the end product;
- minimize "tool" wear.

The conclusions are intended only for cutting operations accompanied by substantial heating effects.

2. GENERAL

According to our present understanding, conventional cutting techniques perform much below their might. The mistakes made are mostly related to the temperature control of both interacting parts, one performing as a "tool", the other performing as a "work-piece".

Nominally, cutting implies shear and efficient cutting requires a large force acting on a very small area which means that the cutting edge must be sharp and remain sharp during action.

From the thermal point of view, the two parts appear now different: the friction work is converted to heat by shear within the work-piece whose temperature increases according to certain laws; the tool - which, by definition, does not yield under the same stress has no internal heat source within its boundaries but being in contact with an external heat source (work-piece) its temperature increases according to different laws.

The temperature distribution in the upper, mechanically affected layer (mal) of he work-piece controls the friction performance and the quality of the process.

The temperature of the tool controls its shear strength, and through it the performance of the cutting process and the wear rate of the tool. Wear is an irrelevant consideration for the work-piece.

The common workshop routine for controlling thermally a cutting process consists in flooding the zone of cutting interaction with some fluid, which acts both as lubricant and coolant. The effect is expected to be sufficient for reaching and maintaining satisfactory thermal steady state conditions. This routine is based on an oversimplified reasoning which, besides passing lightly over the details of the temperature distribution, does not take into consideration the wear rate of the tool. Indeed, when two surfaces are in rubbing contact, both parts undergo wear, in different proportions, function of their shear strength ratio.

On the basis of conventional principles of elasticity and making simple statistical assumptions, it can be shown [1] that the probability of wear of the hard member of a friction couple, "Prob(W)", varies with the shear strength ratio (Hard/Soft) as follows:

Table 1
Probability of wear of the harder member or a friction couple, average conditions.

γ(Hard/Soft	1	1.25	2.0	5.0	10.0	20.0	100.0
Prob (W)	.5	.4	.22	.031	.0042	7E-4	1E-4

The Table shows, for example, that for a strength ratio of 10 the distribution of wear is 99.58% for a work-piece of .42% for a tool. If the cooling fluid would maintain the initial shear strength ratio, the situation would not change. In fact the cutting tool remains continuously in action and its temperature is bound to increase faster, such that the effect of cooling by a stream of fluid is smaller on a tool than on a work-piece, whose each new element enters at about the same temperature in the cutting zone. As a consequence cooling jointly both parts is likely to shift the shear strength ratio to the left (in Table 1) indicating a higher probability of wear for the hard member. It is obvious that in order to reduce the wear rate of the tool the shear strength ratio must be shifted to the right, which requires to cool the tool and to heat the work piece.

The figures given in Table 1 are estimates for average conditions. The surface finish extends the range of probabilities but this and other second order effects are not discussed in the present paper.

3. THE THERMAL STATUS OF THE WORK-PIECE

Consider the part as a unit cube in a semi-infinite slab.

No matter the severity of the mechanical interaction with the tool, the friction shear is confined to a depth of between a few μm to about 200 μm. This is the upper layer where the shear work is converted into heat. At the same time it acts as a heat reservoir the temperature of which can be calculated for steady state conditions knowing the power fed into the system (f•v) and the rate of cooling Jq,out).

Functionally this is the mechanically affected layer (mal), the temperature of which and its distribution to a large extent, control the friction performance and the quality of the cutting action.

The power f•v is fed into the mal through its active interface, S, and drains out, as heat flux, Jq through the lower plane, SS, the mass below (not affected mechanically) and the cold face of the work-piece.

Assuming that the cold face temperature, Tc, is controlled by other factors than those related to friction at the active interface, it follows that at SS, the lower boundary of the mal, the temperature, Tss, at steady state, must be:

$$Tss = (f \cdot v/K) + Tc \qquad (1)$$

where K is the mean thermal conductivity of the material in the mpl. (dimensional terms being unity, do not appear in Eq. (1)).

In these conditions it is evident that Tss is only a function of the power feed (f•v) and the thermal properties of the *mechanically passive layer* (mpl), independent of the events occurring inside the mal for which it acts as an *external constraint.*

@1-At constant power feed (f•v), Tss varies in the same direction and almost by the same amount as Tc (Fig. 2).

There is no need to know the nature of the energy conversion mechanism. At steady state, with no other changes in the system, any adopted conversion process will give the same result. In the present case conversion by "shear" or "flow" seem more convenient. The resistance to shear and to flow are single valued material properties in given conditions, at a given temperature. This makes possible calculation of the temperature T* which would be the temperature inside the mal (the mechanically affected layer), treated as a black box, if the material temperature would be uniform.

The simplest view of the mal is that shown in Fig. 1:

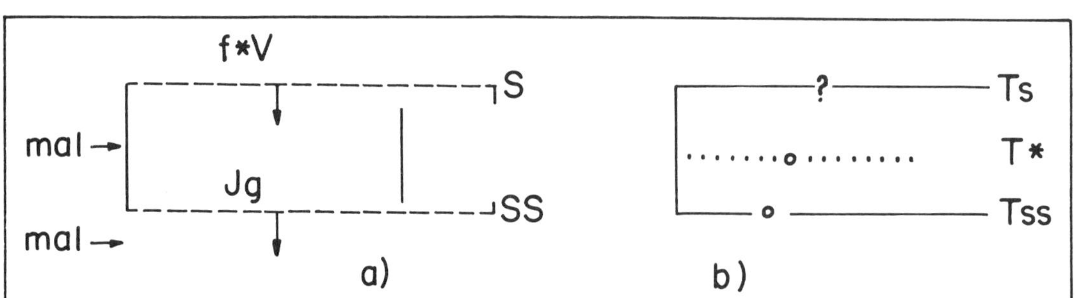

Figure 1 a) The mal-Black Box has input f•v = output Jq; b) If uniform the temperature should be T*. Tss is a constraint to take into consideration.

The restriction of "uniform temperature" in the mechanically affected layer, which should be T*, can be lifted by observing that it corresponds to the shear resistance of the material if it would be uniform throughout the mal. The calculation of T* does not present difficulty [2],[3].

Now the mal must comply to two constraints: T* (function of mal parameters and required by the energy conversion process) and Tss (required by the grad.T for the heat flow through the mpl having Tc as cold face temperature). Because T* is quantitatively related to the mean shear resistance of the material in the mal, if at SS Tss ≤ T*, in order to maintain the mean value of the property, at S Ts ≥ T*. Using the "theorem of means" it is possible to work out a function, F(T), indicating how much should increase Ts for any decrease of Tss relative to T*. In other words:

@2-For given T*, Ts and Tss are varying interdependently but in opposed directions (Fig. 3).

The quantitative relationship between Ts and Tss specified in @2 is presented in Fif. 4: it shows the function F(T)-vs.-T for given T* which has been calculated according to the procedure described previously [2],[3] (for f=9.5 N, v=500 cm/s, Tc=75°C and T*=299°C). Figure 5 explains why the left branch is steeper than the right branch of the curve in Fig. 4.

The general conclusion regarding the work-piece is that Tc can "remote control" Ts and the entire friction behavior dependent on Ts.

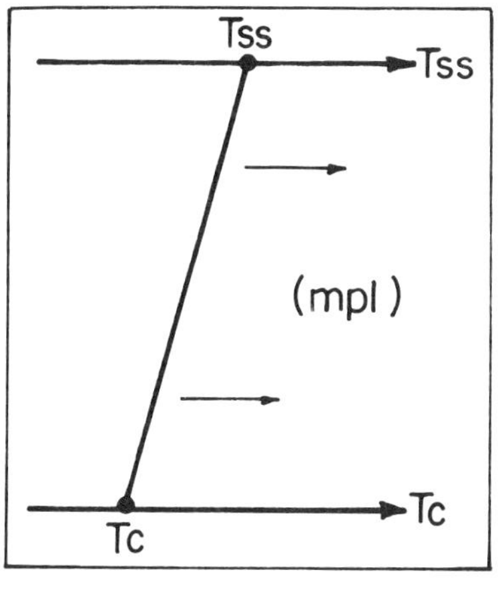

Figure 2: Tss can vary only in the same direction and by the same amount as Tc.

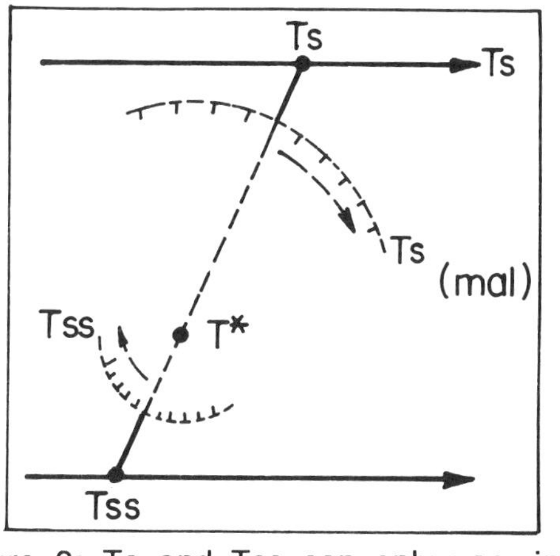

Figure 3: Ts and Tss can only vary interdependently as the ends of a lever with T* as point of rotation.

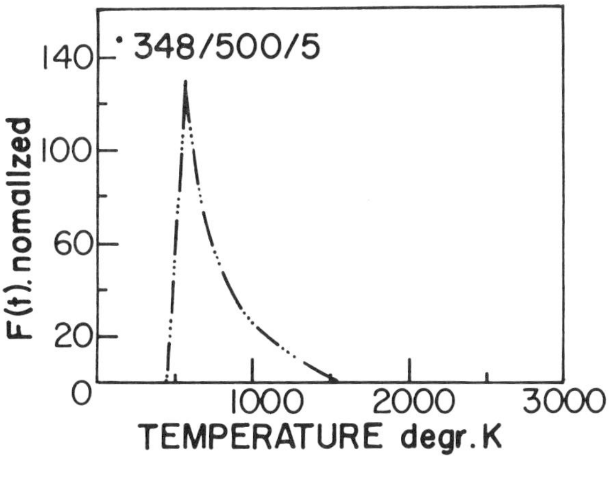

Figure 4: The function F(T)(Ts-vs-Tss) illustrated in Fig. 3 for T*=299°C.

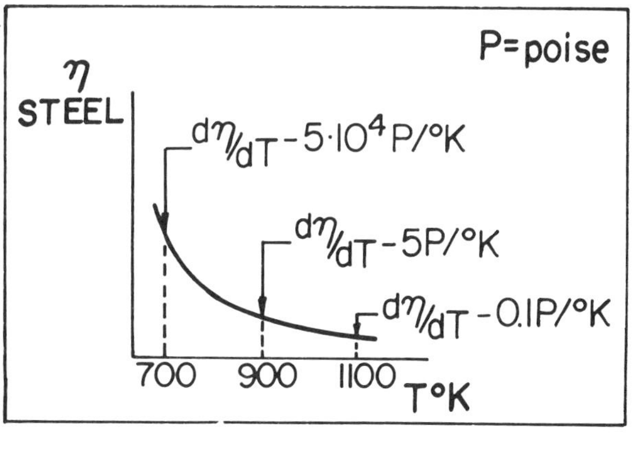

Figure 5: Shape of F(T)-vs-T follows the variation of the resistance to shear with T.

4. THE THERMAL STATUS OF THE TOOL

The tool must exercise on the work-piece a stress exceeding the yield strength in shear of the (work-piece) material. For this, the applied force must be concentrated on a very small area and the tool should have not only a sharp cutting edge but also a sufficiently high strength not to yield under the applied stress.

Therefore, by definition, the tool does not undergo deformation or shear, it has no mechanically affected layer and, implicitly, no internal heat source. Its thermal status is that of the *mechanically passive layer* and depends on its ability to pass on to the environment the amount of heat it receives from the work-piece.

Actually the tool has more difficult cooling conditions than the work-piece, the main problems being:

- The thermal diffusivity of tool materials is rather poor;
- The thermal load on the tool is very high because it is in contact with the work-piece at the S level (Fig. 1) where the temperature is maximum and due to its small size the tool has no room for spreading the heat in its mass. (Observe that the mpl of the work-piece is in contact with the mal at the level SS where the temperature is much smaller.)

Cooling both parts with a fluid coolant is more likely to maintain the thermal steady state of the work-piece than that of the tool where the coolant cannot reach the (hottest) contact interface.

If the tool is more efficiently cooled than the work-piece, no matter where, the flow of heat may be accelerated enough to maintain the material at the nominal strength level. Failure to provide a free path for heat flow will raise the tool temperature, will reduce its strength and will lead to some deformation, producing wear and, eventually, a blunt edge.

Although the term "tool" has been used to identify any cutting tool it should be observed that different rules of cooling may apply to cutting knives or inserts (continuous action) and to saws (periodic action).

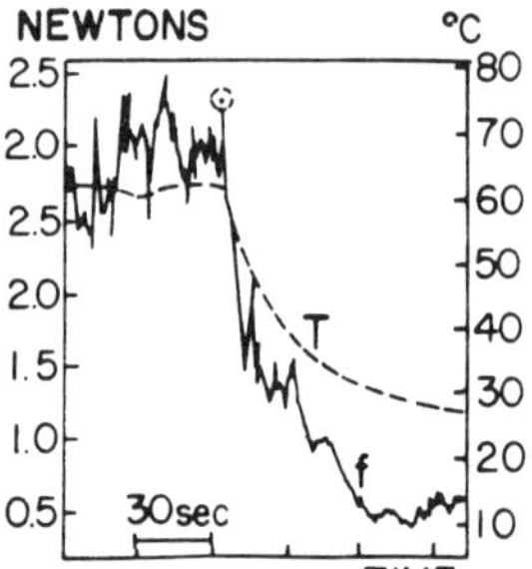

Figure 6: Lowering the cold face temperature improves friction and reduces friction force.

5. THE ROLE OF THE CUTTING FLUID

Usually the cutting process is performed with plenty of fluid around.

From the *lubrication* point of view, in any conceivable conditions only a very thin layer of lubricant is necessary; this can be applied on the work-piece before the process begins and one can select the brand that fits the cutting conditions better.

From the cooling point of view the fluid is always beneficial for the tool wherever it reaches it and of debatable justification for the work-piece.

There are three arguments in favor of work-piece cooling:

- risk of material degradation;
- excessive softening of the material leading to undesired flow (for example instead of a sharp edge the result would be a rounded off edge).
- workshop convenience (simplicity).

There are many arguments against simultaneous tool and work-piece cooling:

- cooled surfaces produce smaller temperature gradients, deeper spreading of surface traces;
- fast cooling may determine reversal of the temperature gradient producing substantial surface damage [4];
- cool work-piece determines fast wear of the tool;
- more energy is required for cutting harder materials.

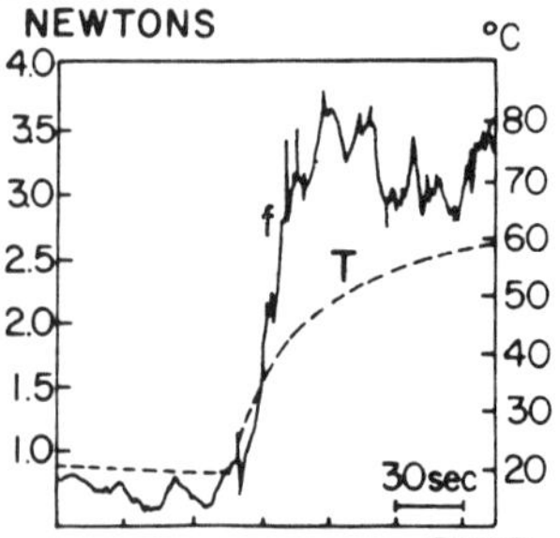

Figure 7: Rising the cold face temperature produces, almost quantitatively, the reverse effect.

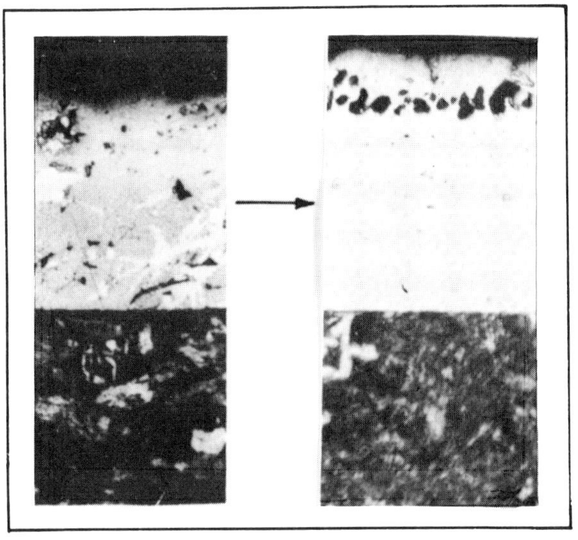

Figure 8: A thick hard coating - although an overlay - acts as a work-piece and develops a TAL instead of a MAL.

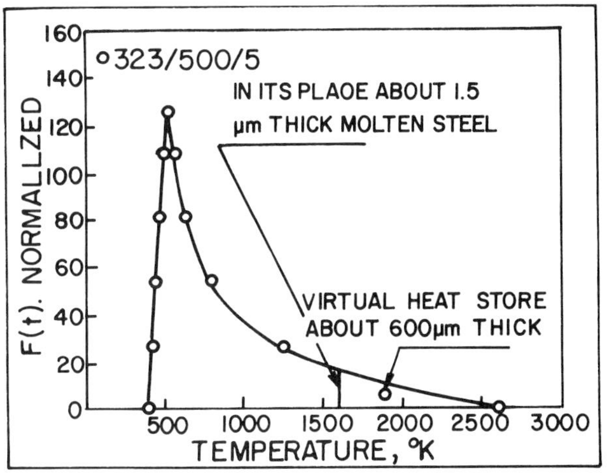

Figure 9: Too low cold face temperature may rise Ts above Tm. A thin mal upper layer melts.

On the contrary, cooling is beneficial [2],[3] if it refers to the the cold face of the work-piece, not always possible. The graph in Fig. 6 illustrates the effect of such cooling on the friction force while grinding Si wafers and Fig. 7 shows that the phenomenon is reversible [5].

6. DISCUSSION

In order to draw practical conclusions regarding the effect of cooling on cutting or grinding operations one has to confront the theoretical projections with the experimental results supporting them. Two aspects are worth considering:

- Benefits for the technological process;
- Tool wear.

Benefits for the technological process and the "mal-third body" analogy . The response of the material to the mechanical action which implies *some form of shear*, depends on the thermal and shear properties of the mechanically affected layer (mal). The mal behaves and must be treated as an independent entity, a transient third body, as conceived by Godet [6]. The attribute "transient" stresses the fact that in the conventional meaning the third body exists as long as the system exists while in the present case it exists as long as the surface shear exists. In other cases the analogy can be extended: for example, an overlay of In or Pb on a harder base, if sufficiently thin, can perform both as mal and third body because of discontinuities of the strength and thermal properties at the interface. More correctly, the overlay is a "mal clone".

The analogy is more elusive if the overlay is thick. For example [7], a thick hard coating (alumina, 100 μm thick) on steel, engaged in high speed sliding friction developed a ~10 μm - this time "thermally affected-layer, or TAL (Fig. 8), absent when the original thickness of the coating was 10 μm; now again the "third body" analogy holds only for the layers undergoing shear or another irreversible mechanical process, the limiting depth being that at which melting or any phase transition occurs.

As a physical entity the mal seldom exceeds a thickness of 200 μm. Shown in Table 2 are the calculated values of some mal parameters, at steady state, under the following conditions: sliding velocity 500 cm/s; load 0.5 MPa; the cold face temperature (Tc) for the first row is 75°C and the results correspond to the F(T)-vs-T graph in Fig. 4. For the second row Tc has been reduced by external means to 50°C and the results correspond to the graph in Fig. 9.

Table 2
Examples of calculated mal parameters reached at steady state in the indicated conditions

Tc	Tss	T*	Ts	Z1,µm	grad	T	Z1,L	Z1,S
348	443	572	1538	71	1.4E05	0.0		7 1
323	394	2609	2609	206	1.1E05		1.5	125

Theoretically, lowering Tc, respectively Tss, raises Ts, Fig. 3). The calculation is unaware of the fact that a change of state can limit the rise of Ts. In the present example Ts exceeds the melting temperature, 1734°K. The difference which the solid state cannot normally accept represents $\Delta H = Cp \bullet (Ts - Tm)$; for metals $Cp \approx 6$ cal/ gatom.degree so that in the second case there would be available 6•(2609-1734)=5250 cal. Virtually this heat should be stored in the upper layer of the mal, about 580 µm thick, above Tm (observe that the melting point of steel is lower than that of Fe). In order to melt one gatom of metal at the melting point one needs the latent heat of melting which as a very rough approximation (Richard rule) equals 2•Tm calories/gatom, in the present case ≈3500 cal. Referring to the same mass it follows that the available heat in the ~580 µm of overheated layer was sufficient to melt ≈5250/3500 = 1.5 µm (Z1,L in Table 2).

For all practical purposes the work-piece has Ts=1734°K, the mechanically affected layer is 125 µm thick (here T=Tm) out of which a layer of molten material 1.5µm thick acts as a third body providing some lubrication. surface melting during friction is not a new idea. For many years it has been advocated and even proved experimentally.

The case discussed brings again into focus the "mal third body analogy". this time only the molten layer acts as an identifiable third body; at the same time it does not have an independent existence being continuously generated by the active mass below. It should be assumed that the third body can undergo phase changes during the shear process.

The unambiguous conclusion reached at this point is that the resistance to shear of the work-piece can be modified by remote control, acting o the cold face temperature.

As a note of warning: the calculations refer to low C steel and use data taken from various handbooks, i.e., one case out of many. The aim of the discussion and of the examples is that of providing an intuitive image of the ways various process variables affect the performance of a cutting process. While for comparative purposes the conclusions are reliable, they cannot lead to quantitative evaluations.

Therefore for each particular case verification and optimization of the remote controlled changes should be made.

When there is no Tc available, like when machining a bar there is still a solution suggested by the function F(T)-vs-T presented in Fig. 3 which offers the option to control Ts directly by surface cooling or heating. Evidently such an action will "bend" the lever, disturbing the steady state. Directing a light stream of air or spreading a boundary layer of lubricant from an oil-soaked sponge will lower the surface temperature. Directing a hot radiation to the surface will raise its temperature just like lowering the cold face temperature. For example a laser beam of controlled width and intensity may improve considerably the chip machining operation. Unlike Tc effects, direct Ts changes distort brutally the temperature distribution in the mal. This is particularly true for cooling. A jet of cold air directed to the surface of a rotating shaft subjected to dry friction may produce severe surface damage [8] as can be seen in Fig. 10. No similar example for the beneficial effect of direct surface increase of Ts is available.

The nearest action producing Ts variation independent of Tc is changing the sliding velocity. An excellent example [5] is shown in Fig. 11 where the most impressive feature is the reversibility of the action-reaction sequence.

There is little to discuss regarding the tool performance by temperature control. There is no doubt that the producers use the best materials and adopt the most appropriate tool design. This does not prevent the prohibitive wear rate of cutting tools; best inserts for high speed cutting (> 250 m/min) have a service life of a few minutes.

Theoretically [1], the wear rate distribution between the soft part (work-piece) and the hard part (tool) of a friction couple is strongly dependent on their shear strength ratio (Table 1). Assume that the temperature for calculating the shear strength ratio is T*. It is easy to show that increasing the shear strength ratio by a factor of 2-2.5, only by lowering the cold face temperature of the work-piece, reduces the probability of wear of the tool about 6 times.

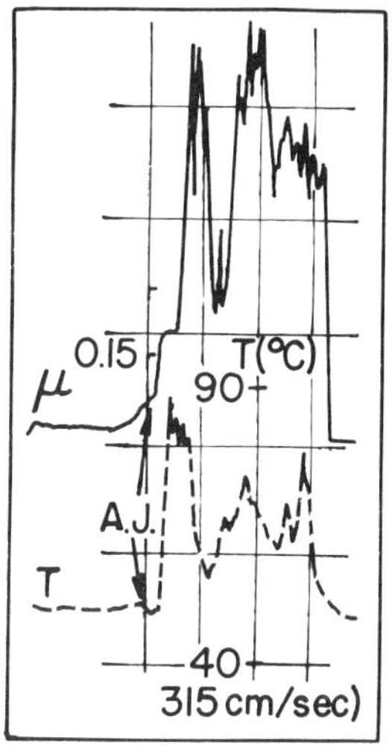

Figure 10: Surface cooling - by cold air - of a steel shaft during friction leads to severe surface damage.

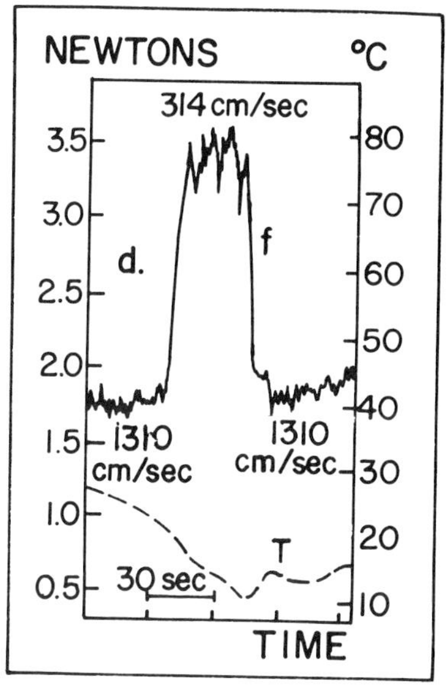

Figure 11: High and low sliding velocities act as high and low, externally controlled, surface temperatures.

An important observation regarding the term **"cooling"**. While referring to the work-piece cooling was meant to be done from outside the mal either acting on Ts or on Tss through Tc. Unlike the work-piece where lowering Tss rises, Ts and the other way round, the tool temperature is lowered throughout no matter which part of the tool is cooled.

The validity of this conclusion is well supported by experiments.

In a study on the wear of non coated hard inserts based on the theoretical background presented in this paper [9], Ber and Goldblatt found that by: eliminating the work-piece **surface** cooling and cooling selectively the cold face of the insert, the tool life increased 60-70%.

In another study Wertheim, Rotberg and Ber [10] used an insert pressure cooled by flushing water through the rake face. The water finally reached the work-piece so that the improvement came only from the more efficient cooling of the tool. They also reported substantial wear reduction.

It seems that the benefit of selectively cooling the tool is well known by specialists.

7. SUMMARY AND CONCLUSIONS

The Tool/Work-Piece friction system appears easy to understand if it is treated as a variant of the so called "Thermal Model" [1-4]. The work-piece is a block of material subjected to shear within a limited depth, which becomes a "heat generator". The heat is passed on to the environment part of it through the remaining body of the work-piece, part of it through the tool.

Cooling is necessary in order to maintain a reasonable thermal steady state. The problem is where to cool. The analysis reaches the conclusion that it is reasonable to assist the heat flow out of the system which means cooling the paths followed by the heat flux, and implicitly, to maintain the temperature of the "heat generator" by efficient drain. On the contrary it is not reasonable to cool specifically the "heat generator" because it will create a decreased temperature difference between work-piece and tool; this will determine an accelerated wear rate of the tool [1].

For some aspects of the analysis direct evidence already exists [9,10]. For other aspects there is plenty of indirect evidence which support qualitatively the conclusions. It is hoped that the paper will provide a clearer picture of the high speed chip machining process.

REFERENCES

[1] Rozeanu, L. and Pnueli, D., "The Wear of the Harder Member in a Friction Couple", Wear, **87** (1983), 317-330.

[2] Rozeanu, L., "A Model for Seizure", ASLE Transactions **16** (1972), 115-120.

[3] Rozeanu, L. and Pnueli, D., "Two Temperature Gradients Model for Friction Failure", ASME Journal Lubrication Technology, **100** (1978), 470-485.

[4] Rozeanu, L. and Pnueli, D., "Hardness Controlled Thermal Friction Failure", ASME Journal Lubrication Technology, **102** (1980), 545-551.

[5] Appelbaum, A., "Friction Properties of Nickel Diamond Composite Coatings", M.Sc. Thesis, Dept. Materials Eng., June 1980.

[6] Godet, M., "Third Bodies in Tribology", Wear, **136** (1990) 39-45.

[7] Rozeanu, L. and Pnueli, D., "Calculations and Experimental Study Thermally Affected Layer Caused by Friction", Journal of Vacuum Science and Technology, A **4**(6) (1986), 2648-2651.

[8] Rozeanu, L., "Friction Transients", ASLE Transactions, **4** (1976), 257-266.

[9] Ber, A. and Goldblatt, M., "The Influence of Temperature Gradient on Cutting Tool Life", CIRP Annals **38/1** (1989), 69-73.

[10] Wertheim, R., Rotberg, J. and Ber, A., "Influence of High Pressure Flushing Through the Rake Face of the Cutting Tool", CIRP Annals **41/1** (1992) 101-106.

SESSION X
INVITED REVIEWS

Chairman :	Professor B Jacobson
Paper X (i)	**Boundary Lubrication and Boundary Films**
Paper X (ii)	**Simulations of Lubricant Behavior at the Interface with Bearing Solids**

Thin Films in Tribology / D. Dowson et al. (Editors)
© 1993 Elsevier Science Publishers B.V. All rights reserved.

Boundary Lubrication and Boundary Films

H A Spikes

Tribology Section
Imperial College of Science, Technology & Medicine
London SW7 2BX

Despite many years of research, boundary lubrication remains the least well understood regime in the Tribological pantheon. This is in part because it is a broad church, spanning a wide range of quite different phenomena which are, in practical terms, very difficult to disentangle. Also the processes involved in boundary lubrication are very localised, hidden within the impenetrable walls of a contact, far away from the gaze of the curious scientist.

This review paper examines our current understanding of boundary lubrication and boundary films and, in particular, the mechanism and nature of friction-reducing, oiliness films. It shows how modern techniques for looking in-situ at boundary films, either directly within contacts or out of contact but whilst still covered with supernatant oil show great promise in advancing our understanding of such films. Up to the present, these techniques have tended to confirm that there are several different mechanisms by which oiliness films are formed and that each of these may have a role to play in boundary lubrication.

1. INTRODUCTION

The regime known as "boundary lubrication" is easier to invoke than it is to define. The earliest use of the term comes from Hardy in 1922 (1):

"*In what is often called complete lubrication, the kind of lubrication investigated by Towers and Osborne Reynolds, the solid surfaces are completely floated apart by the lubricant. There is, however, another kind of lubrication in which the solid faces are near enough to influence directly the physical properties of the lubricant. This is the condition found with 'dry' or 'greasy' surfaces. What Osborne Reynolds calls 'boundary conditions' then operate and the friction depends not only on the lubricant, but also on the chemical nature of the solid boundaries. Boundary lubrication differs so greatly from complete lubrication as to suggest that there is a discontinuity between the two states.*"

Hardy's concept is thus one of a film of liquid lubricant local to the surface, with physical properties so much modified as to be recognisably distinct from the bulk liquid. This picture is perhaps incomplete since, whilst it allows the lubricant to be modified by the rubbing solid, it takes no account of the solid being modified by the lubricant, as appears to occur, for example in extreme pressure lubrication. The definition is valuable, however, in going beyond the merely negative - that boundary lubrication is the regime between hydrodynamic and dry lubrication - to recognise the existence and the protective role of a distinct and characteristic boundary film.

The main problem in creating an adequate

definition of boundary lubrication lies in the very wide range of behaviour that might be included and in deciding whether the definition should be based upon effect, ie friction or wear reduction or upon cause, eg. identifiable film formation. There is also the question as to whether the definition should be confined to liquid lubricants or whether dry surface coatings or additivesdelivered to contacts by a gas flow lie within the realm of boundary lubrication.

This paper will regard boundary lubrication as the regime of lubrication that is found in a liquid-lubricated system where the load is supported primarily, not by hydrodynamic action, but by and through "boundary films". These films must have been formed by the physical or chemical interaction of the solids and the liquid during use and will have structure and thence properties distinguishable both from the bulk liquid and from the rubbing solids.

The literature on boundary lubrication is copious and disparate and this paper does not attempt a comprehensive review. Instead it focuses upon the boundary films which give rise to boundary lubrication and examines our understanding of the nature, origins and properties of these films.

In this review, boundary lubrication is considered in isolation. It must always be recognised, however, that it is rare for real rubbing systems to operate purely in the boundary regime, except at very slow rubbing speeds or with very low viscosity fluids. It is far more common for systems to be in the mixed regime, where some of the load is borne by boundary films and some by hydrodynamically-generated pressure, as illustrated in the well-known Stribeck diagram (figure 1). However, in such conditions, boundary lubrication is still crucial since its efficacy determines the friction, the wear and the likelihood of seizure of rubbing components.

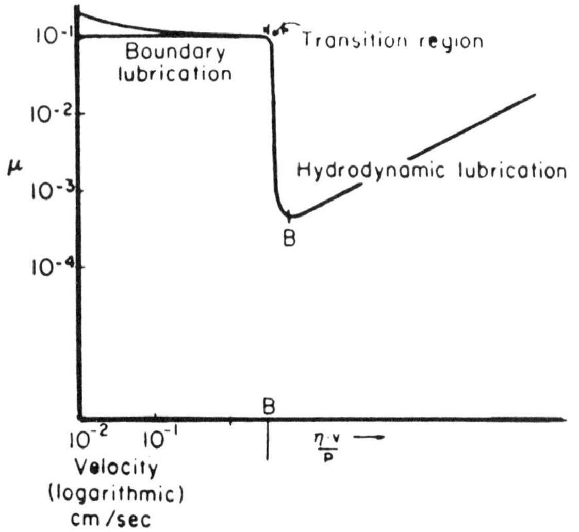

Figure 1. Regions of hydrodynamic and boundary lubrication (from ref. 2).

2. A BRIEF HISTORY

It is now clear that boundary lubrication has always been an important component of practical lubricated systems and many of the empirically-derived lubricant formulations used in the last century included chemicals whose function was solely to enhance the boundary lubricating properties of the oils concerned (3). However the clear recognition of boundary lubrication as a regime separate from hydrodynamic lubrication had to wait until the development and acceptance of modern theories of hydrodynamic film formation.
Once such theories were in place, the phenomenon of "oiliness" could be identified (4)(5). This was described by Stanton in 1923 thus (6):

" . . *in the case of the heating up of the crank-pins of gas-engines it is common practice to substitute castor oil for the mineral oil, with the result that the friction diminishes to its normal value and smooth running results. Now the value of viscosity for castor oil does not differ appreciably from that of a good lubricating mineral oil, so it is evident that the former possesses some lubricating property apart from its viscosity. Although considerable attention has been devoted to the study of this property of good lubricants, no definite relation has yet been found between it and the other physical properties of the oil, and the property itself is somewhat vaguely referred to as 'oiliness'."*

A remarkable explosion of work on this phenomenon started at the beginning of the 1920s. Initially, a range of possible explanations for oiliness were advocated but attention rapidly focused upon the role of adsorbed layers of surfactants. It was shown that single monolayers of adsorbed or reacted long chain surfactants could reduce friction and thus confer oiliness and that the extent of friction reduction depends upon the type of surfactant and its chain length in a manner that implies that oiliness involves the separation of surfaces by a strongly adsorbed film (7)(1).

The next three decades saw a gradual development and confirmation of this adsorbed film concept. Techniques for studying adsorption on solid surfaces were slowly developing and solution concentration measurements (8), radiotracer (9) and contact potential methods (10) were applied to look at liquid adsorption on solid surfaces. By the mid 1960s, the concept of oiliness being due to adsorbed monolayers was widely accepted, although there were still dissenting voices (11).

The first systematic study of what are now recognised to be the other two main classes of boundary lubricant additives, extreme pressure and antiwear agents, began in the 1940s. The late 1920s saw the introduction of hypoid gears, whose high sliding speed and highly loaded contacts could not be satisfactorily protected against scuffing by existing mineral oils, even if fortified by oiliness additives. This led rapidly to two advances, the development of a range of standard scuffing and wear tests (12) and also of extreme-pressure (ep) lubricants (13)(14) based upon sulphur, chlorine and lead-containing additives. For some years there was argument as to whether ep additives acted like oiliness additives but by forming stronger protective film on surfaces (15) or whether they conferred protection by forming an easily worn film which, once worn, reduced contact pressure (16).

The study of additives to reduce wear started in the early 1940s. Initial work by Beeck in 1941 looked at tricresylphosphate and found significant wear reduction, especially when a phosphorus additive was used in combination with a conventional oiliness additive (17). Beeck ascribed the antiwear action of the phosphorus compound purely to smoothing of the surface but later work on a range of phosphorus additives suggested alternative mechanisms such as the formation of iron phosphate soap films (18).

Unfortunately from the point of view of the scientific understanding of extreme pressure and antiwear boundary lubrication, the late 1940s saw the development of a range of very successful additives based upon molecules containing both sulphur and phosphorus. These compounds, including zinc dialkyldithiophosphate, are now recognised to possess both ep and antiwear properties, probably because the sulphur confers extreme pressure properties whilst the phosphorus is able to form thick protective antiwear coatings. This duality of function, however, resulted in the antiwear and ep mechanisms becoming confused and entangled for a number of years.

In the 1950s, researchers started to look carefully at the rubbing contact itself and to appreciate the significance of roughness and asperity contact (19). This lead, in the 1960s and early 1970s, to the development of models of friction and wear based on underlying processes of adhesion, abrasion, corrosion and fatigue. The role of boundary lubricating films in preventing adhesion or enhancing corrosion was fitted naturally into these models (20)(21).

From the 1960s the path of boundary lubrication research has followed, to a large extent, the development of new analytical and experimental techniques. This is unsurprising since the main barrier to understanding boundary films is that they are difficult to study; very thin layers on solid surfaces; compositionally similar to either the bulk liquid or to the substrate; in many cases unable to survive removal from the supernatant lubricant; subject to chemical attack by water or oxygen after rubbing is halted. Prior to the 1960s the techniques available to characterise such films were very limited, being mainly friction and wear tests, adsorption measurements and radiotracers although some valuable X-ray (22) and electron diffraction (23)(24) studies of thin films were also made.

The 1960s saw the development and application to boundary lubrication studies of analytical techniques such as infrared analysis (25) and X-ray fluorescence (26). In the 1970s, specifically surface spectroscopic methods such as XPS, Auger and SIMS became available and were also used to analyse boundary films (27). These techniques significantly advanced understanding of antiwear and ep additives but unfortunately were little help in studying oiliness-type boundary films. This was probably because such films were both too thin for detection by vibrational spectroscopy and unable to survive the washing necessitated by vacuum-based spectroscopic methods. The only significant new technique applicable to studies of oiliness was microcalorimetry, to measure heat of adsorption and to correlate this with friction (28)(29). Thus, despite considerable interest, for fuel lubricity and later for fuel economy reasons, little significant further advance was made over this period concerning the nature of oiliness films.

A major limitation of almost all of the analytical work carried out in the 1960s and 1970s was that it was applied to rubbed surfaces which had been removed from the rubbing contact and then washed, prior to examination, to remove supernatant oil. This meant that the boundary films that were subsequently analysed were, in many cases, only a remnant of the original surface material. Perhaps the biggest advance in the last decade has been the development of techniques for studying boundary films in-situ, either within rubbing contacts or at least submerged in the lubricant within which the film was formed. These include careful observational work to study antiwear films on rubbed tracks (30), the study of contacts between a transparent and a metal surface to permit direct observation into the contact (31), in-contact infrared spectroscopy (32) and the freezing of oil-covered, rubbed surfaces in liquid nitrogen so that surface films can then be examined on a cold-stage using SEM or SIMS (33). Two other important new techniques able to look at surface films immersed in lubricant are force balance techniques (34) and the surface microscopic techniques of scanning tunneling and atomic force microscopy (35).

These and other emerging in-situ techniques hold great promise of advancing our understanding of boundary lubrication and boundary films and will form the focus of this review. As has already been described, boundary lubrication encompasses a number of different processes and chemistries, including phenomena associated with oiliness, antiwear additive action and ep behaviour. For the sake of brevity, the rest of this paper will examine only one of these, boundary lubrication and boundary films that produce the friction reduction effect known as oiliness.

3. OILINESS FILMS

In the beginning there was diversity of opinion concerning the nature of oiliness films. Kingsbury, in 1908, ascribed oiliness to "*an intensified viscosity in that part of the fluid within the region of attraction of the surface molecules of the metal*" (4). In 1920, Wells favoured the effect of surface tension: "*Such a tendency . . . would manifest itself by reducing interfacial tension and would result in better spreading and consequent increased efficiency in lubrication"* (5). Hardy settled upon adsorbed monolayers: *"It is obvious that these relations can be conceived more simply if the limiting lubricating film be composed of two molecular layers, one on the other, with the plane of slip between them"* (36), Deeley however preferred a thicker film. *"It would appear that the unsaturated molecules of the lubricant enter into a firm physicochemical union with the metallic surfaces, thus forming a friction surface which is a compound of oil and metal. This solid surface would appear in the case of metallic surfaces to be much more than one molecule thick, the oil either penetrating into the metal and altering its physical properties or as a result of abrasion forming a paste of metal plus oil"* (6).

During the 1920s a number of workers demonstrated that the effect of oiliness - a reduction in friction coefficient down to a value around 0.1 - could be obtained with just a single monolayer of long chain surfactant between rubbing surfaces (7)(37). It was also shown that the friction value obtained was strongly dependent upon chain length (1) (figure 2).

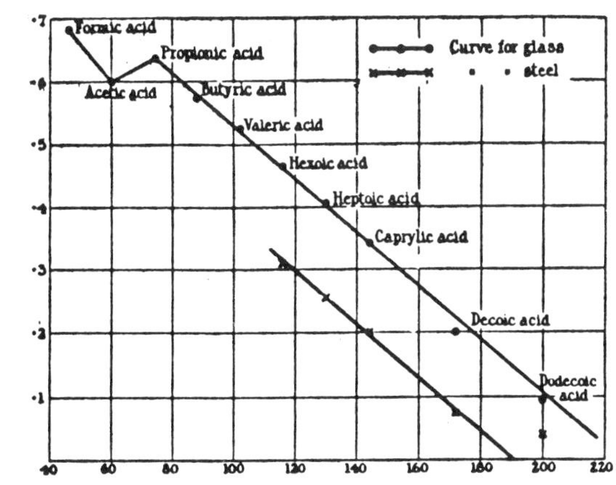

Figure 2. Effect of molecular weight of the friction of fatty acids (from ref. 1). (Abscissa, M.Wt., Ordinate, coefficient of static friction)

Such evidence led to the preeminence of Hardy's monolayer concept of oiliness which became the basis of the well-known monolayer adsorption model of boundary lubrication (figure 3).

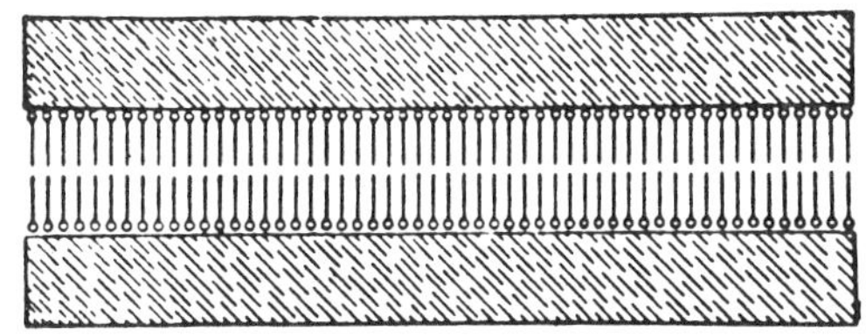

(a) Contact region according to Hardy.

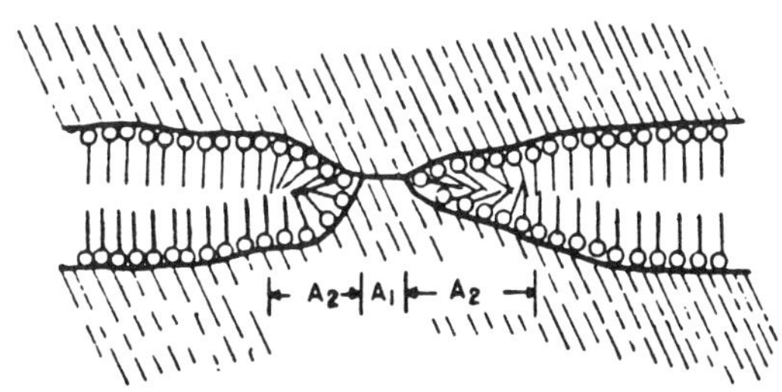

(b) Contact region according to Bowden and Tabor

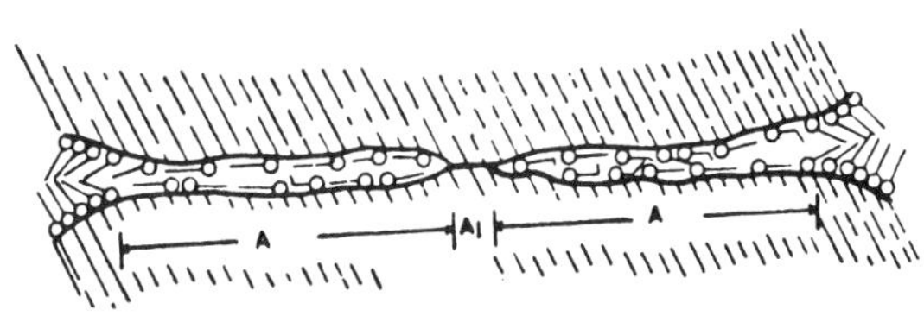

(c) Pressurised film, according to Adamson

Figure 3. Adsorption models of boundary lubrication (from ref. 51).

Further work using deposited monolayers showed that although layers in excess of the first one did not significantly reduce friction they did greatly enhance the film's durability (39) (figure 4). This led to the recognition that effective oiliness requires a self-healing ability - the replenishment of the film during rubbing (39)(40).

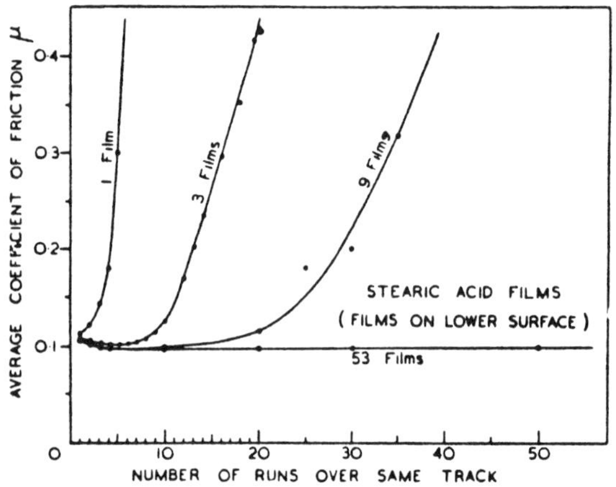

Figure 4. Friction and durability of stearic acid monolayers (from ref. 39).

From the 1940s to 1970s, numerous studies were carried out to study the nature of adsorbed surfactant layers on surfaces (8)(41)(42)(43) and to compare the friction-reducing and adsorption properties of long chain surfactant solutions (44)-(49) although it was rarely possible to carry out friction and adsorption measurements on identical surfaces (50). From all these studies a broad pattern emerged. Oiliness seems to result from near-vertically oriented monomolecular films of long chain polar compounds (51), although at very high pressures these might be squeezed flat (38). To be effective the chain length needs to be at least 12 carbon atoms long and straight rather than branched (52). Normally-refined oils contain natural oiliness surfactants (53). With most metals, including Fe, Ni and Cr, a monolayer is formed on each surface which is partially reversibly and partially irreversibly adsorbed. These two monolayers produce low friction at low temperatures but, as temperature is raised, friction rises sharply at some "critical temperature" which depends upon surfactant type and concentration, suggesting a desorption phenomenon (54)(55). With some additive/metal combinations, notably fatty acids with Cd, Ag, Cu and Zn, a corrosive process occurs to form semi-soluble soaps (8)(46). If alkane solvents are used, solvent is incorporated into the film (56) and the critical temperature tends to be higher if the chain length of the solvent and solute match (57)(58). Recent, clear statements of lubricant-solid adsorption and the monolayer adsorption model are given in (59)(60).

The extent and consistency of all this work has led to the monomolecular adsorption model of boundary lubrication being almost universally accepted for the last thirty years. There has, however, been a consistent thread of doubt.

Although it has been shown that a single monolayer can lubricate surfaces it is much more difficult to prove that such a monolayer is actually the only important boundary film in practical, lubricated systems. Hardy was always careful to stress the possibility of boundary films thicker than monolayers, particularly at low contact pressures (37) and, in his later years, appeared to favour the presence of a quite thick layer between boundary lubricated surfaces (61): "*The simple two-dimensional hypothesis which considers only a monomolecular adsorbed layer in equilibrium with a liquid in mass is not enough, the conception must be three-dimensional and take account of the obvious gross reorganisation of the whole pool of oil which contact with the steel face brings about. This reorganisation must be due to polarisation which spreads from molecule to molecule along the normal from the steel face, it is the kind of polarisation which one of use has called 'diachysis'*" (62).

Trillat, using X-ray diffraction, showed in the 1920s that a lubricant could form a organised layer many molecules thick on metal surfaces (63). In the 1940s and 1950s, work studying the normal approach of lubricated, loaded flats found that there remained a residual, separating film, resistant to high load, typically hundreds of nanometres in thickness with surfactant-containing oils but not with pure, non-polar hydrocarbons (64)(65) (figure 5).

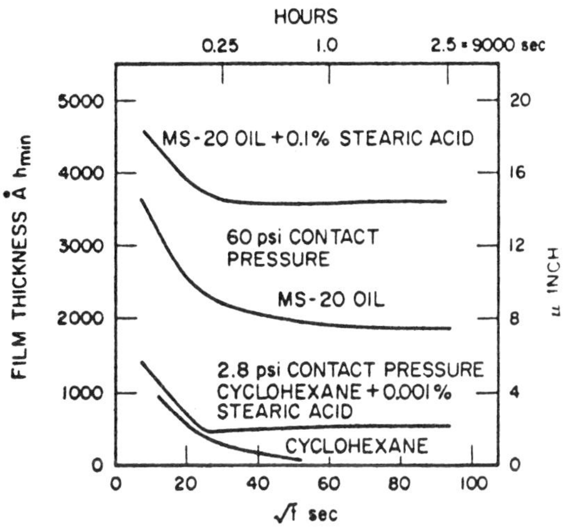

Figure 5. Residual films formed for light mineral oil and cyclohexane solutions (from ref. 65).

Studies by Derjaguin of the meniscus shape adopted by blown droplets of oil on surfaces indicated that liquid layers within a few nanometres of solid surfaces may exhibit anomalous viscosity (66)(67). Using a mercury displacement technique, Smith showed, in the 1970s, the presence of micron thickness viscous layers formed on steel surfaces when heated in fatty acid solutions (68). A significant observation was made by Beeck, who found that fatty acid additives promoted the onset of mixed lubrication so that the transition from boundary to mixed took place at lower speeds (69) (figure 6).

The above work suggests that surfactants or their reaction products with metals may form thick, organised surface layers (10). There have also been indications that lubricants may degrade chemically during rubbing to form polymerised, protective layers (70)(71)(72). Such polymerisation, especially of unsaturated and aromatic components, might be promoted by the high pressures and temperatures at asperity contacts or by exoelectrons emitted during metallic plastic deformation.

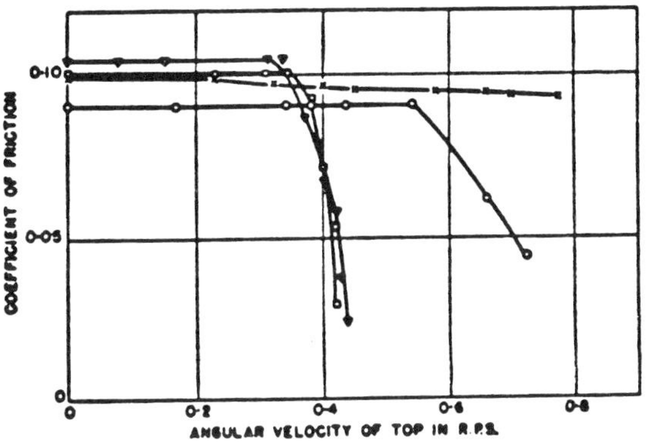

Figure 6. Influence of oiliness additives on speed at which friction falls (from ref. 69).
o castor oil
x ricinoleic acid
▼ fatty acid fraction from castor oil
□ 1% stearic acid in ricinoleic acid

It is worth noting that most surfactant solution, friction studies used to test monolayer adsorption models have used very mild conditions of slow speed and smooth surfaces where such polymerisation effects are least likely to be seen.

This brings us to the mid 1980s and the development of new, in-situ techniques. In this paper, six of these techniques will be considered with respect to the study of oiliness films.

Force Balance Apparatus

Force Balance or Surface Force apparatus were first developed in the late 1960s, primarily to study the attractive forces between surfaces in close proximity (73). They are based upon bringing two very smooth surfaces very close together to produce a known and controllable gap down to fractions of a nanometre. One of the surfaces is mounted so that the normal force exerted on it across the gap by the other surface can be accurately measured. Typically the surfaces are two crossed-cylinders of mica. Since the early 1980s, modern electronics and piezo-electrics has made this technique easier to apply and more sensitive and it has started to be used quite extensively to study both the normal and the shear forces transmitted across a thin film of lubricant between the two surfaces. By using not just the slow approach of two surfaces but also, superimposed upon this, a higher frequency, low amplitude oscillation, both viscous and elastic properties of liquid films down to just one or two molecular layers in thickness can be obtained.

Using this technique, Israelachvili has shown that, for simple molecular organic liquids on very smooth mica surfaces, the normal force is an oscillatory function of distance from the surface. This indicates that the molecules have an effectively layered structure, several molecular diameters out from the surface (74) (figure 7).

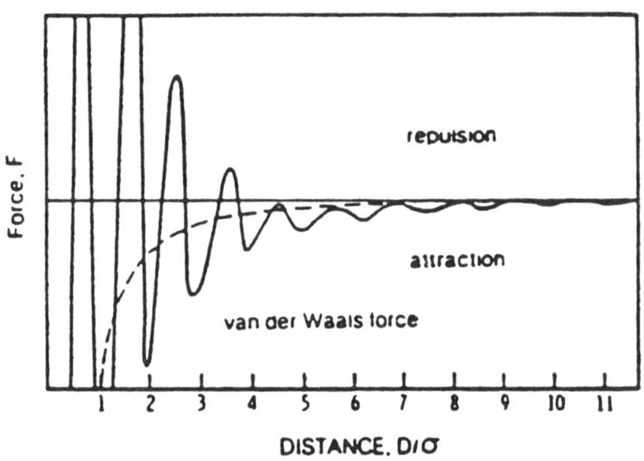

Figure 7. Measured force between two curved molecularly smooth solid surfaces in organic liquids of molecular diameter, d_0 (schematic). If the liquid were to be treated as a structureless continuum, a monotonic attractive van der Waals force law would be expected, as illustrated by the dotted curve (from ref. 74).

Both Israelachvili and Granick have also shown that the shear properties of simple molecular liquids between two mica surfaces retain their bulk properties so long as the separation is equivalent to at least three monolayers. However when the film is thinner than this there are gross deviations from bulk properties with, for one or two layers, effective viscosities up to several orders of magnitude higher than normal (75)(76). It also appears that the film is effectively quantised, tending to jump between one and two or two and three molecular layers (75).

A limitation of Israelachvili's technique from the point of view of lubricant films is that, because his two surfaces consist of mica fastened on glass, the pressures that can be reached are quite low, being limited by the modulus of the adhesive. Recently Georges and co-workers have applied the force balance technique to the contact of a ball on a flat with both bodies made of steel or cobalt-coated glass (77). This has enabled him to reach quite high pressures and also to examine the effect of roughness. Georges has shown that simple molecular liquids on surfaces form "immobile layers" with anomalous viscometric properties typically 1 nm out from surfaces. With low molecular weight polymers, "immobile layers" up to 10 nm thick are seen (78).

Scanning Tunneling and Atomic Force Microscopy

A second pair of techniques that show promise of advancing our understanding of oiliness films are the two scanning microscopy techniques, scanning tunneling and atomic force microscopy. In scanning tunneling microscopy a very sharp tip, with a radius of a few nanometres, is located a fraction of a nanometre above, and moved across, a flat surface. The tip is arranged to move slightly up and down as it traverses the surface, so as to maintain a constant tunneling electric current between the tip and surface (79) (figure 8). The distance the tip moves up and down is measured and this provides a very detailed map of the surface height. The resolution of the technique is such that it can detect the arrangement of individual atoms in molecules on surfaces and figure 9 shows the pattern formed by an adsorbed layer of liquid crystal molecules on a graphite surface (80). The images obtained are not true heights since the tunneling current varies depending upon the film of material present as well as its dimensions, but they are sufficiently detailed in many cases to indicate the alignment and even the mobility of molecular films (35). A particular valuable feature of the method from the point of view of oiliness is that the STM tip appears to pierce through thick, supernatant layers to image only the immediate surface layer and can thus, in principle, be used to look at boundary films submerged in oil (80).

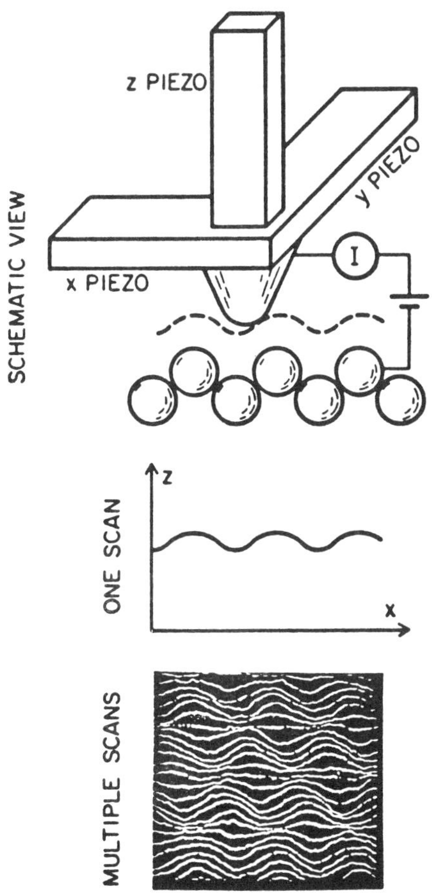

Figure 8. Schematic view of an STM. As the tip follows the dotted trajectory, one scan is recorded. Multiple scans that are plotted together form images (from. ref. 79).

Atomic force microscopy is a similar technique where the tip is positioned above the surface at a distance determined by the reactive force applied on the tip by the surface. Thus the image obtained is of a surface of constant force rather than constant tunneling current.

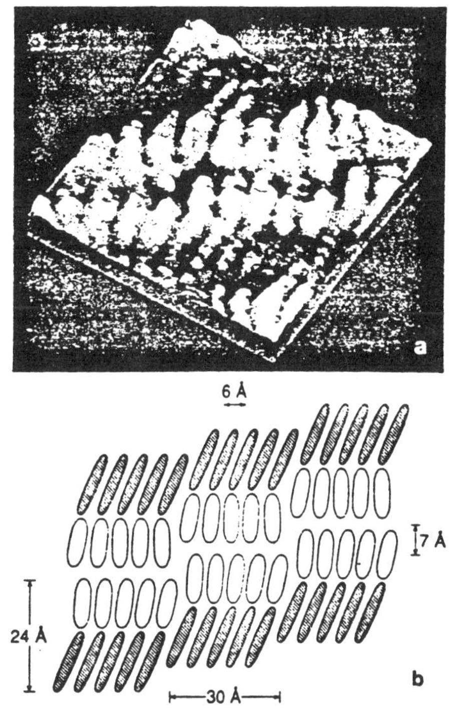

Figure 9. (a) STM image (57 by 57 Å) of 10CB on graphite. (b) Model showing the packing of the 10CB molecules. The shaded and unshaded segments represent the alkyl tails and the cyanobiphenyl head groups respectively. (From ref. 80).

Ultrathin Film Interferometry

Another recently-developed technique able to look directly at oiliness films in contacts is ultrathin film interferometry. Optical interferometry was applied in the 1960s to measure the elastohydrodynamic film-forming properties of lubricants (81). It rolls or slides a steel ball or cylinder on a glass flat and interference fringe patterns are produces which can be used to measure the oil film thickness formed between the two surfaces. This technique traditionally has a lower film thickness limit of around 70 nm but this has recently been overcome and films down to 2 to 5 nm in rolling ball on flat contacts can now be measured (82). Figure 10 shows some preliminary results where the film-forming properties of hexadecane and a solution of stearic acid in hexadecane are compared. Above a contact thickness of 5 nm, both fluids appear to form the same film thickness, whose dependence on rolling speed conforms with elastohydrodynamic behaviour. However at very low speeds, the stearic acid appears to maintain a thin, speed-independent separation which may represent an oiliness film. This technique is still in its infancy and needs further development to produce reliable measurements below 5 nm but holds out considerable promise for studying oiliness film-formation in realistic contacts.

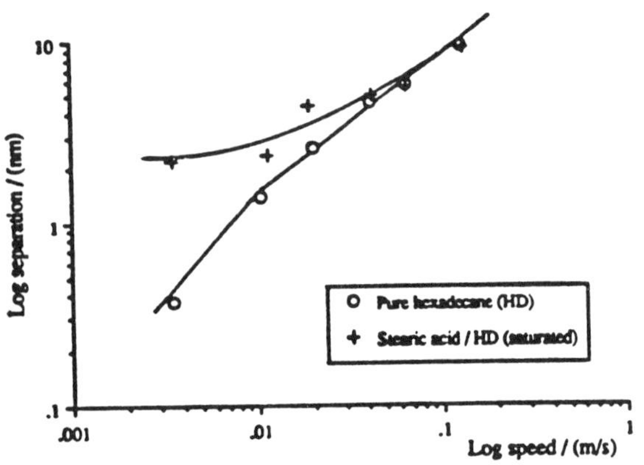

Figure 10. Log. separation vs. log speed for purified hexadecane with and without stearic acid (from ref. 82).

In-track Ellipsometry

The fourth in-situ technique to be described which has been applied to look at oiliness-type boundary lubrication has been developed by Ludema (83)(84). This uses ellipsometry to measure the refractive index and the film thickness of reaction film in the out-of-contact track of a pin on disc-type machine. The ellipsometer is designed with windows which can be submerged in the supernatant oil so that film-thickness measurements can be made directly from the moving disc surface (83) (figure 11).

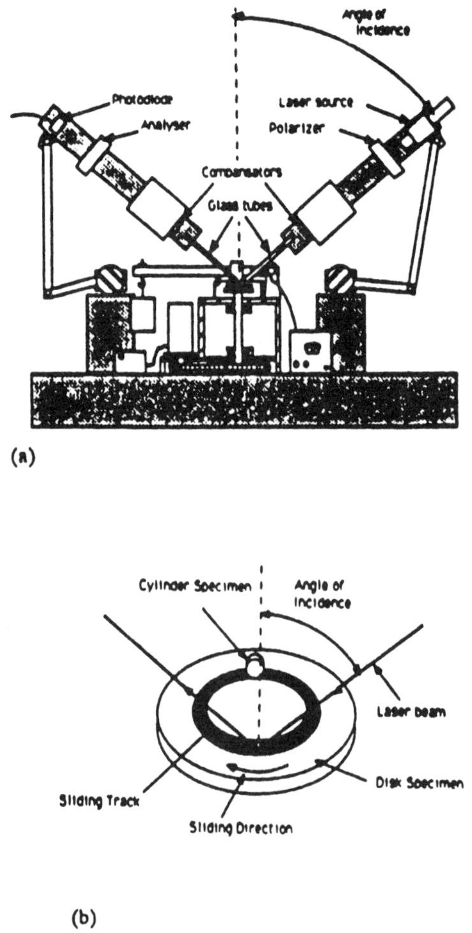

Figure 11. Experimental set-up of in-situ ellipsometry. (a) General view. (b) Expanded view of the sliding surface. (From ref. 83).

Using this technique, Ludema has found that paraffin and mineral oils without additives form films of the order of 5 to 10 nm thick in loaded, sliding, steel on steel contacts. These films take some 10 to 20 minutes to develop and, from ellipsometric evidence, may be two layer structures, with a sublayer of metallic iron/carbon mixture and an overlayer of organo-iron compounds (83) (figure 12).

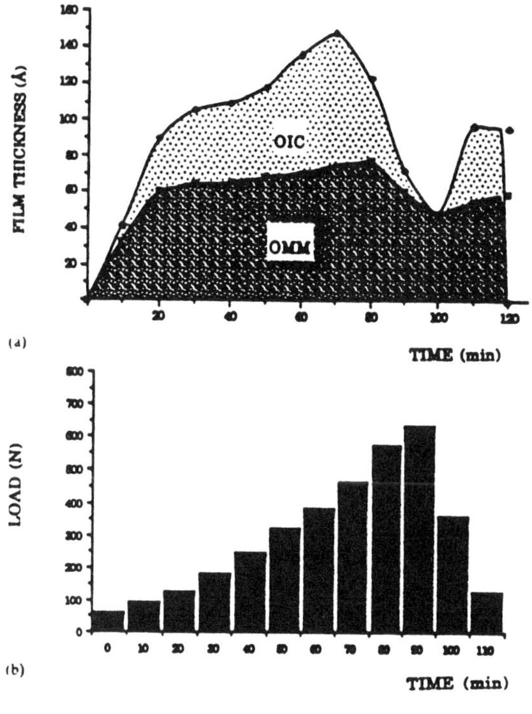

Figure 12. (a) Film thicknesses obtained from two-layer film analysis of ellipsometry measurements from a steel disc. The top film is organo-iron, the bottom is a mixture of iron oxide, metallic iron and iron carbide. (b) Loading history of the specimen. (From ref. 83).

In-Contact Visualisation

This approach consists of rubbing a lubricated steel ball on a transparent flat, typically of sapphire or glass and observing the appearance of the contact zone through a microscope during sliding (31). When white oil or additive-free mineral oil was studied using this technique, a viscous or semi-solid "friction polymer" was formed around the contact with a crescent shape at the latter's leading. The role of the friction polymer in protecting the surfaces was stated to be twofold. It initially prevented wear by virtue of its high viscosity but also formed a protective, solid-like pad between the surfaces (31). When an antiwear additive, zinc dialkyldithiophosphate, (ZDDP) was present, both of these processes were greatly enhanced.

Film Freezing

One of the main limitations of the vacuum-based, surface analytical techniques for studying boundary films on rubbed surfaces is that the surfaces generally have to be washed quite thoroughly before examination. There are two reasons for this. Firstly, volatile oil components from the base fluid would tend to evaporate in the vacuum and deposit on windows and other sensitive parts of the vacuum chamber. Secondly, the supernatant oil layer would obscure the underlying film.

To overcome these problems, a new technique based on fast-freezing has been developed (33). A rubbed metal specimen is taken, still hot, from a wear test rig and plunged into liquid nitrogen. This cools the specimen very rapidly, freezing any adherent oil and also any boundary film. The specimen is then transferred, using conventional frozen tissue-handling equipment, on to the cold stage of an SEM or SIMS. The low temperature prevents sublimation of the frozen oil. It is also possible to scrape away some or all of the frozen oil, exposing the underlying near-surface film prior to insertion into the vacuum, so that details of the variation of composition with depth can be investigated

This technique has so far been used only to study films formed on rubbed iron surfaces by ZDDP additives (33)(85) and has shown the presence of layered films up to 1 μm thick with variations in composition with depth. The technique could however, in principle, be effective for looking at oiliness films.

4. DISCUSSION

All the above techniques show considerable promise in extending our knowledge of oiliness films, although it is too early to say exactly where that knowledge will lead us.

The in-situ work points, so far, to three main mechanisms of oiliness

(i) monolayer adsorption (ultrathin film interferometry and STM)
(ii) viscosity modification to several molecular layers (force balance)
(iii) viscous or pasty reacted layers (ellipsometry, in-situ visualisation)

It is intriguing to see how closely these echo three of the early mechanisms of boundary lubrication proposed at the beginning of the century, Hardy's monolayer model (37), the enhanced viscosity suggestion of Kingsbury (4) and Deeley's concept of a thick, metal-containing layer (6). Only Wells' surface tension hypothesis (5) appears to be unrepresented. However it is now known that surface tension is crucial in controlling oil flow on surfaces and thus in promoting EHD or hydrodynamic film in starved, repeated-pass contacts (86) so his model too has a place in modern lubrication science.

It is not yet possible to assess the relative importance of the three mechanisms outlined above in practical systems. They may well indeed operate under such different conditions of contact severity that they are complementary but not comparable. Thus numerous experiments have shown that adsorbed monolayers do lower friction under relatively mild, low speed, smooth surface, short duration conditions, but it has not been shown if they are significant in longer term, more severe rubbing. Viscous or paste-like reacted films have tended to be found only in quite severe and lengthy rubbing tests and it is not yet clear whether they form in more benign contact conditions. Near-surface viscosity modification, as found by the blow-off technique and by force balances, is likely to be most effective under relatively high speed, low pressure conditions where it might promote hydrodynamic film formation. Here the observations of Beeck, who found that oiliness additives showed just such behaviour in a 4-ball machine are very significant (69) (figure 6).

In principle it should be possible to disentangle the mechanisms, even in practical systems. It would be of interest, for example, to compare the performance of straight and branched chain additives in prolonged wear tests. If the main mechanism of protection is adsorption, then the differences between the structures should be marked. However, if it is decomposition or polymerisation they should not. Further work like Beeck's but using a lower pressure contact would also be valuable, as would freezing work, to investigate the depth and chemistry of any modified layer after different severities of rubbing.

In some ways we have come full circle since the early 1900s. There are clearly several ways that both additive-free and surfactant-containing oils can form boundary films and thus provide boundary lubrication. All are probably significant to a greater or lesser extent, as was recognised sixty years ago (87). However, what has changed is the growth in our capability for studying such films, in particular using new methods that have been developed for looking at surface films in-situ.

5. CONCLUSIONS

This review paper has examined our current understanding of boundary lubrication and boundary films and, especially, the mechanism and nature of friction-reducing, oiliness films. It is concluded that modern techniques for looking, in-situ, at boundary films, either directly within contacts or out of contact but whilst still covered with supernatant oil, show great promise in advancing our understanding of such films. Up to the present, these techniques have tended to confirm that there are several different mechanisms by which oiliness films are formed and that each of these may have a role to play in boundary lubrication.

REFERENCES

1. Hardy, H.B. and Doubleday, I. "Boundary Lubrication, - The Paraffin Series." Proc. Roy. Soc. Lond. A100, pp 550-757, (1922).
2. Bondi, A. Physical Chemistry of Lubricating Oils, Rheinhold, NY. (1951).
3. Musgrave, F.F., "Development and Lubrication of the Automotive Hypoid Gear" J. Inst. Pet. 32, p 32, (1946), citing eg. Brown, R.R. U.S. Patent 22,621, (1859) and Hendrick, E.E., US Patent 90,100 (1869).
4. Kingsbury, A. "A New Oil-Testing Machine and Some of Its Results." Trans. ASME 24, pp 143-160, (1903).
5. Wells, H.M. and Southcombe, J.E. "The Theory and Practice of Lubrication: The 'Germ' Process." J. Soc. of Chem. Ind. 39, pp 51T-60T, (1920).
6. T.E. Stanton, "Friction." Publ. Longmans, Green & Co., London, 1923.
7. Desvaux, H. "Le Frottement des Solides: Lubréfiant." J. de Phys. et la Radium 5, p 845, (1924).
8. Daniel, S.G. "The Adsorption on Metal Surfaces of Long Chain Polar Compounds from Hydrocarbon Solutions." Trans. Farad. Soc. 47, pp 1 345-1359, (1951).
9. Gaines, G.L. Jr. "Some Observations on Monolayers of Carbon-14 Labeled Stearic Acid." J. Coll. Sci. 15, pp 321-339, (1960).
10. Haydon, D.A. "Boundary Potentials and Adsorption at the Oil-Metal Interface." Koll. Zeit. 188, pp 141-147, (1963).
11. Allen, C. M. and Drauglis, E. "Boundary Layer Lubrication: Monolayer or Multilayer." Wear 14, pp 363-384, (1969).
12. van der Minne, J.L. "Comparison of the Behaviour of Various Extreme Pressure Lubricants in Different Testing Machines" I. Mech. E. Proceedings of the General Discussion on Lubrication and Lubricants, vol. 2, pp 429-441, Oct 1937.
13. Mougey, H.C. and Almen, J.O. "Extreme Pressure Lubricants ." Auto. Ind. 65 no 26, pp 758-761, (1931).
14. Miller, F.L. "Extreme Pressure Lubricants and Lubrication" I. Mech. E. Proceedings of the General Discussion on Lubrication and Lubricants, vol. 2, pp 107-117, Oct 1937.
15. Evans, A.E. "Extreme Pressure Lubricants" I. Mech. E. Proceedings of the General Discussion on Lubrication and Lubricants, vol. 2, pp 55-59, Oct 1937.
16. Southcombe, J.E., Wells, J.H. and Waters, J.H. "An Experimental Study of Lubrication Under Conditions of Extreme Pressure" I. Mech. E. Proceedings of the General Discussion on Lubrication and Lubricants, vol. 2, pp 400-411, Oct 1937.
17. Beeck, O., Givens, J.W. and Williams, E.C. "On the Mechanism of Boundary Lubrication. II. Wear Prevention by Addition Agents." Proc. Roy. Soc. Lond, A177, pp 103-118, (1940).
18. Thorpe, R.E. and Larsen, G.G. "Antiseizure Properties of Boundary Lubricants." Ind. Eng. Chem. 41, pp 938-943, (1949).
19. Rabinowicz, E. and Tabor, D. "Metallic Transfer between Sliding Metals: An Autoradiographic Study." Proc. Roy. Soc. Lond. A208, pp 455-475, (1951).2020.
20. Rowe, C.N. "Role of Additive Adsorption in the Mitigation of Wear." ASLE Trans. 13, pp 179-188, (1970).
21. Beerbower, A. "A Critical Survey of Mathematical Models for Boundary Lubrication." ASLE Trans. 14, pp 90-104, (1971).
22. Clark, G.L., Sterrett, R.R. and Lincoln, B.H. "X-ray Diffraction Studies of Lubricants." Ind. Eng. Chem. 28, pp 1318-1328, (1936).
23. Andrew, L.T. "Electron Diffraction Analysis of the Orientation of the

Molecules of Lubricating Oils." Trans. Farad. Soc. 32, pp 607-616, (1936).

24. Brockway, L.O. and Karle, J. "Electron Diffraction Studies of Oleophobic Films on Copper, Iron and Aluminium Surfaces." J. Coll. Sci. 2, pp 277-287, (1947).
25. Francis, S.A. and Ellison, A.H. "Reflection Infrared Studies of Zinc Dialkyldithiophosphate Films Adsorbed on Metal Surfaces." J. Chem. Eng. Data 16, p 83, (1961).
26. Rounds, F.G. "Effect of Additives on the Friction Of Steel on Steel. (I) Surface Topographic and Film Composition Studies." ASLE Trans. 7, pp 11-23, (1964).
27. Ferrante, J. "Practical Applications of Surface Analytical Tools in Tribology." Lub. Eng. 38, pp 223-231, (1981).
28. Groszek, A.J. "Heats of Preferential Adsorption of Boundary Additives at Iron Oxide/Liquid Hydrocarbon Interfaces." ASLE Trans. 13, pp 278-287,
29. Groszek, A.J. "Role of Adsorption in Liquid Lubrication." Proc. Symp. "Interdisciplinary Approach to Liquid Lubricant Technology", Jan. 1972, NASA SP-318.
30. Georges, J.M., Tonck, A,, Mielle, G. and Belin, M. "Chemical Films and Mixed Lubrication." ASLE Trans. 26, pp 293-305, (1983).
31. Sheasby, J.S., Caughlin, T.A. and Habeeb, J.J. "Observation of the Antiwear Activity of Zinc Dialkyldithiophosphate Additives." Wear 150, pp 247-257, (1991).
32. Cann, P.M. and Spikes, H.A. "In Lubro Studies of Lubricants in EHD Contacts Using FTIR Absorption Spectroscopy." STLE Trans. 34, pp 248-256 (1991).
33. Bell, J., Coy, R. and Spikes, H.A. "Cryogenic Studies of Zinc Dialkyl Dithiophosphate Anti-Wear Films." Proc Jap. Int. Tribology Conf, Nagoya, pp 505-510, (1990).
34. Homola, A.M., Israelachvili, J.N., Gee, M.L. and McGuiggan, P.M. "Measurements of and Relation Between the Adhesion and Friction of Two Thin Liquid Films." ASME 89-Trib-3.
35. Andoh, Y., Oguchi, S., Kaneko, R. and Miyamoto, J. "Evaluation of Very Thin Lubricant Films." J. Phys. D 25, pp A71-A75, (1992).
36. Hardy, W. and Bircumshaw, I. "Boundary Lubrication. - Plane Surface and the Limitations of Amonton's Law." Proc. Roy. Soc. Lond. A109, pp 1-27, (1925).
37. Langmuir, I. "The Mechanism of the Surface Phenomena of Flotation." Trans. Farad. Soc. 15, pp 62-74, (1920).
38. A. W. Adamson. Physical Chemistry of Surfaces. Section IX-7B, 2nd Edition. Publ. J. Wiley & Sons, 1967.
39. Bowden, F.P. and Leben, L. "The Friction of Lubricated Metals" Phil. Trans. Roy Soc. A239, pp 1-27, (1940),
40. Adams, N.K. "Molecular Forces in Friction and Boundary Lubrication." I. Mech. E. Proceedings of the General Discussion on Lubrication and Lubricants, vol. 2, pp 197-201, Oct 1937.
41. Bigelow, W.C., Pickett, D.L. and Zisman, W.A., "Oleophobic Monolayers. I. Films Adsorbed from Solution in Non-Polar Liquids." J. Coll. Sci. 1, pp 513-538, (1946).
42. Bigelow, W.C., Glass, E. and Zisman, W.A. "Oleophobic Monolayers. II. Temperature Effects and Energy of Adsorption." J. Coll. Sci. 2, pp 563-591, (1947).
43. Cook, E.L. and Hackerman, N. "Adsorption of Polar Organic J. Phys and Coll. Chem. 55, pp 549-557, (1951).
44. Greenhill, E.B. "The Adsorption of Long Chain Polar Compounds from Solution on Metal Surfaces." Trans. Farad. Soc. 47, pp 625-631, (1951).
45. Greenhill, E.B. "The Lubrication of Metal Surfaces by Mono- and Multi-Layers." Trans. Farad. Soc. 47, pp 631-635, (1951).
46. Bowden F.P. and Moore, A.C. "Physical and Chemical Adsorption of Long Chain Compounds on Radioactive Metals." Trans. Farad. Soc. 47, pp 900-909, (1951).

47. Hutchinson, E. and Rideal, E.K. "Adsorption and Lubrication at Crystal Surfaces. Part I. On the Boundary Lubrication of Sodium Nitrate." Trans. Farad. Soc. 43, pp 435-438, (1947).
48. Hutchinson, E. "Adsorption and Lubrication at Crystal Surfaces. Part II. On the Adsorption of Paraffin Chain Compounds on Sodium Nitrate." Trans. Farad. Soc. 43, pp 439-442, (1947).
49. Hutchinson, E. "Adsorption and Lubrication at Crystal Surfaces. Part III. On the Heats of Wetting and Adsorption on Ionic Crystals." Trans. Farad. Soc. 43, pp 443-447, (1947).
50. Spikes, H.A. and Cameron, A. "A Comparison of Adsorption and Boundary Lubricant Failure." Proc. Roy. Soc. Lond. A336, pp 407-419, (1974)
51. Bailey, A. and Courtenay Pratt, J.S. "The Area of Real Contact and the Shear Strength of Monomolecular Layers of a Boundary Lubricant." Proc. Roy. Soc. Lond. A227, p 500-515, (1955).
52. Levine, O. and Zisman, W.A. "Physical Properties of Monolayers Adsorbed at the Solid Interface. I. Wettability of Aliphatic Polar Compounds and the Effect of Halogenation." J. Chem. Phys. 61, pp 1068-1077, (1957).
53. Groszek, A.J. and Palmer, D.J. "Surface-Active Compounds in Lubricating Oils." J. Inst. Pet. 47, p 295-306, (1961).
54. Frewing, J.J. "The Influence of Temperature on Boundary Lubrication." Proc. Roy. Soc. Lond. A181, pp 23-42, (1942).
55. Frewing, J.J. "The Heat of Adsorption of Long-Chain Compounds and Their Effect on Boundary Lubrication." Proc. Roy. Soc. Lond. A181, pp 270-285, (1944).
56. Cook, H.D. and Ries, H.E. Jr. Radiostearyl Alcohol from n-Hexadecane onto Solid Surfaces." J. Phys. Chem. 63, pp 226-230, (1959).
57. Askwith, T.C., Cameron, A. and Crouch, R.F. "Chain Length of Additives in Relation to Lubricants in Thin Film and Boundary Lubrication." Proc. Roy. Soc. Lond. A291, pp 500-519, (1966).
58. Grew, W. and Cameron, A. "Friction Transition Temperature Effect of Matching Surfactant and Carrier." Nature 214, pp 429-430, (1967).
59. Lockwood, F.E. and Bridger, K. "Liquid Wetting on Metal Surfaces and the Heat of Immersion/Adsorption in Lubrication." ASLE Trans. 30, pp 339-354, (1987).
60. Beltzer, M. and Jahanmir, S. "Effect of Additive Molecular Structure on Friction." Lub. Sci. 1, pp 1-26, (1988).
61. Hardy, Sir W. "Problems of the Boundary State." Phil Trans. Roy. Soc. A203, pp 1-37, (1931).
62. Hardy, Sir W. and Nottage, M. "Adsorption. A Study of Availability and Accessibility." Proc. Roy. Soc. Lond. A136, pp 259-283, (1932).
63. Trillat, J.J. "The Adsorption of Oils in Relation to Lubrication."I. Mech. E. Proceedings of the General Discussion on Lubrication and Lubricants, vol. 2, pp 55-59, Oct 1937.
64. Needs, S.J. "Boundary Film Investigations." Trans. ASME 62, p 331-339, (1940).
65. Fuks, G.I. "The Properties of Solutions of Organic Acids in Liquid "Research in Surface Forces", Vol 1, 1960, ed. B.V. Deryaguin, Transl. by Consultants Bureau, NY, 1964.
66. Deryaguin, B.V., Karassev, V.V., Mechanism of Boundary Lubrication and the Properties of the Lubricating Film. Short- and Long-Range Action in the Theory of Boundary Lubrication." Wear 1, pp 277-290, (1957/58).
67. Akhmatov, A.S. Molecular Physics of Boundary Friction, Israel Prog. for Sci. Transl., Jerusalem (1966), chapt. 15.
68. Smith, A.J. and Cameron, A. "Rigid Surface Films." Proc. Roy. Soc. Lond. A328, pp 541-560, (1972).
69. Beeck, O., Givens, J.W. and Smith, A.E. "On the Mechanism of Boundary Lubrication. I. The Action of Long Chain Polar Compounds." Proc. Roy. Soc. Lond. A177, pp90-102, (1940).

70. Fein, R.S. and Kreuz, K.L. "Chemistry of Boundary Lubrication of Steel by Hydrocarbons" ASLE Trans. 8, pp 29-38, (1965).
71. Zaslavsky, Yu.S., Berlin, A.A., Zaslavsky, R.N., Cherkashin, M.I., Beloserova, K.E. and Rusakova, V.A. "Antiwear, Extreme Pressure and Antifriction Action of Friction Polymer Forming Additives." Wear 20, pp 287-297, (1972).
72. Furey, M.J. "The Formation of Polymeric Films Directly on Rubbing Surfaces to Reduce Wear." Wear 26, pp 369-392, (1973).
73. Israelachvili, J.N. and Tabor, D. "The Measurement of van der Waals Dispersion Forces in the Range 1.5 to 130 nm." Proc. Roy. Soc. Lond. A331, pp 19-38, (1972).
74. Israelachvili, J.N., Fisher, L.R., Horn, R.G. and Christenson, H.K. "Measurement of Adhesion and Short-Range Forces Between Molecularly Smooth Surfaces in Undersaturated Vapours and in Organic Liquids", Microscopic Aspects of Adhesion and Lubrication, Tribol Ser. 7, J.M. Georges, Ed. Elsevier, Amsterdam, pp 55-69, (1982).
75. Israelachvili, J.N., McGuiggan and P.M.Homola, A.M. "Dynamic Properties of Molecularly Thin Liquid Films." Science 240, pp 189-191, (1988).
76. Granick, S. "Motions and Relaxations of Confined Liquids." Science 253, pp 1374-1379, (1991).
77. Tonck, A., Georges, J.M. and Loubet, J.L. and the Rheology of Dodecane between Alumina Surfaces." J. Coll. and Interface Sci. 126, pp 150-163, (1988).
78. Georges, J.M., Millot, S., Loubet, J.L. and Tonck, A. "Drainage of Thin Liquid Films Between Relatively Smooth Surfaces." To be published in J. Chem Phys.
79. Sonnenfeld, R., Schneir, J. and Hansma, P.K. "Scanning Tunneling Microscopy: A Natural for Electrochemistry." Mod. Aspects Electrochem. 21, pp 1-28, (1990).
80. Smith, D.P.E., Hörber, H., Gerber, Ch. and Binnig, G. "Smectic Liquid Crystal Monolayers on Graphite Observed by Scanning Tunneling Microscopy." Science 245, pp 43-45, (1989).
81. Foord, C.A., Hamman, W.C. and Cameron, A. "Evaluation of Lubricants Using Optical Elastohydrodynamics." ASLE Trans. 11, pp 31-43, (1968).
82. Johnston, G.J., Wayte, R. and Spikes, H.A. "The Measurement and Study of Very Thin Lubricant Films in Concentrated Contacts." STLE Trans 34, pp 187-194 (1991).
83. Çavdar, B. and Ludema, K.C. "Dynamics of Dual Film Formation in Boundary Lubrication of Steels. Part I. Functional Nature and Mechanical Properties." Wear 148, pp 305-327, (1991).
84. Çavdar, B. and Ludema, K.C. "Dynamics of Dual Film Formation in Boundary Lubrication of Steels. Part III. Real Time Monitoring and Ellipsometry." Wear 148, pp 347-361, (1991).
85. Bell, J.C. , Delargy, K.M. and Seeney, A.M. "The Removal of Substrate Material Through Thick Zinc Dithiophosphate Anti-Wear Films" Proc. 18th Leeds-Lyon Symposium on Tribology, Lyon, Sept 1991, "Wear Particles" pp 387-396, ed. D. Dowson et al., publ. Elsevier 1992.
86. Guangteng, G., Cann, P.M. and Spikes, H.A. "A Study of Parched Lubrication." Wear 153, pp 91-105, (1992).
87. Hersey, M.D. "The Logic of Oiliness." Mechanical Engineering 55, pp 561-566, (1933).

Thin Films in Tribology / D. Dowson et al. (Editors)
© 1993 Elsevier Science Publishers B.V. All rights reserved.

Simulations of Lubricant Behavior at the Interface with Bearing Solids

P. A. Thompson,[a c] M. O. Robbins,[b *] and G. S. Grest[c]

[a]Department of Mechanical Engineering and Materials Science, Duke University, Durham, NC 27708, U.S.A.

[b]Department of Physics and Astronomy, Johns Hopkins University, Baltimore, MD 21218, U.S.A.

[c]Corporate Research Science Laboratories, Exxon Research and Engineering Company, Route 22 East, Annandale, NJ 08801, U.S.A.

Molecular dynamics simulations are used to study how solid walls affect the structure and dynamics of a contacting fluid. We first describe the changes in structure induced by a single solid wall. Both layering and in-plane order are found within 4 to 10 molecular diameters of the solid surface. The degree of in-plane order is directly correlated with the transfer of shear stress, and thus with the effective boundary condition for fluid flow. A range of boundary conditions from partial slip to entrainment of fluid layers is found. We next consider fluid films confined between two walls. As the film thickness decreases, the ordering influences of the two walls begin to overlap. This may ultimately lead to a phase transition in the film. Spherical molecules crystallize, while chain molecules enter a glassy state. The onset of the glass transition is heralded by dramatic increases in viscosity and by universal non-Newtonian behavior. Crystalline or glassy films have a non-zero yield stress. When this is exceeded the film undergoes a transition to a fluid state and shears easily. This releases stress in the film, which may then refreeze. We show that this is the origin of stick-slip motion at low velocities.

1. INTRODUCTION

Hydrodynamics and elasto-hydrodynamics have been very successful in describing lubrication by micron-thick films [1]. However, these continuum theories begin to break down as film thicknesses become comparable to molecular dimensions. An increasing number of applications require an understanding of lubricants in such severely confined geometries. Examples include lubrication of nanoscale bearings in micromachinery or high-density magnetic disk drives, as well as asperity interactions in macroscopic bearings which operate in the mixed lubrication regime.

Development of new experimental and theoretical techniques for studying thin lubricant films has paralleled the growing interest in their properties. The Surface Force Apparatus (SFA) allows normal and shear forces to be measured between atomically flat solid surfaces while their separation is determined to within 0.1nm using interferometry [2–11] or capacitance [12]. The area of contact in the SFA is typically 100μm which is much larger than the separation between solid walls. The Atomic Force Microscope (AFM) can be used to explore friction in lubricated contacts whose diameter is comparable to the separation (5nm)[13, 14]. This allows spatial resolution of the frictional force on a molecular scale. Quartz crystal oscillators have been used to determine the frictional forces between a surface and an adsorbed film of one or more monolayers [15]. Theoretical advances have been aided by the advent of supercomputers which allow thin films to be simulated at the molecular level using molecular dynamics (MD) [16–29]. These new experimental and theoretical techniques reveal a sequence of dramatic changes in the static and dynamic properties of fluid films as their thickness approaches molecular scales.

In this paper we describe this series of changes, illustrating them with results from our simulations of thin films of spherical and short-chain molecules. We begin by examining the effect of an isolated solid wall on the structure and dynamics of an adjoining fluid. Substantial layering

*Support from the National Science Foundation through Grant no. DMR 9110004 and from the Pittsburgh Supercomputing Center is gratefully acknowledged.

and in-plane order are induced within a few to ten molecular diameters of the solid. The degree of in-plane order is directly correlated with the effective boundary condition for fluid flow. Deviations from the commonly used no-slip boundary condition become substantial at the molecular scale.

We next consider the effects of two confining walls on a fluid film. As the film thickness h decreases, the ordering influences of the two walls begin to overlap and interfere. This leads to oscillations in the energy, normal force and effective viscosity with h [2, 3, 7, 12]. As h decreases further, the entire film may undergo a phase transition to a crystalline or glassy state. These solid-like films resist shear. The onset of the glassy phase is characterized by rapidly increasing relaxation times, which result in reduced diffusion constants and enhanced viscosities. The viscosity develops universal power-law scaling with shear rate that is consistent with recent experiments [10, 11].

Stick-slip motion is observed in both glassy and crystalline films. Transitions between sticking and sliding are accompanied by phase transitions in the film. If the molecules are spherical, static plates induce crystalline order in the film. Stick-slip motion involves periodic shear-melting and recrystallization of the film. For chain-molecules there is a periodic transition between glassy and fluid states. Stick-slip in these systems does not result from a velocity-weakening frictional force like that assumed in many models. Instead it reflects the thermodynamic instability of a shearing solid state.

The following section provides a brief description of our simulation techniques. The subsequent sections consider the effects of single walls and thin films, respectively. Section 5 presents a summary and conclusions.

2. SIMULATION METHOD AND GEOMETRY

Our simulations were performed in a planar, Couette geometry that closely mimics the SFA (Fig. 1). A film of fluid molecules was confined between two solid walls. Periodic boundary conditions were imposed within the plane of the walls

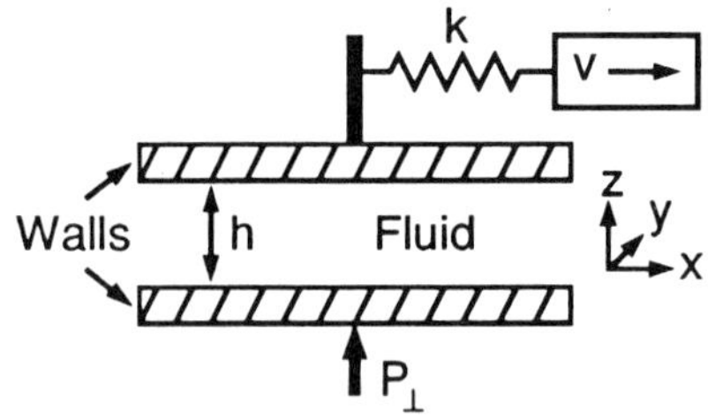

Figure 1. Sketch of the simulation geometry. Solid walls are held together by a constant normal pressure $P_\perp$. The top wall is pulled by a spring connected to a stage moving at constant velocity v. Periodic boundary conditions are imposed in the x and y directions.

to eliminate edge effects. Each wall consisted of $2N_w$ atoms forming two (111) planes of an fcc lattice. Several values of the wall density, ρ^w, were used to explore the effect of commensurability between wall and fluid molecules.

Two types of fluid molecule where considered: spherical molecules and short flexible chains. Interactions between spherical molecules were modeled using a truncated Lennard-Jones (LJ) potential,

$$V(r) = 4\epsilon[(\frac{\sigma}{r})^{12} - (\frac{\sigma}{r_c})^{12} - (\frac{\sigma}{r})^6 + (\frac{\sigma}{r_c})^6], \quad (1)$$

where r is the molecular separation, ϵ and σ are characteristic energy and length scales, and the potential is zero for $r > r_c$.

A well-studied bead-spring model was used for the chain molecules[30]. Each monomer interacted through a purely *repulsive*, truncated LJ potential (Eq. 1 with $r_c = 2^{1/6}$). Adjacent monomers along each chain of length n were also coupled through a strongly attractive potential,

$$V^{CH}(r) = -\frac{1}{2}kR_o^2 \ln[1 - (r/R_o)^2], \quad r < R_o, \quad (2)$$

where $R_o = 1.5\sigma$ and $k = 30\epsilon/\sigma^2$. Previous studies have shown that this potential prevents chain breaking and crossing and yields realistic dynamics for polymer melts[30]. At the densities considered in our study, the mean bond length was $\sim 0.96\sigma$.

The wall-fluid interaction was also modeled with a truncated LJ potential (Eq. 1), but with parameters ϵ^{wf}, σ^{wf}, and r_c^{wf}. This allowed independent variation of the strength and commensurability of wall-fluid interactions. A simple model of the forces within the wall was adopted to speed the calculations: each wall atom was coupled to its lattice site by a stiff spring. The spring constant was always large enough that the root mean squared displacement about lattice sites, δu divided by the nearest-neighbor distance, d, was less than the Lindemann ratio at melting (~ 0.16) [31].

To mimic experiments[5, 6, 8], the simulations were done with a constant pressure $P_\perp$ pushing the walls together. The dynamics of the wall separation h were modeled using a generalization of Andersen's extended system method[32]. Due to the layering described below, the number of fluid molecules, N_f, is nearly constant over a wide range of $P_\perp$ and then decreases suddenly as a layer is squeezed out [2, 7]. Recent simulations show that allowing N_f to vary does not produce significant changes in the dynamics of films of spherical molecules [29]. We thus used a fixed $N_f = m_l N_l$, corresponding to an integral number, m_l, of layers of N_l molecules.

A constant temperature, $T = 1.1\epsilon/k_B$, was maintained by coupling the motion of the wall molecules to a thermal reservoir[30]. The equations of motion were integrated using a fifth-order predictor-corrector algorithm with time step $\Delta t = .005\tau$, where $\tau = (m\sigma^2/\epsilon)^{1/2}$.

We will report all of our results in units of ϵ, σ, and τ. The order of magnitude of these quantities for spherical molecules is 0.03eV, 0.3nm and 3ps, respectively. The unit of pressure, $\epsilon\sigma^{-3}$, corresponds to about a kilobar. Our absolute pressures are somewhat higher than experimental values because the attractive tail of the LJ potential is cutoff. However, changes in pressure are not affected by this shift.

Shear was applied to the system in three different ways. In the SFA, one solid wall is translated by attaching it, through a spring, to a translation stage. Israelachvili's group uses a weak spring and a constant velocity stage [5, 6]. Granick's group imposes a sinusoidal displacement through a stiff spring [8]. However, their results do not depend separately on the frequency and amplitude of the displacement – only the shear rate is important. This indicates that equivalent results would be obtained with a constant displacement velocity and a stiff spring. We have performed simulations with a spring attaching the top wall to a constant velocity stage (see Fig. 1, with a constant force applied to the top wall, and with a constant velocity of the top wall. Constant force simulations correspond to an arbitrarily weak spring and constant velocity simulations to an infinitely strong spring. The registry in the y-direction between top and bottom walls may also be important. We have performed simulations with fixed registry, and with a spring which allows the registry to change. Only minor differences were found between the two algorithms.

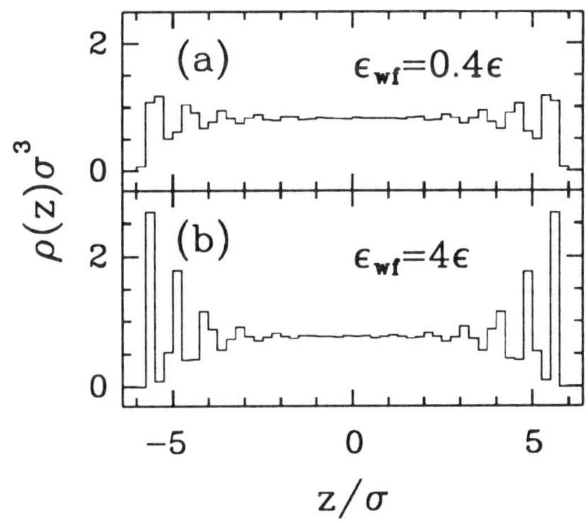

Figure 2. Profiles of the fluid density $\rho(z)$ as a function of z for the indicated values of ϵ^{wf} and $\rho^w = 2.52\bar{\rho}$.

3. INFLUENCE OF A SINGLE WALL

The most well-understood phenomenon observed in the SFA experiments is the layering of fluid molecules parallel to solid surfaces (Fig. 2) [16, 23, 24, 33]. It is induced by structure in the fluid pair-correlation function, and the sharp cut-off in fluid density at the solid wall. An initial fluid layer forms at the preferred wall-fluid spacing. Additional fluid molecules tend to lie in a second layer, at the preferred fluid-fluid spacing. This layer induces a third, and so on. The magnitude of the density oscillations depends on temperature, strength of the wall potential, and on a competition between the solid lattice constant and the molecular spacing of the fluid [16]. For fluids composed of simple spherical molecules, the oscillations typically extend out to a distance of order 5 molecular diameters [16, 23, 24]. For more complex systems containing fluids with chain or branched polymers, the oscillations are negligible beyond ~ 3 molecular diameters [34].

Layering of spherical molecules is illustrated in Fig. 2, which shows the variation of density, ρ, with distance, z, between two walls for $\epsilon^{wf}/\epsilon =$ 0.4 and 1.8 [16]. For this data the wall density was 2.52 times higher than the mean density of the fluid $\bar{\rho}$. Note that increasing the strength of the wall-fluid interaction increases the modulations in the density, or equivalently, the sharpness of the layers. However, the density reaches the constant value expected for a bulk fluid after about four layers for both values of ϵ^{wf}. The rate at which the layering decays depends on the range of the pair-correlation function and is relatively insensitive to ϵ^{wf}.

In addition to normal ordering, solid surfaces induce *in plane* order [16, 19, 27, 25]. Although it has received less attention than layering, in-plane ordering is more important in determining the dynamics of fluid molecules near solid surfaces [16–18]. We have used two measures of the degree of in-plane order in the lth layer away from a solid wall. The first was the spatial probability distribution $P(\vec{r})$ for finding a fluid monomer at a position $\vec{r}$ relative to a unit cell of the solid lattice sites. The second was the two-dimensional static structure factor:

$$S_l(\vec{k}) = 1/N_l \left| \sum_i \exp(i\vec{k}\cdot\vec{r}_i) \right|^2 , \tag{3}$$

where N_l is the number of fluid monomers within the layer.

Figure 3 shows both $P(\vec{r})$ and $S(\vec{k})$ for two fluid layers of spherical molecules immediately below a static wall. The wall and fluid densities are the same and $\epsilon^{wf} = 0.4\epsilon$. Note the strong peaks in P_1 at the centers of the sides of the unit cell, immediately below gaps in the adjacent solid layers. Fluid atoms are 12 times more likely to sit below these gaps than directly below a wall atom. This ordering is also evident in S_1, which shows several Bragg peaks. Atoms in the second layer are much less ordered. The density is highest above the minima in P_1, but P_2 varies by only a factor of two. The structure factor shows weak modulations rather than sharp Bragg peaks.

Figure 4 shows that the degree of in-plane order increases sharply with the strength of the wall fluid coupling. At $\epsilon^{wf}/\epsilon = 1.8$, the first layer is crystallized by the wall. Diffusion within the layer is immeasurably small and the mean-squared displacement about lattice sites is less than the Lindemann criterion for melting[31]. The structure factor of the first layer also satisfies a condition for solid stability found in recent work on bulk systems[35]: The Debye-Waller factor

$$e^{-W} = S_1(\vec{G}_1)/S_1(0) \tag{4}$$

of the shortest reciprocal lattice vector, $\vec{G}_1$, exceeds 0.6. The value of e^{-W} equals unity in an ideal crystal at zero temperature and goes to zero in a bulk fluid. Thus the Debye-Waller factor provides a very convenient measure of the degree of order in a layer, and we will use it throughout the remainder of the text.

In general, increasing the thermal roughness of the solid [16] or the temperature of the fluid reduces the degree of in-plane order [16, 26]. The amount of order is maximized when the wall-fluid coupling is large and there is a high degree of commensurability between wall and fluid densities [16]. At extreme conditions, up to two layers become crystallized by the solid wall [16, 26].

In-plane ordering plays an essential role in the transmission of shear stress across the solid-fluid interface [16]. If the solid wall was completely flat and featureless, it could be moved past the fluid without exerting any force. The appropriate boundary condition for fluid flow would be complete slip. We have investigated the relationship between in-plane order and slip boundary condi-

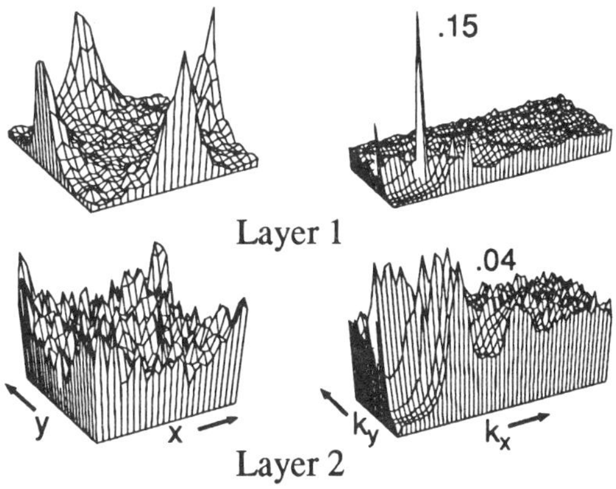

Figure 3. $P(x, y)$ (left) and $S(k_x, k_y)$ (right) in the two fluid layers nearest the walls for spherical molecules, $\epsilon^{wf} = 0.4\epsilon$, $\rho^w = \bar{\rho}$, $r_c^{wf} = r_c = 2.2\sigma$, $N_w = 48$, and $h \simeq 14\sigma$. Note that the scales vary and values for $S(\vec{G}_1)/S(0)$ are given. Wall lattice sites lie at the center and corners of the unit cell for $P(x, y)$.

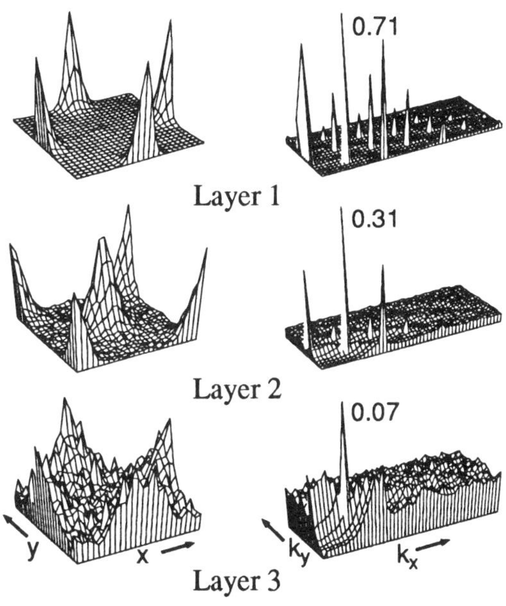

Figure 4. $P(x, y)$ (left) and $S(k_x, k_y)$ (right) in the three fluid layers nearest the walls for $\epsilon^{wf} = 1.8\epsilon$. Other parameters are the same as in Fig. 3.

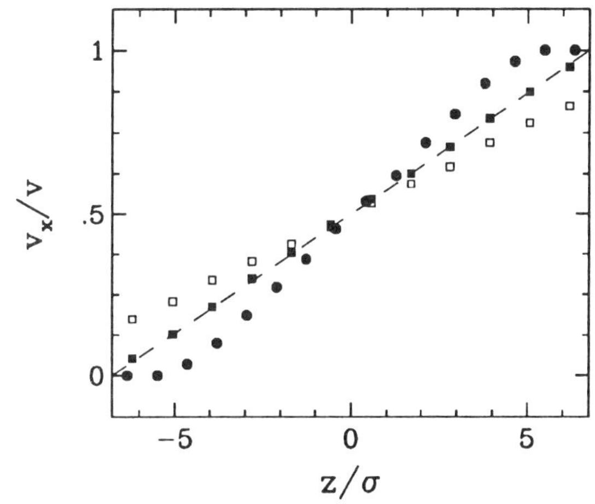

Figure 5. Velocity profiles for spherical molecules, $v = 2\sigma\tau^{-1}$, $r_c^{wf} = r_c = 2.2\sigma$, $\epsilon^{wf} = 0.4\epsilon$ (squares), and $\epsilon^{wf} = 4\epsilon$ (circles). The closed and open symbols indicate $\rho^w/\bar{\rho} = 1$ and 2.52, respectively. The locations of the first layers of solid atoms on each side of the fluid correspond to the vertical borders. The dashed line indicates the flow expected from hydrodynamics with a no-slip BC.

tion for a wide range of wall-fluid couplings, temperatures and wall densities [16]. The fluid was sheared by translating the top wall at constant velocity v in the x direction, and the mean value of v_x in the fluid was calculated as a function of z. Flows with boundary conditions ranging from complete slip to multi-layer entrainment, or "stick", were observed. A similar range of boundary conditions has been found in experiments on thin films [4, 15, 36]

Fig. 5 shows velocity profiles for three systems which exhibit slip, no-slip and "stick" boundary conditions. A no-slip boundary condition implies that v_x/v varies smoothly from zero at the bottom wall to unity at the top wall (Fig. 5). The degree of slip or stick can be determined from the flow in the central region where the film behaves like a bulk fluid. If there is slip, a linear extrapolation of the flow will reach 0 and 1 at distances L_s outside the left and right solid wall, respectively. If there is a stick boundary condition, L_s

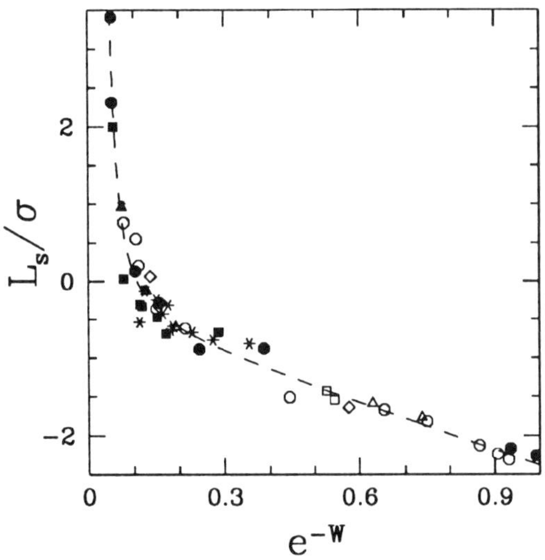

Figure 6. Correlation between L_s and $e^{-W} = S_1(\vec{G}_1)/S_1(0)$. Open and closed symbols indicate results for $\rho^w/\bar{\rho} = 1$ and 2.52, respectively. Circles and triangles are results with wall atoms fixed rigidly to their lattice sites and $k_BT/\epsilon = 1.1$ and 1.4, respectively. Diamonds and squares indicate results for $\delta u/d = 0.1$ at the same temperatures. When the first layer crystallizes, it becomes a solid wall influencing the second layer. Starred points denote $(L_s + 1)$ vs. $S_2(\vec{G}_1)/S_2(0)$ for these systems. Note that they fall on the same curve.

is negative – the fluid velocity reaches that of the walls at a distance $|L_s|$ inside the film. We find that L_s is independent of both h and v [37], and thus provides a unique characterization of stress transfer at the wall-fluid interface.

We have determined L_s as a function of ϵ^{wf}/ϵ, $\rho^w/\bar{\rho}$, T, and thermal roughening of the walls $(\delta u/d)$ [16]. The trends in L_s are exactly opposite to the trends in in-plane order – the slip length decreases as the degree of order increases. A more surprising result is that there seems to be a one-to-one relation between the slip length and Debye-Waller factor for simple spherical molecules. As shown in Fig. 6, the results for a large range of parameters collapse onto a single curve when L_s is plotted against e^{-W}. The transfer of positional order is directly correlated to the transfer of shear stress.

4. CONFINEMENT BY TWO WALLS

4.1. Phase transitions

So far we have focused on the interfacial properties of fluids in systems where the plate separation h is large enough to insure that the fluid exhibits bulk behavior in the central region. At separations below ~ 10 molecular diameters, normal ordering spans the distance across the film. The energy of the film is minimized when the layering effects of the two walls add constructively, and maximized when they add destructively. The resulting oscillations in the force between the walls are readily observed with the SFA [7]. At even smaller separations, $h \sim 5 - 6$ molecular diameters, the in-plane ordering begins to overlap [19–21, 25, 28]. The entire film may then solidify, even though the temperature and pressure are well within the fluid region of the bulk phase diagram [19–21]. The ordering induced by the solid wall lowers the entropy of the fluid state from its bulk value, increasing its free energy relative to the solid phase.

The effect is demonstrated in Fig. 7, which shows the dependence of h, e^{-W}, and the molecular diffusion constant parallel to the walls D, on $P_\perp$ for spherical molecules and chains of length 6. Results for a two-layer system are shown, but similar behavior is found in thicker films.

As expected, the film thickness decreases slightly with increasing $P_\perp$. At the same time, the degree of layering and in-plane order increases, and the diffusion constant decreases. The major difference between spherical and chain molecules is in the form of these changes. There is a sudden drop in D and a transition to an fcc structure with $e^{-W} > 0.6$ in monomer films [35]. The transition pressure, $P_\perp \simeq 7\epsilon\sigma^{-3}$, is substantially lower than the bulk value of $P \simeq 12\epsilon\sigma^{-3}$. In contrast, films of chain molecules remain highly disordered. The steady drop in D relative to bulk values indicates the onset of a glassy phase at a pressure well below the bulk transition pressure.

Thus confinement produces a marked shift in the bulk transition pressures of both types of molecules. The magnitude of the shift depends on the film thickness, the chain length, the strength of the wall-fluid interaction, and the commensurability between wall and fluid. The nature of the transition may also be affected. Large lattice mismatches produce high-order commensurate phases in monomer films[16]. Chains tend to lock in glassy phases, and only crystallize for very large ϵ^{wf} and d close to the equilibrium monomer spacing along chains.

4.2. Universal glassy dynamics

A wall-induced glass transition explains the dramatic increases in the measured relaxation times and viscosities of ultra-thin films [5, 6, 8–11]. To study the dynamic response directly, we sheared our films by translating the top wall at a uniform speed v in the $(11\bar{2})$ direction and found the mean frictional force per unit area, f, exerted on the walls [21]. Figure 8 shows the effective time-averaged viscosity $\mu = fh/v$ as a function of the effective shear rate $\dot{\gamma} = v/h$ and the number of fluid layers m_l for $n = 6$ films at $P_\perp = 4\epsilon\sigma^{-3}$. Note that one may have the same $P_\perp$ at several film thicknesses because of the layer-induced oscillations in the energy of the film with h [7].

The viscous response in Fig. 8 for films with $m_l = 8$ and $m_l = 6$ is nearly the same as that for bulk systems. The films behave like Newtonian fluids with a constant viscosity μ_o until dimensionless shear rates of order 0.1. At higher shear rates the viscosity decreases. This shear-thinning arises because the chains are sheared past each other at a rate which is faster than their relaxation rate. Since strains in the fluid can not relax completely significant changes in fluid structure begin to occur. In our simulations at constant $P_\perp$, the film thickness increases. If h is held fixed, the normal stress begins to increase.

Similar behavior is observed in most bulk polymeric fluids. The shear-thinning regime is commonly fit to a power law, $\mu \propto \dot{\gamma}^{-\alpha}$, and typical values of α range from 0.4 to 0.9 [38]. A maximum relaxation time τ_m can be associated with the shear rate $\dot{\gamma}_c$ where the viscosity begins to fall: $\tau_m = \dot{\gamma}_c^{-1}$. We have determined a value for $\dot{\gamma}_c$ by fitting the high shear rate response to a power law and finding the shear rate at which the fit reaches μ_o.

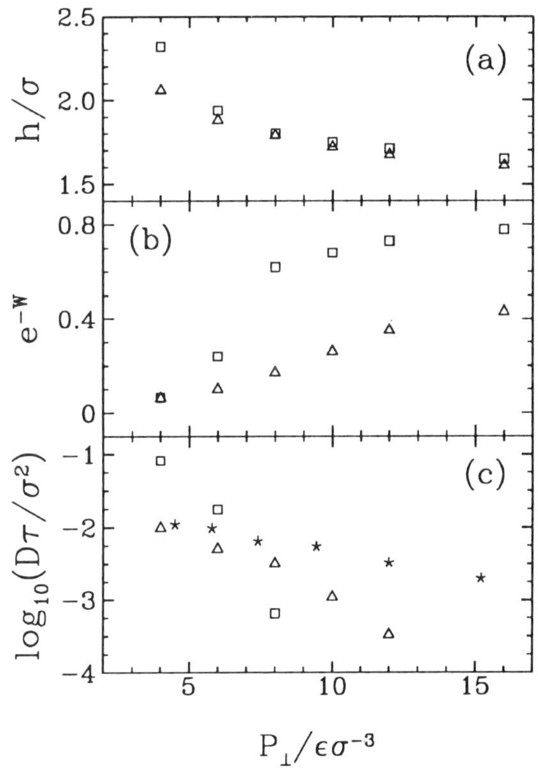

Figure 7. Variation of (a) h, (b) Debye-Waller factor, and (c) diffusion constant with $P_\perp$ in static two layer films with $\epsilon^{wf} = \epsilon$, $r_c^{wf} = r_c = 2^{1/6}\sigma$, $N_w = nN_l = 144$ and $d = 1.20\sigma$. Squares and triangles indicate films with $n = 1$ and $n = 6$, respectively. Stars indicate the bulk diffusion for $n = 6$.

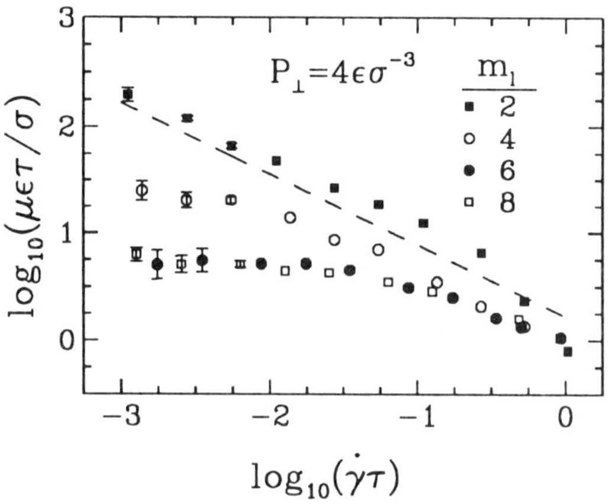

Figure 8. Variation of μ with $\dot{\gamma}$ in films with $n = 6$, $\epsilon^{wf} = 3\epsilon$, $r_c^{wf} = 2.2\sigma$, $N_w = nN_l = 144$, and $d = 1.0\sigma$. The dashed line has slope -2/3.

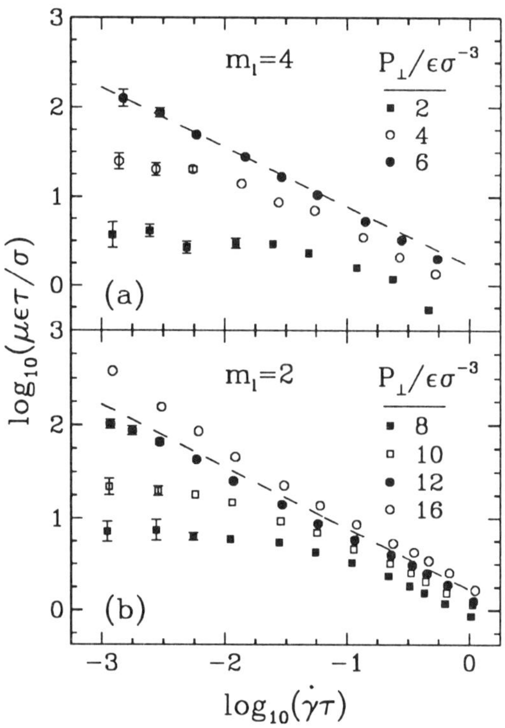

Figure 9. Variation of μ with $\dot{\gamma}$ as a function of $P_\perp$ in films with $n = 6$ and $N_w = nN_l = 144$. In (a) $m_l = 4$, $r_c^{wf} = 2.2\sigma$, $d = 1.0\sigma$, and $\epsilon^{wf} = 3\epsilon$. In (b) $m_l = 2$, $r_c^{wf} = 2^{1/6}\sigma$, $d = 1.20\sigma$, and $\epsilon^{wf} = \epsilon$. The dashed lines have slope -2/3.

As m_l decreases below 6, both τ_m and μ_o increase dramatically (Fig. 8). By $m_l = 2$, they have risen by at least two orders of magnitude relative to their bulk values. The relaxation time exceeds the length of our longest simulations ($\sim 10^4\tau$), and the film has a finite yield stress on these time scales. These changes are completely a result of confinement, since $P_\perp$ is held fixed.

The wall separation can also be varied by fixing m_l and varying $P_\perp$. This is particularly relevant to experiments near the glass transition where the extremely slow dynamics inhibit decreases in the number of molecules in the film. Figures 9(a) and (b) show the dynamic response of $m_l = 4$ and $m_l = 2$ films as a function of $P_\perp$. As expected, both τ_m and μ_o are strong functions of $P_\perp$. By $P_\perp = 6\epsilon\sigma^{-3}$ in Fig. 9(a) the dynamics in 4 layer films are too slow to resolve in our simulations. Hu et al.[10] found similar changes in dodecane films. Increasing $P_\perp$ led to a 5-fold increase in τ_m and μ_o while the number of layers remained fixed. In both their experiments and our simulations the product $\mu_o\dot{\gamma}_c$ is nearly independent of $P_\perp$ and is comparable to the yield stress of a solid-like film (see below).

Increases by more than 5 orders of magnitude in μ_o and 8 orders of magnitude in τ_m have been observed in experiments[9–11]. Based on our simulations we attribute them to a confinement-induced glass transition. A common model of glass transitions is the free volume model which predicts that the relaxation time diverges as the volume decreases to a critical value [39]. Since the area of our films is fixed, this corresponds to a prediction that

$$\tau_m = C\exp(h_o/(h - h_c)), \tag{5}$$

where h_c is the thickness at the glass transition and C and h_o are fitting parameters. Fig. 10 shows a plot of $log_{10}\tau_m$ vs. $h_o/(h - h_c)$ for $h_o = 0.50\sigma$ and $h_c = 1.64\sigma$. The results are clearly consistent with the free-volume expression, although they do not span a large enough range to constitute an accurate test of it. The value of h_c corresponds to $P_\perp = 14\epsilon\sigma^{-3}$. This estimate of the glass transition is also consistent with our calculated values of the diffusion constant which is unmeasurably small for $P_\perp > 12\epsilon\sigma^{-3}$ (Fig. 7).

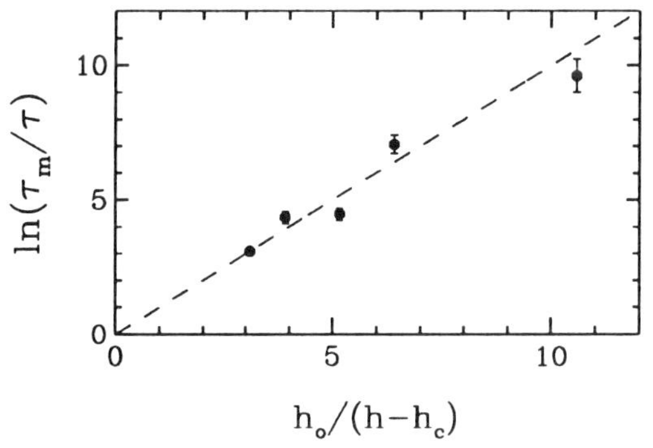

Figure 10. Fit of relaxation time to free volume theory. Dashed line corresponds to $C = 1$.

Bulk diffusion is rapid even for $P_{\perp} > 24\epsilon\sigma^{-3}$.

A striking feature of the non-Newtonian regime in these films ($\dot\gamma > \dot\gamma_c$) is that the viscosity obeys a power law $\mu \sim \dot\gamma^{-\alpha}$ with $\alpha \simeq 2/3$ under constant normal load. The scaling does not depend on many details of the simulations including commensurability between wall and fluid, relative wall orientation, wall stiffness, wall-fluid coupling or chain length [22]. However, we do find a slight increase in α as $P_{\perp}$ is increased towards the glass transition. The same scaling and variation with $P_{\perp}$ has been observed in recent experiments [10, 11]. This is remarkable considering the simplicity of our chain molecules compared to the dodecane, hexadecane, and silicone oil molecules used in the experiments. It suggests that $\mu \sim \dot\gamma^{-2/3}$ may be a universal property of lubricants near a glass transition *under constant normal load* [22]. Simulations at constant h give similar values of γ_c, but the behavior is less universal. The viscosity decreases more slowly with $\dot\gamma$ ($\alpha \sim 1/2$) because the film can not expand to facilitate shear.

Perhaps the most surprising aspect of the power-law response is that the exponent α does not change when the mechanism of shear changes. Consider, for example, the viscous responses shown in Figures 9 (a) and (b). While the power laws characterizing the shear-thinning regime are the same, the nature of the strain is fundamentally different. Strain is uniformly distributed across the film in panel (a), while the film of panel (b) behaves like a plug – strain is localized at the wall-fluid interface.

Plug-like flow produces some counter-intuitive behavior in thin films. The reason is that the apparent viscosity is no longer a property of the film. Instead, it characterizes the dynamics at the weak wall/film interface. Among other things, this means that $\dot\gamma_c^{-1}$ is not necessarily the largest relaxation time *within* the film. In fact, we found that the films in Fig. 9(b) were able to maintain stress over a period much longer than the time $\dot\gamma_c^{-1}$ determined from the effective viscosity. We also observed stick-slip motion (see next subsection) in these systems at $\dot\gamma < \dot\gamma_c$. In contrast, stress in the films of Fig. 9(a) relaxed to zero with a characteristic time constant of $\dot\gamma_c^{-1}$, and there was no evidence of stick-slip motion for $\dot\gamma < \dot\gamma_c$. We also found that the apparent viscosity of plug-like films was independent of chain length, while the bulk value of μ_o increased rapidly with chain length.

Plug-like flow is most readily observed when h is much less than the radius of gyration of a free chain. In this limit, each chain occupies a compact region spanning the distance between walls. This structure reflects the large entropic cost for alignment into a single layer or for sharing a region of space with another molecule [40]. Since any velocity gradient tends to stretch the chains, this topology greatly inhibits flow within the film. We found that shear only extended throughout thin films when the range and strength of the wall-fluid coupling were increased so that it could compete with the multiple interactions between chains (Fig. 9).

The entropic constraints become less important as the walls move apart. As they do, the fraction of the strain which occurs within the film increases. Peculiar behavior may result, such as an *increase* in the *effective* viscosity with increasing film thickness. This behavior has been observed in experiments[8], but seems paradoxical since the film is changing from a glass to a fluid. The explanation is that the apparent viscosity measures the total dissipation at the wall/film interface and within the film. As h increases, there is more strain and dissipation within the film, precisely

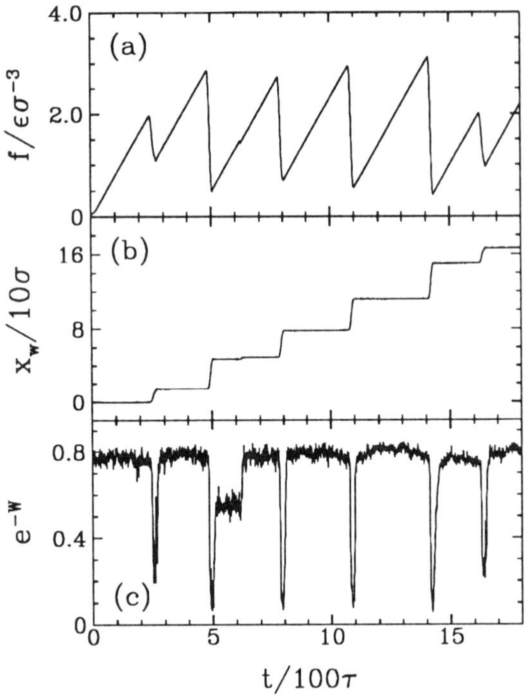

Figure 11. Time profiles of the (a) frictional force per unit area f, (b) displacement of the top wall x_w, and (c) Debye-Waller factor for a system during stick-slip motion with $n = 1$, $m_l = 2$, $P_\perp = 10\epsilon\sigma^{-3}$, $N_w = N_l = 144$ and $v = 0.1\sigma\tau^{-1}$.

because its actual viscosity decreases.

4.3. Stick-slip motion

The longest relaxation time diverges as the films enter a crystalline or glassy state. Such films will not shear until a finite yield stress is exceeded. One of the most intriguing experimental observations is that these films exhibit oscillatory stick-slip motion when they are forced to slide slowly. To probe the molecular origin of this phenomena, we have mimicked the SFA experiments by coupling the top plate through a spring to a stage moving at constant velocity v [19, 20]. Our first studies focused on the behavior of a two layer film of spherical molecules.

Initially, the force per unit area applied by the spring, f, was zero and the top plate was at rest. As the stage moved forward, the spring stretched and f increased. If the film was a liquid, the top plate accelerated until the steady-state f balanced the viscous dissipation. Stick-slip behavior was observed if the film had solidified, as illustrated in Fig. 11(a). Initially, the film responded elastically. The top plate remained stuck, and f increased linearly. Eventually, f exceeded the yield stress of the film f_s, and the top plate began to slide. Since the frictional force was less in the sliding state, the plate accelerated to catch up with the stage and f decreased. At some lower value f_m, the top plate stuck once more, and the process repeated. During the sliding phase, the plate jumped forward by an amount $\Delta x \sim (f_s - f_m)A/k$, where A was the area of the plate. For the case illustrated in Fig. 11(b), Δx ranges from 20 to 30σ. The forward jumps are typically much larger in experiments ($\sim$ 1 μm) because weaker springs are used. Since shearing the film increases the effective repulsion between fluid molecules, sliding plates separate at constant normal load. For the stick-slip motion shown in Fig. 11, h increased by $\sim$ 0.1 to 0.2σ during each slip. The resolution of experiments done thus far has not been sufficient to observe changes of this order ($\lesssim$ 0.1nm) [6].

The relaxation behavior observed in experiments indicates that substantial rearrangement of the molecules occurs [6]. We have calculated the time-dependence of e^{-W} to monitor changes in the degree of order. Figure 11(c) shows that e^{-W} is close to 0.8 in the static state, as expected for a crystal at finite temperature. However, in the sliding state, e^{-W} drops below 0.2, a value consistent with liquid structure and dynamics [16]. Snapshots of instantaneous configurations of atoms also show this loss of order. These observations clearly indicate that the film undergoes solid to liquid phase transitions as it goes from static to sliding states. Analogous transitions, termed shear-melting, occur when bulk solids are sheared [41].

The velocity dependence of the stick-slip behavior is illustrated in Figure 12. In general, we find a unique yield stress f_s in the limit of small v. However, for a system with (111) solid surfaces, two slightly different values corresponding to fcc and hcp ordered phases of the film are observed (Fig. 12(a)). As v increases, the maximum value of f_s remains fixed, but variations in f_s increase substantially. Intermittent periods without stick-slip motion become more frequent (Fig. 12(b-d)). These changes occur because the film is not able to fully order during each stick. Examination of e^{-W} shows a direct correlation between the amount of disorder in the film and the observed decrease in f_s. When v exceeds a certain critical velocity v_c, the film remains in a liquid state and slides smoothly [19, 20]. Similar results are found in experiments[6].

The magnitude of v_c depends on many system parameters, including $P_\perp$, the inertia and crystalline allignment of the walls, and the commensurability of the wall and fluid [20]. In order for the moving wall to stop, its kinetic energy must be converted into potential energy in the film. The maximum potential energy that can be stored in the film scales as the static frictional force, $F_s = Af_s$, times the lattice constant of the wall, σ. Equating this to the kinetic energy at v_c, we find $v_c = c\sqrt{\sigma F_s/M}$, where c is a numerical factor and M is the mass of the moving wall.

We have verified that v_c scales as $M^{-1/2}$ by calculating v_c at several values of M with all other parameters held fixed [20]. Fig. 13 shows that this scaling holds over more than two orders of magnitude in M. In general, the value of c varies from about 0.05 to 0.5, depending on load, commensurability of wall and fluid, and the crystalline alignment of the walls. The largest values of c were only found for epitaxial fluid densities and aligned walls. Neither condition is satisfied in typical experiments. Taking experimental values of $F_s \sim 20mN$, $\sigma \sim 5$Å, $M \sim 0.02$kg, and $v_c \sim 2\mu$m/s, we find $c \approx 0.1$. This is well within the range of theoretical values.

Since the pioneering studies of Bowden and Leben at Cambridge University in the 1930s, stick-slip motion in boundary lubrication has been attributed to a frictional force that decreases with increasing v [42]. However, in our studies, we found that the frictional force increases monotonically with v [19]. This is illustrated by the solid line in Fig. 14 which shows the mean force from runs at constant velocity. Runs at constant

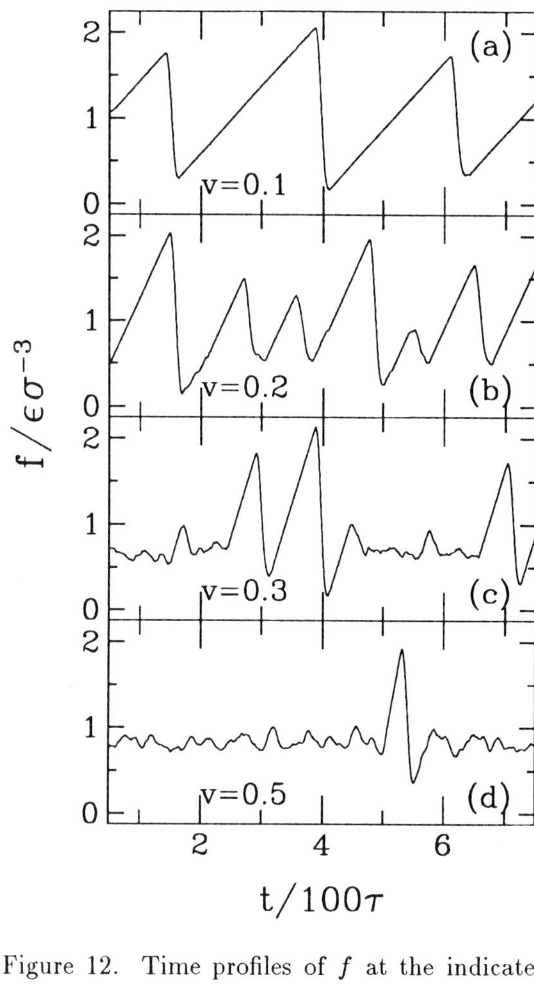

Figure 12. Time profiles of f at the indicated velocities.

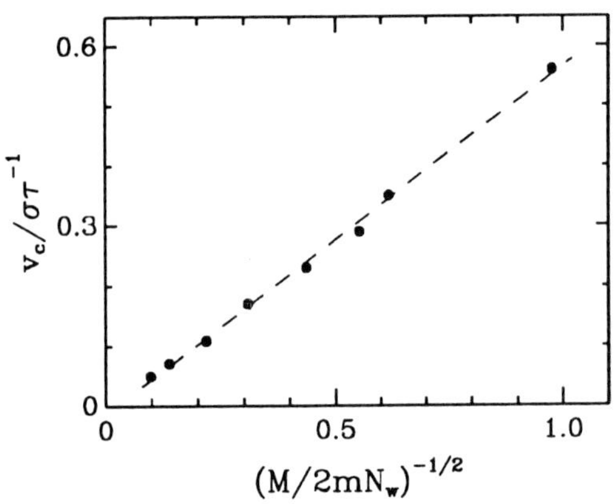

Figure 13. Variation of v_c with $M^{-1/2}$ for the parameters of Fig. 11. The dashed line is a linear fit to the data with $c = 0.3$.

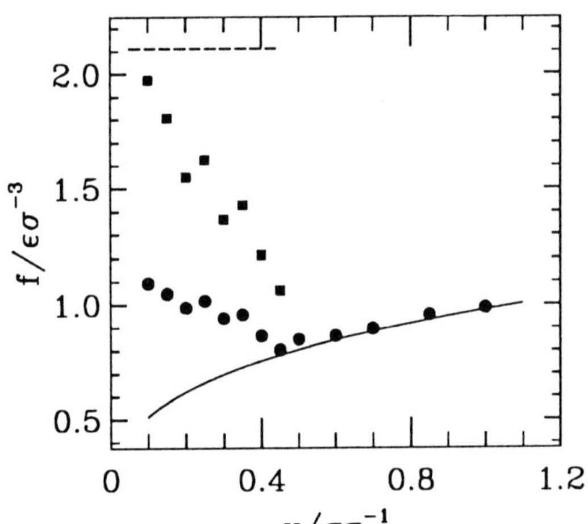

Figure 14. Force vs. v for the system of Fig. 11. The solid line shows the mean force when the top wall is displaced at constant velocity. Circles show the time-averaged force when the wall is pulled through a spring. The two converge above $v_c = 0.45\sigma/\tau$. The dashed line shows the maximum yield stress, and squares indicate the mean yield stress in the stick-slip regime.

force lie on the same curve until the mean velocity drops below v_c. The film then solidifies and v goes to zero. Thus the origin of stick-slip motion in these films can not be due to the dynamic instability associated with negative values of df/dv. Instead it results from the thermodynamic instability of a solid state at finite shear rates and of the disordered sliding state below a critical shear stress [19, 20].

Note that an experiment without sufficient time resolution might observe a frictional force which decreased with v and erroneously conclude that it was responsible for stick-slip. The circles in Fig. 14 show that the time-averaged force decreases with v in the stick-slip regime of our simulations. This time-averaged force represents an average of the force associated with stuck and sliding states. Since the average yield stress of static films decreases with v (see Fig. 12), and the film spends more time in the moving state as v increases, the average force decreases in the stick-slip regime. Similar averaging occurs in a large system where different regions of the surface stick and slip at different times. Interpreting the time or spatially averaged force as an intrinsic $F(v)$ curve, one would conclude that stick-slip motion should occur for $v < v_c$ due to the negative slope. However, this confuses cause and effect. As shown in Fig. 14, the force at constant velocity in the disordered kinetic state is monotonically increasing. It is the stick-slip itself that produces the negative slope.

These observations suggest an experimental method for determining whether stick-slip results from a negative df/dv or from a first order transition between static and sliding states. If there is an intrinsic $f(v)$ curve with a negative slope it will remain unaffected as M increases. In contrast, if stick-slip results from transitions between static and sliding states, v_c will decrease as M increases (Fig. 13). The time-averaged force will only have a negative derivative below v_c (Fig. 14).

Although we have shown results for stick-slip in a single geometry, the behavior is quite universal. Similar results are observed in systems with more complex surfaces and film molecules. What changes is the nature of the stuck state. For example, in systems with amorphous walls [19] or chain-like molecules [21, 22], the static state is glassy rather than crystalline. The disorder in the film leads to greater fluctuations in the yield strength f_s, and reduces v_c [19, 21, 22].

5. Summary and Conclusions

We have described a sequence of changes in the behavior of lubricant molecules which takes place as they become more confined between flat solid walls. Even when the walls are 30 or more molecular diameters apart, deviations from the commonly assumed no-slip boundary condition become apparent [4, 16, 36]. These deviations were found to correlate with the degree of in-plane positional order induced by the wall in the adjacent fluid [16]. As the wall spacing decreases to a few molecular diameters, the structure of the entire film is modified and the film may undergo a phase transition to a crystalline or glassy state [21].

As a film approaches the glassy state, its response exhibits interesting non-Newtonian behavior [19–21]. Shear-thinning begins above a shear rate $\dot{\gamma}_c = \tau_m^{-1}$. The viscosity falls as $\dot{\gamma}^{-\alpha}$, where $\alpha = 2/3$ seems to be a nearly universal exponent for shear at constant $P_\perp$. The divergence of τ_m as $P_\perp$ increases is consistent with a free volume description of the glass transition [39]. The glass transition occurs at much lower pressures than the corresponding bulk transition. The pressure in some bearings may be high enough to induce a bulk phase transition in thick lubricant films. We expect that the viscous response of these vitrified films will be similar to that of the confined glasses considered here.

The solid-like states induced by confining walls are unstable when the film is sheared. A transition to a disordered fluid state occurs. The fluid state itself is only stable above a critical velocity v_c. These two facts lead to stick-slip motion at velocities $v < v_c$. The film transforms periodically between static solid and sliding fluid states.

An open question is whether stick-slip motion has analogous origins in other systems. The ordered static and disordered sliding states need not correspond to solid and fluid structure in an intermediate film. Instead, in the case of solid on solid sliding, they may correspond to states with and without chemical bonds or elastic deformations which increase the coupling between plates. For example, stick-slip motion of a weak solid over a hard solid leaves behind patches of the weak material[42], suggesting periodic welding of the surfaces followed by cracking. Recent simulations show that films that are far below the temperature where they first crystallize tend to crack rather than melt [22]. This also gives rise to stick-slip behavior. Stick-slip is also observed in non-conhesive granular flow. Simulations indicate that its origin is similar to that found here: Periodic dilation and compaction transitions occur [43].

Further work is needed to explore the effects of asperities on the solid surface, of chemical and physical degradation of lubricants, and of more complicated molecular structures. Recent experiments on surfactant coated surfaces reveal transitions between states sliding at different velocities, rather than between a static and sliding state [44]. Molecular dynamics simulations promise to be a powerfull tool in probing these aspects of lubricant behavior.

REFERENCES

1 See for example, D. Dowson and G. R. Higginson, Elastohydrodynamic Lubrication (Pergamon Press, Oxford, 1966).
2 R. G. Horn and J. N. Israelachvili, J. Chem. Phys., 75 (1981) 1400.
3 R. G. Horn, S. J. Hirz, G. Hadziioannou, C. W. Frank and J. M. Catala, J. Chem. Phys., 90 (1989) 6767.
4 J. N. Israelachvili, J. Colloid Interface Sci., 110 (1986) 263.
5 J. N. Israelachvili, P. M. McGuiggan and A. M. Homola, Science, 240 (1988) 189.
6 M. L. Gee, P. M. McGuiggan, J. N. Israelachvili and A. M. Homola, J. Chem. Phys., 93 (1990) 1895.
7 J. N. Israelachvili, Intermolecular and Surface Forces, 2nd ed., (Academic Press, London, 1991).
8 J. Van Alsten and S. Granick, Phys. Rev. Lett., 61 (1988) 2570.

9 J. Van Alsten and S. Granick, Macromolecules, 23 (1990) 4856.
10 H.-W. Hu and G. A. Carson and S. Granick, Phys. Rev. Lett., 66 (1991) 2758.
11 S. Granick, Science, 253 (1992) 1374.
12 J. M. Georges, S. Millot, J. M. Loubet, A. Touck and D. Mazuyer, in this volume.
13 C. M. Mate, G. M. McClelland, R. Erlandsson, and S. Chiang, Phys. Rev. Lett., 59 (1987) 1942.
14 E. Meyer, R. Overney, D. Brodbeck, L. Howald, R. Lüthi, J. Frommer, and J.-J. Güntherodt, Phys. Rev. Lett., 69 (1992) 1777.
15 E. Watts, J. Krim and A. Widom, Phys. Rev. B, 41 (1990) 3466.
16 P. A. Thompson and M. O. Robbins, Phys. Rev. A, 41 (1990) 6830.
17 P. A. Thompson and M. O. Robbins, Physics World, 3 (1990) 35.
18 P. A. Thompson, W. B. Brinckerhoff and M. O. Robbins, J. Adhesion Sci. Technol., in press.
19 P. A. Thompson and M. O. Robbins, Science, 250 (1990) 792.
20 M. O. Robbins and P. A. Thompson, Science, 253 (1991) 916.
21 P. A. Thompson, G. S. Grest and M. O. Robbins, Phys. Rev. Lett., 68 (1992) 3448.
22 P. A. Thompson, G. S. Grest and M. O. Robbins, to be published.
23 J. Magda, M. Tirrell and H. T. Davis, J. Chem. Phys., 83 (1985) 1888.
24 M. Schöen, J. Cushman, D. Diestler and C. Rhykerd, J. Chem. Phys., 88 (1988) 1394.
25 M. Schöen, C. L. Rhykerd, D. Diestler and J. H. Cushman, Science, 245 (1989) 1223.
26 U. Heinbuch and J. Fischer, Phys. Rev. A, 40 (1989) 1144.
27 U. Landman, W. D. Luedtke and M. W. Ribarsky, J. Vac. Sci. Technol. A, 7 (1989) 2829.
28 M. W. Ribarsky and U. Landman, J. Chem. Phys., 97 (1992) 1937.
29 M. Lupowski and F. van Swol, J. Chem. Phys., 95 (1991) 1995.
30 K. Kremer and G. S. Grest, J. Chem. Phys., 92 (1990) 5057.
31 See, for example, J. P. Hansen and I. R. McDonald, Theory of Simple Liquids, 2nd ed., (Academic Press, New York, 1986).
32 H. C. Anderson, J. Chem. Phys., 72 (1980) 2384.
33 S. Toxvaerd, J. Chem. Phys., 74 (1981) 1998.
34 I. Bitsanis and G. Hadziioannou, J. Chem. Phys., 92 (1990) 3827.
35 At the equilibrium melting point $S(\vec{G}_1)/S_0 \sim 0.6$; M. S. Stevens and M. O. Robbins., J. Chem. Phys., in press.
36 D. Y. C. Chan and R. G. Horn, J. Chem. Phys., 83 (1985) 5311.
37 The value of L_s only becomes velocity dependent when the dimensionless shear rate $\dot{\gamma}\tau = \tau v/h$ exceeds 0.1. These shear rates are many orders of magnitude beyond experimental values and the fluid becomes non-Newtonian in this regime.
38 H. A. Barnes, J. F. Hutton and K. Walters, An Introduction to Rheology (Elsevier, Amsterdam, 1989).
39 G. S. Grest and M. H. Cohen, Advances in Chemical Physics, Vol. 48, E. Prigogine and S. A. Rice (eds.), Wiley, New York, 1981, p. 455.
40 S.K. Kumar, M. Vacatello and D. Y. Yoon, J. Chem. Phys., 89 (1988) 5206. A. Yethiraj and C. K. Hall, J. Chem. Phys., 91 (1989) 4827.
41 M. S. Stevens and M. O. Robbins, Phys. Rev. Lett., 66 (1991) 3004.
42 F. P. Bowden and D. Tabor, Friction and Lubrication (Oxford Univ. Press, Oxford, 1958).
43 P. A. Thompson and G. S. Grest, Phys. Rev. Lett., 67 (1991) 1751.
44 H. Yoshizawa, P. M. McGuiggan and J. N. Israelachvili, submitted to Science.

SESSION XI
ELASTOHYDRODYNAMICS (2)

Chairman : Professor H S Cheng

Thin Films in Tribology / D. Dowson et al. (Editors)
© 1993 Elsevier Science Publishers B.V. All rights reserved.

EXPERIMENTAL STUDY TO ANALYSE THE TRIBOLOGICAL BEHAVIOR OF SILICON CARBIDE IN WATER

A. BLOUIN and J. FRENE

Laboratoire de Mécanique des Solides de Poitiers, URA, CNRS.
40, Avenue du Recteur Pineau. 86022 Poitiers FRANCE.

The different properties of ceramics, such as hardness, high temperature stability and anticorrosiveness increase their use in situations of severe service. But their behavior is largely influenced by the atmosphere. In particular, it seems that the presence of water affects the tribological behavior of these materials. For example, for ceramic based upon silicium, the obtained surface after friction in water are very smooth (Ra < 0.01 μm).

But the mode of lubrication is not well understood. Theoritical results have demonstrated that an elastohydrodynamic lubrication could be created in the contact, that is to say one takes into account the deformations of the surfaces and the existence of a fluid film between these elements.

In this study, we present experiments performed at room temperature with silicon carbide (SiC) in water. Thus, the experimental apparatus is composed of a sintered SiC shaft/plane machine. The two specimens are totally immersed in distilled water.

1. INTRODUCTION

Although the specifical properties of ceramics, such as hardness, high temperature and stability, be very interessant in tribological systems, their use are restrictive because of the lack of knowing concerning their variation of properties between different atmospherical conditions.

For example, the presence of water or water vapor affects their tribological behavior. In particular, for ceramics based upon silicium, such as silicon carbide or silicon nitride, authors have found that water improves their properties.

Fischer and Tomizawa [1] in 1985 and 1986 have described the singular reaction of ceramics in water. They are used hot-pressed silicon nitride (Si_3N_4) and sintered silicon carbide (SiC), in a pin/plate contact immersed in water. Indeed, they found after several tests extremely smooth surfaces and small friction coefficients. They assume that during friction there is a formation of a tribological reaction between ceramic and water. They suppose that another phenomena exists in the contact : it seems that hydrodynamic lubrication

Material	Composition	Relative density porosity	Manufacturing process
SiC	SiC (>98%)	d > 3 Residual porosity closed Impermeable materials	Pressing of pur SiC Pressureless sintering at 2000-2200 °C

Table 1 : Manufacturing process of SiC

achieves during friction in water.

Maurin-Perrier and Cartier [2] have performed some tests with a tribometer and Si-ceramics in shaft/bearing and shaft/plane contacts, in several environments with fluids with low viscosity : hexane and water. They likewise obtained low friction coefficients and very smooth surfaces (rugosity R_a about few nanometers) in water. Moreover, the surfaces appear like a "mirror" after tests for SiC material.

These particular results obtained with Si-ceramics in water are not well understood. The mecanisms, which lead to smooth surfaces and low friction, are not well described and explained.

Thus, we have developped theoritical analyses of this problem [3], [4]. The programm is based upon the E.H.D. theory of S.M. Rhode and K.P. Oh [5] and takes into account the high Young's modulus of the SiC and the very low viscosity of water. The studied geometries are shaft/plane or shaft/bearing immersed in water. The calculations shown that deformations of the surfaces are larger than the minimum film thickness and thus they are not neglectible.

For example, for an applied load of about 40 Newton and a rotational speed of the SiC shaft of 600 rpm, the calculated film thickness is equal to 0.008 μm and the deformations are equal to 0.033 μm when the friction force of the SiC plate reaches 8.5 N.

Then, an experimental apparatus has been set up in order to observe the formation of the thin film in a SiC shaft/plane configuration in water.

2. APPARATUS

The experiments were performed on a shaft/plane apparatus consisting of a sintered SiC shaft rotating on a stationary sintered SiC plane. Table 1 shows the manufacturing process of the ceramic [6].

Table 2 shows the corresponding characteristics of this material.

Figure 1 shows the schematical geometry of the apparatus. Table 3 shows the mecanical conditions of the contact. The SiC shaft/plane configuration is installed on a spindle nose of a tribometer as seen on figure 2.

Chemical	Density	Coefficient of linear expansion (20/500 °C)	Vickers hardness	Young Modulus
% weight	kg/m^3	$10^{-6}/°C$	kg/mm^2	GPa
98.5 SiC	3210	4	2400	420
< 2000 ppm : Si, SiO_2		< 1000 ppm : Na,Fe,K ,Ca,Mg		

Table 2 : Characteristics of SiC

The SiC ring is maintained to a conical ring and squeezed by two nuts. The normal load is applied to the shaft with dead weights. The rotational speed of the SiC cylinder is measured with an optical tachymeter and directly read on a frequencemeter. The whole system is immersed in distilled water.

The friction force is measured with a sensor which contains strain gauges. The lengthening of these gauges is proportional to the applied force.

The transducer has two ends with threaded rods. One of this is maintained to the ground with a kneecap and the other sets on a half bronze shaft (to diminish the parasitic friction) which guides the holder of the SiC plate (figure 3).

The film thickness is measured with a capacitive method. It consists in a tungstene probe covered by a teflon dielectric inserted in the SiC flat. Two metallic rings binded on the SiC shaft send the signal emmitted from the electrode (figures 4 and 5).

The electrical characteristics of the interface must be modelized by a capacitor and a resistor in

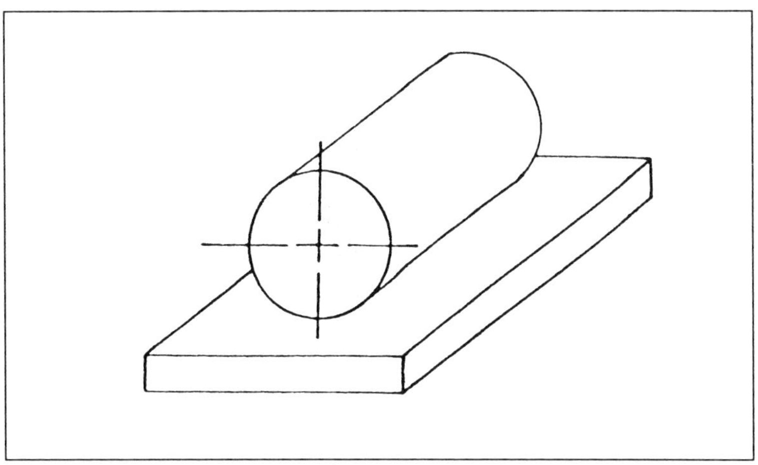

Figure 1 : SiC shaft/plane geometry

Shaft diameter	35 mm
Lenght of contact	20 mm
Plane dimensions	30 × 21 × 8 mm
Rotational speed	20 to 2000 rpm
Applied load	1 to 500 N

Table 3 : Mecanical conditions

parallel.

This method is the modulation of the oscillation frequency of an oscillator. The transducer is connected to an oscillator and the circuit is tuned to the initial capacitance corresponding to an initial reference gap.

Any variation of the gap, and consequently any variation of the capacitance, modulates the oscillation frequency of the oscillator arround 5 MHz. This frequency, in turn, is converted into voltage inversely proportional to the gap.

To calibrate the capacitive sensor, four grooves have been realised on the metallic rings of the SiC cylinder. Their depthes reach 0.8 to 3.5 μm.

Thus, for a given value of depth of one groove, the corresponding value of voltage is obtained with an acquisition system. So a curve could be drawn representing voltage versus film thickness.

Figure 6 shows an example of the gap sensor's response for 100 rpm and a load of 100 N.

A thermocouple has been inserted near the sensor to follow the evolution of the temperature at the interface.

To know the film thickness, a

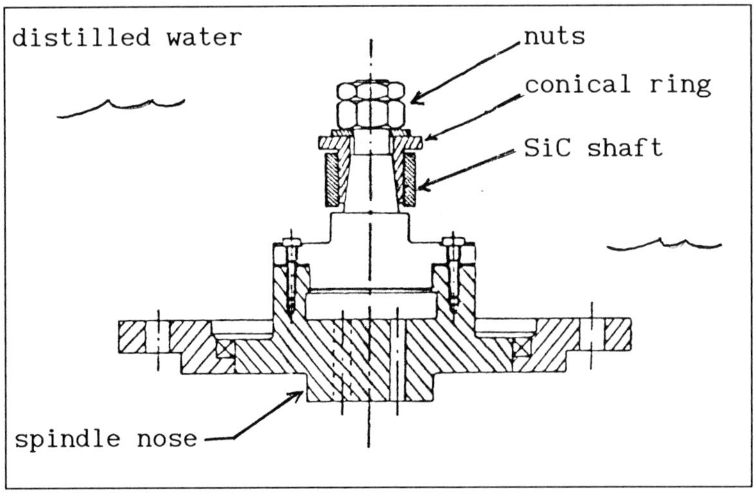

Figure 2 : Experimental apparatus

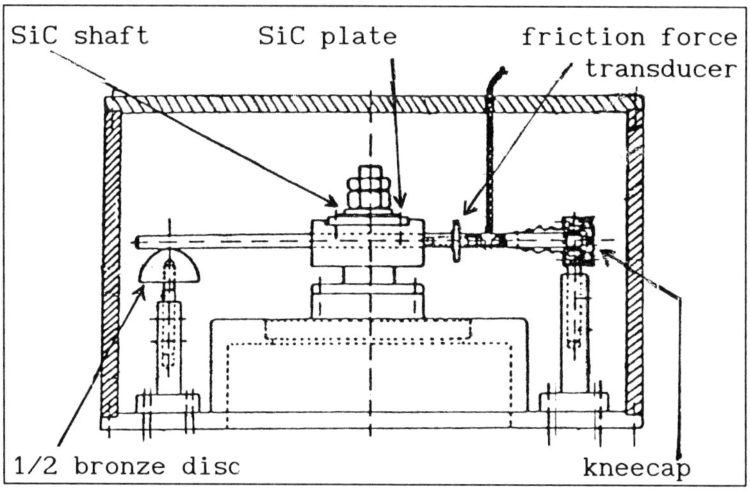

Figure 3 : Apparatus and transducers

load and a rotational speed have been applied. In fact, the film thickness is given by difference of two successive measurements.

Two successive test summaries have been realised : one at high speed and another at 30 rpm. This last speed is our reference speed. We suppose that for this low speed the film thickness is nil in the contact.

To measure the film thickness, the difference between the position of the two graphes has been noted in milliVolt (figure 7).

But, this value must be modify to take into account the drift in temperature of the gap transducer. Then, the milliVolt has been converted into micrometer and

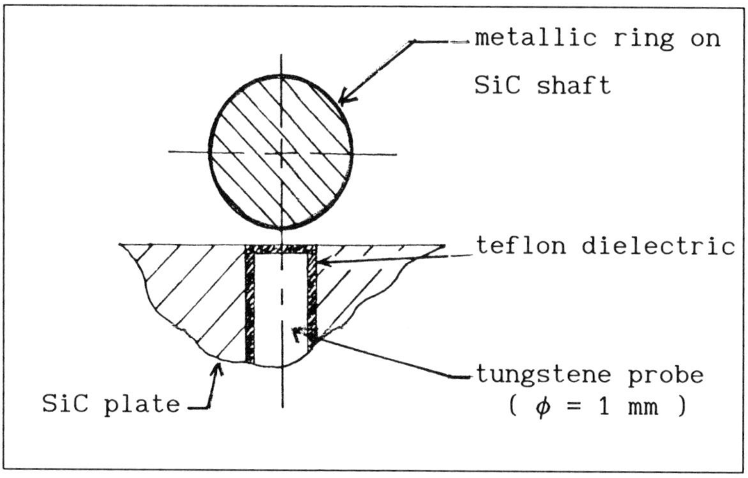

Figure 4 : Capacitive method

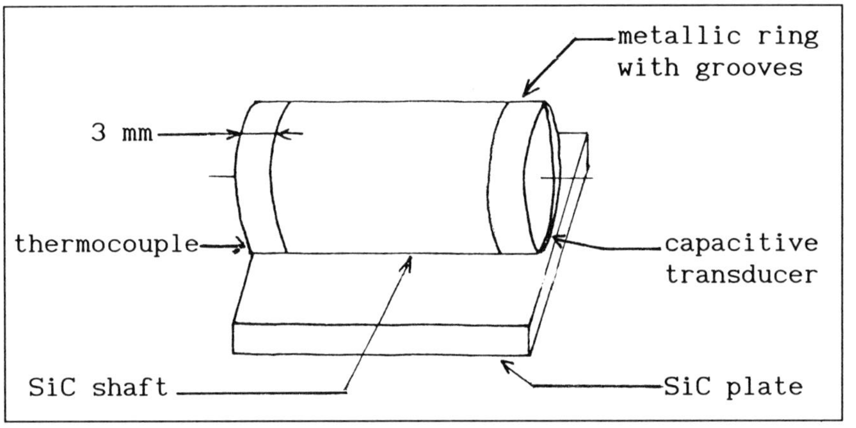

Figure 5 : Geometry of the contact and dimensions

finally, the real film thickness has been obtained.

The sensibility of the probe is about 50mV/μm.

All the samples were cleaned by washing in trichlorethylene before the tests.

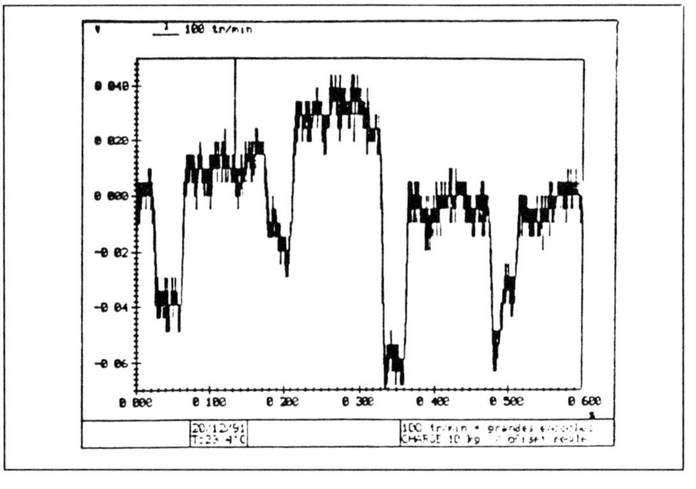

Figure 6 : Gap sensor's response with four grooves realised on the metallic rings of the SiC shaft.
(1000 rpm and 100 N)

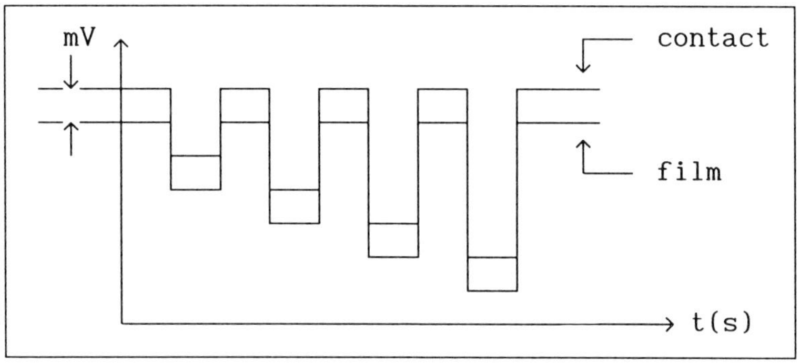

Figure 7 : Measurement of the film thickness

3. RESULTS AND DISCUSSION

The curves given in figure 8 show the evolution of the film thickness versus load for two rotational speeds of the spindle : 600 and 800 rpm. The figure 9 shows the same evolution versus speed for the two different loads : 90 and 100 N.

The film thickness decreases with an increase in load and increases with an increase in rotational speed.

The film thickness varies from 0.4 to 2.5 μm for the whole range of speeds and loads.

The figure 10 shows the effects of increasing speed on the friction force applied to the SiC plate during the test.

At low speed, the friction force reaches 7 N for 90 N and 11 N for 100 N, and it decreases gradually until 400-500 rpm which corresponds with a friction coefficient respectively of 0.08 and 0.11. After that, it remains nearly

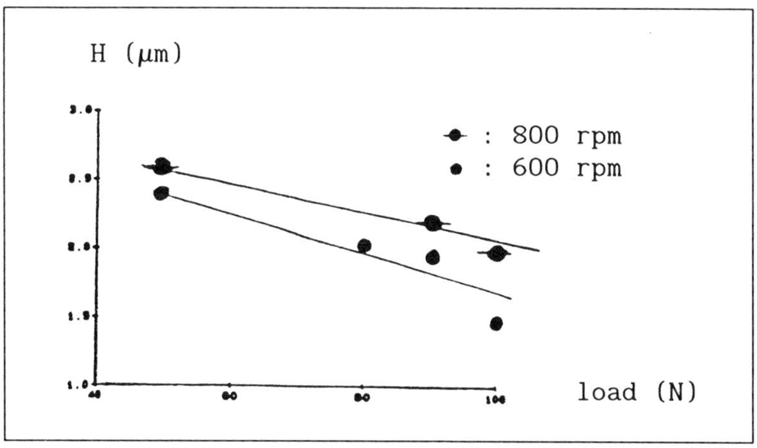

Figure 8 : Film thickness versus load

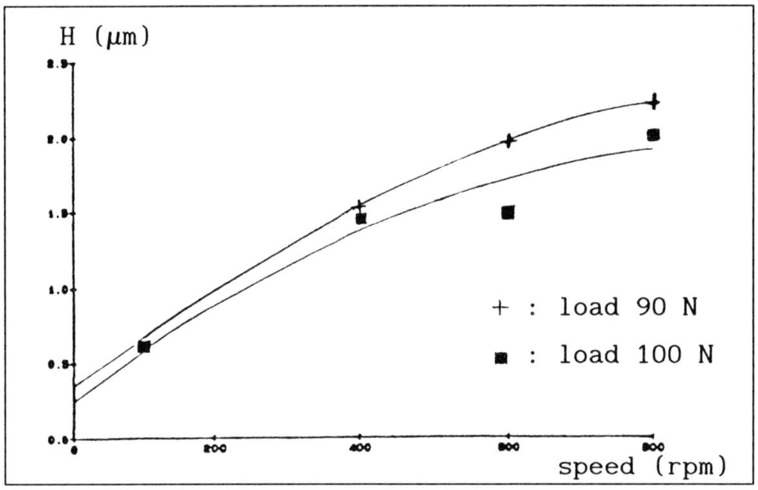

Figure 9 : Film thickness versus speed

constant. The friction force equals to 0.6 N for a load of 90 N and a rotational speed of 1000 rpm. That corresponds to a very low friction coefficient of about 0.007.

Comparing this phenomena with a Stribeck's curve which gives the mode of lubrication versus friction and speed, this nearly constant friction force corresponds with a hydrodynamic lubrication regime in the contact. These results shows that a fluid film is present at the interface and protects the surfaces against wear.

Besides, the observation of the SiC plate's surface shows a shinning part like a "mirror" with

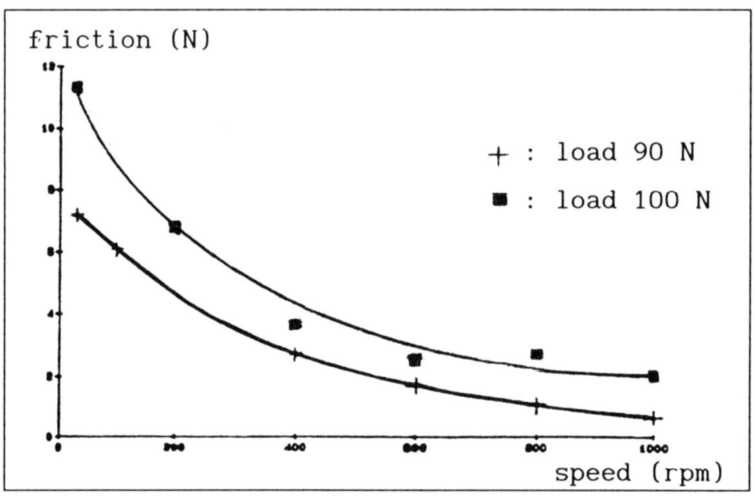

Figure 10 : Friction force versus speed.

an arithmetic rugosity of about few nanometers. This could signify that there is a polishing of the surfaces during tests and consequently the film does't always protect the SiC against wear. We suppose that this is the consequence of the friction at low speed when the film is not again formed. But another fact is very surprising. The results obtained with theoritical programm concerning film thickness are very different from experimental one. The calculated film thicknesses are about 100 times smaller than the experimental one and the calculated friction force is 2 to 5 times higher than the experimental one. We suppose that there is another effect which superposes upon the tribological effect. This could be the consequence of a tribochemical reaction between SiC and water as some authors have already expressed the hypothesis.

Then, we propose the following scheme to explain the obtained results :

- first, at the beginning of the test, at low speed the surfaces rub against themselves and the tribological reaction takes place in the contact. Consequently, adhesive layers are formed near the SiC surfaces, such as colloïdal film [7].

- secondly, at high speed, the convergent surfaces of the contact induce the formation of a hydrodynamic film. This hydrodynamic film is very thin and intercalates between the layers and protects them from wear.

The formation of adhesive layers induces a gradient of mechanical properties in the contact. When we include an increase of the viscosity in the programm, we reduce drastically the difference between theoritical and experimental film thicknesses.

4. CONCLUSION

An experimental apparatus based upon a capacitive method has been set up in order to study the tribological behavior of the SiC material in water. The geometry of a SiC shaft/plane configuration has been analysed.

Some results have been uncovered :
It forms in the contact a film by hydrodynamic effects and by tribochemical reaction. The superposition of these effects protects the SiC surfaces during the test and creates a film which reduces the friction force coefficient.
Its value could be about few hundreath for high rotational speed.

The examination of the surfaces shows that the arithmetic rugosity is about naometers and the surface of the contact appears like "mirror".

The tribochemical mecanism results in extremely smooth surfaces which permits the achievment of elastohydrodynamic lubrication.

ACKNOLEDGMENTS

The authors whish to thank the European Communities Commission for financial support of this

B.R.I.T.E. contract n° RI-1B-295-C and "Ceramiques et Composites" for supplying the ceramic samples and "Hydromécanique Et Frottement" for the help and discussion of results we have had together [8].

REFERENCES

1 FISCHER - TOMISAWA
"Friction and wear of silicon nitride and silicon carbide in water".
ASLE Trans.(1986)
Vol 30,p 41 to 46.

2 M. CARTIER - P. MAURIN-PERRIER
Experimental results from H.E.F.
B.R.I.T.E. contract
N° RI-1B-295-C(CD).

3 A. BLOUIN - J. FRENE
"Etude du comportement d'une céramique en frottement dans l'eau".
Thesis defenced at the University of Poitiers on the 27th January 1992.

4 A. BLOUIN - J. FRENE
"Analyse E.H.D. appliquée au frottement des céramiques dans l'eau".
Matériaux et Techniques.
Spécial Tribologie.Décembre 1991

5 S.M. RHODE - K.P. OH
"Numerical solution of the point contact problem using the finite element method".
International Journal for Numerical methods in Engineering.
Vol 1 - 1977- p 1507 to 1518.

6 SSiC : sintered silicon carbide manufacturing by CERAMIQUES ET COMPOSITES with the name SiC 100.
C - C : BP n°7 - 65460 Bazet - FRANCE.

7 J.M GEORGES - D. MAZUYER
"Pressure effects on the shearing of the colloïdal thin film"
Journal de Physique - Condens. matter 3 - 1991 - p 9545 to 9550

8 B.R.I.T.E. contract.
N° RI-1B-295-C(CD).
"Use of ceramics for advanced tribological systems involving fluids with adverse lubricating properties"
H.E.F. (St Etienne - France)
CERAMIQUES ET COMPOSITES (Bazet France)
A.B.S. PUNPEN (Wolfratshausen - Germany)
BURGMANN (Münich - Germany)
LABORATOIRE DE MECANIQUE DES SOLIDES (Poitiers - France).

Thin Films in Tribology / D. Dowson et al. (Editors)
© 1993 Elsevier Science Publishers B.V. All rights reserved.

Thermal effects approach in elastohydrodynamic contacts by an infrared radiometric technique

L. Bordenet**, J.P. Chaomleffel* and G. Dalmaz*

** Radiall, Laboratoire Central
38509 Voiron Cedex - France

* Laboratoire de Mécanique des Contacts, URA CNRS 856
Institut National des Sciences Appliquées de Lyon
69621 Villeurbanne Cedex - France

Film and contact surfaces temperatures, stresses and traction forces are calculated with a simplified thermal model which takes into account the heat generated by the lubricating film shearing and its dissipation through the film and the solids. The infrared radiometric technique is used to measure temperatures in an EHL contact.

Film and ball surface temperatures are investigated experimentally in a rolling-sliding EHL smooth point contact under steady state and fully flooded conditions, at room temperature. Two filters are used for separating the individual contributions of the ball surface and of the lubricant film from the total radiation received by the detector. Temperature profiles have been obtained by scanning the contact area. With moving surfaces, a fast mode along with a signal treatment is necessary for in-situ measurements in EHL point contacts.

The measured friction forces are closed to those calculated by the simplified thermal model. Experimental and theoretical film and ball surface temperatures are similar in shape and temperature values are in rough agreement.

INTRODUCTION

In elastohydrodynamic contacts, the shearing of thin lubricating film generates heat which is dissipated in the film and the solids and increases temperatures. The study of EHL contacts is mainly concerned with film thickness and traction between two solids in relative motion. Both theoretical and experimental aspects of the contact temperature problem are considered in an EHL rolling-sliding point contact simulator.

The thermo-EHL model used [1] is based on the shear-plane hypothesis proposed by Tevaarwerk [2]. The heat generated by the lubricating film shearing is supposed to be dissipated from the shear-plane located by Blok's theory which gives the heat fluxes associated with each surface [3]. Film and contact surfaces temperatures, stresses and traction forces in the contact are calculated.

The experimental study is concerned by the infrared radiometric technique developed by Turkina, Sanborn and Winer [4] which allows temperature measurements in EHL contacts for very small contact dimensions (film thicknesses from 0.1 to 1 µm and contact areas lower than $1mm^2$).

Wymer and Macpherson [5] used it to study the process of lubricant breakdown and scuffing. Ausherman, Nagaraj, Sanborn and Winer [6] have developed the IR technique and have successfully separated the sliding surface temperature and lubricant film temperature using infrared filters. Winer and co-workers have used this technique to study the influence of speed, load, roughness and lubricant [7] on temperature. Cann and Spikes [8] have used the steady-state contact temperature profiles to obtain information about lubricant rheology under EHL conditions. Shear stresses could be calculated with some assumptions by a thermo-EHL model, from the measured temperature profiles.

The aim of this paper is to try to determine and compare theoretical and experimental temperatures both in the film and on the ball surface in the conjunction zone in order to predict local temperatures in rolling-sliding tribological system operating at room temperature.

1. THEORETICAL APPROACH

It is well known that film thickness can be determined by the flow of an isothermal piezoviscous lubricant between two moving solids elastically deformed in the contact zone and that traction is usually obtained with a thermo-piezo-visco-elasto-plastic shear behaviour.

The isothermal film thickness h(x,z) and pressure p(x,z) distributions analysis is developed for two dimensional newtonian piezoviscous thin film under laminar, steady state conditions for smooth surfaces. The Reynolds equation, Barus piezoviscous law $\mu = \mu_0 \exp(\alpha p)$, compressibility and elastic equation for the contact bodies are considered in the numerical EHL solution detailed in ref [9].

The shear stresses τ_x and τ_z are calculated near pure rolling conditions with the isothermal non-linear Maxwell rheological model proposed by Johnson and Tevaarwerk [10] and based on three lubricant parameters : the elastic shear modulus G, the viscosity μ and the reference Eyring stress τ_0. The limiting shear stress τ_ℓ introduced as a property of the lubricant by Bair and Winer [11 and 12] is considered according to the relation $\tau_o = \tau_\ell / \alpha p$. The shear stresses τ_x and τ_z are calculated with the local values G, μ and τ_0 or τ_ℓ which are functions of the pressure p (x,z).

Thermal effects caused by shearing in the fluid film are investigated using the Tevaarwerk shear-plane hypothesis model (figure 1). The method detailed in ref [1] and [13], is based on solutions for constant flux sources moving on the surface of a semi-infinite half-space solid which allow to find the local temperatures. In this model the velocity distributions throughout the film are not necessary and only the surface velocities are required. The local temperature and the local shear stress of the fluid in this shear-plane are obtained by solving the combined system of the energy equation for the solids and the energy equation for the fluid film.

1.1. Fluid film energy equation

Assuming that the fluid is transported with a mean velocity U in the rolling direction and, as for almost EHL contacts, the film thickness h is smaller than the dimensions a and c of the hertzian contact area, the energy equation can be written :

$$\frac{\partial^2 T_f}{\partial y^2} + \frac{\phi}{\lambda_f} = \alpha_f U \frac{\partial T}{\partial x}$$

with : ϕ, compression heating function,
λ_f, thermal conductivity of the fluid,
α_f, diffusion coefficient for the fluid.

The boundary conditions for fluid film are : at contact entry $T_f = T_e$(x,y,z) and on the solid surfaces $T_f = T_{si}$(x,z) for $y = \pm h/2$. With the shear-plane hypothesis, an additional boundary condition must be satisfied :

$$\left[\frac{\partial T_f}{\partial y}\right]_{y=y_c} = \frac{-\Gamma}{\lambda_f} Q(x, y_c, z)$$

Where Q is the heat generated at the shear-plane which is assumed to be halved in the ratio Γ to Γ-1. For the two solids the Peclet number is defined by Pes = U a α_s where α_s is the diffusivity of the solid material. The Pes are different and Γ coefficient is given by the Blok theory [3] :

$$\Gamma = \frac{\gamma}{\gamma + 1} \quad with\ \gamma = \frac{q_1}{q_2} = \frac{T_{s1}}{T_{s2}} \left[\frac{U_1 \rho_1 \lambda_1 C_1}{U_2 \rho_2 \lambda_2 C_2}\right]^{1/2}$$

The magnitude of the heat generated at the shear-plane is :

$$Q = \tau_x \Delta U + \tau_z \Delta W$$

where τ_x and τ_z are the shear stress calculated previously.

The heat fluxes q_1 and q_2 conducted into the solids are given by :

$$q_{1,2} = -\lambda_f \left[\frac{\partial T_f}{\partial y}\right]_{y=\pm\frac{h}{2}}$$

The energy equation is solved by mean of Laplace transform to give the fluid temperature in the shear plane T_f (x, y_c, z).

1.2. Energy equation for the solids

Under steady-state conditions, and assuming that thermal effects are more important in the rolling direction than in the lateral one, the energy equation for the solids can be written :

$$\frac{\partial^2 T_s}{\partial x^2}+\frac{\partial^2 T_s}{\partial y^2}+\frac{\partial^2 T_s}{\partial z^2}=\alpha_s U\frac{\partial T_s}{\partial x}$$

The solution proposed by Carslaw and Jaeger [14] gives temperature for a solid on which a constant heat flux is applied.

The associated boundary conditions are :

T_s (x,z) = T_i (x,z) for $y = \pm\infty$ where T_i is the solid temperature assumed to be constant,

$$\left[\frac{\partial T_s}{\partial y}\right]_{y=\pm\frac{h}{2}}=0 \qquad \text{outside the source area,}$$

$$\left[\frac{\partial T_s}{\partial y}\right]_{y=\pm\frac{h}{2}}=\frac{-q_i(x,z)}{\lambda_{si}} \qquad \text{inside the source area.}$$

The temperature at the centre of the ith source due to the fluxes at other sources is expressed as :

$$\{T_{si}\}=\left[R_{ij}\right]\{q_j\}$$

where [R_{ij}] is the matrix of influence coefficients.

2. EXPERIMENTS

The measurements of temperature by the infrared radiometric technique with a detector scanning the EHL contact area, film thickness by optical interferometry and traction forces have been performed in a rolling-sliding point contact apparatus under fully flooded conditions at ambient temperature.

The EHL simulator used and the measurements have been described in detail elsewhere [13]. The 10 mm diameter steel ball specimen is smooth with a surface roughness better than Ra = 0.1 µm. The sapphire disc, 10 mm thick, is optically polished and coated with a 200 Å chromium layer. Two lubricants labelled L3 and L6 of different viscosity have been chosen. The characteristics of contact bodies and of the lubricants are respectively given in the table 1 and table 2.

The infrared radiation emission technique has been presented in detail by Winer and co-workers [4], [6], [7]. One contact body must be transparent to infrared radiation. The choice of sapphire is good because its thermal and mechanical properties are similar to steel and it allows high contact pressures up to 3 GPa. In this study, the technique of temperature determination by rapid infrared radiation scanning is used to try to measure local temperature in EHL contacts as a function of time and space.

The RM-50 Infrared Micro Imager of Barnes Engineering Company (RM-50 IMI) consists of an optical scanning system, a liquid nitrogen cooled infrared detector with a very small sensitive field of view. Infrared radiation from each microscopic contact area is converted to an electrical signal that is amplified to a cathode-ray tube and gives thermograms on a screen.

The Indium Antimonide detector at liquid nitrogen temperature (77 K) has a spectral response peak at 4.5 µm in usable range from 1.8 to 5.5 µm in the near infrared and above about 100 Hz to 10 kHz the frequency response is practically flat. The area of the EHL contact is scan by the detector at a frequency of one frame per second (42 elements per line and 64 lines per frame). The temperature sensitivity is 0.1°C near ambient and temperature measuring range of the target is from ambient to 200°C.

The objective magnitude 40x with a spatial resolution of 38 µm and a working distance of 25.4 mm has been used. Note that the objective magnitude 100x with a better spatial resolution of 15 µm and a shorter working distance of 7.6 mm cannot be used in these experiments because the sapphire disc is 10 mm thick.

The output mode of the Barnes RM-50 IMI is a picture which allows to obtain thermograms under quasi-static conditions without enough precision. A electronic output signal is required to allow amplification and fast acquisition data. A specific signal treatment with numerical data have been developed and improved. In practice one line which contains 100 points is obtained within 11.7 ms. Thus the frequency of each line which corresponds to temperature measurements from the inlet to the outlet of the contact is 64 Hz.

The total radiation N received by the detector is :

$$N=\eta_B N_B+\eta_F N_F+\eta_S N_S+\eta_a N_a$$

where N_B, N_F,N_S and N_a are the non attenuated values of radiations emitted by respectively the ball, fluid, sapphire and ambient. Each contribution is associated with an attenuation factor η_B, η_F, η_S and η_a which includes losses due to reflexion and absorption at interfaces between two media. The ball surface only emits and absorbs. Figure 2 shows the transmission characteristics function of the infrared wavelenght for the sapphire and for the two lubricants tested. The emission curves of the lubricants which correspond to a film thickness of 25 µm between two fluorine plates, are very similar with a peak emission at 3.4 µm and a significant emission in the band 3.1 to 3.7 µm at 25°C.

The narrow band filter F (3.29 to 3.54 µm) has been chosen so that essentially all the radiation emitted by the lubricant film is transmitted. The narrow band filter B (5.10 to 5.43 µm) has been chosen so that essentially all the radiation emitted by the ball is transmitted.

The radiation emitted by the ball surface N_B is only a function of ball emissivity ε_B and ball surface temperature T_B since the smooth steel ball is a grey body (constant emissivity over all wavelengths). In the range of ball surface temperature expected the emissivity ε_B is practically constant ($\varepsilon_B \cong 0.1$ to 0.2 for the smooth ball) and is more predominant for wavelenghts of the order of 5 µm. The emissivity of the fluid film ε_F is a function of the film thickness h and of the temperature T_F and a strong function of the wavelenght λ because of the lubricant spectral characteristics. The sapphire emissivity which is expected to be ten times lower than the ball one is neglected.

Instead of analysing and modelling the EHL contact with radiometric equations as suggested by Winer and co-workers, the relation between the radiation measured by the detector and the contact, ball or lubricant film temperature is obtained by calibration. The calibration procedure consists of submerging a stationary unloaded 12.7 mm diameter ball in a constant temperature bath. The specific calibration device shown in figure 3 reproduces geometry, surface polish, lubricant, materials, ambient and environmental temperatures existing in the EHL simulator. The film thickness at contact center is of the order of 0.1 µm. The measured temperature is supposed to be constant in all the small volume including the contact area : ball, lubricant and sapphire have the same temperature. Figures 4, 5 and 6 show the calibration curves obtained respectively without filter, which correspond to a global contact temperature increase ΔT, with filter B for the ball surface temperature increase ΔT_B and with filter F for the lubricant film temperature increase ΔT_F. It is shown that the radiation received by the detector decreases by a factor of 5 with ball filter B and by a factor of 20 with the lubricant filter F. The differences in radiation levels are very small and quantitative temperature variations difficult to measure. The limit in radiance variation sensibility measurement is $\Delta N = 0.01$ W/(m^{2}sr).

The RM-50 IMI is positioned above the contact center with a precision in location of the order of 0.01 mm both for calibration and in EHL measurements. This is possible by a classical optical system which allows direct visual viewing of the contact area. The position of the target spot in the contact area is therefore known.

3. RESULTS

Results are given for a rolling-sliding EHL lubricated smooth point contact under steady state and fully flooded operating conditions at room temperature.

Film thickness, pressure, stress and temperature distributions, load and traction are calculated for a given lubricant.

The infrared radiometric technique developped by Winer and co-workers is adapted to allow in-situ temperature measurements on a steel-ball-sapphire plane disc apparatus.

Calculated values of film thickness, shear stress and traction, ball temperature and lubricant film temperature are compared with experimental measured values.

3.1. Film thickness

The optical film thickness measurements obtained with oil L3 for maximal hertzian pressure of 1, 1.4 and 1.8 GPa for the rolling speed U = 2 m/s and several slide-roll ratios $\Delta U/U$, shows that the computed isothermal central and minimal film thicknesses h_o and h_m are in good agreement with the experimental ones (table 3). A slight decrease in experimental h_o and h_m values due to thermal effects is observed for high slide-roll ratios. Similar results have been obtained with oil L6 with higher film thicknesses.

3.2. Traction

The shear stresses are calculated using the previous thermo-EHL model with isothermal film thickness h(x,z) and pressure p(x,z). For pressures higher than a critical pressure value p*, the mean apparent elastic shear modulus G and the mean lubricant limiting shear stress τ_ℓ increase with local pressure p :

$$G = Ap - B \quad \text{and} \quad \tau_\ell = Cp - D$$

where A, B, C and D are constants (table 4).

Figure 7 shows traction f_x versus slide-roll ratio ΔU/U for lubricant L3, a constant rolling speed and three contact pressures. The proposed thermo-EHL model gives calculed values f_x in good agreement with measured traction forces f_x. Similar agreements have been obtained under different operating conditions and with lubricant L6 for higher film thicknesses.

3.3. Temperatures

The technique of temperature determination by infrared emission has been adapted to obtain ball surface temperature and film temperature distributions :

- the detector scans the center line from inlet to outlet of the EHL contact area at a frequency of 64 lines per second,
- two filters separate the individual contributions of the ball surface and of the lubricant film.

For the lubricant L3, a load W = 36 N, a slide-roll ratio ΔU/U = - 0.78 and a rolling speed U = 2 m/s, figures 8a and 10a show the signal given by the detector respectively with filter B and filter F. A Fast Fourier Transform is applied to these signals to eliminate high frequencies. With the calibration curves (figures 5, 6), the filtered signal (figures 8b and 10b) can be converted into temperature rise (figures 9 and 11) along the rolling axis x. In the same way, the temperature rise on the ball surface and in the lubricant L6 is obtained for a load W = 61 N, a silde-roll ratio ΔU/U = 0.6 and a rolling speed U = 0.6 m/s (figures 12 and 13).

The relation between the radiation measured by the detector and the temperature has been obtained under static conditions by calibration. The use of narrow band filters gives very low radiation values. Differences in radiation of $\Delta N \cong 0.01$ W/(m^2 sr) can be detected. In practice, the ball surface temperature is obtained within 2° C for a temperature ≅ 30° C and 0.5° C for a temperature ≅ 40° C, the lubricant film temperature is obtained within 8° for a temperature ≅ 30° C and 3° C for a temperature ≅ 50°C. The limits of the application of the infrared emission technique in thin elastohydrodynamic films seem to be reached.

Figures 14 and 15 show the experimental and theoretical temperature rise on the ball surface and in the lubricant L3, for a load W = 5.6 N and a rolling speed U = 2 m/s, versus the slide-roll ratio. A rough agreement is observed on these figures. Similar results are obtained for a higher load W = 36 N (figures 16 and 17). Note that the theoretical fluid temperature is the average temperature across the lubricant film thickness.

Experimental and theoretical temperature profiles along the rolling axis x are compared for the lubricant L3 with a rolling speed U = 2 m/s, a load W = 36 N and a slide-roll ratio ΔU/U = - 0.33 (figures 18 and 19). The maximum temperature is located near the contact center and the temperature rise is significant in the Hertz area ($-1 < x/a < 1$).

Similar results have been obtained with oil L6 with higher film thicknesses.

Results obtained with lubricants L3 and L6 also show that :

- at low slide-roll ratios, temperature rise is proportional to slide-roll ratios for both the ball and the film,
- for pressures from 1 to 2 GPa temperatures rise proportionally to pressure for both the ball and the film,
- for the higher viscosity lubricant the ball surface temperature is lower.

These theoretical and experimental results are in general agreement with previous work [7 and 8].

4. CONCLUSION

The proposed thermo-EHL model in which only the surface velocities are considered predicts accurately under our operating conditions :

- stress distributions,
- average temperature distribution in the lubricant film,
- temperature distributions on the surfaces in the conjonction of rolling-sliding EHL contacts.

The infrared radiometric technique developed allows to obtain in-situ quantitative temperature measurements, ball surface and thin lubricant film temperature profiles under the difficult operating conditions occuring in EHL point contacts.

Steady-state ball surface temperature profile and average lubricant film temperature profile could be used to add to the lubricant rheological behaviour knowledge within EHL contacts and to predict thermal effects under conditions approaching those found in rolling element bearings.

REFERENCES

1. CHAOMLEFFEL (J.P.), VERGNE (F.), DALMAZ (G.), "Temperature Calculations and Traction in Rolling-Sliding-Spinning Elastohydrodynamic Points Contacts". Proceedings of the Japan International Tribology Conference Nagoya, 1990, Vol.1, pp.617-622.
2. TEVAARWERK (J.L.), "Traction Calculations using the Shear Plane Hypothesis". Proc. of Leeds-Lyon Symposium on Tribology, 1979, Paper VII(iii), pp.201-215.
3. BLOK (H.), "The Flash Temperature Concept". Wear, 1963, Vol.6, pp.483-494.
4. TURCHINA (V.), SANBORN (D.M.), WINER (W.O.), "Temperature Measurements in Sliding Elastohydrodynamic Point Contacts". Trans. A.S.M.E., july 1974, pp.464-471.
5. WYMER (D.G.), MACPHERSON (P.B), "An Infra-Red Technique for Measurement of Gear Tooth Surface Temperature". Trans. A.S.L.E., 1974, Vol.18, pp.229-238.
6. AUSHERMAN (V.K.), NAGARAJ (H.S), SANBORN (D.M.), WINER (W.O), "Infrared Temperature Mapping in Elastohydrodynamic Lubrication". Trans. A.S.M.E., Jour.Lub.Tech., 1976, Vol.98, pp.236-243.
7. NAGARAJ (H.S.), SANBORN (D.M.), WINER (W.O.), "Effects of Load, Speed and Surface Roughness on Sliding EHD Contact Temperatures". Trans. A.S.M.E., Jour.Lub.Tech., 1977, Vol.99, pp.254-263.
8. CANN (P.M.), SPIKES (H.A.), "The Influence of Sliding Speed and Lubricant Shear Stress on EHD Contact Temperatures". Trans. A.S.M.E., 1990, Vol.33, pp.355-362.
9. VERGNE (F.), "Le contact élasto-hydro-dynamique ponctuel lubrifié par un fluide visqueux linéaire ou non". Thèse de Doctorat : Université Claude Bernard - Lyon I, 1989, 225 p.
10. JOHNSON (K.L.), TEVAARWERK (J.L.), "Shear Behaviour of Elastohydrodynamic Oil Films". Proc. Roy. Soc., London, series A, 1977, Vol.356, pp.215-236.
11. BAIR (S.), WINER (W.O.), "A Rheological Model for Elastohydrodynamic Contacts Based on Primary Laboratory Data", Trans. A.S.M.E., serie F, 1979, Vol.101, pp.258-265.
12. BAIR (S.), WINER (W.O.), "Shear Strength measurements of Lubricants at high pressure", Trans. A.S.M.E., serie F, 1979, Vol.101, pp.251-257.
13. BORDENET (L.), "Approche de la Thermique du Contact Elastohydrodynamique par une Technique de Radiométrie Infrarouge". Thèse de Doctorat : I.N.S.A. de Lyon, 1991, 244 p.
14. CARSLAW (H.S.), JAEGER (J.C.), "Conduction of Heat in Solids". Second edition, Oxford : University Press, 1959, 510 p.

ACKNOWLEDGEMENT

The authors gratefully thank Mr Maurice BOUVIER for his help in the experimental work and for his technical assistance.

Table 1 : Contact bodies characteristics

Material	Steel	Sapphire
Refractive index		1.77
Density ρ (kg/m^3)	7900	3980
Young modulus E (GPa)	210	370
Poisson coefficient ν	0.29	0.34
Specific heat C (J/kg K)	500	700
Thermal conductivity λ (W/m K)	40	30

Table 2 : Lubricants characteristics

Lubricant reference and source	L3 R 620-15 Sun oil Comp	L6 EL 8303 ELF
Viscosity μ_0 at p_0, θ_0 and γ_0 (Pa.s)	0.038	0.328
Temperature-viscosity coefficient δ at p_0 and γ_0 between 20° and 40°C (°C^{-1})	0.0343	0.0635
Pressure-viscosity coefficient α at θ_0 and γ_0 between p_0 and 0.4 GPa (GPa^{-1})	27.8	27.0
Refractive index at θ_0 and P_0	1.505	1.492
Density ρ_0 at θ_0 and p_0 (kg/m^3)	915	895
Specific heat C (J/kg K)	2000	2000
	0.13	0.14

with $\theta_0 = 25°C$, $p_0 = 10^5$ Pa and $\gamma_0 = 10$ to $100\ s^{-1}$
The pressure and temperature coefficients α and δ are defined by the expressions :

$$\mu = \mu_0\, e^{\alpha(p - p_0)} \text{ and } \mu = \mu_0\, e^{-\delta(\theta - \theta_0)}$$

Table 3 :Theoretical and experimental film thicknesses

W N	P_0 GPa	θ f °C	$\frac{U_1 - U_2}{U_1 + U_2}$	h_0(th) µm	h_0(ex) µm	h_m(th) µm	h_m(ex) µm	a µm
5.6	0.967	20.9	0.	0.26	0.32	0.15	0.22	52
		21.1	0.10	0.26	0.30	0.15	0.20	
		21.3	0.20	0.26	0.30	0.15	0.18	
		21.3	0.40	0.26	0.26	0.15	0.17	
16.6	1.39	22.7	0.	0.23	0.24	0.13	0.14	75
		22.7	0.10	0.23	0.23	0.13	0.10	
		23.1	0.20	0.23	0.23	0.13	0.08	
		23.4	0.40	0.23	0.20	0.13	0.08	
36.1	1.81	22.4	0.	0.22	0.23	0.12	0.08	97
		22.6	0.10	0.22	0.21	0.12	0.07	
		23.0	0.20	0.22	0.19	0.12	0.06	
		23.5	0.40	0.21	0.18	0.12	0.05	

Table 4 : Lubricant constants

Fluid	A	B GPa	C	D GPa
L3	0.069	0.021	0.106	0.016
L6	0.098	0.019	0.097	0.018

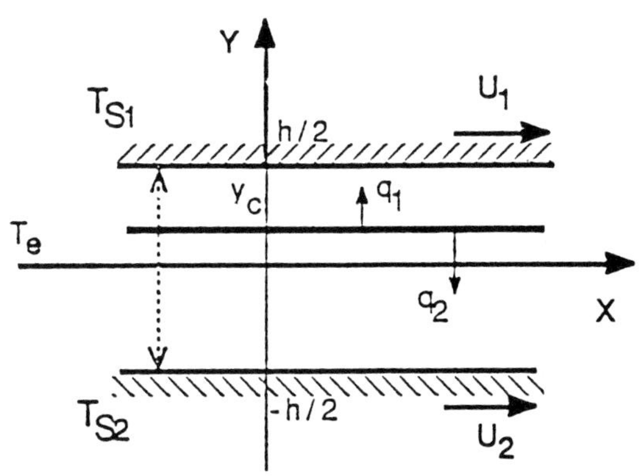

Figure 1. Shear plane model

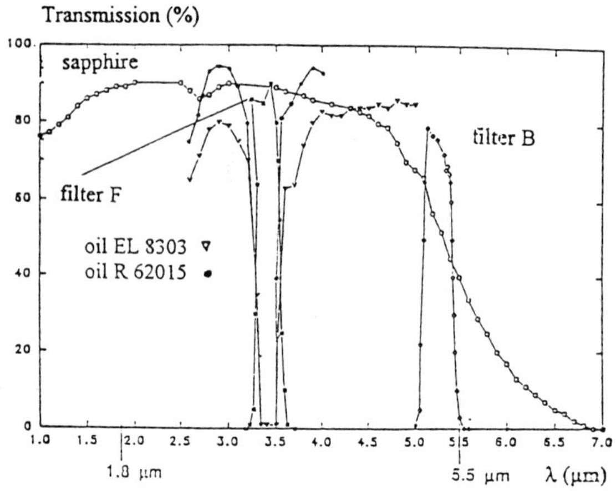

Figure 2. Transmission characteristics

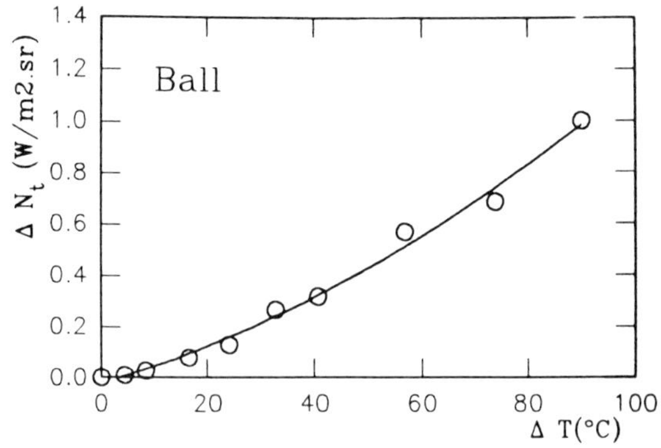

Figure 5. Calibration curve with filter B

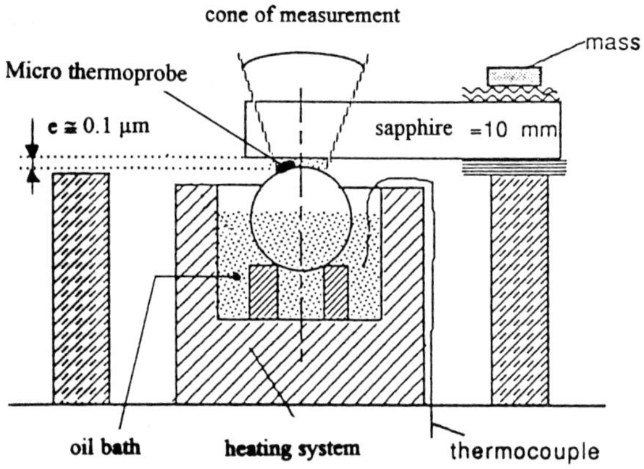

Figure 3. Calibration device

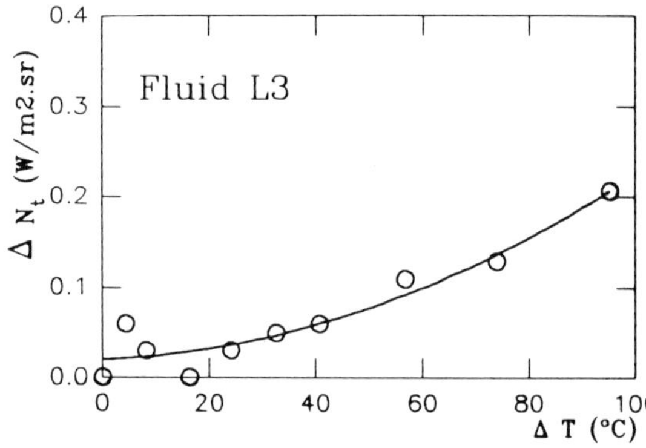

Figure 6. Calibration curve with filter F

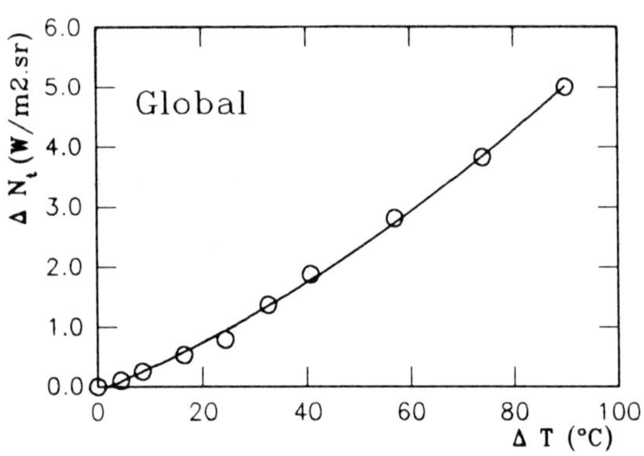

Figure 4. Calibration curve without filter

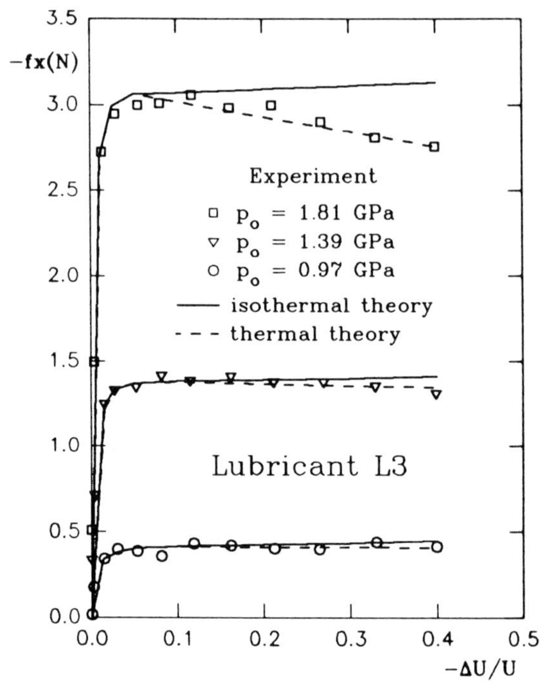

Figure 7. Traction force fx for U = 2 m/s

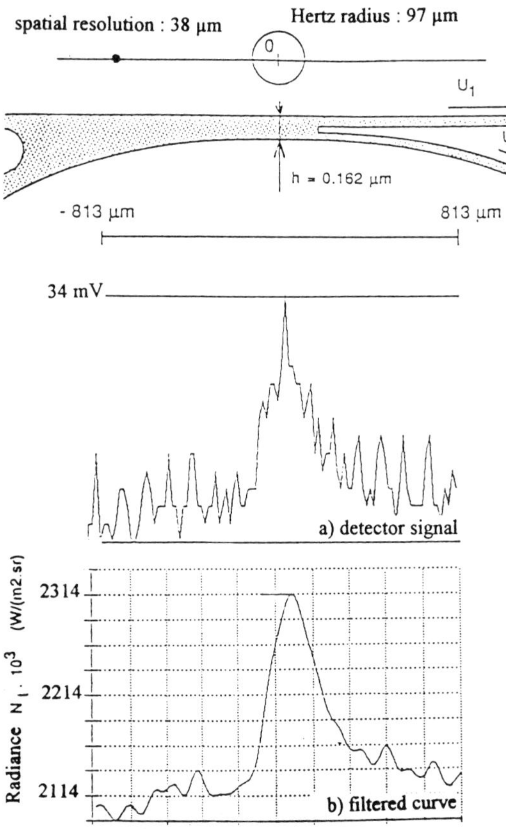

Figure 8. Ball radiation profile along x axis

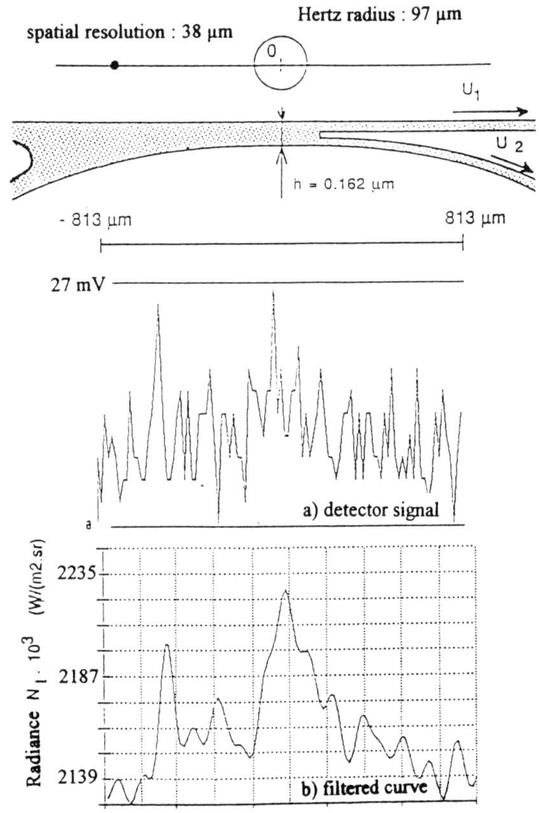

Figure 10. Fluid L3 radiation profile along x axis

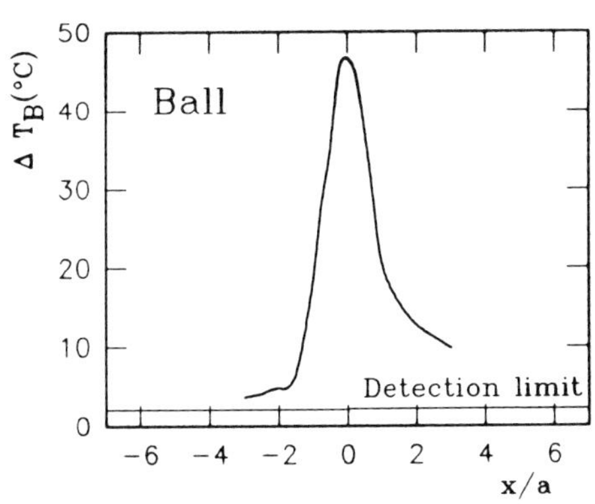

Figure 9. Ball temperature profile along x axis

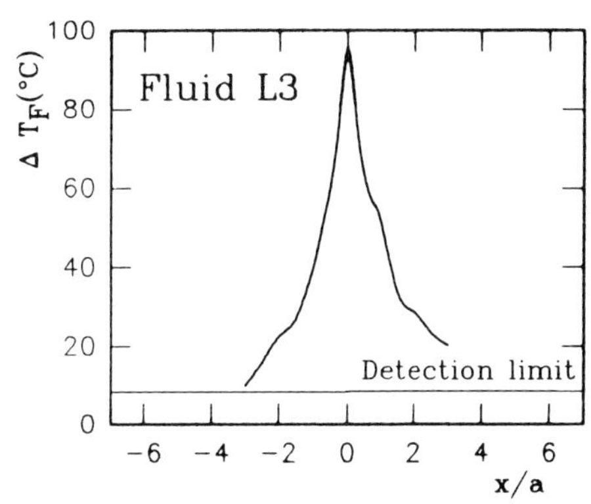

Figure 11. Fluid temperature profile along x axis

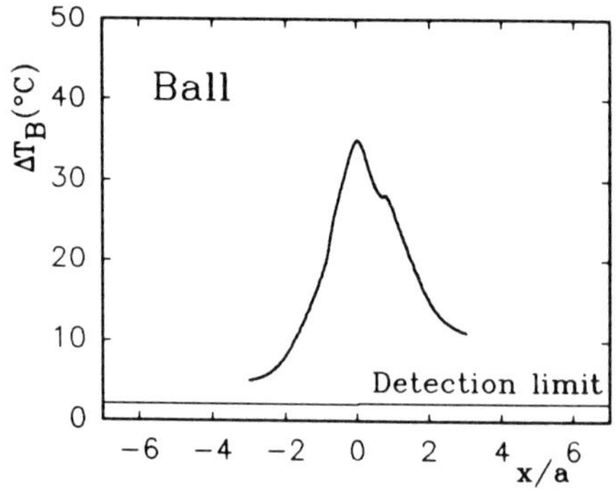

Figure 12. Ball temperature profile along x axis

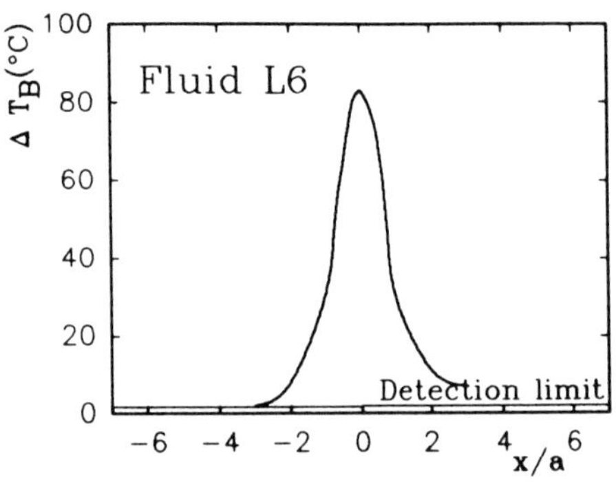

Figure 13. Fluid temperature profile along x axis

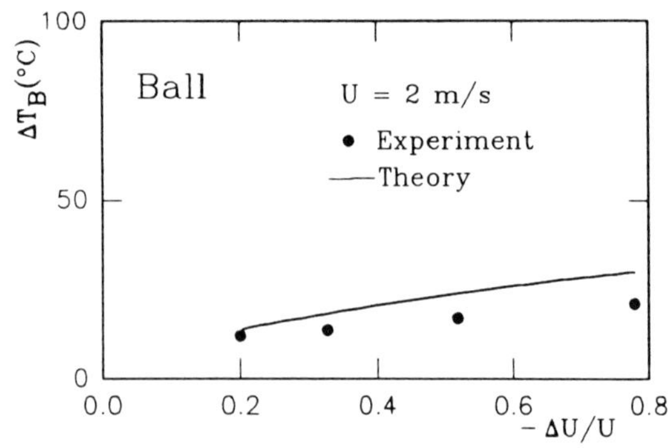

Figure 14. Ball maximum temperature (W = 5.6 N)

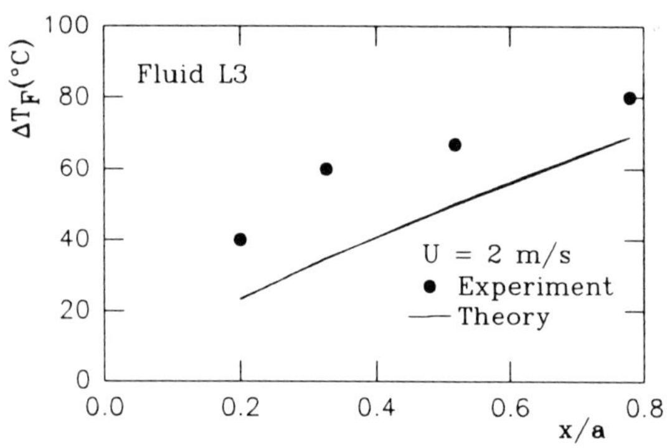

Figure 15. Fluid mean temperature (W = 5.6 N)

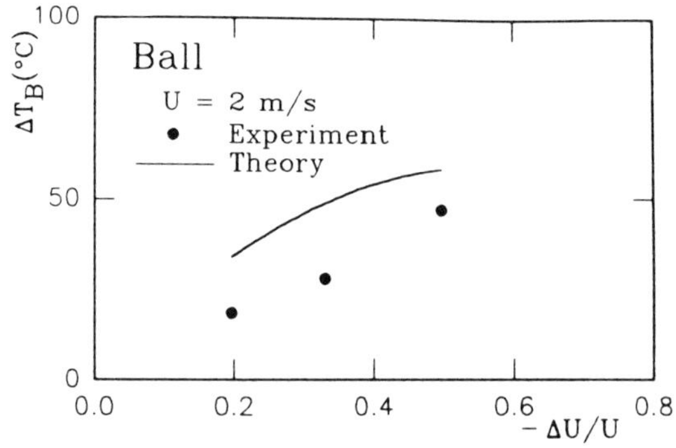

Figure 16. Ball maximum temperature (W = 36 N)

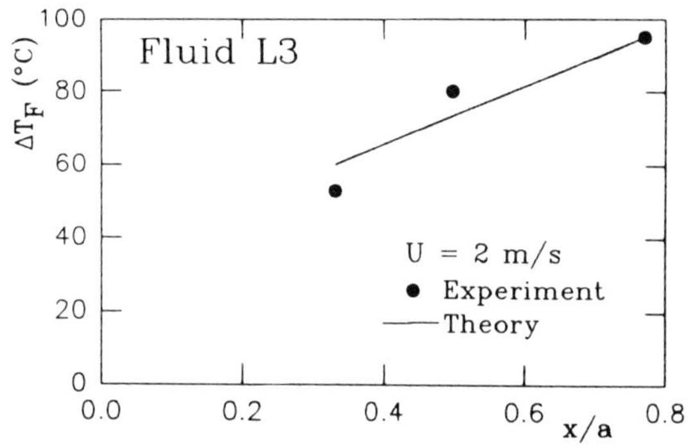

Figure 17. Fluid mean temperature (W = 36 N)

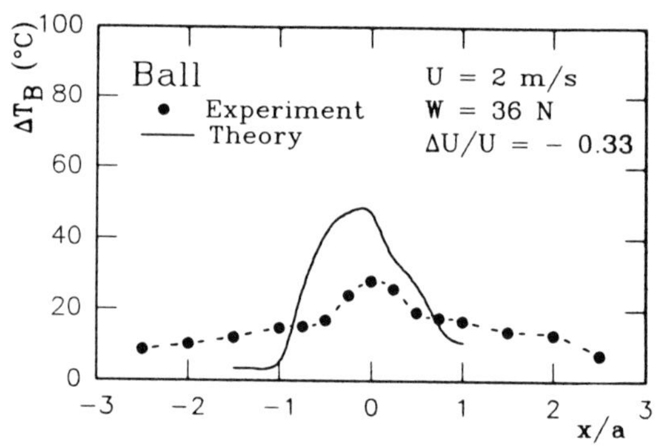

Figure 18. Ball temperature profile

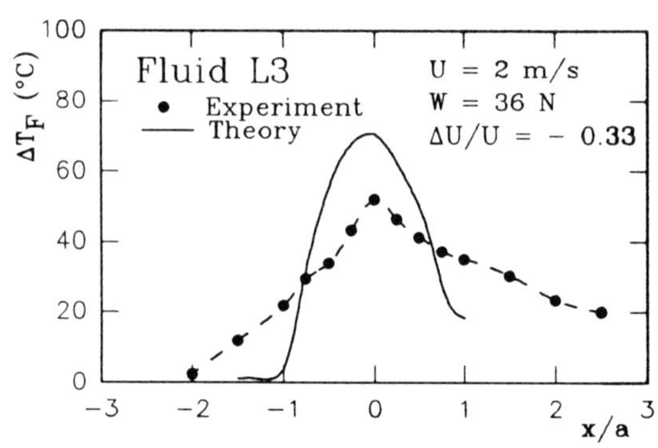

Figure 19. Fluid temperature profile

Thin Films in Tribology / D. Dowson et al. (Editors)
1993 Elsevier Science Publishers B.V.

Experimental Investigations Into Shear Localization in an Operating EHD Contact

Scott Bair[a], Ward O. Winer[a], and Keith W. Distin[b]

[a]Georgia Institute of Technology, School of Mechanical Engineering, Atlanta, GA 30332-0405

[b]Darlington, United Kingdom

Shear localization has been offered as a mechanism for the rate-independent rheology which has been observed for liquid lubricants at high-pressure in concentrated contact as well as in high-pressure rheometers. In recent experiments, shear localization in liquids, in the form of shear bands inclined to parallel boundaries, have been observed in a high-pressure flow visualization cell. Viewing of shear bands in concentrated contact thin films is hampered by the typical geometry of an optical EHD simulator where the line of sight is perpendicular to the plane of the film. However, if light is scattered by reflection from a shear band (plane) there should be a preferred scattering angle which will depend on the direction of liquid shear. A technique is presented which offers evidence of shear bands in EHD films by demonstrating that light is scattered within the film in a direction consistent with the shear direction and previously reported shear band inclination.

1. INTRODUCTION

The relatively thin films which separate machine elements in elastohydrodynamic lubrication operate under a combination of high-pressure and high-shear stress. The study of the shear behavior of these films has been a challenge for the last three decades. An unusual feature of the behavior of these films is that the shear stress saturates at a level (which is approximately proportional to pressure) known as the "limiting shear stress" and becomes relatively rate-independent. This has been observed with disc machines [1,2] and in high pressure rheometers [3,4].

To illustrate, three flow curves for a polyphenyl ether are presented in Figure 1. The two curves representing a temperature of 40°C and

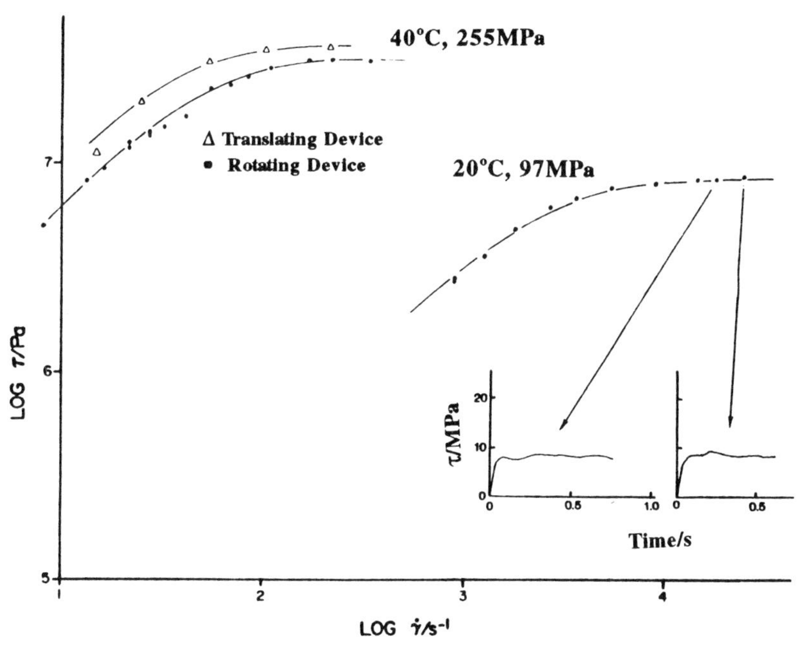

Figure 1. Flow curves for a polyphenyl ether. Shear stress histories inset.

pressure of 255 MPa were obtained with very different rheometers. One rheometer, the translating device, was described in Ref [3] and the data are from that paper. It operates by pulling a wire through the bore of a hollow cylinder while the liquid sample is sheared in the clearance. The second rotating device is a rotating concentric cylinder rheometer [4] which utilizes clearances as small as 1μm to effectively quench the viscous dissipation into the metal cylinders. The two different devices show the same general behavior. However, the rotating device yields a shear stress about 20% lower. Since the calibration for the rotating concentric cylinder rheometer is considerably more straightforward, we consider the data produced by it to be the more reliable. In any event, the similarity in shear behavior for two very different techniques lends credibility to the limiting stress concept. A unique feature of the data obtained with the rotational rheometer in Figure 1 is that the cylinders employed were fabricated from a tungsten-copper composite material which possesses very high thermal conductivity along with the necessary mechanical strength to withstand the shear stress generated. Heating of the cylinders is manifested in a shear stress which decreases linearly with time. The resulting stress histories are inset in figure 1 for the two highest shear rates (1.8 and 2.5 x 10^4 s^{-1}) and are nominally flat for over one-half second.

2. FLOW VISUALIZATION

It had been suggested [2] that shear localization might be the mechanism for the above rate-independent behavior. Recently, shear bands have been reported [5] in liquid lubricants sheared between parallel plates with a hydrostatically imposed pressure. Slip occurred intermittently on planes inclined to the principal shear directions so as to reduce the compressive normal stress on these planes in a manner at least qualitatively consistent with the Mohr-Coulomb criterion. Shear bands are first observed when non-proportional shear behavior occurs as shear stress is increased. A total slip of about 5μm has been measured parallel to the band. Shear bands have been observed in four lubricant types: a mineral oil, cycloaliphatic hydrocarbon, a fluorinated hydrocarbon, and a polyphenyl ether. The orientation of the bands is apparently a property of the material and appears to be independent of temperature, pressure, gap thickness and shear rate.

Shear bands are shown for the polyphenyl ether in Figure 2. There are two types: the first type is most closely aligned with the solid boundaries and the second type is nearly perpendicular to these boundaries. Due to the kinematics of the shear flow, each of the second type rotates with the continuum during the time it is visible. The existence of two types of bands is consistent with an "internal friction" model.

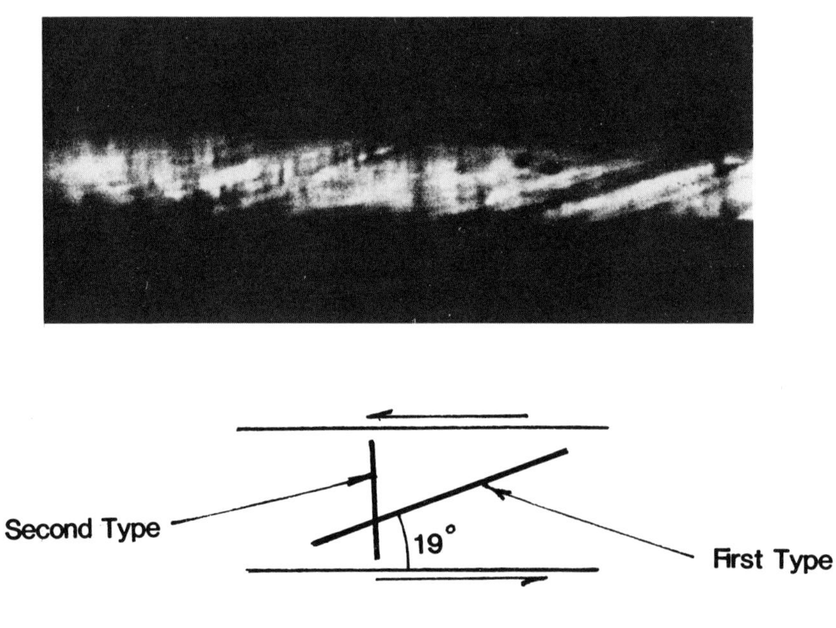

Figure 2. Shear bands in polyphenyl ether at 152 MPa and 5°C.

3. OBSERVATIONS IN AN OPERATING EHD CONTACT

If light is scattered by reflection from a shear band, there should be a preferred scattering direction within the lubricant film which depends upon the direction of shear. The orientation of the second type of band is little changed by the direction of shear and we would expect that the way in which light interacts with this type should be little affected by the shear direction. In the following discussion, then, only the first type of shear band is considered.

An optical EHD simulator was employed in this work as is described in Ref [6]. The contact consisted of a 25.4 mm radius sapphire disc running over a 7.6 mm radius steel roller. The roller had a crown radius to give the contact an ellipticity of 3. The disc was loaded by an 89N weight giving a contact size of 672 x 224 μm and an average pressure of 760 MPa. The polyphenyl ether, 5P4E, was used as a lubricant giving a minimum film thickness of approximately 1.5μm at room temperature.

The original idea was to illuminate the contact with a laser from directly above, through a piece of card with a hole in it, and observe the pattern reflected onto the underside of the card. It was hoped that light would be reflected from the shear bands, expected to be inclined at about 19° to the horizontal, producing a pattern which changed sides when the shearing direction was reversed. This was not observed. Further examination of the optics of the system reveals why. It transpires that any light which may have been reflected from the shear bands regardless of the direction of shear will strike the upper surface of the disc at an incident angle greater than the critical one for that interface, so the light will be totally internally reflected, and could only escape from the disc edges. The refractive index of sapphire is 1.766. The refractive index of the lubricant was measured with a high-pressure refractometer [7] at pressures to 300 MPa and extrapolated to 760 MPa. The value obtained is 1.722.

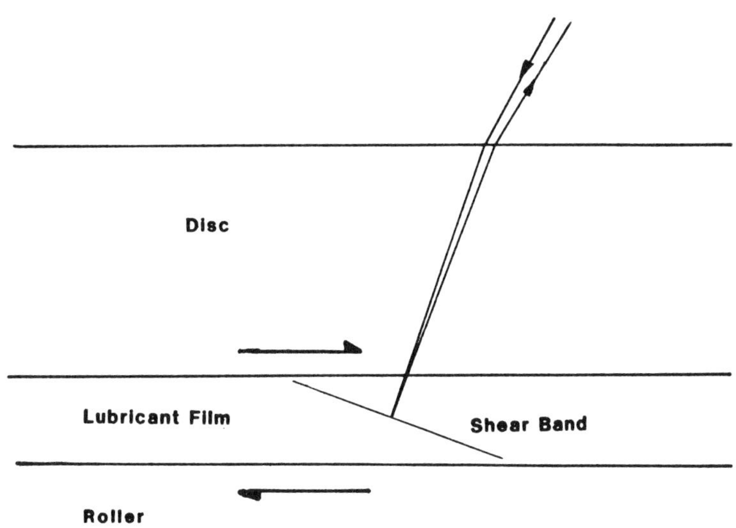

Figure 3. Schematic of light path striking a shear band with negative sliding velocity.

A solution to this problem is to illuminate the contact from an angle (Figure 3). An incident angle of about 34° will be refracted so as to strike a shear band produced by negative sliding velocity approximately normally, and hence be reflected back into the laser. Consequently an angle slightly different was used during the experiments. It was hoped that this would allow the observation of light reflected back toward the source in one direction of shearing as illustrated in Figure 3. Unfortunately, any light that may have been reflected was either too diffuse or coincided with a different brighter reflection. However, it was noticed that the intensity of light escaping from the edge of the disc (see Fig. 4) varied with the direction of rotation and hence shearing. See Table 1. tests A through D. This effect could be caused by shear bands, as explained above or may have been due to reflections from the inlet or outlet regions of the film, as the laser has a 480μm diameter beam some light will strike in front of and behind the contact patch. The inlet meniscus is far enough away to be of no concern. However, the outlet region is much closer to the contact patch. The lubricant film cavitates just beyond the contact and most of the film is thought to adhere to the surfaces. If this is the case it will have little effect on the direction of reflected light.

To illustrate that this is the case, the slide-to-roll ratio was set negative. That is, rotation in the same direction as before results in shear in the opposite one hence changing the inclination of the shear band. The higher intensity light was now observed in the opposite direction of rotation to before (i.e. with the outlet meniscus on the opposite side of the contact). See test F - Table 1. The slide-to-roll ratio was also set to zero and no change in intensity was observed when the direction of rotation was reversed (test E).

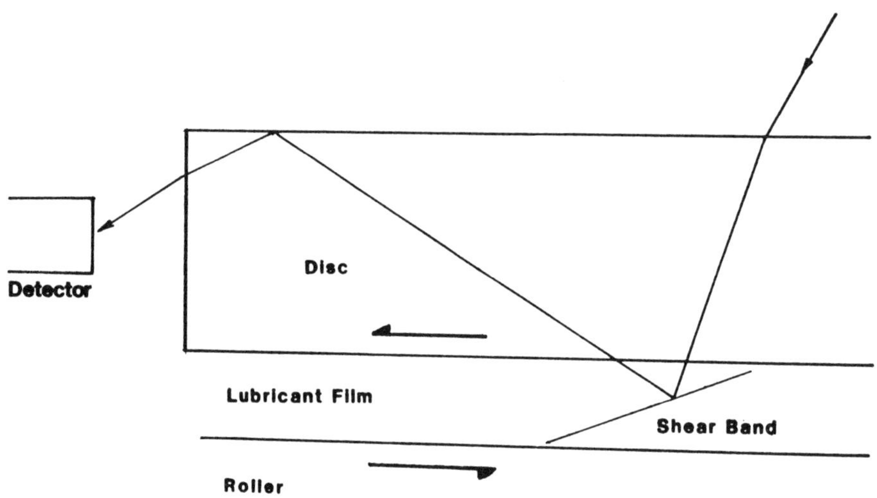

Figure 4. Schematic of light path striking a shear band with positive sliding velocity.

In order to quantify this change in intensity a photovoltaic cell was masked, so that only light from the disc could enter, and placed near the disc edge. The output from the photovoltaic cell is only an indication of the relative intensity for rotation in opposite directions for each absolute value of slide-to-roll ratio, because it proved to be very sensitive to ambient lighting conditions and proximity to the disc edge, which unfortunately changed each time the absolute value of slide-to-roll ratio was changed. The results are listed in Table 1 where light intensity at the edge of the sapphire disc is represented by the photovoltaic output in mV.

These results show that the change in observed intensity is not dependent on the direction of rotation (i.e. meniscus position) but on the direction of shearing. The fact that the intensity change can be seen for both directions of rotation helps to eliminate the possibility that the phenomena may be caused by the geometry of the contact changing, or the position and shape of the inlet and outlet regions changing as the direction is reversed. While this does not conclusively prove that shear bands exist in operating contacts (change in birefringence of sapphire with stress is a possible effect), it does show that there is some mechanism at work within the lubricant which is dependent on the direction of shearing and effects the optical character of the lubricant film.

4. ACKNOWLEDGEMENT

This work was supported in part by a grant from ONR-Tribology Program and this help is greatly appreciated.

Test	Rolling Velocity (m/s)	Sliding Velocity (m/s)	Slide-to Roll Ratio	Intensity Light at Disc Edge (mV)	Change in Intensity (mV)
A	0.228	0.022	0.096	51	-8
	-0.228	-0.022	0.096	43	
B	0.233	0.013	0.056	48	-13
	-0.233	-0.013	0.056	35	
C	0.235	0.008	0.034	54	-17
	-0.235	-0.008	0.034	37	
D	0.237	0.005	0.021	40	-1
	-0.237	-0.005	0.021	39	
E	0.239	0.000	0.000	47	0
	-0.239	0.000	0.000	47	
F	0.245	-0.012	-0.049	52	+7
	-0.245	0.012	-0.049	59	

TABLE 1. Light Intensities for Various Contact Positions

REFERENCES

1. Johnson, K., Tevaarwerk, J., "Shear Behavior of Elastohydrodynamic Oil Films," Proc. Roy. Soc. of London (1977) 365A, 215-236.

2. Evans, C., Johnson, K., "The Rheological Properties of EHD Lubricants," Proc. Instn. Mech. Engrs. (1986) 200, No. C5.

3. Bair, S., Winer, W., "A Rheological Model for Elastohydrodynamic Contacts Based on Primary Laboratory Data," Trans. ASME, Journal of Lubrication Technology (1979) 101, 3, 258-265.

4. Bair, S., Winer, W., "The High Shear Stress Rheology of Liquid Lubricants at Pressures of 2 to 200 MPa," Trans. ASME, Journal of Tribology (1990) 112, 2, 246-253.

5. Bair, S., Qureshi, F., and Winer, W. O., "Observations of Shear Localization in Liquid Lubricants Under Pressure," Trans. ASME, Journal of Tribology, accepted (1992).

6. Bair, S., and Winer, W. O., "Friction/Traction Measurement with Continuously Variable Slide-Roll Ratio and Side Slip at Various Lambda-Ratios," Proceedings of the 1980 Leeds-Lyon Conference (1980), 296-301.

7. Bair, S., "An Experimental Verification of the Significance of the Asymptotic Isoviscous Pressure," STLE, Tribology Trans., accepted (1992).

Thin Films in Tribology / D. Dowson et al. (Editors)
© 1993 Elsevier Science Publishers B.V. All rights reserved.

Fractionated thin film lubrication

A. Jullien, M.-H. Meurisse and Y. Berthier

Laboratoire de Mécanique des Contacts, URA CNRS 856,
Institut National des Sciences Appliquées de Lyon,
69621 Villeurbanne, France.

The analysis of the behaviour of lubricated carbon based composites led to the identification of a new type of lubrication which we called fractionated.
Both visualization on the micro and macroscales and friction measurement were performed during runs; high (0.13-0.15) and stable coefficients of friction were noted. High flat islands with characteristic diameter of about 100 μm are formed during the test; film thicknesses of around 0.1-0.5 micrometers are measured.

Fractionated lubrication is different from mixed lubrication in that, the entire pressure field is broken up into small independent pressurized zones, separated by zero pressure channels.

The questions of load carrying and shear stress in the fractionated film are discussed.

1. INTRODUCTION

This paper presents the tribological analysis of a slip mechanism which uses a porous carbon based composite [1]. The aims of the analysis were to identify :
-the lubrication regime of this mechanism,
-the phenomena governing load carrying capacity, friction and wear.

The operating conditions of the mechanism (a continuously slipping multi-disc clutch) are described in section 2. The tribological analysis was performed in 3 successive steps corresponding to smaller and smaller scales of observation; surfaces of 1cm^2, 1mm^2, and finally 0.05 mm^2. These 3 steps in our tribological analysis and the main results after each step are described in section 3.

The results of these three steps showed that the lubrication regime is non classical hydrodynamic, non mixed and non boundary. We identified a typical surface roughness of the composite material, in flat islands and cavities; we observed activated islands, which at one time were carrying the load, and a mechanism of relay between the activated islands. We called this type of lubrication fractionized [2] or fractionated. Fractionated lubrication is precisely defined in section 4, where questions arising about load carrying generation over the activated islands are discussed.

2. THE MECHANISM

The continuously slipping clutch is a new automotive differential which couples the front axle to the rear axle. This multi-disc contact uses carbon composite discs against steel counterfaces; the lubricant is a fully formulated fluid.

The discs of composite are either plain or grooved; the value of the apparent area of contact is 54 cm^2. The composite is made of carbon fibres in bulk in the carbon matrix, giving a structure consisting in islands and cavities (Figure 1) to the surface of the carbon discs [3].

The specific running conditions of the mechanism imply a constant level of torque, slipping speed and fluid temperature during rather long periods of time (several hours). The torque can be varied by the driven normal load applied to the contact, speed and fluid temperatures are set. Table 1 gives the range of these applied parameters.

A high (0.13-0.15) and stable friction coeffi-

Table 1
Usual running conditions of the clutch

Speed ($m.s^{-1}$)	0.1 to 2
Apparent pressure (MPa)	0.2 to 0.6
Fluid temperature (°C)	70 to 100

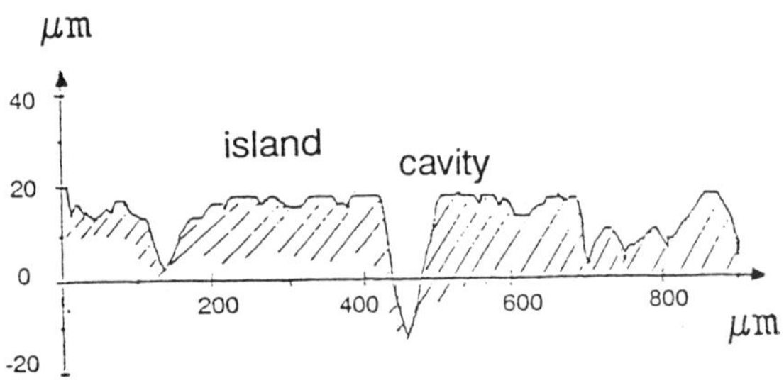

Figure 1. Islands as shown in a two dimensional roughness profile

cient is required throughout the life of the mechanism.

Tests were performed on an industrial SAE type tribometer [3], to study the tribological behaviour of the contact under various running conditions. The friction coefficient did not vary when speed, temperature and apparent applied pressure were doubled from one test to the other [3]. Then, doubts arose about generation of hydrodynamic load capacity in the contact. The tribological analysis of the contact was then undertaken in order to answer that question [3].

3. SUCCESSIVE STEPS IN THE TRIBOLOGICAL ANALYSIS

The analysis was built on both experimental and theoretical investigation, combined into an approach to the mechanism, the first bodies and the third body [4]. Tribological analogies with radial face seals of the same geometry and kinematics allowed us to make the assumption of hydrodynamic lubrication. Thus, in the first step of the analysis we concentrated on hydrodynamic effects, which were investigated at the mechanism scale (1 cm^2).

3.1 Study at the scale of the mechanism ($1 cm^2$).

The industrial SAE tribometer was modified to allow for clutch tests with visualization [2]. The steel counterface was replaced by a pyrex counterface. Experiments were carried out at the mechanism scale ($1cm^2$) : essential results related to the lubricant flow inside the contact were obtained. But, from comparisons between the measurements and classical hydrodynamic theory, we concluded that classical hydrodynamic or mixed lubrication did not occur in this contact.
Furthermore, the absence of a classical hydrodynamic component was explained by the typical structure of the composite [3]. This structure implied that the main phenomena were activated on several islands. We also noticed that the surface of these islands showed morphological changes after friction (Figure 2). These results showed that the scale of analysis had to be reduced. This was done down to the scale of several islands ($1mm^2$) for the second step of the tribological analysis.

3.2 Study at the scale of several islands ($1 mm^2$).

Friction induced phenomena, activated on the islands were investigated. These islands were called "activated islands". As precise simulations were required, the "Pedeba" tribometer of the laboratory was adapted [3] and used. Once we checked that the results of the simulations were in good agreement with the clutch results [3], a parametric study was performed. Simulations including visualizations at this new scale ($1mm^2$) were carried out. At that moment, comparisons with the literature indicated that the lubrication regime of the contact was of the boundary type [3]. The effects on friction, of two counterface materials (pyrex and steel) combined with three different formulated fluids, were examined. Effects of additives induced by boundary lubrication were expected. Then, careful morphological examinations were undertaken with both S.E.M and optical microscopy.

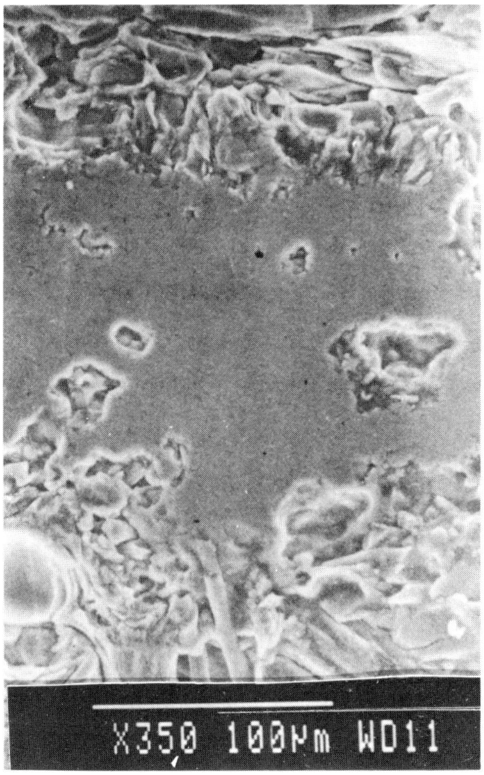

Figure 2. Morphological changes of the surface of the islands (S.E.M.)

The friction coefficients measured under varying fluid temperatures were studied for each contact (one counterface, one fluid). Comparisons of the results obtained with these contacts, of different chemical properties, were carried out.

The comparisons revealed that changing the additives as well as the material of the counterface (pyrex to steel) had negligible effects on friction (Figure 3). But we observed significant morphological differencies in the surface changes of the islands [5]. We noticed that these changes were :

- influenced by the lubricant itself (the morphology of the islands differs strongly in the case of a dry contact),
- influenced by the presence of additives in the fluid,

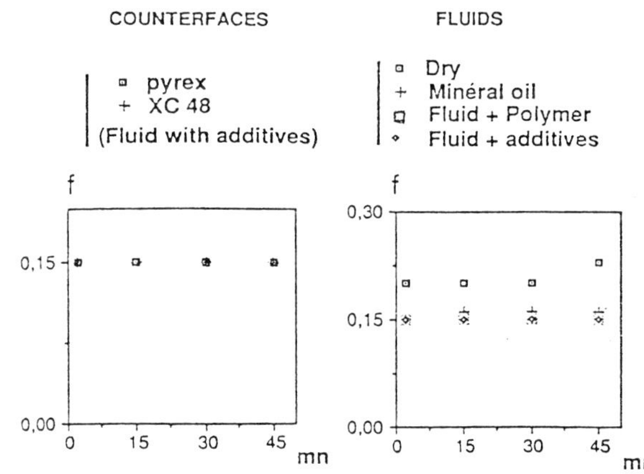

Figure 3. Effect of the additives and the counterface on the coefficient of friction vs time

-not influenced by the material of the counterface.

Doubts about possible boundary lubrication produced over the activated islands, arose. We also felt that direct contact did not govern the tribological behavior of the activated islands either.

So at that time, the phenomena seemed to involve a non-classical lubrication regime. A third step was then undertaken to identify the activated phenomena and to study separation by a fluid film, if any, on the scale of one activated island.

3.3 Study at the scale of one island (0.05 mm^2).

Separation at the scale of one island was studied through simulations.

A new visualisation technique was developed and applied [5]. This technique was partially based on laser fluorescence which allowed for film thickness measurements. Calibration of the technique was carried out by image analysis.

White light was also supplied in order to observe over a period of hours the lubricant supply to the island and the friction-induced morphological changes in the surface. The simulations showed a complete lubricant film over the island. The film thickness was estimated to be in the range 0.1-0.5 micrometers, and quasi constant over an activated island [5].

The morphological surface changes of the island were found to be produced in the presence of lubricant [5]. The island was found to be laterally fed by surrounding cavities and normally by minute interstices (Figures 4.a and 4.b). These simulations also showed the dependance of film thickness on speed and on the normal load.

We concluded that separation over the island was provided by the fluid. We also noticed that this order of magnitude of film thickness was much higher than the boundary values [6]. We thus deduced at the end of this last step that boundary lubrication did not govern the behaviour of this contact.

This complete tribological analysis led to the identification of a new type of lubrication. It strongly differs from the usual lubrication regimes described in literature : we called it fractionized or fractionated lubrication [5]. The main characteristics of fractionated lubrication are discussed below.

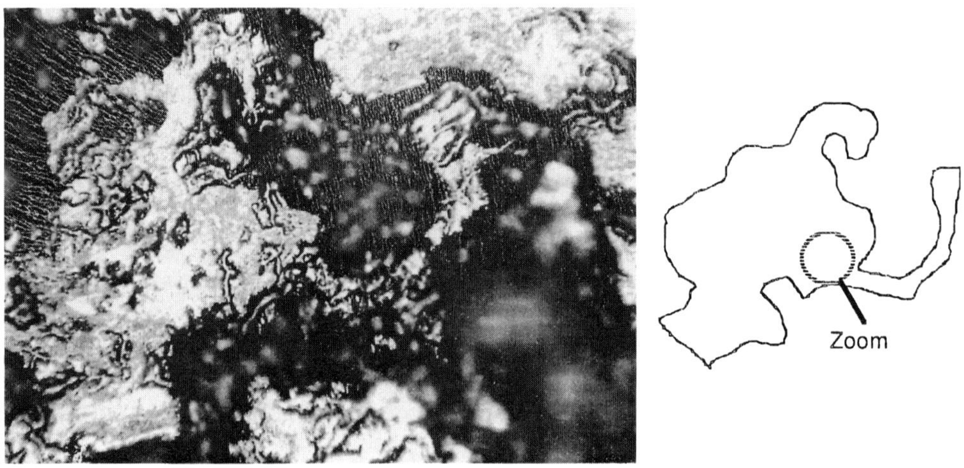

Figure 4.a Optical micrography of an activated island

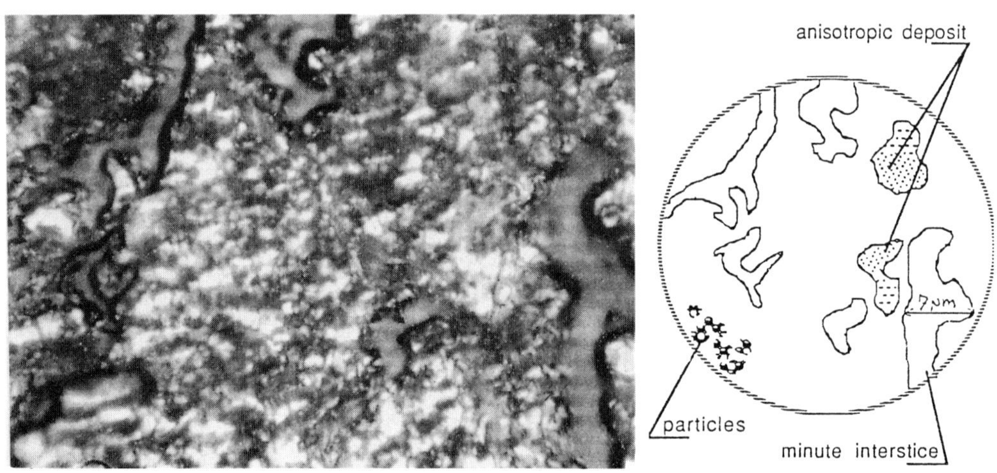

Figure 4.b Zoom of figure 4.a

4. FRACTIONATED LUBRICATION

4.1. Definition

The tribological behaviour of the carbon based material in oil was demonstrated to depend completely on what we called the "islands":

- the normal load is supported by some "activated" islands (1000 to 2000 activated islands exist simultaneously on a carbon disc); the mean pressure in the film on each load carrying island is 0.1 to 0.5 GPa; a mechanism of relay between the islands guarantees the durability of this tribological behaviour;
- the coefficient of friction is governed by the presence, on the activated islands, of a fluid film, the thickness of which was measured by laser fluorescence at between 0.1-0.5μm; additives in the oil influence the evolution during running of the surfaces of the activated islands, but not the friction force itself, which seems to be governed mainly by the rheological properties of the lubricant; due to the low film thickness, the apparent shearing rate is high (5. 10^4 - 10^7 s^{-1}).

As the real lubricated contact zone is split into many independent contact zones, we call this type of lubrication "fractionated lubrication". It is important to note that the independence of the contact zones concerns their location (disjointed and disseminated in the entire apparent area of contact) and especially the decoupling of the flows and pressures in these contact zones, which are separated by a network of channels and deep cavities (around 20 μm depth). In this respect, fractionated lubrication differs fundamentally from mixed lubrication.

We observed bubbles in the cavities of the carbon material and free surfaces of lubricant, which distorted very much but could not go through the gap on the activated islands; these free surfaces hinder the natural feeding of the film by the fluid carried along by the moving solid surface.

The existence of an oil film, completely separating the solid surfaces on the load carrying islands has been proved [5]; the following question then arises: what is the effect does which induces a sufficient pressure in the film to support the applied load (mean pressure of 0.1 to 0.5 GPa in the load carrying zone)?

4.2 The question of the load carrying capacity

Table 2 gives a comparison between the characteristic quantities in the fractionated film on a carrying carbon island and in the hydrodynamic range. It has to be noted that in spite of completely different geometrical charac-

Table 2
Characteristic parameters on an island; comparison with usual hydrodynamic cases.

	Activated island (fractionated film)	Hydrodynamic range from[8]
length or diameter (m)	10^{-4}	10^{-3} to 1
film thickness (m)	10^{-7}	10^{-6} to 10^{-3}
scale factor (length/thickness)	10^{-3}	10^{-3}
mean pressure (Pa)	10^8 to 5.10^8	0 to 10^7
apparent shear rate (s^{-1})	5.10^4 to 10^7	0 to 5.10^5

ristics, the scale factor is maintained. An important consequence is the fact that the basic hypothesis for the Reynolds equation (scale factor much less than 1) still holds.

The 3 elementary load generating effects in hydrodynamic lubrication are examined below:
- the oil wedge: the maximum mean pressure in an inclined plane bearing of 150μm length, 0.1 μm minimum thickness, 1 m/s differential speed,and an optimum upstream/downstream thicknesses ratio of 2.2 is about 2 10^7 Pa when the viscosity is 0.01 Pa.s. The oil wedge effect cannot then explain pressure values greater than 10^8 Pa on the nearly flat activated islands.
- the oil squeeze effect: a normal speed of the order of 1 mm/s would be necessary to generate pressures of 0.1 GPa!
- the stretching effect cannot either be invoked: a negative gradient of tangential speed of 10^4m.s^{-1} along the island is not realistic.

In our attempt to understand the generation of bearing load in the fractionated film, we were led by the idea of the importance of the feeding flow in all cases of starved lubrication [7]. We then studied theoretically the influence of the feeding flow in a plane bearing, sketched in figure 5.

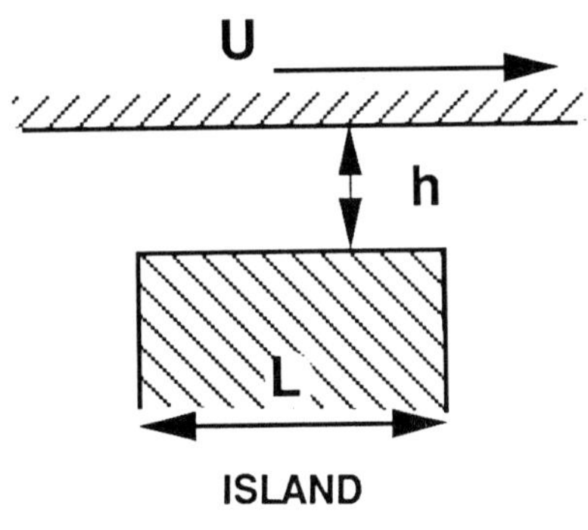

Figure 5. Schematic view of an island (plane bearing)

If one considers the Reynolds model in the simple newtonian 1D case:

$$\frac{d}{dx}\left(\frac{Uh}{2}-\frac{h^3}{12\mu}\frac{dp}{dx}\right)=0 \qquad (1)$$

which implies conservation of flow, it can in the case of parallel surfaces be written:

$$\frac{Uh}{2}-\frac{h^3}{12\mu}\frac{dp}{dx}=k\,\frac{Uh}{2} \qquad (2)$$

where the constant k represents the flow which goes through the contact; if k=1, the flow is the natural Couette flow with a linear gradient of speed, the gradient of pressure is zero and then there is no load. If k>1, the fluid is sucked in, and from (2) the gradient of pressure and then the load, are negative. If k<1, the fluid is decelerated, and the load is positive. Profiles of tangential speed corresponding to the various values of k are shown in figure 6.

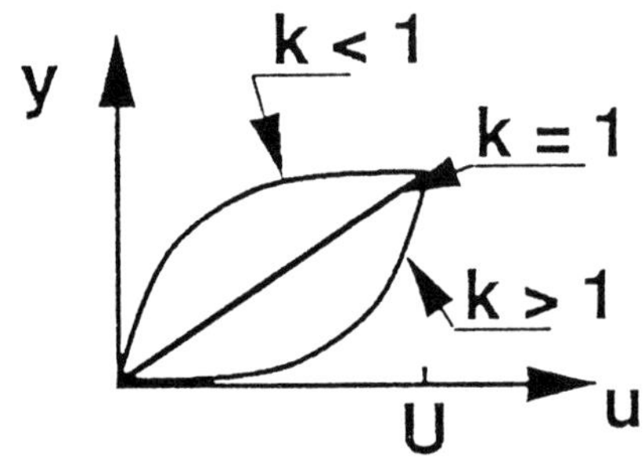

Figure 6. Cross profiles of tangential speed in the plane bearing, following k.

It has to be noted that considering k as an input to equation (2) , e.g. considering the Reynods model with imposed flow does not allow us to write the usual boundary conditions at both the inlet and outlet of the film. One of them has to be relaxed.

The free surfaces of the liquid in the cavities surrounding the islands could act as partial stoppers for the fluid, so that equation (2) with k<1 could explain the generation of bearing load in the fractionated film. Integrating equation (2) over the length of an island gives the pressure increase over the film:

$$\Delta p=\frac{6\ \mu U\ (1-k)\ L}{h^2} \qquad (3)$$

Applying formula (3) with μ = .01 Pa.s, U = 1m/s, L = 150 μm and h = 0.1 μm gives: $\Delta p = 6.(1-k)10^8$ Pa; taking into account the piezoviscosity of the oil would increase this theoretical value yet further. Then, even with a light subambient pressure at the inlet of the film, antagonism between starvation (k<1) and the carrying of the fluid into the contact, due to adherence to the solid surfaces, could explain the load carrying capacity of the fractionated parallel film. The validity of this result is actually purely mathematical, derived from the Reynolds model and has to be considered as a basis for further reflexion...

4.3 The question of the shear rate

Fractionated lubrication allows for very thin films (a few tenths of a micrometer without solid interactions); consequently the apparent shear rate is much higher than the usual range in lubrication (see table 2), and the rheological behaviour of the film can be invoqued to explain the experimental results concerning friction.

5. CONCLUSION

The tribological analysis of a mechanism using a new carbon composite was carried along in order to identify the lubrication regime of the contact. References to literature were made at each step of the analysis to compare the results with classical ones. The behaviour of the contact was studied through global dynamic results (measurements), global and local dynamic results (visualisation of third body) and static results (examination of first bodies). This scheme was reproduced at each scale of analysis (1cm^2, 1mm^2, 0.05mm^2).

The result was that a new lubrication regime was indicated which we called fractionated.

To summarise, the idea of fractionated lubrication includes the following peculiarities:

- the load carrying zone is split into many disconnected zones;
- the regions surrounding the load carrying zones (starved feeding, free surfaces...) -in other words the upstream boundary conditions- have a significative influence on the tribological behavior of the mechanism.
- as many local disconnected carrying zones constitute the entire carrying zone, a high apparent shear rate in the lubricant is possible without any solid to solid contact; the usual "no slip at the wall" boundary conditions have then to be reconsidered.

Fractionated lubrication occurs not only in the case of continuously slipping carbon composites; the ski-snow interface is another example of fractionated lubrication: droplets of water are the disconnected carrying zones, meniscii limiting the droplets influence the dynamics of the interface; also, the shear rate is greater than $10^5 s^{-1}$ (in alpine skiing, the thickness of the droplets is about 15 μm [9]), and the wettability of the ski sole has to be taken into account.

Further experimental investigations concerning the boundaries of the fractionated film - at both the inlet/outlet and the solid walls - are necessary to achieve the understanding of the main phenomena occuring in fractionated lubrication.

REFERENCES

1. D.Benoit, D.Menard, Valeo Z4470 - A new oil friction material generation. Automotive Technology International 90. London : Streling Publication International Ltd,p. 187-190.
2. A.Jullien, M.H. Meurisse, Y. Berthier, Fractionized thin film lubrication : a new regime of lubrication in a carbon based material-oil-glass contact, ASME - STLE Tribology conference, october 18-21, 1992, to be published in Lubrication Engineering.
3. A. Jullien, Y. Berthier, D. Menard, M.-H. Meurisse, Behaviour of wet clutches operating under continuous running conditions with a new carbon based material, presented at the 17th Leeds-Lyon Symposium, Vehicle Tribology, 1990.
4. M.Godet, The third body approach. A mechanical view of wear, Wear, vol. 100, 1984, p.437-452.

5. A. Jullien, M.-H. Meurisse, Y.Berthier, Determination of lubrication regimes and wear through high resolution visualization in contacts using a carbon based composite, submitted for presentation at the 9th International Conference on Wear of Materials, 1993.
6. J.M. Georges, Colloidal behavior of films in boundary lubrication. Microscopic Aspects of adhesion and Lubrication. Ed J.M. Georges, Amsterdam, Elsevier Tribology Series, 1992, vol. 7, p. 729-761.
7. G. Bayada, Variational formulation and associated algorithms, Journal of Lubrication Technology, vol. 105, n° 3, 1983.
8. J. Frêne, D. Nicolas, B. Degueurce, D. Berthe, M. Godet, Lubrification hydrodynamique- paliers et butées, Eyrolles, Paris, 1990.
9. W. Ambach, B. Mayr, ski gliding and water film, Cold Region Science and Technology, 5, 1981.

ACKNOLEDGEMENTS

The authors wish to express their thanks to the Valeo Company and to the french Ministère de la Recherche et de la Technologie for having supported this work, and to Professor Godet for usefull discussions.

SESSION XII
COATINGS AND SURFACE MODIFICATION (2)

Chairman : Professor K L Johnson

Thin Films in Tribology / D. Dowson et al. (Editors)
© 1993 Elsevier Science Publishers B.V. All rights reserved.

WEAR MECHANISMS OF COATED SLIDING SURFACES

K. Holmberg[a], H. Ronkainen[b] and A. Matthews[c]

[a, b] Laboratory of Production Engineering, Technical Research Centre of Finland, PO Box 111, 02151 Espoo, Finland

[c] Research Centre in Surface Engineering, University of Hull, HU6 7RX, UK

The wear mechanisms of coated sliding surfaces are discussed. A new approach to analyse wear mechanisms of coated surfaces is presented. It is based on a classification of the tribological contact process into macromechanical, micromechanical and tribochemical changes, and material transfer. The wear process is dominated by the macromechanical mechanisms, which can be systematically analysed on the basis of four main parameters: the coating, substrate and counterface hardness relationship, the coating thickness, the surface roughnesses and the debris in the contact. This approach is intended to be of assistance in the reliable identification and evaluation of wear mechanisms and for systematic coating design and selection.

1. INTRODUCTION

The deposition of surface coatings on components and tools is today one of the major techniques for reducing wear and inhibiting failure. In the most successful examples wear has been reduced and the lifetime extended by several orders of magnitude.

The results have mainly been achieved by a trial and error approach both concerning the choice of material and the tribological coating parameters, such as coating thickness, hardness and surface roughness. This is due to the lack of a general theoretical concept, that can satisfactorily describe the prevailing friction and wear mechanisms in a tribological contact process with the coated surfaces involved.

The aim of this study is to give a description of the dominating wear mechanisms in dry sliding coated contacts based on experience from the authors' laboratories and from the literature. The approach used is consistent with the earlier published concept for the classification of friction mechanisms of coated surfaces (Holmberg, 1992a and b), which this work extends into.

2. THE TRIBOLOGICAL PROCESS

The tribology of a sliding contact can be understood as a process with certain input data and certain output data as shown in Figure 1. The changes that take place during the tribological process can be suitably divided into macromechanical, micromechanical and tribochemical changes, and material transfer.

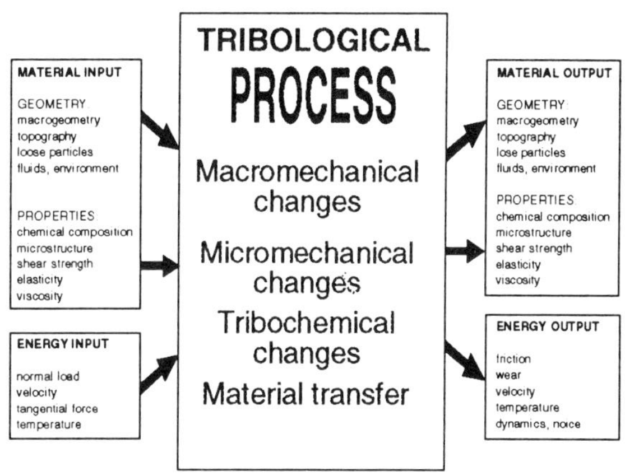

Fig. 1. The tribological process in a contact between two surfaces in relative motion includes mechanical and chemical changes that depend on the input data and result in a changed contact situation determined by the output data at a certain time.

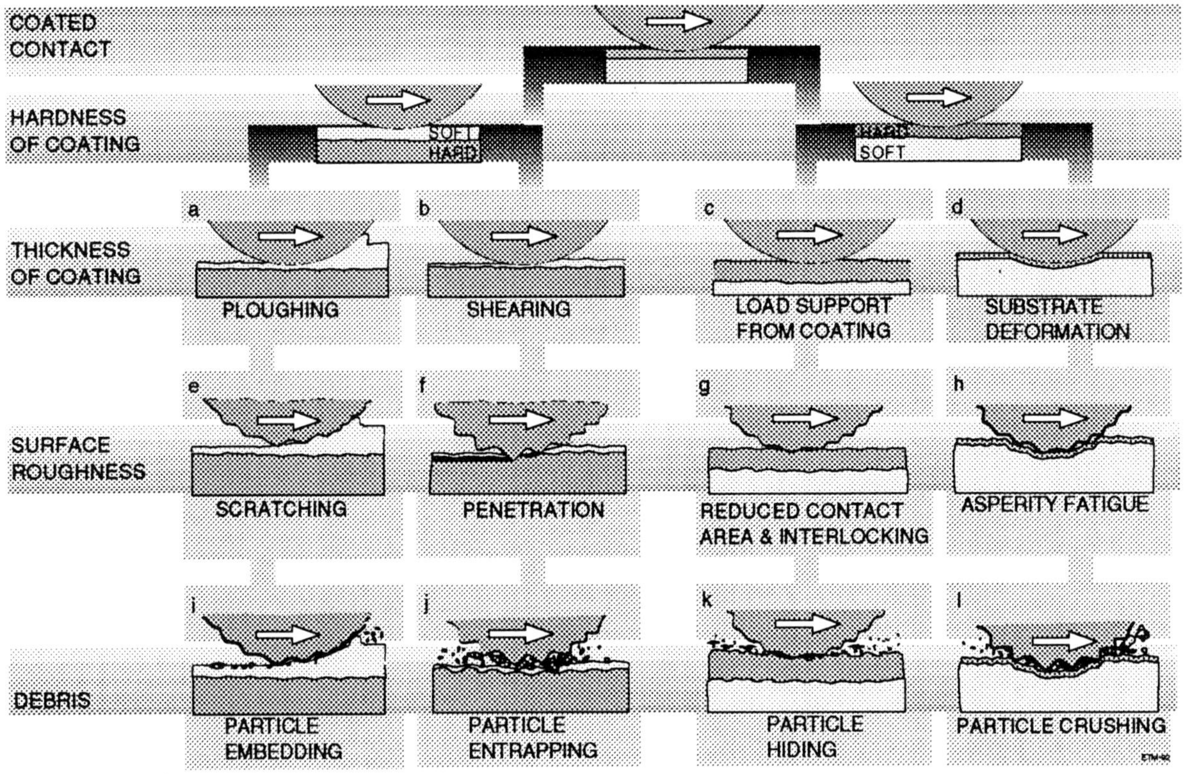

Fig. 2. The macromechanical friction mechanisms in a contact with coated surfaces depend on four main parameters: the hardness relationship, the film thickness, the surface roughnesses and the debris in the contact. Characteristic tribological contact phenomena are shown schematically in the subfigures a-l.

3. MACROMECHANICAL CHANGES

The macromechanical wear mechanisms depend upon the stress and strain distribution in the whole contact, the total elastic and plastic deformations they result in and the total wear particle formation process and its dynamics. In contacts with one or two coated surfaces, four main parameters can be defined that control the tribological contact process. They are

1) the coating-to-substrate hardness relationship,
2) the thickness of the coating,
3) the surface roughnesses and
4) the size and hardness of the debris in the contact.

We will use the schematic presentation for typical tribological contact situations shown in figure 2, which has earlier been used to study the mechanisms of friction (Holmberg, 1992a). The twelve different contact situations can be grouped to represent eight characteristic wear mechanisms as shown in figure 3.

3.1. Plastic deformation

When a hard slider is moving on a soft thick coating which is stiff enough to carry the load, the result will be a groove in the coating, as shown in figure 3a. The removal of material can be estimated from the geometry of the contact. The situation is from a wear point of view very similar to the contact with a hard slider ploughing in a soft bulk material. Plastic deformation was observed as the main wear mechanism when a slider was moving on a plane surface coated with a soft gold layer (Tangena, 1987).

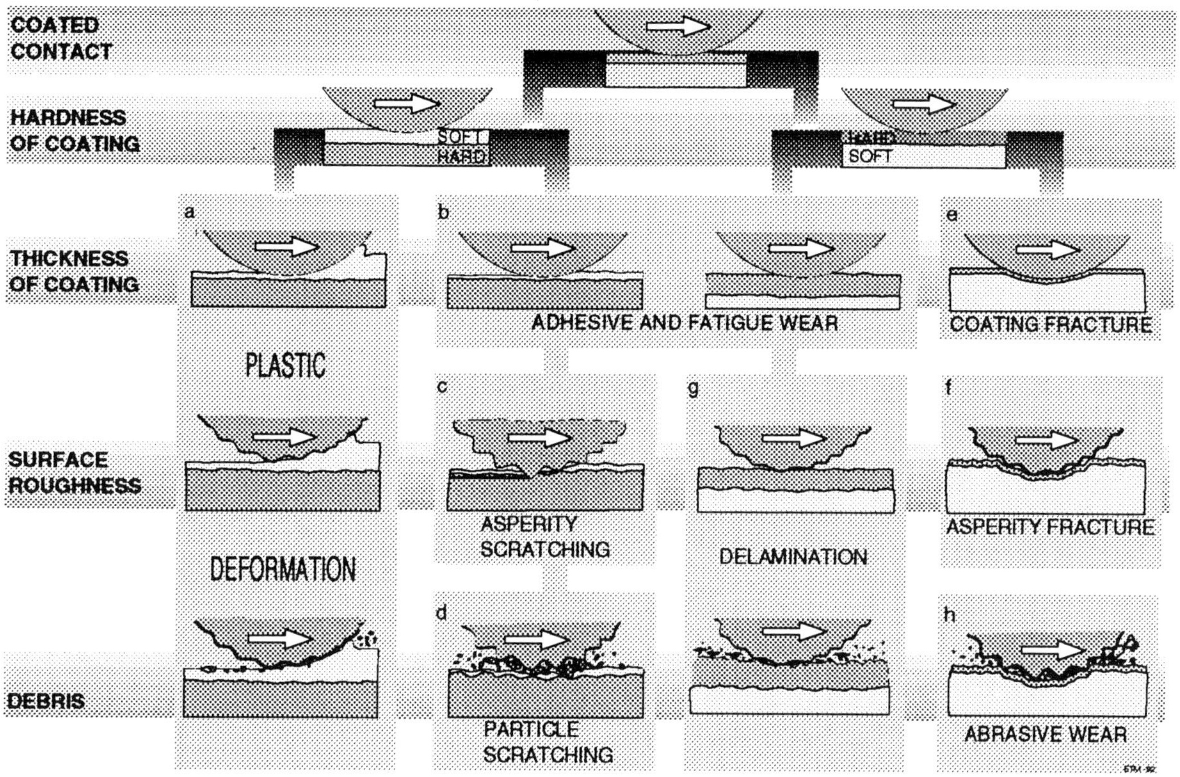

Fig. 3. The macromechanical wear mechanisms in a contact with coated surfaces depend on four main parameters: the hardness relationship, the film thickness, the surface roughnesses and the debris in the contact. Characteristic wear mechanisms are shown schematically in the subfigures a-h.

If the surface of the slider is rough or if small debris are introduced into the contact, the result will be the asperities or the trapped debris causing microploughing and microgrooves in the main groove. The influence of this effect on the total wear is normally only minor.

3.2 Adhesive and fatigue wear

Adhesive wear is present to some extent in several contact situations. Even if all other wear mechanisms have been eliminated by skilful tribological design there will still, in most cases, exist some minor adhesive wear. Adhesive wear is typical of a hard slider moving on a thick hard coating or on a hard substrate covered with a thin soft coating as shown in figure 3b. In addition, fatigue wear may be involved depending on the film material (Fusaro, 1981; and Yamada et al, 1990).

A soft thin film on a hard substrate can be used to reduce delamination wear because of its influence on the dislocation density at the outermost surface layer (Suh, 1986). A thin layer of a soft metal deposited on a hard substrate can retard the wear of the harder substrate by preventing plastic deformation and crack nucleation in the hard substrate. If the soft layer is thicker than a critical value, delamination can occur within this layer, forming loose wear particles which cause ploughing and further plastic deformation.

3.3 Asperity scratching

In contacts with rough surfaces and thin soft films the mechanism of asperity penetration and asperity microploughing, here called scratching, dominates. However, there are interesting examples showing that the surface roughness of the substrate may influence wear even before penetration has taken place (Roberts & Price, 1989). Possible explanations may be an improvement in the load-carrying capacity by the coated substrate asperities or decreased shear strength by increased pressure at the coating between the asperities and the counterface.

Asperities on the surface of a rough slider on a hard substrate coated with a soft coating, as shown in figure 3c, will penetrate the coating and slide directly on the substrate. If the asperities are harder than the substrate, they will also penetrate into the substrate and produce small grooves by scratching both in the coating and in the substrate (El-Sherbiny & Salem, 1984).

When a relatively smooth slider is moving on a rough substrate covered with a thin soft coating, the first stage in the wear process will be a combination of plastic deformation and adhesive wear which decreases the coating thickness until there is contact between the sphere and the substrate asperities through the coating (Fusaro, 1981). After that the wear increases considerably because of both adhesive wear and scratching between the substrate asperities and the surface of the slider.

3.4 Particle scratching

We will now discuss the situation when hard particles of the same size as the film thickness and the surface roughness are introduced into the contact of a rough slider moving on a rough substrate covered with a thin soft coating, as shown in figure 3d. The particles will be caught by the surface roughnesses and cause microploughing both in the soft coating and in the hard substrate in a similar way to the effect of hard asperities penetrating the coating as described earlier. A difference comes from the fact that the particles are free to rotate in the contact and they are often considerably sharper than the asperity angles in engineering surfaces. The wear rate depends on the particle size. It is small for very small particles, increases rapidly with increasing particle size and then stabilises when the particles reach a critical size where a transition from a cutting wear mechanism to a sliding mechanism takes place (Sin et al, 1979).

If the particles in the contact are soft, their tribological effect is quite different. Soft particles with low shear strengths trapped in the contact, can carry part of the load and inhibit direct substrate-to-counterface contact, thus reducing both wear and friction (Yamada et al, 1990).

3.5 Coating fracture

The use of thin hard coatings is often a very favourable solution from the wear point of view. However, if the substrate is not hard enough to carry the load, elastic or plastic deformation will take place in the substrate under the contact. The increased stresses within the coating or at the interface between the coating and the substrate, as the coating deforms along with the surface of the substrate, may be higher than the tensile or shear strength of the material, and result in crack initiation and propagation. Such a contact situation is illustrated in figure 3e.

The harder the substrate, the higher the loading which the coating can resist without failure by fracture, because of increased resistance to deflection (Hedenqvist & al, 1990). With a softer substrate, cracks will occur in the coating both within the contact area and outside at the substrate material pile-up area (Burnett & Rickerby, 1987).

When a highly loaded slider is sliding on a hard coating the friction originating from both shear and ploughing will result in tensile stresses behind the sphere and compressive stresses and material-pile up in front of the slider (Bull, 1991). This can result in several different cracking patterns depending mainly on the geometry and hardness relationships. No complete model for the crack propagation in a coated surface is available, although the theoretical background to it has been discussed by Argon (1980) and Hills et al. (1990).

Even if the coating does not fracture and fail at the first pass, if it is loaded by a sliding counterface repeated loading may result in fatigue cracks and failure. The lifetime of a thin coating may be even considerably longer than that of a thick coating (Erdemir & Hochman, 1988; and Posti & Nieminen, 1989) for several reasons. Firstly, under similar deformation conditions the thicker coating will experience high bending stress levels. Secondly, since the coatings typically have columnar growth morphologies, any crack normal to the surface will be large in the thick coating, and may exceed the critical crack length, whereas in the thin coating this may not be the case.

3.6 Asperity fracture

In the contact between a hard slider and a hard coating, both with surfaces of a high roughness and steep asperity angles as shown in figure 3f, there will be a strong tendency to asperity interlocking with high friction at the beginning of sliding (Hayward & Singer, 1990; and Holiday & Matthews, 1991). Initial wear will take place by the breaking of the asperity peaks and thus smoothening of the surfaces. The hard and sharp wear debris may considerably affect wear during the continuing sliding.

3.7 Delamination

A more common situation is the sliding of a hard rough slider against a hard rough coating with a topography of low asperity angles, typical in engineering surfaces, represented by figure 3g. The sliding action will take place at the tops of the contacting asperities (Bennett et al, 1992). They will be mainly plastically deformed, although the overall contact stress may be less than the yield stress of the materials because the local stresses at the small asperity areas are much higher.

The high stresses at the asperities will thus generate dislocations, pile-up of dislocations and crack nucleation at only some ten micrometres beneath the surface as described by Suh (1973 and 1986) in the delamination theory of wear. Owing to the extremely small plastic deformation of the surface, a large number of cracks must be nucleated before a loose particle can form. In that model the delaminated particles are flake-like and may be some hundred micrometres long (Sue & Troue, 1990; and Yamada et al, 1990).

Wear particles or fragments that are very thin with dimensions less than the magnitude of the surface roughnesses are not expected to markedly influence the wear process, as shown in the lower part of figure 3g, because they can be hidden in the valleys between the asperities while the sliding action takes place at the asperity tops.

3.8 Abrasive wear

Abrasive wear is typical of a hard rough slider sliding on a hard rough coating, with hard particles of sizes larger than the magnitude of the surface roughnesses present in the contact, as shown by figure 3h. The trapped particles will have a scratching effect on both surfaces, and as they are carrying part of the load they will cause concentrated pressure peaks on both surfaces as they try to penetrate them (Rickerby & Burnett, 1987; and Sue & Troue, 1990). The high pressure peaks may well be the origin of crack nucleation in the coating. A rolling movment of the particles, which sometimes may even decrease the friction, can be present.

4. MICROMECHANICAL CHANGES

The origin of the wear phenomena that we observe on a macro level is found in the mechanisms that take place at a micro level. The integration of all the micromechanical mechanisms results in the macromechanical mechanisms discussed above.

The micromechanical tribological mechanisms describe the stress and strain formation at an asperity-to-asperity level, the crack generation and propagation, material liberation and single particle formation. In typical engineering contacts these phenomena are at a size level of about one micrometre or less down to the nanometre level.

Shear and fracture are the two basic mechanisms for the first nucleation of a crack and for its propagation until it results in material liberation and formation of a wear scar and a wear particle. These mechanisms have been discussed by e.g. Argon (1980) and Suh (1986), but there is still today only a very poor understanding of these quite fundamental phenomena.

The tribological micromechanical mechanisms can be studied by using the velocity accommodation mode concept developed by Bertier et al (1989), which seems to be a logical framework for achieving a more fundamental understanding of the mechanical changes in a tribological contact. Velocity accommodation modes are defined as the manner in which the difference in velocity of two surfaces in relative motion is accommodated across the interface.

In a sliding contact with a coated surface, the velocity accommodation can take place at the seven sites shown in figure 4 and by four basic mechanisms: shear, fracture, elastic deformation or rolling. The probability of these mechanisms occurring varies at the different sites in the interface as indicated in figure 4. The mechanisms of elasticity and rolling are mainly related to friction and their influence is relatively minor when considering wear.

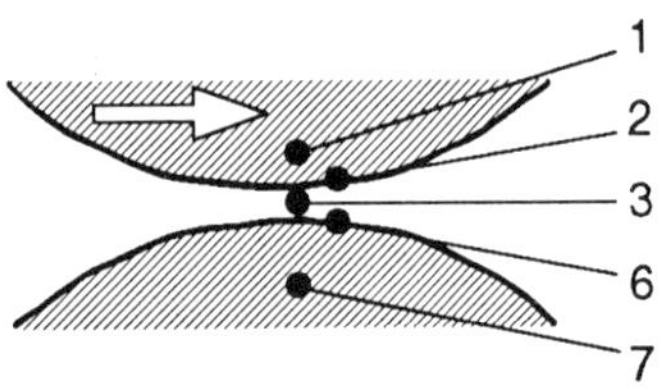

a)

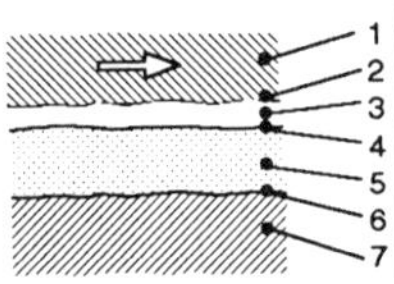

	SHEAR Shear strength	FRACTURE Fracture toughness	ELASTICITY Youngs modulus	ROLLING Rolling resistance
1	-	++	++	-
2	+	-	-	-
3	+++	+	+	+++
4	+++	+	+	+
5	+++	+++	+++	-
6	+	++	+	-
7	-	++	+++	-

+++ = very important; ++ = considerable; + = not important; - = generally negligible

b)

Fig. 4. The velocity accommodation in a coated sliding contact may take place in 1) the counterface, 2) its interface, 3) the intermediate material, 4) the microcoating, 5) the coating, 6) its interface with the substrate or 7) in the substrate. The influence of the velocity accommodation on friction and wear is indicated in the table.

5. TRIBOCHEMICAL CHANGES

The chemical reactions taking place at the surfaces during the moment of sliding contact and also during the periods between repeated contacts change the composition of the outermost surface layer and its mechanical properties. This may have a considerable influence on the wear properties because they are to a great extent determined by the surface properties where phenomena like shear, cracking and asperity ploughing take place. The chemical reactions on the surfaces are strongly influenced by the high local pressures and the flash temperatures, which can be above 1000°C and occur at spots of asperity collisions.

5.1 Formation of thin microfilms on hard coatings

Very low coefficients of friction down to $\mu = 0.1$ have been reported for a hard titanium nitride coating sliding against itself (Mäkelä and Valli, 1985) and even lower values down to $\mu = 0.01$ and 0.05 have been measured for diamond-like hard carbon coatings sliding against silicon nitride (Miyoshi, 1990) and diamond coatings sliding against diamond (Hayward and Singer, 1990).

This can be explained by the formation of low shear strength microfilms on the hard coating or perhaps only on the asperity tips of the coating as shown in figure 5. Thus, if we are studying the contact on a micro scale, we have a soft coating on a hard substrate, although now the coating (e.g. diamond) is playing the role of a hard substrate and the soft microfilm formed is playing the role of a coating. It is obviously favourable if the substrate under the hard coating is hard to avoid fracture of the brittle coating by deformation, to improve the load support and to decrease the real area of contact.

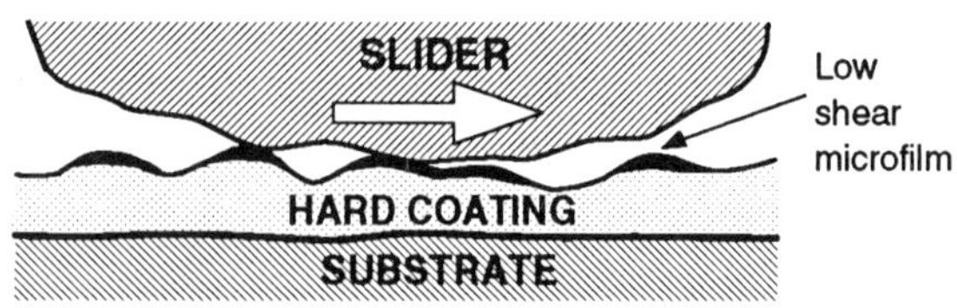

Fig. 5. The formation of a low shear microfilm on the top of a hard coating results in excellent tribological properties and extremely low friction.

The influence of microfilms on friction is obviously considerable in some cases (Holmberg, 1992a), although their influence on wear has not been widely studied. However, it can be assumed that decreased friction which results in decreased shear stresses close to the surface will have a favourable influence on the resistance to wear of the surface.

5.2 Oxidation of soft coatings

In environments including oxygen, such as air, a thin (about 1-10 nm thick) oxide layer is formed very quickly on most metal surfaces. Some oxide layers like copper oxide are sheared more easily than the metal, while others, like aluminium oxide, form a very hard thin layer. Improved lubricity by the oxidation of the top layer of the coating has been observed for lead coatings (Sherbiney & Halling, 1977) and for molybdenum disulphide coatings (Fayelle & al, 1990). Even if the oxide wear particles are hard their abrasive effect is not necessarily important because of their small dimensions compared with the roughnesses of the surfaces or their embedment into soft coatings, as shown in figures 3i and k.

6. MATERIAL TRANSFER

The micro- and macromechanical tribological mechanisms as well as the tribochemical mechanisms explain the phenomena controlling the resistance to motion, i.e. friction, and the liberation of surface material, i.e. wear, to produce wear products. When the material has been liberated from the surface it can still influence the tribological behaviour of the contact in two ways. Loose wear debris in the contact may influence wear as discussed above but the wear debris may also attach itself to the counterface to form a transfer layer on it which considerably changes the tribological properties of the counterface as shown in figure 6. It can be said that a new counterface is formed resulting in a new material pair.

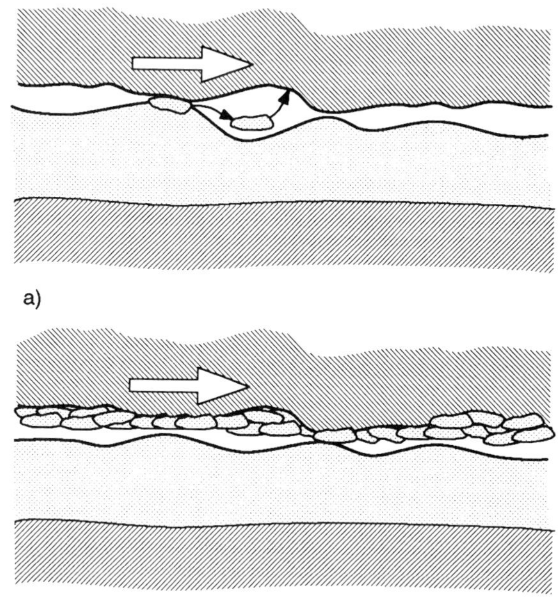

a)

b)

Fig. 6. Wear debris adheres to the counterface and produces a new sliding surface with new tribological properties.

Observations of the considerable influence of transfer films on the tribological behaviour of the contact have been reported e.g. for polytetrafluoroethylene and polyimide coatings (Yamada et al, 1990), titanium nitride coatings (Sue & Troue, 1990; and Ronkainen et al, 1990) and diamond-like hard carbon coatings (Hirvonen et al, 1990; and Ronkainen et al, 1992).

7. CONCLUSIONS

The systematic approach to analysing sliding contacts with coated surfaces based on a classification of the tribological changes into macromechanical, micromechanical and tribo-chemical changes and material transfer, and further to studying the macromechanical changes on the basis of the four main parameters: the hardness relationships, the film thickness, the surface roughnesses and the debris in the contact, is a useful tool for the systematic and reliable identification and evaluation of wear mechanisms and for systematic coating design and selection.

ACKNOWLEDGEMENTS

The authors would like to thank Mr. P. Andersson, Technical Research Centre of Finland, for helpful comments. The support generously provided by the Swedish Academy of Engineering Sciences in Finland and the Technology Development Centre of Finland as well as the support of the UK Science and Engineering Research Council for the work of the Research Centre in Surface Engineering is gratefully acknowledged.

REFERENCES

Argon, A.S., Mechanical properties of near-surface material in: friction and wear. In: Fundamentals of tribology, Suh, N.P. & SakaN.(eds.), MIT Press, London, 1980, 103-125.

Bennett, S., Matthews, A., Valli, J., Perry, A., Bull, S. & Sproul, W., Multipass scratch testing at sub-critical loads. Int. Conf. on Metallurgical Coatings and Thin Films ICMCTF-1992, San Diego 6-10.4.1992, 14 p.

Berthier, Y., Godet, M. & Brendle, M., Velocity accomodation in friction. Tribology Trans., 32 (1989) 4, 490-496.

Bull, S., Failure modes in scratch adhesion testing. Surface and Coatings Technology, 50 (1991) 25-32.

Burnett, P.J. & Rickerby, D.S., The mechanical properties of wear-resistant coatings I: Modelling of hardness behaviour. Thin Solid Films, 148 (1987a) 41-50.

Burnett, P. & Rickerby, D., The relationship between hardness and scratch adhesion. Thin Solid Films, 154 (1987b) 403-416.

El-Sherbiny, M. & Salem, F., A wear equation for solid lubricant films. 3rd Int. Conf. on Solid Lubricants, ASLE, Denver, August 1984, 44-49.

Erdemir, A. & Hochman, R.F., Surface metallurgical and tribological characteristics of TiN-coated bearing steels. Surface and Coatings Technology, 36 (1988) 755-763.

Fayeulle, S., Ehni, P.D., & Singer, I.L., Role of transfer films in wear of MoS_2 coatings. In: Mechanics of coatings. Dowson D. & al eds, Tribology series 17, Elsevier, Amsterdam, 1990, 129-138.

Fusaro, R.L., Mechanisms of lubrication and wear of a bonded solid-lubricant film. ASLE Trans., 24 (1981) 2, 191-204.

Hayward, I.P. & Singer, I.L., The tribological behaviour of diamond coatings. 2nd Int. Conf. on the New Diamond Science and Technology, Sept. 1990, Washington DC.

Hedenquist, P., Olsson, M., Jacobson, S. & Söderberg, S., Failure mode analysis of TiN-coated high speed steel: In situ scratch testing in the scanning electron microcope. Surface and Coatings Technology, 41 (1990a) 31-49.

Hills, D.A., Nowell, D. & Sackfield, A., A survey of cracks in layers propelled by contact loading. In: Mechanics of coatings. Dowson,D. & al (eds), Tribology series 17, Elsevier, Amsterdam, 1990, 203-208.

Hirvonen, J., Koskinen, J., Anttila, A., Lappalainen, R., Toivanen, R., Arminen, E. & Trkula, M., Characterization and unlubricated sliding of ion-beam-deposited hydrogen-free diamond-like carbon films. Wear, 141 (1990) 45-58.

Holmberg, K., A concept for friction mechanisms of coated surfaces. Surface and Coatings Technology, Accepted to be published, 1992, 10 p.

Holmberg, K., A suggestion for the classification of tribological mechanisms in coated surfaces. Finnish Journal of Tribology, 11 (1992b) 1, 1-17.

Holiday, P., Dehbi-Alaoui, A. & Matthews, A., A comparison of techniques for producing diamond-like carbon coatings. Materials Science Forum, 102-104 (1992) 643-654.

Miyoshi, K., Studies of mechanochemical interactions in the tribological behavior of marerials. Surface and Coatings Technology, 43/44 (1990) 799-812.

Mäkelä, U. & Valli, J., Tribological properties of PVD TiN and TiC coatings. Finnish Journal of Tribology, 4 (1985) 2, 74-85.

Posti, E. & Nieminen, I., Coating thickness effects on the life of titanium nitride PVD coated tools. Materials & Manufacturing Processes, 4 (1989) 2, 239-252.

Rickerby, D. & Burnett, P., The wear and erosion resistence of hard PVD coatings. 14th Int. Conf. on Metallurgical Coatings, San Diego March 1987, 29 p.

Roberts, E.W. & Price, W.B., In-vacuo, tribological properties of "high-rate" sputtered MoS_2 applied to metal and ceramic substrates. Mat. Res. Soc. Symp. Proc. Vol. 140 (1989) 251-264.

Ronkainen, H., Holmberg, K., Fancey, K., Matthews, A., Matthes, B. & Broszeit, E., Comparative tribological and adhesion studies of some titanium-based ceramic coatings. Surface and Coatings Technology, 43/44 (1990) 888-897.

Ronkainen, H., Likonen, J. & Koskinen, J., Tribological properties of hard carbon films produced by the pulsed vacuum arc discharge method. Int. Conf. on Metallurgical Coatings and Thin Films ICMCTF-1992, San Diego 6-10.4.1992, 20 p.

Sherbiney, M.A. & Halling, J., Friction and wear of ion-plated soft metallic films. Wear, 45 (1977) 211-220.

Sin, H., Saka, N. & Suh, N.P., Abrasive wear mechanisms and the grit size effect. Wear, 55 (1979) 163-190.

Sue, J.A. & Troue, H.H., Friction and wear properties of titanium nitride coating in sliding contact with AISI 01 steel. Surface and Coatings Technology, 43/44 (1990) 709-720.

Suh, N.P., The delamination theory of wear. Wear, 25 (1973) 111-124.

Suh, N.P., Tribophysics. Prentice-Hall, Englewood Cliffs, New Jersey, U.S.A., 1986. 489 p.

Tangena, A.G., Tribology of thin film systems. Doctoral thesis, Eindhoven Technical University, 1987. 130 p.

Yamada, Y., Tanaka, K. & Saito, K., Friction and damage of coatings formed by sputtering polytetrafluoroethylene and polyimide. Surface and Coatings Technology, 43/44 (1990) 618-628.

Thin Films in Tribology / D. Dowson et al. (Editors)
© 1993 Elsevier Science Publishers B.V. All rights reserved.

The performance of cam mechanism with follower coated by ceramics

S. M. Ghoneam*

Menoufia University, Faculty of Engineering, Dept. of Mech. Design and Prod. Eng., Shebin El-Kom, Egypt

Coating surfaces in cam-follower systems , and gears are designed to reduce the wear ,and especially to avoid scuffing and thus controlling the mode of the failure. The contact temperature is one of the main factors for determining the scuffing phenomena. Moreover, the temperature rise plays a significant role in the behaviour of the thermal stress crack propagation of the coating layer.

In the following paper an attempt is made to analyze the temperature rise in the region of contact of cam-follower surfaces coated with titanium carbide (TiC), and titanium nitride (TiN) ceramics as well as the temperature behaviour of the layer and interface between the layer and substrate material using the finite element technique.The effect of ceramic coating on the temperature distribution is demonstrated from the view-point of the thickness of the ceramic coating layer and surface temperature rise.

1. INTRODUCTION

The known type of wear may occur in cam-follower system, especially scuffing criteria due to high contact temperature and poor lubrication. Scuffing phenomena has been treated by many authors, from different points of view. For example, a review of scuffing emphasizing the importance of the influence of surface asperities on scuffing was done by Ludema [1]. The effect of surface roughness and surface texture on scuffing was investigated and analyzed by Staph et al. [2].While the effect of some materials and operating variables on scuffing was studied by Carper et al. [3]. Experimental analysis concerned with the influence of surface roughness texture on temperature and scuffing in sliding contact was carried out by Wolff [4].While the optimum design of cam system avoiding the scuffing phenomena by eliminating the poor lubrication and high contact temperature was carried out by Ghoneam [5].

But still there are questions concerning, how to improve the wear resistant surfaces of cams and their followers and the possibility of avoiding the scuffing phenomena. Beese and Clarke [6] suggested that the wear resistance of cam and follower surfaces lies with the field of surface coating. Also, many investigators concluded that, coating the surface with ceramic materials provides significant improvement in the tribological behaviour of bearing and machine elements, such as gear system [7,8].

In the field of coating techniques, the thickness of the coating layer is an important factor. The layer should be sufficiently thick to avoid the crack of the layer, and also to avoid fatigue development in the substrate material. The costs of coating should be also taking into consideration. Many surface coating techniques are used. Electrode position method was suggested by Kepple [9]. This method has the additional advantage of being able to use a lower grade steel for the back-up material. The thermal spraying, could be used to deposit a metal coating on cam or follower surfaces [10]. Alternatively, a ceramic coating, could be applied, if brittleness problems are overcome, especially at high temperature operation [11]. Tribology of ceramics were investigated by Kato [12]. The study showed the advantage of ceramic bearings in industrial machines, the main conclusion was that ceramics have a very high potential as a good tribological material. It is clear that, the dominant requirement in the operation of cam systems is to reduce the probability of premature surface failure. An important feature in

* Visiting scholar at Warsaw University of Technology, Institute of Aeronautics and Applied Mechanics, Nowowiejska 22/24, 00-665 Warsaw, Poland.

The follower in contact with the flank:

$$y = (\rho_c - R_2)\,(1 - cos\theta)$$
$$\dot{y} = \omega\,(\rho_c - R_2)\,sin\theta$$
$$\ddot{y} = \omega^2\,(\rho_c - R_2)\,cos\theta$$

The follower in contact with the nose:

$$y = d\,cos(\phi - \theta) + (R_1 - R_2)$$
$$\dot{y} = \omega\,d\,sin(\phi - \theta)$$
$$\ddot{y} = -\omega^2\,d\,cos(\phi - \theta)$$
$$R_c = R_2 + y + \ddot{y}/\omega^2$$
$$V_c = R_c\omega,\quad V_f = \ddot{y}/\omega,\quad V_s = |V_c - V_f|$$
$$\rho_c = 0.0794\text{ m},\; R_1 = 0.0138\text{ m},\; R_2 = 0.0254\text{ m}$$
$$d = 0.037\text{ m},\; b = 0.01\text{ m},\; \omega = 104.72\text{ rad/sec}$$
$$K_f = 7E4\text{ N/m},\; m_f = 2.3\text{ kg}.$$

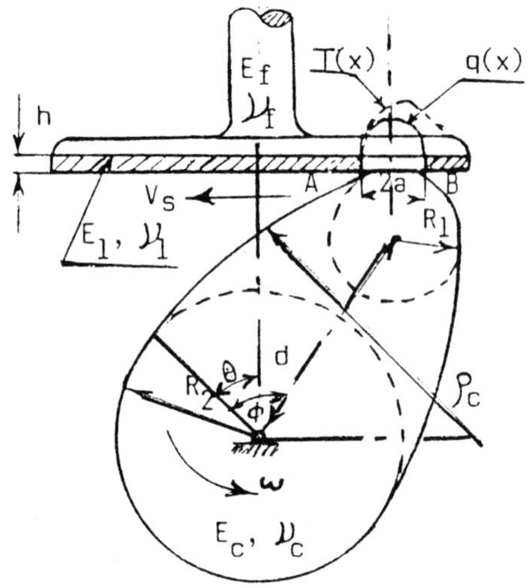

Figure 1. Geometrical model and relations between cam and follower system.

the consideration of how to reduce wear and scuffing is the recognition that the mechanism of coating is essential and that more , analysis are required in this area , such as the temperature performance through the coating layer and substrate materials.

The main purpose of this work is to investigate the role of the coating surface on the contact temperature. Analyze the temperature distribution in the substrate material especially in the region of interface between the coating layer and substrate material was done using the finite element technique . A comparison between the scuffing temperature for uncoated and coated surfaces is presented . To realize the main goal of the present work , the region of contact surface is divided into triangular elements, then, the Laplace's equation is applied. The finite model of the entire surface is derived in the matrix form .The applied boundary conditions based on the contact temperature are calculated according to Blok's theory. A computer program is developed to solve the latter matrix and, then, map the results.

2. ANALYTICAL PROCEDURE

2.1. Finite element contact model

The finite element analysis is presented here to investigate the temperature distribution in the cam follower region of contact, the follower is coated by TiC or TiN ceramics. The nose position for a circular arc cam with a flat - faced follower is selected as shown in figure 1. For this case the maximum Hertz stress as well as the maximum relative velocity exist. Consequently from the theoretical point of view the wear rate may occur. The region of contact between the cam and its follower can be modeled with the finite element mesh shown in the figure 2.

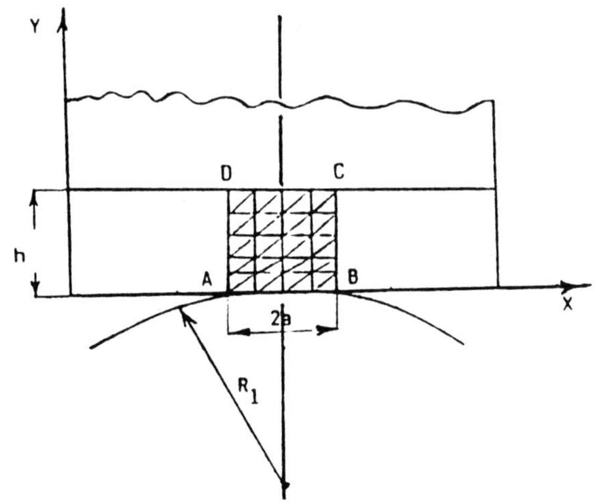

Figure 2. Finite element model.

The follower surface at the region of contact is divided into triangular elements with defined nodes

Table 1
Mechanical and thermal properties for used materials

	Materials		
	Steel	TiN	TiC
Young's modulus (N/m^2)	2.07E11	2.6E11	4.83E11
Poisson's ratio	0.3	0.3	0.3
Thermal conductivity (W/mK)	46.0	20.0	37.0
Specific heat (J/KgK)	544.0	700.0	680.0
Density (Kg/m^3)	7800.0	5440.0	4920.0
Contact conductivity ($Wsec^{1/2}/m^2K$)	13970.94	8726.96	11125.96
Thermal diffusivity (m^2/sec)	12.3E-6	11.7E-6	5.8E-6
Max.Operating temp. (^0C)	427.0	1482.0	1482.0

and coordinates, then Laplace's equation is applied [13]. The finite element model of the entire region is derived in the matrix form with the help of both the known number of nodes and coordinates.

To predict the contact temperature distribution in the interface region between the layer and substrate, a refinement of the mesh near the expected contact region is necessary. The horizontal and vertical dimension of the mesh, **AB** and **BC**, were set equal to the width of contact and the thickness of the layer respectively. The mesh is refined at the center of contact. The closest spacing of nodes is, therefore, immediately beneath the contact surface and the **X**-and **Y** dimensions of the corresponding element are 1 and 0.5 μm respectively.

2.2. Properties of layer and cam–follower material

The material for both the cam and follower are assumed to be the same .The elastic modulus and Poisson's ratio for the layer; E_l, ν_l; and cam–follower; $E_f=E_c$, $\nu_f=\nu_c$; selected in this study are given in Table 1 [14,15]. These elastic properties are appropriate for traditional metallic substrates and wear resistant ceramic coating for machine elements. The friction coefficient values 0.1 and 0.5 were assigned to the properties of the interface elements. These values are representative of lubricated and dry interfaces conditions [16].

2.3. Boundary conditions

2.3.1. The contact temperature concept

For the cases of line contact, Blok's theory is widely used for calculation of the maximum contact temperature [17,18]. According to this theory the maximum contact temperature for the cam follower system can be reformulated as

$$T_{\max} = \frac{AP_m\mu|V_c - V_f|\sqrt{2a}}{\alpha_c\sqrt{V_c} + \alpha_f\sqrt{V_f}} + T_b \tag{1}$$

or

$$T_{\max} = T_f + T_b$$

The first part in equation 1 is called the flash temperature, T_f, and the second part is the average bulk temperature in the cam and follower region, T_b, where:

A is a constant which depends upon the heat flow distribution type,

a is the half width of contact (m),

V_c and V_f are the velocity of cam and follower respectively (m/sec),
α_c and α_f are the contact conductivity for cam and follower respectively i.e. $\alpha = \sqrt{\lambda \rho c}$,
λ is the thermal conductivity (W/m K),
c is the specific heat (J/kg K), and,
ρ is the density (kg/m^3).

For the coating surfaces, the actual pressure, P_m, at the center of contact zone and half width of contact, a, are functions of the properties of the substrate material, coating material, and layer thickness. At first the Hertzian pressure, P_0, and Hertzian half width of contact, a_0, are calculated respectively as follows [19,20]:

$$P_0 = \left(\frac{E_l}{(1-\nu_l^2)} \frac{W}{(1+\beta)\pi R_c b} \right)^{1/2}, \tag{2}$$

and

$$a_0 = 2\left(\frac{W R_c (1+\beta)}{\pi b} \frac{(1-\nu_l^2)}{E_l} \right)^{1/2} \tag{3}$$

where

$$\beta = \frac{(1-\nu_c^2)E_l}{(1-\nu_l^2)E_c},$$

W is the normal load between cam and follower (N), and it is given by: $W = m_f \ddot{y} + K_f y + F_0$, b is the width of cam disc (m), and R_c is the effective radius of cam curvature (m).

Then the actual pressure, and half width of contact are obtained from the diagrams and the tables given in [19] as a function of P_0, γ , and α_0 and a_0, γ and α_0 respectively. The values of γ and α_0 are given respectively as:

$$\gamma = \frac{(1-\nu_f^2)E_l}{(1-\nu_l^2)E_f}, \quad \text{and} \quad \alpha_0 = \frac{a_0}{h}$$

where h is the thickness of the coating layer.

2.3.2. Temp. distribution through the width of contact:

To determine the temp. distribution through the width of contact a numerical method is used to solve the general heat conductance differential equation [18]. To simplify this equation, the displacement speed is assumed sufficiently high to allow heat transfer especially in the normal direction of the surface. Also the heat source is assumed to be one dimensional. The equation for this condition becomes:

$$\frac{\partial^2 T}{\partial x^2} = \frac{\rho\ c}{\lambda} \frac{\partial T}{\partial t} \tag{4}$$

To solve the equation (4), the explicit method using the following relations, is applied [21]:

$$\left. \frac{\partial^2 T}{\partial x^2} \right|_{x=x_i, t=t_i} = \frac{T_{i+1}^j - 2T_i^j + T_{i-1}^j}{(\Delta x)^2} + o(\Delta x)^2 \tag{5}$$

and

$$\left. \frac{\partial T}{\partial t} \right|_{x=x_i, t=t_i} = \frac{T_i^{j+1} - T_i^j}{\Delta t} + o(\Delta t) \tag{6}$$

Subscripts are used to denote position, and superscripts are used for the time. Substituting equations (5) and (6) into (4) and solving for T_i^{j+1} gives the equation for the forward difference method as:

$$T_i^{j+1} = r(T_{i+1}^j + T_{i-1}^j) + (1-2r)T_i^j \tag{7}$$

where

$$r = \frac{\lambda}{\rho\ c} \left[\frac{\Delta\ t}{(\Delta\ x)^2} \right]$$

A computer program is used to solve equation (7) with the help of the end conditions temperature.

3. RESULT AND DISCUSSION

In this section, the contact temperature and temperature distribution through the coating layers of the follower surface are presented in terms of position angle, θ, layer thickness, h, and width of contact, $2a$. To examine the significance of the thickness of the layer and ceramic types in contact temperature, Blok's theory is used. But in the analysis of the effects of the above mentioned parameters on the temperature rise at the interface, the finite element technique is applied. A comparison between the various coatings with layers of different thickness and uncoated follower surfaces are presented, the combination of the follower coating is given in Table 2. In the present analysis the surface roughness is not taken into account, and the bulk temperature is assumed to be 300°C.

Table 2
Coating treatment of cam follower

Cam follower system	follower	cam
U	uncoated	uncoated
NU	TiN	uncoated
CU	TiC	uncoated

Figure 3 shows the flash temperature calculated according to Blok's theory as a function of the position cam angle; increment by one degree; during the rise period. The follower is coated by TiN ceramic with thickness of layer values 5, 50, 100 μm. In this figure there is a comparison between the cam-follower system NU, and U. The contact flash temperature increases with an increase of the thickness of coating layers. The temperature of the coated surface is higher than the un-coated depending on the thickness of layer. This is due to poor contact conductivity of the TiN ceramic with respect to the steel. Also there is significant influence of ceramic type on the temperature rise, this is clear in figure 4. A comparison between the NU, CU, and U coated treatment is clear, the thickness of layer is 10 μm. It can be seen that the flash temperature in NU system is higher than CU and U respectively, this is due to grade of contact conductivity of the TiN, TiC, and steel, respectively. In the above mentioned two figures, the maximum flash temperature occurs at the nose position for high contact pressure and sliding velocity. Therefore, this is region taken into account in the finite element analysis. As mentioned, the bulk temperature for all coating treatments are the same, the flash temperature can be considered an index for the critical surface temperature at the moment of scuffing. Moreover, taken into

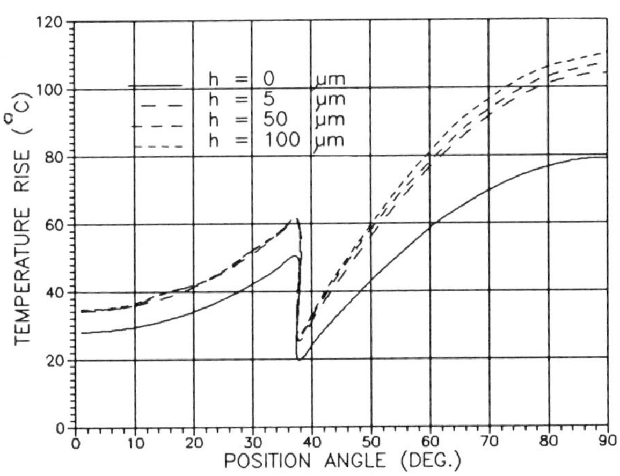

Figure 3. The flash temperature for different film thickness of coating layer (TiN).

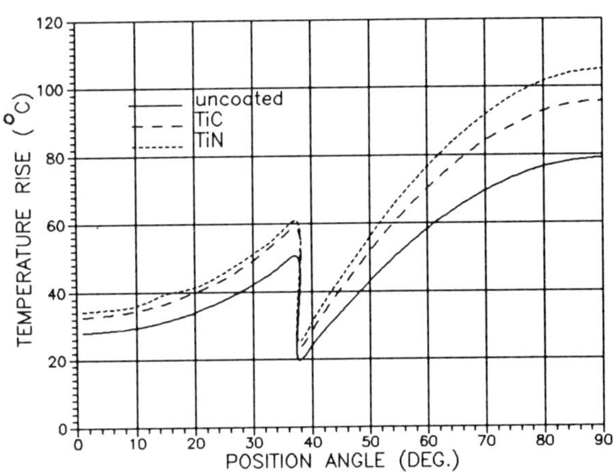

Figure 4. The flash temperature for different type of ceramics coatings at $h = 10\mu$m.

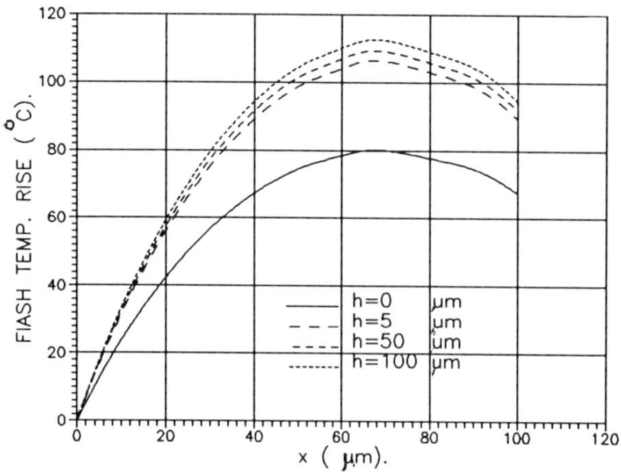

Figure 5. Flash temperature at different position along the width of contact for different thickness of coating layers at $\theta = 90^0$.

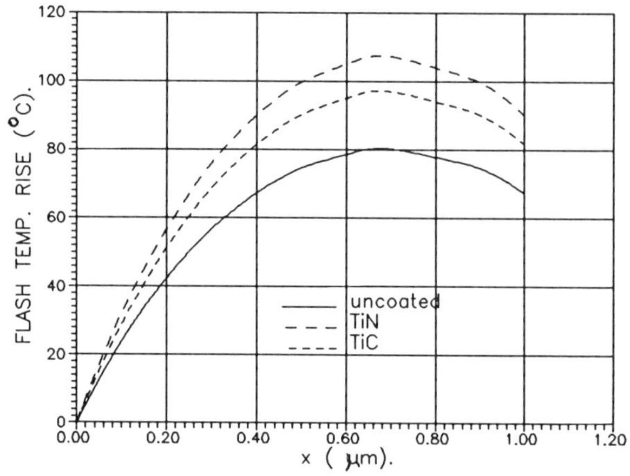

Figure 6. Flash temperature at different position along the width of contact for different ceramics coating at $h = 10\mu$m and $\theta = 90^0$.

account the low maximum operating temperature for the steel with respect to the ceramics materials used; Table 1; it can be conclude that the scuffing phenomena occurs for the case of cam-follower system U rather than cases NU, and CU.

The flash temperature distribution along the width of contact between the cam and its follower at $\theta = 90°$ are shown in figures 5 and 6 for the case of U, and NU. The thickness of layer are 5, 10, 50 and 100 μm, for the cases U, NU, and h=10 μm for CU, respectively. These results are obtained by using equation 7. The maximum temperature occurs between the middle of width of contact and the trailing edge B, approximately at 0.65 of the width of contact. The temperature at the trailing edge is higher than that at the leading edge A.

The temperature distribution for various thickness of layers are shown in figures 7 to 12. The boundary conditions as a function of the thickness of layer and type of ceramic coating are taken from figures 5 and 6, respectively. The diagrams of the temperature distribution are magnified in order make them readily visible.

The effect of TiN ceramic thickness layer on

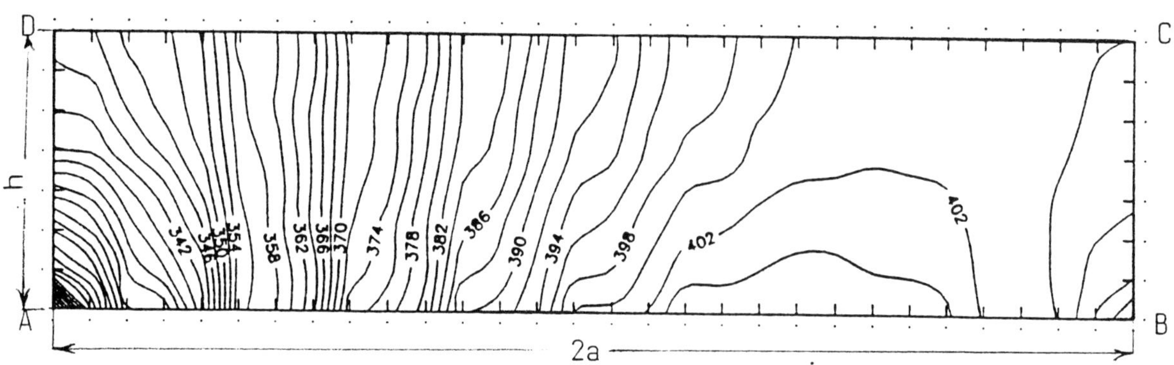

Figure 7. Temperature distribution in coating layer (TiN at $h/a = 0.5$).

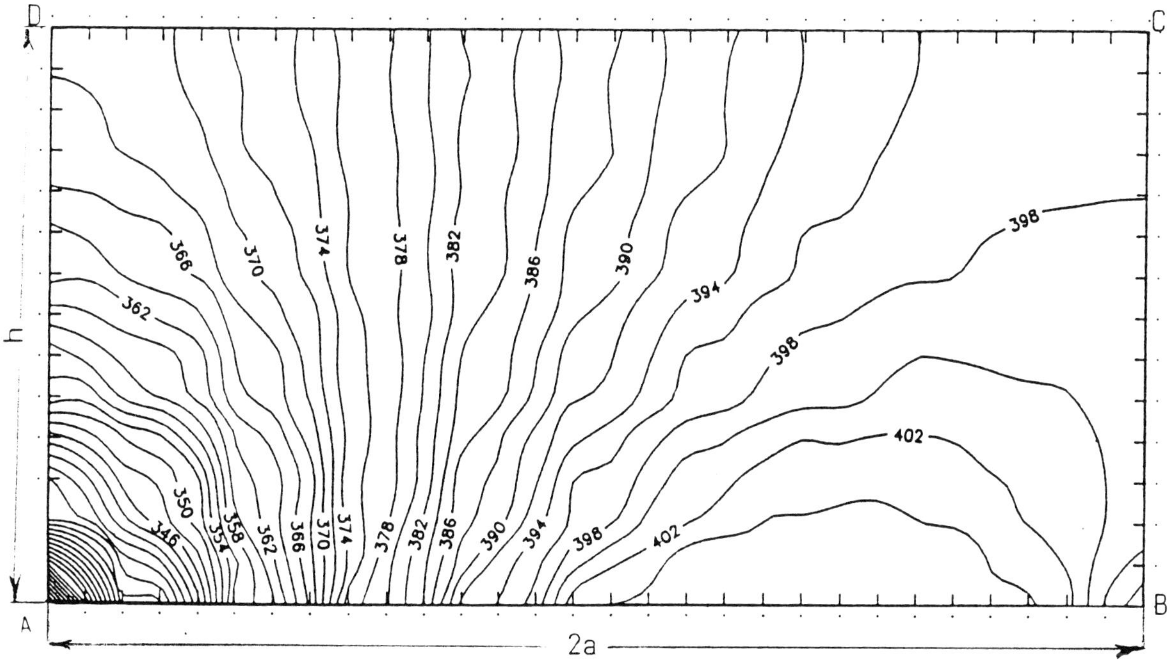

Figure 8. Temperature distribution in coating layer (TiN at $h/a = 1.0$).

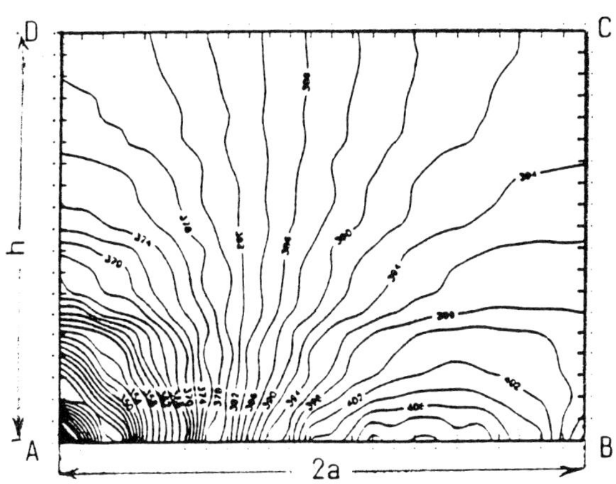

Figure 9. Temperature distribution in coating layer (TiN at $h/a = 1.5$).

temperature distribution of coating followers is examined in figures 7 to 11. A comparison between figure 7 and 11 shows that there is no large difference in the temperature in contact and interface for thin layer of coating $h = 0.1a$. Also the same phenomena exists in the case when the follower

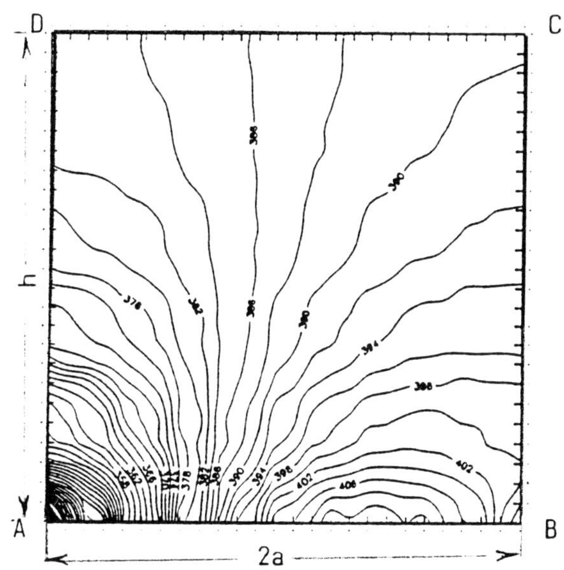

Figure 10. Temperature distribution in coating layer (TiN at $h/a = 2$).

is coated by TiC for the same thickness of layer; figure 12. The variation between the temperature

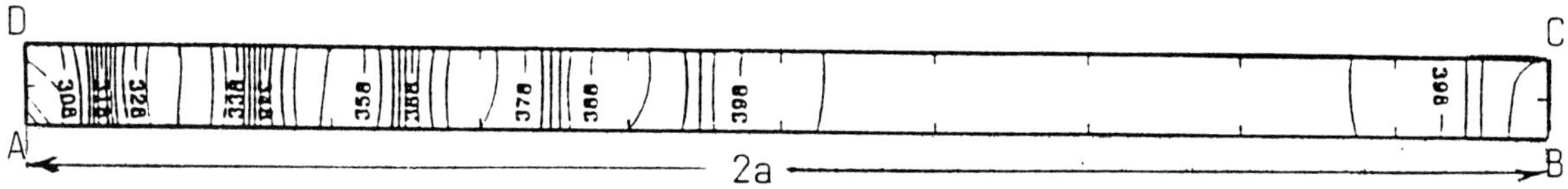

Figure 11. Temperature distribution in coating layer (TiN at $h/a = 0.1$).

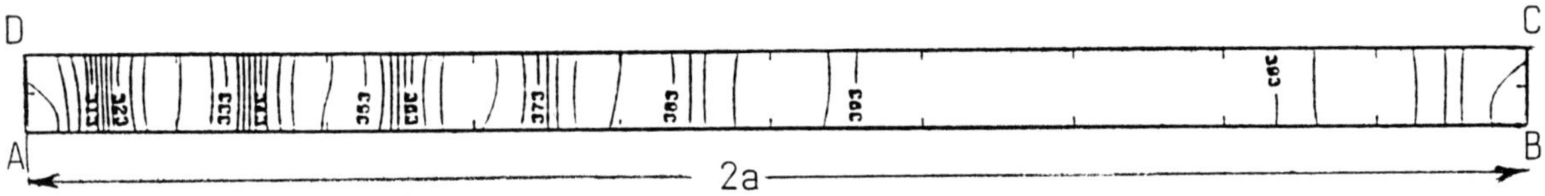

Figure 12. Temperature distribution in coating layer (TiC at $h/a = 0.1$).

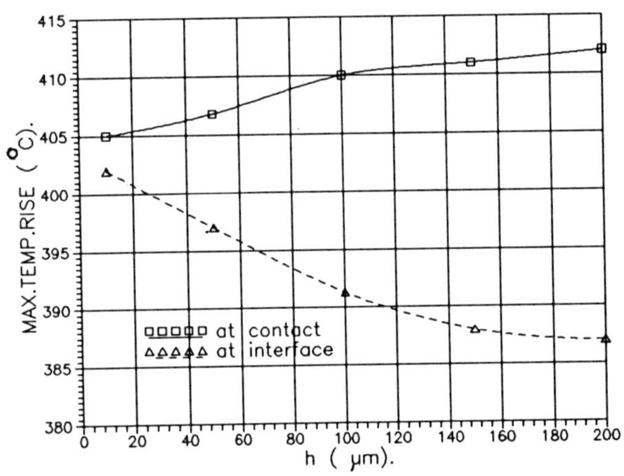

Figure 13. Comparison of max. temperature rise at interface and contact.

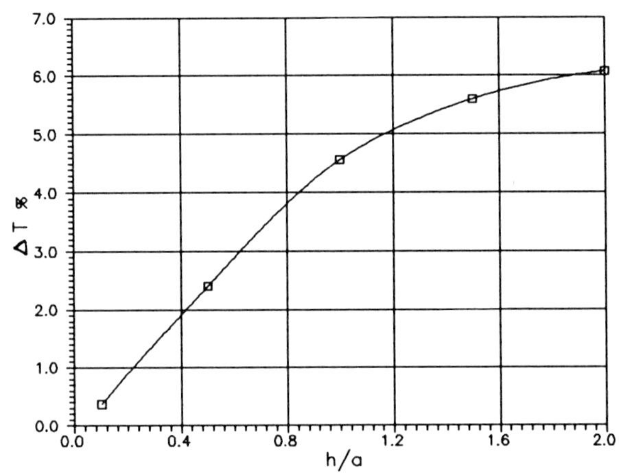

Figure 14. The deviation in max. temperature between contact and interface.

at contact and corresponding once at interface is 0.25 % for the TiC, and 0.37 % for the TiN ceramic. On the contrary, for thick layers of coating values $0.5a$, a, $1.5a$, and $2a$, as shown in figures 7 to 10 respectively, the thickness of layer increase the temperatures of the contact. The increase of thickness of the layer decreases the temperature at the interface, since the ceramic layer reduces the magnitude of heat transfer to the interface. This is due to low thermal conductivities of ceramics. Also, the thickness of the layer has effect on the location of the maximum temperature. The rate of temperature decrement, ΔT, between contact and interface is 6.5% at h/a=2 , 5.6% at h/a=1.5, 4.56% at h/a=1, and 2.4% at h/a=0.5. This is clear in figure 13 and 14. In general it can be noted that the temperature in the interface is higher at the trailing edge B than at the leading edge A.

It is clear that the thickness of layer plays an important role in determining the temperature distribution and its values at the interface. But it must be taken into consideration that the thickness of layer affects on the output deviation error. And costs of ceramic materials and coating technique must be considered. Moreover, the ceramic materials are susceptible to thermal mechanical instability, consequently the thermal stress crack may be exists. This points required more investigation and analysis.

4. CONCLUSION

The temperature distribution of the follower surface with and without a layer of ceramic in contact was investigated using the finite element technique.The major conclusions of this analysis are summarized as follows:

1. The layer thickness, and mechanical and thermal properties of both the substrate and coating material have significant affect for the temperature rise.
2. The finite element analysis has verified the opinion that the temperature at the interface between the coating layer and substrate material depends on the layer thickness and both the mechanical and thermal properties of the coating layer.
3. Coating layers such as TiN and TiC ceramics can improve the wear resistance and avoid severe damage due to scuffing of cam–follower surfaces.
4. Many aspects of the behaviour in cam–follower system such as the influence of ceramic coating on temperature and scuffing, and thermal stress and crack propagation in coating due to thermal-mechanical instability in ceramics are the subject of current research of author program.

Acknowledgement

The author would like to express his thanks to prof. J. Stupnicki of Warsaw University of Technology for his valuable suggestions and advice.

REFERENCES

1. K.C.Ludema, Wear, 100 (1984) 315.
2. H.Staph, et al., J.Mech. Mach. Theory, 8, (1973) 197.
3. H.J.Carper, et al., J. Mech. Mach. Theory, 8 (1973) 209.
4. R. Wolff, Wear, 143 (1991) 99.
5. S.M.Ghoneam, Eng.Research Bull., Faculty of Eng., Menoufia Univ., Vol XI ,Part I (1989) 123.
6. J.G.Beese and H.Clarke, Conf. on Cam Mechanisms, Inst. of Mech Eng., (1978) 95.
7. H.E.Hintermann, Wear, 100 (1984) 381.
8. Y.Terauchi, et al., J. of Tribology International, Vol 20, No 5 (1987) 248.
9. R.K. Kepple, et al., J. Lubrication Technology, 92 (1970) 557.
10. R. F. Smart, Spring Conf. of Inst. of Metall, Majorca (1974).
11. K. T. Scott, Spring Conf. of Inst. of Metall, Majorca (1974).
12. K. Kato , Proc., 5-th Inter. Congr. on Tribology, Vol 1, Helsinki (1989) 75.
13. D.H.Norrie and G. de Vries, An Introduction to Finite Element Analysis, Academic Press (1978).
14. Y. Terauchi, et al., J. Trib. Inter., Vol. 20, No 5 (1987) 248.
15. William A. Glaeser, Materials for Tribology, Tribology Series, 20, Elsevier (1992).
16. K.Komvopoulos, Trans, ASME, J. of Trib., Vol.111 (1989) 430.
17. H.Blok, Wear, 6(1963) 430.
18. I.Iliuc, Tribology of Thin Layers, Tribology Series, 4, Elsevier (1980).
19. P.K.Gupta and J. A. Walowit, ASME, J of Lubr. Tech. (1974) 250.
20. J.A. Walowit and J.N.Anno, Modern Developments in Lubrication Mechanics, App. Science Publishers LTD (1975).
21. C.F. Gerald, Applied Numerical Analysis, second edition, Addison-Wesley (1978).

Thin Films in Tribology / D. Dowson et al. (Editors)
© 1993 Elsevier Science Publishers B.V. All rights reserved.

The Local Yield Map of Hard Coating under Sliding Contact

D.F.Diao*, K.Kato* and K.Hayashi**

* Tribology Laboratory, School of Mechanical Engineering, Tohoku University, Sendai, Japan.

** Institute of Fluid Science, Tohoku University, Sendai, Japan.

The stress distribution in the hard coating and substrate under elastic sliding contact is analyzed with the finite element method. The elliptical distribution of normal contact pressure is supposed and the combinations of coating thickness, friction coefficient and elastic modulus of coating and substrate are changed in wide range. The position of yield initiation is found with the calculated result and the local yield map is introduced in relation to the ratio of the coating yield strength to substrate yield strength and the ratio of the coating thickness to the half-contact width. The local yield map shows that the yield at the coating interface is the most general case in wide range of contact condition and the yield at the coating surface becomes the next general case as the friction coefficient becomes large.

1 Introduction

The hard coating of ceramic or diamond is becoming popurlar in tribological elements to reduce friction and wear[1−2]. Especially wear of substrate is reduced drastically by applying hard coating. Trouble is the spalling or delamination of coating under critical contact condition where crack initiation and its propagation is the dominating mechamsm[3]. In many cases, micro-cracks exist at the surface and coating-substrate interface, and large cracks are propagated from them. Therefore, the critical contact condition for the propagation of a crack becomes the main concern in the design of coating. The first approach for this purpose is to know the critical contact condition for the local yield initiation through the stress distribution in the contact region.

Well established elastic contact theory[4] for continuous elastic solid can not be simply applied for the coated elastic solid. Some attempts have been made to get the contact pressure profile in the case ot cylindrical contact of coated elastic solids[5−11], and the effect of contact pressure profile on the stress distributions in the coating and the coating-substrate interface was analyzed in detail[12]. The indentation size, contact pressure, applied load and amount of plastic deformation in elastic-plastic indentation of a coated solid were obtained by Kennedy and Ling with finite element method[13]. Although mathematical analysis is possible in these cases, analytical expressions are complicated and numerical techniques are generally cumbersome. It has been shown, on the other hand, that the finite element method gives the numerical solution more easily. Komvopolus solved the elastic contact problem of a coated semi-infinite solid compressed by a rigid surface with the finite element method numerically[14]. He also solved the elastic-plastic contact problem in the similar situation, and showed the location, shape and growth of the plastic zone in relation to the critical parameters such as coating thickness and indentation depth[15].

This plastic zone is the source for crack initiation. The usefulness of finite element method is also shown for the analysis of contact of an elastic-plastic two layer half-space[16].

Mathematical stress analysis of sliding contact problem of a coated solid against a cylinder has been studied more recently. King and Sullivan[17] soloved this problem theoretically in 1986 and showed that the equivalent von Mises stress was maximum at the surface when friction coefficient becomes large. Leroy and Villechaiso [18] solved the similar problem by Fourier transform techniques in 1989 and showed the variation of maximum equivalent stress of von Mises under the different contact conditions.

In this paper, therefore, the finite element method is used to obtain stress distributions in the sliding contact region of coated substrste under various different contact conditions and the local yield position is decided with the yield criterion of von Mises. The local yield maps are finally introduced in relation to elastic modulus, yield strength, contact width, coating thickness and friction coefficient.

2 Finite Element Model in Contact

Figure 1 shows the model of externally subjected contact stresses at the contact width 2a and the finite element mesh for the half-space medium. The normal contact stress $P(x)$ and the tangentical contact stress $F(x)$ are given by the following equations;

$$P(x) = P_{max}\sqrt{1-(x/a)^2} \qquad (1)$$

$$F(x) = \mu \cdot P_{max}\sqrt{1-(x/a)^2} \qquad (2)$$

where P_{max} is the maximum contact stress and μ the friction coefficient at the contact interface. x axis is taken on the surface of half-space in the direction of sliding and $x = 0$ at the center of contact. Coating thickness in the half-space is given by t.

An array of 750, 8-node, isoparametric quadrilateral elements was introduced in the contact half-space and the refinement of the mesh near the exact contact region was established. The horizontal and vertical spacing of the nodes at and directly below this contact region were $0.08\mu m$ and $0.16\mu m$ respectively. In order to check the resolution

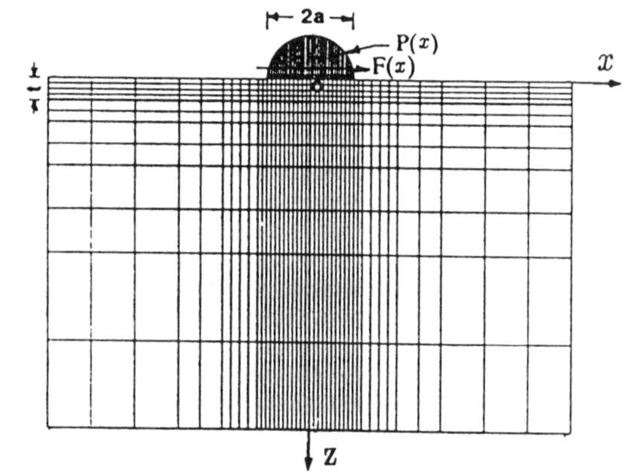

Figure 1 : Finite element mesh for the half-space medium.

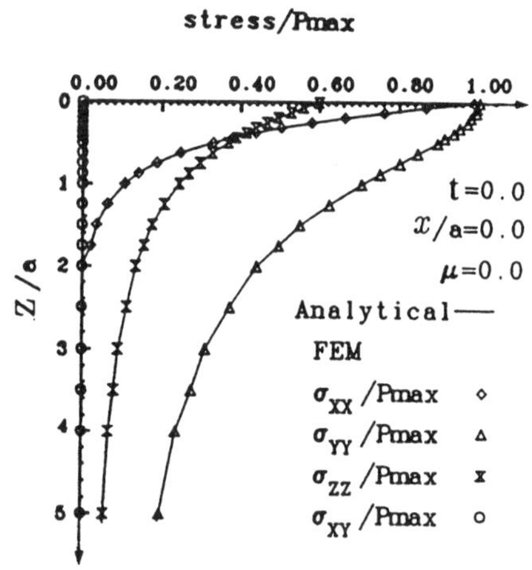

Figure 2 : Analytical and finite element solutions of the elastic stress in a homogeneous half-space medium due to elliptically distributed normal tractions at the surface ($\mu = 0.0, x/a = 0.0$).

and accuracy of this finite sized mesh, the stress values were calculated with FEM and an analytical elastic theory of Smith et al[19] for the simplified model where friction coefficient μ and coating thickness t were taken as zero. Fig.2 shows the calculated stress components of σ_{xx}, σ_{yy}, σ_{zz} and σ_{xy} where ◇, △, ⋈, ○ were obtained with FEM and solid lines with analytical elastic theory. It is obvious from these results that stress values by both methods agree with each other very well. It means that the finite element mesh given in Fig.1 is suitable to give reliable values of stress in the contact zone. After such confirmation about the effectiveness of designed finite element mesh, equivalent stress σ_{vm} of von Mises in the contact zone was calculated with FEM for case of different values of μ, t/a and E_f/E_b, where σ_{vm} was defined as follows,

$$\sigma_{vm} = [\frac{1}{2}\{(\sigma_1-\sigma_2)^2+(\sigma_2-\sigma_3)^2+(\sigma_3-\sigma_1)^2\}]^{1/2} \tag{3}$$

σ_1, σ_2, σ_3 : principal stresses.

3 Calculated Results

3.1 Distribution of equivalent stress of von Mises σ_{vm} /maximum contact pressure P_{max}

Calculated results at x=0 are shown in Fig.3(a),(b),(c),(d),(e) and (f) as relationships between σ_{vm}/P_{max} and z/a, where elastic modulus of coating E_f/elastic modulus of substrate E_b was kept constant at two. t/a was changed from 0.125 to 4.0 and μ was changed from 0.25 to 0.70. The interface between coating and substrate exists where

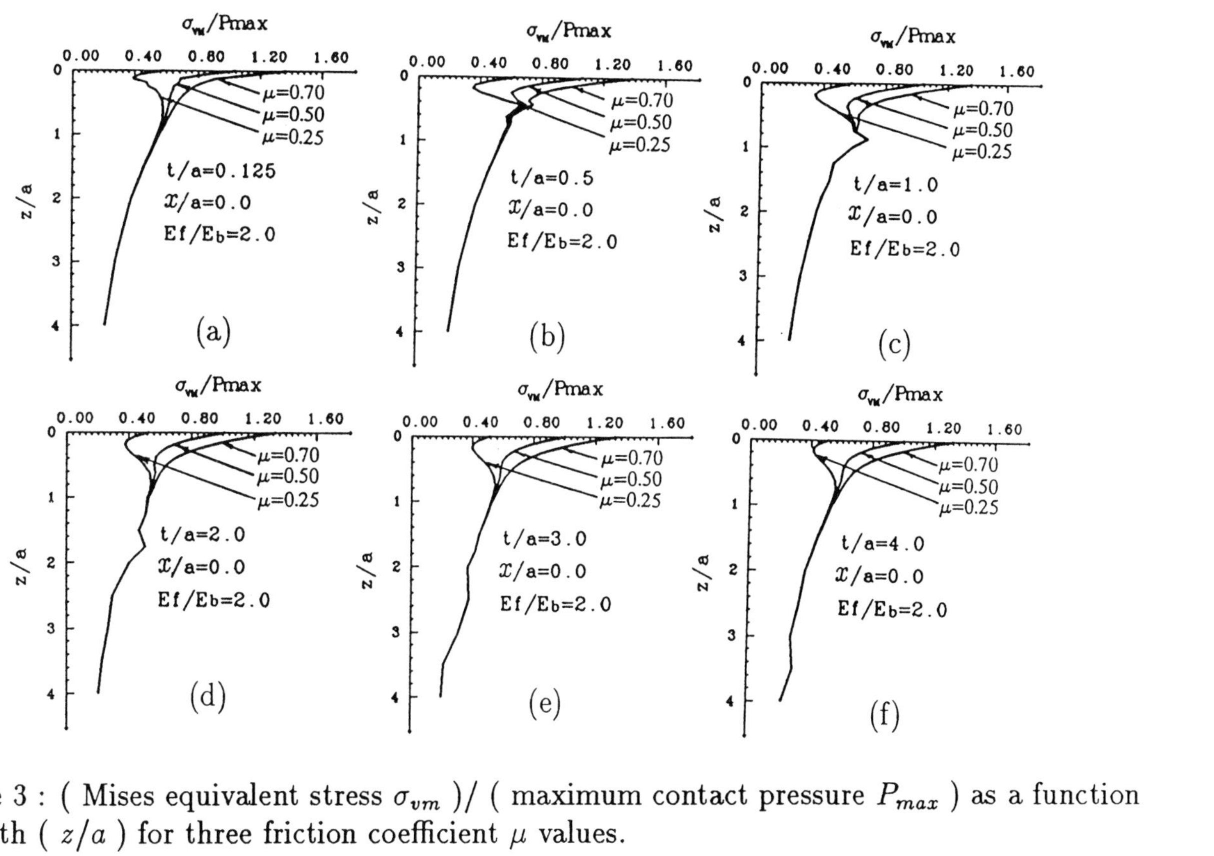

Figure 3 : (Mises equivalent stress σ_{vm})/ (maximum contact pressure P_{max}) as a function of depth (z/a) for three friction coefficient μ values.

$z/a = t/a$. It is generally observed that σ_{vm}/P_{max} has its maximum value at the position below the surface when friction coefficient is small ($\mu = 0.25$) but at the surface when friction coefficient is large($\mu = 0.70$). It is also observed from Fig.3 that there exists a position below which the change of friction coefficient does not cause big difference in values of σ_{vm}/P_{max} and it gradually approaches to zero.

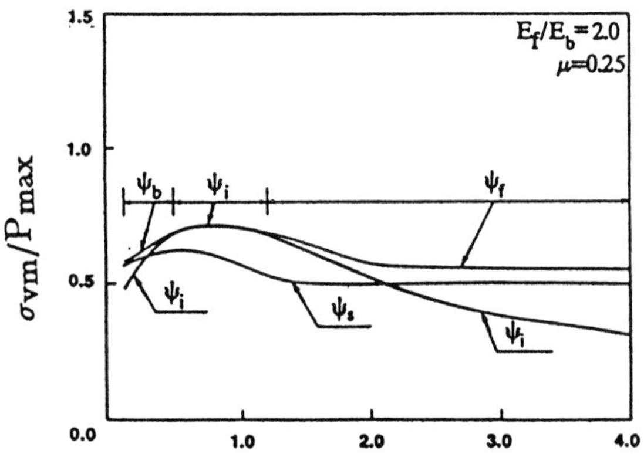

(a)

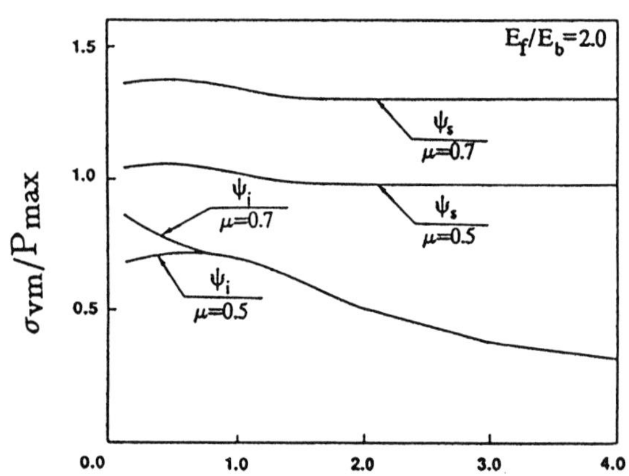

(b)

Figure 4 : Relationship between σ_{vm}/P_{max} and t/a for three friction coefficient μ values.

3.2 The maximum values of σ_{vm}/P_{max} in the coating and the substrate, and the values of σ_{vm}/P_{max} at the coating surface and the interface

From the results of Fig.3, the maximum values of σ_{vm}/P_{max} at the coating and the substrate, and values of σ_{vm}/P_{max} at the coating surface and the interface are obtained and shown Fig.4(a) and (b) as functions of t/a. The meanings of notations in the figure are as follows;

ψ_s : σ_{vm}/P_{max} value at the surface,

ψ_f : the maximum value of σ_{vm}/P_{max} in the coating,

ψ_i : σ_{vm}/P_{max} value at the interface between the coating and the substrate

ψ_b : the maximum value of σ_{vm}/P_{max} in the substrate.

When friction coefficient is small, ψ_s, ψ_f, ψ_i and ψ_b change complicatelly depending on the value of t/a as shown in Fig4(a). But when friction coefficient becomes large, the maximum value does not appear anymore in the coating or in the substrate as shown in Fig.4(b). It means that the value of σ_{vm}/P_{max} at the surface is always larger than that in the coating and the value of σ_{vm}/P_{max} at the interface is always larger than that in the substrate.

3.3 Local Yiled Map of coating and substrate

It is well established that metals follow the yield criterion of von Mises which is given by σ_{vm} =yield strength observed in tension or compression test[20-21]. If we suppose that the yield of ceramics also follows the same criterion, yield at the surface, in the coating

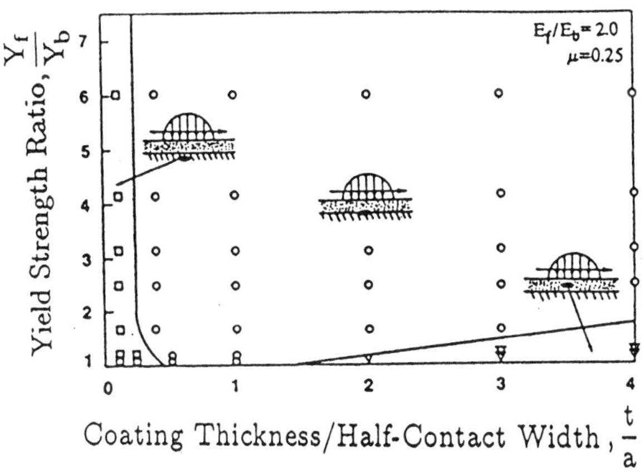

Figure 5 : Local yiled map as a function of the ratio of coating yield strength to the substrate yield strength and the ratio of coating thickness to half-contact width at $\mu = 0.25$.

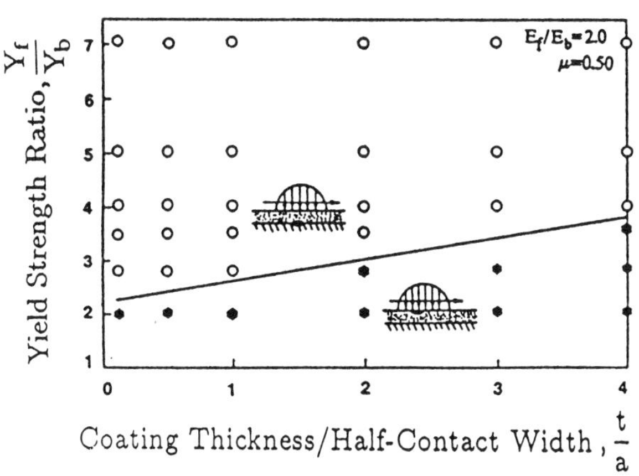

Figure 6 : Local yiled map as a function of the ratio of coating yield strength to the substrate yield strength and the ratio of coating thickness to half-contact width at $\mu = 0.50$.

and in the substrate can occur when following equation are satisfied respectively;

$$\psi_s P_{max} = Y_f \tag{4}$$

$$\psi_f P_{max} = Y_f \tag{5}$$

$$\psi_b P_{max} = Y_b \tag{6}$$

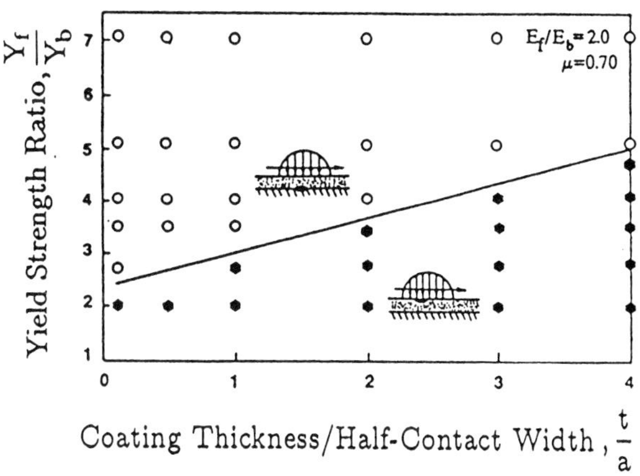

Figure 7 : Local yiled map as a function of the ratio of coating yield strength to the substrate yield strength and the ratio of coating thickness to half-contact width at $\mu = 0.70$.

where Y_f and Y_b are yield strength of materials of coating and substrate respectively. As the yield criterion at the interface between coating and substrate is not known, we suppose that the yield at the interface occurs when σ_{vm} at the interface reaches the smaller yield strength between coating and substrate. In this paper, only the case of $Y_f > Y_b$ was calculated by considering the general case of ceramic coating on the metal. Therefore the yield condition at the interface is given by the following equation;

$$\psi_i P_{max} = Y_b \tag{7}$$

By substituting values of ψ_s,ψ_f,ψ_i and ψ_b at a certain value of t/a in Fig.4 into Eqs.(4),(5),(6) and(7) respectively, we can see which equation gives the smallest value of P_{max} for yield under the given frictional conditions. The position corresponding to the equation for the smallest P_{max} should yield at first under the frictional loading. The positions of yield initiation were decided in this way for the given coating thickness and frictional conditions. The ratio of the coating thickness to the half-contact width t/a was varied between zero and four, and yield strength ratio Y_f/Y_b was varied between one and seven. E_f/E_b was kept constant at two. The position of yield initiation was decided for all combinations of t/a and Y_f/Y_b, and the results are shown in Fig.5, 6 and 7 as local yield maps for cases of $\mu = 0.25$, 0.5 and 0.7 respectively. Solid lines are tentatively drawn as boundaries between two regions of different yield positions. As shown in Fig.5 for relatively low friction, local yield appears in the substrate under wide variety of Y_f/Y_b when t/a is smaller than about 0.3, but it appears at the interface in various combinations of Y_f/Y_b and t/a. Only when Y_f/Y_b is very close unity and t/a is over two, the local yield appears in the coating. If friction coefficient is increased from 0.25 to 0.5, the possibility of yield in the substrate or coating disappears. The yield at the surface dominates in all cases of t/a when Y_f/Y_b is smaller than about three and the yield at the inteface dominates when Y_f/Y_b is larger than about three. The yield at surface dominates at larger values of Y_f/Y_b if friction coefficient is increased from 0.5 to 0.7 as shown in Fig.7. It may be generally concluded from Fig.5, 6 and 7 that the yield initiates most often at the intreface between coating and substrate and it becomes predominant at the surface as μ and t/a become large. The yield in the coating or in the substrate appears only when the friction coefficient is small.

3.4 The effect of E_f/E_b on local yield map

The local yield maps shown in Fig.5, 6 and 7 were obtained under the assumption of constant E_f/E_s value(=2). But in practice, it may change corresponding to the change of Y_f/Y_s. Therefore, it is important to check the effect of E_f/E_s on the decision of the yield position. The effects of E_f/E_b on ψ_s and ψ_i are observed at $t/a = 0.5$ for this purpose in Fig.8 in the case of $\mu = 0.5$. It is obvious from Fig.8 that the relative value between ψ_s and ψ_i does not change so much within the change of E_f/E_b between one and four. This means that the boundary between two regions of different yield positions may not change so much by the change of E_f/E_b and Figs.5,6 and 7 may show the representative local yield maps.

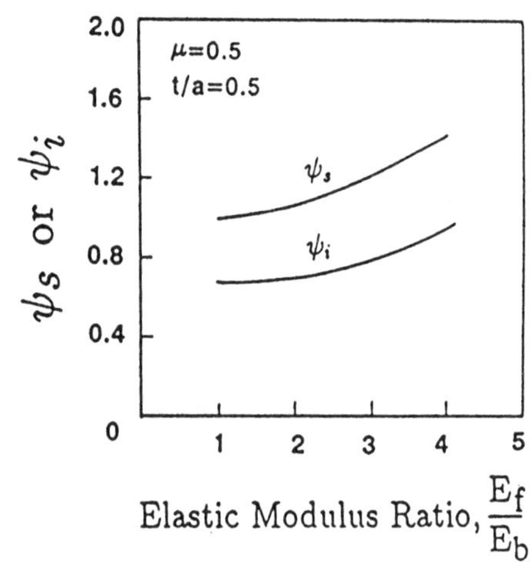

Figure 8 : Influence of the elastic modulus ratio of coating to the substrate on ψ_s and ψ_i.

4 Discussions

Hard coating on soft materials is generally applied to improve the wear resistance of the bulk made of soft material. Ceramic coating on metal is the most representative combination and its wear is mainly caused by crack initiation and propagation in or near-by the coating[3]. Cracks may propagate from defects in brittle behavior when the certain level of stress is attained by the

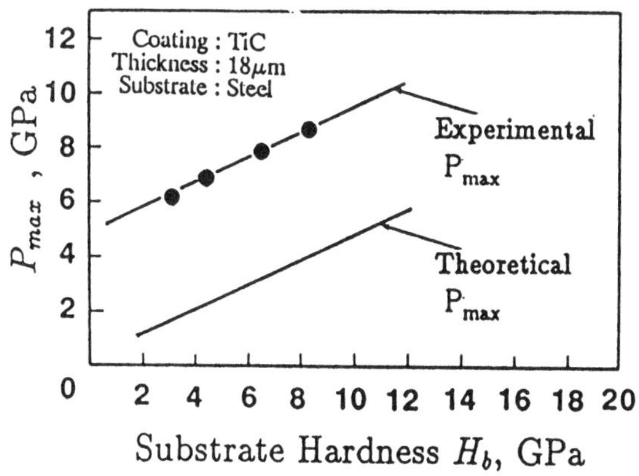

Figure 9 : Comparison between the theoretical value of maximum critical pressure P_{max} of local yield at interface and the experimental value of maximum critical contact pressure P_{max} for spalling of coating from interface at the various substrate hardness.

external force. The equivalent stress of von Mises for the yield initiation should be a good standard of this meaning.

It was shown in local yield maps of Figs.5,6 and 7 that the surface and the interface were two main possible sites for local yield initiation under various frictional conditions. Therefore, the initial defects such as micro-cracks at the surface and the interface become important as stress concentration sources which lead to wear. Depending on combination of Y_f/Y_b, t/a, E_f/E_b and μ, it is clearly determined by Figs.5,6 and 7 that either the surface or the interface should be treated more carefully to avoid the crack propagation. The understanding of wear mechanism of ceramic coating is obtained in this way as the first step. The next step of analysis would be to know the critical value of P_{max} for delamination or spalling of coating started at a certain position decided by the local yield map. The experimental value of P_{max} of this meaning[22] and the theoretical value of P_{max} for the yield initiation at the interface calculated with Eq.(7) are compared in Fig.9 for this purpose. The experimental values of P_{max} were obtained by scratch test on TiC coating of 18μ thickness. Substrate was steel and its hardness was changed by heat treatment between 2 and 8GPa. Theoretical values of P_{max} were calculated by supposing $\mu = 0.5$ and $E_f = 400GPa$, $E_b = 200GPa$, $Y_f = H_f/3(H_f{=}25\text{GPa})$ and $Y_b = H_b/3(H_b{=}2,4,6$ and $8\text{GPa})$, where H_f is coating hardness and H_b is substrate hardness. It is not surprising that the theoretical value of P_{max} is smaller than the experimental one by about 62 percents. Because the value of P_{max} for yield initiation should always give the lower estimation for micro-scopic crack propagation and the experimentally observed value of P_{max} should always give the over estimation for it.

Therefore, the local yield maps introduced and the related equation for P_{max} can be thought useful for the design of tribo-system with coated elements.

5 Conclusions

The local yield maps, which show the positions of yield initiation for hard coating under sliding contact, were introduced in relation to the ratio between yield strength of coating and substrate and the ratio between the coating thickness and half-contact width. The maps show the following conclusions;

1. Yield at the interface between coating and substrate is dominant especially when friction coefficient is small ($\mu \leq 0.25$). It is also dominant at large friction coefficient ($\mu \geq 0.50$) when yield strength ratio Y_f/Y_b is larger than about three.

2. Yield at the surface of coating becomes dominant at large friction coefficient ($\mu \geq 0.5$) when yield strength ratio Y_f/Y_b is smaller than about three.
3. Yield in substrate or the coating occurs only when the ratio of coating thickness to half-contact width is very small ($t/a < 0.3$) or larger than about two under small friction coefficient ($\mu < 0.25$).

References

1. E.Posti and I.Niemiuen,"Influence of coating thickness on the life of TiN-coated high speed cutting tools," Wear, Vol.129, 1989, p.273.
2. H.J.Boving, H.E.Hintermann and G.Stehle, "TiC-coated ball bearing for spin-axis gyro application," ASLE, J. Lubr. Eng., Vol, 37, No.9, 1981, p.534.
3. K.Kato, D.F.Diao and M.Tsutsumi, "The wear mechanism of ceramic coating film in repeated sliding friction," ASME, Wear of Materials, 1991, p.243.
4. K.L.Johnson "Contact Mechanics," Cambridge University Press. 1985.
5. V.M.Aleksandrov, "On the approximate of a certain type of integral equation," Prikladnaya Matematika y Mekhanika, Vol.26, 1962, (in English translation), p.1410.
6. V.M.Aleksandrov, "Some contact problems of elastic layer," Prikladnaya Matematika y Mekhanika, Vol.27, 1963, (in English translation) , p.1164.
7. V.M.Aleksandrov, "Asymptotic methods in contact problems of elasticity theory," Prikladnaya Matematikay y Mekhanika, Vol.32, 1968, (in English translation), p.691.
8. V.M.Aleksandrov, "Asymptotic solution of the contact problem for a thin elastic layer," Prikladnaya Matematika y Mekhanika, Vol.23, 1969, (in English translation) , p.49.
9. P.Meijers, "The contact problem of a rigid cylinder on an elastic layer," Applied Scientific Research, Vol.18, 1968, p.353.
10. T.Wu and Y.P.Chiu, "On the contact problem of layered elastic solids," Quarterly of Applied Mathematics, XXV, 1967, p.233.
11. Y.C.Pao, T.Wu and Y.P.Chiu, "Bounds on maximum contact stress of an indented elastic layer," ASME, J. Appl. Mech., Vol.38, No.3, 1971, p.608.
12. P.K.Gupta, J.A.Walowit and E.F.Finkin, "Stress distributions in plane stress layered elastic solids subjected to arbitrary boundary loading," ASME, J. Lubr. Technol., Vol.48, 1973, p.427.
13. F.E.Kennedy and F.F.Ling, "Elasto-plastic indentation of a layered medium," ASME, J. Eng. Mater. Technol., Vol.96, 1974, p.97.
14. K.Komvopolous, "Finite element analysis of a layered elastic solid in normal contact with a rigid surface," ASME, J. Tribology, Vol.110, 1988, p.477.
15. K.Komvopolous, "Elastic-plastic finite element analysis of indented layered media," ASME, J. Tribology, Vol.111, 1989, p.430.
16. H.Tian and N.Saka, "Finite element analysis of an elastic-plastic two-layer half-space: sliding contact," Wear, Vol.148, 1991, p.261.

17. R.B.King and T.C.O'Sullivan, "Sliding contact stresses in a two dimension layered elastic half-space," Int.J. Solid Structures, Vol.23, No.5, 1987, p.581.

18. J.M.Leroy and B.Villechaise, "Stress determination in elastic coating and substrate under both normal and tangential loads," Proc. of the 16th Leeds-Lyon Symposium on Tribology, 1989, p.195.

19. J.O.Simith and C.K.Liu,"Stress due to tangential and normal loads on an elastic solids with application to some contact stress problems," ASME, J. Appl. Mech., Vol.20, 1953, p.157.

20. W.J.Tomlinson and K.Talbot, "On the relation between hardness and the flow curve of metals," J. Mater. Sci., Vol.3, 1968, p.655.

21. J.R.Cahoon, W.H.Broughton and A.R.Kutzak, "The determination of yield strength from hardness measurements," Metallugical Transaction, Vol.2, 1971, p.1979.

22. P.A.Steinmann and H.E.Hintermann, "Adhesion of TiC and Ti(C,N) coating on steel," J. Vac. Technol. A3(6), 1985, p.3294.

Thin Films in Tribology / D. Dowson et al. (Editors)
1993 Elsevier Science Publishers B.V.

A METHODOLOGY FOR COATING SELECTION

A Matthews[a], K Holmberg[b] and S Franklin[c]

a Research Centre in Surface Engineering, University of Hull, Hull, HU6 7RX, UK.
b Laboratory of Production Engineering, Technical Research Centre of Finland, PO Box 111, SF 02151 Espoo, Finland.
c Philips Centre for Manufacturing Technology, PO Box 218, Building SAQ-2, 5600 MD Eindhoven, The Netherlands.

The wider application of coating technologies requires that design engineers and others who specify coatings should have access to a procedural basis for selection, ideally one which lends itself to computer implementation. Here we report such a basis which involves a nine stage methodology supported by seven selection rules. The approach facilitates the elimination of unsuitable coatings, to arrive as a selection which optimally fulfils the operating needs.

INTRODUCTION

Due to rapid developments in the field of surface engineering, increasing numbers of surface coatings are becoming available to design engineers and others who specify coatings. Simultaneously there has been an increasing recognition of the importance of the surface to the performance of all engineering components. The use of coatings on engineering components and tools has not however increased as much as might be expected; this is due to a number of difficulties:

* There is no systematic analysis and selection procedure which design engineers can use. This is combined with a lack of coating design data in a suitable form for the engineering component designer.
* For many of the newly-developed coatings the optimal application areas have not yet been identified.
* The quality and capabilities of existing coating processes are improving (ie the technology is transient). This means that previous performance data may no longer be applicable. It also means that the knowledge required to make a selection is even greater than previously thought.
* There is still a lack of understanding of the mechanisms involved in determining the tribological behaviour of coated surfaces.
* There is a shortage of coatings experts having a sufficiently broad knowledge of all available techniques to provide a balanced viewpoint.

These are just some of the difficulties, and the result is that coatings are not being used as widely as they should be and that inappropriate (non-optimised) implementation is common.

2. PREVIOUS APPROACHES

Over the past twenty years or so several writers have recognised the importance of coating selection, but no difinitive solution has emerged.

Smart and James (Refs 1,2,3) were two of the first writers to identify the problems of achieving an appropriate selection and they proposed various ways by which this might be achieved. Smart presented a checklist based on two sub-headings: Process Selection and Materials Selection. This is discussed in Ref 4. The approach is rather global and begs many questions. the properties of the same materials deposited by different techniques can, for example, be very different. The lists do not necessarily make the designer's job easier.

James made an effort to define the criteria and their inter-dependencies, and thereby make selection more straightforward. He did this by deviding the selection method into four distinct headings; Requirements, Limitations, Interactions and Economics/design, as described in Ref 4. James defines the ability to redesign as a "resource", although the problem still remains of how to consider, in a methodical way, how the requirements can be matched against limitations and interactions. In essence James established a philosophy for selection, but one which left many questions unanswered. This was developed by Smart, emphasising wear mechanism identification - an approach which has been traditionally used by several writers.

The main drawback with an approach to coating selection which is based on identification of a dominant wear mode is that in almost all cases a number of wear modes operate. This has been succinctly summarised in a sketch produced by Farrow and reproduced in Ref 4. The message from this is that wear is a process which invariably involves at least three or four interdependent mechanisms. This creates a difficulty for many of the idealised approaches to wear control, such as the systems concept (Ref 5).

There thus exists a need for a more robust methodology of selection, one which is built on current knowledge of tribo-contact behaviour, bringing in the latest information about the characteristics of coatings produced by different processes, including their mechanical and chemical properties. It is worth pointing out that until recently it has simply not been possible to produce coatings having the precise properties required to fulfil all the operating needs. Nowadays, especially since the introduction of advance ion- and plasma-based vapour deposition techniques, ceramic, cermet and metallic films can be produced with specifically designed properties. Kramer (Ref 6) has described how ceramic film performance can be predicted for specific wear situations. According to Almond (Ref 7), layered films can provide benefits, eg by arresting crack propagation, a problem which can otherwise increase with thickness. Matthews (Ref 8) proposed mixed phase metal/ceramic coatings produced by layering. The particular laminated structure studied was Ti/TiN which was shown to have beneficial tribological properties in sliding contacts. This composition has now been tested under erosive conditions, confirming the beneficial combination of toughness and hardness.

The availability of suitable coatings, and a considerable improvement in the theoretical understanding of tribological contacts means that the selection problem may now be solved. This is particularly true, given the availability of computer systems which can provide support for such logic-based selection procedures. In this way we shall be able to move away from what James described as the "positive selection approach"; that is, a possible solution will be tried, and if it works it will be used. We may describe this as the "pattern recognition" approach. Rather than the progressive elimination of all feasible solution, the first one to give an improvement is used, often resulting in better solutions being rejected by default. It is far better that a methodical "progressive elimination" approach be adopted. The discipline inherent in this approach encourages the acquisition of information (rather than the assumption of knowledge), and the investigation of alternative ways of solving minor problems, which may be obstructing highly desirable solutions.

3. COATING SELECTION METHODOLOGY

The aim of the approach presented here is to enable the design engineer to effectively minimise the probability of tribological problems occuring during the use of components. The intention is to provide, with a reasonable safety margin, a reliable indication of the material property limitations and characteristics needed.

Because of the complexity of the rules and the need for frequent access to materials data, the methodology lends itself well to incorporation in a computer expert system.

3.1 A systematic framework

The approach presented here is based on matching the requirements of the application (what is needed) with the combinations of properties that can be offered by different coatings (what is possible). The possible solutions are found by progressively eliminating those coatings that do not possess the required combination of properties.

Clearly, in order to achieve a successful marriage between the needs and the possibilities, it is necessary to express the requirements in a way that can be directly compared to the known properties and characteristics of coatings and coating processes. In effect, the component design engineer must translate the requirements into a language that can be understood by the coating process engineer. The complete methodology is represented in Fig 1, in which a number of stages have been differentiated. It is convenient to make a distinction between surface (functional) requirements (stage 4), non-functional requirements (stage 5) and economic and procurement requirements

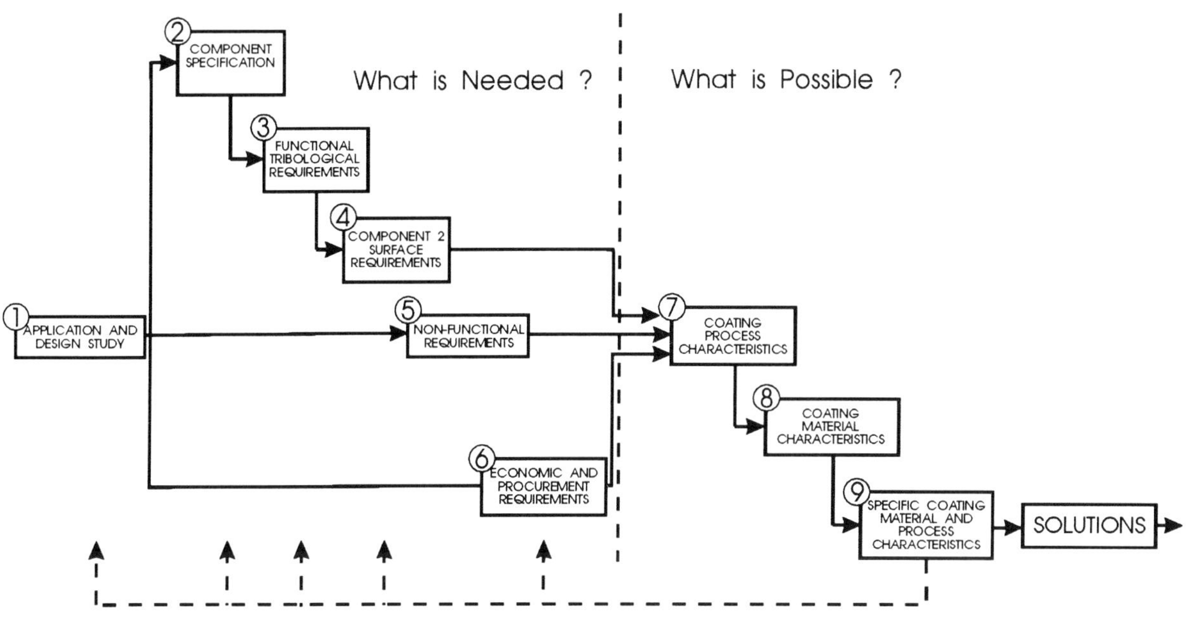

Figure 1. Methodology for Tribological Coating Selection

(stage 6). The purpose of stages 1,2 and 3 is to derive the correct tribological requirements needed for the match by successively refining the design specification. Brief descriptions of the stages are given below. Examples of the contents of each stage are provided in Table 1.

Stage 1: Application and Design Study

This stage forms an early part of all engineering design procedures and involves selection of the global design requirements leading to a general design specification and a definition of the expected working conditions for the complete design (see Pahl and Beitz, Ref 9). The tribologically critical components are identified.

Stage 2: Component Specification

This stage involves a more detailed analysis of the critical components. For each of these, the specific contact and service conditions (contact geometries, forces, velocities, environment etc) and the applicable constraints (overall component dimensions and dimensional accuracies, costs, desired lifetime, etc) are specified. It is necessary to distinguish between the part to be considered for coating (designated part 2) and any counterpart (designated part 1) that is involved in the tribological contact. For simplicity here it will be assumed that the material of part 1 will be pre-specified. The next stage in the methodology is to convert the specifications in stage 2 into requirements of the coatings and coating processes to be considered.

Stage 3: Functional Tribological Requirements

The functional tribological requirements are those that are essential to the tribological performance of the design and will include noise and vibration limitations as well as friction or traction force requirements and allowable wear-rates.

Stage 4: Component 2 Surface Requirements

These are the required limits on the appropriate surface (material) properties, needed in order to fulfill the functional tribological requirements under the applicable service conditions.

Stage 5: Non-Functional Requirements

These requirements are not essential to the tribological function of the component but, nonetheless, represent important constraints on the coating materials and processes that can be applied. Examples are the dimensions required to be coated, maximum allowable dimensional changes associated with the coating or coating process, the need to mask-off of certain parts of the component from coating, desired surface colour, preferred coating materials and processes etc.

Stage 6: Economic and Procurement Requirements

Aspects concerning the allowable costs of the final component and the ease with which a coating can be obtained (either in-house or externally), are covered by these requirements. Also included are legal, environmental and availability factors which may prohibit or favour certain coatings or coating processes.

Stage 7: Coating Process Characteristics

Stages 7, 8 and 9 can be thought of as filters, positively eliminating those coating processes or coatings that are not able to meet the requirements in stages 4, 5 and 6. Stages 7 and 8 can be considered as coarse filters because they are concerned with the invariable characteristics of particular coating processes or coating materials. Stage 7 deals with the process characteristics and property-ranges that are valid for all coatings applied using these processes. Examples of coating processes are high-temperature CVD processes, plasma thermal spraying processes, etc. In this way, if a requirement such as a maximum allowable coating thickness or maximum as-coated surface roughness does not fall within the range that is practically feasible for a particular process, then all coatings applied using this process can be eliminated. If a particular coating material is required, only those processes capable of depositing this material need be considered.

Stage 8: Coating Material Characteristics

Similar to stage 7, stage 8 concerns coating material property-ranges and characteristics that, for practical purposes, can be considered to be independent of the coating process used. For example, the hardness of titanium nitride typically lies between 1500 and 2500 $kgmm^{-2}$, irrespective of whether a CVD or a PVD process is used. Similarly, the elastic modulus of this material will probably not vary greatly with the deposition process utilised.

Stage 9: Specific Coating Material and Process Characteristics

This stage is in some respects a refinement of stages 7 and 8, dealing with the property-ranges and characteristics for specific coating material-process combinations. In this way, a differentiation can be made between the property-ranges of similar coatings deposited by different processes or varied conditions.

3.2 Requriements of Coated Surfaces

The most difficult aspects of the methodology involve the specification of the required tribological behaviour of the surface (stage 3) and the conversion of this information into surface (material) property requirements (stage 4). It is well known that tribological quantities such as friction coefficient and wear-rate are strongly influenced by the tribological system, depending not only on the natures of the interacting surfaces involved in the tribological contact, but also on system conditions including forces, velocities, temperature, atmosphere, etc. The system parameters are often variable and are not mutually independent, adding further complexity to the problem of predicting tribological behaviour. It is therefore not surprising that, in spite of intensive research, the relationships between material surface properties, tribological system conditions and tribological behaviour remain unclear. However, it is still possible to define which material surface properties influence the tribological mechanisms occurring and are therefore important in determining the friction and wear behaviour of coated surfaces. This subject has been reviewed recently by Holmberg (Ref 10). On the basis of this work, the following parameters are of importance:

- Young's modulus and Poisson's constant of the coating and substrate.
- Yield and tensile strength of the coating and substrate.
- Shear strength of the coating and of the coating-substrate interface.
- Thickness of the coating.
- Roughness of the mating surfaces.
- Shear strength and thicknesses of any microfilms present on the coating surface or generated during service.

Ideally the design values for the above properties would be derived from knowledge of the required wear-rate and friction coefficient and the known service conditions. However, for the reasons outlined previously, the general understanding of the tribological phenomena involved, in addition to the theoretical models concerning the contact mechanics of layered surfaces, are not yet at a stage that enables comprehensive relationships to be derived from fundamental principles. Moreover, data on several of the properties listed above, in particular interface shear strengths and properties of microfilms, are not available for the majority of coatings. An alternative approach is to combine some of the more practically-applicable contact mechanics theories with rules of thumb that, although not yet fully explainable on theoretical grounds, are nonetheless well established with tribological design engineers. Such an approach to tribological design has been suggested previously (Ref 11) and is currently being further developed within the framework of an EC BRITE/EURAM Project - 0073/BE-3413.

Table 1 Examples of the Information provided at the different stages in the methodology.

STAGE	EXAMPLES (3 and 4 dependent on the applicable tribological situation)
1	General working conditions, appraisal of design requirements, overall machinery design, identification of critical components.
2.	Geometry of the contact between parts 1 and 2. Material and surface roughness of part 2. Working environment; abrasive particles, liquid, gas, temperature. Forces on the contact. Velocities involved. Nature of the motion involved. Desired lifetime. Maximum allowable tolerance.
3.	Friction or traction force requirements; initial and steady-state. Noise and vibration. Avoidance of localised plastic deformation, fracture, surface fatigue, fretting. Wear-rate.
4.	Yield strength of substrate, in relation to the reduced elastic modulus. Thickness requirement for coating. Wear-factor in relation to the counterpart. Required coating and substrate hardnesses. Preferred coating types.
5.	Dimensions and weights of parts to be considered for coating. Maximum allowable dimensional changes associated with the coating or coating process. Thickness tolerance. Finishing requirements. Requirement for the masking-off of certain parts of the component from coating. Requirement for coating to penetrate holes and recesses. Desired surface colour. Preferred coating materials and processes. Any limitations on the substrate material for the coating.
6.	Allowable costs. Delivery time. Availability in-house, locally, worldwide. Legal and environmental constraints.
7.	Partial-coating ability. Penetration ability. Environmental-friendliness. Ranges of possible thickness, precision of thickness, deposition rate, process temperature and relative distortion of workpiece, maximum coatable dimensions, cost per unit surface area, surface roughness change, relative adhesion, permissible substrates.
8.	Ranges of colour, elastic modulus, Poisson constant, hardness, wear-factor, friction coefficient.
9.	Narrower ranges or specific values for the properties given in 7 and 8.

The intention of this approach is not to calculate the exact property requirements but rather to provide, with a reasonable safety margin, a reliable indication of the material property-limitations and characteristics needed. This enables the design engineer to effectively minimise the probability of tribological problems occuring in service. The design methodology is discussed in greater detail elsewhere (OECD International Research Group on Tribology Meeting, Leeds, September 1992).

3.3 Selection Rules

The design methodology is founded on a number of practical design rules, each of which has one or more materials selection criteria associated with it. Depending on the design situation one, or more commonly several, of these rules are appropriate. By applying the rules successively, the materials selection requirements are progressively refined. The methodology accomodates most of the possible general tribological problems that occur in practical design situations. The selection of which rules are appropriate for a particular design situation is based on the following factors:

- The service conditions under which the component is required to perform (including the type of relative motion between the surfaces).
- The limitations of the rules.
- The constraints imposed by the application on the materials used.

The following rules have been identified using contact conditions illustrated in Fig 2:

Rule 1: Contact stresses

The most important criterion that must be satisfied by the contacting materials in many potential wear situations is that plastic deformation or fracture in the contact region must be avoided under both stationary and moving conditions. In the case of surface-treated or coated materials, this also applies to the substrate material. If plastic deformation or fracture does occur, severe wear is likely to take place from the onset of motion (Ref 12). This does not include failure of roughness peaks, which is inevitable in most contact situations. Surface-treated and coated surfaces require special consideration because the mechanical properties are non-isotropic. The variation in these properties with the depth from the surface must be taken into account. Separate consideration is required for the following 3 situations:

- Untreated and uncoated materials where the yield stength (σy), the elastic modulus (E) and the Poisson constant (ν) are independent of the depth from the outermost surface.
- Surface-treated materials where σy is dependent of the depth from the outermost surface but E and ν can be considered (for practical purposes) to be independent of the depth.
- Surface-coated and surface-treated materials where σy, E and ν are dependent on the depth from the outermost surface.

The rule is based on the theory and analytical solutions developed by Johnson (Ref 13) and Hamilton and Goodman (Ref 14). For coated surfaces, use is made of the analytical solutions provided by Leroy and Villechaise (Ref 15). It is important to note that the minimum allowable strengths for materials calculated in this rule are not intended to be used to provide accurate estimates of the maximum stresses actually occuring. Because the rule is intended for design purposes, the "worst case" situation is used as the basis for the calculations.

Rule 2: Sliding

To practising design engineers, it is often the lifetime of the component (defined by the wear-rate), rather than the physical background of the wear that is important. It has already been pointed out that wear and friction behaviours

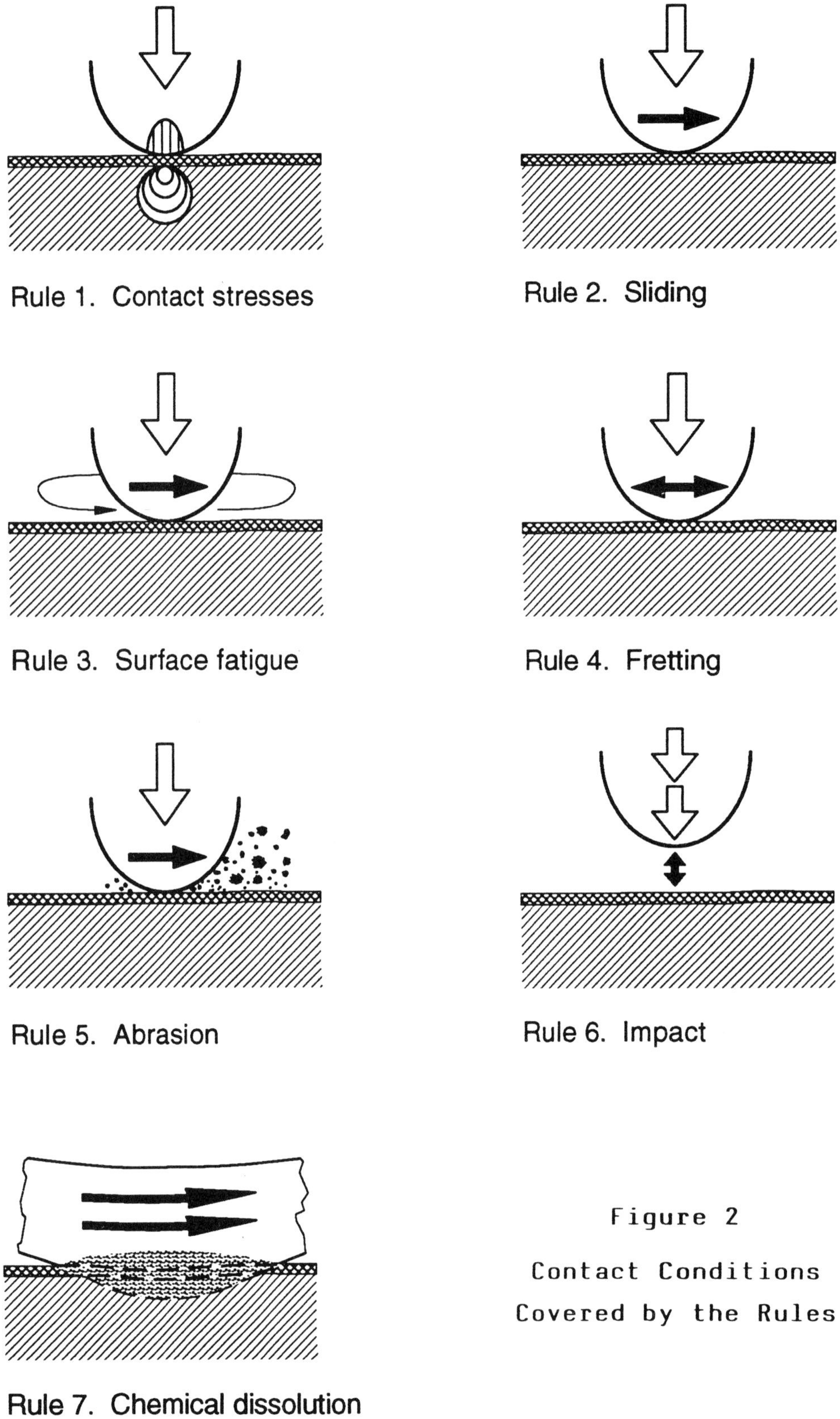

Figure 2

Contact Conditions Covered by the Rules

depend greatly on the specific tribological system of the application. Nevertheless a useful indication of whether or not wear problems are likely to occur in practice (without differentiating between wear mechanisms) can be obtained by using the wear (K) factor proposed by Archard (Ref 16).,

Depending on the test method used, laboratory tribological tests carried out under appropriately consistent conditions can provide the required information. An important prerequisite, valid both for the test data and for the design situation, is that plastic deformation or fracture in the contact region must not take place under conditions of stationary contact. The criteria contained in the contact stresses rule must therefore have been satisfied. Collections of tribological data are available in numerous publications (eg Refs 16,17,18). The values of the K-factors and friction coefficients must, of course, be considered as approximate rather that absolute. However, approximate values are, in practice, sufficient to ensure that a component is designed with a safety factor great enough to avoid the occurrence of tribological problems. It is necessary to compare the maximum permissible wear factor with the probable wear-factors of different material combinations under similar conditions. Although the latter data must be available to the designer, it is possible in some cases to modify the data according to knowledge of the approximate effects of changing different elements of the tribological system.

Rule 3: Surface Fatigue

Surface fatigue is a complex phenomenon influenced by a large number of factors including surface roughness and defects, shape, coarseness and distribution of carbides and inclusions, etc. However, empirical relationships between the contact stress, the lubrication regime, the number of cyclic load changes, the composition and the hardness of materials can be complied from data available in the literature (Refs 16,19,20). Practical experience at Philips Electronics has shown that these relationships can be used successfully as a design tool to minimise the chances of failure occurring due to surface fatigue. The maximum shear stress in a rolling contact is, in the majority of engineering design situations, located ten to a hundred microns below the surface. In addition, it has been shown by Sainsot et al (Ref 21) that for hard thin (10-20 um) coatings on (relatively) soft substrates, the highest stesses tend to be concentrated at the interface between the coating and the substrate. Because the properties change in a discontinuous way at this interface, extra stresses can be present and there is a greater probability of crack initiation. This means that, in order to minimise the probability of surface fatigue, reliable solutions cannot generally be found in applying typical hard wear-resistant surface coatings deposited by PVD and CVD processes, for which the maximum thickness attainable is of the order of ten microns for single layer TiN on steel. Diffusion processes such as carburising and nitriding, which introduce compressive stresses and produce thicker surface layers, are generally more beneficial. Recent research suggests that very thin (ie 0.1um) hard PVD coatings may be advantageous also.

Rule 4: Fretting

If the surfaces of the two parts in a contact oscillate in a tangential direction with a small amplitude compared to the dimensions of the contact, the wear particles formed cannot be released from the contact. Subsequent oxidation of these particles will increase the total volume of material in the contact area, through which there is a possibility that the system will jam. Furthermore, the surface will be roughened by the fretting process, which increases the possibility of fatigue. Fretting wear or fretting corrosion can occur, especially when the contact stress is small, in electrical switch contacts, mechanical (detachable) joints (press-fitted, bolted, rivetted etc. joints) and in precision-engineering constructions which are digitally-controlled and exhibit limit-cycling. The results of fretting are loss of fit in mechanical joints, seizure or high friction in moveable joints, high electrical resistance in electrical contact. The low-amplitude movements necessary for fretting to occur can originate from undamped machine vibrations (also due to electronic position control) or from thermal expansion-contraction effects. In order to minimise the chances of problems occurring in a potential fretting wear situation, use can be made of certain coatings and surface treatments which have been proven to provide effective solutions. Examples are phosphating and sulphiding treatments, MoS_2 coatings and ceramic coatings (in the latter case, both surfaces would normally require coating).

Rule 5: Abrasion

Rules are being developed to indicate the properties required to resist abrasion, which take into account particle velocity, size and shape etc. In PRECEPT this rule deals with the specific case when foreign particles, not originating from wear in the contact, are present in this contact. Based on simple rules of thumb (Refs 1,5,22), which are applied by practising tribological design engineers, the rule provides a minimum surface hardness requirement and, in the case of coatings, a thickness requirement. When this hardness requirement cannot be met by the available coatings, the non-proportionality between hardness and abrasion resistance, that is observed when comparing different material types, is approximated using empirical relationships compiled from published data (Ref 23).

Rule 6 : Impact

The maximum value of the contact pressure occurring in a contact as a result of an elastic impact between 2 massive bodies can be estimated from well-established contact mechanics theory (Refs 13,16).

Rule 7 : Chemical Dissolution

Wear due to the chemical effects of dissolution and diffusion can be accomodated by using the rules developed by Kramer and Suh (eg Refs 6,24). The wear dealt with here is typically that occurring in metal cutting, high temperatures etc., where chemical processes become dominant.

4. EXPERT SYSTEM APPLICATION

The methodology described here has evolved from work in the authors' laboratories over a number of years. In particular at Philips and VTT the emphasis has been on the predictive aspects of tribological theory, whilst at Hull the emphasis has been on expert systems. In earlier expert systems work the aim was generally to encode the experiences of experts in order to eliminate coatings unlikely to fulfil the requirements of the application. The present work, (which encapsulates certain of the philosophies of the PRECEPT EC BRITE project) represents a more rigorous and quantitative treatment which we hope incorporates the best of both approaches (ie case-history information and analysis based on theoretical derivations). The incorporation of this approach into PRECEPT will be covered in later papers, in conjunction with our partners on that project.

5. CONCLUSIONS

An approach to coating selection has been described which involves a nine-stage systematic methodology. Seven selection rules are applied, based on operating conditions. These are intended to avoid the pitfalls of the usual approach of classifying wear mechanisms as such, rather they relate to the type of contact but can be seen as a hybrid approach permitting both case history information and theoretical tribology rules to be applied. The approach lends itself to incorporation into an expert system and should encourage the wider usage of coatings in industry.

6. ACKNOWLEDGEMENTS

This paper includes several ideas and concepts generated under the EC BRITE/EURAM Project PRECEPT (BREU -0073/BE-3413). We gratefully acknowledge financial support for that project, and our collaborators at Lucas Engineering and Systems and CETIM as well as colleagues in Philips, Hull University and VTT for their help and advice.

REFERENCES

1. RF Smart, Tribology Int., April 1978,97.
2. DH James, Surf. J., 9 (1978) 3.
3. DH James, Proceedings 8th Int. Thermal Spraying Conf., Miami, Sept. 1976, Am. Weld. Soc., 1976.
4. A Matthews, KG Swift and P Robinson, Proceedings Surface Engineering and Heat Treatment : Past Present and Future Conference, London, December 1991, Inst. of Metals, London, 1991.
5. H Czichos, Tribology, a Systems Approach to the Science and Technology of Friction, Lubrication and Wear, Elsevier, 1978.
6. BM Kramer and PK Judd, J. Vac. Sci. Technol., A3 (1985) 2439.
7. AE Almond, Vacuum, 34 (1984) 835.
8. A Matthews, PhD Thesis, 1980.
9. C Pahl and W Beitz, Engineering Design, Springer-Verlag, 1984.
10. K Holmberg, Proc. ICMCTF, San Diego, April 1992.
11. CJ Thijsse, in Tribological Design of Machine Elements, Ed. D Dowson et al Elsevier, 1989. Proceedings of 15th Leeds-Lyon Symp. on Tribology, Leeds, Sept. 1988.
12. AG Tangena, PhD Thesis, 1987.
13. KL Johnson, Contact Mechanics, Cambridge University Press, 1985.
14. GM Hamilton and LE Goodman, ASME Journal of Applied Mechanics, 33, (1966), 371.
15. JM Leroy and B Villechaise in Mechanics of Coatings, Ed D Dowson et al, Elsevier, 1990. Proceedings of 16th Leeds-Lyon Symp. on Tribology, Lyon, Sept. 1989.
16. Wear Control Handbook, Eds MB Peterson and WO Winer, ASME, 1980.
17. B Bhushan and BK Gupta, Handbook of Tribology, McGraw-Hill, 1991.
18. SE Franklin, Proc. ICMCTF, San Diego, April 1992.
19. G Neirmann in "Machinenelemente", Zweiter Band "Getriege", Springer Verlag, 1960.
20. CN Rowe and EL Armstrong, Lubrication Engineering, 38, (1982),1.
21. PH Sainsot, JM Leroy, B Villechaise in Mechanics of Coatings, Eds D Dowson et al, Elsevier, 1990. Proceedings of 16th Leeds-Lyon Symp. on Tribology, Lyon, Sept. 1989.
22. TS Eyre, Metals and Materials, March 1992, 143.
23. MM Krushov, Wear, 28 (1974), 69.
24. BM Kramer and NM Suh, J. Eng. Ind., (Trans. ASME), 102 (1980) 303.

SESSION XIII
MOLECULAR SCALE THIN FILMS

Chairman : Dr H A Spikes

Paper XIII (i) **Surface Roughness and Squeezed Films at Molecular Level**

Paper XIII (ii) **The Strain Rate Dependence of the Interface Shear Strength of Poly (methyl methacrylate) Films**

Paper XIII (iii) **Lubrication Regimes for Nanometer-Scale Lubricant Films with Capillary Effects**

Thin Films in Tribology / D. Dowson et al. (Editors)
© 1993 Elsevier Science Publishers B.V. All rights reserved.

Surface roughness and squeezed films at molecular level

J.M. Georges, S. Millot, J.L. Loubet, A. Tonck, and D. Mazuyer.
Laboratoire de Tribologie et de Dynamique des Systèmes.
U.R.A C.N.R.S 855; Ecole Centrale de Lyon, Ecully 69131, France.

Using a new surface force apparatus, static and dynamic measurements have been conducted to investigate the squeeze of n-dodecane and of dicyclohexylmethylpentane from the contact between a relatively smooth sphere and a plane. Seven different solid surfaces (mica, cobalt, gold, platinum and steel) have been tested. It has been found that an "immobile" layer of fluid is present on each solid surface which does not participate in the hydrodynamic flow of the liquid. The thickness of this film is relatively independent of the roughness of the solid, and corresponds to a small number of molecular layers (1 to 4).

I INTRODUCTION

The rheology of simple liquid near a surface or between two surfaces is important for many practical problems, such as the development of an oil film in lubrication conditions. All of these problems require the determination of the position of the "immobile" non slip-plane. In practice measurements suggest that molecules of liquid are not truly "immobile" near a solid surface, but that the component of velocity parallel to the surface is hindered or restricted over the time scale of the experiment in the proximity of the surface. It is also clear that the thickness D_s of this restricted layer varies with the nature of the liquid and the solid.{1}{2}{3}

The aim of the work is to extend measurements of the rheology of near-surface liquid films, to use a surface force apparatus to study the effect of roughness and solid surface type on the "immobile" layer.

II EXPERIMENTAL TECHNIQUES

A -The surface force apparatus

The surface force apparatus has already been partially described in the literature {4}{5}, where it has been mainly used to measure the static force laws between surfaces in close proximity (i.e the force F, as function of the distance separation D between a sphere and a plane). The general principle of the system , is that a macroscopic spherical body can be moved towards and away from a planar one using the expansion and the vibration of a piezoelectric crystal. Generally the specimens are metallic. A sphere of radius R is firmly fixed to a piezoelectric crystal. The plane specimen of mass m is supported by a double-cantilever spring, whose stiffness k° can be adjusted continuously between 4.10^3 and 6.10^6 Nm^{-1}. A drop of liquid is introduced between the plane and the sphere to form a meniscus. A first capacitive sensor C_1 measures the elastic deformation of the cantilever and thus the force transmitted through the liquid to the plane; a second sensor C_2 is designed to measure the relative displacement h between the supports of the two solids. A third capacitive sensor C measures the capacitance between the sphere and the plane. In all the three cases, the capacitance is determined by incorporating the signal in an L.C. oscillator to give a signal-dependent frequency in the range 5 to 12MHz. The resulting fluctuations in oscillation frequency are detected using a low noise frequency discriminator.

The phenomena taking place at the sphere-plane interface (Figure 1), (wetting, surface forces, rheology) are crucially dependent upon the separation between the two surfaces. This distance depends upon both the separation of the undeformed sphere and plane D, and also the extent of solid deformation δ, giving a true separation of H=D-δ. Clearly for negligible solid deformations H=D.

In practice, the true separation is determined as a displacement h, measured by a sensor. In these experiments this is given by $h= H+ h_o$, where h_o is a constant offset. Thus the undeformed separation is given by:

$$D=h-h_o+ \delta \qquad (1)$$

The displacement, h, can be automatically selected under computer control using an input voltage signal. The measured capacitative displacement is then compared with the input signal and the former adjusted using a negative proportional integral (P.I.) feedback loop acting on

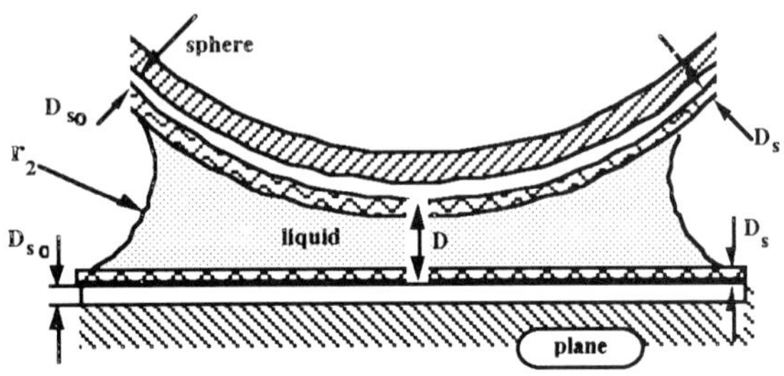

Figure 1 shematic representation of the sphere-plane interface; the roughness is not presented.

the piezoelectric crystal via an high voltage amplifier.

For the work described in this paper, the input control signal has two components. One is a steady ramp which gives a constant sphere-plane approach speed (dD/dt) of 0.1nm/s. Superimposed on this is a small amplitude oscillatory motion of about 0.1nm RMS with a frequency of 0.01 to 500Hz.

The steady approach component allows the quasi-static properties of the separating fluid film to be explored, whilst the oscillatory motion enables determination of dynamic properties of the film.

The question of the location of the origin when measuring the absolute separation of solid surfaces is problematic and requires clarification. Figure 1 shows a shematic contact. The solid surfaces may be covered with an oxide layer of thickness D_{SO}. On top of these there may be immobile liquid layers of thickness D_S. The separation between the solid layers is D.

Three measurements of separation are made in this study, two electrical and one mechanical. The mechanical one derives from the damping function of the interface and is described in more detail in the next section. Broadly, for a fluid film of constant viscosity the inverse of the damping function 1/A is proportional to the mobile layer thickness D- $2D_S$.

The electrical capacitance of the film is also measured as a function of displacement h. The inverse of this function, 1/(dC/dh) is proportional to the separation between the conducting solid surfaces D-$2D_{SO}$. dC/dh is measured as a function of the separation using the synchronised demodulation of an applied electrical oscillation of frequency 4.5 MHz and amplitude 200mV RMS

The third displacement measurement is obtained by applying a constant direct voltage of 1V between the sphere and plane and determining the resultant static force as a function of the sphere-plane displacement with sensor C_1. This electrostatic force F_e is given b

$$F_e = \frac{V^2}{2}\frac{\partial C}{\partial h} = -\frac{\pi \varepsilon_s R V^2}{D} \quad (2)$$

where ε_S is the static dielectric constant of the interface. The inverse of this force showed be proportional to the separation D.

Thus, plots 1/A(h), 1/ (dC/dt), and 1/ F_e(h) against separations show origins at three values of separation corresponding respectively to D-$2D_S$, D and D - $2D_{SO}$. The positions are called O_η, O_e and O_{el} respectively.

Another fact that has to be considered in interpreting capacitance results is the effect of the surface roughness. Theoretical calculations prove that the capitance values Cs and C, for smooth and rough electrodes have derivatives, with respect to the reference distance H=D between ideal smooth and infinitely rigid solids, that are nearly the same, as long as H is at least four to five times the peak-to-valley amplitude of roughness {6}{7}. The expression is:

$$\frac{\partial C}{\partial D} \approx \frac{\partial C_s}{\partial D} \approx -\left(\frac{2 \pi \varepsilon R}{D}\right)_{\left(\frac{D}{R} \ll 1\right)} \quad (3)$$

where ε is the dielectric constant of the interface. As in such conditions H and h only differ by an unknown but constant offset h_0, an extrapolated point such as

$$\left(\frac{\partial C(h)}{\partial h}\right)^{-1} = 0 \quad (4)$$

corresponds to the zero relative to the separation H.

The total force F_T, present between the sphere and the plane can have several origins: a wetting force F_W, a surface force F_S and an hydrodynamic force F_H. The two first can be measured using static experiments, the third with dynamic experiments.

B- Materials

1 Solid Surfaces

Six types of solid surface are used in this study and are listed in table 1. Four of these are metallic cobalt coatings on fused borosilicate glass, whose Poisson's ratio is 0.22 and Young's modulus E= 65 GPa, (glass 732-01, Sovirel Corp.). This cobalt layer was deposited under a low argon pressure (5.10^{-6} Pa), using cathodic sputtering. By

Surfaces	ρ nm	Radius of the sphere (mm) R	thickness of cobalt (nm)	thickness of the second deposit (nm)
(Co_1): Co on borosilicate glass	2.0	3.5	50	------
(Co_2): Co on borosilicate glass	5.2	4.0---5.0	60	------
(Pt) on (Co_1) Pt on Co on borosilicate glass	4	4.8	50	Pt 25
(Au) on (Co_1) Au on Coon borosilicate glass	3 - 4	2.9	50	Au 60
(St_1) Steel 52100 heat treated	12	4.5	-----	-----
(St_2) Steel 52100 heat treated	50	4.5	-----	-----
Mica	0.2	cylinder 15	-----	-----

Table 1 Characteristics of the six types of different solid surfaces used in this study: Peak to valley roughness,ρ measured with a scan length of about 1μm; Radius of the sphere, R or of the cylinders for mica.

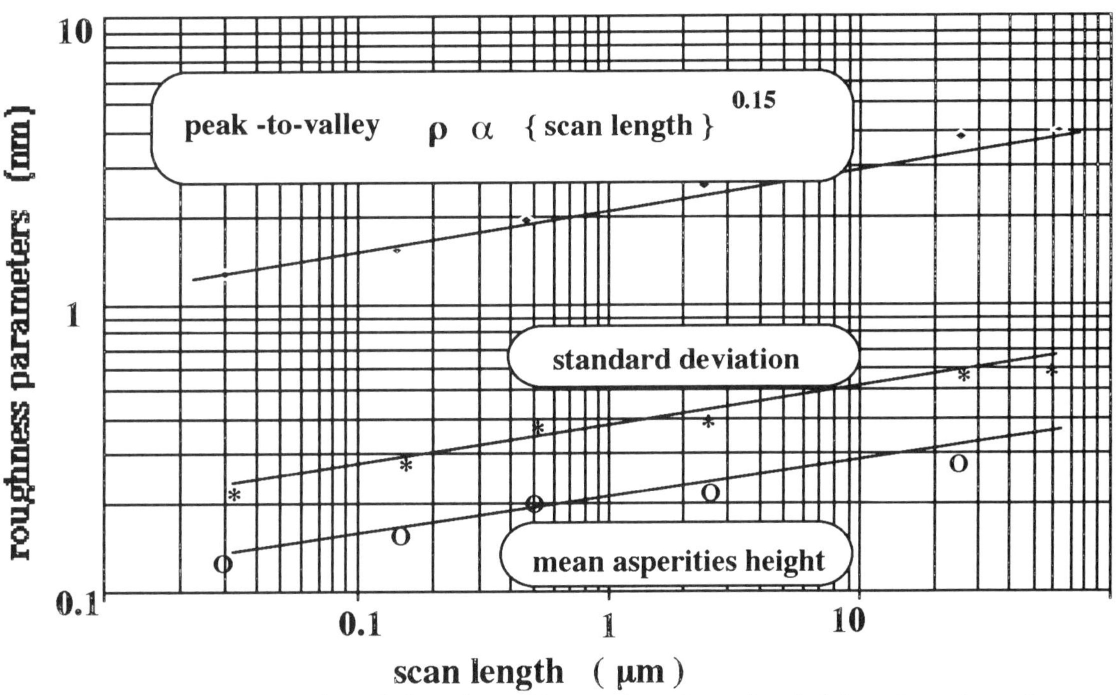

Figure 2 The roughness data for cobalt surfaces (Co_1) such as the peak to height ρ depends on the scanning length. For a scanning length of 1μm, ρ= 2nm.

varying the deposit conditions two types of cobalt deposit were obtained (Co_1 and Co_2). The deposit Co_1, which has lower roughness than the deposit Co_2, was also used as the substrate for two other vacuum deposits one of platinum and one of gold.

A.F.M.and S.T.M examinations of the sputtered surfaces show that, in all the cases, the surfaces consist of irregular connected clusters producing a gently bumpy corrugation with a "blackberry" like roughness. Table 1 summarizes the roughness for each surface (peak to valley ρ measured with a scan length of 1μm). The peak distribution of the A.F.M. image of the cobalt deposit (Co_1) was found to be Gaussian, with the roughness parameters (peak to valley, standard deviation, mean average) are dependent of the scanning length (Figure 2){8}. X.P.S. analysis of these surfaces confirmed the presence of metallic cobalt on the glass, and showed an oxide layer of thickness less than 1nm. Tests were also carried out using 52100 heat-treated steel as both sphere and plane. These surfaces were hand-polished with 0.1μm diamond paste. Finally, test were carried out using two thin sheets of molecularly smooth muscovite mica glued on steel surfaces and mounted as crossed cylinders of radii R=15mm.

Dust minimisation is one key to experimental success. The use of laminar flow bench was sufficient to reduce dust when coupled with a good inspection system, such as optical microscopy with a black field.

2 Liquids

Experiments were carried out with n-dodecane pure grade from Aldrich (packaged under nitrogen in Sure/Seal bottles). 2,2,4,4,6,8,8-heptamethylnonane. This was pure grade and was obtained from Aldrich. It is an isomer of n-hexadecane but by contrast to hexadecane which is flexible, it has a rigid structure.

III INFLUENCE OF THE SURFACE ROUGHNESS ON DRAINAGE AND THE IMMOBILE LAYER.

Using pure n-dodecane as a liquid, drainage experiments were conducted using the solid surfaces indicated in table 1. For each experiment, two measurements were carried out simultaneously on the sphere-plane device during approach: Firstly, the damping function A(h) (measured with a vibration frequency ω=239 rad.s^{-1}), and the electrical capacitance of the sphere-plane interface (except for the non conducting surfaces of mica) were measured; secondly, a constant direct voltage of 1V across the sphere-plane interface was applied and the static force F_e (h) and the damping function measured

The inverse of the damping function 1/A(h), the electrical component $(\delta C/\delta h)^{-1}$; the function $\{F_e(h)\}^{-1}$ were plotted against separation. For all the experiments conducted it was found that for large surface separation, each function varies linearly with the approach displacement h. The extrapolation of each function defines a different origin, as explained earlier.

Figure 3 shows the evolution of these functions in the case of sputtered gold surfaces on fused borosilicate glass (Au). For such surfaces, AFM examination showed irregular connected clusters producing a gently bumpy corrugation with peak-to-valley heights of 3-4 nm and a mean distance between adjacent bumps of about 100nm.

The scales in figure 3 has been calibrated in such a way that the mechanical and electrical functions correspond to distances. For an approach displacement greater than 6.5 nm, each function varies linearly with h. It should be noted that in these experiments, the two electrical origins coincide (O_e- O_{el} =0), that it is not the case for the mechanical origin. The experimental differences between the electrical and the damping functions plotted as a function of the displacement, stay relatively constant to a value of 0.85 nm and the distance O_η-O_e= 0.85nm ±0.5. The fact, that the inverse of the damping function varies linearly, proves that the Reynold's equation is followed, and that the sphere-plane interface is not deformed, so δ=0, or D=H. We conclude as Chan and Horn have already proposed{1}, firstly, that an "immobile" region of n-dodecane is adjacent to each solid surface, and is detected by the measurement of the distance, secondly, that the nature of the gold solid surfaces indicates the absence of an insulating oxide layer so that O_e-O_{el} =0. If the "immobile" layer of dodecane is of thickness D_s at each surface, it appears that the stick boundary conditions are applicable at a distance D_s out from each surface. In this case the hydrodynamic force becomes:

$$\frac{1}{A}=\frac{\dot{D}}{F_H}=\frac{D-2D_s}{6\pi\eta R^2} \quad (5)$$

and therefore O_η-O_e = $2D_s$=0.85nm. This suggest that D_s is comparable to the width of a single hydrocarbon chain lying flat at the surface (the molecular width of n-dodecane is 0.392 nm). Because the radius value R is accurately known, the plot of the function 1/A versus D gives the viscosity of the n-dodecane. For the experimental results at a separation larger than 6.5nm, the viscosity value thus derived is η= 1.44 ± 0.04 mPa.s at temperature of θ=23.5°C, a value in close accordance with those in the literature (η= 1.43 mPa.s) {9}

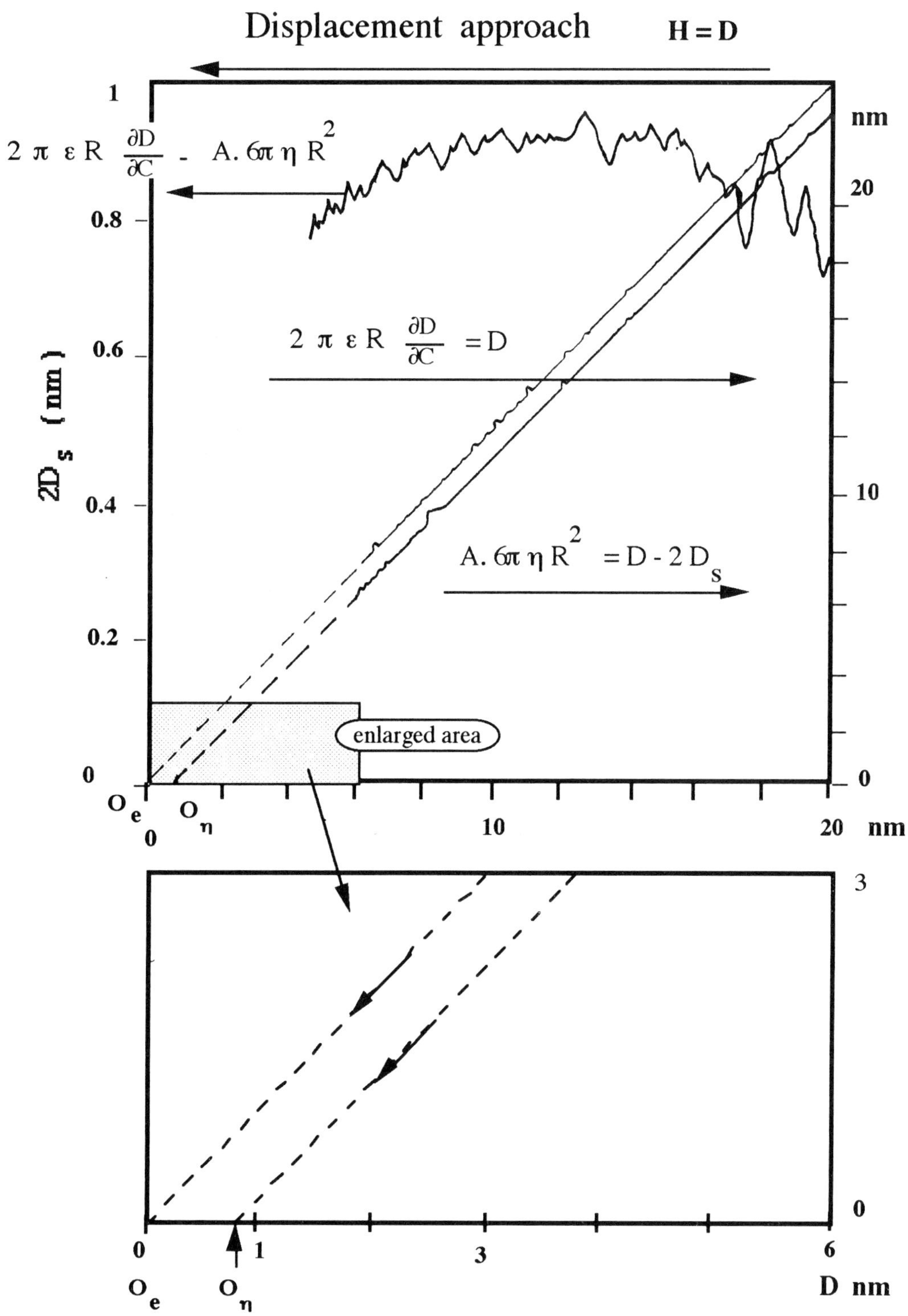

Figure 3 Plot of the three functions, the inverse of the damping function 1/A(h), the electrical component $(\delta C/\delta h)^{-1}$ and the function $\{F_e(h)\}^{-1}$ against the approach displacement for n-dodecane and gold surfaces at 23.5°C. Ideally, if the dielectric constant and the viscosity equal the bulk values at all the distances, the curves should fall on straight lines, passing through the origins. The experimental results agree to these results for D>6.5nm. The two electrical origins are the same (no oxide), and the slope of the inverse of the damping function gives the viscosity η.

A similar experiment was conducted with a rough steel plane surface St_2 against a smooth sphere (St_1) (table1). When D> 100nm, the slope of 1/A function gives a viscosity of the n-dodecane of η= 1.44 ± 0.04 mPa.s at a temperature of θ=23.5°C. This is the same value as for the experiment with gold surfaces. Below a critical separation D_c of 50nm, the measured damping value is higher than the value predicted by hydrodynamics. Also, for D lower than this critical distance D_c, the experimental measurement of the 1/A value decreases much more than linearly with D. This behaviour can be explained by the role played by the surfaces roughness{5}. In this experiment, a least square approximation were used to determine O_η and O_e from the linear part of the function. This gave O_η-O_e = 1nm, a value in accordance with the value found for the gold surfaces, despite of the very high roughness.

Other experiments using the same liquid n-dodecane are conducted for different surfaces. Table 2 summarizes these results. For a solid roughness such that peak to valley for one surface ρ_1 (nm) is less than 50 nm, the thickness of the"immobile" region D_s is relatively independent of the roughness of the solid. In all the case, there is a critical separation above which hydrodynamic theory is obeyed using yhe bulk viscosity of n-dodecane. The viscosity values were always in excellent accordance with the literature. The shear rate in this experiments was typically 1 to 10 s^{-1}.

Surfaces	ρ_1 nm	$2D_s$ ± 0.1 nm	D_s nm	$2D_{so}$ ± 0.1 nm	$\frac{D_s - D_{so}}{\sigma}$
(Co_1): Co on borosilicate glass	2.0	1.5	0.75	0.5-1	1
(Co_2) Co on borosilicate glass	5.2	1.7	0.85	≈1	1
(Pt) Pt on Co on borosilicate glass	4	0.8	0.4	≈0	1
(Au) Au on Co on borosilicate glass	5.5	0.8	0.4	0.0	1.
($St._1$) Steel 52100 heat treated	12	1.5	0.75		1.9
($St._2$) Steel 52100 heat treated	50	1	0.5	?	----
Mica	0.2	0.4 *	0.2		0.5

Table 2 Thickness" of the "immobile" layer of n-dodecane adjacent to each solid surface; this thickness is evaluated as the difference D_s-D_{so} and compared to the molecular width of the n-dodecane(σ=0.392nm).
* The origin zero is obtained by puting a drop of water in the contact at the end of the test.

For some materials that form oxides, such as steel and cobalt, it was found that the distance 2 D_{so}= O_e-O_{el} was not zero as it is the case for the gold surfaces. It is our interpretation that the distance D_{so} determined in this way, is related to the thickness of an oxide layer present on each surface, and in consequence D_s -D_{so} corresponds to an "immobile" layer of dodecane. However, it is conceivable that the relationship is not a simple equality but might depend on the frequency of the oscillating electrical field, i.e. on the frequency-dependence of the dielectric constant of the medium. Clarification of this point requires futher experimentation, which is currently underway. For the present analysis we interpreted D_{so} as being equal to the thickness of the oxide layer. For the experiments with the cobalt (Co_1)(Co_2), it was found that respectively, $2D_s$= 1.5 ± 0.1nm, and 2 D_{so} = 0.5 to 1nm. Therefore, D_S-D_{SO}≈0.4 nm, which can be compared with the width of the molecule s = 0.392 nm, i.e one molecule of n-dodecane is immobile on each surface.

In the case of mica experiments, two mica sheets were glued on steel surfaces, and mounted as crossed cylinders of radii 1.5 cm. Because the surfaces are not conductors, the zero reference (corresponding to O_e) was obtained by putting the surfaces in contact before the insertion of the liquid drop. Again, one molecule of n-dodecane was found to be immobile on each surface.

We do not attach great significance to the precise values of D_s. The important observation is that the "immobile" layer D_S-D_{SO} corresponds to only 1-2 molecular layers; nor do we suggest that these molecules of liquid are truly immobile near the solid surface, but rather that the component of velocity parallel to the surface is hindered or restricted on the time scale of the experiment. The normal component to the surface of the velocity need not be affected, and the molecules next to the surface can be exchanging rapidly with those away from it. A second important inference is that the thickness of the "immobile" layer (D_S-D_{SO}) is almost independent of the surface roughness, even when its value is much lower than the roughness value.

IV GENERAL FEATURES OF FORCES BETWEEN COBALT SURFACES SEPARATED BY DICYCLOHEXYLMETHYLPENTANE.

Using sputtered cobalt surfaces (Co_1) drainage experiments were conducted with 2,4,dicyclohexyl-2-methylpentane (DCMP). This fluid is a semi-rigid hydrocarbon with prominent internal librations and vibrations{9}{10}{11}, and is used as a traction fluid and a lubricant it is also called Santotrac 40.

Figures 4 shows data obtained using the procedure already given for the n-dodecane. It shows the three extrapolated curves corresponding to the electrical component $(\delta C/\delta D)^{-1}$ (which indicates the origin D=0), the function $\{F_e(D)\}^{-1}$ (which gives the distance $2D_{so}= O_{el}-O_e$), and, the inverse of the damping function 1/A(D) (which gives the mechanical origin O_η). It should be noted, first that the two electrical origins do not coincide ($2D_{so}= O_{el}-O_e$=1± 01nm) revealing the presence of a layer attributable to cobalt oxide. Secondly, the distances O_η-O_e and D_S-D_{SO} (O_h-O_e=$2D_s$ =6.5 ± 0.1nm, and D_S-D_{SO} ≈ 2.75 nm), are much greater than those for dodecane with the same cobalt surfaces (table 2). This shows that this distance is much more strongly related to the structure of the molecules than the roughness. The distance D_S-D_{SO} corresponds to about twice the molecular length (1.1-1.3nm) although it is difficult to relate to a particular dimension for such molecule. For the experimental results at a separation larger than 11nm, the slope determination of the function 1/A versus D gives the viscosity of the DCMP, for a known sphere radius (in this case R=4.05 mm). The viscosity is η= 43.5 ± 0.2 mPa.s at a temperature of θ=23.5°C, a value in close accord with those literature (η= 40 mPa.s) {28}. The maximum excess contact pressure, in the case of D>10nm and as indicated by the equation (A4), does not exceed 10Pa. With such pressure, the surfaces are not deformed so that the true displacement is H=D.

For H< $2D_s$= 6.5nm, the contact approach is governed by the elastic deformation of the substrates. So long as the two surfaces are smooth and elastically deform the contact force is related to the total deformation δ by the Hertz equation{12}. In this case, the controlled approach which is H=D-δ, is given by H= $2D_s$-δ, and therefore the function $\delta F/\delta H=-\delta F/\delta(\delta)$ is proportional to $\delta^{0.5}$ (Hertz equation). In consequence, figure 4 shows a plot of the experimental value $(\delta F/\delta H)^2$ versus the displacement H. For H< $2D_s$ the linear relation is verified, with an extrapolated distance such $(\delta F/\delta H)^2$ =0, when H=$2D_s$=6.5nm. Moreover the slope determination gives an elastic modulus of $E/(1-\nu^2)$= 62.5±7 GPa in accordance with a value of the literature (67.7 GPa). This result proves that the mechanical origin of the contact between the smooth sphere and the plane is not H=0, but H=$2D_s$. It also demonstrates that the difference is not totally due to roughness. In this case the Hertz relation (with a radius R=4.05mm) would not be verified. It also confirms the presence of an effectively solid layer of thickness $2D_s$ on the two surfaces .

For D_c> H=D>$2D_s$, the two substrates are not deformed by the hydrodynamic pressure. Indeed the application of elastohydrodynamic theory, which takes into account of the deformation of the solids, indicates that in the conditions of our experiment, the solid deformations are negligible . However for H=D_c=11±0.5nm, the damping function 1/A is no longer a linear function of D. This implies that the radius increases or that the viscosity decreases as the distance decreases. Neither of these two assumptions is probable, but another possibility is that the liquid does not flow in the central region of the contact, and thus does not contribute to the damping. Instead, due to confinement, the liquid behaves elastically. The mechanical transfer function can be interpreted as a damper in parallel with a spring, whose stiffness K_2 results from the central elastic layer The larger

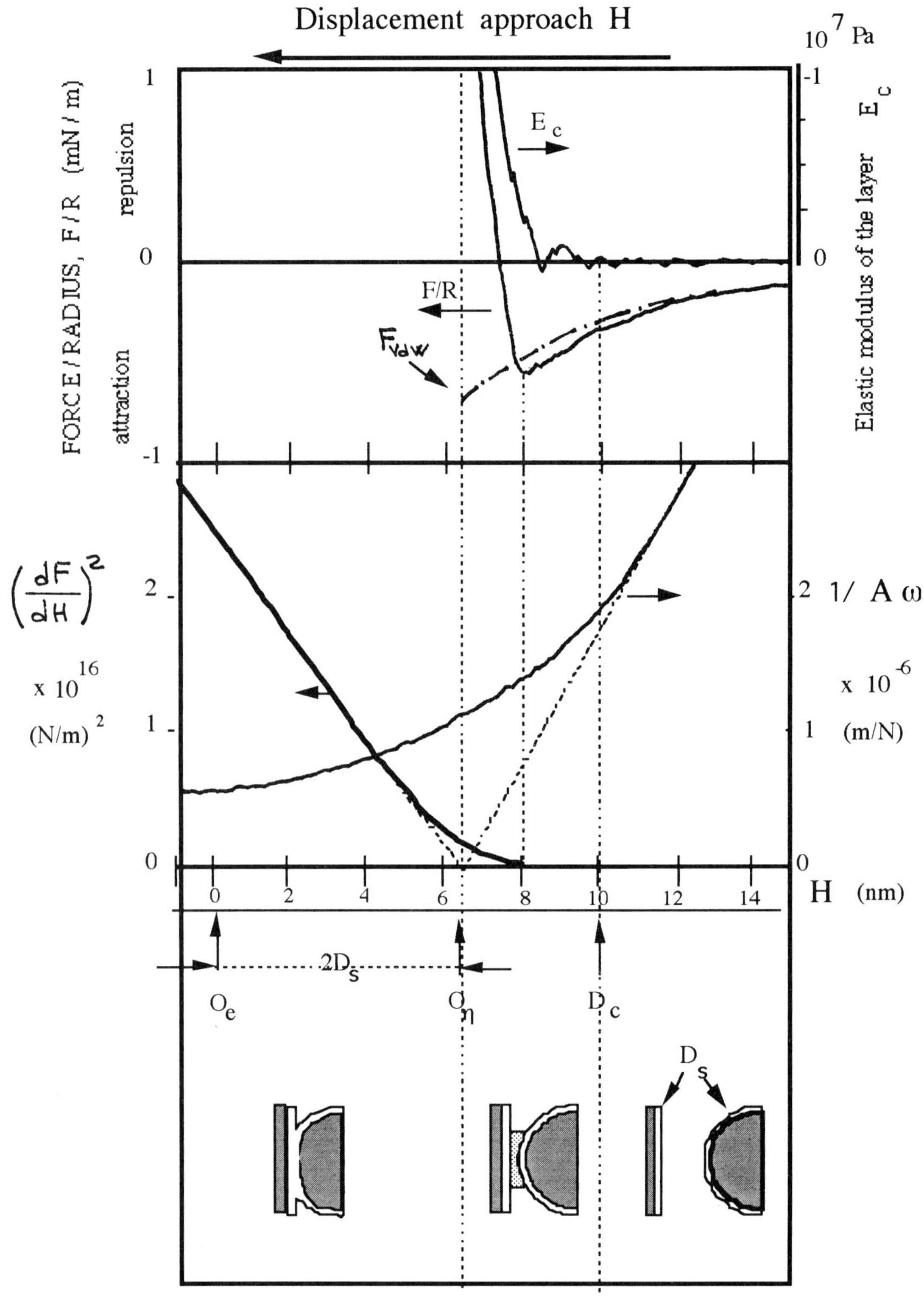

Figure 4: Drainage experiment with 2,4 dicyclohexyl-2-methylpentane (DCMP) between cobalt surfaces (Co_1). The extrapolated curves correspond to the electrical component $(\delta C/\delta D)^{-1}$ (which indicates the origin D=0), the function $\{F_e(D)\}^{-1}$ (which gives the distance $2D_{so}= O_{el}-O_e$). Also shown is the inverse of the damping function 1/A(D).

thickness of the confined layer is: D_c-$2D_{so}$=11-1= 10±0.5nm.

The measured static force can be written as the sum of three components: the Van-der-Waals contribution (F_{vdw}), the entropically driven force (F_{st}) (due to the propensity of the molecules to be oriented parallel to the surface){14}{15}, and the elastic contribution (F_{el}):

$$F_T = F_{vdw} + F_{st} + F_{el} \quad (13)$$

The first two forces are attractive, the last one is repulsive. The Van-der-Waals force, for an undeformed sphere and plane, is given by the relation{3}:

$$F_{vdw} = -\frac{A_{131}R}{6\left(D-2D_{so}\right)^2} \quad (14)$$

where A_{131} is the Hamaker constant of the cobalt-DCPM-cobalt interface, assumed to be close to that of cobalt-alkane-cobalt and then equal to 2 10^{-19}J {13}

The elastic contribution can be computed as follows. It is known {29} that, when a rigid sphere indents a elastic "mattress" deposited on a rigid plane, the elastic force is related to the displacement δ by the following relation {16}:

$$F_{el} = \pi E_c \frac{R\delta^2}{t} \quad (15)$$

where t is the thickness of the mattress, here taken as t= D_c-$2D_{so}$, and E_c is its elastic modulus δ=0 is taken as the displacement at which the sphere begins to touch the "mattress". Therefore the elastic modulus is computed by the use of the second derivative of the force as a function of the deformation dδ=-dH:

$$E_c = -\frac{1}{2\pi R}\left(\frac{d^2F_{el}}{dH^2}\right)\left(D_c - D_{so}\right) \quad (16)$$

The second derivative of F_{vdW} with respect to H is:

$$\frac{d^2F_{vdW}}{dD^2} = -\frac{A_{131}R}{\left(D-2D_{so}\right)^4} \quad (17)$$

Numerical solutions of the equations 16 and 17 have been made and compared with the value of the measurements F(D) and d^2F/dH^2. For D_c-$2D_{so}$=9nm it should be noted that F_{vdw} =-0.40mN/m, d^2F_{vdW}/dH^2= -1.5 10^{11}Pa, F= -0.38mN/m, d^2F/dH^2= +3.5 10^{13}Pa. Therefore it can be concluded that, for H≈D_c the second derivative of the Van-der-Waals force is negligible in comparison with the value of the second derivative of the measured force. The contribution of the entropically driven force is ignored, and it is assumed that the second derivative of this force is negligible in comparison with the value of the second derivative of the measured force. Experimentally we find that d^2F/dH^2>0 for H≈D_c≈ 9.5nm.

In conclusion in the range of H≈D_c, the elasticity of the confined layer is:

$$E_c(H) = -\frac{1}{2\pi R}\left(\frac{d^2F}{dH^2}\right)\left(D_c - D_{so}\right) \quad (18)$$

The plot of this function (figure 4), shows that, as H decreases, the elastic modulus increases progressively, as if the layer is compacted with some detectable instabilities. This behaviour is consistent with a confinement effect, and has already been suggested in the literature{3}{17}{18}. The value of the confined layer is about 3MPa. The elasticity of a molecular network is given by the relation: E_c ≈N kT, where N is the number of links per unit of volume{{19}. The data indicate that N=10^{27} links per cubic meter, or that the average distance between each link is $(1/N)^{0.33}$≈1nm; this distance is very close to that of the length of the molecules concerned(table 4){9}. Therefore the "confined" layer corresponds to a randomly organised "immobile" network of rigid molecules.

The static forces obtained during the outward displacement (figure 7) and plotted as a function of the solid separation D clearly indicate that the maximum elastic deformation carried by the contact has not squeezed out the "immobile" layers ($2D_s$). With complementary experiments it has been found that the "immobile" layers resist to a maximum pressure of 1GPa. The inward approach is evaluated by assuming that the adhesive energy w_i varies linearly during approach between w_1 corresponding to the maximum adhesive force during the approach and w_2 corresponding to the pull-up force.

The interaction between the two cobalt surfaces separated by DCPM now appears clearly. During the approach, an extra attraction in addition to the van-der-Waals force can be detected; then they appear some instabilities at a distance of some 2-3 molecular layers in the interface; finally the "hard" wall appears at H=$2D_s$. The pull-up force corresponds to an extra attraction to the Van-der-Waals force. It should be noted that previous work {14}{15} has shown that for unbranched hydrocarbons and polymer melts, molecules orient preferentially parallel to the surface (which is entropically unfavorable). This type of attractive entropic force, which is similar to the attractive depletion force should be described by a vector (orientational) order parameter. During the squeeze process, the molecules become more aligned, and lose some orientational entropy; so their entropy is lower and we expect an attraction. The extra work of adhesion due to the entropic effect can be obtained by integrating the outward force as a function of the separation and substrating the van-der-Waals contribution. For a contact area estimated to $2\pi(2.7.10^{-6})^2= 2.3\ 10^{-11}\ m^2$, the extra work of adhesion is $1.2\ 10^{-14}$J . The thickness of the interface corresponds to about 5 layers of molecules, so that the number of molecules in the squeezed interface (assuming an area of s.l= $85nm^2$ for each molecule) is:

$(2.3x10^{-11}\ m^2\ x\ 5)\ /\ (0.85x10^{-18}\ m^2)= 1.5x10^8$ molecules. The energy spent by each molecule is then $1.2\ x10^{-14}J\ /1.5x10^8=8\ x10^{-23}J= 0.02$ kT. The orientational entropy loss is generally considered to lie in the range of kT{20} so that, in our crude analysis, we find that the process is therefore probable.

In conclusion, with the semi-rigid hydrocarbon (2,4,dicyclohexyl-2-methylpentane DCMP), a relatively thick "immobile" layer is present near the solid surfaces (compared with a flexible molecule such n-dodecane). During the squeeze process, one confined layer appears due to molecule entanglement. The contact pressure orients the molecule giving a resultant layer subsequently resistant to a relatively high compressive pressure (1GPa).

V CONCLUSIONS

During the squeeze of a thin liquid film between a sphere and a plane, three critical thicknesses are detected. First, when the two surfaces are far from the contact, the flow of the film is hydrodynamic with the bulk viscosity of the liquid. The "immobile" layer (D_s-D_{so}) present on each surface does not partipate in this flow. At, a critical thickness D_c of the sphere-plane interface (such D_c>2(D_s-D_{so})) the separating flow ceases to behave in a classical hydrodynamic fashion. This corresponds to a confined layer. Finally, the ultimate contact corresponds to the elastic deformation of the two solids covered by an immobile layer.

These results suggest also that our measurements, although they concern a thickness layer with a small number of molecules, are explicable by macroscopic theories (as van-der-Waals, Reynold's) because they integrate over a relatively large area (of the order of a micrometer squared) in which a great number of molecules are present. In conclusion the measured thickness represents an averaged value on the measured area.

VI ACKNOWLEDGMENTS

The authors are grateful to H.A. Spikes for this help during the preparation of the manuscript, to R.C. Coy, D. Maugis, J.P. Montfort for many stimulating discussions.The authors are indebted to Shell Research Limited for financial assistance. We also thank the french CNRS and in particular all the members of the GDR 936 "mesures des forces de surfaces en milieu liquides".

VII REFERENCES

{1} D.Y.C. Chan and R.G. Horn, J. Chem. Phys., 83,10, 5211-5324 (1985)

{2} P.Debye and R.L.Cleland, J.Appl. Phys.30,843 (1959)

{3} J.N. Israelachvili,"Intermolecular and Surface Forces", second edition Academic Press, 1992.

{4} A. Tonck, J.M. Georges, and J.L. Loubet, J. of Colloid and Inter. Sci.126,1,1540-1563, (1988)

{5} J.M. Georges, S. Millot, J.L. Loubet, and A. Tonck, submitted to J.Chem. Phys. (1992)

{6} A. Tonck, F. Houze, L. Boyer, J.L. Loubet, and J.M. Georges, J. Phys.:Condens. Matter 3, 5195-5201, (1991).

{7} A. Tonck, F. Houze, L. Boyer, to be published

{8} A. Majumdar and B.Bhushan A.S.M.E trans.1990

{9} D.G. Gillis, S.J. Matthews and L.H. Sutcliffe, Magn. Res. Chem. 28, 371, (1990).

{10} H.A. Spikes, S.T.L.E trans. 33,1, 140-148, (1990)

{11} B.V. Derjaguin and N. V. Churaev, J. Colloid Interface Sci. 49, 249 (1974).

{12} H. Hertz, J. Reine Angew. Math. 92,156,(1882).

{13} J.M. Georges, J.L. Loubet and A. Tonck, C.R. Acad. Paris, La vie des sciences 9,2,1-12, (1992).

{14} J.N. Israelachvili, S.J. Kott, M.L. Gee, and T.A. Witten, Macromolecules, 22,4247,(1989).

{15} J.N. Israelachvili, S.J. Kott, M.L. Gee, and T.A. Witten, Langmuir 5,1111, (1989).

{16} K.L. Johnson, "Contact Mechanics" Cambridge Univ. Press (1985).

{17} R. Horn, S.J. Hirz, G. Hadziioannou, C.W. Frank, J.M. Catala, J. Chem. Phys. 90, 11, 1, 6767-6774 (1989).

{18} R.G. Horn, and J.N. Israelachvili, J.Chem. Phys. 75,3,1400-1411 (1981)

{19} L.R.G. Treloar, "the Physics of the Rubber Elasticity" Clarendon Press Oxford (1975).

{20} P.G. de Gennes, Adv. Colloid Interface Sci. 27,189-209, (1987).

Thin Films in Tribology / D. Dowson et al. (Editors)
© 1993 Elsevier Science Publishers B.V. All rights reserved.

The Strain Rate Dependence of the Interface Shear Strength of Poly(methyl methacrylate) Films

B.J. Briscoe, P.S. Thomas and D.R. Williams, Department of Chemical Engineering and Chemical Technology, Imperial College of Science, Technology and Medicine, London, SW7 2BY, U.K.

The frictional response for a range of thin poly(methyl methacrylate) (PMMA) films has been assessed in terms of the interface shear strength, τ. The dependence of τ on the shear strain rate, the morphology and the molecular structure of the polymer have been experimentally determined. These dependencies are interpreted using a molecular topographical model of the interface shear process. This type of model allows the accommodation of microscopic shear stresses through molecular deformation in the polymer film. In order to identify the molecular nature of these processes, a theoretical stress activated model of molecular flow is employed. The structural properties of the polymer films, and, in particular, the conformation of the pendant ester group, are characterised by FTIR. The current study proposes that the interface rheology of PMMA films is related to the conformation of the pendant ester groups attached to the carbon backbone chain of PMMA.

1. Introduction

In the description of the frictional properties of thin solid lubricant films, a parameter known as the interface shear strength, τ, is often employed. This parameter expresses the numerical value of the frictional force per unit of true contact area or, equivalently, the energy dissipated per unit area of contact per unit of sliding distance. The magnitude of τ is known to be a strong function of several important contact mechanical variables including the contact pressure, the temperature and the nominal strain rate. In addition, τ is also found to be strongly dependent on the morphology and the molecular architecture of polymeric films. This paper reports on a study of the strain rate dependence of the interface shear strength of thin poly(methyl methacrylate) (PMMA) films. The effect of the structural properties of the polymer film, in terms of the molecular conformation, is also discussed. A thermally activated rate limiting model is then applied in order to interpret the interface shear stress response at the molecular level and is subsequently correlated to the molecular conformation of the polymer chains.

1.1 The effect of strain rate

The strain rate dependence of τ has been studied by several authors (Amuzu *et al* (1976), Towle (1971), Briscoe and Smith (1983)). Experimentally, this dependence is generally found to be logarithmic and is expressed in the form:

$$\tau = \tau_0 + \theta \ln\left(\frac{V}{h}\right) \tag{1}$$

where τ_0 and θ are material constants and V is the imposed sliding velocity at constant temperature, contact pressure and film thickness, h. This expression predicts that as the strain rate, approximated by the ratio, V/h, is increased τ is expected to increase logarithmically. This logarithmic dependency is characteristic of retardation in shear in which the imposed shear rates are greater than the lateral molecular relaxation times that may be accommodated by the material. On this basis, the general expectation, as is observed, is that the value of θ is positive. The energy dissipation process within the film for this type of response is supposed to be one of plastic deformation, though some major deviations from this relationship have been observed. Amuzu *et al* (1976) found that, for PMMA, the effect of increasing the strain rate sometimes resulted in a decrease in τ (Figure 1). This trend was also reported for the strain rate dependence of calcium stearate monolayers (Briscoe and Evans (1982)). These deviations have been regarded as arising from retardation in compression (frictional heating is regarded as being a negligible contribution to the decrease in the shear stress). In this retardation mechanism, the relaxation modes normal to the plane of the shear stress become longer than the normal component of the shear strain rate; that is the contact frequency which is approximated by V/d, where d is a contact dimension such as the radius of the contact area. The transient applied contact pressure is, therefore, not accommodated by these normal relaxation modes (the contact pressure dependence of τ is discussed below). The resulting interface shear stress is reduced from the expected value and the value of θ becomes negative. In this case the dissipation process may be one of continuous interfacial brittle fracture.

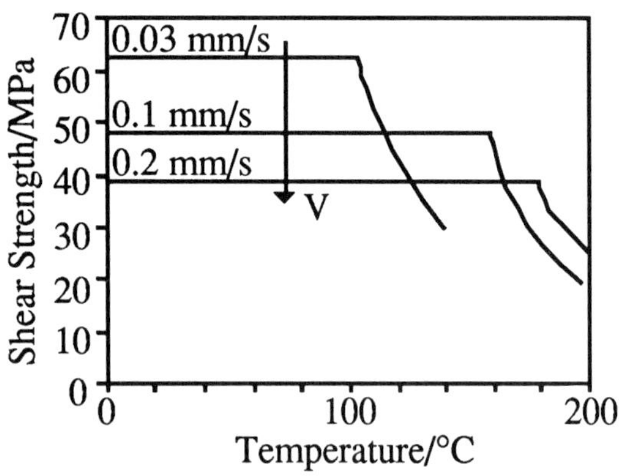

Figure 1. Temperature dependence of the interface shear strength for three sliding velocities (Amuzu *et al* (1976))

The strain rate, or shear retardation, dependence of the τ may be regarded as a function of the intrinsic shear strength, τ_0, in the contact pressure dependence relationship:

$$\tau = \tau_0 + \alpha P \tag{2}$$

where α is the pressure coefficient and P is the contact pressure. The normal shear stress retardation is introduced by a fictive pressure, P', which is less than the true applied pressure, P. Thus:

$$\tau = \tau_0 \ln\left(\frac{V}{h\theta}\right) + \alpha P' \tag{3}$$

where P' (or α) may be expressed as some function of the contact frequency.

1.2 The effect of morphology

τ, and its variation with the contact mechanical variables, is known to be a function of the polymer morphology (relative disposition of the molecular chains and their functional groups). The morphology of a thin polymer film is controlled by the preparation of the film. The effect of morphology on the interface shear strength of isotactic PMMA was studied by Smith (1980) and is shown below. In Figure 2, τ is plotted as a function of temperature for isotactic PMMA cast from two differing solvents. The solvents are described as either 'good' (chloroform $CHCl_3$) or 'poor' (decalin) depending on the compatibility of the solvent with the polymer.

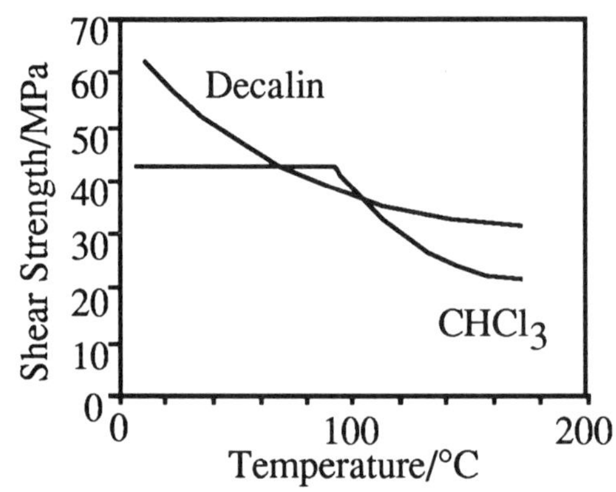

Figure 2. Temperature dependence of τ for isotactic PMMA cast from $CHCl_3$ and decalin (Smith (1980))

Two types of temperature dependence of τ are shown in this Figure. The sample cast from the 'good' solvent shows conventional temperature dependent behaviour of τ for glassy polymers. Two distinct regimes are observed. Below the glass transition temperature, T_g, of the polymer, τ is observed to be essentially constant. Above the T_g ($\approx$105°C for PMMA), a decreasing exponential dependence is generally seen. The sample cast from CCl_4 shows only the latter type of behaviour. The morphology has been thus engineered to extend the ductile region to below that of the normal T_g encountered in monolithic samples. This type of solvent dependence is well established for a variety of polymers (Briscoe and Smith (1983)).

This paper similarly deals with the effect of

solvent quality. In the current study, PMMA films are cast from chloroform ($CHCl_3$) and carbon tetrachloride (CCl_4). Both are effective solvents of PMMA, however, $CHCl_3$ is a much better solvent than CCl_4. This is shown by the relevant theta temperatures: $CHCl_3$, θ = -273°C and CCl_4, θ = +28°C. It is expected that the structure of a film cast from $CHCl_3$ will be more homogeneous and well ordered since the polymer chains are extended forming an affine film with a low density distribution. CCl_4 cast films will have a less homogeneous structure since, due to poorer solvation of the polymer, the polymer chains are not fully extended resulting in a greater distribution of densities and forming a less affine film. It is, therefore, expected that τ will be higher for films cast from the better quality solvents.

1.3 Structural Properties of PMMA

The dependence of the frictional properties on the morphology of the polymer film indicates that the interface shear response of a material is governed by the molecular architecture and the relative molecular conformations. An applied macroscopic shear stress gradient must be accommodated by the relaxation of microstrain by molecular entities. The model adopted in this paper was initially proposed by Pooley and Tabor (1972) and later used by Bowers (1971), Amuzu *et al* (1977) and Briscoe *et al* (1992). The basis of this model is that the deformation processes in an organic polymer film is often controlled by the molecular motion of the substituent groups on the main polymer carbon chain. Bowers (1971) has studied the effect of replacing a hydrogen atom with a fluoro, methyl or trifluoromethyl group on the friction of linear polyethylene chains. The resulting friction was found to increase with the molecular size of the pendant side group. Amuzu *et al* (1977) studied the effect of increasing the length of the side chain of poly(n-alkyl methacrylate) homologues and again found a similar result with the shorter pendant group chain lengths. Basically, the model supposes that the 'roughness' of the main chain is a prime factor in controlling the friction and somehow the main (backbone) chain becomes orientated in the direction of the imposed shear.

In this paper the morphological properties of PMMA films are examined by focusing upon the environment surrounding the pendant methyl ester functional group. This environmental assignment is achieved using vibrational spectroscopy. Fourier Transform Infrared (FTIR) spectroscopy has been shown to be a powerful tool in the analysis of molecular structure and environment of polymeric materials due to the unique conformational sensitivity of the technique. In this paper FTIR spectroscopy is used to detect environmental (intra- and intermolecular) changes in the structure of PMMA films prepared from different sources. Analysis of the vibrational modes within the polymer are used to provide a description of the molecular structure and the side group environment. These deduced structural characteristics are then compared to the observed frictional response.

1.4 Theoretical Molecular Approach to Interface Rheology

For rheological deformation processes to occur, molecular relaxation must also occur concurrently. One popular, but sometimes skeptically viewed, analytical method for describing the plastic deformation of polymeric materials involves the use the Eyring plasticity relationship of a stress modified thermally activated rate limiting process (Eyring 1936). The model produces the following expression:

$$V = A\exp-\left(\frac{\Delta E + \Omega P}{kT}\right)\sinh\left(\frac{\tau v}{kT}\right) \qquad (4)$$

where A is a constant, ΔE is the activation energy for the process, Ω and v are termed the pressure and stress activation volumes, respectively, and k is Boltzman's constant. Here the molecular analysis is, by inference, from the rheological response. The molecular parameters in the Eyring plasticity equation are Ω and v. These parameters are rationalised as being molecular volumes in the flow process. The pressure activation volume, Ω, is thought to be the excess volume required to be created locally in order that a volume of the molecular entity, ν, may flow in the shear stress gradient. Whatever the merits of the model, and the criticisms levelled at the interpretation of Ω and ν values, it has the merit of providing an energy dissipation process, albeit indirectly.

In the high stress limit of Equation 4, which is applicable to the deformation of thin polymer films, where $\tau\nu \gg kT$, the hyperbolic sine function becomes an exponential function:

$$V = \frac{A}{2}\exp\left(-\frac{\Delta E + \Omega P - \tau\nu}{kT}\right) \tag{5}$$

reducing to the logarithmic dependence seen in Equation 1. This theoretical model was initially developed to describe molecular flow in fluids as an example of a thermally activated rate process. It has been successfully used for the deformation of fluids in elastohydrodynamic contacts (Johnson and Tevaarweck (1977)) and for the deformation of bulk materials (Foot *et al* (1987)). It has also been used to interpret τ for thin polymeric films (Briscoe and Smith (1983)) and surfactants (Briscoe and Evans (1982)). Comparing the Eyring plasticity equation with the empirically observed dependence on τ a physical significance may be attributed to the Eyring parameters. By comparing Equations 1 and 5 the coefficient of the velocity dependence may be related to the stress activation volume of the Eyring equation:

$$\nu = \frac{kT}{\theta} \tag{6}$$

By calculating the stress activation volume, the size of the supposed molecular entities involved in the accommodation of a shear stress may, thus, be identified. The pressure coefficient may also be ascribed to the Eyring parameters:

$$\alpha = \frac{\Omega}{\nu} \tag{7}$$

and is found to be the ratio of the two supposed molecular volumes; the local volume which must be created to accommodate the local strain, Ω, to the volume, ν, of the entity which flows.

2. Experimental

2.1. General

The preparation samples of thin polymer films involved the casting of PMMA (Polyscience, MW 100,000) from chloroform ($CHCl_3$) and carbon tetrachloride (CCl_4) onto glass and potassium bromide (KBr) plates. The preparatory details are given in the following sections.

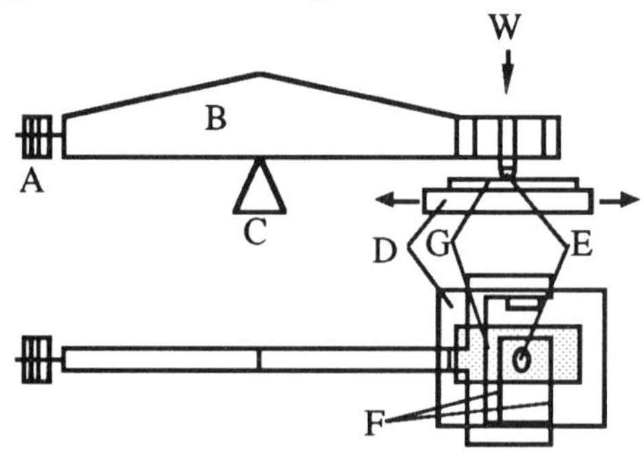

Figure 3. Friction apparatus for the measurement of the interface shear stress as a function of contact pressure, temperature or strain rate. Key: A: counter balance, B: pivoted beam, C: pivot, D: motion stage, E: hemispherical indenter, F: steel spring strain gauges, G: Thin film sample on glass slide.

2.2. Rheology

The friction experiments were carried out using the apparatus shown in Figure 3. The apparatus is based on equipment developed in earlier work by Briscoe, Scruton and Willis (1973) where a hemispherical, flame polished, glass indenter is attached to a pivoted beam which incorporates a force transducer to monitor the frictional force. A platform supporting the flat substrates (glass microscope plates) is raised to the level of the balanced pivoted beam, whereupon a normal load is applied. Relative motion is generated by sliding the platform beneath the indenter. The configuration described affords a spherical contact in which the contact area may be estimated by the use of the classical elasticity equations of Hertz (Timoshenko and Goodier (1951)). The computed Hertzian contact area is used to calculate the interfacial shear stress from the corresponding value of the frictional force. Naturally, there is some uncertainty in the calculated value of τ since the calculation of the contact area assumes that the surfaces are smooth and that the presence of the interposed film does not introduce serious error in the calculation. Previous studies, where the contact area has been directly measured, indicates that this procedure produces reasonably accurate estimates of the shear stress (Briscoe and Evans (1982)). The sample stage is fitted with a heating element where the temperature is controlled by a platinum resistor. The sample temperature was controlled to within ±2°C for low temperatures and ±5°C at temperatures greater that 120°C.

The polymer films were prepared by immersing microscope slides into dilute solutions of PMMA in $CHCl_3$ and CCl_4 at concentrations of 0.02%^{w}/w. Prior to immersion the glass slides were cleaned by a four step process. Initially, the slides were cleansed in acetone, followed by washing in dilute base (1M KOH), dilute acid (1M H_2SO_4) and finally washed in the casting solvent. The slides were then immersed in the polymer solutions for 30 seconds, washed with the casting solvent, in order to avoid variation in the film thickness, and were then allowed to drain vertically. Once dried, the samples were annealed at 160°C for a few hours and then stored in a desiccator until required. The interface shear strength was then measured at room temperature or at 160°C and as a function of the nominal strain rate. Five nominal strain rates were used as defined by the ratio V/h: $8.3 \times 10^4 s^{-1}$, $4.2 \times 10^5 s^{-1}$, $8.3 \times 10^5 s^{-1}$, $1.7 \times 10^6 s^{-1}$ and $5.0 \times 10^6 s^{-1}$. The nominal strain rates were calculated from the average values of the film thickness. The typical film thickness is in the region 100-200nm as determined by linear infrared absorption spectroscopy.

2.3. Spectroscopy

PMMA films were cast onto KBr windows at room temperature from 0.1%^{w}/w solutions in $CHCl_3$ and CCl_4. The solvent was allowed to evaporate to leave a film coating of the polymer on the window. The cast films were annealed at 120°C for 60 hours in air. The samples were then placed in a modified Linkam THMS 600 microscope temperature stage which was inserted in the micro-optics platform of a Bomem Ramspec 152 FTIR spectrometer. The sample cavity was then purged with dry nitrogen and spectra were acquired through a MCT liquid nitrogen cooled detector. The spectra were averaged over 200 scans at a resolution of $4cm^{-1}$. Spectra were recorded from room temperature up to 220°C in intervals of ten degrees.

3. Results

3.1 Rheology

The computed values of τ are plotted as a function of the logarithm of the nominal strain rate in Figures 4 and 5. At high temperature the strain rate dependence of τ shows a conventional response (retardation in shear) for both solvent systems (160°C; $T \gg T_g$). At 30°C, both traces for $CHCl_3$ and CCl_4 show higher values of τ than their 160° analogues. The values of τ for $CHCl_3$ at 30°C are, however, much greater than for the CCl_4 cast sample. They also follow a decreasing strain rate interface shear strength dependence. The sample cast from CCl_4 shows an increasing strain rate dependence and, although the absolute values are greater, the gradient of the curve is less than for those measured at 160°C.

3.2. Spectroscopy

The results shown are the relative intensity of a C-C-O asymmetric stretch of PMMA (Havriliak and Roman (1966)). This mode corresponds to the root of the chain branch attached to the carbon backbone chain, the presumed molecular topographical unit responsible for the main chain 'roughness'. The band is split into two vibrational frequencies. These frequencies are at $1265cm^{-1}$ and $1240cm^{-1}$. The splitting of this asymmetric stretch is associated with the conformation of the ester methyl group relative to the carbonyl band (Figure 6): $1265cm^{-1}$ is associated with structure II and $1240cm^{-1}$ is associated with structure I.

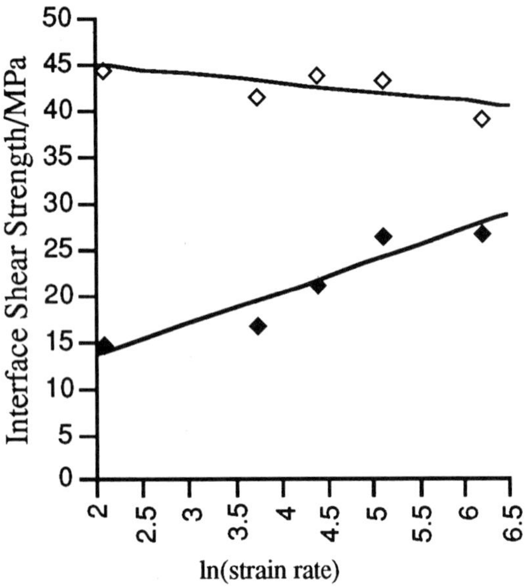

Figure 4. Logarithm of the strain rate dependence of the interface shear strength of PMMA cast from $CHCl_3$ at two ambient temperatures 30°C (◇) and 160°C (◆).

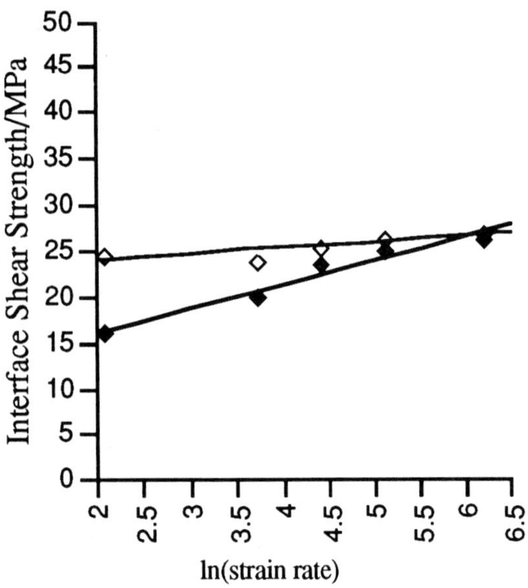

Figure 5. Logarithm of the strain rate dependence of the interface shear strength of PMMA cast from CCl_4 at two ambient temperatures 30°C (◇) and 160°C (◆).

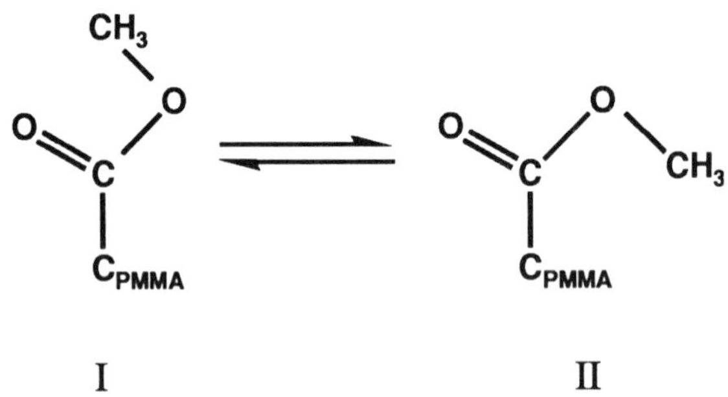

Figure 6. Conformational structures of the substituent ester group attached to the carbon main chain of PMMA. Structure I represents the conformer for the $1240 cm^{-1}$ vibrational mode of the C-C-O asymmetric stretch and structure II represents the conformer for the $1265 cm^{-1}$ vibrational mode.

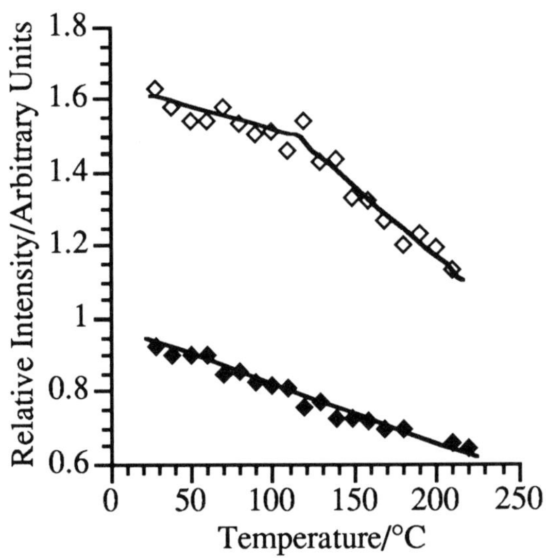

Figure 7. Ratio of the intensities of the $1240 cm^{-1}$ band to the $1265 cm^{-1}$ band plotted as a function of temperature for $CHCl_3$ cast (◆) and CCl_4 cast (◇) PMMA thin films.

The results are plotted as a ratio of the relative intensities of these bands (1240/1265) as a function of temperature (Figure 7). For both the CCl_4 and $CHCl_3$ cast samples the ratio deceases with increasing temperature. The population of the two sites is, however, different for the two solvent cast systems. The CCl_4 cast sample shows predominantly population of the more energetically stable conformer, I. For the $CHCl_3$ cast sample the population distribution is predominantly of conformer II.

4. Discussion

The interface shear strength dependence for PMMA on the strain rate is seen to follow two types of behaviour. The samples cast from both solvents at 160°C show a positive gradient for the strain rate dependence curve. This dependence shows a good correlation to the Eyring stain rate dependence equation and is consistent with plastic deformation within the polymer film. The sample of PMMA cast from CCl_4 at 30°C also shows a positive gradient, however it is much reduced. The $CHCl_3$ sample at 30°C shows a negative gradient and, although it shows a possible logarithmic dependence, it is inconsistent with the process of plastic yield. This type of behaviour has been shown to exist for the contact time dependence by Briscoe and Smith (1983) where retardation in compression is observed.

The calculated values for the Eyring parameters, Ω and v, are shown in Table 1. The values of the stress activation volumes, v, calculated for the interface shear stress data at 160°C are associated with the plastic deformation of the material. These values are 'true' (in as far as the Eyring approach is applicable - the significance of the parameters is notionally still uncertain) values as the deformation mechanism is consistent with the plasticity equation. The data for the samples at 30°C are invalid due to the deformation mechanism comprising partly of retardation in compression for the CCl_4 cast sample and almost entirely retardation in compression for the $CHCl_3$ cast sample. The values of the stress activation volume for the 160°C samples are high. They correspond to the size of several of the basic side group units. If the volume of a methyl methacrylate monomer is taken as $0.14 nm^3$, then the flow volume size is of the order of 16 monomer units for the CCl_4 cast sample.

	α	v/nm^3	Ω/nm^3
$CHCl_3$			
30°C	0.56	-5.9	-
160°C	0.21	1.8	0.38
CCl_4			
30°C	0.24	9.0	-
160°C	0.23	2.3	0.53

Table 1. Calculated values for the Eyring parameters of Equation 5 for PMMA samples cast from $CHCl_3$ and CCl_4 at 30 and 160°C.

The phenomenon of cooperative (multiples of functional units) flow is found in most descriptions of molecular deformation and is not surprising since the created volume required, Ω, to move one molecular entity is probably nearly the same as for many molecular entities. The difference in the size of the stress activation volume for the two solvent cast systems is quite large ($\approx$3 monomer units). This variation may be due either to differences in the number of units in flow or to different sizes of the flow unit. In the following passage the latter premise is pursued.

The molecular model adopted for the accommodation of microscopic shear stresses is a simple topographical model. The model assumes that the macroscopic shear strains are accommodated by the substituent functional groups attached to the carbon backbone chain. In the case of PMMA the substituent functional group in question is the ester functionality; this is the obvious choice if we assume a main chain orientation in the shear plane. FTIR spectroscopy indicates that there is a significant difference in the populations of conformations I and II, as shown in Figure 7, for the different casting solvents used. Specifically, conformation II is more prevalent in the $CHCl_3$ cast films. If the volume of the ester methyl group is taken as $0.034nm^3$ then the size of a flow entity comprising of predominantly structure II (as for the $CHCl_3$ cast sample) is $0.56nm^3$ less than a flow unit comprising of predominantly structure I (as for the CCl_4 cast sample) for 16 monomer units. This value correlates well with the difference in the calculated stress activation volumes ($0.5nm^3$).

The morphological structure of the $CHCl_3$ cast films is described as affine and homogeneous due to the elongation of the chains caused by the effect of solvation. On the application of a shear stress, several molecular units are forced into simultaneous flow. If the polymer chains are well ordered, then the action of the molecular entities sliding past each other cause the ester methyl to conform to the less energetically stable structure due to the even and close packing structure. The CCl_4 cast sample has regions of high density and low density due to the solvent interaction on deposition. An applied shear stress is most likely to be accommodated in the low density regions where the intermolecular interactions are less extreme and therefore the deformation of the ester methyl less pronounced.

5. Conclusions

The interface shear strength of organic polymers is known to be dependent on the molecular structure and overall architecture of the constituent macromolecules. In previous studies, molecular topographical models have been successfully applied to the study of the frictional properties of polymers. These studies have generally investigated the size and rigidity of the substituent functional groups. Analogously, PMMA has a 'rough' molecular profile defined by the pendant ester groups. In this paper a molecular topographical approach has been applied to the interpretation of interface shear strength data for PMMA. The 'roughness' of the substituent group attached to PMMA is examined as a function of the polymer morphology. The morphology of the polymer film is determined by the nature of the casting solvent. The structural conformation of the pendant group in PMMA, investigated using vibrational spectroscopy, has been found to be dependent on the casting solvent and thermal history. The casting solvent determines the molecular order in the film; thermal annealing relaxes the structure. Orientation of the polymer on deposition allows the formation of possible shear planes in which the deformation may occur. During this shear deformation it is assumed that the applied shear stress is accommodated by the relaxation of the substituent pendant group, and associated analyses of both, the rheological and spectroscopic data, provide a reasonable confirmation of this proposition.

Acknowledgements

The authors are pleased to acknowledge the financial support provided by Du Pont de Nemours.

References

Amuzu J.K.A., Briscoe B.J. and Tabor D. (1976) *Proc. JSLE-ALSE Int. Lub. Conf.* p155

Amuzu J.K.A., Briscoe B.J. and Tabor D. (1977) *Trans. ASLE* **20**(2) p152

Bowers R.C. (1971) *J. Applied Phys.* **42**(12) p4961

Briscoe B.J. and Evans D.C.B. (1982) *Proc. Roy. Soc. Lond.* **A380** p389

Briscoe B.J., Scruton B. and Willis R.F. (1973) *Proc. Roy. Soc. Lond.* **A333** p99

Briscoe B.J. and Smith A.C. (1983) *J. Applied Polym. Sci.* **28** p3827

Briscoe B.J., Thomas P.S. and Williams D.R. (1992) *Wear* **153** p263

Eyring H. (1936) *J. Chem. Phys.* **4** p283

Foot J.S., Truss R.W., Ward I.M. and Duckett R.A. (1987) *J. Mat. Sci.* **22** p1437

Havriliak S. and Roman N. (1966) *Polymer* **7** p387

Johnson K.L. and Tevaarweck J.L. (1977) *Proc. Roy. Soc. Lond.* **A356** p215

Pooley and Tabor D. (1972) *Proc. Roy. Soc. Lond.* **A329** p251

Smith A.C. (1980) D.Phil. Thesis, University of Cambridge

Timoshenko S.P. and Goodier J.N. (1951) 'Theory of Elasticity' (New York: McGraw-Hill)

Towle L.C. (1971) *J. Applied Phys.* **42**(6) p2368

Thin Films in Tribology / D. Dowson et al. (Editors)
© 1993 Elsevier Science Publishers B.V. All rights reserved.

Lubrication Regimes for Nanometer-scale Lubricant Films with Capillary Effects

J. L. Streator and J. P. Gerhardstein

George W. Woodruff School of Mechanical Engineering, Georgia Institute of Technology, Atlanta, GA, 30332-0405

Historically, the lubrication of solid surfaces has been characterized as comprising four major regimes: boundary lubrication, mixed lubrication, EHD, and hydrodynamic lubrication. One generally expects that the presence of the lubricant serves to reduce the friction of the interface. However, for very smooth surfaces, the introduction of lubricant can cause a substantial increase in friction over that of unlubricated sliding. While this increase in friction has been attributed to adhesive and/or viscous effects, it has been difficult to predict which of these contributions is most important to a given sliding regime. In the present work, an attempt is made to delineate regimes of lubrication where one or the other of the mechanisms is predominant. Measurements of friction are carried out for several perfluoropolyether lubricants in a thickness range of 0 to 80 nm and at sliding speeds of 0.25 to 250 mm/s. Three distinct regimes of lubrication are observed. The first, an "adhesion" regime, is characterized by large friction at low speeds and a decreasing friction with increasing sliding speed. The second, a hydrodynamic regime, is well described by classical lubrication theory and exists at higher sliding velocities than the first regime. The third regime, identified by an abrupt change in the dependence of friction on velocity, exists at relatively large sliding speeds and indicates substantial shear-thinning of the lubricant. The onset of the third regime is predicted from consideration of the tensile stresses in the lubricant film. It is also found modelling the lubricant as a continuum is valid at least down to 23 nm for the materials studied.

1. INTRODUCTION

Since the classical work of Stribeck (see, for example Czichos [1]) it has been recognized that there are four major regimes of lubrication, namely boundary lubrication, mixed lubrication, elasto-hydrodynamic lubrication (EHD) and hydrodynamic lubrication. It is commonly observed (or assumed) that the interposition of lubricant in the interface achieves a reduction in friction force as compared to unlubricated sliding. The standard argument for this expectation is as follows: Very thin films are able to support only a small fraction of the required load, but shear more readily than the solid/solid junctions at the asperity tips and therefore provide a lower friction force (boundary lubrication). As the film thickness is increased, the load fraction supported by the lubricant becomes comparable to that supported by solid/solid contact and the friction coefficient becomes quite low (mixed lubrication). Appreciably thicker films result in the complete separation of the surfaces and the shear resistance is then contributed solely by the lubricant (hydrodynamic lubrication) The EHD regime is but a special case of the hydrodynamic regime which appears at relatively large loads. In the hydrodynamic regime, the friction increases with sliding velocity, but it is unusual to operate at speeds that result in a friction force greater than for the unlubricated case.

The foregoing scenario applies to countless bearing configurations but there are some engineering applications for which the description is not entirely adequate. In the case of magnetic disk drives, for example, applying too much lubricant can lead to unacceptably large friction forces [2, 3]. The source of this large friction is the adhesive, capillary action provided by a thin liquid film which bridges the gap between to surfaces in close proximity. When large adhesive forces are present the friction has a behavior markedly different from the more common lubrication regimes. Since capillary effects are rarely observed for thick lubricant films, we suspect that there is a well-defined set of operating conditions for which the adhesive forces are important and others for which they are not. The purpose of the present work is to discover transitions from one type of frictional behavior to another and to perhaps add to the general understanding of lubrication phenomena.

2. EXPERIMENTAL

2.1 Materials

A set of four perfluoropolyether lubricants was selected for the study. Some of their properties are identified in Table 1. The temperature at which the viscosities are reported (26°C) corresponds to the ambient temperature during the tests. The reported values of kinematic viscosity were obtained by interpolation, using available viscosity data at other temperatures [4]. In addition to the lubricant, the sliding interface consists of an untextured, carbon overcoated, thin-film rigid magnetic disk, and an Al_2O_3-TiC ceramic slider. The slider is equipped with two rails, or "air-bearing" surfaces which cause it to fly during normal disk drive operation. The relevant feature in the present work is that the combined surface area of the rails is 3.1 mm^2. A sample disk from the test set of disks was found to have a surface roughness of 3-5 nm Ra. The surface of the slider was not measured for roughness, but it is generally found that the slider surfaces are made as smooth or smoother that those of the disks [5]. The use of these very smooth surfaces enabled the study of extremely thin continuous films of lubricant.

The lubricants were applied by a dip-coating technique [6] which consists of dipping the disks in a dilute solution of lubricant, and then raising the disks at a prescribed speed. The solvent chosen was Florinert FC-77, a perfluorochemical. By adjusting the lubricant concentration and the rate of withdrawal, one can obtain a wide range of lubricant film thicknesses [7]. In the present work lubricant concentrations ranged from 0.05 to 4%, and withdrawal rates from 2-8 mm/s, yielding film thickness of 2.3 to 80 nm. For a given withdrawal rate, the resulting film thickness is proportional to the lubricant concentration. After having calibrated the dip-coating process at a given concentration with a set of ESCA (Electron Spectroscopy for Chemical Analysis) measurements, it was a simple matter to prescribe the lubricant film thickness. Verification of the applied lubricant thickness was done via ellipsometry.

2.2 Test Apparatus

The disk is driven by a servo-controlled spindle, which provides a reasonably constant sliding speed range of .25 mm/s to 250 mm/s at a typical sliding track. The slider is attached to a flexible arm known as the slider suspension, which provides a spring load of 150 mN. In order to determine the friction force, we mounted the suspension to a cantilever-beam strain gage transducer. At the gain setting used, the force transducer had a sensitivity of 11.1 mV/mN. The resonant frequency of the transducer was measured and found to be about 800 Hz. Figure 1 schematically illustrates interface under study. Depending on the operating conditions, the slider and disk may or may not be separated by a full film of lubricant.

Table 1.

Lubricant	Kinematic Viscosity (mm^2/s) @26°C	Approx. Density (g/cm^3) @26°C	Surface Tension (dyne/cm) @26°C
143 AZ	29.4	1.9	16
143 AY	104.	1.9	18
143 AX	300.	1.9	18
143 AD	1030.	1.9	19

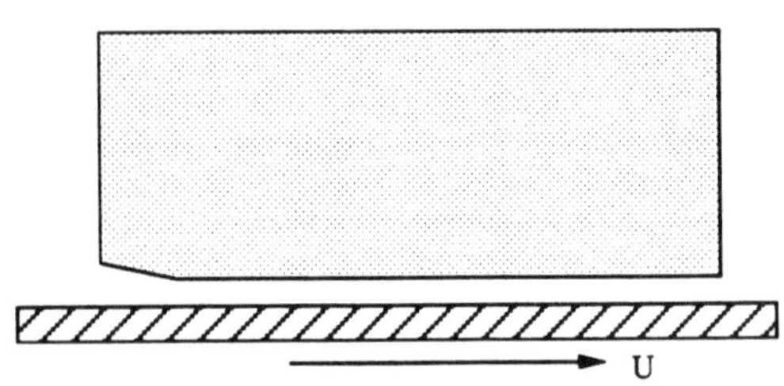

Figure 1. Schematic of interface

2.3 Test Procedure

A disk lubricated to one of the desired thickness (2.3, 4, 10, 23, 40, and 80 nm) and then clamped to the motor spindle. The slider was positioned on the disk at the desired sliding track. The friction was then measured at each of the test speeds (.25, 1.15, 2.5, 5.5, 11.5, 25, 55, 115, and 250 mm/s) for one complete revolution. At least several hundred data points were sampled and the average friction was recorded. To reveal particular friction details a few measurements were done with high-speed sampling (5 kHz) over a portion of the sliding track.

The entire test procedure was conducted beneath a clean air hood to avoid would-be contamination from dust. All tests performed at ambient temperature (26-27°C) and humidity (50-55 %RH).

3.0 RESULTS AND DISCUSSION

3.1 Hydrodynamic Regime

Figure 2 shows friction vs. sliding speed for a lubricant thickness of 80 nm, for each of the four lubricants. We first focus our attention on the middle portion of the graph. For each of the lubricants, the friction force is seen to be approximately linear with sliding speed. In addition, at any given speed within the considered speed range, the relative magnitudes of the friction force correlate reasonably well with the relative viscosities of Table 1. These two features indicate the slider load (150 mN) is borne by a full hydrodynamic film. At the left and right ends of the graph we observe large deviations from linearity, the source of which will be discussed **in later** sections.

Figure 3 shows results for a lubricant thickness of 40 nm. The same qualitative behavior is observed as in Fig. 2. However a close inspection will reveal that the hydrodynamic range has been squeezed and within this range the curves have shifted upward, in accordance with the higher shear rates experienced. Figure 4, which displays results for h = 23 nm, continues in the same pattern. An even smaller hydrodynamic range is observed and,

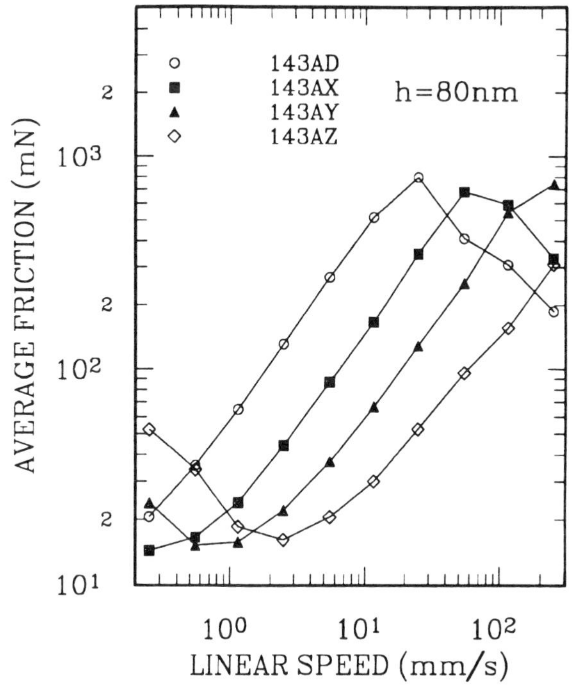

Figure 2: Friction vs. Sliding Speed at 80nm for each lubricant.

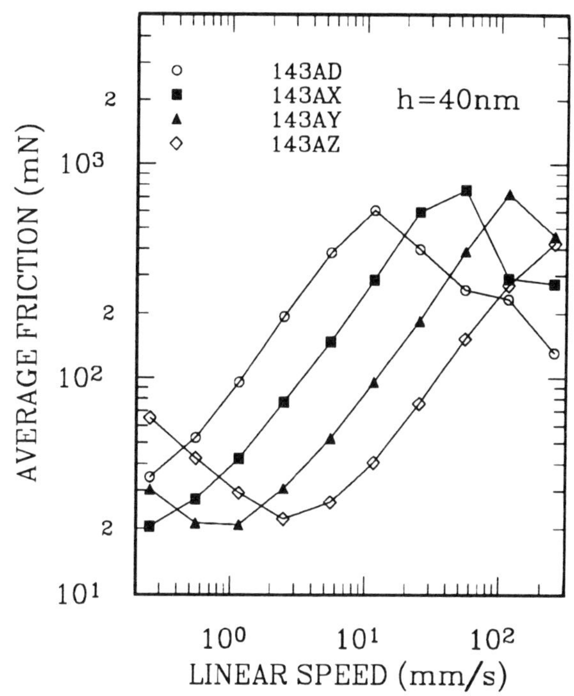

Figure 3: Friction vs. Sliding Speed at 40nm for each lubricant.

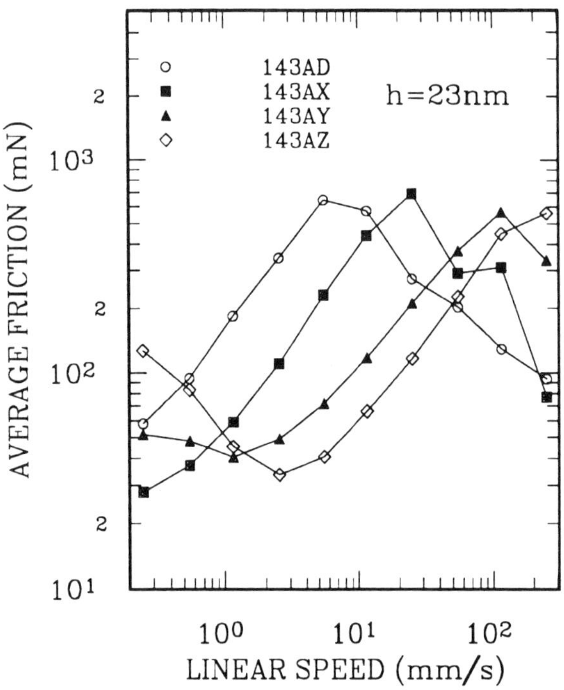

Figure 4: Friction vs. Sliding Speed at 23nm for each lubricant.

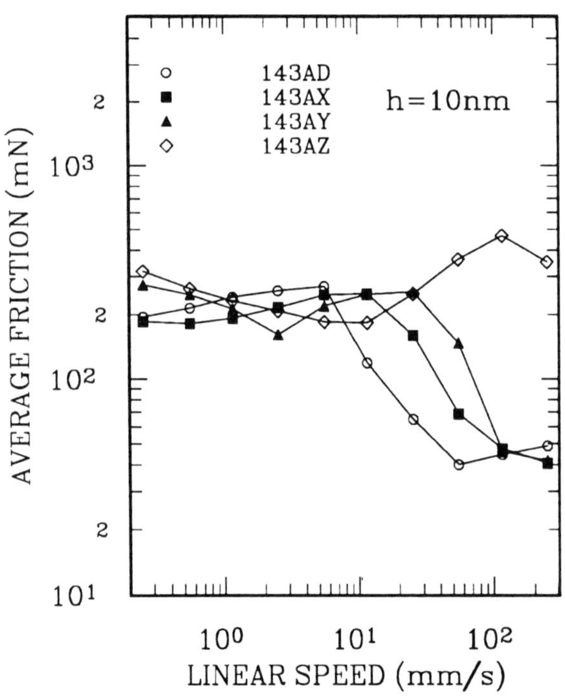

Figure 5: Friction vs. Sliding Speed at 10nm for each lubricant.

at any given speed within the range, the friction is found to be higher than for the thicker films.

3.2 Adhesion Regime

The results for a lubricant thickness of 10 nm, shown in Fig. 5, are quite different from those of the previous thicknesses. Only the lowest viscosity lubricant, 143 AZ, exhibits any friction dependence characteristic of a hydrodynamic film. Even for 143 AZ the friction forces are considerably smaller than would be expected from the increased shear rate. At 4 nm (Fig. 6), none of the friction traces show the behavior expected for a full hydrodynamic film.

Clearly a different mechanism of friction is operative, and it is apparently one of adhesion. Evidence for this conclusion is that the friction levels attained at low speeds in Figs. 5 & 6 are significantly higher than typical unlubricated values (about 50 mN). For example, in the case of 10 nm (Fig. 5), the maximum friction force is 318 mN, more than six times the unlubricated friction level. The large friction observed at low speeds is not consistent with the standard model of boundary lubrication which says that the lubricant will always decrease the friction by replacing solid-solid junctions with the (weaker) liquid film. Moreover, the classic boundary lubrication model suggests that the friction force should decrease as the lubricant film thickness is increased, whereas the measured low-speed friction was found to be larger with the application of a 10 nm lubricant film (Fig. 5) than with a 4 nm film (Fig. 6). (While it is true that the applied lubricant thicknesses is not the operating thickness when a complete film is not present, we would expect, in the absence of adhesion, the operating film thickness to be a monotonically increasing function of the applied thickness.)

The physics of adhesion is complex, and as of yet there is no universally accepted theory for the adhesion of lubricated solid surfaces. Following McFarlane and Tabor (1950) [8], one might attribute the adhesion to the development of meniscus bridges within the contact and the associated Laplace pressure. Alternatively, one could simply consider the attractive van der Waals forces between molecules separated by relatively

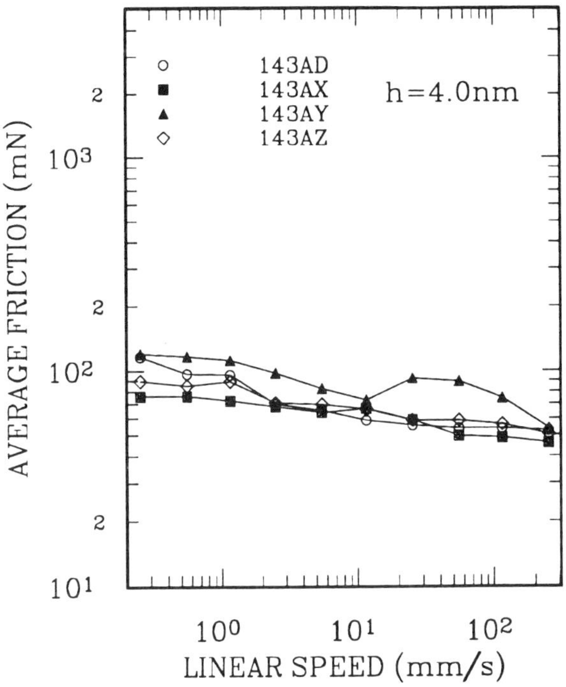

Figure 6: Friction vs. Sliding Speed at 4.0nm for each lubricant.

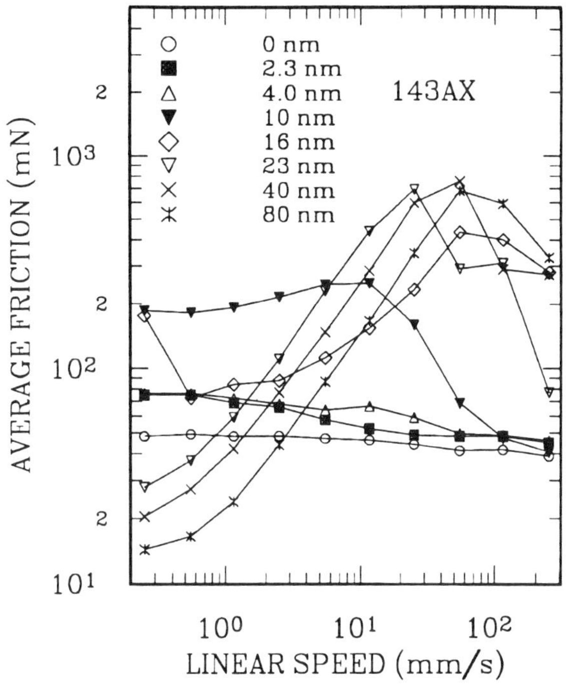

Figure 7: Friction vs. Sliding Speed for 143AX at several lubricant thicknesses.

short distances [9]. In either approach, one may treat the adhesive force as an additional external load on the interface. Since the lubricant is unable to support the increased load, the intensity of the solid-solid asperity contacts must increase, resulting in higher friction.

Figure 7, which shows the friction vs. sliding for lubricant 143AX at eight lubricant thicknesses, better elucidates the transition from the hydrodynamic regime to the adhesion regime. At the three highest thicknesses, the friction traces clearly indicate a complete fluid film at all speeds below the peak friction. At h = 16 nm the beginnings of a transition are evident. While the right side of the curve has similar features to the curves for the higher thicknesses, the friction forces in this region are found to be much lower than would be predicted from the higher shear rates. In addition, the large friction force at .25 mm/s indicates that strong adhesive forces are present. When the film thickness reaches 10 nm, the next transitional stage is reached. At low speeds, the

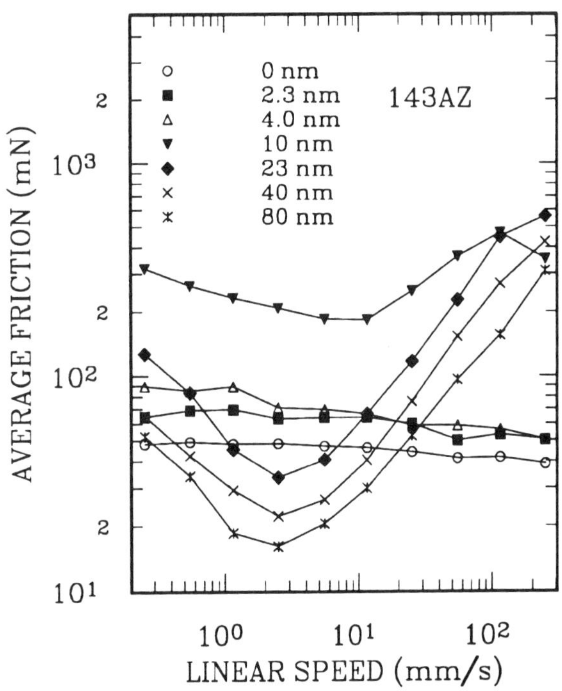

Figure 8: Friction vs. Sliding Speed for 143AZ at several lubricant thicknesses.

friction forces are well above the unlubricated values, while at high speeds, the friction is asymptotic to the unlubricated friction level.

At the lower thicknesses, the friction traces have much the same character of the unlubricated record. However, one can observe a non-negligible adhesive component at low sliding speeds which decreases in a regular way until it eventually disappears at the highest speeds.

Lubricant 143 AZ (Figure 8) exhibits much of the same behavior, although there is a new feature which should be pointed out. For the two highest thicknesses, we are able to observe the classic boundary lubrication regime of the Stribeck curve. At the lowest speed, the friction is very close to the unlubricated value, but as the speed is increased the friction drops significantly, reaching a minimum just before the initiation of a full hydrodynamic film.

Further insight into the nature of the lubrication regimes is obtained with reference to Fig. 9. Here friction versus lubricant thickness is plotted for the lubricant 143AX at the lowest sliding speed (.25 mm/s). The "error" bars represent the ranges of friction values recorded. (No ranges were recorded at 23 nm.) When the lubricant thickness is low, the friction force does not differ much from the unlubricated case. However, a little above this thickness, adhesive forces become large and the friction rises sharply, as observed by previous workers [2, 6]. As the friction is further increase the friction reaches a maximum, after which the interface attains boundary and, eventually, hydrodynamic lubrication.

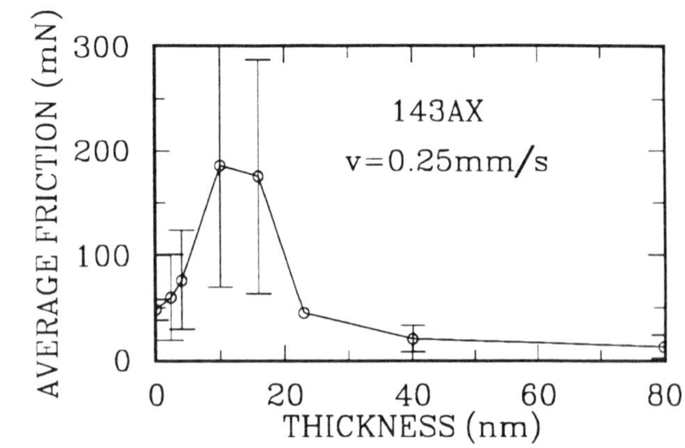

Figure 9: Friction vs. Lubricant Thickness for 143AX at 0.25mm/s.

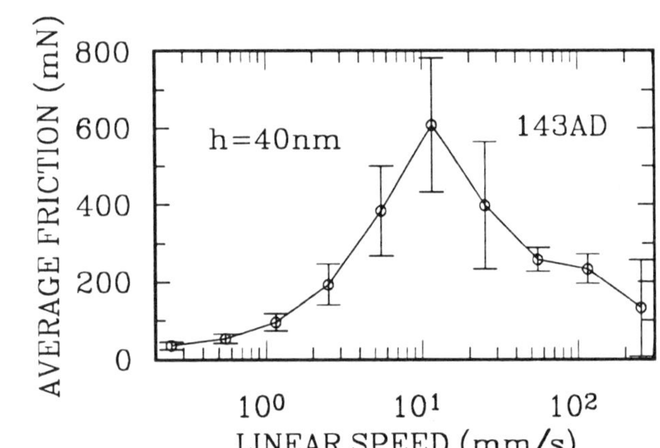

Figure 10: Friction vs. Sliding Speed at 40 nm for 143AD.

3.3 Cavitation Regime

It was observed in many of the previous figures that the friction force at high speed reached a definite peak, after which the friction exhibited a sharp down turn. This transition is highlighted in Figure 10, and the source of it is now explained. To do this we will first establish that the monotonically increasing portion of the curve is in complete quantitative agreement with classic lubrication theory. We first assume that our slider can be modeled as a one-dimensional bearing (this assumption will be justified later). The load and friction are then given by [10]:

$$W = \frac{6\mu UL^2 b}{h_2^2}\frac{\left[\ln a - \frac{2(a-1)}{a+1}\right]}{(a-1)^2} \tag{1a}$$

$$F = \frac{\mu ULb}{h_2}\left(\frac{4}{a-1}\ln a - \frac{6}{a+1}\right) \tag{1b}$$

where W is the slider load, F is the friction force,

h_1 and h_2 are the inlet and outlet film thicknesses, respectively, $a = h_1/h_2$, μ = the dynamic viscosity, U = the sliding speed, L = the length of the bearing (in the direction of sliding), and b = the depth of the bearing. For our slider we have W = .150 N, L = 4.06×10^{-3} m, and b = $.762 \times 10^{-3}$ m. As an example, we take typical values for our bearing as $\mu = 1.9 \times 10^3$ kg/m^3 $\times$ 29.4×10^{-6} m^2/s , U = 2×10^{-3} m/s, and $h_2 = 80 \times 10^{-9}$ m. This gives

$$\frac{\ln a - 2\frac{a-1}{a+1}}{(a-1)^2} = 1.14 \times 10^{-4} \tag{2}$$

It can be shown that (2) will be satisfied for a-1 = 1.37×10^{-3}, and a = 167. We may reject the latter solution as physically unreasonable when we recall the definition of a. Therefore, so far as the hydrodynamic regime is concerned, the film thickness is virtually constant., i.e. $(h_1 - h_2)/h_2 = a - 1 \approx 0$. (See Fig. 1). With this value of a, we obtain via (1b):

$$F = \frac{\mu U L b}{h} \tag{3}$$

Using continuity, it can be shown that the ratio of the flow rate (per unit width) in the transverse direction and the flow rate in the sliding direction is the same order of magnitude as a-1 (i.e. if the gap is constant, there is no need for fluid to flow out of the sides of the bearing). Our one-dimensional flow assumption is therefore validated.

We are now in a position to compare the theoretical prediction (3) to the actual measured values. Since μ is sensitive to temperature it is important to address the issue of viscous heating. By considering the isothermal flow solution for parallel plates, it is readily shown that the predicted fractional change in the viscosity due to heating is given by:

$$\frac{\Delta\mu}{\mu_o} = \frac{\beta\mu_o U^2}{8k} \tag{4}$$

where β is the viscosity-temperature coefficient for the lubricant, μ_o is the nominal viscosity, U is the sliding speed and k is the thermal conductivity. For example, with 143AX, $\beta \approx 0.05$, $\mu_o = 0.57$ Pa-s, k = 0.09 W/m-K. For a sliding speed of 0.1 m/s this gives:

$$\frac{\Delta\mu}{\mu_o} = \frac{(.05)(0.855)(0.1)^2}{8(0.09)} = 6 \times 10^{-4}$$

This result establishes that the viscosity is unaffected by viscous heating. However, if we use the viscosities from Table 1 in equation (3) to predict the friction force, we will not find good agreement with measured values, the predicted forces being substantially higher than measured values. With reference to equation (3) this suggests two possibilities: Either the actual film thickness, h, increases as the speed is increased or the viscosity decreases as the speed is increased, that is, the lubricant is shear-thinning. The former explanation seems physically implausible, since the load remains constant. In contrast some experimental evidence exists to support the latter conclusion.

Cantow et al. (1986) [11], studied the rheology of Fomblin YR at room temperature (25°C) and ambient pressure. (Bhushan, 1990 [12] provides a discussion of their work and presents some of their data in a more readable fashion). Fomblin YR is a branched perfluoropolyether lubricant quite similar to the lubricants used in present tests. They reported the viscosity of YR as a function of shear rate, and found shear-thinning behavior in YR at a range of shear rates and shear stresses that fall within the ranges of the present study. Since it is the shear stress rather than the shear rate [13] that defines the rheological state of a liquid, we make the following assumption in applying their measurements to the lubricants in the current study: At the same values of shear stress each lubricant should experience the same fractional reduction in viscosity. With this hypothesis, the applicability of equation (3) is demonstrated as follows:

1. Fit a parabolic regression to the data of Cantow et al. to relate the fractional reduction of viscosity to the shear stress.

2. For a given sliding speed, compute the actual shear stress in the interface from the measured friction force and the known slider area.
3. For the measured shear stress, predict the ratio of actual viscosity to nominal viscosity based on the regression of part (1).
4. Multiply this ratio by the predicted friction based on the nominal viscosity and compare the result with the measured friction.

Results of the above analysis are applied to lubricant 143AD at 80 nm and to 143AX at 23 nm and are illustrated in Table 2. As indicated the predicted friction force agrees quite well with the measured friction once the shear thinning is accounted for. Similar agreement is found in other cases.

Having established the applicability of (3) we now discuss the source of the peak friction observed in Fig. 10, and in many of the other friction traces. As discussed by Winer and Bair (1987) [13], the ambient pressure influences the maximum sustainable shear stress in a lubricant before it ruptures. Letting p denote the local gage pressure, τ the local shear stress, and p_a the ambient pressure, the analysis proceeds as follows: At any given point in the film, the maximum principal tensile stress is given by τ-p. The lubricant will rupture when the tensile stress exceeds the tensile strength of the lubricant, which is expected to be few atmospheres [13]. Suppose the tensile strength for each lubricant is $3p_a$. Then we have the following inequalities to avoid lubricant rupture:

$$\tau - p < 3p_a$$

$$F = \int_{slider} \tau dA < \int_{slider} p dA + \int_{slider} 3p_a dA$$

$$F < W + 3p_a A = 1.1 \text{ N}$$

This result agrees well the experimental data of Fig. 10 and as well as with the other friction peaks observed throughout the tests.

At speeds beyond the friction peak the friction is seen to drop rather abruptly. This means that the lubricant experiences dramatic shear thinning in this regime. For example, at a speed of 250 mm/s, the friction is 130 mN, which corresponds to a viscosity of 6.7×10^{-3} Pa-s, representing to a 300-fold reduction from its nominal value (1.95 Pa-s). The behavior observed after lubricant rupture may reveal general rheological behavior of the lubricant. Additional study is required to fully understand this regime of material deformation.

4.0 CONCLUSION

The frictional behaviors of several perfluoropolyether lubricants were investigated over a range of speeds (.25 to 250 mm/s) and lubricant thicknesses (0 to 80 nm). Based on analysis of the data we make the following conclusions:

1. In addition to the standard lubrication regimes, two additional regimes of lubrication may exist for smooth surfaces depending on the operating conditions. These are identified as an "adhesion" regime and a "cavitation" regime.

2. The adhesion regime exists at relatively low sliding speeds and is characterized by large friction forces at low speeds and intermediate lubricant thicknesses, and by decreasing friction with increasing sliding speed.

3. The cavitation regime initiates after lubricant rupture and indicates dramatic shear-thinning of the lubricant.

4. The maximum friction observable with any liquid film is ultimately limited by the ambient pressure.

5. The Reynolds equation of lubrication theory accurately predicts the load support and friction force in films as thin as 23 nm for the lubricants studied.

Table 2

SLIDING SPEED (mm/s)	FRICTION FORCE (mN)					
	143 AD @ h = 80 nm			143 AX @ h = 23 nm		
	theoretical using nominal viscosity	experimental	theoretical with shear thinning	theoretical using nominal viscosity	experimental	theoretical with shear thinning
0.25	19.0	20.6	17.7	19.3	27.9	17.2
0.55	46.7	35.6	35.8	47.4	37.1	36.2
1.15	87.3	64.8	67.9	88.7	59.1	70.1
2.5	190.	131.	131.	193.	110.2	137.
5.5	417.	269.2	252.	423.	230.5	263.

ACKNOWLEDGMENT

The authors would like to acknowledge support of this work by the National Science Foundation through Grant No. MSS-9119469. The authors would also like to express appreciation to Dr. Scott Bair for his insightful comments.

REFERENCES

1. Czichos, H., Tribology: A Systems Approach to the Science and Technology of Friction, Lubrication and Wear, Tribology Series 1, Elsevier Scientific Publishing Co., (1978).

2. Yanagisawa, M. "Lubricants on Plated Magnetic Recording Disks," Tribology and Mechanics of Magnetic Recording Systems, ASLE SP-19, pp. 16-20 (1985).

3. Streator, J., Etsion, I. and Bogy, D. B., "The Effect of Lubrication on the Static and Low-Speed Dynamic Friction in Thin Film Disks," Tribology and Mechanics of Magnetic Storage Systems, ASLE SP-25, pp. 24-29 (1988).

4. Product Information, E. I. DuPont & Nemours Inc., Wilmington, Delaware, USA.

5. Gitis, N. V., Volpe, L., and Sonnenfeld, R., "Long-Term Stiction at the Magnetic Thin-Film Disk-Slider Interface," Adv. Info. Storage Syst., Vol. 3, pp. 91-103 (1991).

6. Streator, J. L., Bhushan, B., and Bogy, D. B., "Lubricant Performance in Magnetic Thin Film Disks with Carbon Overcoat--Part I: Dynamic and Static Friction," ASME J. of Trib., Vol. 113, no. 1, pp. 22-31 (1991).

7. Buttafava, P., Bretti, V., Ciardiello, G., Piano, M., Caporiccio, G., and Scarati, A. M., "Lubricant and Wear Problems of perpendicular Recording Thin Flexible Media," IEEE Trans. on Magn., Mag-21, no. 5, pp. 1533-1535 (1985).

8. McFarlane, J. S. and Tabor, D., "Adhesion of Solids and the Effects of Surface Films, " Proc. Mat. Res. Soc. Symp., Vol. 119, pp. 87-92 (1950).

9. Stanley, H. M., Etsion, I. and Bogy, D. B., "Adhesion of Contacting Rough Surfaces in the Presence of Sub-Boundary Lubrication," ASME J. of Trib., Vol. 112, pp. 98-104 (1990).

10. Cameron, A., Basic Lubrication Theory, Longman Group Ltd., (1971).

11. Cantow, M. J. R., Ting, T. Y., Barrall II, E. M., Porter, R. S., and George, E. R., "Shear Dependece of Viscosity for Perfluoropolyether Fluids," Rheol. Acta, Vol. 25, pp. 69-71, (1986).

12. Bhushan, B., Tribology and Mechanics of Magnetic Storage Devices, Springer-Verlag, (1990).

13. Winer, W. O., and Bair, S. S., "The Influence of Ambient Pressure on the Apparent Shear Thinning of Liquid Lubricants--An Overlooked Phenomenon," Paper No. C190/87, Inst. Mech. Engrs. (Lond.), pp. 395-398 (1987).

SESSION XIV
COATINGS AND SURFACE MODIFICATIONS (3)

Chairman : Professor K Kato

Thin Films in Tribology / D. Dowson et al. (Editors)
© 1993 Elsevier Science Publishers B.V. All rights reserved.

Characteristics of Alumina Coatings Made by Machining-Assisted Plasma Spraying

K.Maekawa, I.Ohshima and S.Kajiyama

Department of Mechanical Engineering, Ibaraki University,
Hitachi 316, Japan

Making use of the characteristics of metal cutting, that a surface just after being machined is clean and chemically active, the authors propose a process of machining-assisted plasma spraying (MAPS) in which ceramic particles melted by a plasma heat source are sprayed onto a surface immediately following machining. This paper describes the concept and experimental apparatus, and examines characteristics of alumina coatings deposited by the proposed method. The characterization of bond strength, thermal resistance, wear resistance and grindability shows that MAPS is superior to the conventional technologies where grit blasting is used for a pre-treatment. The mechanisms for improving the interface characteristics are also discussed.

1. INTRODUCTION

Ceramics are used not only as structural components but also as thin films coated over metals and alloys. Developing coating technologies promote the latter use, such as thermal barriers for gas turbines, anti-wear coatings for moving parts and coated cutting tools. In terms of thermal spraying, its industrial applications are still limited since the bond strength of coatings is less than one tenth of the bulk metal. In order to increase the bond strength, various pre-treatments such as cleaning, roughening and heating of the surface to be sprayed play an important role [1].

The purpose of the present study is to develop a new technology to raise the bond strength. A surface just after being machined is chemically active [2, 3] because of its high cleanness and a moderate temperature above 100 °C. The proposed machining-assisted plasma spraying utilizes these characteristics of metal cutting. Characteristics of the alumina coatings are investigated, and mechanisms for governing the bond strength are discussed in the context of the obtained experimental results and thermal analysis by the finite element method.

2. THE CONCEPT

The key point of the proposed method is to spray melted and accelerated particles over the surface just being machined. Figure 1 shows the relationship between the machining point and the spraying position (a) for coating for the outer surface of a cylinder and (b) the end-face of an annular disc. In

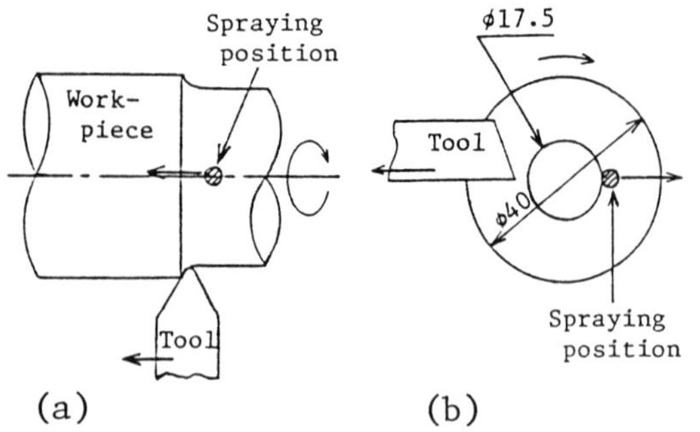

Figure 1. Relationship between cutting point and spraying position: (a) outer surface spraying and (b) cross section spraying.

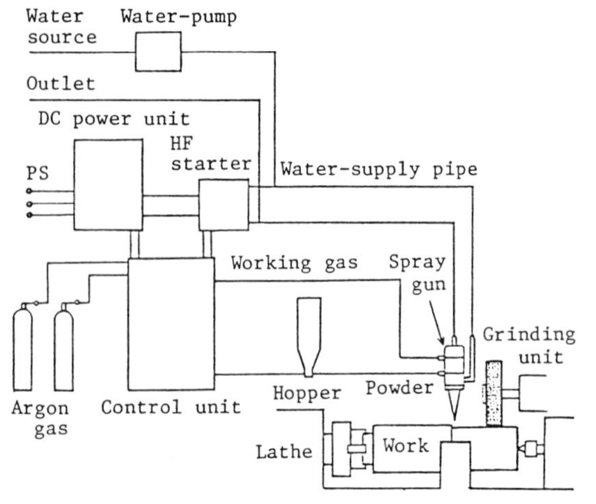

Figure 2. Schematic diagram of the machining-assisted plasma spraying (MAPS) system.

the case of (a) mounting the spray gun on the same carriage that feeds the tool axially maintains the position relationship between them unchanged. A special jig is needed in the case of (b). The integration of machining into spraying causes other factors to take part in the bond strength: residual stresses that develop during machining and heat removed by a chip.

3. EXPERIMENTAL PROCEDURES

3.1. Experimental Apparatus and Spraying Conditions

Figure 2 illustrates the overall structure of the present system. As indicated in Figure 1 (a) a plasma gun is fixed to the carriage of a lathe at the position of 90° behind the cutting point and moves with the cutting tool. The use of the grinding unit is discussed in Section 5.2.

Having performed preliminary experiments, the authors selected the following spraying conditions: an arc current of 200 A, argon as a working and carrier gas, a flow rate of the working gas of 6 l/min, a nozzle diameter of 3.2 mm, a cutting speed of 8 m/min, a feed rate of 0.14 mm/rev in the case of spraying over a cylindrical surface. The spray distance was in the range from 40 to 80 mm. Alumina powder of 10 to 40 μm in diameter was used. The effect of different pre-treatments for a bulk metal of 0.45%C steel was investigated: untreated (surface solvent cleaned), grit blasting and the present machining or clean-cut treatment. The steel has in the annealed state, HV=162 or HR30N=21 (superficial Rockwell hardness).

3.2. Testing and Characterization Methods

The force required for the separation of the coating from the bulk metal was measured by an adhesion test after an ASTM standard, where an annular disc (40 mm in outer diameter and 17.5 mm in inner diameter, the geometry of Figure 1 (b)) with a sprayed coating was glued to a dummy bar, and then was pulled with a special jig. The bond strength can be obtained by dividing the critical force by the glued area. Hardness and thickness of the coating were measured using a superficial Rockwell hardness tester and micrometer, respectively.

As for thermal cycling and thermal shock tests of the coating, the specimen was heated up to specified temperatures in an electric furnace with a heating rate of 23 °C/min. The thermal cycling test was performed between the heating temperatures and 300 °C by air cooling, whereas in the thermal shock test the heated specimen was held 10 min in the furnace, and then was dropped into water at 20 °C. Both tests were repeated until the coating failed.

Wear and scratch tests were performed in the circumferential direction on a bar using the same lathe as MAPS. A constant load was applied using the feed device of the lathe. A pin or an indenter was mounted on the tool holder with a tool dynamometer attached. Both pin and indenter were made of sintered cemented carbide, but the tip configurations were different: a cylindrical pin of a diameter of 2 mm and a tetrahedral indenter of an apex angle of 60°, a radius of 0.8 mm and an attack angle of 83° were used. The sliding velocities of 8 m/min and 2.3 mm/s were used in the wear and scratch tests, respectively. No lubricant was used.

Grindability of the coating was examined with respect to separation from the bulk metal using a cylindrical grinder. The grinding conditions used were as follows: grinder material, GC80H7V4; wheel speed, 2000 m/min; work speed, 8 m/min; feed rate, 0.14 mm/rev; depth of cut, 5 μm/pass.

4. EXPERIMENTAL RESULTS

Figure 3 shows the changes of thickness, hardness and bond strength of alumina coatings produced by MAPS as a function of spray distance. All of them increase with decreasing spray distance. But separation of the coating from the bulk metal occurred at a distance less than 35 mm; therefore 40 mm was determined as the spray distance. The maximum values of thickness, hardness and bond strength obtained are 0.31 mm/pass, 76 (HR30N) and 38 kgf/cm^2, respectively. The conventional spraying yielded the similar thickness and hardness. The bond strength was not measured since the coating was partly segregated from the bulk metal before the adhesion test.

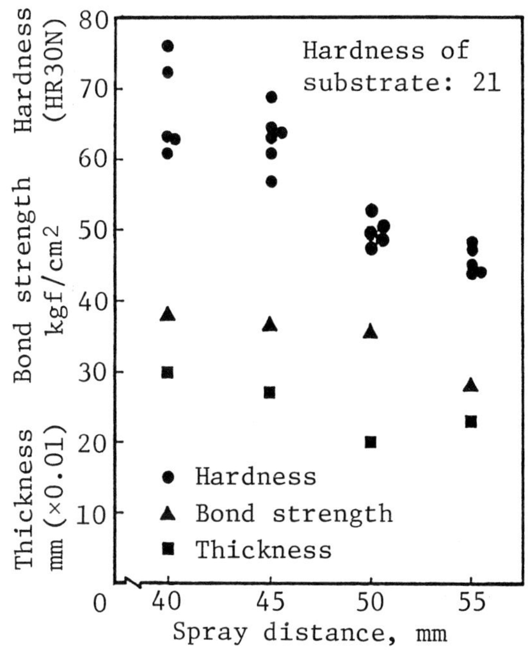

Figure 3. Thickness, hardness and bond strength of alumina coating as a function of spray distance.

Figure 4 shows the results of the thermal cycling test of the coatings produced by MAPS at various testing temperatures. No breakage occurred

(a) 500°C ↔ 300°C (13 cycles)

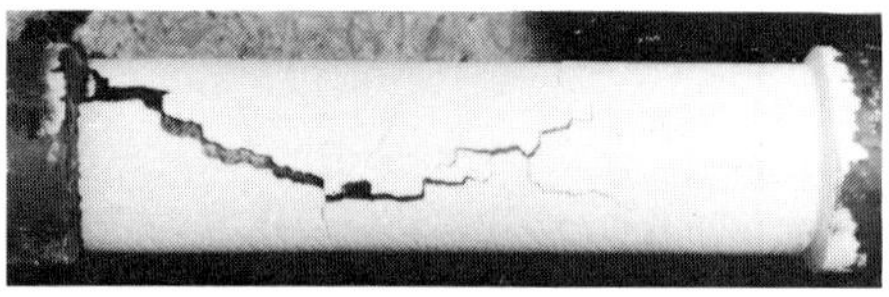

(b) 800°C ↔ 300°C (4 cycles)

(c) 900°C ↔ 300°C (4 cycles)

Figure 4. Coatings in thermal cycling test.

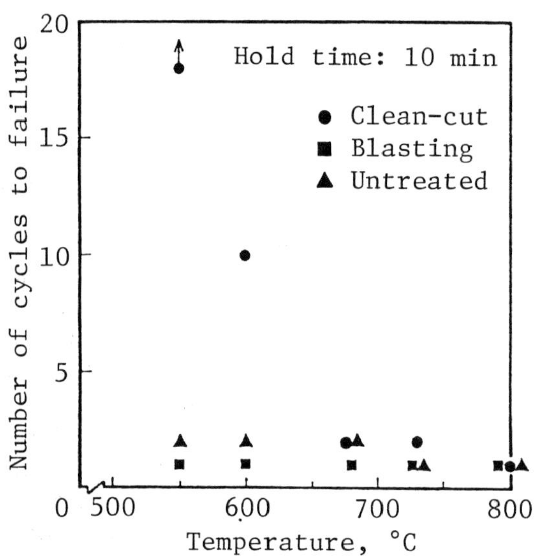

Figure 5. Relationship between temperature and number of cycles to failure in thermal shock test.

at 500 °C after 13 cycles, whereas 4 cycles are critical for failure to take place due to thermal fatigue, as shown in the figure, at the heating temperatures of 800 and 900 °C. Cracks started from an end of the coating and then propagated towards the center. The bulk surfaces exposed due to cracking were severely oxidized. Alumina coatings sprayed with the conventional pre-treatments failed after only one cycle at every heating temperature.

Figure 5 shows for the thermal shock tests the relationship between the heating temperature and the number of cycles to failure of alumina coatings with changing pre-surface treatments, where "clean-cut" means MAPS. No breakage of the coatings produced by MAPS took place at 550 °C after 18 repetitions of the thermal shock. The number of cycles to failure decreases with increasing heating temperature and no more than two was obtained above 680 °C. Spraying over the surfaces after blasting or untreated (as machined and left more than 10 min in air after machining) gives poor thermal shock resistance, regardless of the heating temperatures. Most separation of the coatings occurred during water cooling, and oxidation of the bulk surface was not as severe as that of the thermal cycling test.

Figure 6 shows surfaces worn by the wear test. The surfaces had been finished by a cylindrical grinder before the test. Surface roughness was measured across the sliding direction. Under an applied normal load of 20 N, neither breakage nor wear can be seen in the coating produced by MAPS after 12 hours sliding, as shown in Figure 6 (a), whereas Figure 6 (b) shows the bulk surface after 15 min sliding. The coatings produced by the conventional spraying were separated

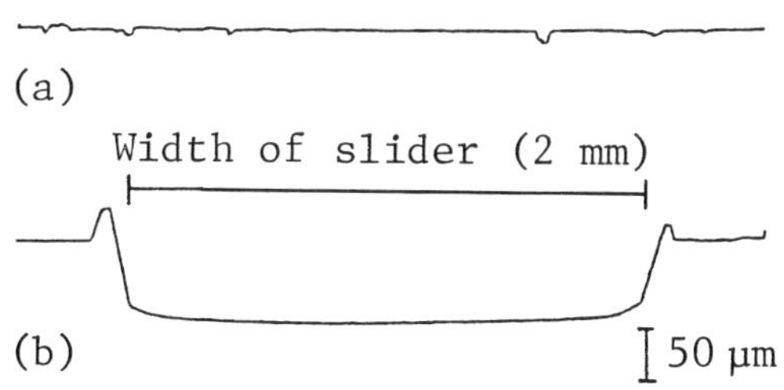

Figure 6. Configuration of wear tracks: (a) alumina with MAPS after 12 hr and (b) bulk surface after 15 min.

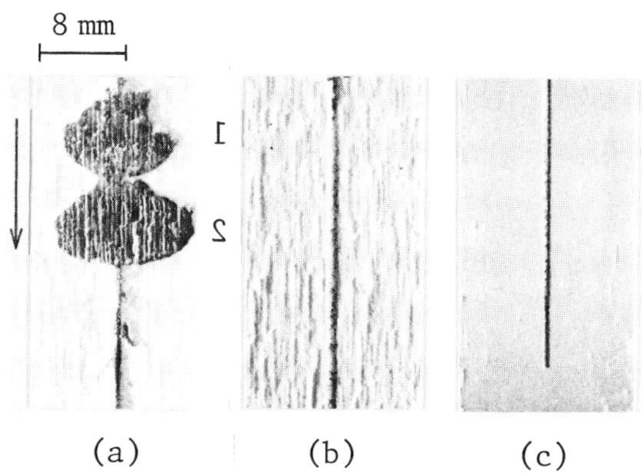

Figure 7. Alumina coatings on 0.45%C steel bar and scratches under applied normal load of 66 N: (a) conventional spraying, (b) MAPS, and (c) MAPS with grinding.

from the bulk metal during the wear test in less than 30 min.

Figure 7 shows the comparison of scratches on alumina coatings produced by (a) the conventional spraying (as machined and left 10

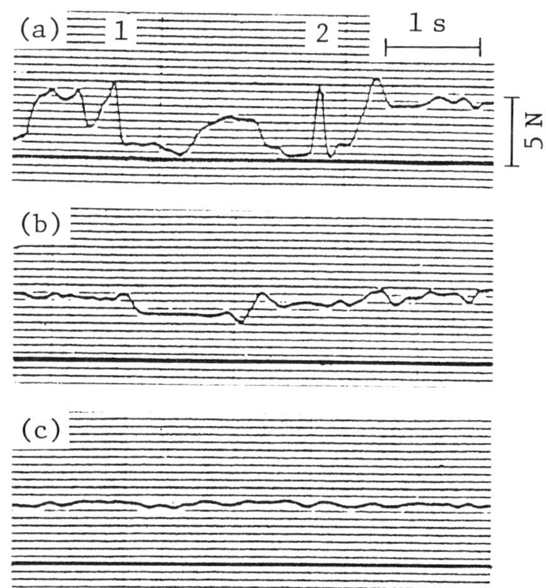

Figure 8. Variation of scratching force: (a) conventional spraying, (b) MAPS, and (c) MAPS with grinding.

Table 1. Scratch characteristics and estimation of bond strength under a normal load of 66 N.

	Thickness, mm	Maximum width, mm	Mean separation area, mm^2	Separation force, N	Shear bond strength, MPa
Conventional spraying	0.16 (0.16)	7.65 (7.13)	41.6 (32.3)	46.6 (47.3)	1.12 (1.46)
MAPS	0.15 (0.18)	NS (7.00)	NS (28.0)	NS (76.5)	NS (2.73)
MAPS with grinding	0.14 (0.14)	4.15 (5.00)	14.5 (16.0)	50.5 (44.8)	3.45 (2.80)

The values in ( ) are under a normal load of 127 N.
NS means no separation.

min in air), (b) MAPS, and (c) MAPS with grinding, when a normal load of 66 N was applied. Figure 8 shows the changes of scratch force or tangential force during scratching. The comparison of both figures follows that a sharp decrease in the force corresponds to large segregation of the coating from the bulk metal. The lower force level is that for scratching the bulk metal. If the strength of the coatings themselves or cohesive bond strength is the same among the three cases, the adhesive bond strength or the interface strength between the bulk metal and the coating produced by the conventional spraying is smallest of them all; large segregation is assumed to take place if the bond strength is low.

Using the area of separation and the scratching force, one can estimate a sort of shear bond strength. Table 1 summarizes the scratch characteristics, in which the force required for cutting the coating and the bulk metal are subtracted from the scratching force, resulting in the separation force. The values in parentheses indicate the results obtained under a normal load of 127 N. It should be also noted that the bond strength calculated here is somewhat underestimated, and it does not correspond to the tensile bond strength obtained in Figure 3. The superiority of MAPS and MAPS with grinding is clearly recognized.

5. DISCUSSION

5.1. Mechanism for Improving Bond Strength

In general, the bond strength of a coating is determined by the velocity on impact of the coating particle, the degree of particle melting and the substrate roughness [4]. With regard to MAPS there seem to be three major factors which govern the bond strength: chemical activity, residual stress and removal of heat by a chip during machining. It is difficult to identify the effect of chemical activity of the surface to be sprayed, but some workers [2, 3] reported that the clean metal surface which is formed by mechanical working or vacuum evaporation of metals has an unexpectedly high reactivity. One of the advantages of low pressure or vacuum plasma spraying is due to the use of a clean surface before deposition [5]. In the present MAPS system under the aforementioned spraying conditions, it takes only 0.236 s for the machined surface to be sprayed. During this time it is plausible that the chemisorption of atmospheric molecules may take place, but it is too short for the surface to be oxidized. Although the strength of chemical bonding between aluminum and iron atoms could be calculated by the molecular orbital method [6], one finds it difficult to distinguish the chemical effect from the mechanical effect or interlocking which depends on the magnitude and geometry of surface roughness [7].

Residual stress distribution for the machining process typically consists of a tensile stress of a high order of magnitude beneath the surface followed by low value of compressive stresses [8]. However, these might be small compared to the thermal stresses caused during plasma spraying. In order to clarify the influence of the residual stresses, three-dimensional thermal analysis by the finite element method has been employed. Postulating that an arbitrary

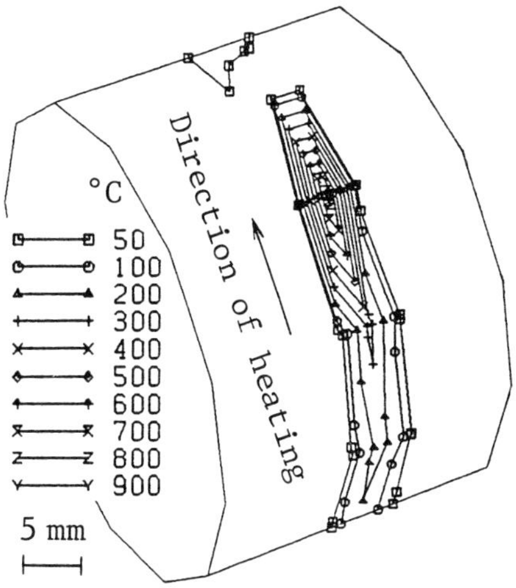

Figure 9. Distribution of surface temperature (°C) after one rotation from start of heating.

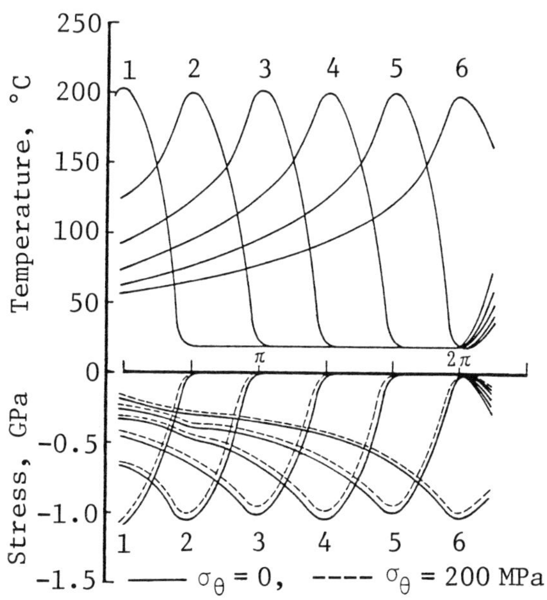

Figure 10. Time evolution of temperature and hoop stress at 0.5 mm beneath surface.

incident power of 1.26 kW with heating spot of 29.3 mm^2 moves on the surface accompanied by a convective heat loss of 1 $Wm^{-2}K^{-1}$, and neglecting the particle-plasma (melting) and particle-substrate (solidification) interactions [4], temperature rise and thermal stresses of a steel bar (Young's modulus: 206 GPa, Poisson's ratio: 0.3, yield stress: 300 MPa, thermal expansion: 10^{-5} K^{-1}) were calculated. Figure 9 shows the distribution of surface temperature just after one revolution or at 0.942 s from the start of heating. Figure 10 shows the time evolution of temperature and hoop stress at 0.5 mm beneath the surface. The broken lines indicate the hoop stress when an initial hoop stress of 200 MPa is taken part in the surface elements of 1 mm thickness. The assumed residual stress was tensile, being applied to the elements between the machining point and the spraying position. The solid lines represent the hoop stress without initial stress. The difference of the hoop stress is only 5 pct during the heating; consequently, little effect of the residual stress after the cooling to room temperature was obtained.

In the MAPS process, removal of heat by a chip must be taken into account. Assuming that a chip volume of 9.33 mm^3/s is heated up to 500 °C, a rough estimation gives only a heat loss of 20 W which is 1.6 pct of the input heat in the thermal analysis.

5.2. Integration of Grinding into MAPS

Surface finish by grinding is indispensable to the sprayed coating, for application to moving parts such as cams or bearing shafts. Whether the coating quality is good or bad depends not only on the spraying conditions but also on centering operations when a cylindrical grinder is used; even the coating by

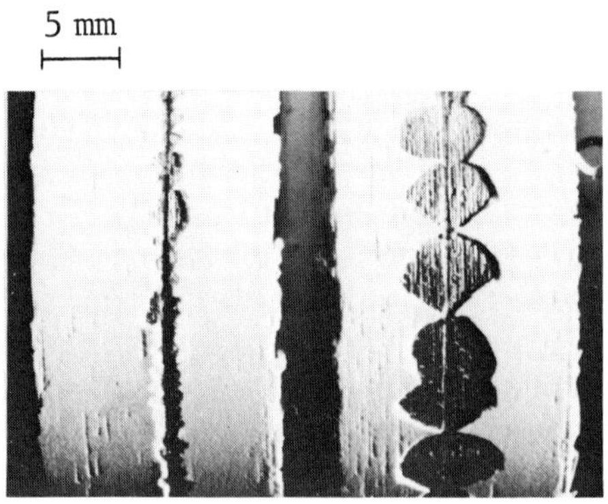

(a) (b)

Figure 11. Scratches under applied normal load of 165 N: (a) MAPS with simultaneous grinding and (b) MAPS with batch grinding.

MAPS was sometimes fractured or ground unevenly, resulting in exposure of the bulk metal. A candidate to solve the problem is to use a centerless grinder; an alternative is to integrate the grinding process into the MAPS system, as shown in Figure 2.

Figure 11 shows an advantage of the integration of grinding into MAPS, where typical scratched grooves under an applied load of 165 N are compared between the combined methods: (a) MAPS with simultaneous grinding and (b) MAPS with a batch of grinding more than 10 min after MAPS. The grinding conditions used were the same as those of the grindability test. No defects were seen on the ground surface by visual inspection before the scratch test. The combination of grinding with MAPS yields a higher bond strength of the coating to which the applied grinding force makes a contribution. On the contrary, the batch grinding induced degradation of the coating followed by periodic and large separation from the bulk metal. Under an applied load of 66 N no separation occurred in both cases.

6. CONCLUSIONS

A process of plasma spraying combined with machining has been proposed and successfully employed to produce a high quality of alumina coatings. Making use of a clean and active bulk surface which has just been machined is essential in the proposed machining-assisted plasma spraying (MAPS). Integration of grinding into the MAPS system, in which the grinding is conducted simultaneously with MAPS, makes a contribution to raising the bond strength between the coating and the bulk metal. The characterization of the coatings showed that MAPS was superior to the conventional spraying technologies where various surface pre-treatments such as cleaning, roughening and heating are fundamental.

The mechanisms for governing the interface characteristics in the MAPS process have been discussed in the context of the obtained experimental results and thermal analysis by the finite element method. It is concluded that the primary role of raising the bond strength is due to chemical activity of the bulk surface to be sprayed; of course, spraying conditions surely affect the bond strength as well.

ACKNOWLEDGMENTS

The authors gratefully acknowledge the support of the Japanese Ministry of Education (Grant-in-Aid for Scientific Research, Nos. 01460097, 03650093) and the helpful cooperation of Daiichi Metco Co., Ltd. and Hitachi Koki Co., Ltd.

REFERENCES

1. B.J.Gill and R.C.Tucker Jr, Mater. Sci. Technol., **2** (1986) 207.
2. G.W.Rowe, E.F.Smart, K.C. Tripathi, ASLE Trans., **20** (1977) 347.
3. S.Mori, M.Suginoya and Y.Tamai, ibid., **25** (1982) 261.
4. D.Apelian, M.Paliwal, R.W.Smith and W.F.Schilling, Int. Metals Rev., **28** (1983) 271.
5. A.R.Nicoll, H.Gruner, G.Wuest and S.Keller, Mater. Sci. Technol, **2** (1986) 214.
6. A.B.Anderson and S.P.Mehandru, J. Electrochem. Soc., **132** (1985) 1695.
7. R.L.Mullen, R.C.Hendricks and G.McDonald, Ceram. Eng. Sci. Proc., **8** (1987) 559.
8. E.Brinksmeier, J.T.Cammett, W.König, P.Leskovar, J.Peters and H.K.Tönshoff, Annals of CIRP, **31** (1982) 491.

Thin Films in Tribology / D. Dowson et al. (Editors)
© 1993 Elsevier Science Publishers B.V. All rights reserved.

Friction and wear behaviour of PVD-coated tribosystems

H. Sander[a] and D. Petersohn[b]

[a]Federal Institute of Materials Research and Testing (BAM)
Laboratory 5.22, Festkörperreibung und Schwingungsverschleiß
Unter den Eichen 87, 1000 Berlin 45, Germany

[b]Federal Institute of Materials Research and Testing (BAM)
Laboratory 5.31, Physikalische Oberflächentechnologien; Vakuumbeschichtungstechnik
Unter den Eichen 87, 1000 Berlin 45, Germany

ABSTRACT

The purpose of this investigation is to determine the friction and wear behaviour of several PVD-coatings on steel against steel counterbodies under lubricated and unlubricated sliding conditions in dry and humid air and in vacuum. The aim is to minimise friction and wear by selecting suitable coatings and to optimise them in order to produce a self-lubricating system which needs no external source of lubricant.

The experiments were performed by means of a model test system with a ball on disc tribometer with oscillating motion. Both friction and wear results are described as a function of coated sliding pair combination and test conditions. A wide variety of coating systems has been tested: TiN, TiC, TiCN, WN, WCN, TaN, TiAlN, CrN, soft metal and chalcogenide coatings and amorphous carbon (Me-C:H).

The results showed that the friction and wear behaviour of PVD-coated steel sliding pairs is governed by the deposition variables, coating structure and the surrounding conditions (atmospheric humidity). The best results with respect to low friction coefficient and high wear resistance were achieved with carbon coatings.

1. INTRODUCTION

The deposition of hard coatings on cutting and forming tools by physical vapour deposition (PVD) to increase their endurance and efficiency is nowadays widely practised, whereas the coating of engineering components is still in the development phase.

The subject of the present investigation is to apply this deposition technique to tribological machine elements, with two common objectives:

- to increase the wear resistance of the surface material,
- and to modify its frictional behaviour.

The aim of the work was to minimize friction and wear by optimization of the coatings, in order to produce a self-lubricating system which needs no external source of lubricant during its lifetime.

The investigation is related to conveyor chains which are widely used in production lines, for example in the automobile industry. Improvements in the wear resistance of these chain elements, which must work without fluid lubrication to avoid contamination by oil, especially in paint shop areas, are of great interest. Other important areas where the tribological applications of surface coatings can be helpful are, for example, in food-processing machinery where contamination of the product by a liquid lubricant must be avoided, or in space where a liquid lubricant would evaporate.

2. EXPERIMENTAL DETAILS

2.1. Coating process

Coatings of about 5 μm thickness were deposited by physical vapour deposition (PVD), which allows mechanically resistant materials to be deposited at substrate temperatures of 200-500 C. The fact that PVD processes are carried out at relatively low temperatures makes them very attractive, since the coating process may then not influence the microstructure and properties of the underlying substrate material. Consequently, coatings can be applied to previously quenched and drawn steels without loss of hardness or distortion.

Quality control starts with the arrival of the specimens to be coated at the coating plant. The specimens are examined for surface quality. Hardness measurements and surface roughness determinations are carried out prior to coating.

After coating additional measurements are performed to obtain information on the thickness, hardness and adhesion of the coatings. The adhesion is established by a scratch test in which the so-called critical load is determined.

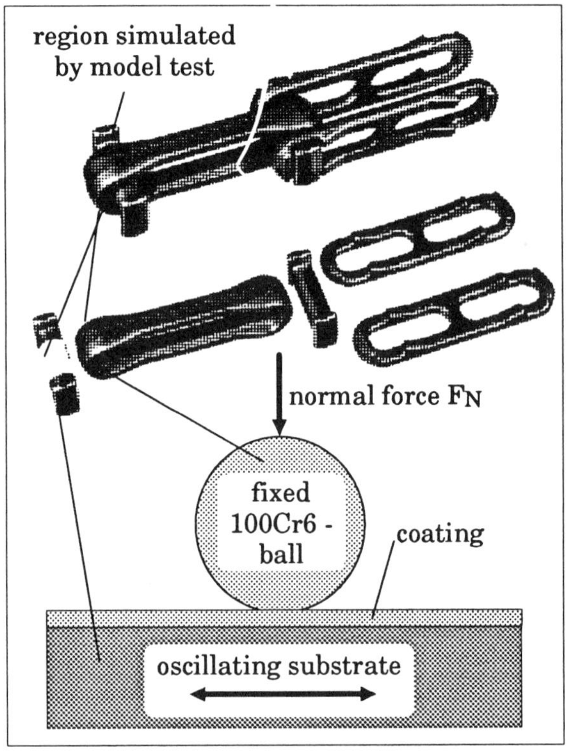

Figure 1. Elements of a conveyor chain and geometry of model test arrangement.

2.2. Experimental approach

The upper part of Figure 1 shows the contact regions in the conveyor chain links. Improvement of the friction and wear behaviour by deposition of hard coatings onto the contact surfaces was investigated by model tests, in which the contact conditions occurring in practice were simulated. The geometry of the model test is shown in the lower part of Figure 1. It contains only those parts of the practical system which take part in the wear and friction process, separated from the rest of the system. The plane corresponds to the link pin whereas the ball, which is pressed against the specimen by a normal force, corresponds to the chain link of the simulated system. The normal force applied to the contact simulates the pressure caused by loading the conveyor chains.

2.3. Apparatus

The experiments were carried out using a specially designed ball-on-disc tribometer for reciprocating sliding, which was capable of measuring the frictional force, the ball-disc distance, the electrical resistance of the contact and the relative position of the two counterbodies on a continuous basis. The test rig used to achieve these objectives is shown schematically in Figure 2.

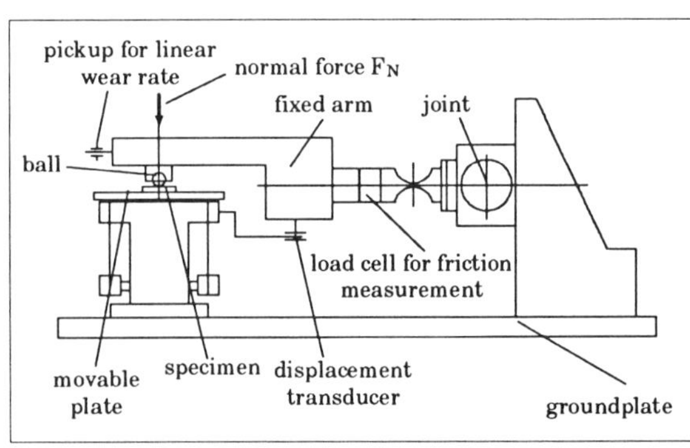

Figure 2. Schematic diagram of the experimental apparatus.

The top specimen (a sphere of 10 mm diameter) is located in a holder which is clamped to a fixed arm with transducers for friction force and wear rate measurement. The lower specimen, a coated disc, is mounted on a movable horizontal plate, supported by two vertical elastic beams. Contact is achieved by pressing the ball against the flat surface of the coated specimen under a fixed normal force F_N, applied by a dead weight. An electromagnetic vibrator under closed loop control was used to produce a reciprocating motion of the support plate. The heart of the test equipment is contained in a vacuum chamber which allows the surrounding conditions to be changed. The tangential friction force F_t, resulting from the oscillating movement, can be measured with an inductive load cell, mounted on the fixed arm. The linear wear rate is determined continuously by measuring the distance perpendicular to the plane of the specimen by strain gauges attached to a spring which contacts the upper ball holder. This method of measurement gives the sum -for both ball and specimen- of all material lost below the original datum surface. The slip amplitude is detected by strain gauges attached to a spring which contacts the oscillating plate. The experiments were carried out under normal loads between 1 and 50 N, at a slip amplitude of 100 μm (i.e. a stroke of 200 μm) and a frequency of 5 Hz. The triangular driving waveform was used to give a constant sliding speed (2 mm/s). The number of sliding cycles was up to 10^6.

After ultrasonic cleaning with acetone, the specimens were mounted on the holders and brought precisely into contact without misalignment. The advantage of using bearing steel balls (100Cr6 German standard) as one counterbody in all the tests is, that they are manufactured in large batches, so that little variation occurs in the size, finish and mechanical properties of the balls within one batch. A further advantage is that alignment of the two counterbodies is straightforward, so that the initial contact pressure is precisely controllable by the normal force.

2.4. Specimen and material

A basic objective of the experimental work was to use materials similar to those found in practice. Thus the following steels were selected as coating substrates for the model tests:

- SAE 1552, structural steel
- 100 Cr 6, ball bearing steel
- S 6-5-2, high speed steel
- C 45, C 60 and 42 CrMo 4, tempered steel
- X 155 CrVMo 12 1, cold working steel

These specimens were cylindrical in shape, with a diameter of 25 mm, a thickness of 4 mm and a central hole 7 mm in diameter for mounting. After machining, the specimens were first heat treated to the hardnesses shown in Table 1 and then polished to a surface roughness of 0.01 μm Ra. With one pair of specimens about 10 tests can be performed by rotating both the ball and the disk through a small angle.

Table 1
Hardness HK 0.075 of the substrate materials.

Material	initial state	after hardening	after coating
100Cr6	260	870	570
C45	280	720	-
42CrMo4	400	740	600
X155CrVMo	320	800	790
S6-5-2	370	730	1020
SAE1552	320	740	480

3. RESULTS

3.1. Lubricated contact system

As many contact systems operate in the presence of a lubricant, lubricated systems were investigated first. Figure 3 shows the progressive optimisation of the friction and wear coefficient of the lubricated system, achieved by variation of the material and TiN coating,.

In the first stage of improvement, a distinct decrease of friction and wear is achieved by

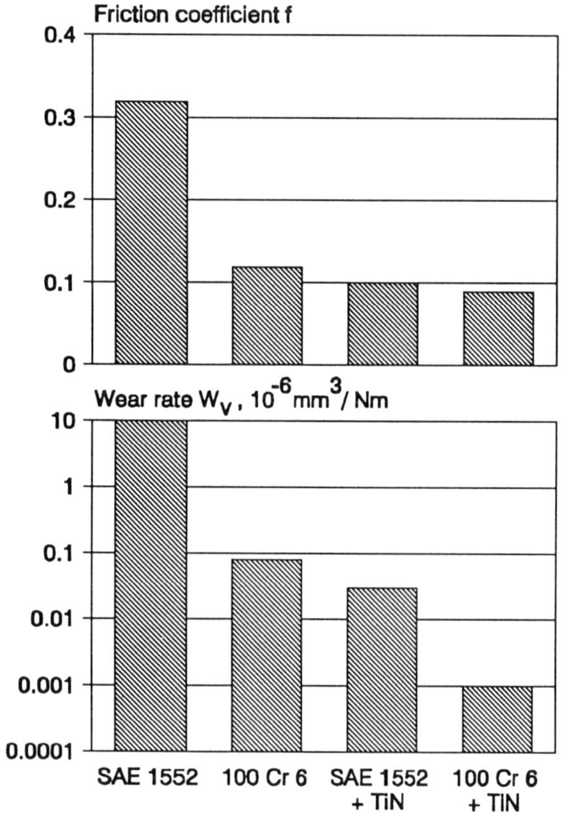

Figure 3. Friction and wear behaviour of the lubricated tribosystem; normal force F_N= 20 N, initial pressure p_o= 1226 MPa, sliding amplitude s_o= 100 μm, number of cycles N_{max}= 10^6, lubricant= ISO VG 100 (Optimol).

replacing the conveyor chain material SAE1552 by the high quality bearing steel 100Cr6. In the second step, about the same improvement is achieved by coating the chain material SAE1552. The best result is reached in the third step, by both changing the material and coating it with TiN. The friction coefficient is then lower than 0.1 and the wear is reduced by 4 orders of magnitude compared to the initial state. Transferring these result to the practical chain element or similar systems the following improvements are achieved:

- low friction reduces dissipated energy,
- very low wear rate, so that almost no wear particles are produced,
- single lubrication for the whole lifetime, because the function of the lubricant is not disturbed by wear products,

and from the economical point of view:

- reduction of maintenance and inspection costs,
- no cost for the disposal of drained off lubricant.

The replacement of TiN by the hard- coatings TiC or Ti(C,N) leads under the same conditions to nearly the same results:

- friction coefficient f= 0.1,
- wear rate k< 0.05 $10^{-6} mm^3/Nm$

It is remarkable that this behaviour is achieved in spite of the relatively high contact pressure (p_o= 1228 MPa) which remains during the whole duration of test because there is nearly no wear. This is caused by the low shear stress which results from the low friction.

3.2. Unlubricated system

3.2.1. Hard coatings

Values of the friction coefficient f and wear rate k for the optimised hard coatings TiN, TiC and Ti(C,N) deposited on the alloy 100Cr6, compared to the uncoated system, are listed in Table 2 for two different atmospheric conditions: moist air of 25 % and dry air of <1 % relative humidity (RH). The results are averaged over test durations of 250000 cycles.

The optimization was achieved by varying the PVD process conditions under the criterion of minimising friction and wear.

In dry air the hard coatings decreased the wear by more than one order of magnitude, whereas in moist air only a small improvement was achieved. The friction was relatively high for TiN, still quite high for TiC and low for Ti(C,N), but in all cases the frictional work was still too high, so that all coatings were defective at the end of the tests. Because of the relatively high friction, the pure hard coatings are unsuitable for use in unlubricated systems.

Table 2
Friction and wear of TiN, TiC and Ti(C,N) coatings on 100Cr6 in dry and moist air in comparison with the uncoated system; normal force F_N= 5 N (initial pressure p_o= 770 MPa), sliding amplitude s_o= 100 μm, sliding speed v_s= 2 mm/s, number of cycles N_{max}= 250000.

coating	f		k, $10^{-6}mm^3/Nm$	
	RH<1%	RH=22%	RH<1%	RH=22%
TiN	0.78	0.38	2.3	5.3
TiC	0.44	0.36	3.7	4.5
Ti(C,N)	0.28	0.36	1.6	3.9
none	0.61	0.59	40	6.0

Similar high friction in moist air was shown by the hard coatings WN, W(C,N), TaN and TiN+CuN, as indicated in Table 3.

Table 3
Friction and wear of WN, W(C,N), TaN and TiN+CuN coatings on 100Cr6 in moist air of 22% relative humidity; normal force F_N= 5 N (initial pressure p_o= 770 MPa), sliding amplitude s_o= 100 μm, sliding speed v_s= 2 mm/s, number of cycles N_{max}= 250000.

coating	WN	W(C,N)	TaN	TiN+CuN
f	0.6	0.59	0.69	0.55
k,$10^{-6}mm^3/Nm$	1	3	43	8

As hard coatings which are characterised by high oxidation stability, TiAlN and CrN were thoroughly studied. In addition, CrN coatings are characterised by low internal stresses, which permits high coating thicknesses. The tribological behaviour of (TiAl)N and CrN coatings under three different atmospheres is shown in Figure 4.

All the hard coatings investigated reduced wear, but the friction was so high that the contact surfaces were damaged by the high shear stress. It was therefore attempted to reduce the friction by additionally depositing low friction films on to the hard coatings. Since the lubricated, hard coated systems had shown that below a friction coefficient of about 0.1 no damage occurs, additional self lubricating films were added in an attempt to reach this

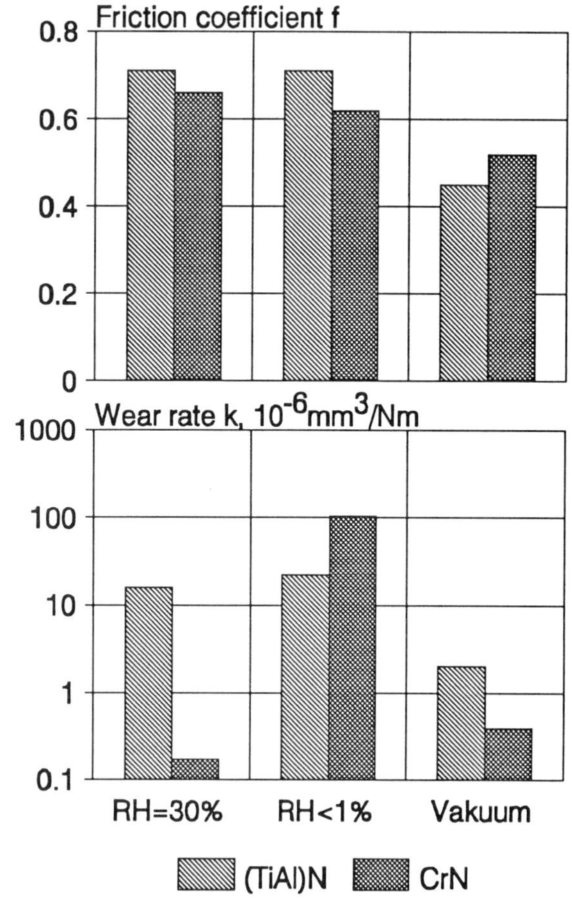

Figure 4. Friction and wear of (TiAl)N and CrN coatings on 100Cr6 in moist air of 30%, dry air of <1% relative humidity and vacuum of p= 0.1 mbar residual pressure; normal force F_N= 5 N (initial pressure p_o= 770 MPa), sliding amplitude s_o= 100 μm, sliding speed v_s= 2 mm/s, number of cycles N_{max}= 50000.

value. The hypothesis underlying the behaviour of such complex coatings is the principle of function sharing. The hard coating serves as a wear resistant, supporting foundation and a low friction coating over that serves as a solid lubricant.

3.2.2. Soft metallic coatings

At first soft metallic coatings were used as a friction reducer on TiN hard coatings. The minimum friction coefficient for the materials

Au, In, Cu and Pb on TiN against steel 100Cr6 is listed in Table 4.

Table 4
Minimum friction values of soft metallic vapour deposited films on TiN on 100Cr6 in dry and humid air; normal force F_N= 5 N (initial pressure p_o= 770 MPa), sliding amplitude s_o= 100 μm.

coating	friction coefficient f_{min}	
	rel.h.= 0.1 %	rel. h.= 20 %
TiN+Au	0.17	0.19
TiN+In	0.24	0.15
TiN+Cu	0.27	0.27
TiN+Pb	0.28	0.32

Table 5 shows friction coefficient f, wear rate k and life time N_L for some soft metallic films in dry air. The life time is defined by the number of cycles at which the friction increased rapidly. As a typical example, the tribological behaviour of the coating TiN+Pt is shown in Figure 5.

Table 5
Friction coefficient f, wear rate k and lifetime N_L of Au, Pt on TiN as well as In films in dry air of <1 % relative humidity (material: S 6-5-2 for In, 100Cr6 else).

coating	TiN+Au		TiN+Pt		In
F_N, N	1	5	1	5	5
f	0.57	0.44	0.7	0.18	0.63
k, $10^{-6}mm^3/Nm$	0.84	2.15	35	~0	79
N_L, 10^3	11	2	7	18	~0

The relatively high average friction values can be ascribed to the fact that the high pressure and the reciprocating sliding displaces the soft metal out of the contact zone by plastic strain, so that the high value friction of the hard coating returns.

3.2.3. Chalcogenide coatings

From the group of chalcogenides the selenide coatings WSe_2 and $NbSe_2$ and the sulfide coating MoS_2, which is known as a self-lubricating coating [1], were investigated. These coatings were directly deposited on the specimens without a hard coating as support. The resulting friction and wear behaviour of these coatings on different materials in dry and humid air is summarized in Table 6.

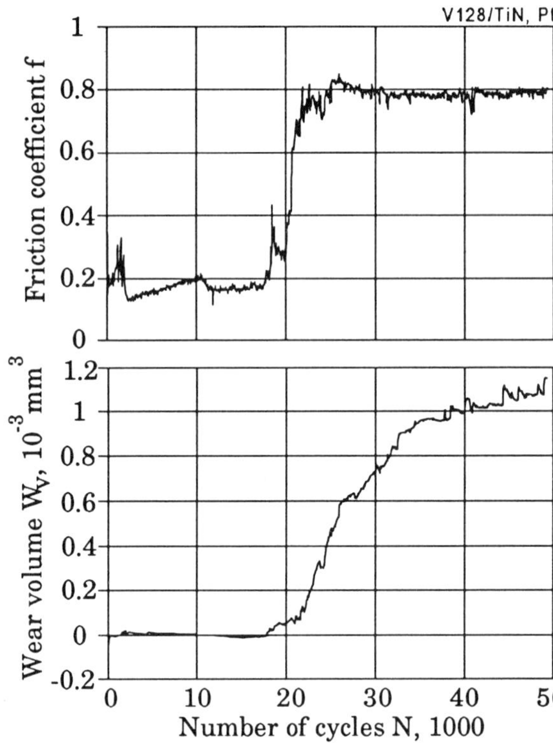

Figure 5. Friction coefficient f and wear volume W_v versus number of cycles of the TiN+Pt coated tribosystem at 5 N normal force and 100 μm sliding amplitude in dry air of <1 % relative humidity.

f_{min} is the minimum friction coefficient, f(250.000) and k(250.000) are the coefficients averaged over 250000 cycles and N_L is the life time in cycles. The function of the selenide coatings as a solid lubricant is too short, like that of the soft metallic coatings. The longest lifetime is shown by WSe_2 in humid air; however in dry air this coating fails immediately. Relative to the surrounding conditions, the two coatings show opposite behaviour: WSe_2 has the better tribological quality (low friction and wear) in humid air, whereas $NbSe_2$ acts better in dry air. The MoS_2 coating yielded better results. The dependence of the tribological behaviour on the atmospheric conditions is the same as that of $NbSe_2$, but the values of friction and wear are much lower. In dry air the friction is as low as that of a lubricated system.

Table 6
Friction and wear coefficient of dichalcogenide coatings; normal force F_N= 5 N.

coating	WSe_2			$NbSe_2$	MoS_2	
target	C 45			42CrMo4	100Cr6	
rel. humidity, %	<1	22	<1	22	<1	22
f_{min}	0.6	0.09	0.24	0.42	0.03	0.11
f(250000)	0.63	0.58	0.62	0.63	0.07	0.52
k, $10^{-6}mm^3/Nm$	46	0.63	0.02	15	~0	~0
k(250000)	77	13.3	71	10.8	0.88	13.3
N_L, 10^3	-	20	5	5	50	5

Because MoS_2 coatings directly deposited on the target material showed good, and in dry air excellent, tribological behaviour, the next step was to test MoS_2 deposited on to a supporting hard coating. On account of the strong influence of humidity on the friction and wear behaviour of MoS_2 coatings, the tests were additionally carried out in vacuum. As substrate material the ball bearing steel 100Cr6 was chosen, and WC was used as the hard coating. The results are shown in Figure 6. In vacuum and dry air the friction is lower than that of the lubricated system (f< 0.1), but the wear rate is one order of magnitude higher (see Figure 3). The relatively high wear rate is

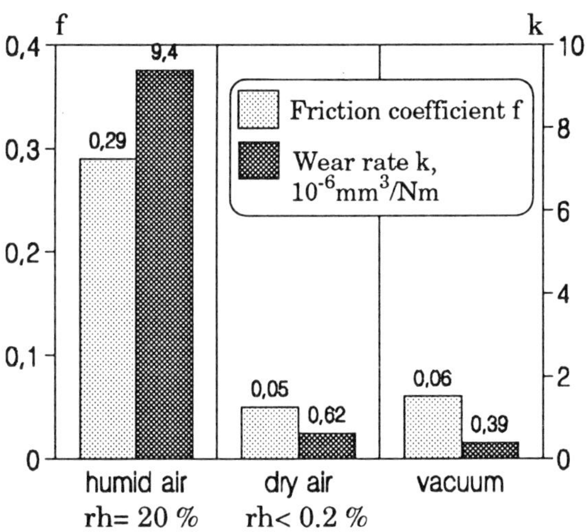

Figure 6. Friction and wear coefficient of the coating system WC+MoS_2 on steel 100Cr6; normal force F_N= 5 N, number of cycles N_{max}= 250000.

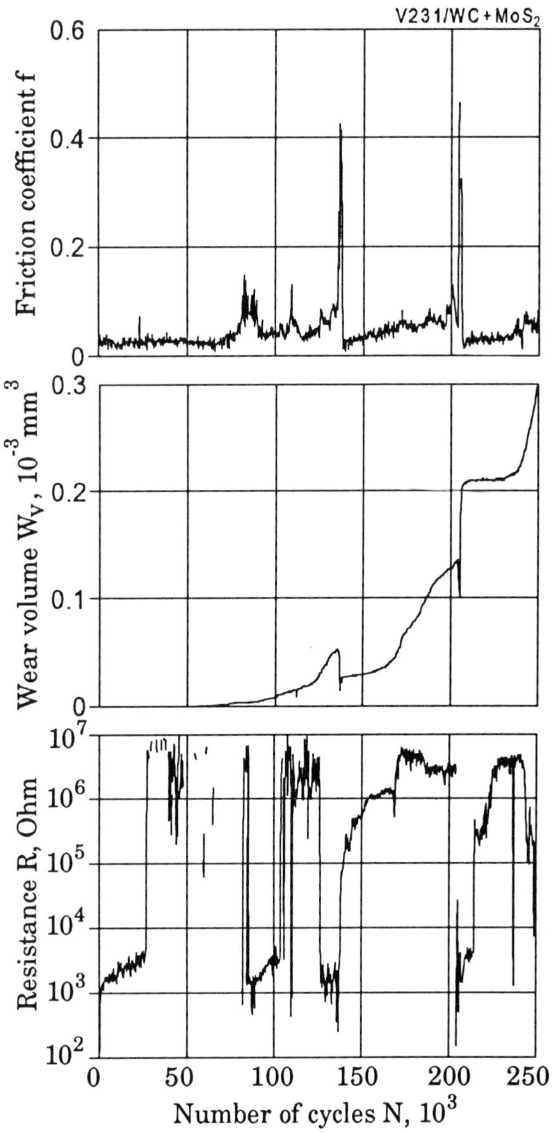

Figure 7. Friction coefficient, wear volume and resistance versus number of cycles of WC+MoS_2 on steel 100Cr6 at 5 N normal force and 100 μm sliding amplitude in dry air of < 1 % relative humidity.

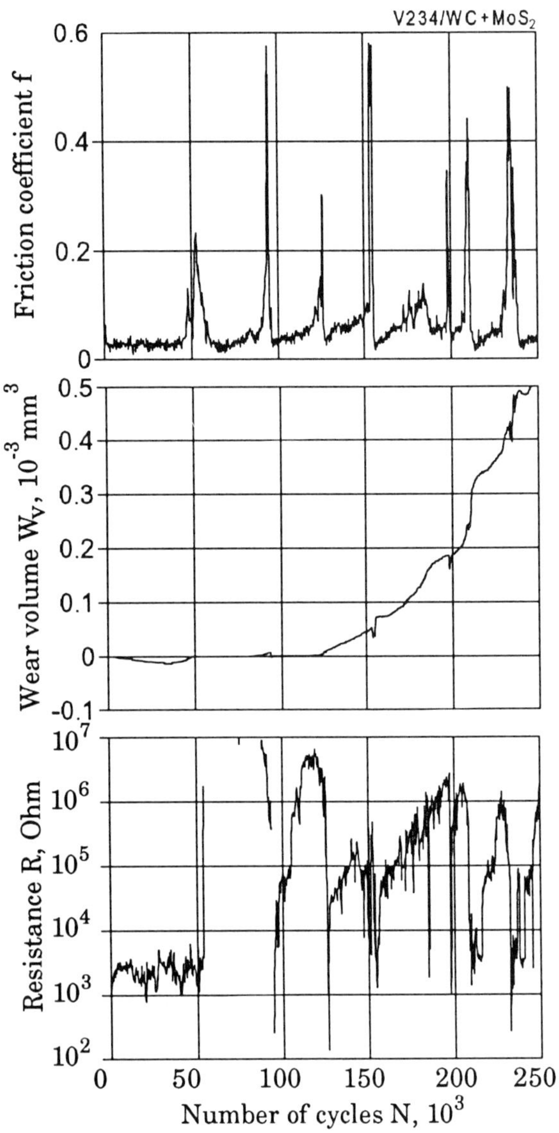

Figure 8. Friction coefficient, wear volume and resistance versus number of cycles of MoS_2 on steel 100Cr6 at 5 N normal force and 100 µm sliding amplitude in dry air of < 1 % relative humidity.

caused by breakdown of the solid lubricating coating so that metallic contact and adhesion occurred. The peaks of the friction curves in Figure 7, associated with an increase of wear, confirm this suggestion. The relatively high resistance in dry air indicates the presence of oxidised hard wear particles, with high resistance, which act as abrasive particles. Comparison of Figures 7 and 8 shows that the supporting WC hard coating lowers the number of adhesive contacts so that friction and wear are decreased.

3.2.4. Carbon coatings

Carbon coatings were investigated because of their advantageous tribological behaviour [2, 3]. The friction and wear behaviour of the carbon coatings Ti-C, W-C, Ta-C and Si-C on the bearing steel 100Cr6, averaged over 50000 cycles in humid air of 30 % relative humidity, are shown in Figure 9.

The friction coefficient is equal to or lower than 0.3, and comparable with that of MoS_2 under the same atmospheric conditions. However the wear rate, compared with that of

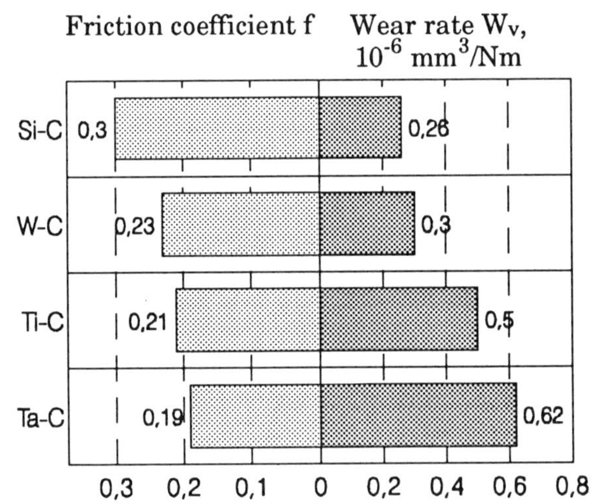

Figure 9. Friction and wear coefficient of carbon coatings on steel 100Cr6; normal force F_N= 5 N, number of cycles N_{max}= 50000.

MoS_2, is clearly lower. The carbide building material has no great influence on friction and wear because the upper lubricating film of about 1-2 µm thickness consists of pure carbon. The influence of the surrounding conditions is therefore represented in Table 7 as an example by the coating Ta-C. It can be seen that the atmosphere has relatively little influence on the behaviour of the optimised Ta-C coating.

Table 7
Friction and wear coefficient of Ta-C on bearing steel 100Cr6 under the surrounding conditions humid air, dry air and vacuum; normal force F_N= 5 N, number of cycles N_{max}= 50000.

Ta-C	Air		Vacuum
	rel.h.=30 %	rel.h.<1%	po~0.1mbar
f	0.19	0.16	0.1
k, $10^{-6}mm^3/Nm$	0.62	0.31	1.3

Friction coefficient f, wear rate k and electrical resistance R of the coating W-C on tempered steel C60 in humid air as a function of the number of cycles are shown in Figure 10.

The resistance at first stays at 2 kOhms for nearly 10^5 cycles, and then suddenly decreases to approach a value below 10 Ohms. Subsequently, large fluctuations appear in the resistance trace. The first reduction in resistance can be ascribed to perforation of the carbon coating, leading to metallic contact followed by adhesion.

The friction coefficient behaves in a parallel manner to the resistance. The friction curve starts at a stable value, less than 0.2. After the drop in resistance which indicates sudden metallic contact between the slider and the substrate, the friction coefficient rises and large fluctuations appear. This transition between low and high values of friction coefficient is due to the gradual contact of the slider and the substrate, leading to local adhesion, and to the abrasive effect of the wear particles.

Typical of the carbon coatings in humid air is the tendency for material to be transferred into the contact area before the wear increases, so that the wear becomes negative over a period of cycles.

The rapid decrease in resistance indicating perforation of the lubricating carbon film permits the definition of a lifetime.

Figure 11 shows as an example the result of a test which was stopped when the resistance had fallen to 50 Ohms. The friction coefficient is nearly constant at 0.15 over this number of cycles. The wear volume increases monotonically with a low rate. The lifetime of the coated tribosystem can be defined by the number of cycles at the end of the test: in this particular case N_L(50 Ohms)= 37400 cycles. The entire wear track of the coated contact surface is shown in Figure 12. Only a small mark parallel to the sliding direction can be observed, which is probably due to the abrasive action of a debonded hard coating particle.

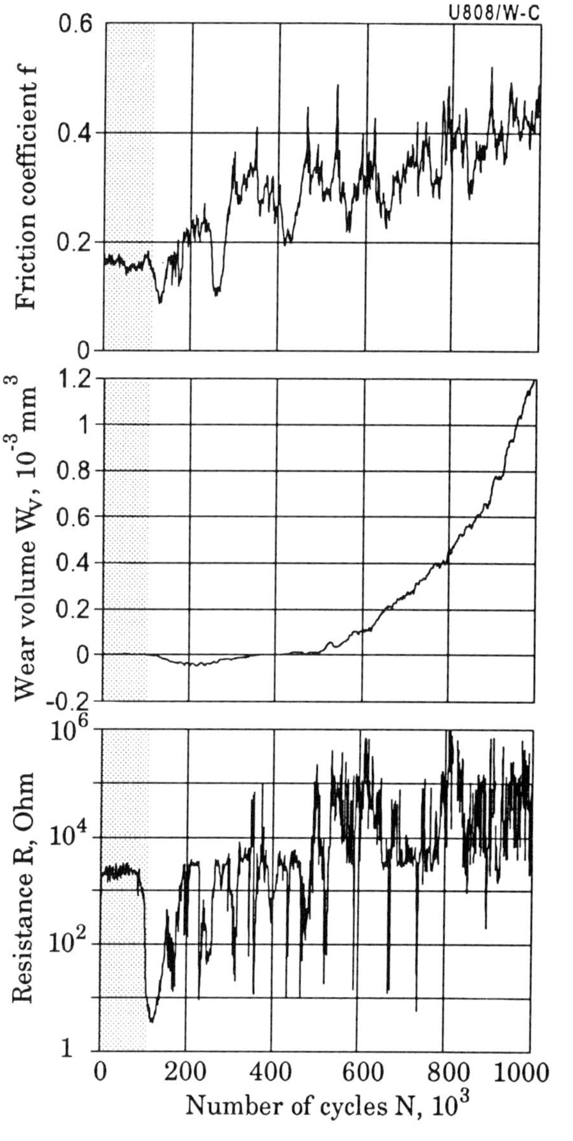

Figure 10. Friction coefficient, wear rate and electrical resistance of W-C coating on steel C60 in humid air of 49 % relative humidity; normal force F_N= 5 N, number of cycles N_{max}= 10^6.

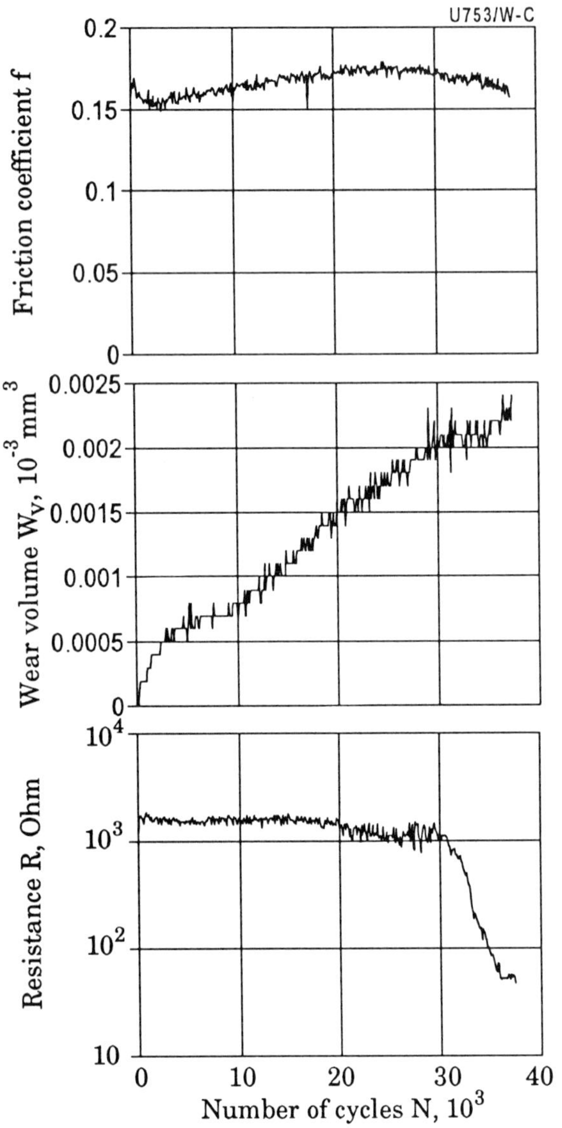

Figure 11. Friction coefficient, wear rate and electrical resistance of W-C coating on steel X20Cr13 in humid air of 36 % relative humidity; normal force F_N= 10 N, life time N_L(50 Ohms)= 37400 cycles; test stopped at R= 50 Ohms.

Subsequent development of the wear volume after this lifetime is shown in Figure 10. It can be seen that a region with a low wear rate follows the lifetime. Conditions during this regime represent an emergency running period from the practical point of view.

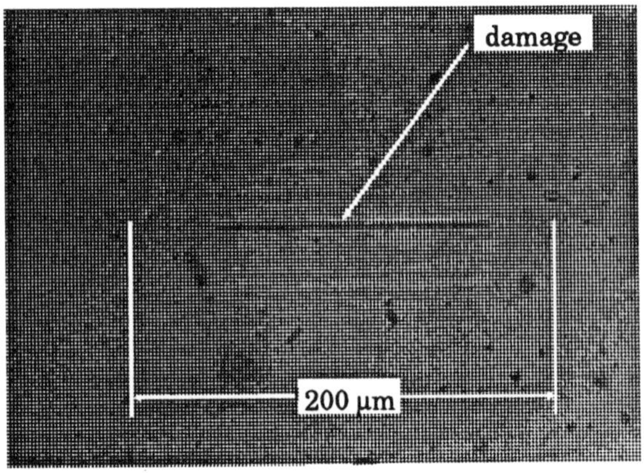

Figure 12. Surface damage of W-C coating on steel X20Cr13 after a life time of N_L(50 Ohms)= 37400 cycles; normal force F_N= 10 N, relative humidity 36 %.

Electrical resistance can be used to detect the end of the lifetime for all of the carbon coatings produced and investigated in this work in humid air. The method can also be used with commercial low conducting CVD-carbon coatings, but a higher value of resistance must be chosen. The extent of damage to the lubri-

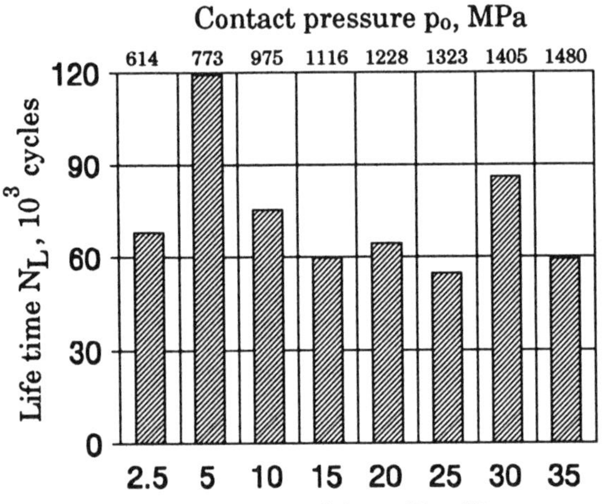

Figure 13. Life time N_L(50 Ohms) of W-C on steel 100Cr6 as a function of normal force in humid air of 35 % relative humidity.

cating film at the end of lifetime is related to the choice of the cut-off resistance.

The lifetime of a W-C coating on bearing steel 100Cr6 as a function of normal force is shown in Figure 13. It can be seen that the lifetime decreases weakly with rising normal force and that the scatter is relatively high. This scatter is related to the stochastic character of the wear process. For all values of normal force the wear scars showed the same characteristics as seen in Figure 12.

4. CONCLUSION

An investigation of PVD-coatings on steel surfaces subjected to reciprocating sliding conditions has yielded the following results:

- In lubricated systems every hard coating which prevents adhesion reduces wear. The friction coefficient is then about 0.1.
- Unlubricated hard coated systems have high friction coefficients above 0.5, with the result that the dissipated frictional energy damages the contact surfaces.
- Soft metallic and chalcogenide films, as friction-reducing coatings on supporting hard coatings, lower friction but the lifetime is short. From this group the best results were achieved with MoS_2 on WC.
- Lowest friction and wear coefficients were produced with amorphous carbon coatings of about 1-2 μm thickness. The life times were greater than those of the other self lubricating coatings investigated, even at high contact pressures. Because of the low conductivity of the carbon films it is possible to detect the end of life time by resistance measurement, which may be important for practical applications. The dependence of the tribological behaviour on humidity can be reduced by optimisation of the PVD-process.

ACKNOWLEDGEMENT

This study was sponsored by the Deutsche Forschungsgesellschaft (DFG) and the authors are grateful for this support.

REFERENCES

1. V. Buck, Preparation and Properties of different Types of sputtered MoS_2 Films, Wear, 114 (1987) 263 - 274
2. R. Messier, A. R. Badzian, T. Badzian, K. E. Spear, P. Bachmann and R. Roy, From Diamond-like Carbon to Diamond Coatings, Thin Solid Films, 153 (1987) 1 - 9
3. J. Franks, K. Enke and A. Richardt, Diamond-like carbon-properties and applications, Metals and Materials (November 1990), 695 - 700

Thin Films in Tribology / D. Dowson et al. (Editors)
© 1993 Elsevier Science Publishers B.V. All rights reserved.

Electroviscous Effect of Liquid Crystals

Shin Morishita[a] and Yoshitsugu Kimura[b]

[a]Department of Naval Architecture and Ocean Engineering, Yokohama National University
156 Tokiwadai, Hodogaya-ku, Yokohama 240 Japan

[b]Institute of Industrial Science, University of Tokyo
7-22-1 Roppongi, Minato-ku, Tokyo 106 Japan

This paper describes the electroviscous effect of a thermotropic liquid crystal in nematic phase. Liquid crystals are characterized by orientation of their own molecules, whose direction can be controlled by the applied electric field or magnetic field. Because of the controllability of this orientation of molecules, liquid crystals are being used in displays for computers and television sets, in which optical anisotropy is controlled by applying electric field. From the tribological point of view, the controllability of orientation produces two possibilities; low friction and viscosity variation in a prescribed manner. In the present paper, the viscosity variation caused by the applied electric field strength, which is known as the electroviscous effect, is studied at varying electric field strength, shear field strength or temperature. The test sample is a thermotropic liquid crystal which shows nematic phase. By controlling the degree of orientation of the molecules, the possibility of using liquid crystals in tribological applications is shown where accurate control of lubrication characteristics is of critical importance.

1. INTRODUCTION

A liquid crystal is characterized in its phase which is observed as an intermediate state between a solid crystal and an isotropic liquid. This intermediate phase, liquid crystalline phase, is caused by the orientation of the molecules, and because of the orientation, it shows optical anisotropy. Displays for computers and television sets are typical applications of liquid crystals, in which the orientation of the molecules is controlled by the applied electric field strength.

The orientation of the molecules produces other characteristic features, which attract the interest of engineers as well as scientists. One of them is low friction when liquid crystals are used as a lubricant in tribological field, which is being studied by several authors[1,2] Another interesting feature of liquid crystals is apparent viscosity variation when they are subjected to magnetic field, the Freedericksz transition, or to electric field, the electroviscous effect.

One of the authors has studied ER (Electro-rheological) fluids whose apparent viscosity can be varied by the applied electric field strength. An ER fluid is a colloidal suspension of fine semiconducting particles in a dielectric liquid. The wide range of, and the quick response in, the variation of their viscosity provides ER fluids with wide applications to mechanical components. One of the application studies is using ER fluids as a lubricant. However, the ER effect, which is defined as the apparent viscosity variation of an ER fluid with the applied electric field strength, becomes very small under high shear field. This deficiency is fatal in applying ER fluids to the areas where they are suffered from high shear rates as is the case with lubrication.

There have been a few researches on the electroviscous effect of liquid crystals. Bjornstahl and Snellman [3] first reported the apparent viscosity variation of a liquid crystal when an electric field is applied. Nomura and Ogawa[4] reported the electroviscous effect of a liquid crystal under rather low electric field. Honda et al. [5] compared the electroviscous effect of different kinds of liquid crystals, and showed remarkable apparent viscosity variation at high electric field and strong dependence of the effect

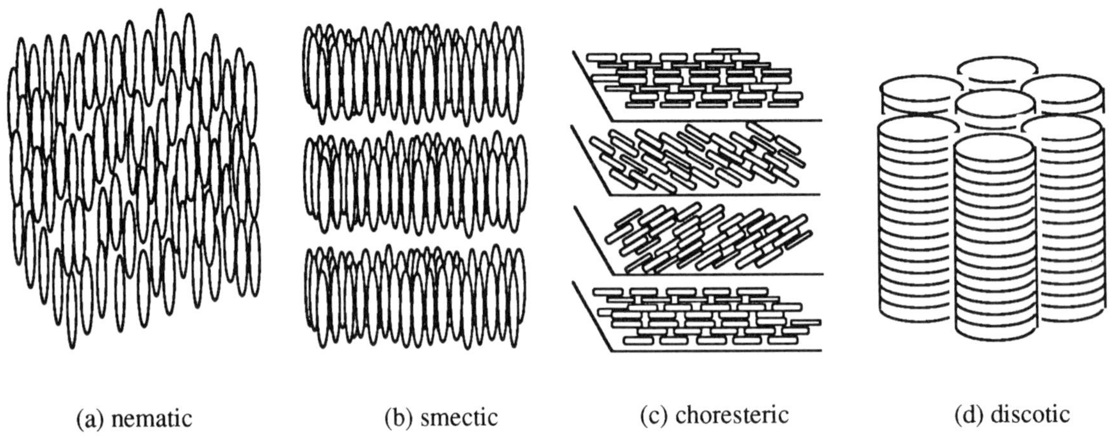

Figure 1. Molecular models of liquid crystals

on the frequency of the applied field.

In the present paper, the electroviscous effect of a nematic type liquid crystal is studied under various electric field strength, shear field strength and temperature in comparison with that of ER fluids.

2. LIQUID CRYSTALS

A liquid crystal is defined as a liquid which has the nature of a crystal in showing optical or other physical anisotropy caused by the structure [1,7,8]. Not a few liquids show liquid crystalline states which can be classified into four phases; nematic, smectic, cholesteric, and discotic phases, according to the shape and the orientation of the molecules. Figure 1 shows the idealized model of the molecules of liquid crystals.

The liquid crystalline phase is also called the mesophase between the solid state and the liquid state [7]. This phase can appear in either of the two ways, one in a given temperature range, while the other in a given concentration range; they are called the thermotropic and lyotropic liquid crystals, respectively. In the case of a thermotropic liquid crystal, a solid changes into an optically opaque liquid crystalline phase above the melting point. Being heated further, the liquid becomes an optically clear isotropic phase at a temperature, which is called the clearing point. In the present paper, a thermotropic liquid crystal in the class of nematic phase is studied.

Viscosity of liquid crystals is usually measured by a rotational type viscometer, which determines apparent Newtonian viscosity assuming a uniform shear rate. A nematic liquid crystal flows like a normal organic liquid, but the orientation causes its anomalous viscous behavior.

The viscosity of liquid crystals depends on the orientational order and the degree of orientation of the molecules. A vector, the director, is introduced in order to represent the direction of the orientational order of nematic or smectic liquid crystals whose molecules have a cylindrical shape. The director is a unit vector whose direction corresponds to that of the major axis of the molecules.

The three limiting viscosities can be defined if the director is fixed in the external electric or magnetic field [7]:

η_1: director parallel to velocity gradient (Fig.2-a)
η_2: director parallel to flow direction (Fig.2-b)
η_3: director normal to shear plane (Fig.2-c)

The three viscosities, η_1, η_2 and η_3, the Miesowicz

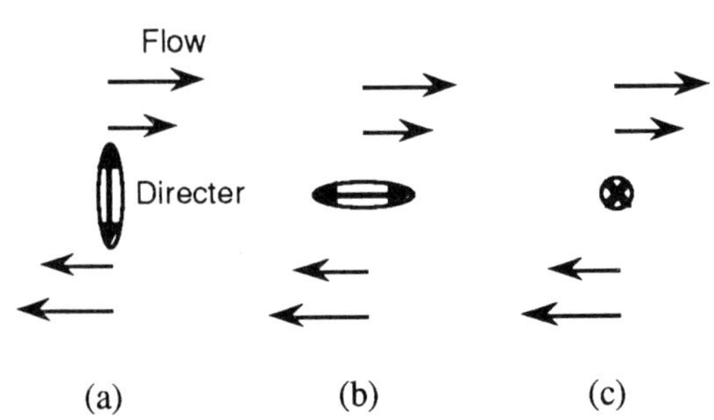

Figure 2. Miesowicz viscosity

viscosities, generally have a sequence of η1>η3>η2 in nematic liquid crystals, where η3 behaves like the extrapolated isotropic viscosity [8]. It was reported that a rotational type viscometer measures η3 with nematic liquid crystals [8].

Application of external electric field can control the director, and in this way it affects the viscosity realizing the electroviscous effect. This effect depends on the dielectric constant of the liquid crystal under an applied electric field. Two dielectric constants are defined for the liquid crystal molecules having anisotropy; one is parallel to the director, $\varepsilon_{\parallel}$, and the other normal to the director, $\varepsilon_{\perp}$. The subtraction $\Delta\varepsilon = \varepsilon_{\parallel} - \varepsilon_{\perp}$ is called the dielectric anisotropy, which can be positive or negative. When the dielectric anisotropy is positive (p-type liquid crystals), the director becomes parallel to the applied electric field, and when negative (n-type liquid crystals), the director has the direction normal to the electric field.

It is expected that a p-type nematic liquid crystal has a larger electroviscous effect than an n-type one, and a large dielectric constant and a large dielectric anisotropy causes a larger electroviscous effect.

3. EXPERIMENTAL DETAILS

3.1 Test Sample

The physical properties of the test sample of liquid crystal are shown in Table 1. A nematic liquid crystal on the market which was developed for displays was tested because of easy procurement. This sample was composed of several liquid crystals to obtain suitable viscosity, clearing point, or other properties. The main components of the present sample were biphenyls and esters.

Table 1. Properties of test sample

Property	Value
Melting point	- 20.0 °C
Clearing point	65.8 °C
Resistivity	2.4×10^{10} Ω cm
Viscosity (20 °C)	0.0537 Pa·s
Ordinary axis	1.508
Refractive anisotropy	0.156
Threshold voltage	1.3 V
Dielectric constant	23.0 F/m
Dielectric anisotropy	14.6 F/m

3.2 Experimental Apparatus

The experimental apparatus, in Fig.3, is composed of a rotational type viscometer, a high voltage source, a temperature controller, and a microcomputer. The rotational type viscometer had a pair of concentric cylinders, between which the test sample was filled.

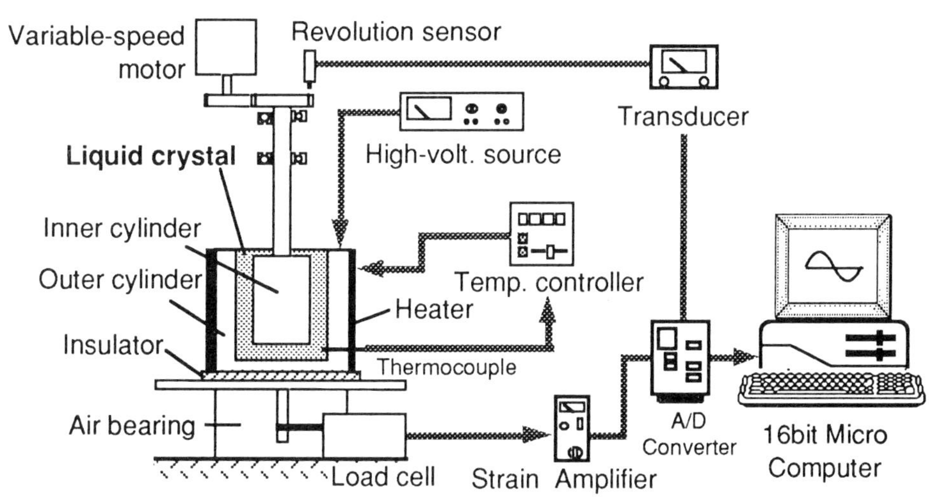

Figure 3. Experimental apparatus

The inner cylinder was rotated by a variable speed motor to apply a shear field to the test sample. The torque transmitted through the test sample was measured by a load cell which was pressed by a torque arm fixed to the outer cylinder. The outer cylinder was supported by an air bearing to minimize the friction. The inner cylinder had the outer diameter of 50 mm and the length of 50 mm, around which the clearance of 0.5 mm was formed. The test sample was poured into the clearance, and was subjected to shear ranging 100 to 600 sec^{-1} when the inner cylinder was rotated at 20 to 120 rpm. The clearance at the bottom of the inner cylinder was 5 mm, which was much greater minimizing the contribution of viscous shear on the bottom surface Both cylinders are made of copper.

Between the cylinders, a DC field was applied except for the case where the frequency contribution to the electroviscous effect was studied. The capacity of the DC source was 3 kV/mm in field strength, and 30 mA in electric current, while for the AC source, 1 kHz in frequency, 2 kV/mm in field strength of peak-to-peak, and 2 mA in electric current; square wave was used for the AC field.

Temperature of the test sample was kept with an accuracy of ±0.5 °C when detected by a thermocouple equipped at the bottom of the inner cylinder, with a temperature controller.

All the signals from the load cell, the voltage source, and the revolution counter were put into a microcomputer through a 12 bit A/D converter in a usual technique.

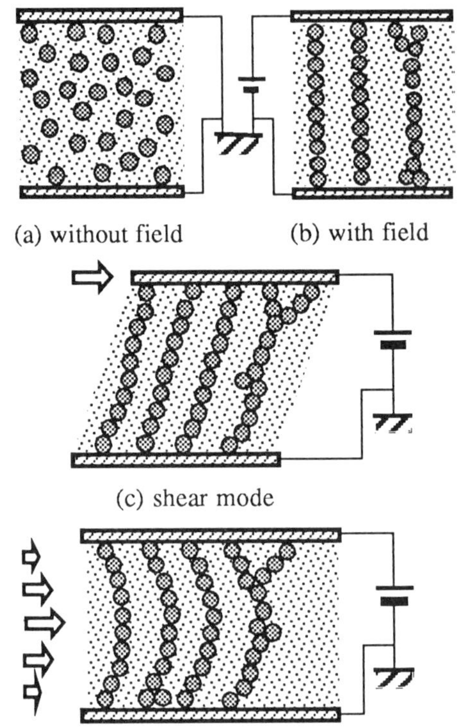

Figure 4. Mechanism of ER effect

4. RESULTS

4.1 Results with ER Fluids

For comparison with the present results, some results for an ER fluid obtained in a previous experiment is described. An ER fluid is a colloidal suspension of fine semiconducting particles in a dielectric liquid [9-12], as shown in Fig.4(a), and this special colloidal suspension was first introduced by Bjornstahl in 1939 [3]. In 1947 Winslow proposed several applications of ER fluids to mechanical components as well as their basic characteristic behavior [12]. The semiconducting particles dispersed in the ER fluids are usually cornstarch, cellulose, silica-gel or alumina powder, whose particle size is from several tens of nanometers to 10 micrometers. Application of electric filed forms lots of chains of the particles, bridging the electrodes, Fig.4 (b), which resist the

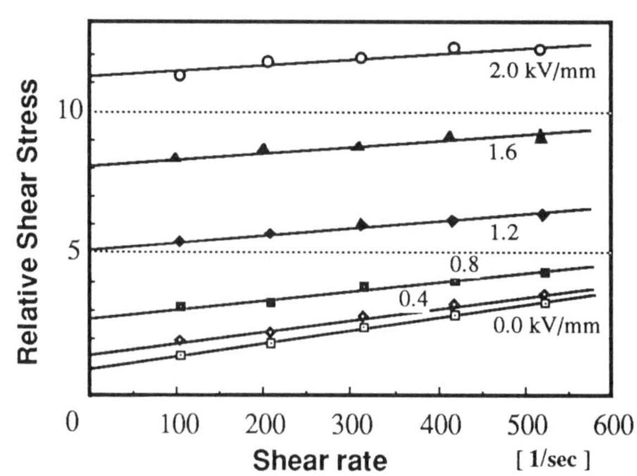

Figure 5. Typical characteristics of ER fluid

shear movement (Fig.4(c)) or the flow of the fluids (Fig.4(d)), resulting in an increase in the apparent viscosity.

Typical characteristics of an ER fluid are shown in Fig.5, where the relative shear stress is defined as the ratio of shear stress with electric field to that without electric field. ER fluids behave like a Bingham plastic having a yield stress at zero shear rate, which increases as the field strength increases. The gradient of each line gives the Newtonian viscosity of the fluid. It should be noted that the viscosity variation with the applied electric field is reversible.

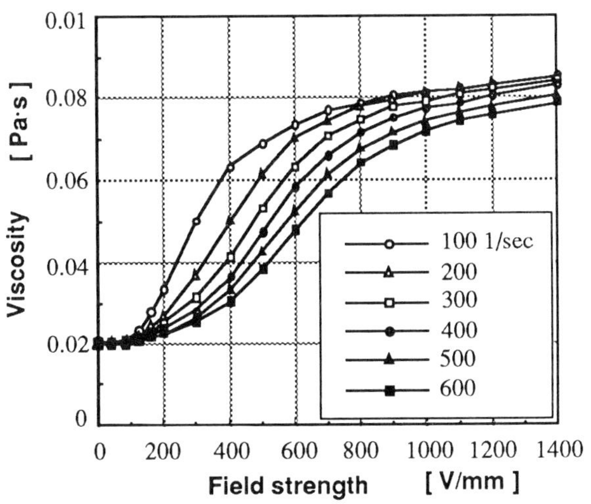

Figure 6. Electroviscous effect

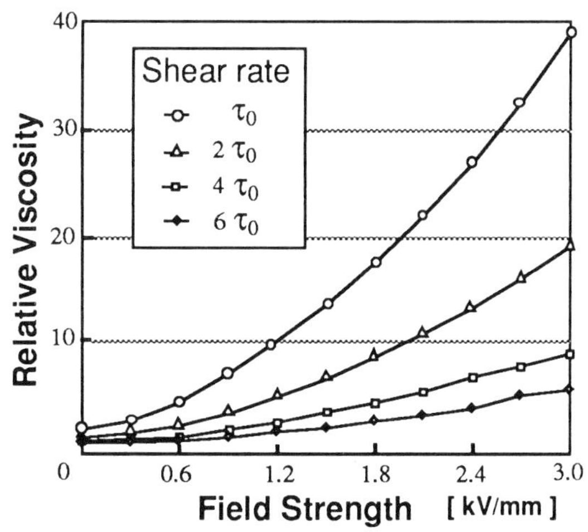

Figure 7 Viscosity variation of ER fluid

The electroviscous effect in ER fluids, which is called as ER effect, observed as the increase of apparent viscosity is very large; the apparent viscosity becomes tens or hundreds times larger than that in the absence of electric field. One of the authors has applied the ER fluid to vibration control actuators, and successful results have been obtained [13-15]. Unfortunately, however, the ER effect greatly depends on the shear field as the small gradients of the lines in Fig.5 suggests, which limits application of ER fluids to tribological purposes.

4.2 Field strength and Shear Rate Effects

Figure 6 shows the electroviscous effect at various shear rates at 40 °C. The effect appears from the field strength of 100 V/mm, and the viscosity increases drastically with the applied field strength thereafter. As the field strength is increased further, the electroviscous effect becomes gradual, and the viscosity seems to approach a maximum value regardless of the applied shear field strength. Higher shear rates cause similar eventual increase of the viscosity, but at higher electric field strength. This forms a sharp contrast with the ER effect, shown in Fig.7, which is a rearrangement of Fig.5, where the relative viscosity is the ratio between the apparent viscosities with and without electric field.

The electric current across the film at 40 °C is shown in Fig.8, as a function of the applied electric field strength and the shear field strength. The electric current was increased almost linearly with the

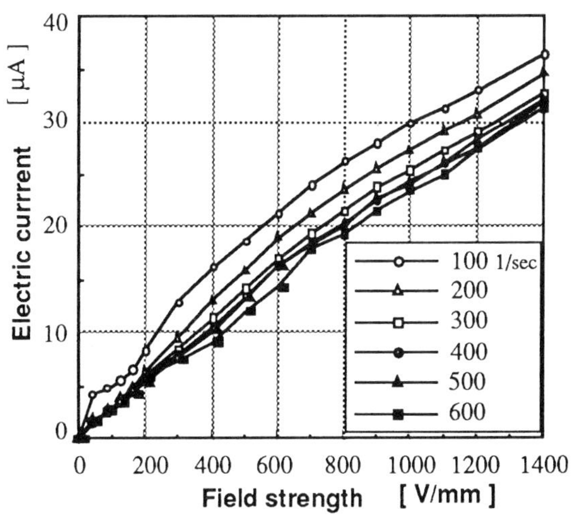

Figure 8. Electric current

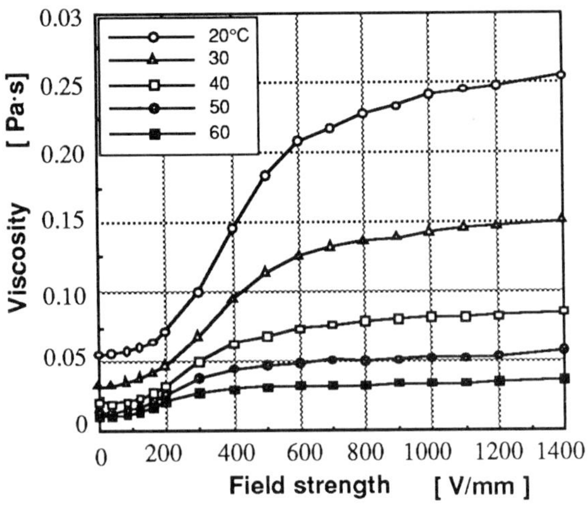

Figure 9. Temperature effect on electroviscosity

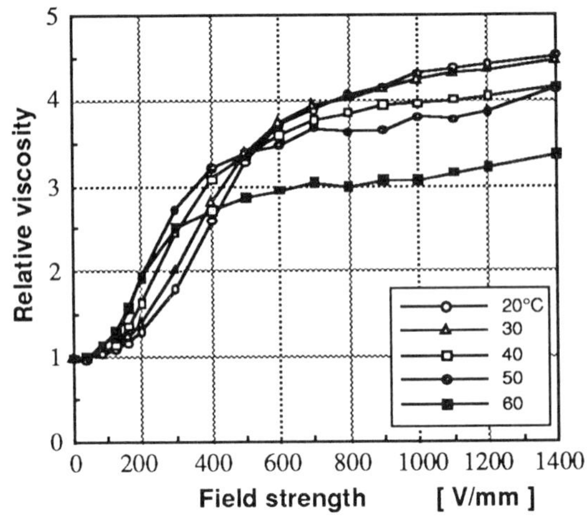

Figure 10. Temperature effect on electroviscosity (dimensionless form)

electric field strength, while shear field strength had less effects.

4.3 Temperature Effects

The electroviscous effect of the test sample depends on the temperature, as shown in Fig.9, between its melting point and the clearing point. At higher temperatures, no electroviscous effect is observed.

As the temperature is increased, the viscosity of the test sample becomes low, and the electroviscous effect seems to be lowered. However, when replotted in the relative viscosity defined in 5.1, Fig.10, it turns out that a higher temperature causes the electroviscous effect to appear at a lower electric field strength and lowers the attainable relative viscosity but moderately. The ER effect is also dependent on the temperature, but, the effect if said to become larger as the temperature is increased [11] by a mechanism not yet clarified.

Figure 11 shows the dependence of electric current on the temperature. The electric current becomes larger with the temperature; it also increases with the electric field strength as in Fig.11, but non linearly.

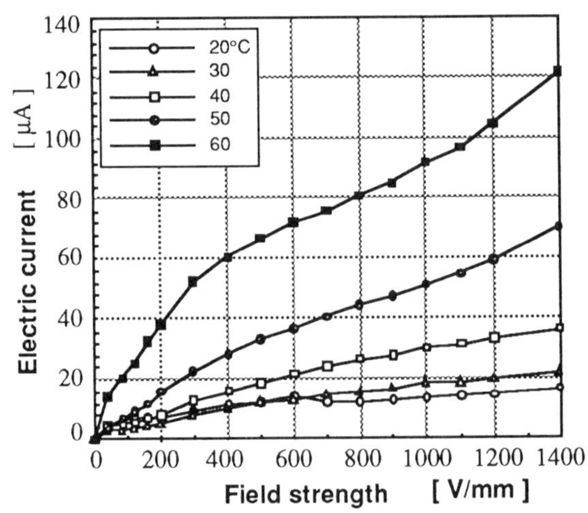

Figure11. Temperature effect on electric current

4.4 Frequency Effects

Frequency of the applied electric field has little effect on the apparent viscosity, as shown in Fig.12, in the range studied. This behavior is common to the results with ER fluids reported by Klass and Matinek [9]. However, the apparent viscosity of the ER fluid decreased suddenly when the frequency exceeds 200 Hz, which is a feature not found with the present test sample. With liquid crystals, Honda et al. [5] obtained a result which showed a marked increase in the apparent viscosity in a high frequency range. The reason of the disagreement with the present result might be found in the difference in the liquid crystals.

5. CONCLUSIONS

The electroviscous effect is studied of a nematic liquid crystal composed of biphenyls and esters, which was developed for displays. The conclusions obtained are as follows:

(1) The viscosity of the liquid crystal can be varied by the applied electric field. In contrast to ER fluids, the electroviscous effect is not weaken even by high shear field, if the applied electric field is strong enough.

(2) When compared in the relative viscosity, temperature exerts moderate effects in lowering the threshold electric field strength for the electroviscous effect to appear, and in lowering the attainable valued at a higher electric field strength.

(3) The frequency of applied electric field contributes little to the electroviscous effect of liquid crystals in the range up to 1 kHz.

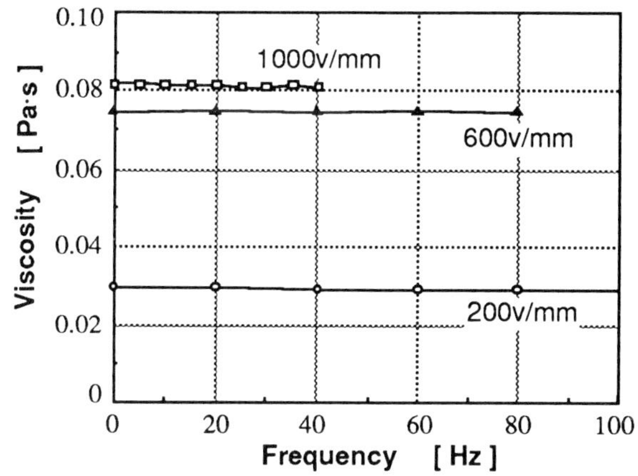

Figure 12. Electroviscous effect in AC field

ACKNOWLEDGEMENT

The authors thank to Mr. Yutaka Fujita (RODIC Co.,Ltd.) for providing the test sample and helpful discussion.

REFERENCES

1. Biresaq,G.: Tribology and the Liquid-Crystalline State, ACS Symposium Series No.441 (1990)
2. Tichy,J.A. and Rhin,Y.: Trans ASME, J. Tribology, 111, 1 (1989) 169
3. Bjornstahl,Y. and Snellman,O.: Kolloid Zeitshrift, 86 (1939) 223
4. Nomura,A and Ogawa,S.: Japanese J. Applied Physics, 16, 4 (1977) 639
5. Honda,T., Sasada,T, and Kurosawa,K.: Japanese J. Applied Physics, 17, 9 (1978) 1525
6. Yun,G.Y. and Eredrickson,A.G.: Phys. Fluid, 16 (1973) 1
7. W.H.de Jeu: Physical Properties of Liquid Crystalline Materials, Gordon and Breach (1980)
8. Benicewicz,B.C., Johnson,J.F. and Shaw,M.T.: Mol. Crystals and Liquid Crystals, 65 (1981) 111
9. Klass D.L. and Martinek T.W.: J. of Applied Physics, 38, 1 (1967) 67
10. Block,H. and Kelly,J.P.,,"Electro-rheology",J. of Physics D: Applied Physics, 21, 12 (1988) 1661
11. Jordan T.C. and Shaw, M.T.: IEEE Trans. on Electrical Insulation, 24, 5 (1989) 849
12. Winslow,W.M.: J. of Applied Physics, 20 (1949) 1137
13. Morishita,S. and Mitsui,J.: Trans. ASME, J. Acoustics and Vibration 114,3 (1992) 354
14. Morishita,S., Mitsui,J., and Kuroda,Y.: Proce. PVP Conf., ASME, PVP-211 (1991) 1
15. Morishita,S. and Ura,T.: Proc. Recent Advances in Adaptive and Sensory Materials and Their Applications, Virginia Polytech., (1992) 537

Thin Films in Tribology / D. Dowson et al. (Editors)
© 1993 Elsevier Science Publishers B.V. All rights reserved.

The effect of oxygen adsorption on field stimulated exoelectron emission from high purity aluminum

N.Ohmae, M.Mori, M.Tagawa and M.Umeno

Department of Precision Engineering, Faculty of Engineering, Osaka University, 2-1 Yamadaoka, Suita, Osaka 565, Japan.

Emission properties of field stimulated exoelectron emission upon exposure to oxygen were studied. Three emission parameters, i.e., rate of excitation, rate of emission, and number of trap sites were calculated by means of a new method which is based upon the two-process model. The enhancement of emission by oxygen exposure was due to the increment of electron traps. The traps formed by the oxygen adsorption located at -3.62eV and -4.00eV measured from the vacuum level.

1.INTRODUCTION

It has been known that exoelectron emission depends strongly upon surface conditions. Therefore, exoelectron emission has been recognized as a useful tool for detecting change in the surface properties. In the field of tribology, *in-situ* measurement of lubricant failure, detection of surface damage etc. by exoelctron emission have been expected [1-5]. Exoelectron emission from metals also were widely studied, and high purity aluminum is the one of the typical materials.[6-8] It has been argued that exoelectrons are emitted from electron traps in the aluminum oxides or at the interface. Therefore, many investigations were carried out in conjunction with the growth of oxide films.[9,10] Generally, ultraviolet (UV) light is used as a stimulus for exoelectron (photo stimulated exoelectron emission, PSEE). However, UV light is effective not only for the process of electron emission from trap levels, but also for the excitation of electrons to trap levels. It is not easy for PSEE to separate the both processes, and other useful methods are desired.

Field stimulated exoelectron emission (FSEE) from aluminum was first reported by Tagawa et al. in 1986 [11] and FSEE has a unique feature for studying exoelectrons. In our previous study, storage effect of exoelectrons was clearly detected for FSEE, and it becomes clear that the two-process model, proposed by Shigekawa and Hyodo [12], also was applicable to FSEE [11].

In the case of FSEE, excitation is accomplished by a thermal process, and emission is provided by the tunnel effect. Stimulus for FSEE is the strong electric field which does not give any influences for electron excitation. This is completely different from PSEE. In the two-process model, three emission parameters exist, i.e., the rate of excitation, the rate of emission and the number of trap sites. It is necessary to evaluate these all parameters in order to characterize the emission property of FSEE. However, interval of applying pulse and increment of emission are only measured in conventional experiments of storage effect. So that it is not easy to evaluate each emission parameter. Though, we proposed the method of estimating the rate of emission in the previous work [13], the other two parameters are still unknown.

In this study we propose the improved method for estimating three emission parameters individualy. Variations of these parameters during oxygen exposure were studied. Moreover, total energy distributions (TED) of FSEE were successfully measured to determine the energies of trap levels.

2.THEORY

A schematic diagram of the storage effect is indicated in Figure 1. The storage effect is the phenomenon that emission current increases temporarily after the pause of stimuli (irradiation of UV light for PSEE or applied voltage for FSEE), and is well explained by the two process model that electrons are storaged at the trap level during the pause of stimuli. In the two process model, it is assumed that there are exoelectron traps of S_0 above the Fermi level. The excitation of electrons from the Fermi level to trap level occurs at a constant rate β. Exoelectron emission from the occupied traps takes place with a rate of α, during the stimuli. The increment of emission by storage effect, I_p, and its saturation value I_{ps} are expressed by [12]

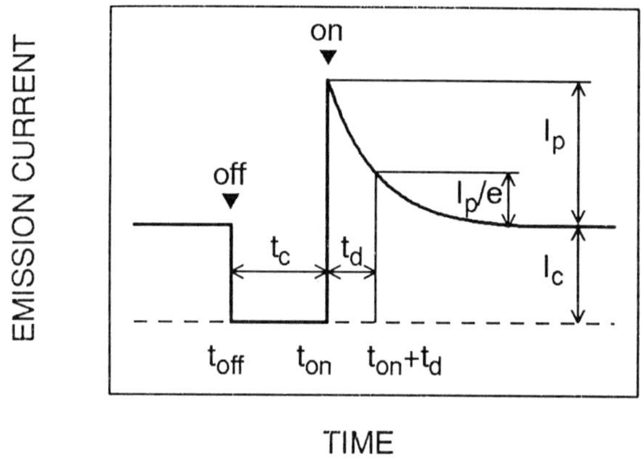

Figure 1. Storage effect of PSEE and FSEE.

$$I_p=(e\alpha^2S_0)/(\alpha+\beta)\,[1\text{-}\exp(\text{-}\beta t_c)] \qquad (1),$$

$$I_{ps}=(e\alpha^2S_0)/(\alpha+\beta) \qquad (2).$$

Equations (1) and (2) lead to

$$I_p=I_{ps}[1\text{-}\exp(\text{-}\beta t_c)] \qquad (3).$$

The rate of excitation β is calculated by a slope of $[\ln(1\text{-}I_p/I_{ps})]$ vs. $[t_c]$ plot as proposed in the previous work [13]. The decay of emission current after re-applying voltage is expressed by

$$I(t)=e\alpha\beta S_0/(\alpha+\beta)+e\alpha^2S_0/\ (\alpha+\beta)\ [1\text{-}\exp(\text{-}\beta t_c)]\exp[\text{-}(\alpha+\beta)(t\text{-}t_{on})] \qquad (4).$$

where t_{on} denotes the time when voltage pulse is re-applied. The continuous emission of FSEE, I_c, is calculated from equation (4) when $t \rightarrow \infty$,

$$I_c=e\alpha\beta S_0/(\alpha+\beta) \qquad (5).$$

By solving equations (1), (4) and (5), equation (6) is given.

$$I(t)=I_c+I_p\exp[(\text{-}(\alpha+\beta)(t\text{-}t_{on})] \qquad (6).$$

Equation (6) indicates that the FSEE current is devided into continuous emission and temporary increment as shown in Figure 1. The component of field emission (I_f) will be added to the equation (6) in a practical emission mode. Since these emission could not be separated, we have to focus the decay of emission which is caused only by the storage effect. The second term of the equation (6) is rewritten as

$$I_d(t)=I_p\exp[(-(\alpha+\beta)(t-t_{on})] \quad (7).$$

Therefore,

$$\log[I_d(t)/I_p] = (\alpha+\beta)(t-t_{on}) \quad (8).$$

Assuming that the emission decays to 1/e of the peak value at time [t_{on} + t_d], next relation is obtained.

$$\log[I_d(t_{on}+t_d)/I_p] = -1 \quad (9).$$

From equations (8) and (9),we obtain

$$\alpha+\beta = 1/t_d \quad (10).$$

One can evaluate the value of α from t_d and β obtained before. Moreover, the number of emission sites S_0 is also calculated by using equation (2).

This method is not limited to FSEE, but can be extended to exoelectron emission with other stimuli.

3.EXPERIMENTAL PROCEDURE

Extremely high purity aluminum (99. 9999%) was used to avoid influences of inpurities. The samples is machined into

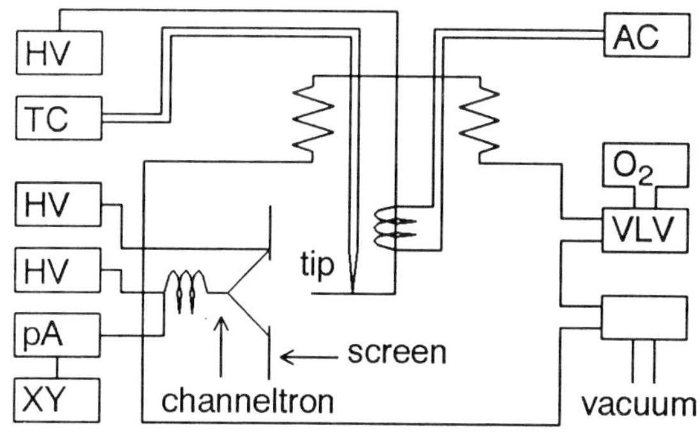

Figure 2. Schematic diagram showing FSEE apparatus.

a tip and finished by electropolishing using an electrolyte (10%$HClO_4$ and 90% C_2H_5OH) at 0~5V DC. Experimental apparatus used in this study is illustrated in Figure 2. The aluminum tip was located at the bottom of a manipulator. Tip temperature was controlled by a resistive heating and monitored with CA thermocouple (room temperature to 600K). Anode electrode was placed opposite to the tip which served as a fluorescent screen for visual obserbation of the emission pattern. Electrons emitted from the tip were detected by the channeltron electron multiplier equipped at the center of the screen. The ultimate vacuum pressure of the aparutus was lower than 4×10^{-10}Torr. Annealing at 420K for 30 min. was carried out for each tip in order to reduce lattice imperfections. High purity oxygen gas was introduced through a valiable leak valve. The experimental method for measuring TED spectrum were described elsewhere [14].

4.RESULTS AND DISCUSSION

Figure 3 shows storage effect of FSEE

from 99.9999% pure aluminum before oxygen exposure, and the dependence of increments of emission on the interval of pulsed voltages can be clearly seen. The increment of emission, I_p, saturates when plotted against interval in Figure 4. This feature is the same as that found for PSEE. The relationships between interval of pulsed voltage and increment of emission were measured at oxygen exposures of 0.01, 0.02, 0.05, 0.1, 0.2, 0.5, 1.0, 2.0, 5.0 and 10.0L, as shown in Figures 5 (a)~(j), respectively. One can find similarities in the shape of the graph, i.e., I_p increases at an early stage and then saturates. However, the saturation current becomes large at increased oxygen exposures. The saturation current I_{ps} against oxygen exposure is indicated in Figure 6. I_{ps} increases rapidly until 0.02L, and hardly increases

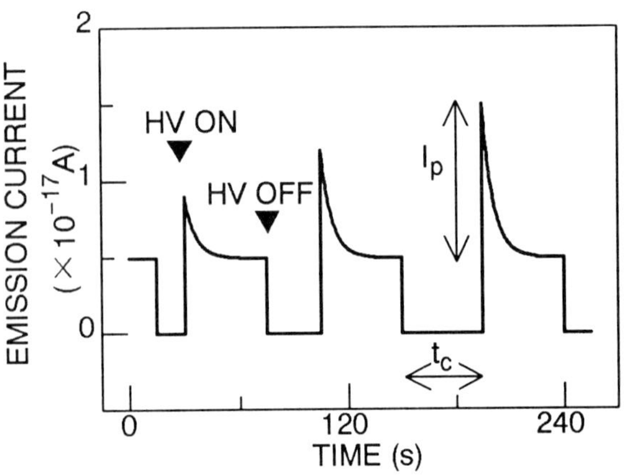

Figure 3. Storage effect of FSEE observed for 99.9999% pure Al.

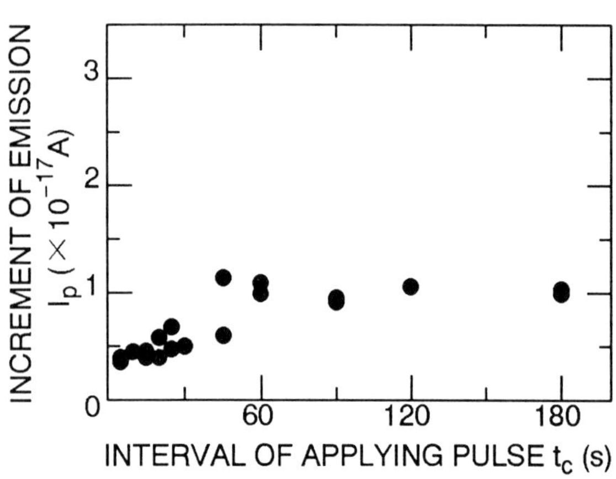

Figure 4. Relationship between interval of pulses and increment of emission before oxygen exposure.

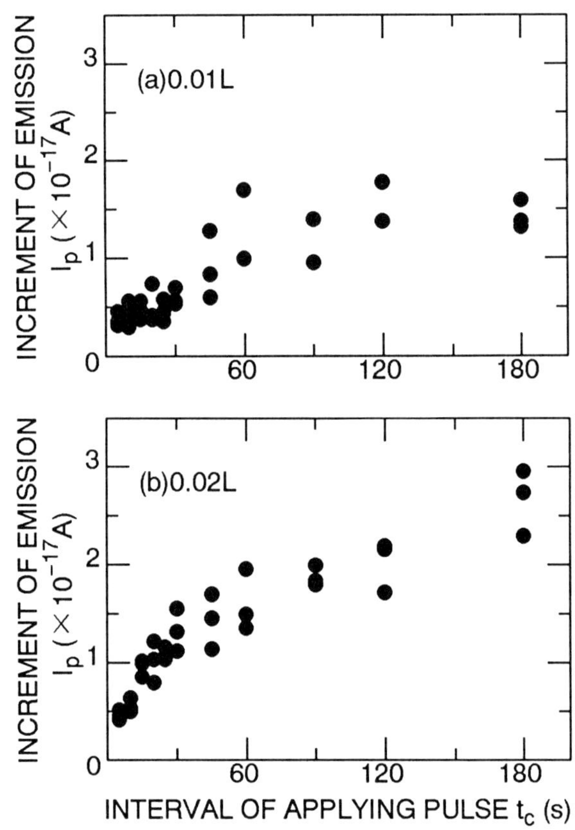

Figure 5. Relationship between interval of pulses and increment of emission after oxygen exposures at; (a)0.01L, (b)0.02L, (c)0.05L, (d)0.1L, (e)0.2L, (f)0.5L, (g)1.0L, (h)2.0L, (i)5.0L, and (j)10.0L.

- CONTINUE TO NEXT PAGE -

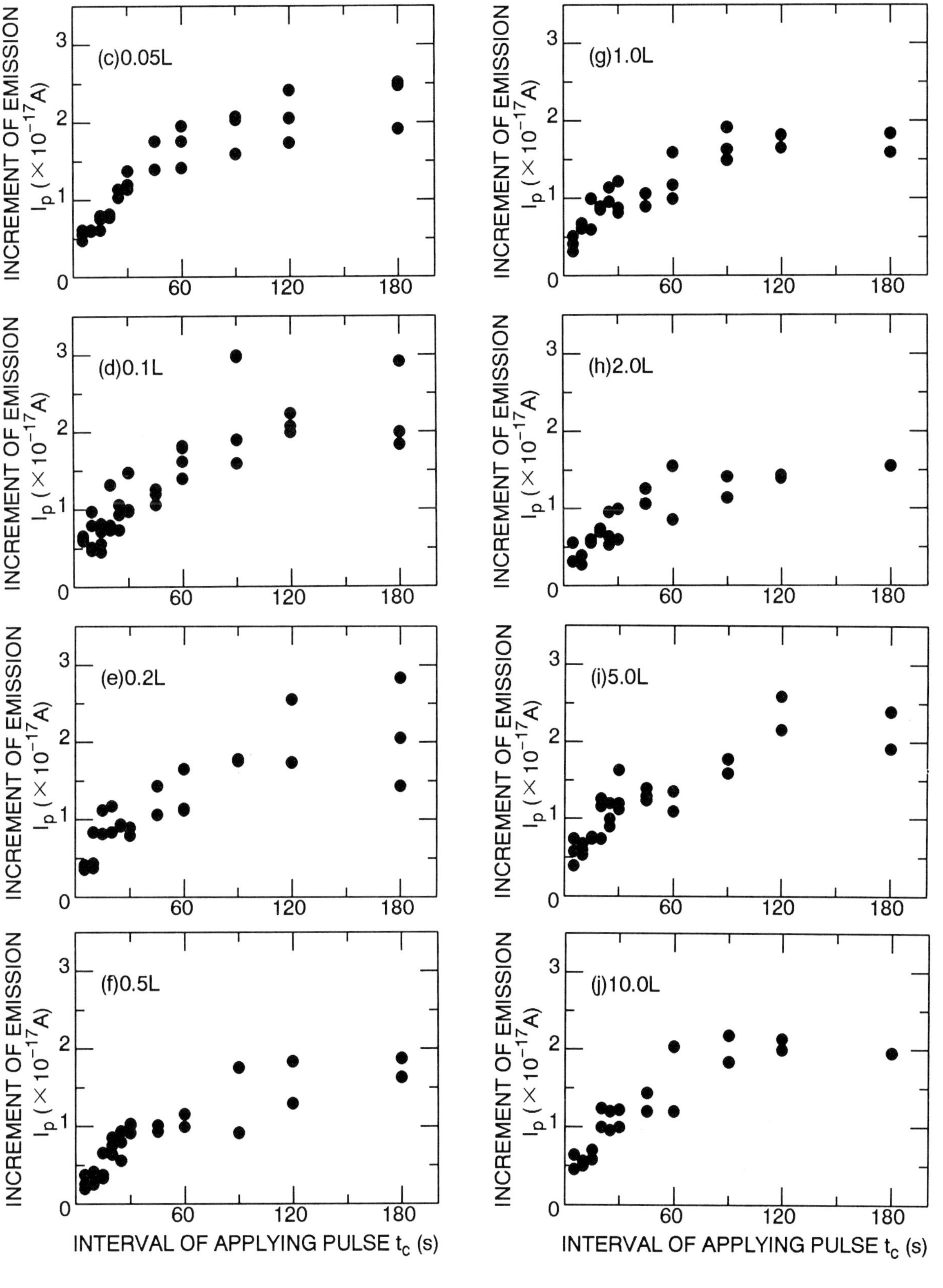
(c)0.05L
(d)0.1L
(e)0.2L
(f)0.5L
(g)1.0L
(h)2.0L
(i)5.0L
(j)10.0L
INCREMENT OF EMISSION
I_p ($\times 10^{-17}$A)
0
1
2
3
60
120
180
INTERVAL OF APPLYING PULSE t_c (s)

over 0.02L. The rate of excitaion, β, is plotted in Figure 7. β decreases from 3.7×10^{-2}/s (unexposed) to $2.1\sim3.0\times10^{-2}$/s after oxygen exposures. The decreased β means that a mean energy of trap levels shifted to higher energies, namely, the energy difference between the trap level and the Fermi level becomes large. In contrast, the rate of emission, α, is constant as indicated in Figure 8. This suggests that the emission mechanism of FSEE does not change after the exposure to oxygen. Change in the number of traps, S_0, with exposure to oxygen is shown in Figure 9. It is obvious that S_0 increases rapidly until 0.02L, and becomes almost constant at 1.2×10^4 over 0.02L. The tendency of Figure 9 correlates closely to that in Figure 6. The above analyses depict that the change in emission current of FSEE due to oxy-

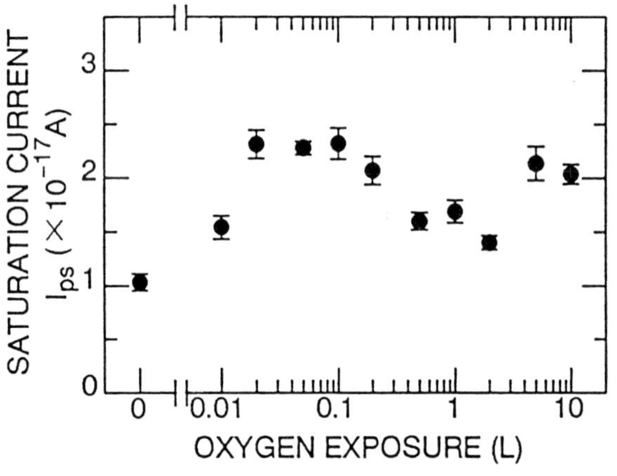

Figure 6. Change in saturation current of FSEE with respect to oxygen exposures.

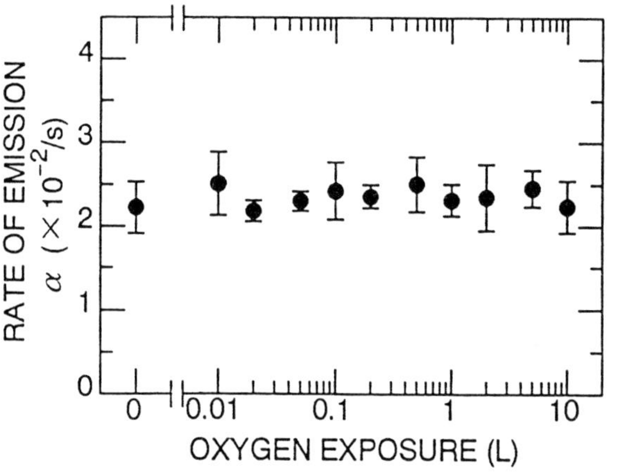

Figure 8. Change in emission probability (α) with respect to oxygen exposures.

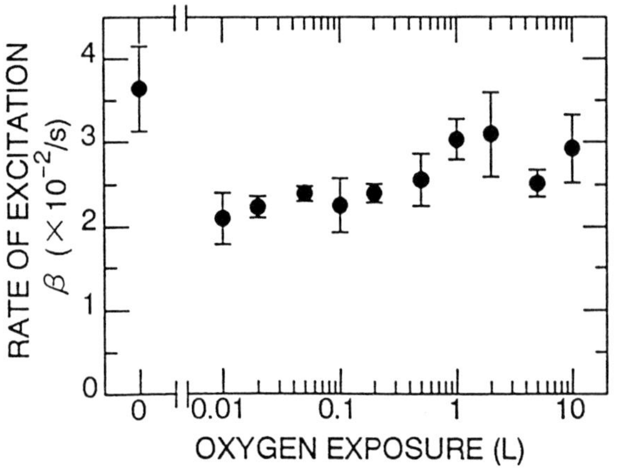

Figure 7. Change in excitaion probability (β) with respect to oxygen exposures.

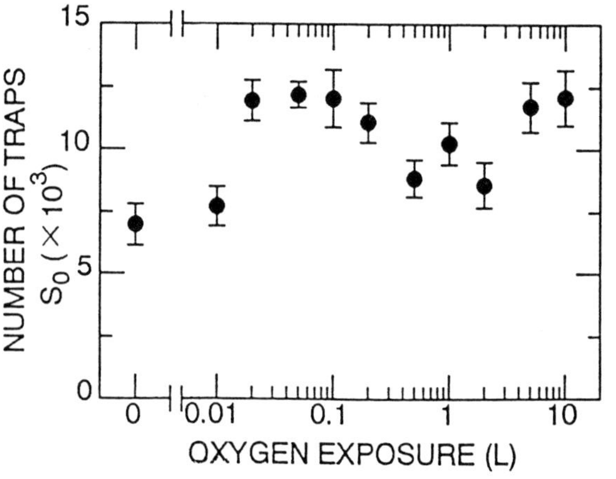

Figure 9. Change in number of electron traps (S_0) with respect to oxygen exposures.

gen exposure is due to the change in the number of traps.

The energy levels of the traps formed by oxygen exposure were examined by TED of FSEE. TED spectra before and after exposure to oxygen (900L) were shown in Figures 10 (a) and (b), respectively. Broken lines in Fig. 10 show the peak synthesis using computer. Two peaks at -4.00eV and -3.80eV were observed before exposure. Additional new peak at -3.62eV appears after oxygen exposure, locating at a level shallower than the former two peaks. The emission current of each peak

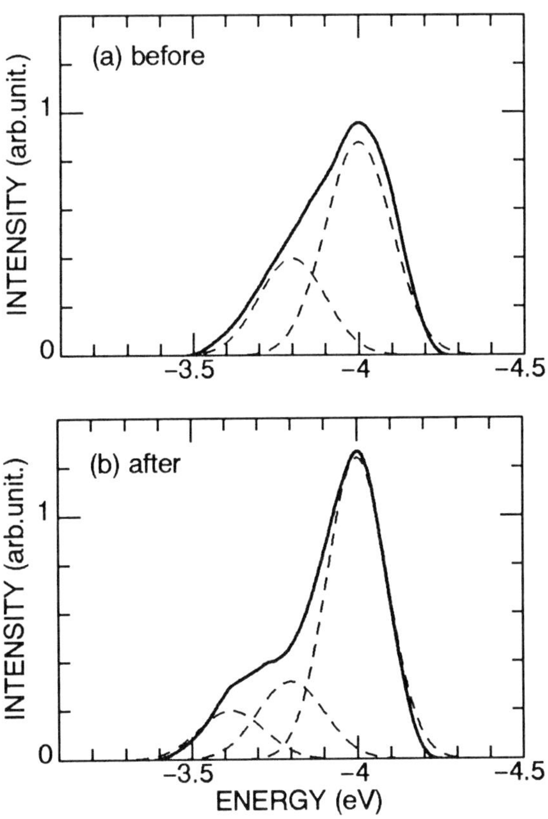

Figure 10. Total energy distribution of FSEE, (a) before and (b) after oxygen exposure at 900L.

Table 1. Peak synthesis of energy distribution of FSEE before and after oxygen exposure.

Peaks	before exposure	after exposure
-4.00eV	20.7 $\times 10^{-17}$A	26.0 $\times 10^{-17}$A
-3.80eV	9.3 $\times 10^{-17}$A	7.4 $\times 10^{-17}$A
-3.62eV	0.0 $\times 10^{-17}$A	6.6 $\times 10^{-17}$A
TOTAL	30.0 $\times 10^{-17}$A	38.0 $\times 10^{-17}$A

is summarized in Table 1. Peaks at -4.00 eV and -3.62eV were enhanced by the oxygen exposure, while the peak at -3.80eV lowered. It is, therefore, concluded that the former two peaks are the trap levels due to oxygen adsorption and that the level at -3.80eV is different. The peak at -4.00eV relates to aluminum surface, since this peak was observed before oxygen exposure. The change in work function due to oxygen exposure at 10L was calculated to be -0.2eV in this experiment. The value of -0.2eV is much smaller than that reported previously [15]. In fact, atomic concentration of 60% was detected by Auger electron spectroscopy even before oxygen exposure. Therefore, FSEE peak at -4.00eV before exposure is liberated from native oxide with a stoichiometory of Al_2O_3. The new trap level at -3.62eV implies that the oxygen atoms with different binding state from native oxide layer exist after oxygen exposure. The former appears to be chemisorbed, whereas the latter physisorbed. The trap level at -3.80eV may be caused by small impurities at the surface, since the intensity of this level decreases with oxygen exposure. However, further research is necessary in order to clarify details of this emission trap.

5.CONCLUSIONS

A new method for estimating the emission parameters studied. Three emission parameters, i.e., the rate of excitation, emission, and the number of sites were evaluated for high purity Al before and after oxygen exposures. It was found that enhancement of emission by oxygen exposure was due to the increment of emission sites. Total energy distibution of FSEE agrees well with the change in the pramenters. It became clear that peaks at -4.00 and -3.62eV were enhanced by oxygen exposure. The enhancement of peak intensities at -4.00 and -3.62eV by oxygen exposure signifies that these levels are due to aluminum oxides with different binding states.

REFERENCES

[1] E.Rabinowicz, D.M.Boyd and N. Ohmae, Proc. 2nd Int. Conf. on Solid Lubrication, Denver, CO (1978) p9.

[2] W.J.Baxter, Metallugical Trans., 6A. (1975) 749.

[3] E.Rabinowicz, Sci. Am. 236 (1977) 74.

[4] W.J.Baxter, J. Appl. Phys., 44. (1973) 608.

[5] P.A.March and E.Rabinowicz, ASLE Trans., 20 (1976) 315.

[6] H.Shigekawa, R.Iwatsu, M.Okada and S.Hyodo, Jpn. J. Appl. Phys., 22 (1983) 42.

[7] J.Ferrante, ASLE Trans., 20 (1977) 328.

[8] D.L.Doering, T.Oda, J.T.Dickinson and P.Bräunlich, Appl. Surf. Sci., 3 (1979) 196.

[9] B.Z.Rosenblum, P.Bräunlich and L. Himmel, J. Appl. Phys., 48 (1977) 5262.

[10] B.Z.Rosenblum, J.P.Carrico, P. Bräunlich and L.Himmel, J. Phys. E, 10 (1977) 1056.

[11] M.Tagawa, S.Takenobu, N.Ohmae and M.Umeno, Appl. Phys. Lett., 53 (1988) 626.

[12] H.Shigekawa and S.Hyodo, Jpn. J. Appl. Phys., 21 (1982) 1278.

[13] M.Tagawa, M.Umeno, N.Ohmae and S.Takenobu, Jpn. J. Tribology, 34 (1988) 1175.

[14] M.Tagawa, K.Tsuzuki, M.Mori, N. Ohmae and M.Umeno, Appl. Phys. Lett., 61 (1992) in press.

[15] U.Memmert and P.R.Norton, Surf. Sci., 203 (1988) L689.

SESSION XV
BIO-TRIBOLOGY (1)

Chairman : Professor K Ludema

Thin Films in Tribology / D. Dowson et al. (Editors)
© 1993 Elsevier Science Publishers B.V. All rights reserved.

A Model analysis of squeeze film effect in a hip joint

K. Ikeuchi and M. Oka
Research Center for Biomedical Engineering, Kyoto University,
53 Kawahara-cho, Shogoin, Sakyo-ku, Kyoto 606-01, Japan

A model for squeeze film effect in a hip joint is presented. Gap flow, Flow in the cartilage, solid contact of the cartilage are considered. The result shows that most of applied load is supported hydrodynamically by synovial fluid, while contact arises in a short time. The contact pressure is kept low by fluid exuding from the cartilage to the gap.

1. INTRODUCTION

The major lubrication mechanism in a synovial joint is elastohydrodynamic action between porous surfaces with boundary lubrication providing surface protection in case of severe loading [1]. Low friction of synovial joint is confirmed by numerical analysis based on elastohydrodynamic lubrication (EHL) theory [2-3], while quantitative study of other lubrication mechanisms has scarcely reported.

This paper presents a model for squeeze film effect in a hip joint in consideration of the following items:

(1) Non-Newtonian viscosity of gap flow
(2) Flow in cartilage
(3) Elastic deformation of subchondral bone
(4) Change of cartilage thickness
(5) Asperity contacts

2. MODELING

A hip joint is approximated by an axially symmetric pair of bone and cartilage (Fig. 1). Every strain is assumed to be small enough so that linear deformation theory may be valid.

2.1. Deformation

Articular cartilage is approximated by 3 element model (Fig. 2) [4]. Instantaneous elastic deformation arises at spring A, while creep deformation arises at the parallel combination of spring B and the dashpot. The former results from simultaneous movement of solid and liquid phases, while the latter results from their relative movement.

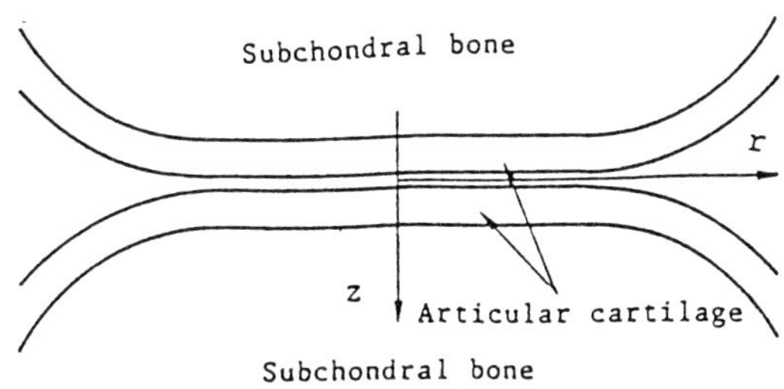

Figure 1. Axially symmetric model of hip joint.

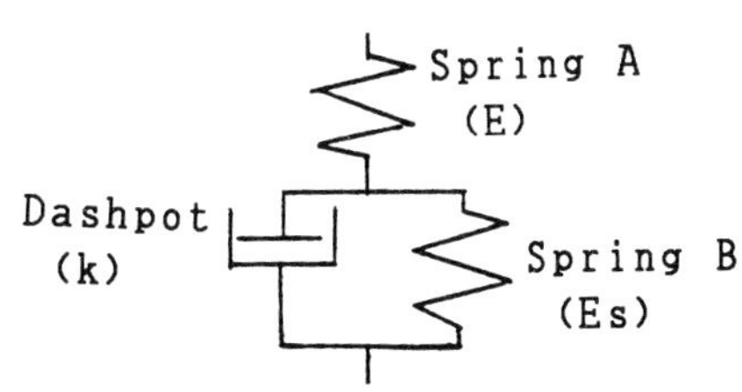

Figure 2. 3-element model.

Under short term movement like walking, since elastic deformation is predominant over creep deformation in cartilage, the former is calculated by an exact theory, while the latter is estimated by an approximate model.

Decrease of cartilage thickness due to elastic deformation is calculated with the coordinate system shown in Fig. 3 by the following equations [5,6].

$$d_{g_j}=\sum_{j'}p_{g_{j'}}G_{j,j'}\delta r_{j'} \tag{1}$$

where p_g is fluid pressure in the gap. The Green function is given by the following equations.

$$G_{j,j'}=\frac{(1-\nu^2)r_j'}{Ea}\int_0^{\infty}f(\zeta,\nu)J_0\left(\frac{r_j\zeta}{2a}\right)J_0\left(\frac{r_{j'}\zeta}{2a}\right)d\zeta \tag{2}$$

$$f(\zeta,\nu)=\frac{(3-4\nu)\sinh(\zeta)-\zeta}{(3-4\nu)\cosh(\zeta)+5-12\nu+8\nu^2+\zeta^2/2} \tag{3}$$

Where a, E and ν are thickness, modulus of elasticity and Poisson's ratio of the cartilage respectively. Jo is Bessel function of 1st kind and 0th order.

Since elastic modulus of subchondral bone (Es) is much higher than that of cartilage (E), the effect of surface deflection of the bone on cartilage deformation is neglected.

The decrease of cartilage thickness (d_c) due to elastic deformation by contact pressure (p_c) is expressed in the same manner as follows:

$$d_{c_j}=\sum_{j'}p_{c_{j'}}G_{j,j'}\delta r_{j'} \tag{4}$$

Since articular cartilage is thin and it is attached to bone, the decrease of cartilage thickness due to creep deformation is approximated by a rod model, and it is expressed with the coordinate system shown in Fig. 4, as

$$d_{f_j}=\frac{1}{E_m}\sum_i(p_{g_j}+p_{c_j}-p_{f_{i,j}})\delta z_i \tag{5}$$

Where Em is elastic modulus of solid matrix in cartilage. The elastic deformation and the creep deformation can be added after individual calculation because of linear deformation assumption.

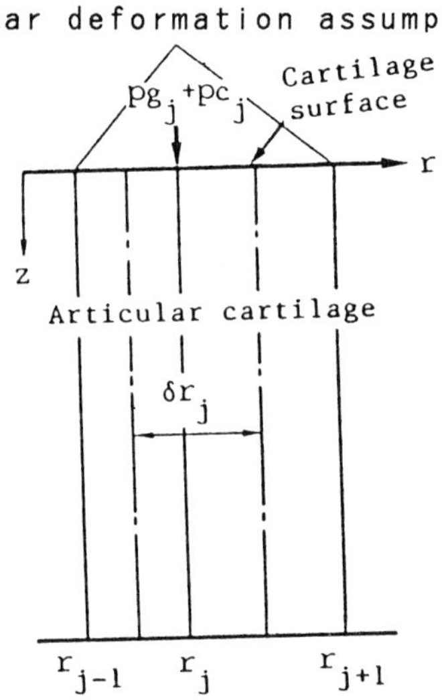

Figure 3. Coordinate system for elastic deformation.

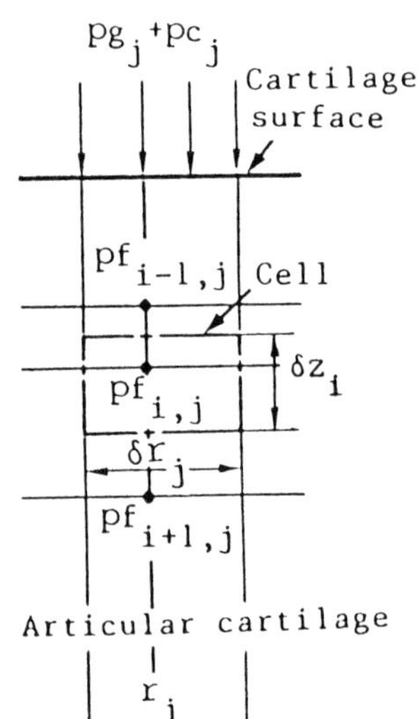

Figure 4. Coordinate system for creep deformation.

The bone composed of cancellus bone and subchondral bone plate is approximated by half space of isotropic elastic material. Since overlaying cartilage is thin and its elastic modulus is much lower than that of the bone, the effect of stresses due to bending and elongation of the cartilage is neglected and only normal stress transmitted by the cartilage is considered. Accordingly, it is approximated that the subchondral bone deforms only by pressure applied to the cartilage surface. the surface deflection [7] is expressed with Green functions as

$$d_{s_j}=\sum_{j'}(p_{g_{j'}}+p_{c_{j'}})G'_{j,j'}\delta r_{j'} \qquad (6)$$

for $r_j < r_{j'}$.

$$G'_{j,j'}=\frac{4(1-\nu_s^2)}{\pi E_s}K\left(\frac{r_j}{r_{j'}}\right) \qquad (7)$$

for $r_j > r_{j'}$.

$$G'_{j,j'}=\frac{4(1-\nu_s^2)}{\pi E_s}\frac{r_{j'}}{r_j}K\left(\frac{r_{j'}}{r_j}\right) \qquad (8)$$

where Es is elastic modulus and ν_s is Poisson's ratio of the bone.

The gap width is given from geometric relation as

$$h_j=h_{0_j}+h_r+2d_{s_j}+2d_{g_j}+2d_{c_j}+2d_{f_j} \qquad (9)$$

where $h_o(r)$ is surface profile function without deformation, and h_r is the relative position.

2.2 Flow in cartilage

Figure 5 shows the coordinate system in the cartilage. Fluid pressure (p_f) is defined at each grid point, and cell is defined by the dotted lines which pass the center of four surrounding grid points. The conservation of incompressible fluid in a cell is expressed as

$$q_1+q_2+q_3+q_4=r_j\delta r_j d(\delta z_i)/dt \qquad (10)$$

where q_1-q_4 are the volumetric flow rates per unit angle, and they are given by

$$q_1=\frac{r_j\delta r_j k_{i-(1/2)}}{\delta z_{i-(1/2)}}(p_{f\,i-1,j}-p_{f\,i,j}) \qquad (11)$$

$$q_2=\frac{r_j\delta r_j k_{i+(1/2)}}{\delta z_{i+(1/2)}}(p_{f\,i+1,j}-p_{f\,i,j}) \qquad (12)$$

$$q_3=\frac{r_{j-(1/2)}\delta z_i k_i}{\delta r_{j-(1/2)}}(p_{f\,i,j-1}-p_{f\,i,j}) \qquad (13)$$

$$q_4=\frac{r_{j+(1/2)}\delta z_i k_i}{\delta r_{j+(1/2)}}(p_{f\,i,j+1}-p_{f\,i,j}) \qquad (14)$$

where k is hydraulic permeability. The calcified cartilage in the deep zone is assumed to be impermeable.

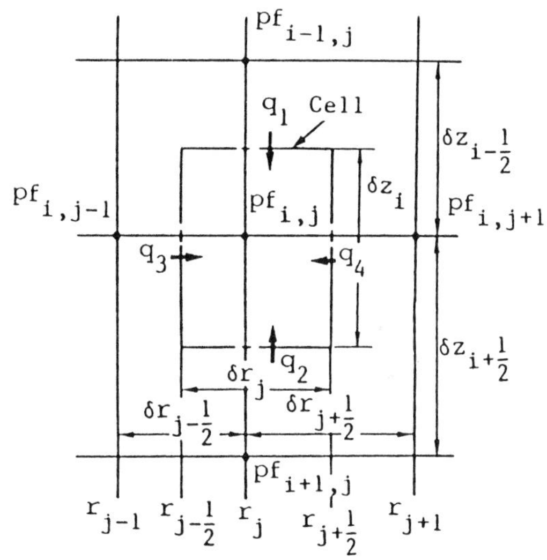

Figure 5. Coordinate system for flow in cartilage.

2.3. Flow in cartilage with gap

Figure 6 shows the coordinate system for cartilage with gap. The mass conservation in the cell is expressed as follows:

$$2q_1+q_3+q_4+q_5+q_6=r_j\delta r_j d(h_j)/dt \tag{15}$$

If synovial fluid flows according to the power law, its shear stress is expressed in terms of shear rate as

$$\tau^m=\mu_0(du/dz) \tag{16}$$

the flow rate per unit depth for negative pressure gradient is led with a similar method as in a conventional hydrodynamic lubrication problem as follows:

$$q=\frac{h^{2+m}}{(2+m)2^{1+m}\mu_0}\left(-\frac{dp_g}{dx}\right)^m \tag{17}$$

For axially symmetric problem, q_5 and q_6 are expressed by

$$q_5=\frac{h_{j-(1/2)}^{2+m}r_{j-(1/2)}}{(2+m)2^{1+m}\mu_0}\left(\frac{p_{gj-1}-p_{gj}}{\delta r_{j-(1/2)}}\right)^m \tag{18}$$

$$q_6=-\frac{h_{j+(1/2)}^{2+m}r_{j+(1/2)}}{(2+m)2^{1+m}\mu_0}\left(\frac{p_{gj}-p_{gj+1}}{\delta r_{j+(1/2)}}\right)^m \tag{19}$$

2.4. Asperity contacts

Since surface irregularity with several scales exists on cartilage surface, solid contacts may take place on the surface. As shown in Fig. 7, elastic deformation takes place with contact of asperities. The asperities are modeled to be springs of h_c in length. Locally averaged contact pressure (p_c) is expressed with gap width as follows:

for $h_i>h_c$

$$p_{cj}=0 \tag{20}$$

for $h_i<h_c$

$$p_{cj}=e(h_c-h_j) \tag{21}$$

where e is the stiffness of asperities per unit area.

2.5. Method of calculation

The above equations of deformation and flow are solved for unknowns h_r, p_f, p_g, and p_c. The calculation was performed successively with Newton-Raphson's iterative scheme. Initial condition is $p_f=0$, $p_g=0$, $p_c=0$ and $h_r=1$mm.

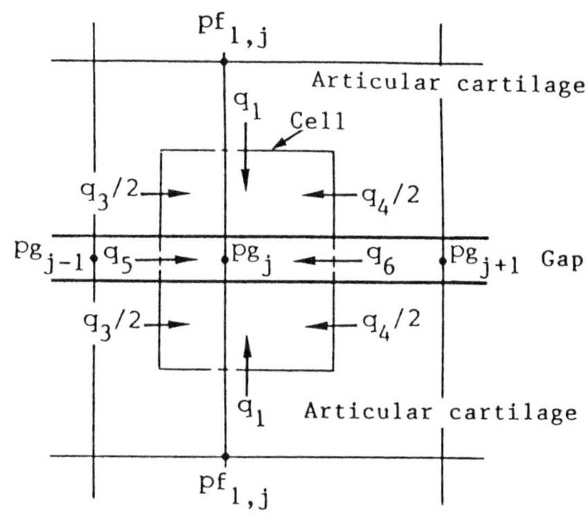

Figure 6. Coordinate system for gap flow and interstitial flow.

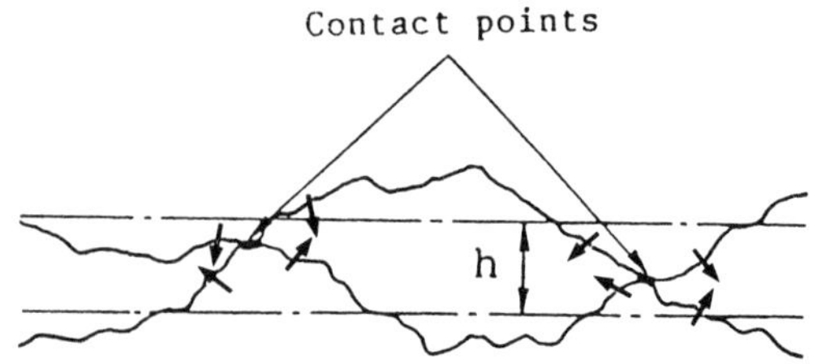

Figure 7. Magnified view of contact.

2.6. Characteristic values

Material and shape parameters are adopted from a normal joint. For articular cartilage, a=3mm, E=3MPa, ν =0.5, Em=0.7MPa and k=$3\times10^{-16}m^4/(Ns)$. For surface asperities, hc=10 μm and e=$-10^{10}N/m^3$. For subchondral bone, Es=300MPa and ν_s=0.25. For non-Newtonian viscosity of synovial fluid m=3 and μ_0 =$8\times10^3Pa^3.s$. Applied load is 2kN. Initial profile is given in mm as follows:

for r < 15 mm

$$h_0=0 \tag{22}$$

for r > 15 mm

$$h_0=(r-15)^2/80 \tag{23}$$

3. RESULTS

Figure 8 shows that film thickness continues to decrease at the central region with outward flow from the squeeze film. The flow is restricted by the peripheral zone, where film thickness is smallest. Figure 9 shows that fluid pressure scarcely changes with time. Figure 10 shows that the cartilage contacts at the peripheral zone in 0.19s after loading. The contact pressure at the peripheral zone increases at first and then approaches a certain value.

Figure 11 shows flow velocity from the cartilage to the gap. Fluid weeps from the cartilage to the gap [8] at the contact region, while flow in reverse direction is not presented.

Figure 12 shows that the thickness of the cartilage decreases by creep deformation at the peripheral zone and inside of it, while it increases at the out side of the peripheral zone due to radial flow in the cartilage.

Figure 13 shows the ratio of load supported by contact pressure to total load. If complete hydrodynamic lubrication condition is kept, the ratio is zero. The ratio is 100% without any hydrodynamic lubrication effect.

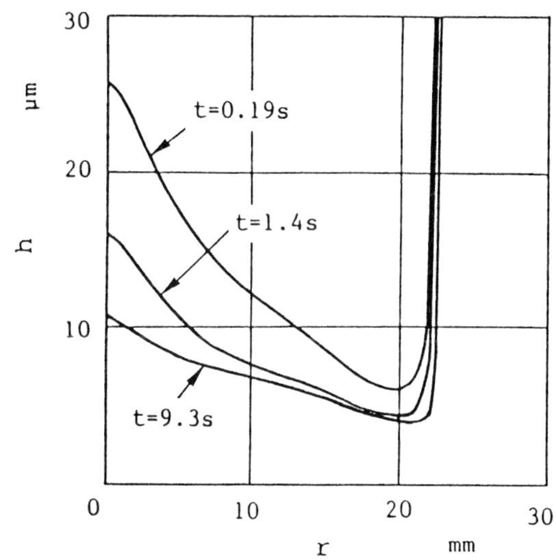

Figure 8. Film thickness.

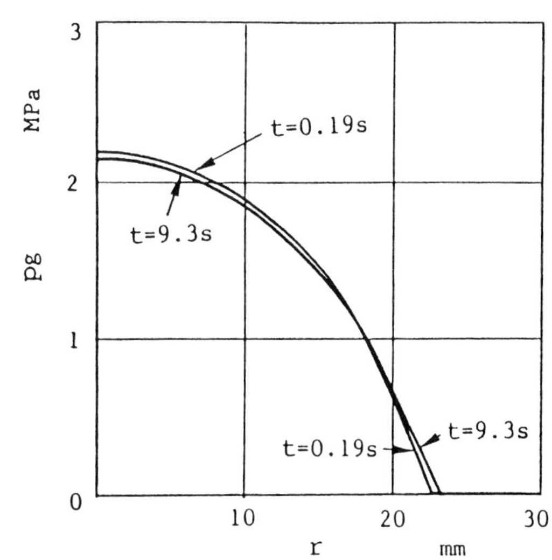

Figure 9. Fluid pressure in gap.

It is shown that contact arises within 0.1s for non-Newtonian fluid, while noncontacting condition is kept for more than 0.3s for Newtonian fluid of which viscosity is 0.1Pa.s. The effect of interstitial flow on contact pressure, which is small at the early stage, increases after 1s with increase of contact pressure.

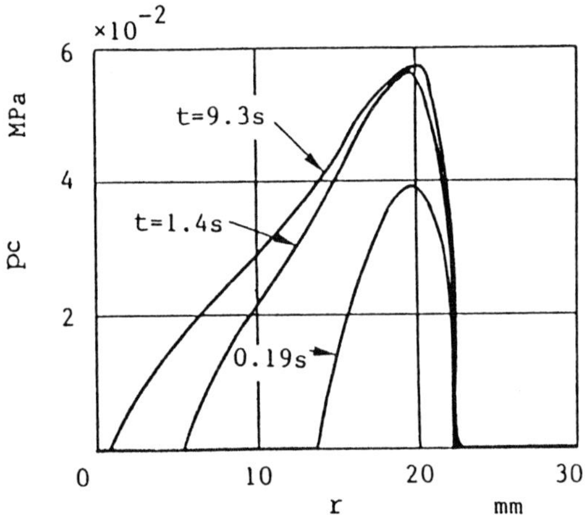

Figure 10. Contact pressure on cartilage surface.

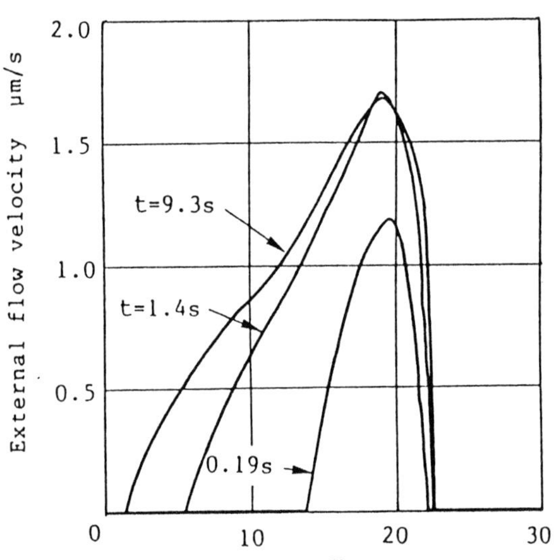

Figure 11. Flow velocity exuding from cartilage to gap.

4. DISCUSSION

The minimum film thickness decreases from 1mm to 6 μm in 0.19s (Fig. 8). Since the surface asperities are modeled to be linear springs of which length (h_c) is 10μm, contact arises in a short time after loading when film thickness becomes less than 10μm.

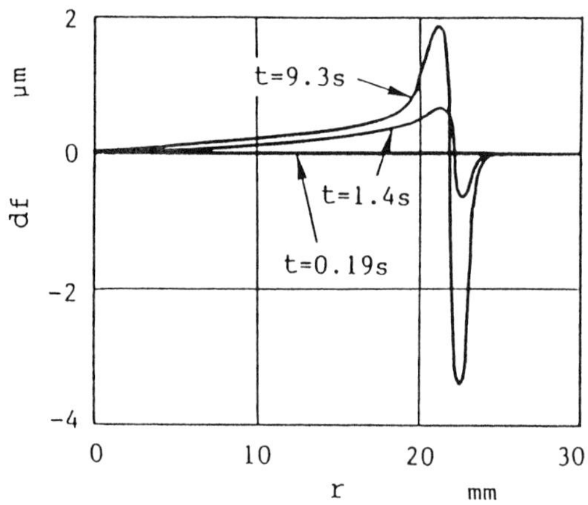

Figure 12. Decrease of cartilage thickness by creep deformation.

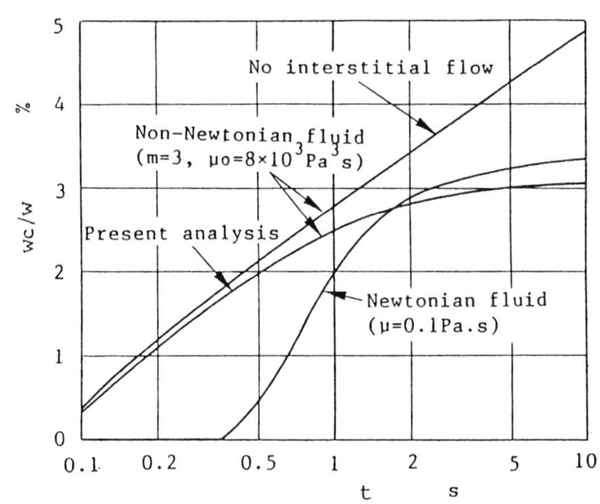

Figure 13. Ratio of load supported by contact pressure to total load.

Since gap is almost 1mm just after loading, shear rate of the gap flow is high, and hence equivalent viscosity of the synovial fluid is low due to its non-Newtonian character. At the early stage, since resistance to outward flow is low because of low equivalent viscosity, film thickness decreases radically and contact arises in a short time. Thereafter, with decrease of film thickness, the equivalent viscosity rises and resistance to outward flow increases rapidly. Squeeze film effect is thus maintained for longer than that estimated with the Newtonian fluid of which viscosity is 0.1Pa.s.

The fluid pressure is kept almost unchanged and the interface is in good lubrication condition for a long time, while solid contact arises in between cartilage surfaces in a short time.

In the contacting region, fluid which exudes from the cartilage to the gap restricts the decrease of minimum film thickness (Fig. 8) and the increase of maximum contact pressure (Fig. 10), from 1.4s after loading. Without interstitial flow, since no fluid exudes, contact pressure and load support by contact pressure continues to increase (Fig. 11). Thus the flow from the cartilage to the gap plays an important role in keeping good lubrication condition.

Fluid exudes in the contacting region, while scarcely any flow arises in the noncontacting region. According to Darcy's law, water flows in proportional to pressure gradient. At the moment of contact between cartilages, the fluid in the cartilage is pressuraized and pressure gradient arises across a layer beneath the surface [9]. Since this layer is very thin at first, the pressure gradient is steep enough even if the contact pressure is not high. Fluid exudes to gap by this pressure gradient. While in noncontacting region, since no steep pressure gradient arise and permeability of cartilage is low, flow between the cartilage and the gap is negligibly small.

The film thickness in a noncontacting squeeze film with porous material [10] decreases faster than that with impermeable material, because interstitial outward flow arises in parallel with the gap flow. However, since the permeability of articular cartilage is very low, the effect of the additional flow is negligibly small in a noncontacting region.

According to the present simulation, flow direction is always from the cartilage to the gap, and hence water content of the cartilage continues to decrease. This is because new contact arises at initially noncontacting place in the present model. On the contrary, if surfaces are separated after a long term contact, fluid would be imbibed by the cartilage.

5. CONCLUSION

The result from a model analysis of a hip joint shows that most of the applied load is supported by fluid pressure, while solid contact arises in a short time due to non-Newtonian viscosity of synovial fluid. The fluid which exudes from the cartilage to the gap acts to restrict the increase of contact pressure. A hip joint thus maintains good lubrication condition under various operating conditions.

NOMENCLATURE

d_c: decrease of cartilage thickness due to elastic deformation by contact pressure
d_f: decrease of cartilage thickness due to creep deformation
d_g: decrease of cartilage thickness due to elastic deformation by fluid pressure
d_B: deflection of subchondral bone
E: instantaneous elastic modulus of cartilage
Em: elastic modulus of solid matrix in cartilage
Es: elastic modulus of bone
e: stiffness of asperities per unit area
h: gap width
h_c: height of asperity
h_o: surface profile without deformation
h_r: relative position of mating faces
k: permeability of cartilage
m: viscosity index of synovial fluid
p_c: contact pressure
p_f: fluid pressure in cartilage
p_g: fluid pressure in gap
q: flow rate
r: radial coordinate
t: time
w: joint load ($= 2\pi \int_0 (p_g+p_c) rdr$)
w_c: load supported by solid contact pressure ($= 2\pi \int_0 p_c rdr$)
z: coordinate normal to surface
μ_o: coefficient of viscosity
ν: Poisson's ratio of cartilage for instantaneous deformation
ν_B: Poisson's ratio of bone

REFERENCES

1. D. Dowson, Proc. Instn Mech. Engrs, Vol. 181, Pt. 3J (1966-1967) 45.
2. J.B. Medley and D. Dowson, ASLE Trans., Vol. 27, No.3 (1984) 243.
3. E.F. Rybicki, J.S. Strenkowski, M.A. Tamm and W.A. Glaeser, Wear, No. 47 (1978) 279.
4. J.R. Parsons and J. Black, J. Biomechanics, Vol. 10 (1977) 21.
5. V. Castelli, G.K. Rightmire and D.D. Fuller, Trans. ASME, J. Lub. Technol., Vol. 89, No.4 (1967) 510.
6. K. Ikeuchi, H. Mori, T. Ohkubo and S. Ichi, Bulletin JSME, Vol. 25, No. 202 (1982) 646.
7. H. Christensen, Trans. ASME, J. Lub. Technol., Vol. 92, No. 1 (1970) 145.
8. C.W. McCutchen, Proc. Instn Mech. Engrs, Vol. 181, Pt3J (1966-1967) 55.
9. V.C. Mow, S.C. Kuei, W.M. Lai and C.G. Armstrong, Trans ASME, J. Biomechanical Engng, Vol. 102, No. 1 (1980) 73.
10. J.S. Hou, V.C. Mow, W.M. Lai and M.H. Holmes, J. Biomechanics, Vol. 25, No. 3 (1992) 247.

Thin Films in Tribology / D. Dowson et al. (Editors)
© 1993 Elsevier Science Publishers B.V. All rights reserved.

An experimental study of contact condition and friction in hip prostheses

K. Ikeuchi, S. Goto, K. Isobe and M. Oka

Research Center for Biomedical Engineering, Kyoto University, 53 Kawahara-cho, Shogoin, Sakyo-ku, Kyoto 606, Japan

A femoral head of quartz glass was attached to a hip joint simulator to observe the contact region. The area of contact changes for an ultra high molecular weight polyethylene (UHMWPE) socket. Squeeze film effect, which reduces friction at heel strike, diminishes during step phase. Frictional torque is lower for a compliant socket of silicon rubber. However, if extremely compliant socket (E=1.7 MPa) is used, friction which was low at the start of walking increases and becomes even higher than the case with UHMWPE socket. This transition is late with more viscous fluid.

1. INTRODUCTION

A tribological study of an artificial joint is important to reduce wear particles which result tissue reaction, bone absorption and loosening.

Since load applied to a hip joint changes radically and sliding velocity is low in a walking cycle, the lubrication condition in a hip prosthesis is more complicated than that in an engineering bearing [1]. An experiment with a joint simulator is hence necessary to investigate lubrication condition in a hip prosthesis. However, it is not easy to detect what really happens in a hip prosthesis.

In order to get more direct information in detail, we observed the interface between a femoral head and a socket with a transparent femoral head [2]. The purpose of this study was to investigate the effects of socket compliance and fluid viscosity on the contact area and lubrication condition through visual study and friction measurement.

2. APPARATUS AND METHOD

Figure 1 shows the cross sectional view of a hip prosthesis attached to the simulator. A femoral head of 28.1 mm in diameter is made of quartz glass. The contact area is observed through a flat surface at the bottom with a micro video camera. Dynamic load is applied to the head via a lower beam, while flexion-extension movement is applied by oscillating the socket. The simulator is driven with a variable speed electric motor and two cams.

Three sockets of different materials are used to investigate the effect of socket compliance. The inside diameters are 28.2 mm and radial clearances are 0.05 mm.

One of the sockets is made of UHMWPE (modulus of elasticity is 450 MPa), and its outside diameter is 42 mm. It has no backing and attached to the simulator with bone cement (PMMA resin).

Other sockets (Fig. 2) are made of silicon rubber. Elastic modulus of silicon rubber A is 11 MPa, and

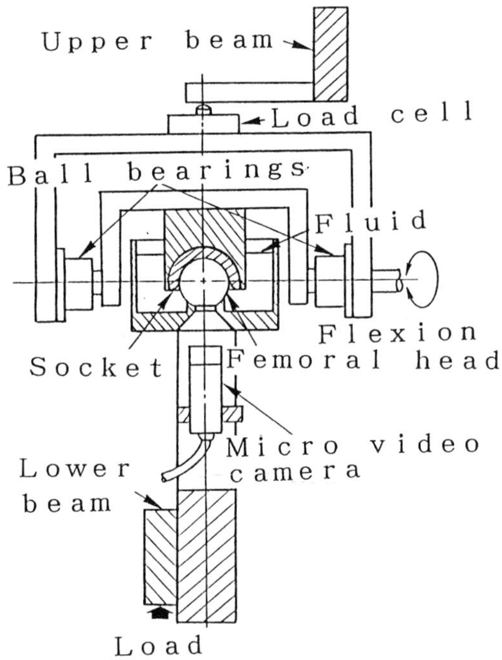

Figure 1. Cross sectional view of apparatus.

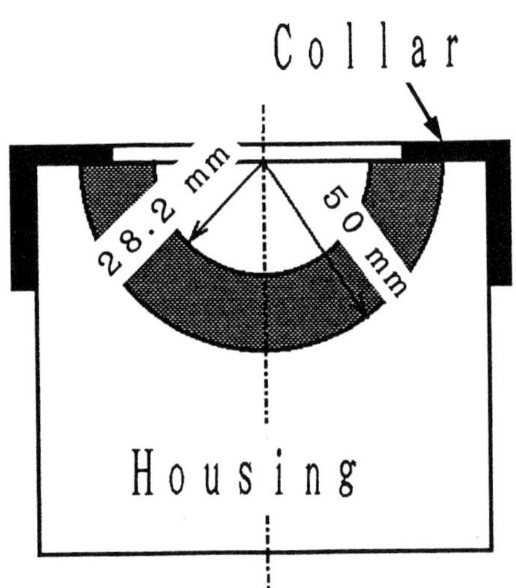

Figure 2. Compliant socket.

that of silicon rubber B is 1.7 MPa. Their outside diameters are 50 mm. They are cast in metal housings and fixed by epoxy resin and collars as shown in Fig. 2.

The joint is lubricated with Indian ink thickened with carboxymethyl cellulose. The viscosity of the solutions was measured with a capillary tube visco-meter. The measured values were 0.01 Pa.s, 1.0 Pa.s and 10Pa.s at 25°c.

Figure 3 shows the change in flexion-extension angle and applied load in a walking cycle.

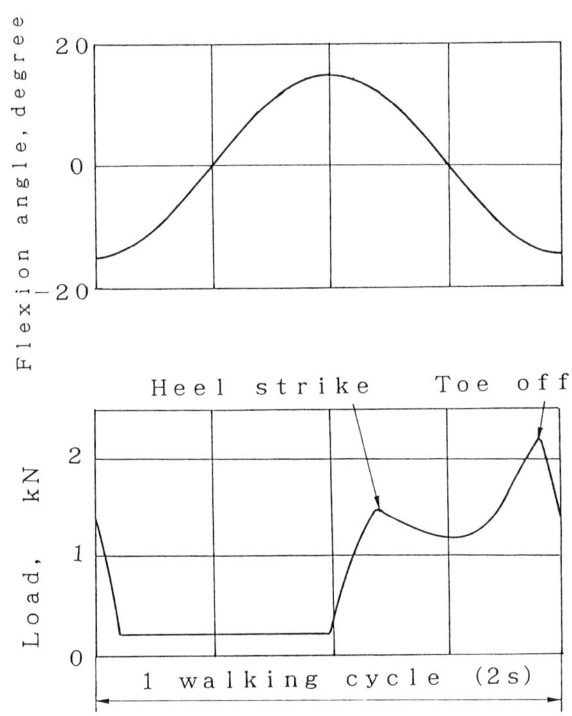

Figure 3. Flexion angle and load.

3. RESULTS

Figure 4 shows the interface between the femoral head and the UHMWPE socket in a walking cycle. A bright contact region can be distinguished from circumferential dark region. The area of contact region which was small in stance phase becomes larger at heel strike and it takes maximum value at toe off. When the area of contact region decreases with the beginning of swing phase, cavitation is observed.

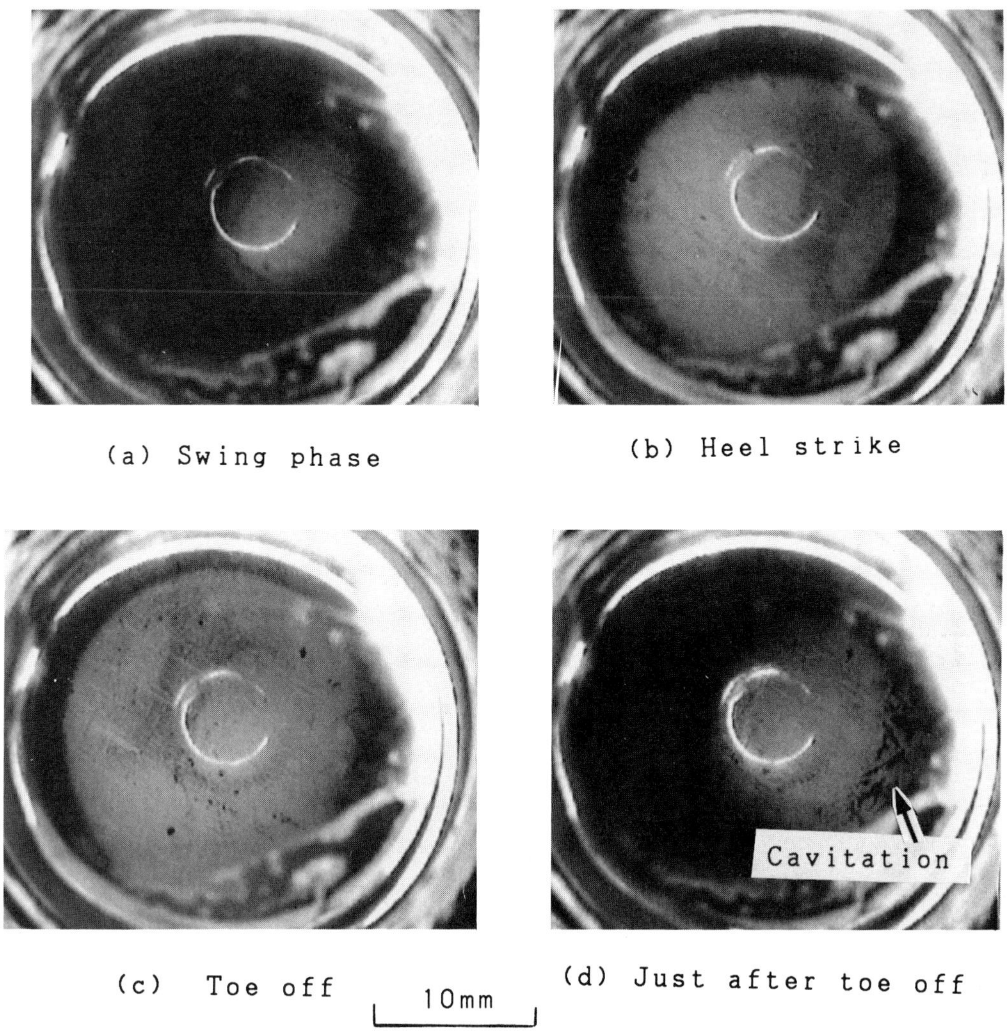

(a) Swing phase (b) Heel strike

(c) Toe off (d) Just after toe off

Figure 4. Interface with UHMWPE socket.

Figure 5 shows the interface between the femoral head and the socket of silicone rubber B in an early walking cycle. The contact region always extends over the range of vision. The difference of lubrication condition between swing phase and stance phase is not remarkable, because of positive and negative squeeze film effects. Many dark spots represent the fluid entrapped in the dents on the silicon rubber surface. Those spots diminishes with time and the interface becomes brighter at 75th walking cycle (Fig. 6).

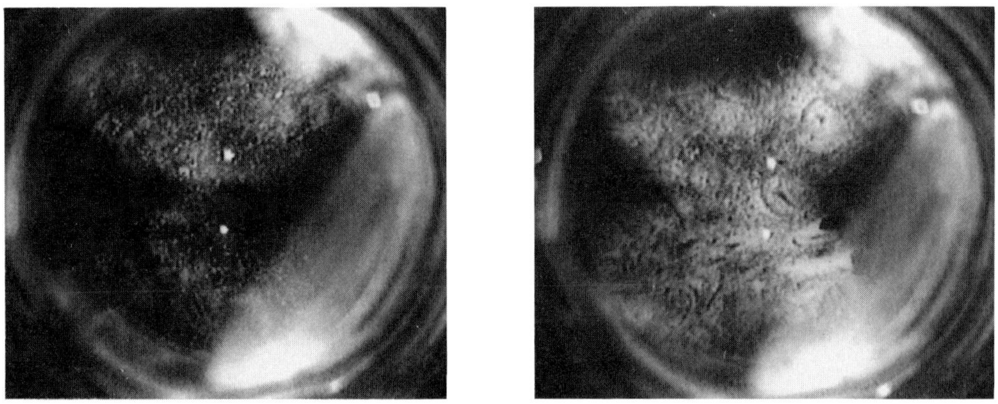

(a) Swing phase 10mm (b) Stance phase

Figure 5. Interface with socket of silicon rubber B before transition.

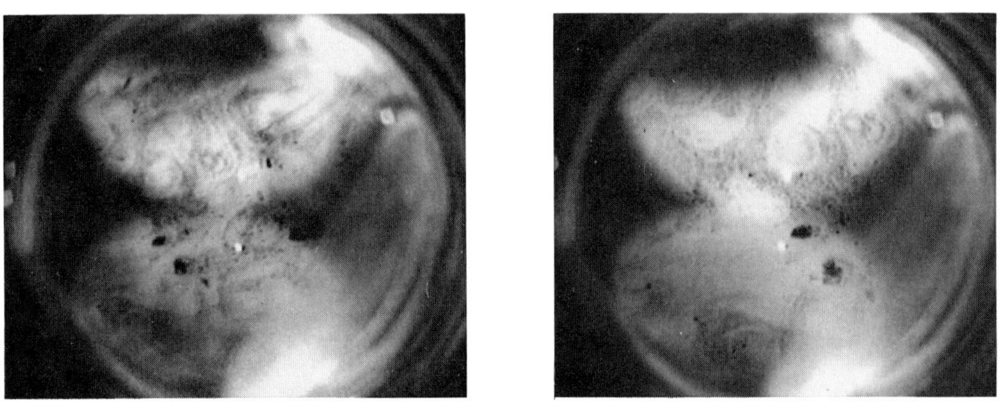

(a) Swing phase 10mm (b) Stance phase

Figure 6. Interface with socket of silicon rubber B after transition to high friction mode.

Figure 7 shows the frictional torque for UHMWPE socket. Friction is higher at toe off than that at heel strike. Figure 8 shows that frictional torque for the socket of silicon rubber A is lower than that for the socket of UHMWPE. At the early stage, frictional torque is also low for the socket of silicon rubber B, but it increases and becomes even higher than the case with UHMWPE after 150s (Fig. 9).

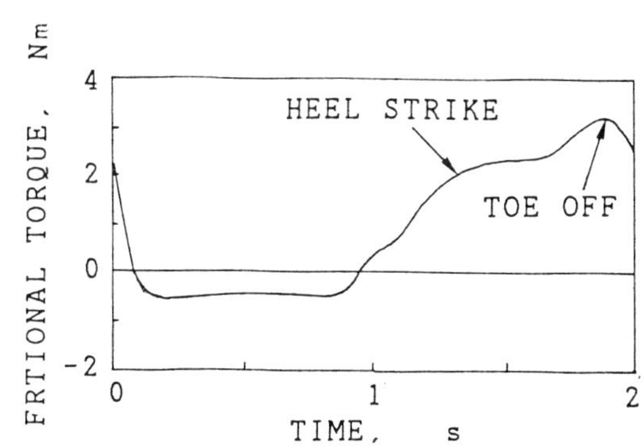

Figure 7. Frictional torque in one cycle for UHMWPE socket.

Figure 10 shows the effect of fluid viscosity on frictional torque at the moment of toe off. A fluid with higher viscosity results lower frictional torque. The frictional torque scarcely changes with time and hence transition does not appear.

Figure 11 shows that frictional torque increases with time for silicon A socket. The time of transition from low frictional torque mode to high torque mode is late for a highly viscous fluid.

Figure 12 shows more radical transition to high friction mode for the socket of silicone rubber B. For a highly viscous fluid, the time of transition is late, and the torque at toe off is low after the transition.

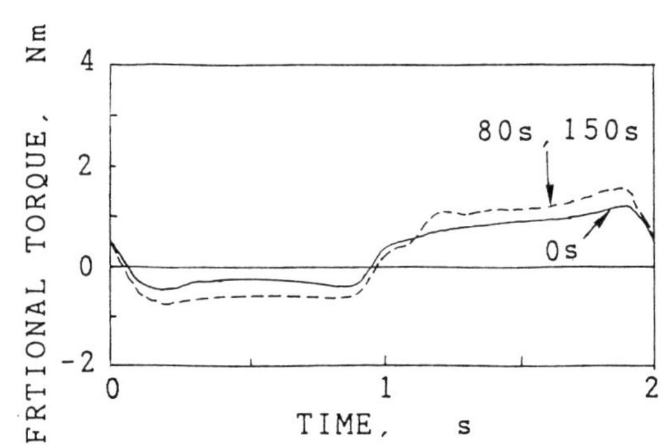

Figure 8. Frictional torque in one cycle for socket of silicon rubber A.

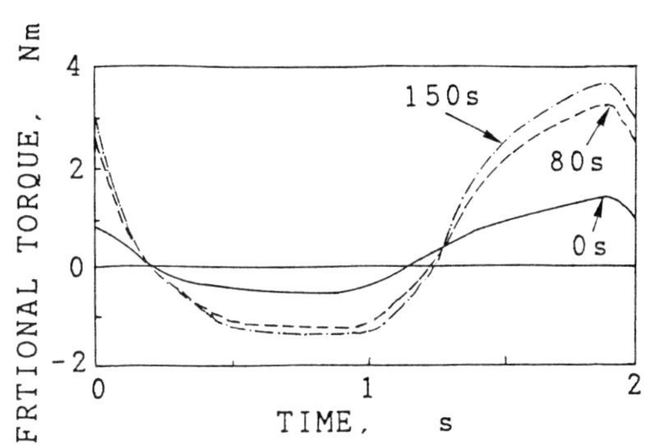

Figure 9. Frictional torque in one cycle for socket of silicon rubber B.

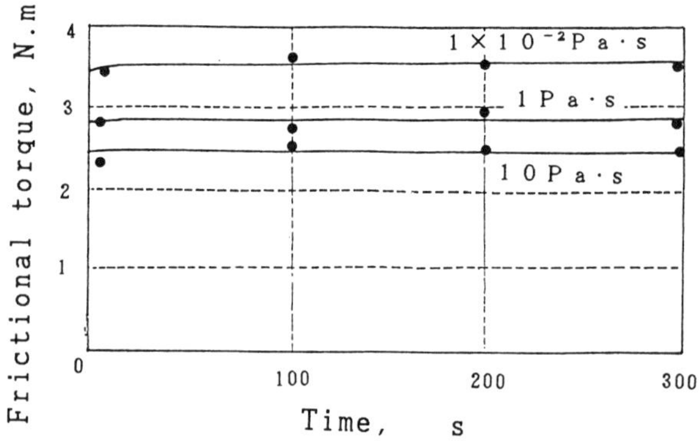

Figure 10. Frictional torque at toe off for UHMWPE socket.

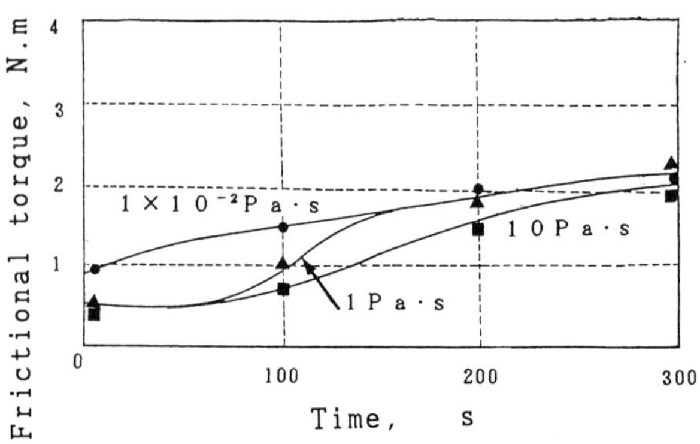

Figure 11. Frictional torque at toe off for socket of silicon rubber A.

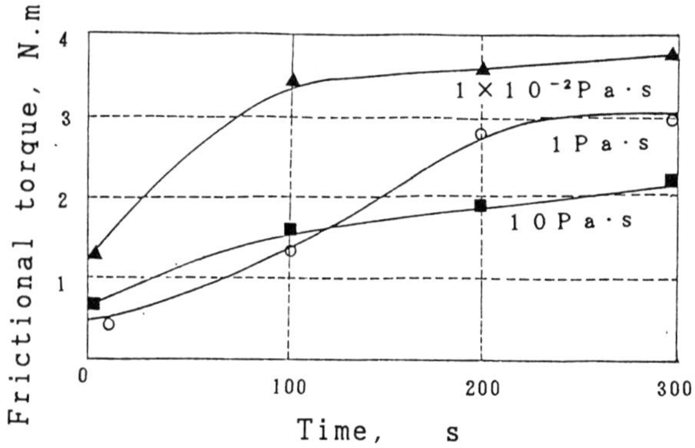

Figure 12. Frictional torque at toe off for socket of silicon rubber B.

4. DISCUSSION

For an UHMWPE socket, since gap is thin in the contact region, fluid can be pressurized and hence it can support load, while the gap is thick in the outside dark region. Most of the joint load is supported by the contact region. The contact area changes in response to load change (Fig.4), while the lubrication condition in the contact region changes in response to load change (squeeze film effect) and sliding motion (hydrodynamic wedge effect). From the measured values of frictional torque and their continuous increase during step phase(Fig. 7), it is concluded that contact region is in elasto-mixed lubrication condition [3,4]. Hydrodynamic load support continues to decrease in step phase and it recovers in swing phase. Judging from the fact that frictional torque at toe off keeps approximately constant value (Fig. 9), squeeze film effect [5] diminishes during each stance phase.

For a socket of silicon rubber, contact region is always large. At the early cycles, many spots of trapped fluid are observed and frictional torque is low. So elastohydrodynamic squeeze film [5] and micro-elastohydrodynamic lubrication [6] seem to be effective. However, if socket is too compliant, fluid continues to exude from the gap until the high frictional torque mode takes place.

In order to realize a low friction prosthesis, compliance of a socket should be optimized by material selection and design. This increase of friction might be minimized to make radial clearance (difference in radii between a ball and a socket) larger than the present value (0.05 mm).

5. CONCLUSION

For a socket of UHMWPE, area of contact region changes in response to joint load. Most of joint load is supported at the contact region. Though squeeze film is effective at heel strike, it diminishes during stance phase.

For a compliant socket, contact region is always wide, lubrication condition is better and frictional torque is lower. However, if compliance of a socket is too high, fluid continues to exude from the gap and the lubrication condition finally changes to high friction mode.

REFERENCES

1. J. O'Kelly, A. Unsworth, D. Dowson and V. Wright, Proc. Instn Mech. Engrs, Vol. 8, No. 3 (1979) 153.
2. K. Ikeuchi, H. Mori, H. Inoue and M. Oka, Proc. Japan International Tribology Conf. Nagoya (1990) 1.
3. K. Ikeuchi, H. Mori and Y. Murai, Japanese Journal of Tribology, Vol. 34, No. 9 (1989) 1065 (Allerton Press).
4. K. Ikeuchi and M. Oka, Frontiers Med. Biol. Engng, Vol. 2, No. 3 (1990) 167 (VSP).
5. K. Mabuchi and T. Sasada, Wear, 140 (1990) 1.
6. K. P. Baglin, Proc. Instn Mech. Engrs, Vol. 200, No. C6 (1986) 415.

Thin Films in Tribology / D. Dowson et al. (Editors)
© 1993 Elsevier Science Publishers B.V. All rights reserved.

Friction and Mixed Lubrication in Soft Layer Contacts

L. Caravia[a], D. Dowson[a], J. Fisher[a] , P.H. Corkhill[b], B.J. Tighe[b]

[a]Department of Mechanical Engineering,
University of Leeds, Leeds, LS2 9JT, United Kingdom.

[b]Department of Chemical Engineering and Applied Chemistry,
Aston University, Birmingham, B4 7ET, United Kingdom.

The use of thin layers of a compliant material to form the bearing surface in human total replacement joints has been found to promote fluid film lubrication. There may be periods,however, such as at start up of motion, when the fluid film breaks down. The present study investigated the friction of compliant layers in conditions of mixed or boundary lubrication. The friction of three materials - two polyurethanes of different elastic modulii and a hydrogel terpolymer - was measured when sliding against hard indenters of varying surface roughness. The measured friction was found to be low and decreased as the polyurethane modulus and indenter surface roughness decreased. The hydrogel gave the lowest friction values.

1. INTRODUCTION

The natural synovial joint operates with low friction and enjoys fluid film lubrication in normal articulating conditions [1]. Theoretical studies have shown that the soft, compliant nature of the articular cartilage which forms the bearing surfaces of the joint allows the formation of a fluid film through micro-elastohydrodynamic action [2]. Micro-elastohydrodynamic action occurs when pressure perturbations in the fluid film caused by the surface roughness causes the surface asperities themselves to be flattened, resulting in a much smoother surface which can maintain a fluid film. In the harder materials used in current replacement joint designs the asperities are not easily flattened and micro-EHL does not occur so readily . As a result these joints do not operate in the fluid film regime [3]. There is now much interest in the use of soft material layers as one, or both, of the bearing surfaces in total human replacement joints [4]. It is thought that these 'cushion form' replacement joints will function in the fluid film regime through micro-EHL action and, in addition, in knee prostheses through the generation of a thicker fluid film. The superior lubrication compared to existing forms of replacement joints will help to reduce wear and frictional torques (which can lead to loosening of the joint) and will thus prolong the useful life of the replacement. This has been demonstrated experimentally with both prototype hip and knee prostheses in joint simulator tests with continuous motion and cyclic loading [5, 6]. It has to be remembered that artificial joints are not always in continuous motion and periods when there is limited movement and a high level of loading may well prevent full fluid lubrication being maintained in these compliant joints. The long term success of cushion form bearings may well be dependent upon the tribological performance when the fluid film breaks down.

A number of different candidate materials have been considered for cushion form bearings, these include silicone rubber [7] silicone copolymers [8], and polyurethanes [9, 10]. More recently hydrogels have also been proposed as possible materials [11, 12]. These hydrogels have a high water content and are microporous and can release water when deformed in a similar manner to natural synovial joints.It is possible with appropriate tribological design to produce an equivalent optimum elastohydrodynamic lubricating film thickness with all these materials [4] for a particular cycle of movement for any joint configuration. However, it is not known at present if particular types of materials have preferential micro-elastohydodynamic lubrication, or are able to trap or retain fluid by micro deformations which can help preserve thin lubricating films and reduce the level of friction in a mixed lubrication regime when there is a limited amount of movement at the articulating surfaces.

Three materials have been examined in this study: two polyurethanes with different elastic modulii and a hydrogel terpolymer. The frictional properties of these materials have been studied in conditions of fluid film and mixed lubrication. In comparing the friction of the elastomeric polyurethanes with the hydrogel the aim was to study whether there was a frictional advantage in using a high water content hydrogel polymer, when the fluid film was low and the contact entered a mixed lubrication regime.

2. MATERIALS

Tecoflex aliphatic, linear 93A and 80A injection grade polyurethanes with nominal elastic modulii of 20MPa and 6MPa respectively were used. Flat layers (0.8mm thick) were injection moulded onto flat polyethersulphone substrates. The surface roughness (Ra) of the polyurethane layers was on average $0.08\mu m$.

The hydrogel used was an NVP-CHexMA-HexMA terpolymer [12] flat layer which was bonded onto a PVC substrate. The layer was 0.8mm thick and had a nominal elastic modulus of 20MPa. The nature of the hydrogel material did not allow the surface roughness to be measured using normal surface profilometry techniques. However, the layer was prepared by copolymerisation in situ between two glass plates with an Ra of $\ll 0.01\mu m$. The surface roughness of the layer might be expected to be of the same order as that of the glass plates it was formed against but, in practice, probably had an increased waviness due to shrinkage. This effect was noted with the polyurethane layers which were injection moulded against a counterface of Ra $0.01\mu m$. It was found that the polyurethane layers had a higher Ra value than the counterface and an increased waviness compared to the counterface.

Medical grade spherical stainless steel indenters of surface roughnesses $0.20\mu m$ and $0.05\mu m$ Ra and a spherical glass indenter of $\ll 0.01\mu m$ Ra were used to slide against the compliant material layer. The variation in the indenter roughness was used to change the lubrication regime in tests that were carried out at constant velocity and load and hence with virtually the same elastohydrodynamic film thickness.

All tests were carried out with the layer and indenter completely submerged in deionised water.

3. APPARATUS AND METHODS

The friction experiments were carried out on a sliding friction apparatus (Figure 1) adapted from a sledge microtome [13]. The indenter was held

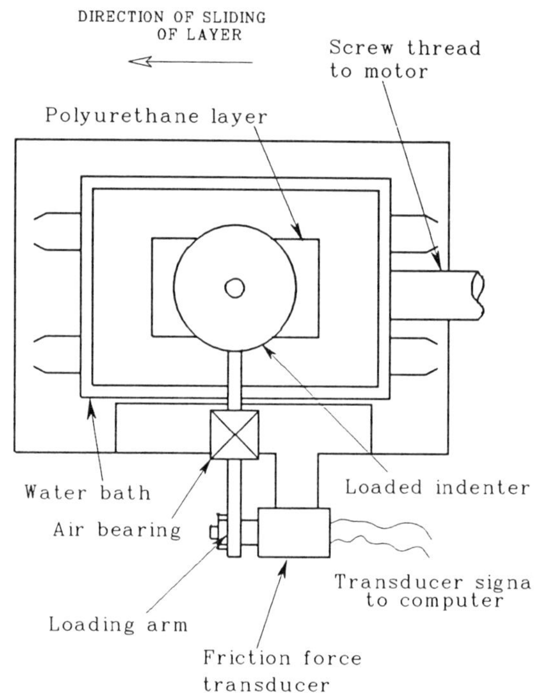

Figure 1. Sliding friction apparatus

at one end of a pivoted arm and loaded onto the flat material layer. The arm was pivoted on an air bearing to keep apparatus friction to a minimum. The extreme end of the arm was restrained by a piezo- electric force transducer. The voltage signal from the transducer's charge amplifier was passed into an analogue to digital converter connected to a microcomputer. The signal was analysed using Unkelscope software. The software used calibration data to convert the transducer signal into a friction force. The friction was measured by recording the frictional force on the indenter as it slid over the soft layer at a constant velocity.

The polyurethane and hydrogel layers were fixed into position in a water bath which was driven by a motor to slide in one direction at constant sliding speed. The sliding distance was measured using a linear variable differential transducer (LVDT) and the signal from this was fed directly into the analogue to digital converter and then into the Unkelscope software. Using previously determined calibration data, as for the friction force, the transducer signal was converted into displacement.

The sliding velocity was 8mm/s. This was chosen to fall within the physiological range when the joint is moving at a slow velocity such as at start up of motion. In addition it has been shown previously that the friction was almost independent of sliding speed at this velocity [14]. The contact stress between the material layers and the indenter was constrained at 2MPa. In order to keep the stress constant for materials of different elastic modulus, the indenter radius was varied. Constraining the contact stress at 2MPa at a constant velocity and load ensured that a similar elastohydrodynamic film thickness was achieved in each contact.

Thirty five sliding tests were carried out for each material with each indenter and the mean and sample standard deviation was calculated. Statistical comparison was carried out using a one way analysis of variance and Student's t test.

4. RESULTS

After the initial start up of sliding the measured friction settled to a constant steady state value as shown in Figure 2. The shape of the curve shown is explained in a recent paper (14). In general, the friction was low and in all tests the coefficient of friction was less than 0.5. Figure 3 shows the coefficient of friction values for the three material layers sliding against the three different surface roughness indenters. Each bar represents the average coefficient of friction value for the 35 tests carried out for each of the nine material/indenter combinations. Figure 3 shows clearly that:

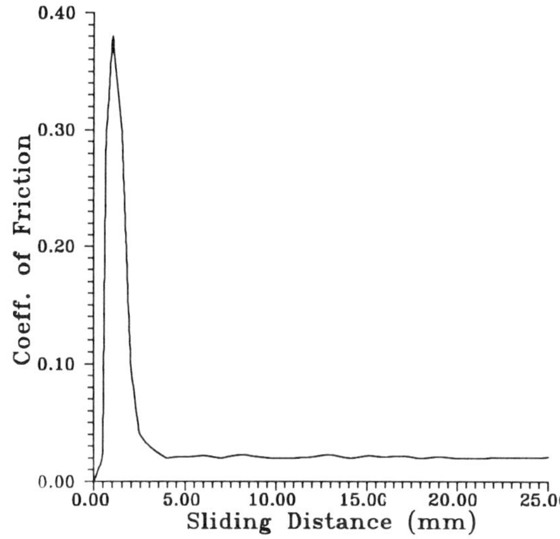

Figure 2. Typical friction trace from a single sliding traverse.

- increasing the indenter surface roughness caused the friction to increase
- reducing the elastic modulus whilst otherwise keeping the same material - polyurethane - resulted in lower friction being measured
- the lowest friction values were obtained with the hydrogel material.

The softer, lower elastic modulus polyurethane (E=6MPa) gave lower friction than the harder, 20MPa elastic modulus polyurethane ($p<0.01$ for all values). When sliding against the two smoother indenters (Ra's of $\ll 0.01\mu m$ and $0.05\mu m$) the softer polyurethane gave very similar friction values for both indenters of $\mu = 0.04$. Increasing the indenter surface roughness to $0.20\mu m$ Ra caused a large increase in the friction to

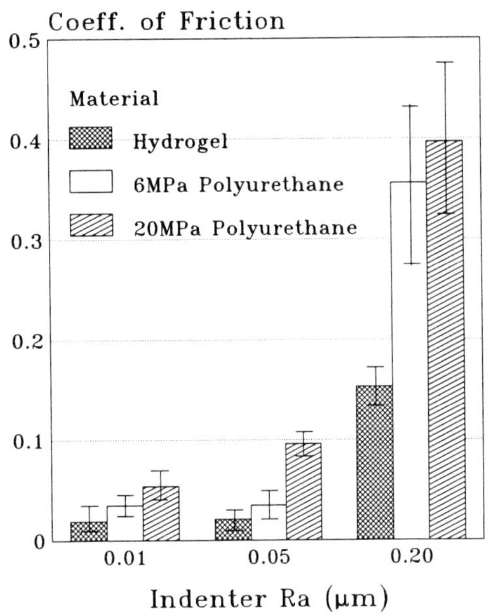

Figure 3. The effect of indenter roughness on the coefficient of friction values for all three compliant materials.

μ = 0.36 (p<0.001). The higher modulus polyurethane showed a bigger difference between the two smoothest indenters with friction values of μ = 0.05 when sliding with the smoothest indenter and μ = 0.09 when sliding with the 0.05μm Ra indenter. The roughest indenter again gave much higher friction coefficient values of 0.4 (p<0.001).

The hydrogel gave the lowest friction values, giving lower values than the polyurethanes for all three indenters (p<0.01). Like the softer polyurethane the hydrogel gave little difference in the friction values when sliding with the two smoothest indenters, with friction coefficient values of around 0.02 for both indenters. When sliding with the roughest indenter the friction values increased to μ = 0.15 (p<0.001), which was still lower than for the polyurethanes.

5. DISCUSSION

The ability of compliant materials to maintain a thin fluid film through micro- elastohydrodynamic action, or indeed other microscopic deformations, is likely to be extremely important in prolonging the life of cushion form types of human replacement joints. However, the compliant material must also be able to withstand conditions when the fluid film has broken down. This study has looked at the frictional properties of two different types of materials when sliding in the mixed lubrication regime or in the presence of a very thin fluid film.

Table 1
Pressure required to flatten the surface asperities of the compliant material

Modulus (MPa)	Pressure (MPa)
6	0.061
20	0.205

The calculated theoretical film thickness [4] for these compliant materials sliding under the experimental conditions was 0.052μm for the two 20MPa materials and 0.057μm for the 6MPa material. These film thicknesses were approximately one tenth of the fluid film thickness calculated during continuous motion, high velocities and dynamic loading during a normal walking cycle [4]. However, the film thickness was established to be at a much lower level in this study in order to investigate the effect of film breakdown and mixed lubrication in different material combinations. Analysis of the combined roughness of the indenters and the compliant layers showed that a full fluid film was not expected in any of the tests, unless micro-EHL or flattening of the asperities was achieved. The elastic modulus of a material had an effect on the magnitude of the pressure perturbation that was required to flatten the surface asperities [4]. The lower the elastic modulus the easier it was to deform the material and so lower pressures were required to produce an effectively smoother surface which could maintain a thinner fluid film (Table 1).

This was shown by the differences in the friction results for the two polyurethanes, with the lower elastic modulus material giving lower friction. For this softer polyurethane there was little difference in the friction measured when sliding with the two smoother indenters suggesting that with these two indenters the material/indenter conjunction was mostly operating in the fluid film regime. The level of friction detected (0.04) could be attributed to hystereses losses in the macroscopic deformation of the polyurethane. In contrast the higher modulus polyurethane showed a bigger difference in friction between the two indenters indicating that, because a fluid film was harder to maintain with this harder material there was a greater percentage of mixed lubrication in the conjunction when sliding with the $0.05\mu m$ Ra indenter than with the softer polyurethane. When sliding with the roughest indenter both polyurethanes gave much higher friction corresponding to a much greater degree of direct surface contact between the polyurethane and the indenter.

The use of glass for the smoothest indenter was for experimental ease only in that a very smooth surface was easily generated on glass. However, the different surface energies of stainless steel and glass may well affect the friction measured by affecting the strength of adhesion bonds formed when the indenter was in direct contact with the layers.

The hydrogel gave significantly lower friction than the polyurethanes with all three indenters and the low values recorded indicate that there was a large percentage of fluid film lubrication in the layer/indenter conjunction. Hydrogels are water swollen polymers and in this case the equilibrium water content of the NVP-CHexMA-HexMA hydrogel was 57%. A large percentage of this water was free to move in the polymer network [12] and could be squeezed out when pressure was applied. This water could then be available to form a fluid film when the external lubricating water had been squeezed out of the conjunction. Recent work by Jin et al [15] has shown that under normal squeeze film action, when loading porous materials, fluid in the material flowed into the material in the centre of the contact where the pressure was high and out of the material at the edge where the pressure was low. The net effect was that the fluid film was depleted by the porosity. However, this only takes effect after a period of approximately one second and the analysis was carried out for perfectly smooth surfaces. In this experiment the hard indenter was continuously deforming new hydrogel material as it moved across the surface of the soft material at a velocity of 8mm/s and each element of hydrogel was depressed for less than 0.5s. More importantly the hydrogel was clearly a rough surface and it is not known at present whether the local deformation of the porous hydrogel asperities could produce a local enhancement of the fluid film, what might be called a micro-squeeze effect. Perhaps equally important is that there was a large percentage of water in the hydrogel and when the surfaces of the hydrogel and indenter were in contact when film breakdown occurs in the mixed lubrication regime, there will be less area for the adhesive force to develop and hence friction should be lower. Clearly the mechanisms associated with micro deformation and fluid movements in the porous asperities which may help maintain thin fluid films and reduce friction in the mixed lubrication regime are not fully understood. However, the hydrogel did give lower values of friction than polyurethanes when sliding with all three indenters indicating that there was a higher proportion of fluid film in the contact.

Hydrogels have been previously considered for use in human replacement joints but in the past these materials have proved to have little structural integrity and have thus been considered unsuitable. The authors are currently engaged in a program of research which aims to produce hydrogels with the necessary strength, wear and durability properties to survive in the stringent operating conditions in total replacement joints.

6. CONCLUSIONS

- The coefficient of friction increased by a factor of ten with high modulus polyurethane when fluid film breakdown occured.

- Micro-elastohydrodynamic lubrication mechanisms were more effective in the lower modulus polyurethane and helped to maintain the fluid film when the ratio of the film thickness to the asperity heights was reduced.

- NVP-CHexMA-HexMA hydrogel showed lower levels of friction than both polyurethanes, in sliding conditions where the fluid film was thin and the contact approached a mixed lubrication regime. This may be explained by the free water content of the hydrogel and microscopic deformations which release water into the contact.

REFERENCES

1 D. Dowson, Ch. 29 in 'Biomechanics of Diarthrodial Joints Vol.2' (eds. V.C. Mow, A. Ratcliffe, S. Woo),Springer-Verlag, London, 1990, 305-345.
2 D. Dowson, Z.M Jin, Engng. in Med.(1986), 15, 63-65.
3 J. Fisher, D. Dowson, Proc. Instn. Mech. Engrs.,Part H (1991), 205, 73-79.
4 D. Dowson, J. Fisher, Z.M. Jin, D.D. Auger, B. Jobbins, Proc. Instn. Mech. Engrs., Part H (1991), 205, 59-63.
5 D.D. Auger, D. Dowson, J. Fisher, Proc. 8th Meeting of European Society of Biomechanics, Rome, 1992, 115.
6 D.D. Auger, D. Dowson, J. Fisher, Proc 19th Leeds-Lyon Symposium on Tribology, 'Thin Films in Tribology' , Elsevier, Amsterdam, 1993.
7 D.D. Auger, J.B. Medley, J. Fisher, D. Dowson, Proc 16th Leeds-Lyon Symposium on Tribology, 'Mechanics of Coatings' , Elsevier, Amsterdam, 1990, 251-260.
8 A.J. Yeadon, M. LaBerge, J.B. Medley, G.M. McNeice, Development and Design with Advanced Materials (eds. G.H. Sih, S.V. Hoa, J.T Pindera), Elsevier, 1990, 219.
9 A. Unsworth, M.J. Pearcy, E.F.T. White, G. White, Tribology, lubrication and wear - 50 years on, C219/87, MEP, London, 1987, 715-724.
10 J.B. Medley, R.M. Pilliar, E.W. Wong, A.B. Strong, Engng. in Med. (1980), 9, 59-65.
11 M. Oka, T. Noguchi, P. Kumar, K. Ikeuchi, T Yamamuro, S.H. Hyon, Y. Ikada, Clinical Materials (1990), 6, 361-381.
12 P.H. Corkhill, A.S. Trevett, B.J. Tighe, Proc. Instn. Mech. Engrs., Part H (1990), 204, 147-155.
13 P.S. Walker, D. Dowson, M.D. Longfield, V. Wright, Rheological Acta (1969), Bad 8, Heft 2, 234-239.
14 L. Caravia, D. Dowson, J. Fisher, Wear (1993), (in press).
15 Z.M. Jin, D. Dowson, J. Fisher, Proc. Instn. Mech. Engrs., Part H, (in press)

Thin Films in Tribology / D. Dowson et al. (Editors)
© 1993 Elsevier Science Publishers B.V. All rights reserved.

Measurements of Elastic Modulus and Hardness of Biocompatible Polyurethanes and the effects on Contact Area within the Hip Prosthesis

J.M.Blamey

Centre for Biomedical Engineering, School of Engineering and Computer Science

University of Durham, Durham DH1 3LE, U.K.

SYNOPSIS

The use of soft layers on the bearing surfaces of artificial hip joints has shown large advantages over current devices. However, the Hardness and Elastic Modulus of the layers are critical for good tribological properties. Thus, these properties must be monitored during optimisation of materials for this application. This paper reveals the interaction between the two mechanical properties and indicates how Elastic Modulus can be calculated from a simple Hardness test. The advantage of the Hardness test is that it may be performed on any size or thickness of sample. This removes processing differences between components and test pieces. The results from Hardness measurements were found to be highly dependent on the backing material and the extent of bonding to it as well as the thickness of the material. In addition, it was noted that the 93 and 100A grades from the Tecoflex range of biocompatible polyurethanes exhibited substantial visco-elastic properties. The work on Hardness / Elastic Modulus interactions has allowed the effective modulus of an acetabular component to be estimated, which can then be applied to the lubrication equations. These theoretical studies support current experimental findings.

1. INTRODUCTION

Artificial hip joints formed from metal cups lined with soft layers have provided excellent tribological properties when lubricated with fluids of similar viscosities to synovial fluid. However, the tribological properties of elastomeric layers are highly dependent on their elastic properties. Unsworth *et al* [1] [2] and more recent work by the author has concluded that hardness values of 3.5 - 9.0 N/mm^2 (H20/30 DIN 53456) are required for the best tribological results. The reason for this is that lubrication is highly dependent on the contact area. If the layer is made harder the area of the contact decreases and contact pressures rise. This can cause breakdown of the fluid film between the articulating surfaces and increase friction substantially. Moving to a softer material can increase the contact area but dimensional changes under load will tend to be greater, requiring larger clearances to eliminate grabbing. In addition, the fatigue strength of the softer polymers tends to be lower. This means that for a commercial hip prosthesis a compromise must be reached with mechanical strength and tribological properties.

Hardness is a measurement of the ability of an indenter to penetrate a surface (DIN 53456, Brinell or Vickers) or the loss of energy of a ball dropped onto the surface (Shore). It can be used to estimate the elastic modulus of a material [3], [4]. However, hardness is dependent on the thickness of the layer as well as its surface properties. In addition, the interface between indentor and layer and the layer and the rigid substrate are important [5]. Changing the applied load can lead to changes in the measured hardness values. As such it is only possible to use hardness measurements as a comparison with other materials tested under identical conditions. In addition materials exhibiting visco-elastic properties will tend to show different values if the speed of loading and time at maximum load are altered.

The aim of this study was to assess the effect of changing layer parameters on the hardness values obtained from tests based on DIN 53456 with a holding load of 5N and a maximum load of 25N. A comparison was also made between hardness and the elastic modulus measurements made on strip samples in order to make comparisons between the experimental and theoretical results. This enabled a calibration to be made from the results for strip samples which could be applied to metal or polyethylene backed cups.

2. THEORY

2.1. Theory of Hardness Measurements

The hardness test detailed in DIN 53456 seemed the most appropriate to use for an elastic material since the penetration is monitored during the test. This is unlike the Brinell test, which relies on measurements of the size of the indentation in the surface following removal of the load (irreversible plastic deformation).

The DIN 53456 hardness test relies on a loaded steel ball (Vicker's Hardness > 800) penetrating the surface of a layer and defines

$$\text{Hardness} = \frac{\text{Major Load}}{\text{Area of Indentation}} \quad (1)$$

which can be rewritten as

$$H\ (20/30) = \left[\frac{0.21}{0.21 - e_r + e}\right]\left[\frac{F}{\pi\ d\ e_r}\right] \quad (2)$$

A minor load is applied to the specimen to allow accurate determination of the starting penetration. A plot of load against penetration for Estane 57, one of the biocompatible polyurethanes, is shown in figure 1. It can be seen that at low loads (below 3-4N) the initial contact is still being formed and results would be inaccurate if a load below this were used as the minor load.

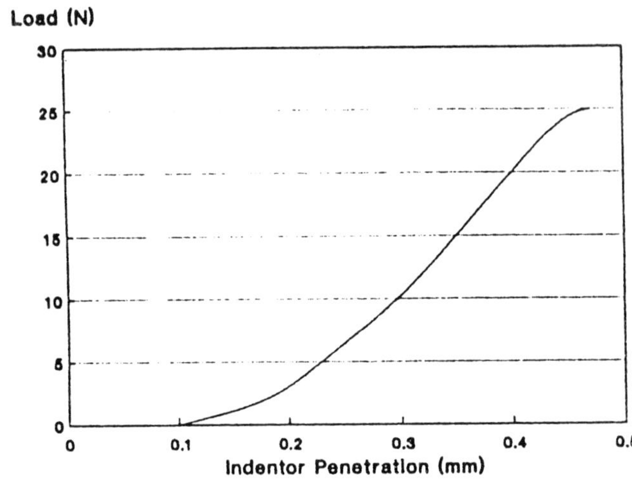

Figure 1. A Plot of Load / Penetration for the Hardness Test

Tests were done whereby the minimum load was altered and the change in final hardness values could be seen.

Table 1
Hardness when Minimum Load is Altered

Minor Load (N)	Major Load (N)	Hardness (N/mm^2)
2.5	22.5	2.99
5.0	25.0	3.28
10.0	25.0	3.32

The minor load was held for 5 seconds and then quickly ramped to the major load. The major load could be chosen to give a penetration between 0.150 and 0.350mm after thirty seconds at that value, as detailed in DIN 53456.

A major load of 20N was chosen to give a range of Hardness values of 9.5 - 3.4 N/mm^2, corresponding to the acceptable range of penetrations (figure 2).

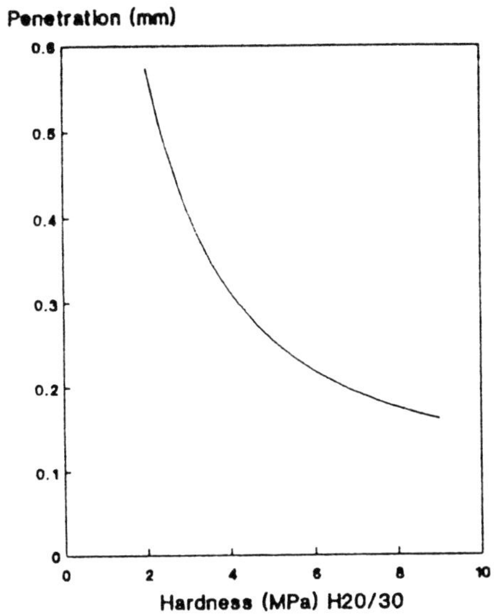

Figure 2. A Plot of Indentor Penetration Versus Hardness for the DIN 53456 Hardness Test

At low penetrations or with harder samples the results from equation 2 tended to drift away from the load/indentation area plot (figure 3), indicating the problem of using too small an indentation. At low values of hardness the curves are very similar, thus it is likely that penetrations greater than 0.350 mm are acceptable (providing the samples are thick enough), increasing the range of hardness values that can be accurately measured down to 2.0 N/mm^2.

The DIN test specified a minimum material thickness of 4 mm to eliminate the effects of the backing material. Unsworth *et al* [1], [2] have specified that layer thicknesses of 2 mm gave good tribological results and by using a thin layer the shear

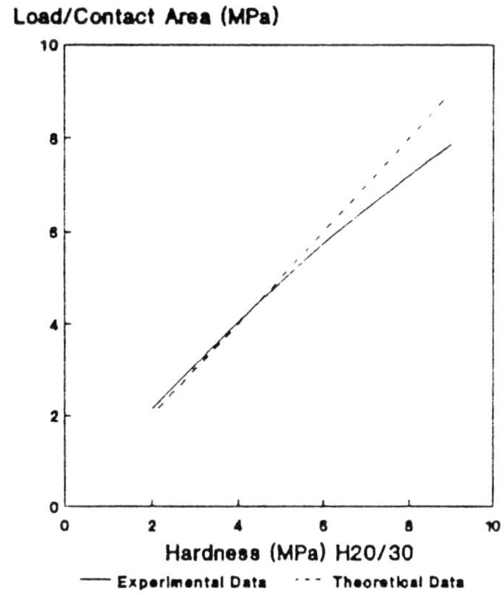

Figure 3. A Plot of the Calibration of the DIN 53456 Hardness Test

stresses at the interface between this layer and its backing material will be reduced [6].

2.2. Theory of Hardness and Elastic Modulus Interactions

The load / extension curves of a polymer shows many different regions which represent activities within the matrix. Urethanes consist of many chains containing both hard and soft segments, and variations in the quantity of each alters the mechanical properties of the polymers.

A typical load / deflection curve is shown in figure 4. It can be seen that at low loads the curve is quite steep, representing a stiff sample. This corresponds to untangling of the coils of the chains which make up the polymer and a re-orientation of these chains. This part of the curve can be explained theoretically with a good degree of accuracy. The second part of the curve, which shows a constant gradient, involves stretching of the coils of polymer to produce straight chains of material. The upturn in the modulus of the material prior to failure represents the behaviour of the chains which have been stretched near to their maximum lengths. The gradient and extent of the final region can indicate crystallization induced by molecular orientation.

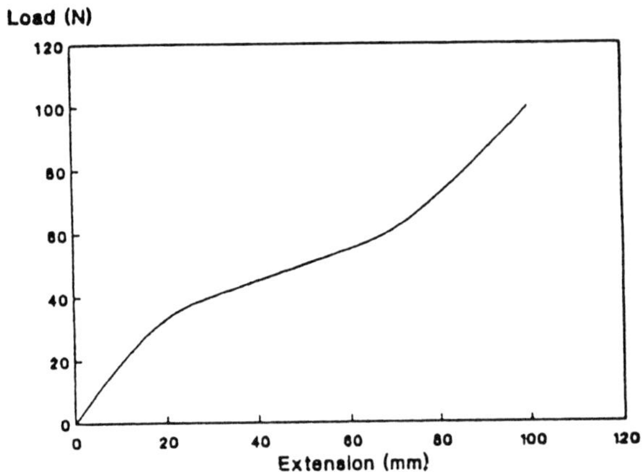

Figure 4. A Load / Extension Plot for a Tensile Polyurethane Specimen

Therefore, the value of elastic modulus measured is highly dependent on the part of the curve used. In use as a compliant layer in an artificial prosthesis it has previously been calculated that the maximum value of compressive stresses caused by the pressures generated is 3.5 MPa [6]. This value of stress corresponds to a point on the first region of the load / extension curve, so the modulus during this region is most important.

Hertzian Theory predicts that elastic modulus may be calculated directly from parameters obtained from indentation of a ball into the surface of the solid. This is the same arrangement as the hardness test used in this study.

$$E = \left(\frac{9}{16}\right)\left(\frac{F\,d}{2\,a^3}\right) \tag{3}$$

a, the contact radius is calculated from the penetration of the ball between the minimum load of 5N and the maximum load of 25N used for the hardness test.

$$a = (d\,e - e^2)^{1/2} \tag{4}$$

This analysis does not take into account the effects of sample thickness and as such relies on an infinite solid. From the previous hardness tests it can be seen that for thickness values greater than 4mm the effect of further changes in thickness is minimal. Therefore, this analysis should be limited to thickness at or above 4mm.

Waters [4] used results obtained by Timoshenko and Goodier [7] to obtain expressions for elastic modulus from indentations on thin sheets, where interaction with the backing material was apparent. He used a number of ball sizes from 3 to 10 mm in diameter pressed into thin rubber sheet which were from 1.5 to 18.5 mm in thickness. They were supported on a rigid metal block. His experimental results allowed the following equation to be formed:

$$E = \left(\frac{9}{16}\right)\left(\frac{2\,F}{d}\right)\left(\frac{\{1 - (\exp)^{-At/a}\}}{e}\right)^{3/2} \tag{5}$$

where A = 0.67 for a lubricated lower boundary and 0.41 for a non-lubricated lower boundary

In all of the tests done at Durham a PTFE spray was used to ensure lubricated conditions when the polyurethane sheets were tested. Thus, a value of 0.67 was used for A in the theoretical analysis.

3. MATERIALS AND METHODS

A number of commercially available elastomers were used covering a range of specified hardness values, measured on the Shore scale.

The thermosetting Castomer system was produced in three grades, 75, 80 and 85A, as a three part system and was cast onto flat plates to form the samples. The upper surface was not constrained by a plate as a better surface finish could be obtained by leaving it to cure in the air. The biologically compatible urethanes were formed using two methods. The thicker samples (4mm) were produced by injection moulding while the thinner samples were made by compression moulding between flat plates. In this way thickness could be specified by shims placed between the plates. Tecoflex EG80, 85, 93 & 100A materials (Aliphatic Polyether Urethane) were all investigated as was Pellethane 2363-80A (Aromatic Polyether Urethane) and Estane '5714F1' (Polyether Urethane) and '58271' (Polyester Urethane). The Tecoflex polyurethanes seemed softer than the other materials of the same Shore hardness, which is why so many grades of material were investigated.

DIN 53456 specifies that samples should be smooth, flat and the faces parallel, but does not quantify this. Samples were checked for flatness and parallelism by taking measurements of thickness over the surface. Surface roughness was measured on a number of samples using an Alpha-Step as 0.6-1.1μm (Ra). The samples were placed on a lubricated flat steel plate of 12 mm thickness. A five millimeter hardened steel ball was used as the indentor, with a Houndsfield testing machine providing the loading. A further investigation was made by bonding samples of Tecoflex to a polyethylene backing with cyanoacrylate cement.

4. RESULTS

As expected, the hardness values for all of the elastomers increased as the sample thickness was reduced. Figure 5 shows the results for the castomer samples and figure 6 shows a similar plot for the biocompatible urethanes.

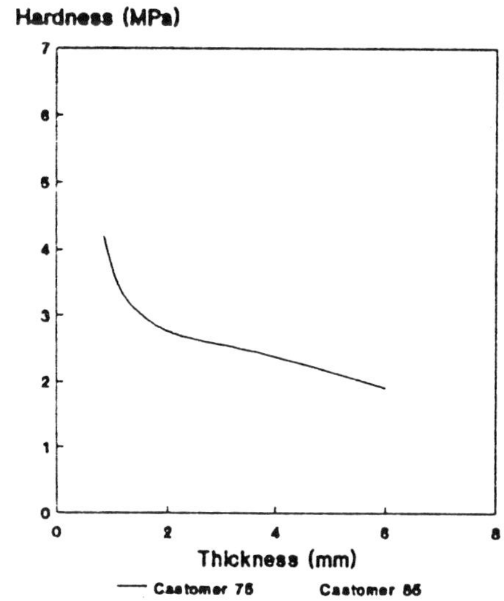

Figure 5. The Effect of Sample thickness on Hardness Values (DIN 53456)

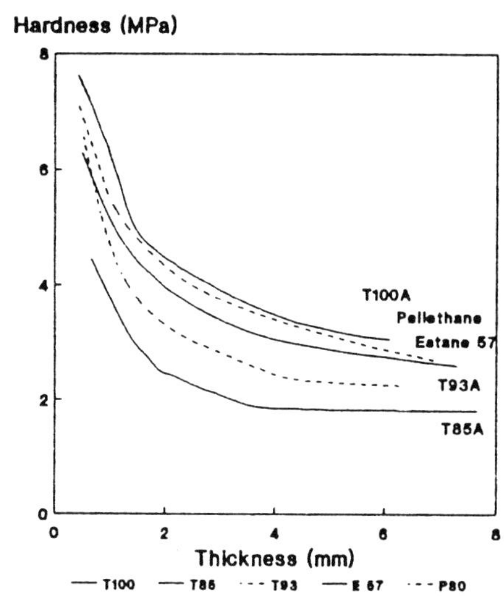

Figure 6. The Effect of Sample thickness on Hardness Values (DIN 53456)

The general shape of the curves is the same with a lower limit of hardness at thicknesses greater than about 4.0mm. This represents the stage at which the stressed region under the ball is not affected by the lower surface of the material. At very low thickness values it would be expected that the curves for all of the materials would coincide as the effect of the steel support becomes dominant. This is suggested by the shape of the curves in figures 5 and 6. During the period at maximum loading it was noted that the Tecoflex elastomers, in particular the 93 and 100A grades, consistantly showed extensive 'creep' with the penetration increasing by as much as 0.15 mm. This is much greater than the other biocompatible elastomers where increases were only about 0.03 mm. This was investigated by measuring the penetration of the indentor (and hence calculating hardness) immediately on reaching the major load and comparing values. Figure 7 shows that this accounted for 25 and 40% of the total penetration for Tecoflex 93 and 100A respectively and 5-10% for the other biocompatible urethanes.

The bonded samples of Tecoflex showed a similar shape of curve (figure 8) to figures 5 and 6, but with higher hardness values, especially with the thinner samples. Typical increases were in the region of 30%. At higher thickness values the curves for the 85 and 93A grades coincided with those for the unbonded samples.

Results from the hardness / elastic modulus interactions were collated on three graphs. Figures 9a and 9b show the accuracy of the basic Hertzian theory for sheets of 4 mm thickness. It can be seen that the slope of the experimental data, which comprises a number of different urethanes, fits well with the theoretical line. Differences in the magnitude of the hardness values from experiment and theory for measured elastic moduli are also small. Figures 5 & 6 showed that the variation of hardness with thickness was similar for all of the elastomers.

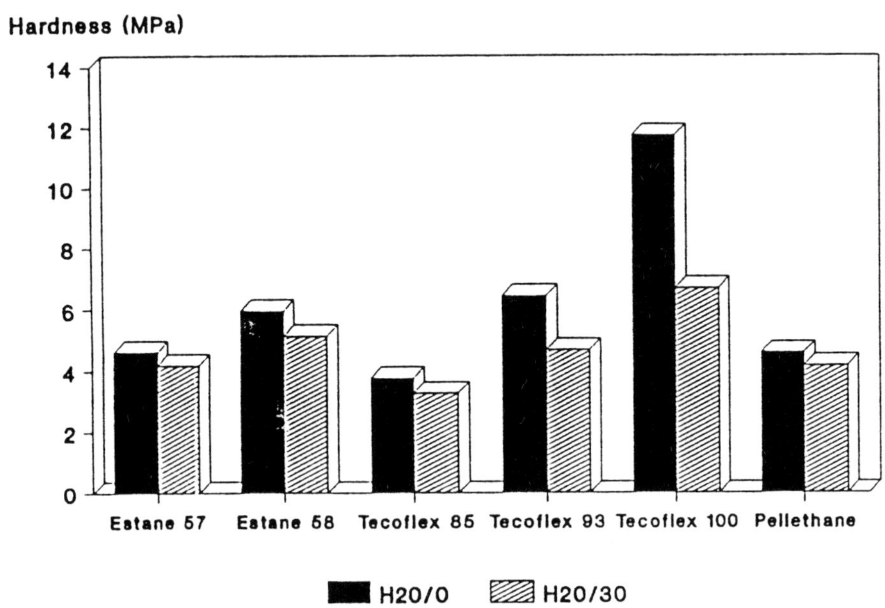

Figure 7. The Visco-elasticity of thin samples of Biocompatible Polyurethanes undergoing Hardness Testing

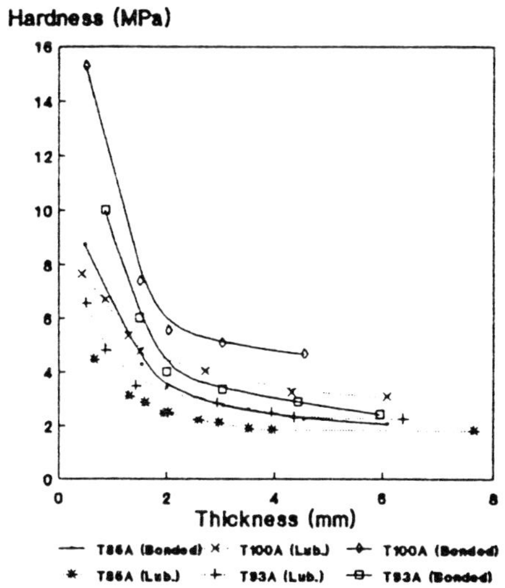

Figure 8. The effect of sample thickness and bond with backing material on hardness for Tecoflex urethanes

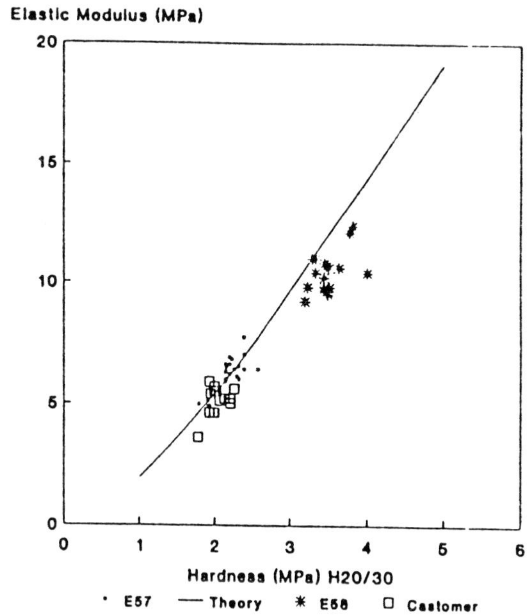

Figure 9b. A Plot of Elastic Modulus against results from the DIN 53456 Hardness Test, showing Hertzian Theory

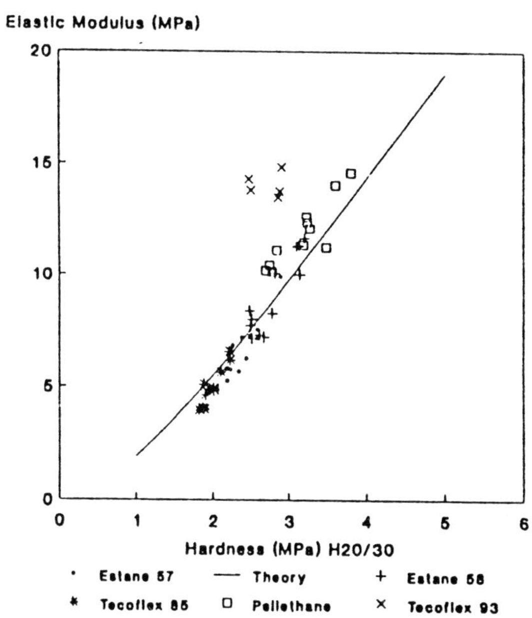

Figure 9a. A Plot of Elastic Modulus against results from the DIN 53456 Hardness Test, showing Hertzian Theory

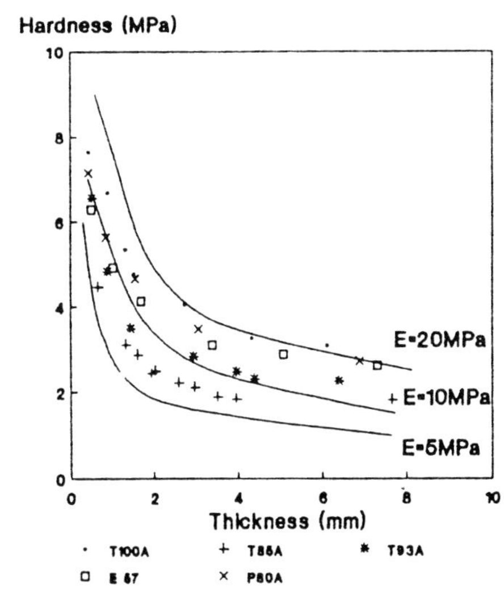

Figure 10. Waters' Theoretical Results and the comparisons with practical data from the range of biocompatible polymers

Therefore for simplicity only the biocompatible polyurethanes were used to investigate Waters' theoretical results. His data was expressed as lines of constant elastic modulus on figure 10. It can be seen that there is good agreement between the theoretical curve and the measured data.

5. DISCUSSION

The DIN 53456 hardness test specifies that "the maximum penetration of the ball into the samples should not exceed 10% of the sample thickness". To investigate the changes that occur at large relative penetrations, the data from figures 5 and 6 was modified to produce figures 11 & 12 where hardness is plotted against the percentage penetration of the indentor.

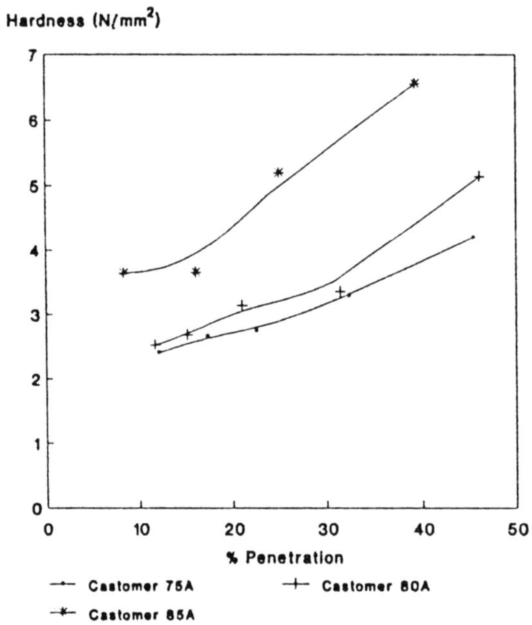

Figure 11. The Change in Hardness with Penetration of the Indentor as a percentage of the sample thickness

This showed that hardness was approximately constant at penetrations of up to 20% of the sample thickness. Therefore, if this test is being performed on samples where the material properties are important, penetration values must be kept below this proportion of the sample thickness. However, with samples destined for tribological testing, a bond is formed between the urethane and the backing material.

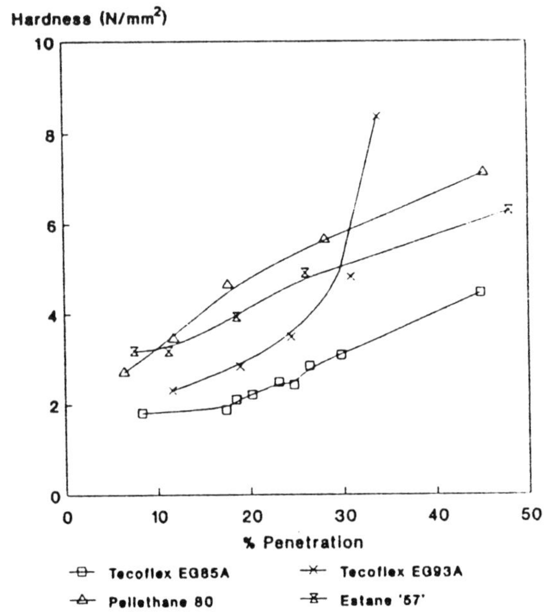

Figure 12. The Change in Hardness with Penetration of the Indentor as a percentage of the sample thickness

The effect of this bond on the hardness value was also assessed. A 13.3% reduction in hardness was obtained when the bond to the metal backing was destroyed. This was with a 2 mm layer, which is the optimum thickness for the cup, and indicates the effect of the backing material even though the penetration of the indentor was less than 20%. It is thought that this is due to the reduction in the flow of the elastomer with a 100% bond (figure 13). From figures 6 & 8 it can be seen that a bond with the backing material seems to reduce the effective layer thickness by about 15%.

Therefore this must be taken into account when estimating the hardness of the bonded layers.

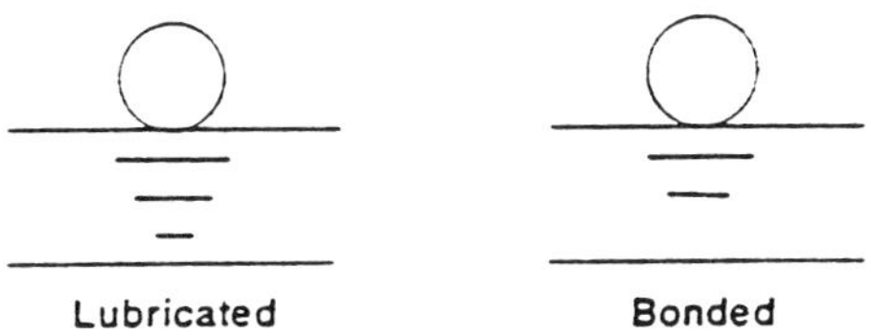

Figure 13. The effect of the bond with the backing material on the Indentor Penetration

Measurement of the elastic modulus of a material is not easy to accomplish. It relies on testing samples which can rarely be produced from the actual manufactured component. Tensile samples (dumbells) have to be produced, leading to possible inaccuracies if processing has been accomplished differently. The tensile specimens have to be cut and checked for stress concentrations and then accurately measured dimensionally. The testing machine must have grips which provide good transference of stress with minimal slippage and the forces developed can be quite high neccessitating a stiff loading frame on the testing machine.

Figures 9a & b and 10 have shown that the elastic modulus can be determined easily and simply from hardness tests on any thickness of sample. Account can even be made for the backing material, such as in the case of cup samples used for tribological assessment. Using figure 10 the effective elastic modulus of a 2mm bonded layer can be estimated at > 20MPa.

6. CONCLUSIONS

The effects of backing materials and bonding of the lower face of polyurethane layers have been quantified. Information has been presented which will allow the elastic modulus of elastomeric materials to be obtained from the results of a simple hardness test on any sample thickness. In addition the effective elastic modulus of acetabular cups has been estimated. This will allow greater accuracy in the tribological assessment of soft layer prostheses. The test is both simple and quick to perform and the spread of results from samples of the same material is normally small. The use of this method will certainly speed up research in this area and will help *in-vivo* testing as smaller samples could be implanted. Work on other polymers of different modulii would allow this technique to be expanded.

Acknowledgements

The author wishes to thank Professor A. Unsworth and Dr. T.V.Parry for their comments. The Wolfson Foundation funded this work and Action Research are now funding its application. Thanks are due to them

REFERENCES

1. Unsworth, A., Pearcy, M.J., White, E.F.T. and White, G. (1987) "Soft Layer Lubrication of Artificial Hip Joints" *Proc. Instn. Mech. Engrs.* **Tribology Vol.,** 715-724.

2. Unsworth, A., Pearcy, M.J., White, E.F.T. and White, G. (1988) "Frictional Properties of artificial hip joints" *Engng. in Med.* **17(3)**, 101-104.

3. Taylor, D.J. and Kragh, A.M. (1970) "Determination of the rigidity modulus of thin soft coatings by indentation methods" *J. Physics D: Applied Physics*, **3** 29-32.

4. Waters, N.E. (1965) "The indentation of thin rubber sheets by spherical indentors" *Brit. J. Appl. Phys.* **Vol.16**, 557-563.

5. Jarrar, M.J. (1989) "Asymptotic behavior of thin elastic layers bonded and unbonded to a rigid foundation" *Int. J. Mech. Sci.,* **31(3)**, 229-235.

6. Armstrong, C.G., Lai, W.M. and Mow, V.C. (1984) "An analysis of the unconfined compression of articular cartilage" *J. Biomech. Engng.* **106** 165-173.

7. Timoshenko, S.P. and Goodier, J.N. "Theory of Elasticity, Edition 3" (1970), McGraw-Hill

NOMENCLATURE

- a contact radius
- A Waters' constant
- d diameter of indentor
- e penetration
- e_r reduced penetration
- E elatic modulus of polyurethane
- F load applied during hardness test
- t thickness of polyurethane layer

Thin Films in Tribology / D. Dowson et al. (Editors)
1993 Elsevier Science Publishers B.V.

Fluid film lubrication in natural hip joints

Z.M. Jin, D. Dowson and J. Fisher

Department of Mechanical Engineering, The University of Leeds, Leeds, LS2 9JT, United Kingdom

The film thickness in the normal human hip joint has been predicted under steady-state entraining motion, pure dynamic squeeze-film motion and the transient condition involving both the entraining and squeeze-film motions. The simplified method adopted to predict the film thickness has been compared with the full numerical method under both steady-state entraining and transient conditions for the normal human ankle joint and good agreement has been found. The transient film thickness predicted for the hip joint is almost constant throughout one walking cycle due to the combined effect of entraining during the swing phase, generating a sufficiently large lubricant film and squeeze-film action preserving this lubricant film during the stance phase. However, the lubricant film thickness predicted is only of the order of 1.5 (μm) and thus it seems clear that micro-elastohydrodynamic action also plays as important a part in the lubrication of hip joints as it was found in a previous study to have in the ankle joint.

1. INTRODUCTION

Studies of the lubrication of natural synovial joints have advanced considerably in the last thirty years (Mow and Mak, 1986; Dowson, 1990). In particular, there have been two predominant fluid film lubrication mechanisms proposed for synovial joints: elasto-hydrodynamic lubrication in which the elasticity of cartilage is assumed to be important (Dowson, 1967) and biphasic lubrication, in which the porosity of articular cartilage is assumed to be mainly responsible for fluid film generation (Mansour and Mow, 1977). Recently Hou et al (1992) and Jin et al (1992) examined the squeeze-film lubrication of normal human hip joints incorporating the biphasic elasticity equation for articular cartilage. It was found that under physiological walking conditions, as experienced in hip joints, the film thickness was slightly reduced, but the elasticity was not significantly affected. In other words, articular cartilage can be treated as a single phase material for the purpose of elasticity analysis. This assumption has been adopted in elastohydrodynamic lubrication analysis of synovial joints by Dowson and his colleagues and has also been demonstrated in an experimental study (Johnson et al, 1977).

Elastohydrodynamic lubrication analysis of synovial joints is mainly involved with predictions of the lubricant film thickness. Early attempts were based on the existing film thickness formulae for semi-infinite solids (Dowson, 1967). The effects of the finite layer thickness of articular cartilage and the cyclic variation of the load and speed experienced in a normal human ankle joint were later considered by Medley et al (1984). They found that an almost constant film thickness was generated during continuous normal walking by the combined entraining and squeeze-film motions. However, the film thickness predicted was less than 1 (μm) and was smaller than the surface roughness of articular cartilage. This conflict was later removed by Dowson and Jin (1986, 1987) who introduced the roughness of articular cartilage into the elastohydrodynamic lubrication analysis. It was found that the original surface roughness of cartilage was largely smoothed out by micro-elastohydrodynamic action, thus enabling a continuous lubricant film to be developed during the normal walking cycle. Recently the elastohydrodynamic lubrication analysis has been extended to hip joints by Dowson and Yao (1990). The finite difference method was adopted to solve the governing equations and a considerable amount of computing time was required. The purpose of the present study is to develop a film thickness formula under the conditions of entraining motion in an alternative approach and to perform the transient film thickness analysis for human hip joints under dynamic conditions.

1.1 Notation

- $a(x)$ cavitation boundary in the side-leakage direction.
- a_o half width of dry contact.
- A_j, B_j coefficients for finite difference approximation to derivatives.
- d total layer thickness of articular cartilage.
- E elastic modulus of articular cartilage.
- E' equivalent elastic modulus, $E' = \frac{2E}{1-\upsilon^2}$
- h film thickness and also the centre line film thickness in the entraining direction.
- h_c central film thickness.
- h_m minimum film thickness.
- h_o minimum undeformed film thickness.
- p pressure distribution and also the centre line pressure distribution in the entraining direction.
- p_j pressure values at node j.
- R equivalent radius
- r,z polar coordinates
- u entraining velocity
- u_1 sliding velocity, 2u
- v_o squeeze-film velocity, $\frac{dh_c}{dt}$
- w total load
- w_e load due to entraining motion
- w_s load due to squeeze-film motion
- x,y,z coordinates
- x_e inlet boundary
- x_o cavitation boundary in the entraining direction
- x_i mesh point at node i
- $\alpha,\beta,\Gamma,\varsigma$ coefficients in the film thickness formulae
- δ elastic deformation
- η viscosity of synovial fluid
- ν Poisson's ratio

2. FORMULATION

The normal human hip joint was modelled as a sphere-on-plane configuration with an equivalent radius (R) as shown in Figure 1.

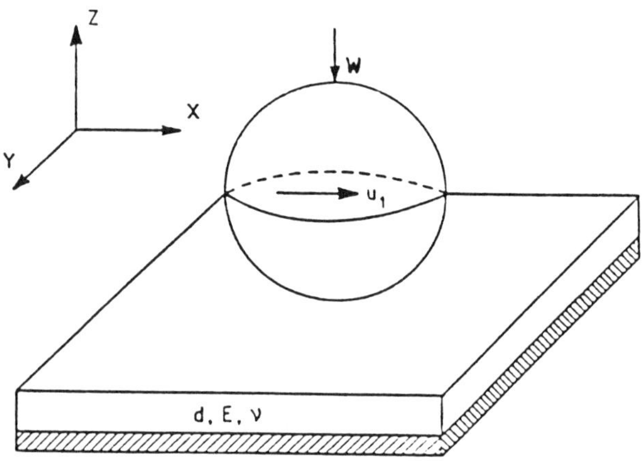

Fig.1. A Model of a Rigid Sphere Sliding against a Compliant Layered Surface Firmly Bonded to a Rigid Substrate Representing Normal Human Hip Joints

Articular cartilage was modelled as a single phase elastic layer and the porosity was neglected (Jin et al 1992). Thus its main property was described by an elastic modulus (E) and Poisson's ratio (ν). The value of Poisson's ratio was chosen to be 0.4 in order to compensate the fluid loss due to deformation (Hayes and Mockros, 1971). This also enabled the constrained column model to be applied to predict elastic deformation (Medley et al 1984). Furthermore, the articular cartilage was assumed to be firmly bonded to the 'rigid' sub-chondral bone. Synovial fluid was modelled as a Newtonian, isoviscous and incompressible lubricant and thus its main parameter was viscosity (η). The dynamic load cycle was chosen from Seireg and Arviker (1975) and the entraining velocity was calculated from Graham and Walker (1973) as shown in Figure 2. The details can be found in Jin (1988).

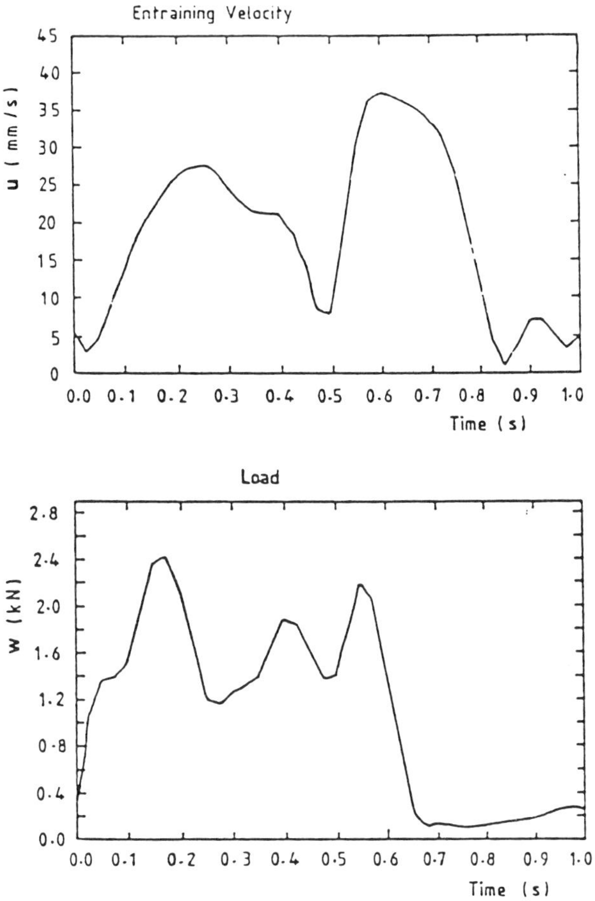

Fig. 2 Load and Speed in Normal Human Hip Joint During Walking

3. GOVERNING EQUATIONS AND NUMERICAL ANALYSIS

The transient film thickness analysis was based upon the superposition method from Holland (1978). Both the film thickness formulae under the entraining and squeeze-film motions were required and the superposition method was adopted to combine these two equations to predict the transient film. This simplified approach appeared to be reasonable under heavily loaded conditions as in human synovial joints. The film profiles under both entraining and squeeze-film motions appear to be almost constant throughout the contact region and both the film rupture boundary due to cavitation under entraining motion and the periphery boundary under squeeze-film motion approached the edge of the dry contact region.

3.1 Steady-State Film Thickness Formula Under Entraining Motion

The governing equations for this case consists of the following equations:

- Reynolds' equation,

$$\frac{\partial}{\partial x}\left(h^3 \frac{\partial p}{\partial x}\right) + \frac{\partial}{\partial y}\left(h^3 \frac{\partial p}{\partial y}\right) = 12\eta u \frac{\partial h}{\partial x} \qquad (1)$$

- Elasticity equation based on the constrained column model

$$\delta = \frac{d}{E}\left[1 - \frac{2\nu^2}{1-\nu}\right] p \qquad (2)$$

- Film thickness equation based on the parabolic approximation:

$$h = h_o + \frac{x^2 + y^2}{2R} + \delta \qquad (3)$$

- load equation:

$$w_e = \iint p \, dx \, dy \qquad (4)$$

The boundary conditions are

- at the edge of the computation zone

$$p = 0 \qquad (5)$$

- at the film rupture boundary due to cavitation,

$$p = 0 \qquad (6)$$

$$\frac{\partial p}{\partial x} = \frac{\partial p}{\partial y} = 0 \qquad (7)$$

The full numerical solution to the above equations was obtained by Dowson and Yao (1990) based on the finite difference method for typical conditions experienced in normal human hip joints.

Some interesting features of the film profile and the pressure distribution were observed.

- in the entraining direction, the film profile appeared to be similar to a plane inclined surface bearing assumed by Medley (1981) for line contact conditions.

- in the side-leakage direction, the film profile appeared to be essentially constant within the main contact conjunction.

- the pressure distributions in both the entraining and the side-leakage directions were found to be similar to the dry contact pressure distribution of a parabolic type.

Thus it was possible to make the following assumptions within the contact region:

- h(x,y) was constant in the side-leakage direction,

$$h(x,y) = h(x) \tag{8}$$

- p(x,y) was parabolic in the side-leakage direction.

$$p(x,y) = p(x)\left[1 - \left(\frac{y}{a(x)}\right)^2\right] \tag{9}$$

where a(x) is the cavitation boundary in the side-leakage direction. Reynolds' equation under these assumptions was reduced to,

$$\frac{\partial}{\partial x}\left(h^3 \frac{\partial p}{\partial x}\right) - \frac{2h^3 p}{a^2(x)} = 12\eta\, u \frac{\partial h}{\partial x} \tag{10}$$

where the pressure (p) and the film thickness (h) in equation (10) were used to represent the centre line pressure distribution and the film thickness in the entraining direction respectively.

The cavitation boundary in the side-leakage direction was assumed as following (see Figure 3).

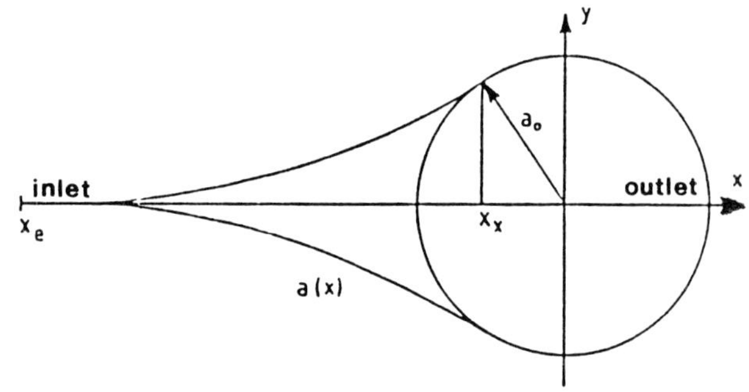

Fig. 3. The Boundary in the Side-Leakage Direction Adopted in the Present Simplified Solution Method

- for $x > x_x$, the dry contact boundary

$$a(x) = a_o \sqrt{\left[1 - (x/a_o)^2\right]} \tag{11}$$

- for $x \le x_x$, a parabolic curve connected to the inlet boundary (x_e)

$$a(x) = a_o \sqrt{\left[1 - \left(\frac{x_x}{a_o}\right)^2\right]} \left[\frac{\left(x_e - x\right)}{\left(x_e - x_x\right)}\right]^2 \tag{12}$$

The load equation (4) became

$$w_e = \frac{4}{3} \int_{x_e}^{x_o} p(x)\, a(x)\, dx \tag{13}$$

where x_o is the film rupture boundary due to cavitation in the entraining direction. Thus, the present three-dimensional lubrication problem was reduced to two-dimensions and was readily solved with less computing effort and more nodal points.

The finite difference method was adopted to solve the reduced two-dimensional governing equation in pressure (10). The solution domain was chosen from the inlet to a sufficiently large outlet position where the cavitation boundary could be included and the whole region was divided into a number of non-uniform meshes $(x_1, x_2 \ldots, x_n)$. Equation (10) was discretised as follows

$$p_{i-1}\left[h_{i-1}^3 A_{i-1} + \left(3\left(\frac{\partial h}{\partial x}\right)_i h_i^2 - 12\eta u C_s\right) B_{i-1}\right] + p_i\left[h_i^3 A_i + \left(3\left(\frac{\partial h}{\partial x}\right)_i h_i^2 - 12\eta u C_s\right) B_i - \frac{2h_i^3}{a^2(x_i)}\right] + p_{i+1}\left[h_{i+1}^3 A_{i+1} + \left(3\left(\frac{\partial h}{\partial x}\right)_{i+1} h_i^2 - 12\eta u C_s\right) B_{i+1}\right] = 12\eta u \frac{x_i}{R} \tag{14}$$

where the coefficients $(A_{i-1}, A_i, A_{i+1}, B_{i-1}, B_i, B_{i+1})$ were given by Dowson and Jin (1992).

Equation (14) was subjected to the boundary conditions given in equation (5).

$$p_1 = p_n = 0 \tag{15}$$

It was solved in the domain $(x_2, x_3, \ldots, x_{n-1})$ by a matrix inversion method after assuming an initial pressure distribution. The cavitation boundary (x_o) was determined according to the following algorithm.

- find the minimum pressure value at node (m).
- if $p_m < 0$ replace the negative pressure values by zero.
- resolve equation (14) in the domain (2, m-1).
- repeat until $p_2, p_3, \ldots, p_{m-1}, > 0$ and the cavitation boundary was found at $x_o = x_m$

The pressure iteration was repeated until the difference between the calculated and the assumed pressure was sufficiently small. The convergent pressure distribution obtained enabled the load capacity to be calculated.

$$w_c = \frac{4}{3}\sum_{i=1}^{m-1} 0.5\left[p_i\, a(x_i) + p_{i+1}\, a(x_{i+1})\right](x_{i+1} - x_i) \tag{16}$$

and if this was different from the load capacity specified, iteration was performed by modifying the undeformed minimum film thickness (h_o).

The inlet boundary was located at a value of (x) 2.5 times the half width of the dry contact and the outlet boundary 1.5 times the half zone width downstream of the centre of the conjunction. The effect of (x_x) on the predicted film thickness was investigated and it was found that as the load increased it had only a small effect provided that $\left(\frac{x_x}{x_o}\right)$ was chosen between -0.98 and -0.707. A relative difference of 0.001 was ensured for both the pressure and load iterations and the total number of nodes was chosen between 2000 and 8000 depending upon the case considered.

3.2 Dynamic film thickness formula under pure squeeze-film motion

The Reynold's equation under this condition is conveniently expressed in polar coordinates.

$$\frac{1}{r}\frac{\partial}{\partial r}\left(h^3 r \frac{\partial p}{\partial r}\right) = 12\eta \frac{\partial h}{\partial t} \tag{17}$$

and the boundary condition are as follows:

- at periphery p = 0 (18)
- at the centre of the contact due to symmetry

$$\frac{\partial p}{\partial r} = 0 \tag{19}$$

Following Higginson (1978), the film profile was assumed to be constant (h_c) within the contact region and the squeeze-film velocity $\left(\frac{\partial h}{\partial t}\right)$ was also assumed to be constant $v_o = \frac{dh_c}{dt}$ The local variation of film profile with radial coordinate and time was thus not simulated. Equation (17) was reduced to,

$$\frac{\partial p}{\partial r} = 6\eta v_o \frac{r}{h_c^3}$$

or $$p = -\frac{3\eta v_o}{h_c^3}\left[a_o^2 - r^2\right] \tag{20}$$

where the periphery boundary was assumed to be equal to the half width of dry contact (a_o)

The load equation for the pure squeeze-film motion was,

$$w_s = 2\pi \int_0^{a_o} r\, p\, d r \tag{21}$$

Thus by combining equations (20) and (21), the following equation was obtained,

$$w_s = -\frac{3\pi\eta\ a_o^4\ v_o}{2h_c^3} \tag{22}$$

3.3 Governing equation under transient conditions

Following Holland (1978), the total instantaneous load imposed on the joint (w) was assumed to be carried by the pressure produced by both entraining motion and squeeze-film motion,

$$w = w_e + w_s \tag{23}$$

where (w_e) and (w_s) were assumed to be the loads respectively carried by entraining and squeeze-film action alone.

Thus,

$$w_e = f_e\left(h_c, u, \eta, E, \upsilon, d, R\right) \tag{24}$$

$$w_s = f_s\left(h_c, \frac{dh_c}{dt}, \eta, E, \upsilon, d, R\right) \tag{25}$$

and equation (23) became,

$$w = f_e\left(h_c, u, \eta, E, \upsilon, d, R\right) + f_s\left(h_c, \frac{dh_c}{dt}, \eta, E, \upsilon, d, R\right) \tag{26}$$

where f_e and f_s are functions relating to the load and the other parameters for entraining motion and squeeze-film motion respectively.

Equation (26) is an ordinary first order differential equation which could readily be solved once the film thickness expression for entraining motion was obtained as described in the next section. NAG routine DO2BAF was used for this purpose (Numerical Algorithm Group Ltd., Wilkinson House, Jordan Hill Road, Oxford, OX2 8DR). Furthermore it was noted that the solution procedure outlined above was also applicable to human ankle joints provided that the corresponding film thickness formula was adopted.

4. RESULTS

The following parameters representing normal human hip joints were chosen to perform the transient film analysis:

R = 1 m
E = 16 MPa
υ = 0.4
d = 4 mm

and the load and the speed are shown in Figure 2. Furthermore, under physiological shear rate, the viscosity of healthy synovial fluid is about 0.01 (Pas) and the viscosity of degenerated synovial fluid

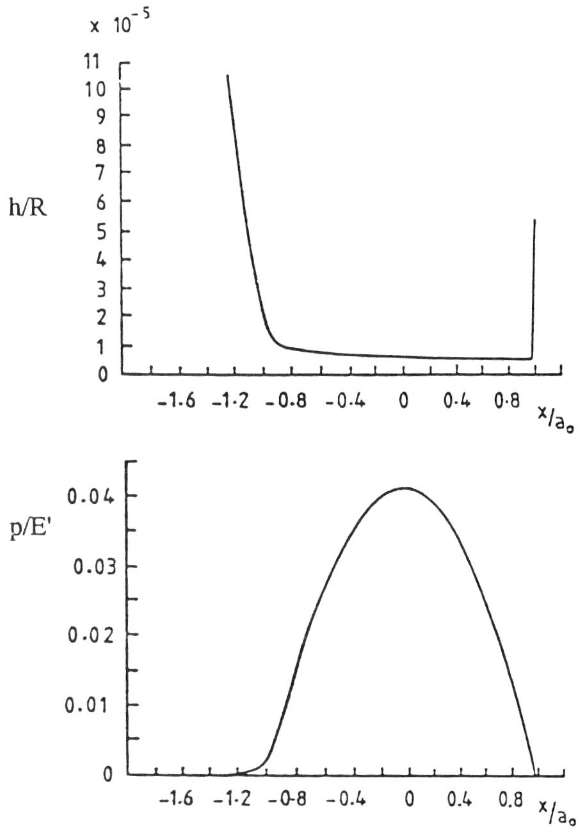

$\frac{\eta u}{E'R} = 10^{-10}$; $\frac{w_e}{E'R^2} = 2.5 \times 10^{-5}$;

$\frac{d}{R} = 4 \times 10^{-3}$

Fig. 4 Prediction of the Centre Line Film Thickness and Pressure Distribution in the Entraining Direction

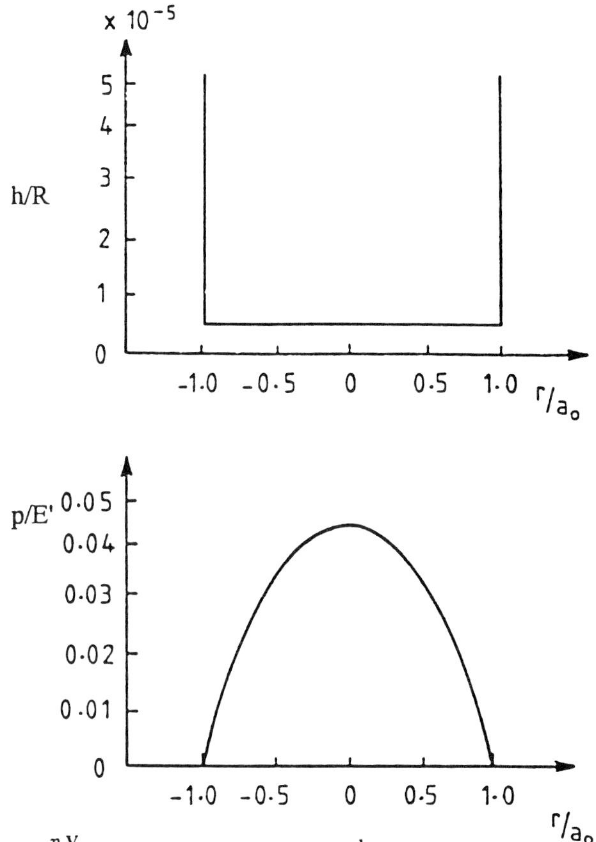

$\frac{\eta V_o}{E'R} = -4.682 \times 10^{-15}$; $\frac{d}{R} = 4 \times 10^{-3}$;

$\frac{w_s}{E'R^2} = 2.5 \times 10^{-5}$

Fig. 5 Prediction of Film Thickness and and Pressure Distribution under Squeeze-film Motion

is no better than the viscosity of water (Cooke et al, 1978).

A typical film profile and pressure distribution on the centre line in the entraining direction are shown in Figure 4 for steady-state entraining conditions. Figure 5 shows a typical film profile and pressure distribution under pure squeeze-film motion.

A series of results have been obtained for the minimum and central film thickness under entraining motion, covering a wide range of load

$$\left(10^{-5} \leq \frac{w_e}{E'R^2} \leq 5 \times 10^{-5}\right),$$ speed

$$\left(10^{-11} \leq \frac{\eta u}{E'R} \leq 10^{-9}\right)$$ and layer thickness

$$\left(5 \times 10^{-4} \leq d/R \leq 4 \times 10^{-3}\right).$$

This enabled the minimum and central film thickness formulae to be obtained. There were expressed as equations (27) and (28) for convenience and Tables 1 and 2 show the present results as well as those from Dowson and Yao (1992)

$$\frac{h_c}{R} = \alpha_c \left(\frac{\eta u}{E'R}\right)^{\beta_c} \left(\frac{w_c}{E'R^2}\right)^{r_c} \left(\frac{d}{R}\right)^{\zeta_c} \qquad (27)$$

	α_c	β_c	r_c	ζ_c
Present	1.50	0.52	-0.16	0.37
Dowson and Yao (1992)	2.10	0.56	-0.18	0.37

Table 1 Comparison of the coefficients in the central film thickness formula between the present study and Dowson and Yao (1992).

$$\frac{h_m}{R} = \alpha_m \left(\frac{\eta u}{E'R}\right)^{\beta m} \left(\frac{w_c}{E'R^2}\right)^{r_m} \left(\frac{d}{R}\right)^{\zeta_m} \qquad (28)$$

	α_m	β_m	r_m	ζ_m
Present	1.25	0.52	-0.15	0.37
Dowson and Yao (1992)	1.77	0.56	-0.19	0.37

Table 2 Comparison of the coefficients in the minimum film thickness formula between the present study and Dowson and Yao (1992).

Substituting equations (27) and (22) into equation (26), the transient film thickness formula was readily obtained:

$$\frac{dh_c}{dt} = \frac{2h_c^3}{3\pi\eta a_o^4}\left[13.4\ E'R^2\left(\frac{\eta u}{E'R}\right)^{3.36}\left(\frac{d}{R}\right)^{2.36}\left(\frac{h_c}{R}\right)^{-6.40} - w\right] \qquad (29)$$

which is an ordinary differential equation which was readily solved for cyclic variations of load and speed experienced in synovial joints. Figure 6 shows a comparison of the transient film thickness predictions between the present simple method and the full numerical solution (Jin, 1988) for conditions representative of the normal human ankle joint. Figure 7 shows comparisons of the transient film thickness, the film thickness based upon entraining alone and the film thickness based upon pure squeeze-film motion assuming a similar film thickness as the transient film thickness at the start of the cycle. Figure 8 shows the effect of degenerated synovial fluid on the transient film thickness predictions.

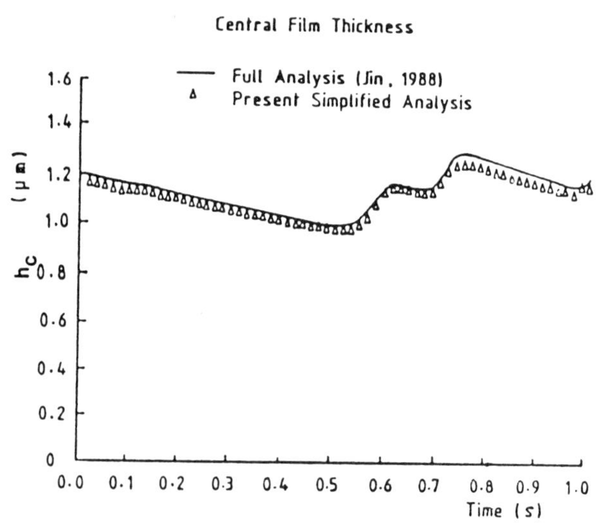

Fig. 6 Comparison of the Transient Film Thickness Prediction Between the Full and Simplified Analyses for Normal Human Ankle Joint

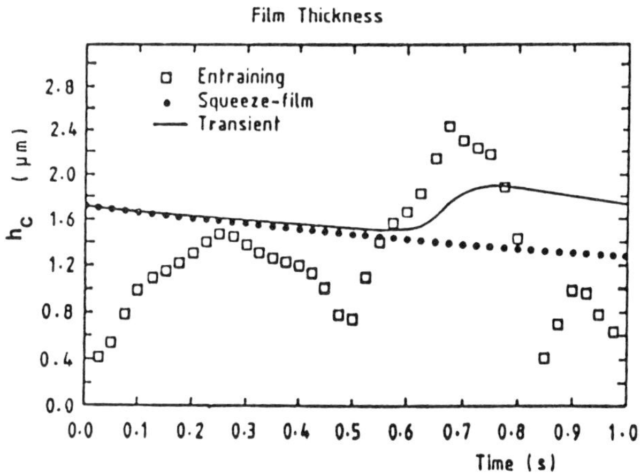

Fig. 7 Prediction of Film Thickness During One Walking Cycle for Three Different Lubrication Models

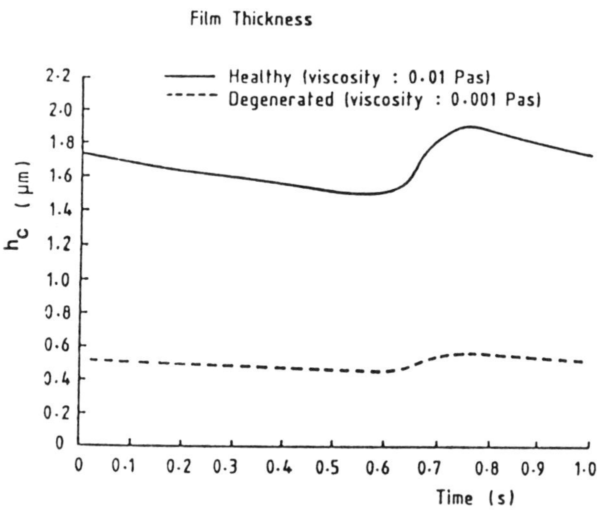

Fig. 8 Effect of Degenerated Synovial Fluid on the Prediction of Transient Film Thickness

5. DISCUSSION

Typical features of the film profile and pressure distribution can be seen from Figure 4 under steady-state entraining motion. The film profile within the loaded conjunction is slightly inclined, which was described as a plane-inclined surface bearing by Medley (1981). The hydrodynamic pressure was close to a parabolic distribution and approached dry contact conditions. Some differences were observed in Figure 5 under pure squeeze-film motion. The film profile within the loaded conjunction was assumed to be constant and the corresponding pressure distribution was parabolic as expressed in equation (20).

Good agreement between the steady-state film thickness formulae was obtained under entraining motion between the present method and that of Dowson and Yao (1992).

Good agreement was achieved between the present simple solution for the transient film thickness predictions and the full numerical solution by Jin (1988) for the normal human ankle joint as shown in Figure 6. Thus a similar conclusion was expected when applied to normal human hip joints.

Figure 7 shows a similar variation of the transient film thickness with time as in the human ankle joint. It is noted that during the first half cycle, the lubrication mechanism is predominantly squeeze-film, which preserves the lubricant film thickness during this period of time. On the other hand, entraining action is seen to be important during the second half of the cycle, which generates a sufficiently large lubricant film thickness. Thus the combination of these two lubrication mechanisms generates an almost constant, continuous lubricant film throughout walking cycle. Furthermore, it is also noted that the transient film thickness predicted is of the order of 1.7 (μm) and still considerably smaller than the surface roughness of articular cartilage (Jin, 1988). Thus micro-elastohydrodynamic action is also likely to be important in the human hip joint.

The effect of degenerated synovial fluid is to reduce the transient film thickness predicted. It is noted that a decrease of viscosity by 10 times results in a decrease of the film thickness prediction by a factor of 3. Furthermore, the film thickness variation is also reduced, implying that squeeze-film action is more important when the film thickness is reduced.

6. CONCLUSIONS

- A simplified numerical procedure has been developed to predict film thickness under steady-state and pure squeeze film motions.

Good agreement has been found between the simple method and a full numerical solution (Dowson and Yao, 1992).

- A simplified transient film thickness prediction procedure has been developed and good agreement has been found between this simple method and the full numerical solution for the human ankle joints (Jin, 1988).

- The variation of the transient film thickness during one walking cycle for normal human hip joints shows that in the first half cycle squeeze-film action plays an important role in preserving the lubricant film thickness, while in the second half cycle, entraining action is essential to generate a sufficiently large lubricant film. Thus the combination of these two lubrication mechanisms gives an almost constant, continuous lubricant film thickness throughout the walking cycle.

- The transient film thickness predicted for normal human hip joints is still much smaller than the surface roughness of articular cartilage. Thus micro-elastohydrodynamic lubrication action must also be important in the lubrication of hip joints.

- The effect of degenerated synovial fluid results in a decrease of the transient film thickness predictions.

REFERENCES

1. A.V. Cooke, D. Dowson and V. Wright. The rheology of synovial fluid and some potential synthetic lubricants for degenerate synovial joints. Engng. Med., 7, pp 66-72, 1978.
2. D. Dowson. Models of lubrication in human joints. In Lubrication and Wear in Living and Artificial Human Joints. Proc. Inst. Mech. Engrs. 181, 3J : pp 45-54, 1967.
3. D. Dowson. Biotribology of natural and replacement synovial joints. In Biomechanics of Diarthrodial Joints (Vol.2.), eds V.C. Mow, A. Ratcliffe and S.L.Y. Woo, Springer-Verlag 1990.
4. D. Dowson and Z.M. Jin. Micro-elastohydrodynamic lubrication of synovial joints. Engineering in Medicine, 15(2), pp 63-65, 1986.
5. D. Dowson and Z.M. Jin An analysis of micro-elastohydrodynamic lubrication in synovial joints considering cyclic loading and entraining velocities. Proc. 13th Leeds-Lyon Symposium on Tribology, Elsevier, Amsterdam, pp. 375-386, 1987.
6. D. Dowson and Z.M. Jin. A full numerical solution to the problem of micro-elastohydrodynamic lubrication of a stationary compliant wavy layered surface firmly bonded to a rigid substrate. Submitted to Engineering in Medicine, 1992.
7. D. Dowson and J.Q. Yao. A full solution to the problem of film thickness prediction in natural synovial joints, in Mechanics of Coating, Proc. 16th Leeds-Lyon Symposium on Tribology, eds. D. Dowson, C.M. Taylor and M. Godet, Elsevier, Amsterdam, pp 91-102, 1990.
8. D. Dowson and J.Q. Yao. A study of the deformation and lubrication of synovial joints, to be published, 1992.
9. J.D. Graham and T.W. Walker. Motion in the hip: the relationship of split line patterns to surface velocities. In Perspectives in Biomedical Engineering, ed. R.M. Kenedi, The Macmillan Press Ltd., London and Basingstoke, pp. 161-164, 1973.
10. W.C. Hayes and L.F. Mockros. Viscoelastic properties of human articular cartilage. J. Appl. Physiol., 41, pp 562-568, 1971.
11. G.R. Higginson. Squeeze films between compliant solids. WEAR, 46. pp. 387-395, 1978.
12. J. Holland. Die instantionäre elasto-hydrodynamik, Konstruktion, 30 H.9, pp. 363-369, 1978.
13. Z.M. Jin, D. Dowson and J. Fisher. The effect of porosity of articular cartilage on the lubrication of normal human hip joint, Engineering in Medicine, in press.
14. G.R. Johnson, D. Dowson and V. Wright. The elastic behaviour of articular cartilage under sinusoidally varying compressive stress. Int. J. Mech. Sci., 19, pp. 301-308, 1977.

15. J.M. Mansour and V.C. Mow. On the natural lubrication of synovial joints : normal and degenerate. Trans ASME, J Lub Technol, 99 (2), pp.163-173, 1977.
16. J.B. Medley. The lubrication of normal human ankle jonts. PhD Thesis, Department of Mechanical Engineering, University of Leeds, 1981.
17. J.B. Medley, D. Dowson and V. Wright. Transient elasto-hydrodynamic lubrication models for the human ankle joint. Engineering in Medicine, 13, pp.137-151, 1984.
18. V.C. Mow and F. Mak. Lubrication of diarthrodial joints. In Bioengineering Handbook, eds. H. Skalak and S. Chien, McGraw-Hill, 1986.
19. A. Seireg and R.J. Arvikar. The prediction of musclar load sharing and joint forces in the lower extremities. J. Biomechanics, 8, pp. 89-102, 1975.

SESSION XVI
COMPLIANT SURFACES

Chairman : Dr J C Bell

Paper XVI (i)	**Compliant Foil Bearings Technology and Their Applications to High Speed Turbomachinery**
Paper XVI (ii)	**Thickness Measurement Methods for Thin Lubricant Films : Application to a Radial Face Seal**
Paper XVI (iii)	**Lubricating of O-Ring Seals in Pneumatic Applications with Ultra Thin Grease Films**
Paper XVI (iv)	**Thin Film Tribology and Rubber Bearings**

Thin Films in Tribology / D. Dowson et al. (Editors)
© 1993 Elsevier Science Publishers B.V. All rights reserved.

Compliant Foil Bearings Technology and Their Application to High Speed Turbomachinery

H. Heshmat[a] and P. Hermel[b]

[a] **Mechanical Technology Incorporated, 968 Albany-Shaker Road, Latham, New York 12110, U.S.A.**

[b] **ABG-SEMCA, 408, Avenue Des Etats-Unis, BP 2010 Toulouse Cedex, France**

ABSTRACT

The results of recent design and development work on compliant air lubricated foil journal bearings to meet the requirements of new commercial aircraft (FALCON-2000) cooling turbo-compressors are presented. Stringent requirements are imposed on compliant foil bearings in these applications to assure safe and durable operation. Compliant foil bearings required to operate at high speed with an insignificant steady state rotor weight while dynamic forces (internal and/or external shock and vibration) exceeding steady loads by an order of magnitude. Bearing design analysis, bearing performance and system rotor dynamics, including critical speeds and rotor imbalance response are discussed. Full rotor bearing development and operational procedures consisting of test results of steady state and random vibrations obtained from both a rotor bearing simulator and in an actual system are presented.

INTRODUCTION

The continued growth at MTI in machinery utilizing foil bearings is illustrated in Figure 1 which shows applications defined in terms of operating speeds and journal bearing diameters. The data on Figure 1 is suitable as a general design guide when considering the potential for a foil bearing solution in advanced machinery. Not illustrated or discussed in this paper is the potential for foil bearings with liquid or particular cryogens [1, 2, 3] or mixed-phase lubrication where the bearing characteristics such as compliance and potential low cost offer advantages.

The use of the foil bearing in aircraft for cabin cooling turbines (ACMs) has undoubtedly been the most meaningful and successful application since the late 1970s. In these application foil bearings, clearly demonstrating the advantages of long service life with minimum maintenance and the avoidance of cabin contamination by the elimination of the conventional ball bearing--oil lubrication systems. The forgiveness of the foil bearing relative to the rolling element bearing in minimizing damage to the aerodynamic components of this type of machinery in the event of a system malfunction has also been demonstrated. A three-wheel ACM is in use on Boeing 747 aircraft in transatlantic service which utilizes MTI's typical foil bearings with a journal bearing of 51 mm (2.0 in.) in diameter and a thrust bearing with 101 mm (4 in.) diameter. The ACM applications now have been extended to military aircraft such as fighters where the bearings must meet the additional requirements of very high speed and severe gyroscopic moments. A recently enacted program for the development of a high speed ACM for a European commercial aircraft FALCON 2000 provided the opportunity for developing a lightweight rotor/bearing system. The overall system description that follows will serve to point out the particular requirements for a high-speed rotor/bearing in a turbocompressor.

AIR CYCLE MACHINE DESCRIPTION

Figure 2 shows a cross-section of the basic ACM turbocompressor elements. The rotor assembly consists of a single-stage turbine driving a single-stage centrifugal impeller through a hollow shaft assembly. The bearing arrangement used to support the high-speed rotor of the turbocompressor is shown schematically in Figure 2. It consists of a pair of inboard foil journal bearings and a set of double-acting thrust foil bearings at the mid-span of the rotor.

Considerable effort has been devoted to developing stable and relatively quiet bearing shaft systems. For example, the bearing-shaft assembly has undergone several analytical iterations to optimize the bearing configuration, clearance for axial hydrodynamic unbalance load, rotor unbalance in each bearing plane, and external shock and vibration. The overall rotor assembly weight is about 772 gram (1.7 lb), with a journal bearing length of 31 mm (1.22 in.), a diameter of 35 mm (1.375 in.), and a nominal diametral clearance of 0.076 mm (0.003 in.). The thrust foil bearing outer diameter is 72 mm (2.85 in.) and the inner diameter is 38 mm (1.5 in.).

ROTOR BEARING SIMULATOR

A finite element model of the ACM turbo systems rotor was developed for use in predicting the rotor system dynamic performance parameters such as critical speeds, stability, and response in the process of completing the foil bearing design iterations. Based on rotordynamic analysis the system tradeoff studies to evaluate potential compliant foil bearing locations, stiffness range and sizes were determined.

Bearing stiffness is primarily a function of the elastic properties of the foils which are under the control of the designer. This provides great flexibility for tuning the stiffness to meet overall system dynamic performance requirements. The elastic structure also provides high shock load capacity. In foil bearings, the clearance geometry required to generate load-carrying hydrodynamic films is provided by the elastic deflection of the foils. This allows foil bearings to operate with much larger bearing clearances than rigid bearings. Operating with gases, foil bearing clearances can be almost an order of magnitude greater than rigid bearings, which is a distinct advantage. The larger clearances and the flexibility of the foils makes foil journal bearings much more tolerant to rotor unbalance. The combination of smooth foils and bump foils also provides great flexibility for the designer to customize the bearing properties. As the bump width decreases, the bump stiffness increases, which enhances the formation of a convergent clearance between the smooth foils and the journal. Compliant foil bearings are not limited to the basic combination of a single smooth foil and a single bump foil, as shown in Figure 1. One or more smooth and bump foils can be added to stiffen the structural top foil as indicated in Figure 3. This figure shows how two bump foils may be used to modify the stiffness. Under no load, there is a clearance between the peaks of the two bumps as shown in Figure 3. As load is applied, it initially only deflects the upper bump foil until the load is sufficient to bring the two bump foils into contact. As the load is further increased, both bump foils have to deflect, which increases the stiffness.

In the design development of compliant foil bearings (CFBs) for specific applications, it is most economical and a sound practice to build a simulator to evaluate the bearing/rotor system performance. The simulator rotor was designed to have a mass distribution and inertia as close as possible to the real rotor and with correct bearing locations. Figure 4 shows a simulator which was developed to evaluate journal bearings for the ACM. The rotor was driven by an air turbine and its lateral motion was measured by four displacement probes. Two probes were mounted orthogonally in a plane between the bearings, and the other two were located towards the end of the rotor, over the simulated turbine. A tachometer was used to pick up the speed from the simulated compressor. Thermocouples were located near the bearings to monitor the bearing ambient temperature.

Once a range of bearing coefficients had been determined based on a system critical speed map [4], a tradeoff study was conducted to evaluate system sensitivities to changes in bearing position, bearing stiffness and damping requirements, and rotor imbalance. The speed dependent stiffness characteristics for the final bearing design were predicted for speeds up to 100,000 rpm and temperatures up to 150°C (300°F), based on the expected static load to be supported by the bearings.

Figure 5 is representative of the predicted foil bearing dynamic characteristics for the ACM turbo-compressor systems using a foil bearing computer code [5, 6, 7]. Figure 6 shows plots of the first two predicted rigid body natural frequencies as a function of rotor speed and bearing damping coefficients of 0 to 350 N.s/m (0 to 2 lb.s/in.

Frequency, log decrement and precession direction for the conical and translator modes were predicted for various speeds, using the estimated damping values and the calculated stiffness coefficients. Figure 7 shows the rotor bearing system stability results for a spin speed up to 100,000 rpm. This figure indicates the level of damping required to achieve the necessary positive value for log decrement (stable operation) for the two most troublesome rigid body modes. As part of the design study, the impact of the static and dynamic loading on the compliant foil bearing design was assessed. Figure 8 shows sample pressure profiles vs. extent of pad, indicating a uniqueness of this bearing with respect to a conventional foil bearing. Figure 9 summarizes the foil bearing predicted performance in terms of load, eccentricity ratio, and bearing power loss as a function of rotor speed.

SIMULATOR DEVELOPMENT AND TESTING

The simulator is shown schematically in Figure 4; Figure 10 is a photograph of the test rotor/bearings; Figure 11 is a photo of the test installation mounted on a shaker.

During the tests, all probe outputs were recorded on magnetic tap; the rotor orbit was displayed on oscilloscopes; and an FFT signal analyzer was used to identify resonant frequencies. Figure 12 shows the test facility setup, recording, monitoring and control units.

TEST RESULTS

Figure 13 shows the signal analyzer output for the two transducers at the turbine end when running at 80,000 rpm. The major resonant frequency was subsynchronous at 14,700 rpm and had a peak amplitude of approximately 14.7 μm (0.58 mil), while having translatory mode shape (second rigid body natural frequency). At the end of the test, the drive air was shut off at 80,000 rpm. Figure 14 shows the signal analyzer output during the coastdown phase. This figure shows the growth of the synchronous component at frequencies ranging from 11,550 rpm to 15,000 rpm, the first and second rigid body critical speeds range. The peak amplitudes of these frequencies were only of the order of 5 μm (2×10^{-4} in.).

A significant amount of testing was completed with this rotor/bearing arrangement and successful foil gas bearing performance at steady state operating conditions was demonstrated. The most important result of this series of tests was confirmation of analytical predictions as shown in Figure 15. The predicted damped natural frequencies (first and second rigid body critical speeds) based on $B_{xx} = B_{yy} =$ 2 lb-s/in is plotted against the spin speed of the rotor (Figure 15), along with corresponding experimental data.

The rotor/bearing system was also subjected to vibration tests by mounting the simulator on a shaker (Figure 11). An acceleration sweep of 1 g peak-to-peak was applied in a vertical plane at a spin speed of 69,000 rpm to determine the bearing damping level. The acceleration sweep range was from 100 Hz up to 2 Khz. Figures 16(a) and 16(b) show a typical rotor response where the rotor spin speed was 69,000 rpm and corresponding elapse time for acceleration sweep lasted about 300 sec. As shown in Figure 16(a), the damped natural frequency was determined to be 13,500 rpm with a translatory mode (i.e., second rigid body critical speed). Results shown in Figure 16(c) depict a time response plot from the sweep data at 69,000 rpm. Experimental data shown in Figure 16, such as log decrement, amplitude and corresponding frequency, along with predicted bearing stiffness coefficients at 69,000 rpm, were used to extract the bearing damping coefficient. For a single degree of freedom system this corresponds to a bearing damping level of 324 N.S/m (1.85 lb-s/in).

The above test procedure at various g levels at different spin speeds was conducted to identify bearing damping levels as well as maximum g levels that the system could satisfactorily perform. The bearings were subjected to a form of shock test by suddenly imposing a large acceleration lasting 10 seconds. Figure 17 shows the rotor motion when a total of 100 g peak-to-peak acceleration was applied with a rotor speed of 55,000 rpm. The upper plot in Figure 17 shows the vertical component of the rotor motion measured at the turbine end. The peak amplitude was only 9.5 μm (0.374 mil) and the main rotor motion was at a frequency of 200 Hz (12,000 rpm).

Results of these experiments clearly indicated (a) the validity of the predicted dynamic characteristic performance, and, (b) CFBs have higher tolerance to the external shock and vibration with a significant levels of damping.

VALIDATION TESTING

This section presents results of validation testing of selected foil bearings in an actual prototype ACM system. These tests also consisted of steady-state and dynamic test operations.

Steady-State Operation - Based on the preliminary simulator evaluation, the selected foil bearings were installed in the ACM system as part of the final development process. The unit shown in Figure 18 was successfully operated, achieving full operation speed (i.e., an idle speed of 6,000 rpm and a nominal operating speed of 66,000 rpm). This unit was instrumented for general data acquisition, however, displacement probes placed in a similar fashion as described earlier were utilized here too, with the exception of their locations. A set of probes was mounted orthogonally at either end of the rotor, outboard of the turbine and compressor. The following displacement output data conveniently refers to their locations.

Figure 19 shows the signal analyzer output for the vertical outboard displacement probe at the compressor end running at 66,000 rpm. The major resonant frequency was synchronous due to mechanical runout and had a peak amplitude of 16.01 μm. The minor resonant frequencies were subsynchronous related to two rigid body critical speeds. The subsynchronous frequency at 245 Hz (translatory mode) correlated extremely well with the results obtained from the simulator test (see also Figure 15 for comparison). The results of steady state operation in general showed good agreement with the preliminary simulator tests,including predicted performance data.

DYNAMIC TEST

The ACM unit was also subjected to vibration tests by mounting the units on a shaker. While the ACM was rigidly mounted on the shaker, low and high frequency vibrations in the form of a sine wave were applied. At any given operating speed the g level was increased linearly over a designated frequency range. An example of such tests, including input vibration and output data is given in Figure 20.

The upper plot in Figure 20 shows an applied g level in the vertical direction over the frequency range of 20 to 30 Hz. The vertical displacement probes reading a typical rotor response for the rotor speeds of 25,000 and 55,000 rpm are shown in Figure 20. The largest peak-to-peak amplitude of the rotor was about 100 μm at the turbine end.

Additional required tests were to be carried out with a random vibration input. Figure 21 shows the applied power spectral density that the ACM was subjected to.

A set of typical rotor responses at two spin speeds of 25,000 and 55,000 rpm, subject to random vibration, are shown in Figure 22. The peak amplitudes of the rotor were 57 μm and 66 μm respectively, which occurred at subsynchronous frequencies of 150 Hz at speeds of 417 Hz (25,000 rpm) and 203 Hz at speeds of 917 Hz (55,000 rpm). These subsynchronous frequencies are associated with rotor/bearing critical speeds which were predicted and demonstrated in earlier tests (for comparison see Figure 15). From these results it seems that the applied external vibration has an insignificant effect on the magnitude of the subsynchronous frequency. However, its impact was on the growth of subsynchronous amplitude.

As part of the development, the ACM unit was subjected to various levels of shock at all operating speeds, including zero rpm. The ACM rotor speed was first stabilized at any given test run,and an external shock was then applied to the unit. A sample shock wave is shown in composite plots in Figure 23. The elapse time for a shock wave was 11 msec; while peak-to-peak amplitudes of 7.7 g had a period of 1.6 msec. Figure 23 shows the rotor response when a total 7.7 g peak-to-peak shock was suddenly applied with rotor speeds of 25,000 and 55,000 rpm. The amplitude and subsynchronous frequency as a result of the applied shock wave remained below 11 μm at 55,000 rpm and below 3 μm at 25,000 rpm, as given in the plots of Figure 23.

CONCLUSIONS

The analytical and experimental studies conducted on the application of compliant, air-lubricated foil bearings for an Air Cycle Machine for commercial aircraft applications confirmed their potential in terms of load-carrying capability, suppression of rotor unbalance and whirl-free characteristics.

Successful transition and development of foil bearings for ACMs were achieved and it was demonstrated that a systematic approach toward advancement and application of foil bearings with an extremely reliable analytical tool is possible, yet it can be risk free. Thus, the approach taken here was most economical.

The analytical design tools were successfully used to assist foil bearing development for ACMs. Predicted bearing dynamic characteristics (bearing stiffness and damping coefficients) were in good agreement with experimental results.

The high speed ACM rotor bearing system was subjected to acceleration levels up to 100 g peak-to-peak, demonstrating their shock resistance at high speed on air film capability.

ACKNOWLEDGMENTS

The authors are grateful for the support of the development effort and for permission for publication from both ABG-SEMCA and Mechanical Technology Incorporated. Our thanks for the help received from our numerous colleagues during the overall systems development and preparation of this paper.

REFERENCES

1. Heshmat, H. "Analysis of Compliant Foil Bearings with Spatially Variable Sitffness", presented at *AIAA/SAE/ASME/ASEE 27th Joint Propulsion Conference,* June 24-26, 1991, Sacramento, CA, Paper No. AIAA-91-2102.

2. Heshmat, H. "A Feasibility Study on the Use of Foil Bearings in Cryogenic Turbopumps", presented at *AIAA/SAE/ASME/ASEE 27th Joint Propulsion Conference*, June 24-26, 1991, Sacramento, CA, Paper No. AIAA-91-2103.

3. Heshmat, H. and Shapiro, W. "Application of Compliant Fluid-Film Bearings to the High-Pressure Oxygen Turbopump of the SSME," submitted for

presentation and publication at the *5th ETO Propulsion Technology Conference*, May 19-22, 1992, NASA, George C. Marshall Space Flight Center, Huntsville, AL.

4. Heshmat, H. "Gas-Lubricated Foil Bearings for Heat Pumps with Counter-Rotating Heat Exchangers." ASME Paper 82-DET-139 (September 1982).

5. Heshmat, H., Walowit, J., and Pinkus, O. "Analysis of Gas-Lubricated Compliant Journal bearings." *J. Lubr. Tech. Trans.* ASME 105, No. 4 (1983): 647-655.

6. Ku, C.-P.R. and Heshmat, H. "Compliant Foil Bearing Structural Stiffness Analysis: Part I: Theoretical Model Including Strip and Variable Bump Foil Geometry," J. of Tribology, Trans. ASME Vol. 113, no. 2 (1992): 394-400.

7. Ku, C.-P.R. and Heshmat, H. "Compliant Foil Bearing Structural Stiffness Analysis, Part II: Experimental Investigation." Accepted for publication in the J. of Tribo., Trans. ASME, ASME Paper No. 92-Trib-6, to be presented at STLE/ASME Joint Trib. Conf., San Diego, CA, October 18-21, 1992.

Foil Journal Bearing*

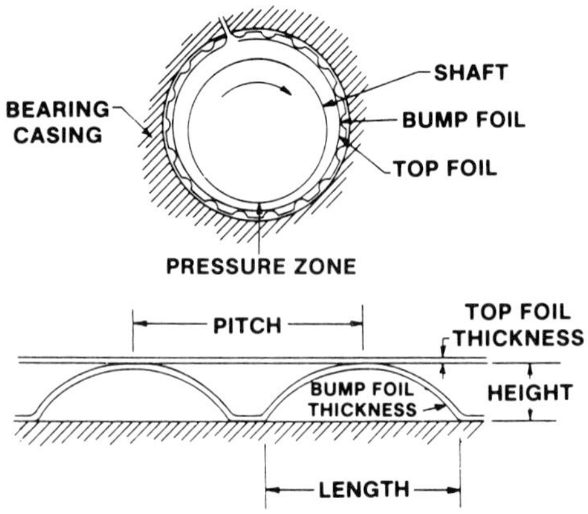

TYPICAL DETAILS OF BEARING

Foil Thrust Bearing*

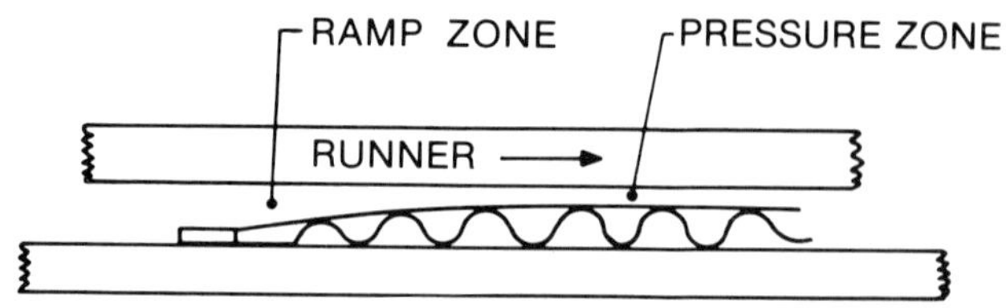

PAD OPERATION

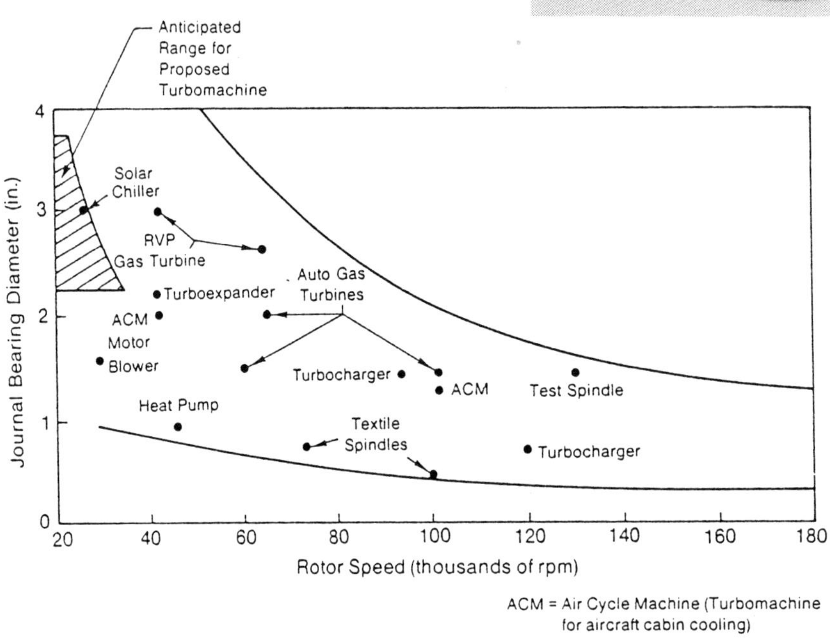

Fig. 1 Examples of Foil Bearing Applications

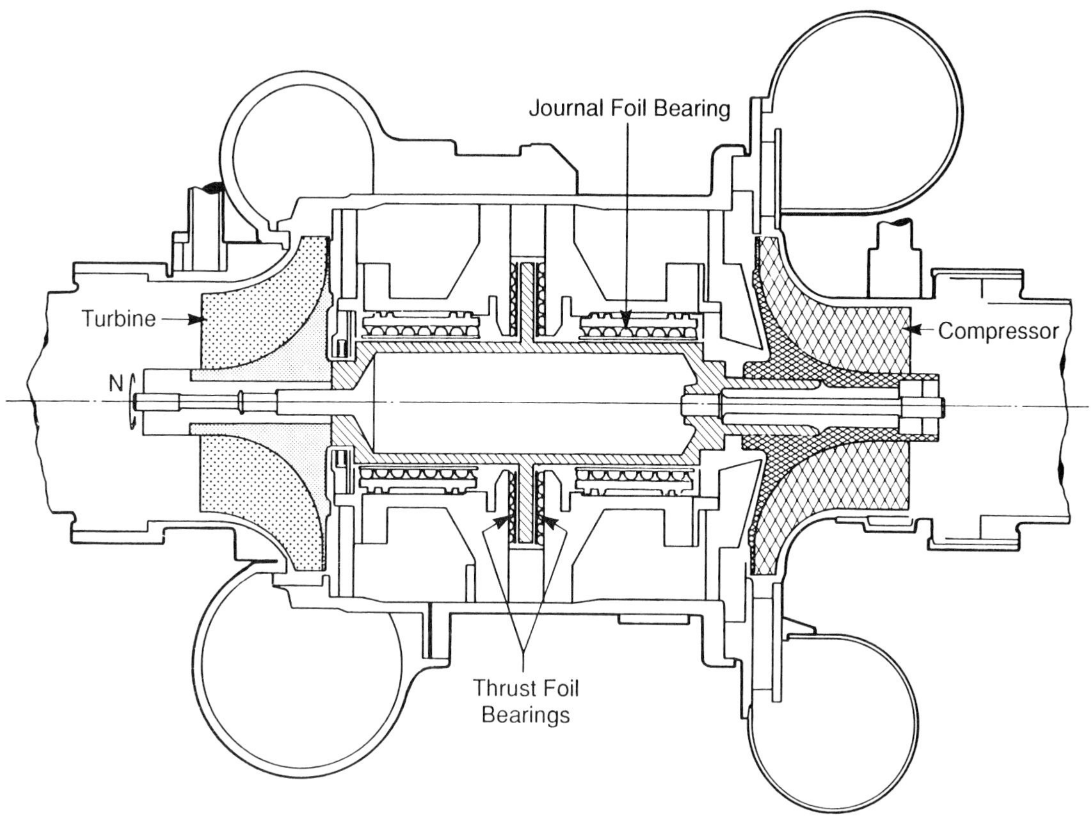

Fig. 2 Application of Compliant Foil Bearings in Aircraft Air-Cycle Machines; N ≈ 90,000 rpm

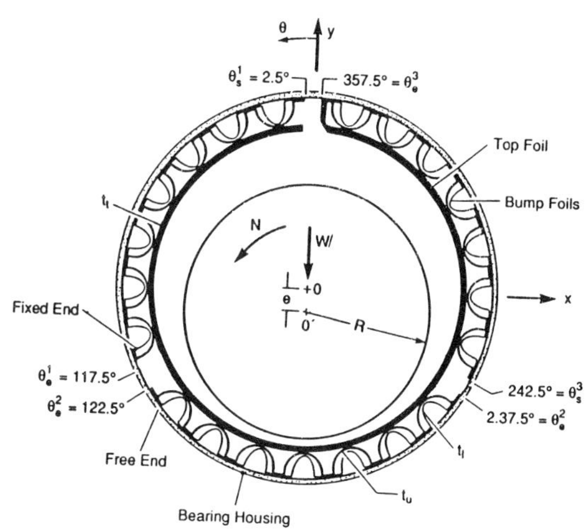

Fig. 3 Foil Bearing for High Speed and High Temperature Application

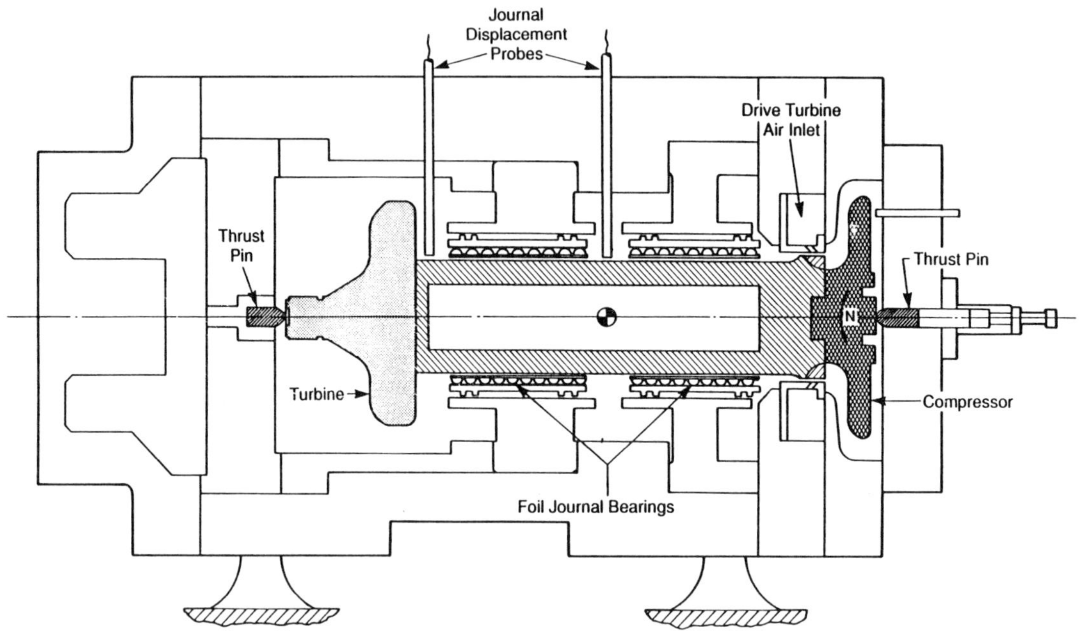

Fig. 4 High Speed Compliant Foil Journal Bearing Test Rig (Air-Cycle Machine Simulato

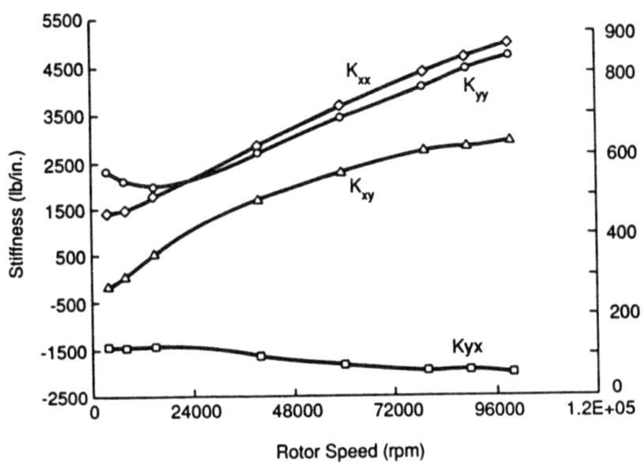

Fig. 5 Stiffness Characteristics of Foil Journal Gas Bearings

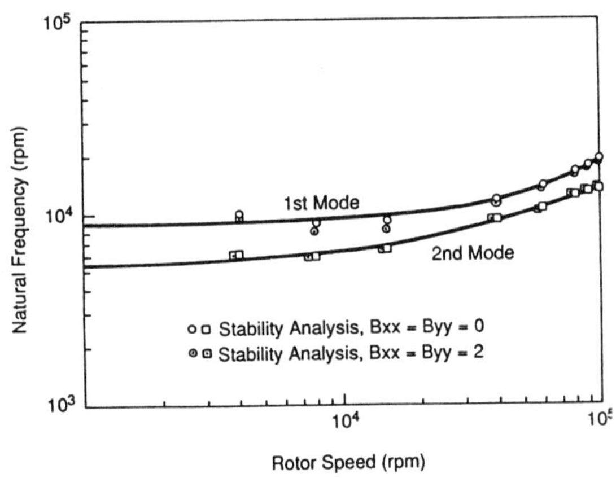

Fig. 6 Rotor/Bearing Critical Speeds as Function of Rotor Speed ai Damping Coefficients

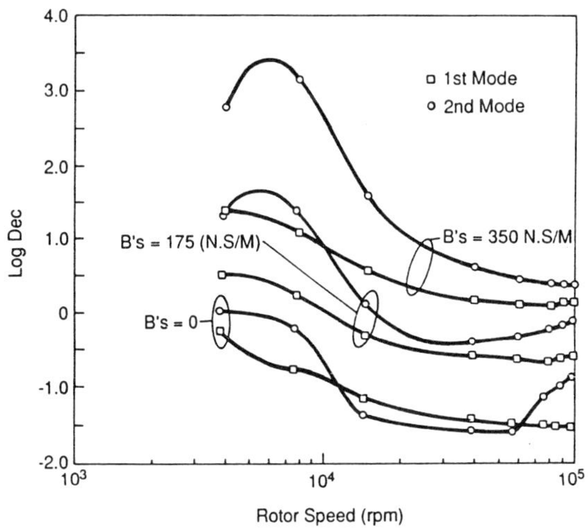

Fig. 7 Log Decrement vs. Rotor Speed

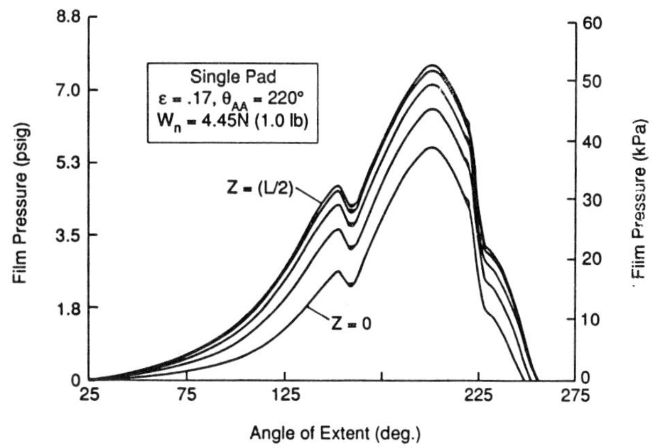

Fig. 8 Pressure Profile vs. Extent of Pad

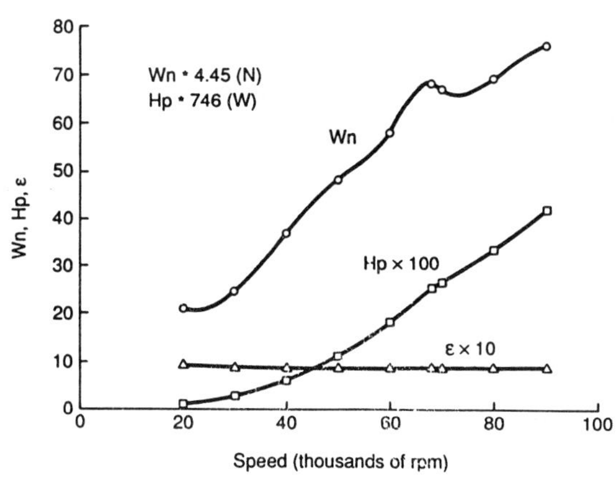

Fig. 9 Foil Bearing Performance: Load Wn, Eccentricity Ratio ϵ, and Power Loss Hp

Fig. 10 Photograph of the Simulator Rotor and Foil Journal Bearings

Fig. 11 Photo of ACM Simulator Rig Mounted on a Shaker

Fig. 12 Test Facility Setup, Recording, Monitoring, and Control Units

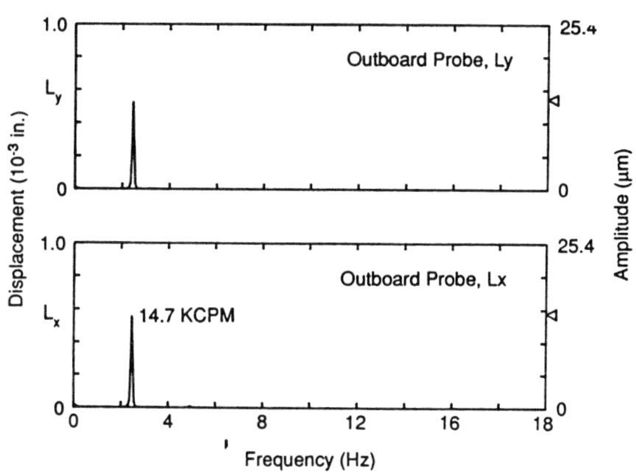

Fig 13 Analyzer Output, Turbine End at 80,000 rpm, Vertical = L_y, Horizontal = L_x

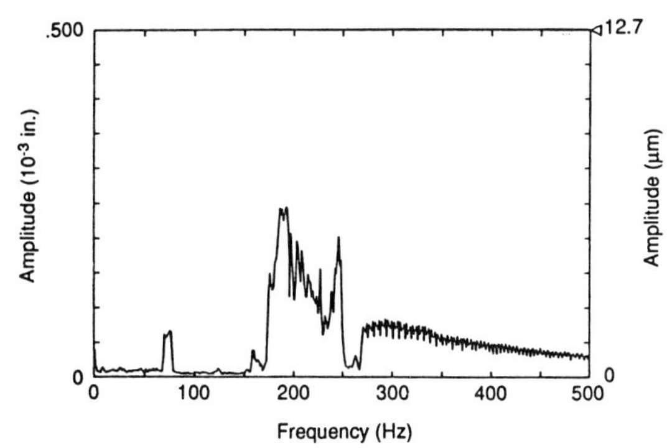

Fig. 14 Typical Tracking Plot of Coastdown from 80,000 rpm at Compressor End

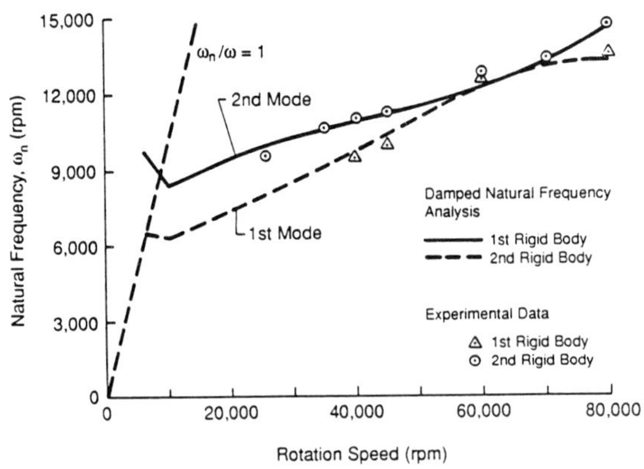

Fig. 15 Turbo Compressor Critical Speed Map; With Compliant Foil Bearing, (ACM)

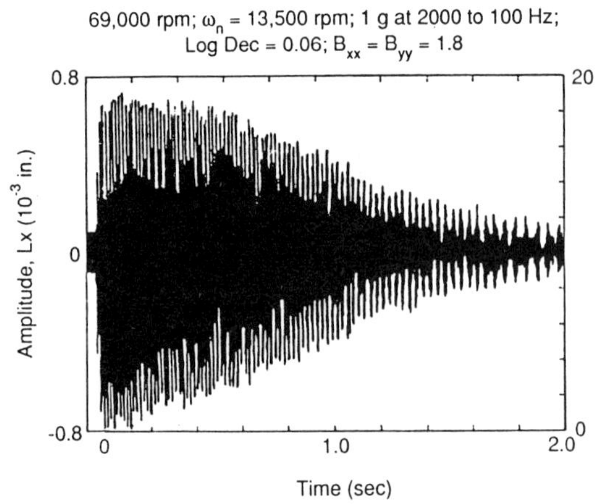

Fig. 16(c) Vibration Sweep: Descending

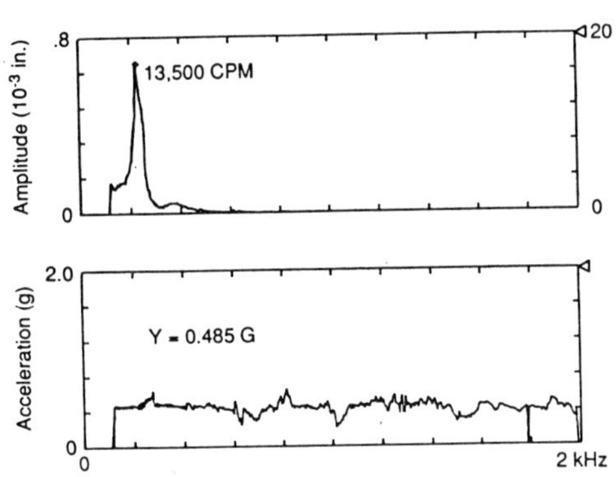

Fig. 16 (a&b) Rotor Response at 69,000 rpm with 1.0 g Acceleration Sweep

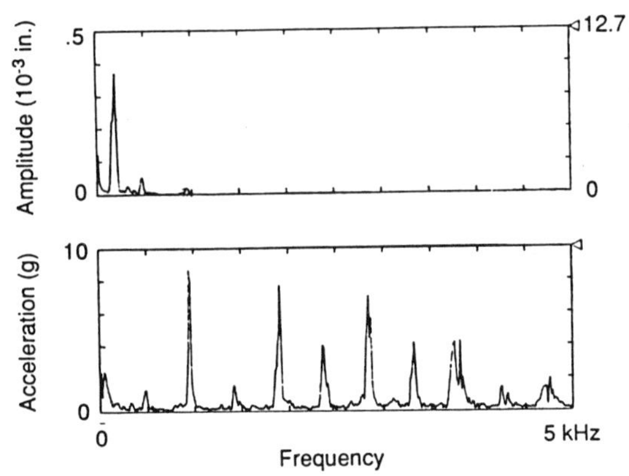

Fig. 17 Rotor Response at 55,000 rpm with 100 g Peak-to-Peak Acceleration

Fig. 18 Photo of FALCON 900 Foil Bearing Air Cycle Machine

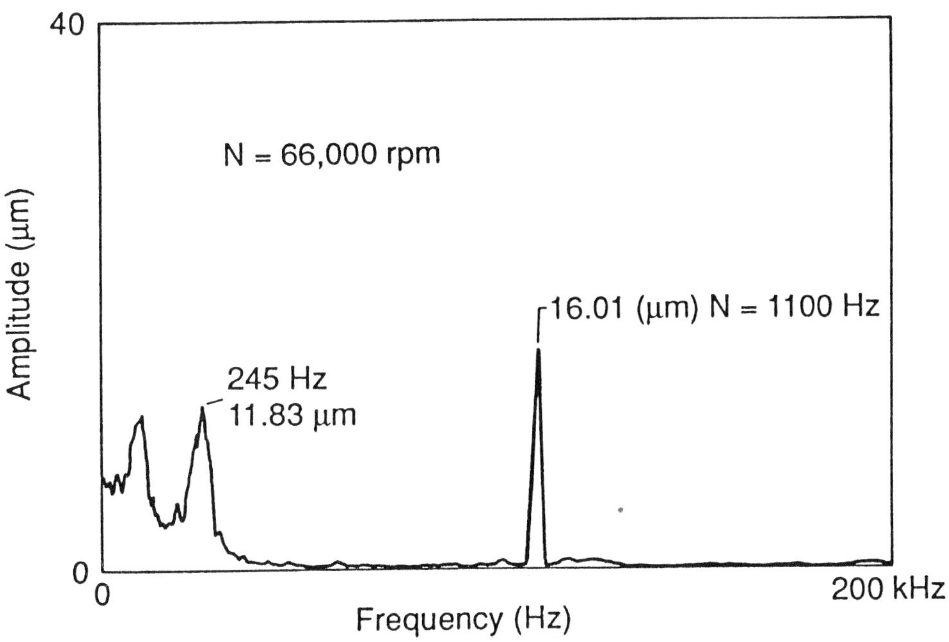

Fig. 19 Analyzer Output for Displacement Probe at Compressor End at 66,000 rpm

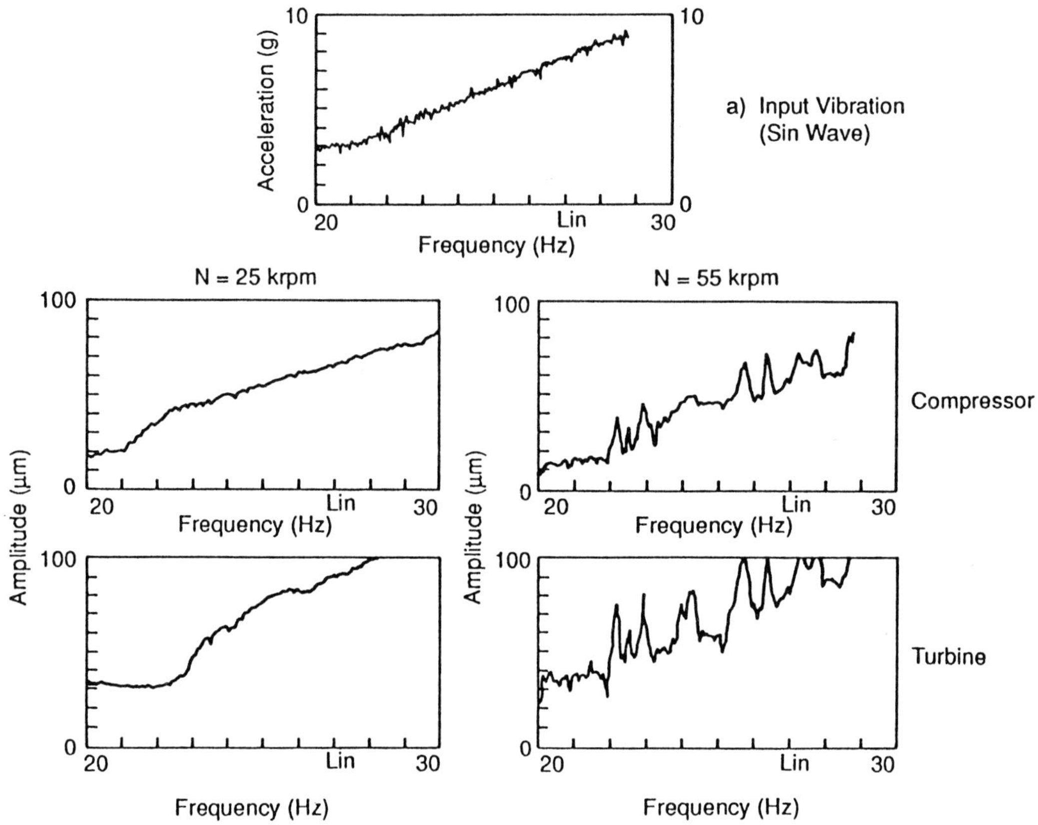

Fig. 20 ACM Rotor Response at 25,000 rpm and 55,000 rpm with 3.0 to 9.0 g Peak Acceleration at Low Frequency

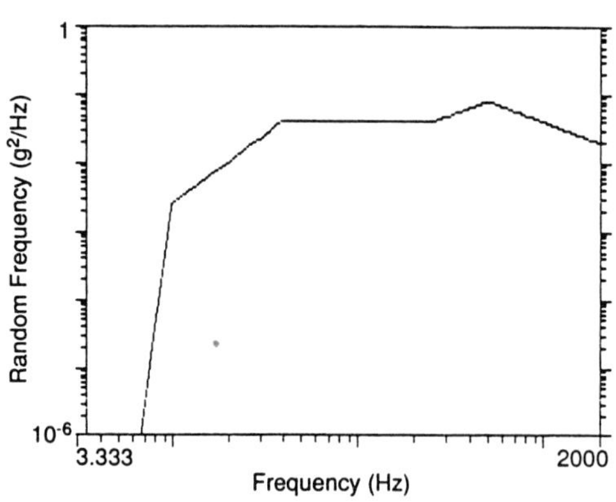

Fig. 21 Spectrum of Applied Random Vibration to ACM

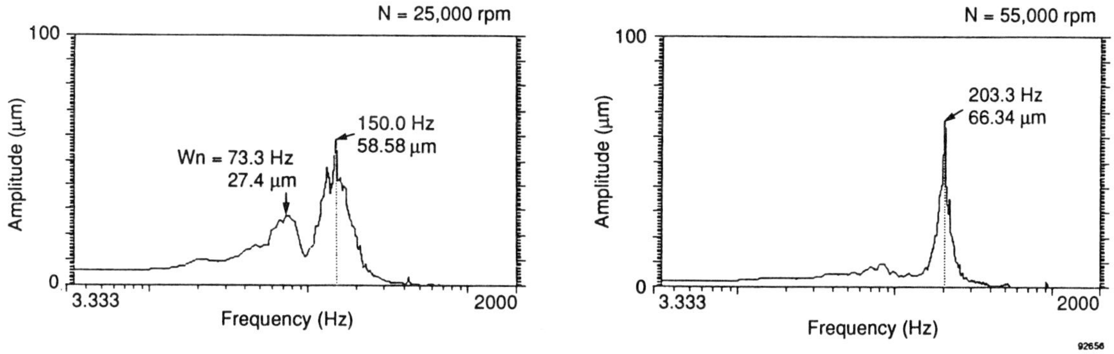

Fig. 22 ACM Rotor Response to External Random Vibration (Amplitude vs. Frequency)

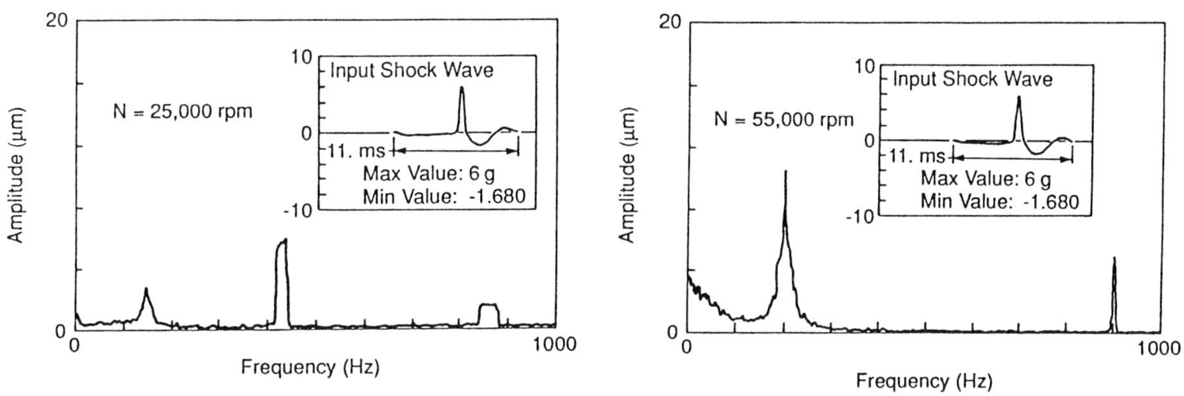

Fig. 23 ACM Response to External Shock Wave (+6 to -1.7 g Shock)

Thin Films in Tribology / D. Dowson et al. (Editors)
1993 Elsevier Science Publishers B.V.

THICKNESS MEASUREMENT METHODS FOR THIN LUBRICANT FILMS. APPLICATION TO A RADIAL FACE SEAL.

B. Tournerie, M. Lépine(*) and J. Frêne

Laboratoire de Mécanique des Solides -U.R.A. CNRS- Université de Poitiers

((*) Now engineer at Jeumont Schneider Industrie)

This paper presents the measurement techniques of thin lubricant film thickness which can be used in experimental research or in industrial applications in the field of lubrication and more particularly in the case of dynamic sealing. An application to an experimental study of a low pressure mechanical seal is then reported. The measurement technique and the calibration are described in details. Examples of results are presented and analysed.

1. INTRODUCTION

The knowledge of the lubricant film thickness (or of the interfacial gap) is essential to understand the lubrication mechanisms of a contact. However, the thickness measurement of very thin films is quite difficult and generally the analysis of the lubrication mode is deduced from other parameters easier to measure. The main method consists in the analysis of the variation of the friction coefficient f versus the duty parameter $G=\mu V/W$, where μ is the dynamic viscosity of the lubricant, V the linear speed of the contact and W the load applied to the contact. The f-G curves, often encountered in literature are supposed to permit delimitation of the different lubrication modes ; solid contact, mixed friction and hydrostatic or hydrodynamic lubrication. These curves must be interpreted in function of parameters which are not taken into account in the G factor such as ; the pressure of the lubricant, the geometry and the roughness of the surfaces, the nature of the materials and in particular the influence of the temperature on their mechanical properties. In addition, the curves present a very large scatter [1-2] because they are affected by evolutions of the contact during the tests ; distortions and wear of the contact surfaces, presence of wear particles and debris, fatigue of the materials,... etc. Another method consists in measuring the relative displacements of the surfaces of the lubricated contact. But, to be precise, this technique is often as difficult to carry out as a direct measurement of thickness. In fact the measurement of the displacements of the two elements of the contact integrates also the distortions of both these elements and their supports and it is generally difficult to distinguish displacements and distortions. The leakage flow measurement can be used, in some cases, in indirect methods of mean thickness determination. However, these methods need a good knowlegde of the geometry and kinematic of the lubricated surfaces and this condition is seldom satisfied. In the case of dynamic seals the occurence of a leakage indicates a partial or a total surface separation and the presence of hydrodynamic or hydrostatic effects, but does not permit to deduce the shape and the value of the sealing gap.

The direct measurement of the interface thickness, along the zone of contact and in dynamic conditions allows to determine the lubrication mode, the interface geometry and the kinematic of the solids in contact. This paper presents the measurement techniques of thin lubricant film thickness through a review of experimental works and industrial tests.

An application of a capacitive measurement technique to an experimental study of a low pressure mechanical seal is then reported.

2. REVIEW AND ANALYSIS OF THE MEASUREMENT METHODS

Generally, the fluid film thickness is obtained by the measurement of the distance between the two surfaces separated by the film. The measurement methods can be divided into two main categories [3]

- optical techniques which are principally fluorescence and interferometric methods
- electrical techniques which mainly include resistance, inductance, capacitance and impedance methods.

2.1. Optical Techniques

2.1.1. Fluorescence method

The lubricated surfaces are illuminated by an UV or blue light. The lubricant is doped with fluorescentdye. As the lubricant is lightened, the fluorescent dye emits a light of longer wavelength (generally green) whose intensity is dependent on the film thickness. The measurements are made by means of optical detectors and oscilloscope. Light sources and detectors must have great stability. Output signals are calibrated by comparison with known thicknesses of the lubricant film in the same conditions.

This technique has been developed for measuring oil film thickness on a piston through a transparent window, by using a mercury lamp [4] or HeCd laser providing blue light [5,6]. Visible light allows the use of standard optical components.

The calibration of such measuring device is difficult because of their non-linear response in function of the film thickness. It is also complicated by multiple reflections of the different surfaces [6]. Moreover, when the lubricant is oil, blue light must be used because UV light produces oil degradation and non stable fluorescent activity [4,5,6].

Recent developments of the technique have been carried out [7] for the study of lip seals. The novelty of this work consist mainly of a very precise optimisation and calibration methods and of the use of CCD camera and PC-based image processing for 3D mapping of the film thickness.

2.1.2. Interferometry

The variation of film thickness between two surfaces, one is reflective and the second is semi-transparent, can give interference pattern under the effect of an appropriate incident light. Interference fringes give curves of equal thicknesses.

Appropriate reflectivity of the two surfaces is necessary to have a good contrast of the interferograms. It can be obtained with glasses of refractive index higher than those of lubricants and by depositing thin reflective layers of metal oxydes on the face in contact with the lubricant film. The external face of the transparent element must also be coated to reduce spurious reflections. It is also necessary that surfaces present high quality of finish ; rough surfaces do not allow the use of interference technique [8].

Two types of interferograms can be obtained depending on the nature of the light source ; white or monochromatic light.

With white light, given by low voltage Tungsten filament lamp [8-12] or halogen lamp [7, 11], the interference fringes are coloured and easily identified [10, 14]. Absolute thickness can be measured with a good accuracy of about 0.02 μm. Nevertheless for thickness greater than 1 μm the interferogram becomes muddy white and can no longer be interpreted.

With monochromatic light sources such as Mercury discharge lamp with band filter or Sodium lamp [8, 10-12] or CW He-Ne laser [12, 15-20] a pattern of bright and dark fringes is obtained. Two successive bright or dark fringes correspond to a thickness variation of about $\lambda/4$, where λ is the wavelength of the light source. The thickness variation can be measured with an accuracy greater than 0.1 μm, but the interferogram does not indicates neither the increasing or the decreasing variation nor the absolute thickness. However, techniques allow to overcome this obstacle in some cases.

For thin thicknesses h, it is possible to choose wavelengths in visible or infrared spectrum in order to have $h < \lambda/4$. Thus the intensity of the first order fringe is an univocal function of the thickness [12].

In some cases the thickness increases, for an increasing speed from the contact. By counting the fringes from the contact, the absolute thickness can be found [8, 10, 11, 14, 19-21].

If two light wavelengths or more are used, by means of different sources or by means of a monochromator, the comparison of the interferograms leads to the determination of the fringe order and thus to the thickness [17, 18, 22, 23].

Some interferometric studies have been performed in the case of radial face seals [21, 24-26] or in similar case [8].

For piston ring lubrication experiments, it has been noted that simultaneous occurence of liquid phase disturb the interference observation [16].

A large number of studies have been devoted to ball-plane E.H.D. contact with very thin lubricant films [9-11, 15, 22, 27, 28].

As interferometric methods are particularly accurate for very low thicknesses of gas, they have been largely employed in the case of magnetic disk-heads [12, 13, 15, 17-20, 22, 23].

2.2. Electrical Techniques

2.2.1. Resistance Method

The resistance method involves the use of electrical conductor materials and the insulation of some elements of the apparatus. In the literature, no real direct resistance method of thickness measurement has been encountered, but interfacial electrical resistance in lubricated contacts has been studied [29]. It is very difficult to evaluate the relationship between electrical resistance and thickness because the resistivity of both lubricant and metal-lubricant interface is extremely sensitive to impurities or to lubricant additives [30, 31, 32]. Furthermore short circuits, due to impurities or chemical compound formed by additives and surfaces reaction, can occur between the surfaces [33]. In some cases, however electrical breakdown indicates asperity contact which is characteristic of boundary lubrication [34, 35].

2.2.2. Sparkling voltage method

Studies were carried out to measure minimum film thickness between two plane surface [29, 36] by means of sparkling voltage method. It was found that breakdown voltage is higly dependent on the purity of the lubricant, temperature, and frequency of the applied AC current. No recent development of this technique has been found in the literature.

2.2.3. Inductance methods

The inductance of a wire coil varies as its separation from a magnetic material changes. This phenomenon is the basis of a displacement transducer used in some works. In the case of very thin films, the commercially available inductive transducers have limitation on the minimum distance that they can distinguish from zero. Mutual-inductance transducers must be used [37] to reduce the temperature-dependance of the sensitivity. These transducers are breakable and quite large. For these reasons these transducers are particularity employed to measure the relative displacements of two elements. Thickness variations of the lubricant film between these element's surfaces can be indirectly obtained [38-40] from the displacements.

In the case of bearings, where the film thickness may be important, inductance transducers are generally used [41-43]. In this case, the problems caused by material inhomogeneity may be solved by filters included in the measurement circuit.

Miniature-transducers have been developped for specific applications ;bearings of diesel engine [44], lip seals by using magnetic fluids [45].

2.2.4. Impedance and capacitance methods

The basic principle is that two surfaces, separated by the lubricant which is a dielectric film, form a capacitor whose capacitance is inversely proportional to the film thickness. Practical methods can be distinguished by the design of the capacitive probe and by the associated capacitive measurement equipment.

For cylindrical bearings, in the absence of cavitation, the minimum film thickness is calculated from the total capacitance of the oil film between the shaft and the big-end bearing ([46], [47]) A similar technique has been used for face seals [29] For one of the first measurements of film thickness in mechanical seals, Denny [48] used two segments stuck on the stationnary ring as electrodes. Some special-type of electrodes were developed for particular applications [49] but generally a wire, forming the central electrode, is inserted in a tubing used as electrical screen. The central electrode is insulated from the screen by means of expoxy resin or insulating sleeve. The probe is bounded in one of the element of the lubricated contact through an insulating sleeve or a layer of resin. The extremity of the probe is then lapped flush to the surface [46,50-57].

In order to prevent short circuit when the surfaces are in contact and to have in this case a reference capacitance, the probe can be fixed slightly below the surface [58] and covered by a thin and sheet of dielectric on its extremity [24, 59-62].

Methods using such transducers are often named capacitive methods although they are methods of reactance or impedance measurement. Two main principles of measurement are generally encountered :

- modulation of the oscillation frequency of an oscillator
- use of an universal bridge driven by an AC source.

In the first method [18, 24, 29, 40, 46, 48, 53-64], the transducer is connected to an HF oscillator and the circuit is tuned to the initial capacitance corresponding to the initial reference gap. Any variation of the gap, and consequently any variations of the capacitance, modulates the oscillation frequency of the oscillator which, in turn, is converted into voltage inversely proportional to the gap.

In the second method [46, 47, 50-57] a three terminal bridge is generally used. The central electrode is connected to the imput and the screen to the neutral of the bridge. The counterface element is connected to the earth of the bridge but insulated from the main earth of the test rig. The ouput signal is an amplitude modulated wave whose frequency is the one of the source driving the bridge generally of about 20 to 50 kHz. This signal can be demodulated producing an output voltage proportional to the probe capacitance.

Some authors calculate the thickness from capacitance measurement values. In order to determine these values, the probe is replaced by external calibrated capacitors or included in the decade controls of the equipments [29, 46, 48, 63]. Other authors directly calibrate the output voltage in terms of capacitance or thickness. They generally use auxiliary calibration equipment to compare output voltage to known capacitances or to known thicknesses in conditions more or less close to the experimental ones [39, 50, 54, 62, 63]. It is generally found that changes in calibration due to temperature and pressure in the test range remain very small.

2.3. Field of application

Interferometric methods are particularly accurate for very low thicknesses, specially with white light for submicron measurements. If monochromatic light basically permits to measure only the variations of thickness, superposition methods allow to get absolute values with a very good accuracy. The problem encountered with fringe contrast render the use of these techniques easier with gaseous lubricants.

For thicknesses of the order of the micrometer, capacitance methods are the most employed. The accuracy of these methods is limited by the difficulties encountered for the calibration.

When the thickness is very important -in the case of bearings for example- inductance methods are quite efficient.

3. APPLICATION TO A MECHANICAL SEAL

Mechanical seals are components which must satisfy the two following antagonistics requirements :

- low leakage ; that implies very thin lubricant film or face contact
- low friction and low wear ; that requires the presence of a sufficiently thick lubricant film.

The gap clearance measurement in the interface is the best way to know if a given face seal, operate under hydrodynamic of mixed friction lubrication conditions and allows the analysis of the factors influencing the transition between these two modes.

3.1. Description of the test rig

Major features are :

- a vertical precision shaft driven by a 3.7 kw motor through a variable speed drive
- a hydraulic power pack which supplies fluid at controlled flowrate and temperature, and controlled pressure in the range 0 to 0.1 MPa
- a test housing equiped with electrical transducers
- a computerized data acquisition system

A drawing of the test seal and housing is presented in figure 1. The rotor consists of a ring of metal-impregnated carbon assembled in a copper alloy carrier. The stator is made of steel. The mean diameter of the seal faces is 70 mm and the face width 4 mm.

The oil is fed to the inside of the seal, the outside of the seal is in the atmospheric air. The secondary seal between the floating stator and its carrier is an elastometric O-ring whose active diameter is 74 mm. The balancing ratio is B = 1. The closing force of the seal is applied by springs placed in blind holes drilled in the seat holder. The holder is fixed to the frame of the test cell through the fluid collector and the torquemeter.

Local film temperature is measured by a thermocouple embedded in a blind hole under the stator surface.

Friction torque is transmitted from the stator to the holder by the antirotation locks and thence to the torquemeter.

The displacements of the floating seal seat are measured by three equidistant proximity probes mounted on the holder in front of a shoulder on the ring.

The film thickess is measured by capacitance method.

3.2. Principle of the film thickness measurement

The measurement method is capacitance based, using a frequency modulation of a carrier wave. The capacitance probe, of about 1 mm diameter, bounded in an insulator tube, is embedded in the seal ring. In order to prevent thermal differential expansion the probe is made of steel like the stator and it is fixed on the extremity near the sensing end. In order to avoid short circuits when the thickness is very thin the extremity of the probe is covered by a thin layer of dielectric before to being made flush to the surface. The probe is connected to a circuit formed by a tuning plug, an oscillator and a capacitance converter. The capacitance converter is initially tuned to the frequence f_o for the capacitance C_o, corresponding to the initial film thickness h_o, in order to deliver a zero output voltage . Any change Δh of the thickness and thence of the transducer capacitance ΔC produces a change of the resonance frequency of the circuit, which in turn is converted into an output voltage V theoretically given by :

$$V=K.\varepsilon_\ell.a\left[\frac{1}{d\,\frac{\varepsilon\ell}{\varepsilon d}+(h_o+\Delta h)}-\frac{1}{d\,\frac{\varepsilon\ell}{\varepsilon d}+h_o}\right] \quad (1)$$

where : K is the converter constant, ε_ℓ and ε_d are the dielectric constant of the lubricant and of the dielectric sheet, a is the area of transducer sensing end, d is the thickness of the dielectric sheet, $h = h_o + \Delta h$ is the lubricant film thickness.

If the probe is not insulated on its extremity, d = 0, when face contact occurs ($h_o = 0$ or $\Delta h =-h_o$), V increases beyond the converter range. If $d \neq 0$ when face contact occurs, we have V = 0.

If we note $h_c = \varepsilon_\ell d/\varepsilon_d$, h_c can be considered as equivalent to a constant thickness to which the converter is initially tuned. The sensivity of the measurement is then :

$$\frac{dV}{dh}=-k\varepsilon_\ell a\frac{1}{(h_c + h)^2} \quad (2)$$

where h_c contributes to reduce the sensitivity but also attenuates the non-linearity of the measurement.

3.3. Calibration on a calibration cell

A special apparatus has been built for calibration purposes. The cell is enclosed by the experimental stator equiped with the transducer to be calibrated and can be filled with various fluids (oil, air,...etc) under controled pressure. A micrometric screw allows to define a fluid film of known thickness in front of the sensing end of the transducer. The test cell can be placed in a thermostatic chamber for temperature influence evaluation.

Figure 2 shows a V-h calibration curve obtained at the ambiant temperature in the oil. Similar curves were obtained in the air. No influence of the pressure has been observed up to 0.5 MPa. These curves have been established with a good repeatability. For large variations of the thickness h these curves are non linear, but for variations less than 10 μm the effect of non linearity remain small. In this range of small thicknesses the output voltage V can be expressed approximately by the linear function

$$V = Ah + B \quad (3)$$

where B depends on initial conditions of the tuning.

Figure 3 shows an example of calibration curves, established at various temperature in the air. Similar curves were obtained in the oil. The sensitivity A is less affected by the temperature than the factor B which is responsible for noticeable temperature drift. The capacitance of the transducer can be considered as equivalent to the circuit shown on figure 4 where :

- C_ℓ is the capacitance due to the film thickness
- C_d is the capacitance due to the dielectric sheet
- C_p, named "parasitic" capacitance is due to insulator between the electrode and the stator, to connexions, to connecting wire,

...etc.

The capacitance C is then :

$$C = C_a + C_t = C_a + \frac{C_\ell \, C_d}{C_\ell + C_d} \qquad (4)$$

By means of the calibration cell, adjusting the film thickness on three particular values of the thickness (Δh = 0, Δh = 10 μm, Δh ∞ the values of C_ℓ, C_d and C_p have been evaluated for various temperatures as shown on figure 5. It is found that the "parasitic" capacitance C_p is larger than C_d and C_ℓ but that their relative variations are of the same order of about 2 or 3 percent for a temperature variation of about 30°C. This can be attributed to dielectric constant variations in function of the temperature and also to differential thermal expansion of insulator and steel probe. Unless the transducer design is modified this inconvenience cannot be avoided but it must be taken into account in calibration procedure.

3.4. Calibration on the test cell and measurements

3.4.1. Calibration

The electrical environment of the test cell is not exactly the same as that of the calibration cell, in particular due to the earth rotating contact. So V-h curves have also been established in situ, in test conditions. For this, on the face of one of the carbon rotors of the studied seals, slots of known depth, from 1.5 to 15 μm, have been machined. The V-h calibration curves have been obtained, with a good repeatability by this means, showing linearity in conformity with relation (3) in the thickness range from 0 to 10 μm. By heating the hydraulic circuit it has been verified that the sensitivity A is not sensibly affected in oil by the temperature T and the variation of B in function of T have been established. Depending mainly on the probe diameter the value of A is of the order of -0.20 V/μm to -0.35 V/μm with a relative variation less than 5 percent in the range from 20°C to 60°C and the temperature drift of B is linearly dependant on T and of the order of 0.05 V/°C to 0.1 V/°C. For each transducer the values of A and B are then included in the microcomputer programs of measurement acquisition and treatment.

3.4.2. Measurements and results

The thickness measurements have been performed for experimental studies which have been the subject of several papers [40, 61, 65]. Other measurements were simultaneously carried out ; vibrations of the floating stator, friction torque, face temperature, leakage. Wear of the rotor face was examined after each series of tests. Only few examples of thickness measurements are presented and discussed in this paper in order to illustrate the capabilities of the measurement technique. These results concern with tests of seals equiped with three different carbon rotors presenting different face misalignments (characterized by the parameter z) and different face radial tapers (described by the angle β) (figure 6 - Table 1). The same closing force was applied to the three seals ;

$$F_o = (37 + 833 p_s) \text{ in N}$$

where p_s is the sealant pressure in MPa. Test were conducted for various rotational speeds N and pressures p_s.

For low misalignment conditions, case 1 and 2, the measured displacements of the floatting stator show a typical "dynamic tracking" ; the stator misalignment follows the rotor's one with the same amplitude at the shaft rotational frequency. Figure 7 shows an example of the thickness measurement for case 2. All the measurements performed for cases 1 and 2 are similar. Figure 8 shows the varitions of the measured film thickness for different values of the rotational speed and of the sealant pressure. It is found that :

- As the capacitive probe is set at the mean radius and the faces are tapered, the measured thickness is greater than that at the outer radius. Thickness values on the figure 8-a are small, of the same order of the thickness variation due to the rotor radial taper between outer radius and probe location.
- The signal of figure 7 indicates that the film thickness remains constant during shaft rotation and that the faces are parallel.
- The measured friction torque is quasi independ of both the rotational speed and the sealant pressure.
- For case 1 and 2 no leakage was observed.

From these observations we can conclude that faces are in contact on the outer circumference. Thickness variations are due to thermal distortion of the carbon face leading to an accentuated taper angle β when speed and thence face temperature increases. For case 1, where the misalignment is very small, the influence of the sealant pressure is very slight. At a speed greater than 1200 rpm noise characteristic of rubbing contact appears. After test the rotor face is worn on the outer radius.

For case 2, where the misalignment is greater than for case 2 the influence of the sealant pressure is more sensible and for speed higher than 1200 rpm no noise neither rubbing contact occur. Although the faces are in contact on the external circumference, fluid pressure generate opening force in the tapered gap on the inner side filled with oil. This hydrodynamic pressure is opposed to the closing force and contributes to a softer asperity contact on the outer zone.

For high misalignment condition, case 3, the amplitude of the stator vibrations is twice the rotor deflection and their frequency is half that of shaft rotation. In this case the signal delivered by the capacitive measurement system presents typical peaks as shown on figure 9. By comparing the floating stator oscillations with the rotating misalignment of the rotor these peaks have been found to occur in zone where the seal faces are divergent and where the interfacial gap is increasing ; these are zones of phase changing. The recorded peaks are due to sudden change of the fluid dielectric constant when gas zones periodically replace the oil in front of the capacitive transducer. Such signal cannot be interpreted in terms of film thickness and does not give reliable measurements.

4. CONCLUSION

The review of the main thickness measurement techniques for lubricant film has shown that, for submicrometer range optical methods have to be employed, for thicknesses of the order of the micrometer impedance or capacitance methods provide good results, and for higher thicknesses inductance methods are quite efficient.

An application of a capacitive based technique of thickness measurement to a low pressure face seal has shown that the technique may be accurate under condition of careful calibration and permits to analyse in detail the lubrication mechanisms. In addition, capacitive transducer allows to localise cavitation zones.

REFERENCES

1. A.O. Lebeck, Parallel sliding load support in the mixed friction regime. Part 1 - The experimental data, Trans. of the ASME, Journal of Tribology, Vol. 109, p. 189-195, January 1987.
2. R.K. Flitney and B. Nau, Performance testing of mechanical seals, 13th Int. Conf. on Fluid Sealing (BHR Group), Brugge, Belgium, April 1992.
3 I. Sherrington and E.H. Smith, Experimental methods for measuring the oil-film thickness between the piston-rings and cylinder-wall of internal combustion engines, Triboloy International, Vol. 18, n° 6, p. 315-320, December 1985.
4. A.E. Smart and R.A.J. Ford, Measurement of thin liquid films by a fluorescent technique, Wear, 29, 41-47, 1979.
5. R.A.J. Ford and C.A. Foord, Laser based fluorescence techniques for measuring the liquid films, Wear, 51, p. 289-297, 1978.
6. L.L. Ting, Development of a laser fluorescence technique for measuring piston ring oil film thickness, Transactions of the ASME, Jour. of Lub. Tech., Vol. 102, p. 165-171, April 1980.
7. G. Poll, A. Gabelli, P.G. Binnington, and J. Qu, Dynamic mapping of rotary lip seal lubricant film by fluorescent image processing, 13th Int. Conf. on Fluid Sealing (BHR Group), Brugge, Belgium, April 1992.
8. C.R. Mc Clune and D. Tabor, An interferometric study of lubricated rotary face seals, Tribology International, 219-227, August 1978.
9. C.A. Foord, C. Hammann and A. Cameron, Evaluation of lubricants using optical elastohydrodynamics, ASLE Trans., 11, 31-43, 1968.
10. G. Dalmaz, L'hydrodynamique du contact sphère-plan, Thèse de Docteur-Ingénieur, Université C. Bernard, Lyon, Mai 1971.
11. A. Cameron and R. Gohar, Theoretical and experimental studies of the oil film in lubricated point contact, Proc. Roy. Soc., A 291, 520, 1966.
12. J.M. Fleisher and C. Lin, Infrared laser interferometer for measuring air-bearing separation, IBM Journal of research and development, Vol. 18, n° 6, p. 529-533, 1974.
13. Y. Mitsuya and T. Ohkubo, High kundsen number molecular rarefaction effects in gas-lubricated slider bearings for computer flying heads, Transactions of the ASME, Journal of Tribology, Vol. 109, 276-282, April 1987.
14. C. Lin and R.F. Sullivan, An application of white light interferometry in thin film measurements, I.B.M. J. of Res. Develp., May 1972.
15. F.J. Westlake and A. Cameron, A study of ultra-thin lubricant films using an optical technique, Proc. Instn. Mech., Vol. 182, p. 75-78, 1967-68.
16. Y. Wakuri, S. Ono, M. Soejima and K. Masuda, Oil-film behaviour of reciprocating slider with circular profile (optical measurement of oil-film separation boundary), Bulletin of the JSME, Vol. 24,

n° 194, August 1981.
17. A. Sereny and V. Castelli, Experimental investigation of slider gas bearings with ultra-thin films, Transactions of the ASME, Journal of Lub. Tech., Vol. 101, p. 510-515, October 1979.
18. G.A. Domoto and Y.T. Hsia, An experimental investigation of molecular rarefaction effects in gas lubricated bearings at ultra-low clearances, Transactions of the ASME, Journal of Lub. Tech., Vol. 105, p. 120-129, January 1983.
19. T. Ohkubo and J. Kishigami, Accurate measurement of gas-lubricated slider bearing separation using visible laser interferometry, Transactions of the ASME Journal of Tribology, Vol. 110, p. 148-155, January 1988.
20. A. Nigam, A visible laser interferometer for air bearing separation measurement to submicron accuracy, Transactions of the ASME, Journal of Lub. Tech., Vol. 104, p. 60-65, January 1982.
21. F.K. Orcutt, An investigation of the operation and failure of mechanical face seals, Transactions of the ASME, Journal of Lub. Tech., p. 713-725, October 1969.
22. V.M. Lieskovsky, Optical interference techniques for spacing measurements in the 1 to 6 micron range, Mesucora 63. Congres. International, Paris, France 1963.
23. K. Tanaka, T. Odaka, Y. Takeuchi, and H. Daito, Experimental examination of numerically calculated flying height considering second-order slip flow at ultra thin spacing and low ambient pressure for magnetic disk head slider, Proc. 5th Int. Cong. on Tribology. Eurotrib 89, Vol. 3, p. 240-245, 1989.
24. M. Lépine, Etude des conditions de lubrification d'un joint d'étanchéité à faces radiales : détermination de la pression interfaciale et de l'épaisseur du film, Thèse de Doctorat de l'Université de Poitiers, Mai 1990.
25. B.S. Nau, Film cavitation observations in face seals, 4th Int. Conf. on fluid sealing, BHRA, session 5, p. 190-228, 1969.
26. B.S. Nau, An investigation into the nature of the interface film, the pressure generation mechanism and centripetal pumping in mechanical seals, BHRA, report RR 754, January 1963.
27. L.N. Lacey, G.H. Kelsall, and H.A. Spires, Thick antiwear films in elastohydrodynamic contacts, Part. 1 : film growth in rolling/sliding. EHD contacts, ASME Trans. Vol. 29, 3, p. 299-305, 1986.
28. O. Isaksson, Measurement of the influence of sliding velocity on oil film thickness in an elastohydrodynamic point contact, Proc. 5th Int. Cong. on Tribology. Eurotrib 89, Vol. 2, p. 403-408, 1989.
29. J.N. Anno, Hydrodynamic film-thickness determination in face-seal, Lubrication Engineering, Jour. of ASLE, March 1975, p. 132-135.
30. M.S. Filowitz, W.H. King and J.K. Appeldoorn, Oil film thickness in a bearing of a fired engine, SAE, Trans., 82.05.11, 1982.
31. J.A. Spearot, C.K. Murphy and R.C. Rosenberg, Measuring the effect of oil viscosity on oil film thickness in engine Journal Bearings, SAE Trans., 83.16.89, 1983.
32. T. Ishii, S. Miyagawa and I. Taguma, Research and development on the hydrodynamic pressure generation between contacting flat surfaces. ASLE Trans., Vol. 29, 3, 339-346, 1986.
33. M. Suzuki, K.C. Ludema, The wear

process during "running-in" of steel in lubricated sliding, Trans. ASME, Journ. of Tribology, October 1987, Vol. 109, 587.
34. S. Furuhama and T. Sumi, A dynamic theory of piston-ring lubrication, 3rd report : measurement of oil-film thickness, Bull. of ASME, 1961, 4(16), 744.
35. T.A. Dow, C.A. Schiele and R.D. Stockwell, Tecnique for experimental evaluation of piston-ring/cylinder film thickness, Trans. ASME, J. Lub. Technology, 1983, 105, 353.
36. A. Dyson, Investigation of the discharge-voltage method of measuring the thickness of oil films formed in a disc machine under conditions of elasto-hydrodynamic lubrication, Proc. Inst. Engrs., Vol. 181, part 1, n° 23, p. 633-649, 1966-67.
37. M.G. Lewis, A miniature mutual inductive proximity transducer, J. Phys. E., 1974, 7, 269.
38. R.D. Wing and O. Saunders, Oil-film temperature and thickness measurements on the piston-rings of a Diesel Engine, Proc. I. Mech. E., 1972, 186, 1.
39. R. Haardt, Les joints d'étanchéité à faces radiales : les effets transitoires introduits en lubrification hydrodynamique par leur mésalignement, Thèse de Docteur Ingénieur, Université C. Bernard, Lyon, 1975.
40. J. Digard de Cuisard, M. Gentile, M. Bouchon, B. Tournerie and J. Frêne, Experimental study on lubrication modes in low pressure mechanical face seals, Proc. 11th International Conference on fluid sealing, BHRA., paper F2, 1987.
41. J. Frêne, Contribution à l'étude des régimes de transition dans les paliers lisses, Thèse Docteur-Ingénieur. Faculté des Sciences, Université Lyon, 1970.
42. D. Nicolas, Les régimes non laminaires en lubrification. Réduction du frottement par addition de polymères, Thèse d'Etat, INSA et Université Claude Bernard, Lyon I, 1979.
43. J. Ferron, Contribution à l'étude des phénomènes thermiques dans les paliers hydrodynamiques, Thèse de Docteur de 3ème cycle. Université Poitiers, 1982.
44. J.L. Charron, Measurement of oil film thickness in diesel motor main bearings : implementation of the method and experimental results, Proc. of the 9th Leeds-Lyon Symposium, Sept. 1982.
45. G. Poll and A. Gabelli, Formation of lubricant film in rotary sealing contacts-part II : A new measuring principle for lubricant film thickness, ASME/STLE Tribology Conference, Toronto, Canada, October 1990.
46. T.W. Bates and P.G. Evans, Effects of oil rheology on Journal bearing performances : Part 1 - Instrumentation of the big-end bearing of a fired engine, Proc. JSLE, Int. Tribology Conference, Tokyo, July 1985.
47. T.W. Bates, B. Fantino, L. Launay and J. Frêne, Oil film thickness in an elastic connecting-rod bearing : comparison between theory and experiment, STLE Tribology Transactions, 33, 2, p. 254-266, 1990.
48. D.F.Denny, Some measurement offluid pressures between plane parallel thrust surfaces with special reference to radial-face seals, Wear, 4, 64-83, 1961.
49. S. Furuhama, C. Asahi and M. Hiruma, Measurement of piston ring oil film thickness in an operating engine, STLE Trans., Vol. 26, 3, 325-332, 1983.
50. G.M. Hamilton and S.L. Moore,

The lubrication of piston rings-First paper-Measurement of the oil-film thickness between the piston rings and liner of a small diesel engine, Proc. Instn. Mech. Engrs, Vol. 188, 20/74, 1974.
51. S.R. Brown and G.M. Hamilton, The partially lubricated piston ring, Journal of Mechanical Engineering Science, Vol. 19, n° 2, p. 81-89, 1977.
52. G.M. Hamilton, The hydrodynamics of a cam follower, Tribology Int., June 1980, p. 113-119.
53. S.L. Moore, G.M. Hamilton, The starved lubrication of piston rings in a diesel engine, Journal of Mechanical Engineering Science, Vol. 20,n° 6, p. 345-352, 1978.
54. T.G. Doust and A. Parmar, An experimental and theoretical study of pressure and thermal distortions in a Mechanical Seal, ASLE Trans., Vol. 29, 2, 151-159, 1981.
55. T.G. Doust and A. Parmar, Hydrostatic effects in a mechanical face seal, ASLE Trans., Vol. 29, 4, 467-472, 1986.
56. T.G. Doust and Parmar A., Transient thermoelastic effects in a mechanical face seal, Proc. 11th, Int. Conf. on Fluid sealing BHRA, paper F4, p. 407-422, 1987.
57. M.S. Ozogan and C.M. Taylor, Running in of cylindrical rollen bearings with axial loading, Proc. of Leeds-Lyon Symposium on Tribology, Leeds, England, Sept. 1984.
58. Y. Haiqing, A. Xiaolan and D. Fupin, Experimental measurement of unsteady elastohydrodynamic lubrication film thickness between cam and tappet, Proceeding of 5th Int. Cong. on Tribology, Eurotrib 89, Vol. 2, Helsinki 1989.
59. K. Shin, Y. Tateishi and S. Furuhama, Measurement of oil-film thickness between piston-ring and cylinder, SAE Paper 83.00.68 (1983).
60. N. Grice, I. Sherrington and E.H. Smith, Measurements of the circumferential distribution of oil-film thickness around the piston-rings of a reciprocating piston apparatus, Proc. 5th Int. Cong. on Tribology Eurotrib 89, Vol. 4, p. 192-197, 1989.
61. B. Tournerie and M. Lépine, Experimental study on Factors influencing the Lubrication mode of radial face seals, Proc. 5th Int. Cong. on Tribology, Eurotrib 89, Vol. 4, p. 353-358, 1989.
62. M. Lépine and B. Tournerie, Contribution à l'étude du comportement des films minces lubrifiants ; pression et épaisseur dans les cas des joints à faces radiales, Actes 9ème Congrès Français de Mécanique. Tome 1, p. 222-223, 1989.
63. B.A. Batch and E.H. Iny, Pressure generation in radial face-seals, 2nd Int. Conf. on fluid sealing, BHRA, paper F4, April 1964.
64. B.A. Stanghan-Batch, Face lubrication in mechanical seal, Proc. 9th Tribol. Conv., paper C59/71, Inst. Mech. Engrs, 1971.
65. B. Tournerie, M. Lépine, and J. Frêne, Experimental analysis of the optimum lubrication conditions for mechanical seals, 13th Int. Conf. on Fluid Sealing, BHR Group, Brugge, Belgium, April 1992.

Cases	Misalignment Z (μm)	Taper β (μm/mm)
1	1	2,5
2	4,5	2,5
3	25	4,5

Table 1 : Misalignments and taper angle of the carbon rotors at an homogeneous temperature of 80° C

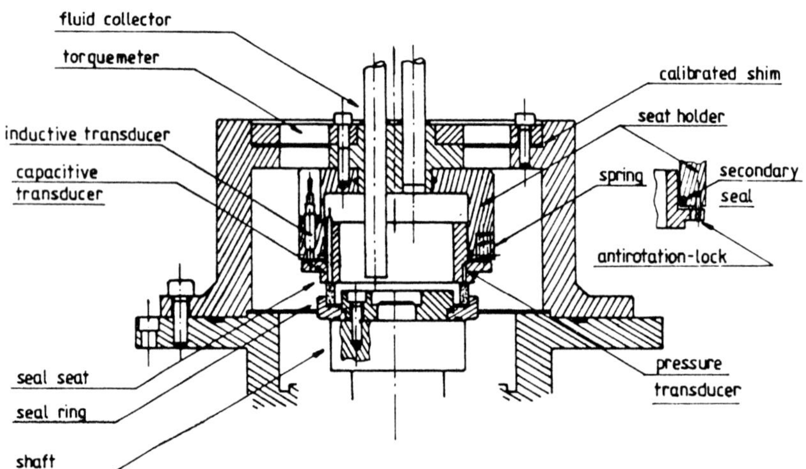

Figure 1 : Experimental seal equiped with electrical transducers

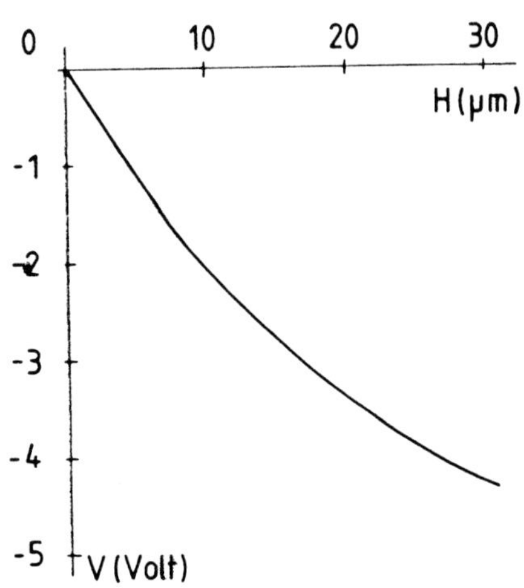

Figure 2 : Calibration curve at the ambiant temperature in oil.

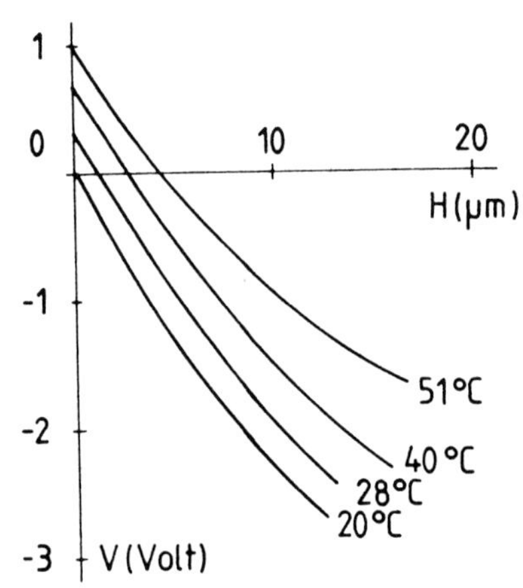

Figure 3 : Calibration curves for various temperatures in air.

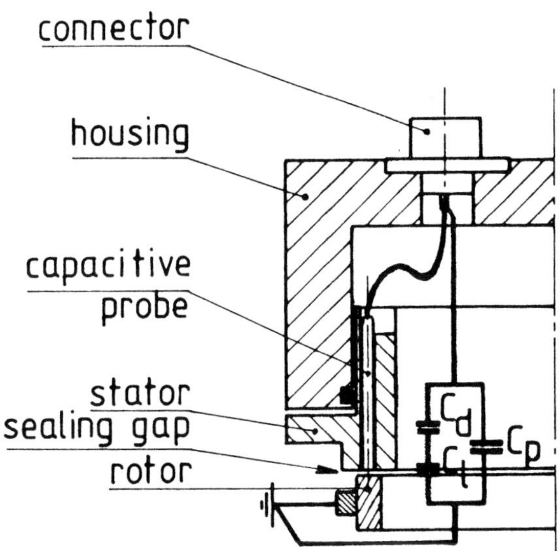

Figure 4 : Equivalent circuit of the capacitance transducer.

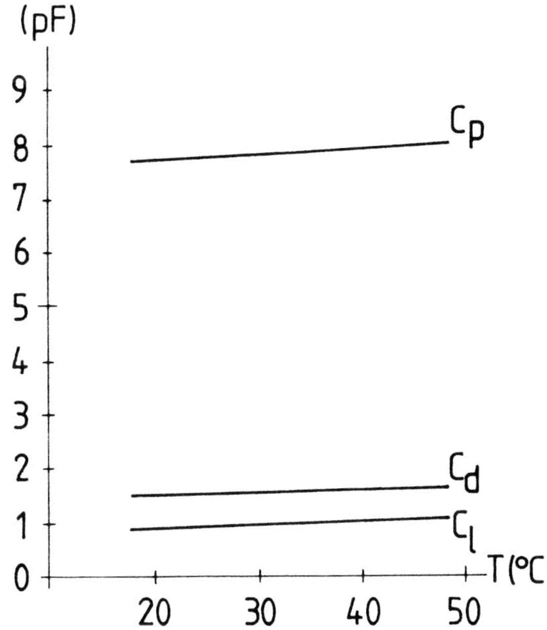

Figure 5 :Measured values of the capacitances C_ℓ , C_p and C_d for various temperatures.

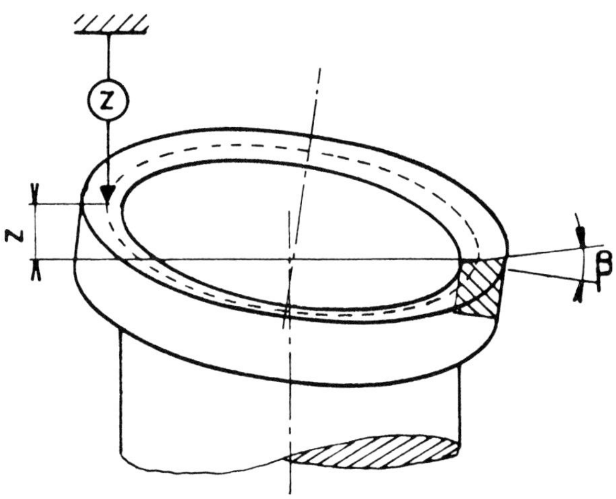

Figure 6 : Misalignment and cone angle of the rotor face.

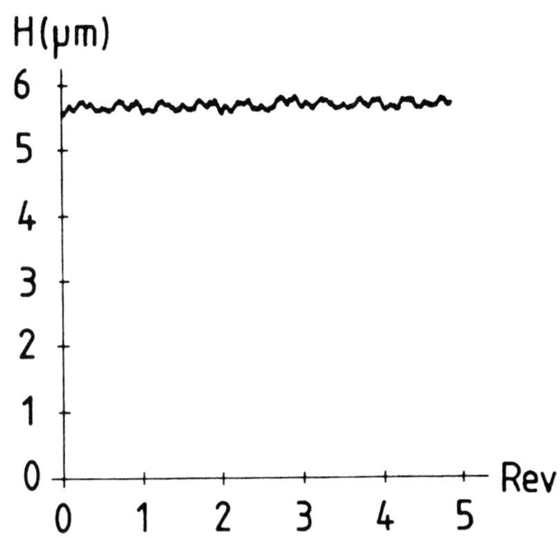

Figure 7 : Film thickness measurement (case 2).

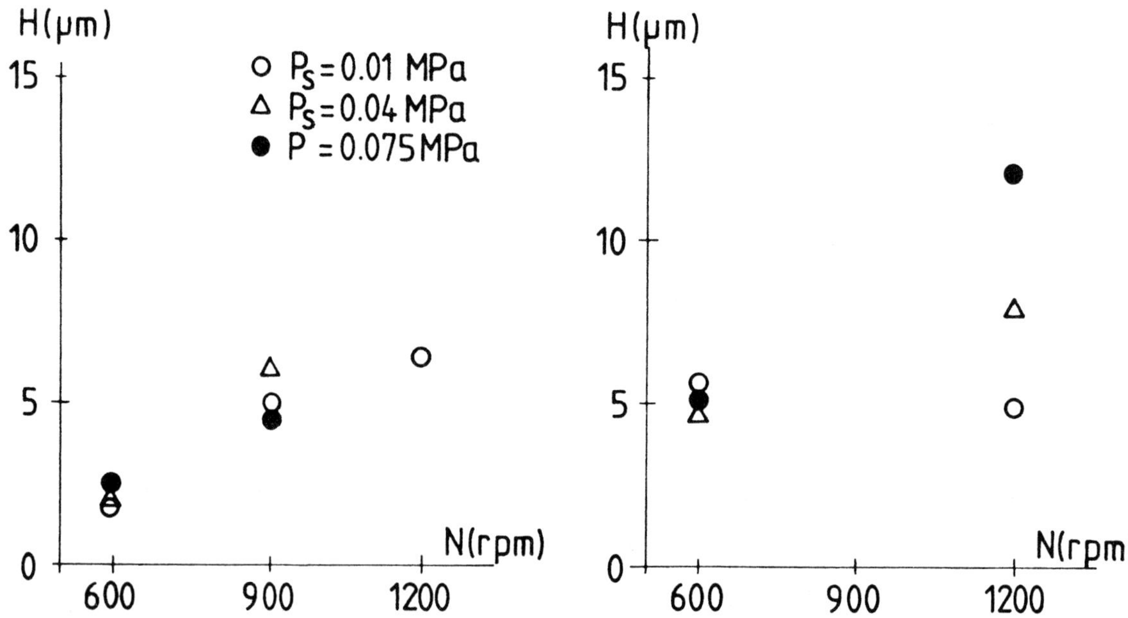

8-a ; case 1

8-b ; case 2

Figure 8 : Film thickness variation.

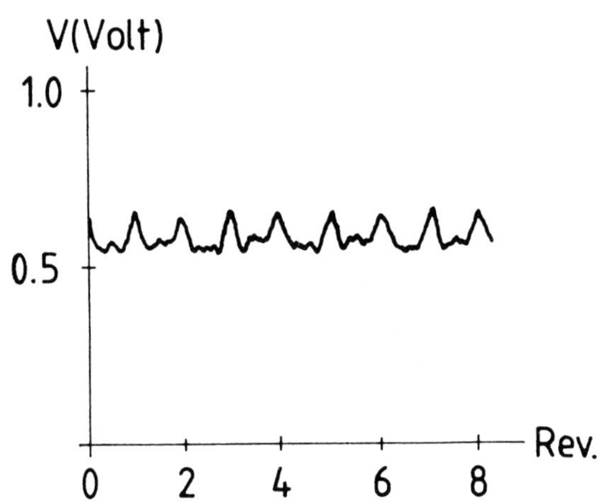

Figure 9 : Signal delivered by the measuring system for case 3.

Thin Films in Tribology / D. Dowson et al. (Editors)
1993 Elsevier Science Publishers B.V.

Lubricating of O-Ring seals in pneumatic applications with ultra thin grease films

V. PICHON*, P. BOUCHAYER** and M. GODET***

* Chercheur CIFRE / CPOAC détaché au L.M.C
** Ingénieur CPOAC
*** Laboratoire de Mécanique des Contacts
URA CNRS 866
Institut National des Sciences Appliquées de Lyon
69621 Villeurbanne

Thin grease films are known to be long lived in pneumatic applications. However little quantitive data concerning film thickness and life exists.

Experimental results on grease feeding conditions, film thickness and life are given in this paper. Grease film thickness beyond 0.3 micrometer are shown to be very effective.

1. Introduction

This paper presents the results of a study on the performance of an O-ring mounted on the piston of a pneumatic valve with 5 openings and two static positions. The piston is driven forth by air pressure which acts on the left end of the piston (fig.1) and is recalled by a spring.

The valve operating conditions depend on the application :

- the pressure varies between 0.2 and 1.2 MPa,
- the piston frequency varies between 0 and 2 Hz.
- piston velocity which depends also on pressure varies between 0 and 2 m/s.
- life should exceed 20 million cycles

As the pressure which acts on the piston is variable, the friction induced by the O-ring must be as low and as constant as possible to guarantee satisfactory valve performance. Low friction values are obtained by greasing the seals and optimising the seal housing geometry.

O-rings are mounted either (fig.2) :

- in extension: the seal is stretched as the piston groove diameter is larger than the seal internal diameter,
- floating: the seal external diameter is larger than the chamber or sleeve internal diameter but there is no contact between seal and piston groove.

Satisfactory valve operation is thus a compromise between sealing, friction and life. The internal and external dimensions of the O-ring studied here are respectively 15.1 and 20.5 mm. The tore diameter is thus 2.7 mm. An experimental simulation (fig.3) was set-up to visualise seal operation.

2. Experiments

2.1. The Taguchi method

Complete seal life tests take 4 months as 20 million cycles must be run at a nominal frequency of 2 Hz. Shorter tests must therefore be set up to:

- identify the parameters which govern the sealing and friction compromise,
- determine the minimum grease film thickness necessary to guarantee low friction and low wear.

As the number of parameters which govern the compromise is large, the method of experimental design, developed by Taguchi, was used.

The Taguchi method limits the number of tests necessary to identify the governing parameters, the role of each of these parameters, and the interaction between parameters in the control of the phenomenon studied (fig.4). The method consists in developing a model of the phenomenon under study (friction, wear or life in this case) which is later validated by comparing model predictions to experimental results with a minimum amount of tests. A detailed description of the method is found in references [1] [2].

A global model is set up by first establishing empirically a list of possible governing parameters, and a number of conditions required to describe each parameter. A table of tests to be run is established from that list of parameters and conditions. The mean effect of each parameter on the phenomenon studied is calculated from the test results and, once combined according to the Taguchi model, these effects determine the global model response for each test. The residue, which is equal to the difference between the test result and the predicted model response, can be calculated, and a variance analysis set up to identify the governing parameters of the phenomenon. This analysis compares the variance of the phenomenon (due to uncontrolled factors) calculated from the residue, and the variance introduced by the individual parameters themselves. If the variance introduced by a given parameter is larger than the variance of the phenomenon, that parameter is considered to be influent.

The table of tests to be run can be generated :

- along classical lines (complete plan or matrix approach): tests are run for all possible parameter combinations. Individual or combined effects of parameters are determined but the time and effort necessary is considerable.
- following Taguchi (partial plan): here, only specific tests are run. Tests are limited in number if certain interactions between parameters are assumed , à priori, to be negligible.

The definition of a Taguchi, or partial plan, must verify a certain number of conditions including that of orthogonality which supposes that, during the variation of one parameter, the number of variations of the other parameters are equal. For example, figure 7 which is representative of a partial plan, shows that when the contact parameter is "dry" (not lubricated) or "greased" (lubricated) 8 tests are run for both sleeve materials (anodised aluminium and P.P.M.A.). This guarantees that the respective influence of each factor is correctly estimated and also that the test plan can be broken up, if necessary, into sub-plans.

2.2. Test devices and measurements

Two experimental devices were built :

- a quasi-static bench to measure grease film thickness,
- a dynamic bench to measure friction.

Film thickness is measured by weighing the amount of grease on the sleeve. Reproducible results were obtained with a high precision scale (± 0,1 mg). Uniform grease distribution was achieved using a short piston drawn manually once down the sleeve (fig.5).

Friction is measured on a bench (fig.6) which simulates valve operation. A dynamometer is placed between the hydraulic actuator and the piston. Variable speed and variable length tests were conducted. Transparent sleeves can be mounted on the bench for visualisation.

2.3. First test program

The list of possible governing parameters established empirically, and their variations, includes :

- contact nature : dry and grease lubricated,
- sleeve material : anodised aluminium and P.P.M.A.,
- seal mounting : minimum and maximum extension, minimum and maximum floating,

- seal material : soft elastomer with a Shore hardness of 55 and standard elastomer with a Shore hardness of 70,
- sliding speed : 0.01 and 0.1 m/s,
- pressure : 0.2 and 0.8 MPa.

The phenomena studied are:
- static sealing
- dynamic sealing
- friction

As the Taguchi method exploitation is based on numbers and not on words, the values 1 and 0 are given respectively for effective sealing and for leaking seals. Friction was measured after 100 cycles when steady-state conditions were reached. The value retained is an average of the values measured during this stroke which vary by ± 10% with respect to that average.

2.4. Methodology and Results

A partial experimental plan, limited to 32 tests instead of the 128 that the matrix method would have required, was set up using the Taguchi standard table L_{32} (fig.7). This plan presupposes that interactions other than the following are negligible : contact nature and mounting, seal material and mounting, seal material and pressure.

Tests results (fig.7) indicate that the three phenomena studied (static sealing, dynamic sealing and friction) behave quite differently under dry and lubricated conditions. Columns 2 and 3 show that under both static and dynamic conditions no leakage was noted under greased conditions while leaks were observed under dry dynamic running conditions. Visualisation tests showed that, in these runs, leakage followed O-ring torsion in its groove (fig.8) which resulted in local strictions and sealing loss. The change in friction behaviour from dry to greased conditions was so important that it overshadowed other effects which could not be analysed properly. Figure 9 shows that for an average value of 130 N, friction ranges from 170 to 20 N from dry to lubricated conditions.

The Taguchi partial plan was thus divided into two sub-plans, one for the dry, the other for the greased conditions. The analysis of the "dry" sub-plan leads to a high value of the residue which lessened the impact of the conclusions. This was related to the dispersion (fig.10) in the experimental results of the "dry" sub-plan. Dispersion indicates that one or more factors, which are not identified, have a weight equal to, or larger than, those included in the Taguchi analysis. Work on the "dry" conditions was therefore abandoned, friction forces of over 150 N were too large and the cases of sealing loss under dynamic conditions (70 %) too frequent.

The analysis of the "greased" sub-plan gave low residue and conclusions could therefore be drawn from it. Three "governing" parameters were identified: pressure, seal material and mounting (fig.11). A supplementary test was run to validate the model , friction results were within 3% of the model predicted value which was satisfactory. The tendencies predicted by the model agree with those commonly accepted in practice (fig.12).

3. Grease film thickness

3.1. Governing parameters

The results of the first series of tests confirmed that pneumatic valves could not operate satisfactorily without grease. They also identified three governing parameters: pressure, material seal and mounting, suggesting that in future tests the other parameters could be kept constant without loss of generality. It was therefore possible to investigate the role of grease film thickness, and pressure, on friction and sealing by varying only these three parameters. These parameters were reduced to one: the 70 Sh elastomer and the floating mount were chosen as being closer to applications The other fixed parameters were:

- sleeve material: P.P.M.A. was chosen for visualisation
- velocity: the low velocity of 0,01 m/s was chosen to facilitate visualisation

The second test program was therefore limited to the investigation of the effects of pressure and initial grease film thickness on friction and "working" grease film thickness. The "working" film thickness is the thickness of the grease film left on the sleeve after having been partly scraped off during one stroke of the piston equipped with O-rings. It is called "working thickness", as it serves to lubricate the next piston (or seal) stroke. Working thicknesses were measured on the quasi-static bench described above. The initial film thickness could be controlled in the same way. Early tests run without pressure show (fig.13) that the grease film thickness drops rapidly during the first few cycles.

Equivalent tests cannot be run on the dynamic bench, on which friction can be measured, as grease rims (fig.14) are built up at each end of the stroke so the working thickness cannot be determined by weighing.

3.2. Test programs and results

Two test programs were run :

1. the friction force was determined on the dynamic bench for different pressure and initial film thickness values,
2. the effect of pressure and initial film thickness on the working thickness were determined on the quasi-static bench.

This breakdown, which is imposed by the measuring techniques used, assumes that the results of both plans can be related and that it will be possible to understand the influence of the working film thickness on friction. Figure 15 shows these plans and the tests results.

Results

Friction data analysis shows that both pressure and the interaction between pressure and initial film thickness are governing parameters. It also shows that between 0.6 and 0.9 micrometers, the influence of the original film thickness is negligible on friction (fig.16-2). The interaction between pressure and initial film thickness is important only when these parameters are respectively 0.8 MPa and 0.3 micrometers (fig.16-4).

Working film thickness data analysis show that both pressure and initial film thickness are governing parameters but that the interaction between the two is negligible (fig. 17).

The liaison between the two sets of results is presented in figure 18. In figure 18-1, the solid line gives the variation of working film thickness with pressure and the dashed line gives the variation of friction with pressure. These variations are understandable because :

- when the pressure is low, the working film thickness is maximum (0.4 µm) and the friction minimal (7 N).
- when the pressure increases gradually, the working film thickness drops and friction increases.

Figure 18-2 can be explained in the same manner :

- when the initial film thickness is minimal (0.3 µm), the working thickness is minimal (0.2 µm) and friction is maximum (18 N).
- when the initial film thickness increases, the working film thickness increases slightly and friction drops slightly also.

Figure 18-3 is more difficult to explain even though the reasoning is basically similar. The solid lines give the variation of working thickness with initial film thickness for different pressure values and the dashed lines give the variation of friction with initial film thickness for the same pressure values. That figure summarises clearly the analyses of both experimental plans and can be explained as follows:

- when the pressure is zero (lines with open circles), the working film thickness which varies between 0.3 and 0.4 µm is maximum, and friction which varies between 5 and 10 N is fairly constant and low.
- when the pressure increases and reaches 2 bars (lines with triangular symbols) the working film thickness which varies between 0.3 and 0.2 µm drops, and friction which varies between 10 and 12 N, increases slightly.

- when the pressure reaches its maximum of 8 bars (lines with + signs), the working thickness continues to drop slightly between 0.2 and 0.15 µm, and then friction increases suddenly. Friction is fairly constant (20 N) for initial film thicknesses between 0.5 and 0.9 µm which lead to "working" film thicknesses larger than 0.2 µm but it rises sharply when for initial film thicknesses of 0.3 µm, "working" film thicknesses of 0.15 µm are achieved.

Results show that a threshold is reached when the "working" film thickness reaches 0.2 µm when friction goes up significantly by as much as 75% to values beyond 20 N, for two seals, which is not acceptable industrially.

This study has identified a limiting running condition (0.8 MPa, 0.3 µm) which will serve for the study of the seal life.

3.3. Seal life

Seal life was studied on the dynamic bench operating under the following conditions :

- floating mount
- P.P.M.A. sleeve
- 70 Sh elastomeric seal
- pressure: 0.8 MPa
- initial film thickness : 0.3 µm.

Friction is measured and sealing was controlled every 200 cycles. During the test which lasted for 5000 cycles, loss of sealing efficiency was observed on the lower seal when it twisted in its groove. This generated a non symmetrical friction cycle as it only happened on the piston down stroke (fig.19)

This shows that as the grease film is scraped off, the conditions are close to those observed under dry friction. Further tests are needed to explain the variation of friction with time and to understand the role played by the grease rims created at both ends of the stroke even with such low initial film thickness.

Conclusion

Conditions, which control O-ring sealing efficiency and friction in pneumatic valves, was studied experimentally. Because of the great number of parameters which control the phenomena, the Taguchi experimental design method was used. Results showed that :

- O-rings cannot be used "dry" as leakage was observed in 70 % of the cases and friction reached 150 N for two seals.
- greased seals gave a better performance. No leakage was observed and friction was maintained around 15 N.
- For greased seals, friction was governed by :
  - pressure : an increase from 0.2 to 0.8 MPa causes an increase in friction of 35 %,
  - seal material : an increase in hardness from 55 to 70 Sh causes an increase in friction of 20 %,
  - seal mount: a change from an extension mount to a floating mount causes a drop in friction of 20 %.

Sleeve material and speed, within the conditions tested, did not affect friction.

This first set of experiments was followed by a second in which the effect of grease thickness on both sealing efficiency and friction was studied. In this program, the sleeve was made out of P.P.M.A. to able visualisation, the velocity was 0.01 m/s, 70 Sh seals were mounted floating. The relation between friction and "working" film thickness (or film thickness left after most of the original grease has been scraped off in the first few cycles) was studied as a function of original grease film thickness and pressure which were varied respectively from 0.3 to 0.9 µm and from 0 to 0.8 MPa. Satisfactory results for both sealing efficiency and friction were obtained with very thin "working" films (0.2/0.4 µm) but a threshold value of 0.2 µm was noted below which friction increased significantly.

Life tests were run under severe operating conditions: a "working" film thickness of 0.2 µm and a pressure of 0.8 MPa. In time, friction increased and seal twisting, observed only under dry conditions, was also noted here after 5000 cycles. No steady state conditions were reached in this test.

Acknowledgements

The authors wish to thank the Compagnie Parisienne d'Outillage à Air Comprimé groupe BOSCH for their financial support , and for the interest shown all along this program. They also wish to thank Mr. PERROT, technician in the Laboratoire de Mécanique des Contacts of INSA, and Mr. DERRIEN, C.P.O.A.C. technician, for their technical assistance particularly in setting up the experimental devices.

References

1. G. TAGUCHI, Introduction to quality engineering, Tokyo : Asian Productivity Organization, 1986, 189 p.
2. M. PILLET, Introduction aux plans d'experiences par la methode Taguchi, Paris : Les Editions d'Organisation, 1992, 224 p.

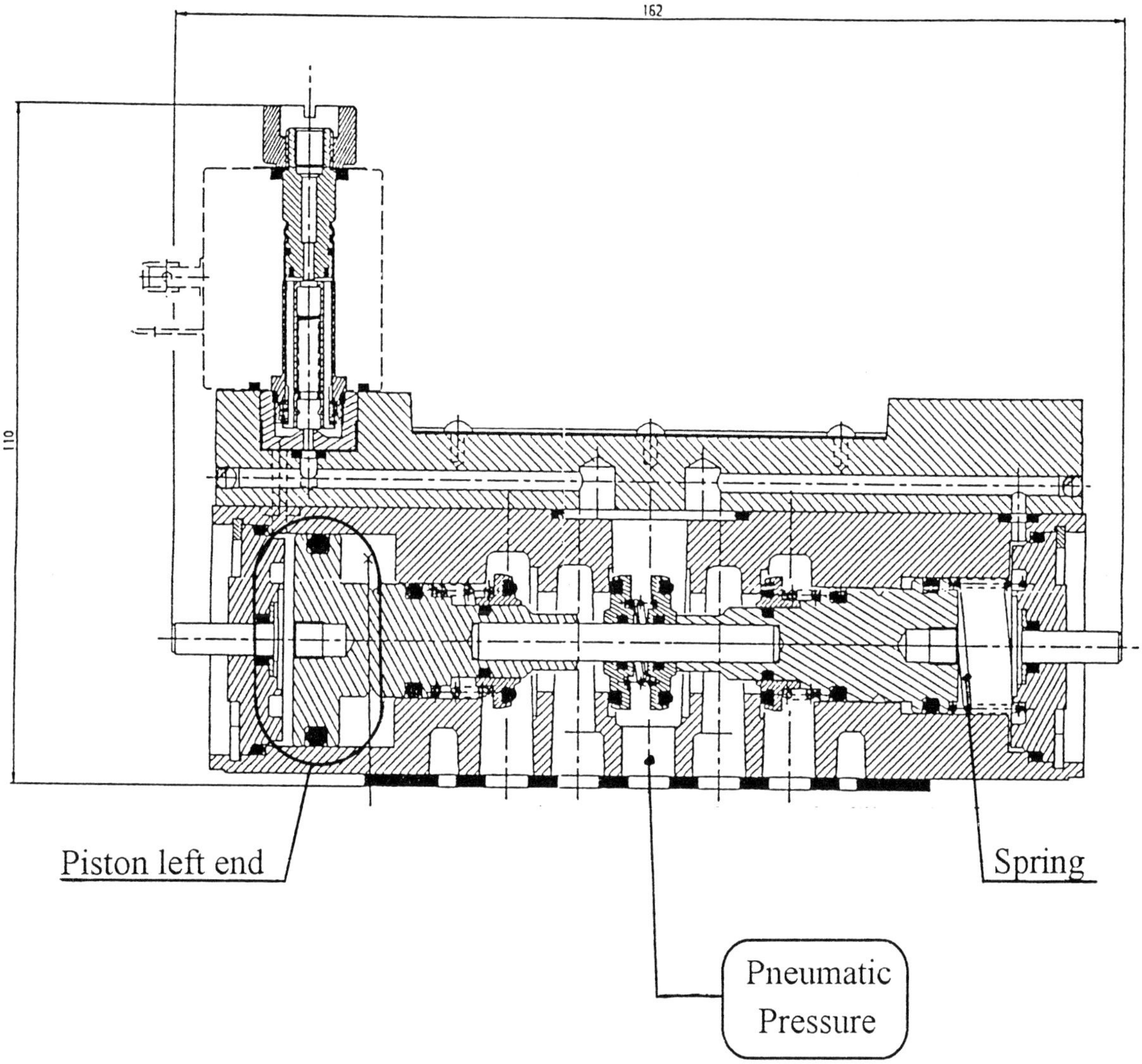

Figure 1 : Pneumatic valve

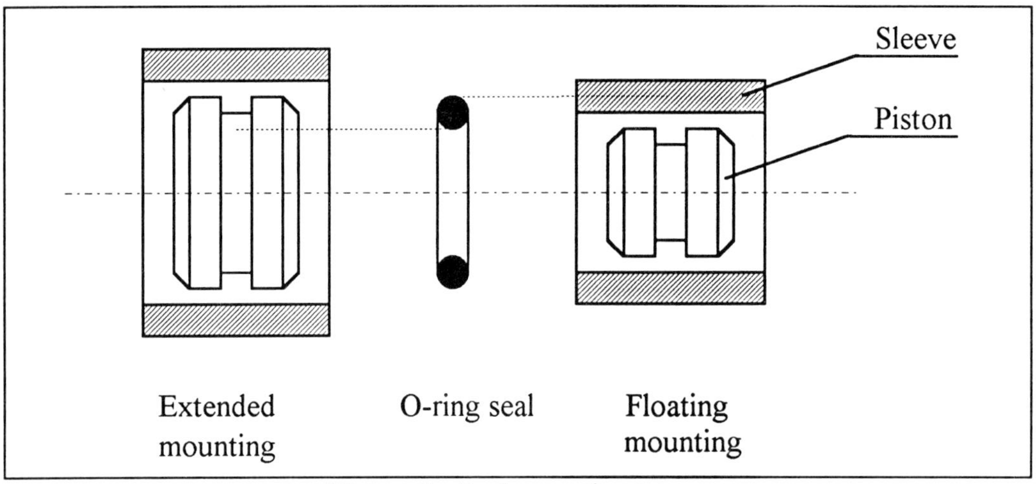

Figure 2 : Seals mouting types

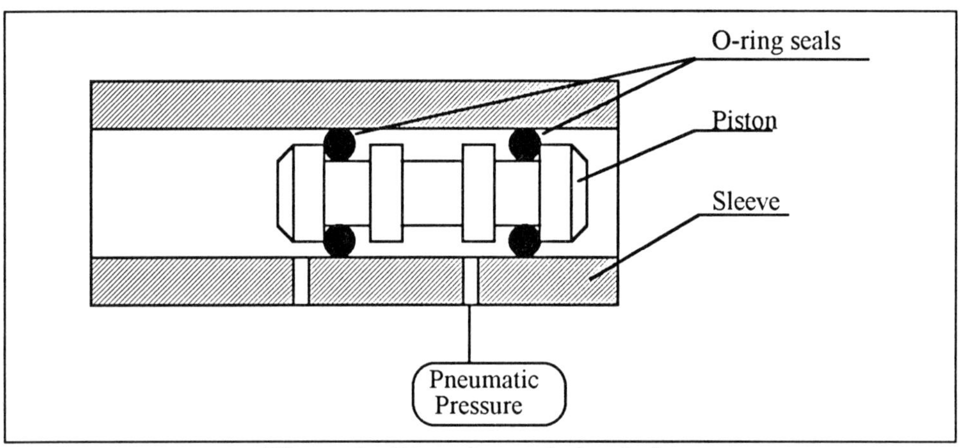

Figure 3 : Experimental simulation

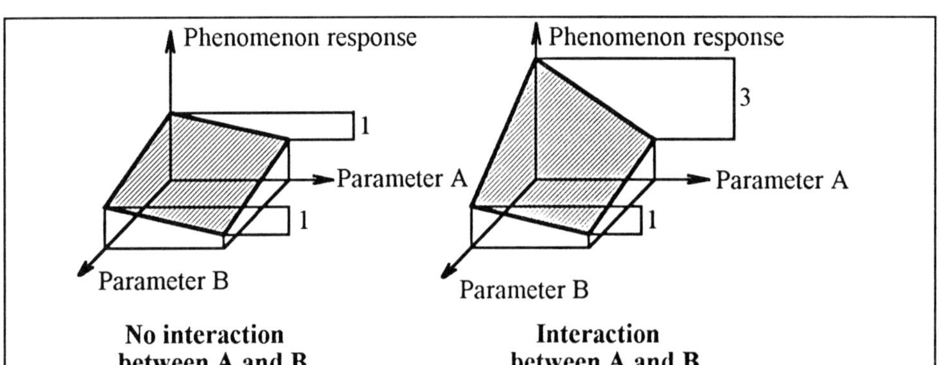

Figure 4 : Interaction between parameters A and B

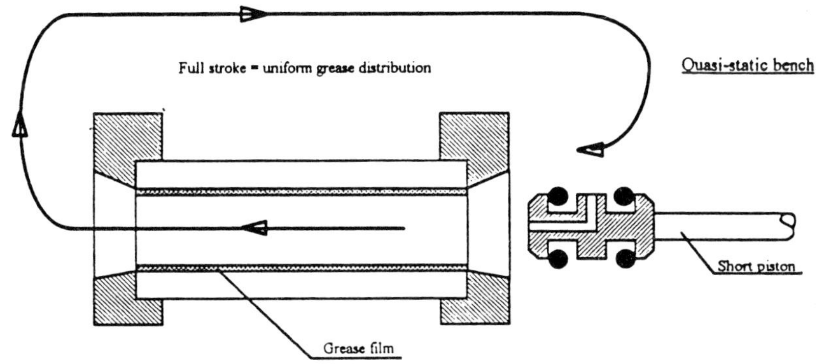

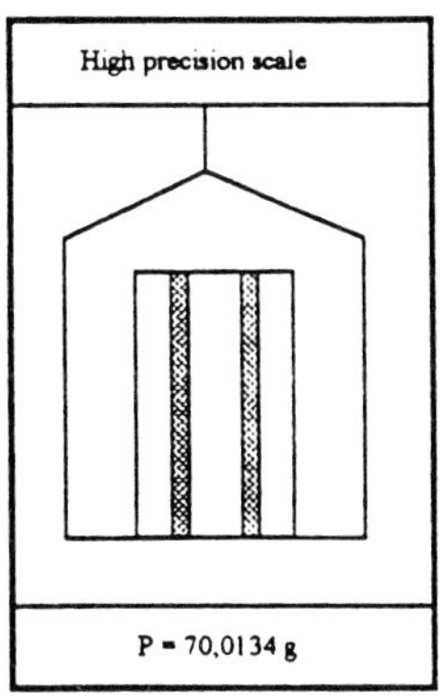

Formula :

$$h = \frac{\Delta p}{\rho S}$$

Δp : grease weight
ρ : grease volumic mass
S : internal sleeve surface

Figure 5 : Grease film thickness measurement

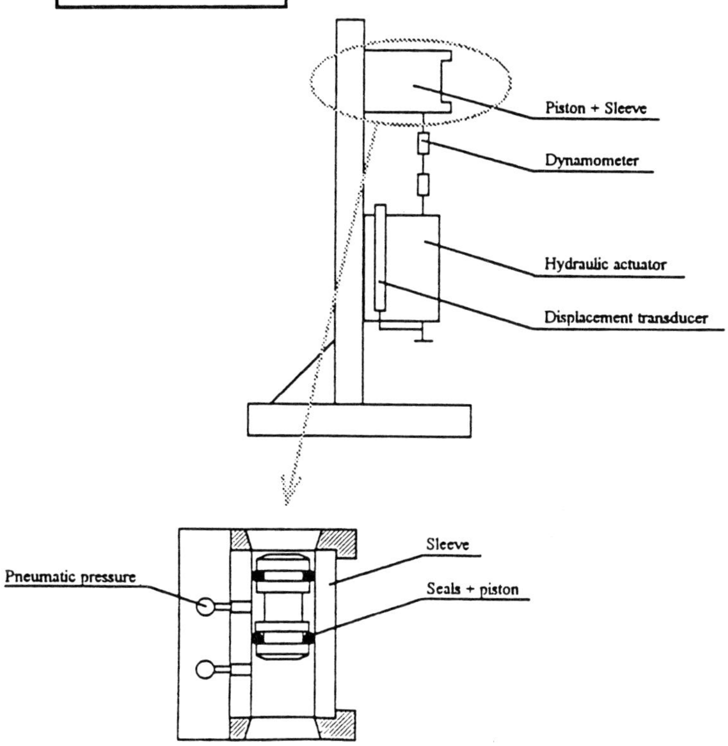

Figure 6 : Dynamic bench

Test	Parameters						Tests results		
	Contact nature	Sleeve material	Seal mounting	Seal material	Pressure	Sliding speed	Friction (N)	Static Sealing	Dyna Sealing
1	dry	Anod alu	Min ext	70 Sh	0,2 MPa	0,01 m/s	90	1	1
2	dry	P.P.M.A.	Min ext	70 Sh	0,8 MPa	0,1 m/s	150	1	1
3	dry	P.P.M.A.	Min ext	55 Sh	0,2 MPa	0,01 m/s	90	1	0
4	dry	Anod alu	Min ext	55 Sh	0,8 MPa	0,1 m/s	260	1	0
5	dry	Anod alu	Max ext	70 Sh	0,2 MPa	0,01 m/s	145	1	1
6	dry	P.P.M.A.	Max ext	70 Sh	0,8 MPa	0,1 m/s	300	1	1
7	dry	P.P.M.A.	Max ext	55 Sh	0,2 MPa	0,01 m/s	110	1	0
8	dry	Anod alu	Max ext	55 Sh	0,8 MPa	0,1 m/s	260	1	0
9	dry	Anod alu	Min float	70 Sh	0,2 MPa	0,1 m/s	120	1	0
10	dry	P.P.M.A.	Min float	70 Sh	0,8 MPa	0,01 m/s	110	1	1
11	dry	P.P.M.A.	Min float	55 Sh	0,2 MPa	0,1 m/s	200	1	0
12	dry	Anod alu	Min float	55 Sh	0,8 MPa	0,01 m/s	140	1	0
13	dry	Anod alu	Max float	70 Sh	0,2 MPa	0,1 m/s	260	1	0
14	dry	P.P.M.A.	Max float	70 Sh	0,8 MPa	0,01 m/s	140	1	0
15	dry	P.P.M.A.	Max float	55 Sh	0,2 MPa	0,1 m/s	145	1	0
16	dry	Anod alu	Max float	55 Sh	0,8 MPa	0,01 m/s	125	1	0
17	greased	P.P.M.A.	Min ext	70 Sh	0,2 MPa	0,01 m/s	16,5	1	1
18	greased	Anod alu	Min ext	70 Sh	0,8 MPa	0,1 m/s	21	1	1
19	greased	Anod alu	Min ext	55 Sh	0,2 MPa	0,01 m/s	9,5	1	1
20	greased	P.P.M.A.	Min ext	55 Sh	0,8 MPa	0,1 m/s	18	1	1
21	greased	P.P.M.A.	Max ext	70 Sh	0,2 MPa	0,01 m/s	20	1	1
22	greased	Anod alu	Max ext	70 Sh	0,8 MPa	0,1 m/s	22	1	1
23	greased	Anod alu	Max ext	55 Sh	0,2 MPa	0,01 m/s	10	1	1
24	greased	P.P.M.A.	Max ext	55 Sh	0,8 MPa	0,1 m/s	17	1	1
25	greased	P.P.M.A.	Min float	70 Sh	0,2 MPa	0,1 m/s	10	1	1
26	greased	Anod alu	Min float	70 Sh	0,8 MPa	0,01 m/s	15	1	1
27	greased	Anod alu	Min float	55 Sh	0,2 MPa	0,1 m/s	9,5	1	1
28	greased	P.P.M.A.	Min float	55 Sh	0,8 MPa	0,01 m/s	20	1	1
29	greased	P.P.M.A.	Max float	70 Sh	0,2 MPa	0,1 m/s	15	1	1
30	greased	Anod alu	Max float	70 Sh	0,8 MPa	0,01 m/s	14	1	1
31	greased	Anod alu	Max float	55 Sh	0,2 MPa	0,1 m/s	10	1	1
32	greased	P.P.M.A.	Max float	55 Sh	0,8 MPa	0,01 m/s	15	1	1

Figure 7 : Partial plan

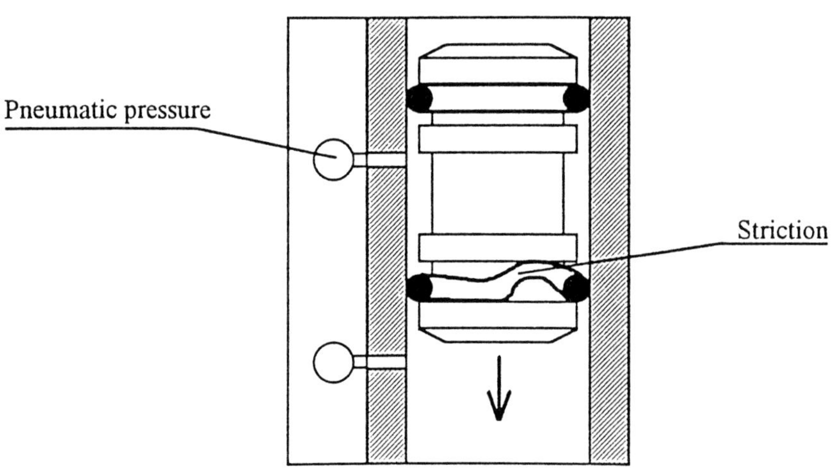

Figure 8 : Seal twist in it's groove

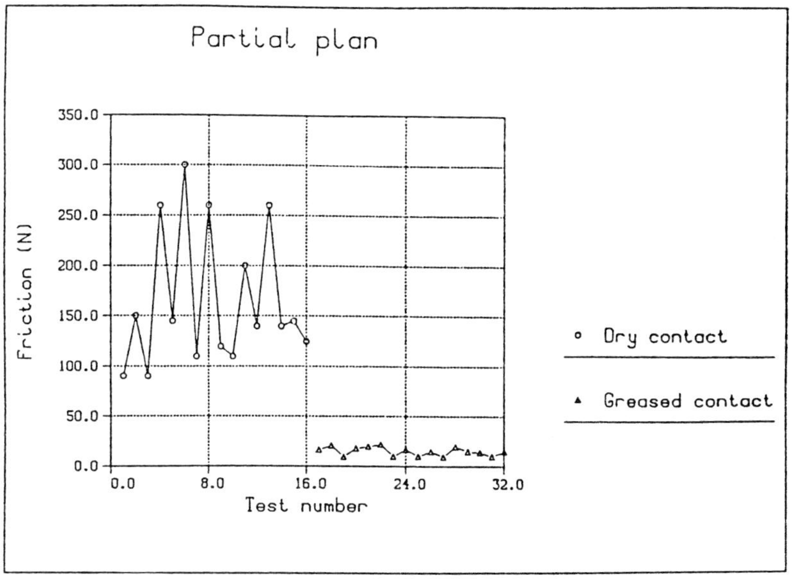

Figure 9 : Friction variation during tests

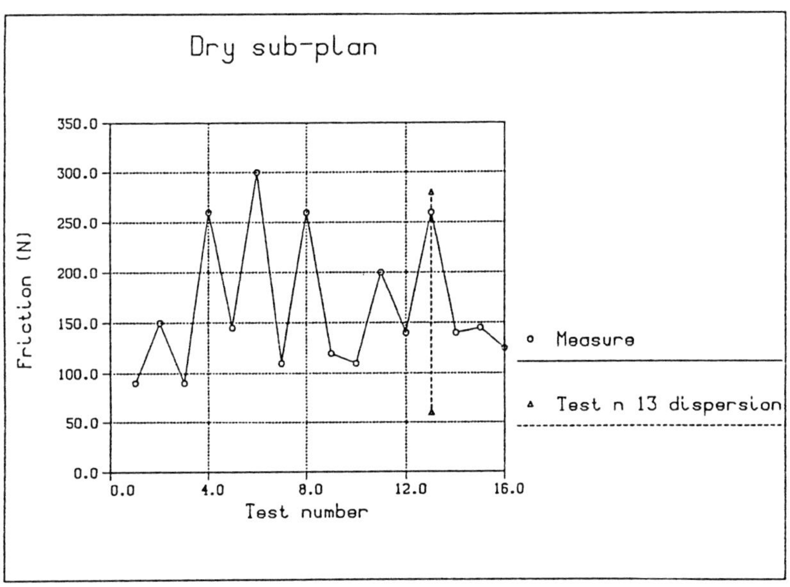

Figure 10 : Measurement dispersion during test n°13

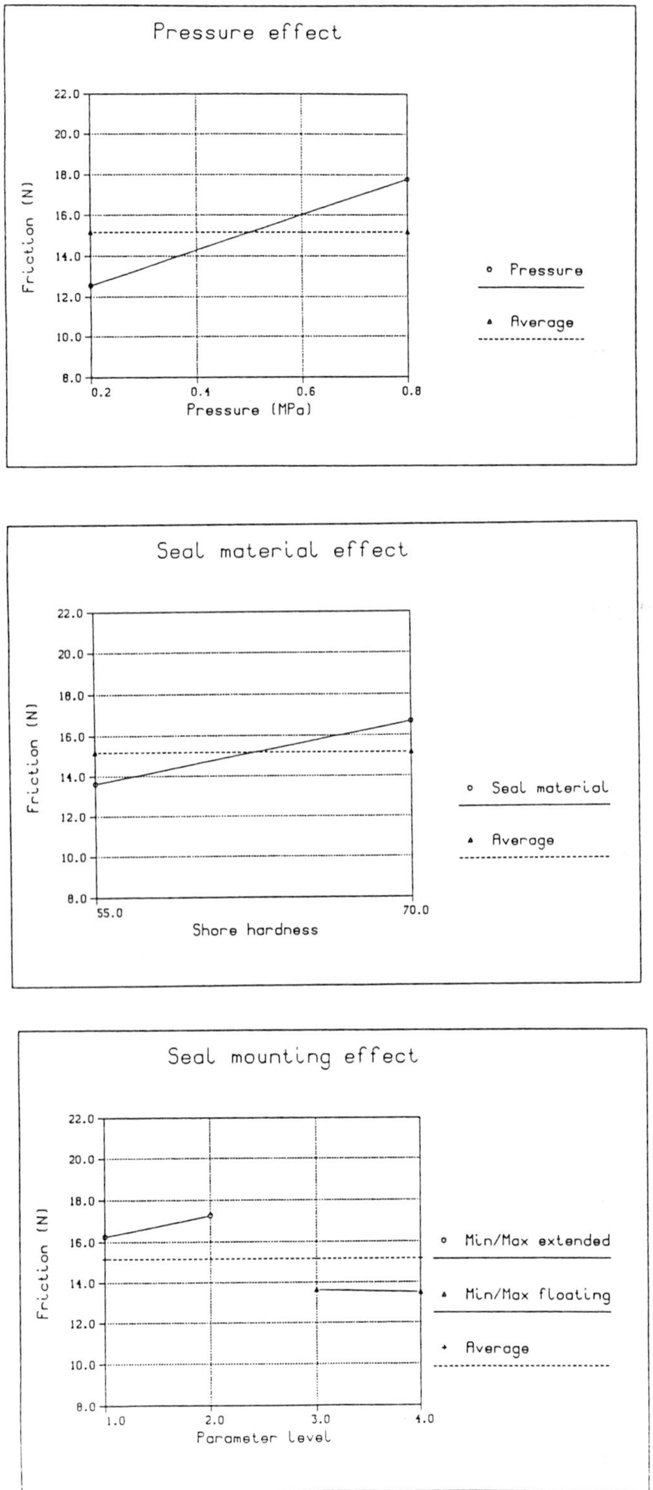

Figure 11 : Greased sub plan analysis

Parameter variation	Physical variation	Friction tendency
Pressure ↗	σ_N ↗	F ↗
Hardness ↗	Contact area between seal and sleeve ↘	F ↘
Seal mounting Extended → Floating	σ_N ↘	F ↘

with : σ_N : normal stress from seal to sleeve

Figure 12 : Practical tendencies

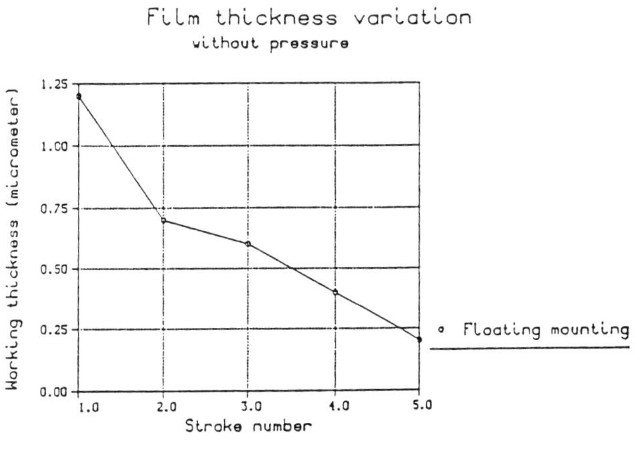

Figure 13 : Film thickness variation

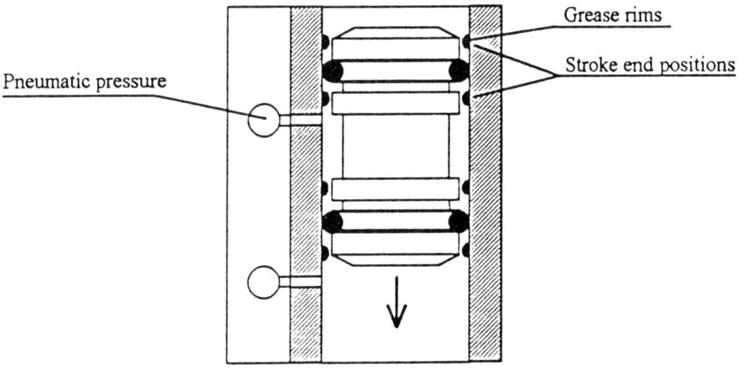

Figure 14 : Grease rims

Parameters studied

Parameters	Variations		
Pressure	0 MPa	2 MPa	8 MPa
Initial thickness	0,3 μm	0,6 μm	0,9 μm

Plans

Test	Pressure	Initial thickness
1	0 MPa	0.3 μm
2	0 MPa	0.6 μm
3	0 MPa	0.9 μm
4	2 MPa	0.3 μm
5	2 MPa	0.6 μm
6	2 MPa	0.9 μm
7	8 MPa	0.3 μm
8	8 MPa	0.6 μm
9	8 MPa	0.9 μm

Friction (N)
8
6,4
5,8
12,3
12,4
10,7
34
20,9
23

Working thickness (μm)
0,3
0,45
0,4
0,2
0,25
0,26
0,14
0,26
0,20

Tests results

Figure 15 : Two tests programs

Friction = f (Pressure, Initial thickness)

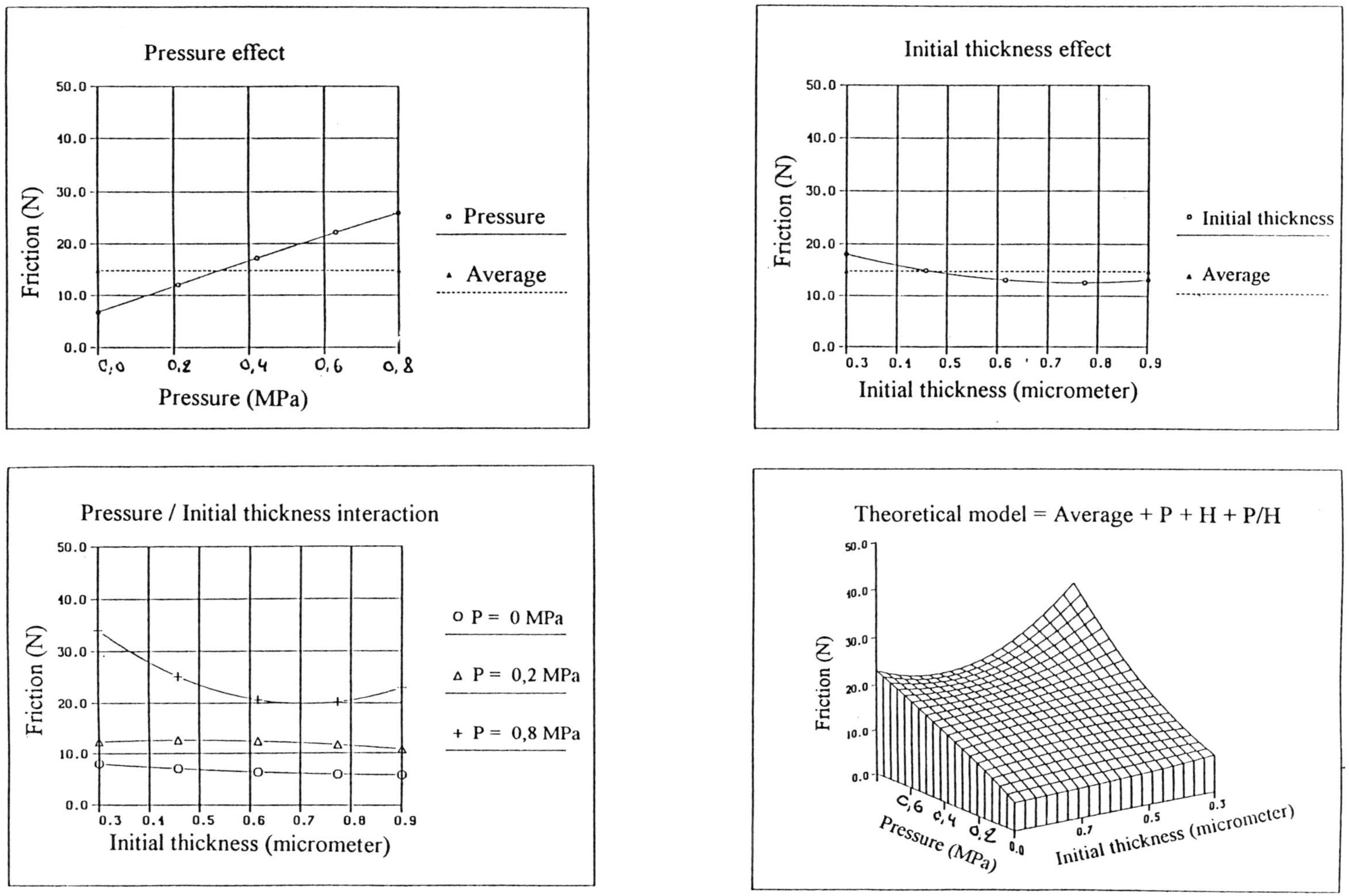

Figure 16 : Friction plan analysis

Working thickness = f (Pressure, Initial thickness)

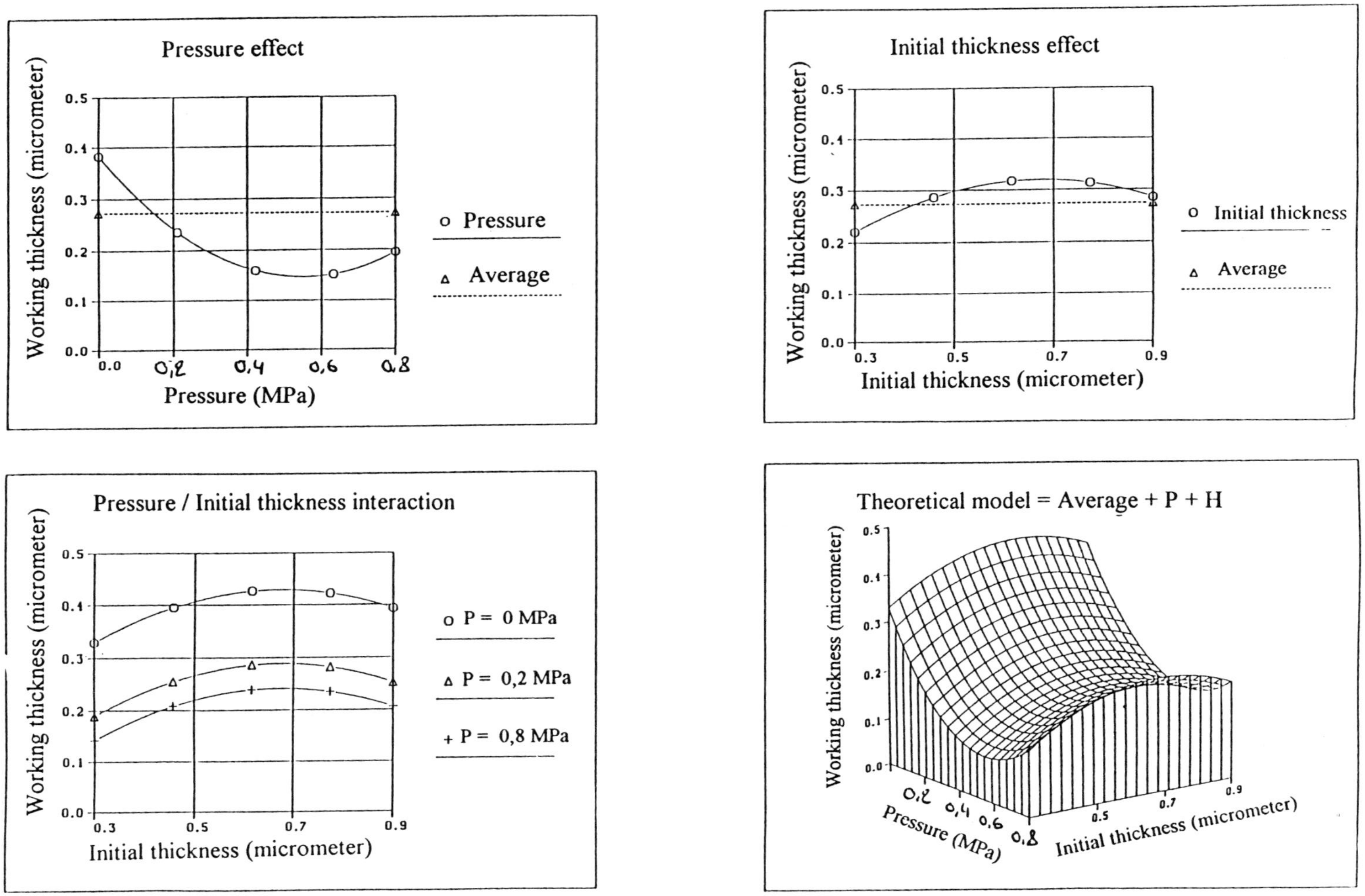

Figure 17 : Working thickness plan analysis

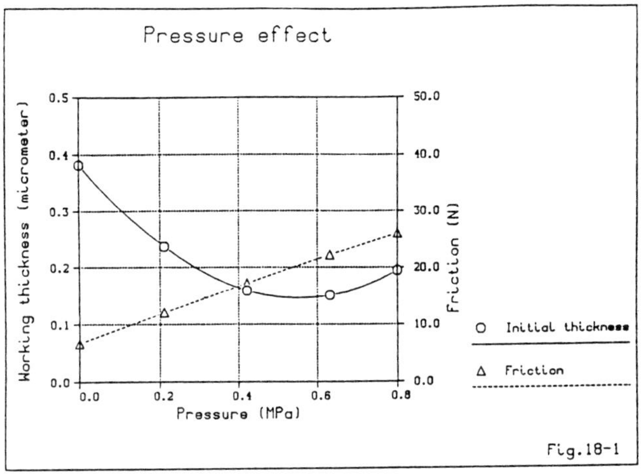

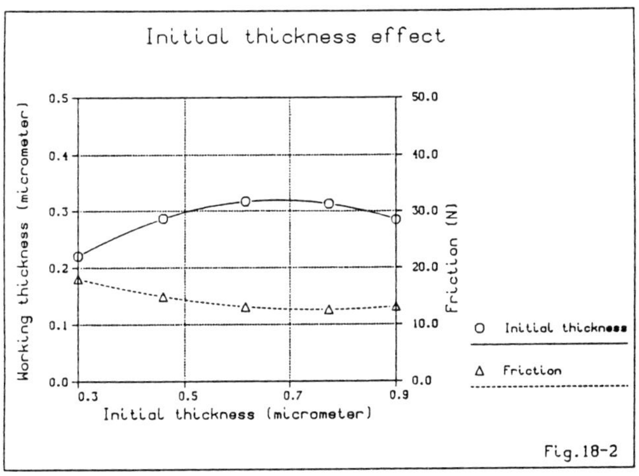

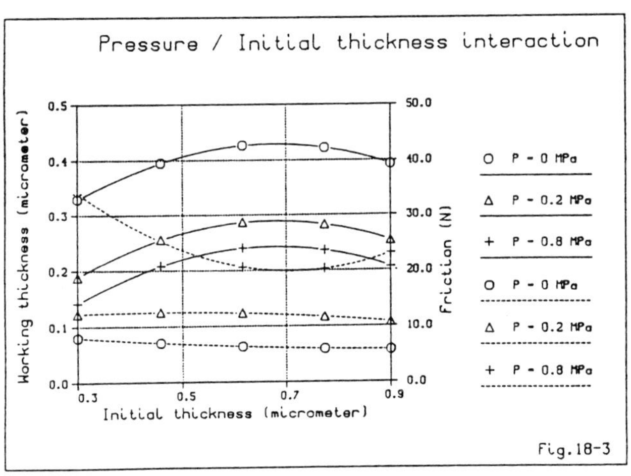

Figure 18 : Liaison between friction and working thickness

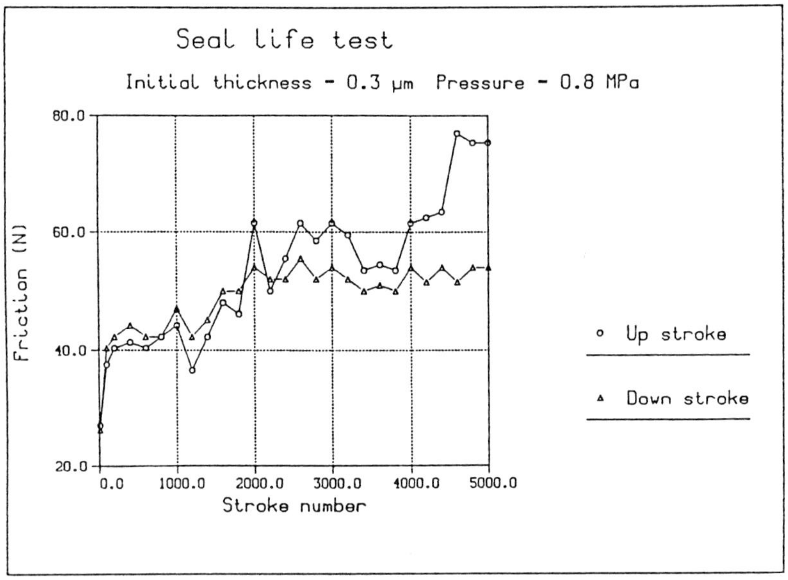

Figure 19 : Friction variation

Thin Films in Tribology / D. Dowson et al. (Editors)
1993 Elsevier Science Publishers B.V.

THIN FILM TRIBOLOGY AND RUBBER BEARINGS

Roy L. Orndorff, Jr.[a] and John C. Holzheimer[b]

[a]Research and Development Department, [b]Engineering Department, BFGoodrich Company, Engineered Polymer Products, 150 Division Drive, Wilmington, NC 28401, USA

Plasto-Elastohydrodynamics is the name given to the interaction of elastic and hydrodynamic forces in properly designed rubber bearings. Resulting design features include reduced rubber thickness (forms a lubricant-trapping pocket), glass-smooth bearing surface (maintains thin film to very low shaft speeds), wide staves (eliminates edge interaction), reduced number of staves or lands (lower friction) and shorter bearings. The rubber bearing designs now in use originated in the 1920's. These new improvements will enable rubber bearings to perform in a manner superior to other water-lubricated bearing materials and possibly even to oil-lubricated metal bearings in marine applications. Retained will be unique advantages such as the ability to handle grit-laden lubricant.

1. INTRODUCTION

Rubber bearings have been used for seventy years in many different and difficult applications. These usually involve operations where water or process fluid lubricants are used and the lubricant is often filled with grit or sand. In fact, the term "gravel grinders" is sometimes used in a derogatory manner to describe such bearings. It is usually assumed in the traditional tribological literature that such bearings can never perform beyond the boundary and mixed-film regimes (1), and that they will inherently be inefficient with unacceptably high wear rates.

The experimental work reported here is intended to show that rubber bearings can in fact be designed to operate very efficiently from a hydrodynamic standpoint with thin film lubrication. The bearings will have very low wear rates and superior performance at low journal speeds.

2. HISTORICAL AND TECHNICAL BACKGROUND

2.1 History

The first recorded use of rubber as a bearing material was by Charles Frederic Sherwood who was faced with an emergency mine pump bearing failure on a weekend early in this century. He had no brass bearing spares and he was desperate. So he cut off a piece of rubber hose and installed it. Much to his surprise, it worked very well and the wear rate was less than that of the metal bearing. That was the start. His patent was issued in May 1922 and showed various designs having molded rubber lugs supporting the journal. In December of 1922 BFGoodrich purchased the patent and began the worldwide business.

2.2 Technical Background

The oldest research report in the rubber bearing files is dated December 10, 1925, and was entitled "Report of Cutless Bearing Tests" by Burton B. Annis. It was initiated by the need to analyze and solve some field failures of spiral groove bearings. The conclusion was that spiral groove bearings were not good performers because the edges of the grooves consistently scraped the lubricant off of the journal. In fact the wear rate in the field and the lab appeared to be radically reduced once the bearings were worn down to the bottom of the groove (to form a round bore bearing). Annis concluded that the round bore "plain" bearing was the best as far as clean water lubrication was concerned, but that they tended to overheat because the clearance did not admit enough lubricant for cooling purposes. Fluted (landed) bearings gave favorable friction and wear results and were recommended by Annis as the solution. He also recommended the study of differently shaped lands and different rubber thicknesses. (This R & D work was not started until the late 1960's and then without the knowledge of the existence of Annis' report.) Annis recommended that a harder shaft sleeve should be used in gritty water. This report appears to be the origin of the rubber bearing design as it is known today (Fig. 1).

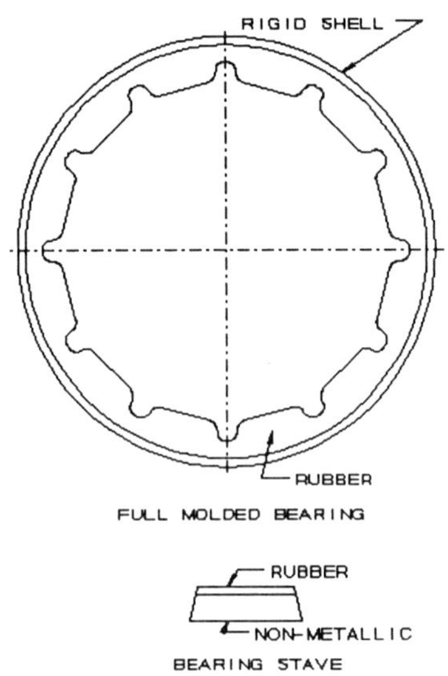

Figure 1. Types of Rubber Bearings

Other historical data tracing the evolutionary development of rubber bearings is included in references 2 and 3. One interesting point was the non-sensitivity of rubber bearings to large clearance changes noted by a number of researchers. This was because each land works as an independent bearing support pad. Another interesting point noted by early researchers is that rubber bearings do not wipe and suddenly fail as do metal bearings. The failure mode is very soft and gradual. Invariably the eventual cause of failure is overheating due to fluid lubricant supply restrictions. Rubber bearings allow a shaft to turn on it's center of gyration even if this is not the geometrical center of rotation. This reduces stresses and vibration at high speeds in both the bearing and the surrounding supporting structure.

A well known application in the 1930's illustrating high speed performance was the use of rubber bearings in Miss England III which set a world speed record. (They were also used in Miss England I and II.) Shaft speed was 200 s^{-1} ($22ms^{-1}$).

A point of disagreement among researchers in the 1930's was whether or not landed rubber bearings could function hydrodynamically. Shannikov (4), in 1939, experimentally proved that the lubricating action was hydrodynamic even though he used thick wall bearings. This was proven by correlating the coefficient of friction vs circumferential shaft velocity performance curve with the measured fluid pressure distribution in both the circumferential and axial directions. (He did not note the formation of a pocket.) There is no theoretical tribological reason that a properly-designed landed rubber bearing cannot function as well as a round-bore (cylindrical) bearing from a hydrodynamic standpoint.

A materials study was started at BFG with the objective of improving the rubber compound. At the end of this study limited testing was done with samples having reduced rubber thickness. Friction was much less and it was quite obvious significant performance improvements were possible by design changes. On the strength of this work, further research was done. As a result it was discovered that wear rate and coefficient of friction could both be reduced by over ninety percent, compared to the thicker rubber stave designs then in use. For the first time, hydrodynamic lubrication could be developed with rubber bearings across most of the shaft speed range (2).

It was found that once the rubber thickness was in the range of 3.18mm to 7.95mm, a unique action took place. The rubber deformed and shifted to form a very favorable lubricant trapping pocket (Fig. 2). The deforming action was initially elastic. However, during the run-in period of the bearing, non-elastic plastic creep occurred and then set took place to permanently shape the pocket. This action is called Plasto-Elastohydrodynamic Lubrication. The pocket stays in place after the bearing is unloaded. It is readily apparent end-to-end and side-to-side when a straight edge is laid across it. The pocket is too irregular in shape and depth to form by molding or machining.

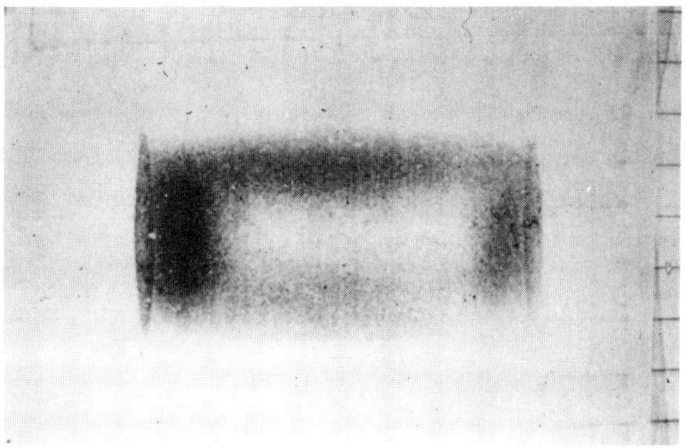

Figure 2. Rubber Pocket

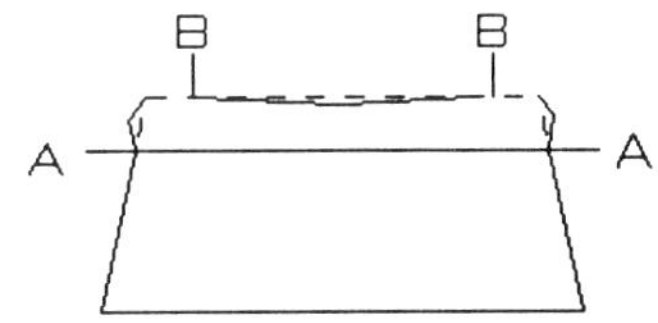

AA = Rubber Width,
BB = Shaft Contact Patch Width
Dashed lines are outline of new strip.

Figure 3. Actual Stave Tracing

The formation of the pocket is not a wear function and rubber is essentially incompressible. The rubber is redistributing itself to a new shape. In the lab we have proven this by comparing dry weight loss on the wear tester with sample arm deflection. Field results have also indicated the favorable benefits of the pocket formation. Fig. 3 is a before and after tracing of the cross-section of a thin rubber stave after five years of successful service in a large naval ship (2). The ship reported a significant operational noise reduction after installation of these staves. Dry weight loss measurement indicated no wear had occurred in the five years. The rubber had simply shifted to form the pocket. This does raise one problem. It is customary in marine maintenance to measure wear-down by measuring the increase in clearance between the bearing and the top of the journal. The deflection required to form the pocket is thus falsely charged to wear using this customary method.

3. RECENT RESEARCH

The initial work was done with a limited number of elastomer thickness values. They were chosen to be representative of the entire range of possible thickness values. A second investigation was performed on a complete range of thickness values in an organized manner. This work was done with individual sample blocks, as was the original research. Finally, a limited number of rubber thickness values were investigated in a full bearing configuration. Also investigated in the full bearing configuration were the effects of very smooth elastomer surfaces and the elimination of the lightly-loaded side staves (lands) of the bearing.

3.1 Thickness Study

<u>Test Conditions:</u> The friction force tests were performed on the single-station test machine. Total test time was two weeks. Test runs were made each working day at three different radial loadings. Nominal pressure is the load divided by the product of the bore multiplied by the bearing length. (Present design nominal pressure is 276 Nominal kPa.) Quoted nominal pressure values are those that the test sample would experience if it was the heaviest-loaded stave in a bearing having that nominal pressure. Journal rotation was varied from 0 to 4.17 s^{-1}. Special emphasis was used to measure the starting or zero shaft speed friction force. The

journal was slowed down under load to zero shaft speed and restarted a minimum of three different times in order to obtain an accurate average value. Start-up conditions were very severe because the journal was not fully immersed. In order to take the judgement factor completely out of the data reduction process, the lowest friction force value was used out of the ten daily runs for each velocity step in each load range. Most of the data points came from the last several runs after the sample was well run-in.

The friction torque measuring system used in the single arm and full bearing coefficient of friction test machines are differential dynamometers that were originally invented in the 18TH century. They are rugged, easily calibrated, and give very accurate and repeatable results (2).

Wear tests were run on the three-station loaded arm machine. Test period for all samples was 28 days of continuous operation. Test conditions were severe (1530 kPa nominal bearing pressure, 0.333 s^{-1} shaft speed). Journal diameter was 254mm and sample lengths were 50.8mm for both the friction and the wear testing.

Results: Wear results are shown in Table 1. The coefficient of friction results are shown in Table 2. The columns are in order of increasing rubber thickness, reading left to right.

Discussion of Results: The coefficient of friction results in Table 2 show an optimum range of rubber thickness values. Within this range, 4.76mm to 7.95mm, the friction force at all speeds and loads appears to be a minimum. The two thinnest rubber samples have friction coefficient values approximating values for the harder and thicker rubber standard design before 1980. The hardness reading for the thinnest samples show a hardness higher than the hard rubber stave (No. 1, Table 1). This is the effect of the closeness of the hard plastic backing material to the rubbing surface. The tip of the Shore A gage needle approximates the shape of a piece of grit or sand, and therefore reflects actual sandy water operation. Because of the increase in effective hardness, the thinnest samples would not be good designs from a friction standpoint and would cause excessive journal wear in sandy water. The wear samples showed scratches on the rubber surface for the two thinnest rubber samples and therefore had the worse surface conditions and the wear rates were also out of proportion for the thicknesses. The coefficient of friction appeared to decrease with increasing load in the optimum range. This was especially true at the slower shaft speeds.

TABLE 1
WEAR TEST RESULTS

NO.	SAMPLE	WEIGHT LOSS GRAMS	APPARENT HARDNESS DURO-METER	CONTACT PATCH WIDTH, MM
1	12.7MM RUBBER THICKNESS STANDARD BRASS BACKED HARDER COMPOUND, 87 DUROMETER	0.2504	82	39.1
2	9.53MM RUBBER THICKNESS SOFTER COMPOUND, 70 DUROMETER	0.1687	71	36.6
3	7.95MM RUBBER THICKNESS SOFTER COMPOUND 70 DUROMETER	0.0392	71	34.8
4	6.35MM RUBBER THICKNESS SOFTER COMPOUND, 70 DUROMETER	−0.013	76	34.0
5	4.60MM RUBBER THICKNESS SOFTER COMPOUND, 70 DUROMETER	0.107	74	34.5
6	3.18MM RUBBER THICKNESS SOFTER COMPOUND, 70 DUROMETER	0.1445	76	34.3
7	2.38MM RUBBER THICKNESS SOFTER COMPOUND, 70 DUROMETER	0.0472	77	30.5
8	1.59MM RUBBER THICKNESS SOFTER COMPOUND, 70 DUROMETER	0.1296	83	24.9
9	0.79MM RUBBER THICKNESS SOFTER COMPOUND, 70 DUROMETER	0.0746	89	19.8

TABLE 2
COEFFICIENT OF FRICTION VALUES, (Value in Table x 10^{-4}, Ex: 2.0 = 0.0002 = Coefficient of Friction)
Sample Length = 50.8mm, Journal Diameter = 254mm

SHAFT SPEED (s^{-1})	Rubber Thickness, mm 0.79	1.59	2.38	3.18	4.76	6.35	7.95	9.53
				Nominal kPa = 310 (780N)				
0.000	856.0	614.0	456.0	514.0	194.0	357.0	194.0	400.0
0.167	776.0	137.0	262.0	285.0	46.0	39.2	74.0	314.0
0.417	97.0	32.9	57.0	61.2	10.3	10.3	16.9	57.0
0.833	27.5	12.5	16.2	21.9	3.1	6.8	7.8	18.1
1.250	21.8	10.5	11.4	16.1	2.0	6.7	6.7	13.3
1.667	22.9	10.7	8.8	18.2	3.1	6.0	8.8	9.7
2.083	20.9	8.6	9.6	13.5	3.9	6.7	7.7	11.4
2.500	22.7	11.4	10.5	13.5	5.8	8.6	9.1	11.4
2.917	22.7	14.3	11.4	22.7	6.7	10.5	13.3	14.3
3.330	19.9	14.3	10.5	19.9	3.9	7.7	8.6	11.4
3.750	20.9	14.3	12.4	21.8	4.8	9.6	11.4	12.4
4.167	19.9	13.3	10.5	22.7	5.8	9.6	8.6	10.5
				Nominal kPa = 489 (1223N)				
0.000	879.0	391.0	391.0	443.0	124.0	254.0	124.0	313.0
0.167	1272.0	124.0	227.0	182.0	33.0	36.0	55.0	200.0
0.417	327.0	26.7	51.0	37.7	8.1	8.1	10.6	58.0
0.833	87.0	8.2	13.0	13.6	3.4	4.6	7.6	14.2
1.250	62.0	8.2	8.8	13.0	2.8	5.2	7.0	10.6
1.667	62.0	8.9	8.3	14.9	2.9	5.9	7.1	8.9
2.083	32.8	10.0	7.6	14.8	2.2	5.8	7.0	10.0
2.500	27.4	10.0	8.8	16.0	5.2	7.0	7.0	10.6
2.917	26.2	12.4	11.2	19.6	5.2	8.8	9.4	11.2
3.330	23.2	11.2	9.4	18.4	4.0	7.0	8.8	10.0
3.750	25.0	11.8	8.8	20.2	6.4	8.8	9.4	11.2
4.167	24.0	11.8	8.8	20.8	6.4	7.6	8.8	11.2
				Nominal kPa = 668 (1670N)				
0.000	933.0	453.0	325.0	307.0	91.0	211.0	91.0	240.0
0.167	933.0	133.0	325.0	187.0	19.0	45.0	64.0	453.0
0.417	393.0	43.0	45.0	43.0	6.1	9.2	13.0	53.0
0.833	187.0	9.8	20.8	26.3	2.8	5.4	8.0	35.0
1.250	91.0	7.1	8.4	13.7	2.3	4.9	7.1	7.1
1.667	80.0	6.3	9.8	15.1	2.4	5.4	7.4	8.1
2.083	53.0	7.5	9.7	14.6	2.3	5.3	7.5	9.3
2.500	43.0	8.9	9.3	15.9	4.0	6.7	8.9	9.3
2.917	43.0	8.9	10.2	18.1	5.3	8.0	10.2	10.2
3.330	43.0	8.9	8.4	17.2	4.5	6.7	9.3	8.9
3.750	43.0	8.9	8.0	16.8	5.3	8.0	9.7	10.2
4.167	43.0	9.7	8.9	15.5	4.9	8.0	9.3	9.7

3.2 **Full Bearing Tests**

Test Conditions: The full bearing friction test machine is a horizontal shaft unit with a variable speed drive system. Shaft speed can be varied from 0 to 8.33 s^{-1} by means of a three speed transmission system. This enables the machine to develop the higher torque required at start-up under load (breakaway) and for low speed operation. Friction force measurements are made at breakaway and in steps from 0.008 to 6.67 s^{-1}. Shaft diameter is 171mm and bearing length is 146mm, giving and L/D ratio of 0.85. Standard test run-in period was 24 hours and the running time at each shaft speed step was 15 minutes. Static breakaway tests were made after 3 hours under load. Radial load for all tests was 6911 N (276 nominal kPa).

Four rubber thickness values were investigated. They were 3.18mm, 4.76mm, 6.35mm and 7.95mm. They were molded with a standard finish (0.8μm) and also with a glass-smooth finish. The latter was obtained by molding the rubber surface against a glass plate. The bearing utilized twelve staves. To evaluate the effect of using a lesser number of load-carrying staves, tests were run using four staves from the unloaded six staves in the upper half of each bearing. Figure 4 is a cross sectional view of the test machine housing showing the four-stave arrangement.

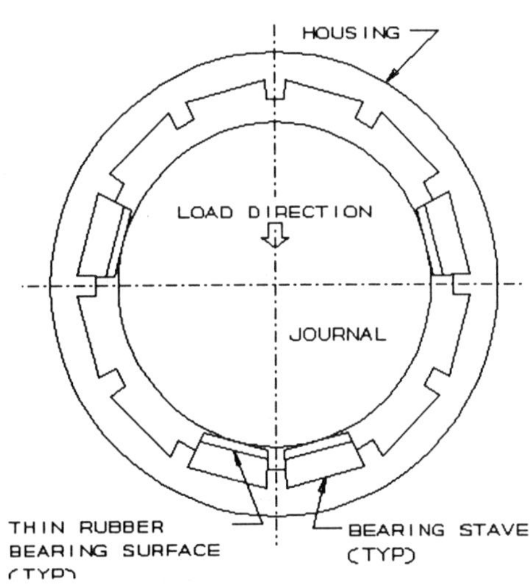

Figure 4. Four stave Lab Bearings

Results: Figure (5) and Figure (6) are plots of coefficient of friction versus shaft speed for twelve staves standard finish, and four stave-glass smooth surface.

Discussion of Results: The curves show a reduction in coefficient of friction with rubber thickness. The effect is more pronounced with the twelve stave standard finish version. All of the plots are of the characteristic hydrodynamic shape. The coefficient values proceed with increasing shaft speed through the steadily decreasing mixed-film region to a minimum value range at which full hydrodynamic lubrication is attained. Further increases in shaft speed result in increased friction coefficients because the fluid film is being sheared at a faster rate. The only curve to show much indication of a boundary lubricated (flat) regime is the highest friction sample: 7.95mm rubber thickness, twelve-stave standard finish. This sample also had the minimum friction coefficient (f = 0.0070) point occur at the highest shaft speed (6.67 s^{-1}). Other thickness values of standard finish twelve stave bearings had their minimum friction points occur in the shaft speed range 2.33 to 3.33 s^{-1} (f = 0.0019). The minimum coefficient (f = 0.0007) was much lower for the glass smooth four-stave bearings. This indicates the importance of a very smooth surface in developing low friction even with the formation of a lubricant trapping pocket. The very smooth surface does not have the usual surface asperities present to penetrate the water film and contact the journal, thereby increasing the friction.

Another advantage of the glass-smooth stave surface is that it eliminates the run-in period. This period can last for one hundred hours in the laboratory, much longer in actual practice. The 24 hour run-in was not long enough for most of the non-glass smooth samples. It was long enough to show trends. (The thinner rubber samples tend to have a more favorable decrease in coefficient of friction value during the run-in period than do the thicker rubber samples. This is based on further testing with this rig and the single stave tester.)

There is always the question of how realistic it is to recommend a glass smooth surface in the real world.

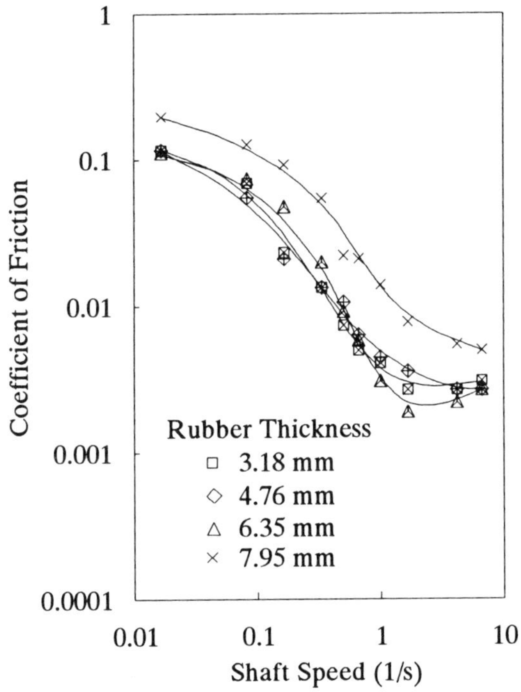

Figure 5. Twelve Stave Standard Finish

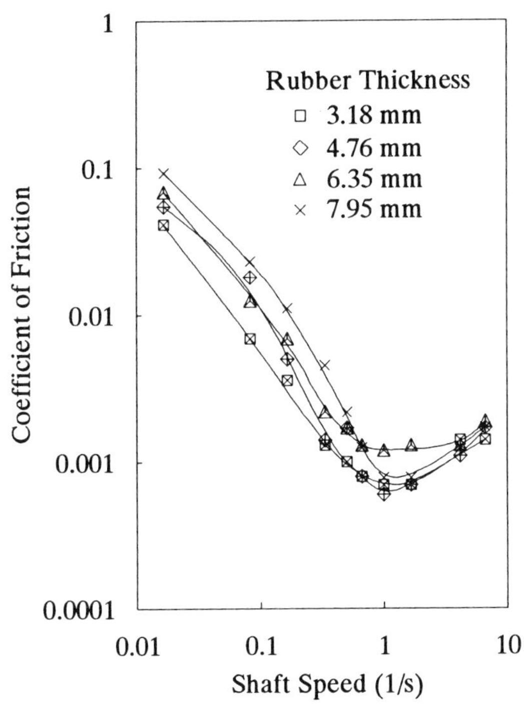

Figure 6. Four Stave Glass Smooth Finish

A newly constructed ship often remains dockside for months after leaving the drydock while fitting-out is completed. Barnacles and other marine growth soon make the clearance spaces and channels between the bearing and the lands their home. Then when the ship's propeller shaft starts turning the barnacles are ground up so that they tend to act as a very rough abrasive during the first few revolutions of the shaft. The abrasive particles are much too numerous and the shaft speed too slow for the usual rejection process to take place. The shaft is then damaged which in turn damages the bearing. Some of the damage is smoothed out during subsequent operation, but the low coefficient of friction values of the original glass-smooth rubber surfaces are probably never reattained. The answer is to protect the bearing by pumping water through it at dockside and/or using a shroud at the aft end. In the case of a pump, the bearing should be covered with a tarpaulin during final assembly in the shop to keep out debris and dirt.

Two present disadvantages indicate that a lesser number of wider staves would improve performance. First, It was theorized that the lightly-loaded side staves of a bearing contribute a disproportionate share of the total friction torque of the bearing. Secondly, narrower staves contribute to bearing noise (squeal) due to edge interaction with the journal, scraping off the lubricant.

An extended program was completed evaluating short test blocks on the single-arm coefficient of friction test machine. The lighter loads present in the side staves of an actual bearing were emulated. It was discovered that in the range of optimum rubber thickness values, the higher the load the lower the friction coefficient. Table 3 shows a typical comparison for one rubber thickness value. Improved bearing performance was borne out by the data shown in Figure 6. Stave locations are shown in Figure 4.

TABLE 3
COEFFICIENT OF FRICTION VS LOAD/LENGTH

JOURNAL SPEED, (1/s)	RADIAL LOAD 2.37 N/mm	RADIAL LOAD 34.2 N/mm
0.000	0.2407	0.0474
0.417	0.0128	0.0076
0.833	0.0058	0.0031
1.670	0.0039	0.0015
3.330	0.0024	0.0011

SHAFT DIAMETER = 254 mm

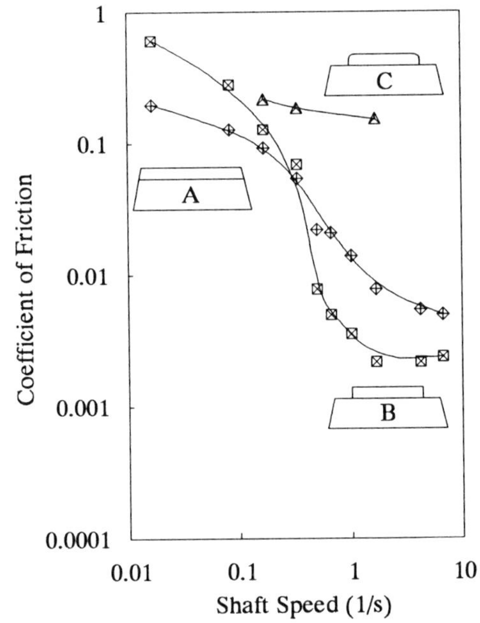

Figure 7. Narrow Stave Friction Comparison

We had noticed in the field and in the lab that with thicker rubber staves considerable edge interaction could occur. This results in squeal as the stave leading edge scrapes off the lubricant. The trend in modern large ships has been for the journal (shaft) diameter-to-land width ratio (JDL) to increase. It is now frequently in the range of 7 to 10. The full bearing test machine has a JDL ratio of 4. An experiment was run in the lab to duplicate the field experience. We reduced the stave effective bearing surface circumferential length to obtain a JDL of 7.2. As shown in Fig. 7B, the friction at higher shaft speeds was less than that for the standard stave (Fig. 7A). This is probably because of the favorable angle of attack formed as the rubber bulged to form the pocket. At slower shaft speeds, however, the situation reversed as the increasing friction force sheared the rubber in the circumferential direction resulting in edge interaction. Next, the edges of the special staves were broken (lightly ground) in an attempt to reduce the sharp lip. The result surprisingly was a large increase in friction at the higher shaft speeds (Fig. 7C). Later tests with machined and molded edges having various curvatures all showed significant increases in friction probably due to the approach and departure angles being too great. The answer is to make the staves wider and the rubber thinner in order to remove any possibility of edge interaction.

Squeal noise and stick-slip noise are different. Squeal is due to an edge interaction and occurs at higher shaft speeds. Slip-stick is caused by a breakdown of the lubricant film in the boundary regime and occurs at very low shaft speeds. Squeal problems are solved by design changes and/or stiffness increases. Stick-slip problems are solved by design (thin, smooth surface) and the basic hydrophobic nature of the elastomer compound. All potential non-metallic water-lubricated bearing materials exhibit stick-slip. The engineering challenge is to drive the onset of stick-slip to as low a shaft speed as possible (essentially zero effect) by means of design and material choices. For example thin rubber smooth surface staves operate routinely on the full bearing test machine at rotational speeds of 0.0083 s^{-1} (4.49mm s^{-1}) without any stick/slip.

4. THEORY OF OPERATION

4.1 Plasto-Elastohydrodynamic Lubrication

Prior research results along with recent results reported and discussed in this paper indicate that elastomer bearings do develop efficient lubrication. How they accomplish this is uniquely and significantly different from how metal and other rigid surface bearings function.

The main points are that Plasto-Elastohydrodynamic Lubrication involves the formation of a pocket in the rubber layer. Pocket formation is aided by the thickness of the rubber and its stiffness. A very smooth bearing surface reduces run-in time to a short period and also reduces the coefficient of friction across the whole shaft speed range. Elimination of the inefficient side staves further reduces friction. Wider staves eliminate the possibility of edge interaction. Reduction of the bearing length from the customary length-to-diameter value of four would further improve friction performance. Inherent in this discussion is the requirement that the rubber bearing compound possess the wear resistance required to operate at higher actual contact pressures.

There is some experimental indication that reduced area of contact between journal and bearing surface can be beneficial to the development of hydrodynamic lubrication even though the contact pressure is higher. This is because only a limited amount of water can be pumped into the contact area. Too much area of contact and the already thin film will not be able to maintain a thickness adequate to separate the surfaces. Control of the approach angle of the journal at the bearing surface is very important.

4.2 Gritty Water Operation

The flat land/approach angle relationship functions in the process of grit rejection when the lubricating film has sand or dirt in it. The mechanism of bearing wear for any bearing material has a number of progressive steps. First, a piece of grit enters the contact patch and becomes embedded in the bearing material. It then machines the journal (shaft) as a crude cutting tool. Finally, the rough grooves in the journal machines the bearing. A rubber bearing, especially one with hardness of around 70 Shore A, interrupts the progressive wear process. Instead of being embedded and retained in the bearing surface the grit particles are instead depressed and rolled through the contact patch. They exit into the adjacent lubricant groove and are then washed axially out of the bearing. Because of the elasticity of the rubber most of the grit particles leave no trace of their passage. The wear process is effectively short circuited. The grit particles do not enter the next land in line because they are carried out of the bearing in the excess lubricant flow. The adjacent land cannot accept the grit particles due to an efficient low approach angle. The hydrodynamically pumped fluid is carried into the land entrance by the boundary layer attached to the journal. The grit and the excess fluid flows axially out of the area defined by the angle of approach. If there is lubricant being pumped into the forward end of a bearing by ship motion or a pump the particle ejection process out the rear end of the bearing will be more rapid.

The grit ejection process was vividly illustrated when we started the full bearing test machine for the first time in our new plant location. Not realizing the water was full of sand, our first coefficient of friction test turned into a wear test. There was no journal wear. There were no wear circumferential marks on the bottom, most heavily loaded staves. The flat land/approach angle grit ejection process was working. However the process did not work as well on the lightly-loaded side lands both on the approach and departure sides. Faint scratches were visible on the staves where the grit particles had been engulfed by the hard polymer alloy material and rolled through the contact patch with no damage to the journal.

5. CONCLUSIONS AND FUTURE WORK

Rubber bearings correctly designed will efficiently compete with other bearing materials. This is important as oil lubrication becomes expensive because of supply and pollution control problems.

Water is the nearest thing to a universal lubricant. It is readily available almost everywhere. The cost is low because it does not come out of a well in the Middle East. It does not pollute and stores twice as much heat as compared to oil. Water and rubber bearings are a natural match.

A designer of rubber bearings has more decisions to make than do the designers of most other bearings. This can be a strength offering important advantages to the end users. Up to this point in time designers have had to use mostly empirical methods based on experimental data reported in this and prior technical papers. From time to time various theoreticians have attacked the problem (5). However, no one has been successful in developing and solving the equations that would fully and mathematically describe the complex interaction of hydrodynamic and elastic forces involved in the successful operation of rubber bearings. Such an effort is needed to turn the experimentally-based empirical methods currently used into a science.

For example, calculation of the minimum film thickness (h_0) is very important in rigid bearing design. But how relevant is it when a pocket is formed (Fig. 2 and Fig. 3)? One reference (6) states that all compliant surface bearings have the minimum film thickness value at the trailing edge of the contact patch because the film pressure at that point drops below ambient. We have seen the result in harder rubber bearing staves returned after long service in the field. It is evidenced by a scalloped wear pattern at the trailing edge caused by cavitation erosion resulting from the below-ambient pressure.

Future work will involve learning more about how to control the approach angle/contact area relationship. Included in this effort will be new hard polymer alloys that promise to extend the load capabilities of water-lubricated thin film bearings into a range significantly beyond the capability of any other materials.

ACKNOWLEDGEMENTS: The authors wish to thank Kenneth Helmsin for his support and Donald Hatch, Harold Bridges, Robert Will, Gena Hamby and Greg Marney for their help.

REFERENCES

1. P.J. Lidgitt, D.W.F. Goslin, G. Rockwell, G.S. Ritchie, Hard-on-hard water lubricated bearings for marine applications, Proceedings of the 13TH Leeds-Lyon Symposium on Tribology, Elsevier, Amsterdam (1986) 129.

2. R.L. Orndorff, Jr., Water-Lubricated Rubber Bearings, History and New Developments, Naval Engineers Journal, (Nov. 1985) 39-52.

3. J.M. Foulk, Water-Lubricated Rubber Bearings; A Brief Overview of History Application Guidelines and Recent Advancements, The Society of Naval Architects and Marine Engineers, (1986) Gulf Section.

4. V.M. Shannikov, Rubber Bearings, Their Use, Construction and Lubrication, Kauchuik-i-Rezina, (1939) No. 12, 18-22.

5. C. Giannikos and R.H. Buckholz, Hydrodynamically Lubricated Plane Slider Bearings Using Elastic Surfaces, Proceeding of the 13TH Leeds-Lyon Symposium on Tribology, Elsevier, Amsterdam (1986) 165.

6. D.D. Fuller, Theory and Practice of Lubrication for Engineers, John Wiley & Sons, New York (1984) 449.

SESSION XVII
WEAR / COATINGS

Chairman : Professor B J Hamrock

Paper XVII (i)	**Micro-Raman Spectroscopy for the Characterization of Wear Induced Surface Modifications on Hard Coatings**
Paper XVII (ii)	**Application of the Thermomechanical Wear Transition Model to Layered Media**
Paper XVII (iii)	**Tribodynamic Modeling**
Paper XVII (iv)	**Modelling of Discontinuous Transfer Films : Role of Local Shear Stress in the Build Up of Load Carrying Capacity**

Thin Films in Tribology / D. Dowson et al. (Editors)
© 1993 Elsevier Science Publishers B.V. All rights reserved.

Micro-Raman spectroscopy for the characterization of wear induced surface modifications on hard coatings

B. Blanpain, M. Franck, H. Mohrbacher, E. Vancoille, J.P. Celis and J.R. Roos

Departement Metaalkunde en Toegepaste Materiaalkunde, Katholieke Universiteit Leuven, de Croylaan 2, B-3001 Leuven, Belgium

Micro-Raman spectroscopy can be a helpful tool in determining wear induced surface modifications due to its ability to identify compounds and measure internal stress in an area with a diameter as low as 1 μm. Micro-Raman spectroscopy is used here to characterize surface reactions, transfer layers and internal stress relaxation induced by wear processes for four different hard coatings: PVD TiN, solar beam oxidized TiN, CVD diamond and diamond-like carbon coatings. The Raman spectroscopic investigation of partially worn coatings has resulted in very specific information complementary to results that can be obtained with commonly used surface characterization techniques.

1. INTRODUCTION

The friction and wear properties of unlubricated tribo-pairs are governed by wear induced surface modifications. Techniques able to monitor or characterize these modifications are thus of value to understand the performance of tribo-pairs.

Micro-Raman spectroscopy has a large potentiality in this area due to the specific information it can provide: compound identification including the distinction between polymorphs and polytypes, the evaluation of crystallinity and the determination of internal stress. In addition, the technique is non-destructive and the sampling area can be restricted to an area of a few microns in diameter.

Raman spectroscopy has been used previously to determine wear products in the sliding of steel against silicon nitride and graphite [1]. In this paper some examples are presented of the use of Raman spectroscopy to characterize surface modifications after laboratory wear testing. The different cases that will be discussed are: TiN, solar beam oxidized TiN and diamond-like carbon coatings all worn against corundum, and a diamond coating worn against diamond and a cemented carbide material. With these examples the wide applicability of this characterization technique in the study of the tribological behavior of hard coatings is illustrated.

2. EXPERIMENTAL PROCEDURE

The Raman spectra were acquired with a Jobin-Yvon Mole S3000 spectrograph equipped with a triple monochromator, a 512 multidiode array and a microscope. The laser source is a Spectra Physics Ar-ion laser, which was operated at a wavelength of 514.5 nm. For most of the experiments the spectral slit width was set around 8 cm^{-1}, with exception of the spectra taken of the diamond samples where the spectrograph was operated at a spectral slit width of 2 cm^{-1}. The laser power and the magnification of the lens of the microscope are mentioned separately for each spectrum. At a lens magnification of a 100x the laser spot size is around 1 to 2 μm in diameter.

The wear testing equipment used are a ball-on-disk apparatus [2] for the study of the wear of TiN versus corundum and a fretting apparatus [3] for the other studies.

Experimental details on the materials and other experimental conditions used will be given as the different case studies are presented.

3. CASE STUDY 1: WEAR DEBRIS ANALYSIS IN THE SYSTEM TiN/CORUNDUM

TiN coatings are used extensively in all types of cutting tools. These coatings have the capability to enhance tool life, improve surface finish and increase productivity. But they are subject to abrasion by hard inclusions in the machined steel, and prone to oxidation at elevated temperatures induced by high cutting speeds. An understanding of the dependence of the coating wear on sliding velocity necessitates a detailed analysis of wear debris formed on TiN.

PVD TiN (thickness 2 µm) was deposited in a Balzers type equipment on polished high speed steel (AISI M2) disks (0.3 µm R_a). The coated disks were tested at a load of 5 N in a ball-on-disk apparatus against a polished corundum ball (10 mm diameter, 0.2 µm R_a). The coating wear volume versus the sliding speed is given in Figure 1. The wear increases to a maximum at 0.2 m/s and then decreases as the speed increases to 1.0 m/s. This dependency can be modeled by a mild oxidational model relating the flash temperature at the contact points with the wear rate of the coating [4].

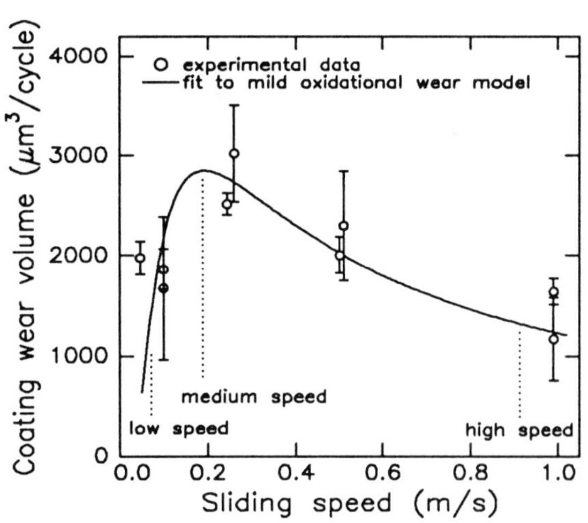

Figure 1: Coating wear as a function of sliding velocity for a corundum 10 mm diameter ball sliding against a PVD TiN coated HSS steel in a ball-on-disk experiment. The tests were run for 15,000 cycles at 5 N load.

Raman spectroscopy was performed on the white powder-like debris that accumulated along the wear track on the disk. The debris analyzed was produced at three different sliding speeds: low (0.01 m/s), medium (0.1m/s) and high (1 m/s). The Raman spectra were acquired immediately after the production of the debris to minimize contamination from the environment.

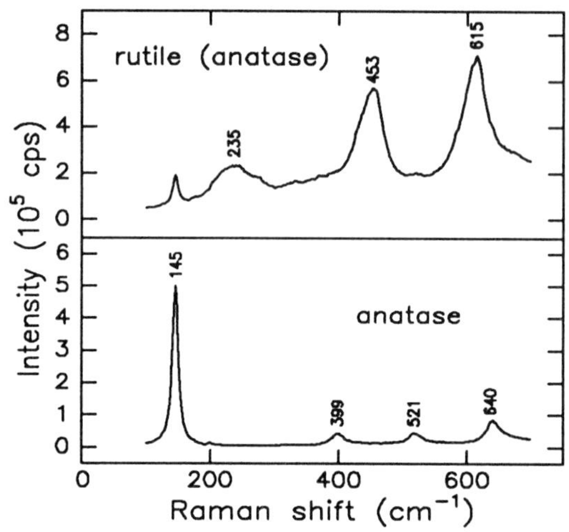

Figure 2: Raman spectra of rutile and anatase standard powders. (laser power 15 mW, lens 20X).

Figure 2 shows the Raman spectrum for reference rutile and anatase powders, two common polymorphs of TiO_2. The distinct features observed have also been reported in literature [5]. The rutile standard powder contained some anatase contamination. Figure 3 shows the spectrum taken at 15 mW for the debris produced at three different speeds. The peaks of the different spectra are broad. All samples show the rutile features around 440 cm^{-1} and 610 cm^{-1}, while the spectra of the low and medium speed debris have an additional distinctive broad peak around the position of the most intense anatase peak. The broad features of Figure 3 are typical for highly defected submicron sized particles. The spectra obtained by exposing the particles to a higher laser power of 150 mW are presented in Figure 4. The rutile peaks have sharpened, and for the low and medium speed debris a sharpened anatase peak is evident. The anatase peak is not present in the spectrum for the debris made at the highest speed (Figure 4c). Obviously the Raman activity of the wear debris is enhanced by the laser irradiation at 150 mW. Apparently the as-formed debris is heated to such an extent that defect annealing and grain growth processes of the rutile and anatase debris particles are coming into effect. From peak-fitting and comparison with a mixture of anatase and rutile it was found that the wear debris produced at low sliding speed consisted approximately of 25 vol% anatase. The metastable anatase readily converts to rutile form at 700°C [6]. It was observed that the anatase peak can be annealed out at a laser power of 1W. Electron diffraction patterns of the debris particles showed that all types of debris could be indexed as rutile while no evidence of the anatase phase was found. Raman spectroscopy appears to be more sensitive to the presence of the anatase phase in a rutile matrix than electron diffraction techniques .

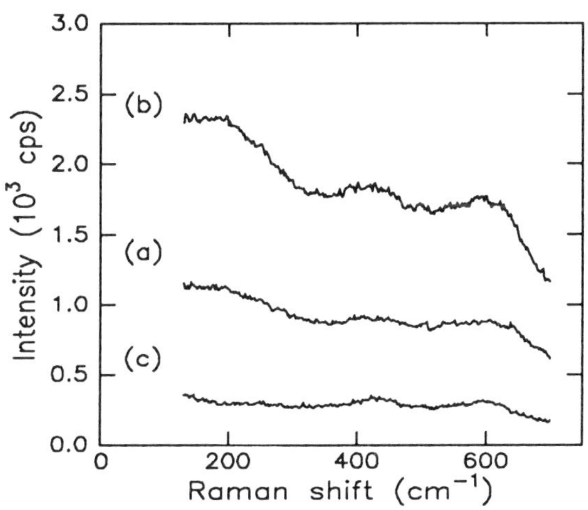

Figure 3: Raman spectra of debris formed at low speed (a), medium speed (b) and high speed (c) (laser power 15 mW, lens 20X).

The white wear debris created from a TiN coated disk sliding against an corundum ball was investigated by Raman spectroscopy and found to be TiO_2. In the debris produced at the highest speed, only rutile was found, whereas for the debris formed at low and medium sliding speed a mixture of anatase and rutile could be identified. Because of their extremely small size, these debris particles were very sensitive to heating effects under the laser beam.

4. CASE STUDY 2: SOLAR BEAM OXIDIZED TiN AGAINST CORUNDUM IN A FRETTING TEST.

It has been reported that slightly oxygen-deficient rutile (TiO_{2-x}) can behave like a low shear strength, lubricious oxide [7]. In the previous case study it was shown that different polymorphs of TiO_2 can be formed during wear of TiN. Here, the friction and wear behavior of a TiN coating partially oxidized by surface treatment with solar radiation, is investigated. Micro-Raman spectroscopy was used to analyze wear-induced modifications of the oxidized TiN surface.

PVD TiN (5.6 µm thick, R_a = 0.3 µm) was deposited on a polished high speed steel (ASP23) disk using a Balzers ion-plating equipment. The coated samples were treated in ambient air with concentrated solar radiation in the Central Receiver System of the Almeria Solar Plant (Spain). Heating the sample to 800 °C and holding this temperature during 10 s, induced an oxidation of the TiN. A uniform oxide layer of 1.1 µm thick was formed. The oxidized TiN coating was tested against a corundum counterbody ball of 10 mm diameter under fretting conditions (displacement stroke 50 µm, load 2 N). The frequency was fixed at 8 Hz, and the tests were run for 50,000 cycles.

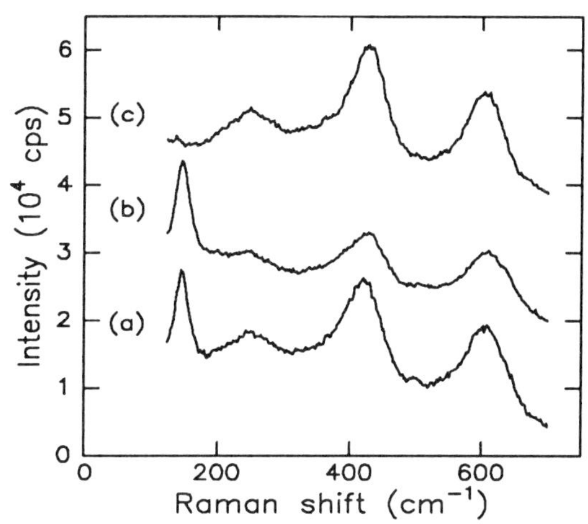

Figure 4: Raman spectra of debris formed at low speed (a), medium speed (b) and high speed (c) (laser power 150 mW, lens 20X). Spectrum (b) and spectrum (c) were shifted up by 15,000 and 30,000 cps respectively.

Figure 5 displays the evolution of the coefficient of friction for the oxidized TiN coating against a corundum ball during a test of 50,000 fretting cycles. From the start of the test, the coefficient of friction rapidly increases until a maximum value of about 0.7 is reached after 5,000 cycles. At this point, an abrupt drop of the coefficient of friction to 0.2 occurs. The shear stresses caused by the high frictional forces acting on the contacting bodies were so high that the interface strength was exceeded in the central contact zone, resulting in a failure of the titanium oxide/TiN interface. A thick third body layer is created which almost entirely fills the central zone of the contact area. This layer originates from the local detachment of the oxide layer from the TiN coating, and the entrapment of the debris in between the contacting bodies.

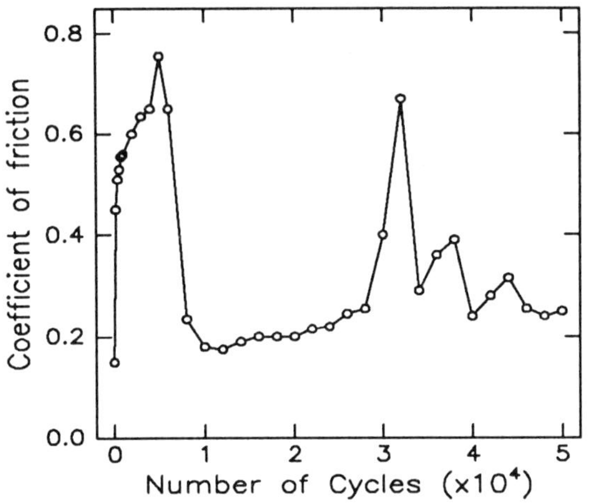

Figure 5: Coefficient of friction versus number of cycles during fretting of an oxidized TiN coating against a corundum ball.

The fluctuations of the friction force after 30,000 cycles are probably due to an extension of the area of oxide layer detachment.

The Raman spectrum of the original surface oxide layer on the TiN coating (Figure 6a) shows the typical features of rutile. The characteristic phonon bands at 448 cm^{-1} and 611 cm^{-1} closely match the values reported for single crystal rutile [5]. Raman spectra were taken from the peripheral contact zone (Figure 6b), from the wear debris scattered in and outside the contact area (Figure 6c), and from the layer covering the central zone of the contact area (Figure 6d). Although the surface morphology of the peripheral contact zone is severely modified after 50,000 fretting cycles, no distinctive change of the Raman spectrum compared to the original state is observed. The Raman spectrum of the powder-like wear debris (Figure 6(c)) exhibits the rutile features as well, but peak broadening together with peak shifts to 444 cm^{-1} and 600 cm^{-1} has occurred. Further peak shifting to 440 cm^{-1} and 592 cm^{-1} is found on the Raman spectrum of the layer covering the central contact zone (Figure 6(d)). The peak shifting of the 447 cm^{-1} line can be linked to the oxygen deficiency of the rutile compound [5]. Significantly broader bands were reported in case of an amorphous TiO_2 phase that generally exhibits a spectrum similar to that of rutile [8].

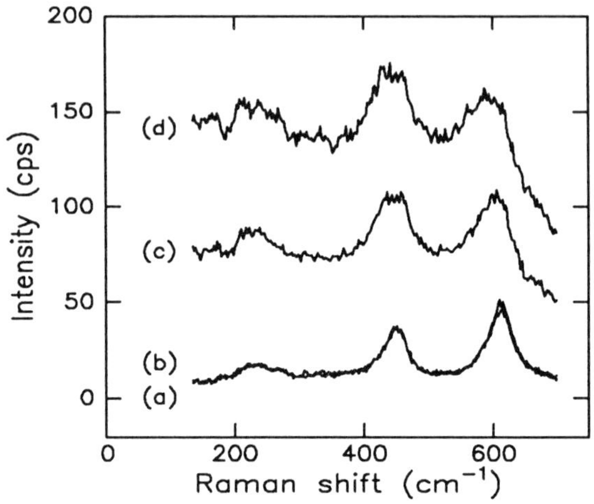

Figure 6: Raman spectra of (a) the original oxidized TiN coating, and the wear-induced surface modifications: (b) the peripheral contact zone, (c) the scattered wear debris, and (d) the layer covering the central contact zone (laser power 5 mW, lens 100x, slit width 2 cm^{-1}).

Raman spectroscopy has proven to be an interesting technique to determine surface modifications induced by fretting on an oxidized TiN coating. Due to a strong Raman activity and structure-related sensitivity of the line shapes and positions of the rutile spectrum, the use of this technique may provide a better understanding of the role of titanium oxides. A distinctive peak shifting and broadening of the rutile spectrum was observed as a result of wear-induced modifications in this work. For an improved understanding of structural modifications of rutile additional studies of the effect of specific defects (e.g. vacancies, nano-size particles, ...) on the Raman spectrum are necessary.

5. CASE STUDY 3: FRETTING WEAR OF CVD DIAMOND COATINGS

CVD diamond films are an important new material for tribological applications. To fathom the applicability of these diamond coatings, their tribological behavior needs to be understood.

Diamond/diamond and diamond/cermet tribo-pairs were subjected to fretting wear tests. Diamond coated silicon nitride balls were used as counterbodies. The flat test samples were WC-Co (grade K10) inserts and diamond coated WC-Co inserts. Diamond coated silicon nitride balls were used as counterbodies. The evolution of the coefficient of friction for such fretting contacts is shown in Figure 7. In the case of diamond wearing against diamond, a sharp drop of the coefficient of friction is observed during the first 1,000 cycles from 0.5 to values below 0.1. In diamond versus WC-Co contacts, the coefficient of friction gradually increases from values at around 0.4 reaching values of about 0.7 after 0.5×10^6 cycles.

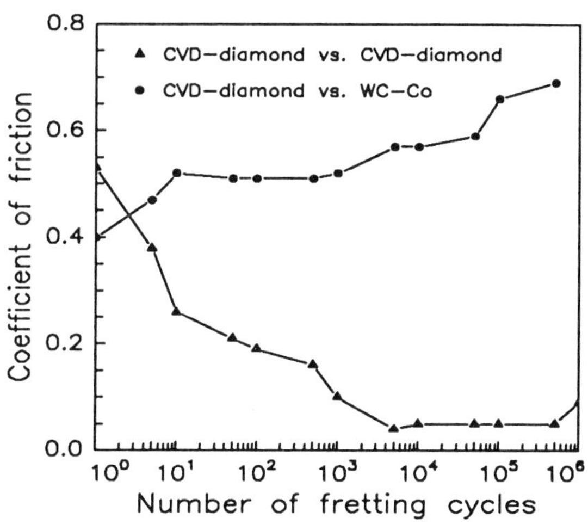

Figure 7: Coefficient of friction of diamond versus diamond and diamond versus WC-Co tribo-pairs determined in ball-on-flat fretting tests (displacement 100 µm, normal load 1N, vibrational frequency 8 Hz, relative humidity 60%).

Analysis of the wear damage in diamond-diamond tribo-pairs by SEM revealed flat worn areas on both of the originally rough diamond surfaces. Raman spectroscopy was used to extricate information about the surface modification of the diamond coated flat specimen. The spectrum of a pristine diamond surface is displayed in Figure 8a exhibiting a sharp diamond peak at 1338 cm^{-1} and a broad peak feature around 1570 cm^{-1}. This feature indicates the presence of some sp^2-bonded carbon in the coating [9]. The peak of the CVD diamond coating is shifted by 5.5 cm^{-1} compared to that of a single crystalline diamond specimen shown in Figure 8c. This shift can be attributed to compressive stresses present in the diamond coating resulting either from the mismatch in the coefficient of thermal expansion between diamond and the WC-Co substrate or from the growth of the CVD diamond coating. A relationship between peak shifts of the diamond line and stresses were given by Sharma et al. [10] suggesting a shift of 5 cm^{-1} corresponds to stresses of approximately 2 GPa. Peak broadening (9 cm^{-1} FWHM) is observed for the polycrystalline diamond coating compared to 3.3 cm^{-1} in the single crystalline specimen. Such broadening has been explained as the effect of an increased defect density [11]. A Raman spectrum representing the center of the flat worn area is shown in Figure 8b. A lower scattering intensity of the diamond peak is observed. This can be associated with a reduced coating thickness due to wear. A linear increase of the Raman scattering

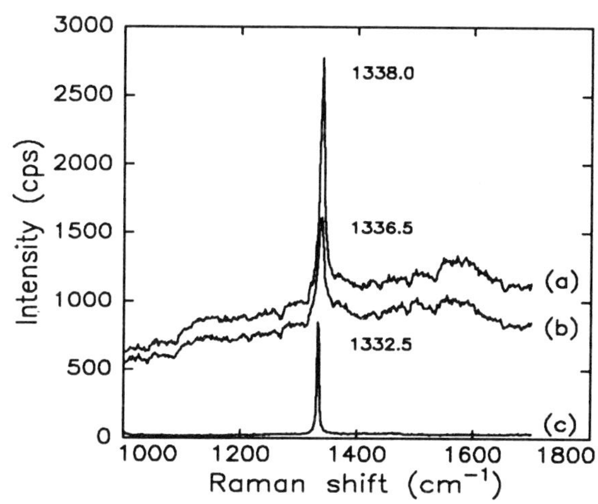

Figure 8: Raman spectra of (a) CVD-diamond as deposited, (b) CVD-diamond after fretting wear against CVD-diamond, (c) and single crystalline high pressure diamond; (laser power 100 mW, lens 100X).

intensity with the diamond film thickness within the focusing depth of the laser has already been indicated [12]. The peak shift in the worn area is smaller than in the original surface indicating a lower residual stress state. No significant change in the scattering contribution of the broad peak feature at 1570 cm^{-1} could be found. This leads to the conclusion that the fraction of sp^2-bonded carbon was not altered during the fretting test. The temperature-shear-induced phase transformation from diamond to graphite, which has been suggested as a self-lubricating mechanism [13] could not be verified. Graphite, whose scattering cross-section is about 50 times larger than that of diamond has a sharp peak at 1580 cm-1 [12]. Within the sensitivity of the present measurement no such peak could be observed. The low energy input by fretting wear and the high thermal conductivity of diamond seem not to promote the conditions for graphitization.

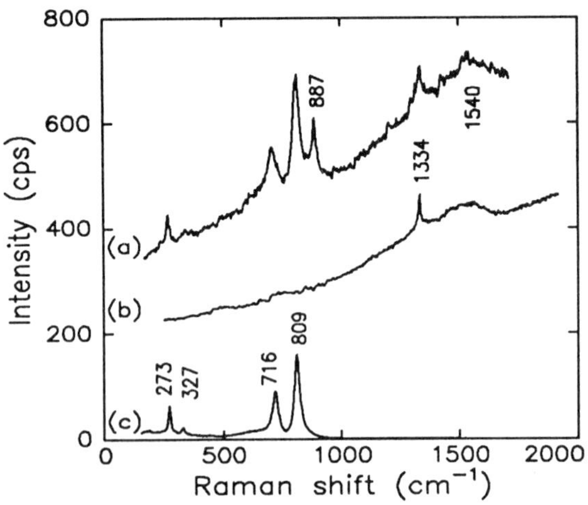

Figure 9: (a) Transfer layer on CVD-diamond after fretting wear against WC-Co, (b) CVD-diamond as deposited, and (c) tungsten oxide (WO_3) reference powder (laser power 20 mW, lens 100X).

A different form of surface modification is found in case a diamond coating was tested against WC-Co. SEM-analysis showed the formation of a transfer layer on top of the diamond coating. The sharp asperities of the diamond coating were masked and seemed to remain undamaged. The chemical composition of the transfer layer was determined qualitatively with an energy dispersive X-ray detector (Be-window) in the SEM. This indicated the presence of tungsten and cobalt substantiating the transfer of material from the WC-Co flat to the diamond coated counterbody. A Raman spectrum taken on the transfer layer is displayed in Figure 9a. The characteristic peaks of the as-deposited diamond coating (Figure 9b) are apparent notwithstanding the transfer layer is covering the diamond coating. Five additional peaks emerge due to the presence of the transfer layer. The first four peaks of a WO_3 reference powder (Figure 9c) are also present in the spectrum taken of the transfer film. Tribo-oxidation of WC to WO_3 was also observed by Fayeulle et al. in wear experiments between WC-Co and MoS_2 coated steel [14]. The peak at 887 cm-1 indicating the presence of another Raman active compound in the transfer layer could not yet be identified. Raman spectra taken on WC and Co_2O_3 reference powders do not exhibit Raman peaks and can therefore not be the origin of this peak.

The Raman spectroscopy investigation demonstrated that the residual stress state of diamond films can be altered during wear tests and that a transfer film containing WO_3 is formed on the diamond coating in the wear of diamond against a cemented carbide.

6. CASE STUDY 4: FRETTING WEAR OF DIAMOND-LIKE CARBON AGAINST CORUNDUM

During the last decade diamond-like carbon (DLC) coatings have received a substantial interest as they exhibit interesting physical, chemical and mechanical properties [15] and can be produced by a wide range of techniques at reasonably low temperatures. DLC materials also have favorable tribological properties and it was recently shown that they could be promising as protective coatings in applications limited by fretting wear [16]. DLC coatings have been intensively studied by Raman spectroscopy. The investigation of debris formed during the wear of DLC coatings however has up to now not been investigated extensively.

DLC coatings were deposited on a polished M2 steel substrate by arc ion plating with a graphite target in an Ar atmosphere[17]. A 2 N loaded polished corundum ball of 10 mm diameter was oscillated against the DLC coated flat at a stroke of 100 µm and a frequency of 8 Hz under ambient conditions (23°C, 60% RH).

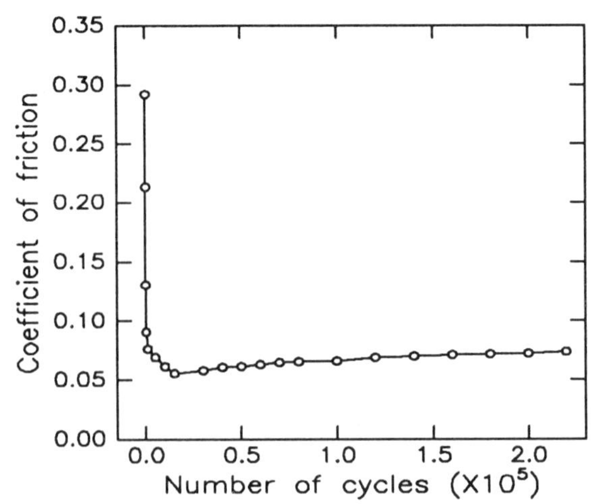

Figure 10: Coefficient of friction as a function of the number of fretting cycles for a diamond-like carbon coating against a corundum ball.

During the fretting test the coefficient of friction dropped from an initial value of 0.3 to a low value between 0.05 and 0.10 (Figure 10). It was observed that when the corundum ball was shifted after a run of several thousand cycles to an unworn area of the coating, the coefficient of friction in this second test attained the low value of 0.1 instantaneously. To elucidate this effect a fretting test was run for 2000 cycles and the worn areas of the DLC coating as well as the corundum counterbody were analyzed with Raman spectroscopy.

The Raman spectrum of the as-deposited coating is shown in Figure 11a. The spectrum consists of the overlap of two broad peaks at around 1340 cm^{-1} and around 1580 cm^{-1} often referred to as the D and G peaks [18]. A spectrum taken of the

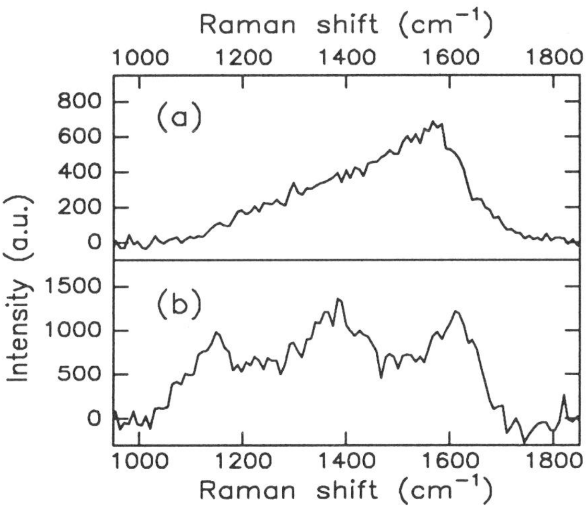

Figure 11: (a) Raman spectrum of the as-deposited DLC coatings. (b) Raman spectrum of the wear debris ejected from the fretting contact zone in the direction of the vibrational movement after fretting wear against a corundum counterbody (2000 cycles).

debris at the edge of the fretting pit on the DLC coating clearly contains three peaks (Figure 12). Two of the peaks of the spectrum can be assigned to sharpened D and G peaks of the as-deposited material. This sharpening has also been noticed during thermal annealing of DLC coatings [19] and is due to a partial graphitization of the diamond-like carbon structure. The interpretation of the peak around 1150 cm^{-1} is not yet totally accepted. Nemanich et al. have suggested that this feature may be due to the presence of nanocrystalline or amorphous diamond [20]. The hertzian contact pressure maximum for the fretting test conditions equals 700 MPa, which is in the stability regime of graphite. At this time we can only suggest that this feature at 1150 cm^{-1} is due to a defected form of graphite, which needs to be further identified. In Figure 12 a spectrum taken from material transferred to the corundum ball is presented. When compared to the spectrum of the debris ejected out of the contact zone on the DLC flat (Figure 11b) and that of the as-deposited coating we can notice that it is intermediary between these two: the D and G peaks have sharpened, and the peak feature around 1150 cm^{-1} is emerging. This can be understood assuming that the debris gradually transforms during the ejection from the contact zone to the state of the spectrum in Figure 11b.

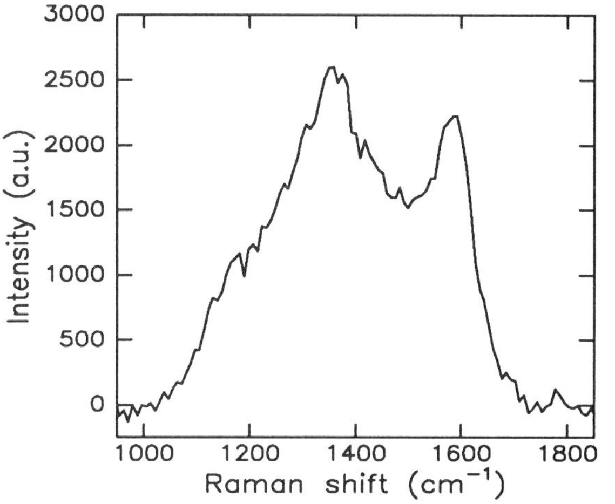

Figure 12: Raman spectrum of debris at the edge of the contact zone on the corundum ball after fretting wear (2000 cycles) against the DLC coating.

These results indicate that the friction and wear behavior of the arc evaporated DLC coating discussed is strongly influenced by the formation of debris showing the characteristics of disordered graphitic particles, which act as a lubricating body between the DLC coating and the corundum counterbody. A peak feature at 1150 cm^{-1} has been observed in the spectra of the debris and is speculated to be a disorder type in graphite that needs to be identified further.

CONCLUSIONS

Micro Raman spectroscopy has been shown to supply useful information on the wear induced surface modification of hard coatings as a result of its ability to detect local phase reactions and internal stress changes. This was demonstrated with four particular cases of laboratory wear testing on hard coatings.

It was found that in the wear of TiN coatings against corundum the debris consists of anatase and rutile components, where the anatase component is reduced as the sliding speed is increased. Solar beam oxidized TiN coatings exhibit Raman spectra with distinct rutile peaks, which are broadened and shifted considerably as a result of wear testing. For diamond coatings wearing against diamond it was observed that the internal stress is modified during a wear test; diamond wearing against a cemented carbide resulted in a WO_3 containing transfer film on the diamond. In the wear of diamond-like carbon coatings against corundum the formation of a disordered graphitic debris was established and its relevance to the frictional behavior of this tribo-pair was indicated.

ACKNOWLEDGMENTS

We are thankful to M. Van Stappen (WTCM, Diepenbeek, Belgium), T. Leyendecker (Cemecoat GmbH, Aachen, Germany) and F. Löffler (RWTH, Aachen, Germany) for providing the TiN, diamond and diamond-like carbon samples respectively.
One of the authors (BB) is grateful to the National Fund for Scientific Research (NFWO, Belgium) for a senior research fellowship. This work has been financed in part by IUAP and IWONL contracts.

REFERENCES

1. A. Gangopadhyay, S. Jahanmir and B.E. Hegemann, Proceedings of the 16th Leeds-Lyon Symposium on Tribology, eds. D. Dowson, C.M. Taylor and M. Godet, Tribology series **17** (1990) 63.
2. E. Vancoille, J.R. Roos, L.M. Stals, Proceedings of the 17th Leeds-Lyon Symposium on Tribology, eds. D. Dowson et al., Tribology series **18** (1992) 193.
3. J.R. Roos, J.P. Celis, M. Franck and H. Pattyn, Surf. Coat. Techn. **45** (1989) 89.
4. E. Vancoille, J.P. Celis, J.R. Roos, to be presented at Wear of Materials Conference, San Fransisco (1993) and accepted for publication in Wear.
5. J.C. Parker, R.W. Siegel, J. Mater. Res. **5** (1990) 1246.
6. T.B. Massalski, Binary Alloy Phase Diagrams, American Society for Metals (1982), 1793.
7. M.N. Gardos: MRS Symp. Proc. 140 (1989) 325.
8. P.L. White, G.J. Exarhos, M. Bowden, N.M. Dixon, D.J. Gardiner: J. Mater. Res..**6** (1991) 126.
9. L.H. Robins, E.N. Farabaugh, and A. Feldman, J. Mater. Res., **11** (1990) 2456.
10. S.K. Sharma, H.K. Mao, P.M. Bell, and J.Z. Xu, J. Raman Spectros. **16** (1985) 350.
11. R.G. Buckley, T.D. Moustakas, L. Ye, and J. Varon, J. Appl. Phys. **66** (1989) 3595.
12. D.S. Knight and W.B. White, J. Mater. Res. **2** (1989) 385.
13. M.N. Gardos and B.L. Soriano, J. Mater. Res. **5** (1990) 2599.
14. S. Fayeulle, P.D. Ehni, and I.L. Singer, Surf. Coat. Techn. **41** (1990) 93.
15. J. Robertson, Surf. Coat. Techn. **50** (1992)185.
16. B. Blanpain, J.P. Celis, J.R. Roos, J. Smeets and J. Ebberink, Thin Solid Films **223** (1993)65.
17. O. Knotek, F. Löffler, J. Brand and W. Burgmer, Surf. Coat. Techn. **49** (1991) 370.
18. J. Wagner, M. Ramsteiner, Ch. Wild and P. Koidl, Phys. Rev. B. **40** (1989) 1817.
19. N. Wada, P.J. Gaczi and S.A. Solin, J. Non-Cryst. Sol. **35** (1980) 543
20. R.J. Nemanich, J.T. Glass, G. Lucovsky, and R.E. Shroder, J. Vac. Sci. Technol. **A6** (1988) 1783.

Thin Films in Tribology / D. Dowson et al. (Editors)
© 1993 Elsevier Science Publishers B.V. All rights reserved.

Application of the Thermomechanical Wear Transition Model to Layered Media

R.S. Cowan and W.O. Winer

Tribology and Rheology Laboratory, Georgia Institute of Technology
George W. Woodruff School of Mechanical Engineering
Atlanta, GA 30332-0405 USA

Sliding surfaces in contact produce a stress field and frictional heat source which may induce severe wear from material yielding. The initiation of such thermomechanical wear will be influenced by a surface film, possessing different thermal and mechanical properties than those of the substrate. Conditions for layered systems that can be treated as monolithic, and therefore amenable to the application of the current thermomechanical wear transition model, are discussed considering thermal and isothermal effects. A range of layer thickness and material combinations are presented, such that when coupled with the conditions of operation can be used for design guidance. As an example, steel coated with titanium nitride is examined by comparing calculated data, based on the thermomechanical wear model, with experimental results.

1. INTRODUCTION

When components in relative motion are mechanically engaged, the region of contact experiences a temperature rise which may affect the performance and longevity of the respective surfaces. This common observation has been associated with the dissipation of thermal energy, produced in overcoming frictional resistance as sliding at the contact interface occurs. At the micro-level, this temperature increase can be substantial. A localized change in material properties, an enhancement in chemical reactivity, and ultimately, failure of the mechanical system can result.

By considering the thermal stresses generated from frictional heating, Ting and Winer [1] analyzed the complete stress field of the sliding contact by superimposing its effects on the isothermal stresses induced from surface tractions. Results were integrated with material yield criteria to promote the postulate that plastic deformation on the surface is a wear mechanism. A model for determining the dominant factors which would reduce or control such thermomechanical wear was developed.

Expressed in the form of wear equations, the thermomechanical wear theory provides a means for determining a critical temperature and a non-dimensional critical stress, for use in identifying the transition between mild and severe thermomechanical wear. The relationships are governed by fundamental material and thermal properties, forming three non-dimensional parameters which are measures of time, material property dependence on temperature, and thermal stress intensity. Good agreement between the theoretical expectations and experimental results for dry and lubricated contacts under load has been demonstrated [2,3].

A surface film, possessing different thermal and mechanical properties than those of the substrate, will influence the behavior of frictional heating and plastic deformation at the loaded contact [4]. It will not only affect the coefficient of friction, but also influence the dissipation of thermal energy. As a result, layer thickness and material combinations exist that will deny the onset of thermomechanical wear.

This review explores the conditions in which layered systems can be treated for effective application of the current thermomechanical wear transition model. Such criteria should be of value to design engineers, concerned with the likelihood that layered media will severely wear in an application of sliding contact.

2. THERMOMECHANICAL WEAR ANALYSIS

At the microlevel, it is observed that a seemingly smooth surface is composed of a series of asperities. Thus, when contact between surfaces is made under low pressure, the interface is not one coherent area, but is made up of several small regions where respective surface peaks touch. Assuming these real areas of contact are sparsely distributed, they may be treated as isolated spherical asperities, which can be locally represented by a spherical asperity in contact with the flat surface of a semi-infinite body [5].

In the thermomechanical wear model, a circular contact patch of radius, a, and a maximum Hertzian pressure, P_0, occurs when a total normal load, F_N, is applied between the stationary spherical asperity and the moving semi-infinite flat surface. Serious thermal effects are expected at the asperity, as it receives more of the generated frictional heat than any particular point of contact on the surface moving with velocity V.

The total stress field of the sliding contact, σ_{ij}, is comprised of isothermal and thermal stresses. The former may be calculated from the analyses of Hamilton and Goodman [6,7]. The latter may obtained via the Navier's equation with a thermoelastic displacement potential as demonstrated in [1]. Expressed as:

$$\sigma_{ij} = P_0 \{ \bar{\sigma}_{ij}^{i} (\underset{\sim}{x}, \nu, f) + G_t \cdot \bar{\sigma}_{ij}^{t} (\underset{\sim}{x}, \nu, Fo) \} \quad (1)$$

the total stress field is a function of contact location, $\underset{\sim}{x}$, Poisson's ratio, ν, friction coefficient, f, and Fourier number, Fo. The parameters $\bar{\sigma}_{ij}^{i}$ and $\bar{\sigma}_{ij}^{t}$ represent non-dimensional isothermal and thermal stresses. The variable G_t, signifying the relative importance of the thermal stress contribution, can be used as a thermomechanical wear-control parameter. It is defined as:

$$G_t = \frac{E\,\alpha_t}{k\,(1-\nu)} \gamma f V a \quad (2)$$

where E, α_t, and k are the stationary material's elastic modulus, thermal expansion coefficient, and thermal conductivity. The fraction of total frictional heat input to the stationary body is γ.

The response of a material to an applied stress may be either brittle or ductile depending on the temperature. Ceramics, for instance, exhibit brittle failure at low temperature and ductile failure at high temperature [8]. Accordingly, the fracture criterion used in assessing thermomechanical wear differs. By utilizing von Mises theory as a wear criterion for ductile material, the second invariant, J_2, of the total stress field σ_{ij} is related to the material's tensile yield strength, σ_y, by

$$\sqrt{3 \cdot J_2} \geq \sigma_y \quad (3.1)$$

For brittle material, the maximum stress theory is used to relate the maximum principal stress, σ_1, to the yield strength by

$$\sigma_1 \geq \sigma_y \quad (3.2)$$

Recognizing that the tensile yield strength can be typically expressed by a linear function of temperature as

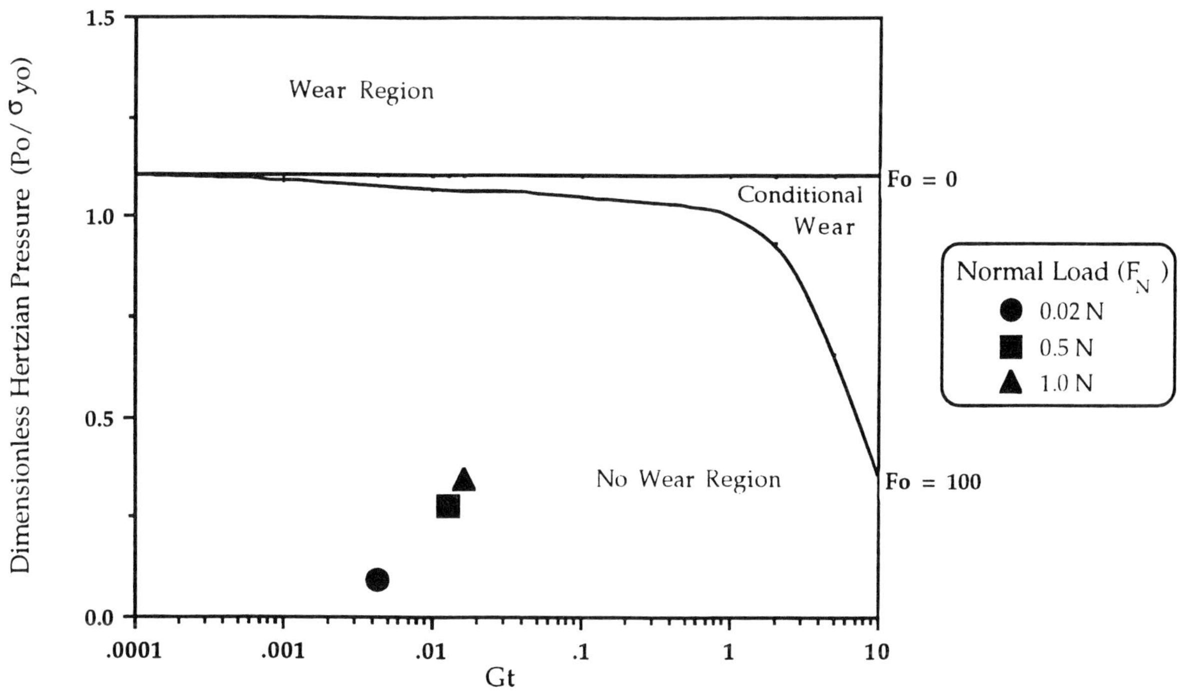

Figure 1. Wear Map for TiN (f = 0.46)

$$\sigma_y = \sigma_{yo} - n \cdot T \tag{4}$$

Ting and Winer [1] hypothesize that thermomechanical wear is initiated when the Hertzian contact pressure, P_o , satisfies the following basic relationship:

$$P_o \geq C_o \cdot \sigma_{yo} \tag{5}$$

The value C_o is dependent on material and thermal properties, the total stress field, and the selected failure criterion, σ_{yo} is a reference tensile yield strength at a reference bulk temperature, T is the temperature rise above the bulk temperature, and n is a constant coefficient. By definition, the dimensionless Hertzian pressure on the contact is P_o/σ_{yo}, which is used to develop thermomechanical wear maps for designer use [2].

Figures 1 and 2 are examples of wear maps with an operating point represented by P_o/σ_{yo} and G_t, for a specific Fourier value, Fo, and friction coefficient, f. The operating condition relative to the dimensionless Hertzian pressure, P_o/σ_{yo}, will determine whether thermomechanical wear occurs as suggested by equation (5).

The wear map consists of three regions: wear, no-wear, and conditional wear. The wear region is above the Fo = 0 line, the no-wear region is the area below the curve of an Fo value associated with thermal steady-state, and the conditional wear region is between the Fo = 0 line and the steady-state curve. Operation in the no-wear region means no thermomechanical wear will result, because the Hertzian pressure is less than the value required to cause yielding. Likewise, an operating point in the immediate wear region causes prompt wear as the Hertzian contact

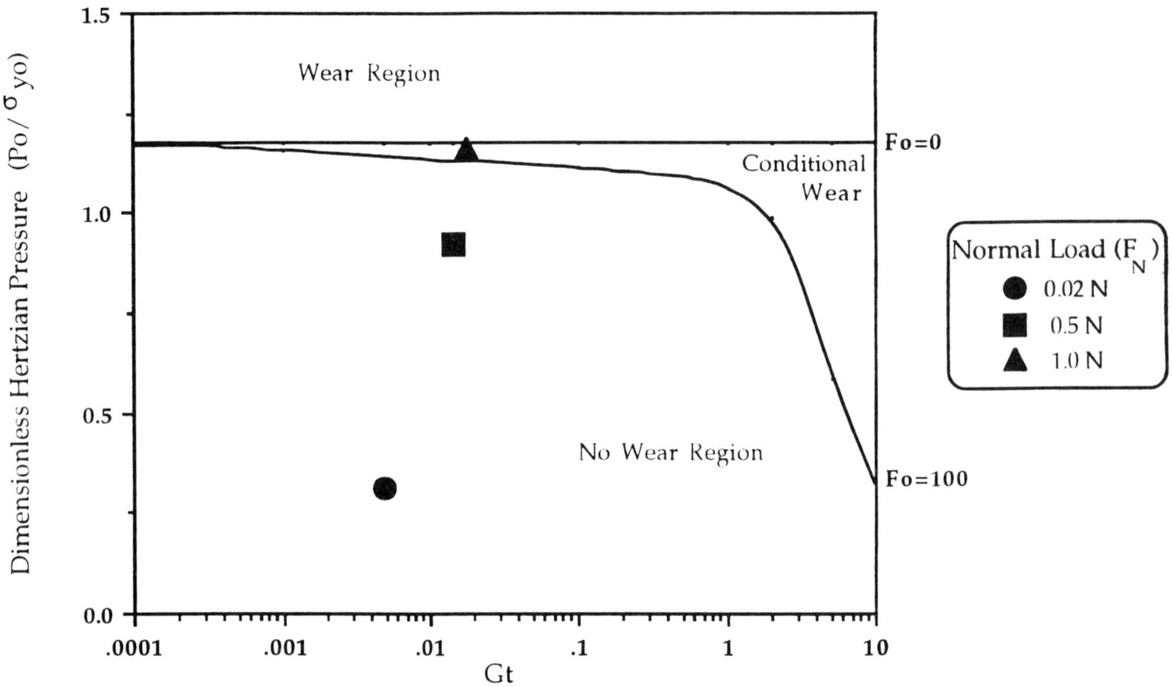

Figure 2. Wear Map for AISI 1095 Steel (f = 0.46)

pressure exceeds that required for yielding. If an operating point is located in the conditional wear region, the occurrence of wear depends on the magnitude of *Fo*, that is, how long the frictional heat source exists.

3. MONOLITHIC TREATMENT OF LAYERED MEDIA

The quality of surfaces in sliding contact is often enhanced by protective coatings, layered composites, and thin films. As expected, it has been observed that for hard films on soft substrates, load capacity increases as the layer thickness increases, while the converse is true for soft films on hard substrates [9,10]. Yet, detailed knowledge of the frictional contact stress field is required for understanding the impact of the layer and substrate on such mechanical reliability issues as fatigue, fracture, interface delamination, and plastic deformation. Generally, such studies are presented for applications possessing extremely thin or thick surface layers, thereby permitting the body to be treated as monolithic [11,12].

For successful application of the current thermomechanical wear model, the layered system must emulate a monolith. Thus, film thickness and material combinations shall be identified for consistent prediction of thermomechanical wear, recognizing their impact on the contact's friction coefficient and the resulting isothermal and thermal stress fields.

3.1. Sliding Contact Model

Given the thermomechanical wear criteria of Section 2, the sliding contact

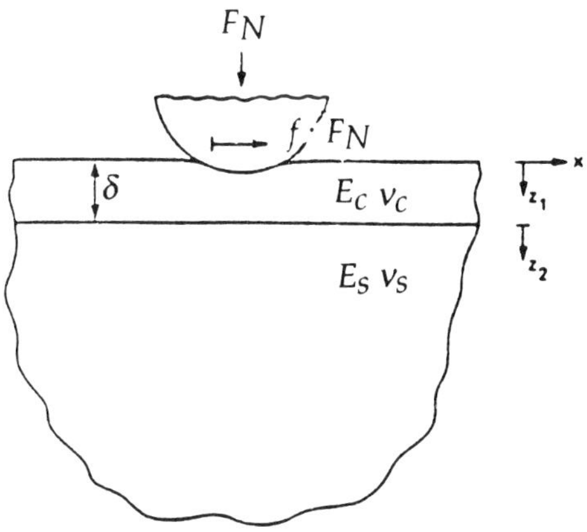

Figure 3. Contact problem notation and cross-sectional geometry

configuration is altered to reflect the conditions and results cited in the literature for layered media. Figure 3 depicts the cross-sectional geometry for the sliding contact between a spherical indenter and an elastic layered half-space.

Potential material property differences between the film layer (coating), substrate, and indenter are noted through the use of lower-case subscripts *c*, *s*, and *i*, respectively. The layer thickness is represented by δ.

It is assumed that the film layer is perfectly bonded to the substrate. The radius of the loaded region is a, F_N is the total normal load, and $f \cdot F_N$ is the x-directed total tangential force. When the film and substrate materials are identical, the uncoated or Hertzian case results, where a_0 and P_0 represent the half-contact width and maximum pressure used for normalizing results. Residual stress and inertial effects are neglected.

3.2. Thermomechanical Wear Factors

3.2.1. Coefficient of Friction

The coefficient of friction, f, is sensitive to a variety of factors. These include the material composition, surface finish, sliding velocity, and temperature, as well as the presence of contamination, lubrication, and humidity. As a result, for any two surfaces, f may fluctuate over several orders of magnitude varying with time and location. Since it directly enters the isothermal and thermal stress calculations to the first power, the coefficient of friction provides a major source of analysis uncertainty.

Komvopoulos [13] investigated the mechanical aspects of boundary-lubricated layered surfaces and concludes that when the layer is not disrupted, the friction coefficient is on the order of 0.1 and the wear tracks are indistinguishable from the virgin surface. Fragmentation and removal of the surface layer is found to result with relatively high friction coefficients. The reported presence of adhesion and plowing under such circumstances will likely mask the initiation of thermomechanical wear.

3.2.2. Isothermal Considerations

Studies of a Hertzian contact as shown in Figure 3 provide an indication as to how the contact size can vary because of layer and substrate material interaction. Figures 4 [14] and 5 [15] summarize such results, where:

$$1/E_s^* = (1 - \nu_s^2)/E_s + (1 - \nu_i^2)/E_i \tag{6.1}$$

$$1/E_c^* = (1 - \nu_c^2)/E_c + (1 - \nu_i^2)/E_i \tag{6.2}$$

The actual footprint radius is a, with a_0 representing the contact radius from only bulk film material. The pressure P is similarly denoted. Observe that as the film thickness (δ) increases, a tends to a_0. For any constant values of load and indenter radius, a is approximately equal to a_0 for $\delta \geq a_0$.

Through the use of contour plots, O'Sullivan and King [15] illustrate the impacts of layer stiffness and friction coefficient on the isothermal stress field. Using a least squares iterative approach with Popkovich-Neuber elastic potentials, the layered-surface sliding contact problem was

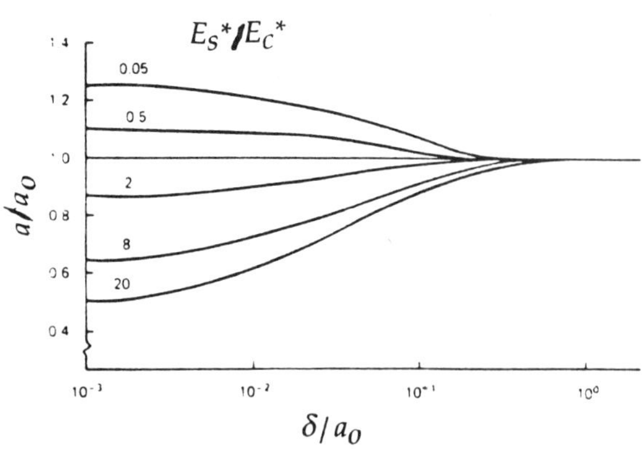

Figure 4. Effect of film thickness on Hertzian contact of coated surfaces. Ref. source [14]

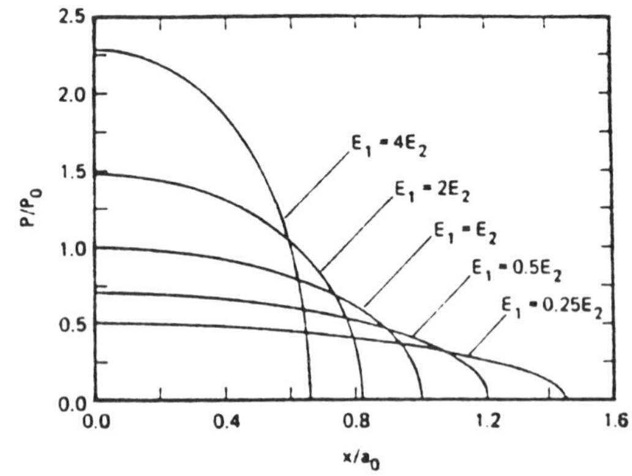

Figure 5. Normal contact pressure profiles beneath indenter along x-axis. Ref. source [15]

examined. Consideration was given to the von Mises and maximum normal stress results for ductile and brittle failure, assuming a film thickness (δ) of a_0, where a_0 represents the contact radius for the non-layered case ($E_C = E_S$). For a stiffer layer ($E_C = 2 \cdot E_S$) of thickness δ, the constant load yields a reduced contact size ($a = 0.819 \cdot \delta$) and von Mises stresses significantly higher than the non-layered case for various friction coefficients. The effect of a stiffer layer amplifies the expected maximum tensile stress increases associated with rising friction coefficients. In contrast, for a compliant layer ($E_C = 0.5 \cdot E_S$) of thickness δ with a range of friction coefficients, an increased contact size ($a = 1.207 \cdot \delta$) occurs exhibiting lower von Mises and tensile stresses than the non-layered case.

From [15], such trends suggest that for values of a_0 where $2 \cdot a_0 < \delta$, the maximum stresses are in the layer, similar to the homogeneous case with modulus E_C. For values of a_0 where $(a_0/6) > \delta$, the isothermal stress field is dominated by the substrate. In between ($a_0/6 \leq \delta \leq 2 \cdot a_0$), the stress field is heavily dependent on the value of E_C relative to E_S.

3.2.3. Thermal Considerations

The impact that a layered media has on the thermal stress field is directly related to its influence in dissipating thermal energy. This would be reflected by a change in the surface temperature increase (or flash temperature) expected for a sliding contact.

Ling solved the heat transfer problem of a layered half-space with a moving heat source [4]. The result is presented in the form of an integrated infinite series. For ease of analysis, Gecim and Winer [16] closely approximated the thermal problem by modeling a quickly rotating layered cylinder, subjected to a small heat source.

From the governing differential equations [16], Figure 6 was developed, illustrating how the surface temperature may behave as a function of the thermal contact coefficients of the substrate (b_S) and layer (b_C). The thermal contact coefficient, b, may be computed from

$$b = k/\sqrt{D} \tag{7}$$

where k and D are the thermal conductivity and thermal diffusivity of the appropriate material. Presented in non-dimensional form, the plot gives an indication of the surface

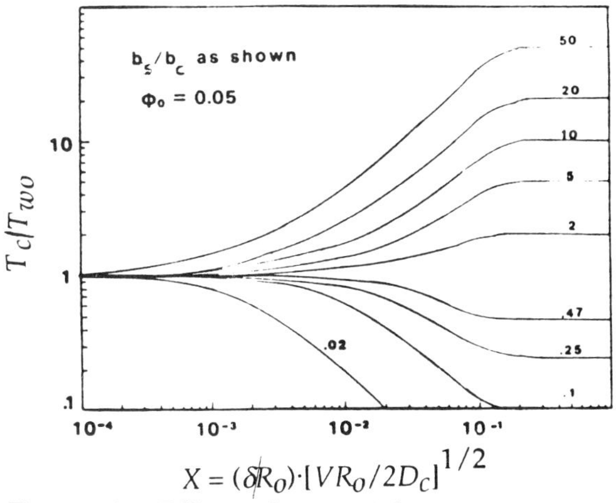

Figure 6. Effect of material characteristics and operating conditions on maximum surface temperature rise with film to that without. Ref. source [16]

temperature rise with the film, T_c, to that without, T_{wo}. The horizontal axis is represented by X, where

$$X = (\delta/R_o)\cdot[VR_o/2D_c]^{1/2} \qquad (8)$$

The variable R_o represents the surface radius (with film), V the surface velocity, D_c the thermal diffusivity of the layer, and δ the film depth.

For an angular contact width of $\Phi = 0.05$ radian, typical of a gear, cam, or rolling element bearing contact, Figure 6 suggests that for the thin film/low speed condition, $X < 10^{-3}$ thermal penetration goes well below the film into the substrate; thus, film thermal resistance becomes insignificant, and the temperature ratio approaches unity. For the thick film/high speed case, $X > 10^{-1}$, the temperature rise is confined within the film, such that the temperature ratio approaches the ratio of the thermal contact coefficients. In between ($10^{-3} \le X \le 10^{-1}$), thermal penetration is heavily dependent on the value of b_s relative to b_c. Note that a heat source larger than 0.05 radian would shift the curves to the right, implying that the region where the film dominates would require thicker film/higher speed.

4. EXPERIMENTAL ASSESSMENT

Using the current thermomechanical wear transition model, wear maps were created for TiN and AISI 1095 steel, as shown in Figures 1 and 2. An operating point in the wear map is represented by P_o/σ_{yo} and G_t, for a specific Fourier value and friction coefficient. Material and thermal properties [17,18] used to generate these graphs are given in Table 1.

Very little data exist with regard to the wear failure of layers in sliding contact. Komvopoulos, Saka, and Suh [19] examined the significance of layer thickness, normal and tangential surface tractions, and mechanical properties of layer and substrate on wear, through pin-on-disk experiments. Titanium and steel materials, with and without TiN coatings of various thicknesses, were used.

Some of their findings are presented in Figures 1 and 2, following data conversion to a dimensionless Hertzian pressure and thermal stress parameter, G_t. Assuming the thermomechanical wear stress field is confined to the TiN layer (Figure 5):

$$G_t = \frac{E\,\alpha_t}{k\,(1-\nu)}\gamma f\,V\,a = 4.44 \times 10^4 \cdot f\,V\,a_o \qquad (9.1)$$

where γ is the heat partition factor (=0.50, a maximum value for low Fourier number) [20]. The Hertzian contact radius, a_o, for a hemispherical pin-tip and flat disk subjected to a normal load, F_N, is given in [14] as:

$$a_o = \left[\frac{3\,F_N\,R_t\,(1 - \nu^2)}{2\cdot E}\right]^{1/3} \qquad (9.2)$$

Table 1
Representative Mechanical and Thermal Properties of Selected Materials

	E (GPa)	α_t (m/m-°C)	k (W/m-°C)	ν	σ_{yo} (MPa)	n (MPa/°C)
Titanium Nitride	250	8×10^{-6}	30	0.25*	2000.*	0*
AISI 1095 Steel	200	13×10^{-6}	40	0.30	525.	0.2

*Assumed
Ref. source: [17, 18]

where Rt is the pin-tip radius. Rt was reported to be 0.0032 m. Substituting the material properties for TiN:

$$a_o = 2.62 \times 10^{-5} \cdot (F_N)^{1/3} \qquad (9.3)$$

The ratio of the peak normal pressure, P_o, to the reference tensile yield stress can be expressed as

$$P_o/\sigma_{yo} = 0.348 \cdot (F_N)^{1/3} \qquad (9.4)$$

Similarly, assuming the thermomechanical wear stress field is confined to the AISI 1095 steel substrate (Figure 2):

$$G_t = \frac{E\,\alpha_t}{k\,(1-\nu)} \gamma f\, V\, a = 4.64 \times 10^4 \cdot f\, V\, a_o \qquad (10.1)$$

$$a_o = 2.80 \times 10^{-5} \cdot (F_N)^{1/3} \qquad (10.2)$$

$$P_o/\sigma_{yo} = 0.870 \cdot (F_N)^{1/3} \qquad (10.3)$$

The experimental data presented were obtained using steel substrates with a 0.2 µm thick TiN coating. A steady-state friction coefficient of 0.46 was reported. For each load condition, a contact radius exceeding 6·δ or 1.2 µm resulted, suggesting that the isothermal stress state is heavily dependent on the properties of the steel substrate. With a thin film and slow sliding velocity of 3. cm/sec, equation (8) yields an X value << 10^{-3}, indicating thermal penetration goes well below the film into the substrate. This is reflected in the values of G_t. Using the guidelines of Section 3, the thermomechanical wear model can be employed, treating the layered system as a monolith of steel.

It was reported that the 0.02 N load case showed no evidence of wear. Burnishing occurred at 0.5 N, with severe damage present from a normal load of 1.0 N. This is consistent with the thermomechanical wear predictions of Figure 2. As expected, treating the TiN as a monolith to detect thermomechanical wear yielded poor results (Figure 1).

5.0 CONCLUSION

A method has been presented for determining the conditions in which a layered media can be treated as monolithic. Dependent on isothermal and thermal factors, layer thickness and material combinations can be evaluated so as to minimize the likelihood of thermomechanical wear.

Applications, however, do exist in which the layered system cannot be treated as monolithic. For such situations, successful utilization of the current thermomechanical wear model is suspect.

As mechanical and thermal demands on machine elements increase, the ability to identify the transition between mild and severe wear becomes increasingly important. Coupled with the promising surface engineering developments associated with layered media, the concepts addressed in this review should be of increasing value to design engineers, concerned with product longevity and performance.

REFERENCES

[1] B. Y. Ting and W. O. Winer, *Journal of Tribology*, **111**, 1989, pp. 315-322.

[2] J. Yang and W. O. Winer, *Journal of Tribology*, **113**, 1991, pp. 262-269.

[3] J. Yang, R. S. Cowan, and W. O. Winer, "Prediction of Failure Transitions in Sliding Contacts by a Thermomechanical Wear Theory", *Journal of Tribology*, To be published.

[4] F. F. Ling, *Surface Mechanics*, (New York: Wiley Interscience), 1973.

[5] E. Ioannides and J. C. Kuijpers, *Journal of Tribology*, **108**, 1986, pp.394-402.

[6] G. M. Hamilton and L. E. Goodman, *ASME Journal of Applied Mechanics*, **33**, 1966, pp. 394-402.

[7] G. M. Hamilton, *Proc. Inst. Mech. Engrs.*, **197C**, 1983, pp. 53-59.

[8] L. E. Toth, *Transition Metal Carbides and Nitrides*, (New York: Academic Press), 1971, p. 169.

[9] S. Bair, S. Ramalingam, and W. O. Winer, *Wear*, **60**, 1980, pp.171-180.

[10] J. M. Georges and E. Rabinowicz, *Wear*, **14**, 1969, pp. 171-180.

[11] P. K. Gupta and J. A. Walowit, *Journal of Lubrication Technology*, 1974, pp. 250-257.

[12] Y. C. Pao, T. Wu, and Y. P. Chui, *Journal of Applied Mechanics*, 1971, pp. 608-614.

[13] K. Komvopoulos, *Tribology Transactions*, **34**, 1991, pp. 266-280.

[14] R. D. Arnell, P. B. Davies, J. Halling, and T. L. Whomes, *Tribology: Principles and Design Applications*, (New York: Springer-Verlag), 1991, pp 19-20.

[15] T. C. O'Sullivan and R. B. King, *Journal of Tribology*, **110**, 1988, pp. 235-240.

[16] B. Gecim and W. O. Winer, *Journal of Tribology*, **108**, 1986, pp. 92-97.

[17] W. J. Lackey et al., *Proc. International Symposium on Ceramic Components for Heat Engines*, Japan, 1983.

[18] *Metals Handbook*, 9th Edition, Vol. 1, (ASM: Metals Park), 1978.

[19] K. Komvopoulos, N. Saka, N. P. Suh, *Journal of Tribology*, **109**, 1987, pp. 223-231.

[20] B. Y. Ting, "Thermomechanical Wear Theory", PhD Dissertation, Georgia Tech, 1988.

Thin Films in Tribology / D. Dowson et al. (Editors)
© 1993 Elsevier Science Publishers B.V. All rights reserved.

TRIBODYNAMIC MODELING

S. L. Rice, F. A. Moslehy, and S. Elmi

This paper concerns the dynamic modeling of a generic tribosystem involving solid/solid contact, consisting of two mechanical subsystems and an interface, i.e., any wear test apparatus such as a pin-on-disc machine. Typically in such machines, one mechanical subsystem is comprised of shafts and gears and bearings and a power source which provides motion. This motion is resisted at the tribological interface by a second mechanical subsystem comprised of "stationary" elements. Both subsystems are capable of vibration in various modes. The vibrations are excited by tribological interactions, including processes of debris formation and attrition, and time-varying relative motion of asperities and debris within the interface. The vibrations are important because they result in fluctuating normal and frictional forces which experimentalists measure and attempt to interpret. The vibrational displacements are also significant and not widely appreciated in terms of their influences on crack initiation and propagation in the interface region. Moreover, designers of equipment having planar joints should be aware of the magnitudes and frequencies of vibrations which occur due to solid/solid interaction at tribological interfaces.

This paper presents modeling techniques being developed to deal with an entire tribosystem, including both mechanical "sides" of a generic machine, as well as the interface. Prior work has focussed only on the dynamic response of "one side" of such a system. However, all elements of a triboapparatus, including the frictional interface, have stiffness and damping attributes which must be considered. This paper describes methodology by which such elements can be included in a tribodynamic model. The methodology is applicable to any system having solid/solid tribocontact across an interface comprised of asperities and/or debris.

1. INTRODUCTION

Except for idealized special cases, friction and wear are complex phenomena. The complexity arises from the fact that there are many factors which operate either alone or in combination to yield a friction coefficient or wear rate for a system. Among the significant factors are operating parameters (such as relative surface velocity, nominal contact stress, system stiffness, and load cycling factors), ambient conditions, and numerous surface and near-surface physical and chemical material state variables.

Because friction and wear involve so many factors, the associated theory has remained relatively undeveloped. Simple friction "laws" are well known, but cannot be relied upon. Similarly, various predictive wear equations exist, but these are based upon empirical factors which mask complexity in "experimentally determined" coefficients.

This paper explores the portion of this complexity which is due to dynamic phenomena associated with the mechanics of tribocontact. Such "tribodynamic modeling" seeks to determine time variations in surface displacements as well as in the resultant frictional (and normal) forces which are transmitted across and through the interface. Data resulting from such modeling are particularly important in understanding the behavior of systems where tribocontact occurs through asperities and/or debris acting across thin interfacial films, with or without a lubricant. Indeed, vibrational amplitudes in the normal direction may be large compared to film dimension, and therefore significant in modifying local contact conditions. Similarly, transverse and/or torsional oscillations influence asperity and debris interaction, and therefore are important in both initiation and propagation of surface and subsurface cracks.

Quite simply, contact mechanics must consider

dynamic as well as quasi-static interactions between specimen and counterface. In actual systems, neither the resultant frictional force, **F**, nor the applied normal load, **N**, is constant. Fluctuations in these forces can excite vibrations in the triboapparatus, as can variations in debris/asperity interaction, and non-uniform distribution of transfer material. In addition to fundamental mechanistic effects, the vibrations of a tribosystem lead to significant variations in its friction and wear characteristics.

For example, it has been shown that simply by changing the specimen stiffness in pin-on-disc tests with metals, not only are wear rates dramatically altered, both the amount and the direction of transfer between specimen and counterface materials is significantly different [01]. A similar finding occurs when composite dental restorative materials are tested against enamel [02]. Thus, what some investigators would erroneously interpret as behavior due to material factors is instead attributable to tribodynamic phenomena.

Various researchers have studied dynamic aspects of parts of a triboapparatus [03-09]. Usually the apparatus is "separated" at the frictional interface, and one "side" of the machine is excited by perturbations in the normal (**N**) and/or frictional (**F**) forces. Meanwhile, the dynamic behavior of the other "side" of the apparatus is neglected. This approach is satisfactory for mechanical systems such as vibratory feeders where a good model can be obtained with reference to motion which occurs only on one side of the frictional interface.

For tribotest machines, however, the above approach is not adequate since "both sides" of the apparatus must be considered. This is necessary because the force and motion fluctuations occurring on one side of the machine are transmitted to the other side through the frictional interface. Moreover, tribological events arising at the interface (such as production of wear debris or asperity interactions) contribute to the excitation of (both sides of) the overall system. In addition, vibration itself influences the transmission of forces and displacements, so there is "coupling" or "feedback" which affects dynamic behavior, and therefore, frictional and wear outcomes.

For example, asperities interacting with one another in the direction of sliding will do so more (or, alternatingly, less) frequently when a transverse oscillation is superposed on a nominally steady relative velocity. Similarly, vibrations in the direction of the normal load affect the nominal "separation" of specimen and counterface surfaces. As this separation changes, so does the (non-linear) stiffness of the interface, and this in turn affects the dynamic behavior of the system. Thus, tribological variables affect the dynamic behavior of the apparatus, and the dynamics of the machine affect tribological measurements/outcomes. Clearly, both "sides" of the apparatus, as well as the interface, must be incorporated into the tribodynamic model.

2. PIN/DISC MACHINE MODELING

For actual tribosystems, each must be modeled individually, since the particular mechanical configuration determines vibrational modes and response to given excitation. However, the major modeling issues are methodological and therefore, applicable to the generic system. Such a system is comprised of a "specimen side", an interface, and a "counterface side".

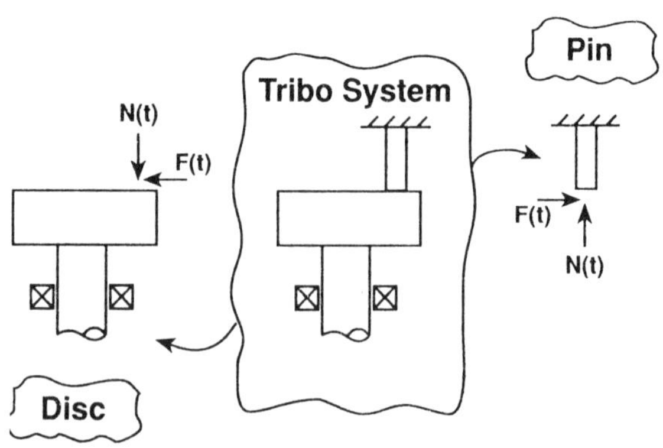

Dynamic Assemblies

Figure 1

Figure 1 shows a particular configuration of the pin-on-disc apparatus which is discussed herein. In this representative machine, the "specimen side" (pin subassembly) is nominally fixed, but obviously is capable of vibrating in various principal directions. Such motions are suggested in Figure 2, where fluctuations in the normal force (**N**) tend to excite longitudinal pin vibrations, and variations in the frictional force (**Fl**) cause transverse vibrations.

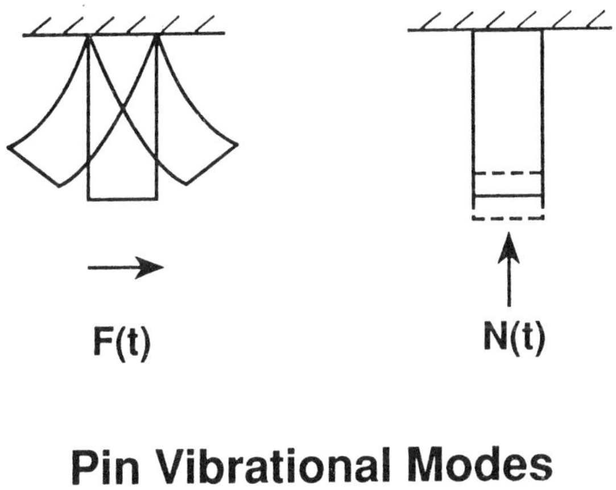

Figure 2

The "counterface side" (disc subassembly) of this machine includes the mechanical drivetrain as well as structural supports. As shown in Figure 3, the disc subassembly is subject to normal vibrations [excited by **N(t)**], transverse vibrations [excited by **F(t)** and a bending moment, **M(t)** due to the normal force, **N(t)**, acting through a moment arm about the disc centerline], and torsional vibrations [excited by **F(t)** acting at a radial distance from the disc centerline].

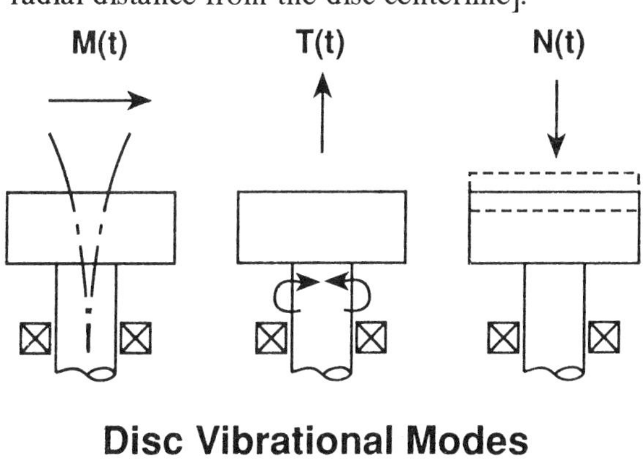

Figure 3

The frictional interface is shown in Figure 4, comprised of asperities and wear debris, both of which carry a portion of the applied normal (quasi-static) load. The portion carried by asperities is determined by the Greenwood/Williamson model [10], and that carried by debris is determined by equilibrium considerations [11]. The applied load and the nominal separation of the specimen and counterface surfaces, **d**, are related as shown in Figure 5 for various asperity and debris characteristics.

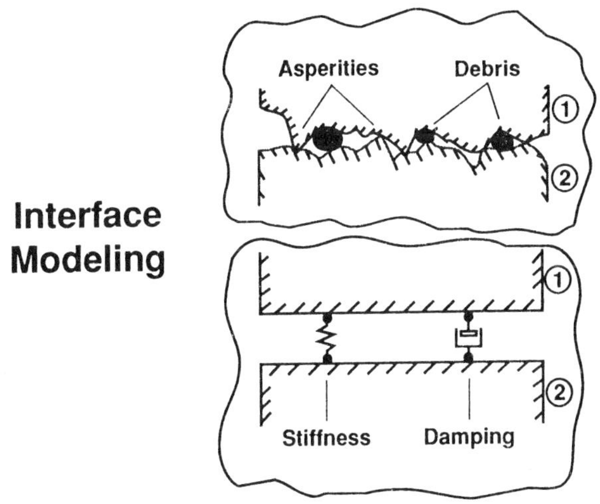

Figure 4

The load-"deflection" behavior shown in Figure 5 constitutes a non-linear stiffness which is represented by the lumped parameter interface "spring". It is clear that changes in asperity topography and in debris size or quantity affect the interface stiffness, as does the nominal separation between contacting surfaces. Thus, vibrations in the normal direction (which cause variation in nominal separation), **d**, and/or any events which alter interface geometry or composition lead directly to changes in the instantaneous stiffness of this spring.

Also represented in the interface model is a lumped parameter "dashpot" to account for energy losses. Equivalent viscous damping is assumed throughout the machine, with coefficients estimated by proportional damping [12].

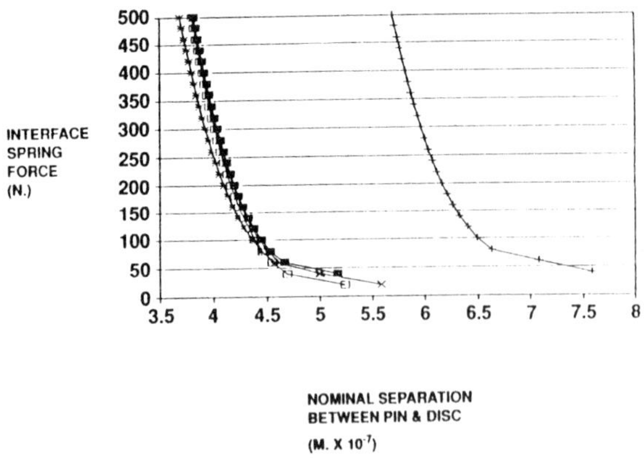

Figure 5

As suggested in Figure 4, the instantaneous normal load, **N(t)**, is comprised of terms due to interface stiffness and damping. The damping term is added to the stiffness term only when the velocities of the specimen and counterface elements are such that their surfaces are in relative "approach" (not when they are moving apart). The coefficient of friction, $\boldsymbol{\mu}$, is calculated based upon current conditions at each time step, and depends upon the instantaneous value of the nominal separation of specimen and counterface surfaces, **d** [11]. The current value of the friction force, **F(t)**, is obtained from the product of $\boldsymbol{\mu}$ and **N(t)**.

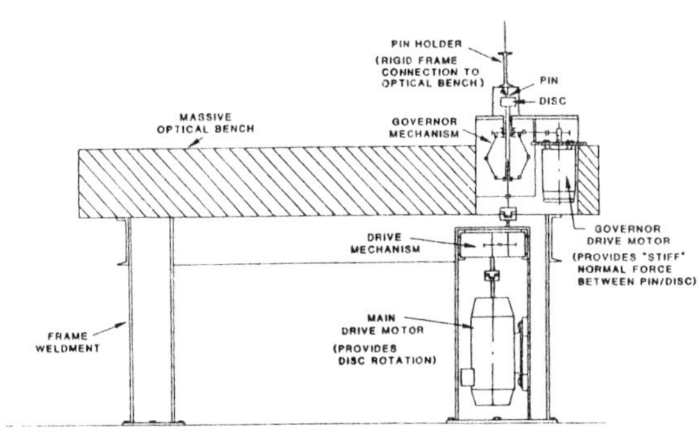

Figure 6

Figure 6 shows an overall schematic of the physical pin-on-disc system under consideration. This apparatus was developed for laser speckle studies of deformations in pin specimens in real-time wear tests [13]. Accordingly, the pin is designed to be (nominally) stationary and is fixed to a rigid frame which is mounted onto a massive optical bench. The counterface disc is loaded against the pin by a governor mechanism. The disc is rotated independently by a main drive motor which provides torque through gears in a drive mechanism, as seen in more detail in Figure 7.

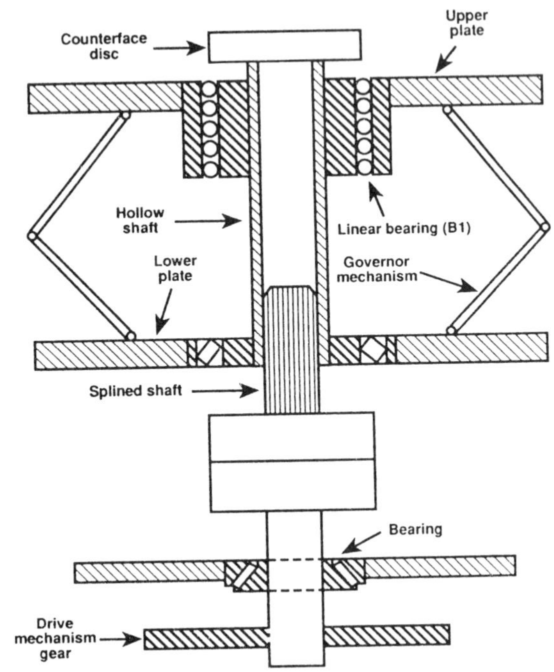

Figure 7

This figure shows the gear which rotates the splined shaft which drives the hollow shaft to which the counterface disc is attached. Because of the spline and bearing connections shown, this disc rotation occurs independently from motion of the governor mechanism, which is used to apply the normal load. This loading is due to inertial forces which depend upon the governor rotational speed (driven through the upper plate shown in Figure 7). The upper plate is constrained from vertical movement, so inertial forces of the governor act to pull up the lower plate. When the counterface disc is in contact with the pin specimen (not shown in Figure 7), the governor inertial forces are transmitted through the hollow shaft to create the normal loading between pin and disc, and the faster the governor speed, the greater the applied load, **N**.

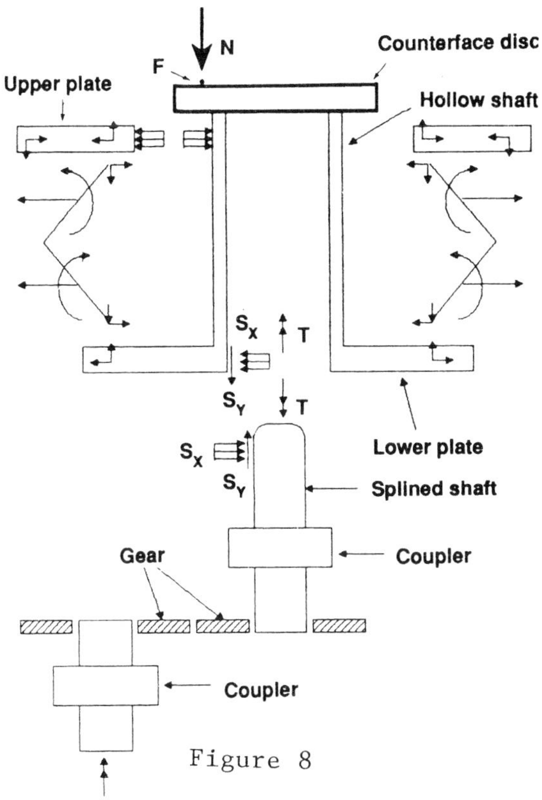

Figure 8

Figure 8 shows a free body diagram for elements of the drive mechanism of Figure 7. Because of the governor action, the normal load, **N**, is created between pin and disc, and when the disc is rotated, the frictional force, **F**, arises. Since the normal force acts through a moment arm to the disc centerline, **r**, a bending moment, **M**, with magnitude given by the product of **N** and **r**, is applied to the disc. Similarly, the friction force gives rise to a torque, **T**, with magnitude given by the product of **F** and **r**. Further reference to Figure 8 shows that the torque, **T**, is transmitted through the hollow and splined shafts, and thence through couplers and gears, to the main drive motor. In contrast, the bending moment, **M**, is equilibrated at the linear bearing, B1, where the slope of the hollow shaft is constrained to be vertical. Thus, different parts of the system are subject to different vibrational constraints.

Consider next the motion of the system in the normal direction (along the line of action of the normal contact force, **N**). The mass elements of the subassembly on the "counterface side" of the machine which are subject to vibration in this direction are shown schematically in Figure 9, as are stiffness components. The governor mechanism has its own effective mass and a single degree of freedom [14]. It also has a particular force/configuration or stiffness characteristic [14], which can be combined with that of the hollow shaft into an equivalent spring which physically has its fixed end at the upper plate, since this element is constrained from vertical motion.

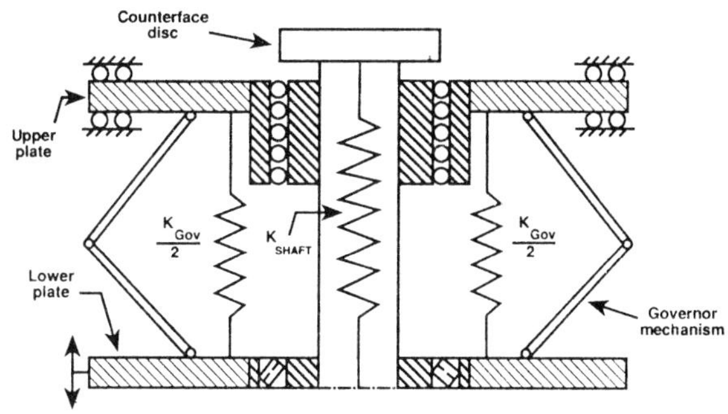

Figure 9

As noted earlier, the counterface disc is subject to a bending moment, **M**, so it deflects angularly as does a part of the hollow shaft to which it is attached. This deflection is constrained by the flexural stiffness of the shaft as well as the linear bearing which forces the deflection slope to zero, as mentioned above. This can be represented as a torsional spring which acts to return the disc surface to a horizontal position. Thus, the counterface disc has two degrees of freedom, each with associated stiffness and damping elements, as shown on the bottom part of Figure 10.

This figure also shows the stiffness and damping elements of the interface, as discussed previously. Note that there is not a fixed connection between the interface and the mass elements which it most directly influences. Thus, dynamic separation of specimen and counterface is possible, in which case the interface forces cease to act. Indeed, these forces are present only when there is effective compression across the interface, as represented by a minimum nominal separation, **d**, of specimen and counterface surfaces.

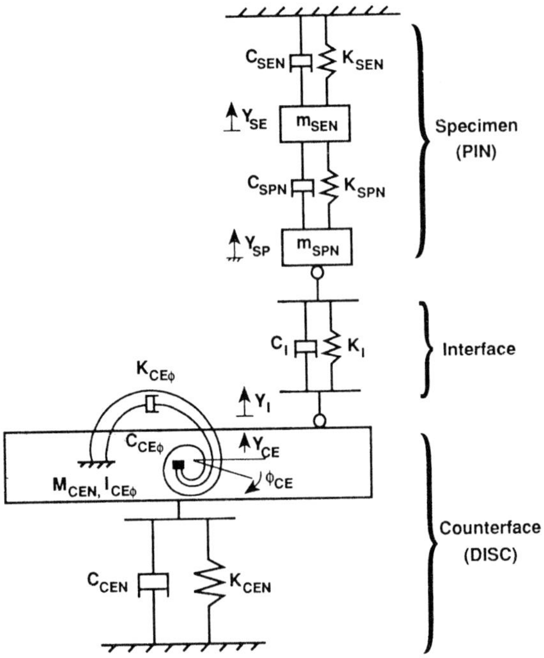

Figure 10

Figure 10 also shows two mass elements for the specimen, as well as associated stiffnesses and dampers. Two each of such elements were chosen in order to allow evaluation of the role of near-surface plastic deformation as has been documented in wear studies on metals [15]. Because plastically deformed zones have significantly reduced stiffness, it is of interest to explore this effect on the dynamic behavior of the system. Thus, the specimen mass is represented in two elements. The mass nearest the interface is very small, representing a plastically deformed zone of roughly 500 microns in thickness. When the specimen is being loaded (compressed), the stiffness of this zone is estimated from the tangent modulus from a representative stress-strain curve for a plastically deforming metal. When the machine dynamics are such that the specimen is being unloaded, the elastic stiffness is used in the model. Finite element analyses are used to determine stiffness values for specimens of various and geometries, of different materials, and having plastically deformed zones of varying depth.

Models for motion in the transverse and torsional directions are more straightforward, as are the equations of motion. Due to space limitations, details are not presented here. Excitation to the system is due to interactions of debris and asperities at the frictional interface. Because of the model used to represent the interface, several parameters can be changed to simulate changes in specimen and counterface surfaces which occur, say, during run-in. For example, asperity radius or height or density; or debris size or density; or near-surface material stiffnesses can be changed. For purposes of initial simulations, random variations in debris and asperity height have been chosen. Changes in these parameters do not necessarily occur at the same time, but rather depend on the average times between interactions. These times can be obtained by dividing the average distance between asperities (or between debris particles) by their relative speed of approach.

In the computer simulation, the time intervals between excitation events are monitored. When an interval expires, or when a computer run is initiated, a normal displacement pulse is calculated as the product of the average asperity height (or debris size) and a random number selected from the interval between 0 and 2. This displacement pulse is partitioned as it would be accommodated if the system were in static equilibrium; i.e., the more compliant side of the system absorbs more of the pulse than does the stiffer side. When the pulse is applied, the forces in the interface spring and dashpot are calculated, and are set equal to zero if necessary (e.g., if surfaces are separating, it is not physically realistic for a damping force to act to maintain their compression; similarly, if the surfaces have separated to the extent that the force in the interface spring has become zero, then the normal force, **N**, is set equal to zero).

Input parameters for computer implementation for dynamic simulations include those necessary to describe surface topography (such as required for Greenwood/Williamson calculations [10,11]), as well as nominal debris size and density, material properties and operating speed. For a given apparatus, preliminary calculations will have been completed to determine values for the stiffness and mass matrices of the system. A general Jacobi routine is used to calculate eigenvalues, from which damping coefficients are determined. Initial conditions for displacements and velocities are specified, and an initial nominal surface separation, **d**, at the interface, is calculated. The interface spring and damping forces yield the normal force, and the friction force is calculated as the product of μ and **N**.

A fourth-order Runge-Kutta technique is used to solve the equations of motion. Because system eigenvalues are relatively large (of the order of 10^14), a very small time step is required (of the order of 10^-8). Consequently, solutions are implemented on a Cray Y-MP Supercomputer.

3. CONCLUSIONS

Tribodynamic modeling techniques are described which incorporate both specimen and counterface sides of a pin-on-disc apparatus. While details are developed for a particular machine, the techniques are generally applicable. When computer results are available from a variety of cases, it will be possible to assess the importance of various parameters in terms of their roles in influencing fluctuations of normal and frictional forces. It will also be possible to quantitatively evaluate the dynamic oscillations in various vibrational modes, and thereby, to determine the roles of these oscillations in contributing to microfracture of asperities, to velocity accomodation mechanisms [e.g., 16], and to other tribological processes occurring within the interface zone.

REFERENCES

[01] Rice, S. L., Nowotny, H., and Wayne, S. F., the Role of Specimen Stiffness in Sliding and Impact Wear. Wear, 77(1): 13-28. 1982.

[02] Rice, S. L., Bailey, W. F., Pacelli III, P. F. T., and Blanck, W. R., Influence of Variation in Enamel Stylus Stiffness on the Sliding Wear Behavior of a Composite Dental Restorative. J. Dental Research, 61(3): 493-496. 1982.

[03] Ko, P. L., and Brockley, C. A., The Measurement of Friction and Friction-Induced Vibrations. J. Lubr. Tech., 92: 543-549. 1970.

[04] Tolstoi, D. M., Significance of the Normal Degree of Freedom and Natural Normal Vibrations in Contact Friction. Wear, 10: 199-213. 1976.

[05] Elder, J. A., and Eiss, N.S., A Study of the Effect of Normal Stiffness on Kinetic Friction Forces between Two Bodies in Sliding Contact. ASLE Trans., 12: 234-241. 1969.

[06] Broszeit, E., Kloos, K. H., Hess, F. J., and Wagner, E., Influence of Vibrations on the Wear Properties of a Tribological System, Wear, 28: 395-401. 1974.

[07] Soom, A., and Chen, J. W., Simulation of Random Surface Roughness-Induced Contact Vibrations at Hertzian Contacts during Steady Sliding, J. Tribology, 108: 123-127. 1986.

[08] Hess, D. P., and Soom, A., Normal and Angular Motions at Rough Planar Contacts During Sliding with Friction, J. Tribology, 114(3): 567-578. 1992.

[09] Suk, M., Ishii, T., and Bogy, D., The Influence of Air-Bearing Surface Geometry on the Dynamics of Sliders, J. Tribology, 114(1): 26-31. 1992.

[10] Greenwood, J. A., and Williamson, J. B. P., Contact of Nominally Flat Surfaces, Proc. R. Soc. Lond. A, 295: 300-310. 1966.

[11] Zhang, J., Moslehy, F. A., and Rice, S. L., A Model for Friction in Quasi-Steady-State Sliding, Part I: Derivation and Part II: Numerical Results and Discussion. Wear, 149: 1-26. 1991.

[12] Meirovitch, L., Elements of Vibration Analysis, McGraw-Hill, Inc. 1986.

[13] Seif, M. A., Mohr, P. J., Moslehy, F. A., and Rice, S. L., Deformation and Strain Fields in Pin Specimens in Sliding Contact by Laser Speckle and Metallographic Techniques, J. Tribology, 122: 506-513. 1990.

[14] Seif, M. A., Moslehy, F. A., and Rice, S. L., Dynamic Stiffness Analysis in Tribocontact, Wear, 119: 353-368. 1987.

[15] Rice, S. L., Nowotny, H., and Wayne, S. F., Characteristics of Metallic Subsurface Zones in Sliding and Impact Wear, Wear, 74(2): 131-142. 1982.

[16] Berthier, Y., Vincent, L., and Godet, M., Velocity Accomodation in Fretting, Wear, 125: 25-38. 1988.

Thin Films in Tribology / D. Dowson et al. (Editors)
1993 Elsevier Science Publishers B.V.

MODELLING OF DISCONTINUOUS TRANSFER FILMS

ROLE OF LOCAL SHEAR STRESS IN THE BUILD UP OF LOAD CARRYING CAPACITY

by M. BRENDLE P. TURGIS and R. GILMORE

Centre de Recherches sur la Physico-Chimie des Surfaces Solides (C.N.R.S.)
24, Avenue du President Kennedy F - 68200 MULHOUSE (France)

Abstract

Transfer on thoroughly polished surfaces frequently occurs as discontinuous films, consisting of isolated, discrete, fixed particles, able to play the role of asperities. In this work, assuming particles are dome shaped, a simple model is proposed to describe the transfer films, in which the two dimensional size is the same as that measured by image analysis, and to calculate its contribution to load carrying capacity. This model includes the various shear stresses existing along the transfer particles and analyses the mechanisms of particle growth. This latter is shown to be a determinant parameter in the build up of L.C.C. A change of the actual shear stress with temperature (indirectly assessed from the size of the fragments detached during the frictional process) accounts for most of the experimental results obtained previously, and supports the present model.

1. INTRODUCTION

While the role of transfer films, for instance in dry friction, is generally considered as most important in controlling the friction and wear of a given system, the mechanisms by which this actually occurs are still not well understood. There is general agreement that many parameters are involved, such as adhesive interactions, physical and chemical adsorption, mechanical properties and surface topography, but until recently, the links between these parameters, the extent of transfer and the frictional force were missing.

Transfer often occurs as incomplete films and consists of isolated discrete particles. Image analysis was revealed to be an interesting tool to study quantitatively the influence of some important parameters on the characteristics of transfer. This was for instance applied to graphite, when a pin made of compacted graphite powder is rubbed against thoroughly polished discs (1)(2). Furthermore, the transfer particles on a steel counterface have the important characteristic of displaying no no mobility and a progressive growth around fixed nucleation sites (2). This allowed us to show for the first time that the frictional behaviour of such transfer films can be understood by considering each particle as an asperity, either grooving, ploughing or carrying elastically the load.

Simultaneously, during the progressive build up of these transfer particles with increasing rubbing time, the variation of the frictional force often displays a maximum. Since the beginning, we suggested that this maximum could be the result of the onset of load carrying capacity (L.C.C.) on the transfer particles. In order to make this concept clearer, it is sufficient to consider the extreme situations: obviously, one particle alone is unable to carry the load, since in general the pressure which would then be developed on such a small area would largely exceed the hardness of the material. Consequently, if the bonding of this particle is strong enough, this particle can only groove the counterface. In contrast, L.C.C. can be achieved on the whole population of transfer particles existing within the nominal area of contact, provided they are in sufficient number and size for the mean applied pressure to be equivalent to the hardness.

A simple analysis reveals that growth only becomes possible if the shear stress developped between the particle and the pin surface is high enough to induce a local transfer, in general by a fracture process. Moreover, this shear stress is a function of the particle shape, the applied normal pressure and the adhesive interactions. This may explain the complexity of the system and the large variety of possible situations. For instance, some experimental results suggest that the growth may stop before full L.C.C. is reached on the transfer particles. Hence, the experimental observation of limiting coefficients of friction (as a function of time) associated with a limiting extent of transfer is not a criteria of full L.C.C. on the transfer particles.

The aim of this work is therefore to develop a model for transfer films, consisting of isolated, fixed discrete particles (as is actually the case for the graphite steel system) and to analyse the mechanisms of L.C.C. both in terms of topography and of interfacial interactions. It will also take into account the actual value of the shear strength existing at the particle surface. Finally, we will show how this model is supported by the experimental results obtained by image analysis, for instance by the influence of temperature on the actual interfacial shear resistance.

2. MODELLING OF TRANSFER FILMS

2.1. Modelling of individual particles.

Particle shape

In order to apply the basic concepts of tribology, is useful to consider each particle as being a simple hemispherical dome.

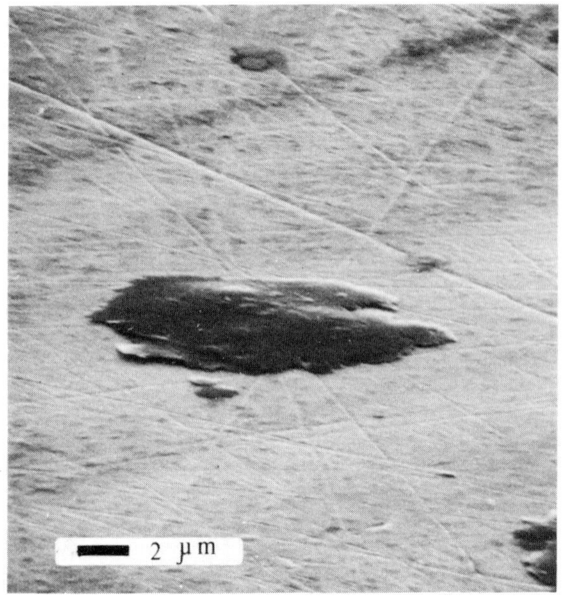

Fig.1: Typical view of a transfer particle of graphite on steel, as seen by S.E.M.

In Fig. 1 is shown an actual transfer particle as seen by S.E.M. with a steel surface tilt angle of 30° : it clearly shows that the particle has a certain height and displays indeed the shape of a spherical dome.

This led us to adopt the following model: the transfer particles will be considered as spherical domes having a height $\mathbf{h_i}$ and a radius $\mathbf{a_i}$ as illustrated in Fig. 2a :

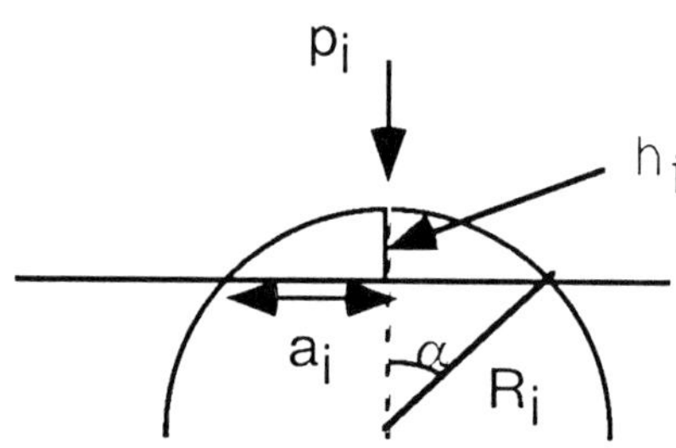

Fig . 2a : Schematic representation of transfer particle cross section.

Shear stresses

The completion of our transfer model requires the introduction of the shear stresses developed along the various interfaces of the particles and the limiting shear stress of graphite τ , as illustrated in Fig. 2b .

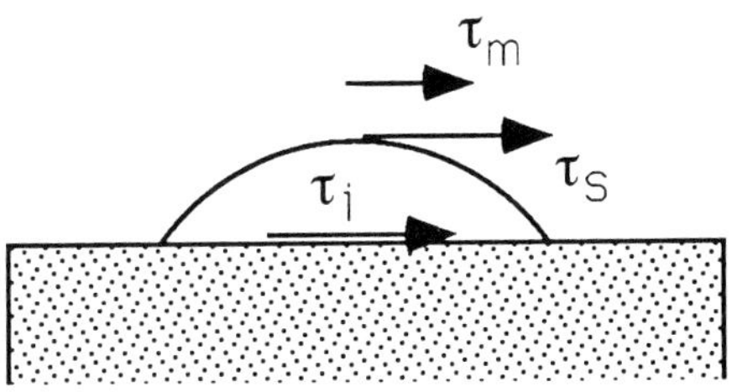

Fig. 2b : Schematic representation of the shear stresses existing at the various interfaces of a transfer particle.

where :

τ_i : resolved shear stress at the graphite steel interface;

τ_s : resolved shear stress at the graphite-graphite interface;

τ_m : maximal hertzian shear stress underneath the asperity;

τ_g : limiting shear stress of graphite

Referring to the work of Y. BERTHIER; M. GODET et al (3) the sites where nucleation will occur are then simply those where the following condition is fulfilled:

$$\tau_i > \tau_g$$

and similarly, the condition of particle growth by the following condition:

$$\tau_i > \tau_s > \tau_m > \tau_g$$

Of course, the actual values of these stresses are dependent on pressure, temperature and environment. Referring to B.J. Briscoe et al (4) the pressure dependance is of the form:

$$t = t_0 (1 + k.P)$$

In particular, this pressure dependance results in nucleation occuring at the points where the pressure is the highest, (areas of actual contact), and furthermore at the points displaying the highest adhesive interactions.

In the same way, the actual growth of a given particle is influenced by a large number of parameters and will be considered in more detail below.

2.2. Modelling of transfer films

A transfer film consists of many isolated particles, as illustrated in Fig. 3 .

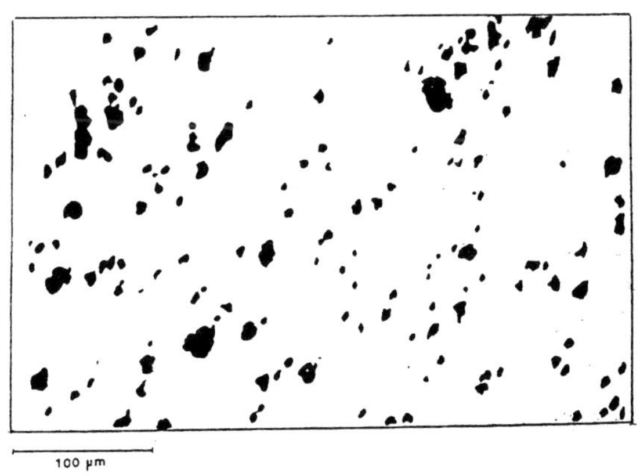

Fig. 3 : Typical view of a transfer film of graphite in highly contrasted S.E.M.

In this two dimensional view, the 3 dimensional shape of the particles and, for instance, the similarity in this 3 dimensional shape, which seems to characterise the transfer particles formed on a given surface (as characterised by a given surface energy and possibly a given surface contamination) when the particles are observed either by S.E.M. or with an optical stereomicroscope is not visible. In our model, the same projected area, i.e. πa_i^2 can correspond to transfer particles having different radius of curvature R_i and/or different height h_i.

The similarity in shape previously described leads us to assume that this similarity may be expressed by the constancy of the angle α formed by the tangent to the dome and the rubbing plane:

$$\sin \alpha = a_i / R_i = \text{constant}$$

Thus, within a given transfer film, the transfer particles differ from each other essentially by their radius of curvature. Taking n_i to be the number of particles having a projected area:

$$s_i = \pi a_i^2$$

the total area covered by the particles actually active within the nominal contact area, **S**, is:

$$S \sum n_i a_i^2 \pi = S X$$

where **X** is the area fraction covered by the transfer particles, and is measurable by image analysis.

3. LOAD CARRYING MECHANISMS

In order to clarify the evolution of the load carrying mechanism during the build up of a transfer film, it may be useful to recall that we are considering the case where a pin of soft material is rubbed on a hard, thoroughly polished disc. Thus, the actual area of contact a_r is determined by the hardness **H** of the pin by:

$$a_r = W / H.$$

In the absence of a transfer film, this area may possibly result from the roughness of the pin, and the sliding will necessarily occur at the graphite metal interface i.e. with a shear stress τ_i.

In the presence of a transfer film, the topography changes, and sliding will at least partly occur at the graphite-graphite interface along the transfer particles with a shear stress τ_s. While in the first situation the zones of high normal stress were stationary with respect to the pin, in the last situation they have to move across the pin by inducing either a purely elastic or elastoplastic deformation. In each case the mechanical response is different, as described below.

3.1. 1st Mechanism: Elastic load carrying

Again based on the assumption that each particle plays the role of a hemispherical indenter, it is possible to calculate the normal force which may be carried by the elastic deformation induced by the penetration of the particles into the "pin". Referring to HERTZ's law (5), the load, w_i, carried by one particle having a radius a_i is such that:

$$a_i = (3w_i R_i / 4E^*)$$

where E^* is defined by:

$$\frac{1}{E^*} = \frac{1-\gamma_1^2}{E_1} + \frac{1-\gamma_2^2}{E_2}$$

with E_1/E_2 and n_1/n_2 being respectively the Young's modulus and the coefficient of Poisson of the antagonist materials (It is interesting to note, that while both the pin and the asperities are made of the same material, the asperities appear in general to be tougher than the original material).

For a given particle, this load is limited by the penetration depth which cannot exceed the height of the particle, h_i. Nevertheless, the total normal load applied, **W**, can be carried by this mechanism, provided the number and size of the transfer particles are large enough. For instance, full LCC on the transfer particles is reached if:

$$W = S \sum n_i w_i = S \sum n_i a_i^3 \, 4 E^* / 3 R_i$$

$$\text{i.e.} = (4/3\pi) \, . \, E^* . \, S.\sum (a_i / R_i) \, . \, \pi a_i^2$$

$$= A S X \sin \alpha$$

$$\text{where } A = 4/3\pi \, . \, E^*$$

This relationship predicts that even in static conditions L.C.C. may be reached for different values of **X**; the other controlling parameters being the shape of the particles expressed here by the ratio:

$$a_i \,/\, R_i = \sin\alpha$$

The flatter the particles, the larger the area fraction needed to carry the load by elastic deformation.

Another parameter which may vary from one powder to another is **E*** ; and the stiffer the pin (high **E***) the smaller will be the area **X**.

3.2. 2nd Load Carrying Mechanism

During tangential displacement, each particle has to cross the contact by inducing elastic or plastic deformation of the pin. Each particle is then working more or less in the same way as an inclined plane and the driving force exerted along these surfaces necessarily induces a normal component contributing to LCC. In an elastic contact, the symetry gives a net result equal to zero. In a plastic contact, however, the rear side is either not in contact, or at least submitted to lower contact pressures (and hence lower shear stresses); this leads to a positive contribution to L.C.C.

The theoretical analysis of the elastic or plastic contact of a hemispherical indenter with interfacial shear stress is not yet completed. Therefore, in a simplified approach and to a first approximation, we will consider each half dome to be equivalent to an inclined plane, having the same projected area $\pi a_i^2/2$, but with a smaller angle α' .

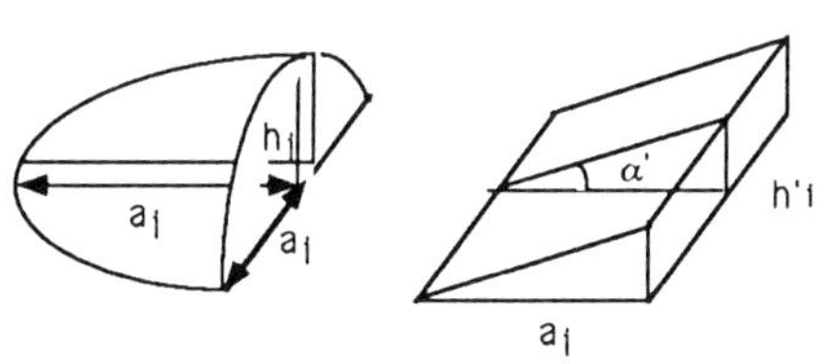

Fig. 4 Schematic representation of transfer particle contact as half dome and its rectangular inclined plane equivalent.

The resolved shear stress being τ , the shear resistance along the plane is:

$$f_i = \pi\, a_i^2 \cdot \tau_s \,/\, 2\cos\alpha'$$

and the corresponding normal component is:

$$f_i \sin\alpha' = \pi\, a_i^2 .\ \tau_s .\ \sin\alpha' \,/\, 2\cos\alpha'$$

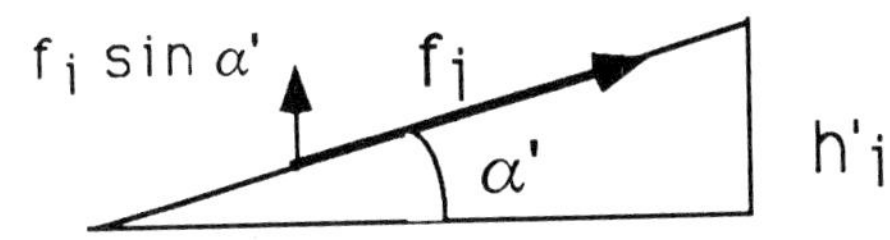

Fig. 5 : Schematic representation of the components of the frictional force along the transfer particle leading edge.

$$f_i \sin\alpha' = \pi\, a_i^2 .\ \tau_s .\ tg\,\alpha' \,/\, 2$$

$$= \pi\, a_i^2 .\ \tau_s .\ h' \,/\, 2\, a_i$$

where **h'** is the height of the equivalent inclined plane. The latter being the height of a rectangle having a base identical to the diameter of the particle and the area equivalent to that of the section becomes:

$$k\,\pi\, R_i^2 = 2\, a_i\, h'_i = 2\, a_i^2 .\ h'_i \,/\, a_i$$

hence:

$$h'_i \,/\, a_i = k\,\pi\, R_i^2 \,/\, 2\, a_i^2$$

And for the whole population of transfer particles it becomes:

$$\Sigma\, f_i \sin\alpha' = S\, \Sigma\, n_i.\pi\, a_i^2.\ tg\,\alpha' .\tau_s / 2$$

and hence the dynamic contribution of the frictional force to L.C.C.

$$W_D = S .\ tg\,\alpha' .\ \tau_s .\ X \,/\, 2$$

$$W_D = B .\ S .\ X.\ \tau_s$$

where $B = 1/2\ \ tg\,\alpha'$

This relation shows that the contribution of

frictional force to L.C.C. is again proportional to the actual shear stress along the transfer particles and to a geometrical parameter, which on the basis of our hypothesis is expressed here by $tg.\alpha'$.

For each of the mechanisms described above taken individually, L.C.C. appears to be proportional to the area fraction, **X**, covered by the transfer particles i.e. a parameter measurable by image analysis.However, each relation can only be actually applied to some extreme situation.

3.3. Actual mechanisms of L.C.C.

In a real intermediate situation, the models described above have to be completed by the following remarks:

- both mechanisms probably always coexist within a given population of transfer particles: the smallest particles interacting purely elastically, the largest inducing plastic deformation. The critical size, s_c, characterising the transition between these two behaviours has even been estimated experimentally and shown to be in the range of 10 to 15 μm^2 (2).

- whatever the activated mechanism, it is probable that the normal load is not necessarily entirely carried on the transfer particles. Even if the first mechanism of elastic L.C.C. is prevailing, a given proportion of the load may rest on the initial surface, i.e. without volumic interaction with the transfer particles.

Hence, for a real situation, the total normal load is carried by 3 different mechanisms, each on a different fraction of the total actual area of contact:

$$a_r = S\ X_h = W/H$$

If X_c is the area fraction corresponding to the particles larger than s_c, it becomes:

$$W = SA(X - X_c) + SB\ X_c\ \tau_s + SC(X_h - X)$$

Full L.C.C. on the transfer particles can only be achieved if the particle growth is not stopped before $X = X_h$. This points to the determinant influence of the parameters controlling the particle growth considered below.

4. CRITERIA OF PARTICLE GROWTH

Within a given population of transfer particles, each particle normally ignores the existence of its neighbours (at least until some critical particle density is reached) and the event of growth or non growth is only determined by the surrounding local physical and physico-chemical parameters.

It should therefore be possible to apply the information obtained from previous studies on model single asperities interacting with a flat surface. On an individual scale the growth of a transfer particle is characterised by the irreversible attachment of a small fragment previously detached from the counterface. This process is similar to the mechanism of wedge formation. In their systematic investigation, K. HOKKIRIGAWA and K. KATO (6) show that among the possible types of interactions, such as ploughing, wedge formation or cutting, the wedge formation is restricted to a limited domain, depending on the degree of penetration, the relative shear strength ratio of the two materials and the asperity geometry (see Fig. 6)

In our model, the degree of penetration:

$$h_i / R_i = a_i / tg\alpha\ R_i$$

is constant for a given population. On the other hand, the relative shear strength ratio is normally also kept constant for a given experiment.

Hence all the particles usually display the same probability of growth, and the experimentally observed decrease in particle growth with increasing duration of rubbing can only result from a progressive change in the shape of the particles, i.e. in the angle α (7).

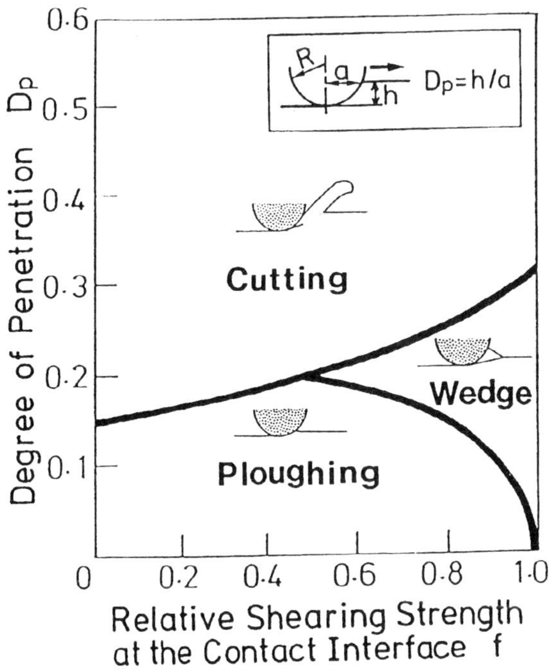

Fig. 6 Wear mode diagramm in abrasive wear according to K. HOKKIRIGAWA and K. KATO (4)

A change in the actual value of the shear stress, τ_s, is expected from one experiment to the other, for instance if they are performed at various temperatures. It should account for the differences observed in the extent of transfer and the variations in the frictional forces.

It would therefore be very important that the actual shear stress existing at the top of the transfer particles be assessed experimentally. Such a possibility is offered indirectly by the determination of the size of the elemental particles detached during one step as described below.

4.1. Indirect assessment of the interfacial shear stress τ_s

From our previous work we know that each transfer particle grows progressively by the agglomeration of small elemental particles around fixed nucleation sites. The elemental particles transferred at each step being much smaller than the initial graphite powder particle, they have to be detached by a fracture process. As mentioned previously, these processes are adsorption sensitive, and hence dependent on environment and temperature. The work necessary for this fracture process has to be transmitted via the sliding interface, and it is therefore not surprising that it is controlled by the actual interfacial shear stress τ_s.

Referring to O.AYASHI and K. LUDEMA (8), the diameter of the elemental particles detached during such a process is indeed related to the surface energy, χ_s, and the mechanical strength of the material by the relation:

$$2R = k \cdot 4E / \chi_s$$

where:
2R = minimal particle diameter
k = function of particle shape
E = Young's modulus

Owing to the additional complexity introduced by the anisotropic structure of graphite, we will simply retain that the size of the elemental particle is inversely proportional to τ_s.

Fortunately, this same anisotropy of graphite makes it relatively easy to discriminate the elemental particles, for instance in optical microscopy with polarised light and additional retarder plate (9)(10).

Consequently, the simple comparison of the size of the elemental particles within the agglomerates forming the transfer particles allows us to have some idea of the respective interfacial shear stresses which prevailed during their formation: the smaller the size of the elemental particles, the larger was the shear stress.

5. RESULTS

5.1 Previous results

All the experimental results described in previous papers support and form the background of the present model, and are more or less supporting it. Therefore, we will restrict ourself to a few new experimental results supporting the hypothesis that the actual shear stress is a determinant parameter for the comprehension of the frictional behaviour of transfer films.

Since our first investigations in the field of graphite transfer (11)(12), we could show that when rubbing on thoroughly polished discs, the variations of the frictional force with rubbing time belong to one of two typical curve types, depending on the disc temperature. Below 200°C, friction continuously increases towards a limiting value, this latter being itself an increasing function of temperature. Above 200°C the variations of the frictional force display a maximum systematically, while the extent of transfer is visibly higher than in the previous case. Hence, the existence of a maximum appeared to be associated with the build up of a significant transfer film, and to the simultaneous decrease in the actual area of contact.

The quantification of transfer by image analysis later allowed us to confirm the previous hypothesis and to show that below 200°C the area fraction covered by the transfer particles is indeed lower than 2%, while above 200°C it may reach values as high as 10%.

5.2. Temperature dependance of L.C.C.

Whatever the exact relationship, it is possible to assume in first approximation that the maximum in friction corresponds to the transition from one prevailing mechanism to another, and is associated with a critical extent of transfer.

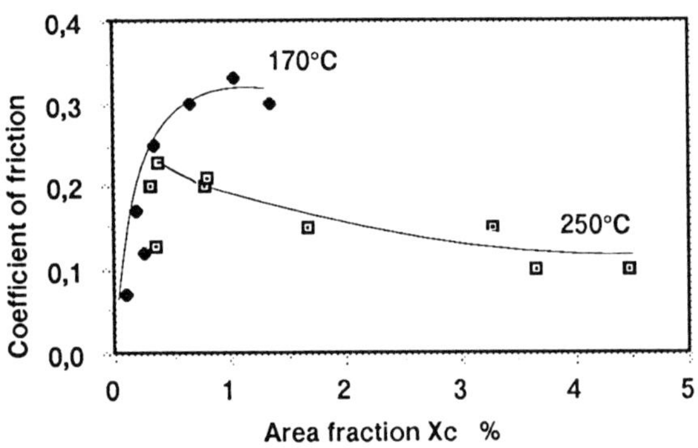

Fig. 7: Comparison of the evolution of the coefficient of friction with the extent of transfer at 170 and 250 °C respectively.

In Fig. 7 are compared the variations of the frictional force F_T as a function of the area fraction, **X**, covered by the transfer particles as observed with increasing rubbing time (up to 150 mins.) respectively at 170°C and 250°C. It is visible that the maximum in friction is not reached for the same area fractions at 170°C as at 250°C. But a closer examination also reveals, that for identical area fractions, the maximum in friction is already reached at 250°C, while at 170°C, no decrease in friction is displayed at even larger extents of transfer. This behaviour is the experimental evidence that the area fraction alone is not sufficient to characterise the friction transition. Referring to our model, the other parameters could be the shape and volume of the particles, and/or the local shear stress, τ_S. Since shear stress determines also volume and shape, we would normally predict that the shear stress at 170°C is lower than at 250°C. An experimental assessment of this statement would bring more support to our model.

5.3. Experimental assessment of the elemental particle size

Optical microscopy

Referring to the indirect method described above, we examined systematically the size of the elemental particles constituting the transfer particles formed at various temperatures. Those formed at 170°C did indeed appear to be much larger (1-2 μm) in comparison to those formed at 250°C (0.3 μm). In other words, the interfacial shear stress existing at 170°C is indeed smaller than at 250°C.

Thus, the method consisting of deriving the actual shear stress from the size of the elemental particles seems to be valid in principle and very promising. However, the use of optical microscopy appears unsuitable for obtaining accurate values for a large number of particles, i.e. having a statistical weight.

Granulometry

The use of a laser diffraction particle analyser normally gives the particle size distribution and would therefore be best adapted to our problem. However, it appeared to be very difficult to disaggregate the elemental particles in order to obtain a suspension of isolated elemental particles.

Gas adsorption

In the same line of thought, it should be possible to obtain some statistical information on the mean particle size, by measuring the specific area of the particles by gas adsorption. However, this method again is still not directly applicable to our particles, since the amount of available transfer particles is very small and difficult to collect. Therefore we attempted to solve the problem by a different approach: instead of studying the influence of temperature on the size of elemental transfer particles, i.e. of fragments detached during the frictional process, we studied the influence of temperature on the size of the wear particles formed in slightly different conditions. The latter were simply obtained by the interaction of a rasp rubbing against a graphitic disc at a controlled temperature. This easily allowed us to obtain some 100 mg of wear debris enabling the determination of active surface area by using a method described elsewere (13). This method consists of measuring the amount of CO and CO2 thermally desorbed from the edge or prismatic surfaces of graphite by heating the sample to 900°C in a vacuum. More precisely, these species resulted themselves from previous adsorption of oxygen under well defined conditions (300°C, 300 torr) after all the previously present chemisorbed species were desorbed by heating under vacuum up to 900°C.

The results, illustrated in Fig. 8 reveal that the active surface is indeed a function of temperature and that, in particular, a maximum is displayed at 200°C. Referring to previous considerations, a maximum in specific area corresponds to a minimal particle size and hence to a maximal shear stress during the generation of the particles. The subsequent decrease in specific area reflects either an

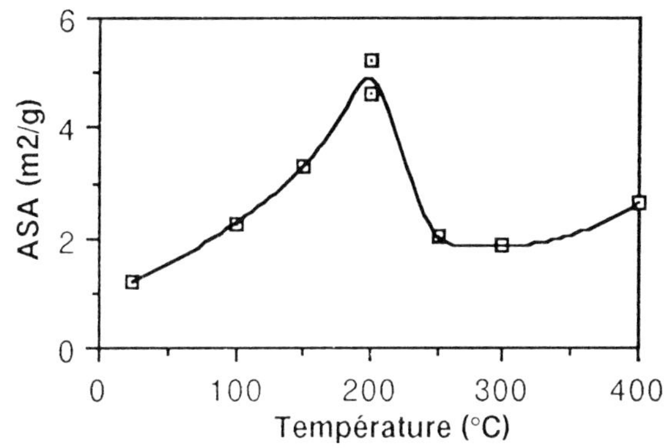

Fig. 8 Evolution of the active surface area (A.S.A.) with the temperature prevailing during the abrasive process.

actual decrease in the shear stress, or the disappearance of the smallest particles by burning. The oxidation of graphite especially on mechanical activated surfaces may indeed start at relatively low temperatures. The hypothesis that this may already occur during the frictional process of transfer particle formation is also consistant with a decrease in local shear stress and the decrease in the coefficient of friction. But in that case, it does not support the mechanical lift effect.

In every case, the shear stress developed along the interface of the transfer particles appears to be actually a function of the temperature, and at least below 200°C, is an increasing function of temperature.

This change in τ_s may well partly result from the influence of the physical water adsorption. However, since any fracture of graphite creates fresh highly reactive surfaces which necessarily stabilise by chemisorption of molecules from the environment, the possibility that the nature and proportion of these chemical groups may change with the temperature prevailing during their formation cannot be discarded. This last possibility has been confirmed by XPS analyses (14).

7. CONCLUSION

Assuming simple dome shaped individual particles, we have shown that discontinuous transfer films can be described using the same parameters as those used in image analysis, such as the area fraction covered by the transfer particles, the particle density and the mean area of the particles.

The 3rd dimension of the particles is however also important and accounts for load carrying capacity by two different mechanisms: by elastic indentation of the dome into the counterface, and by the mechanical reaction resulting from the driving force along the inclined plane formed by the transfer particles. In both cases the model of transfer particles adopted allows us to calculate the contribution to L.C.C. as a function of the area fraction and a parameter characterising the aspect ratio: the angle α .

In a dynamic situation i.e. during friction , the particles may continue their growth. Referring to the work of HOKKIRIGAWA and KATO, this process was shown to be a function of the shear stress developed along the particle interface and again of the particle shape. Hence, in the final analysis, the shear stress appears to be one of the most determinant parameters, since it controls all the mechanisms of L.C.C., either by the possibility of particle growth, or by the mechanical response of friction along the particle surface.

On the base of previous experimental results we already know that the same graphite, rubbing on the same steel can give rise to a large variety of frictional and transfer behaviours, by changing only the temperature. Our model accounts for these behaviours, provided that the actual shear stress is a function of temperature.

The actual change of the value of this shear stress, assessed indirectly by measuring the size of the elemental particles making up the agglomerates forming the transfer particles by 2 different methods validate the model developed.

REFERENCES

1. M. Brendlé, J. Fatkin, P. Turgis and R. Gilmore, Tribology Transactions **33** (4) pp 471-480, (1990)
2. Brendlé, M., Turgis, P. 18th Leeds Lyon Symposium on Tribology "Wear particles - from craddle to grave",Ed. D. Dowson et al.,Elsevier pp 313 - 321 1992
3. Y. Berthier, M. Godet and M. Brendle Tribology Transactions **32** (4) 490-496 (1989)
4. B.J. Briscoe and A.C. Smit, A.S.L.E Trans., **25** (3) pp 349-354
5. K.L. Johnson "Contact mechanics"; Cambridge University Press, pp 90-93 (1985).
6. K.Hokkirigawa and K. Kato, Tribology International 21 pp 51- (1988)
7. K. Hokkirigawa and K. Kato, Wear of Materials Vol.1, pp 1-8 (1989).
8. O.O. Ajayashi. and K.C. Ludema K. Wear of Materials Vol.1, pp 349-359 (1989).
9. R.A. Forrest and H. Marsh, Carbon, **15**, pp 349-350 (1977).
10. R.A. Forrest and H. Marsh Proceedings of the "5th London International on Carbon and Graphite Conference"pp 328-333 (1978)
11. Brendlé, M. and Colin, G., 7th Leeds Lyon Symposium on Tribology, in "Friction and Traction" Ed. D. Dowson et al., Butterworths (London) pp 124-132 (1981)
12. M Brendlé, and J. Fatkin, 12th Leeds Lyon Symposium on Tribology, "Mechanisms and Surface Distress Ed. D. Dowson et al Butterworths pp 93-99 (1986)
13. P. Ehrburger Carbon **29** pp 763-768 (1991)
14. M. Brendlé, G. Nancé and R.Gilmore (in progress)

ACKNOWLEDGMENTS

The authors wish to express their thanks to LONZA G+T Ltd. (Switzerland) for having supported this work over several years and for the constant interest and helpful discussion.

The authors are indebted to LONZA - ALUSUISSE and especially to Dr. H. HOTZ for having allowed them to perform the image Analysis.

Question

Can you comment on the shape factor of the particles as a function of temperature, speed......Is it possible to determine that by the image analysis system?

Answer

This question concerns a very important and wide field which alone would need a paper. As we tried to describe in previous papers and to explain in the above, the shape and size of the transfer particles depend on a large number of parameters which simultaneously affect:

- the particle density,
- the particle size distribution
- the size of the elemental particles constituting the agglomerates.

Their variations with temperature, surface polishing method ,etc... were already described in previous papers.

The shape factor itself contains different aspects :

- a two-dimensional one, describing the contour of the particles;
-and a three-dimensional one, encompassing the height - diameter ratio, and more precisely the shape, i.e. dome shaped as assumed in this work, or of different cross section such as gaussian;

Concerning the two-dimensional aspect, it is quite clear that the contour is never a perfect circle. However, the shape factor "roundness" as measured by image analysis is in general close to one. For a long time we measured systematically this parameter, but we could not derive any general statement from it.

Also related to the contour of the particle is the fractal number of the contour line. Some measurements achieved by Rocques Carmes et al (a) on our samples, revealed that the fractal number associated with particles formed at 250°C on non preoxidised steel discs (1.12) is smaller than that corresponding to those formed at the same temperature on preoxidised steel discs : (1.24). Whether this change is related to the size of the elemental particles, or to the contamination by iron oxide particles mixed with the elemental graphite particles is still not established.

Concerning the 3 dimensional aspect, until yet it could not be assessed quantitatively, but only estimated from optical stereomicroscopy and even better from S.E.M. observations. Indeed, the shape then appears to vary from spherical dome shaped particles, to something still having axial symmetry but of Gaussian cross section or even to a volume displaying only longitudinal symmetry. For a given sliding speed, these changes are apparently less influenced by the temperature itself, than by some other unidentified parameter.

Pictures illustrating the large variety of particle shape were already given in previous papers (12) and are even better illustrated by colour photographs in the dissertation (b).

In contrast, an increase in sliding speed dramatically modifies the shape factor. Obviously, higher speeds enhance the tendancy of forming long elongated particles, suggesting that the particles which normally build up at low speed cannot mechanically withstand the increased shear stresses and/or chocks associated with higher speeds.

Additional references

a) C. ROCQUES CARMES, D. WEHBI, A. LE MEHAUTE and C. TRICOT Proceedings of the 7th International Congress on Stereology 2-9th September 1987 CAEN (France)

b) P. TURGIS Dissertation - University of Haute Alsace 1992

SESSION XVIII
BIO-TRIBOLOGY (2)

Chairman : Professor J-M Georges

Thin Films in Tribology / D. Dowson et al. (Editors)
© 1993 Elsevier Science Publishers B.V. All rights reserved.

The friction and lubrication of human *stratum corneum*.

S.A. Johnson[a], D.M. Gorman[a], M.J. Adams[a], and B.J. Briscoe[b]

[a]Unilever Research Port Sunlight Laboratory,
Wirral, Bebington, Merseyside, L63 3JW, United Kingdom.

[b]Department of Chemical Engineering and Chemical Technology, Imperial College,
Prince Consort Road, London, SW7 2BY, United Kingdom.

In vivo measurements are reported of the lubrication imparted by water, various aqueous media, and some hydrophobic oils, to a glass probe sliding on the inner human forearm. The primary aim of the paper is to show that the tribology of the human skin surface may be satisfactorily interpreted on the basis of precedents developed for the rationalization of frictional data for elastomers and thermoplastics, and selected data are interpreted on this basis. Water itself plasticises the *stratum corneum* and, consequently, has profound effects on the frictional behaviour. The effective lubrication observed in surfactant solutions can be explained in terms of electrostatic double-layer repulsion between charged layers adsorbed on the contacting surfaces. Viscosity effects associated with conventional hydrodynamic lubrication are demonstrated using topically applied silicone oils.

1. INTRODUCTION

The *stratum corneum* is the surface layer of mammalian skin and, for humans, is about 10 to 15 μm in thickness over much of the body. It is comprised of dead cells that are formed mainly from the protein keratin. The cells are closely packed in a lipid matrix to form a thin membrane that is supported and constrained by a bed of connective tissue. This subsurface material behaves like a very soft elastomer of much lower elastic modulus than the *stratum corneum*. The interpretation of the friction and lubrication characteristics of the *stratum corneum* is complicated both by the intrinsic variability of biomaterials, and also by the high moisture sensitivity of its surface and bulk mechanical properties. *In vitro* studies have shown that its elastic modulus, at low moisture contents, is typical of a glassy polymer [1]. However, moisture-induced plasticisation reduces the modulus to values comparable to that of a soft elastomer [1].

Previous studies on the tribology of human skin have attempted to demonstrate a correlation between certain tactile sensations and the friction between the skin surface and a variety of probes [2-7]. In addition, friction measurements have been used to provide *in vivo* information about the effects of age [8,9], hydration [2,4,7-15], dermatitis [16], and cosmetic products [2,4,6,7,10-13,17] on both the interfacial and bulk properties of *stratum corneum.* Measurements have also been made as a function of anatomical site [8,9], and probe type [4,10,12-15,18].

Most previous studies on the friction of human skin, using sliding probes, have involved torque measurements on discs or flat cones rotating at constant velocity. Typical instruments have either employed adapted viscometers [12,13,17], rotating wheels [4,7-9,11,19], or oscillating discs [6,16,20] pressed against the skin surface. Measurements have also been reported for the friction between a moving finger and a variety of fixed contact surfaces [14,18].

The equipment used in the current study employs a more conventional tribological configuration based on a linear, reciprocating, sliding action. This system allows the control of both the applied normal load and the sliding velocity, coupled with an accurate measurement of the frictional response. A smooth, convex, glass lens was selected as a probe to facilitate theoretical modelling, and to aid comparison with extensive earlier results on the friction and lubrication of glass/elastomer contacts.

The current paper describes measurements of the sliding friction of this rigid contact member on an inner human forearm, for a range of clean and lubricated conditions. During sliding, some energy is dissipated in deforming the subsurface tissue. The quantitative modelling of the resistive force associated with this process requires a knowledge of the viscoelastic properties of this material. In the present case, these data were obtained *in vivo* by cyclic normal indentation on the same inner forearm.

2. EQUIPMENT AND MATERIALS

2.1. Friction apparatus

Friction measurements were made *in vivo* using an instrument based on the Tabor-Eldredge design [21]. A smooth glass lens was used as the contact member for reasons discussed above. Dead loads of 0.05 to 1 N were applied at the contact by adding weights to the end of the balance arm to which the lens was attached. The lens had a radius of curvature of 8 mm, and was sufficiently large to produce a contact diameter, over this load range, that was less than its actual diameter (10 mm).

The skin surfaces used were regions on the inner forearms of two male subjects. The limbs were positioned such that an approximately horizontal surface was presented to the glass contact. The contact was moved repeatedly along the arm over a length of about 50 mm, with imposed sliding velocities in the range 0.25 to 50 mm/s.

The frictional force was measured by strain gauges included within the lens holder. After suitable conditioning, the signal from the strain gauges was recorded either by a chart recorder, or a data logger and a computer.

2.2. Normal indentation apparatus

An Instron Universal Testing Machine was used to measure the mechanical response of the skin surface to normal loading by the glass lens. The lens and region of the inner arm were those employed in the friction experiments. Loading/unloading cycles were carried out at velocities between 0.1 and 2.5 mm/s, with the point of strain reversal set at a maximum force of 2 N. The force profiles during cycling were displayed on a chart recorder. Experiments were carried out on both 'dry' and wet skin.

2.3. Skin and probe preparation

In order to generate reproducible results, a standard conditioning and cleaning procedure was adopted to prepare the *stratum corneum*. This involved rubbing the arm for around fifteen seconds with a mild anionic detergent (sodium lauryl ether-3EO sulphate) dispersed in deionized water, before rinsing and rubbing with large quantities of deionized water. Subsequently, the cleaned region was only allowed to come in contact with either the test materials, or deionized water. After a few days, natural shedding of the skin cells produced a surface which was apparently free of, otherwise ubiquitous, insoluble calcium soap contamination.

Before each experiment, the lens probe was cleaned with an acetone-moistened tissue. For measurements on 'dry' skin, the substrate was always allowed to equilibrate for several minutes to the ambient temperature (294 K) and relative humidity (50 %) of the environmentally-controlled room in which the friction apparatus was situated.

2.4. Materials

Sodium and calcium chlorides, sodium dodecyl sulphate (SDS), and pH-buffer solutions were obtained from BDH. Solutions were prepared with deionized water. DC200-series polydimethylsiloxanes with viscosities between 0.8 and 59000 mPas were supplied by Fluka, Hopkins and Williams, and also by Dow Corning. All the fluids were used at 294 K, and applied in amounts to give optimal lubrication.

3. RESULTS

3.1. Normal indentation

Typical loading/unloading data for the glass lens on the 'dry' inner forearm, obtained at a velocity of 2.5 mm/s, are shown in Fig. 1.

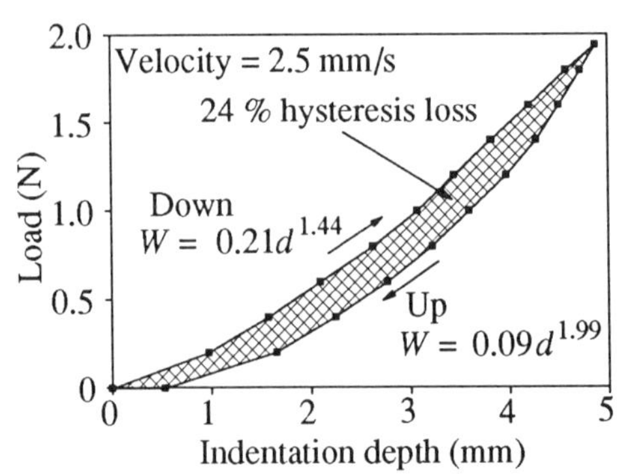

Figure 1. Indentation of the human inner forearm.

No major differences in these data were observed for slower speeds down to a velocity of 0.1 mm/s. The data show a hysteresis energy loss of 24 % of the elastic input energy. A temporary set of *ca.* 0.5 mm is also apparent which is recoverable on a longer timescale. No differences in the indentation behaviour were observed either when the arm was wet, or when the associated fist was clenched or unclenched. The loading and unloading data may be represented by power-law expressions of the form shown on the figure, where W is the applied load and d is the imposed penetration depth.

3.2. Friction

In the current work, either continuous or intermittent ('stick-slip') motion was observed depending on the conditions. For continuous sliding, the mean value of the measured force was taken as the mean dynamic friction. In cases of intermittent sliding, the average value of the peaks and troughs was taken as the mean value of the dynamic frictional force at the imposed sliding velocity, and the average value of the peaks was taken as the mean value of the static frictional force. These are the conventional definitions of the static and dynamic frictional force [22]. The means were computed from data obtained in both sliding directions. In this way, it was generally possible to obtain a discrimination within ± 3 %.

The coefficient of friction, μ, is defined as being the ratio of the frictional force, F, to the normal load, W; thus $\mu = F/W$. For the present system, the capillary forces resulting from any liquid bridges between the lens and the skin surface were found to be effectively negligible. Hence W is given by the applied dead load, even in the presence of fluids. The dynamic frictional force, F, was found to vary as some power of W, where the load index, n, was always less than unity; *ie.* $F = kW^n$ ($n < 1$), where k is a constant. Under these circumstances, the coefficient of friction is load dependent, designated by $\mu(W)$, and given by $\mu(W) = kW^{n-1}$.

3.2.1. Lubrication by water

Figure 2 shows a representative trace of the friction between the glass lens and 'dry', clean, skin for a load of 0.2 N and a sliding velocity of 4 mm/s. The corresponding data for wet skin that had been immersed in water for 120 seconds are also shown; 'stick-slip' friction is evident in this case. The slopes of the 'stick' phases of the motion provide information about the stiffness of the *stratum corneum*.

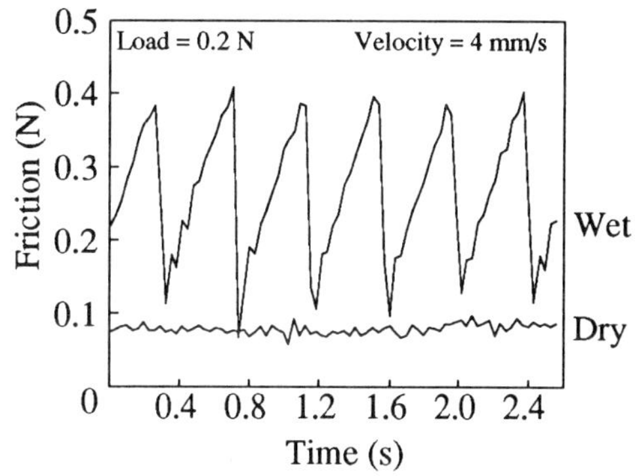

Figure 2. Wet and 'dry' skin/glass friction.

Table 1
Dynamic friction power-laws†

	Value of fixed parameter		
	W = 0.2 N	V = 3 mm/s	V = 8 mm/s
Dry‡	$0.063V^{0.05}$	$0.25W^{0.95}$	$0.30W^{0.93}$
Wet	$0.36V^{-0.29*}$	$1.4W^{0.88}$	$1.2W^{0.83}$
SDS#	$0.011V^{0.19}$	----------	$0.063W^{0.83}$
DC200§	$0.035V^{0.45}$	----------	$0.22W^{0.59}$

† Friction and load, W, in Newtons; imposed velocity, V, in mm/s.
‡ Equilibrated at 50 % R.H. and 294 K.
* For $V > 4$ mm/s.
# 0.1 M in deionized water.
§ 12200 mPas silicone oil.

Fitted power-law expressions for the load dependence of the friction, for various sliding conditions, are given in Table 1 for two imposed sliding velocities, V. These data indicate that both the indices and coefficients are sensitive to the sliding velocity. The variation in the frictional force with velocity (for W = 0.2 N) for 'dry' and wet skin is shown in Fig. 3. The occurrence of 'stick-slip' motion is associated with a dynamic frictional force that reduces with velocity, as is the case for wet skin at velocities above 3.5 mm/s, or a static frictional force that increases with dwell time [22]. The frictional force increase with velocity for 'dry' skin is consistent with the observed continuous sliding. The velocity dependencies of the friction may also be represented by power-law expressions analogous to those for the load dependence, and these are also given in Table 1. The corresponding fitted-curves are superimposed on the experimental data in Fig. 3.

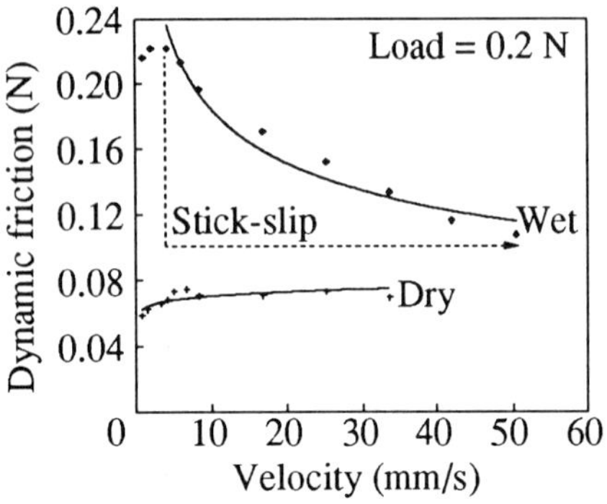

Figure 3. Velocity dependence of wet and 'dry' skin/glass friction.

3.2.2. Lubrication by salt and pH-buffer solutions

A small decrease in the wet friction was observed as the concentration of a simple electrolyte (NaCl) was increased in deionized water - for a 1 M solution this reduction was *ca.* 15 %. In contrast, pH had a major influence on the wet friction (Fig. 4), with the values decreasing markedly with decreasing hydrogen ion concentration. Intermittent motion was only observed with the pH 4 buffer solution. These pH effects were reversible by rinsing both arm and contact with deionized water.

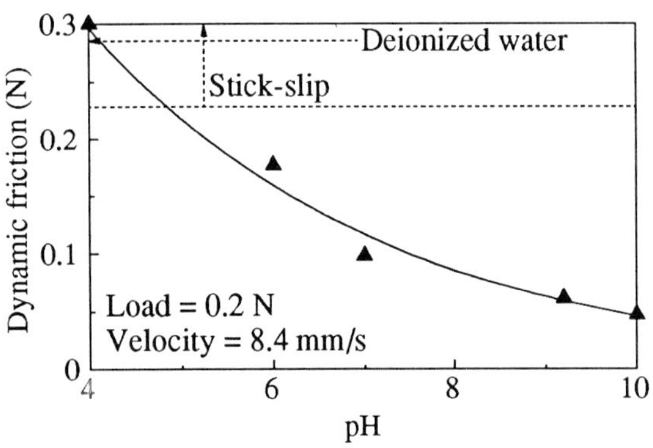

Figure 4. The effect of pH on skin/glass friction.

3.2.3. Lubrication by apolar fluids and aq. SDS

Figure 5 shows the velocity dependence of the friction of the 12 and 12200 mPas silicone oils. For a load of 0.2 N, the latter produced the lowest friction of the oils tested at the lowest sliding velocity (0.25 mm/s). The least viscous oils tended to exhibit a rise in friction at the lowest velocities (Fig. 5), accompanied by a tendency towards 'stick-slip' motion. Figure 5 also shows the velocity dependence of the lubrication imparted to wet skin by a 0.1 M SDS dispersion. This concentration resulted in the lowest friction of the materials investigated over the whole velocity range available. Table 1 gives the appropriate fitted power-law expressions for both the 12200 mPas silicone oil and the 0.1 M SDS dispersion, and Fig. 5 shows the corresponding fitted-lines.

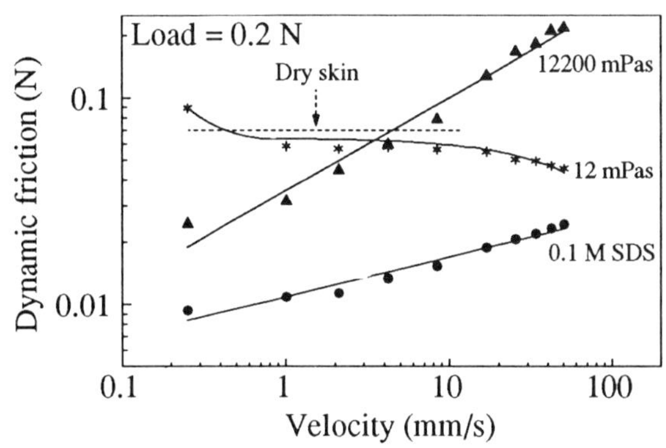

Figure 5. Velocity dependence of skin/glass friction for two silicone oils and aqueous SDS.

4. DISCUSSION

Friction may be usefully regarded as the energy dissipated per unit sliding distance. For a deformable substrate, which is unlubricated or boundary lubricated, work is expended at both the interface and in the bulk of the material when a rigid sphere slides over it [22]. To a first approximation, the total frictional force, F_{TOT}, may be given by the sum of two non-interacting terms - a deformation term, F_{DEF}, and an interfacial adhesion term, F_{INT}; thus

$$F_{TOT} = F_{DEF} + F_{INT} \tag{1}$$

It is also useful to ascribe two components to the frictional force for perfectly hydrodynamically-lubricated deformable contacts. For small contact pressures, the interfacial component is that appropriate to isoviscous elastohydrodynamic lubrication (IEHL) and designated by F_{IEHL}; hence

$$F_{TOT} = F_{DEF} + F_{IEHL} \tag{2}$$

In the 'mixed' lubrication regime all three components of friction, from eqns. 1 and 2, have to be considered.

4.1. Deformation component of friction: an assessment of its contribution

This component arises from the work dissipated by the nominally rigid slider in deforming the bulk of the compliant contacting material. The deformation volume is confined to a region beneath the slider that has a size of the order of the contact diameter. For viscoelastic materials, such as skin, the deformation-recovery cycle is associated with viscous energy losses. This process has been modelled, and examined experimentally, by Greenwood and Tabor [23]. The following expression was obtained for the resistive force in the current contact configuration:

$$F_{DEF} = \beta \left(\frac{9}{128r}\right)^{2/3} \left(\frac{1-\nu^2}{E}\right)^{1/3} W^{4/3} \tag{3}$$

where β is the viscoelastic hysteresis loss fraction, r is the radius of curvature of the rigid slider (or roller), and E and ν are the real part of the Young's modulus and the Poisson ratio of the compliant substrate respectively. For sinusoidally-driven systems, β is equivalent to $\pi \tan \delta$, where $\tan \delta$ is the loss tangent of the viscoelastic substrate and given by the ratio of the loss and storage moduli. For the more complex deformation path involved with sliding (or rolling), $\beta = k_1(\pi \tan \delta)$ where k_1 is found to be between 1 and 3 [23,24].

The bulk deformation parameters required to estimate the deformation component of friction from eqn. 3 were obtained in the present work by normal indentation with a sphere. That this type of experiment only senses subsurface bulk properties was confirmed by the insensitivity of the response to the presence of moisture, which only acts to markedly plasticise the *stratum corneum.* Assuming that the skin deforms in a Hertzian manner, the indentation depth, d, as a function of load, slider radius and material parameters is given by [25]

$$d = \frac{a^2}{r} = \left(\frac{9}{16r}\right)^{1/3} \left(\frac{1-\nu^2}{E}\right)^{2/3} W^{2/3} \tag{4}$$

where a is the Hertzian contact radius. This equation predicts that the displacement dependence of the load should be a power-law of the form $W = k_2 d^{1.5}$, where k_2 is a constant. The experimentally observed dependence (see Fig. 1) is close to this form, particularly for the loading part of the cycle. Exact agreement would not be expected given the inhomogeneous structure of the skin, and the effects of viscoelastic hysteresis loss and temporary set. Nevertheless, using the data in Fig. 1 an approximate modulus can be calculated by substitution of the following parameters into eqn. 4: d = 0.003 m, W = 1 N, r = 0.008 m and ν(skin) = 0.49 (assumed). This leads to an elastic modulus with a value of 40 kPa. The loss fraction, β, can be estimated from the area of the hysteresis loop in Fig. 1. Adopting a value of 0.24 from the figure (*ie.* k_1 = 1), eqn. 3 predicts that F_{DEF} = 0.009 N and hence μ_{DEF}(0.2 N) = 0.05 for a load of 0.2 N.

The elastic modulus and loss tangent of viscoelastic materials generally exhibit frequency, temperature, and strain dependence. This means that the friction parameters estimated above are only approximate, since the deformation frequencies in the indentation experiments did not correspond exactly to those that were induced during sliding. The mean deformation frequency in sliding is *ca.* $(3V/2a)$; *ie.* 0.1 - 30 Hz, for V = 0.25 - 50 mm/s and W = 0.2 N ($2a$ = 5.6 mm from eqn. 4). The saw-tooth displacement-time profile employed for indentation comprises a convolution of many frequencies, some of which are in this regime. In spite of these inadequacies, the measured mechanical parameters enable a first-order estimation of the deformation component of friction.

The predicted values of F_{DEF} and μ_{DEF} using eqn. 3 may be treated as the lower limits for F_{TOT} and μ_{TOT} that can be achieved through perfect lubrication. Clearly, for the case of the glass contact sliding on clean 'dry' skin, and clean wet skin at velocities below 50 mm/s, the contribution from the deformation component of friction is relatively small (Fig. 3). However, in systems well-lubricated with low viscosity fluids, such as sliding in water containing 0.1 M SDS (Fig. 5), or for sliding on wet skin at high velocities, it is clear that the hysteresis loss component dissipated in the skin will be much more important [23]. Experimentally, the 0.1 M SDS dispersion resulted in a coefficient of friction, $\mu_{TOT}(0.2\ N)$, of 0.05 at low sliding velocities (Fig. 5). This agrees with the calculation made above. However such agreement must be treated as fortuitous since the experimental load index, $n = 0.83$, is quite different from the theoretical value of 4/3 predicted by eqn. 3. This discrepancy may be associated with the complex mechanical gradients associated with the skin system. One further important point to note is that the relatively low composite skin modulus, in engineering terms, leads to very low contact pressures. For a load of 0.2 N the mean Hertzian pressure under the glass lens is calculated to be only 8 kPa - clearly any fluid lubrication will be of an isoviscous nature.

4.2. Lubrication by fluids

4.2.1. IEHL - computation of fluid film traction

For a soft elastomeric sphere sliding in a fluid on a rigid flat surface, the minimum film thickness, h, is given, for smooth surfaces, by [26]

$$h = 0.8\ (\eta V r)^{0.6} \left(\frac{W}{2a}\right)^{-0.2} E^{-0.4} \tag{5}$$

where η is the viscosity of the fluid at the characteristic shear rate beneath the contact of V/h. The contact diameter, $2a$, required to calculate h can be obtained from eqn.4.

Ignoring the small contribution from viscous drag outside the contact region, the fluid film traction can then be estimated from the calculated shear stress, S, and the Hertzian contact area, A (from eqn. 4), over which it is generated:

$$F_{IEHL} = SA = \left(\frac{\eta V}{h}\right) A \tag{6}$$

Assuming that eqn. 5 applies to the inverted geometry of the present experiments, this model predicts that, in perfectly lubricated contacts, μ_{IEHL} should be proportional to $(\eta Vr)^{0.4}$, $(WE)^{-0.2}$ and $h^{0.67}$. The data in Table 1 show that the velocity dependence of the 12200 mPas silicone fluid is comparable to that which would be expected on this basis. The rising dependence of friction with velocity ensures continuous, rather than intermittent, motion. However, the 12 mPas oil provided very little lubrication. Table 2 gives calculated film thicknesses and coefficients of friction, at $V = 10$ mm/s, for these two silicone fluids and for water ($\eta = 1$ mPas). The friction in the fluids measured experimentally is generally much higher than that predicted, even if the subsurface bulk hysteresis loss components are included. This is probably indicative of skin roughness effects being important in the experimental regime studied. For comparison, the (undeformed) average roughness (R_a) and the mean peak-to-valley height of the inner forearm are known to be *ca.* 20 µm [27] and 100 µm [28] respectively. It can be seen from the data in Table 2 that sliding in low viscosity fluids will be in the 'mixed' lubrication regime for $V = 10$ mm/s.

In the case of lubricated sliding on rough surfaces, where the film thickness is of the same order as the roughness, it is necessary to consider the effect of asperity-flattening and the EHL provided by the thin fluid film interposed at each asperity tip. This micro-EHL approach is becoming well-developed and has been adopted with some success to explain the low friction observed in human and artificial synovial joints [29].

Table 2
Calculated[†] IEHL film thicknesses and coefficients of friction

	Fluid viscosity (mPas)		
	1	12	12200
h (µm)	0.31	1.4	89
$\mu_{IEHL}(0.2\ N)$	0.0041	0.011	0.17

[†] For $V = 0.01$ m/s, $W = 0.2$ N, $E = 40$ kPa, $\nu = 0.49$, $r = 0.008$ m and $a = 0.0028$ m.

4.2.2. Lubrication in aqueous systems

Fluid film lubrication in water is promoted by surfactants that aid wetting, and also cause the glass and skin to separate at asperity tips by electrical double-layer repulsion [30] between charged adsorbed layers. Comparison of Fig. 5 with Fig. 3 demonstrates the reduction in the wet friction that may be achieved by the presence of ionic surfactants.

There is considerable indirect evidence that surface electrical effects are important in skin friction under wet conditions generally. Firstly, extensive studies [26,31] on the lubrication of smooth glass/rubber contacts have shown the importance of double-layer phenomena in governing both friction and squeeze-film behaviour. In particular, Richards and Roberts [31] have shown that a 0.01 M SDS dispersion produces stable films of thickness 10 - 30 nm under static loading conditions in the pressure range 10 - 60 kPa, and that these films are maintained during sliding. The mean pressure in the present experiments was of a similar magnitude, implying that skin/glass lubrication could also result from similar effects.

Secondly, the reduction in skin/glass friction produced simply by increasing the pH (Fig. 4) can be explained in terms of the different electrical characteristics of the glass and skin surfaces. Glass has a negative surface charge over the whole pH range tested [32], whereas the isoelectric point for the *stratum corneum* is at *ca.* pH 5 [33], which is about the pH of normal skin [28]. At smaller pH values than this, net attraction occurs at asperity tips leading to fluid film collapse. At pH values higher than the isoelectric point of skin, the zeta potentials of both the glass and skin will become more negative leading to greater surface repulsion, and hence thicker films. Deionized water (pH 7) produces results similar to pH 4 - 5 buffer probably because of the buffering action provided by the skin itself.

Thirdly, the skin/glass friction remained small on rinsing the 0.1 M SDS dispersion from the contact surfaces with deionized water, implying a major role for surface-adsorbed surfactant. If, instead, the surfactant was rinsed with 0.01 M $CaCl_2$ (very hard water) then the friction increased irreversibly due to the deposition of insoluble calcium dodecyl sulphate. In contrast, rinsing off the surfactant solution with NaCl of the same strength had little effect. More concentrated brine solutions did increase the friction, but the effect was readily reversed by rinsing with deionized water. This is to be expected because ionic shielding of the potential in the double-layers will lead to thinner films and higher friction [26]. That brine alone had little effect on the friction is because neither Na^+ or Cl^- are expected to interact strongly with glass or skin. Nevertheless, the slight reduction in friction with concentration could be a weak double-layer effect [26].

In summary, IEHL of skin by fluids is complicated both by roughness and double-layer effects. That the wet friction of clean skin against glass shows intermittent motion and finite static frictional forces implies that, at relatively low velocities, water is excluded from at least some of the asperity tip contacts. The origins of the resistive forces associated with shear of 'dry' asperity contacts are discussed in the next section.

4.3. The interfacial adhesive component of friction

This friction component can be expressed as the product of an interfacial shear stress, S, and a real area of contact, A, thus

$$F_{INT} = S\,A \tag{7}$$

In order to implement this model for skin, **both** S and A have to be identified. The remainder of this section is devoted to the origins of the shear stress, with the real area of contact being addressed in the next section.

It has been remarked previously that, in the dry state, the *stratum corneum* is a glassy polymer while in the wet state it has an elastic modulus which is typical of a soft elastomer. The mechanisms responsible for the interfacial frictional energy dissipation can be quite different for these two classes of material. These mechanisms are described below to indicate which material properties are thought to be important.

For glassy polymers, the shear process generally involves plastic flow or fracture of material in a region about 100 nm thick adjacent to the surface [34]. This mechanism leads to the adhesion model of friction [22] with S identified as the interfacial shear strength of the polymer, τ. The flow process appears to be stress-aided and thermally-activated for those organic systems where data are available. The following empirical equation [35] is found to describe the variation of τ with temperature, T, velocity, and mean pressure, $P=(W/A)$:

$$S = \tau = \tau_o \exp\left(\frac{Q}{RT}\right)\ln\left(\frac{V}{h\phi}\right) + P\alpha_o \exp\left(\frac{-V}{2a\theta}\right) \tag{8}$$

where R is the gas constant, h the thickness of the shear band, and a is the Hertzian contact radius. τ_o is an intrinsic shear strength and α_o, Q, ϕ and θ are also material constants that, to first order, are independent of V, T and P. $1/\phi$ and $1/\theta$ are the characteristic times of the shear and compression processes respectively.

The interesting point is that τ has a functionality which is consistent with the expected yield or fracture behaviour in the analogous bulk material [34,35].

Elastomers can accommodate larger elastic strains than glassy polymers. Consequently, models of elastomeric friction based on a molecular adhesive 'stick-slip' process are also credible. Grosch [36] measured the friction of a number of elastomers under dry conditions as a function of sliding velocity, and found maxima in the data. In addition, for each elastomer the friction as a function of strain rate at different temperatures could be reduced to a single master curve using the WLF transform. The position of the maximum in the master curve seemed to correlate with that for the bulk loss modulus, which suggested that hysteresis losses were important in governing the magnitude of the shear stress. Moore [37] has described a number of models for elastomeric friction where the interfacial component is proportional to tan δ. On the other hand, Ludema and Tabor [38] have shown how the observed behaviour can be explained by considering the effects of viscoelasticity on the shear strength and the contact area in the glassy polymer model. On this basis, the increase in friction with velocity, at low velocities, reflects a similar increase in shear strength with shear rate to that of glassy polymers (eqn. 8). The decrease in friction at higher velocities results from a decrease in contact area, brought about by an increase in storage modulus with contact frequency.

More recently it has been shown that, under certain circumstances, the dry and wet friction of soft elastomers may be ascribed to a reversible peeling (adhesive) process involving the propagation of Schallamach waves [39-41]. This process does not involve true sliding. Instead gross motion occurs through wave travel, and energy is dissipated during a local detachment process. Briggs and Briscoe [40] have derived the following expression for the interfacial shear stress in such cases:

$$S = \frac{1}{2\pi\rho}\frac{G}{(1-\nu)} \tag{9}$$

where ρ is the ratio of the width of a detachment wave to the interfacial displacement associated with the passing of the wave, and G is the shear modulus of the elastomer. It is not yet clear whether detachment waves are an important energy dissipation source for sliding on human skin.

Of some interest is the characteristic change in the friction that occurs when certain glassy polymers (and probably elastomers) are plasticised by a suitable fluid. This process has been studied in detail by Cohen and Tabor [42] for Nylon plasticised by water. For Nylon/Nylon contacts the friction decreases immediately on addition of water due to a reduction in the interfacial shear strength, but then increases slowly with time, to a level slightly less than that for the dry contact, due to an increase in the real area of contact (see the next section). Thus, under steady state conditions, the reduction in the interfacial shear strength is more important than the increase in the real area of contact. No increase in friction with water immersion time was observed for glass/Nylon contacts because, apparently, a lubricating water film became adsorbed on the glass.

In summary, there are a number of ways of describing the origins of the interfacial shear stress for elastomers and thermoplastics that can be applied to skin.

4.3.1. Skin - the real area of contact

The current data show that human *stratum corneum* behaves differently from a polyamide (Nylon) in contact with water. The addition of water significantly increases the friction of 'dry' skin. Furthermore, the increase in friction is asymptotic after the addition of water, reaching steady state after a period of 1 - 2 minutes with no indication of a reduction during this time. This implies that, for friction, the dominant effect of water is most probably to increase the contact area.

Assuming that a smooth Hertzian contact is formed in the current experiments, it is possible to estimate the real area of contact using eqn.4. This predicts that the area and thus the friction, for a constant interfacial shear stress, should be proportional to $(W/E)^{2/3}$. In cases where smooth contacts are not formed, then the use of the Hertz equation will over-estimate the real area of contact, and hence also the friction. Multiple-asperity elastic contacts typically exhibit a load index of unity. However, values between 2/3 and unity are theoretically possible for special cases of the type considered by Archard [43]. For example, Adams et al [44] have shown that the friction for an elastic sphere having hemispherical asperities of equal radii should be proportional to $(W/E)^{8/9}$. It should be noted that the pressure dependence of the shear strengths of glassy polymers (eqn. 8) can also lead to load indices for friction in the range 2/3 to unity, even for single-asperity Hertzian contacts [44]. However, a pressure-sensitivity would not be manifested by this route for soft elastomers and skin because of the small pressures. All the experiments reported here that were not hydrodynamically-lubricated showed load indices in the range 0.83 to 0.95 (Table 1). It may be significant that the smaller load index was associated with the wet case. This suggests that the asperities of the *stratum corneum* are sufficiently compliant to partially conform to the surface of the glass slider. Asperity flattening will be greater in the wet case where the elastic modulus of the *stratum corneum* is reduced by plasticisation, as discussed below, and hence the contact will be more closely approximate to Hertzian conditions.

The reduction in the elastic modulus of the *stratum corneum*, as the moisture content is increased, has been observed both *in vitro* [1,45,46], and *in vivo* [47-49]. Table 1 shows that the wet friction at low velocities was about a factor of 4 greater than the 'dry' value. However, the observed value was sometimes appreciably less than this (*ca.* a factor of 2) if either the skin was not sufficiently clean before the wet measurement, or if the skin was not properly equilibrated before the 'dry' measurement. On the basis of eqns. 4 and 7, this reduction in friction would correspond to a factor of 8 reduction in the elastic modulus. This factor would be expected to lie in the range 4 - 8 for rough elastic contacts of the type considered by Archard. According to the literature [1,46], water reduces the elastic modulus of *stratum corneum* by factors in the range 2 - 2000. Therefore, it is reasonable to conclude that the increase in friction for skin in water is due to the greater compliance of the surface asperities, and hence an increase in the real area of contact. This is consistent with the corresponding reduction in the load index as described above.

Under 'dry' conditions, it would appear that the velocity dependencies of the contact area and shear stress terms are either small or self-compensate to some extent. This leads to little variation in the friction with velocity under 'dry' conditions (Fig.3). The modulus of the *stratum corneum* would be expected to be considerably more strain-rate dependent in the wet than in the dry glassy state. This is probably a contributory factor in the reduction in friction with increasing velocity that leads to intermittent motion for sliding in water. An additional factor is likely to be the development of partial IEHL at higher velocities as discussed in Section 4.2.1.

5. CONCLUSIONS

The present data indicate that the tribology of skin may be interpreted in terms of precedents derived from studies on the friction of elastomers and thermoplastics.

The large increase in friction in the presence of water is apparently a result of the moisture-dependent mechanical properties of the *stratum corneum*. In the present system 'stick-slip' friction is an attribute of wet, clean, skin above a threshold velocity. The lubricating effects of surfactants and high pH are probably a result of films produced at asperity tips by repulsion between electrical double-layers on the contacting surfaces. The low frictional forces observed in surfactant dispersions are close to those predicted from a consideration of hysteresis deformation losses in the skin.

Low viscosity silicone oils are poor lubricants because the surface roughness is sufficiently large to penetrate any films developed by elasto-hydrodynamic lubrication. A 12200 mPas silicone fluid apparently produced a fully-developed hydrodynamic contact over the full range of velocities studied.

ACKNOWLEDGEMENT

The work described here was undertaken as part of a DTI initiative in Colloid Technology.

REFERENCES

1. A.C. Park and C.B. Baddiel, J. Soc. Cosmet. Chem., **23** 3 (1972).
2. J.K. Prall, J. Soc. Cosmet. Chem., **24** 693 (1973).
3. E.L. Cussler, S.J. Zlotnick and M.C. Shaw, Percep. Psycho., **21** 504 (1977).

4. L.J. Wolfram, J. Soc. Cosmet. Chem., **34** 465 (1983).
5. E.L. Cussler, Cosmet. and Toilet., **93** 17 (1978).
6. M. Lodén, H. Olsson, L. Skare and T. Axéll, J. Soc. Cosmet. Chem., **43** 13 (1992).
7. S. Nacht, J.-A. Close, D. Yeung and E.H. Gans, J. Soc. Cosmet. Chem., **32** 55 (1981).
8. A.B. Cua, K.-P. Wilhelm and H.I. Maibach, Br. J. Derm., **123** 473 (1990).
9. P. Elsner, D. Wilhelm and H.I. Maibach, Dermatologica, **181** 88 (1990).
10. S.Comaish and E. Bottoms, Br. J. Derm., **84** 37 (1971).
11. D.R. Highley, M. Coomey, M. DenBeste and L.J. Wolfram, J. Invest. Derm., **69** 303 (1977).
12. A.F. El-Shimi, J. Soc. Cosmet. Chem., **28** 37 (1977).
13. W.A. Gerrard, Bioeng. Skin, **3** 123 (1987).
14. O.S. Dinc, C.M. Ettles, S.J. Calabrese and H.A. Scarton, J. Tribology, **113** 512 (1991).
15. B. Buchholz, L.J. Frederick and T.J. Armstrong, Ergonomics, **31** 317 (1988).
16. M. Lodén, H. Olsson, T. Axéll and Y.W. Linde, Br. J. Derm., **126** 137 (1992).
17. M.D. Batt, W.B. Davis, E. Fairhurst, W.A. Gerrard and B.D. Ridge, J. Soc. Cosmet. Chem., **39** 367 (1988).
18. A.D. Roberts and C.A. Brackley, J. Phys. D, **25** A28 (1992).
19. J.S. Comaish, P.R.H. Harborow and D.A. Hofman, Br. J. Derm., **89** 33 (1973).
20. V. Henricsson, A. Svensson, H. Olsson and T. Axéll, Scan. J. Dent. Res., **98** 529 (1990).
21. K.R. Eldredge and D. Tabor, Proc. Roy. Soc. A, **229** 181 (1955).
22. F.P. Bowden and D. Tabor, **The Friction and Lubrication of Solids Part II**, OUP, London, 1964.
23. J.A. Greenwood and D. Tabor, Proc. Phys. Soc., **71** 989 (1958).
24. B.J. Briscoe and D. Tabor, in **Polymer Surfaces**, p1 ed. D.T. Clark and W.J. Feast, John Wiley & Sons, Chichester, 1978.
25. S.P. Timoshenko and J.N. Goodier, **Theory of Elasticity**, McGraw-Hill, Tokyo, 1970.
26. A.D. Roberts and D. Tabor, Proc. R. Soc. Lond. A, **325** 323 (1971).
27. U. Hoppe, G. Sauermann and R. Lunderstädt, J. Soc. Cosmet. Chem., **36** 105 (1985).
28. H.C. Korting, M. Megele, L. Mehringer, D. Vieluf, H. Zienicke, G. Hamm and O. Braun-Falco, Int. J. Cosmet. Sci., **13** 91 (1991).
29. D. Dowson and Z.-M. Jin, J. Phys. D, **25** A116 (1992).
30. J.N. Israelachvili and H. Wennerström, J. Phys. Chem., **96** 520 (1992).
31. S.C. Richards and A.D. Roberts, J. Phys. D, **25** A76 (1992).
32. J.N. Groves and A.R. Sears, J. Coll. Int. Sci., **53** 83 (1975).
33. S. Rothman, **Physiology and Biochemistry of the Skin**, The Univ. of Chicago Press, Chicago, 1954.
34. B.J. Briscoe, in **Friction and Wear of Polymer Composites**, p25 ed. K. Friedrich, Elsevier, Amsterdam, 1986.
35. B.J. Briscoe and A.C. Smith, in **Reviews on the Deformation Behaviour of Materials III**, p151 ed. P Feltham, Freund, Tel-Aviv, 1980.
36. K.A. Grosch, Proc. Roy. Soc. A, **274** 21 (1963).
37. D.F. Moore, **The Friction and Lubrication of Elastomers**, Pergamon Press, Oxford, 1972.
38. K.C. Ludema and D. Tabor, Wear, **9** 329 (1966).
39. A. Schallamach, Wear, **17** 301 (1971).
40. G.A.D. Briggs and B.J. Briscoe, Phil. Mag. A, **38** 387 (1978).
41. M. Barquins, Mat. Sci. and Eng., **73** 45 (1985).
42. S.C. Cohen and D. Tabor, Proc. Roy. Soc. A, **291** 186 (1966).
43. J.F. Archard, Proc. Roy. Soc. A, **243** 190 (1957).
44. M.J. Adams, B.J. Briscoe and L. Pope, in **Tribology in Particulate Technology**, p8 eds. B.J. Briscoe and M.J. Adams, Adam Hilger, Bristol, 1987.
45. K.J. Hall and J.C. Hill, J. Soc. Cosmet. Chem., **37** 397 (1986).
46. M. Takahashi, M. Yamada, Y. Machida and Y. Tsuda, J. Soc. Cosmet. Chem., **35** 171 (1984).
47. P.L. Dorogi, G.M. Dewitt, B.R. Stone and E.M. Buras, Bioeng. Skin, **2** 59 (1986).
48. E.R. Cooper, P.J. Missel, D.P. Hannon and G.B. Albright, J. Soc. Cosmet. Chem., **36** 335 (1985).
49. G.B.E. Jemec, B. Jemec, B.I.E. Jemec and J. Serup, Plast. Reconstr. Surg., **85** 100 (1990).

Thin Films in Tribology / D. Dowson et al. (Editors)
© 1993 Elsevier Science Publishers B.V. All rights reserved.

The adaptive multimode lubrication in knee prostheses with compliant layer during walking motion

T. Murakami [a], N. Ohtsuki [b] and H. Higaki [a]

[a] Department of Mechanical Engineering, Faculty of Engineering, Kyushu University, Hakozaki, Higashi-ku, Fukuoka 812, Japan

[b] College of General Education, Kyushu University, Ropponmatsu, Chuo-ku, Fukuoka 810, Japan

In the present paper, the lubrication modes in knee prostheses different in geometric design and materials under walking conditions are evaluated on the basis of the experimental results in knee joint simulator tests. The fluid film formation is investigated by applying the electric resistance method for conductive femoral and tibial components. The frictional behaviour also is examined to clarify the tribological severity.

To improve the lubricating performance during walking, the design concept of the adaptive multimode lubrication mechanism including weeping, gel film, adsorbed film etc. is proposed to protect articulating surfaces under severe conditions. In this study, frictional behaviour of sliding pair of gelatin gel models lubricated with gelatin solution is observed to investigate the effect of gel film. Furthermore, the application of polyvinylalcohol (PVA) hydrogel as weeping articular surface is evaluated.

1. INTRODUCTION

A great number of total joint replacements particularly for diseased hip and knee joints have been executed all over the world since the successful clinical application of total hip prosthesis by Charnley in 1962 (1). However, in certain cases, serious tribological problems such as loosening and wear occured. This is attributable to the poorness of the fluid film formation in joint prostheses, most of which are composed of ultrahigh molecular weight polyethylene (UHMWPE) and anticorrosive metals or bioceramics. Wear debris attacks the living tissue inducing the loosening of joint prosthesis (2, 3). High friction also accelerates the loosening. To attain long endurance of joint prostheses, it is required to improve the lubricating film formation.

In total hip prostheses as spherical conforming bearings consisted of polyethylene and metal, Sasada and Mabuchi (4, 5) pointed out that the optimum geometric design enables them to be operated under fluid film lubrication by an elastohydrodynamic squeeze action under walking condition. They suggest minimum film thickness of about 2 μm is preserved for lubricant viscosity of 0.015 Pa·s during level walking. Besides, it was recently reported (6-10) that the application of compliant artificial cartilage or soft layer to rubbing surface of total hip or ankle joint is capable of improving its lubricating characteristics.

On the other hand, in many kinds of knee prostheses with low geometrical congruity, the sufficient fluid film is unlikely to be formed to prevent the direct contact between femoral and tibial components, as suggested by the authors (11-14). Furthermore, they pointed out that the elasto-hydrodynamic film formation in a knee prosthesis with an elastomeric tibial component is considerably excellent compared with the prosthesis with a polyethylene tibial one. However, further studies are required to realize a knee joint with a compliant artificial cartilage.

In the present paper, firstly the recent idears on the lubrication modes in natural synovial joints are shortly reviewed from the experimental and theoretical investigations [13]. The functions of natural synovial joints are the final targets to be attained with high performance joint prostheses.

Next, the lubrication modes in knee prostheses under walking conditions are evaluated on the basis of the experimental results in knee joint simulator tests, and then the design concept of the adaptive multimode lubrication for knee prostheses is proposed.

To investigate the lubricating phenomena of gel film, frictional behaviour of sliding pair of gelatin gel models lubricated with gelatin solution is observed. Furthermore, the frictional behaviour of cylindrical knee prosthesis with weeping articular surface of PVA hydrogel is evaluated.

2. ADAPTIVE MULTIMODE LUBRICATION IN NATURAL SYNOVIAL JOINTS

The natural synovial joint is an ideal joint or bearing which has a very low friction coefficient of 0.005 - 0.02, high load-carrying capacity, high mobility accompanied with stability, and a long life of 70 - 80 years. Concerning the excellent lubrication mechanism of the natural synovial joints, various proposals (13) have been offered from the experimental or theoretical studies, since Reynolds (15) suggested that the fluid film formation in natural joints may be mainly due to squeeze action. The lubrication mechanisms such as hydrodynamic (rigid, isoviscous), elastohydrodynamic, microelastohydrodynamic, weeping, biphasic or triphasic, boosted, mixed, elastomixed, gel film and boundary lubrication have been proposed.

As discussed in detail by Murakami (13), the elastohydrodynamic film formation, which is enhanced by the elastic deformation of articular cartilage with subchondral bone and viscous effect of synovial fluid, appears to play the main role in protecting the compliant articular surfaces, supported by the various supplemental lubrication mechanisms when local contact of cartilages occurs during normal activity. This lubrication mode was denoted as the adaptive multimode lubrication by Murakami in 1990 (13), as suggested by Dowson in 1967 (16) and Sasada in 1978 (17).

The authors' experimental simulator tests on knee joint model with compliant rubbing surface(11-14) suggests that the lubrication modes change during walking as shown in Fig.1, which is modified from the proposal by Unsworth et al. (18). It exhibits that the fluid film thickness slightly decreases at peak loads during the loading stance phase, and then recovers to the previous level by the entrainment of fluid during the swing phase. For thin film phases under severe conditions, supplemental lubrication mechanisms appear to protect articular surfaces.

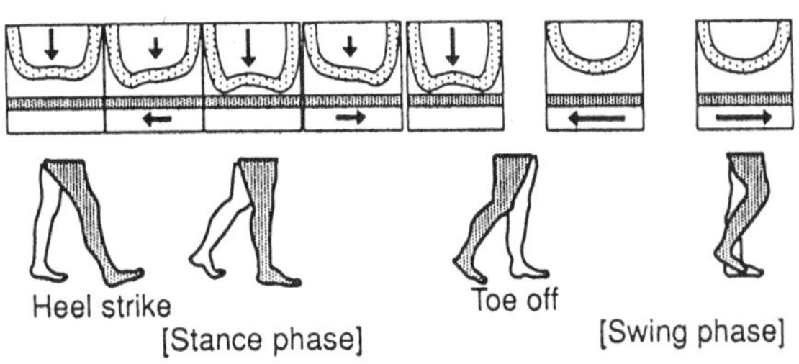

Figure 1. Lubricating film profiles in simplified knee joint model during walking motion

The microelastohydrodynamic lubrication concepts were applied to ankle synovial joints under walking conditions by Dowson and Jin (19-21). By using the cylinder-near-a-plane model, in which the former is covered with a soft elastic layer of sinusoidal surface roughness, they found that the surface roughness itself causes local pressure perturbations to occur which flatten the initial asperities to a significant extent. Thus, microelastohydrodynamic lubrication contributes to the effective lubricating films in synovial joints. However, the flow in articular cartilage indicated by McCutchen (22), and Mansour and Mow (23) is excluded in their analysis.

Ikeuchi et al.(24) considered the flows in both the articular cartilage and the gap between the cartilages in a hip joint under step load. They pointed out that a circumferential zone with local contact points appears just after step loading while most of the applied load is supported by pressurized synovial fluid entrapped at the center. The creep deformation follows in the contact zone due to an outward flow from the cartilage caused by a significant pressure gradient near the surface, improving the conformity and the lubrication. It was further found that the non-Newtonian viscosity of the synovial fluid enhances the squeeze film effect in smaller gap, and the deformation of the subchondral bone also plays a significant role in the entrapment of fluid in a hip joint.

Furthermore, in their experimental study (25), no creep deformation is detected when the cartilage is pressed by fluid alone. This suggests that the cartilage can be simplified as a nonporous incompressible elastic material under full fluid film conditions.

As described above, the elastohydrodynamic fluid film and various supplementary lubrication mechanisms appears to contribute in protecting the synovial articular surfaces and reducing friction, depending on the severity of operating conditions.

Consequently, the lubrication mode in natural synovial joints is regarded as the "adaptive multimode lubrication".

3. SIMULATOR TESTS

3.1 Experimental for simulator tests

A walking motion was simulated by using the same knee joint simulator in Fig.2 as reported in the previous studies[11-14]. Flexion-extension motion of a femoral component was achieved by cam driven rack-and-pinion system. Time-dependent tibial axis load was applied by the personal computor controlled hydraulic system. In these tests, the walking period

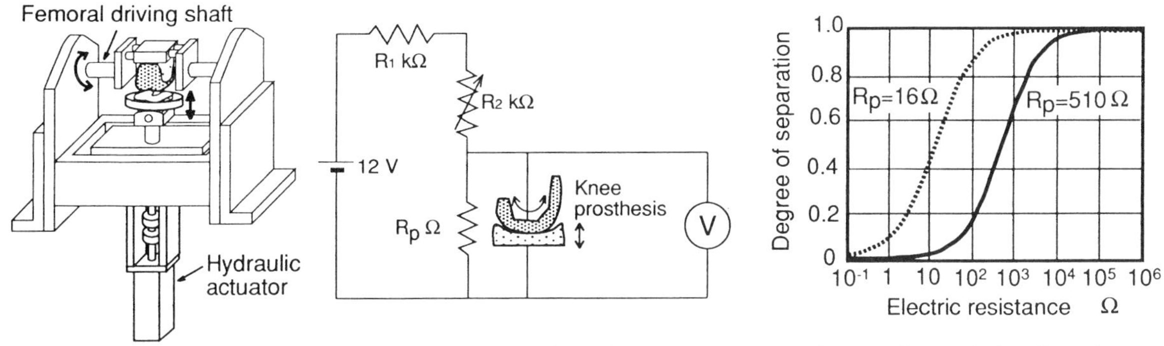

Figure 2. Knee joint simulator, electric circuit and relationship between degree of separation and electric resistance

is 2 s, and the anterior-posterior movement of tibial component is restricted within narrow limits.

The lubricating film formation during walking motion was examined by the electric resistance method. The polyethylene surface was conducted by Pt or Au coating. The extent of fluid film formation was evaluated by the degree of separation, which is defined as the ratio of measured voltage to the applied voltage of 100 mV (degree of separation is 1 : full separation, degree of sepatation is 0 : contact). This value depends on the electric resistance value Rp in parallel to rubbing specimens.

And, the degree of separation = 0 does not mean intimate contact in all conjunction but local intimate contact. So, the fluid film formation in mixed lubrication mode should be investigated by other methods. To examin in detail the tribological severity, the torque of upper driving axis was measured by using strain gauges attached to the shaft.

The following knee prosthesis models as shown in Fig.3 were tested.

[A] Anatomical knee prosthesis with metallic(Co-Cr-Ni alloy steel COP) femoral and polyethylene tibial components: Okayama University Mark II (11,12).

[B] Cylindrical knee prosthesis with metallic femoral and polyethylene tibial components. Radius of femoral steel component is 30 mm, and radial clearance is 0.3 mm. Tibial polyethylene component has sliding mobility on the holder base plate to adjust the load centering.

[C] Anatomical metallic femoral component and a compliant tibial component. The femoral component is the same one with type [A]. Tibial conductive silicone rubber plate of 3 mm thickness is adhered to acrylic plate of 5 mm thickness. The elastic modulus of silicone rubber is 9.1 MPa, similar to cartilage.

Type A represents most of existing anatomical knee prostheses. Type B represents the knee prostheses with better geometric congruity such as Kitasato University Cylindric type (26) and Oxford Knee (27).

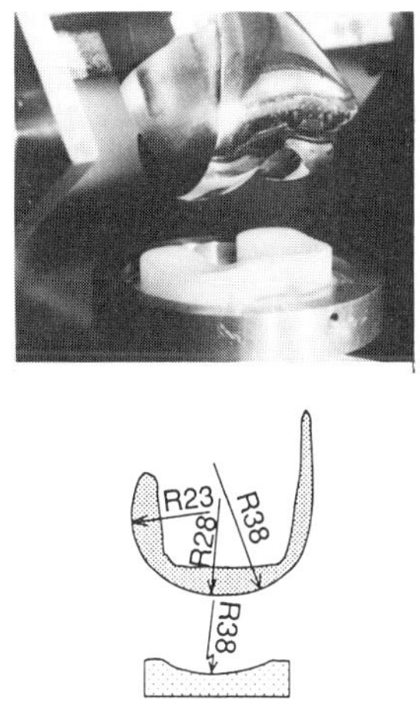

COP/UHMWPE
Ra=0.03μm/0.34μm
(a) Anatomical type A

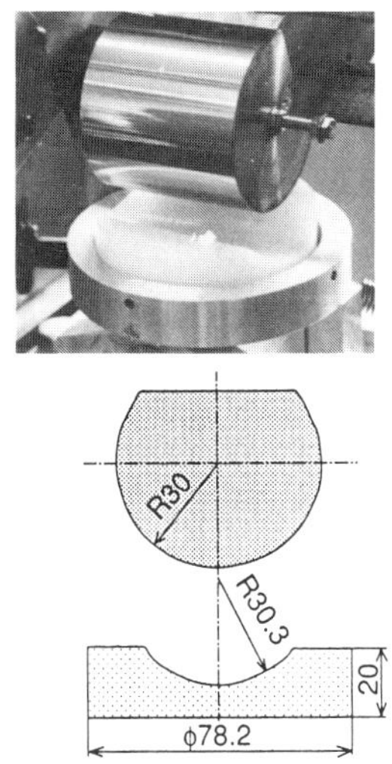

S45C/UHMWPE
0.02μm/0.5μm
(b) Cylindrical type B

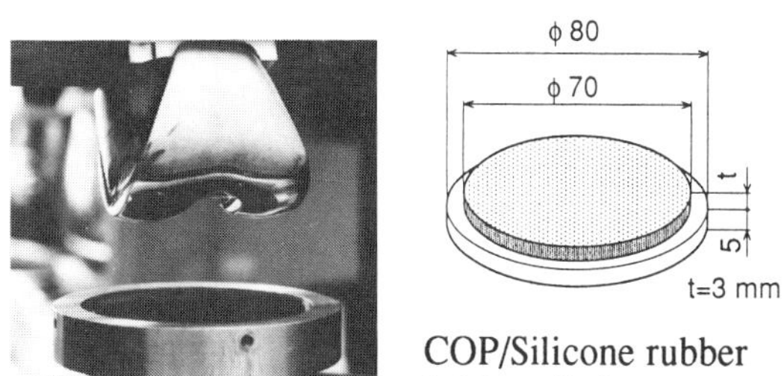

COP/Silicone rubber
0.03μm/0.4μm
(c) Type C with compliant tibial component

Figure 3. Three types of knee prostheses, material combinations and surface roughness Ra

In contrast, type C is new modified type (11). As lubricants, silicone oils (S-1, S-10, ..., S-10000, Number means a kinetic viscosity mm^2/s at 25 °C) and distilled water were used. All tests were carried out at room temperature (13 to 15 °C).

3. 2 Results and discussions on simulator tests

3.2.1 Fluid film formation in electric resistance method

In walking or swinging motion, the mutual relationship between wedge or entraining, and squeeze film actions becomes important for fluid film formation. Under usual walking condition, fluid film formation in natural knee joint is likely to be enhanced by the swinging motion as described before by the authors (11-13).

The examples of fluid film measurement for type A during the fifth walking cycle at different loading patterns are shown in Fig. 4. For this type, the degree of separation with Rp = 510 Ω is 0 except for a part of swing phase under an extremely light load condition, even when a high viscous lubricant (S-10000 of about 10 Pa·s) is used. However, the results with Rp = 16 Ω exhibit the existence of very thin film corresponding to the electric resistance of order of 10^{-1} Ω at some phases under mixed lubrication during walking. It indicates significant contact occurs at the second and third peak loads during stance phase. With decreasing viscosity, these signals disappeared.

In contrast, perfect separation is almost established with high viscosity oil S-10000 during walking in a knee prosthesis B with better geometrical congruity. Furthermore, with medium viscosity S-100 of about 0.1 Pa·s, considerable elastohydrodynamic film is formed even during the loading stance phase (Fig.5, Rp = 510 Ω). The degree of separation decreases near stroke ends at the second and third peak loads during stance phase. At the first peak load, considerable film is preserved, which is attributed to squeeze film action after film recovery during preceding swing phase. However, under lubricated conditions with low viscosity lubricants, it showed the significant direct contact between rubbing surfaces particularly during loading stance phase.

So, the effect of compliant tibial component on fluid film formation was investigated. In this prosthesis C, the fluid film formation is enhanced and better separation is confirmed except the slight

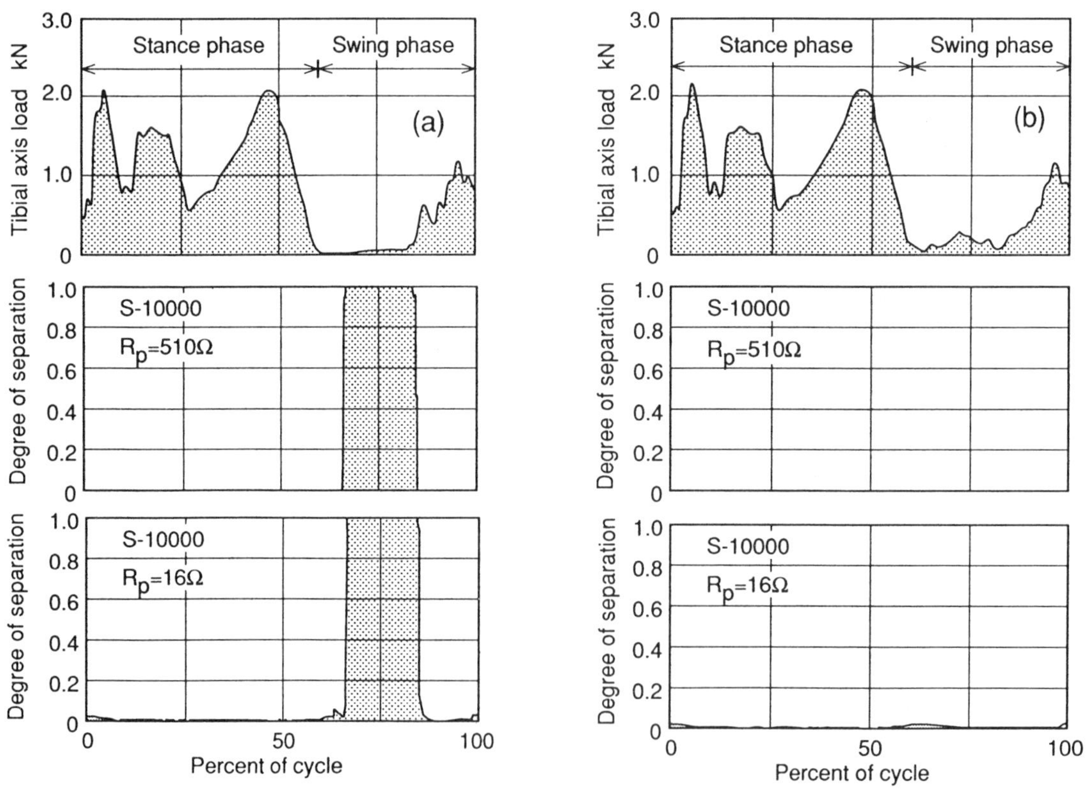

Figure 4. Fluid film formation in knee prosthesis A with polyethylene tibial component during walking under different load patterns

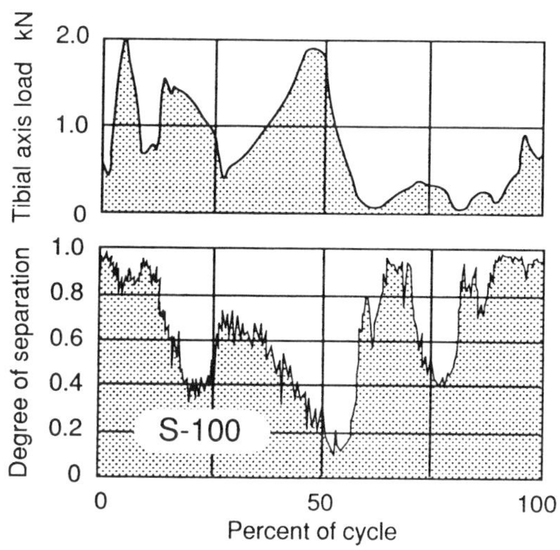

Figure 5. Fluid film formation in cylindrical knee B

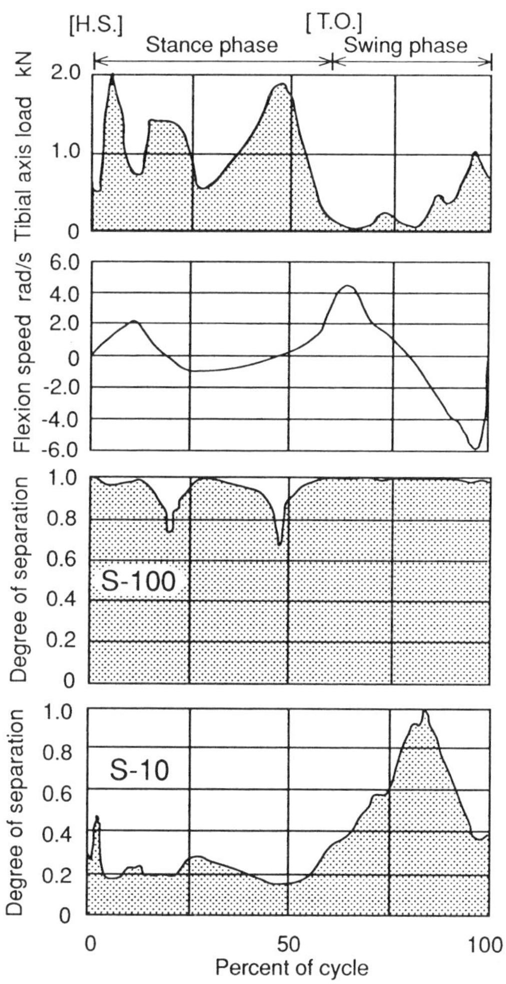

Figure 6. Fluid film formation in knee prosthesis C with silicone rubber tibial component

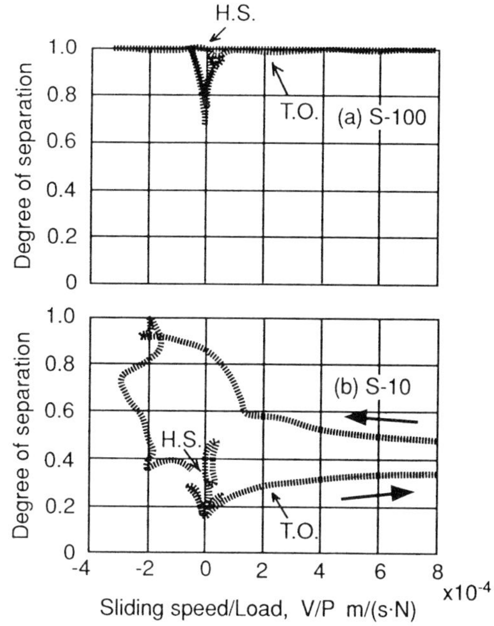

Figure 7. Fluid film changes during walking motion

drops with medium viscosity of S-100 during walking motion as shown in Fig.6 with Rp = 510 Ω. The film profile in knee prostheses with a compliant layer lubricated with medium or high viscosity lubricant appears to change in similar modes to natural knee joint. During the stance phase, the degree of separation decreases slightly at stroke ends at the second and third peak loads, and then recovers to the perfect separation level during the swing phase. With decreasing viscosity, however, the direct contact occurs in this type, too.

To consider the effect of entrainment, the degree of separation is plotted against the operating parameter of [Sliding speed (V)] / Load (P) in Fig. 7. With S-100, the separation slightly drops like V-shape near V/P=0 during stance phase from heel strike (H.S.) to toe off (T.O.). With S-10, the degree of separation changes complicately with V/P depending on its history during stance and swing phase. During stance phase, the separation changes like normal squeeze action and V-shape near V/P = 0. In contrast, at stroke end of V/P = 0 during swing phase, the degree of separation is high, which is brought about by the preceeding recovery of fluid film.

In this electric resistance method, it is difficult to evaluate the difference of lubrication state under mixed lubrication with low viscosity conditions. Therefore, the fluid film formation was evaluated from the viewpoint of frictional behaviour.

3.2.2 Frictional characteristics

The frictional behaviour was evaluated by measuring the variation in the torque of the femoral driving shaft (12, 13, 28). These values include the torque caused by frictional force and eccentric loading, and inertia effect. So, relative values of frictional force ΔF for different lubricants (silicone oil and distilled water) to those of S-100 for A and B or S-1000 for C, which showed minimum torque, are plotted in Fig.8. Positive frictional value means posterior to anterior direction to femoral component.

For all types, the friction increases with a decrease in viscosity except high viscosity range. It is noted that distilled water exhibits lower friction than S-1 with similar viscosity. It is further noticed that the frictional severity is increased near stroke ends during stance phase with low viscosity lubricants. For cylindrical type B, high viscosity condition in full fluid film lubrication mode (Fig.8(b')) shows excessive high friction at high speed phases.

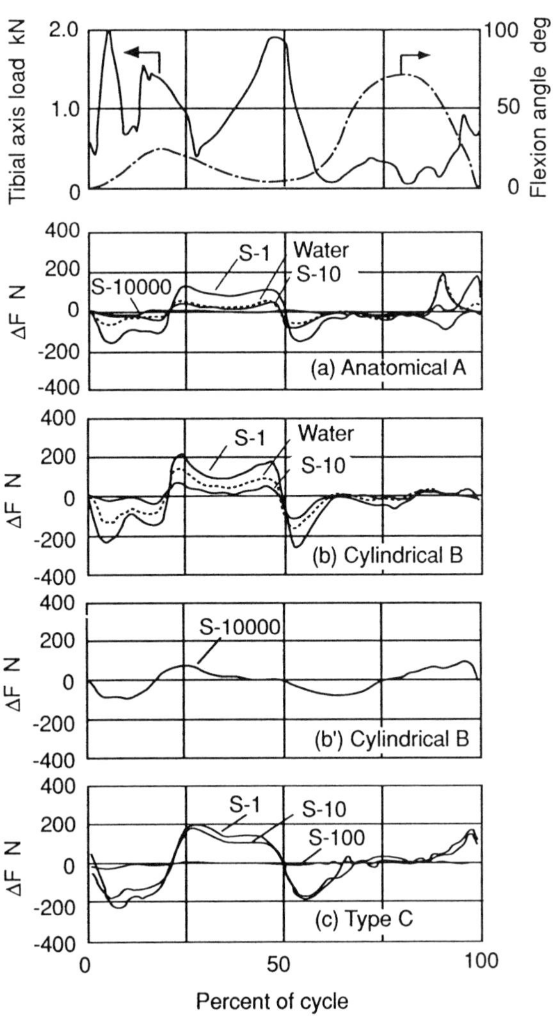

Figure 8. Frictional behaviour during walking

The elastohydrodynamic film formation in cylindrical knee prosthesis B and an anatomical one with compliant layer C are considerably excellent compared with the anatomical prosthesis A with a polyethylene tibial one. From the viewpoint of design, larger contact areas and lower contact pressure are needed for thicker fluid film as pointed out by Dowson et al. for cushion form hip prosthesis (10). The apparent contact areas and the average pressures which were examined by the pressure sensitive paper (p>0.5 MPa, Fuji Film) under the condition of extended and loaded at 2.0 kN are shown in Fig. 9.

The knee prosthesis C composed of anatomical femoral component and silicone rubber flat tibial one shows higher contact pressure than conformal cylindrical type B. With medium viscosity lubricant S-100, however, type C exhibits better separation than type B. This is attributable to microelastohydrodymnamic effect in low elastic modulus material with appropriate asperity of type C. The high electric sensitivity of coated layer of cylindrical type B in local slight contact may affect the observed separation.

As described above, the knee prosthesis with larger contact areas B or C appears to be operated under fluid film formation with medium viscosity lubricant of 0.1 Pa·s. For artificial joints in human body, however, the viscosity of body fluid or secondary synovial fluid is about 0.01 Pa·s or less. So, the adaptive multimode lubrication mechanism including microelastohydrodynamic, weeping, gel film etc. is required to protect articulating surfaces under severe conditions with low viscosity lubricants. The microelastohydrodynamic lubrication is likely to enhance effective fluid film lubrication in artificial joints with compliant layer as discussed in detail by Dowson et al. (10,21). So, in this study, the effect of gel film (26) on lubrication is discussed besed on the reciprocating tests of gelatin models.

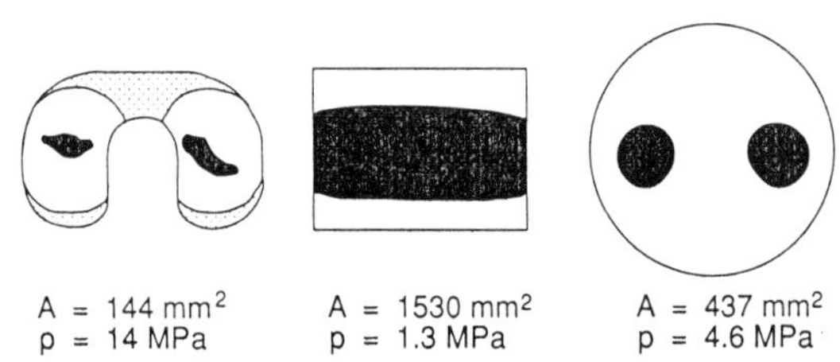

(a) A (b) B (c) C

Figure 9. Apparent contact areas at 2 kN at extended

4. RECIPROCATING TEST OF SLIDING PAIR OF GELATIN HYDROGEL

4.1 Experimental for gelatin models

Frictional behaviour of sliding pair of gelatin gel models lubricated with dilute gelatin solution was observed to evaluate the effect of gel film.

The reciprocating tests were carried out for upper gelatin gel disc of 20 mm radius and 15 mm thickness, and lower gelatin gel flat plate supported with acrylic block inside 5 mm below rubbing surface (Fig.10). Both gelatin hydrogel specimens were prepared by gelatin 20 wt.% water solution. The upper surface roughness is Ra=0.50 μm, and the lower is 0.44 μm.

As lubricant, gelatin 1.5 wt.% water solution was used, which has sol-gel transition around 15 - 21 °C, depending time and pressure. To standardize the gel condition, firstly the solution was warmed up to 30°C corresponding to about 0.03 Pa·s, and then cooled to definite temperature whitin 10 min as shown in Fig.11(a). The viscosity was measured by cone/plate viscometer. When the gelatin solution was set aside, it showed time-dependent viscosity, resulting in gel state for appropriate time, even under constant temperature (Fig.11(b)).

In this test, four kinds of conditions were compared, i.e., (G-1) immediately after cooling, (G-2) after keeping at constant temperature for 30 min in outer bath, and (G-3) after keeping at constant temperature for 30 min in immersed condition in liquid bath of reciprocating apparatus, (G-4) the same with G-3, but initial temperature of test specimen was cooled 10°C lower than gelatin solution.

Reciprocating test was run under the conditions of stroke L= 20 mm and frequency ω = 1.0 Hz, immediately after loading of weight w = 130 N/m. Contact width in sliding direction is about 15 mm. Contact stress condition in gelatin gel was directly observed by photoelastic method.

4.2 Results and discussions on gelatin models

Representative frictional behaviour (f : coefficient of friction) during initial ten cycles at 18°C, where the lowest friction is observed, is shown in Fig.12. In G-1, friction is low only at the initial stages and

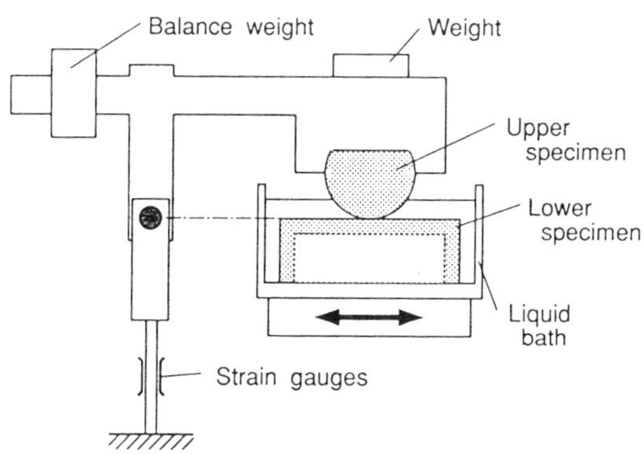

Figure 10. Reciprocating test apparatus

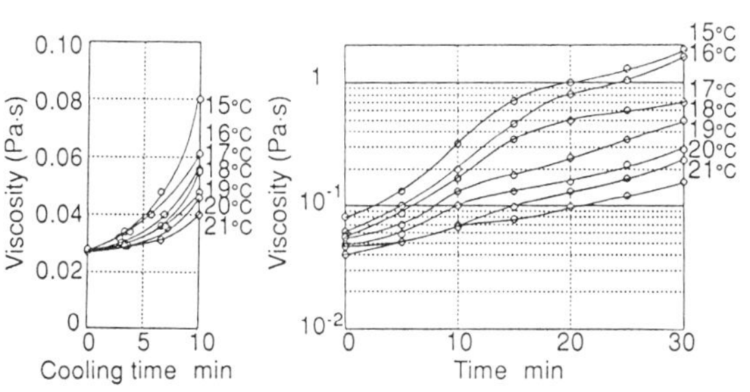

Figure 11. Time dependent viscosity of gelatin solution at shear rate of 100 1/s

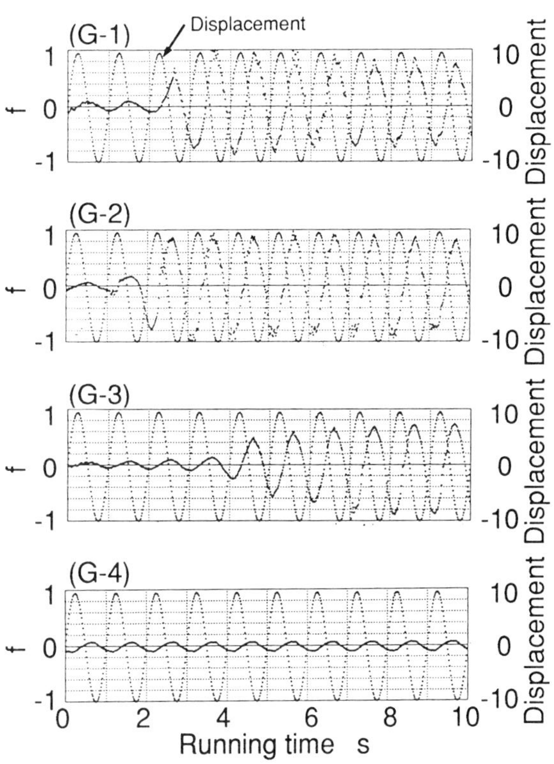

Figure 12. Frictional behaviour in reciprocating test

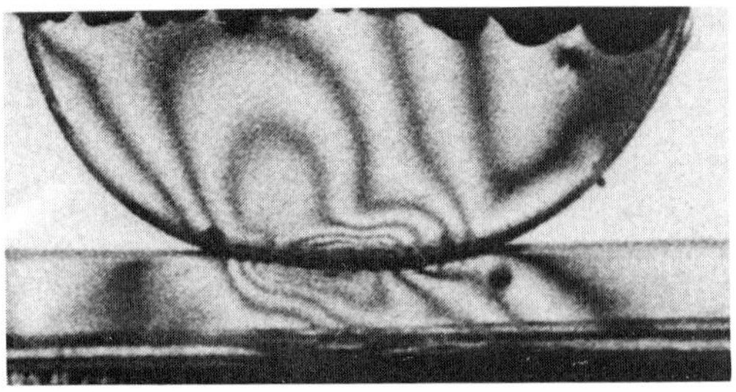

(G-2)

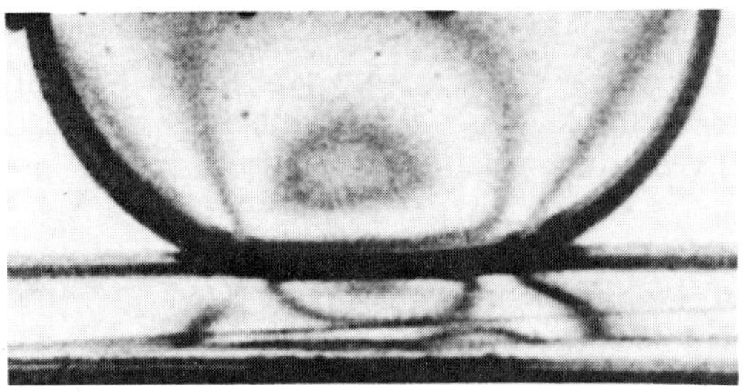

(G-4)

Figure 13. Photoelastic patterns at stroke centre

then abruptly increases. In G-2 with higher viscosity, the similar behaviour is observed. In G-3, the transition from low friction to high occurs moderately. In contrast, in G-4, stable low friction condition (f<0.1) is preserved. It is noted that the maximum friction in G-4 is observed at deceleration phase, while the maximum occurs at stroke centre for fluid film lubrication region at initial stages for G-1 to G-3. This fact suggests that in G-4 gel film plays main role in frictional behaviour.

Fig.13 shows photoelastic patterns corresponding to principal shear stress distributions at the stroke centre, for left moving stroke of lower specimen, in the 5th cycle for G-2 and G-4. Contact stresses are very different depending on friction condition.

As described above, the conditions apprppriate for gel film formation on the surfaces improve the friction condition. In G-4 condition, the deposition of gel film on the gelatin surfaces appears to be enhanced by temperature gradient. It is suggested that the solid-like behaviour of gelatin solution is effective in reducing friction at the critical gelled condition.

However, the coefficient of friction is not so much low. This may be due to the operating conditions in this reciprocating test under shorter stroke and constant load, which suppress the fluid film formation, resulting in larger load support by gel film. Furthermore, gelatin hydrogel has very week weeping behaviour because of scarcity of exudative free water, which facilitates the adhesion. So, another simulator test was carried out using cylindrical knee prosthesis with PVA hydrogel as weeping surface.

5. CYLINDRICAL KNEE PROSTHESIS WITH PVA HYDROGEL

The cylindrical knee prosthesis B was compared with the cylindrical knee with concave silicone rubber or PVA hydrogel layer of 3 mm thickness, where the latters are denoted as B-SR and B-PVA, respectively. PVA hydrogel was prepared with PVA 15 wt.% water solution by repeated freezing-thawing method by Nanbu (6, 29). In these simulator tests, stainless steel femoral component of 30 mm radius was used. Concave tibial components for B-SR and B-PVA have about 30 mm radius. These prostheses have larger apparent contact areas. As lubricants, silicone oils, distilled water and sodium hyaluronate (molecular weight = 1.96×10^{6}) 0.5 wt% water solution were used. Sodium hyaluronate solution behaves as non-Newtonian fluid : 0.430 Pa·s at shear rate of 10 1/s, 0.074 Pa·s at 400 1/s at 15 °C.

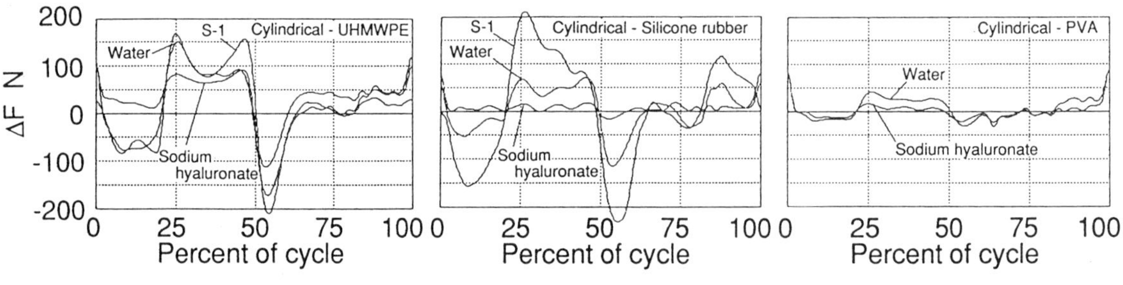

(a) B (b) B-SR (c) B-PVA

Figure 14. Frictional behaviour for three kinds of cylindrical knee prostheses during walking

Frictional behaviour for three kinds of cylindrical knee in walking simulator tests is shown in Fig. 14. Compliant layer of silicone rubber or PVA hydrogel improves fluid film formation compared with type B with polyethylene tibial component. With distilled water, B-PVA exhibits lower friction than B-SR. With sodium hyaluronate solution, both B-PVA and B-SR show low friction. At the present stage, it is difficult to discriminate between them for hyaluronate solution. Further researches shall clarify detailed lubrication mechanisms.

6. DISCUSSIONS

As shown in the first half of this paper, the considerable enlargement of apparent contact areas in knee prosthesis design by the improvement of geometrical congruity for B or the application of compliant rubber material for C is likely to significantly promote fluid film formation with medium viscosity lubricant of 0.1 Pa·s.

However, the lubricant viscosity in replacement joints is about 0.01 Pa·s or less. The knee prostheses B and C show considerable contact between rubbing surfaces by thinning of fluid film with decreasing viscosity. Therefore, the authors propose the design concept of the adaptive multi-mode lubrication mechanism to protect articulating surfaces and reduce friction under severe conditions with low viscosity lubricants.

Conformal cylindrical knee prosthesis with compliant surfaces such as silicone rubber or PVA hydrogel is a feasible design to realize low friction and low wear (10, 30).

One of the unsolved problems on the clinical application of compliant materials or artificial cartilage is their short durability. So, it is better to reduce the contact pressure in knee prosthesis by improving the geometrical congruity.

Furthermore, it is shown in the simulator tests that the knee prosthesis C composed of the anatomical femoral component and silicone rubber flat tibial one exhibits abrupt friction increase with decreasing viscosity (S-10 in Fig.8(c)). The appropriate geometrical congruity is required to preserve the lubricating film formation particularly under low viscosity conditions.

As discussed above, cylindrical knee prostheses with compliant layer exhibit good frictional behaviour even with low viscosity lubricants. On the lubricating effect of sodium hyaluronate solution as polymer solution, the cylindrical knee joints with compliant layer exhibit its effectiveness. This means the thicker fluid film is formed in these prostheses, which allows the polymer to be entrained into the conjunction (31).

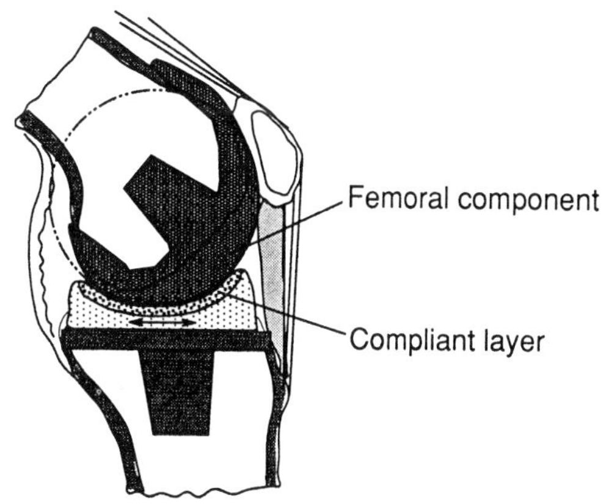

Figure 15. An example of design of cylindrical knee prosthesis with compliant layer

On the lubrication mechanism in joint prosthesis with PVA hydrogel, weeping or exudation of fluid from PVA seems to make lubrication better. The gel film derived from bulk PVA or hyaluronate solution may play a role to protect rubbing surfaces. In this case, the similar phenomena to those in gelatin tests above may occur, although considerable fluid film contributes to lowering friction. Detailed researches are required to elucidate these lubrication mechanisms.

To compromize the reqirements for the lubrication performance and mobility, the adequate freedom of sliding motion (27) are required. An example of design is shown in Fig.15.

7. CONCLUSIONS

In anatomical knee prosthesis with low geometric congruity composed of metallic femoral and polyethylene tibial components, the mixed lubrication appears to prevail during usual walking motion.

In the cylindrical knee prosthesis with polyethylene tibial component or anatomical knee prosthesis with compliant tibial layer, the elastohydrodynamic film is considerably established with medium viscosity of about 0.1 Pa·s during walking motion.

Therefore, the design concept of the adaptive multimode lubrication was proposed to improve lubricating performance of joint prostheses even under low viscosity conditions. The gel-like behaviour of lubricant is likely to be effective in reducing friction at the critical gelled condition.

Cylindrical knee with PVA hydrogel exhibits excellent frictional behaviour under low viscosity condition.

However, further studies are required to realize a knee prosthesis with a compliant artificial cartilage for clinical use.

ACKNOWLEDGEMENTS

The authors wish to express their appreciation to the former graduate students at Kyushu University, Messrs. K.Kariya, A.Hirashima and H.Shimanouchi for thier cooperation, to Dr. M. Nanbu of Nippon Oil Co., Ltd. for his advice on PVA hydrogel, and to Prof. Y.Sugioka of Dept. of Orthopaedic Surgery at Kyushu University and Prof. K.Ogata at Fukuoka University for their comments from orthopaedic viewpoint. Knee prosthesis was provided by Mizuho Ika Co., Ltd., and sodium hyaluronate was supplied by Roman Industries Co., Ltd. Financial supports were given by the Grant-in-Aid for Scientific Research of The Ministry of Education, Science and Culture,and Takeda Science Foundation.

T. Murakami thanks Prof. D. Dowson for his kindful and valuable advices for this study since the author's stay at the University of Leeds in 1981- 82.

REFERENCES

1. J. Charnley, Low Friction Arthroplasty of the Hip, Springer-Verlag, (1979)
2. H.G. Willert and M. Semlitsch, J. Biomed. Mater. Res., 11 (1977) 157.
3. I. C. Clarke and P. Cambell, Progress in Bioengineering, ed. by J.P. Paul, et al., Adam Hilger (1989) 104.
4. T. Sasada and K. Mabuchi, Proc. JSLE Int. Trib. Conf., (1985), 949.
5. K. Mabuchi and T. Sasada, Wear, 140 (1990) 1.
6. T. Sasada, M. Takahashi, M. Watakabe, K.Mabuchi, Y. Tsukamoto and M. Nanbu, J. Jpn. Soc. Biomaterials (in Japanese), 3, 3, (1985) 151.
7. M. Oka, T. Noguchi, P. Kumar, K. Ikeuchi, T. Yamamuro, S.H. Hyon and Y. Ikada, Clin. Mater., 6 (1990) 361.
8. A. Unsworth, M.J. Pearcy, E.F.T. White and G. White, Proc. Inst. Mech. Eng., C219/87 (1987) 715.
9. D.D.Auger, J.B.Medley,J.Fisher and D. Dowson, Mechanics of Coatings, ed. by D. Dowson, C.M. Taylor and M. Godet, Elsevier (1990) 264.
10. D. Dowson, J. Fisher, Z.M. Jin, D.D. Auger and B.Jobbins, Proc. Instn. Mech. Engrs., 205, Part H, (1991) 59.
11. T. Murakami and N. Ohtsuki, Fluid Film Lubrication - Osborne Reynolds Centenary, ed. by D. Dowson, C.M.Taylor, M. Godet, and D. Berthe, Elsevier, (1987) 387.
12. T. Murakami and N. Ohtsuki, Progress and New Directions of Biomechanics, ed. by Y.C. Fung, K. Hayashi, and Y. Seguchi, Mita Press (1989) 403.
13. T. Murakami, JSME International Journal, Series III, 33, 4 (1990) 465.
14. T. Murakami and N. Ohtsuki, Proc. Third U.S.A.-China-Japan Conf. Biomechanics, (1991) 78.
15. O. Reynolds, O., Phil. Trans. Roy. Soc. London, 177 (1886) 157.
16. D.Dowson, Proc. Instn. Mech. Engrs., 181, Pt 3J, (1966-67) 45.
17. T. Sasada, J. Jpn. Soc. Lubr. Eng., (in Japanese), 23, 2 (1978) 79.
18. A. Unsworth, D. Dowson and V. Wright, Ann. Rheum. Dis., 34 (1975) 277.
19. D. Dowson and Z-M. Jin, Engng. Med., 15 (1986) 63.
20. D. Dowson and Z-M. Jin, Fluid Film Lubrication - Osborne Reynolds Centenary, ed. by Dowson, D. et al, Elsevier Sci. Pub., (1987) 375.
21. D. Dowson and Z.M. Jin, J. Phys. D : Appl. Phys., 25 (1992) A116.
22. C.W. McCutchen, Nature, 184 (1959) 1284.
23. J.M. Mansour and M.C. Mow, Trans. Am.Soc. Mech. Engrs., J. Lub. Tech., 99, (1977) 163.
24. K. Ikeuchi, M. Oka and H. Mori, Trans. Jpn. Soc. Mech. Eng., (in Japanese), 55, 510, C (1989) 508.
25. K. Ikeuchi, M.Oka and K. Gi, Trans. Jpn. Soc. Mech. Eng., (in Japanese), 55, 516, C (1989) 2123.
26. T. Sasada, Y. Tsukamoto and K. Mabuchi, Biotribology, (in Japanese), Sangyo-Tosyo, (1988) 60.
27. J.W. Goodfellow and J. O'Cornnor, Clin. Orthp. & Rel. Res., 205 (1986) 21.
28. N. Ohtsuki and T. Murakami, Proc. Jpn. Int. Trib. Conf. Nagoya, 2 (1990) 713.
29. M. Nanbu, Kobunshi-kako (J. Polymer Processing), 32 (1983) 32.
30. D. Dowson, Proc. Instn. Mech. Engrs., Int. Conf. The Changing Role of Engineering in Orthopaedics, (1989) 1.
31. T. Murakami, A. Hirashima and N. Ohtsuki, to be published in Proc. JAST Trib. Conf. Morioka, Oct.1992 (in Japanese) (1992).

Thin Films in Tribology / D. Dowson et al. (Editors)
© 1993 Elsevier Science Publishers B.V. All rights reserved.

Friction of Cylindrical Cushion Form Bearings for Artificial Joints – A Comparison of Theory and Experiment –

D.D. Auger *, D. Dowson, J. Fisher
Department of Mechanical Engineering,
University of Leeds, Leeds, U.K., LS2 9JT

Cylindrical cushion bearings have been tested under conditions of dynamic loading and variable velocity. Newtonian lubricants were used and friction was measured throughout the cycle. Stribeck assessment of each bearing was performed using the point of maximum load and velocity in the cycle. This experimental data of friction coefficient versus Sommerfeld number was compared to theoretical predictions and showed good agreement. Detailed agreement of friction throughout the cycle between theory and experiment was also obtained.

1. INTRODUCTION

During the past decade, interest has been growing in cushion form bearings designed for total joint replacement. These cushion bearings utilise thin layers of compliant material in order to enhance the lubricating conditions at the articulating surfaces [1]. If a continuous lubricating film can be maintained within a cushion joint, then these prostheses have several potential advantages over conventional implants. Cushion joints would benefit from lower frictional torque and reduced generation of wear debris. Current artificial joints have relatively stiff metallic or ceramic on polymer bearing surfaces [2] that operate in a mixed or boundary lubrication regime [3]. This results in the generation of wear debris which has been shown to cause adverse tissue reactions, bone resorption and loosening of the prostheses [4–6]. Cushion joints which operated on a full fluid film would eliminate wear, reduce the incidence of prosthetic loosening and prolong implant life.

The lubrication regime in which cushion implants operate can be determined experimentally by measuring the friction and carrying out Stribeck assessment of the results [7, 8]. Experimental studies have shown that cushion bearing design will require careful optimization of both the materials and the geometry [7, 9, 10]. Theoretical studies can help to identify and optimize the more important parameters of design [1]. However, no one has successfully obtained agreement between theory and experiment for low modulus layered contacts under dynamic loading and sliding conditions. This verification is required before confident use can be made of theory in the design process.

Cushion bearings have proved difficult to study under conditions of dynamic loading and variable velocity as found in artificial joints. For cushion line contacts, good detailed agreement between theory and experiment has been found under conditions of constant load and sinusoidal velocity [11]. However, agreement between experiment and theory under conditions of dynamic loading and sinusoidal velocity was not found to be good in a recent comparison [8]. In the present study, a cylindrical bearing design, and a modified experimental method, were adopted and a comparison between theoretical predictions and experimental observations of friction for variable loading and velocities as found in artificial joints was carried out.

1.1. Nomenclature

$b, b(t)$ line dry contact half width

E_1 modulus of elasticity of the femoral component

E_2 modulus of elasticity of the layer

$F, F(t)$ applied load

*Author supported by Overseas Research Students Awards scheme, Tetley and Lupton Scholarship and Johnson and Johnson Orthopaedics. The work was also supported by the Science and Engineering Research Council and the Arthritis and Rheumatism Council. Thanks to Mr. H.D. Darby and Mr. P. Wood for providing technical support.

h_{cen} central film thickness

L length of the line contact

R reduced radius ($\frac{1}{R} = \frac{1}{R_1} - \frac{1}{R_2}$)

S Sommerfeld number: ($S = \frac{\eta UL}{2F}$)

t_h thickness of the low modulus layer

$U, U(t)$ sliding velocity

α percent flattening of deforming asperities

η lubricant viscosity

λ surface separation ratio: ($\lambda = \frac{h_{cen}}{\sigma}$)

λ_e effective surface separation ratio

$\mu, \mu(t)$ coefficient of friction

σ combined RMS roughness: ($\sigma = \sqrt{\sigma_1^2 + \sigma_2^2}$)

σ_1 RMS roughness of layer

σ_2 RMS roughness of metal counterface

SSD sample standard deviation

2. THEORY

A complete analysis of the lubrication of a cushion bearing would be computationally prohibitive [12, 13] so a simplified approach was adopted for the present study. A plane inclined surface bearing model was used by previous researchers [8, 11] but even this required a significant computational effort. Therefore, for the current study, an estimate of the friction coefficient throughout the dynamic cycle was derived from the shear stress relation for a Newtonian fluid. The film shape was assumed to be the same as the dry contact shape:

$$\mu(t) = \frac{2Lb(t)\eta\frac{U(t)}{hcen}}{F(t)} \tag{1}$$

The load, $F(t)$ and the sliding velocity, $U(t)$ were measured throughout the cycle. The contact half width, $b(t)$ during the cycle was calculated using the full elasticity solution [14]. A steady state value of central film thickness, h_{cen} was calculated to estimate the thickness of the film throughout the dynamic cycle. Previous theoretical studies [12, 13] have shown that the film thickness is not expected to vary a great deal during a dynamic cycle. Therefore, a reasonable estimate of the film thickness can be obtained by using the average cycle conditions in a steady state formula. A previous investigator [15] developed formulae which were suitable for the present study:

$$\begin{aligned} \frac{0.8h_{cen}}{R} &= 1.335\exp^{-0.2396\frac{b}{t_h}} K \qquad (2) \\ \text{for } \frac{b}{t_h} &< 2.0 \\ \frac{0.8h_{cen}}{R} &= 1.159\left(\frac{b}{t_h}\right)^{-0.4875} K \\ \text{for } \frac{b}{t_h} &\geq 2.0 \\ \text{where } K &= \left(\frac{\eta U}{E_2 R}\right)^{0.6}\left(\frac{F}{E_2 RL}\right)^{-0.2} \end{aligned}$$

Equation 1 was used to estimate the friction in the cylindrical cushion bearings throughout the dynamic cycle. The data from the region which corresponded to peak loading and velocity in the cycle was also used in a similiar fashion to that of previous studies [7, 8], to construct theoretical Stribeck curves for comparison with experimental results. The initiation of film breakdown was predicted using a surface separation ratio [16] defined as the ratio of the central film thickness to the combined RMS roughness of the two articulating surfaces:

$$\begin{aligned} \lambda &= \frac{h_{cen}}{\sigma} \qquad (3) \\ \text{where } \sigma &= \sqrt{\sigma_1^2 + \sigma_2^2} \end{aligned}$$

Normally with hard bearing surfaces, asperity contact and film breakdown occurs when $\lambda \leq 3.0$ and this results in a steep increase in friction with a reducing Sommerfeld number. This criterion however, ignores microelastohydrodynamic effects and assumes that there will be no deformation of the asperities and effective smoothing of the surfaces. Cushion bearings benefit from microelastohydrodynamic effects [1]. Flattening of the asperities reduces the RMS roughness of the surfaces and allows thinner lubricating films to be preserved. For the present study, the surface separation ratio was calculated assuming that no deformation of the surface asperities occurred. Hence, a comparison of theory with the experimental results was expected to show the benefits achieved by microelastohydrodynamic lubrication.

3. MATERIALS AND METHODS

3.1. The Contact

Four flat cushion bearings were constructed using two different low modulus medical grade polyurethanes[2] and two different layer thicknesses. Table 1 shows the details of each layer. The modulii were estimated from Shore A hardnesses measurements and the technical information supplied with the material. The Poisson's ratio was assumed to be 0.5. These flat cushioned components articulated against a metallic cylinder which had a radius of 20 mm. The bearing was 62 mm long, which was sufficient to allow it to be confidently treated as a line contact [17].

Table 1
Cushioned components

Flat	t_h (mm)	E_2 (MPa)	σ_2 (μm)	σ (μm)
1	2.5	20	0.113	0.119
2	5.0	20	0.080	0.088
3	2.5	6	0.102	0.111
4	5.0	6	0.078	0.090

A Talysurf 5 profilometer was used to measure the roughnesses of the components. Six traverses were carried out on each layered flat and sixteen on the cylinder. The results were averaged and the combined RMS roughnesses for each set of components was tabulated (table 1).

3.2. Lubricants

Testing was always performed with the cushion bearing fully flooded with lubricant. Table 2 shows the lubricants which were used. Water, two low viscosity carboxymethylcellulose sodium salt solutions (LV) and two mineral oils were used. The LV and water solutions were named according to the percentage by weight ratios of carboxymethylcellulose sodium salt to total weight. The HVI 60 mineral oil was prepared by Shell and the HS 2:1 oil was a mixture of Shell HVI 650 and a polybutene thickening agent (BP: HVIS 10). This mixture was prepared in the ratio of 2:1 by volume and was thus named HS 2:1.

[2]Tecoflex: Courtesy of Thermedics Inc. Woburn, MA 01888 USA

Table 2
Lubricant viscosities

Lubricant	Viscosity (Pa.s)
Water	0.000845
LV 0.05%	0.00136
LV 0.25%	0.00312
HVI 60	0.0254
HS 2:1	0.486

The viscosities of the lubricants were measured using a Ferranti–Shirley cone–on–plate viscometer over a range of temperatures (20–50°C) and shear rates (0–15500 s^{-1}). Over the range of shear rates employed, the lubricants were judged to be Newtonian [18]. Table 2 gives the viscosities of the lubricants at the average operating temperature of 33.8°C for the pendulum simulator apparatus over all of the tests which were carried out.

3.3. Pendulum Simulator Apparatus

The friction tests were carried out using the pendulum simulator apparatus (figure 1) which has been recently modified [8]. The cylindrical femoral component was mounted on the moving frame, at the fulcrum of a driven pendulum. A variable load pattern was imposed by a complex cam and follower hydraulic system and the component was oscillated back and forth at 1 Hz by a Scotch yoke mechanism. Figure 2 shows the average load and position cycles for all the tests which were carried out. The vertical dashed lines show the region of data which corresponded to maximum load and maximum sliding velocity that was used to produce Stribeck plots of the results.

The cushioned tibial component was mounted on the fixed frame suspended on a tray which floated on hydrostatic bearings. The tray was free to rotate on the hydrostatic bearings but was restrained by a transducer which measured the torque between the articulating surfaces as the

Figure 1. Pendulum simulator apparatus

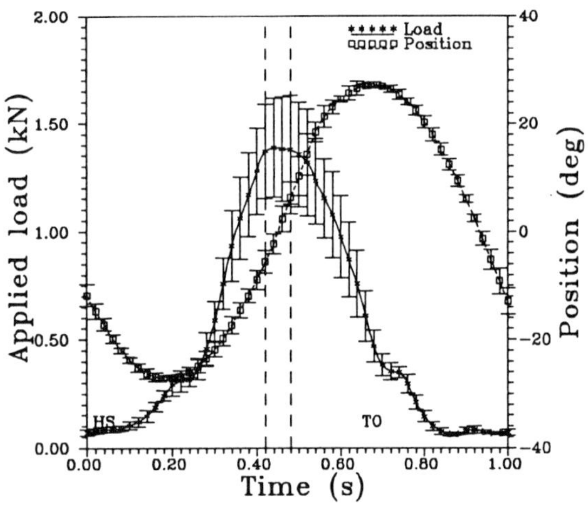

Figure 2. The dynamic cycle

femoral component oscillated forwards and backwards. Load and position were also measured throughout the cycle.

3.4. Test Protocol

Fluid film lubrication and very low levels of friction were expected to occur in the cushion bearings, so for this study a new test protocol was developed. It was found that uncertainties in the measurements of frictional torque were substantially reduced by combining the results of two measurements. Firstly, the torque was recorded with the femoral head travelling in the forward direction during the stance phase of loading. Secondly, the torque was measured with the components moving in the reverse direction during stance. These two measurements were combined by assuming that the frictional torques during the forward and reverse cases were equal in magnitude but opposite in sign. In addition, any measurement uncertainty was assumed to be independent of the direction of bearing movement during the stance loading phase of the cycle. Previous researchers [19, 20] have used similiar techniques in order to determine the zero levels in their measurements.

For a given set of components and lubricant, the forward and reverse procedure was repeated three times and two cycles of data were collected during each test. The six complete cycles of data were averaged and the sample standard deviations at each point in time were calculated. The resulting cycle was compared with its corresponding theoretical cycle and the data from the peak region was also used to plot Stribeck curves of the results for comparison with theory.

4. RESULTS AND DISCUSSION

The average load of 1380 N (SSD = 236) and average sliding velocity of 67.3 mm/s (SSD = 15.3) in the peak region were used with the lubricant viscosities shown in table 2 to calculate the theoretical friction coefficient from equation 1 for each component. The estimated friction coefficients were plotted against Sommerfeld number and compared with the experimental Stribeck curves for each flat.

4.1. Stribeck Comparisons

Figures 3 and 4 show the results for the 2.5 and 5.0 mm 20 MPa flats, 1 and 2. Figures 5 and 6 show the results for the 2.5 and 5.0 mm 6 MPa flats, 3 and 4. The Sommerfeld numbers at which the surface separation ratio $\lambda = 3.0$ are shown in figures 3–6 as vertical dashed lines. The onset of film breakdown, surface to surface contact and increasing friction was expected to occur at decreasing Sommerfeld numbers to the left of $\lambda = 3.0$. The results for the 20 MPa layers shown in figures 3 and 4 agreed with this theory in general terms and a rise in the measured friction was observed for decreasing Sommerfeld numbers. The 6 MPa

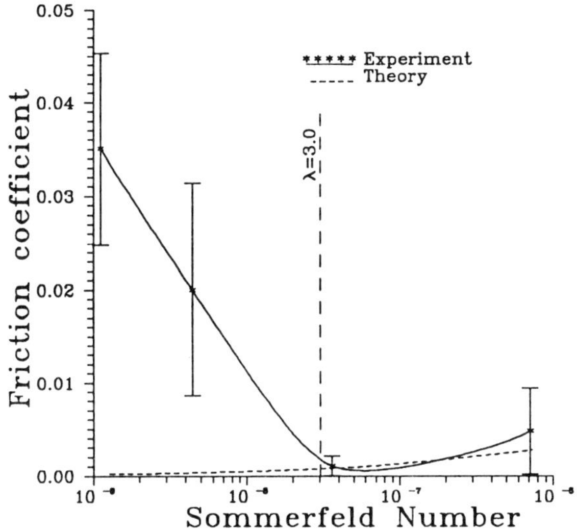

Figure 3. Stribeck results for flat 1

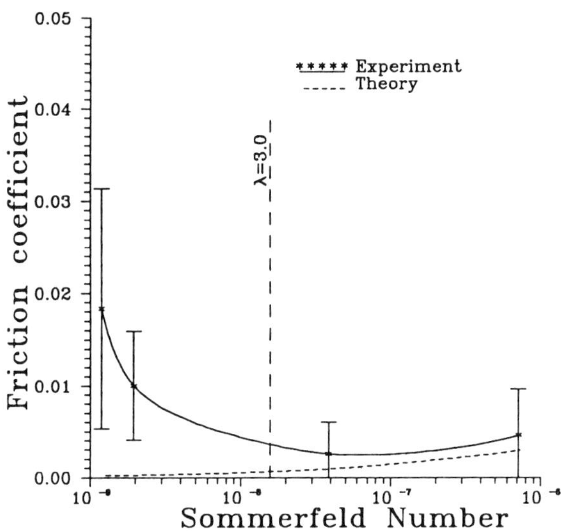

Figure 4. Stribeck results for flat 2

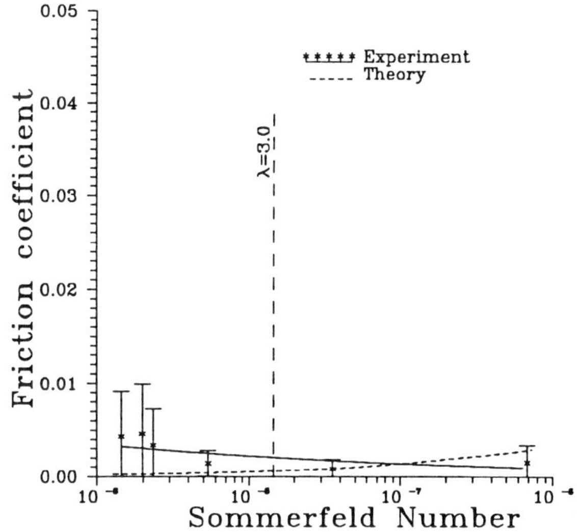

Figure 5. Stribeck results for flat 3

layers in figures 5 and 6 demonstrated much flatter Stribeck curves and a rise in measured friction was not as obvious. This was evidence that microelastohydrodynamic effects were enhancing the lubrication of the softer components in the mixed lubrication regime to the left of $\lambda = 3.0$.

The data from figures 3–6 is tabulated in table 3. This table also shows the 95% confidence intervals of the mean measured coefficients of friction. For the 20 MPa layers, at $\lambda \geq 3.0$, the predicted average friction coefficients were within the 95% confidence limits of the mean experimental values. One value was within the 99% confidence limit. This was very good agreement between theory and experiment. The 6 MPa layers also showed good agreement with theory. The predicted values of friction for the HVI 60 lubricant were within the 95% confidence limits of the mean measured friction coefficients. The theory, however, overestimated the friction for the thicker HS 2:1 lubricant for flats 3 and 4. These values were outside the 95% confidence limits of the experimental means, but, the estimated friction coefficient for flat 4 with HS 2:1 as the lubricant was within the 99% confidence limit which was still good agreement with theory. The final theoretical point for flat 3 with HS 2:1 as the lubricant fell just outside the 99% confidence limit of the experimental mean value. The deviation was in the fourth decimal place which clearly with our

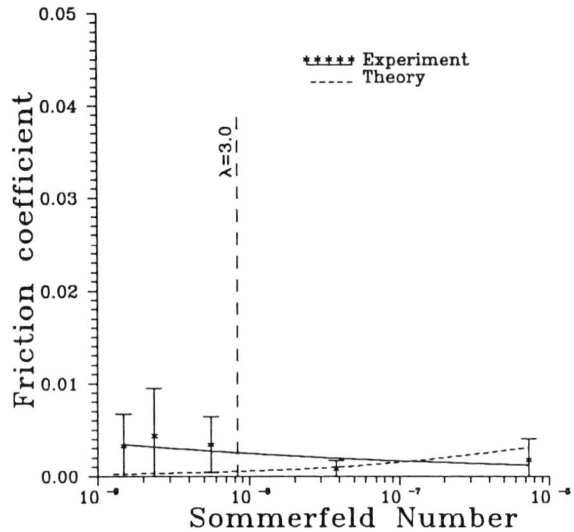

Figure 6. Stribeck results for flat 4

friction apparatus was not a measurable difference. Agreement between theory and experiment was exceptionally good for the first time.

At Sommerfeld numbers to the left of $\lambda = 3.0$, surface to surface contact is normally expected to begin and friction to rise. In this region of the Stribeck curves, the theoretical values of the mean friction for the 6 MPa layers were outside the 95% confidence interval of the measured mean values. However, the theoretical values were within a single sample standard deviation of the measured mean friction coefficients. This was evidence to suggest that microelastohydrodynamic effects were operating to smooth the surfaces and preserve thinner lubricating films. Such experimental evidence of microelastohydrodynamic lubrication has not previously been reported.

In previous work [8], very low levels of friction ($\mu \leq 0.005$) were also measured, but it was not possible to make a quantitative comparison with theory. Apparatus uncertainties, such as bearing misalignment and vibrations prevented such a comparison. The test protocol employed in this investigation, along with the greater number of experiments, has eliminated these difficulties and allowed the experimental results to be compared with theory.

4.2. Comparison throughout the cycle

A detailed comparison between theory and experiment throughout the cycle verifies the Sommerfeld comparison already made. The friction results for the lubricants HVI 60 and HS 2:1 which produced surface separation (λ) ratios greater than 3 for all of the layers, showed excellent agreement with theory throughout the entire cycle. The results for flat 1 and flat 4 are shown in figures 7 and 8 respectively and compared with theory. Even during the swing phase, from $t = 0.66$ s to $t = 1.0$ s, when the applied load approached zero causing a rise in the coefficient of friction, the agreement was good. Previous results [8], were only in close agreement during one small portion of the stance phase of the cycle. The agreement found in this investigation between theory and experiment has not previously been shown for layered contacts under full dynamic conditions.

4.3. Microelastohydrodynamic effects

The goal of cushion bearing design is to promote fluid film lubrication. Fluid film lubrication can be achieved in two ways. A very thin film can be preserved through improvements in surface quality and/or increasing the likelihood of surface smoothing due to microelastohydrodynamic lubrication. Alternatively, fluid film lubrication can be achieved by promoting thicker lubricating films sufficient to separate the surface asperities in the first place.

In the thick fluid film lubrication regime at λ values greater than 3.0, the experimental results agreed well with the theoretical predictions. Elastohydrodynamic lubrication theory alone was sufficient to explain the results in the fluid film region. This study has demonstrated for the first time agreement between theory and experiment for low modulus layered line contacts under conditions of dynamic load and variable velocity.

In addition, in the mixed lubrication regime, the friction remained low especially for the 6 MPa layers, when it was expected to have risen, according to Stribeck assessment. The friction coefficient for the 6 MPa layers was less than 0.005 for all the lubricants tested. Even with water as the lubricant, friction remained low. Water produced estimated elastohydrodynamic central film thicknesses of 59–96 nanometers (table 4) and when compared to the RMS roughnesses of the layers

Table 3
Tabulated experimental results compared with theory.

Flat # 1: 2.5 mm 20 MPa

	Experiment			Theory	
Lubricant	μ	SSD	95% Confidence interval† of μ	μ	λ
Water	0.0351	0.0102	$0.0309 \leq \mu \leq 0.0393$	0.0002	0.45
LV 0.25%	0.0200	0.0114	$0.0152 \leq \mu \leq 0.0314$	0.0004	0.98
HVI 60	0.0011	0.0011	$0.0006 \leq \mu \leq 0.0015$	0.0008 *	3.46
HS 2:1	0.0048	0.0046	$0.0028 \leq \mu \leq 0.0068$	0.0028 *	20.32

Flat # 2: 5.0 mm 20 MPa

	Experiment			Theory	
Lubricant	μ	SSD	95% Confidence interval† of μ	μ	λ
Water	0.0183	0.0130	$0.0128 \leq \mu \leq 0.0238$	0.0002	0.66
LV 0.05%	0.0100	0.0059	$0.0075 \leq \mu \leq 0.0125$	0.0003	0.88
HVI 60	0.0026	0.0035	$0.0011 \leq \mu \leq 0.0040$	0.0009 **	5.12
HS 2:1	0.0046	0.0050	$0.0024 \leq \mu \leq 0.0067$	0.0030 *	30.07

Flat # 3: 2.5 mm 6 MPa

	Experiment			Theory	
Lubricant	μ	SSD	95% Confidence interval† of μ	μ	λ
Water	0.0043	0.0048	$0.0023 \leq \mu \leq 0.0063$	0.0002	0.70
LV 0.05%	0.0034	0.0038	$0.0018 \leq \mu \leq 0.0054$	0.0003	0.93
LV 0.05%	0.0046	0.0053	$0.0024 \leq \mu \leq 0.0068$	0.0003	0.93
LV 0.25%	0.0014	0.0014	$0.0008 \leq \mu \leq 0.0020$	0.0004	1.53
HVI 60	0.0008	0.0010	$0.0004 \leq \mu \leq 0.0013$	0.0009 *	5.37
HS 2:1	0.0015	0.0019	$0.0007 \leq \mu \leq 0.0023$	0.0029	31.57

Flat # 4: 5.0 mm 6 MPa

	Experiment			Theory	
Lubricant	μ	SSD	95% Confidence interval† of μ	μ	λ
Water	0.0033	0.0034	$0.0019 \leq \mu \leq 0.0047$	0.0002	0.97
LV 0.05%	0.0044	0.0051	$0.0022 \leq \mu \leq 0.0066$	0.0003	1.30
LV 0.25%	0.0034	0.0030	$0.0021 \leq \mu \leq 0.0047$	0.0004	2.13
HVI 60	0.0008	0.0009	$0.0004 \leq \mu \leq 0.0012$	0.0010 *	7.51
HS 2:1	0.0018	0.0023	$0.0008 \leq \mu \leq 0.0027$	0.0031 **	44.11

† $\mu \pm \frac{t_{n-1} SSD}{\sqrt{n}}$ where $t_{n-1} = 2.069$ *and* $n = 24$

* Values within the 95% confidence limits.

** Values within the 99% confidence limits ($t_{n-1} = 2.807$).

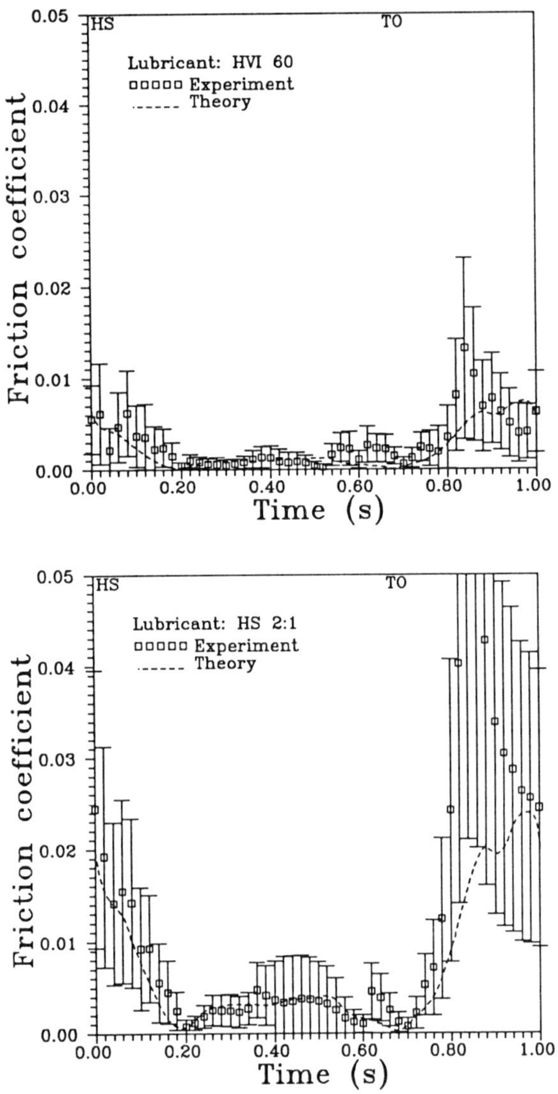

Figure 7. Comparison with theory throughout the cycle for flat 1 lubricated with HVI 60 (top) and HS 2:1 (bottom).

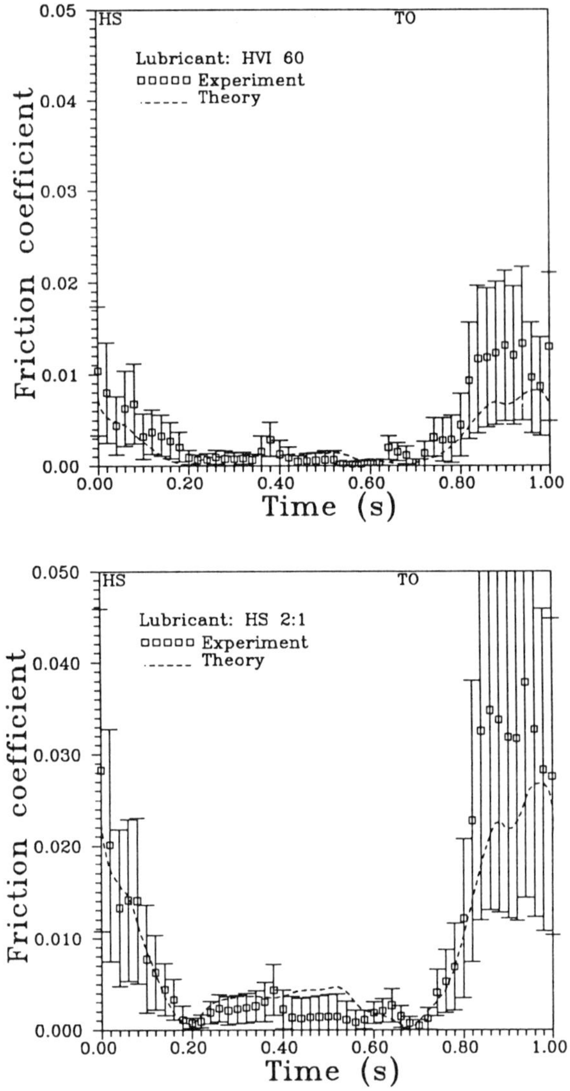

Figure 8. Comparison with theory throughout the cycle for flat 4 lubricated with HVI 60 (top) and HS 2:1 (bottom).

(table 1) some asperity deformation microelastohydrodynamic lubrication must have occured preserving a thin lubricating film and maintaining low friction in the contact.

Table 4
Estimated film thicknesses when lubricated with water

Flat	t_h (mm)	E_2 (MPa)	h_{cen} (μm)
1	2.5	20	0.059
2	5.0	20	0.066
3	2.5	6	0.081
4	5.0	6	0.096

For the purposes of bearing design, the amount of asperity deformation which can be expected to occur in a cushion bearing needs to be known. The surface separation criteria of $\lambda = 3.0$ has been successfully used in the past to estimate the onset of film breakdown in engineering bearings [16]. Therefore, for each material modulus used in this study, the percentage flattening of the deforming asperities was calculated that was necessary to maintain an effective surface separation ratio of 3.0:

$$\alpha = 1.0 - \sqrt{\frac{\left(\frac{h_{cen}}{\lambda_e}\right)^2 - \sigma_1^2}{\sigma_2^2}} \qquad (4)$$

where $\lambda_e = 3.0$

The measured friction has been plotted against the percent asperity deformation required to maintain $\lambda_e = 3.0$ in figure 9. The required deformation of the asperities depends upon the estimated central film thickness and the surface finishes of the components. As the percentage deformation increases, low friction indicates that fluid film lubrication was being maintained by deformation of the surface asperities. This benefit of microelastohydrodynamic lubrication has not previously been shown experimentally. A large number of data points appear at 100% deformation because of the very thin films present in the layered flat bearings. The thin estimated lubricating films meant that full flattening of the asperities was required to achieve a surface separation ratio of 3.0. In some cases, even with

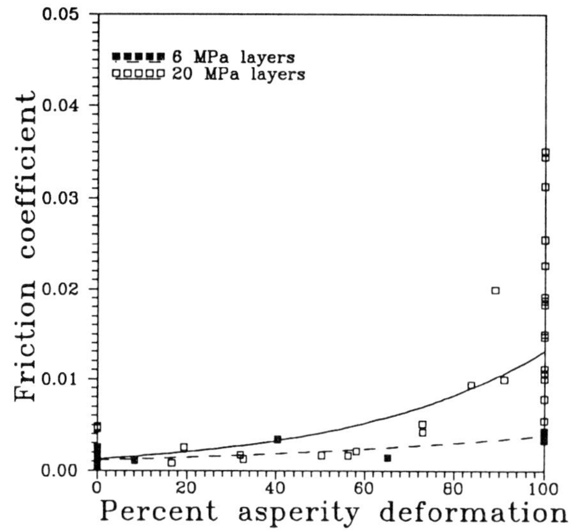

Figure 9. Surface smoothing effects. A graph of friction coefficient plotted against percentage asperity deformation needed to achieve and effective surface separation ratio of 3.0.

100% flattening of the polyurethane asperities, a surface separation ratio of 3.0 was not achieved because, depending upon the lubricant viscosity and the estimated film thickness, the asperities of the metallic counterface which were not deforming may have produced surface to surface contact anyway. At deformations less than 100%, $\lambda = 3.0$ was always achieved.

The friction of the 6 MPa material remained low up to very high asperity deformations. This result demonstrated that microelastohydrodynamic lubrication was effective in preserving very thin lubricating films. The rising characteristic of the friction at larger percentage deformations for the 20 MPa material suggests that a limit of asperity deformation may be present at which microelastohydrodynamic effects can no longer preserve a thin lubricating film. For design purposes, friction levels of 0.005 or less are thought to be appropriate. Asperity deformations of up to 60% can be expected from the 20 MPa material and complete asperity flattening can be expected from the 6 MPa material. Hence an effective surface roughness which takes account of the percentage flattening can then be calculated for soft layer contacts. These results show that there are limits to the effectiveness of microelastohydrodynamic lubrication in preserving fluid film lubrication. The effect of material modulus has been shown experimentally for the first time. Of course, other factors such as surface features and geometry need to be investigated further in order to fully understand the effects of microelastohydrodynamic lubrication on cushion bearing design.

5. CONCLUSIONS

The frictional torque throughout a dynamic loading cycle was measured for a cylindrical cushion knee joint articulating against flat layered tibial components. The friction measurements produced by the new test protocol demonstrated reduced bearing misalignment effects and force vector offsets and apparatus vibrations were also reduced.

Comparisons were made between theory and experiment. Good agreement was found for the Stribeck curves and for friction throughout the cycle. Such agreement under dynamic conditions has not previously been demonstrated. These results give greater confidence in the use of elastohydrodynamic theory for the further design of cushion bearings for total joint applications.

In addition, the softer 6 MPa layers showed the added benefits of microelastohydrodynamic lubrication, producing a relatively flat Stribeck curve and low levels of friction for the lubricants tested which produced Sommerfeld numbers that were below the surface separation threshold of $\lambda = 3.0$. Deformation of the surface asperities was more easily obtained for the lower modulus 6 MPa material. This caused thinner films to be preserved to lower Sommerfeld numbers. These results showed the enhanced lubrication of cushion bearings over conventional implants. A comparison of the two materials used in this study of friction has shown the benefits of microelastohydrodynamic lubrication in preserving fluid film lubrication as well as the potential limits of effectiveness of microelastohydrodynamic action.

REFERENCES

1 D. Dowson, J. Fisher, Z.M. Jin, D.D. Auger, B. Jobbins, J. Eng. in Med. H205 (1991), 59–68.

2 J. Fisher, D. Dowson, J. Eng. in Med. H205 (1991), 73–79.

3 A. Unsworth, Phys in Med. Biol. 23 (1978), 253–268.

4 D.W. Howie, B. Vernon-Roberts, R. Oakeshott, B. Manthey, JBJS, 70–A (1988), 257–263.

5 D.W. Murray, N. Rushton, JBJS, 72–B (1990), 988–992.

6 J. Livermore, D. Ilstrup, B. Morrey, JBJS, 72–A (1990), 518–528.

7 A. Unsworth, M.J. Pearcy, E.F.T. White, G. White, Tribology – friction, lubrication and wear, 50 years on, C219/87 (1987), 715–724, MEP, London.

8 D.D. Auger, J.B. Medley, J. Fisher, D. Dowson, 'Mechanics of coatings', 16th Leeds-Lyon, (1990), 251–260, Elsevier, Amsterdam.

9 J. Blamey, S. Rajan, A. Unsworth, R. Dawber, J. Biomed. Eng. 13 (1991), 180–184.

10 A.J. Yeadon, M. LaBerg, J.B. Medley, G.M. McNeice, 'Development and design with advanced materials', (1990), 219–226, Elsevier, Oxford.

11 J.R. Gladstone, J.B. Medley, 'Mechanics of coatings', 16th Leeds-Lyon, (1990), 241–250, Elsevier, Amsterdam.

12 T.J. Smith, J.B. Medley, 'Fluid film lubrication – Osborne Reynolds centenary', 13th Leeds-Lyon, (1987), Elsevier, Amsterdam.

13 D. Dowson, Z.M. Jin, 'Fluid film lubrication – Osborne Reynolds centenary', 13th Leeds-Lyon, (1987), 375–386, Elsevier, Amsterdam.

14 P. Meijers, Applied Sci. Res. 18 (1968), 353–383.

15 J.B. Medley, D. Dowson, V. Wright, Eng. in Med. 13 (1984), 137–151.

16 K.L. Johnson, J.A. Greenwood, S.Y. Poon, Wear, 19 (1972), 91–108.

17 J. Halling, Principles of tribology, (1975) Macmillian Press, New York.

18 D.D. Auger, Tribology of cushion bearings used for total joint replacement, PhD thesis, University of Leeds, Leeds, U.K.

19 J. O'Kelly, A. Unsworth, D. Dowson, A. Hall, V. Wright, Eng. in Med. 7 (1978), 73–83.

20 B.J. Roberts, A. Unsworth, N. Mian, Ann. Rheum. Dis. 41 (1982), 217–224.

Thin Films in Tribology / D. Dowson et al. (Editors)
1993 Elsevier Science Publishers B.V.

BONDING OF SOFT LAYERS TO RIGID BACKINGS

J.M.Blamey, P.J.Mullin, J.Seaton, A.Unsworth & T.V.Parry
Centre for Biomedical Engineering
School of Engineering and Computer Science,
University of Durham, Durham City, DH1 3LE, U.K.

SYNOPSIS

Successful bonding of polyurethane layers to polyethylene backings has been achieved using a combination of mechanical and chemical mechanisms. These joints have been analysed using peel and blister testing as well as microscopy. Strength was found to be high, but largely as a result of the mechanical interlock between the two materials rather than chemical bonding and this is currently being addressed to improve strength still further. Bonded samples have been soaking in Mamalian Ringers solution at 37°C for 750 days with no visible signs of delamination. Long term fatigue studies of the bond in Mamalian Ringers have recently started, using loaded femoral heads to introduce representative stresses in the bond.

1. INTRODUCTION

The use of soft elastomeric layers in hip prostheses poses problems with fixation. In the same way as cartilage they must be supported by a rigid backing to maintain dimensional stability under high loading. This backing material should be bonded to the elastomer to eliminate problems with grabbing owing to the high poisson's ratio of the material. When no bond is present the layer tends to be pushed from the surface of the backing onto the ball leading to increased friction (figure 1).

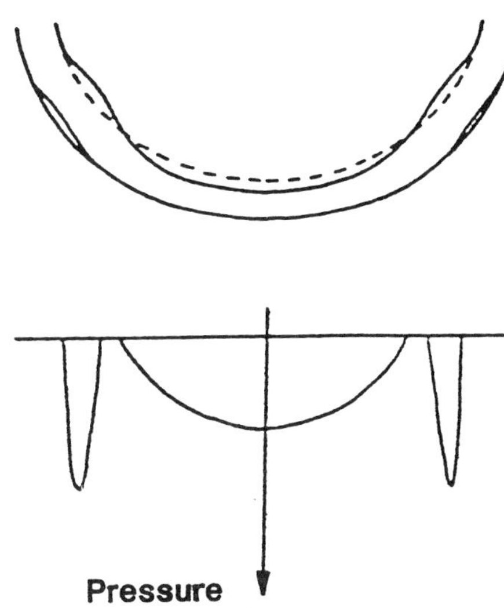

Figure 1. Grabbing of the Femoral Head onto the Acetabular Cup

When correctly bonded, the rigid backing will reduce the effective Poisson's ratio of the elastomeric layer from its initial value of 0.48-0.50 allowing closer clearances, without the risk of grabbing. The tribological properties are improved because the larger contact area reduces pressures on the articulating surfaces and promotes fluid film lubrication [1].

Use of adhesives in this application poses problems with biocompatibility, biostability, hydrolysis, fatigue and creep response. In addition it requires a relatively flexible adhesive due to the low elastic modulus of the adherands.

2. THEORY

2.1. Theory of Joint Design

Joints should be specifically designed to facilitate several bonding criteria:

1. Allow adhesive application
2. Allow adhesive to cure
3. Put the adhesive under acceptable loading

Bonding of the elastomeric layer to the rigid backing would take place in a controlled environment (i.e. not *in-vivo*) and so the first two criteria are easily achieved. However ensuring that the loading is not too high for the adhesive requires thorough stress analysis and possible redesigning of the interface. Forces must be transmitted between the substrates through the adhesive bond which consists of many minute adhesion sites. If enough of these sites are sharing the load, a successful joint can be formed. A number of loading types may be applied to joints:

2.1.1. Tensile/Compressive Loading

In their idealised form these joints have the highest strength with stress being distributed over the entire bond area (figure 2a). However deflection of the joint components can lead to non-uniform loading and give rise to cleavage stresses (Figure 2b) reducing the load carrying capacity of the joint with stresses concentrated over a small cross-section of adhesive.

Figure 2a. Tensile/Compressive Loading of Adhesive Joints

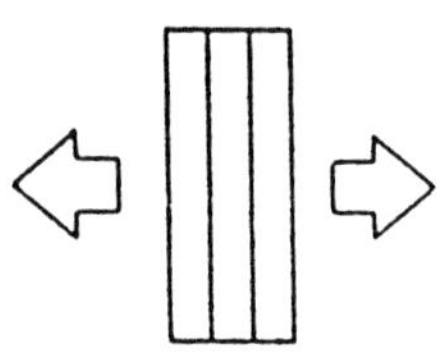

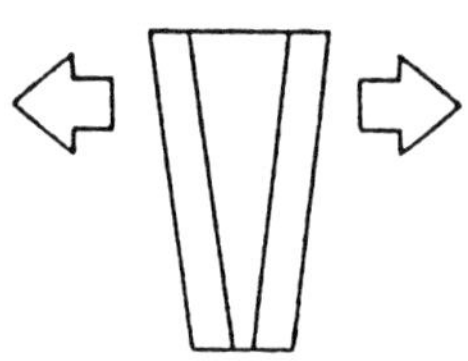

Figure 2b. Cleavage in Adhesives

2.1.2. Shear Loading

Joints which include shear loading of the adhesive apply stress over the entire bond area, but higher stresses are concentrated at the ends of the bond in this simple model (figure 3). The strength of a joint under shear loading is typically 10% of that of a compressively (with no cleavage) loaded bond. In-service conditions are normally limited to situations where shear stresses are less than 20% of the ultimate shear stress of the adhesive [2].

2.1.3. Peel Loading

The peel strength is usually the weakest property of a joint, with only toughened adhesives capable of maintaining reasonable peel strengths. This property is normally assessed with a similar arrangement to that shown with a peel angle of 90 to 180° (figure 4). Bond Strengths of only 0.1% of those obtained under compressive loading and 1% of the strengths obtained in shear loading are typically measured under peel loading.

Figure 3. Shear Loading of Adhesives

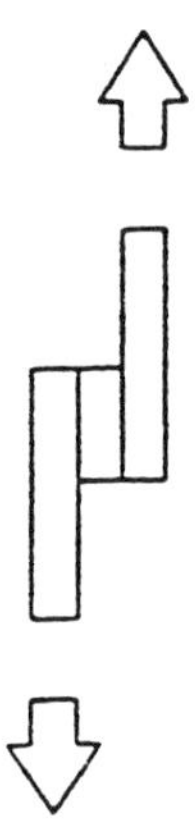

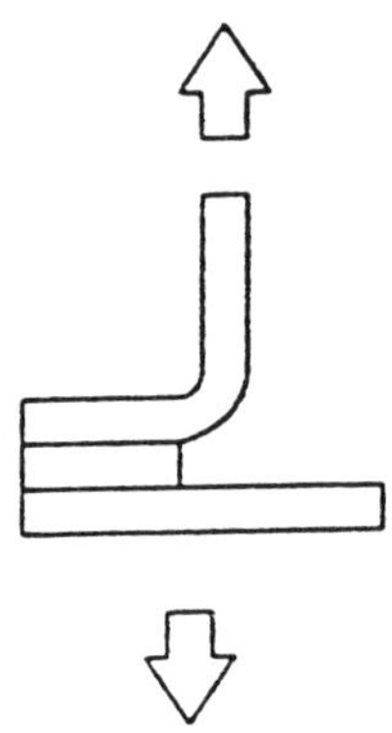

Figure 4. Peel of Adhesive Joints

2.2. Materials

Initially, metal backed cups were used in the assessment of soft layers with the layer bonded directly to the backing under compression moulding. The bond was found to be extremely weak and holes had to be drilled in the stainless steel backing to provide a mechanical interlock and allow the layers to be tribologically tested [3] (figure 5).

Figure 5. Stainlees Steel backed Acetabular Cups

This would be unsuitable for an *in-vivo* model and does not maximise the potential of the layers, requiring large clearances to limit grabbing during operation. Also if there were problems with the elastomeric layer during operation and failure of the lugs occurred (as has been noted during long term *in-vitro* studies) the resulting metal on metal contact would tend to require immediate revision surgery.

Polyethylene in its ultra high molecular weight (UHMW) form has been widely used in knee and hip prostheses since Charnley optimised his hip prosthesis to make use of it. Since this time its biological properties have been well documented, with little tissue response as well as minimal polymer degradation [4], [5]. Therefore it would seem to be a good material to act as a backing for the elastomers. Failure of the elastomeric layer would lead to a large clearance UHMWPE/metal contact which could maintain hip mobility for a reasonable time, reducing the need for emergency revision surgery.

A further reason for using UHMWPE can be understood by considering the fixing arrangements for many currently available prostheses. Porous and hydroxyapatite coatings on metallic surfaces have become popular for improving bone fixation with the polyethylene acetabular component a "snap-fit" into a metallic ring. The Durham prosthesis could easily be incorporated into many of the currently available systems. This would allow a direct comparison to be made between the performance of soft layers and UHMWPE in clinical prostheses.

2.3. Adhesive Theory

The performance of an adhesive is highly dependent on the adherands. Surface cleanliness is often important with degreasing often mandatory prior to application of the cement. However, some adherands are generally considered to produce poor bonds and polyethylene is amongst these. This is because of its waxy hydrocarbon surface which does not allow the polar molecules of the adhesive to bond well [6]. This material requires oxidation of the surface, replacing the hydrogen (H^+) ions with oxygen ions (O^{2-}) to exhibit a polar surface for improved adhesion. Even with such modification there are only a few adhesives which are suitable.

2.3.1. Hot-melt Adhesives

Thermoplastic polymers are used in a liquid form at an elevated temperature to cover the surfaces prior to pressure and cooling to complete the bond. Problems can occur because full wetting of the surface may be poor due to the low surface energy of polyethylene. In addition elevated (in-service) temperatures will tend to lead to weakening of the bond through adhesive softening. This may be a problem in the body (37°C) if a low temperature adhesive were to be used. Biocompatibility should be good, assuming a compatible thermoplastic was used. Bond strength is highly dependent on suitable adherands and a polar surface which may also be roughened.

2.3.2. Polyurethane Adhesives

Base materials of isocyanate and an amine or glycol are mixed and form a liquid material with quite a high viscosity which is cast onto the surfaces to effect a bond. The castomer resin (non-biocompatible) which was used during the early phases of soft layer work [7] was moulded in-situ through this reaction. The bonds produced can be of good strength especially in peel and impact loading but they are often the result of cross-linking of the urethane which can render it non-biocompatible. Also problems can occur in hot wet environments due to hydrolysis. This was observed with the castomer material following studies at 80°C in saline, with cracks appearing on the surface (figure 6)

Figure 6. Hydrolysis Damage to a Castomer (Polyurethane) Cup

2.3.3. Solvent Cast Systems

These range from low viscosity liquids to pastes and represent a thermoplastic polymer (or urethane) carried by a suitable solvent. These can be highy biocompatible, for example Tecoflex resins in Tetrahydrofuran have been successfully used in the body [8]. Cross linked materials tend to give the greatest strength with only light loading possible with non-cross linked adhesives. Precise placement is mandatory as there is an instantaneous "grab" with curing often required to remove all of the solvent. The adhesive layers can be quite thick and so surface roughening can be used. Because of their thickness, adhesives are suitable for fatigue applications.

2.3.4. Alternative Systems

Other possible adhesives include Nitrile and Silicone rubbers. Although both are highly resistant to biodegradation and they are obtainable in a variety of hardnesses, the strength of bonds obtained with the two adherands will be questionable. Toughened acrylics, a family which includes bone cement, show good biocompatibility, but are often of high elastic modulus and as such are not suitable for bonding such a flexible material as polyurethane.

2.3.5. Contributing Considerations

Biocompatibility and minimal degradation in a hostile liquid environment at an elevated temperature are necessary for all biological adhesives. However, to improve the adhesive process the adherands should have elastic modulli close to or straddling that of the adhesive. Thus the adhesive acts as a form of stress reliever at the interface. Also if a large amount of motion is required, high values of ultimate elongation for the adhesive are essential.

Roughening of a surface is a good method of improving a bond and many successful bonds rely on mechanical interlock. In this application this would be an important avenue to explore as many of the biocompatible adhesives are only capable of withstanding nominal loads with all but the polyurethane systems incapable of supporting shear stresses above 3.5MPa for prolonged periods.

2.4. Bond Analysis

Tests by Unsworth *et al* [3], [7] indicated that layer thicknesses of 2 mm or larger were required for optimum tribological properties. Armstrong [9] studied the interface between cartilage and bone and obtained the following simplified expression for shear stress:

$$\tau\ max = \left(\frac{\delta P}{\delta x}\right) \ y \qquad (1)$$

This analysis assumed an incompressible layer i.e. ν = 0.500 and a flexible layer on a rigid backing material. The maximum shear stress of the bond can be seen to occur at the interface between the two layers beneath the point of maximum pressure gradient. Reductions in shear stress thus require a thinner layer and lower value of the pressure gradient.

Eberhardt *et al* [10] expanded on this work and obtained similar expressions with normal stresses (representing compressive loads) reaching 1.92 $P/\pi a^2$ at the interface between soft and hard layers. For the soft elastomeric joints this represents 3.5-6.3MPa. Shear stresses reach a maximum at a distance of 14mm from the central point, measured along the bond line. Previously layer thicknesses of 2mm had exhibited good tribological properties, however mechanical interlock has been suggested as being paramount to a successful bond. This would require an average layer thickness of 3-4mm to allow at least 2mm of uncontaminated (with UHMWPE) elastomeric layer. This corresponds to $^a/_t$ values of 6.1-4.6. Shear stresses of 0.68-0.90MPa can be calculated from Armstrong [9].

Lees [2] stated that for fatigue applications the proof stress of the adhesive in the mode of expected stress should be five times that expected under the loading to give a good bond life. This means that strength of the bond must be above 18MPa in compression and 4.5MPa in shear. Although this value of compressive stress should not cause a problem to many adhesives, because of the non-uniform loading cleavage stresses will be introduced, which will probably be a maximum at the edge of the contact area.

In any bond regions of low strength or pores within the adhesive layer will be present. The ability of an adhesive to limit crack growth from regions of high stress concentration such as pores is an inexact science which can depend on many inherent properties. However, most joints fail because they cannot meet peel and cleavage overloads with shear overloads being rare. The measurement of this property which may be related to peel strength is also a problem with many potential tests. Peel tests have been previously introduced and a range of specimen geometries exist as no peel test can fulfil the requirements of all cases (BS 5350 Parts C9-14, 1978). However, blister tests can also be used. The interface is expanded using a pressurised fluid over an area where the bonding has been interrupted (figure 7).

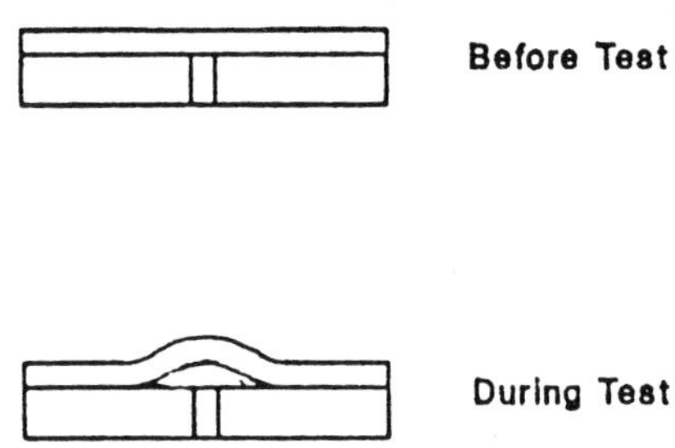

Figure 7. The Blister Test Arrangement

This test indicates the resistance of an adhesive to crack growth in a cleavage configuration and the fluid used (which should be relatively incompressible) can mimic a degrading in-service environment.

Fracture mechanics for the blister test arrangement can be simplified by a number of assumptions [11]:

Opening Loading: Tensile loads are applied perpendicular to the plane of the crack, leading to opening. This is Mode I loading as there are no shear stress components parallel to the crack.

Plane Strain: conditions also apply, as the crack is internal

For adhesive cases, the following equation was derived [12]:

$$\theta = \frac{P_c^2\, b}{E\,(1-\nu^2)} \left[\frac{3}{32} \left[\left(\frac{b}{t}\right)^3 + \left(\frac{c}{t}\right) \frac{4}{(1-\nu)} \right] + \frac{2}{\pi} \right]^{-1} \quad (2)$$

A full finite element analysis of the cup is planned to confirm the loading applied to the bond during in-service conditions. This is likely to reveal some degree of peel, cleavage, compressive and tensile forces on the bond. Problems arise because of the high Poissons' ratio of the elastomer and it has been noticed from early fatigue testing that there are cleavage forces operating around the upper edge of the cup under loading, leading to fracture of the bond in this region. This confirms that it is not only the contact area which has to be rigidly bonded and indicates the need for a full analysis.

3. Materials and Methods

3.1. Preparation Methods

Much of the research concentrated on the preparation of hot melt processed components. These consisted of test pieces formed in dies which were compressed between the heated plattens of a compression moulding machine. Temperature control allowed the plattens to be adjusted independently and pressures of up to 45 tonnes could be applied. Three moulds were used for the majority of specimens and they are shown schematically in figure 8.

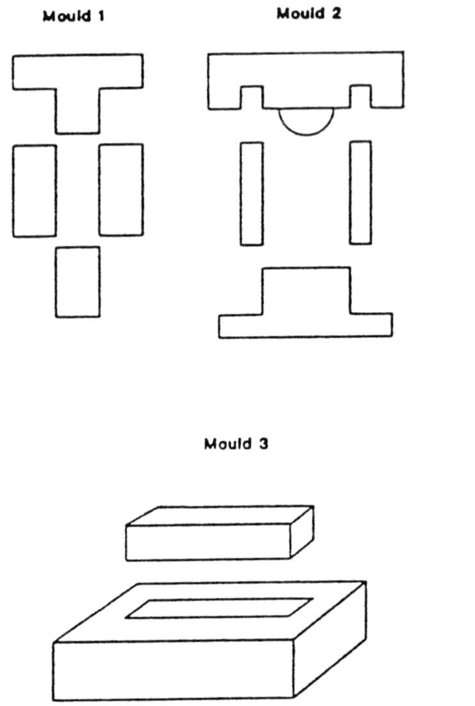

Figure 8. The Moulds used for the Bonded Specimens

Figure 9. The Specimens Produced

Mould 1 prepared disc specimens of 31mm diameter and thickness of 8mm (figure 9). These could be easily prepared because of the simplicity of the mould, with trapped air not causing a problem. Blister test specimens could be constructed by incorporating a disc of PTFE (to stop bonding) over the central portion of the PE/PU interface. The discs could also be sliced to yield peel test specimens although these had a relatively short bond length which could lead to inaccuracies (BS 5350 Parts C9-14, 1978).

Mould 2 addressed this problem and yielded specimens of 70mm in length and 20mm wide. PTFE could be used in the moulding process to yield an unbonded region but it was found that slicing along the bond line was adequate to start the "crack". Mould 3 allowed prototype acetabular cups to be constructed. These required a layer thickness of at least 2mm to give the required cup hardness for acceptable tribological properties. The internal radius of the component and the bone line could be adjusted by the use of a range of acetabular moulds.

Extensive experimental results concluded that the best components were formed when the component was made on one heat cycle. UHMWPE which was in powdered form was compacted with a load varying between 10 and 30 tonnes depending on the mould. Polyurethane granules (circa 3mm in diameter) were then added and compressed into the polyethylene surface and the mould was heated. Thermal degradation can be a problem with urethanes so it is preferable to spend as little time at the higher temperatures as possible. A higher pressure tends to allow better conduction between mould and sample but can adversely affect the removal of trapped air from the sample. [The acetabular cup specimens proved to be the most difficult samples to produce without bubbles]. A compromise pressure was found by experimentation, and this was used at the higher temperatures during the moulding process. A graph of platten and mould temperatures and pressures for a bubble free acetabular cup is shown in figure 10. The temperature measured in the centre of the mould can be seen to lag behind the platten temperature due to the length and large thermal mass of the die. The temperature of the centre of the component will also lag behind that of the mould because of the low thermal conductivities of the polymers. Although the melting temperature of the polyethylene (130°C) was lower than the urethanes no degradation was noticed, possibly due to the reduced conduction of the polyethylene powder compared with the urethane pellets.

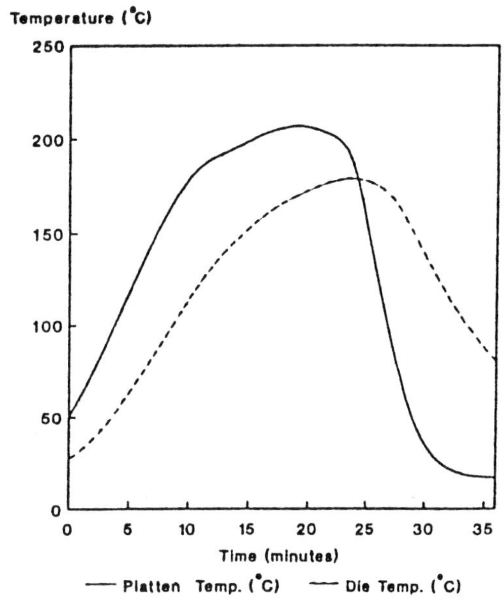

Figure 10. The Temperatures of Plattens and Mould 2 during moulding

3.2. Materials

Three biocompatible polyurethanes were investigated. All had been previously assessed for degradation resistance [13] and all are currently being used in medical applications. The processing temperatures of the materials were in the range 155-170°C, although lower temperatures could be employed if high compaction pressures were used. The use of very high temperatures led to bubble formation, whilst lower temperatures meant incomplete melting and inter-granular pores. Estane 5714F1 (77°A), Pellethane 2363 (80°A) and Tecoflex EG (85 & 93°A) were used, all polyetherurethanes. The materials were in the form of pellets of typically 3-4mm diameter. Pellethane was of regular angular particles and tecoflex of spheroids. Estane was in the form of irregular chips. The reliance of the bond on the mechanical interlock meant that the shape of the urethane particles was likely to have a substantial effect on the mechanical strength of the bond.

3.3. Testing Methods

3.3.1. Peel tests

A floating roller arrangement was used as it produced more constant numerical data than other methods (BS 5350) with a bond formed mainly on mechanical interlock. [The chemical bond was found to be relatively weak as specimens with little or no interlock could easily be pulled apart by hand] it was likely that a large spread of results would be apparent and so the test had to produce as consistent results as possible. The geometry of the arrangement (figure 11) was as BS 5350 Part C9 (1978).

A computer controlled Houndsfield tensile testing machine was used to apply the loading at a constant strain rate of 0.4 mm/second.

The effect of soaking in saline at 37°C for 48 hours prior to testing was investigated. The urethane would be fully hydrated at this stage [13].

Specimens were prepared in widths of 10 and 20 mm and a cut was prepared along the bond line using a 0.5mm width saw blade. At least 15mm of the bond line was failed and a graph of load/time obtained which allowed the bond strength to be calculated as load/unit width of bond line.

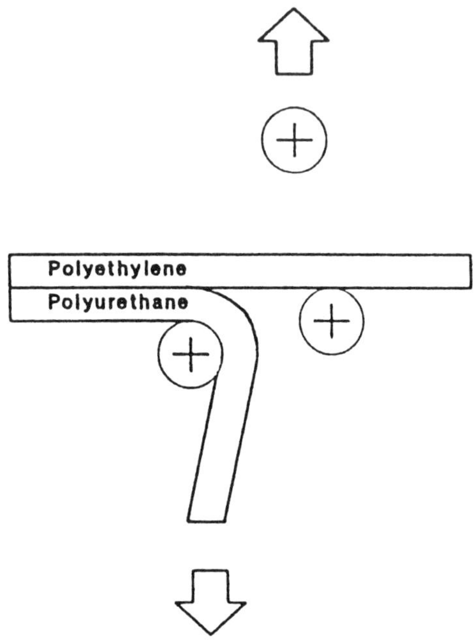

Figure 11. The Floating Roller Peel Test Arrangement

The specimens prepared with mould 3 were also assessed for shear strength through tensile testing of the polyurethane component and assssing failure if there was delamination along the bond line.

3.3.2. Blister Tests

Tecoflex 85°A/polyethylene specimens were prepared in mould 1 with a PTFE disc incorporated into the interface. In fact the disc was often above the interface because of the rough nature of the bond with granules up to 5mm in diameter forming the interface. This moves the geometry to one which favours cohesive rather than adhesive failure [12].

A PTFE disc of 11 mm and a nominal layer thickness of 4 mm were used. The disc was cut from a sheet using a simple punch. Although this method can produce imperfect cracks, with sharper edges than specified [12] in this case this was thought to be unimportant because of the irregular nature of the bond. A countersunk hole was drilled in the specimens to transfer the pressurised oil to the crack thus forming the blister. The sample was mounted in a brass fitting which contained a perspex window so that failure could be seen as the crack extended. A dye was added to the oil to allow easier identification of the crack front.

4. Results

4.1. Peel Tests

The results of peel testing are displayed in Table 1.

Table I
Peel Strengths of PE/PU Bond

Polymer	Hardness (MPa)	Bond Strength (N/mm)	Sample Width (mm)
Estane 57	2.85	9.75	10
Estane 57 (saline)	2.42	7.25	10
Pell 2363	3.21	13.9	10
Teco 85	2.15	15.0	10
Teco 93	3.35	12.2	19
Teco 93 (vacuum)	3.26	6.8	19

These strength values compare favourably with results for adhesives. For example the peel strength of a toughened epoxy is 5.2N/mm. However, this is a pure chemical bond however with no mechanical interlock. The reduction in strength of 25% when the samples have been hydrated in saline is acceptable and will be due to the softening of urethanes when hydrated [13] and the increased deformation and hence failure at the mechanical interlock.

The improvements in strength with Pellethane and Tecoflex may be due to the higher elastic modulus or more regular granules which form the interlock. However, a larger sample group is required before conclusions may be drawn, as these results were drawn from testing 6-8 samples of each material.

It can be seen that vacuum formed tecoflex layers (which are then compression moulded to the polyethylene backing in a two stage process) are much less strong. This is because of the inferior mechanical interlock.

4.2. Blister Tests

The movement of the crack front was monitored and the failure energy was calculated with a range of values of 2.9-8.5kJ/m^2. These values compared favourably with results obtained from epoxy/brass bonds of 0.4kJ/m^2. Failure was adhesive at the Polyurethane/ Polyethylene bond in the vast majority of cases.

For the blister test the thickness of the elastomeric layer and the initial crack size both govern the failure pressures. At low values of $^{t}/_{a}$ the elastomeric material will undergo plate like deflection. This causes larger destructive forces than for a high value of $^{t}/_{a}$ which acts as an infinite medium with deflection within the material giving a "near field" only analysis. This can be seen from calculations of the adhesive fracture energy which increases as $^{t}/_{a}$ increases (figure 12) and the results of Andrews and Stevenson [12] which showed a many fold increase in the failure pressure as the thickness of specimen was increased.

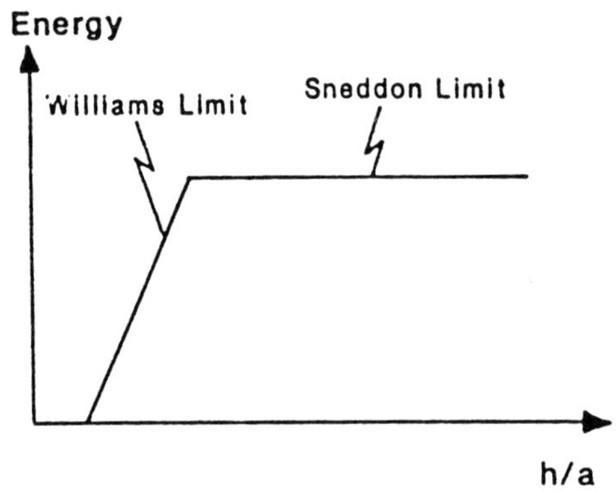

Figure 12. Graph of Adhesive Fracture Energy for the Blister Test

5. Conclusions

The results from the peel tests showed a low scatter, even with such low failure lengths, which was surprising with the irregular nature of the bond. However, much longer specimens should improve the accuracy of the results.

The specimens which had the bond tested in shear by tensile loading of the polyurethane showed very high bond strengths as failure was always of the polyurethane, and not at the bond. This means that the failure shear stress of the bond is in the region required for the application.

The use of the blister test in this application is still at an early stage and much work is required in this area. Widely scattered results of other researchers have been found to be the result of imperfect PTFE discs leading to very sharp cracks. This should not be so critical in this application owing to the compliant nature of the materials blunting the crack tips and the very irregular bond.

A full analysis of the acetabular cup using finite elements techniques is planned to accurately display the damaging interface forces and this may necessitate the testing of the bonds in shear.

Long term tests on the bond strength have begun on a Dartec Fatigue Testing Machine. The cups are loading in compression to 4000 N in an attempt to fail the bond. This should given an indication of the most likely crack propagation sites so these can be improved if necessary.

REFERENCES

1. O'Carroll, S., Jin, Z.M., Dowson, D., Fisher, J. and Jobbins, B. (1990) "Determination of contact area in 'cushion form' bearings for artificial hip joints" *Proc. Instn. Mech. Engrs.: Part H*, **Vol. 204**, 217-223.
2. Lees, W.A. "Adhesives in Engineering Design" (1984), The Design Council, Springer-Verlag, Berlin and New York.
3. Unsworth, A., Pearcy, M.J., White, E.F.T. and White, G. (1988) "Frictional Properties of Artificial Joints" *J. Engng. Med.* **17**: 101-104.
4. Gibbons, D.F. (1978) "Polymeric Biomedical Materials" *The British Polymer Journal,* **Vol.10**, December, 232-237.
5. Hastings, G.W. (1978) "Load Bearing Polymers used in Orthopaedic Surgery" *The British Polymer Journal,* **Vol.10**, Dec. 251-255.
6. Thomas, D.R.K. (1991) Private Communication
7. Unsworth, A., Pearcy, M.J., White, E.F.T. and White, G. "Soft Layer Lubrication of Artificial Hip Joints" (1987) *Proc. Instn. Mech. Engrs.* **Tribology Volume**, 715-724.

8. Brown, D.L. (1988) "Custom made aliphatic polyuretanes for medical use" presented at "Medical Plastics '88 Conference", Oxford, U.K., Sept 8th.
9. Armstrong, C.G., (1986) "Analysis of the stresses in a thin layer of articular cartilage in a synovial joint" *Engng. in Med.* **Vol.15 No.2,** 55-62.
10. Eberhardt, A.W., Keer, L.M., Lewis, J.L. and Vithoontien, V. (1990) "An Analytical Model of Joint Contact" *ASME J. Biomechanical Engineering,* **Vol.112,** 407-411.
11. Timoshenko, S.P. and Goodier, J.N. "Theory of Elasticity, Edition 3" (1970), McGraw-Hill
12. Andrews, E.H., and Stevenson, A. (1978) "Fracture Energy of Epoxy Resin under Plane Strain Conditions" *J. Mater. Sci.* **13,** 1680-1688.
13. Blamey, J.M., Rajan, S., Unsworth, A. and Dawber, W.R.K. (1991) "Soft layered prostheses for arthritic hip joints: a study of materials degradation" *J. Biomed. Eng.,* **13(5),** 179-184.
BS 5350 Parts C9-C14 (1978): Peel Testing Methods.

NOMENCLATURE

a	Contact Radius
b	Blister Radius
c	Initial Crack Length for Blister Test
p	Pressure in the Contact Area
P_c	Critical Pressure in Blister Test
R_1	Radius of the Femoral Component
R_2	Radius of the Acebabular Comp.
t	Thickness of the Urethane Layer
W	Load Applied in the Prosthesis
x	Distance along Contact from Pole
y	Distance through Soft Layer
ν_1	Poisson's Ratio of the Femoral Comp.
ν_2	Poisson's Ratio of the Acetabular Component
θ	Adhesive Fracture Energy
τ_{max}	Maximum Shear Stress at the Layer interface

Thin Films in Tribology / D. Dowson et al. (Editors)
© 1993 Elsevier Science Publishers B.V. All rights reserved.

THE REAL AREA OF CONTACT MEASURED ON ELASTOMERS

M. Visscher, C.P. Hendriks and K.G. Struik

Eindhoven University of Technology
Department of Mechanical Engineering

In this paper optical profilometry is used to measure the roughness deformation of a rough elastomer in contact with a smooth glass plate. Two conditions are considered: a wet and a dry contact.

In the wet contact, the deformed roughness texture is hardly influenced by the contact load. This is probably caused by a high stiffness of the liquid film, compared with the asperity and bulk stiffness.

In the dry contact, increased asperity flattening is observed with increasing load. Besides, different deformations occur at different length scales: Small scale asperities still persist on extensively flattened larger scale asperities. This is probably by asperity interaction, which will be more significant at the smaller length scales, where the distance between the asperities is smaller compared to the asperity height.

1. INTRODUCTION

The tribology of rough surfaces has received much attention during the last decades. One of the items studied is the roughness deformation under static load, which is e.g. important in the case of boundary and mixed lubrication. Up till now several models have been developed to predict the real area of contact, but experimental verification appears to be difficult.

Recently a new method for measurement of the deformed roughness texture of elastomers under static load has been developed and presented by Visscher and Struik (1992a) and by Visscher (1992). It uses an optical profilometer, which is able to scan the surface roughness through a (possibly loaded) glass plate on it.

This paper provides a brief literature review on the real area of contact, followed by description of the experimental method used by the authors. Next preliminary measurements, both with and without a liquid in the contact area, will be presented and discussed. Finally a brief discussion will be given on the determination of the real area of contact and on the used method for the measurement.

2. LITERATURE REVIEW

The literature will be briefly reviewed here. More elaborate reviews can be found in Thomas and King (1977), Woo and Thomas (1980), Bhushan (1985a), McCool (1986), Visscher and Struik (1992a), Hendriks (1992b, 1993) and Visscher (1992 pp. 69ff.).

2.1. Theoretical work

In general, theoretical work concerns the real area of contact in dry contacts. Different modeling of the roughness texture is applied. Greenwood and Williamson (1966) described the asperity summits with spheres of equal radius but of different height and Onions and Archard (1973) considered the different length scales in the roughness texture, modeling the roughness by small spheres superimposed on larger spheres. Majumdar and Bhushan (1991) also accounted for the different length scales, but used fractals to describe the roughness texture. Numerical calculations of the roughness deformation under static load are presented by Lai and Cheng (1985), Lubrecht and Ioannides (1991), Xian and Zheng (1991) and Lee and Cheng (1992). Different predictions are obtained for the real area of contact, due to the different modeling of the roughness texture.

In most papers, asperity interaction is not considered. Interaction is caused by deformation of the bulk material and can have a significant influence. According to Vergne et al. (1985), roughness valleys can possibly persist due to this interaction, even at high loads. Podbevsek (1992 section 5.3) showed that a non-contacting asperity can be lowered by the load on a neighbouring higher, contacting asperity. Therefore, negligence of interaction probably yields over-estimation of the real area of contact.

2.2. Experimental work

A large number of experimental methods have been applied to study the real area of contact. Bhushan (1985a), investigating magnetic tapes, discussed different methods possibly suitable for polymeric materials. The most important methods are the electrical and the optical, but most are not suitable for polymers. Bhushan (1985b) chose interferometry and found that the real area of contact can be seriously over-estimated (e.g. by 30 percent) due to the small slopes in the roughness texture of magnetic tapes, which make the boundary between the contact area and the surrounding non-contacting area unclear. Bhushan stated, that the accuracy would increase with increasing roughness slopes. However, the interference pattern is disturbed by light scatter at higher slopes, which makes interpretation of the picture difficult or even impossible (Visscher, 1992 pp. 126-127). Furthermore, present methods often provide an idea of the real area of contact only, but do not yield an indication of the total deformation of the roughness texture. This latter point, however, appears to be important, as will be discussed in section 7 below.

3. THE METHOD

The real area of contact is measured using the optical profilometer shown in fig. 1. It is derived from a compact disc transducer, described by Bouwhuis

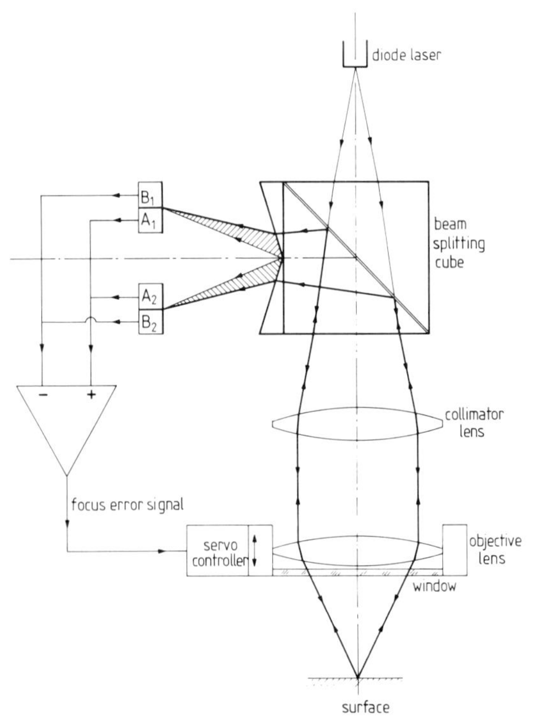

Figure 1 Optical profilometer.

and Braat (1978) and by Bouwhuis et al. (1987), and used for contactless shape and roughness measurements (Struik and Chang, 1987). The principle of the method is, that four photodiodes (A_1, A_2, B_1 and B_2) receive the same amount of light when the surface is in focus and the so-called **focus error signal** is zero. If the surface is out of focus, the outer diodes (B_1 and B_2) receive a different amount of light than the inner diodes (A_1 and A_2) and the focus error signal is positive or negative, depending on the surface position (above or below the focal plane of the objective lens). When out of focus occurs, the objective lens is moved by the servo controller to a position where the focus error signal is zero again, focusing the lens onto the surface. Measurement of the lens position during scanning then yields the shape and the roughness profile of the surface.

The accuracy in the height measurement is about 0.01 μm and the resolving power, determined by the focus spot diameter, is about 1 μm.

The same method can be applied to measure the deformed roughness profile when a glass plate is pressed onto the scanned surface (fig. 2). This additional glass plate influences the measurement, due to additional reflecting surfaces and due to refraction, which causes spherical aberration. These factors are investigated and measures are taken to reduce or eliminate their influence (see Visscher, 1992 section 5.3; Visscher and Struik, 1992a and 1992b). The most important factor is the reflection on the contacting glass surface, which can disturb the measured profile significantly. This influence is not well understood at the moment and this reflectance should therefore be eliminated, when the deformed roughness profile is to be measured quantitatively.

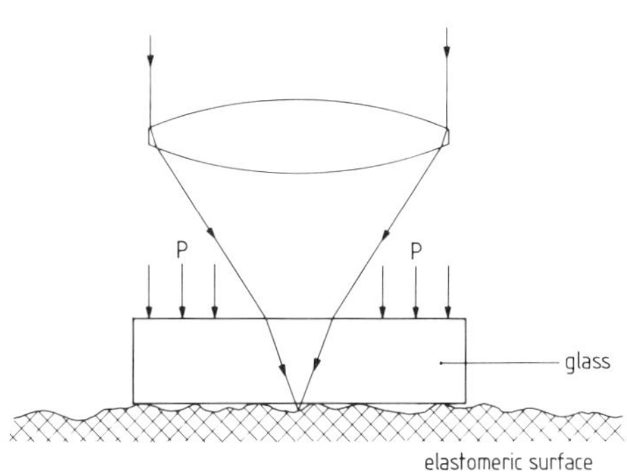

Figure 2 Scanning through a glass plate loaded onto the surface.

4. MEASUREMENTS WITH LIQUID IN THE CONTACT AREA

The liquid used in the contact is an oil mixture of 75 percent Shell Ondina 15 and 25 percent of Shell ondina 68. This mixture has an equal index of refraction as the glass plate and reflection on the contacting glass surface is therefore avoided.

The measurements are performed on a flat piece of polyurethane. The undeformed roughness texture and the derived height distribution are shown in fig. 3.

4.1. Results

Measurements were performed at five loads and the results of all five measurements are presented by Hendriks (1992a and 1993) and by Visscher (1992). Two results, one at the lowest and one at the highest applied load, are given here in fig. 4 and 5.

Fig. 4a shows a number of higher, flattened asperities at the lower load. The larger areas are indicated by A, B and C and are also shown in fig. 4b, while fig. 4c shows the height distribution curve of the deformed rough-

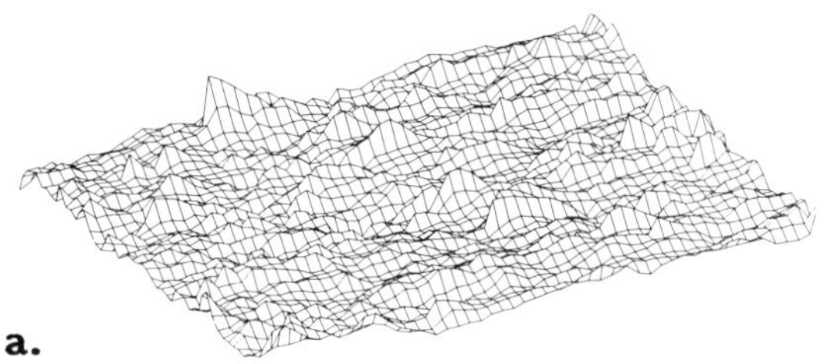

a.

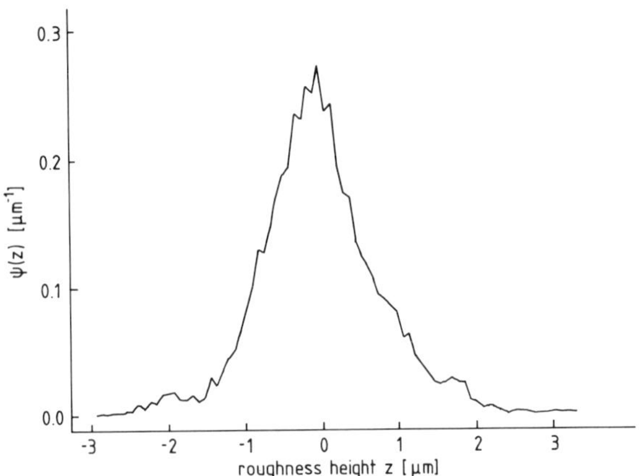

b.

Figure 3 Surface plot (a) and height distribution (b) of the undeformed roughness texture of the polyurethane plate. ($\psi(z)$ = probability density function of the roughness height z)
Measured area = 200 μm $\times$ 200 μm;
Sample distance = 2 μm
(in both directions);

ness texture.

Fig. 5 shows the flattened areas and the height distribution curve at the higher load. This measurement is not performed on exactly the same part of the surface, but equal areas (A and B) can be recognized. Besides, ripples are present in the height contour lines, due to some kind of clearance in the test rig. These problems are caused by the fact that the present test rig was initially not designed for three-dimensional measurements.

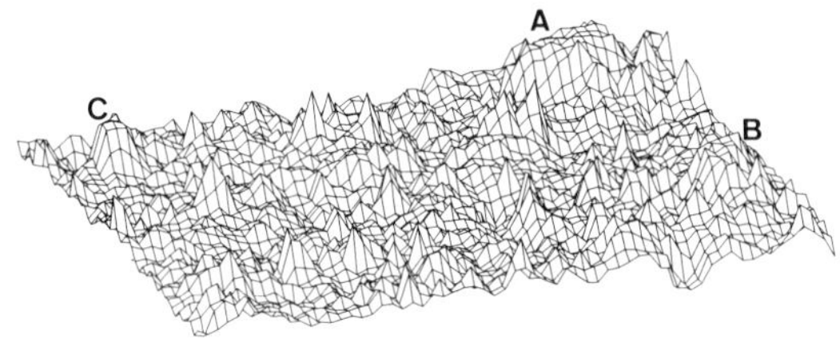

a.

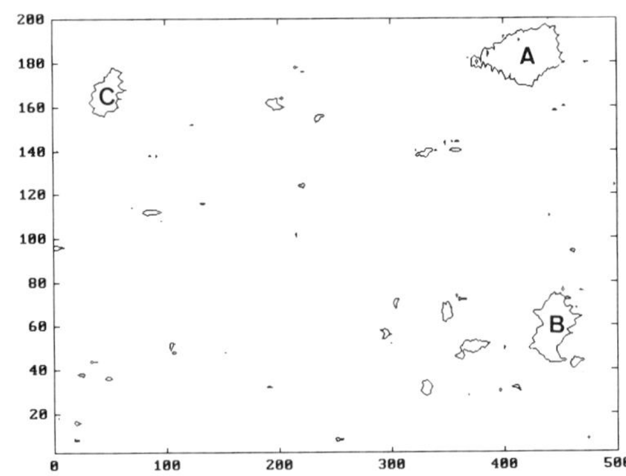

b.

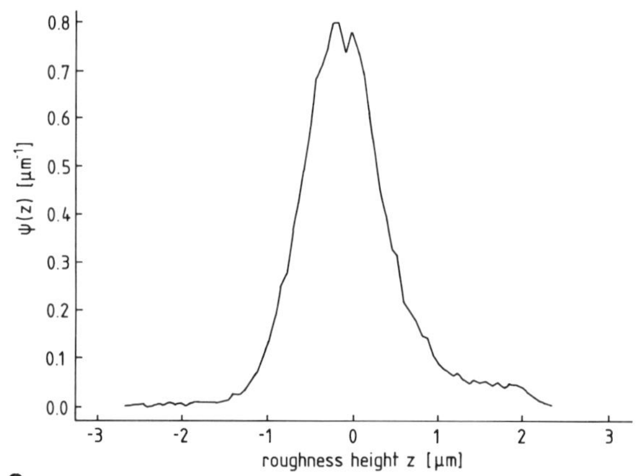

c.

Figure 4 Surface plot (a), highest height contour lines (b) and height distribution (c) of the deformed roughness texture of the polyurethane plate at an average contact pressure of 0.032 MPa and with liquid.
Measured area = 500 μm $\times$ 200 μm;
Sample distance = 2 μm
(in both directions);

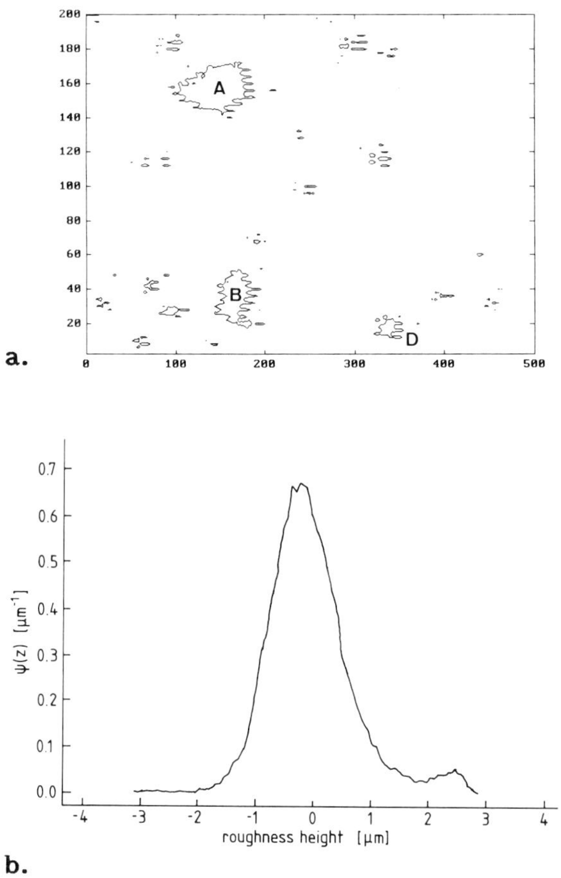

Figure 5 Highest height contour lines (a) and height distribution (b) of the deformed roughness texture of the polyurethane plate at an average contact pressure of 0.15 MPa and with liquid.
Measured area = 500 μm $\times$ 200 μm;
Sample distance = 2 μm (in both directions);

4.2. Discussion

Only a few asperities appears to be flattened at both loads. Also, the differences between the measurement at the higher load and the measurement at the lower load are small. This is illustrated in fig. 6, where the standard deviation σ of the roughness height distribution as well as the size of the areas A and B are shown for the different loads. These parameters are clearly not load-dependent. The differences found in σ must be subscribed to the fact that the measurements are not performed on exactly the same part of the surface. The differences in derived dimension of the areas A and B are probably caused by uncertainty in the determination of these dimensions (see section 6 for further discussion).

Apparently, increase of contact load does hardly or not affect the roughness deformation. This can be explained, when the contact load is largely supported by the liquid and not by contacting asperities. Then increase in contact load yields impression of the bulk material and not of the asperities. Evidence for this idea is given by a measurement presented by Visscher (1992 fig. 5.4 page 81) and by Hendriks (1993), where flattened but non-contacting asperities are shown. Also, Johnson et al. (1972) found for EHL contacts that the film stiffness is often much larger than the asperity and bulk stiffness. The same can be true for the present contact situation, in which the average thickness of the "squeezing" film is much less then the contact dimension (roughly 1 μm and 20 mm respectively). Therefore the liquid can not squeeze out of the contact and the film stiffness will be very large due to the liquid's low compressibility.

5. MEASUREMENTS WITHOUT LIQUID IN THE CONTACT AREA

Measurements were also performed on the dry contact situation. Then no liquid is present in the contact area and the roughness profile is measured inaccurately due to reflection on the glass (see section 3 above). However, only one reflecting surface (the glass to elastomer interface) is present in the

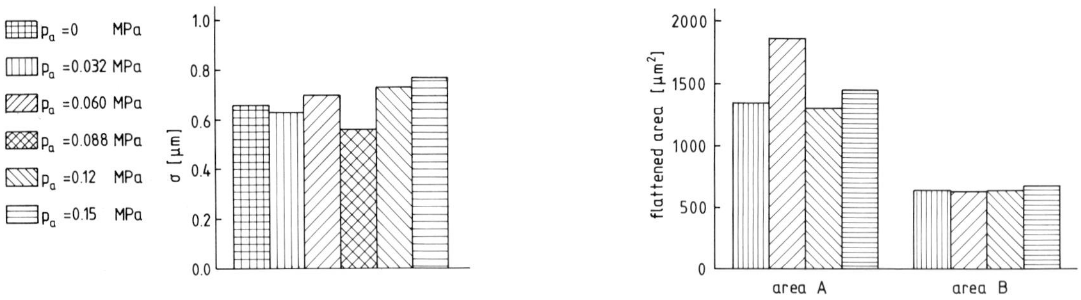

Figure 6 Standard deviation σ of the roughness height distribution and the derived dimensions of the flattened areas A and B, measured with liquid in the contact area at different values of the average contact pressure p_a. (The areas A and B were not present in the measurement at 0.088 MPa, due to horizontal shift between the measurements)

real contacts and the measurements are therefore reliable within these real contacts. These measurements can thus be used to determine the real area of contact, enabling experimental verification of the contact models like reviewed in section 2.1.

5.1. Results

Measurements were performed at six loads on the same piece of polyurethane as in section 4 (see fig. 3 for the undeformed roughness texture). The results of all six measurements are presented by Hendriks (1993) and by Visscher (1992) and two results are shown here in fig. 7. The roughness height distribution is not derived, because of the unreliability in the height measurement outside the real contacts.

5.2. Discussion

Fig. 7 clearly shows the increased flattening at higher loads and one can expect that the real area of contact can easily be derived from these plots. However, this is not as easy, due to an interesting feature found when looking more closely to the results, as will be discussed now.

Fig. 8 shows a cross section of fig. 7b, in which flattened areas are found on a length scale of the order of 10 μm. Within these flattened areas, smaller scale height variations are found. These height variations are

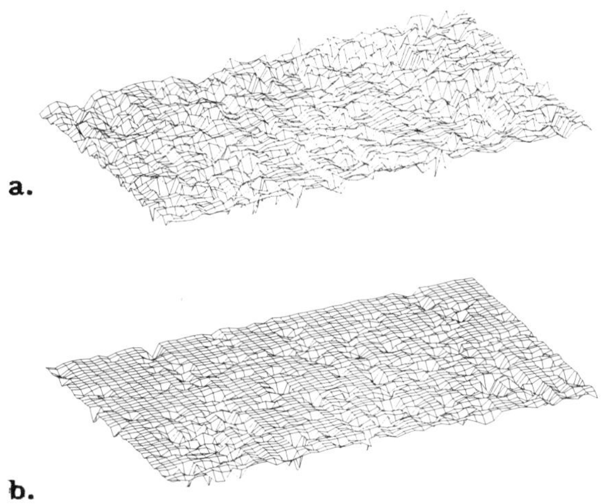

Figure 7 Measured surface plot of the deformed roughness texture of the polyurethane plate in dry contact with the glass plate.
Measured area = 512 μm × 256 μm;
Sample distance = 2 μm (in both directions);
a. Average contact pressure = 0.06 MPa
b. Average contact pressure = 0.12 MPa

significantly larger than the glass plate's roughness (of the order of 0.1 μm and 0.01 μm respectively) and can therefore not be explained by the glass plate's roughness. Instead, smaller scale "real contacts" appear to exist within a larger scale flattened area. A 1 μm scale "contact" is perhaps also not a real area of contact, but may also have smaller scale contacts within its area. This compares to the idea of describing the roughness texture by fractals, but smaller scale contacts (smaller than some micrometer) can not be resolved by the profilometer due to the focus spot diameter, which is about 1 μm.

A possible explanation for the existence of small scale roughness on large scale flattened areas is the larger asperity height to wavelength ratio at smaller length scales (see Hendriks, 1993). Interaction is then more significant at the smaller length scales than at the larger length scales and the smaller scale asperities still persist when the larger scale asperities are largely flattened.

A consequence of this feature is, that the real area of contact can in principle not be determined. Nevertheless, the "real area of contact" can in practice be derived for a particular length scale, ignoring the effects occurring at smaller scales.

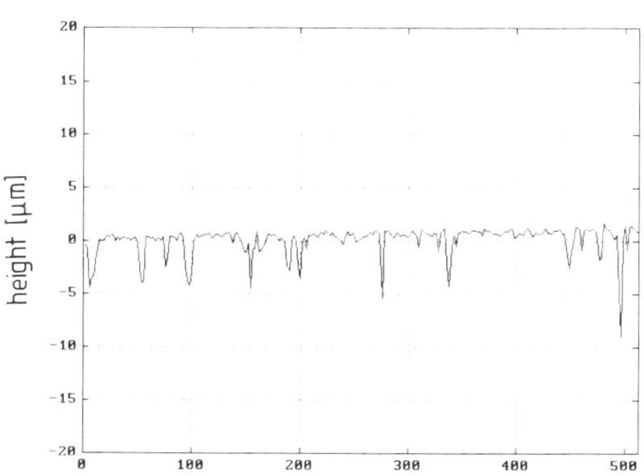

Figure 8 Cross section of fig. 7b.

Fig. 9 shows a first rough estimation of the real area of contact at a length scale of the order of 10 μm (Hendriks 1993). This curve is not a straight line, perhaps due to the fact that the different measurements were not performed on exactly the same part of the surface, but the real area of contact on this length scale is more or less proportional to the contact load.

Verification of contact models has not been performed yet, because more research is required (see section 6 and 7 below). Further investigation is in progress and will also include measurements at higher loads.

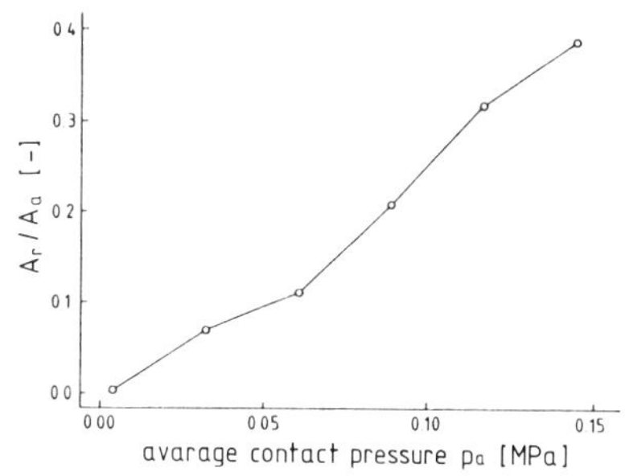

Figure 9 Ratio of the real area of contact A_r and the apparent area of contact A_a versus contact load.

6. DERIVATION OF THE REAL AREA OF CONTACT

Determining the real area of contact from a measurement, the following two points must be considered:

- The different length scales in the roughness texture;
- The reproducibility.

The former item is already discussed in section 5.2 above. The real area of contact can only be determined on a particular length scale. Consequently, one must first consider which length scale(s) is/are important for the problem considered and the measurement parameters (resolving power and measurement length) must be chosen accordingly.

How the real area of contact on a particular length scale can be determined accurately is still subjected to investigation and will be discussed in more detail by Hendriks (1993).

The latter item, dealing with the reproducibility of the measurements, also needs more investigation. Considering e.g. fig. 4b and 5a, a particular flattened area can emerge or disappear when a following measurement is performed on a slightly different part of the surface: e.g. area C in fig. 4 and area D in fig. 5. Consequently, the derived real area of contact can be different when measured on a different part of the surface.

Besides, a waviness with a wavelength of e.g. 1000 μm can yield differences in the real area of contact between parts of the surface in the valleys of the waviness and parts on the summits of the waviness: It is possible that contact only occurs on the waviness summits, when its amplitude is large compared with the roughness height variations. Then measurements must in principle be performed on a part of the surface larger than the wavelength of the waviness, but an objection can be, that the sample distance must be increased to keep the number of measurement points restricted to a practical level.

Finally, the reproducibility in determination of the dimension of one single area appears to be somewhat questionable, as it appears in fig. 6 that the dimension of area A is significantly larger at an average contact pressure of 0.060 MPa than at the other pressures, smaller as well as larger.

Further investigation on the determination of the real area of contact is necessary and in progress. More details will be presented by Hendriks (1993).

7. DISCUSSION ON THE EXPERIMENTAL METHOD

Finally some remarks will be made on the characteristics of the experimental method used.

As already mentioned in section 2.2 above, it is preferred to measure the whole deformed roughness texture instead of the real area of contact only. The reason is, that the different deformations at different length scales can then be studied, yielding better physical insight into the contact problem as illustrated in section 5.2. Using a method which only measures the real area of contact, only the larger scale "flattened areas", as e.g. shown in fig. 8, are found or only the smaller scale contacts, present on the larger scale flattened area, are determined, depending on the resolving power of the used device.

For the dry contact situation, the (deformed) roughness texture is not measured quantitatively well with the method used in this paper. However, the differences in deformation at the different length scales are shown qualitatively, which enables the derivation of the "real area of contact" (or: the flattening) at different length scales. Therefore this method is in essence suitable for further, more detailed investigation of the contact problem.

8. CONCLUSIONS

Optical profilometry is helpful to study both the wet and the dry contact of a rough elastomer and a glass plate. The conclusions of the measurements presented above will be listed below.

Measurements with liquid in the contact area

Measurements on a polyurethane surface show that the roughness height distribution is not affected by the load, after initial deformation has occurred at lower loads. Also, the size of flattened areas found in the measurements remains constant at increasing load.

An explanation for these results is probably, that a large part of the contact load is supported by the liquid and not by the contacting asperities.

Measurements on the dry contact

In the dry contact situation, contacting flattened areas are found and their size (of the order of 10 μm) is more or less proportional with the load. However, these "large scale" flattened areas are not smooth and therefore, different "real areas of contact" can apparently be found on different length scales.

The importance of this result is, that the "real area of contact" is only defined for a particular length scale and one must first consider which length scale(s) is (are) important, before the real area of contact is determined.

ACKNOWLEDGEMENT

We thank Mr. A.C.M.E. van Kalmthout, Mr. J.W. Versteeg and Mr. J.A. Peels for their technical support and Dr. J.J. Baalbergen and Prof. J.J.M. Braat (both of Philips Research Laboratories, Eindhoven) for their advise. Financial support was given by the Technology Foundation STW.

REFERENCES

Bhushan, B., **1985a**, "The real area of contact in polymeric magnetic media - I: Critical assessment of experimental techniques", *ASLE Trans.*, Vol. 28, No. 1, pp. 75-86

Bhushan, B., **1985b**, "The real area of contact in polymeric magnetic media - II: Experimental data and analysis", *ASLE Trans.*, Vol. 28, No. 2, pp. 181-197

Bouwhuis, G., and **Braat, J.J.M.**, **1978**, "Video disk player optics", *Applied Optics*, Vol. 17, No. 13, pp. 1993-2000

Bouwhuis, G., **Braat, J.J.M.**, **Huijser, A.**, **Pasman, J.**, **Rosmalen, G. van**, and **Schouhamer Immink, K.**, **1987**, "*Principles of optical disc systems*", Adam Hilger Ltd., Bristol and Boston, 3rd ed. (1st ed. 1985), ISBN 0-85274- 785-3

Greenwood, J.A., and **Williamson, J.B.P.**, **1966**, "Contact of nominally flat surfaces", *Proc. Royal Soc. London*, Vol. A295, pp. 300-319

Hendriks, C.P., **1992a**, "*Meting van de vervormde ruwheid van elastomeren onder statische kontaktbelasting - deel 1: Dataverwerking en interpretatie*", Eindhoven Univ. Tech., The Netherlands, Report no. T&M D53

Hendriks, C.P., **1992b**, "*Modellen van het statisch belaste, droge kontakt - Literatuuroverzicht*", Eindhoven Univ. Tech., The Netherlands, Report no. T&M D54

Hendriks, C.P., **1993**, M.Sc. Thesis, Eindhoven Univ. Tech., The Netherlands (to be published)

Johnson, K.L., **Greenwood, J.A.**, and **Poon, S.Y.**, **1972**, "A simple theory of asperity contact in elastohydrodynamic lubrication", *Wear*, Vol. 19, pp. 91-108

Lai, W.T., and **Cheng, H.S.**, **1985**, "Computer simulation of elastic rough contacts" *ASLE Trans.*, Vol. 28, No. 2, pp. 172-180

Lee, S.C., and **Cheng, H.S.**, **1992**, "On the relation of load to average gap in the contact between surfaces with longitudinal roughness" *Trib. Trans.*, Vol. 35, No. 3, pp. 523-529

Lubrecht, A.A., and **Ioannides, E.**, **1991**, "A fast solution of the dry contact problem and the associated sub-surface stress field, using multilevel techniques", *J. Trib. (Trans. ASME F)*, Vol. 113, No. 1, pp. 128-133

Majumdar, A., and **Bhushan, B.**, **1991**, "Fractal model of elastic-plastic contact between rough surfaces", *J. Trib. (Trans. ASME F)*, Vol. 113, No. 1, pp. 1-11

McCool, J.I., **1986**, "Comparison of models for the contact of rough surfaces.", *Wear*, Vol. 107, No. 1, pp. 37-60

Onions, R.A., and **Archard, J.F.**, **1973**, "The contact of surfaces having a random structure", *J. Phys. D: Appl. Phys.*, Vol. 6, pp. 289-304

Podbevsek, F.P., **1992**, "*On the roughness deformation of elastomers under static load: Asperity interaction*", Report Eindhoven Univ. of Tech.

Struik, K.G., and **Chang, F.M.**, **1987**, "Measurement of shape and roughness by a modified compact disc sensor coupled to a personal computer", *Proc. 4th Int. Seminar on Precision Engineering*, Cranfield Inst. of Technology, UK, pp. 81-90

Thomas, T.R., and **King, M.**, **1977**, "*Surface topography in engineering: a state of the art review and bibliography*", BHRA Fluid Engng., Cranfield (UK), ISBN 0 900983 66 3

Vergne, Ph., **Villechaise, B.**, and **Berthe, D.**, **1985**, "Elastic behavior of multiple contacts: Asperity interaction", *J. Trib. (Trans. ASME F)*, Vol. 107, No. 2, pp. 224-228

Visscher, M., **1992**, "*The measurement of the lubricant film thickness and the roughness deformation of lubricated elastomers*", Ph.D. Thesis, Eindhoven Univ. of Tech., the Netherlands

Visscher, M., and **Struik, K.G.**, **1992a**, "The measurement of the roughness deformation of elastomers under static load", *Proc. 8th. Int. Coll. "Tribology 2000"* (Organized by the Technische Akademie Esslingen, Germany), paper 9.6

Visscher, M., and **Struik, K.G.**, **1992b**, "Optical profilometry of mechanically inaccessible surfaces", paper presented at the ASPE 1992 Annual meeting (to be published in *Prec. Engng.*)

Woo, K.L., and **Thomas, T.R.**, **1980**, "Contact of rough surfaces: A review of experimental work", *Wear*, Vol. 58, pp. 331-340

Xian Ling, and **Zheng Linqing**, **1991**, "A numerical model for the elastic contact of three-dimensional real rough surfaces", *Wear*, Vol. 148, pp. 91-100

SESSION XIX
PANEL DISCUSSION

Chairman :	Professor D Dowson	
Panel	Dr R C Coy	(Shell Research, UK)
	Professor M Godet	(I.N.S.A., Lyon, France)
	Professor B Jacobson	(S.K.F., Netherlands)

CHAIRMAN'S INTRODUCTION

This Panel Discussion represents an innovation in the structure of the Leeds-Lyon Symposia. There has once again been a very substantial number of papers presented at the Symposium and with so many important isues emerging in the subject, the organizers felt that it would be appropriate to invite three distinguished tribologists to record their impressions of developments recorded at the meeting.

This innovation appeared to be well received.

Dr R C Coy (Shell, Thornton Research Centre, Chester, UK).

Following my participation in the Symposium I would like to present my personal view on the issues facting us as we develop our understanding of the factors dominating thin film lubrication. By thin film lubrication I mean around 10 nm or less, although for some systems this may be an order of magnitude greater.

Firstly, it is very difficult to measure such thin films, however several techniques have now been developed using both optical and electrical methods. These allow us to examine much more closely what is happening under such thin film conditions. What they demonstrate is quite remarkable in that a purely physical model - Elastohydrodynamic lubrication - predicts correctly oil film thicknesses down to a few nanometers. Below this thickness there is a surface modified lubricant layer which is relatively immobile and of the order of 1 to 4 molecules thick and which does not participate in the hydrodynamic flow of the liquid. This film depends on the nature of the surfaces and on the fluid molecular structure. The film properties are more solid-like than liquid and thus the bulk rheological properties no longer apply and hence "classical" EHD theory breaks down.

Of course under extreme conditions of pressure, temperature and shear rate the bulk behaviour of fluids can also change and as we have seen over the past few days there are theoretical approaches that attempt to take these non-Newtonian effects into account. To incorporate both surface and non-Newtonian effects will require a more general approach where lubricant properties are P, T, $\dot{\gamma}$, directionally and surface dependent. Table 1 compares a "classical" and a more "general" approach.

Table 1:

Comparison of "classical" and a more "general" approach.

Classical	General
Isotropic Fluids	An-isotropic Fluids (ordering, orientation etc.)
Newtonian Bulk Rheology	Non-Newtonian and Surface Dominated Rheology
Non-Directional Transport Properties, η, K, D	Directional Transport Properties, $\eta_{x,y,z}$ - $K_{x,y,z}$, $D_{x.y.z}$
Steady State Conditions	Transient Conditions
Energy Input to Fluid Molecules <kT	Energy Input to Fluid Molecules ≥ kT

A further complication relates to the "solid" components of a tribological system. Metal and ceramic components have a wide range of surface modifications which alter their properties. These solids may be modified prior to (or during) operation to give surface films or coatings with properties which vary with depth below the surface. Lubricants contain additives such as zinc dialkyl dithiophosphates which form surface reacted films that are themselves of a complex layered nature here the outermost layer will be the surface modified fluid. Thus any tribological system can be represented schematically as shown in Figure 1.

The significance of each of these layers will depend very much on the prevailing conditions. Under thick film, less severe conditions "classical" Newtonian Hydrodynamic and Elastohydrodynamic solutions are sufficient whereas the thinner the films, and the higher the pressures and shear rates, the greater the need to incorporate a more general solution. It is interesting to note that no one has yet managed to combine physical and chemical models to predict lubrication breakdown and failure. Thus the challenge for the future must surely be to develop models that incorporate these thin film effects, the layered solid properties and the chemistry so that failure can be predicted and more importantly better strategies developed for its prevention.

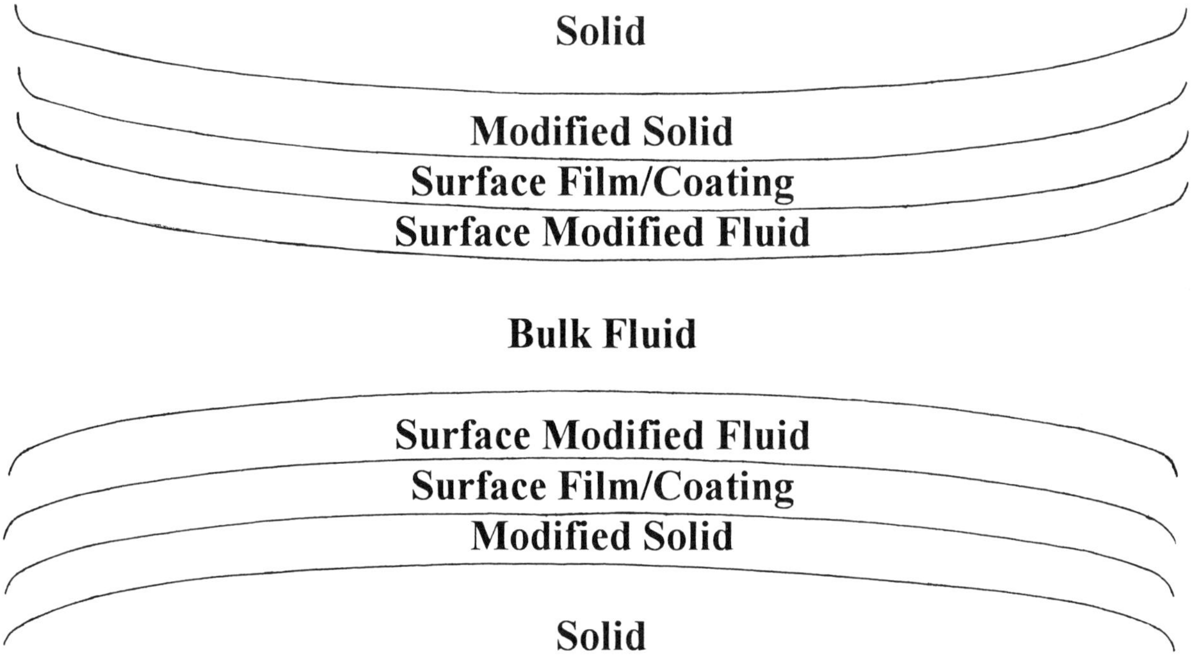

Figure 1. Schematic Contact

<u>**Professor Bo Jacobson**</u> **(SKF Engineering & Research Centre BV, The Netherlands)**

Before the Symposium, the members of the Panel, Dr R C Coy, Professor M Godet and myself, were asked to listen to the various presentations and try to find important trends in the science of tribology as discussed during the Symposium. The panel members visited the lectures according to their own interests, so as there were three parallel sessions most of the time, and only three panel members, some lectures were not attended by any of the panel members if two of us happened to go to the same lecture.

From my personal point of view, the most important trend discussed during the Symposium was the need for much better and much more detailed information about the rheological behaviour of lubricants. Our knowledge and understanding of the relationships between the stress tensor, the strain tensor and the strain rate tensor for lubricants in the transient loading of elastohydrodynamic lubricant films is very limited. No clear and concise experimental measurements are available regarding the coupling of stresses and strains in perpendicular directions, so it is not known how a sliding speed in one direction influences the lubricant flow on an asperity level perpendicular to that direction.

Another related lack of our understanding is the detailed behaviour of the lubricant at and close to the lubricted surfaces. The models used for numerical calculations normally rely on either a stress boundary condition or a non-slip boundary condition between the oil and the bearing surface. Depending on which one is used, the resulting calculated oil film thickness between rough surface asperities is very different.

The forces acting between the lubricant molecules and the atoms in the lubricated surfaces reach much further than just one atomic layer. This means that the flow properties of the lubricant (and of the lubricated surfaces!) are influenced by the presence of the other material. This normally leads to a lower mobility of the lubricant molecules close to the surface, but it is not known if the limiting shear strength of the lubricant is influenced by the presence of the surface. Some lubricants together with some surfaces may behave according to the stress boundary condition while other systems behave according to non-slip boundary conditions.

The chemical-physical influence of the lubricant molecules on the lubricated surfaces lead to another important topic discussed during the Symposium: surface engineering. By optimizing the combinations of layered coatings with the substrate on one side and the lubricant on the other, it should be possible to reach lower friction, lower wear rate, longer endurance life and lower cost at the same time as the influence on the environment is decreased.

The last point I should like to put forward for future investigation is the running-in phenomena and their influence on surface properties and surface roughness parameters. The surface roughness parameters used today are more relted to specification and control of surface manufacturing than to the functioning of the surfaces after manufacture. Depending on the specific load level, the slide roll ratio, the cleanliness of the lubricant, the temperature and so on, identical surfaces will run in to different final surface structures or fail. The surface roughness parameters which need to be controlled are thus different for different applications, and how to choose the correct surface roughness parameter for a specific application is not well understood today. The roughness parameter chosen for control of the manufacturing of the surface should thus take into consideration the safe running-in to final surface, the possibility to build up a thick enough oil film for the final surface, and the stress field under the run-in surface under operating conditions. It should also be remembered that failures are normally caused by extreme events and not by statistical mean values, and extreme events are by definition not possible to measure using standard surface roughness measurements.

<u>Professor M Godet</u> **(Laboratoire de Mécanique des Contacts, INSA, Lyon, France).**

A definition

Surface engineering is a new word, coined by material experts to include the various industrial operations (coatings, tratments, paints etc.) performed on the surfaces of different materials. These operations are intended to embellish these surfaces or protect them from different forms of aggression such as oxidation, wear, and other forms of surface damage. Recently, the meaning of the word was extended to include the **design** of coated machine elements. Tools for dsign of coated elements, which are similar in essence to those used in the design of common massive parts, are needed when machine elements are subjected to the often severe running conditions encountered in tribological applications. This discussion is limited to hard coatings. A lot more work must be done before a similar approach can be proposed for soft coatings.

As design is an old subject, it has little attraction and has proved difficult to defend in the face of such a fashionable subject as surface engineering, which, even when limited to its original definition, is capable of keeping armadas of scientists busy with large research budgets. Competition is difficult.

The design process

The design process in surface engineering is broken down into 3 steps:

(1) the machine element (or mechanism),
(2) the rubbing surfaces (or first-bodies),
(3) the interface (or third bodies).

This discussion will concentrate on points 1 and 2.

Machine element (or mechanism)

The term "mechanism" is used here in its original sense as representing a mechanical structure, such as a bearing, and not a fundamental physical or chemical process, adhesion for instance. Classical design calls for the identification of:

- loads,
- velocities,
- temperature,
- environment;
- ......

imposed by the mechanism studied. Note that some of these parameters can vary by several orders of magnitude when one goes from one mechanism to another, from automotive gears to computer recording heads for instance. The conditions met in practice are so different that a surface engineering problem must begin with a thorough analysis of the operating conditions of the mechanism under study.

Rubbing element (or first-bodies)

First-bodies are the individual rubbing parts themselves. Gear teeth, bearing balls, cable strands are first-bodies. It is convenient to identify, even schematically, with the view of illustrating the difference in approach between them, the material and mechanical aspects of design in surface engineering:

Material tasks are reduced to:

- the setting up of a list of potential material combinations (coating + substrate) drawn essentially from experience,
- the qualification from a material point of view (hardness residual stresses, coating to substrate adhesion,...) of these combinations,
- the identification of thermomechanical characteristics (E, ν, k, ρ, c_p, α) and limits of both substrate and coating, where E, ν, and k, ρ, c_0, α are respectively the classical elasticity and heat transfer coefficients.

Mechanical tasks are reduced to the determination of the:

- stress,
- stress intensity factors,
- temperature

fields imposed by the mechanism to the first-bodies. This is particularly important in coated systems as coating to substrate thermomechanical compatibility problems are added to the classical strength of materials substrate and/or coating problems. It has been shown that compatibility problems are governed essentially by modulus, conductivity, and dilation coating substrate ratios E_c/E_s. k_c/k_s α_c/α_s and each of these three ratios can determine whether for given running conditions, the coating/substrate combination retained is adequate from a first-body point of view and this independently of any specific tribological aspects.

In other words, it is necessary, once the material requirements are met, to establish whether from a thermomechanical point of view the combination retained for the first-body can withstand the running conditions imposed by the mechanism. Only then is it necessary to assess again for representative running conditions, the tribological performance of that combination.

That assessment can be approached theoretically if the mechanism runs under thick viscous film conditions or experimentally in all other circumstances. Remember that the same exacting control must be exercised in choosing the experimental operating conditions as it is impossible to extrapolate friction and wear results from classical laboratory test benches to industrial applications. Close simulation must be achieved to produce meaningful results.

Difficulties experienced today

The difficulties experienced today in setting up the design phase of surface engineering are many, and most of them come from the lack of dialogue between material and design (or mechanical) experts. Again schematically, the first consider their work done when they have performed the tasks listed above, the second have not yet included the design of coated elements in their obligations, and are led to make decisions based on commercial claims and summary tests. Further if the mechanicals looked at the problem in more detail they would realise that they could not go ahead with their tasks as most of the thermomechanical coefficients and limits needed for their analyses are not available today. Progress in surface engineering, as opposed to surface science, requires the setting up of research programs in which both material and mechanical experts are included. The rest is waste.

The XIXth Leeds-Lyon Symposium

The Symposium was rich in surface engineering papers, even though its title suggested a more fundamental approach. This illustrates once more that it is difficult in Tribology to define a clear frontier between fundamental and applied aspects.

- Session 1 gave an overview of thin film tribology which touched upon surface engineering.
- Session 2 concentrated on coatings, and stressed the material aspects of the problem. There was however an opening towards the mechanics of coatings which was most encouraging.
- Session 4 discussed coating action.
- Chemical and corrosion effects were discussed in Session 8.
- Session 11 presented a very strong plea to bring methodology in coating evaluation and distinguish in so doing between the various forms of coated surface damage including substrate yielding.
- Session 14 discussed performance evaluation of various coatings.

Conclusion

The XIXth Leeds-Lyon Symposium brought a significant contribution to surface engineering and all attendees are thankful for this. Continued progress requires:

- multidisciplinary (materials, mechanics, tribology) programs and meetings at both national and European levels.
- favouring existing coating characterisation rather than the definition of new coatings.

Recommended reading

Mechanics of Coatings, (Proceedings of the 16th Leeds-Lyon Symposium) Ed: D Dowson, C M Taylor and M Godet, Elsevier Tribology Series No. 17, Amsterdam 1990.

Written Discussion - Contributions

DISCUSSION

SESSION V - ANALYSIS AND DESIGN OF HEAVILY STRESSED COMPONENTS

'Micro-Wear of Tapered Roller Bearings Lubricated by Oil With and Without Foreign Particles' by S Noguchi, T Sawemoto and S Aihara.

Mr D Parker (Glacier Vandervell Ltd)

The authors presented some interesting results derived from the comparisons of surface profiles. This implied the ability to precisely relocate the measuring device on the component under study. This relocation must have been to an accuracy much finer than the footprint of the probe.

Are the authors able to present details of the technique adopted to derive quantitative data from direct comparisons of surface profiles?

Reply by Dr S Aihara (NSK Ltd, Fujisawa, Kanagawa, Japan)

The method of re-location used in these studies was explained elsewhere (1), and by the time of publication of the proceedings, an English translation of the mentioned paper will be available (A1). The principle of re-location method adopted was based on that by Williamson and Hunt (7). A combination of a goniometer stage, a linear scale and a digital computing system enabled us to re-locate the initial position precisely. The system is illustrated in Fig. A1. However, re-locating plates shown in Fig. 2 are essential.

(Ref. A1) S Noguchi and T Sawamoto, Japanese Journal of Tribology, to be published, Allerton Press. (English Translation of Ref. (6)).

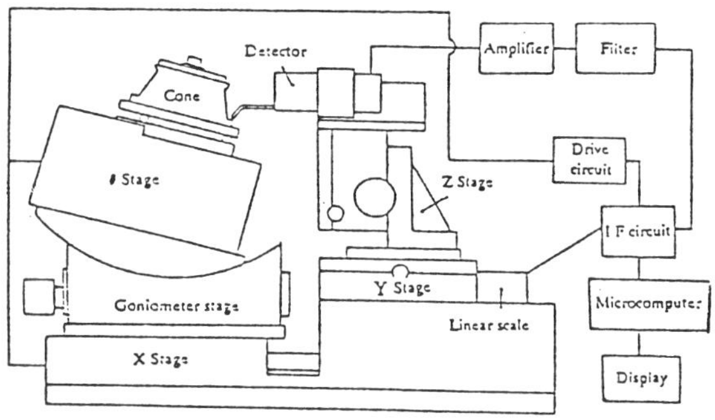

Fig. A1 Schematic Diagram of the Surface Profile Measuring Device

Dr J A Williams (Cambridge University Engineering Department, England)

Even the smallest of your particles is many times larger than the minimum film thickness in the contact. Do you think it possible that a significant fracture of the particles occurs, leading to the production of abrasives with dimensions more like the minimum film thickness, and that these are responsible for the increased wear rate. Are there perhaps particles of a particular size range that are especially destructive to the bearing surfaces? In our own work on three body abrasive wear [1,2] we have given some thought to just how individual particles pass through such a gap and bring about surface damage. Do the authors have any views - perhaps deduced from examination of the worn surfaces, or of the debris isolated by Ferrography, on the actual mechanisms of the damage process?

(Ref. 1) J A Williams and A M Hyncica, 'Abrasive Wear in Lubricated Contacts', J Physics, D, 25, (1992), A81-A90.

(Ref. 2) J A Williams and A M Hyncica, 'Mechanisms of Abrasive Wear in Lubricated Contacts', WEAR, 152, (1992), 57-74.

Reply by Dr S Aihara (NSK Ltd., Fujisawa, Kanagawa, Japan).

The difference between a sliding bearing and a rolling bearing is as follows, and causes the difference in wear. The lubricated contact of a rolling bearing is an Elasto-Hydrodynamic contact, where oil film thickness is small, almost comparable to the surface roughness, with very high contact pressures and a very short duration of each contact. As we mentioned in the paper, the wear rate with contaminated oil is still much smaller than that for a lubricated sliding contact. This may be related to the short contact time.

In this study, we put only a minute amount of hard particles into the oil. Therefore, most of the trapped particles detected by Ferrography should be worn particles. But we could not find any large particles left and so we think that fracture of the particles occurred during rotation.

Though our interest was in the amount of wear, we think that the wear was caused by three body abrasive wear, as the entire raceway was covered with very short scratches.

'The Effect of Compliance on Peak Oil Film Pressure in Connecting Rod Bearings', by D Fenner, C McIvor, J M Conway-Jones and H Xu.

Professor J F Booker (School of Mechanical and Aerospace Engineering, Cornell University, USA).

The present work represents a step forward on the long path towards rigorous extrapolation of test rig data to design.

The powerful analysis tool developed and reported previously by the first two authors [1-3] has an excellent chance of bridging the considerable distance from a relatively rigid fatigue test rig to a relatively flexible modern connecting rod. It won't be easy, however.

The system (Sapphire fatigue test rig) chosen for study here has an unusually massive housing, allowing remarkable agreement with classic EHL (modified Hertzian) contact results [13]. This system is also unusual in that journal and sleeve probably make comparable contributions to its structural compliance (though this was apparently not considered in the present study).

In contrast, our own 'model EHL' work reported previously at the 15th Leeds-Lyon Symposium [D1] and elsewhere [D2] (based on a relatively non-iterative EHL transient analysis) studies a generic connecting rod bearing similar to modern General Motors practice. Previous structural optimisation work by others [D3] with similar connecting rods shows that their high compliance can either increase or decrease global minimum film thickness while producing multiple local maxima of film pressure.

Such modern applications are very different from the classic (but relatively rigid) Ruston-Hornsby VEB III widely studied by previous workers (including the present authors and discusser).

In this context could the authors comment further on the implications of the present work for design of modern bearings?

Additional References

D1 Kumar, A, Booker, J F and Goenka, P K. 'Dynamically Loaded Journal Bearings: A Modal Approach to EHL Design Analysis', Tribological Design of Machine Elements, (Proceedings 15th Leeds-Lyon Symposium on Tribology, Leeds, England, September 6-9, 1988), Eds. D Dowson, C M Taylor, M Godet and D Berthe, Elsevier, Amsterdam, The Netherlands, 1989, 305-315, 508-510.

D2 Kumar, A, Goenka, P K and Booker, J F, 'Modal Analysis of Elastohydrodynamic Lubrication: A Connecting Rod Application', Trans. ASME, J Tribology, 1990, 112, 524-534.

D3 Goenka, P K and Oh, K P, 'An Optimum Connecting Rod Design Study - a Lubrication Viewpoint', Trans. ASME, J Tribology, 1986,

108, 487-493.

Reply by Dr D Fenner (Kings College, London, UK) **and Mr J M Conway-Jones** (Glacier Vandervell, UK).

The authors are appreciative of the comments of Professor J F Booker. Our paper draws attention to the dependence of the axial variation in the oil film thickness and pressure on the elasticity of the bearing support, and indicates the direction for future work to link basic material fatigue test data with more realistic estimates of oil film pressure variation.

Work completed recently by the authors confirms the existence of multiple local maxima of oil film pressure in automotive type connecting rods when the load is acting on the more compliant cap half of the bearing. Future work will aim to assess this effect more fully.

'A Transient Elastohydrodynmamic Lubrication Analysis for Spiral Bevel Gears' by S G Lim, D E Brewe and R F Handschuh.

ir Harry van Leeuwen (Eindhoven University of Technology, Netherlands).

From the presentation I conclude that the authors have investigated the effect of variable load and radius of curvature of the film thickness profile. The resulting values for the transient central and minimum film thickness differ from the stationary values by some ± 17% and ± 7%, respectively, and show an appreciable phase lag.

The authors refer to a classical paper by Gu [2], who suggested a 'quasi-steady state criterion'. This criterion boils down to: how transient is the lubricated contact? It is defined by the ratio between the contact duration time, (td), and the contact zone flow time, (tf). A difference between Gu's paper and this one is the high frequency variation of the contact load during the contact, which decreases (td), and hence the value of the criterion, (td/tf), and the high value of the contact load.

Note that in Figure 1 of the Synopsis ($\partial h/\partial t$) is almost constant in the contact zone, except a small deviation at the outlet.

I would like to see an answer to the following questions:

1. What is the origin of the high frequency oscillations in the contact force? How does the radius of curvature vary with time? What is the average value for the quasi-steady state criterion, (td/tf)?

2. Can the authors give average numerical values of the rolling and sliding speed, of the radius of curvature, of the contact load frequency, and of the contact length?

3. Have the authors tried Vichard's curve fits? If so, do they differ appreciably from the results for the full numerical solution?

Reply by Dr S G Lim, Mr D E Brewe and Dr R F Handschah (NASA Lewis, USA).

The authors wish to thank Professor van Leeuwen for his thoughtful remarks. The following responses are given in the order of questions.

1. The impact of the transient effect can be explained using equation (9). The squeeze number S in the right hand side plays an important role. When ω_D is large in comparison with μ_m/R, a large squeeze effect is expected. The origin of ω_D is the fluctuation of the operating parameters at the contact. In the global point of view (assuming the surface is smooth), the most notable source of the transient effect is the dynamic loading. The high frequency may come from the vibrations of the gear system or gear profile error. The order of magnitude of the amount of load fluctuation is based on the experimental results in reference (10). Unlike spur gears, the expression t_d/t_f is not available for the spiral bevel gear. As mentioned in reference (2), this ratio is small for a large load since t_f is increased due to the enlargement of the Hertzian contact length. In fact, according to the right hand side of equation (9), we can see that a large dimensionless load, W, causes a large squeeze

effect. In conclusion, the transient effect is important not only for the highly loaded case (small t_d/t_f) but for a large squeeze parameter S.

2. Much of the information to this question was not a part of the synopsis, but appears in the full paper (see Fig. 5 and 6). Since this is an isothermal investigation we only required the average surface speeds to solve the Reynolds equation.

3. Indeed, we tried the squeeze velocity distribution, $\partial h/\partial t$, for Vichard's approximation provided in the accompanying figure. This can be generated using equation (16). The operating parameters are the same as in Fig. 5. The curve 1 is matched with (a) in Fig. r, 2 with (b), etc. Even though the curves 2 and 4 look straight, they do in fact contain a small amount of curvature on a finer scale. The plot range is $-3b \le x \le b$, where b is semi-Hertzian length. There is a very close qualitative similarity between the full numerical solution and Vichard's approximation. But, there is a notable quantitative difference. Also, we notice a small spike in $\partial h/\partial t$ near the exit region, while that in the Vichard's approximation is constant in the Hertzian contact region.

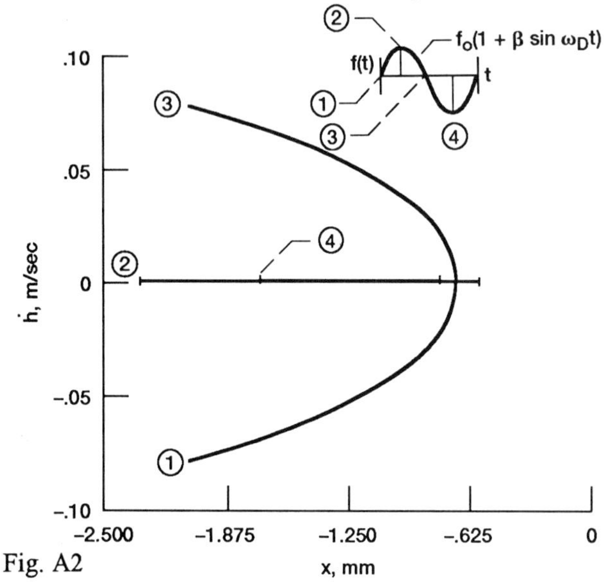

Fig. A2

SESSION VII - ELASTOHYDRODYNAMICS 1

'Non-Dimensional Groups in Elastohydrodynamic Lubrication' by D Dowson, E S Song and C M Taylor.

ir Harry van Leeuwen (Eindhoven University of Technology, Netherlands).

Dimensionless groups in EHL have been a favourite subject for discussions for more than 30 years now. For the Barus viscosity relationship and linear elastic solids, K L Johnson's (1970) paper has made an end to much rumour. This presentation left me behind with some questions, which I would like to discuss.

In my opinion, if the Barus relationship is adopted, the same numerical values for dimensionless groups as suggested by, e.g. Dowson and Higginson, or Moes, or Johnson, will all lead to the same result, i.e. the same numerical value for the non-dimensional film thickness (Note: the transformation from Dowson and Higginson's to the other's groups is not unambiguous).

But if the Roelands' model is employed, a different set of non-dimensional groups results, e.g. the groups H, U, W and Z (which is Roelands' parameter Z_R) in Dowson and Higginson's notation. Maybe a smaller set can be derived, but I have not given it a try.

For reasons of comparison with results from the Barus era, the results obtained for the Roelands equation have been transformed in the 'Barus' - set. Roelands (1966) suggests αi (see Roelands (1966) p. 163) and $\alpha^* = 1/p_{iv}$, as (see Roelands 1966) p. 451):

$$\alpha_1 = \alpha_i (Z, \eta_o), \text{ and} \quad \ldots (a)$$
$$\alpha^* = \alpha^* (Z, \eta_o), \text{ respectively} \quad \ldots (b)$$

In the first relationship it is easy to swap from α_i to Z (given η_o) and vice versa; in the second this is troublesome because of the resulting incomplete gamma function.

From the synopsis I conclude that the authors keep G, W, and U constant, and in addition have fixed values for F, Rx, Ry, Er, α and Z. Hence, the parameters in G and W are fixed, and U is kept constant by keeping the product of η_o and u_e (the entrainment or rolling speed) constant.

If η_0 is varied, so does u_e. But, according to eqns. (a) and (b), η_0 is determined by Z and α_i or α^*. If I assume that α_i or α^* is set equal to the given value of α, then η_o is fixed, so u_e can not be varied.

I have a problem in understanding what is fixed and what is varied in the numerical experiments. Can the authors appreciate my problem?

Additional References:

K L Johnson, 1970, "Regimes of Elastohydrodynamic Lubrication", Journal of Mechanical Engineering Science, Vol. 12, No. 1, pp. 9-16.

C J A Roelands, 1966, "Correlational Aspects of the Viscosity-Temperature-Pressure Relationship of Lubricating Oils", PhD Thesis, Delft University of Technology, 1966, 495 pp.

Reply by Mr E S Song (Department of Mechanical Engineering, The University of Leeds, UK).

We have illustrated in this paper that two additional dimensionless groups are required for the adoption of the Roelands relationship. When the Barus relationship (eq. 1) is adopted replacing equation (4), the same procedure taken before equation (17) gives;

$$H_{cen} = f\,(G, U, W, E'\,\beta_1) \qquad (22)$$

for circular or line contact EHL problems, since $E'\beta_2 = (E'\beta_1)(\beta_2/\beta_1)$ and (β_1 and β_2) are assumed to be constants. When we further assume an incompressible fluid taking equation (5) out, the corresponding relation becomes;

$$H_{cen} = f\,(G, U, W) \qquad (23)$$

Therefore, it may be true that fixed values of (G, U, W) for the Barus model lead to the same result if only the analyses concern incompressible fluids. And this was the case which Johnson or others addressed. However, Table 3 shows that, for a compressible fluid, fixed (G, U, W) for employing the Barus relation gives different results according to the values of (E') taken.

Any exclusive effect of adopting the alternative viscosity-pressure model can be observed when, for example, the circular or line contact EHL results are compared in terms of (G, U, W, η_o/η_∞, E'/γ). This is because all the input variables which influence the result will be uniquely transformed by fixing these five dimensionless groups and using equation (7).

The dimensionless film thickness relations presented in this paper comprise the maximum number of separate independent non-dimensional groups. Even if a smaller set could be derived in some cases, such a set should be separate from that built upon the maximum number of the groups.

Table 3. Variation of the results for fixed (G = 5713, U = 4.69 x 10^{-13}, W = 1.82 x 10^{-8}) - adopting the Barus equation (1) and equation (5) for the compressibility. ($R_x = R_y$ = 0.0254 (m), and Pure Sliding)

Data No.	1	2
η_o	319.0	2.95e-4
β	5.71e-8	0.57e-8
E'	1.00e11	10.0e11
F	1.174	11.74
u_e	3.73e-6	40.33
$H_{cen} \times 10^5$	0.359	0.334
$H_{min} \times 10^5$	0.230	0.230
$\bar{p}$ at $p=p_h$	1.05	1.22
$p_h \times 10^9$	0.096	0.96

SESSION XV - BIO-TRIBOLOGY (1)

'A Model Analysis of Squeeze Film Effect in a Hip Joint' by K Ikeuchi and M Oka

Professor D Dowson (Department of Mechanical Engineering, The University of Leeds, UK).

Professor Ikeuchi has presented a very nice and comprehensive analysis of the squeeze-film lubrication of a model of synovial joints.

The flow of fluid within the cartilage is a key element in the analysis and I would like to learn a bit more about the modelling of the cartilage structure. Is the present model compatible with the bi-phasic structure analysed by Van Mow et al?

Does the author assume isotropic conditions within the cartilage? - or is account taken of the variation in effective elasticity and porosity normally and transversely to the cartilage surface?

Reply by Professor K Ikeuchi (Research Center for Biomedical Engineering, Kyoto University, Japan).

We wish to thank Professor Dowson for his valuable comments. It is always important to examine the inevitable assumptions and approximations in any model.

The present model is comparable with the more strict bi-phasic model in that articular cartilage is composed of a porous elastic solid and a liquid absorbed by the pores. Therefore, we expect agreement between the results from the two models if appropriate material parameters are used.

An isotropic condition is assumed in the EHL theory for the instantaneous deformation of the cartilage. However, there are no such assumptions on the time dependent deformation due, in part, to interstitial flow. Though we used uniform values for the elastic modulus and permeability constant in the cartilage for simplicity, it is possible to change them at each mesh point, as the characteristics values vary in both normal and tranverse directions. We would like to investigate the effect of anisotropy of the cartilage on the lubrication condition in the future.

'Friction and Mixed Lubrication of Soft Layer Contacts' by L Caravia, D Dowson, J Fisher, P H Corkhill and B J Tighe.

ir Prof. K Ikeuchi (Kyoto University, Japan).

It is exciting that a cushion form bearing shows high potential not only in hydrodynamic lubrication conditions but also in mixed lubrication conditions. The discusser thinks that both high water content and relatively low permeability is necessary for a porous layer to keep the squeezing from the layer for a long time. Have you measured the permeability of the hydrogel? How can we control this value?

Reply by Ms L Caravia, Professor D Dowson, Dr J Fisher, Dr P H Corkhill, Professor B J Tighe.

Permeability is an important material property and is likely to affect both the lubrication regime and the deformation and stress distribution in the contact. However, it was not investigated as a variable in this initial study. It is possible to make hydrogels with different permeabilities and the effect of permeability on the tribological properties will be investigated in future work.

Mr J M Blamey (University of Durham, UK).

Do the hydrogels have problems with fatigue wear under the applied loads? Are they mechanically strong enough?

Reply by Ms L Caravia, Professor D Dowson, Dr J Fisher, Dr P H Corkhill, Professor B J Tighe.

It is important to draw attention to the fatigue strength of hydrogels. Poor mechanical properties of traditional hydrogels has been a major constraint on their use in artificial joints. We have also recently demonstrated that low elastic modulus polyurethanes have unacceptable fatigue strength and durability (1) in cushion knee prostheses. Over the past three years, our colleagues at Aston University have made significant advances in improving the mechanical properties of hydrogels by introducing novel copolymer and terpolymer compositions and by developing semi-interpenetrating networks (2,3,4). We are currently assessing the long term wear and durability of a range of novel hydrogel materials and hope to be able to publish these results in the near future. However, the importance of the high level of start up friction found with smooth, non-porous polyurethane materials (5) must be emphasised - coefficient of friction values of between 0.5 and 1.0

cannot be tolerated in total artificial joints. We feel that the identification of high start up friction as a potential problem and the development of hydrogels as a possible solution is a significant step in the development of the 'cushion bearing concept' at Leeds University.

References:

1. D D Auger, D Dowson and J Fisher, Proc. 8th Meeting of European Society of Biomechanics, Rome, 1992, 115.

2. D A Baker, P H Corkhill, C O Ng, P J Skelly, B J Tighe, Polymer, 29, 1988, 691-700.

3. P H Corkhill, B J Tighe, Polymer, 31, 1990, 1526-1537.

4. P H Corkhill, A S Trevett, B J Tighe, Proc. Instn. Mech. Engrs., Part H (1990), 204, 147-155.

5. L Caravia, D Dowson, J Fisher, Wear (1993), (in press).

'Measuerements of Elastic Modulus and Hardness of Biocompatible Polyurethanes and the Effects on Contact Area Within the Hip Prosthesis', by J M Blamey.

Dr J Fisher (Department of Mechanical Engineering, University of Leeds, UK).

Elastic Modulus and Hardness are traditionally used to define fundamental properties of materials.

In this study the terms have been used to describe the properties of a composite structure (the soft layer and the rigid substrate). The changes in the parameters which are described as elastic modulus and hardness in this study may be due to changes in the design of the structural composite (in particularly the thickness of the layer). They are not necessarily representative of changes in the fundamental properties of the material.

Is it not therefore more appropriate to use alternative terms to describe the parameters measured in this study.

Reply by Mr J M Blamey (School of Engineering and Computer Science, University of Durham UK).

In no way did I mean to suggest that the fundamental properties of the material are altered. Comments I made relate to the softening of the material as conditions change from laboratory to in-vivo.

I do not think that either elastic modulus or hardness define fundamental properties. Hardness, in particular, is a comparative test, dependent on the thickness of the sample, the backing material and the bond to it, test used, indenter maximum load, time under load, rate of loading and so on. As such hardness is designated H20/30 for urethanes and H132/30 for polyethylene. The measure of elastic modulus is dependent on the part of the stress-strain curve which is considered, particularly for rubbers/ polyurethanes.

The effective elastic modulus, E', used in calculations of EHL and squeeze film thickness values is not a fundamental property of the material, but a combination of values for the two surfaces. The main aim of the study was to calculate, through experimental means, this effective modulus to allow more representative calculations of EHL and squeeze films for layered surfaces. This has been successfully done through hardness testing and experimental contact area assessments and values of effective elastic moduli of the acetabular component of 150-250 MPa have been obtained for bonded samples. These are much closer to the bulk modulus of the material (~ 1500 MPa) than the measured tensile modulus (10-20 MPa). This figure can then be used to calculate the effective modulus of the joint (E') for use in tribological assessment.

'Fluid Film Lubrication of Natural Hip Joints' by Z Jin, D Dowson and J Fisher.

ir Harry van Leeuwen (Eindhoven University of Technology, The Netherlands).

The authors state in their study on the transient EHD film thickness in elliptical soft contacts that the so-called Holland method, which was conceived in the sixties for calculations on dynamically loaded bearings, yields accurate numerical results. I am surprised, and want to explain this below.

In his 'Verfahren der summierten Traganteile' (= Method of superposed load parts) Holland solves the steady state entrainment effect and the non steady state squeeze effect separately, and superposes the resulting loads afterwards. He extended his method to nonconformal contacts in 1977 (referenced below), and used it in the analysis of heavily loaded EHD line contacts as gears and cam mechanisms.

As the superposition principle is allowed only if the differential equation is linear, including in its boundary conditions, Holland's method is questionable in the case of rigid bearings. At high journal eccentricities this method is in error (Polish investigators published on this in about 1986 in Wear - I cannot trace down the correct reference at this moment). The reason is that the free boundary conditions of the two load carrying effects are always different, hence inconsistent.

By applying this method to EHD problems, where film thickness (and with hard contacts, viscosity too) is pressure dependent, it is once more not allowed to split the full problem into two separate problems. Apart from numerical errors (see Haiqing et al (1984)), Holland's approach to the hard EHD problem is even more questionable because his empirical transient film thickness formula is not based on numerical data, but on false reasoning ("Faktorenvergleich" = comparison of factors). By comparing the empirical film thickness formulae obtained by Herrebrugh (1968) (steady state, isoviscous elastic, H1), Dowson and Higginson (1977) (steady state, piezoviscous elastic, D1) and Herrebrugh (1970) (pure squeeze, isoviscous elastic, H2), he arrives at his new film thickness formula (D2) by stating that the ratios of factors and exponents (of arbitrarily chosen non-dimensional groups!) in these equations are like:

$$H1 : D1 = H2 : D2$$

I wonder why the results obtained in the present study can be considered as being accurate. Are they compared against results of a full analysis? If so, how can my doubts be taken away?

Additional References.

Dowson, D and Higginson, G R. (1977), "Elastohydrodynamic Lubrication", SI (2nd) Edition, Pergamon, Oxford, p. 96.

Herrebrugh, 1968, "Solving the Incompressible and Isothermal Problem in Elastohydrodynamic Lubrication Through an Integral Equation" J of Lubrication Technology, Trans. ASME, Vol. 90, No. 1, pp 262-270.

Herrebrugh, 1970, "Elastohydrodynamic Squeeze Films Between Two Cylinders in Normal Approach", J of Lubrication Technology, Trans ASME, Vol. 92, No. 2, pp 292-302.

Holland, J, 1977, "Die Instationäre Elastohydrodynamik - eine Möglichkeit zur Erfassung der Beansprunchung in Maschinenelementen (The instationary elastohydrodynamics - a possibility for the description of loads in machine elements)", Proceedings Eurotrib '77, 2nd European Tribology Congress (Oct. 1977), Vol. I, pp 38/1-38/6.

Haiqing, Y, Xiaze, Z and Yuzian, H, 1984, "The computation of unsteadily loaded EHL film thickness and other lubrication parameters of cam-tappet pairs of IC engines and analysis of their performance", in: Developments in Numerical and Experimental methods Applied to Tribology by D Dowson, C M Taylor, M Godet and D Berthe (eds), Proceedings 10th Leeds-Lyon Symposium on Tribology (Lyon, 1983), Paper VII (iii), Butterworths, London, pp 171-181.

Reply by Dr Z M Jin, D Dowson and J Fisher
(Department of Mechanical Engineering, The University of Leeds, UK).

The authors would like to thank Harry van Leeuwen for his interesting questions. As he rightly mentions, the superposition method proposed by Holland is, in general, not applicable to rigid bearings and hard EHD problems since the Reynolds' equation including boundary conditions is not linear under these conditions. However, under heavily loaded soft EHD problems, as we considered for human synovial joints, the following characteristics have been found:

- The film profile is almost constant within the contact region and the hydrodynamic pressure is almost identical to the dry contact pressure under steady-state entraining motion, squeeze-film motion and their combination.

- The cavitation boundary due to film rupture under steady-state entraining and combined entraining and squeeze-film motions and the boundary pressure under squeeze-film motion are all very close to the edge of the dry contact region.

Under these two conditions, Reynolds' equation becomes linear and the superposition principle can be utilized.

The validity of this simple approach was examined by running the full transient numerical analysis, where both the entraining term and the squeeze-film term were included in Reynolds' equation for normal human ankle joints (Dowson and Jin, 1987). The predicted average film thickness throughout one walking cycle was shown in Figure 6 of the full paper and good agreement was found. Thus it is expected that the present simple superposition approach can also be applied to human hip joints.

SESSION XVII - WEAR/COATINGS

'Micro-Raman Spectroscopy as a Tool for the Characterization of Wear Induced Surface Modifications on Hard Coatings' by B Blanpain, M Franck, H Mohrbacher, E Vancoille, J P Celis and J R Roos (Katholieke Universiteit Leuven, Belguim).

Dr E Rosset (EPFL-DMX-Tribology Group. Ecole Polytechnic, Switzerland).

What is the effect of surface roughness or debris distribution size in the Micro-Raman Spectroscopy?

Reply by Dipl. Ing. H Mohrbacher (Katholieke Universiteit Leuven, Belgium).

The effect is seen in the scattering intensity registered by the detector. The scattering intensity depends on the scattering cross-section which is a material property, the experimental setup (laser wavelength and power, collection angle of microscope front lens, detector efficiency.), and the number of scattering units encountered by the laser spot size. With increasing surface roughness the specific surface area will be increased and more scattering units are excited within the same laser spot size. Thus, the scattering intensity will be enhanced. If the roughness is very high, part of the scattered light is reflected away from the collection cone of the microscope front lens resulting in a reduced detected scattering intensity. For the debris distribution, a similar argument applies, i.e., a more compact debris comprises a higher area density of scattering units and, thus, a higher scattering intensity. The peak position itself being solely a material property is not affected by surface roughness.

SESSION XVIII - BIO-TRIBOLOGY (2)

'Friction of Cylindrical Cushion Form Bearings for Artificial Joints - A Comparison of Theory and Experiment', by D D Auger, D Dowson and J Fisher (Department of Mechanical Engineering, The University of Leeds, UK).

ir Harry van Leeuwen (|Eindhoven University of Technology, The Netherlands).

It is not an easy task to perform measurements in dynamically loaded lubricated systems, like in this joint simulator. The authors should be commended for their valuable work. I have the following comments and questions:

(a) Being a novice in the bio-tribology world, I wondered if the choice of the point of maximum load and velocity is an arbitrary one or not. Will the conclusions change if a different point in time is chosen?

(b) In addition, I conclude that the authors assume that friction due to pressure induced flow is less important than due to shear flow. I agree with this opinion.

(c) The 99.9% confidence limits are quite wide. I forgot to pay attention to the scales (linear?). The authors conclude good correspondence between theoretical and experimental Stribeck curves. However, in my opinion the tendency is different: the theoretical curve shows a positive gradient (full film conditions), while the experimental curve shows a negative gradient, unless the limits of the confidence intervals are used.

Would the authors please comment on these remarks?

Reply by D D Auger, D Dowson and J Fisher (Department of Mechanical Engineering, The University of Leeds, UK).

(a) The point of maximum load and velocity was chosen because it is the centre of the stroke and therefore the point in the cycle which is least affected by vibrations and artefacts associated with the reversals of direction of the femoral head at the ends of each stroke. Furthermore, the frictional force is greatest at the point of peak load and therefore the accuracy of the measurement is best at the point in the cycle. Finally, for the data in the fluid film regime, the conclusions would not change if a different point in the cycle were chosen. A detailed comparison of the experimental results with the theory was carried out for the friction throughout the cycle and good agreement was obtained. The choice of one portion of the cycle to represent each test was a convenient way of displaying a lot of results concisely.

(b) In a full fluid film lubrication regime, with a near parallel film and a near symmetric pressure distribution representative of elasto-hydrodynamic conjunctions the shear of the lubricating film is considered to be the major factor contributing to the friction force.

(c) It is appropriate to draw attention to the relatively widespread variation of the experimental results. However, it has to be recognised that the levels of friction are very low ($\mu < 0.005$) and small artefacts introduced by the dynamic loading, cyclic motion and any misalignment factors can have a marked effect on the experimental values presented. This is why such a large number of tests were performed for each case. The means of the samples and the population standard deviations presented are in fact estimates of the true friction in the contact. If a further set of samples were taken then the mean values could fall anywhere within the confidence limits of the previously measured set of means. Therefore a mean value predicted from theory which falls within these same confidence limits must be concluded to be a reasonable estimate of the experimentally measured mean. Furthermore, this agreement was only achieved in the fluid film lubrication regime at surface separation ratios greater than three. The experimental data spanned the mixed lubrication regime through to the fluid film lubrication regime and to produce a readable graph the results were presented with a linear scale on the y-axis for friction coefficient and a logarithmic scale on the x-axis for the Sommerfeld number. There were, in fact, only two lubricants tested for each case which produced a sufficiently large Sommerfeld number. There were, in fact, only two lubricants tested for each case which produced a sufficiently large Sommerfeld number to achieve a surface separation ratio greater than three. Therefore, the slope of the experimental data in the fluid film lubrication regime could not be determined nor compared with the slope of the theoretical curve. Finally, it should be remembered that this is the first reported good correlation between theory and experiment in the fluid film lubrication regime under physiological conditions. In addition, in the mixed lubrication regime, it was evident for the lower modulus layers that micro-elasto-hydrodynamic effects were dominating the contact and enhancing the lubrication, producing much lower levels of friction than expected.

SESSION XIX - PANEL DISCUSSION

Dr A Kapoor (University of Cambridge, UK).

In response to Professor B O Jacobson's remarks about changing surface topography and its importance I would like to point to my work with Professor K L Johnson, which was presented in the IIIrd session. In this theoretical investigation we found surface topography to change drastically during sliding. The asperity height distribution changes from Gaussian to one which is bounded and the asperity curvature changes such that the highest asperities are almost flat (curvature = 0). However, since the separation for practical load ranges between 1 to 3 σ, only a small top portion of the total surface plays a part in the contact process and consequently undergoes a change. The problem is the profilometer which considers all of the surface and so changes in the top (less than 25%) are appreciably watered down. What we need is a method of separating out the useful portion of the surface which the contact sees from the whole of the surface which the profilometer sees and then evaluate the steady state surface roughness parameters.

Consider λ-ratio which relates film thickness to the roughness of the whole of the virgin surface. As a result of sliding and shakedown of asperities the roughness of the "profile seen by contact" is appreciably reduced and the steady state λ-ratio is much higher. No wonder the bearing keeps running well despite low λ-values obtained by using roughness of the "surface seen by the profilometer".

Reply by Professor B Jacobson (SKF, The Netherlands).

I agree with Dr Kapoor that the surface topography of sliding contacts between rough surfaces change rapidly during running in and that the run-in surfaces later determine the contact stresses and the lubricant needed to keep the surfaces separated by a lubricant film. Dr Kapoor also suggests that a special type of profilometry should be used which considers only the worn high tops in the surface structure.

There are at least two problems here which have to be considered. One is the totally different behaviour of asperities in pure rolling motion and in different degrees of sliding; the other is the definition of the λ ratio itself.

For pure sliding experiments at high loads, the lubricant can be expected to behave strongly non-Newtonian, which puts a limit to the pressure gradients possible to maintain in the contact. This means that the asperities will not be very much deformed elastically and the roughness in the lubricated contact is almost the same as that outside the contact. If the roughness is larger than the "mean" film thickness, the asperities will be worn on the tops and become flatter until a steady state is reached.

In a pure rolling situation, the shear stresses in the oil only come from the pressure gradients in the oil film. This means that the total shear strength of the oil can be used to build up pressure gradients, so the pressure gradients possible to maintain in a pure rolling situation are much higher than those possible to maintain in a sliding contact with the same film thickness.

This leads to the conclusion that the degree of elastic deformation of the asperities is a function of the kinematics of the contact. This in its turn means that the active surface roughness in a contact is not just a function of the roughness as measured outside the contact but also depends on the kinematics of the contact.

As I see it, this is the main problem when it comes to determining an actual λ value. I agree that the calculated oil film thickness should not be compared with the surface roughness of the new surfaces, but to determine how the run-in surfaces work for a certain oil film thickness it is also necessary to know the degree of sliding motion because this, together with the load levels, determines how much the asperities are elastically deformed in the lubricated contact.

19th Leeds-Lyon Symposium on Tribology

'Thin Films in Tribology'

Leeds - 8th-11th September 1992

List of Attending Authors

Title	Name	Affiliation
Dr	M J Adams	Unilever Research Port Sunlight Lab. Quarry Road East Bebington Wirral L63 3JW UK
Dr	S Aihara	NSK Ltd Tribology Res. & Dev. 1-5-50 Kugenuma-Shinmei Fujisawa Kanagawa 251 Japan
Mr	D D Auger	The University of Leeds Dept. of Mech. Eng. Leeds LS2 9JT UK
Prof	T Bell	University of Birmingham School of Metallurgy & Materials, Edgbaston Birmingham B15 2TT UK
Dr	Y Berthier	Institut National des Sciences Appliquees de Lyon, Laboratoire de Mecanique des Contacts, 20 Avenue Albert Einstein, 69621 Villeurbanne Lyon France
Mr	J Blamey	University of Durham S.E.C.S. South Road Durham DH1 3LE UK
Dr	M C C Brendle	C.N.R.S. 24 Avenue du President Kennedy 68200. Mulhouse France
Prof	B J Briscoe	Imperial College Department of Chemical Engineering London SW7 2BY UK
Ms	L Caravia	The University of Leeds Dept. of Mech. Eng. Leeds LS2 9JT UK
Mr	T A Caughlin	The University of Western Ontario Materials Engineering London Ontario Canada N6A 5B9
Prof	H S Cheng	Northwestern University 219 Catalysis, Evanston Il 60208 USA
Prof	T H C Childs	The University of Leeds Dept. of Mech. Eng. Leeds LS2 9JT UK

Title	Name	Affiliation
Dr	J M Conway-Jones	Glacier Meetal Co Ltd. Argyle House, Joel Street, Northwood Hill, Middlesex HA6 1LN, UK.
Dr	R C Coy	Shell Research Ltd. Thornton Research Centre, Lubrication & Special Products Dept. P O Box 1 Chester CH1 3SH, UK
Prof	G Dalmaz	Institut National des Sciences Appliquees de Lyon, Laboratoire de Mecanique des Contacts, 20 Avenue Albert Einstein 69621 Villeurbanne Lyon France
Prof	D Dowson	The University of Leeds Dept. of Mech. Eng. Leeds LS2 9JT UK
Dr	D N Fenner	King's College London Dept. of Mech. Eng. Strand London WC2R 2LS UK
Dr	J Fisher	The University of Leeds Dept. of Mech. Eng. Leeds LS2 9JT West Yorkshire, U.K.
Prof	J Frêne	Université de Poitiers Laboratoire de Mécanique des Solides, 40 Ave. du Recteur Pineau 86022 Poitiers Cedex France
Prof	J M Georges	Ecole Centrale de Lyon, BP 163, 69131 Ecully Cedex France
Ass Prof	S M Ghoneam	Menoufia University Faculty of Engineering Shebin-El-Kom Menoufia University Shebin-El-Kom Egypt
Prof	M Godet	Institut National des Sciences Appliquees de Lyon, Laboratoire de Mecanique des Contacts, 20 Avenue Albert Einstein 69621 Villeurbanne Lyon France
Dr	J A Greenwood	University of Cambridge, Dept. of Engrg. Trumpington St. Cambridge CB2 1PZUK
Prof	B J Hamrock	Ohio State University Mech. Eng. Dept. 206 West 18th Ave. Columbus Ohio 43210 1107, USA
Dr	H Heshmat	Mechanical Technology Inc 968 Albany-Shaker Road Latham New York 12110 USA
Dr	K Hiratsuka	Tokyo Institute of Tech. Dept. Mech. Eng. for Production 2-12-1 Ookayama Meguroku Tokyo 152 Japan
Dr	K Holmberg	VTT Laboratory of Production Eng. Metallimiehenkuja 6, P O Box 111 SF-02150 Espoo Finland

Title	Name	Affiliation
Prof	K Ikeuchi	Kyoto University Research Center for Biomedical Engrg. 53 Kawahara-cho, Shogoin Sakyo-Ku Kyoto 606 Japan
Dr	Y Iordanoff	Universite Paul Sabatier Laboratoire de Genie Mecanique IUT A 50 Chemin des Maraichers 31062 Toulouse France
Prof	B O Jacobson	SKF Engrg & Research Centre, Postbus 2350, 3430 DT Nieuwegein The Netherlands
Dr	Z M Jin	The University of Leeds Dept. of Mech. Eng. Leeds LS2 9JT West Yorkshire, UK
Prof	K L Johnson	University of Cambridge Dept. of Engrg. Trumpington St. Cambridge CB2 1PZUK
Dr	S A Johnson	Unilever Research Port Sunlight Laboratory Quarry Road East Bebington Wirral L63 3JW UK
Dr	A Kapoor	University of Cambridge Dept. of Engrg. Trumpington St. Cambridge CB2 1PZUK
Dr	E Kassfeldt	Luleå Univ. of Technology Dept. of Mech. Eng. S-95187 Luleå Sweden

Title	Name	Affiliation
Prof	K Kato	Tohoku University Laboratory of Tribology School of Mech. Eng. Sendai Japan
Prof	S C Laroiya	Technical Teachers' Training Institute Dept of Mechanical Eng. Sector 26 Chandigarh India 160019
Dr	S G Lim	NASA Lewis Research Center MS 23-2, 21000 Brookpark Road Cleveland Ohio 44135 USA
Prof	J-J Liu	Tsinghua University Dept. of Mech. Eng. Beijing 100084, China
Dr	A A Lubrecht	SKF Engineering & Research Centre P O Box 2350, 3430 DT Niuwegein The Netherlands
Dr	K Maekawa	Ibaraki University 4-12-1 Nakanarusawa Hitatchi 316 Japan
Prof.	K Matsubara	Tokai University Dept. of Production Mechanical Eng. Kirakaname, Hitastsuka-shi, Kanagawa 259-12. Japan.
Dr	M-H Meurisse	Institut National des Sciences Appliquees de Lyon, Laboratoire de Mecanique des Contacts, 20 Avenue Albert Einstein 69621 Villeurbanne Lyon France

Title	Name	Affiliation
Mr	S Mischler	Ecole Polytechnique Fédérale de Lausanne Laboratoire de Metallurgie Chimique/ Tribologie MX-C Ecublens CH-1015 Lausanne, Switzerland
Mr	G E Morales-Espejel	University of Cambridge Churchill College Cambridge CB3 0DS, UK
Prof	S Morishita	Yokohama National Univ. Dept. of Naval Architecture & Ocean Engeering,156 Tokiwadai, Hodogaya ku, Yokohama 240, Japan
Prof	T Murakami	Kyushu University Dept. of Mech Eng. Fac. of Eng. 6-10-1 Hakozaki Higashi-ku, Fukuoka 812 Japan
Mr	C M Myllerup	Technical University of Denmark, Department of Machine Elements, Build 403, 2800 Lyngby Denmark
Prof	N Ohmae	Osaka University Dept. of Precision Eng. Fac. of Eng. 2-1 Yamadoaka Suita Osaka 565 Japan
Mr	I Ohshima	Ibaraki University 4-12-1 Nakanarusawa Hitachi 316 Japan
Dr	A Olver	Imperial College Dept. of Mech. Eng. Exhibition Road London SW7 2BX UK

Title	Name	Affiliation
Mr	R L Orndorff, Jr	BF Goodrich Aerospace Division, 150 Division Drive Wilmington NC 28401 USA
Mr	V Pichon	Institut National des Sciences Appliquees de Lyon, Laboratoire de Mecanique des Contacts, 20 Avenue Albert Einstein 69621 Villeurbanne Lyon France
Prof	S L Rice	University of Central Florida, P O Box 25000, Coll. of Eng. Orlando Florida 32816, USA
Prof	M O Robbins	John Hopkins University Dept. of Physics & Astronomy Baltimore MD 21218 USA
Mr	E Rosset	Ecole Polytechnique Fédérale de Lausanne Laboratoire de Metallurgie Chimique/ Tribologie MX-C Ecublens, CH-1015 Lausanne, Switzerland
Dr	H Sander	Bundesanstalt für Materialforschung und-prüfung (BAM) Unter den Eichen 87, 1000 Berlin 45 Germany
Dr	J Seabra	Porto University Faculdade de Engenharia Rua dos Bragas 4099 Porto Codex Portugal

Title	Name	Affiliation
Dr	**J S Sheasby**	The University of Western Ontario Materials Engineering London Ontario Canada N6A 5B9
Mr	**E S Song**	The University of Leeds Dept. of Mech. Eng. Leeds LS2 9JT UK
Dr	**H A Spikes**	Imperial College Dept. of Mech. Eng. Exhibition Road London SW7 2BX UK
Dr	**T A Stolarski**	Brunel University Dept. of Mech. Eng. Uxbridge Middlesex UB8 3PH UK
Prof	**J L Streator**	Georgia Institute of Technology ,George W Woodruff School of Mech. Eng. Atlanta GA 30332-0405 USA
Prof	**C M Taylor**	The University of Leeds Dept. of Mech. Eng. Leeds LS2 9JT UK
Mr	**P S Thomas**	Imperial College Department of Chemical Engineering London SW7 2BY UK
Dr	**A A Torrance**	Trinity College Parsons Building. Dublin 2 Ireland
Mr	**B Tournerie**	Universite de Poitiers Laboratoire de Mecanique des Solides, 40 Av. du Recteur Pineau 86022 Poitiers Cedex France
Mr	**E Vancoille**	Dept. MTM KU-Leuven de Croylaan 2, B3001 Leuven Belgium
Mr	**M Visscher**	Eindhoven Univ. of Tech. Meierijsingel 16 NL 5262 JV Vught The Netherlands
Dr	**J A Williams**	University of Cambridge University Engineering Dept. Trumpington Street Cambridge CB2 1PZ, U.K.
Prof	**W O Winer**	Georgia Institute of Technology ,George W Woodruff School of Mech. Eng. Atlanta GA 30332-0405 USA
Dr	**H Xu**	T & N Technology Ltd Cawston House Cawston Lane Rugby Warks CV22 7SA UK

19th Leeds-Lyon Symposium on Tribology

'Thin Films in Tribology'

Leeds - 8th-11th September 1992

List of Delegates

Title	Name	Affiliation
Dr	M J Adams	Unilever Research Port Sunlight Lab. Quarry Road East Bebington Wirral L63 3JW UK
Dr	S Aihara	NSK Ltd Tribology Res. & Dev. 1-5-50 Kugenuma-Shinmei Fujisawa Kanagawa 251 Japan
Mr	D D Auger	The University of Leeds Dept. of Mech. Eng. Leeds LS2 9JT UK
Mr	de Baets	University of Ghent Laboratori Machine Bouw, St Pietersnieuwstraat 41, B-9000 Gent Belgium
Dr	T W Bates	Shell Research Ltd Thornton Research Centre P O Box 1 Chester CH1 3SH UK
Dr	J C Bell	Shell Research Ltd. Thornton Research Centre Lubrication & Special Products Dept P O Box 1 Chester CH1 3SH, UK.
Prof	T Bell	University of Birmingham School of Metallurgy & Materials, Edgbaston Birmingham B15 2TT UK
Dr	Y Berthier	Institut National des Sciences Appliquees de Lyon, Laboratoire de Mecanique des Contacts, 20 Avenue Albert Einstein 69621 Villeurbanne Lyon France
Mr	J Blamey	University of Durham S.E.C.S. South Road Durham DH1 3LE UK
Prof	H Blok	Dr H Colijnlaan 4, Flat 19 2283 XM Rijswijk The Netherlands
Prof	J F Booker	Cornell University (Upson Hall), School of Mech. & Aerospace Eng. Ithaca New York 14853 USA
Dr	J Borde	CNRS Representative in the UK 6 Cromwell Place London SW7 2JN UK
Mr	J Bos	University of Twente Fac of Mech. Eng. P O Box 217, Enschede The Netherlands

Title	Name	Affiliation
Mr	**M Breen**	King's College London Dept. of Mech. Eng. Strand London WC2R 2LS, UK.
Ir	**F Bremer**	Nederlandse Philips Bedrÿven BV, Centre for Manuf. Tech. Building SAQ-2100, 5600 MD Eindhoven The Netherlands
Dr	**M C C Brendle**	C.N.R.S. 24 Avenue du President Kennedy 68200 Mulhouse France
Prof	**B J Briscoe**	Imperial College Department of Chemical Engineering London SW7 2BY UK
Mr	**K de Bruyn**	WTCM - Oppervlakteveredeling Universitaire Campus B-3590 Diepenbeek Belgium
Dr	**P M Cann**	Imperial College Tribology Section Dept. of Mech. Eng. Exhibition Road London SW7 2BX UK
Ms	**L Caravia**	The University of Leeds Dept. of Mech. Eng. Leeds LS2 9JT UK
Mr	**T A Caughlin**	The University of Western Ontario Materials Engineering London Ontario Canada N6A 5B9
Prof	**H S Cheng**	Northwestern University 219 Catalysis, Evanston Il 60208 USA
Prof	**T H C Childs**	The University of Leeds Dept. of Mech. Eng. Leeds LS2 9JT UK
Dr	**R J Chittenden**	The University of Leeds, Industrial Unit of Tribology, Department of Mechanical Engineering Leeds LS2 9JT UK
Dr	**K J Chugg**	Ethyl Petroleum Additives Ltd. London Road, Bracknell, Berkshire RG12 2UW, UK
Mr	**S Cole**	University of Birmingham School of Metallurgy & Materials, Edgbaston Birmingham B15 2TT, UK.
Dr	**J M Conway-Jones**	Glacier Meetal Co Ltd. Argyle House, Joel Street, Northwood Hill, Middlesex HA6 1LN, UK.
Dr	**D Cooper**	BP International Ltd Sunbury Research Centre Sunbury-on-Thames Middlesex TW16 7AZ UK
Dr	**R C Coy**	Shell Research Ltd. Thornton Research Centre Lubrication & Special Products Dept. P O Box 1 Chester CH1 3SH, UK
Prof	**G Dalmaz**	Institut National des Sciences Appliquees de Lyon, Laboratoire de Mecanique des Contacts, 20 Avenue Albert Einstein 69621 Villeurbanne Lyon France

Title	Name	Affiliation
Ing.	J-M Dewulf	Societe Europeenne de Propulsion (SEP) DGPL, Foret de Vernon, BP 802 27200 Vernon, Cedex France
Prof	D Dowson	The University of Leeds Dept. of Mech. Eng. Leeds LS2 9JT UK
Mr	D Eastbury	Group Editor - Science & Tech. Butterworth Heinemann, Linacre House, Jordan Hill Oxford OX2 8DP UK
Mr	D Enwright	Commissioning Editor Butterworth Heinemann Linacre House Jordan Hill Oxford OX2 8DP UK
Dr	D N Fenner	King's College London Dept. of Mech. Eng. Strand London WC2R 2LS UK
Dr	J Fisher	The University of Leeds Dept. of Mech. Eng. Leeds LS2 9JT West Yorkshire, U.K.
Prof	L Flamand	Institut National des Sciences Appliquees de Lyon, Laboratoire de Mecanique des Contacts, 20 Avenue Albert Einstein 69621 Villeurbanne Lyon France
Prof	J Frêne	Université de Poitiers Laboratoire de Mécanique des Solides, 40 Ave. du Recteur Pineau 86022 Poitiers Cedex France

Title	Name	Affiliation
Prof	J M Georges	Ecole Centrale de Lyon, BP 163, 69131 Ecully Cedex France
Mr	S M Ghoneam	Warsaw University Of Tech. Institute of Aircraft Tech. & Applied Mechanics ul Nowowiejska 24 00-665 Warsaw Poland
Mr	C Godeau	Institut National des Sciences Appliquees de Lyon, Laboratoire de Mecanique des Contacts, 20 Avenue Albert Einstein 69621 Villeurbanne Lyon France
Prof	M Godet	Institut National des Sciences Appliquees de Lyon, Laboratoire de Mecanique des Contacts, 20 Avenue Albert Einstein 69621 Villeurbanne Lyon France
Mr	R J M Gomes	University of Minho Escola de Engenharia - Azurem, 4800 Guimarãs Portugal
Dr	J A Greenwood	University of Cambridge, Dept. of Engrg. Trumpington St. Cambridge CB2 1PZ UK
Mr	J W Hadley	Liverpool Polytechnic SETM, Byrom Street Liverpool L3 3AF, UK
Dr	G M Hamilton	The University of Reading Dept. of Engeering Whitknights Reading RG6 2AY UK

Title	Name	Affiliation
Prof	**B J Hamrock**	Ohio State University Mech. Eng. Dept. 206 West 18th Ave. Columbus Ohio 43210-1107, USA
Mr	**D Han**	The University of Leeds Dept. of Mech. Eng. Leeds LS2 9JT UK
Mr	**R T Harding**	The University of Leeds Dept. of Mech. Eng. Leeds LS2 9JT UK
Dr	**H Heshmat**	Mechanical Technology Inc. 968 Albany-Shaker Road Latham New York 12110 USA
Dr	**K Hiratsuka**	Tokyo Institute of Tech. Dept. Mech. Eng. for Production 2-12-1 Ookayama Meguroku Tokyo 152 Japan
Dr	**K Holmberg**	VTT Laboratory of Production Eng. Metallimiehenkuja 6, P O Box 111 SF-02150 Espoo Finland
Dr	**C P Hunter**	IRSID, BP 50, 42702 Firminy France
Prof	**K Ikeuchi**	Kyoto University Research Center for Biomedical Engrg. 53 Kawahara-cho, Shogoin Sakyo-Ku Kyoto 606 Japan
Mr	**G E Innes** The	University of Leeds Dept. of Mech. Eng. Leeds LS2 9JT UK
Dr	**Y Iordanoff**	Universite Paul Sabatier Laboratoire de Genie Mecanique IUT A 50 Chemin des Maraichers 31062 Toulouse France
Prof	**B O Jacobson**	SKF Engrg & Research Centre, Postbus 2350, 3430 DT Nieuwegein The Netherlands
Mr	**F Ji**	The University of Leeds Dept. of Mech. Eng. Leeds LS2 9JT West Yorkshire, UK
Dr	**Z M Jin**	The University of Leeds Dept. of Mech. Eng. Leeds LS2 9JT West Yorkshire, UK
Mr	**B Jobbins**	The University of Leeds Dept. of Mech Eng. Leeds LS2 9JT West Yorkshire, UK
Prof	**K L Johnson**	University of Cambridge Dept. of Engrg. Trumpington St. Cambridge CB2 1PZ, UK
Dr	**S A Johnson**	Unilever Research Port Sunlight Laboratory Quarry Road East Bebington Wirral L63 3JW UK
Mr	**D A Jones**	The University of Leeds Dept. of Mech Eng. Leeds LS2 9JT West Yorkshire, UK
Mr	**G J Jones**	Glacier Vandervell Ltd Argyle House, Joel Street Northwood Hills Middlesex HA6 1LN UK

Title	Name	Affiliation
Dr	**A Kapoor**	University of Cambridge Dept. of Engrg. Trumpington St. Cambridge CB2 1PZ UK
Dr	**E Kassfeldt**	Luleå Univ. of Technology Dept. of Mech. Eng. S-95187 Luleå Sweden
Prof	**K Kato**	Tohoku University Laboratory of Tribology School of Mech. Eng. Sendai Japan
Mr	**R Kumagai**	Petroleum Energy Center (PEC), 4-10 Ohnodai 1-chome, Midori-ku, Chiba 267 Japan
Mr	**E Kvistgaard**	Man B&W Diesel A/S Engrg. Stamholmen 161 DK-2650, Hvidovre Copenhagen Denmark
Mr	**C K K Lai**	The University of Leeds Dept. of Mech. Eng. Leeds LS2 9JT UK
Dr	**J K Lancaster**	'Windfalls', Marton-cum-Grafton York YO5 9QY UK
Prof	**S C Laroiya**	Technical Teachers' Training Institute Dept of Mechanical Eng. Sector 26 Chandigarh India 160019
Mr	**H J van Leeuwen**	Eindhoven Univ. of Tech. Dept. of Mech. Eng. WH04-102 P O Box 513 NL-5600 MB Eindhoven The Netherlands
Dr	**J L Ligier**	Glacier Sic 4 Rue de la Liberte 4 BP 2073 74089 Annecy Cedex France
Mr	**C H Lim**	University College Swansea, Mech. Eng. Dept. Singleton Park Swansea SA2 8PP UK
Dr	**S G Lim**	NASA Lewis Research Center MS 23-2, 21000 Brookpark Road Cleveland Ohio 44135 USA
Prof	**J-J Liu**	Tsinghua University Dept. of Mech. Eng. Beijing 100084, China
Dr	**A A Lubrecht**	SKF Engineering & Research Centre P O Box 2350, 3430 DT Niuwegein The Netherlands
Dr	**J Leopold**	TU Chemnitz Am Bahrehang 35 0-9081 Chemnitz Germany
Mr	**J Liu**	The University of Leeds Dept. of Mech. Eng. Leeds LS2 9JT UK
Prof	**K Ludema**	University of Michigan Dept. of Mech. Eng. 2250 G G Brown Building Ann Arbor MI 48109-2125 USA
Mr	**J McDonald**	Superintendent, Quality Engrg. Test Est. Ottawa, Ontario, Canada, KIA 0K2
Mr	**A McNicol**	The University of Leeds Dept. of Mech. Eng. Leeds LS2 9JT UK

Title	Name	Affiliation
Mr	M-T Ma	The University of Leeds Dept. of Mech. Eng. Leeds LS2 9JT UK
Mr	Y Maeda	Petroleum Energy Center (PEC) Advanced Tech. & Res. Institute, 4-10 Ohnodai 1-chome, Midori-ku, Chiba 267 Japan
Dr	K Maekawa	Ibaraki University 4-12-1 Nakanarusawa Hitatchi 316 Japan
Dr	S Mahmood	The University of Leeds Dept. of Mech. Eng. Leeds LS2 9JT UK
Dr	C N March	The University of Leeds Industrial Unit of Tribology Department of Mechanical Engineering Leeds LS2 9JT UK
Prof.	K Matsubara	Tokai University Dept. of Production Mechanical Eng. Kirakaname, Hitastsuka-shi, Kanagawa 259-12. Japan.
Dr	M-H Meurisse	Institut National des Sciences Appliquees de Lyon, Laboratoire de Mecanique des Contacts, 20 Avenue Albert Einstein 69621 Villeurbanne Lyon France
Mr	S Mischler	Ecole Polytechnique Fédérale de Lausanne Laboratoire de Metallurgie Chimique/Tribologie MX-C Ecublens CH-1015 Lausanne, Switzerland
Ing.	D Moes	Thomassen International BV, P O Box 70 6990 AB Rheden The Netherlands
Dr	A J Moore	B P International Ltd BP Research Centre Chertsey Road Sunbury-on-Thames Middlesex TW16 7LN UK
Mr	G E Morales-Espejel	University of Cambridge Churchill College Cambridge CB3 0DS, UK
Prof	S Morishita	Yokohama National Univ. Dept. of Naval Architecture & Ocean Engeering,156 Tokiwadai, Hodogaya ku, Yokohama 240, Japan
Prof	T Murakami	Kyushu University Dept. of Mech Eng. Fac. of Eng. 6-10-1 Hakozaki Higashi-ku, Fukuoka 812 Japan
Mr	C M Myllerup	Technical University of Denmark, Department of Machine Elements, Build 403, 2800 Lyngby Denmark
Dr	D Nelias	Institut National des Sciences Appliquees de Lyon, Laboratoire de Mecanique des Contacts, 20 Avenue Albert Einstein 69621 Villeurbanne Lyon France
Prof	N Ohmae	Osaka University Dept. of Precision Eng. Fac. of Eng. 2-1 Yamadoaka Suita Osaka 565 Japan

Title	Name	Affiliation
Mr	I Ohshima	Ibaraki University 4-12-1 Nakanarusawa Hitachi 316 Japan
Dr	A Olver	Imperial College Dept. of Mech. Eng. Exhibition Road London SW7 2BX UK
Dr	E Onsøyen	SINTEF Production Eng N-7034 Trondheim Norway
Mr	R L Orndorff, Jr	BF Goodrich Aerospace Division, 150 Division Drive Wilmington NC 28401 USA
Mr	D D Parker	Glacier Vandervell Ltd Riccarton, Kilmarnock Scotland KA1 3NA UK
Mr	V Pichon	Institut National des Sciences Appliquees de Lyon, Laboratoire de Mecanique des Contacts, 20 Avenue Albert Einstein 69621 Villeurbanne Lyon France
Dr	S Y Poon	Inter Technology Research (Newark) Ltd., Potterdike House, Lombard St. Newark Notts NG24 1XG, UK
Mr	M Priest	The University of Leeds Industrial Unit of Tribology Department of Mechanical Engineering Leeds LS2 9JT UK
Mr	S Raadnui	Flat 5C, Vivian House Roman Bridge Close Blackpill Swansea SA3 5BG UK
Mr	C Radcliffe	The University of Leeds Dept. of Mech. Eng. Leeds LS2 9JT UK
Prof	S L Rice	University of Central Florida, P O Box 25000, Coll. of Eng. Orlando Florida 32816, USA
Dr	J W Richmond	T & N Technology Ltd, Cawston House Cawston Lane Rugby Warks CV22 7SA UK
Prof	M O Robbins	John Hopkins University Dept. of Physics & Astronomy Baltimore MD 21218 USA
Mr	G Roche	Institut National des Sciences Appliquees de Lyon, Laboratoire de Mecanique des Contacts, 20 Avenue Albert Einstein 69621 Villeurbanne Lyon France
Mr	E Rosset	Ecole Polytechnique Fédérale de Lausanne Laboratoire de Metallurgie Chimique/Tribologie MX-C Ecublens, CH-1015 Lausanne, Switzerland
Dr	H Sander	Bundesanstalt für Material forschung und-prüfung (BAM) Unter den Eichen 87, 1000 Berlin 45 Germany
Dr	J Seabra	Porto University Faculdade de Engenharia Rua dos Bragas 4099 Porto Codex Portugal

Title	Name	Affiliation
Dr	**J S Sheasby**	The University of Western Ontario Materials Engineering London Ontario Canada N6A 5B9
Mr	**A J Smith**	NSK-RHP Bearing Res Ltd Ruddington Fields Business Park, Ruddington Nottingham NG11 6JZ
Mr	**E S Song**	The University of Leeds Dept. of Mech. Eng. Leeds LS2 9JT UK
Dr	**H A Spikes**	Imperial College Dept. of Mech. Eng. Exhibition Road London SW7 2BX UK
Mr	**P Stephan**	Universite Paul Sabatier Laboratoire de Genie Mecanique IUT A 50 Chemin des Maraichers 31062 Toulouse France
Dr	**T A Stolarski**	Brunel University Dept. of Mech. Eng. Uxbridge Middlesex UB8 3PH UK
Prof	**J L Streator**	Georgia Institute of Technology ,George W Woodruff School of Mech. Eng. Atlanta GA 30332-0405 USA
Mr	**M Taniguchi**	The University of Leeds Dept. of Mech. Eng. Leeds LS2 9JT UK
Prof	**C M Taylor**	The University of Leeds Dept. of Mech. Eng. Leeds LS2 9JT UK
Dr	**R I Taylor**	Shell Research Ltd. Thornton Research Centre P O Box 1 Chester CH1 3SH UK
Mr	**P S Thomas**	Imperial College Department of Chemical Engineering London SW7 2BY UK
Dr	**A A Torrance**	Trinity College Parsons Building. Dublin 2 Ireland
Mr	**B Tournerie**	Universite de Poitiers Laboratoire de Mecanique des Solides, 40 Av. du Recteur Pineau 86022 Poitiers Cedex France
Dr	**Y D Tridimas**	Liverpool Polytechnic School of Engrg. & Tech. Mangt. Byrom Street Liverpool L3 3AF, UK.
Dr	**P J Tweedale**	National Centre of Tribology, AEA Industrial Tech. Risley Warrington WA3 6AT UK
Mr	**E Vancoille**	Dept. MTM KU-Leuven de Croylaan 2, B3001 Leuven Belgium
Dr	**C H Venner**	University of Twente Fac. of Mech. Eng. P O Box 217 Enschede The Netherlands
Mr	**P Vergne**	Institut National des Sciences Appliquees de Lyon, Laboratoire de Mecanique des Contacts, 20 Avenue Albert Einstein 69621 Villeurbanne Lyon France

Title	Name	Affiliation
Prof	**L Vincent**	Ecole Centrale de Lyon Dept. of Materiaux-Mecanique Physique 69131 Ecully, Cedex France
Ir	**H Visscher**	University of Twente Faculty of Engineering Tribology Group Postbox 217 7500 AE Enschede The Netherlands
Mr	**M Visscher**	Eindhoven Univ. of Tech. Meierijsingel 16 NL 5262 JV Vught The Netherlands
Prof	**J Vizintin**	Fac. of Mech. Eng. Murnikova 2, 61000 Lujubljana pob. 394, Slovenia
Mr	**D Wang**	The University of Leeds Dept. of Mech. Eng. Leeds LS2 9JT UK
Dr	**J A Williams**	University of Cambridge University Engineering Dept. Trumpington Street Cambridge CB2 1PZ, U.K.
Mr	**W H Wilson**	MCB University Press 18 Adel Park Gardens Leeds LS16 8BN UK
Prof	**W O Winer**	Georgia Institute of Technology ,George W Woodruff School of Mech. Eng. Atlanta GA 30332-0405 USA
Mr	**Y Xie**	University of Cambridge University Engineering Dept. Trumpington Street Cambridge CB2 1PZ, U.K.
Dr	**H Xu**	T & N Technology Ltd Cawston House Cawston Lane Rugby Warks CV22 7SA UK
Mr	**R Yamashita**	The University of Leeds Dept. of Mech. Eng. Leeds LS2 9JT UK
Mr	**L S Yang**	The University of Leeds Dept. of Mech. Eng. Leeds LS2 9JT UK